$x^2 = a$ is equivalent to $x = -\sqrt{a}$ or $x = \sqrt{a}$

$x^2 > a$ is equivalent to $x < -\sqrt{a}$ or $x > \sqrt{a}$

$x^2 < a$ is equivalent to $-\sqrt{a} < x < \sqrt{a}$

Exponents and Radicals

$a^0 = 1 \qquad a^1 = a$

$a^{-1} = \dfrac{1}{a} \qquad a^{-n} = \dfrac{1}{a^n}$

$a^m a^n = a^{m+n} \qquad \dfrac{a^m}{a^n} = a^{m-n}$

$(a^m)^n = a^{mn}$

$a^n b^n = (ab)^n \qquad \dfrac{a^n}{b^n} = \left(\dfrac{a}{b}\right)^n$

$a^{1/n} = \sqrt[n]{a} \qquad a^{1/2} = \sqrt{a}$

$a^{m/n} = \sqrt[n]{a^m} = \left(\sqrt[n]{a}\right)^m$

$\sqrt[n]{ab} = \sqrt[n]{a}\sqrt[n]{b} \qquad \sqrt[n]{\dfrac{a}{b}} = \dfrac{\sqrt[n]{a}}{\sqrt[n]{b}}$

Factorials

$0! = 1 \qquad 1! = 1$

$n! = 1 \cdot 2 \cdot 3 \cdots (n-1) \cdot n$

GEOMETRY

Area (A), Circumference (C), Surface Area (S), and Volume (V)

Triangle, base b and height h: $A = \dfrac{1}{2}bh$

Rectangle, base b and height h: $A = bh$; $C = 2b + 2h$

Circle, radius r: $A = \pi r^2$; $C = 2\pi r$

Sector of a circle, radius r and angle θ (in radians): $A = \dfrac{1}{2}r^2\theta$; $C = r\theta$

Sphere, radius r: $V = \dfrac{4}{3}\pi r^3$; $S = 4\pi r^2$

Cylinder, base radius r and height h: $V = \pi r^2 h$

Rectangular box with dimensions a, b, c: $V = abc$; $S = 2ab + 2ac + 2bc$

Cube of side a: $V = a^3$; $S = 6a^2$

Analytic Geometry

Distance between points x_1 and x_2 on a number line: $d = |x_2 - x_1|$

Distance between points (x_1, y_1) and (x_2, y_2) in a plane: $d = \sqrt{(x_2 - x_1)^2 + (y_2 - y_1)^2}$

Midpoint of a line segment joining (x_1, y_1) and (x_2, y_2): $M = \left(\dfrac{x_1 + x_2}{2}, \dfrac{y_1 + y_2}{2}\right)$

Slope of a line through points (x_1, y_1) and (x_2, y_2): $m = \dfrac{y_2 - y_1}{x_2 - x_1}$

Equation of a line through (x_1, y_1) of slope m (point-slope equation): $y - y_1 = m(x - x_1)$

Line of slope m and y-intercept b (slope-intercept equation): $y = mx + b$

Equation of a circle of radius r centred at the origin: $x^2 + y^2 = r^2$

TRIGONOMETRIC RATIOS AND TRIGONOMETRIC FUNCTIONS

π radians = 180°; i.e., 1 radian = $\dfrac{180^\circ}{\pi}$ and $1^\circ = \dfrac{\pi}{180}$ radians

Right-angle triangle

$\sin\theta = \dfrac{\text{opposite}}{\text{hypotenuse}}$ $\qquad$ $\csc\theta = \dfrac{1}{\sin\theta} = \dfrac{\text{hypotenuse}}{\text{opposite}}$

$\cos\theta = \dfrac{\text{adjacent}}{\text{hypotenuse}}$ $\qquad$ $\sec\theta = \dfrac{1}{\cos\theta} = \dfrac{\text{hypotenuse}}{\text{adjacent}}$

$\tan\theta = \dfrac{\text{opposite}}{\text{adjacent}}$ $\qquad$ $\cot\theta = \dfrac{1}{\tan\theta} = \dfrac{\text{adjacent}}{\text{opposite}}$

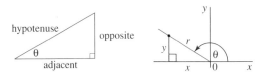

General angles

$r = \sqrt{x^2 + y^2}$

$\sin\theta = \dfrac{y}{r}$ $\qquad$ $\csc\theta = \dfrac{1}{\sin\theta} = \dfrac{r}{y}$

$\cos\theta = \dfrac{x}{r}$ $\qquad$ $\sec\theta = \dfrac{1}{\cos\theta} = \dfrac{r}{x}$

$\tan\theta = \dfrac{y}{x}$ $\qquad$ $\cot\theta = \dfrac{1}{\tan\theta} = \dfrac{x}{y}$

Calculus for the Life Sciences: Modelling the Dynamics of Life

Second Canadian Edition

Frederick R. Adler
University of Utah

Miroslav Lovrić
McMaster University

1914–2014
Nelson Education celebrates 100 years of Canadian publishing

Calculus for the Life Sciences: Modelling the Dynamics of Life, Second Canadian Edition

by Frederick R. Adler and Miroslav Lovrić

Vice President, Editorial, Higher Education:
Anne Williams

Publisher:
Paul Fam

Executive Editor:
Jackie Wood

Marketing Manager:
Leanne Newell

Developmental Editor:
Suzanne Simpson Millar

Permissions Coordinator:
Lynn McLeod

Content Production Manager:
Claire Horsnell

Production Service:
Integra Software Services Pvt. Ltd.

Copy Editor:
Julia Cochrane

Proofreader:
Integra

Indexer:
Ursula Acton

Design Director:
Ken Phipps

Managing Designer:
Franca Amore

Interior Design:
Cheryl Carrington

Cover Design:
Martyn Schmoll

Cover Image:
iLexx/iStockphoto.com

Compositor:
Integra Software Services Pvt. Ltd.

Copyright © 2015, 2012 by Nelson Education Ltd.

Printed and bound in the United States of America
1 2 3 4 17 16 15 14

Adapted from *Modeling the Dynamics of Life: Calculus and Probability for Life Scientists,* Second Edition, by Frederick R. Adler, published by Brooks/Cole, a division of Thomson Learning, Inc. Copyright © 2005 by Brooks/Cole, a division of Thomson Learning, Inc.

For more information contact Nelson Education Ltd., 1120 Birchmount Road, Toronto, Ontario, M1K 5G4. Or you can visit our Internet site at http://www.nelson.com

Statistics Canada information is used with the permission of Statistics Canada. Users are forbidden to copy this material and/or redisseminate the data, in an original or modified form, for commercial purposes, without the expressed permissions of Statistics Canada. Information on the availability of the wide range of data from Statistics Canada can be obtained from Statistics Canada's Regional Offices, its World Wide Web site at <http://www.statcan.gc.ca>, and its toll-free access number 1-800-263-1136.

ALL RIGHTS RESERVED. No part of this work covered by the copyright herein may be reproduced, transcribed, or used in any form or by any means| graphic, electronic, or mechanical, including photocopying, recording, taping, Web distribution, or information storage and retrieval systems| without the written permission of the publisher.

For permission to use material from this text or product, submit all requests online at www.cengage.com/permissions. Further questions about permissions can be emailed to permissionrequest@cengage.com

Every effort has been made to trace ownership of all copyrighted material and to secure permission from copyright holders. In the event of any question arising as to the use of any material, we will be pleased to make the necessary corrections in future printings.

Library and Archives Canada Cataloguing in Publication Data

Adler, Frederick R., author
Calculus for the life sciences : modeling the dynamics of life / Frederick R. Adler, University of Utah, Miroslav Lovrić McMaster University. – Second Canadian edition.

Includes index.
ISBN 978-0-17-653078-5 (bound)

1. Calculus–Textbooks.
2. Life sciences–Mathematics–Textbooks.
I. Lovrić, Miroslav author
II. Title.

QA303.2.A354 2014   570.15118
C 2013-904680-1

ISBN-13: 978-0-17-653078-5
ISBN-10: 0-17-653078-9

Contents

Chapter 1 — Introduction to Models and Functions — 1

1.1 Why Mathematics Matters 1

1.2 Models in Life Sciences 4
Types of Dynamical Systems 7

1.3 Variables, Parameters, and Functions 10
Describing Measurements with Variables, Parameters, and Graphs 11
Describing Relations between Measurements with Functions 13
Catalogue of Important Functions 19
Exercises 22

1.4 Working with Functions 25
Inverse Functions 29
Creating New Functions 36
Transformations of Graphs 37
Exercises 44

1.5 Logical Reasoning and Language in Math and Life Sciences 47
Building Blocks: Definitions and Theorems 48
Implications 50
Math Results and Their Interpretation 51

Chapter Summary: Key Terms and Concepts 52
Concept Check: True/False Quiz 53
Supplementary Problems 53
Project 54

Chapter 2 — Modelling Using Elementary Functions — 55

2.1 Elementary Models 55
Proportional Relations 55
Linear Functions 58
Power Functions 63
Exercises 69

2.2 Exponential and Logarithmic Functions; Exponential Models 72
Exponential Functions 72
Logarithmic Functions 75
Exponential Models 78
Semilog and Double-Log Graphs 86
Exercises 89

2.3 Trigonometric and Inverse Trigonometric Functions 91
Angles and Trigonometric Ratios 92
Trigonometric Functions 94
Describing Oscillations 98

Inverse Trigonometric Functions   102
Exercises   107

Chapter Summary: Key Terms and Concepts   110
Concept Check: True/False Quiz   110
Supplementary Problems   110

Chapter 3 Discrete-Time Dynamical Systems   113

3.1 Introduction to Discrete-Time Dynamical Systems   113
Discrete-Time Dynamical Systems and Updating Functions   113
Solutions   116
Manipulating Updating Functions   122
Units and Dimensions   125
Exercises   126

3.2 Analysis of Discrete-Time Dynamical Systems   129
Cobwebbing: A Graphical Solution Technique   130
Equilibria: Geometric Approach   132
Equilibria: Algebraic Approach   135
Exercises   139

3.3 Modelling with Discrete-Time Dynamical Systems   141
Absorption of Caffeine   141
Elementary Population Models   142
Dynamics of Alcohol Use   147
Exercises   151

3.4 Nonlinear Dynamics Model of Selection   153
A Model of Selection   153
The Discrete-Time Dynamical System and Equilibria   156
Exercises   160

3.5 A Model of Gas Exchange in the Lung   164
A Model of the Lungs   164
The Lung System in General   167
Lung Dynamics with Absorption   170
Exercises   172

Chapter Summary: Key Terms and Concepts   174
Concept Check: True/False Quiz   175
Supplementary Problems   175
Project   177

Chapter 4 Limits, Continuity, and Derivatives   179

4.1 Investigating Change   179
The Average Rate of Change   179
Instantaneous Rate of Change   182
Exercises   185

4.2 Limit of a Function   187
Limit   187
Calculating Limits   192
Calculating Limits Algebraically—Limit Laws   195
Exercises   201

4.3 Infinite Limits and Limits at Infinity   205
Infinite Limits   205
Limits at Infinity   209
Infinite Limits at Infinity   215

Comparing Functions at Infinity   219
Application to Absorption Functions   222
Limits of Sequences   224
Exercises   226

4.4 Continuity   229
Continuous Functions   230
Input and Output Precision   238
Hysteresis   240
Exercises   241

4.5 Derivatives and Differentiability   243
Differentiable Functions   243
Graphs and Derivatives   247
Approximating Derivatives   252
Exercises   254

Chapter Summary: Key Terms and Concepts   257
Concept Check: True/False Quiz   257
Supplementary Problems   258
Project   259

Chapter 5 | Working with Derivatives   261

5.1 Derivatives of Powers, Polynomials, and Exponential Functions   261
Derivatives of Constant and Power Functions   261
The Sum (Difference) Rule for Derivatives   265
Derivatives of Polynomials   265
The Exponential Function   268
Exercises   271

5.2 Derivatives of Products and Quotients   273
The Product Rule   273
The Quotient Rule   277
Exercises   280

5.3 The Chain Rule and the Derivatives of Logarithmic Functions   283
The Derivative of a Composite Function   283
Derivatives of Logarithmic Functions   288
Exponential Measurements and Relative Rate of Change   290
Proof of the Chain Rule   292
Exercises   294

5.4 Derivatives of Trigonometric and Inverse Trigonometric Functions   296
Derivatives of Sine and Cosine   296
Other Trigonometric Functions   300
Derivation of the Key Limits   301
Derivatives of Inverse Trigonometric Functions   303
Exercises   305

5.5 Implicit Differentiation, Logarithmic Differentiation, and Related Rates   307
Implicit Differentiation   308
Logarithmic Differentiation   310
Related Rates   312
Exercises   315

5.6 The Second Derivative, Curvature, and Concavity   316
The Second Derivative   316
Using the Second Derivative for Graphing   320
Applications   322
Approximating the Second Derivative   325

Acceleration   325
Exercises   326

5.7 Approximating Functions with Polynomials   330
The Tangent and Secant Lines   330
Quadratic Approximation   333
Taylor Polynomials   338
Exercises   340

Chapter Summary: Key Terms and Concepts   342
Concept Check: True/False Quiz   342
Supplementary Problems   342
Project   343

Chapter 6 Applications of Derivatives   345

6.1 Extreme Values of a Function   345
Minima and Maxima   345
Absolute Extreme Values   355
Exercises   363

6.2 Three Case Studies in Optimization   365
Strength of Bones   365
Maximizing the Rate of Food Intake   368
Optimizing Area and the Shape of a Honeycomb   372
Exercises   375

6.3 Reasoning about Functions: Continuity and Differentiability   377
Continuous Functions: The Intermediate Value Theorem   377
Rolle's Theorem and the Mean Value Theorem   381
Continuity and Differentiability   384
Exercises   386

6.4 Leading Behaviour and L'Hôpital's Rule   388
Leading Behaviour of Functions at Infinity   388
Leading Behaviour of Functions at Zero   392
The Method of Matched Leading Behaviours   393
L'Hôpital's Rule   396
Exercises   403

6.5 Graphing Functions: A Summary   405
Exercises   413

6.6 Newton's Method   414
Solving Equations   414
Newton's Method   414
Exercises   422

6.7 Stability of Discrete-Time Dynamical Systems   424
Motivation   424
Stability and the Slope of the Updating Function   427
Evaluating Stability with the Derivative   429
Exercises   434

6.8 The Logistic Dynamical System and More Complex Dynamics   437
The Logistic Dynamical System   437
Qualitative Dynamical Systems   439
Analysis of the Logistic Dynamical System   444
Ricker Model   446
Exercises   448

6.9 Case Study: Panting and Deep Breathing Online

Chapter Summary: Key Terms and Concepts   450
Concept Check: True/False Quiz   451
Supplementary Problems   451
Projects   453

Chapter 7 Integrals and Applications   455

7.1 Differential Equations   455
Differential Equations: Examples and Terminology   458
Graphical Solution of Pure-Time Differential Equations   462
Euler's Method for Solving Differential Equations   464
Exercises   470

7.2 Antiderivatives   473
Antiderivatives   473
Rules for Antiderivatives   476
Solving Simple Differential Equations   480
Exercises   484

7.3 Definite Integral and Area   486
Area   487
Sigma Notation   492
Area and Definite Integral   493
The Definite Integral   497
Properties of Integrals   502
Exercises   507

7.4 Definite and Indefinite Integrals   510
The Fundamental Theorem of Calculus:
    Computing Definite Integrals with Indefinite Integrals   510
The Integral Function and the Proof of the Fundamental Theorem of Calculus   519
Exercises   526

7.5 Techniques of Integration: Substitution and Integration by Parts   529
A Useful Shortcut   529
The Chain Rule and Integration by Substitution   530
Integration by Parts   538
Integration Using Taylor Polynomials   543
Exercises   545

7.6 Applications   548
Integrals and Areas   549
Integrals and Volumes   555
Integrals and Lengths   560
Integrals and Averages   564
Integrals and Mass   565
Exercises   568

7.7 Improper Integrals   571
Infinite Limits of Integration   571
Improper Integrals: Examples   573
Infinite Integrands   579
Exercises   583

Chapter Summary: Key Terms and Concepts   584
Concept Check: True/False Quiz   584
Supplementary Problems   585
Projects   586

viii   Contents

Chapter 8 — Differential Equations — 589

8.1 Basic Models with Differential Equations 589
Autonomous Differential Equations 589
Exercises 598

8.2 Equilibria and Display of Autonomous Differential Equations 602
Equilibria 602
Graphical Display of Autonomous Differential Equations 605
Exercises 607

8.3 Stability of Equilibria 609
Recognizing Stable and Unstable Equilibria 610
Applications of the Stability Theorem 612
A Model for a Disease 614
Outline of the Proof of Theorem 8.3.1 615
Exercises 617

8.4 Separable Differential Equations 619
Separation of Variables 619
Exercises 628

8.5 Systems of Differential Equations; Predator–Prey Model 630
Predator–Prey Dynamics 630
Dynamics of Competition 632
Newton's Law of Cooling 633
Applying Euler's Method to Systems of Autonomous Differential Equations 634
Exercises 638

8.6 The Phase Plane 640
Equilibria and Nullclines: Predator–Prey Equations 640
Equilibria and Nullclines: Competition Equations 643
Equilibria and Nullclines: Newton's Law of Cooling 644
Exercises 645

8.7 Solutions in the Phase Plane 648
Euler's Method in the Phase Plane 648
Direction Arrows: Predator–Prey Equations 651
Direction Arrows: Competition Equations 654
Direction Arrows: Newton's Law of Cooling 656
Exercises 657

8.8 Dynamics of a Neuron   Online

Chapter Summary: Key Terms and Concepts 659
Concept Check: True/False Quiz 660
Supplementary Problems 660
Projects 662

Answers to Odd-Numbered Exercises 665
Index I-1

Preface

Calculus for the Life Sciences: Modelling the Dynamics of Life is truly a *calculus* book for *life sciences* students.

This book covers limits, continuity, derivatives, integrals, and differential equations, which are standard topics in introductory calculus courses in universities. All concepts, definitions, and theorems are explained in detail and illustrated in a number of fully solved examples. A variety of approaches—algebraic, geometric, numerical, and verbal—facilitate the understanding of the material and will be of great help to any students who may read the book on their own.

As new mathematics concepts and ideas are introduced, applications illustrating their use are presented. Questions arising from life sciences situations are employed to motivate the construction of several mathematical objects (such as derivatives and integrals). A narrative introduces the context of an application and shows its relevance and importance in life sciences.

The major purpose of this book is to build *quantitative skills* and to introduce its readers to the *insights* that mathematics can provide into all branches of life sciences.

Although the importance of quantitative skills in the life sciences is a widely accepted fact, realities—in many cases—conceal their vital role. In part, this is because mathematics tends to be hidden, working in the background: once something has been figured out and becomes a standard, the mathematics that was used is no longer needed, as practitioners use software, diagrams, and charts instead. For instance, doctors and nurses do not look at half-life information to calculate a dosage every time they need to administer a medication. An ultrasound technician can "calculate" the volume of blood in a patient's heart chamber by selecting a few items from the pull-down menus in a computer program, completely unaware of the fairly sophisticated mathematics (calculation of volume by approximate integration) used to design the software.

As for the insights, a simple mathematical model explains why the heart of a mouse beats about 15 times as fast as the heart of an elephant. With mathematical formulas that describe human growth, we can understand why body mass index does not give reliable estimates of body fat in children and tall people.

Furthermore, mathematics has the power to *reveal otherwise invisible worlds* in the vast universe of numerical data that has been gathered about almost every single phenomenon related to life. Joel E. Cohen writes:[1]

> For example, computed tomography can reveal a cross-section of a human head from the density of X-ray beams without ever opening the head, by using the Radon transform to infer the densities of materials at each location within the head. Charles Darwin was right when he wrote that people with an understanding "of the great leading principles of mathematics ... seem to have an extra sense." Today's biologists increasingly recognize that appropriate mathematics can help interpret any kind of data.

---

[1] J. E. Cohen. Mathematics is biology's next microscope, only better; biology is mathematics' next physics, only better. PLoS Biol 2(12): e439, 2004. Published online December 14, 2004. doi: 10.1371/journal.pbio.0020439. PMCID: PMC535574.

Although a great deal of biology can be done without mathematics, the powerful new technologies that are transforming fields of biology—from genetics to physiology to ecology—are increasingly quantitative, as are many questions at the frontiers of knowledge. Along with genetics, mathematics is one of two unifying factors in the life sciences. And as biology becomes more important in society, mathematical literacy becomes as vital for an informed citizen as it is for a researcher.

Modelling and the Dynamics of Life

The goal of this book is to teach the mathematical ideas that will help us understand various phenomena in life sciences. These are the same ideas that researchers use in their work, as well as in collaborations with colleagues engaged in more empirical activities. They are not specific techniques, such as differentiation or integration by parts, but rather they revolve around building *mathematical models*.

A mathematical model is that crucial link between a life sciences phenomenon and its description in terms of mathematical objects. We will gain the skills needed to construct a model, make sure it works, and understand what it implies—we will learn how to *translate* appropriate aspects of a life sciences problem and its assumptions into formulas, equations, and diagrams; how to *solve* the equations involved; and how to *interpret* the results in terms of the original problem. For instance:

- We build several models in an attempt to predict what the population of Canada will be in the near (and not so near) future. In each case, we critically examine the assumptions that we made, as well as comment on the results of the model.

- Regular clinical breast examinations are key to early detection of breast cancer. Do these exams suffice, or should one also consider mammography? By using an exponential model to understand how cancer cells grow, we conclude that in some cases mammography gives a significant lead time over clinical examination in early detection of cancer.

- To study the interaction between populations, we take a detailed look at the predator-prey model, which could be used, for instance, to describe the interaction between foxes and rabbits in an ecosystem. In addition, a model of selection will help us understand the behaviour of two different bacterial cultures sharing the same space and resources.

- We investigate limited population growth models, such as the logistic model and the Allee effect.

- Using discrete-time dynamical systems, we build our understanding of aspects of the consumption and absorption of drugs such as alcohol and caffeine.

Content

The organization and presentation of the material in this book mirror the relationship between mathematics and the life sciences. Mathematics enables us to describe, explain, and understand biological processes. In turn, the questions life scientists would like to know the answers to stimulate the development of mathematical ideas and techniques.

A quick overview of the content of the book is contained in the following table:

	An Overview of the Book
Chapter 1 **Introduction**	We illustrate how math is used in the life sciences (Section 1.1); introduce the idea of a mathematical model (Section 1.2); review basics about functions, such as graphs and transformations of graphs, composition, and inverse functions; and use the opportunity to discuss elementary models (Sections 1.3 and 1.4). In the final section, we talk about building blocks of math (definitions, theorems), math reasoning patterns (how to think of an implication), and interpreting math results in the context of applications in life sciences.
Chapter 2 **Modelling Using Elementary Functions**	We review properties of linear, power, and transcendental functions, and, at the same time, introduce important models (many of which we revisit as we learn more math). For instance, using power functions, we study heartbeat frequency in mammals and derive an important relation between the size of an animal and the size of its body cover, we predict the changes in Canadian population, we study cell growth using exponential models, and we describe various oscillations and learn how forensic technicians use blood splatters to identify the events that led to their formation.
Chapter 3 **Discrete-Time Dynamical Systems**	We study algebraic and geometric properties of discrete-time dynamical systems (recurrence relations) in Sections 3.1 and 3.2. We discuss a whole spectrum of models, including limited population and consumption of coffee and alcohol (Section 3.3) and a nonlinear population model of selection (Section 3.4). As we learn more math, we revisit these models and enrich our understanding. The chapter closes with a more advanced model on gas exchange in the lung (Section 3.5).
Chapter 4 **Limits, Continuity, and Derivatives**	This chapter contains an in-depth discussion of major calculus concepts: limits (Sections 4.2 and 4.3), continuity (Section 4.4), and derivatives (Sections 4.1 and 4.5). With applications never far from sight, we introduce several absorption functions (Section 4.3). In order to understand how practitioners reason and speak about absorption, we compare these functions in terms of how quickly they approach zero or infinity.
Chapter 5 **Working with Derivatives**	In this fairly technical chapter, we cover all differentiation rules (Sections 5.1 to 5.4), including implicit and logarithmic differentiation and related rates (Section 5.5). Besides traditional questions (found in all calculus texts), the reader will find exciting related rates situations from biology. We learn how to calculate higher-order derivatives and use them to study graphs (Section 5.6) and build various approximations of functions, including Taylor polynomials (Section 5.7).
Chapter 6 **Applications of Derivatives**	This chapter offers a wide range of applications of derivatives: computing absolute and relative extreme values (Section 6.1), identifying leading behaviour and calculating limits using L'Hôpital's rule (Section 6.4), drawing graphs of functions (Section 6.5), and approximating solutions of equations using Newton's method (Section 6.6). To illustrate how optimization works, we study the strength of bones and the feeding patterns of animals and analyze the way bees build their honeycombs (Section 6.2). Section 6.3 makes important connections between continuity and differentiability. The derivatives help us understand the stability of a dynamical system (Section 6.7) as well as gain an insight into complex dynamic behaviour such as chaos (Section 6.8). An advanced study of breathing patterns (Section 6.9) can be downloaded from the textbook's companion website.
Chapter 7 **Integrals and Applications**	We introduce differential equations, an important ingredient in many mathematical models in biology (Section 7.1), as motivation for a need to reverse differentiation. We proceed as usual: antiderivatives (Section 7.2) and area and the definite integral (Section 7.3) meet in the statement of the Fundamental Theorem of Calculus (Section 7.4). We work through standard integration methods (Section 7.5) and discuss improper integrals (Section 7.7). We go through a wide range of applications in Section 7.6 (area, volume, length, mass, etc.), including estimating the surface area of Lake Ontario and the volume of a heart chamber.
Chapter 8 **Differential Equations**	The focus of this chapter is on autonomous differential equations and their applications. We introduce and investigate several important models, such as the logistic model, the Allee effect, the law of cooling, diffusion, and the continuous model of selection (Section 8.1). We develop tools for a qualitative analysis: equilibria and display (Section 8.2), stability (Section 8.3), and then solve some equations using separation of variables (Section 8.4). The remaining sections are devoted to models based on systems of differential equations, such as predator–prey and competition (Section 8.5), and their analysis using phase-plane techniques (Sections 8.6 and 8.7). A more challenging section on the dynamics of a neuron (Section 8.8) can be downloaded from the textbook's companion website.

Presentation

Easy-to-read and easy-to-follow narratives, carefully drawn pictures and diagrams, clear explanations, large numbers of fully solved examples, and broad-spectrum applications make the material suitable for a variety of audiences with a wide range of interests and backgrounds.

In creating this book, we were guided by several important principles:

- Convey excitement about the material.
- Convey the relevance and importance of mathematics and its applications.
- Use a variety of approaches—algebraic, numeric, geometric, and verbal.
- Motivate the introduction of new mathematical objects, concepts, and algorithms.
- Review background material so it does not become an obstacle in explaining advanced material.
- Provide more examples and illustrations for more difficult material.
- Revisit important concepts as often as possible, in a variety of contexts.
- Revisit applications and enrich models as new mathematics ideas are introduced.

Conceptual Understanding and Mathematical Thinking

This text provides the reader with an opportunity to build a clear understanding of a relatively small number of ideas and concepts from the calculus of functions of one variable. It offers exhaustive discussions and carefully crafted examples; clear and crisp statements of theorems and definitions; and—acknowledging that many of us are visual learners—numerous illustrations, graphs, and diagrams. Important concepts are revisited as often as possible, placed in a variety of contexts, and applied in solving life sciences problems.

Development of Mathematical (Quantitative) Skills

It is impossible to fully master almost any topic in mathematics without adequate skills in symbolic (algebraic) manipulation. This book contains a large number of fully solved examples designed to illustrate formulas, algebraic methods, and algorithms, and to provide an ideal opportunity for students to improve on their routine in the technical intricacies of calculations.

Topics that some students may find challenging (such as Riemann sums and integration methods, working with Taylor polynomials, and graphing using leading behaviour) are accompanied by a large number of solved examples. Use of technology (graphing calculator or mathematical software such as Maple) is strongly encouraged, since it might provide insights that will enhance understanding of the material.

In-Depth Explorations of Particular Models

This book includes several extended applications. Early on, we study the phenomenon of the interaction between two types of bacteria, mutant and wild, in Section 3.4, and certain aspects of the dynamics of gas exchange in the lungs, in Section 3.5. Applications of optimization to feeding patterns of animals, calculations of the strength of bones, and a study of the ways bees build their honeycombs are found in Chapter 6.

Phase-plane methods are used in Chapter 8 to analyze two-dimensional systems representing interactions between two species. Some of these models can be assigned as individual or group study projects.

End-of-Section and End-of-Chapter Problems, Computer Exercises, and Projects

In addition to routine skills problems, each section includes a wide variety of modelling problems to emphasize consistently the importance of *interpretation*. As well, more challenging mathematics questions will be found here, including requests to construct proofs that are omitted from the text. Each chapter includes supplementary problems that introduce a variety of new applications and can be used for review or practice sessions.

The book contains more than 50 exercises designed to be explored on a graphing calculator (preferably programmable) or on a computer using software such as Maple. These exercises emphasize *visualization, experimentation,* and *simulation*—all of which are important aspects of conducting research in life sciences, as well as many other areas.

Most chapters include projects suitable for individual or group exploration. Examples include the following:

- modelling the balance between selection and mutation (Chapter 3)
- studying periodic hematopoiesis using a discrete-time dynamical system (Chapter 4)
- experimenting with different numerical schemes for solving differential equations (Chapter 7)
- carefully studying models of adaptation by cells (Chapter 8)

Teaching

Although this is a book for *life sciences,* it is a *calculus textbook* as well. It can be used to teach a variety of courses, spanning a wide range of flavours and levels of difficulty. The coverage of math theory, techniques, and algorithms, as well as applications, allows an instructor to tune the course to an appropriate balance of math rigour and applications in life sciences.

Several suggestions for courses are outlined in the chart on page xiv.

Great care has been taken to make the level of exposition—both mathematical and applications—adequate for first-year students. A course instructor can assign, with confidence, parts of the material as homework or as optional reading.

No matter which course is taught from this text, the benefits are obvious. *The modelling approach is naturally a problem-solving approach.* Students will not remember every technique they have learned. This book emphasizes understanding what a model is and recognizing what a model says. To be able to recognize a differential equation, interpret the terms, and use the solution is far more important than knowing how to find a solution algebraically. These reasoning skills, in addition to familiarity with the models in general, are what will stay with the motivated student and what will matter most in the end.

Suggestions for Courses

One-semester, first-year calculus for life sciences	Start with Chapter 1 (introduce the subject, justify using math to model life sciences phenomena, warm-up); Chapter 2 (review linear, power, and transcendental functions, discuss numerous models); Chapter 4; Chapter 5 (might not require much time); Chapter 6 (pick among the topics and applications presented); Chapter 7 (selection of topics and applications); if time permits, discuss Section 8.1. Discrete-time dynamical systems (in this context, recurrence relations) in Chapter 3 can be skipped, or covered to a desired depth. For instance: extract basics from Sections 3.1 and 3.2 to cover models in Section 3.3 (such as limited population or alcohol consumption), or cover Sections 3.1 and 3.2 in detail, discuss many interesting models in Sections 3.3 and 3.4, and then discuss stability and further models in Sections 6.7 and 6.8.
One-semester, first-year calculus for life sciences (advanced)	Taking advantage of students' background, discuss features of models in Chapters 1, 2, and 3, perhaps including Section 3.5; Chapters 4 and 5 might not require much time; possibly include case studies in Section 6.2, or discussion of stability of equilibria and dynamics of chaos in Sections 6.7 and 6.8; cover Chapter 7 and the first four sections of Chapter 8.
Two-semester, first-year calculus for life sciences	First semester as above. Use the opportunity to finish the material in Chapters 7 and 8 to a desired depth. The remaining time could be spent discussing topics from one or more of the Modules: Several Variables, Probability and Statistics, and Linear Algebra, which are available as separate books (see page xv for a Table of Contents of each module).
One-semester calculus course of a more theoretical nature	Start with Chapters 1 and 2 (to review properties of elementary functions; some applications can be used to discuss properties of functions that involve parameters); Chapter 4 and Chapter 5, including some proofs; Chapter 6, at least Sections 6.1 and 6.3; Chapter 7 (with details regarding Riemann sums); if time permits, Sections 8.1 to 8.3.
Second-year modelling course for students who took traditional calculus	Focus on models in Chapters 1 and 2 (also a good review of functions); Chapter 3, perhaps all of it; pick themes from Chapters 4 and 5 (for instance, absorption functions in Section 4.3 or related rates with life sciences context in Section 5.5); cover Section 5.7; selection of material from Chapter 6, such as Sections 6.2, 6.7, and 6.8; focus on applications in Chapter 7 (Section 7.6); Chapter 8, including systems of autonomous differential equations. Sections 3.5, 6.9, and 8.8 contain some challenging material but are appropriate for second-year students.
One-semester introductory course on dynamics of growth, or population modelling	Start with a selection of topics in Chapter 2 (exponential model, allometry); then Chapter 3, in particular Sections 3.3 and 3.4, with application exercises and computer exercises in Section 3.4 introducing additional models, including chaotic behaviour; necessary topics in Chapters 4 and 5, using models as examples (logistic, gamma distributions, Hill functions, von Bertalanffy limited growth, etc.); Sections 6.7, 6.8 (*exercises* in these two sections and *projects* at the end of the chapter provide many opportunities for extensions [chaos, Ricker model, etc.]); Chapter 7; major focus on autonomous differential equations, systems, and phase planes in Chapter 8.
Project-driven, problem-based course (learning math through applications)	Consider questions such as the following: Why can't we use radiocarbon dating to determine the time when dinosaurs died (Section 2.2)? What happens to the amount of alcohol in our body if we consume one drink per hour (Section 3.3)? How do bones grow and what makes them strong (Sections 2.1 and 6.2)? Why do bees' honeycombs have hexagonal cross-sections (Section 6.2)? How do we approximate the area of a lake, or the volume of a heart chamber (Section 7.6)? How do we model interactions of two species (Sections 8.5, 8.6, 8.7)? This is just a sample of starting points suggested in this book—students realize that they need to learn more math if they wish to understand these life sciences problems and find good answers.

Life Sciences Modules to Accompany This Text

This book covers the calculus of functions of one variable. Quite often, longer (two-semester) life sciences courses include topics from other disciplines. To allow for flexibility in planning courses and resources, additional material has been organized in three separate modules. Intertwining math foundations and applications, the modules cover functions of several variables, basics of probability and statistics, and elementary linear algebra. These modules can be bundled with the text, or alternatively, a custom package can be compiled, including material from the textbook and the modules. Contact your sales representative for more information.

Functions of Several Variables (144 pages), ISBN: 978-0-17-657136-8

1. Introduction
2. Graph of a Function of Several Variables
3. Limits and Continuity
4. Partial Derivatives
5. Tangent Plane, Linearization, and Differentiability
6. The Chain Rule
7. Second-Order Partial Derivatives and Applications
8. Partial Differential Equations
9. Directional Derivative and Gradient
10. Extreme Values
11. Optimization with Constraints

Probability and Statistics (195 pages), ISBN: 978-0-17-657135-1

1. Introduction: Why Probability and Statistics
2. Stochastic Models
3. Basics of Probability Theory
4. Conditional Probability and the Law of Total Probability
5. Independence
6. Discrete Random Variables
7. The Mean, the Median, and the Mode
8. The Spread of a Distribution
9. Joint Distributions
10. The Binomial Distribution
11. The Multinomial and the Geometric Distributions
12. The Poisson Distribution
13. Continuous Random Variables
14. The Normal Distribution
15. The Uniform and the Exponential Distributions

Linear Algebra (138 pages), ISBN: 978-0-17-657137-5

1. Identifying Location in a Plane and in Space
2. Vectors
3. The Dot Product
4. Equations of Lines and Planes
5. Systems of Linear Equations
6. Gaussian Elimination
7. Linear Systems in Medical Imaging
8. Matrices
9. Matrices and Linear Systems
10. Linear Transformations
11. Eigenvalues and Eigenvectors
12. The Leslie Model: Age-Structured Population Dynamics

About the Nelson Education Teaching Advantage (NETA)

The **Nelson Education Teaching Advantage (NETA)** program delivers research-based instructor resources that promote student engagement and higher-order thinking to enable the success of Canadian students and educators. To ensure the high quality of these materials, all Nelson ancillaries have been professionally copy-edited.

Be sure to visit Nelson Education's **Inspired Instruction** website at http://www.nelson.com/inspired/ to find out more about NETA. Don't miss the testimonials of instructors who have used NETA supplements and seen student engagement increase!

Assessing Your Students: *NETA Assessment* relates to testing materials. **NETA Test Bank** authors create multiple-choice questions that reflect research-based best practices for constructing effective questions and testing not just recall but also higher-order thinking. Our guidelines were developed by David DiBattista, psychology professor at Brock University and 3M National Teaching Fellow, whose research has focused on multiple-choice testing. All Test Bank authors receive training, as do the copy-editors assigned to each Test Bank. A copy of *Multiple Choice Tests: Getting Beyond Remembering*, Prof. DiBattista's guide to writing effective tests, is included with every Nelson Test Bank.

Technology in Teaching: *NETA Digital* is a framework based on Arthur Chickering and Zelda Gamson's seminal work "Seven Principles of Good Practice In Undergraduate Education" (AAHE Bulletin, 1987) and the follow-up work by Chickering and Stephen C. Ehrmann, "Implementing the Seven Principles: Technology as Lever" (AAHE Bulletin, 1996). This aspect of the NETA program guides the writing and development of our **digital products** to ensure that they appropriately reflect the core goals of contact, collaboration, multimodal learning, time on task, prompt feedback, active learning, and high expectations. The resulting focus on pedagogical utility, rather than technological wizardry, ensures that all of our technology supports better outcomes for students.

Instructor Resources

All NETA and other key instructor ancillaries are provided on the Instructor's Resources Companion Site at www.nelson.com/site/calculusforlifesciences, giving instructors the ultimate tool for customizing lectures and presentations.

NETA Test Bank: This resource was written by Andrijana Burazin and reviewed by Miroslav Lovrić, McMaster University. It includes more than 275 multiple-choice questions written according to NETA guidelines for effective construction and development of higher-order questions. The Test Bank was copy-edited by a NETA-trained editor and underwent a full technical check. Also included are almost 100 true/false questions.

The NETA Test Bank is available in a new, cloud-based platform. **Testing Powered by Cognero®** is a secure online testing system that allows you to author, edit, and manage test bank content from any place you have Internet access. No special installations or downloads are needed, and the desktop-inspired interface, with its drop-down menus and familiar, intuitive tools, allows you to create and manage tests with ease. You can create multiple test versions in an instant, and import or export content into other systems. Tests can be delivered from your learning management system, your classroom, or wherever you want.

Instructor's Solutions Manual: This manual contains complete worked solutions to exercises in the text. Prepared by text author Miroslav Lovrić, McMaster University, it has been independently checked for accuracy by Caroline Purdy, University of New Brunswick.

Image Library: This resource consists of digital copies of figures, short tables, and graphs used in the book. Instructors may use these jpegs to create their own PowerPoint presentations.

DayOne: Day One—Prof InClass is a PowerPoint presentation that instructors can customize to orient students to the class and their text at the beginning of the course.

Enhanced WebAssign Cengage Learning's **Enhanced WebAssign**™, the leading homework system for math and science, has been used by more than 2.2 million students. Created by instructors for instructors, **EWA** is easy to use and works with all major operating systems and browsers. **EWA** adds interactive features to go far beyond simply duplicating text problems online. Students can watch solution videos, see problems solved step by step, and receive feedback as they complete their homework.

Enhanced WebAssign™ allows instructors to easily assign, collect, grade, and record homework assignments via the Internet. This proven and reliable homework system uses pedagogy and content from Nelson Education's best-selling Calculus textbooks, then enhances it to help students visualize the problem-solving process and reinforce concepts more effectively. **EWA** encourages active learning and time on task and respects diverse ways of learning.

Visit www.cengage.com/ewa for more information and to access **EWA** today!

Student Ancillaries

Student Solutions Manual

The Student Solutions Manual contains detailed worked solutions to the odd-numbered exercises in the book. It was prepared by Miroslav Lovrić, McMaster University, and technically checked by Caroline Purdy, University of New Brunswick.

Enhanced WebAssign

Cengage Learning's **Enhanced WebAssign**™ is a groundbreaking homework management system that combines our exceptional Calculus content with the most flexible online homework solution. **EWA** will engage you with immediate feedback, rich tutorial content, and interactive features that go far beyond simply duplicating text problems online. Visit www.cengage.com/ewa for more information and to access **EWA** today!

Companion Website

Access additional material, including online-only advanced material not included in the text, in the **Companion Website** for *Calculus for the Life Sciences*. Visit www.nelson.com/site/calculusforlifesciences.com.

Acknowledgments

Thanks to Paul Fam for getting this project started. To the team at Nelson—Jackie Wood, Sean Chamberland, Leanne Newell, Paulina Kedzior, and Lindsay Bradac—thank you for your understanding and guidance and for helping me think through many details of the project. To Suzanne Simpson Millar and Katherine Goodes, thank you for your many words of advice and encouragement, for your generosity, and for your patience in dealing with me. To the production team—Julia Cochrane, Claire Horsnell, and Indumathy Gunasekaran—a big thank-you for making it all happen. Many thanks to others who have contributed to the project in various ways and whose names or involvement are not included here.

I wish to acknowledge the reviewers who offered encouragement and guidance throughout the development of this book:

Alan Ableson, *Queen's University*
Troy Day, *Queen's University*
Shay Fuchs, *University of Toronto*
Thomas Hillen, *University of Alberta*
David Iron, *Dalhousie University*
Robert Israel, *University of British Columbia*
Merzik Kamel, *University of New Brunswick*
Petra Menz, *Simon Fraser University*

Wes Maciejewski, *University of British Columbia*
Caroline Purdy, *University of New Brunswick*
Norman Purzitsky, *York University*
Joe Repka, *University of Toronto*
Linda Wahl, *University of Western Ontario*

    I would especially like to send a big thank-you to my family and friends for their support and for not asking too many questions ("Dude, when will you finally finish this?!").

Miroslav Lovrić
Hamilton, 2014

Chapter 1

Introduction to Models and Functions

This chapter opens by answering an obvious and easy question—why do we need mathematics in the life sciences? We list some (of many!) questions from biology, health sciences, and elsewhere that are answered—using mathematics—in this book. As well, we mention a few present-day research problems that have very little chance of being solved or fully understood without mathematics.

We introduce the main tools needed to study biology using mathematics: **models** and **functions.** A **model** is a collection of mathematical objects (such as functions and equations) that allows us to interpret biological problems in the language of mathematics. Biological phenomena are often described by measurements: a set of numeric values with units (such as kilograms or metres). Many relations between measurements are described by **functions,** which assign to each input value a unique output value.

After talking about what constitutes a mathematical model and presenting a few examples, we briefly **review functions** and their properties. In this chapter, we discuss the **domain and range** and the **graph of a function, algebraic operations with functions, composition of functions,** and **inverse functions.** We build new functions from old using shifting, scaling, and reflections. We catalogue important elementary functions and note their properties. The reader who is familiar with functions might skip this material and move to the next chapter.

We introduce the four approaches—algebraic, numeric, geometric, and verbal—that we use throughout the book to discuss functions, their properties, and their applications.

In the last section we discuss aspects of logical reasoning that we need to follow and to understand mathematical expositions. As well, we contrast the language used in mathematics and in the life sciences, in order to motivate learning an important skill—communicating scientific facts, ideas, and results across disciplines.

1.1 Why Mathematics Matters

FIGURE 1.1.1
Complete skeleton of a triceratops, Royal Tyrrell Museum
Photo courtesy of the Royal Tyrrell Museum, Drumheller, Alberta

In the summer of 2012, a team of palaeontologists from the Royal Tyrrell Museum ("Canada's dinosaur museum") in Drumheller, Alberta, unearthed the skeleton of a large triceratops, a herbivorous dinosaur that lived in what is now North America some time between 68 and 65 million years ago. An adult triceratops measured 8–9 m in length and about 3 m in height, and weighed between 6,000 and 12,000 kg (Figure 1.1.1).

No human has ever seen a living dinosaur, so how do we know all this?

Estimates of age, size, weight, and many other quantities are obtained *using mathematics*, based on the data collected from the bones and from the site where they were found. Among other techniques, potassium-argon dating (see the note following Example 2.2.13) is used to compute the time interval when triceratops lived on Earth. By counting the growth lines in the MRI scan of certain bones (unlike an X-ray, an MRI image is *calculated*), researchers can determine how old the dinosaur was when it died. Then, using the formula (adapted from G. M. Erickson, K. C. Rogers, and S. A. Yerby, Dinosaurian growth patterns and rapid avian growth rates. *Nature*, 412 (429–433), 2001)

$$M = \frac{12{,}000}{1 + 2.9e^{-0.87(t-7.24)}} \tag{1.1.1}$$

we can find an approximation of the body mass, M (in kilograms), of the triceratops based on the age at death, t (in years).

Allometry—a branch of life sciences—is the study of numeric relationships between quantities associated with human or animal organisms. For instance, the allometric formula (adapted from M. Benton, and D. Harper, *Basic Palaeontology*, Harlow, U.K.: Addison Wesley Longman, 1997)

$$\text{Sk} = 0.49 \text{Sp}^{0.84} \tag{1.1.2}$$

relates the skull length, Sk, of a larger dinosaur to its spine length, Sp (both measured in metres). From the triceratops' vertebrae found at the Drumheller site, researchers could figure out its spine length and then use formula (1.1.2) to calculate the size of its skull. (Further examples of allometric relationships can be found in Examples 2.1.12 to 2.1.16 and Example 5.5.10.)

This example, and many more that we will encounter in this book, echo this important message:

> Mathematics is an indispensable tool for studying life sciences. It deepens our understanding of life science phenomena and helps us to figure out the answers to questions that would otherwise be hard (or impossible) to find.

To further emphasize this message, we give a sample of questions that we will discuss and answer—using mathematics—in this book. Needless to say, we view mathematics in its broadest sense, i.e., including probability, statistics, numeric techniques and simulations, and computer programming.

- My body mass index is 27 (above normal range). How much weight should I lose to lower it to 24 (healthy weight)? My body mass index is 16 (below normal range). How much weight should I gain to bring it to 18 (healthy weight)? (See Example 2.1.11.)

- According to the Statistics Canada 2011 Census, about 33.5 million people lived in Canada in May 2011 (exactly 33,476,688 people were enumerated in the census). How many people will live in Canada in 2021? (See Examples 2.1.10 and 2.2.14.)

- If a student consumes one alcoholic drink (12 oz of beer, or 5 oz of white wine, or 1.5 oz of tequila or vodka) every hour, how much alcohol will be in that student's body after five hours? How long will it take the student to sober up? (See Section 3.3, in particular Examples 3.3.6–3.3.8.)

- How do forensic pathologists identify the location of impact (say, from a bullet) by analyzing blood splatters on the floor? (Read Example 2.3.15.)

- Which part of a skeleton grows faster: the skull or the spine? (See Example 5.5.10.)

- What is the surface area of Lake Ontario? (This information is needed, for instance, when scientists study the impact of pollutants on lake fauna; see Example 7.6.6.)

- Scientists believe that the fossils found in the Burgess Shale Formation in British Columbia are about 505 million years old. The Joggins Fossil Cliffs in Nova Scotia contain fossils from the so-called coal age of Earth's history, about 310 million years ago. How were these estimates obtained? A tree trunk (Figure 1.1.2) was unearthed near the city of Kaitaia in New Zealand. How long ago did the tree die? (See the note following Example 2.2.13.)

FIGURE 1.1.2

Ancient kauri tree trunk (unearthed and placed upright for display)
Miroslav Lovric

- An MRI (magnetic resonance imaging) scan shows that a smaller blood vessel branches off the right coronary artery at an angle of 50 degrees. Is there a reason for medical concern? (Read Example 6.1.18.)

- Long bones in mammals (such as the femur) are hollow, filled with blood cell–producing marrow. Although lightweight, they are strong enough to support the entire body, enabling it to move in various ways. However, under continuous stress (often identified in athletes) or due to an acute event (such as a fall or a car crash), a femur can break. Can we somehow grow a stronger femur by making its walls thicker? (See Section 6.2.)

- How much valuable time is saved if a breast cancer is detected in a mammogram compared to a clinical breast examination detection? (See Example 2.2.15.)

Formulas (1.1.1) and (1.1.2), which describe life science quantities and phenomena using mathematical objects (in this case formulas involving exponential and power functions) are said to constitute a **mathematical model** (or just a **model,** as is common in practice). For instance, formula (1.1.1) **models** the relationship between age at death and body mass for larger dinosaurs. Formula (1.1.2) is a **model** for a relationship between skull length and spine length for large dinosaurs. We will say more about models in the next section.

In this book we study numerous ways of describing how populations (of cells, bacteria, animals, or humans) change. The simplest model, which uses elementary mathematics, assumes that the birth and the death rates are constant. If the birth rate is larger than the death rate, the model implies that the population will grow exponentially. Of course, beyond a certain point, this is neither realistic nor possible for any population.

To make this model better mimic reality, we introduce various modifications: for instance, we can make the birth and the death rates change over time, we can include the carrying capacity (carrying capacity is the largest number of individuals that can live in an ecosystem), or we might need to include a term that accounts for the minimum number of individuals needed for the population to avoid extinction. Having made some or all of these modifications to our model, we realize that *we need to know more mathematics* in order to work with it.

This is not all—we might need to add terms that account for harvesting and seasonal changes in the population size. As well, it might be necessary to include the effects of a disease, a natural disaster, or another random event that might affect the population. For all this, we need to know *even more mathematics*. Hence another important message:

> As the model—the description of a life sciences phenomenon using mathematics—moves closer to reality, it also becomes more complex, and more mathematics is needed to work with it.

In other words, as we learn more math, we are able to probe deeper into a problem, understand it better, gain new insights, and obtain more meaningful results and answers. To further stimulate interest in studying life sciences and mathematics together, we list several problems that are the topics of present research:

- What are the risks to the indigenous fish populations in the Great Lakes from the new species of fish brought in the tanks of large cargo ships?

- What are the distinct features of the trafficking of eosinophils as they migrate from bone marrow to the blood and, ultimately, to the lungs? How can this enhance our understanding of certain aspects of the development of allergic asthma? (Eosinophils are white blood cells, important components of our immune system, defending it against parasites and certain infections. Allergic asthma is a disease of the airways that develops as a consequence of an immune-inflammatory response to allergen exposure (such as dust, pollen, or various drugs), causing inflammation in the lungs.)

- According to the Director of Biodiversity Programs at the Royal Botanical Gardens (RBG) in Hamilton, Ontario, there is a need to cut "an apparent overpopulation of deer wintering on its lands." An aerial survey of the part of the RBG lands in 2010 identified 267 deer, which is deemed (by the Ontario Ministry of Natural Resources) to be six to nine times the desired number of deer in the area. Is the ministry right? What is the carrying capacity of the RBG lands, i.e., how many deer should be removed (culled or relocated) from there?

- Due to long exposures to zero or near-zero gravity, astronauts working in the International Space Station suffer from spaceflight osteopenia (bone loss; on average, they lose about 1% of their bone mass per month spent in space). On the other hand, sea urchins are known to continue the production of calcium, unaffected by the lack of gravity. By modelling the growth and development of calcium plates in the skeletons of the urchins, researchers are trying to shed more light on the dynamics of calcium recycling, hoping to reduce the effects of osteopenia in humans.

- In order to better understand the pathophysiology of hydrocephalus (potentially brain-damaging buildup of fluid in the skull), researchers are studying the interaction between the cerebrospinal fluid and the brain tissue. Present efforts are focused on using partial differential equations (we will study differential equations in this book) to gain new insights into this interaction.

Dear Student:

This book gives you an opportunity to learn mathematics and to see how it is used in the wide spectrum of applications in the life sciences. To see applications in action (and to get more from them!) you need to understand the underlying math concepts, formulas, and algorithms. This is why you will find a large number of fully solved examples, as well as exercises, ranging from easy and routine to more complex, theoretical, and challenging. Work on as many of them as you can.

Learning mathematics is not easy. Like everything you really care for, it requires seriousness, dedication, a significant amount of time, and lots of hard work. But in the end, it will be worth it!

> No subject teaches logical thinking, develops analytic and problem-solving skills, and demonstrates how to deal with complex problems better or more effectively than math.

Have you ever wondered why math majors score at or near the top in all standardized tests, including MCAT, GMAT, and LSAT?

1.2 Models in Life Sciences

Living systems, from cells to organisms to ecosystems, are characterized by change and dynamics. Living things grow, maintain themselves, and reproduce. Even remaining the same requires dynamical responses to a changing environment. Understanding the mechanisms behind these dynamics and deducing their consequences is crucial to understanding biology, biochemistry, ecology, epidemiology, physiology, population genetics, and many other life sciences.

This dynamical approach is necessarily mathematical because describing dynamics requires quantifying measurements. What is changing? How quickly is it changing? What is it changing into?

In this book, we use the language of mathematics to describe quantitatively how living systems work and to develop the mathematical tools needed to compute how they change. From measurements describing the initial state of a system and a set of rules describing how change occurs, we will attempt to predict what will happen to the system.

For example, by knowing the initial amount of a drug taken (caffeine, Tylenol, alcohol, etc.) and how it is processed by the liver or kidneys (dynamical rules), we can predict how long the drug will stay in the body and what effects it might have. By knowing how many elephants live in a certain area and quantifying the factors that influence how they reproduce, we can predict what will happen to their population in the future.

To study a life sciences phenomenon using mathematics, we build a **model.** How does a model work?

First, we identify a **problem** we need to study, or a **question** we need to answer. Assume that a virus (say, H1N1) appears within a population. Will the virus spread? How many people will get infected? How many will die? Will our hospitals have adequate resources to treat increasing numbers of patients? These are just a handful of the questions that we would like to know the answers to. Underlying all of them is the basic question: we know (approximately) how many people are infected today. How many will be infected tomorrow? in three days? in a week? in a month?

Next, we need to **quantify the measurements** that we will use to investigate the problem. If we wish to use mathematics, we need numerical data. In our case, the basic measurement is the number of people infected with the H1N1 virus at any given time. Given the tools we have at our disposal, it is very useful to think of the number of people infected as a **function** (dependent variable). We might wish to use a symbol such as $I(t)$ to denote that function and to say that we will investigate how it changes with respect to time. As well, we have some data—we might know (or be able to estimate) how many people are infected at the moment. This is called an **initial condition;** we will denote it by $I(0)$, relabelling the time so that time = zero corresponds to the start of our investigation.

We also need to define the **rules that govern the dynamics** of the model. How does $I(t)$ change, and what affects it? What makes it grow, and what makes it decrease? Because we are studying change, we need a mathematical tool that is designed to do that—the derivative! That is why calculus is so relevant. If we wish to study change, we must use calculus.

Using functions and derivatives, we translate data, observations, measurements, and dynamic rules into **formulas, equations,** and whichever other mathematical tools we think will be useful. Two tools to which we devote a large part of this book, **discrete-time dynamical systems** and **differential equations,** are essential components of many mathematical models.

By doing all this we have just built a **model for the spread of a virus.** Other models are created in more or less the same way (Figure 1.2.3).

To obtain **solution(s),** we use a variety of mathematical techniques (differentiation, integration, numerical algorithms, solving differential equations, limits of sequences, etc.). In rare cases we obtain exact solutions; usually we can obtain an approximation from an algorithm, a calculator, or a computer.

Finally, we **interpret the mathematical results.** Do they make sense? Do they answer our question(s)? Do they help us better understand what is going on? Quite often, the answer to these questions is not an enthusiastic yes, but we can usually say yes to the following questions—can we get more out of this? can we do better? can we make it more accurate? Thus, mathematical modelling is itself a dynamical process—in light of new data that we might have obtained, or new developments or new insights, we refine or modify our model and run it again, in hopes of obtaining better answers and gaining a better understanding of the problem.

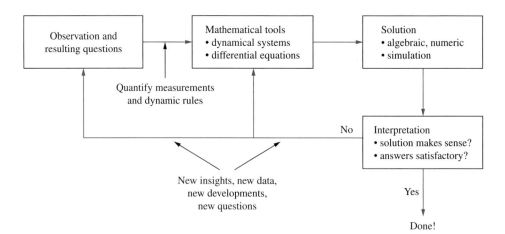

FIGURE 1.2.3
Mathematical model

Example 1.2.1 Models for the Population of Canada

In Example 2.1.10 we ask the following question about the population of Canada: based on the 2001 and 2006 Census data, what will the population be in 2011? (We picked the year 2011 so that we can compare our results with the actual 2011 Census data.)

We build a linear model (i.e., using a linear function) and solve it, obtaining the estimate of 33,219,000 for the year 2011. Not finding the answer satisfactory (following the "No" direction in the flow chart in Figure 1.2.3), we decide to try another model.

In Example 2.2.14 we use exponential growth as a model for the Canadian population, arriving at the estimate of 33,305,000. Comparing to the actual population in 2011 (33,477,000), we see that the exponential model is a bit more accurate than the linear model. However, since both models give underestimates of 170 thousand or more, we find neither truly satisfactory.

So it is back to the drawing board—but we need to learn more mathematics. To build richer, more appropriate models, we need to learn about derivatives, integrals, and differential equations.

Example 1.2.2 Models of Malaria

Early in this century, Sir Ronald Ross discovered that malaria is transmitted by certain types of mosquitoes. Because the disease was (and remains) difficult to treat, one promising strategy for control seemed to be reducing the number of mosquitoes. Many people thought that all the mosquitoes would have to be killed to eradicate the disease. Because killing every single mosquito was impossible, it was feared that malaria might be impossible to control in this way.

Ross decided to use mathematics to convince people that mosquito control could be effective. The problem can be formulated dynamically as a problem in population growth. Ross knew that an uninfected person can become infected upon being bitten by an infected mosquito and that an uninfected mosquito can be infected when it bites an infected person (Figure 1.2.4). From these assumptions, he built a **mathematical model** describing the population dynamics of malaria. With this model he proved that the disease *could* be eradicated without killing every single mosquito. We see evidence of this today in the United States, where malaria has been virtually eliminated even though the mosquitoes capable of transmitting the disease persist in many regions.

Example 1.2.3 Models of Neurons

Neurons are cells that transmit information throughout the brain and body. Even the simplest neuron faces a challenging task. It must be able to amplify an appropriate incoming stimulus, transmit it to neighbouring neurons, and then turn off and be ready

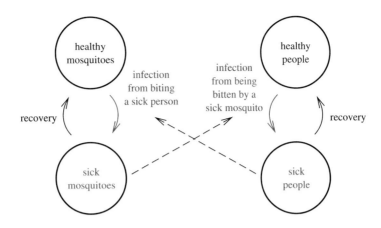

FIGURE 1.2.4
The dynamics of malaria

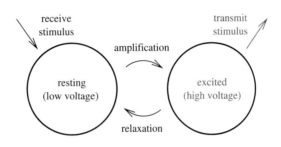

FIGURE 1.2.5
Mathematics description of the dynamics of a neuron

for the next stimulus. This task is not as simple as it might seem. If we imagine the stimulus to be an input of electrical charge, a plausible rule is "If electrical charge is raised above a certain level, increase it further." Such a rule works well for the first stimulus but provides no way for the cell to turn itself off. How does a neuron maintain functionality?

In the early 1950s, Hodgkin and Huxley used their own measurements of neurons to develop a mathematical model of dynamics to explain the behaviour of neurons. The idea is that the neuron has fast and slow mechanisms to open and close specialized ion channels in response to electrical charge (Figure 1.2.5). Hodgkin and Huxley measured the dynamical behaviour of these channels and showed mathematically that their mechanism explained many aspects of the functioning of neurons. They received the Nobel Prize in Physiology or Medicine for this work in 1963 and, perhaps even more impressive, developed a model that is still used today to study neurons and other types of cells.

Types of Dynamical Systems

Biological phenomena can be studied using three types of dynamical systems: discrete-time, continuous-time, and probabilistic systems. The first two types, which are covered in this book, are **deterministic,** meaning that the dynamics include no chance factors. In this case, the values of the basic measurements can be predicted exactly at all future times. Probabilistic dynamical systems include chance factors, and values can be predicted only on average. (We do not study probabilistic dynamical systems in this book.)

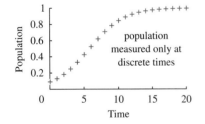

FIGURE 1.2.6
Measurements described by a discrete-time dynamical system

Discrete-Time Dynamical Systems

Discrete-time dynamical systems describe a sequence of measurements made at equally spaced intervals (Figure 1.2.6). These dynamical systems are described mathematically by a rule that gives the value at one time as a function of the value at a previous time. For example, a discrete-time dynamical system describing population growth is a rule that gives the population in one year as a function of the population in the previous year. A discrete-time dynamical system describing the concentration of oxygen in the lung is a rule that gives the concentration of oxygen in a lung after one breath as a function of the concentration after the previous breath. Mathematical analysis of the rule can provide scientific predictions, such as the maximum population size or the average concentration of oxygen in the lung. The study of these systems requires **differential calculus** (Chapters 4, 5, and 6).

Continuous-Time Dynamical Systems

Continuous-time dynamical systems, usually known as **differential equations,** describe measurements that change continuously (Figure 1.2.7). A differential equation consists of a rule that gives the **instantaneous rate of change** of a set of measurements. The beauty of differential equations is that information about a system at one time is sufficient to predict the state of a system at all future times. For example, a continuous-time dynamical system describing the growth of a population is a rule that gives the rate of change of population size as a function of the population size itself. The study of these systems requires the mathematical methods of **integral calculus** (Chapters 7 and 8).

Quantifying Measurements

If we wish to use mathematics, we need numbers—we have to quantify all data, measurements, and relations between measurements that we plan to use.

In this book we work with the data provided and are not concerned about how they were collected, nor do we worry about their accuracy or validity.

The most important thing about numbers is that we can compare them and draw important conclusions from these comparisons. There are many possible relationships between quantities; in Table 1.2.1 we list those most commonly used.

To describe the range of (real-number) values, we use inequalities or interval notation; see Table 1.2.2 (it is assumed that $a < b$).

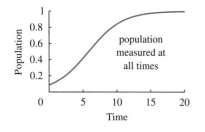

FIGURE 1.2.7
Measurements described by a continuous-time dynamical system

Table 1.2.1

Symbol	Relationship between Quantities, Meaning, and Example
$A = B$	Quantities A and B are equal, for example, $1/2 = 0.5$. In mathematics and to mathematicians, this means that the quantities are absolutely identical. In most modelling examples, however, rarely (if ever) are two phenomena equal or identical. Many times we use the equals sign when we actually mean to say "approximately equal."
$A \approx B$	A is approximately equal to B, as in $\pi \approx 3.14$. The exact meaning of "approximately equal" depends on the context (even within a context, its meaning might not be fixed). Due to a measurement error, every instrument we use returns an approximate (and not true) reading of the quantity measured.
$A \neq B$	A and B are neither equal nor approximately equal.
$A > B$	A is greater than B, or B is smaller than A.
$A \gg B$	A is much greater than B, or B is much smaller than A. The effects of B can be ignored; for instance, if $A \gg B$ then $A \pm B \approx A$ and $B/A \approx 0$.
$A \propto B$	A is proportional to B; if B triples, so does A (we will discuss examples in forthcoming sections).
$A \propto 1/B$	A is inversely proportional to B; if B doubles, then A halves (we will discuss examples in forthcoming sections).
order of magnitude (powers of 10)	One power of 10 is one order of magnitude. If A is about 10 times as large as B, then A is larger by one order of magnitude. A is three orders of magnitude larger than B if A is about a thousand times as large as B. The mass of a dog is two orders of magnitude less than the mass of an elephant.

Table 1.2.2

Interval	Inequality or Symbols	Picture	Represents All Real Numbers between a and b
(a, b)	$a < x < b$		excluding both a and b
$(a, b]$	$a < x \leq b$		excluding a and including b
$[a, b)$	$a \leq x < b$		including a and excluding b
$[a, b]$	$a \leq x \leq b$		including both a and b

			Represents All Real Numbers
(a, ∞)	$x > a$		greater than a
$[a, \infty)$	$x \geq a$		greater than or equal to a
$(-\infty, b)$	$x < b$		less than b
$(-\infty, b]$	$x \leq b$		less than or equal to b

			Represents
$(-\infty, \infty)$	$\mathbb{R}$		all real numbers

The intervals (a, b), (a, ∞), and $(-\infty, b)$ are called **open intervals,** whereas $[a, b]$, $[a, \infty)$, and $(-\infty, b]$ are **closed intervals.**

In many cases, we need to **estimate** the numbers that we need to use in our model. For instance, it is difficult (in many cases impossible) to know exactly how many bacteria are in a Petri dish, or exactly how many people are infected with a virus, or exactly how much sodium enters a cell in an hour through the process of diffusion.

Example 1.2.4 Estimating the Number of Bacteria in a Petri Dish

Bacteria might be too small to count or, for other reasons, counting is dismissed (as impractical or time-consuming or impossible). Here is one possible approach. We can estimate the area that a colony of bacteria occupies in a Petri dish. Then, knowing how thick the colony is, i.e., how many layers of bacteria there are, we can calculate the volume occupied by the colony (volume = area times thickness).

The key measurement is density, in this case defined as

$$\text{density} = \frac{\text{number of bacteria}}{\text{volume}}$$

If we know the density, then the formula

$$\text{total number of bacteria} = \text{density} \times \text{volume}$$

approximates the total number of bacteria in the colony. Note that none of the numbers involved in this calculation could possibly be exact.

Example 1.2.5 Counting Elephants

How many elephants are there in the Southern African highlands, a large ecosystem that spans several countries? It is impossible to track down every elephant and come up with an exact number.

What we can do is pick a small representative subregion of the whole region, one that—in all significant ways—mimics the whole region. Using a variety of means (air surveillance, acoustic monitoring, seismic sensors, etc.) we figure out how many elephants are in that small subregion, and based on it compute the density

$$\text{density} = \frac{\text{number of elephants}}{\text{area}}$$

Table 1.2.3

Quantity	Dimension	Sample Units and Conversion
length	length	metre, centimetre, kilometre, mile, foot, inch
		1 m = 3.28084 ft, 1 in. = 2.54 cm, 1 mile = 1.60934 km
area	length^2	square metre, square inch
volume	length^3	cubic metre, cubic foot, litre, gallon
duration	time	second, hour, day
speed	$\frac{\text{length}}{\text{time}}$	metres per second, miles per hour
acceleration	$\frac{\text{length}}{\text{time}^2}$	metres per square second
mass	mass	kilogram, gram, pound, ounce
		1 kg = 2.20426 lb, 1 oz = 28.34952 g
mass density	$\frac{\text{mass}}{\text{length}^3}$	kilograms per cubic centimetre, grams per litre

To estimate the number of elephants in the whole ecosystem, we multiply this density by the area of the ecosystem.

Data and measurements come with **units.** In Table 1.2.3 we list the ones most commonly used.

Whenever necessary, we convert from one set of units to another using conversion factors (some of which are given in Table 1.2.3).

Example 1.2.6 Converting between Units

To convert 120 km/h into miles per minute, we note that 1 mile = 1.60934 km. Thus,

$$1 \text{ km} = \frac{1}{1.60934} \text{ miles}$$

and so

$$120 \frac{\text{km}}{\text{h}} = 120 \left(\frac{\frac{1}{1.60934} \text{ miles}}{60 \text{ min}} \right) = 1.24275 \frac{\text{miles}}{\text{min}}$$

To express the density of 2.7 kg/cm^3 in pounds per cubic inch, we need the conversion factors

$$1 \text{ kg} = 2.20462 \text{ lb} \quad \text{and} \quad 1 \text{ cm} = \frac{1}{2.54} \text{ in.}$$

Thus

$$2.7 \frac{\text{kg}}{\text{cm}^3} = 2.7 \left(\frac{2.20462 \text{ lb}}{\left(\frac{1}{2.54} \text{ in.}\right)^3} \right) = 2.7(2.20462)\left(2.54^3 \frac{\text{lb}}{\text{in.}^3}\right) = 97.54357 \frac{\text{lb}}{\text{in.}^3}$$

Note that most conversion factors we have used here are approximations.

1.3 Variables, Parameters, and Functions

Quantitative science is built on measurements, often referred to as data. Mathematics provides tools—notation, language, concepts, algorithms, and calculations—necessary for describing, analyzing, and thinking about measurements and relations between them. In this section, we develop the algebraic notation needed to describe measurements, introducing **variables** to describe measurements that change *during* the course of an experiment and **parameters** that remain constant during an experiment but can change *between* different experiments. The most important types of relations between measurements are described with **functions,** where the value of one can be computed from the value of the other.

Describing Measurements with Variables, Parameters, and Graphs

Algebra uses letters or other symbols to represent numerical quantities.

Definition 1.3.1 A **variable** is a symbol that represents a measurement that can change during the course of an experiment.

A **dependent variable** represents a measurement that depends on another measurement, called an **independent variable.** In general, an experiment might involve several dependent and independent variables. For instance, designing an experiment whose aim is to study the growth of puppies, we declare their mass and body length as dependent variables and take time and quality of nutrients as independent variables.

Let us mention another example: an experiment measures how the population of bacteria in a culture changes over time. Because two changing quantities are being measured (time and bacterial population), we need two variables to represent them. We choose variables that remind us of the measurements they represent. In this case, we can use t to represent time and b to represent the population of bacteria. Thus, t is the independent variable and b is the dependent variable.

In a different experiment, we determine that the rate (call it r) at which the human liver metabolizes alcohol is based on the amount of alcohol (call it a) present in the body. So, in this case, a is the independent variable and r is the dependent variable.

Because there are fewer letters in the alphabet than quantities measured, the same letter will be used to denote different quantities in different experiments. For this reason we need to define the variables explicitly, indicating the symbols representing them, every time we build a model. As well, we need to check the definitions and symbols when we are using someone else's model.

Thinking about data is often facilitated by using visual interpretations, such as sketches, diagrams, and graphs.

We draw graphs in the **Cartesian coordinate system,** i.e., by using two perpendicular number lines that we call **axes.** The point of intersection of the two axes is called the **origin.** The independent variable (we will also refer to it as the input, or the argument) is placed on the horizontal axis, and the dependent variable (or output, or value) is placed on the vertical axis (Figure 1.3.8). The axes are labelled with (all or some of) the variable name, the measurement it represents, and the units of measurement (Figure 1.3.9). We never draw a graph without labelling the axes.

If we do not wish to assign special symbols to variables, or when we are dealing with general or abstract contexts, we usually use x for the independent variable and y for the dependent variable. In that case, the axes are called the x-axis and the y-axis.

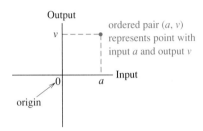

Figure 1.3.8

The components of a graph in Cartesian coordinates

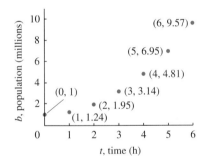

Figure 1.3.9

Results of bacterial growth experiment

Example 1.3.1 Describing Bacterial Population Growth

Table 1.3.1

t	b
0	1.00
1	1.24
2	1.95
3	3.14
4	4.81
5	6.95
6	9.57

Table 1.3.1 lists measurements of bacterial population size (in millions), denoted by the variable b, at different times t after the beginning of an experiment.

To graph the data, we label the horizontal axis by t, denoting time (in hours), and the vertical axis by b, denoting population (in millions).

The six data points from the table are shown in Figure 1.3.9. Each row in the table represents one data point: for instance, $t = 0$ and $b = 1$ are shown in the graph as the point with coordinates (0, 1). Likewise, $t = 4$ and $b = 4.81$ generate the point (4, 4.81) on the graph.

Example 1.3.2 Describing the Dynamics of a Bacterial Population

Suppose several bacterial cultures with different initial population sizes are grown in controlled conditions for one hour and then carefully counted. The population size acts as the basic measurement at both times. We must use different variables to represent these values, and we choose to use **subscripts** to distinguish them. In particular, we let b_i (for the **initial** population) represent the population at the beginning of the experiment, and we let b_f (for the **final** population) represent the population at the end. The following table and Figure 1.3.10 present the results for six colonies.

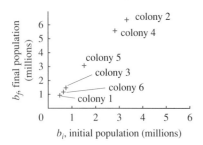

FIGURE 1.3.10

Results of alternative bacterial growth experiment

Colony	Initial Population, b_i	Final Population, b_f
1	0.47	0.94
2	3.30	6.60
3	0.73	1.46
4	2.80	5.60
5	1.50	3.00
6	0.62	1.24

Experiments of this sort form the basis of discrete-time dynamical systems and are the central topic of Chapter 3.

Experiments are done under a particular set of controlled conditions that remain constant during the experiment. However, these conditions might differ from experiment to experiment.

Definition 1.3.2

A **parameter** is a symbol that represents a measurement that does not change during the course of an experiment.

Tracking the growth of bacterial populations over time might take place at temperatures that are constant during an experiment but differ between experiments. The temperature, in this case, is represented by a parameter. Parameters, like variables, are represented by symbols that recall the measurement. We can use T to represent temperature.

Example 1.3.3 Variables and Parameters

Suppose a biologist measures growing bacterial populations at three different temperatures: 25°C, 35°C, and 45°C. During the course of each experiment, the temperature is held constant, while the population changes.

t	b when $T=25°C$	b when $T=35°C$	b when $T=45°C$
0	1.00	1.00	1.00
1	1.14	1.45	0.93
2	1.30	2.10	0.87
3	1.48	3.03	0.81
4	1.68	4.39	0.76
5	1.92	6.36	0.70
6	2.18	9.21	0.66

Figure 1.3.11 compares the population sizes of the three populations, which all started at the same value. The population grows most quickly at the intermediate temperature of 35°C and declines at the high temperature of 45°C.

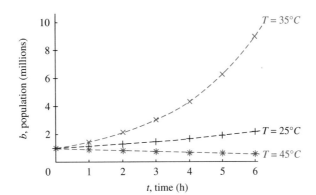

FIGURE 1.3.11

Bacterial growth experiment at three temperatures

Describing Relations between Measurements with Functions

The ways in which measurements in an experiment depend on each other can be described using **relations** or **functions.**

A **relation** between two variables is the set of all pairs of values that occur. Let us consider an example.

Example 1.3.4 A Relation between Temperature and Population Size

T	P
25	2.18
25	2.45
25	2.10
25	4.25
35	9.21
35	7.39
35	6.36
45	0.66
45	0.93

Suppose the temperature, T, and final population size, P, are measured for nine populations, with the results shown in the table and Figure 1.3.12. These values could result from repeating the experiment in Example 1.3.3 several times and measuring the population at time $t = 6$.

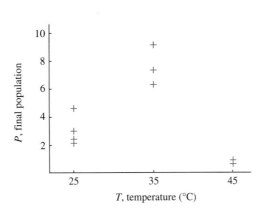

FIGURE 1.3.12

Final population size at three temperatures

Formally, the relation between the final population, P, and the temperature, T, consists of the following ordered pairs, recorded in the form (T, P): (25, 2.18), (25, 2.45), (25, 2.10), (25, 4.25), (35, 9.21), (35, 7.39), (35, 6.36), (45, 0.66), and (45, 0.93).

Different values of the population P are related to each temperature, perhaps because of differences in experimental conditions.

We will not be studying relations in detail in this book. Instead, we focus on a special type of relations called functions.

Definition 1.3.3 A **function** f is a rule that assigns to each real number x in some set D a unique real number $f(x)$ in a set R.

Thus, a function is a relation in which each value x appears *exactly once* as the first coordinate in the listing of all ordered pairs $(x, f(x))$ that represent it. Clearly, the relation we studied in Example 1.3.4 is not a function.

14   **Chapter 1**   Introduction to Models and Functions

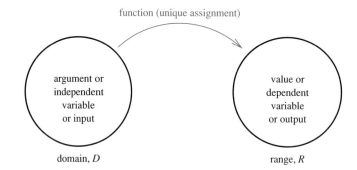

FIGURE 1.3.13

The basic terminology for describing a function

The set D is called the **domain** of the function f, and the set R of all real numbers $f(x)$ for all x in D is its **range**. The symbol x is called the **independent variable** (or input, or argument) of f, and the symbol y, or $f(x)$, is called the **dependent variable** (or output, or value); see Figure 1.3.13.

Thus, we can say that a function returns a unique output for each input that is allowed (i.e., for each input that belongs to its domain). We often use the notation $y = f(x)$ to denote a function.

If the domain of a function f is not given explicitly, then it is assumed that it consists of all real numbers for which the formula for f makes sense (that is, gives a real number). For instance, if $f(x) = \frac{1}{x}$ and no other information is given, then we say that the domain of f consists of all real numbers x such that $x \neq 0$. This domain is referred to as a **natural domain.**

However, if the domain of the function is given explicitly (and is, appropriately, called a **given domain**), then that is what we work with. For example, consider $g(x) = \frac{1}{x}$, where $x > 0$. Then $g(2) = \frac{1}{2}$ and $g(19.62) = \frac{1}{19.62} = 0.051$, but $g(-4)$ is not defined.

In applications, the particular context will determine the domain. For instance, if we try to use the function $b(m) = 3m - 6$ to model the amount of a certain enzyme in our body that depends on some variable m, then the domain of $b(m)$ will consist of numbers $m \geq 2$ (since the amount of enzyme cannot be a negative number).

Likewise, the domain of the function $s = f(l)$ that relates the speed, s, of an ant to its body length, l, is an interval $[L_m, L_M]$, where L_m and L_M are, respectively, the minimum and the maximum possible body length of an ant.

Example 1.3.5   Data That Can and Cannot Be Described by a Function

The data in the bacterial growth example (Example 1.3.1) can be described by a function. Each value of the input, t, is associated with only a single value of the output, b. Using functional notation, we can describe this dependence as $b = f(t)$.

The data in Example 1.3.4 cannot be represented as a function. For instance, to the value $T = 35$ of the independent variable, three values of the dependent variable, $P = 9.21$, $P = 7.39$, and $P = 6.36$, have been assigned.

Example 1.3.6   A Function Describing Bacterial Population Growth

The bacteria population in Example 1.3.1 obeys the formula

$$b(t) = \frac{t^2}{4.2} + 1$$

The population size, b, is a function of the time, t. The argument (independent variable) of the function b is t, the time after the beginning of the experiment. The value of the function is the population of bacteria. The formula summarizes the relation between these two measurements. The output is found by squaring the input, dividing by 4.2, and then adding 1.

The function b takes time after the beginning of the experiment as its input. Because negative time does not make sense in this case, the domain of this function consists of all positive numbers and zero. We write that

b is defined on the domain $t \geq 0$

The square of any real number is zero or positive, and thus

$$\frac{t^2}{4.2} \geq 0$$

Adding 1 to both sides, we get

$$b(t) = \frac{t^2}{4.2} + 1 \geq 1$$

Thus, the range of $b(t)$ is *contained* in the interval $[1, \infty)$. To prove that it is *equal to* $[1, \infty)$, we need to pick any number β in $[1, \infty)$ and show that there is a number t such that $b(t) = \beta$. Indeed,

$$\frac{t^2}{4.2} + 1 = \beta$$
$$t^2 = 4.2(\beta - 1)$$
$$t = \sqrt{4.2(\beta - 1)}$$

After discussing modifications to graphs (shifts and scaling), we will be able to verify our claim about the range by drawing the graph of $b(t)$. ▲

Example 1.3.7 A Function with Non-numeric Domain

Consider the data in Table 1.3.2. These data describe a relation between two observations: the identity of the species and the number of legs. We can express this as the function L (to remind us of legs). According to the table,

$$L(\text{ant}) = 6, \quad L(\text{crab}) = 10$$

and so forth. The domain of this function is "types of animals," and the range is the non-negative integers $(0, 1, 2, 3, \ldots)$. We plot the input ("animal") along the horizontal axis and the output ("number of legs") on the vertical axis (Figure 1.3.14).

Note that in this example we used non-numeric values (animals), so—strictly speaking—L is not a function (according to our definition, functions operate on real numbers). Nevertheless, it is sometimes convenient to use functions to describe certain phenomena that involve non-numeric data. ▲

Functions can be described in four ways: (1) numerically (by means of a table of values), (2) algebraically (as a formula), (3) pictorially (as a graph), and (4) verbally. Life scientists and applied mathematicians need to learn to use all four methods and translate fluently between them (see Section 1.4). In particular, we must know how to translate graphical information into words that communicate key observations to colleagues and the public.

We often represent functions visually in the form of a **graph.**

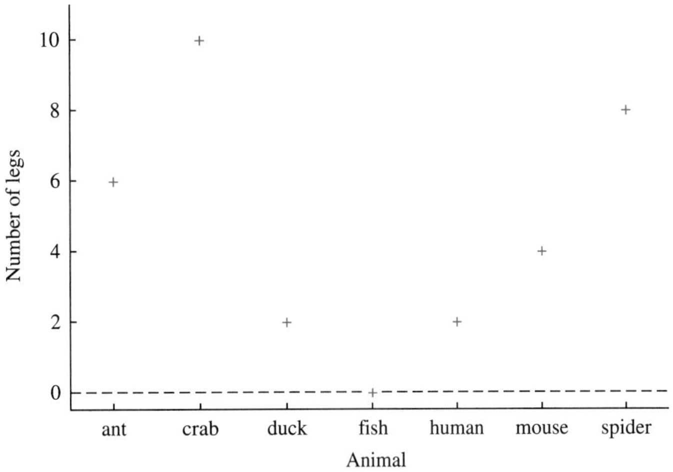

Table 1.3.2

Animal	Number of Legs
ant	6
crab	10
duck	2
fish	0
human	2
mouse	4
spider	8

FIGURE 1.3.14

Numbers of legs of various organisms, plotted on a graph

Definition 1.3.4 The **graph** of a function $y = f(x)$ is a curve that consists of all points $(x, f(x))$ with x in the domain of f; see Figure 1.3.15.

We can also say that the graph of f contains all points (x, y) for which x is in the domain and $y = f(x)$ is the corresponding value of f. In some situations, it will be useful to think of a point on the graph as an ordered pair (input, corresponding output).

Given x, we think of $f(x)$ as the height of the graph above the number x or as the **signed vertical distance** from the number x on the x-axis to the graph of the function. Because the distance is a positive number or zero, we have to add the word "signed." "Signed" means that

$f(x)$ = vertical distance if the graph is above the x-axis

(for example $f(1) = 4$ in Figure 1.3.15) and

$f(x)$ = negative vertical distance if the graph is below the x-axis

(such as $f(5) = -1$ in Figure 1.3.15).

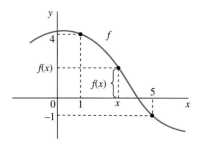

FIGURE 1.3.15
Graph of a function

Example 1.3.8 Constructing the Graph of a Function

A common way to draw the graph of a function is by plotting points: we construct a table of values, transfer the points into the coordinate system, and then connect them with a smooth curve.

Suppose we wish to graph the function

$$f(x) = 0.6x^3 - 2x^2 - x + 4$$

Taking several values of x, we first obtain the table of values (Table 1.3.3) and then sketch the graph by smoothly connecting the points we obtain, as shown in Figure 1.3.16.

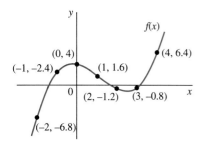

FIGURE 1.3.16
The graph of the function
$f(x) = 0.6x^3 - 2x^2 - x + 4$

Table 1.3.3

x	y = f(x)	Point (x, y) on the Graph
−2	−6.8	(−2, −6.8)
−1	2.4	(−1, 2.4)
0	4	(0, 4)
1	1.6	(1, 1.6)
2	−1.2	(2, −1.2)
3	−0.8	(3, −0.8)
4	6.4	(4, 6.4)

For convenience, we use small integer values for x in the calculations needed for the table of values.

In many situations, however, connecting points with a simple smooth curve might not produce a satisfactory graph. As a matter of fact, it could produce an incorrect graph—for instance, our choices for numeric values of x might miss important features of the graph, such as asymptotes or extreme values; see Example 1.3.9. In general, we have to use a variety of tools—many of which will be discussed in this book—to draw accurate graphs of functions.

Plotting points, besides being the most elementary way of graphing functions, is also the most widely used and implemented method, for instance in graphing calculators and in mathematical software.

Example 1.3.9 Sketching by Plotting Points Can Produce Erroneous Graphs

Sketch the graph of the function

$$f(x) = x^6 - 3x^5 - 5x^4 + 15x^3 + 4x^2 - 12x + 3$$

for x in the interval $[-2, 3]$.

We create a table of values by picking integer values for x; see Table 1.3.4.

Table 1.3.4

x	y = f(x)	Point (x, y) on the Graph
−2	3	(−2, 3)
−1	3	(−1, 3)
0	3	(0, 3)
1	3	(1, 3)
2	3	(2, 3)
3	3	(3, 3)

The points from Table 1.3.4 are plotted in Figure 1.3.17a. How do we connect them with a smooth curve to obtain the graph of $f(x)$? Do we draw a straight line?

FIGURE 1.3.17

Sketching the graph of a function

It turns out that a straight line is not even close to the actual graph. Using a graphing device, we plot an accurate graph of $f(x)$ in Figure 1.3.17b. Clearly, if we wish to draw the graph by hand, or obtain more—or more precise—information (such as intervals where the function is increasing and where it is decreasing, or the x-values where the function assumes the largest value) we need tools beyond calculating entries in a table of values.

Example 1.3.10 Describing Results in Graphs and Words

A more complicated pattern of change in population size is presented in the adjacent table.

Time	Population Size
0	0.86
2	1.69
4	2.98
6	4.49
8	5.69
10	6.17
12	5.95
14	5.29
16	4.41
18	3.50
20	2.67
22	1.96
24	1.41

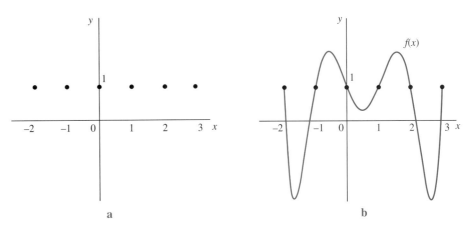

FIGURE 1.3.18

The population of bacteria in a culture

We can see (more easily from the graph in Figure 1.3.18 than from the table) that the bacterial population grew during the first 10 hours and declined thereafter. The population reached a maximum at time $t = 10$ h. This graph and its description can be used to understand the results even without a mathematical formula.

Example 1.3.11 Sketching a Graph from a Verbal Description

Conversely, it can be useful to sketch a graph of a function from a verbal description. Suppose we are told that a population increases between time $t=0$ and time $t=5$, decreases nearly to 0 by time $t=12$, increases to a higher maximum at time $t=20$, and goes extinct at time $t=30$. A graph (Figure 1.3.19) translates this information into pictorial form. Because we were not given exact values, the graph is not exact. It instead gives a **qualitative** picture of the results.

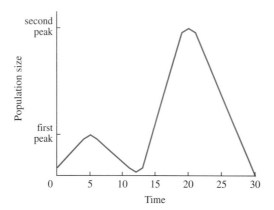

Figure 1.3.19
A bacterial population plotted from a verbal description

The graph in Figure 1.3.18 consists of a set of isolated points. It describes an experiment with a discrete set of data (or measurements), and it represents a **discrete variable.** An experiment with continuous data drawn as a curve, representing a **continuous variable,** is shown in Figure 1.3.19.

Both graphs consist of points (either a finite or an infinite number of them), and each point carries two pieces of information: the value of the independent variable as its first coordinate, and the corresponding value of the function (dependent variable) as its second coordinate.

The fact that a function assigns a unique value y to each value x in its domain means that its graph cannot contain two or more points with the same x-coordinate. This observation provides a useful geometric test that allows us to recognize functions.

Vertical Line Test A set of points represents a function if every vertical line crosses the set at most once.

The reason that we said "set of points" is to include both discrete graphs (such as in Figures 1.3.11, 1.3.12, and 1.3.18) and continuous graphs (Figures 1.3.16, 1.3.17b, and 1.3.19). "At most once" means "zero or once"; i.e., a vertical line does not have to cross the graph at all (Figure 1.3.20a). If even one line crosses a curve more than once, then we know that the curve cannot represent a function (Figure 1.3.20b; see also Example 1.3.12). The discrete set of points in Figure 1.3.20c cannot represent a function.

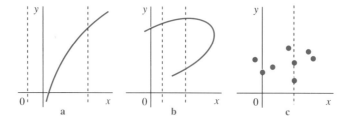

Figure 1.3.20
Vertical line test

Example 1.3.12 A Mathematical Formula Describing a Relation That Is Not a Function

The set of solutions for x and y satisfying the equation

$$x^2 + y^2 = 1$$

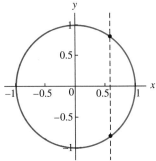

the vertical line $x = 0.6$ intersects the curve at two points

FIGURE 1.3.21
The circle is not the graph of a function

FIGURE 1.3.22
Constant function

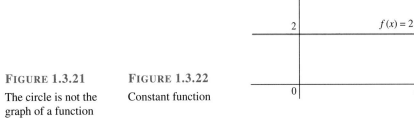

is the circle of radius 1 centred at the origin (Figure 1.3.21). Each value of x between $x = -1$ and $x = 1$ is associated with two different values of y. For example, the value $x = 0.6$ is associated with both $y = 0.8$ and $y = -0.8$.

Catalogue of Important Functions

We list several important functions, draw their graphs, and indicate their domains and ranges. In the following sections, we will study their properties and use them in various applications. As well, we will add more functions to our catalogue.

(1) The function $f(x) = c$, where c is a real number, is a **constant function.** It assigns the same number, c, to all values of the independent variable, x. Its domain is the set of real numbers, and its range consists of the single value c. The graph in Figure 1.3.22 shows the constant function $f(x) = 2$.

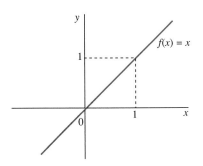

FIGURE 1.3.23
Linear function $f(x) = x$

(2) The function $f(x) = mx + b$, where m and b are real numbers, is called a **linear function.** Its graph is a line with slope m and y-intercept b. When $m \neq 0$, both the domain and the range consist of all real numbers. If $m = 0$, the line is horizontal, so it represents the constant function $f(x) = b$. The special case $f(x) = x$ is a line of slope 1 through the origin; see Figure 1.3.23. Sometimes we calculate the linear function from the point-slope form

$$y = m(x - x_0) + y_0$$

where m is the slope of the line and (x_0, y_0) is a point on the line. Using $f(x)$ instead of y, we write

$$f(x) = m(x - x_0) + y_0$$

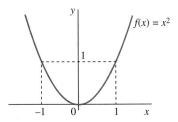

FIGURE 1.3.24
Parabola $f(x) = x^2$

(3) The graph of a **quadratic function**, $f(x) = x^2$, is a parabola with the vertex located at the origin. The graph is symmetric with respect to the y-axis (Figure 1.3.24). The domain of $f(x)$ is the set of all real numbers, and the range is $[0, \infty)$. The graphs of $f(x) = x^n$, where $n = 4, 6, 8, \ldots$, look similar to the graph of $f(x) = x^2$.

(4) The graph of a **cubic function**, $f(x) = x^3$, is shown in Figure 1.3.25. It is symmetric with respect to the origin; its domain and range consist of the set of real numbers. The graphs of $f(x) = x^n$, where $n = 5, 7, 9, \ldots$, look similar to the graph of $f(x) = x^3$.

(5) The domain of the function $f(x) = \dfrac{1}{x}$ is the set of all real numbers x such that $x \neq 0$. Its graph is a hyperbola, symmetric with respect to the origin, whose asymptotes are the coordinate axes (Figure 1.3.26). From the graph, we see that the range of

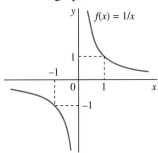

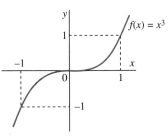

FIGURE 1.3.25
The graph of $f(x) = x^3$

FIGURE 1.3.26
Hyperbola $f(x) = 1/x$

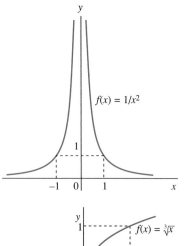

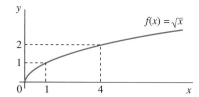

FIGURE 1.3.27
The graph of $f(x) = 1/x^2$

FIGURE 1.3.28
Square root function $f(x) = \sqrt{x}$

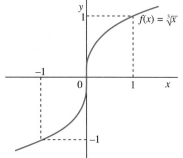

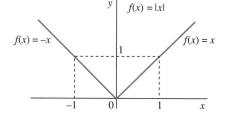

FIGURE 1.3.29
Cube root function $f(x) = \sqrt[3]{x}$

FIGURE 1.3.30
Absolute value function $f(x) = |x|$

$f(x) = \frac{1}{x}$ is $f(x) \neq 0$ (we usually say $y \neq 0$). Using interval notation, we write the range as $(-\infty, 0)$ and $(0, \infty)$.

(6) The graph of $f(x) = \dfrac{1}{x^2}$ is symmetric with respect to the y-axis; the coordinate axes are its asymptotes (Figure 1.3.27). Its domain is the set of all real numbers x such that $x \neq 0$, and the range consists of positive numbers, i.e., $(0, \infty)$.

(7) The domain of the **square root function**, $f(x) = \sqrt{x}$, is the set $[0, \infty)$, and the range is $[0, \infty)$. The graph has the shape of a parabola; see Figure 1.3.28.

Note that—by definition—the square root function is *positive or zero;* for example, $\sqrt{25} = 5$. This is why, when solving the equation $x^2 = 25$ (i.e., when finding *all x* whose square is 25), we need to include the negative root, and write $x = \pm\sqrt{25} = \pm 5$.

(8) The graph of the **cube root function**, $f(x) = \sqrt[3]{x}$, is symmetric with respect to the origin; see Figure 1.3.29. Both domain and range are the set of all real numbers.

(9) The **absolute value function**, $f(x) = |x|$, is defined by

$$|x| = \begin{cases} x & \text{if } x \geq 0 \\ -x & \text{if } x \leq 0 \end{cases}$$

From the definition, we see that we can calculate the absolute value of any number, so the domain consists of all real numbers. The graph (Figure 1.3.30) shows that the range is the set $[0, \infty)$.

In the next section we will add polynomials, rational functions, and algebraic and transcendental functions to our list of important functions (see subsection Creating New Functions).

Functions such as $f(x) = |x|$, where different formulas are used in different parts of the domain, are said to be **defined piecewise.** Another piecewise-defined function is discussed in the next example.

1.3 Variables, Parameters, and Functions

Example 1.3.13 Function Defined Piecewise

Sketch the graph of the function

$$y = f(x) = \begin{cases} 2x+1 & \text{if } x < 1 \\ -x+2 & \text{if } x \geq 1 \end{cases}$$

If $x < 1$, then $y = 2x + 1$, which represents the line with slope 2 and y-intercept 1. Alternatively, we could have computed two points, say, $(0, 1)$ and $(-1, -1)$, and joined them with a straight line.

For $x \geq 1$, we get the line $y = -x + 2$ (slope -1 and y-intercept 2).

We sketch the graph of the function $f(x)$ in Figure 1.3.31. The filled dot on the graph indicates that $f(1) = 1$. The empty dot indicates that the point does not belong to the graph. (Note that two filled dots, one above the other, would violate the vertical line test.)

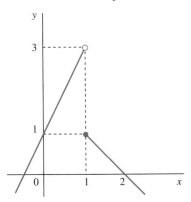

FIGURE 1.3.31
Graph of a function defined piecewise

In Example 4.4.3 we use piecewise-defined functions to describe how the number of cancer cells changes due to a combination of chemotherapy and surgery.

Example 1.3.14 Computing the Domain of a Function I

Find the domain of the function $f(x) = \dfrac{1}{x\sqrt{x+4}}$.

Since division by zero is not defined (i.e., is not a real number), we see that $f(x)$ is not defined when $x = 0$ and $x = -4$. As well, the square root of a negative number is not defined, so we need to make sure that $x + 4 \geq 0$, i.e., $x \geq -4$.

We conclude that the numbers x that are in the domain of $f(x)$ satisfy $x \geq -4$, $x \neq -4$ (i.e., $x > -4$), and $x \neq 0$. So, the domain is the interval $(-4, \infty)$ from which the number 0 has been removed. Said differently, the domain consists of two intervals, $(-4, 0)$ and $(0, \infty)$.

Example 1.3.15 Computing the Domain of a Function II

Find the domain of the function $f(x) = \sqrt{9 - x^2}$.

Since the square root is defined for positive numbers and zero only, the numbers x that belong to the domain of $f(x)$ must satisfy $9 - x^2 \geq 0$, i.e.,

$$x^2 \leq 9$$

Therefore, $-3 \leq x \leq 3$, and the domain is the interval $[-3, 3]$.

How did we find the solution to the inequality $x^2 \leq 9$ in the previous example? In Table 1.3.5 we recall solutions to basic quadratic inequalities.

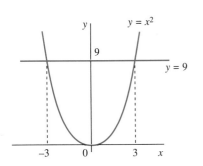

FIGURE 1.3.32
Solving $x^2 \leq 9$ geometrically

Table 1.3.5

Inequality (assume that $a > 0$)	Solution as an Inequality	Solution in Interval Form
$x^2 < a$	$-\sqrt{a} < x < \sqrt{a}$	$(-\sqrt{a}, \sqrt{a})$
$x^2 \leq a$	$-\sqrt{a} \leq x \leq \sqrt{a}$	$[-\sqrt{a}, \sqrt{a}]$
$x^2 > a$	$x < -\sqrt{a}$ or $x > \sqrt{a}$	$(-\infty, -\sqrt{a})$ or $(\sqrt{a}, \infty)$
$x^2 \geq a$	$x \leq -\sqrt{a}$ or $x \geq \sqrt{a}$	$(-\infty, -\sqrt{a}]$ or $[\sqrt{a}, \infty)$

Alternatively—we can solve the inequality $x^2 \leq 9$ without having to memorize the facts given in Table 1.3.5—we reason geometrically.

The graph of $y = x^2$ is a parabola with its vertex located at the origin. The graph of $y = 9$ is a horizontal line crossing the y-axis at 9; see Figure 1.3.32. The two graphs intersect when $x^2 = 9$, i.e., when $x = \pm 3$. To solve $x^2 \leq 9$ means to identify all x-values for which the parabola $y = x^2$ lies below the line $y = 9$, or where it touches the line. From the graph, we see that the x-values must belong to the interval $[-3, 3]$.

Arguing in a similar way, we can verify all entries in Table 1.3.5.

We have already used the words "increasing" and "decreasing" to describe the behaviour of functions.

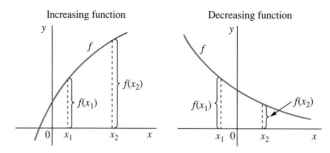

FIGURE 1.3.33
Increasing and decreasing functions

Definition 1.3.5 A function $f(x)$ is said to be **increasing** on an interval I if
$$f(x_1) < f(x_2) \text{ for all } x_1 \text{ and } x_2 \text{ in } I \text{ such that } x_1 < x_2$$
It is called **decreasing** if
$$f(x_1) > f(x_2) \text{ for all } x_1 \text{ and } x_2 \text{ in } I \text{ such that } x_1 < x_2$$
See Figure 1.3.33.

Note that the terms "increasing" and "decreasing" are defined on an interval because we are comparing the values of f at two points. It does not matter whether the interval I is open or closed. As well, it might include $-\infty$ or ∞ (or both).

If we decide to use closed intervals, then we must accept the fact that there will be points that belong to both an interval where a function is increasing and an interval where it is decreasing. For instance, $f(x) = x^2$ is decreasing on $(-\infty, 0]$ and increasing on $[0, \infty)$. Alternatively, if we use open intervals, we say that $f(x) = x^2$ is decreasing on $(-\infty, 0)$ and increasing on $(0, \infty)$. This time, 0 does not belong to either interval.

Perhaps the best way to visualize the statements in the definition is to think of $x_1 < x_2$ as "x_1 is earlier than x_2." In that case, $f(x_1) > f(x_2)$ means that f is greater for earlier values of x, so f must be decreasing.

Looking at our catalogue of functions, we see that the functions $f(x) = x$, $f(x) = x^3$, and $f(x) = \sqrt[3]{x}$ are increasing on $(-\infty, \infty)$. The function $f(x) = \sqrt{x}$ is increasing on $[0, \infty)$, and the function $f(x) = \dfrac{1}{x}$ is decreasing on $(-\infty, 0)$ and $(0, \infty)$. The function $f(x) = \dfrac{1}{x^2}$ is increasing on $(-\infty, 0)$ and decreasing on $(0, \infty)$. The functions $f(x) = x^2$ and $f(x) = |x|$ are decreasing on $(-\infty, 0]$ and increasing on $[0, \infty)$. The function of Example 1.3.13 is increasing on $(-\infty, 1)$ and decreasing on $(1, \infty)$.

Summary Quantitative science is built upon measurements, and mathematics provides the methods for describing and thinking about measurements and the relations between them. **Variables** describe measurements that change during the course of an experiment, and **parameters** describe measurements that remain constant during an experiment but might change between different experiments. **Functions** describe relations between different measurements when a single output is associated with each output; functions can be recognized graphically with the **vertical line test.** The elementary functions we studied in this section will help us understand how more complicated functions work.

1.3 Exercises

Mathematical Techniques

▼1–2 ▪ Give mathematical names to the measurements in the following situations, and identify the variables and parameters.

1. A scientist measures the density of wombats at three altitudes: 500 m, 750 m, and 1000 m. He repeats the experiment in three different years, with rainfall of 30 cm in the first year, 50 cm in the second, and 60 cm in the third.

2. A scientist measures the density of bandicoots at three altitudes: 500 m, 750 m, and 1000 m. She repeats the experiment in three different years that have different densities of wombats, which compete with bandicoots. The density is 10 wombats per square kilometre in the first year, 20 wombats per square kilometre in the second, and 15 wombats per square kilometre in the third.

3–8 ▪ Sketch the graph of each function. Identify the intervals where the function is increasing and the intervals where it is decreasing.

3. $f(x) = -4.2$
4. $f(x) = 3x - 6$
5. $y = \dfrac{2}{x} + 4$
6. $f(x) = 2\sqrt[3]{x}$
7. $f(x) = \dfrac{1}{x^2} + 1$
8. $f(x) = |x - 5|$

9–12 ▪ Graph the given points and say which point does not seem to fall on the graph of a simple function that describes the other four.

9. $(0, -1), (1, 1), (2, 2), (3, 5), (4, 7)$
10. $(0, 8), (1, 10), (2, 8), (3, 6), (4, 4)$
11. $(0, 2), (1, 3), (2, 6), (3, 11), (4, 12)$
12. $(0, 30), (1, 25), (2, 15), (3, 12), (4, 10)$

13–16 ▪ Evaluate each function at the given algebraic arguments.

13. $f(x) = x + 5$ at $x = a$, $x = a + 1$, and $x = 4a$
14. $g(y) = 5y$ at $y = x^2$, $y = 2x + 1$, and $y = 2 - x$
15. $h(z) = \dfrac{1}{5z}$ at $z = \dfrac{c}{5}$, $z = \dfrac{5}{c}$, and $z = c + 1$
16. $F(r) = r^2 + 5$ at $r = x + 1$, $r = 3x$, and $r = \dfrac{1}{x}$

17–23 ▪ Find the domain of each function.

17. $f(x) = 3 - x + x^2$
18. $g(x) = \dfrac{2.3}{x - 4}$
19. $f(x) = \dfrac{x - 2}{x^2 + 1}$
20. $f(x) = \sqrt{2x - 7}$
21. $y = |x - 1| + \dfrac{1}{x}$
22. $f(x) = \dfrac{x - 2}{x^2 - 1}$
23. $f(x) = \sqrt{\dfrac{1}{x - 2}}$

24–27 ▪ Find the domain of each function, and sketch its graph.

24. $f(x) = \begin{cases} 2x + 1 & \text{if } x < 1 \\ \sqrt{x} & \text{if } x \geq 1 \end{cases}$

25. $f(x) = \begin{cases} \dfrac{1}{x} & \text{if } x < 4 \\ \sqrt{x} & \text{if } x \geq 4 \end{cases}$

26. $y = \begin{cases} 2 & \text{if } x \leq -1 \\ x - 1 & \text{if } -1 < x < 0 \\ \sqrt{x} & \text{if } x \geq 0 \end{cases}$

27. $g(x) = \begin{cases} x^2 & \text{if } x > 0 \\ -x^2 & \text{if } x \leq 0 \end{cases}$

28–35 ▪ Determine the range of each function.

28. $f(x) = 3$
29. $f(x) = -x$
30. $y = -x^2$
31. $f(x) = x^2 + 3$
32. $f(x) = -x^2 + 3$
33. $g(x) = \sqrt{2x}$
34. $f(x) = \sqrt{4 - x}$
35. $g(x) = \sqrt[3]{x^2}$

36–37 ▪ Sketch graphs of the following relations. Is there a more convenient order for the arguments?

36. A function whose argument is the name of a province and whose value is the highest altitude in that province.

Province	Highest Altitude (m)
British Columbia	4663
Alberta	3747
Ontario	693
Québec	1651
New Brunswick	817
Newfoundland and Labrador	1652

37. A function whose argument is the name of a bird and whose value is the average length of that bird.

Bird	Length (cm)
Cooper's hawk	50
Goshawk	66
Sharp-shinned hawk	35

Applications

38–41 ■ Describe what is happening in the graphs shown.

38. A plot of cell volume against time in days.

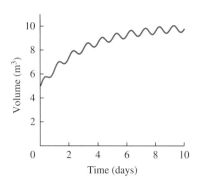

39. A plot of a Pacific salmon population against time in years.

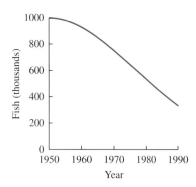

40. A plot of the average height of a population of trees plotted against age in years.

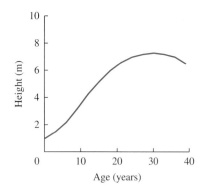

41. A plot of an Internet stock price against time.

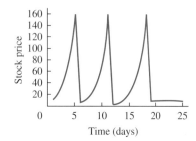

42–45 ■ Draw graphs based on the following descriptions.

42. A population of birds begins at a large value, decreases to a tiny value, and then increases again to an intermediate value.

43. The amount of DNA in an experiment increases rapidly from a very small value and then levels out at a large value before declining rapidly to zero.

44. Body temperature oscillates between high values during the day and low values at night.

45. Soil is wet at dawn, quickly dries out and stays dry during the day, and then becomes gradually wetter again during the night.

46–49 ■ Evaluate the following functions over the suggested range, sketch a graph of the function, and answer the biological question.

46. The number of bees, b, found on a plant is given by $b = 2f + 1$, where f is the number of flowers, ranging from 0 to about 20. Explain what might be happening when $f = 0$.

47. The number of cancerous cells, c, as a function of radiation dose, r (measured in rads), is

$$c = r - 4$$

for r greater than or equal to 5, and is zero for r less than 5. Suppose r ranges from 0 to 10. What is happening at $r = 5$ rads?

48. Insect development time, A (in days), obeys $A = 40 - \frac{T}{2}$, where T represents temperature in degrees Celsius for $10 \leq T \leq 40$. Which temperature leads to the most rapid development?

49. Tree height, h (in metres), follows the formula

$$h = \frac{100a}{100 + a}$$

where a represents the age of the tree in years. The formula is valid for any positive value of a, which ranges from 0 to 1000. How tall would this tree get if it lived forever?

50–55 ■ Use a graphing device to plot the following functions. How would you describe them in words?

50. $f(x) = x^5 - 12x^4 + 51x^3 - 92x^2 + 60x + 4$ for $-0.3 \leq x \leq 5.2$

51. $g(x) = 8 - \frac{3.5 + 3x^2}{x^2 + 0.4}$ for $-7 \leq x \leq 5$

52. $h(x) = \sqrt{|x - 1| + |3 - x| + |x - 5|}$ for $0 \leq x \leq 6$

53. $|g(x)|$ for $-7 \leq x \leq 5$ (using the function in Exercise 51)

54. $k(x) = x^4 - 2x^3 - x^2 + 2x$ for $-1.5 \leq x \leq 2.5$

55. $k(x) + |k(x)|$ for $-1.5 \leq x \leq 2.5$ (using the function in Exercise 54)

1.4 Working with Functions

In this section, we study various ways of combining functions in order to create new ones. Algebraically, we can do so by applying elementary **algebraic operations,** computing an **inverse function,** or forming a **composition** of two functions. Geometrically, we can **shift, scale,** or **reflect** existing graphs to create new ones.

Elementary Algebraic Operations Using the sum, difference, product, and quotient, we can generate new functions.

The height of the graph of the sum of two functions is the height of the first plus the height of the second. Geometrically, we can graph each of the pieces and add them together point by point.

Algebraically, the value of the function $f + g$ is computed as the sum of the values of the functions f and g.

Definition 1.4.1 The sum, $f + g$, of the functions f and g is the function defined by

$$(f + g)(x) = f(x) + g(x)$$

Their difference, $f - g$, is defined by

$$(f - g)(x) = f(x) - g(x)$$

The value of the product, $f \cdot g$, is computed as the product of the values of the functions f and g.

Definition 1.4.2 The product, $f \cdot g$, of the functions f and g is the function defined by

$$(f \cdot g)(x) = f(x) \cdot g(x)$$

Their quotient, f/g, is defined by

$$\left(\frac{f}{g}\right)(x) = \frac{f(x)}{g(x)}$$

If the domain of f is D_f and the domain of g is D_g, then the domain of $f + g$, $f - g$, and $f \cdot g$ is the intersection $D_f \cap D_g$, since both functions, f and g, need to be defined. (The intersection, $A \cap B$, of two sets, A and B, is the set that contains all numbers that belong to both A and B.) The domain of f/g is the intersection $D_f \cap D_g$ from which the values of x such that $g(x) = 0$ (if any) have been removed, to avoid dividing by zero.

Example 1.4.1 Adding Biological Functions

If two bacterial populations are separately counted, the total population is the sum of the two individual populations. Suppose a growing population is described by the function

$$b_1(t) = t^2 + 1$$

and a declining population is described by the function

$$b_2(t) = \frac{5}{1 + 2t}$$

The individual population sizes and their sum are computed in the following table and graphed in Figure 1.4.34.

t	$b_1(t)$	$b_2(t)$	$(b_1 + b_2)(t)$
0	1	5	6
0.5	1.25	2.5	3.75
1	2	1.67	3.67
1.5	3.25	1.25	4.5
2	5	1	6
2.5	7.25	0.83	8.08
3	10	0.71	10.71

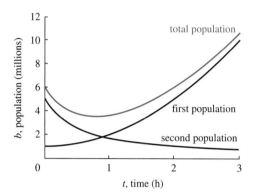

FIGURE 1.4.34
Adding biological functions

We can draw the sum $b_1(t) + b_2(t)$ from the graphs of b_1 and b_2 without calculating anything. Take some value for the variable t; the graph of $b_1 + b_2$ at t is the sum of the height of b_1 at t and the height of b_2 at t; thus, all we need to do is to stack the two heights, one on top of the other, to obtain the height that corresponds to their sum at t.

Example 1.4.2   Multiplying Biological Functions

Many quantities in science are built as products of simpler quantities. For example, the mass of a population is the product of the mass of each individual and the number of individuals. Consider a population growing according to

$$b(t) = \frac{t^2}{4.2} + 1$$

t	b	μ	$\mu \cdot b$
0	1	1	1
1	1.24	0.5	0.62
2	1.95	0.33	0.65
3	3.14	0.25	0.79
4	4.81	0.2	0.96
5	6.95	0.17	1.16
6	9.57	0.14	1.37

(Example 1.3.6). Suppose that as the population gets larger, the individuals become smaller. Let $\mu(t)$ (the Greek letter mu)[1] represent the mass of an individual at time t, and suppose that

$$\mu(t) = \frac{1}{1+t}$$

We can find the total mass $m(t)$ (also called the biomass) at time t by multiplying the mass per individual by the number of individuals:

$$m(t) = b(t)\mu(t) = \left(\frac{t^2}{4.2} + 1\right)\left(\frac{1}{1+t}\right) = \frac{t^2 + 4.2}{4.2(1+t)}$$

See the table, and also Figure 1.4.35.

---

[1] Quite often, we use Greek letters to represent variables and parameters.

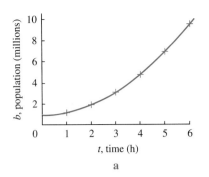
a

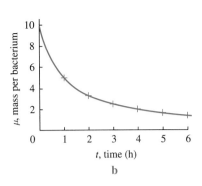
b

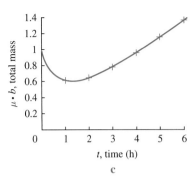
c

FIGURE 1.4.35
Multiplying biological functions

From the graph in Figure 1.4.35c, we conclude that the total mass of this population initially declines, and then starts increasing somewhere between one hour and two hours.

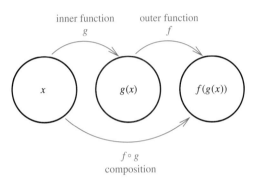

FIGURE 1.4.36
Composition of functions

Composition of Functions One of the important ways to combine functions is through **composition**, where the output of one function acts as the input of another.

Definition 1.4.3 The composition, $f \circ g$, of functions f and g is the function defined by

$$(f \circ g)(x) = f(g(x)) \tag{1.4.1}$$

We say "f composed with g evaluated at x" or "f of g of x." In calculations, we sometimes refer to f as the **outer function** and to g as the **inner function.** See Figure 1.4.36.

Example 1.4.3 Computing the Value of a Functional Composition

Consider the functions

$$f(x) = 4 + x - x^2$$
$$g(x) = 2x$$

To find the value of the composition $f \circ g$ at $x = 2$, we compute

$$\begin{aligned}(f \circ g)(2) &= f(g(2))\\ &= f(4)\\ &= 4 + 4 - 4^2\\ &= -8\end{aligned}$$

Similarly, to find the value of the composition $g \circ f$ at $x = 2$, we compute

$$\begin{aligned}(g \circ f)(2) &= g(f(2))\\ &= g(4 + 2 - 2^2)\\ &= g(2)\\ &= 4\end{aligned}$$

Example 1.4.4 Computing the Formula of a Functional Composition

Consider again the functions $f(x) = 4 + x - x^2$ and $g(x) = 2x$ from Example 1.4.3. The formula for the composition $f \circ g$ is given by

$$(f \circ g)(x) = f(g(x))$$
$$= f(2x)$$
$$= 4 + 2x - (2x)^2$$
$$= 4 + 2x - 4x^2$$

In Example 1.4.3 we computed that $(f \circ g)(2) = -8$. If we evaluate by substituting into the formula $(f \circ g)(x) = 4 + 2x - 4x^2$, we find

$$(f \circ g)(2) = 4 + 2 \cdot 2 - 4 \cdot 2^2 = -8$$

matching our earlier result.

Likewise, we find the composition $g \circ f$ as follows:

$$(g \circ f)(x) = g(f(x))$$
$$= g(4 + x - x^2)$$
$$= 2(4 + x - x^2)$$
$$= 8 + 2x - 2x^2$$

To find the composition $(g \circ f)(x)$, the key step is substituting the output of the function $f(x)$ into the function $g(x)$. In Example 1.4.3 we computed that $(g \circ f)(2) = 4$. If we evaluate by substituting into the formula $(g \circ f)(x) = 8 + 2x - 2x^2$, we find

$$(g \circ f)(2) = 8 + 2 \cdot 2 - 2 \cdot 2^2 = 4$$

again matching our earlier result. ▲

Example 1.4.4 illustrates an important point about the composition of functions: the answer is generally different when the functions are composed in a different order. If it happens that $f \circ g = g \circ f$, we say that the two functions **commute.**

When the two compositions do not match, we say that the two functions do not commute. If you think of functions as operations, this makes sense; sterilizing the scalpel and then making an incision produces a quite different result than making an incision and then sterilizing the scalpel.

Example 1.4.5 Composition of Functions in Biology

Numbers of bacteria are usually measured indirectly, for instance by measuring the optical density of the medium (such as water) populated by the bacteria. Water allows less light through as the population becomes larger. Suppose that the optical density, ρ, is a function of the bacterial population of size b with formula

$$\rho(b) = \frac{2.3}{2 + b}$$

illustrated in Figure 1.4.37a. Since the size, $b(t)$, of the bacterial population changes with time, the optical density as a function of time is the composition of the function $\rho(b)$ with the function $b(t)$.

Suppose that

$$b(t) = 0.24t^2 + 1$$

(Figure 1.4.37b). Then

$$\rho(b(t)) = \rho(0.24t^2 + 1) = \frac{2.3}{2 + 0.24t^2 + 1} = \frac{2.3}{3 + 0.24t^2}$$

(Figure 1.4.37c).

The composition $b \circ \rho$ is not merely different from the composition $\rho \circ b$, it does not even make sense. The function b accepts as input only the time t, not the optical density returned as output by the function ρ. ▲

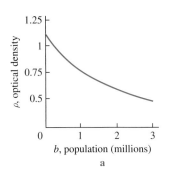
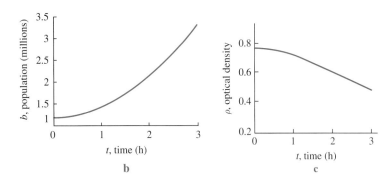

FIGURE 1.4.37

Composing biological functions

Inverse Functions

Example 1.4.6   Census Data on the Population of Canada

The population of Canada has been steadily increasing since the 1950s. Table 1.4.1 shows the census data, obtained from the Statistics Canada website (http://www.statcan.gc.ca).

We think of the population as the function $P = f(t)$, where P is measured in thousands and t denotes time (year in C.E. counting).

Looking at Table 1.4.1, we see that in 1981 the population was $P(1981) = 24,343$ thousands (or 24,343,000); in 2006, the population was $P(2006) = 31,613$ thousands (i.e., just over 31.6 million).

For various reasons we might need to use these data in a different way, and ask questions such as, "When did the population of Canada reach 20 million? 30 million?"

Table 1.4.1

Year, t	Population, $P = f(t)$, in Thousands
1951	14,009
1956	16,081
1961	18,238
1966	20,015
1971	21,568
1976	22,992
1981	24,343
1986	25,309
1991	27,297
1996	28,847
2001	30,007
2006	31,613
2011	33,477

Wikipedia, Population of Canada by year. Found at: http://en.wikipedia.org/wiki/Population_of_Canada_by_year; Statistics Canada. 2011. *Population and Dwelling Counts, for Canada, Provinces and Territories, 2011 and 2006 Censuses*, "Population and dwelling count highlight tables, 2011 Census." Statistics Canada Catalogue no. 98-310-XWE2011002. Ottawa, Ontario. http://www12.statcan.gc.ca/census-recensement/2011/dp-pd/hlt-fst/pd-pl/Table-Tableau.cfm?LANG=Eng&T=101&S=50&O=A; Statistics Canada. 2007. *Population and Dwelling Counts, for Canada, Provinces and Territories, 2006 and 2001 Censuses, 100% Data* (table). "Population and dwelling count highlight tables, 2006 Census." "2006 Census: Release topics." Census. Statistics Canada Catalogue no. 97-550-XWE2006002. Ottawa, Ontario. March 13. http://www12.statcan.ca/english/census06/data/popdwell/Table.cfm?T=101; Statistics Canada. 2001. *Population and Dwelling Counts, for Canada, Provinces and Territories, 2001 and 1996 Censuses, 100% Data* (table). "Population and dwelling count highlight tables, 2001 Census." "2001 Census: Release topics." Census. Statistics Canada Catalogue no. 92-377-XIE02001. Ottawa, Ontario. http://www12.statcan.ca/english/census01/products/standard/popdwell/Table-PR.cfm.

Table 1.4.2

Population P, in Thousands	Year $t = f^{-1}(P)$
14,009	1951
16,081	1956
18,238	1961
20,015	1966
21,568	1971
22,992	1976
24,343	1981
25,309	1986
27,297	1991
28,847	1996
30,007	2001
31,613	2006
33,477	2011

Wikipedia, Population of Canada by year. Found at: http://en.wikipedia.org/wiki/Population_of_Canada_by_year; Statistics Canada. 2011. *Population and Dwelling Counts, for Canada, Provinces and Territories, 2011 and 2006 Censuses*, "Population and dwelling count highlight tables, 2011 Census." Statistics Canada Catalogue no. 98-310-XWE2011002. Ottawa, Ontario. http://www12.statcan.gc.ca/census-recensement/2011/dp-pd/hlt-fst/pd-pl/Table-Tableau.cfm?LANG=Eng&T=101&S=50&O=A; Statistics Canada. 2007. *Population and Dwelling Counts, for Canada, Provinces and Territories, 2006 and 2001 Censuses, 100% Data* (table). "Population and dwelling count highlight tables, 2006 Census." "2006 Census: Release topics." Census. Statistics Canada Catalogue no. 97-550-XWE2006002. Ottawa, Ontario. March 13. http://www12.statcan.ca/english/census06/data/popdwell/Table.cfm?T=101; Statistics Canada. 2001. *Population and Dwelling Counts, for Canada, Provinces and Territories, 2001 and 1996 Censuses, 100% Data* (table). "Population and dwelling count highlight tables, 2001 Census." "2001 Census: Release topics." Census. Statistics Canada Catalogue no. 92-377-XIE02001. Ottawa, Ontario. http://www12.statcan.ca/english/census01/products/standard/popdwell/Table-PR.cfm.

Answering these questions requires that we change the way we think of the population function—we need to view time as a function of population. By reversing the roles of the variables, we have created the **inverse function,** which we denote by f^{-1}; i.e.,

$$t = f^{-1}(P)$$

So the value of $t(P)$ is the time (year) when the population was equal to P. For instance, $t(20,015) = 1966$. Thus, it was in 1966 (or perhaps late in 1965) that the population of Canada reached 20 million. From $t(30,007) = 2001$ we conclude that the population of Canada crossed the 30 million mark in 2001.

To produce the table of values for f^{-1}, we do not need new data—all we need is to switch the two columns in the table that we already have. See Table 1.4.2.

Example 1.4.7 Conversion between Degrees Celsius and Degrees Fahrenheit

The function

$$F = f(C) = \frac{9}{5}C + 32$$

gives the conversion from degrees Celsius to degrees Fahrenheit. The temperature in degrees Celsius is the independent variable, and the temperature in degrees Fahrenheit is the dependent variable. For example, the Fahrenheit equivalent of 25°C can be calculated as follows:

$$F = f(25) = \frac{9}{5}(25) + 32 = 78$$

Thus, 25°C = 78°F.

Now suppose that a thermometer shows a temperature of 14°F. How many degrees Celsius is that? Clearly, we need to undo the function that converts degrees Celsius into degrees Fahrenheit. In symbols, we need the inverse function to express C as a function of F.

Starting with
$$F = \frac{9}{5}C + 32$$
we get
$$\frac{9}{5}C = F - 32$$
$$C = \frac{5}{9}(F - 32)$$

Thus,
$$C = f^{-1}(F) = \frac{5}{9}(F - 32)$$

For $F = 14°F$,
$$C = f^{-1}(14) = \frac{5}{9}(14 - 32) = \frac{5}{9}(-18) = -10$$

Thus, $14°F$ is the same as $-10°C$.

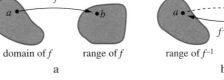

FIGURE 1.4.38
A function and its inverse

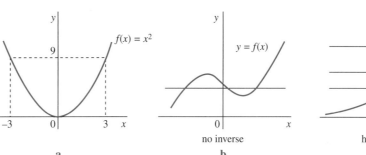

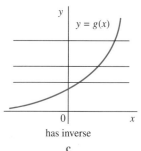

FIGURE 1.4.39
Horizontal line test

Recall that a function f is a rule that assigns to each number a in the domain of f a unique number $b = f(a)$ in the range of f. We can visualize this using an arrow diagram; see Figure 1.4.38a. The arrow starts at the domain of f, points at its range, and indicates that b is the unique number assigned to a.

The idea of the inverse function is to reverse (or undo) this process. Given f, we denote its inverse (as we have done already) by f^{-1}. The solid arrow in Figure 1.4.38b represents f^{-1}. It indicates that the domain of f^{-1} is the same as the range of f, and the range of f^{-1} is the same as the domain of f. As well, $f^{-1}(b) = a$.

For instance, if f takes $a = 2$ and returns 11, i.e., $f(2) = 11$, then f^{-1} takes $b = 11$ and returns 2, i.e., $f^{-1}(11) = 2$. Likewise,

$$\text{if } f(-5) = 18 \text{ then } f^{-1}(18) = -5$$

and

$$\text{if } f^{-1}(42) = 13 \text{ then } f(13) = 42$$

Not every function has an inverse function. Take, for instance, $f(x) = x^2$. Since $f(3) = 9$ and $f(-3) = 9$, what number should f^{-1} assign to 9? 3 or -3? Recall that in order to define a function, we need *uniqueness* of the assignment; in those situations where we cannot achieve that uniqueness, we say that the function does not have an inverse.

Looking at the graph of $f(x) = x^2$ in Figure 1.4.39a, we see that $f(3)$ and $f(-3)$ have the same height; that is, the horizontal line through 9 on the y-axis crosses the graph at two points, $(3, 9)$ and $(-3, 9)$. We conclude that if any horizontal line crosses the graph of a function twice (or three times, or more), then that function cannot have an inverse (another example is provided in Figure 1.4.39b).

However, if every horizontal line crosses the graph of a function exactly once or not at all, then the function has an inverse (such as the function $g(x)$ in Figure 1.4.39c).

Horizontal Line Test If every horizontal line crosses the graph of a function $f(x)$ at most once, then $f(x)$ has an inverse function.

Let us look at some graphs we drew in the previous section and apply the test: we conclude that the functions

$$f(x) = x, \; f(x) = x^3, \; f(x) = 1/x, \; f(x) = \sqrt{x}, \text{ and } f(x) = \sqrt[3]{x}$$

have an inverse, whereas

$$f(x) = x^2, \; f(x) = x^4, \; f(x) = 1/x^2, \text{ and } f(x) = |x|$$

do not.

A function that satisfies the horizontal line test is also called **one-to-one**, since it never assumes the same value twice.

Now that we know and understand how to determine which functions have an inverse, we can formally define inverse functions.

Definition 1.4.4 Assume that a function, f, satisfies the horizontal line test and that its domain is A and its range is B. The **inverse function**, f^{-1}, of f is the function with domain B and range A such that

$$f^{-1}(b) = a$$

whenever $f(a) = b$.

From the definition, we calculate the compositions

$$(f^{-1} \circ f)(a) = f^{-1}(f(a)) = f^{-1}(b) = a$$

and

$$(f \circ f^{-1})(b) = f(f^{-1}(b)) = f(a) = b$$

In words, composing a function and its inverse (in either order) does not change the value of the number the composition is applied to. This sounds reasonable: if we take a number, apply f to it, and then undo f (i.e., apply f^{-1}), we should get our number back.

Note that a function and its inverse come in pairs: f^{-1} is the inverse of f, and f is the inverse of f^{-1}.

Why did we use a and b in the above, and not x and y? We are used to the variable x representing the independent variable and y representing the function (dependent variable). In defining the inverse function, we had to switch the roles of x and y (i.e., y becomes the independent variable and x becomes the dependent variable), so, to avoid confusion, we decided to use "neutral" symbols for the variables.

But now that we understand how inverse functions work, we will resume the usual practice of using x to denote the independent variable for *any* function we work with, and we will write

$$f(f^{-1}(x)) = x \text{ and } f^{-1}(f(x)) = x$$

keeping in mind that x in $f(f^{-1}(x)) = x$ belongs to the domain of f^{-1}, and x in $f^{-1}(f(x)) = x$ represents a number in the domain of f.

The above two formulas are called the **cancellation formulas**, for obvious reasons.

Example 1.4.8 Cancellation Formulas I

Consider the functions $f(x) = x^3$ and $g(x) = \sqrt[3]{x}$. The fact that

$$f(g(x)) = f(\sqrt[3]{x}) = (\sqrt[3]{x})^3 = x$$

1.4 Working with Functions

and
$$g(f(x)) = g(x^3) = \sqrt[3]{x^3} = x$$
proves that $f(x)$ and $g(x)$ are inverses of each other.

Example 1.4.9 Cancellation Formulas II

Let us check that the conversion formulas
$$F = f(C) = \frac{9}{5}C + 32 \quad \text{and} \quad C = f^{-1}(F) = \frac{5}{9}(F - 32)$$
from the beginning of this subsection are indeed inverses of each other:
$$\begin{aligned}
f^{-1}(f(C)) &= f^{-1}\left(\frac{9}{5}C + 32\right) \\
&= \frac{5}{9}\left(\left(\frac{9}{5}C + 32\right) - 32\right) \\
&= \frac{5}{9}\left(\frac{9}{5}C\right) \\
&= C
\end{aligned}$$

As well,
$$\begin{aligned}
f(f^{-1}(F)) &= f\left(\frac{5}{9}(F - 32)\right) \\
&= \frac{9}{5}\left(\frac{5}{9}(F - 32)\right) + 32 \\
&= F - 32 + 32 \\
&= F
\end{aligned}$$

Recall that a function assigns a unique number, $y = f(x)$, to each x in its domain. Algebraically, a function is a formula that expresses y in terms of x. To find a formula for the inverse function, we have to find the rule that undoes this; i.e., we need to express x in terms of y.

▶▶ **Algorithm 1.4.1** Finding the Inverse, f^{-1}, of a Function f

1. Write $y = f(x)$.
2. Solve for x in terms of y, if possible.
3. Express f^{-1} as a function of x by interchanging x and y.

If the variables involved have a meaning (i.e., if we are working in the context of an application), we skip step 3. Note that in the conversion formulas in Example 1.4.7 (see also Example 1.4.9), we solved $F = (9/5)C + 32$ for C, got $C = (5/9)(F - 32)$, and left it in that form. Switching F and C would not make sense at all, as it would produce an erroneous conversion formula.

So we do step 3 only when we are working with abstract variables x and y, i.e., when there is no context.

Example 1.4.10 Finding an Inverse

Consider the population that changes in accordance with the equation
$$b(t) = \frac{t^2}{4.2} + 1$$

(Figure 1.4.40a). If we wish to find the time, t, from the population, b, we must solve for t:

$$\frac{t^2}{4.2} + 1 = b$$

$$\frac{t^2}{4.2} = b - 1$$

$$t^2 = 4.2(b - 1)$$

$$t = \sqrt{4.2(b - 1)}$$

Since t is time, we took the positive square root.

This function is graphed in Figure 1.4.40b. The last step requires that $b \geq 1$ because we cannot take the square root of a negative number. For example, the time associated with a population of 5 is

$$t = \sqrt{4.2(5 - 1)} \approx 4.1$$

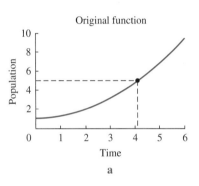
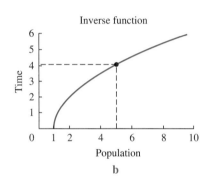

Figure 1.4.40

Going backwards with the inverse function

Example 1.4.11 Finding the Inverse of a Function

To find the inverse of the function $f(x) = \dfrac{x+1}{x-1}$, we start by writing it in the form

$$y = \frac{x+1}{x-1}$$

which we solve for x. Cross-multiplying and gathering all terms with x on the left, we get

$$(x - 1)y = x + 1$$
$$xy - y = x + 1$$
$$xy - x = y + 1$$
$$x(y - 1) = y + 1$$
$$x = \frac{y+1}{y-1}$$

We are done—the inverse function is

$$f^{-1}(y) = \frac{y+1}{y-1}$$

To write f^{-1} in the usual form, with x representing the independent variable (there is no context here, so we do step 3 from Algorithm 1.4.1), we replace y with x and obtain

$$f^{-1}(x) = \frac{x+1}{x-1}$$

Note that, in Example 1.4.11, the inverse function, $f^{-1}(x)$, is equal to the original function, $f(x)$; in other words, the function $f(x)$ is its own inverse. Convince yourself that the function $f(x) = 1/x$ has the same property.

Example 1.4.12   A Function with No Inverse

Consider the data in the following table.

Initial Mass (g)	Final Mass (g)	Initial Mass (g)	Final Mass (g)
1	7	9	20
2	12	10	18
3	16	11	15
4	19	12	12
5	22	13	9
6	23	14	6
7	23	15	3
8	22	16	1

Suppose you were told that the mass at the end of the experiment was 12 g. Initial masses of 2 g and 12 g both produce a final mass of 12 g. You cannot tell whether the input was 2 or 12. So this function has no inverse (Figure 1.4.41).

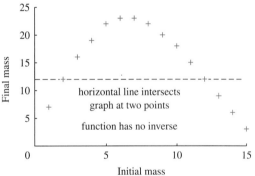

FIGURE 1.4.41
A function with no inverse

Algorithm 1.4.1 might fail because the algebra is impossible. Step 2 requires solving an equation, and many equations cannot be solved algebraically.

Example 1.4.13   A Case When Finding the Inverse Explicitly Is Not Possible

The function $f(x) = x^5 + x + 1$ is an increasing function (see Figure 1.4.42), and therefore it satisfies the horizontal line test. (After learning about derivatives, we will be able to prove that f is indeed increasing for all x.)

So f has an inverse function. In order to find it, we need to solve the equation

$$y = x^5 + x + 1$$

for x. However, this is not possible. French mathematician Evariste Galois proved that there is no formula for solutions of polynomial equations of degree five or higher.

But it is not all bad news. It is possible to compute the approximate inverse of this function (and others) using a numerical algorithm and a calculator or computer.

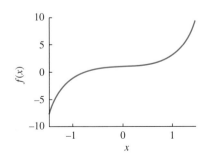

FIGURE 1.4.42
A function with an inverse that is impossible to compute algebraically

How do we find the inverse f^{-1} if f is given as graph?

Assume that for some function, $f(3) = 0$; this means that the point $(3, 0)$ belongs to its graph. But then $f^{-1}(0) = 3$, which means that the point $(0, 3)$ belongs to the graph of f^{-1}. Likewise, if the point $(5, 1)$ belongs to the graph of f, then the point $(1, 5)$ must belong to the graph of f^{-1}. What is the geometric relation between the points $(3, 0)$ and

(0, 3) and between the points (5, 1) and (1, 5)? Looking at Figure 1.4.43a, we see that these pairs of points are symmetric with respect to the line $y = x$.

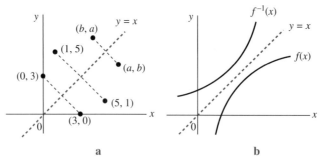

FIGURE 1.4.43

Finding the inverse geometrically

Thus, to draw the inverse, we reflect the given graph across the diagonal line $y = x$, as shown in Figure 1.4.43b.

Finding the Inverse of a Function

If the function is given as a formula $y = f(x)$, then solve for x in terms of y, if possible (see Algorithm 1.4.1). If the function is given as graph, then reflect it with respect to the diagonal $y = x$.

Example 1.4.14 Finding an Inverse Algebraically and Geometrically

Let us try to find the inverse function of the function $f(x) = x^2$, where $x \geq 0$.

It is true that $y = x^2$ does *not* have an inverse when we think of it as defined on the set of all real numbers. However, here we are dealing with the function $y = x^2$ defined for $x \geq 0$ only. Its graph is the part of the parabola $y = x^2$ in the first quadrant, together with the origin; see Figure 1.4.44a.

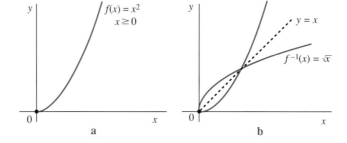

FIGURE 1.4.44

Finding the inverse geometrically

Clearly, this function satisfies the horizontal line test, so it must have an inverse. Solving $y = x^2$ for x, we get

$$x = \pm\sqrt{y}$$

Because we chose to limit ourselves only to $x \geq 0$, we must choose the plus sign, and thus the inverse is

$$f^{-1}(x) = \sqrt{x}$$

To obtain the graph of $f^{-1}(x)$, we reflect the graph of $f(x)$ for $x \geq 0$ with respect to the line $y = x$; see Figure 1.4.44b.

Creating New Functions

Using elementary algebraic operations and compositions, we are able to form important families of functions.

A function of the form

$$p(x) = a_n x^n + a_{n-1} x^{n-1} + \cdots + a_1 x + a_0$$

where n is an integer, $n \geq 0$, and $a_1, a_2, \ldots, a_n$ are real numbers, is called a **polynomial.** If $a_n \neq 0$, then $p(x)$ is said to be a polynomial of degree n. The domain of any polynomial is the set of real numbers. The functions

$$f(x) = x^3 - 1.3x^2 + 23,\ f(x) = 3 - x^5,\ f(x) = x - \pi,\ \text{and}\ f(x) = 2$$

are polynomials of degree 3, 5, 1, and 0, respectively. As well, the functions listed in (1)–(4) in the Catalogue of Important Functions in Section 1.3 are polynomials.

The quotient of two polynomials is called a **rational function.** For example,

$$f(x) = \frac{2x^3 - 13x - 4}{x + 4},\ f(x) = \frac{2.5}{x},\ \text{and}\ f(x) = \frac{x + 1}{x^2 - x - 1}$$

and functions (5) and (6) from the Catalogue of Important Functions in Section 1.3 are rational functions. Because every polynomial $p(x)$ can be written as the quotient $p(x)/1$, we conclude that polynomials are rational functions. The domain of a rational function consists of all real numbers for which the polynomial in the denominator is not zero.

Starting with polynomials, and combining them using composition, elementary algebraic operations (addition, subtraction, multiplication, division), and power and root functions ($\sqrt{x}, \sqrt[3]{x}, \sqrt[4]{x}$, etc.), we obtain **algebraic functions.** The following are examples of algebraic functions:

$$f(x) = \sqrt{\frac{x^3 - 13x}{x - 1}},\ f(x) = \sqrt[4]{1 - x + 2x^2} - 3x,$$

$$f(x) = \frac{1}{\sqrt[3]{x^4 - 2 + x}} + \frac{2.5}{x},\ \text{and}\ f(x) = (\sqrt{x} + 3)^6$$

Functions that are not algebraic are called **transcendental functions.** Examples of transcendental functions are trigonometric and inverse trigonometric, exponential, and logarithm functions. We will study these functions in detail in forthcoming sections.

Transformations of Graphs

We now examine ways of obtaining new graphs from existing graphs.

Shifts Assume that $c > 0$ is a constant. The fact that the function $y = f(x) + c$ is obtained by adding c to every value of $f(x)$ means that the points on the graph of $y = f(x) + c$ are obtained by moving points on the graph of $y = f(x)$ up c units. Similarly, the graph of $y = f(x) - c$ is obtained by moving the graph of $y = f(x)$ down c units; see Figure 1.4.45.

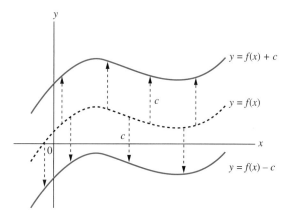

FIGURE 1.4.45
Vertical shifts

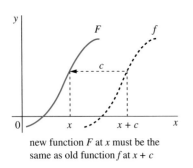

FIGURE 1.4.46
Horizontal shifts

new function F at x must be the same as old function f at $x+c$

Now consider a new function, F, defined by $F(x) = f(x+c)$, where the graph of f is known. What is the relation between the graphs of f and F?

The formula $F(x) = f(x+c)$ says that the value of the new function F at x is the same as the value of f at $x+c$ (note that $x+c$ is c units to the *right* of x); see Figure 1.4.46. So, starting with f, we obtain the graph of F by moving f to the *left* c units.

The case $F(x) = f(x-c)$ is argued analogously.

Let $c > 0$. To Obtain the Graph of	
$y = f(x) + c$	shift the graph of $f(x)$ up c units
$y = f(x) - c$	shift the graph of $f(x)$ down c units
$y = f(x+c)$	shift the graph of $f(x)$ left c units
$y = f(x-c)$	shift the graph of $f(x)$ right c units

Figures 1.4.47 and 1.4.48 illustrate the rules presented in the table.

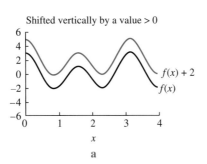
a

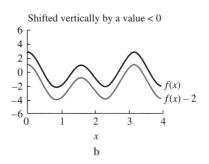
b

FIGURE 1.4.47
Vertically shifting a function

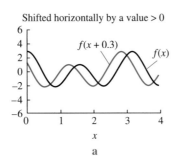
a

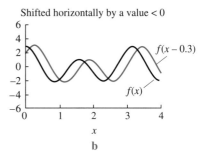
b

FIGURE 1.4.48
Horizontally shifting a function

Example 1.4.15 Vertically and Horizontally Shifting a Function

Consider the function

$$f(x) = \frac{1}{1+x^2}$$

defined on $-5 \leq x \leq 5$, drawn in Figure 1.4.49. We will shift the value and the argument of this function by values both greater than and less than 0.

Shifting the function vertically corresponds to adding a constant to $f(x)$.

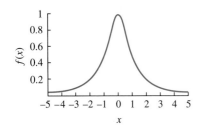

FIGURE 1.4.49
The original function

1.4 Working with Functions

Argument	Original Function	Vertically Shifted by a Value > 0	Vertically Shifted by a Value < 0	Horizontally Shifted by a Value > 0	Horizontally Shifted by a Value < 0
x	$f(x)$	$f(x)+2$	$f(x)-2$	$f(x+2)$	$f(x-2)$
−5	0.04	2.04	−1.96	0.1	0.02
−4	0.06	2.06	−1.94	0.2	0.03
−3	0.1	2.1	−1.9	0.5	0.04
−2	0.2	2.2	−1.8	1	0.06
−1	0.5	2.5	−1.5	0.5	0.1
0	1	3.0	−1	0.2	0.2
1	0.5	2.5	−1.5	0.1	0.5
2	0.2	2.2	−1.8	0.06	1
3	0.1	2.1	−1.9	0.04	0.5
4	0.06	2.06	−1.94	0.03	0.2
5	0.04	2.04	−1.96	0.02	0.1

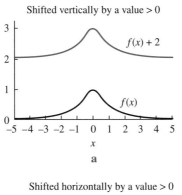
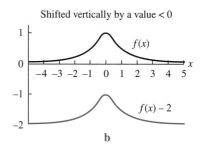
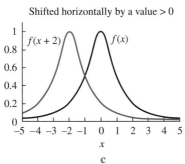
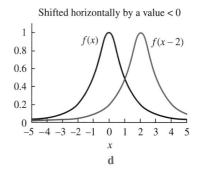

FIGURE 1.4.50
Vertically and horizontally shifting a function

Thus, shifting vertically moves the function up if it is shifted by a value greater than 0 or down if it is shifted by a value less than 0. Shifting horizontally moves the function to the left if it is shifted by a value greater than 0 or to the right if it is shifted by a value less than 0. See Figure 1.4.50.

Scaling Take a function $f(x)$ and form the function $F(x) = cf(x)$ for $c > 0$. To obtain the value of F at x, we take the value of f at x and multiply it by c. In other words, we scale the graph vertically (which amounts to stretching if $c > 1$ and compression if $c < 1$).

In Figure 1.4.51 we show the graphs of $f(x)$, $2f(x)$, and $f(x)/3$. We say that the graph of $f(x)$ needs to be scaled vertically by a factor of 2 (or stretched vertically by a factor of 2) to produce the graph of $2f(x)$. As well, the graph of $f(x)$ has to be scaled vertically by a factor of $1/3$ (or compressed vertically by a factor of 3) to obtain the graph of $f(x)/3$.

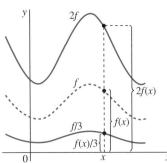

FIGURE 1.4.51
Vertical scaling

FIGURE 1.4.52
Horizontal scaling

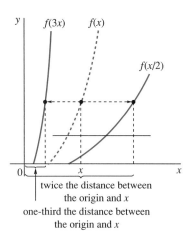

Given the graph of a function f, we define a new function, F, by $F(x) = f(cx)$, where c is a positive constant. What is the relation between the graphs of f and F?

Assume, for the sake of argument, that $c = 3$; i.e., consider $F(x) = f(3x)$. The value of F at x is the same as the value of f at $3x$ (which is three times as far away from the origin as x). So we obtain the graph of F by compressing f horizontally by a factor of 3.

Likewise, to obtain $f(x/2)$, we stretch the graph of f horizontally by a factor of 2, as shown in Figure 1.4.52.

To Obtain the Graph of	
$y = cf(x), c > 1$	stretch the graph of $f(x)$ vertically by a factor of c
$y = cf(x), 0 < c < 1$	compress the graph of $f(x)$ vertically by a factor of $1/c$
$y = f(cx), c > 1$	compress the graph of $f(x)$ horizontally by a factor of c
$y = f(cx), 0 < c < 1$	stretch the graph of $f(x)$ horizontally by a factor of $1/c$

See Figures 1.4.53 and 1.4.54 for an illustration of how these rules work.

We note that when $0 < c < 1$, the transformation (compression or stretching) is by a factor of $1/c$ and not c. It is a matter of language, i.e., what the words we use suggest in their usual, daily meaning. Consider the case of $f(x/2)$ (i.e., $c = 1/2$) that we discussed above; we argued that the graph of $f(x/2)$ is an expansion of the graph of $f(x)$. How do we quantify the size of the expansion? Saying "expansion by a factor of $c = 1/2$" actually suggests compression, so we say "expansion by a factor of $1/c = 2$."

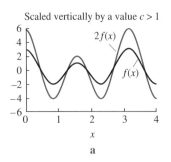

a

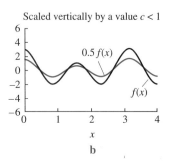

b

FIGURE 1.4.53
Vertically scaling a function

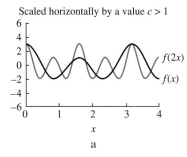
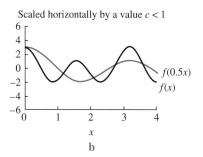

FIGURE 1.4.54

Horizontally scaling a function

Example 1.4.16 Vertically and Horizontally Scaling a Function

Consider again the function

$$f(x) = \frac{1}{1+x^2}$$

shown plotted for $-5 \leq x \leq 5$ in Figure 1.4.49 above. We will scale the value and the argument of this function by values both greater than and less than 1.

Consider the input $x = -1$. Then

$$2f(-1) = 2 \cdot \frac{1}{1+(-1)^2} = 1$$

$$0.5f(-1) = 0.5 \cdot \frac{1}{1+(-1)^2} = 0.25$$

$$f(2 \cdot (-1)) = f(-2) = \frac{1}{1+(-2)^2} = 0.2$$

$$f(0.5 \cdot (-1)) = f(-0.5) = \frac{1}{1+(-0.5)^2} = 0.8$$

The following table gives several values of the scaled functions.

Argument	Original Function	Vertically Scaled by a Value > 1	Vertically Scaled by a Value < 1	Horizontally Scaled by a Value > 1	Horizontally Scaled by a Value < 1
x	$f(x)$	$2f(x)$	$0.5f(x)$	$f(2x)$	$f(0.5x)$
−5	0.04	0.08	0.02	0.01	0.14
−4	0.06	0.12	0.03	0.02	0.2
−3	0.1	0.2	0.05	0.03	0.31
−2	0.2	0.4	0.1	0.06	0.5
−1	0.5	1	0.25	0.2	0.8
0	1	2	0.5	1	1
1	0.5	1	0.25	0.2	0.8
2	0.2	0.4	0.1	0.06	0.5
3	0.1	0.2	0.05	0.03	0.31
4	0.06	0.12	0.03	0.02	0.2
5	0.04	0.08	0.02	0.01	0.14

Scaling vertically makes the graph of the function taller if it is scaled by a value greater than 1 and shorter if it is scaled by a value less than 1. Scaling horizontally makes the graph of the function thinner if it is scaled by a value greater than 1 and wider if it is scaled by a value less than 1. See Figures 1.4.53 to 1.4.55.

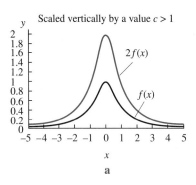
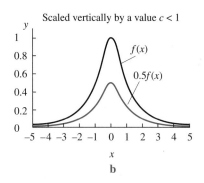
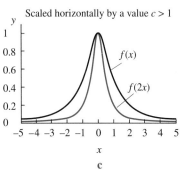
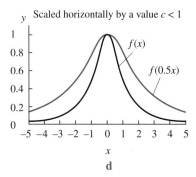

FIGURE 1.4.55

Vertically and horizontally scaling a function

Reflections How do we get the graph of $-f$ from the graph of f?

Since $f(x)$ and $-f(x)$ lie across the x-axis from one another (see Figure 1.4.56a), we conclude that the graph of the function $-f$ is the mirror image of the graph of f with respect to the x-axis.

Since x and $-x$ lie symmetrically across the y-axis, it follows that the graph of $f(-x)$ is the mirror image of the graph of $f(x)$ with respect to the y-axis (Figure 1.4.56b).

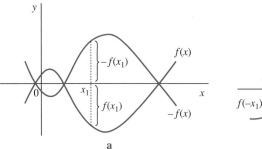
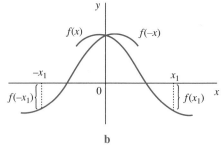

FIGURE 1.4.56

Reflections

To Obtain the Graph of	
$y = -f(x)$	reflect the graph of $f(x)$ with respect to the x-axis
$y = f(-x)$	reflect the graph of $f(x)$ with respect to the y-axis

Example 1.4.17 Reflecting a Graph

Based on the graph of the function $y = f(x)$ given in Figure 1.4.57a, we will draw the graph of the function $y = |f(x)|$.

We write the definition of absolute value:

$$|f(x)| = \begin{cases} f(x) & \text{if } f(x) \geq 0 \\ -f(x) & \text{if } f(x) < 0 \end{cases}$$

Now we interpret this. The first line says that whenever $f(x) \geq 0$, i.e., when the graph of $y = f(x)$ lies above the x-axis or touches it, the function $y = |f(x)|$ is the same as $y = f(x)$.

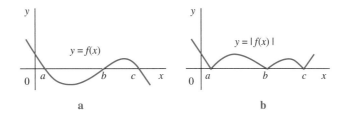

FIGURE 1.4.57

Obtaining the graph of $y = |f(x)|$ from the graph of $y = f(x)$

The second line says that whenever $f(x) < 0$, i.e., when the graph of $y = f(x)$ lies below the x-axis, the function $y = |f(x)|$ is the same as $y = -f(x)$.

So, to obtain the graph of $y = |f(x)|$ from $y = f(x)$, we keep the parts of the graph of $y = f(x)$ that are above the x-axis (or touch the x-axis), and reflect (across the x-axis) those parts of the graph of $y = f(x)$ that lie below the x-axis; see Figure 1.4.57b.

In this book we will study models for various populations (cells, bacteria, animals, humans, etc.). The changes in population size are often described by specifying the growth rate or the per capita growth rate. The **growth rate** tells us how a population changes in small time intervals (we will make this definition precise after we learn about the derivatives); it is expressed as the number of individuals per unit of time. The **per capita growth rate** (or the **specific growth rate** or the **relative growth rate**) is the growth rate divided by the population size. Assume that at the moment when the population of bacteria reaches 120,000, its growth rate is 15,000 bacteria/hour. The per capita growth rate at that moment is 15,000/120,000 = 0.125 bacteria/hour per bacterium.

Example 1.4.18 Monod Growth Function

In some situations, the per capita growth rate of a population depends on the concentration of the nutrient. To model these populations, we assume that with no nutrient present, the per capita growth rate is zero; when the nutrient is available in unlimited quantities, the per capita growth rate reaches a certain level (sometimes we say that it *stabilizes* or *plateaus* at a certain level). This behaviour is described by the **Monod growth function**

$$r(c) = \frac{Ac}{k+c} \tag{1.4.2}$$

where c is the concentration of the nutrient ($c \geq 0$), $r(c)$ is the per capita growth rate, and A and k are positive constants. The graph of $r(c)$ is shown in Figure 1.4.58 (in Exercises 78 and 79 we suggest the use of transformations to obtain this graph). We see that $r(0) = 0$, and, as c increases, the values of $r(c)$ approach the **saturation level** A.

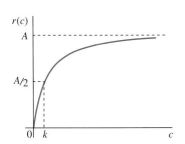

FIGURE 1.4.58

Monod growth curve

Substituting $c = k$ into Equation 1.4.2, we obtain

$$r(k) = \frac{Ak}{k+k} = \frac{Ak}{2k} = \frac{A}{2}$$

In words, the parameter k represents the concentration required for the per capita rate to reach one half of its saturation value, A (k is called the **half-saturation constant**).

Note an important feature of the graph of the Monod function, related to doubling the values of the concentration: doubling a small concentration produces a larger change in the per capita rate (Figure 1.4.59a) than doubling a large concentration (Figure 1.4.59b).

In a simplified case of enzyme kinetics, formula (1.4.2), written as

$$v = \frac{Ac}{k_M + c} \tag{1.4.3}$$

models the speed of the reaction (i.e., the rate of production of an enzyme from its substrate). The variable c is the concentration of the substrate, A is the maximum speed of the reaction, and k_M is called the **Michaelis constant.** Formula (1.4.3) is referred to as the **Michaelis-Menten model.**

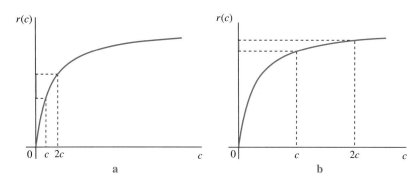

FIGURE 1.4.59

Reactions of the per capita rate to doubling the concentration

Summary New functions are built by combining functions through **addition, subtraction, multiplication, division,** and **composition**. In functional composition, the output of the **inner function** is used as the input of the **outer function.** The composition of functions is not commutative. A function that passes the horizontal line test has an **inverse function.** Geometrically, we create new graphs by **shifting, scaling,** and **reflecting** given graphs.

1.4 Exercises

Mathematical Techniques

1–4 ■ Graph each pair of functions on the same coordinate axes. Compute the value of the sum at $x = -2$, $x = -1$, $x = 0$, $x = 1$, and $x = 2$, and plot the result.

1. $f(x) = 2x + 3$ and $g(x) = 3x - 5$
2. $f(x) = 2x + 3$ and $h(x) = -3x - 12$
3. $F(x) = x^2 + 1$ and $G(x) = x + 1$
4. $F(x) = x^2 + 1$ and $H(x) = -x + 1$

5–8 ■ Graph each pair of functions on the same coordinate axes. Compute the value of the product at $x = -2$, $x = -1$, $x = 0$, $x = 1$, and $x = 2$, and graph the result.

5. $f(x) = 2x + 3$ and $g(x) = 3x - 5$
6. $f(x) = 2x + 3$ and $h(x) = -3x - 12$
7. $F(x) = x^2 + 1$ and $G(x) = x + 1$
8. $F(x) = x^2 + 1$ and $H(x) = -x + 1$

9–12 ■ Find the formulas for the product $f \cdot g$ and the quotient f/g, and identify their domain.

9. $f(x) = x^3$, $g(x) = 1 - x$
10. $f(x) = 1$, $g(x) = \dfrac{2}{x}$
11. $f(x) = \sqrt{x}$, $g(x) = \sqrt{x - 1}$
12. $f(x) = \dfrac{x^3}{2}$, $g(x) = \dfrac{2}{x}$

13–20 ■ Find the formulas for the compositions $f \circ g$ and $g \circ f$. Simplify where possible.

13. $f(x) = x^3$, $g(x) = 4 - 2x$
14. $f(x) = 12 - x^2$, $g(x) = 4$
15. $f(x) = \dfrac{1}{x}$, $g(x) = x - 3$
16. $f(x) = \dfrac{x - 1}{2}$, $g(x) = \dfrac{1 - x}{2}$
17. $f(x) = \dfrac{x - 1}{x + 1}$, $g(x) = \dfrac{1}{x}$
18. $f(x) = \sqrt{x - 2}$, $g(x) = \sqrt{2 - x}$
19. $f(x) = \dfrac{x}{1 + x^2}$, $g(x) = \sqrt{x}$
20. $f(x) = |x - 2|$, $g(x) = x^3 - 1$

21–24 ■ Find both compositions of each pair of functions. Which pairs of functions commute?

21. $f(x) = 2x + 3$ and $g(x) = 3x - 5$
22. $f(x) = 2x + 3$ and $h(x) = -3x - 12$
23. $F(x) = x^2 + 1$ and $G(x) = x + 1$
24. $F(x) = x^2 + 1$ and $H(x) = -x + 1$

25–28 ■ Find the inverse of each function when an inverse exists. In each case, compute the output at an input of 1, and show that the inverse undoes the action of the function.

25. $f(x) = 2x + 3$

26. $g(x) = 3x - 5$

27. $F(x) = x^2 + 1$

28. $F(x) = x^2 + 1$ for $x \geq 0$

29–32 ▪ Graph each function and its inverse if it exists. Mark the given point on the graph of each function.

29. $f(x) = 2x + 3$. Mark the point $(1, f(1))$ on the graph of f and the corresponding point on the graph of f^{-1} (based on Exercise 25).

30. $g(x) = 3x - 5$. Mark the point $(1, g(1))$ on the graph of g and the corresponding point on the graph of g^{-1} (based on Exercise 26).

31. $F(x) = x^2 + 1$. Mark the point $(1, F(1))$ on the graph of F and the corresponding point on the graph of F^{-1} (based on Exercise 27).

32. $F(x) = x^2 + 1$ for $x \geq 0$. Mark the point $(1, F(1))$ on the graph of F and the corresponding point on the graph of F^{-1} (based on Exercise 28).

33–38 ▪ Sketch the graph and state the domain and the range of each function. Determine whether or not the function has an inverse. If it does, find it.

33. $f(x) = \sqrt[3]{x} + 4$

34. $g(x) = \sqrt{x} + 4$

35. $y = \dfrac{1}{x^2}, -1 \leq x \leq 1$

36. $f(x) = |x - 2|, -6 \leq x \leq 10$

37. $y = \dfrac{13}{x - 3}$

38. $y = \dfrac{3x - 5}{7}$

39–42 ▪ Find the inverse of each function.

39. $f(x) = \sqrt[3]{2x - 11} - 1$

40. $y = (x - 2)^5$

41. $g(x) = \dfrac{1 - x}{2x}$

42. $f(x) = \dfrac{1 - x}{2 + x}$

43–45 ▪ Find the inverse of each function without interchanging the variable labels.

43. The height of a gorilla is given by $h(m) = a\sqrt[3]{m}$, where m is the mass of the gorilla and a is a positive constant.

44. The quantity v depends on the quantity a according to $v(a) = \dfrac{3a}{a + 2}$.

45. The quantity p depends on the quantity m according to $p(m) = \dfrac{2}{1 + 3/m}$.

46–49 ▪ Using the graph of the function $g(x)$, sketch the graph of the shifted or scaled function, say which kind of shift or scale it is, and compare with the original function.

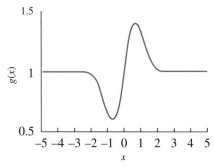

46. $4g(x)$

47. $g(x) - 1$

48. $g(x/3)$

49. $g(x + 1)$

50–59 ▪ Starting with the graph of a known function (see the subsection Catalogue of Important Functions in Section 1.3), use transformations to sketch the graph of each function.

50. $f(x) = -\dfrac{1}{x^2}$

51. $f(x) = \dfrac{1}{(x - 3)^2}$

52. $f(x) = |x - 7|$

53. $f(x) = \sqrt{x - 3}$

54. $f(x) = 4\sqrt{x}$

55. $f(x) = 3 - \sqrt{x - 2}$

56. $f(x) = 1 - \dfrac{1}{x}$

57. $f(x) = x^2 + x + 6$ (Hint: Complete the square.)

58. $f(x) = x^2 - 3x + 1$

59. $f(x) = |x - 2| - 3$

Applications

60–63 ▪ Consider the following data describing the growth of a tadpole.

Age, a (days)	Length, L (cm)	Tail Length, T (cm)	Mass M (g)
0.5	1.5	1.0	1.5
1	3.0	0.9	3.0
1.5	4.5	0.8	6.0
2	6.0	0.7	12.0
2.5	7.5	0.6	24.0
3	9.0	0.5	48.0

60. Graph length as a function of age.

61. Graph tail length as a function of age.

62. Graph tail length as a function of length.

63. Graph mass as a function of length, and then graph length as a function of mass. How do the two graphs compare?

64–67 ▪ The following series of functional compositions describe connections between several measurements.

64. The number of mosquitoes (M) that end up in a room is a function of how much the window is open (W, in square centimetres) according to $M(W) = 5W + 2$. The number of bites (B) depends on the number of mosquitoes according to $B(M) = 0.5M$. Find the number of bites as a function of how much the window is open. How many bites would you get if the window were 10 cm² open?

65. The temperature of a room (T, in degrees Celsius) is a function of how much the window is open (W, in square centimetres) according to $T(W) = 40 - 0.2W$. How long you sleep (S, measured in hours) is a function of the temperature according to $S(T) = 14 - \frac{T}{5}$. Find how long you sleep as a function of how much the window is open. How long would you sleep if the window were 10 cm² open?

66. The number of viruses (V, measured in trillions) that infect a person is a function of the degree of immunosuppression (I, the fraction of the immune system that is turned off by medication) according to $V(I) = 5I^2$. The fever (F, measured in degrees Celsius) associated with an infection is a function of the number of viruses according to $F(V) = 37 + 0.4V$. Find fever as a function of immunosuppression. How high will the fever be if immunosuppression is complete ($I = 1$)?

67. The length of an insect (L, in millimetres) is a function of the temperature during development (T, measured in degrees Celsius) according to $L(T) = 10 + \frac{T}{10}$. The volume of the insect (V, in cubic millimetres) is a function of the length according to $V(L) = 2L^3$. The mass (M, in milligrams) depends on volume according to $M(V) = 1.3V$. Find mass as a function of temperature. How much would an insect weigh that developed at 25°C? Would you be frightened to meet this insect?

68–69 ▪ Each of the following measurements is the sum of two components. Find the formula for the sum. Sketch a graph of each component and the total as functions of time for $0 \leq t \leq 3$. Describe each component and the sum in words.

68. A population of bacteria consists of two types, a and b. The first follows $a(t) = 1 + t^2$, and the second follows $b(t) = 1 - 2t + t^2$, where populations are measured in millions and time is measured in hours. The total population is $P(t) = a(t) + b(t)$.

69. The above-ground volume (stem and leaves) of a plant is $V_a(t) = 3t + 20 + \frac{t^2}{2}$, and the below-ground volume (roots) is $V_b(t) = -t + 40$, where t is measured in days after seed germination and volumes are measured in cubic centimetres. The domain is $0 \leq t \leq 40$. The total volume is $V(t) = V_a(t) + V_b(t)$.

70–73 ▪ Consider the following data describing a plant.

Age, a (days)	Mass, M (g)	Volume, V (cm³)	Glucose Production, G (mg)
0.5	1.5	5.1	0.0
1	3.0	6.2	3.4
1.5	4.3	7.2	6.8
2	5.1	8.1	8.2
2.5	5.6	8.9	9.4
3	5.6	9.6	8.2

70. Graph M as a function of a. Does this function have an inverse? Could we use mass to figure out the age of the plant?

71. Graph V as a function of a. Does this function have an inverse? Could we use volume to figure out the age of the plant?

72. Graph G as a function of a. Does this function have an inverse? Could we use glucose production to figure out the age of the plant?

73. Graph G as a function of M. Does this function have an inverse? What is strange about it? Could we use glucose production to figure out the mass of the plant?

74–77 ▪ The total mass of a population (in kilograms) as a function of the number of years, t, is the product of the number of individuals, $P(t)$, and the mass per person, $M(t)$ (in kilograms). In each of the following exercises, find the formula for the total mass; sketch graphs of $P(t)$, $M(t)$, and the total mass as functions of time for $0 \leq t \leq 100$; and describe the results in words.

74. The population of people, P, is $P(t) = 2 \times 10^6 + 2 \times 10^4 t$, and the mass per person, $M(t)$, is $M(t) = 80 - 0.5t$.

75. The population, P, is $P(t) = 2 \times 10^6 - 2 \times 10^4 t$, and the mass per person, $M(t)$, is $M(t) = 80 + 0.5t$.

76. The population, P, is $P(t) = 2 \times 10^6 + 1000t^2$, and the mass per person, $M(t)$, is $M(t) = 80 - 0.5t$.

77. The population, P, is $P(t) = 2 \times 10^6 + 2 \times 10^4 t$, and the mass per person, $M(t)$, is $M(t) = 80 - 0.005t^2$.

78. Define the function
$$f(x) = \frac{2x}{x+1}$$
Using long division, or otherwise, show that
$$f(x) = 2 - \frac{2}{x+1}$$
Starting with the graph of $y = 1/x$, sketch the graph of $f(x)$ for $x \geq 0$.

79. Define the function
$$r(c) = \frac{Ac}{k+c}$$
where A and k are positive constants. Using long division, or otherwise, show that
$$r(c) = A - \frac{Ak}{k+c}$$
Starting with the graph of $y = 1/c$, sketch the graph of $r(c)$ for $c \geq 0$.

Computer Exercises

80. Have a graphing device plot the following functions for $-2 \leq x \leq 2$. Do they have inverses?
    a. $h_1(x) = x + 2x$
    b. $h_2(x) = x^2 + 2x$
    c. $h_3(x) = x^3 + 2x$
    d. $h_4(x) = x^4 + 2x$
    e. $h_5(x) = x^5 + 2x$

    Have the device try to find the formula for the inverses of these functions and plot the results. Does the device always succeed in finding an inverse when there is one? Does it sometimes find an inverse when there is none?

81–86 ■ ● Have a graphing device plot the following functions. How would you describe them in words?

81. $f(x) = x^2 e^{-x}$ for $0 \leq x \leq 20$.
82. $g(x) = 1.5 + e^{-0.1x} \sin x$ for $0 \leq x \leq 20$.
83. $h(x) = \sin 5x - \cos 7x$ for $0 \leq x \leq 20$ for x measured in radians.
84. $f(x) + g(x)$ for $0 \leq x \leq 20$ (using the functions in Exercises 81 and 82).
85. $g(x) \cdot f(x)$ for $0 \leq x \leq 20$ (using the functions in Exercises 81 and 82).
86. $h(x) \cdot h(x)$ for $0 \leq x \leq 20$ (using the function in Exercise 83).

1.5 Logical Reasoning and Language in Math and Life Sciences

To further our understanding of interactions between mathematics and life sciences we compare the language of the two disciplines and illustrate how life scientists communicate their mathematical ideas and results.

Mathematics studies abstract ideas and concepts and operates with abstract objects. Its language is precise, clear, and unambiguous, and it is to a great extent communicated through symbols and formulas. On the other hand, the narratives in life sciences are closer to the language used in our everyday communication. Symbols and formulas are often replaced by verbal descriptions, metaphors, and analogies. Sometimes—for this reason—a life sciences text might appear to be unclear or ambiguous. Although there is a some truth in this, the reality reveals a more complex picture. Let us examine a few examples.

In the summary of a study of an infectious disease, we read, "the number of infected people will climb until it eventually plateaus at 1200." A mathematician prefers "will increase" to "will climb." The precise math meaning of "plateau" is given by the concept of the horizontal asymptote, which is defined and calculated using limits. The word "eventually" is (mathematically) ambiguous and cannot be precisely quantified; however, it does have a meaning to a specialist.

A population biologist might say that, "according to our model, the population will blow up in finite time" (see Example 8.4.9). Nothing will explode! The phrase "blow up" means that the model predicts that in finite time, the population will surpass any number, no matter how large (one million, one billion, 100 billion, etc.). In math, this situation is described using limits (vertical asymptote), and we say that "the population approaches $+\infty$ as the time approaches a certain (finite) value."

The meaning of the sentence "Moose calves experience a large initial growth rate" is clear to a life scientist familiar with the context, but leaves a mathematician puzzled. How large is large? Is 10 large, or is 100 large? What does "initial" refer to, five days or three weeks? What does the term "growth" refer to—height, weight, volume, or something else?

Whereas a life scientist plots "relative surface area versus volume," a mathematician sketches the graph of the relative surface area as a function of volume (and so places the volume on the horizontal axis and the relative surface area on the vertical axis; see Figure 2.1.17).

We will encounter similar situations as we work our way through this book. They will help us understand how life scientists translate mathematical ideas and results into their language, and vice versa—how we translate life sciences information and problems into math ideas, formulas, and procedures.

Next, we introduce several building blocks of mathematical exposition (such as definitions and theorems) and discuss logical constructions, i.e., the ways we reason in mathematics. This context provides us with a good opportunity to further contrast mathematics and life sciences.

Building Blocks: Definitions and Theorems

A **definition** in mathematics is a statement that introduces something new—a new object, concept, or property of a mathematics object—based on the previously established objects, concepts, and/or properties.

For instance, we define the natural logarithm function $y = \ln x$ (see Definition 2.2.1 in Section 2.2) as the inverse function (the concept of the inverse function was previously introduced in Section 1.4) of the natural exponential function $y = e^x$ (introduced at the start of Section 2.2). We define the derivative of a function (Section 4.5) based on the previously established concept of the limit (Sections 4.2 and 4.3).

Definitions in mathematics are clear, concise, and unambiguous. Once established, they rarely change. The definition of $\sqrt{A}$ ($\sqrt{A}$ is the positive number whose square is A) has not changed for thousands of years (the narratives or symbols have changed many times, but not the meaning). The definition of the derivative has not changed since the times of Newton and Leibniz.

The reason that math can afford to work with clearly and precisely defined notions lies in its abstractness. On the other hand, biology does not study abstract objects. It is for this reason that defining objects or ideas in life sciences can be challenging (or even impossible). How do we define "life"? After years of research and scientific (and not only scientific) debate, we still do not have a workable, generally acceptable definition of climate change. Even well-established definitions (such as the definition of a mammal) have "grey zones," i.e., situations in which we are not sure how the definition applies (the platypus lays eggs but is a mammal). In spite of many attempts, the term "species" has not been clearly (precisely) defined in biology.

Definitions in life sciences are not etched in stone. For instance, the definitions of obesity and attention deficit hyperactivity disorder (ADHD) have been modified fairly recently. As well, they do not apply universally; for instance, in Japan, a person whose BMI (body mass index; see Example 2.1.11 for the definition) is higher than 25 is considered obese. In China, that threshold is 28, and in North America it is 30.

In certain situations (think of marketing and public relations), definitions are purposely kept vague. What is "organic"? What does it mean to say a certain product is "green"? What amount of fat is considered "low fat"? Worse yet—definitions can be completely misleading. For instance, the "no sodium" declaration on bottled water does not mean that there is zero sodium. The precise definition (not easy to find!) states that "no sodium" means "less than 5 mg sodium per 100 mL."

Sometimes the same word carries different meanings in math and in life sciences. For a mathematician, the function in Figure 1.5.60a is periodic: all highs are equal, all lows are equal, and the values of the function repeat at points that are a fixed distance apart from each other. A life scientist will call the behaviour depicted in Figure 1.5.60b periodic. Although this function is not (mathematically) periodic, it does explain the same idea—alternate highs and lows.

A **theorem** establishes a relationship between previously defined mathematics objects and/or previously proved properties of mathematics objects. It consists of two parts: assumption (or assumptions) and conclusion (or conclusions). The language used in the statement of a theorem identifies which is which. Very often, we write

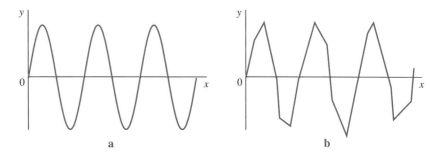

FIGURE 1.5.60

Periodic behaviour: in mathematics (a) and in life sciences (b)

a theorem in one of these two forms (or their linguistic variations): "If (list of assumptions), then (list of conclusions)" or "Assume that (list of assumptions). Then (list of conclusions)." The logical form "if ... then ..." is called an **implication.** More about this soon.

Consider the following theorem (the context is real numbers, and the term "positive" is defined as "greater than zero"): "Assume that a and b are positive numbers. Then their sum, $a + b$, and their product, ab, are positive." This theorem has two assumptions ("a is positive" and "b is positive") and two conclusions ("the sum $a + b$ is positive" and "the product ab is positive"). As we gain more experience, we might use phrases that are closer to everyday language (as is often done in math); for instance, we might say "The sum, $a + b$, and the product, ab, of positive numbers a and b are positive."

How do we use a theorem? First, we must check that *all* assumptions are satisfied. Only then can we state that the conclusions are true. For instance, let $a = 2.47^3$ and $b = e^{-0.3}$. Since both a and b are positive, we conclude that the sum, $a + b = 2.47^3 + e^{-0.3}$, and the product, $ab = 2.47^3 e^{-0.3}$, are positive. If just one of the assumptions of the theorem does not hold, we *cannot use* the theorem—even though its conclusions might still be true. For instance, if we pick $a = 12$ and $b = -4$, then the assumption "b is positive" does not hold. However, the sum $a + b = 8$ is positive. As well, if $a = -12$ and $b = -4$ (both assumptions fail to hold), it is still true that $ab = 48$ is positive.

This is really important, so we repeat:

> A theorem can be applied (i.e., its conclusions are true) only if all assumptions are satisfied. If one or more assumptions fail to hold, then we cannot use the theorem.

One more example. The statement "A cat is a mammal" can be expressed in math language as "Assume that an animal is a cat. Then it is a mammal." This theorem consists of one assumption ("an animal is a cat") and one conclusion ("an animal is a mammal"). How do we use this theorem? We take an animal and check: if it is a cat, then the theorem says that it is a mammal. If the animal we're holding is not a cat, the theorem does not apply. Its conclusion ("an animal is a mammal") could still be true (we might be holding a mouse) or not (we might be holding a bird).

A theorem comes with a proof. A **proof** is a sequence of steps that starts at the assumption(s), advances by using previously proven mathematical techniques and deductive reasoning, and arrives at each conclusion. We cannot prove a theorem in mathematics by demonstrating that it holds for specific examples. Picking two positive numbers (say, $a = 15$ and $b = 220$) and showing that their sum and product are positive does not suffice to prove the theorem we introduced a bit earlier. Even if we repeat this experiment (picking positive numbers a and b and verifying that $a + b$ and ab are positive) a hundred million times—we still have not proven the theorem.

Reasoning in life sciences is quite different. Starting with results of repeated experiments, scientists try to formulate a general principle (that's what inductive reasoning is about). How many repeated experiments (with the same or similar results) are needed depends on the specific situation. For instance, a step in the approval process for new drugs in Canada consists of clinical trials on humans. If the drug demonstrates its effectiveness and safety (i.e., if its therapeutic value outweighs the side-effects) on a *small* sample (could be one hundred humans; so, one hundred repeated experiments), then it is cleared for approval, i.e., it is deemed effective and safe for *everyone* in Canada.

Implications

We mentioned that an **implication** is a statement of the form "If A then B," where A and B are math statements; it is denoted by $A \Rightarrow B$. Sometimes it is useful to visualize "If A then B" as "A is a subset of B." For instance, the implication "If an animal is a cat, then it is a mammal" is visualized in Figure 1.5.61: all cats are placed in the smaller box; the smaller box is a subset of the larger box, which contains all mammals.

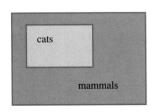

FIGURE 1.5.61

Visualizing implication as a subset relation

How do we use this diagram? Look at the animal labelled x in Figure 1.5.62a: x is in the small box, so x is a cat. Since the smaller box is inside the larger box, x is also in the larger box, and so we conclude that x is a mammal. Thus, "animal is a cat" implies "animal is a mammal."

The direction of an implication is not (in general) reversible; i.e., $A \Rightarrow B$ and $B \Rightarrow A$ are two different statements, and neither follows from the other. (The statement $B \Rightarrow A$ is called the **converse** of the statement $A \Rightarrow B$.) In our case, $B \Rightarrow A$ reads "If an animal is a mammal, then it is a cat," which is obviously not true. This is clear from the diagram in Figure 1.5.62b: the animal marked x is in the larger box (it is a mammal), but it is not in the smaller box (it is not a cat). So, we need to keep in mind the following:

> The direction of an implication matters! Knowing that "if A then B" is true does not tell us whether "if B then A" is true or false.

Once we understand what implication is and how to work with it in this obvious context of cats and mammals, we apply it in abstract math situations. For instance, Theorem 6.3.4 in Section 6.3 proves that the implication "If a function is differentiable, then it is continuous" is true. However, the converse statement, "If a function is continuous, then it is differentiable," is shown to be false in Section 4.5.

Not distinguishing between an implication and its converse is a common logical fallacy. The fact that a person who has meningitis suffers from fever, chills, nausea, and vomiting can be expressed as the implication "If meningitis, then fever, chills, nausea, and vomiting." A person experiencing fever, chills, nausea, and vomiting, thinking of the worst possible scenario, might conclude that she has meningitis. However, this thinking is incorrect—there is no reason to believe that the converse "If fever, chills, nausea, and vomiting, then meningitis" is true. And it is not—fever, chills, nausea, and vomiting could be symptoms of food poisoning (as a matter of fact, entering the four symptoms into the internet self-diagnosing tool WebMD (http://symptoms.webmd.com) returns 99 possible medical conditions).

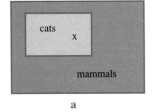
a
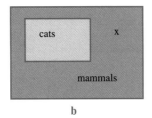
b
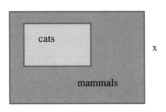
c

FIGURE 1.5.62

Arguing about implication using subset language

So, we cannot just reverse the direction of the implication and hope to keep the validity of the statement. However, there is a way to correctly reverse the direction of an implication.

Let us go back to the implication "If A then B" in the context "If an animal is a cat, then it is a mammal." Consider the animal labelled x in the diagram in Figure 1.5.62c: since x is not in the larger box, it is not a mammal; but then x is not in the smaller box either, so x is not a cat. Thus, we conclude that "If an animal is *not* a mammal, then it is *not* a cat." In general, from the implication $A \Rightarrow B$ we obtain the **contrapositive** statement not $B \Rightarrow$ not A.

For instance, from the theorem "If a function is differentiable, then it is continuous" (Theorem 6.3.4, Section 6.3) we derive the true statement "If a function is not continuous, then it is not differentiable" (Example 4.5.4 in Section 4.5). Another example: the contrapositive of "If you drink, then do not drive" is the equivalent statement (i.e., it preserves the meaning) "If you drive, then do not drink."

In reality, things are more complicated. Consider the implication "If disease, then symptoms" we just discussed. Its converse, "If symptoms, then disease," does not hold. But, it could happen that the contrapositive "If no symptoms, then no disease" does not hold either! There could be a rare form of some disease that does not show any of the known symptoms.

Math Results and Their Interpretation

The final step in solving a life sciences problem using math consists of translating math results (in the form of graphs, procedures, and/or numeric results) back into the context of the application. That this is not always an obvious task is illustrated in the following examples.

In Example 2.1.13, we study the relationship between the heartbeat frequency, h, and the body mass, B, of a mammal, given by the function $h = 241B^{-1/4} = 241/\sqrt[4]{B}$. For a mathematician, the domain of the function h consists of all positive values, and its graph is shown in Figure 1.5.63a. For a life scientist, the domain is the interval from the lightest mammal, B_L, to the heaviest mammal, B_H, as shown in Figure 1.5.63b.

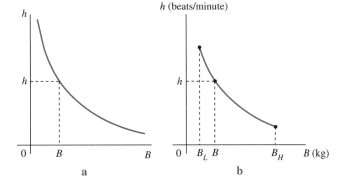

FIGURE 1.5.63

Abstract mathematics graph (a) and its interpretation (b).

For a mathematician, the value of h for a given value of B is unique (think of the vertical line test). The graph in Figure 1.5.63b needs to be read differently: given a mammal of body mass B, the corresponding value of h is at best an approximate value of its heartbeat rate (not all animals with the same body mass, B, have exactly the same heartbeat rates); alternatively, the value h could be interpreted as an average of heartbeat rates of all mammals of the same body mass, B.

The graph in Figure 1.5.64a shows how a certain quantity (in this case population, denoted by P) changes over time. A mathematician will say that the values of P approach zero as t approaches infinity (and that P remains positive, i.e., it is never equal to zero). In reality, the population will go extinct (i.e., P will be zero) in some finite time. In this context, the "tail" of the graph in Figure 1.5.64a (i.e., the part for large values of t) makes no sense.

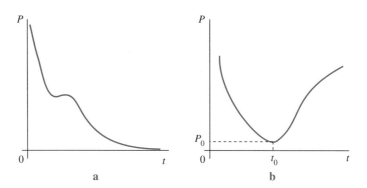

FIGURE 1.5.64
Interpreting graphs

A mathematician will say that the function $P(t)$ in Figure 1.5.64b reaches its minimum value at t_0 and then starts increasing. A population biologist will see a different picture: it might happen that the minimum population, P_0, is below the minimum needed for survival (Allee effect; see Example 8.1.3) and that the population will go extinct (rather than recover as suggested by the graph).

We will meet many more situations where we need to interpret mathematical results as we continue reading this book.

Our final remark is about rounding off decimal numbers. There is no convention telling us how many decimal places to use in our calculations. Sometimes we will round off to three, and sometimes to five or six decimal places. We need to keep this in mind, as rounding off to a different number of decimal places will produce different numeric results. Consider the function $P(t) = 30,007 e^{kt}$ of Example 2.2.14. We calculated that $k = 0.010428$ and, using that value, obtained $P(10) = 33,305.105$. If we round off the value of k to three decimal places ($k \approx 0.010$), then $P(10) = 33,162.864$; rounding to four decimal places ($k \approx 0.0104$), we obtain $P(10) = 33,295.781$, and so on.

The usual round-off technique (based on the relevant decimal being smaller than, equal to, or larger than 5) gives $4.2 \approx 4$, $4.5 \approx 4$ (sometimes $4.5 \approx 5$), and $4.7 \approx 5$. In reality, this routine might not make sense, or the way we round off (up or down) might not matter. Consider the following two examples.

How many 50-seat buses do we need so that each of 210 students has a seat on a bus? Dividing 210 by 50 we obtain 4.2. But this time $4.2 \approx 4$ makes no sense—we need 5 buses. Modelling a population of moose, we might arrive at the number 328.4. So, it is either 328 or 329 moose, and in practice it makes no difference which number we pick.

Chapter Summary: Key Terms and Concepts

Define or explain the meaning of each term and concept.
Mathematics model: discrete-time dynamical system, continuous-time dynamical system
Functions: relation, function, vertical line test; dependent variable, independent variable, parameter; discrete variable, continuous variable; domain, range; graph
Types of functions: constant, linear, quadratic, cubic, square root, cube root, absolute value; function defined piecewise; increasing, decreasing; polynomial, rational function, algebraic function, transcendental function
Operations on functions and transformations: algebraic operations, composition, horizontal line test, inverse function; shifts, scaling, reflections

Concept Check: True/False Quiz

Determine whether each statement is true or false. Give a reason for your choice.

1. The composition of $f(x) = x^2$ and $g(x) = 7$ is $(g \circ f)(x) = 49$.
2. The function $f(x) = x^3$ is the only polynomial of degree three that has an inverse function.
3. Reflect the graph of $f(x)$ with respect to the x-axis, and then compress it horizontally by a factor of 3. The resulting function is $-f(x/3)$.
4. Every increasing function passes the horizontal line test.
5. The function $f(x) = 2x + 1$ is the inverse of the function $g(x) = 2x - 1$.
6. The domain of the function $f(x) = \sqrt{|x|}$ consists of all real numbers.
7. The range of the function $g(x) = \sqrt{5x - 2}$ is the set $[0, \infty)$.
8. If $h(x) = x^2 + 1$, then $h(z + 1) = z^2 + 2$.
9. If $f^{-1}(3) = 2$, then $f(2) = 3$.
10. The inverse of an increasing function is a decreasing function.

Supplementary Problems

1. The function
$$A(c) = \frac{nc}{k + c^2}$$
(where k and n are positive constants) describes the amount of a chemical absorbed as a function of concentration, c. What happens to $A(c)$ if the concentration, c, is allowed to grow larger and larger?

2. Explain how to transform the graph of $f(x) = 1/(1 + x^2)$ drawn in Figure 1.4.49 to obtain the graph of the function
$$g(x) = \frac{4x^2}{1 + x^2}$$
(Hint: rewrite $g(x)$ using long division.) Sketch the graph of $g(x)$.

3. Sketch the graph of the function $f(x) = \Big|||x| - 2| - 3\Big|$.

4. Estimate the mass of Earth in kilograms, assuming that Earth is a sphere of radius 6,371 km and its density is 5.5 g/cm^3.

5. Find the domain of the function
$$f(x) = \frac{13x^2 - 4}{1 - \sqrt{7 - 2x}}$$

6. The density of soil forming a forest floor is given by
$$d(x) = \frac{0.7}{0.8 + \sqrt{0.1 + \frac{1}{x+1}}}$$
where x is the depth in metres (so $x = 0$ labels the surface, and $x = 3$ is 3 m below the surface). State what question is answered by finding the inverse function. Find the inverse function of $d(x)$.

7. In a rectangular region 15 km long and 3 km wide within a national park, researchers identified (say, by aerial surveillance) 187 deer. Estimate the number of deer in a circular region of radius 12 km within the park.

8. Given $f(x) = x^3 + x$, find
$$\frac{f(x+h) - f(x)}{h}$$
Simplify your answer.

9. Given $f(x) = 2x^2 + 7$, find and simplify the expression
$$\frac{f(x+h) - f(x-h)}{2h}$$

10. Define the function
$$g(x) = \frac{1}{2}(f(x) + |f(x)|)$$
Describe how to obtain the graph of $g(x)$ from the graph of $f(x)$.

11. A 65 kg swimmer burns about 400 kcal (kilocalories) per hour. Knowing that 1 kcal is approximately 4200 joules and 1 watt = 1 joule/second, about how many watts does the swimmer burn?

12. Consider a colony of 100,000 bacteria, arranged in a circular region on a Petri dish.

    a. Assuming that each bacterium weighs $3 \cdot 10^{-9}$ g, find the mass of the colony.

    b. Take the density of bacteria to be that of water (at 4°C), i.e., 1 g/cm^3. Find the volume occupied by the colony.

    c. Assume that the colony is $2 \cdot 10^{-3}$ cm thick (this is roughly the thickness of a cell). Find the area occupied by the colony, and from there the radius of the region occupied.

Project

1. Consider the function

$$f(x) = \frac{ax^2 + b}{x^2 + 4}$$

where $a, b \geq 0$. Use a graphing device to answer the following questions.

a. Take $a = 1$ and plot the graph of $f(x)$ on $[0, 25]$ with $b = 1$, $b = 3$, $b = 5$, and $b = 8$. What do all four graphs have in common? Which are increasing, and which are decreasing? Find the value of b that separates increasing graphs from decreasing graphs.

b. Take $b = 0$ and plot the graph of $f(x)$ on $[0, 25]$ for several values of a. What do all of the graphs have in common? How does the value of a affect the graph?

c. Take $b = 4$ and plot the graph of $f(x)$ on $[0, 25]$ with $a = 0$, $a = 3$, $a = 8$, and $a = 20$. Which are increasing, and which are decreasing? Find the value of a that separates increasing graphs from decreasing graphs. How does the value of a affect the graph in the long term (i.e., for large values of x)?

Chapter 2
Modelling Using Elementary Functions

Modelling consists of taking a description of a biological phenomenon and converting it into mathematical form. Living things are characterized by change, and one goal of modelling is to **quantify these dynamics with an appropriate function.**

We study a whole variety of phenomena, including bacterial and tumour cell growth, the dynamics of the population of Canada, blood circulation time, heartbeat frequency, the growth of bones, and radiocarbon dating. Whenever we use a model, we keep the following questions in mind:

- What biological process are we trying to describe?
- What biological questions are we seeking to answer?
- What are the basic measurements and their units?
- What are the relationships between these measurements?
- What do the results mean biologically?

In this chapter we model phenomena using **linear, power, exponential, logarithmic, trigonometric,** and **inverse trigonometric** functions. As we develop new mathematical tools, we enrich existing models and build new ones. For instance, in Chapter 3, we will further the study of exponential models that we start here. As well, we will learn about discrete-time dynamical systems and apply them to studying gas exchange in the lungs and the behaviour of competing populations.

2.1 Elementary Models

In this section, we study **linear** and **power functions** and investigate models that can be built using these functions. The graph of a linear function is a line; it is characterized by the fact the rate of change (the change in output over the change in input) is constant. For example, the relationship between degrees Fahrenheit and degrees Celsius is linear. A line that goes through the origin represents a **proportional relationship.** In this case, if the input is scaled by some factor, then the output is scaled by the same factor. We also investigate properties of power functions and use them to understand biological phenomena such as blood circulation time, growth of bones, and heartbeat frequency.

Proportional Relations

The simplest relations are **proportional** relations, meaning that the output is **proportional** to the input. Mathematically, this means that the ratio of the output to the input is constant. The general formula for a proportional relation is

$$f(x) = mx$$

where m is a real number. The ratio of the output mx to the input x is

$$\frac{\text{output}}{\text{input}} = \frac{mx}{x} = m$$

as long as $x \neq 0$. The real number m is called the **constant of proportionality.** Constants of proportionality, like all measurements, have units and dimensions.

To describe proportional relations geometrically, we recall the definition of the slope of a line.

Definition 2.1.1 **Slope of a Line**

The slope of the line passing through the points (x_1, y_1) and (x_2, y_2) with $x_1 \neq x_2$ is given by

$$\text{slope} = \frac{\text{change in output}}{\text{change in input}} = \frac{y_2 - y_1}{x_2 - x_1} \quad (2.1.1)$$

See Figure 2.1.1. The assumption $x_1 \neq x_2$ guarantees that the line is not vertical (the slope of a vertical line is not defined). The changes in the input x and the output y are often written as

$$\Delta x = x_2 - x_1$$
$$\Delta y = y_2 - y_1$$

where Δ (the Greek letter delta) means "change in." Then

$$\text{slope} = \frac{\Delta y}{\Delta x} \quad (2.1.2)$$

This notation will prove very useful when we study derivatives later in this book.

A line has a constant slope. This means that no matter where we measure the slope (i.e., no matter which points we use), we obtain the same value.

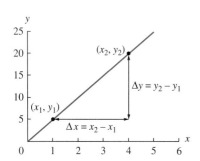

FIGURE 2.1.1
Slope and Δ notation

Example 2.1.1 Finding the Slope between Two Data Points

In Figure 2.1.2, the data points are $(x_1, y_1) = (1, 5)$ and $(x_2, y_2) = (4, 20)$. The slope is

$$\text{slope} = \frac{\Delta y}{\Delta x} = \frac{y_2 - y_1}{x_2 - x_1} = \frac{20 - 5}{4 - 1} = \frac{15}{3} = 5$$

Recall the general formula for a proportional relation:

$$f(x) = mx$$

where m is a constant. Taking two points, $(x_1, f(x_1))$ and $(x_2, f(x_2))$, we compute

$$\text{slope} = \frac{f(x_2) - f(x_1)}{x_2 - x_1} = \frac{mx_2 - mx_1}{x_2 - x_1} = \frac{m(x_2 - x_1)}{x_2 - x_1} = m$$

Thus, the graph of the proportional relationship $f(x) = mx$ is the line through the origin with slope equal to the constant of proportionality.

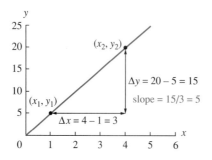

FIGURE 2.1.2
Finding the slope

2.1 Elementary Models

Example 2.1.2 Proportional Relation between Mass and Volume

The fundamental relation between the mass, M, and the volume, V, of an object, or a living organism, states that

$$M = \rho V$$

where ρ is the density. From $\rho = M/V$ we see that a unit of density is a unit of mass divided by a unit of volume, such as kilograms per cubic metre (kg/m^3) or grams per cubic centimetre (g/cm^3).

In the case when the density, ρ, is *constant* (e.g., if the entire object is built of the same material), the relation $M = \rho V$ establishes a *proportional relation between mass and volume*. (When we study living organisms, we often assume that their density is constant.) ▲

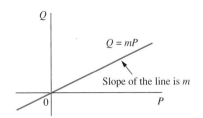

FIGURE 2.1.3
Q is proportional to P

We use the symbol $\propto$ to denote the proportional relationship. Thus, $Q \propto P$ means that the quantity Q is proportional to the quantity P; i.e., there is a constant, m (called the constant of proportionality), such that

$$Q = mP$$

Geometrically, in the PQ coordinate system (P is on the horizontal axis and Q is on the vertical axis; see Figure 2.1.3), the relation $Q = mP$ is the line that goes through the origin (if $P = 0$, then $Q = 0$) with slope m.

Thus, we use the expression $M \propto V$ to say that mass is proportional to volume. The proportional relationship $y = mx$ can be expressed as $y \propto x$.

Assume that the input, x, in the proportional relationship $f(x) = mx$ is scaled by a factor of a, i.e., that it is equal to ax. Then

$$f(ax) = m(ax) = a(mx) = af(x)$$

i.e., the *output is scaled by the same factor*.

So, for example, if the input triples, the output triples. If the input decreases by 40%, the output decreases by 40%. The fact that the output scales by the same factor as the input characterizes the proportional relationship (thus distinguishing it from all other relationships between input and output).

> **Proportional Relationship $f(x) = mx$ or $f(x) \propto x$**
>
> The ratio of the output to the input is constant and equal to m.
> $f(x)$ is the line of slope m going through the origin.
> If the input is scaled by some factor, then the output is scaled by the same factor.

Example 2.1.3 Proportional Relation between Heart Mass and Body Mass

Researchers have discovered that for smaller mammals, the mass of the heart is roughly proportional to the mass of the body. If the heart of a 4.7-kg dog weighs about 33 g (0.033 kg), determine the mass of the heart of a cat if the cat has a mass of 1.8 kg.

Denoting the body mass by B and the heart mass by H (both measured in kilograms), we write $H \propto B$; i.e., there is a constant, m, such that

$$H = mB$$

Using the given data for a dog, $H = 0.033$ when $B = 4.7$, we obtain

$$0.033 = 4.7m$$

and

$$m = \frac{0.033}{4.7} \approx 0.007$$

Therefore, $H \approx 0.007B$. In words, the heart mass of a smaller mammal is about 0.7% of its body mass. In particular, the heart mass of a cat that weighs 1.8 kg is

$$H \approx 0.007(1.8 \text{ kg}) \approx 0.0126 \text{ kg}$$

i.e., about 12.6 g.

Linear Functions

Proportional relations are described by functions that perform a single operation on their input: multiplication by a constant. The graphs of such functions are lines with slope equal to the constant of proportionality. Furthermore, these lines pass through the point (0, 0) because an input of 0 produces an output of 0.

Many other functions also have linear graphs but do not pass through the origin.

Definition 2.1.2 A function of the form $f(x) = mx + b$, where m and b are real numbers, is called a **linear function**.

The domain of a linear function consists of all real numbers. If $m \neq 0$, its range consists of all real numbers as well. When $m = 0$, the line is horizontal and the range is a single value, b.

Note that a linear function represents a proportional relationship *only if $b = 0$* (see Example 2.1.4).

Example 2.1.4 A Linear Function That Is Not a Proportional Relation

The graph of the function

$$y = f(x) = x + 1$$

is a line (Figure 2.1.4). But the relation between the input, x, and the output, y, is *not* a proportional relation. Two points on this line are (1, 2) and (2, 3). At the first, the ratio of output to input is $2/1 = 2$. At the second, the ratio of output to input is $3/2 = 1.5$. Clearly, the ratio of output to input is not constant.

For linear functions, it is the ratio of the **change in output**, Δy, to the **change in input**, Δx, that is constant. Suppose we start at the point (0, 1) on the graph. The ratio of change in output to change in input between this point and (1, 2) is

$$\frac{\Delta y}{\Delta x} = \frac{2 - 1}{1 - 0} = \frac{1}{1} = 1$$

The ratio of change in output to change in input between (0, 1) and (2, 3) is

$$\frac{\Delta y}{\Delta x} = \frac{3 - 1}{2 - 0} = \frac{2}{2} = 1$$

Recall that a **line** is characterized by a **constant slope,** like a constant grade on a road. We use this fact to find a formula for a line. First, choose any point that lies on the graph of the function and call it the **base point** (Figure 2.1.5). If the base point has

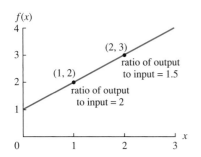

FIGURE 2.1.4

A linear function that is not a proportional relation

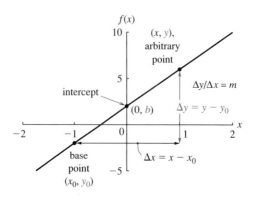

FIGURE 2.1.5

The elements of the general linear graph

coordinates (x_0, y_0), the slope between it and an arbitrary point (x, y) on the line is

$$\text{slope} = \frac{\Delta y}{\Delta x} = \frac{y - y_0}{x - x_0}$$

Because the slope between any two points on the graph is constant,

$$\frac{y - y_0}{x - x_0} = m$$

for some fixed value of m. Multiplying both sides by $(x - x_0)$, we find

$$y - y_0 = m(x - x_0)$$

We have thus obtained the **point-slope form** of the equation of a line.

Rearranging the point-slope equation, we obtain

$$y - y_0 = m(x - x_0)$$
$$y = m(x - x_0) + y_0$$
$$y = mx - mx_0 + y_0$$
$$y = mx + (-mx_0 + y_0)$$

Denoting $-mx_0 + y_0$ by b, we obtain the **slope-intercept form** of the equation of the line

$$y = mx + b$$

Note that when $x = 0$, then $y = b$; i.e., b represents the y-intercept.

For example, the graph of the linear function $f(x) = -4x + 5$ is a line of slope -4 that crosses the y-axis at the point $(0, 5)$. The graph of the constant function $f(x) = -2$ (which is a special case of the linear function $f(x) = mx + b$ for which $m = 0$ and $b = -2$) is a horizontal line (since its slope is zero), crossing the y-axis at $(0, -2)$.

Example 2.1.5 Recognizing a Linear Function

We can easily recognize linear functions by the operations done to the input variable. If the formula involves only adding, subtracting, and multiplying and/or dividing by constants, the equation describes a linear function. For example, the formulas

$$M = 0.4N - 1, \quad y = \frac{2x - 7}{4}, \quad \text{and} \quad p = -t + 0.66$$

describe linear functions (i.e., a linear relationship between the output and the input). Note that the second formula can be written as

$$y = \frac{2x - 7}{4} = \frac{2x}{4} - \frac{7}{4} = \frac{1}{2}x - \frac{7}{4}$$

so it represents the line with slope $1/2$ and y-intercept $-7/4$. ▲

Example 2.1.6 Recognizing a Nonlinear Function

The function

$$b(t) = \frac{5.0}{1 + 2t}$$

is not a linear function because the input variable, t, appears in the denominator. The function

$$b(t) = t^2 + 3t + 2$$

is not linear because the input variable, t, is squared. ▲

Example 2.1.7 Finding the Equation of a Line

We find the equation of the line in Figure 2.1.6. From $(x_1, y_1) = (-0.5, 2.7)$ and $(x_2, y_2) = (2.5, 0.6)$, we compute the slope

$$m = \frac{\Delta y}{\Delta x} = \frac{y_2 - y_1}{x_2 - x_1} = \frac{0.6 - 2.7}{2.5 - (-0.5)} = \frac{-2.1}{3} = -0.7$$

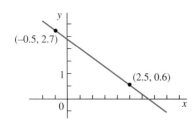

FIGURE 2.1.6
A line defined by two points

Thus, the point-slope equation is

$$y - 2.7 = -0.7(x - (-0.5))$$

(We used the point $(-0.5, 2.7)$ as the base point, but we could have used either point.) Rearranging the terms, we obtain

$$y = -0.7(x + 0.5) + 2.7$$
$$= -0.7x + 2.35$$

Thus, the given line has slope -0.7 and y-intercept 2.35.

Example 2.1.8 Plotting a Line from an Equation

Suppose we wish to plot the linear function $F(x)$ given by

$$F(x) = -2x + 30$$

Plugging in $x = 0$ gives $F(0) = 30$, and $x = 10$ gives $F(10) = 10$ (Figure 2.1.7). The graph of the line connects the points $(0, 30)$ and $(10, 10)$. This line goes down and to the right with a slope of -2.

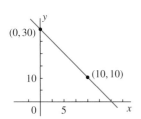

FIGURE 2.1.7
Graphing a line from its equation

When the graph goes down when viewed from left to right (as in Example 2.1.8), we say that the function is a **decreasing function.** A larger input produces a smaller output. Linear functions with negative slopes are decreasing functions. In contrast, a positive slope corresponds to an **increasing function.** Larger inputs produce larger outputs. A slope of exactly zero corresponds to a function with equation

$$f(x) = 0 \cdot x + b = b$$

Such a function always takes on the constant value b, the y-intercept, and has as its graph a horizontal line.

Slope	Graph	Function
positive	goes up	increasing
negative	goes down	decreasing
zero	horizontal	constant

An example of each type is shown in Figure 2.1.8.

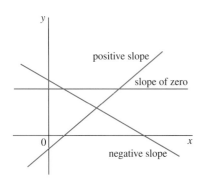

FIGURE 2.1.8
Linear functions with positive, negative, and zero slopes

Linear Relationship $f(x) = mx + b, b \neq 0$

The ratio of output to input *is not constant*.
The rate of change (the change in output over the change in input) *is constant* and equal to m.
$f(x)$ is the line of slope m and y-intercept b.

Example 2.1.9 Linear Relation between Degrees Celsius and Degrees Fahrenheit

Water freezes at a temperature of $0°C$ and boils at a temperature of $100°C$. The Fahrenheit temperature scale has been defined so that the freezing temperature of water is $32°F$ and the boiling temperature is $180°$ higher, i.e., $212°F$.

Knowing that their relationship is linear, we can now derive the formula that converts degrees Celsius into degrees Fahrenheit. Thus, $C =$ temperature in degrees Celsius is the independent variable, and $F =$ temperature in degrees Fahrenheit is the dependent variable. In other words, we want to find the relationship in the form $F = f(C)$.

In Figure 2.1.9 we plot the freezing and boiling points on the respective scales as well as the line whose equation we are looking for.

The slope is

$$m = \frac{\Delta F}{\Delta C} = \frac{212 - 32}{100 - 0} = \frac{180}{100} = \frac{9}{5}$$

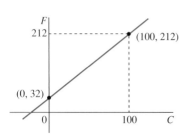

FIGURE 2.1.9
Conversion between Celsius and Fahrenheit

Using $(0, 32)$ as the base point, we get
$$F - 32 = \frac{9}{5}(C - 0)$$
and
$$F = \frac{9}{5}C + 32$$

This is the formula that we used in Example 1.4.7. Note that the slope (this time written with units)
$$m = \frac{9°\text{F}}{5°\text{C}} = 1.8\frac{°\text{F}}{°\text{C}}$$
says that it takes $1.8°\text{F}$ to make up $1°\text{C}$.

An interesting thing happens when we replace F by C in the conversion formula $F = \frac{9}{5}C + 32$. We get
$$C = \frac{9}{5}C + 32$$
$$-\frac{4}{5}C = 32$$
and therefore
$$C = 32\left(-\frac{5}{4}\right) = -40$$

What does this calculation show?

Clearly, the Celsius and Fahrenheit scales are different—for instance, the measurements of $65°\text{C}$ and $65°\text{F}$ or $-2°\text{C}$ and $-2°\text{F}$ do not represent the same temperature. However, there is a place where the two scales meet, and that is at -40; i.e., $-40°\text{C}$ and $-40°\text{F}$ represent the same temperature.

Example 2.1.10 Linear Model for the Population of Canada

The table below shows the population of Canada (see Example 1.4.6), according to the three most recent government censuses.

Year	Population, in Thousands
2001	30,007
2006	31,613
2011	33,477

Statistics Canada. 2011. *Population, urban and rural, by province and territory (Canada)*, "Population estimates and projections, 2011 Census." Statistics Canada Catalogue no. 98-310-XWE2011002. Ottawa, Ontario. http://www.statcan.gc.ca/tables-tableaux/sum-som/l01/cst01/demo62a-eng.htm.

Based on the data for the years 2001 and 2006, we will build a linear model for the increase in population. Then we will use the 2011 data to test our model, i.e., to see how accurately it predicts the population in 2011.

To simplify the calculation, we relabel the time (common practice!) by declaring that $t = 0$ corresponds to the year 2001; the data are given in Table 2.1.1.

Table 2.1.1

Time, t	Population, $P(t)$, in Thousands
0	30,007
5	31,613
10	33,477

FIGURE 2.1.10

Linear model for the population of Canada

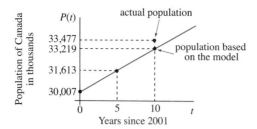

See Figure 2.1.10. From the data points (0, 30,007) and (5, 31,613), we compute the slope

$$m = \frac{\Delta P}{\Delta t} = \frac{31{,}613 - 30{,}007}{5 - 0} = \frac{1606}{5} = 321.2$$

To get the point-slope form, we use (0, 30,007) as the base point:

$$P(t) - 30{,}007 = 321.2(t - 0)$$

We obtain the formula

$$P(t) = 321.2t + 30{,}007$$

which represents the *linear model* for $P(t)$, the population of Canada (in thousands), as a function of time, t (in years, since 2001), based on the 2001 and 2006 census data.

The slope of $P(t)$ is 321.2. This means that the rate of increase in Canadian population was 321,200 new people per year. Note that we assumed that the population changed linearly between 2001 and 2006. We use the term "new people" to include all contributors to the increase in population, such as newborn children and new immigrants.

Our model predicts the population of Canada in 2011 to be

$$P(10) = 321.2(10) + 30{,}007 = 33{,}219$$

thousands (or 33,219,000). Comparing with the census data for 2011, where the real population is 33,477 thousands (or 33,477,000), we see that our model gives an underestimate. We conclude that between 2006 and 2011 the Canadian population increased faster than between 2001 and 2006; had it increased at the same rate, the population in 2011 would have been 33,219,000.

As we can see, the linear model was not really successful in predicting the population in 2011 (it underestimated it by 258,000). In forthcoming sections we will build models based on functions that are not linear and will better model the dynamics of the change in Canadian population. ▲

Note Why was the linear model not appropriate in predicting the population change? Let us look at the data again, but with a new column, in which we calculate successive differences between population counts:

Year	Time, t	Population, in Thousands	Difference, in Thousands
2001	0	30,007	
2006	5	31,613	31,613 − 30,007 = 1606
2011	10	33,477	33,477 − 31,613 = 1864

If the population of Canada were changing linearly (or nearly linearly) between 2001 and 2011, then the differences would have been the same (or nearly the same). Obviously that is not the case—thus, we cannot accurately predict the change in Canadian population by applying a linear model.

While the linear model might not be appropriate for predicting the future, it likely produces reasonably accurate population estimates *between* census years. Based on the known data for 2001 and 2006, we derive the formula

$$P(t) = 321.2t + 30{,}007$$

to model the Canadian population. We do not have census data for 2002 or 2005. However, using our model we calculate the population in 2002 to be

$$P(1) = 321.2(1) + 30{,}007 = 30{,}328.2$$

thousands (or 30,328,200), and in 2005 to be

$$P(4) = 321.2(4) + 30{,}007 = 31{,}291.8$$

thousands (or 31,291,800).

This process of predicting values (such as the population in 2002 or 2005) between known values (population in 2001 and 2006) is called **interpolation;** we say that we **interpolated** a prediction between known values.

Power Functions

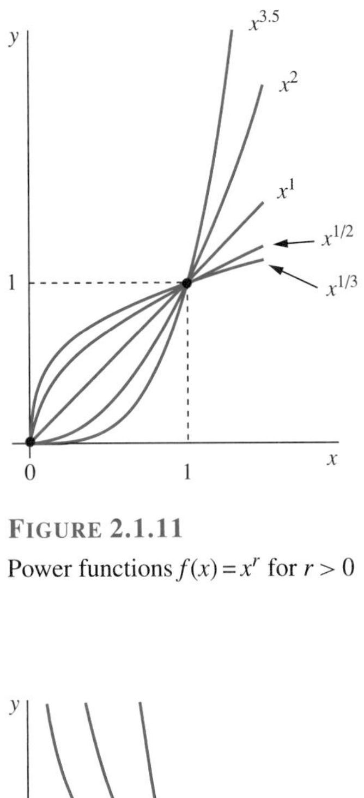

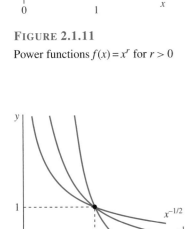

FIGURE 2.1.11

Power functions $f(x) = x^r$ for $r > 0$

Recall that a power function is of the form $f(x) = x^r$, where r is a real number. It is usually assumed that $r \neq 0$; when $r = 0$, then $f(x) = x^0 = 1$ is a constant function, which does not have all of the basic properties that general power functions do.

The graphs in Figure 2.1.11 show several functions $f(x) = x^r$, where $r > 0$, plotted for $x \geq 0$. Note that they all intersect at $(0, 0)$ and at $(1, 1)$. As well, their mutual relationship (in terms of which is smaller and which is larger) reverses as x passes through $x = 1$. For instance, $x^2 < x$ if $0 < x < 1$, but $x^2 > x$ if $x > 1$. Likewise, $x^{1/2} = \sqrt{x} < x^{1/3} = \sqrt[3]{x}$ for $0 < x < 1$, but $\sqrt{x} > \sqrt[3]{x}$ for $x > 1$. As x increases, all functions $f(x) = x^r$, $r > 0$, increase as well. All functions $f(x) = x^r$ for $r > 0$ are defined on $[0, \infty)$, and their range is $[0, \infty)$. Some functions are also defined for negative values of x (such as $f(x) = x^5$ and $f(x) = x^{1/3}$), but some are not (such as $f(x) = x^{1/4}$ and $f(x) = x^{9/2}$).

In Figure 2.1.12 we plot $f(x) = x^r$ for several *negative* values of r. All of the functions intersect at $(1, 1)$, and, as before, their mutual relationship reverses at $x = 1$; e.g., $x^{-1/2} < x^{-1}$ for $0 < x < 1$, but $x^{-1/2} > x^{-1}$ for $x > 1$, etc. All of the functions seem to approach zero as x increases and grow larger and larger as x approaches 0. All functions $f(x) = x^r$ for $r < 0$ are defined on $(0, \infty)$, and their range is $(0, \infty)$. Some functions are also defined for negative values of x (such as $f(x) = x^{-1}$ and $f(x) = x^{-2}$), but some are not (such as $f(x) = x^{-1/2}$ and $f(x) = x^{-3/4}$).

We will study power functions and their properties (such as asymptotes and limits) in detail in Chapter 4.

Next, we investigate several models that involve power functions. First, recall the following:

FIGURE 2.1.12

Power functions $f(x) = x^r$ for $r < 0$

Two Quantities P and Q Are
proportional if there is a constant $a \neq 0$ such that $Q = aP$
inversely proportional if there is a constant $a \neq 0$ such that $Q = \dfrac{a}{P}$

For instance, the formula $V = \frac{4}{3}\pi r^3$ says that the volume, V, of a sphere is proportional to the cube of its radius, r. In the formula (taken from M. Silva, Allometric scaling of body length: Elastic or geometric similarity in mammalian design, *Journal of Mammalogy*, 79:1(20–32), 1998)

$$L = 5.328 B^{2.516}$$

L is the body length (in millimetres) of a red fox and B is its mass (in kilograms). Thus, the length of a fox is proportional to $B^{2.516}$, i.e., to its mass B raised to the power of 2.516.

It has been determined that the number of insect species, N, is inversely proportional to the square of the length, L, of the species (see May, R., *Diversity of Insect Faunas*. Blackwell Scientific Publications, London (1978), 188–203). Thus, we can write

$$N = \frac{a}{L^2} = aL^{-2}$$

for some nonzero constant a.

Example 2.1.11 Body Mass Index

The body mass index (BMI) is defined as the quotient

$$\text{BMI} = \frac{m}{h^2}$$

where m is a person's mass in kilograms and h is that person's height in metres. How do we make sense of this formula?

BMI depends on two variables, the height and the mass. The trick to understanding what BMI means is to keep one of the variables fixed at some value, while changing the other. In terms of definitions from the last chapter, this means that we *declare one independent variable to be a parameter*.

First, we let height be the parameter. If $h = 1.55$ m, then

$$\text{BMI} = \frac{m}{1.55^2} \approx 0.416m$$

So BMI is proportional to m, and its graph is a line of slope 0.416 (Figure 2.1.13). Similarly, by taking $h = 1.65$ m we get

$$\text{BMI} = \frac{m}{1.65^2} \approx 0.367m$$

(line of slope 0.367), and if $h = 1.75$ m, then

$$\text{BMI} = \frac{m}{1.75^2} \approx 0.327m$$

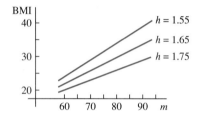

Figure 2.1.13
Body mass index, height as parameter

See Table 2.1.2 for the particular values of BMI.

Table 2.1.2

Mass (kg)	BMI When $h = 1.55$	BMI When $h = 1.65$	BMI When $h = 1.75$
60	24.97	22.04	19.59
65	27.06	23.88	21.22
70	29.14	25.71	22.86
75	31.22	27.55	24.49
80	33.30	29.38	26.12
90	37.46	33.06	29.39

Because BMI is proportional to m, a 10% increase in mass (keeping height fixed) results in a 10% increase in BMI.

Next, we consider body mass as the parameter. If we take the body mass to be 60 kg, the corresponding formula is

$$\text{BMI} = \frac{60}{h^2}$$

Similarly, we obtain the formulas for the BMI for those whose mass is 65 kg:

$$\text{BMI} = \frac{65}{h^2}$$

and those whose mass is 80 kg:

$$\text{BMI} = \frac{80}{h^2}$$

2.1 Elementary Models

Table 2.1.3

Height (m)	BMI When m = 60	BMI When m = 65	BMI When m = 80
1.45	28.54	30.92	38.05
1.50	26.67	28.89	35.56
1.55	24.97	27.06	33.30
1.60	23.44	25.39	31.25
1.65	22.04	23.88	29.38
1.70	20.76	22.49	27.68

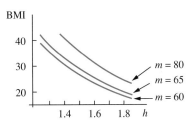

FIGURE 2.1.14

Body mass index, mass as parameter

The values of BMI are given in Table 2.1.3, and the three curves, representing BMI for the three chosen values of the parameter, are shown in Figure 2.1.14.

Looking at the three formulas above, we see that BMI is inversely proportional to the square of the height. Thus, a 10% increase in height *does not* produce a 10% increase in BMI. As a matter of fact, because the relationship is inversely proportional, *an increase in height causes a decrease in BMI*.

Let us figure out what the decrease is. Suppose that the person's mass is m kilograms; their original BMI (call it Old BMI) is

$$\text{Old BMI} = \frac{m}{h^2}$$

If their height increases by 10% (so that it is $h + 0.1h = 1.1h$), and their mass remains at m kilograms, their new BMI is

$$\text{New BMI} = \frac{m}{(1.1h)^2} = \frac{m}{1.1^2 h^2} = \frac{1}{1.1^2} \frac{m}{h^2} \approx 0.826 \frac{m}{h^2} = 0.826(\text{Old BMI})$$

i.e., it is 82.6% of their previous BMI.

Note that this calculation is independent of the particular choice of m. So, no matter what our mass, if we grow in height by 10% and keep our mass unchanged, our BMI will drop to 82.6% of its previous value. ▲

Note The ratio mass/height^2 was first used in the nineteenth century as a way of estimating healthy body mass based on a person's height. Originally given the name Quetelet Index (after its Belgian inventor, A. Quetelet), it was renamed the body mass index (BMI) in the early 1970s. Since then, BMI has gained popularity among health professionals as an aid in discussing problems with their over- and underweight patients.

It has been determined that BMI does not give adequate estimates of percentage of body fat in a number of cases (such as children and tall people). Other measurements, such as the ratio of mass to the *cube* of height, or the ratios

$$\frac{\text{weight}}{\text{height}^r}$$

with powers of r between 2.3 and 2.7, have been proposed and used.

Example 2.1.12 Blood Circulation Time in Mammals

Blood circulation time is the average time needed for blood to reach a site in the body (of a human or an animal) and come back to the heart. It has been determined that for mammals, the blood circulation time is proportional to the fourth root of the body mass.

Denote the blood circulation time by T and the body mass by B. We measure T in seconds and B in kilograms. The relationship between T and B can be expressed as

$$T(B) = a\sqrt[4]{B}$$

for some constant a.

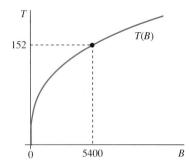

FIGURE 2.1.15

Blood circulation time as a function of body mass

The graph of $T(B)$ is shown in Figure 2.1.15. The algebraic domain of $T(B)$ is $B \geq 0$; in reality, the domain of $T(B)$ is an interval from the mass of the lightest mammal on Earth to the mass of the heaviest mammal on Earth.

If the body mass doubles, how does the blood circulation time change?

We know that if T were proportional to B, then doubling B would cause T to double. But that is not the case here, since T is proportional to the *fourth root of B*.

Let us compute it—replacing B by $2B$, we get

$$T(2B) = a\sqrt[4]{2B} = a\sqrt[4]{2}\sqrt[4]{B} = \sqrt[4]{2}(a\sqrt[4]{B})$$

i.e., since $a\sqrt[4]{B} = T(B)$ we obtain that

$$T(2B) = \sqrt[4]{2}\, T(B)$$

Thus, as the body mass of an animal doubles, its blood circulation time increases by a factor of $\sqrt[4]{2} \approx 1.19$. Likewise, if B increases tenfold, then

$$T(10B) = a\sqrt[4]{10B} = a\sqrt[4]{10}\sqrt[4]{B} = \sqrt[4]{10}(a\sqrt[4]{B}) = \sqrt[4]{10}\, T(B)$$

shows that T scales by a factor of $\sqrt[4]{10} \approx 1.78$. This means that the blood circulation time of an elephant weighing 5400 kg is about 1.78 times as long as the blood circulation time of a cow that weighs 540 kg.

The way to find the value of the constant a in

$$T(B) = a\sqrt[4]{B}$$

is to measure T and B for one animal. Suppose that a 5400-kg elephant has a blood circulation time of 152 s. Then

$$152 = a\sqrt[4]{5400}$$

and

$$a = \frac{152}{\sqrt[4]{5400}} \approx 17.73$$

We have fully recovered the formula:

$$T(B) = 17.73\sqrt[4]{B}$$

Thus, the blood circulation time for a mouse that weighs 100 g = 0.1 kg is about

$$T(0.1) = 17.73\sqrt[4]{0.1} \approx 9.97 \text{ s}$$

and for a sperm whale whose mass is 38 tonnes = 38,000 kg, it is equal to

$$T(38{,}000) = 17.73\sqrt[4]{38{,}000} \approx 247.5 \text{ s}$$

Example 2.1.13 Heartbeat Frequency in Mammals

For mammals at rest, the heartbeat frequency (number of heartbeats per minute), h, has been determined to depend on the body mass according to

$$h = 241 B^{-0.25} = 241\left(\frac{1}{B^{0.25}}\right)$$

In words, the heartbeat frequency is *inversely* proportional to $B^{0.25} = B^{1/4}$, i.e., to the fourth root of the body mass. B is measured in kilograms, and the unit of h is 1/min.

The graph of h is shown in Figure 2.1.16. It shows that the hearts of smaller animals beat much more quickly than the hearts of larger animals. For instance, the heart rate of a mouse with mass $B = 100$ g (0.1 kg) is

$$h(0.1) = 241\left(\frac{1}{0.1^{0.25}}\right) \approx 428.6$$

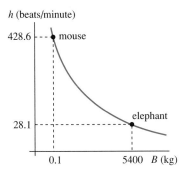

FIGURE 2.1.16

Heartbeat frequency as function of body mass

times per minute. The heartbeat frequency of an elephant with mass 5400 kg is

$$h(5400) = 241 \left(\frac{1}{5400^{0.25}}\right) \approx 28.1$$

beats per minute.

How does h scale as B increases? For example, if the body mass increases tenfold, how does the heart rate change? From

$$h(B) = 241 \left(\frac{1}{B^{0.25}}\right)$$

we get

$$h(10B) = 241 \left(\frac{1}{(10B)^{0.25}}\right) = 241 \left(\frac{1}{10^{0.25} B^{0.25}}\right) = \frac{1}{10^{0.25}} \left(241 \left(\frac{1}{B^{0.25}}\right)\right) \approx 0.56 h(B)$$

So the heart rate of an animal is about 56% of the heart rate of another animal that is one tenth its mass.

Note Organisms (humans, animals, plants, cells, etc.) do not grow isometrically—different parts and organs grow at different rates. For instance, the human skull grows faster when a child is 2 years old than when she is 12 years old. Long bones grow faster in diameter than in length (Example 2.1.16). Mathematical formulas that capture these patterns of change are called **allometric formulas** or **allometric relations.** The formulas relating blood circulation time (Example 2.1.12) and heartbeat frequency (Example 2.1.13) to body mass of mammals are examples of such relations.

Allometric relations are usually obtained by fitting a curve to the data that have been collected. Quite often they are expressed in the form $y = ax^r$, where a is a positive constant. (Further examples of allometric formulas can be found in Exercises 42 and 43.)

Example 2.1.14 Relationship between Surface Area and Volume

Recall that the surface area, S, and the volume, V, of a sphere of radius r are given by

$$S = 4\pi r^2 \text{ and } V = \frac{4}{3}\pi r^3$$

We are going to figure out how the surface area depends on the volume. Although an animal is not sphere-shaped, this relationship will help us understand how animal growth affects mass (related to volume) and heat regulation (related to surface area). For instance, if an animal grows to twice its volume, how does its surface area (skin area) change?

We start from

$$V = \frac{4}{3}\pi r^3$$

isolate r

$$4\pi r^3 = 3V$$
$$r^3 = \frac{3V}{4\pi}$$
$$r = \left(\frac{3V}{4\pi}\right)^{1/3}$$

and substitute into the formula $S = 4\pi r^2$ for the surface area:

$$S(V) = 4\pi \left(\left(\frac{3V}{4\pi}\right)^{1/3}\right)^2$$
$$= 4\pi \left(\frac{3V}{4\pi}\right)^{2/3}$$
$$= 4\pi \frac{3^{2/3} V^{2/3}}{(4\pi)^{2/3}}$$
$$= (4\pi)^{1/3} 3^{2/3} V^{2/3}$$
$$\approx 4.84 V^{2/3}$$

We conclude that the surface area is not proportional to the volume, but proportional to the volume raised to the power of 2/3. (So scaling V will not produce the same scaling of S.)

The constant of proportionality, 4.84, characterizes the sphere, and it changes if we consider other solids. For instance, the relationship between the surface area and the volume of a cube is $S = 6V^{2/3}$. However, the fact that *the surface area is proportional to $V^{2/3}$ holds no matter what three-dimensional solid (or animal) is considered.* So, in general,

$$S(V) = aV^{2/3}$$

for a positive constant, a.

We can now answer our question: if an animal grows to twice its volume (thus twice its mass, since volume and mass are proportional), its surface area will change according to

$$S(2V) = a(2V)^{2/3} = 2^{2/3}(aV^{2/3}) = 2^{2/3}S(V) \approx 1.59 S(V)$$

i.e., it will increase by a factor of about 1.6. ▲

Example 2.1.15 Surface Area to Volume Ratio and What It Means

Continuing with the previous example, we compare surface area to volume for animals of varying sizes. Using $S = aV^{2/3}$, we calculate

$$\frac{S}{V} = \frac{aV^{2/3}}{V} = \frac{a}{V^{1/3}} = aV^{-1/3}$$

which is the *relative surface area* (see Figure 2.1.17). We conclude that the relative surface area decreases as the animal size increases; i.e., smaller animals tend to have relatively more surface cover (skin or fur, for instance) than larger animals. Since heat loss is proportional to surface area, smaller animals tend to lose heat relatively faster than larger animals. That is one of the reasons why the hearts of smaller animals beat faster (Example 2.1.13), or why their blood circulates faster (Example 2.1.12). ▲

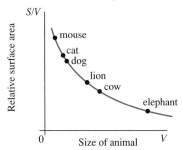

FIGURE 2.1.17
Relative surface area of animals

Example 2.1.16 How Bones Grow

Human legs need to be strong enough to support the body mass. As we grow, our legs (bones as well as muscles) must grow in a way to support the increase in body mass.

Suppose that a baby girl grows to twice her size. Because volume is length times width times height, as each of the length, the width, and the height doubles, the volume of the baby increases eightfold. Since volume is density times mass (i.e., mass and volume are proportional), the mass of the baby increases eightfold as well.

It is known that the strength of a bone is proportional to its cross-sectional area. With a reasonable approximation that the cross-section of a bone is a disk of radius r, we conclude that the strength of a bone is proportional to πr^2.

So if a bone in the baby's body grows so that its radius doubles, the cross-sectional area will be $\pi(2r)^2 = 4\pi r^2$, i.e., four times the original cross-sectional area—and so her body strength will increase fourfold as well. But this is no match for the eightfold increase in mass. In other words, to support an eightfold increase in mass, brought on by the baby's dimensions all doubling, her bones must increase their radius by *more* than a factor of 2. How much more? We need to find the radius of the cross-sectional disk (call it R) that will guarantee the eightfold increase in strength, i.e.,

$$\pi R^2 = 8\pi r^2$$

Thus,

$$R^2 = 8r^2$$
$$R = \sqrt{8r^2} \approx 2.83r$$

We come to an important conclusion: the *way we grow is not by mere scaling* (i.e., an adult is not a scaled version of a baby). In our case, if the bone in the baby's body

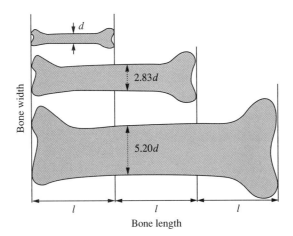

FIGURE 2.1.18
How bones grow

grows to twice its length, its width (the diameter of the cross-section) grows by a factor of 2.83. See Figure 2.1.18.

Likewise, if a body grows to three times its original size (so the length of a bone triples), the width of the bone will have to grow by a factor of $\sqrt{27} \approx 5.20$ to maintain its strength.

Summary The graphs of many important functions in biology are lines. We derived the link between lines and **linear functions**. A **proportional relation** is a special type of linear function in which the ratio of the output to the input is always the same. This constant ratio is the **slope** of the graph of the relation. Lines can be expressed in **point-slope form** or **slope-intercept form.** The slope can be found as the change in output divided by the change in input. Equations of linear functions can be used to **interpolate**—that is, to estimate outputs from untested inputs. The behaviour of **power functions,** $f(x) = x^r$, depends on the sign of r. Assume that the variable is positive (such as time, volume, or mass). Negative values of r describe phenomena that decrease as the variable increases. Power functions $f(x) = x^r$ with positive values of r increase as the variable increases. Many phenomena, such as blood circulation time, growth of bones, and heartbeat frequency, can be described and understood using **allometric relations**.

2.1 Exercises

Mathematical Techniques

1–4 ▪ For each line, find the slope between the two given points by finding the change in output divided by the change in input. What is the ratio of the output to the input at each of the points? Which are proportional relations? Which are increasing and which are decreasing? Sketch a graph.

1. $y = 2x + 3$, using points with $x = 1$ and $x = 3$
2. $z = -5w$, using points with $w = 1$ and $w = 3$
3. $z = 5(w - 2) + 8$, using points with $w = 1$ and $w = 3$
4. $y - 5 = -3(x + 2) - 6$, using points with $x = 1$ and $x = 3$

5–6 ▪ Check that the point indicated lies on the line, and find the equation of the line in point-slope form using the given point. Multiply out to check that the point-slope form matches the original equation.

5. The line $f(x) = 2x + 3$ and the point $(2, 7)$.
6. The line $g(y) = -2y + 7$ and the point $(3, 1)$.

7–12 ▪ Find the equation in slope-intercept form for each line. Sketch a graph indicating the original point from the point-slope form.

7. The line $f(x) = 2(x - 1) + 3$.
8. The line $g(z) = -3(z + 1) - 3$.
9. A line passing through the point $(1, 6)$ with slope -2.
10. A line passing through the point $(-1, 6)$ with slope 4.
11. A line passing through the points $(1, 6)$ and $(4, 3)$.
12. A line passing through the points $(6, 1)$ and $(3, 4)$.

13–16 ▪ Check whether each function is linear.

13. $h(z) = \dfrac{1}{5z}$
14. $F(r) = r^2 + 5$
15. $P(q) = 8(3q + 2) - 6$
16. $Q(w) = 8(3w + 2) - 6(w + 4)$

17–18 ■ Check that each curve does not have constant slope by computing the slopes between the points indicated.

17. $h(z) = \dfrac{1}{5z}$ at $z = 1$, $z = 2$, and $z = 4$. Find the slope between $z = 1$ and $z = 2$, and the slope between $z = 2$ and $z = 4$.

18. $F(r) = r^2 + 5$ at $r = 0$, $r = 1$, and $r = 4$. Find the slope between $r = 0$ and $r = 1$, and the slope between $r = 1$ and $r = 4$.

19–24 ■ Solve each equation. Check your answer by plugging in the value you found.

19. $2x + 3 = 7$
20. $\dfrac{1}{2}z - 3 = 7$
21. $2x + 3 = 3x + 7$
22. $-3y + 5 = 8 + 2y$
23. $2(5(x - 1) + 3) = 5(2(x - 2) + 5)$
24. $2(4(x - 1) + 3) = 5(2(x - 2) + 5)$

25–28 ■ Solve each equation for the given variable, treating the other letters as constant parameters.

25. Solve $2x + b = 7$ for x.
26. Solve $mx + 3 = 7$ for x.
27. Solve $2x + b = mx + 7$ for x. Are there any values of b or m for which the equation has no solution?
28. Solve $mx + b = 3x + 7$ for x. Are there any values of b or m for which the equation has no solution?

29–32 ■ Most unit conversions are proportional relations. Find the slope and graph the relation between each pair of units.

29. Place inches on the horizontal axis and centimetres on the vertical axis. Use the fact that 1 in. = 2.54 cm. Mark the point corresponding to 1 in. on your graph.

30. Place centimetres on the horizontal axis and inches on the vertical axis. Use the fact that 1 in. = 2.54 cm. Mark the point corresponding to 1 in. on your graph.

31. Place grams on the horizontal axis and pounds on the vertical axis. Use the fact that 1 lb ≈ 453.6 g. Mark the point corresponding to 1 lb on your graph.

32. Place pounds on the horizontal axis and grams on the vertical axis. Use the fact that 1 lb ≈ 453.6 g. Mark the point corresponding to 1 lb on your graph.

33–34 ■ Not very many functions commute with each other (Section 1.4). The following problems ask you to find all linear functions that commute with the given linear function.

33. Find all functions of the form $g(x) = mx + b$ that commute with the function $f(x) = x + 1$. Can you explain your answer in words?

34. Find all functions of the form $g(x) = mx + b$ that commute with the function $f(x) = 2x$. Can you explain your answer in words?

35–43 ■ Express the relationship between the quantities algebraically, i.e., by writing a formula.

35. y is proportional to x; when $x = 6$, $y = 7$.
36. a is inversely proportional to b; when $b = 1$, $a = -4$.
37. y is proportional to the square root of x; when $x = 4$, $y = 8$.
38. y is proportional to the square of x; when $x = 2$, $y = 12$.
39. a is inversely proportional to the cube of b; when $b = 1$, $a = 3$.
40. p is proportional to the cube root of q; when $q = 1000$, $p = 11$.
41. y is inversely proportional to the square of x; when $x = 7$, $y = 2$.
42. The lung volume, V (in millilitres), of a mammal is proportional to $B^{1.07}$, where B is the body mass (in kilograms). Find the formula for V as a function of B if you know that a lung of a 12-kg dog has a volume of 764 mL.
43. The mass, M, of a fish skeleton is proportional to $B^{1.03}$, where B is the body mass (M and B are measured in grams). Express M as a function of B given that the skeleton of a 1-kg fish has a mass of 40.5 g.

44–47 ■ Sketch the pair of functions, f and g, for $x > 0$ and identify which one increases faster for $x > 1$. If necessary, make a table of values for f and g.

44. $f(x) = x^3$, $g(x) = x^4$
45. $f(x) = x$, $g(x) = \sqrt{x}$
46. $f(x) = x^{3.2}$, $g(x) = x^{2.4}$
47. $f(x) = x^{1.1}$, $g(x) = x$

48–51 ■ Sketch the pair of functions, f and g, for $x > 0$ and identify which one decreases faster for $x > 1$. If necessary, make a table of values for f and g.

48. $f(x) = x^{-3}$, $g(x) = x^{-4}$
49. $f(x) = \dfrac{1}{x}$, $g(x) = \dfrac{1}{\sqrt{x}}$
50. $f(x) = x^{-0.1}$, $g(x) = x^{-0.5}$
51. $f(x) = x^{-1.1}$, $g(x) = x^{-0.5}$

Applications

52–55 ■ Many fundamental relations express a proportional relation between two measurements with different dimensions. Find the slope and the equation of the relation between each pair of quantities.

52. Volume = area × thickness. Find the volume, V, as a function of the area, A, if the thickness is 1.0 cm.

53. Volume = area × thickness. Find the volume, V, as a function of the thickness, T, if the area is 7.0 cm^2.

54. Total mass = mass per bacterium × number of bacteria. Find the total mass, M, as a function of the number of bacteria, b, if the mass per bacterium is 5.0×10^{-9} g.

55. Total mass = mass per bacterium × number of bacteria. Find the total mass, M, as a function of mass per bacterium, m, if the number of bacteria is 10^6.

56–59 ■ A ski hill has a slope of -0.2. You start at an altitude of 3000 m.

56. Write the equation giving altitude, a, as a function of horizontal distance moved, d.

57. Write the equation of the line in feet (1 ft = 0.3048 m).

58. What will your altitude be when you have gone 700 m horizontally?

59. The ski run ends at an altitude of 2600 m. How far will you have gone horizontally?

60–63 ▪ The following data give the elevation of the surface of Lake Louise in Alberta.

Year, y	Elevation, E (m)
1980	1749.2
1985	1750.0
1990	1750.8
1995	1751.9
2000	1751.6
2005	1750.8
2010	1750.0

60. Graph these data.

61. During which 10-year periods is the surface elevation changing linearly?

62. What was the slope between 1980 and 1990? What would the surface elevation have been in 2010 if things had continued as they began? How different is this from the actual elevation?

63. What was the slope during the period between 2000 and 2010? What would the surface elevation have been in 1980 if things had always followed this trend? How different is this from the actual elevation?

64–67 ▪ Graph the following relations between measurements of a growing plant, checking that the points lie on a line. Find the equations in both point-slope and slope-intercept form. What do the y-intercepts mean?

Age, a (days)	Mass, M (g)	Volume, V (cm³)	Glucose Production, G (mg)
0.5	2.5	5.1	0.0
1.0	4.0	6.2	3.4
1.5	5.5	7.3	6.8
2.0	7.0	8.4	10.2
2.5	8.5	9.5	13.6
3.0	10.0	10.6	17.0

64. Mass as a function of age. Find the mass on day 1.75.

65. Volume as a function of age. Find the volume on day 2.75.

66. Glucose production as a function of mass. Estimate glucose production when the mass reaches 20.0 g.

67. Volume as a function of mass. Estimate the volume when the mass reaches 30.0 g. How will the density at that time compare with the density when $a = 0.5$?

68–71 ▪ Answer the following questions for each Atlantic province.

a. Build a linear model for the population based on the census data for 2001 and 2006.

b. What does the value of the slope say about the change in the population?

c. According to your model in (a), what is the predicted population in 2011? Compare with the census data.

Province	Population in 2001	Population in 2006	Population in 2011
Newfoundland and Labrador	512,930	505,469	514,536
Prince Edward Island	135,294	135,851	140,204
Nova Scotia	908,007	913,462	921,727
New Brunswick	729,498	729,997	751,171

Statistics Canada. 2011. *Population and Dwelling Counts, for Canada, Provinces and Territories, 2011 and 2006 Censuses*, "Population and dwelling count highlight tables, 2011 Census." Statistics Canada Catalogue no. 98-310-XWE2011002. Ottawa, Ontario. http://www12.statcan.gc.ca/census-recensement/2011/dp-pd/hlt-fst/pd-pl/Table-Tableau.cfm?LANG=Eng&T=101&S=50&O=A; Statistics Canada. 2007. *Population and Dwelling Counts, for Canada, Provinces and Territories, 2006 and 2001 Censuses, 100% Data* (table). "Population and dwelling count highlight tables, 2006 Census." "2006 Census: Release topics." Census. Statistics Canada Catalogue no. 97-550-XWE2006002. Ottawa, Ontario. March 13. http://www12.statcan.ca/english/census06/data/popdwell/Table.cfm?T=10.

68. Newfoundland and Labrador.

69. Prince Edward Island.

70. Nova Scotia.

71. New Brunswick.

72–75 ▪ Consider the data in the following table (adapted from *Parasitoids* by H. C. F. Godfray), describing the number of wasps that can develop inside caterpillars of different masses.

Mass of Caterpillar (g)	Number of Wasps
0.5	80
1	115
1.5	150
2	175

72. Graph these data. Which data point does not lie on the same line as the others?

73. Find the equation of the line connecting the first two points.

74. How many wasps does the function predict would develop in a caterpillar with a mass of 0.72 g?

75. How many wasps does the function predict would develop in a caterpillar with a mass of 0 g? Does this make sense? How many would you really expect?

Computer Exercise

76. Consider the function

$$f(x) = \frac{x^r}{a + x^r}$$

where a and r are positive constants. Use a graphing device to answer the following questions.

a. Take $a = 1$ and plot the graph of $f(x)$ on $[0, 20]$ with $r = 0.5$, $r = 1$, $r = 2$, and $r = 3$. What do all of the graphs have in common for increasing values of x?

b. Keep $a = 1$ and plot the graph of $f(x)$ on $[0, 1]$ with $r = 0.5$, $r = 1$, $r = 2$, and $r = 3$. What is common to all four graphs? Describe how the graphs differ.

c. Take $r = 2$ and plot the graph of $f(x)$ on $[0, 20]$ with $a = 1$, $a = 5$, $a = 10$, and $a = 20$. What is common to all four graphs? Describe how the graphs differ.

2.2 Exponential and Logarithmic Functions; Exponential Models

In this section we introduce **exponential functions** and their inverses, **logarithmic functions.** We draw their graphs and explore their numeric features and a number of other properties. The basic algebraic rules for working with these functions are contained in the laws of exponents and the laws of logarithms. In the second half of the section we explore models based on **exponential growth** and **exponential decay,** such as radiocarbon dating and the growth of a tumour. As well, we further investigate the population of Canada based on recent census data and the exponential growth model. **Semilog** and **double-log** graphs help us visualize functions with a wide range of values.

Exponential Functions

An exponential function is a function of the form $y = a^x$, where the base, a, is positive and the exponent, x, is any real number. Although we can sometimes compute a power of a negative number, such as $(-3)^4$, the exponential function is defined for positive bases only (we comment on this a bit later).

The domain of $y = a^x$ consists of all real numbers. Since $a^x > 0$ for all real numbers x (remember that $a > 0$), it follows that the range of the exponential function $y = a^x$ consists of positive numbers; i.e., it is the interval $(0, \infty)$.

In Figure 2.2.19 we show graphs of exponential functions $y = a^x$ for several values of a. To compare their behaviour, we show the table of values as well (Table 2.2.1).

The graph of $y = a^x$ is increasing if $a > 1$. Moreover, the larger the value of the base, a, the faster the graph grows (compare $y = 2^x$, $y = 3^x$, and $y = 5^x$ in Figure 2.2.19).

If $0 < a < 1$, the graph of $y = a^x$ is decreasing. As the base, a, decreases, the graph of $y = a^x$ decreases at a faster rate (compare $y = (1/2)^x$ and $y = (1/10)^x$ in Figure 2.2.19).

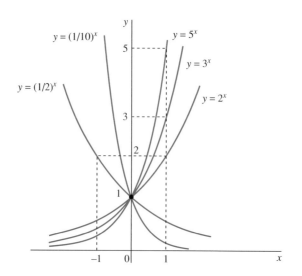

FIGURE 2.2.19

Exponential functions $y = a^x$

2.2 Exponential and Logarithmic Functions; Exponential Models

Table 2.2.1

x	$(1/2)^x$	$(1/10)^x$	2^x	3^x	5^x
−2	4	100	1/4	1/9	1/25
−1	2	10	1/2	1/3	1/5
0	1	1	1	1	1
1	1/2	1/10	2	3	5
2	1/4	1/100	4	9	25
3	1/8	1/1000	8	27	125
...	...	...	...	...	...

When $a = 1$, then $y = 1^x = 1$ is a constant function.

Since $a^0 = 1$ for all $a \neq 0$, the graph of $y = a^x$ goes through the point $(0, 1)$ on the y-axis.

What about $y = a^x$ for *negative* values of a? Take, for instance, $a = -4$, so that $y = (-4)^x$. If $x = 1/2$, we get

$$y = (-4)^{1/2} = \sqrt{-4}$$

which is not a real number, so $x = 1/2$ is not in the domain of $y = (-4)^x$. Likewise, when $x = 1/4$, then

$$y = (-4)^{1/4} = \sqrt[4]{-4}$$

which is again not real. We can identify many more values for x that are not in the domain of $y = (-4)^x$; see Exercise 52. The same is true for $y = a^x$ if a is any negative number.

This fact about the domain makes the function $y = a^x$ for $a < 0$ quite complicated (and in a way useless), and thus we do not study it in calculus. (However, such functions make a lot of sense for complex numbers.)

From

$$\left(\frac{1}{a}\right)^x = \frac{1}{a^x} = a^{-x}$$

we conclude that the graph of $y = (1/a)^x$ is the mirror image of the graph of $y = a^x$ with respect to the y-axis; compare the graphs of $y = 2^x$ and $y = (1/2)^x$ in Figure 2.2.19.

The key to using exponential functions is knowing the **laws of exponents,** summarized in the following table. This table also includes examples using $a = 2$ that can help in remembering when to add and when to multiply.

Laws of Exponents (for $a > 0$)

	General Formula	Example with $a = 2$, $x = 2$, and $y = 3$
Law 1	$a^x \cdot a^y = a^{x+y}$	$2^2 \cdot 2^3 = 2^5 = 32$
Law 2	$(a^x)^y = a^{xy}$	$(2^2)^3 = 2^6 = 64$
Law 3	$a^{-x} = \dfrac{1}{a^x}$	$2^{-2} = \dfrac{1}{2^2} = \dfrac{1}{4}$
Law 4	$\dfrac{a^y}{a^x} = a^{y-x}$	$\dfrac{2^3}{2^2} = 2^{3-2} = 2$
Law 5	$a^1 = a$	$2^1 = 2$
Law 6	$a^0 = 1$	$2^0 = 1$

Table 2.2.2

x	e^x
0	1
5	148.4
10	22,026.5
15	3.27×10^6
20	4.85×10^8
50	5.18×10^{21}
100	2.69×10^{43}

In the case when $a = e \approx 2.71828$, we obtain the **natural exponential function** (or just the **exponential function**) $y = e^x$. This function appears in a number of applications, such as population modelling, drug absorption, and radioactive decay.

The reasons we work with this particular exponential function will be revealed once we learn about limits and derivatives in Chapter 5.

Since $e^x > 0$ for all x, the range of $y = e^x$ is $(0, \infty)$. Its domain is the set of all real numbers. As with any exponential function, $y = e^x$ crosses the y-axis at $(0, 1)$; i.e., it satisfies $e^0 = 1$.

The table of values in Table 2.2.2 shows that $y = e^x$ increases very rapidly. For instance, e^{15} is greater than 3 million, and e^{20} exceeds 485 million.

The graphs of the functions $y = e^x$ and $y = e^{-x}$ are shown in Figure 2.2.20.

The graphs of functions of the form $y = e^{\alpha x}$ (commonly seen in applications) look like $y = e^x$ if $\alpha > 0$ and $y = e^{-x}$ if $\alpha < 0$. Figure 2.2.21 shows several functions of the form $y = e^{\alpha x}$.

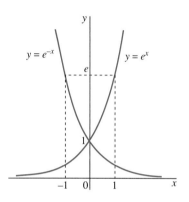

FIGURE 2.2.20

Exponential functions $y = e^x$ and $y = e^{-x}$

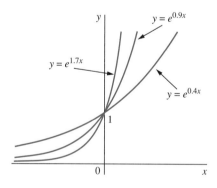 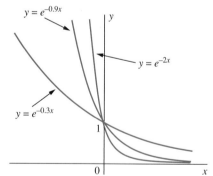

FIGURE 2.2.21

Exponential functions $y = e^{\alpha x}$

Example 2.2.1 The Laws of Exponents for the Base e

- $e^3 \cdot e^4 = e^{3+4} = e^7$ (law 1).
- $e^3 + e^4$ cannot be simplified with a law of exponents.
- $(e^3)^4 = e^{3 \cdot 4} = e^{12}$ (law 2).
- $e^{-2} = \dfrac{1}{e^2}$ (law 3).
- $\dfrac{e^4}{e^3} = e^{4-3} = e^1 = e$ (laws 4 and 5).
- $e^0 = 1$ (law 6).

Table 2.2.3

x	f(x)	Successive Quotients
0	c	
1	ce^α	$\frac{ce^\alpha}{c} = e^\alpha$
2	$ce^{2\alpha}$	$\frac{ce^{2\alpha}}{ce^\alpha} = e^\alpha$
3	$ce^{3\alpha}$	$\frac{ce^{3\alpha}}{ce^{2\alpha}} = e^\alpha$
4	$ce^{4\alpha}$	$\frac{ce^{4\alpha}}{ce^{3\alpha}} = e^\alpha$

We say that a quantity f **changes exponentially** if

$$f(x) = ce^{\alpha x}$$

where c and α are nonzero constants.

How do we tell whether or not a quantity changes exponentially? In other words, what is special about the data that come from exponentially changing quantities? In Table 2.2.3 we calculate the values of $f(x) = ce^{\alpha x}$ for equal increments of x, as well as the successive quotients of the values of f.

All quotients are equal—and that is what characterizes exponential functions. (Note that we have not made any assumptions: α could be positive or negative.)

> **Exponential Model $f(x) = ce^{\alpha x}$ with $c \neq 0$, $\alpha \neq 0$**
>
> The ratio between values of $f(x)$ for equally spaced values of x is constant.
> If $c > 0$ and $\alpha > 0$, the function $f(x)$ is increasing.
> If $c > 0$ and $\alpha < 0$, the function $f(x)$ is decreasing.

Logarithmic Functions

The graph of the natural exponential function passes the horizontal line test, and thus the function $y = e^x$ has an inverse function. In the arrow diagram in Figure 2.2.22a we recall the domain and the range of $y = e^x$.

Definition 2.2.1 The inverse function of the natural exponential function, $y = e^x$, is called the **natural logarithmic** (or **natural log**) **function** and is denoted by $y = \ln x$.

Due to the inverse relationship between $y = e^x$ and $y = \ln x$, the domain of $y = \ln x$ is $(0, \infty)$, and the range is the set of all real numbers; see Figure 2.2.22b.

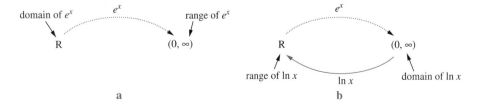

FIGURE 2.2.22
Exponential and logarithmic functions

Remember that

$$\ln A = B \text{ is equivalent to } e^B = A$$

See the diagram in Figure 2.2.23. In particular, $\ln 1 = 0$ since $e^0 = 1$.

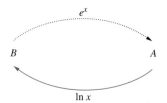

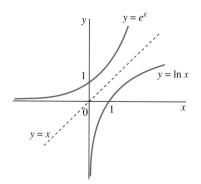

FIGURE 2.2.23
Exponential and logarithmic functions as inverses

FIGURE 2.2.24
Natural logarithm function, $y = \ln x$

Example 2.2.2 Exponential and Logarithmic Functions

- From $\ln 100 \approx 4.605$ it follows that $e^{4.605} \approx 100$.
- From $e^5 \approx 148.41$ it follows that $\ln 148.41 \approx 5$.
- From $\ln 0.1 \approx -2.303$ it follows that $e^{-2.303} \approx 0.1$.
- From $e^{-3} \approx 0.04979$ it follows that $\ln 0.04979 \approx -3$.

The following two formulas will be quite useful.

Cancellation Laws for Natural Exponential and Logarithm Functions

$e^{\ln x} = x$ if $x > 0$.

$\ln(e^x) = x$ for all real numbers x.

For instance, $e^{\ln 6.7} = 6.7$, $\ln e^{-3} = -3$, and $\ln e = 1$.

The graph of $y = \ln x$ is obtained (as with any inverse function) by reflecting the graph of $y = e^x$ with respect to the diagonal $y = x$; see Figure 2.2.24. The graph increases from negative infinity (the meaning of negative infinity will be made clear once we introduce limits) through 0 at $x = 1$ and rises more and more slowly as x becomes larger.

As a matter of fact, $y = \ln x$ is among the slowest growing functions; see the table of values (Table 2.2.4). For instance, the natural log of one billion is about 20.72, and $\ln 10^{21}$, which is the natural log of the diameter of our Milky Way Galaxy in metres, is about 48.35.

An important use of logarithms is the application of the laws of logarithms to functions and numbers. The laws of logarithms presented in the accompanying table are the laws of exponents in reverse.

Table 2.2.4

x	ln x
0.001	−6.90776
0.01	−4.60517
0.1	−2.30259
1	0
10	2.30259
100	4.60517
1000	6.90776

Laws of Logarithms (for $x, y > 0$, and any number p)

	General Formula	Example with $x = 5$, $y = 2$, and $p = -4$
Law 1	$\ln(xy) = \ln x + \ln y$	$\ln 10 = \ln(5 \cdot 2) = \ln 5 + \ln 2$
Law 2	$\ln x^p = p \ln x$	$\ln 5^{-4} = -4 \ln 5$
Law 3	$\ln(1/x) = -\ln x$	$\ln(1/5) = -\ln 5$
Law 4	$\ln(x/y) = \ln x - \ln y$	$\ln(5/2) = \ln 5 - \ln 2$
Law 5	$\ln e = 1$	
Law 6	$\ln 1 = 0$	

Example 2.2.3 The Laws of Logarithms in Action

- $\ln 3 + \ln 4 = \ln(3 \cdot 4) = \ln 12$, using law 1.
- $\ln 3 \cdot \ln 4$, $\ln(3 + \pi)$, and $\ln(\sqrt{2} - 3)$ cannot be simplified with laws of logs.

- $\ln(3^4) = 4\ln 3$, using law 2.
- $\ln(1/3) = -\ln 3$, using law 3.
- $\ln(4/3) = \ln 4 - \ln 3$, using law 4.

The formula $e^{\ln x} = x$ can be used to express a general exponential function in terms of e^x. As an example, consider $y = 5^x$. Since $5 = e^{\ln 5}$, we write

$$5^x = (e^{\ln 5})^x = e^{(\ln 5)x}$$

Using a calculator we find that $\ln 5 \approx 1.60944$, so

$$5^x \approx e^{1.60944x}$$

In general,

$$a^x = (e^{\ln a})^x = e^{(\ln a)x}$$

We conclude that the graph of $y = a^x$ is obtained by horizontally stretching or compressing the graph of $y = e^x$. If $\ln a < 0$, the resulting graph has to be reflected across the y-axis.

Besides $y = e^x$, another very commonly used exponential function is $y = 10^x$. Its inverse is the **logarithm to the base 10,** which is written

$$y = \log_{10} x$$

and is read "log base 10 of x." Thus,

$$\log_{10} A = B \text{ is equivalent to } A = 10^B$$

In particular,

- $\log_{10} 1000 = 3$, since $10^3 = 1000$
- $\log_{10} 0.01 = -2$, since $10^{-2} = \frac{1}{10^2} = \frac{1}{100} = 0.01$
- from $10^6 = 1{,}000{,}000$ we conclude that $\log_{10} 1{,}000{,}000 = 6$
- from $10^{2.7} \approx 501.187$ we conclude that $\log_{10} 501.187 \approx 2.7$

In most ways, the exponential function with base 10 and the log base 10 work much like the exponential function with base e and the natural logarithm. All laws of exponents and logarithms are the same except law 5, which becomes

$$\text{Law 5 of exponents:} \quad 10^1 = 10$$
$$\text{Law 5 of logarithms:} \quad \log_{10} 10 = 1$$

Example 2.2.4 Converting Logarithms

We find a conversion formula that expresses $\log_{10} x$ in terms of $\ln x$.

Let $\log_{10} x = A$. Then $x = 10^A$, and, after applying the natural logarithm to both sides, we have

$$\ln x = \ln 10^A$$
$$\ln x = A \ln 10$$
$$A = \frac{\ln x}{\ln 10}$$

Thus,

$$\log_{10} x = \frac{\ln x}{\ln 10}$$

i.e.,

$$\log_{10} x \approx \frac{\ln x}{2.30259} \approx 0.43429 \ln x$$

We conclude that the graph of $y = \log_{10} x$ is a scaled version of the graph of $y = \ln x$.

Likewise, if $\ln x = B$, then $x = e^B$, and
$$\log_{10} x = \log_{10} e^B = B \log_{10} e$$
i.e.,
$$B = \frac{\log_{10} x}{\log_{10} e}$$
and finally we obtain the conversion formula
$$\ln x = \frac{\log_{10} x}{\log_{10} e}$$

Exponential Models

Numerous measurements, experiments, and methods across a wide spectrum of applications in sciences can be described using exponential functions. In such cases, we write the measurement $S(t)$ as a function of t as
$$S(t) = S(0)e^{\alpha t}$$

The number $S(0)$ represents the value (initial value) of the measurement at time $t = 0$. The parameter α describes at what rate the measurement changes; α has dimensions of 1/time.

When $\alpha > 0$, the function $S(t)$ is increasing, describing **exponential growth** (Figure 2.2.25a). When $\alpha < 0$, the function $S(t)$ is decreasing, showing **exponential decay** (Figure 2.2.25b). The function $S(t)$ increases most quickly with large positive values of α, and it decreases most quickly with large negative values of α.

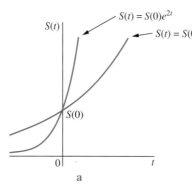
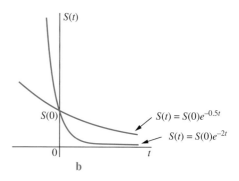

FIGURE 2.2.25
The function $S(t)$ shown for several values of α

One important number describing measurements in exponential models is the **doubling time.** When $\alpha > 0$, the measurement is increasing. A convenient measure of the speed of increase is the time it takes the initial value to double.

Example 2.2.5   Computing a Doubling Time from Scratch

Suppose
$$S(t) = 150 e^{1.2t}$$
with t measured in hours. This measurement starts at $S(0) = 150$ and doubles when $S(t) = 300$, or
$$150 e^{1.2t} = 300$$
$$e^{1.2t} = 2$$
$$1.2t = \ln 2$$
$$t = \frac{\ln 2}{1.2} \approx 0.5776$$

2.2 Exponential and Logarithmic Functions; Exponential Models

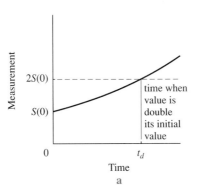
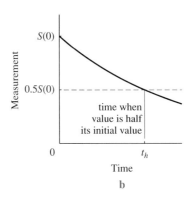

FIGURE 2.2.26
Doubling times and half-lives

As a check, we compute

$$S(0.5776) = 150e^{1.2 \cdot 0.5776} \approx 300$$

Thus, the doubling time for $S(t)$ is approximately 0.5776 hours (Figure 2.2.26a). ▲

We can solve for the doubling time for a measurement modelled by $S(t) = S(0)e^{\alpha t}$ by finding the time, t_d, when $S(t_d) = 2S(0)$:

$$S(t_d) = S(0)e^{\alpha t_d} = 2S(0)$$
$$e^{\alpha t_d} = 2$$
$$\alpha t_d = \ln 2$$
$$t_d = \frac{\ln 2}{\alpha} \approx \frac{0.6931}{\alpha}$$

The **general formula for the doubling time** is

$$t_d = \frac{\ln 2}{\alpha}$$

The doubling time becomes smaller as α becomes larger, consistent with the fact that measurements with larger values of α increase more quickly.

Example 2.2.6 Computing a Doubling Time with the Formula

Suppose $S(t) = 150e^{1.2t}$ as in Example 2.2.5. Then $\alpha = 1.2$, and the doubling time is

$$t_d \approx \frac{0.6931}{1.2} \approx 0.5776 \text{ h}$$ ▲

When $\alpha < 0$, the measurement is decreasing, and we can ask how long it will take to become half as large. This time, denoted t_h, is called the **half-life** and can be found by setting $S(t)$ equal to $1/2$ of $S(0)$ instead of $2S(0)$ as we did in the doubling-time calculations:

$$S(t_h) = S(0)e^{\alpha t_h} = 0.5S(0)$$
$$e^{\alpha t_h} = 0.5$$
$$\alpha t_h = \ln 0.5$$
$$t_h = \frac{\ln 0.5}{\alpha} \approx -\frac{0.6931}{\alpha}$$

Therefore, the **general formula for the half-life** is

$$t_h = \frac{\ln 0.5}{\alpha} = \frac{\ln 1/2}{\alpha} = -\frac{\ln 2}{\alpha} \approx -\frac{0.6931}{\alpha}$$

The half-life becomes smaller as α grows more and more negative. Remember to apply this equation only when $\alpha < 0$.

Example 2.2.7 Computing the Half-Life

If a measurement follows the equation
$$M(t) = 240e^{-2.3t}$$
with t measured in seconds, then $\alpha = -2.3$ and the half-life is
$$t_h \approx \frac{-0.6931}{-2.3} \approx 0.3013 \text{ s}$$
(see Figure 2.2.26b). ▲

Conversely, if we are told the initial value and the doubling time or half-life of some measurement, we can find the formula.

Example 2.2.8 Finding the Formula from the Doubling Time

Suppose $t_d = 26{,}200$ years is the doubling time for some measurement m. Because
$$t_d \approx \frac{0.6931}{\alpha}$$
we can solve for α as
$$\alpha \approx \frac{0.6931}{t_d} = \frac{0.6931}{26{,}200} = 2.645 \times 10^{-5}$$
If we are also told that $m(0) = 0.031$, then the formula for $m(t)$ is
$$m(t) = 0.031 e^{2.645 \times 10^{-5} t}$$
▲

Example 2.2.9 Finding the Formula from the Half-Life

Suppose $t_h = 6.8$ years is the half-life for some measurement V. Because
$$t_h \approx -\frac{0.6931}{\alpha}$$
we can solve for α as
$$\alpha \approx -\frac{0.6931}{t_h} = -\frac{0.6931}{6.8} \approx -0.1019$$
If we are given that $V(0) = 23.1$, then the formula for $V(t)$ is
$$V(t) = 23.1 e^{-0.1019 t}$$
▲

Example 2.2.10 Thinking in Half-Lives

Assume that $M(t)$ describes the amount of some drug t hours after it was administered to a patient (thus, $M(0)$ is the amount of the drug initially given to the patient). The half-life of the drug is known to be t_h hours.

One half-life (i.e., t_h hours) after the initial dosage of $M(0)$ units was given to the patient, one half of it is absorbed, so the amount of drug still present in the body is $0.5M(0)$; another half-life later, it decreases to one half of $0.5M(0)$, i.e., to $0.5(0.5M(0)) = 0.5^2 M(0)$, etc.

In Table 2.2.5 we show the dynamics of exponential decrease for six half-lives.

As we can see, after four half-lives, only about 6% of the initial drug is still left unabsorbed by the body. After another half-life, it is down to about 3%. Since many drugs are not effective when they reach less than 5% of their original level, practitioners use this fact as a ball-park—after four to five half-lives, the drug that was administered is no longer effective. ▲

Table 2.2.5

Number of Half-Lives	Amount Left in the Body	Percent Amount Left in the Body
0	$M(0)$	100
1	$0.5M(0)$	50
2	$0.5^2 M(0)$	25
3	$0.5^3 M(0)$	12.5
4	$0.5^4 M(0)$	6.25
5	$0.5^5 M(0)$	3.125
6	$0.5^6 M(0)$	1.5625

Example 2.2.11 Thinking in Half-Lives

Consider the measurement $M(t)$ given in Example 2.2.7, with a half-life of 0.3013 s. To figure out how much the value will have decreased in 2 s, we could plug into the original formula, finding

$$M(2) = 240 e^{-2.3 \cdot 2} \approx 2.41$$

The value decreased by a factor of nearly 100. Alternatively, 2 s is

$$\frac{2}{0.3013} \approx 6.638$$

half-lives. After this many half-lives, the value will have decreased by a factor of $2^{6.638} \approx 99.59$, so that $M(2) \approx \frac{240}{99.59} \approx 2.41$. We can think of using half-lives as converting the exponential to base 2. ▲

Example 2.2.12 Radiocarbon Dating: Determining the Age of Artifacts

Carbon-14 (or radiocarbon) dating is a common method of determining the age of a wide range of organic material. It is based on measuring the amount of a radioactive carbon isotope, ^{14}C, left in the sample under investigation and comparing it with the known amount that is present in the same, or a similar, living sample.

Cosmic radiation breaks down nitrogen molecules (in the upper layers of Earth's atmosphere, in particular at altitudes of 9 km to 15 km) into an unstable isotope of carbon, namely ^{14}C. Rain, winds, and other atmospheric activities bring ^{14}C down to Earth's surface, where, through photosynthesis, it gets introduced into living plants. Animals and humans absorb ^{14}C through food chains. ^{14}C decays into stable carbon isotopes; however, through absorption of new ^{14}C, a living organism maintains a constant level of ^{14}C radioactivity. Once an organism dies, the levels of ^{14}C decrease through radioactive decay. It has been determined that ^{14}C has a half-life of 5730 years.

Radiocarbon analysis of several wooden artifacts determined that they contain about 29.7% of the level of ^{14}C in a living tree today. Find the approximate age of the artifacts knowing that the half-life of ^{14}C is 5730 years.

Let $C(t)$ denote the level of ^{14}C present in an artifact; by t we denote the time—but in this case, pointing into the past. So $t = 2500$ means 2500 years ago. Substituting the half-life information into the radioactive decay formula

$$C(t) = C(0) e^{kt}$$

we get

$$0.5 C(0) = C(0) e^{k(5730)}$$

and thus
$$e^{5730k} = 0.5$$
$$5730k = \ln 0.5$$
$$k = \frac{\ln 0.5}{5730} \approx -0.00012097$$

So we can describe the radioactive decay of ^{14}C with the formula
$$C(t) = C(0)e^{-0.00012097t}$$

To find the age of the artifacts, we replace $C(t)$ by $0.297C(0)$ and compute t:
$$0.297C(0) = C(0)e^{-0.00012097t}$$
$$e^{-0.00012097t} = 0.297$$
$$-0.00012097t = \ln 0.297$$
$$t = \frac{\ln 0.297}{-0.00012097} \approx 10{,}035.74$$

Thus, the artifacts are about 10,000 years old. Note that this is an estimate, as a number of factors have not been included in this calculation. For instance, contamination of the sample with certain chemicals (such as limestone or various acids from soil) can skew the estimate. As well, we assumed (but had no support to do so) that the wood from which the artifacts were made contained the same levels of ^{14}C as a living tree (today). ▲

Example 2.2.13 Limits of Radiocarbon Dating

The smallest level of ^{14}C radioactivity that can be detected (using present technology) is about one-tenth to one-twentieth of a percent (i.e., between 0.1% and 0.05%).

The age of an object that has 0.1% of ^{14}C left can be calculated as in Example 2.2.12 (by replacing 0.297 with 0.001):
$$C(t) = 0.001C(0) = C(0)e^{-0.00012097t}$$
i.e.,
$$e^{-0.00012097t} = 0.001$$
$$t = \frac{\ln 0.001}{-0.00012097} \approx 57{,}103.04$$

Similarly, replacing 0.1% by 0.05% we calculate
$$t = \frac{\ln 0.0005}{-0.00012097} \approx 62{,}832.95$$

Thus—at this time—radiocarbon dating can be used to date objects no older than 57,000 to 63,000 years. In reality, objects this old cannot be dated reliably using radiocarbon dating, one reason being the accuracy of determining the amount of ^{14}C left. Suppose that a sample contains 0.05% of ^{14}C, but we measure it to be 0.1% (thus, we made an error of 0.05%). The above calculations show that our estimate is off by over five thousand years. ▲

Note Dinosaurs died millions of years ago, so clearly their age was not determined using radiocarbon dating. How were they dated? One method consists of dating the rocks, clay minerals, and other materials that surround the remains (usually fossilized bones) of a dinosaur. Again, the idea of radioactive decay is used, but we need a substance whose half-life is much longer than 5730 years. Such substances exist—for instance, K-Ar dating (potassium-argon dating) is based on the fact that the half-life of radioactive potassium, ^{40}K, is about 1.248 billion years.

So when did the dinosaurs die? About 65 million years ago (see Exercise 69).

2.2 Exponential and Logarithmic Functions; Exponential Models

Note Radiocarbon dating and potassium-argon dating are **radiometric dating** techniques. Radiometric dating is based on observing the level of a certain isotope (which occurs naturally in various objects) and the products of its decay.

Numerous alternative methods have been used for dating organic and inorganic matter. For instance, the tree trunk from the Introduction (Figure 1.1.2 in Section 1.1) was determined to have existed through six MIS (marine isotope stages, i.e., alternate warm and cool periods in Earth's climate), which makes it about 200 thousand years old.

Example 2.2.14 Modelling the Population of Canada Using Exponential Growth

In Example 2.1.10 we built a model for the increase in Canadian population using a linear function. As well, we explained why such a model is not really appropriate. Let us look at the data for the population between 1986 and 2011. We have added differences, as well as ratios, of successive population counts.

Year	Population, in Thousands	Difference of Successive Population Counts	Ratio of Successive Population Counts
1986	25,309		
1991	27,297	27,297 − 25,309 = 1988	27,297/25,309 ≈ 1.079
1996	28,847	28,847 − 27,297 = 1550	28,847/27,297 ≈ 1.057
2001	30,007	30,007 − 28,847 = 1160	30,007/28,847 ≈ 1.040
2006	31,613	31,613 − 30,007 = 1606	31,613/30,007 ≈ 1.054
2011	33,477	33,477 − 31,613 = 1864	33,477/31,613 ≈ 1.059

Statistics Canada. 2011. *Population, urban and rural, by province and territory (Canada)*, "Population estimates and projections, 2011 Census." Statistics Canada Catalogue no. 98-310-XWE2011002. Ottawa, Ontario. http://www.statcan.gc.ca/tables-tableaux/sum-som/l01/cst01/demo62a-eng.htm.

As we noted earlier, equal, or nearly equal, successive differences suggest linear behaviour—which is not the case here. However, the fact that successive quotients are nearly the same means that the behaviour of the population could be close to exponential.

So let us model the population of Canada using exponential growth, $P(t) = P(0)e^{rt}$, where t is time in years relabelled so that $t = 0$ represents 2001. As in the linear case, we will use the 2001 and 2006 data to build the model and then test it against the 2011 data.

Year	Time, t	Population $P(t)$, in Thousands
2001	0	30,007
2006	5	31,613
2011	10	33,477

Recall that $P(0)$ represents the initial population; thus

$$P(t) = 30{,}007 e^{rt}$$

and all we need to do is to calculate r. Substituting the data for $t = 5$ (year 2006), we get

$$31{,}613 = 30{,}007 e^{r(5)}$$
$$e^{5r} = \frac{31{,}613}{30{,}007}$$
$$5r = \ln \frac{31{,}613}{30{,}007}$$
$$r = \frac{1}{5} \ln \frac{31{,}613}{30{,}007} \approx 0.010428$$

Thus—based on the assumption of exponential growth, and using the known data for 2001 and 2006—we obtain the formula

$$P(t) = 30{,}007 e^{0.010428t}$$

The population in 2011 is calculated to be

$$P(10) = 30{,}007 e^{0.010428(10)} \approx 33{,}305 \text{ thousands}$$

This prediction is closer to the actual population (33,477 thousands) than the one obtained from the linear model. However, it is still not very good. ▲

Example 2.2.15 Benefits of Mammography over Clinical Breast Examination

Cancer is the name for a range of diseases characterized by the uncontrolled growth of cells (i.e., beyond the limits of what is considered normal or healthy), their invasion and/or destruction of surrounding tissue, and their movement (through blood or lymph nodes) toward other locations in the body. In this example, we focus on the dynamics of cell growth forming tumour tissue in the case of breast cancer.

By $C(t)$ we denote the number of cancer cells, where $C(0) = 1$ (cancer starts with a single cell showing certain genetic abnormalities). As the time unit we choose the doubling time and assume that it is constant throughout the process of tumour growth in an individual.

The doubling time depends on many factors. In Table 2.2.6 we show how doubling time depends on age: we give minimum and maximum doubling times for certain age groups, as well as the average.[1]

After the first doubling time, the number of cancer cells is $C(1) = 2C(0) = 2^1$; after another doubling, $C(2) = 2C(1) = 2^2$; etc. In general, after t doubling times have elapsed, there are $C(t) = 2^t$ cancer cells. For instance, $C(10) = 2^{10} = 1024$ and $C(20)$ is just over one million.

After 27 doublings, the cancer is about 5 mm in diameter and contains $C(27) = 2^{27} \approx 134$ million cells. In practice, tumours of this size can be detected in a mammogram. When the tumour grows to about 10 mm in diameter, it can be detected in a clinical breast exam.

Cancer is a serious problem and should be dealt with as early as possible. What is the lead time gained by mammography over clinical breast examination? That is, how much sooner can breast cancer be detected if mammography is used?

During one doubling-time interval, the number of cancer cells doubles, and so does the volume of the cancer. But how much does the cancer grow in size; i.e., by how much does its diameter increase? If the diameter doubles, then the volume (being proportional to the cube of the diameter) increases eightfold. So the diameter needs to increase by a smaller value. Write

$$V = kd^3$$

Table 2.2.6

Age	Doubling-Time Interval in Days	Average Doubling Time in Days
younger than 50	44–147	80
between 50 and 70	121–204	157
older than 70	120–295	188

---

[1] Peer, P. G., J. A. van Dijck, J. H. Hendriks, R. Holland, and A. L. Verbeek. Age-dependent growth rate of primary breast cancer. *Cancer* 71(11):3547–3551, 1993.

where V is the volume, d is the diameter, and k is a constant. We are looking for the value of the diameter (call it d_d) such that

$$2V = k\, d_d^3$$

Therefore,

$$2kd^3 = kd_d^3$$
$$d_d^3 = 2d^3$$
$$d_d = \sqrt[3]{2d^3} = \sqrt[3]{2}\, d \approx 1.26d$$

So, if the volume doubles, the diameter increases by about 26%. Thus, after the 28th doubling, the diameter of the tumour is $1.26(5) \approx 6.30$ mm. The next doubling increases the diameter to $1.26(6.30) \approx 7.94$ mm. After the 30th doubling the tumour reaches a size (diameter) of $1.26(7.94) \approx 10.0$ mm. So it takes three doublings for the tumour to grow from 5 mm to 10 mm (in diameter); see Figure 2.2.27. Thus, for a woman between 50 and 70 years of age, taking average doubling time (see Table 2.2.6), the lead time is $3(157) = 471$ days, i.e., about a year and 3.5 months. ▲

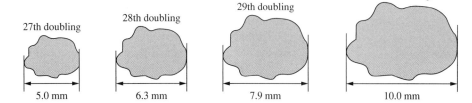

FIGURE 2.2.27

Dynamics of growth of cancer

We will introduce further models involving exponential functions as we learn more math. In Example 7.5.8 we investigate the formula

$$L(t) = M\left(1 - e^{-rt}\right)$$

for the length, $L(t)$, of a fish as a function of time (M and r are positive constants). An important population model is based on the **logistic function**

$$P(t) = \frac{L}{1 + \frac{L-P_0}{P_0} e^{-kt}}$$

where $P(t)$ is the population at time t, L represents the carrying capacity, k is the growth rate, and P_0 is the initial population (Example 8.4.5). The function

$$S(D) = e^{-\alpha D - \beta D^2}$$

(where $\alpha, \beta > 0$) models the fraction of cancer cells surviving a treatment with a radiation dose D (Example 2.2.17). A commonly used model for reproduction is the **Ricker model**

$$P(x) = rxe^{-x}$$

where $r > 0$ (see Section 6.8). Concentration of a pollutant (say, in the air) at a location x units away from the source changes according to the formula

$$c(t) = \frac{N}{\sqrt{4\pi kt}} e^{-x^2/4kt}$$

where the quantities N and k characterize the pollutant (see Exercise 81). The same model is used to describe the movement of molecules in the process of osmosis, the way the heat flows away from a source, or the concentration of a drug diffusing in blood.

Semilog and Double-Log Graphs

In some cases, the values of certain functions (such as exponential and power functions) spread over a wide range (see, for instance, Table 2.2.2 or Table 4.3.6 in Section 4.3). In that case, or when the independent variable assumes a wide range of values as well, we can use logarithms to construct graphs that are more useful than the graphs drawn in the usual xy-coordinate system. Let us consider an example.

Suppose we are to graph the function $f(x)$ defined by the table of values in the margin.

Note that the values of $f(x)$ range from 0.056 to values that are several orders of magnitude larger (24.34 and 120.12). The graph of the original data is difficult to read because the large vertical scale makes the small values almost indistinguishable (Figure 2.2.28a). If we take the natural logarithm of the data (Table 2.2.7), however, the values are much easier to compare (Figure 2.2.28b).

When we plot the values of $\ln f(x)$ against the values of x we say that we are drawing a **semilog graph** of $f(x)$. In other words, a semilog graph plots the natural logarithm of the output against the input. Needless to say, we cannot use semilog graphs for functions whose range includes zero or negative numbers.

Instead of taking the natural logarithm we could take the logarithm to the base 10 to obtain a semilog graph (see Table 2.2.7 and Figure 2.2.29a). In that case, it is common practice to place the values of $f(x)$ on the vertical axis, as shown in Figure 2.2.29b. Note that the value 1 on the vertical axis in Figure 2.2.29a corresponds to the value 10^1 on the vertical axis in Figure 2.2.29b, the value 2 corresponds to the value $10^2 = 100$, the value -1 corresponds to the value $10^{-1} = 0.1$, and so on.

x	f(x)
0	120.12
1	24.34
2	2.19
3	0.89
4	0.056
5	0.078
6	0.125
7	0.346
8	1.128

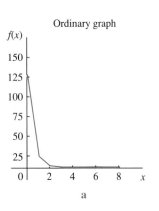
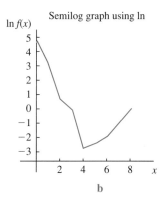

FIGURE 2.2.28
Ordinary graph and semilog graph

Table 2.2.7

x	f(x)	ln f(x)	log₁₀ f(x)
0	120.12	4.79	2.08
1	24.34	3.19	1.39
2	2.19	0.78	0.34
3	0.89	−0.12	−0.05
4	0.056	−2.88	−1.25
5	0.078	−2.55	−1.11
6	0.125	−2.08	−0.90
7	0.346	−1.06	−0.46
8	1.128	0.12	0.05

2.2 Exponential and Logarithmic Functions; Exponential Models

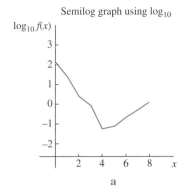
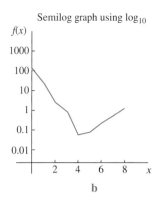

FIGURE 2.2.29
Semilog graph

Looking at semilog graphs (Figure 2.2.28b and Figure 2.2.29), we see that $f(x)$ reaches its minimum when $x = 4$ and increases steadily after that (this cannot be easily deduced from the graph in Figure 2.2.28a.)

Taking the natural logarithm of the exponentially changing quantity $S(t) = S(0)e^{\alpha t}$, we obtain

$$\ln S(t) = \ln\left(S(0)e^{\alpha t}\right) = \ln S(0) + \ln e^{\alpha t} = \ln S(0) + \alpha t$$

Thus, the semilog graph of $S(t)$ is a line of slope α and vertical intercept $\ln S(0)$.

Example 2.2.16 A Semilog Graph of an Exponentially Growing Value

Suppose

$$S(t) = 150 e^{1.2t}$$

with t measured in hours (Example 2.2.6 and Figure 2.2.30a). To plot a semilog graph of $S(t)$ against t we find the natural logarithm of $S(t)$:

$$\begin{aligned} \ln S(t) &= \ln\left(150 e^{1.2t}\right) \\ &= \ln 150 + \ln e^{1.2t} \\ &\approx 5.01 + 1.2t \end{aligned}$$

Therefore, the semilog graph is a line with intercept 5.01 and slope 1.2 (Figure 2.2.30b).

Example 2.2.17 Linear-Quadratic Survival Model

In radiotherapy, several models are used to describe the survival rate, $S(D)$, of clonogenic cells (such as cancer cells) as a function of the applied radiation dose, D. The linear-quadratic model states that

$$S(D) = e^{-\alpha D - \beta D^2}$$

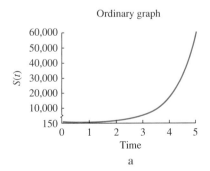
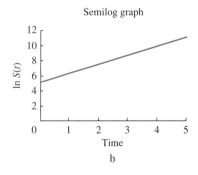

FIGURE 2.2.30
Original graph and semilog graph

FIGURE 2.2.31

Semilog graph of the survival curve

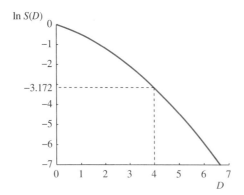

where D is the radiation dose (measured in grays (Gy), units of absorbed dose) and α and β are the parameters that characterize how the radiation beam neutralizes a cell. For instance, in the case of breast cancer cells (adapted from Qi, X. S., et al., Is α/β for breast cancer really low? *Radiotherapy and Oncology* 100: 282–288, 2011)

$$S(D) = e^{-0.401D - 0.098D^2}$$

For instance, $S(4) = e^{-0.401(4) - 0.098(16)} \approx 0.0419$ says that about 4.2 percent of cells will survive the radiation dose of 4 Gy. To draw the semilog graph of $S(D)$ (as is common practice!) we compute

$$\ln S(D) = -0.401D - 0.098D^2$$

The semilog graph is a parabola through the origin (Figure 2.2.31). We keep in mind that the values on the vertical axis are the natural logarithms of the survival rates. Thus, when $D = 4$, then (from the graph) $\ln S(D) \approx -3.172$ and (computing) $S(D) \approx e^{-3.172} \approx 0.0419$. ▲

In Section 2.1, we investigated several allometric relations in the form $y = ax^r$, where a is a positive constant (see Examples 2.1.12–2.1.14 and the note following Example 2.1.13). Computing the natural logarithm of y, we obtain

$$\ln y = \ln(ax^r) = \ln a + r \ln x$$

So the semilog graph of y is a vertically scaled (also reflected across the x-axis if $r < 0$) and shifted version of $\ln x$. This isn't of much use; however, in the coordinate system where $Y = \ln y$ (that's what was done for a semilog graph) and $X = \ln x$ (we now apply the same idea to the horizontal axis), we obtain the line

$$Y = \ln a + rX$$

of slope r and Y-intercept $\ln a$. In this construction we can replace the natural logarithm by the logarithm to the base 10 (as is done in the next example).

When we apply the logarithm to the values in both the domain and the range of a function, we obtain a **log-log plot** or a **double-log plot.** That is, a double-log plot is the plot of the logarithm of the output versus the logarithm of the input.

Example 2.2.18 Double-Log Plot of the Blood Circulation Time in Mammals

In Example 2.1.12 we obtained the relation

$$T(B) = 17.73\sqrt[4]{B} = 17.73B^{1/4}$$

between the body mass, B, of a mammal (in kilograms) and its blood circulation time, $T(B)$ (in seconds); we graphed it in Figure 2.1.15. To draw the log-log plot we compute

$$\log_{10} T(B) = \log_{10}\left(17.73B^{1/4}\right) = \log_{10} 17.73 + \frac{1}{4}\log_{10} B$$

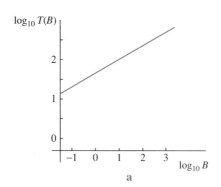
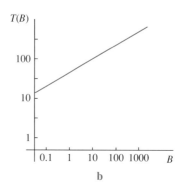

FIGURE 2.2.32
Double-log plots of the blood circulation time

Thus, the log-log graph is a line of slope 1/4 with vertical intercept at $\log_{10} 17.73 \approx 1.24$. In Figure 2.2.32a the labels on the axes represent the logarithms of the values, and in Figure 2.2.32b the actual values of B and $T(B)$ are shown.

Summary   We introduced **exponential functions** and their inverses, **logarithmic functions.** Using exponential functions, we studied various applications based on exponential growth and decay. The **natural logarithm** function can be used to solve equations involving the exponential function, including finding **doubling times** and **half-lives.** Measurements that cover a large range of positive values can be conveniently displayed on a **semilog graph,** which reduces the range and which produces a linear graph if the measurements follow an exponential function. In some cases we can use a **log-log plot.**

2.2 Exercises

Mathematical Techniques

1–10 ▪ Use the laws of exponents to rewrite each expression, if possible. If no law of exponents applies, say so.

1. 43.2^0
2. 43.2^1
3. 43.2^{-1}
4. $43.2^{-0.5} + 43.2^{0.5}$
5. $43.2^{7.2}/43.2^{6.2}$
6. $43.2^{0.23} \cdot 43.2^{0.77}$
7. $(3^4)^{0.5}$
8. $(43.2^{-1/8})^{16}$
9. $2^{2^3} \cdot 2^{2^2}$
10. $4^2 \cdot 2^4$

11–20 ▪ Use the laws of logarithms to rewrite each expression, if possible. If no law of logarithms applies or the quantity is not defined, say so.

11. $\ln 1$
12. $\ln(-6.5)$
13. $\log_{43.2} 43.2$
14. $\log_{10}(3.5 + 6.5)$
15. $\log_{10} 5 + \log_{10} 20$
16. $\log_{10} 0.5 + \log_{10} 0.2$
17. $\log_{10} 500 - \log_{10} 50$
18. $\log_{43.2}(5 \cdot 43.2^2) - \log_{43.2} 5$
19. $\log_{43.2}(43.2^7)$
20. $\log_{43.2}(43.2^7)^4$

21–22 ▪ Apply the laws of logarithms with base equal to 7 to compute the following.

21. Using the fact that $\log_7 43.2 \approx 1.935$, find $\log_7\left(\dfrac{1}{43.2}\right)$.
22. Using the fact that $\log_7 43.2 \approx 1.935$, find $\log_7[(43.2)^3]$.

23–26 ▪ Solve the following equations for x. Plug in your answer to check.

23. $7e^{3x} = 21$
24. $4e^{2x+1} = 20$
25. $4e^{-2x+1} = 7e^{3x}$
26. $4e^{2x+3} = 7e^{3x-2}$

27–30 ▪ Sketch a graph of each exponential function. For each, find the value of x where the function is equal to 7. For the increasing functions, find the doubling time, and for the decreasing functions, find the half-life. For which value of x is the value of the function 3.5? For which value of x is the value of the function 14?

27. $y = e^{2x}$

28. $y = e^{-3x}$

29. $y = 5e^{0.2x}$

30. $y = 0.1e^{-0.2x}$

31. Express $y = 1.4^x$ in the form $y = e^{ax}$ (i.e., find a). Explain how to transform the graph of $y = e^x$ to obtain the graph of $y = 1.4^x$.

32. Express $y = 0.27^x$ in the form $y = e^{ax}$ (i.e., find a). Explain how to transform the graph of $y = e^x$ to obtain the graph of $y = 0.27^x$.

33–37 ▪ Calculate the composition $f(g(x))$. Simplify if possible.

33. $f(x) = e^{2x}, g(x) = \ln 2x$

34. $f(x) = e^{2x}, g(x) = 3x - 4$

35. $f(x) = e^{2x}, g(x) = 3x - \ln 4$

36. $f(x) = \ln x, g(x) = 3x^3 e^{-4x}$

37. $f(x) = \log_{10} x, g(x) = 4x^{-2} 10^{-x}$

38–43 ▪ Solve each equation for x.

38. $\ln(2x - 1) = -1$

39. $\dfrac{1}{\ln(x - 1)} = 4$

40. $\ln x + \ln(x - 1) = \ln 2$

41. $\log_{10}(x - 99) = 3$

42. $\ln(\ln x) = 0$

43. $\log_{10}(\log_{10} x) = 1$

44–47 ▪ In each case find the inverse function and say what it represents.

44. The heartbeat frequency, $h = 241B^{-0.25}$, from Example 2.1.13.

45. The blood circulation time, $T(B) = 17.73 \sqrt[4]{B}$, from Example 2.1.12.

46. The density of soil forming a forest floor is given by $d(x) = 0.7 \left(0.8 + e^{-0.2x}\right)^{-1}$, where x is the depth in metres (so $x = 0$ labels the surface, and $x = 3$ is 3 m below the surface).

47. The function $S(D) = e^{-\alpha D - \beta D^2}$ (where $\alpha, \beta > 0$) represents the percent of cancer cells surviving a treatment with a radiation dose D.

48–51 ▪ In the same coordinate system, sketch each pair of functions.

48. $y = \ln x$ and $y = \ln 2x$

49. $y = \ln x$ and $y = \ln x^2$

50. $y = \ln x$ and $y = \log_{10} x$

51. $y = \log_{10} x$ and $y = -\log_{10}(x - 2)$

52. Show that $x = 3/8, x = 5/8, x = 7/8, x = -1/2$, and $x = -3/2$ are not in the domain of the function $y = (-4)^x$. Find a few more numbers, x, that are not in the domain of $y = (-4)^x$. Can you find a hundred more? A thousand more? How many are there?

Applications

53–56 ▪ Suppose that the size of an organism at time t is given by

$$S(t) = S(0)e^{\alpha t}$$

where $S(0)$ is the initial size. Find the time it takes for the organism to double and to quadruple in size in the following circumstances.

53. $S(0) = 1$ cm and $\alpha = 1$/day

54. $S(0) = 2$ cm and $\alpha = 1$/day

55. $S(0) = 2$ cm and $\alpha = 0.1$/hour

56. $S(0) = 2$ cm and $\alpha = 0$/hour

57–58 ▪ Suppose that the size of an organism at time t is given by

$$S(t) = S(0)10^{\alpha t}$$

where $S(0)$ is the initial size and t is measured in days. Find the time it takes for the organism to double in size by converting to base e. How long will it take to increase by a factor of 10?

57. $S(0) = 2.34$ and $\alpha = 0.5$

58. $S(0) = 2.34$ and $\alpha = 0.693$

59–62 ▪ The amount of carbon-14 (^{14}C) left t years after the death of an organism is given by

$$Q(t) = Q(0)e^{-0.000122t}$$

where $Q(0)$ is the amount left at the time of death. Suppose $Q(0) = 6 \times 10^{10}$ ^{14}C atoms.

59. How much is left after 50,000 years? What fraction is this of the original amount?

60. How much is left after 100,000 years? What fraction is this of the original amount?

61. Find the half-life of ^{14}C.

62. About how many half-lives will occur in 50,000 years? Roughly what fraction will be left? How does this compare with the answer to 59?

63–65 ▪ Suppose a population has a doubling time of 24 years and an initial size of 500.

63. What is the population in 48 years?

64. What is the population in 12 years?

65. Find the equation for population size, $P(t)$, as a function of time.

66–68 ▪ Suppose a population is dying with a half-life of 43 years. The initial size is 1600.

66. How long will it take for the population to reach 200?

67. Find the population in 86 years.

68. Find the equation for population size, P(t), as a function of time.

69. The clay material found near the fossilized bone of a dinosaur has been determined to contain about 96.45% of its original amount of the isotope ^{40}K. Knowing that the half-life of ^{40}K is 1.248 billion years, find the time when the clay was formed, and thus, roughly, the age of the fossilized dinosaur bone.

70–73 ▪ Plot semilog graphs of the values from the indicated earlier problems.

70. The growing organism in Exercise 53 for $0 \leq t \leq 10$. Mark where the organism has doubled in size and where it has quadrupled in size.

71. The carbon-14 in Exercise 59 for $0 \leq t \leq 20,000$. Mark where the amount of carbon has halved in size.

72. The population in Exercise 63 for $0 \leq t \leq 100$. Mark where the population has doubled.

73. The population in Exercise 66 for $0 \leq t \leq 100$. Mark where the population has halved in size.

74–77 ▪ Sketch, or describe in words, the log-log plot of each function.

74. $f(t) = 3\sqrt{t}$ where $t > 0$ (use ln).

75. $f(t) = 11\, t^{-1/3}$ where $t > 0$ (use ln).

76. The heartbeat frequency, $h = 241 B^{-0.25}$, from Example 2.1.13 (use $\log_{10}$).

77. The relationship $S = 4.84 V^{2/3}$ from Example 2.1.14 (use $\log_{10}$).

Computer Exercises

78. Use a computer to find the following. Plot the graphs to check.

   a. The doubling time of $S_1(t) = 3.4 e^{0.2t}$.

   b. The doubling time of $S_2(t) = 0.2 e^{3.4t}$.

   c. The half-life of $H_1(t) = 3.4 e^{-0.2t}$.

   d. The half-life of $H_2(t) = 0.2 e^{-3.4t}$.

79. Have a computer solve for the times when the following hold. Plot the graphs to check your answers.

   a. $S_1(t) = S_2(t)$ with S_1 and S_2 from Exercise 78.

   b. $H_1(t) = 2 H_2(t)$ with H_1 and H_2 from Exercise 78.

   c. $H_1(t) = 0.5 H_2(t)$ with H_1 and H_2 from Exercise 78.

80. Have a computer plot the function

$$h(x) = e^{-x^2} - e^{-1000(x - 0.13)^2} - 0.2$$

for values of x between -10 and 10.

   a. Describe the result in words.

   b. Blow up the graph by changing the range to find all points where the value of the function is 0. For example, one such value is between 1 and 2. Plot the function again for x between 1 and 2 to zoom in.

   c. If you found only two points where $h(x) = 0$, blow up the region between 0 and 1 to try to find two more.

81. The formula

$$c(t) = \frac{N}{\sqrt{4\pi k t}} e^{-x^2/4kt}$$

models the concentration of a pollutant in the air at a location x units away from the source at time $t > 0$. The source is located at $x = 0$. Assume that $N = 10$ and $k = 1$.

   a. Take $x = 0$ and plot $c(t)$ on the interval $[0.5, 30]$. What does the model say about the change in concentration over time?

   b. Consider the location $x = 2$ and plot $c(t)$ on the interval $[0.5, 30]$. What does the model say about the change of concentration over time? How does it compare to (a)?

   c. Consider the location $x = 4$ and plot $c(t)$ on the interval $[0.5, 30]$. What does the model say about the change of concentration over time? How does it compare to (b)?

   d. Determine what happens to the graphs in (a)–(c) as t keeps increasing (plot the three graphs over the intervals $[0.5, 50]$, $[0.5, 100]$, etc.). Does it make sense?

2.3 Trigonometric and Inverse Trigonometric Functions

We have used linear and exponential functions to describe several types of relations between measurements. Important as they are, these functions cannot describe **oscillations,** processes that repeat in cycles. Heartbeats and breathing are examples of biological oscillations. In addition, the daily and seasonal cycles imposed by the movements of Earth drive sleep-wake cycles, seasonal population cycles, and the tides. In this section we investigate **trigonometric functions,** discovering a variety of their important properties. We use sine and cosine functions to describe simple oscillations, such as daily and monthly temperature cycles. Finally, we construct inverse trigonometric functions for the sine and tangent functions.

Angles and Trigonometric Ratios

Recall that a **positive angle** is measured counterclockwise from the direction of the positive *x*-axis. If it is measured clockwise, it is **negative.** The units commonly used to measure angles are **degrees** (°) and **radians** (rad). Unless stated otherwise, in calculus and elsewhere in mathematics, we use radians. For example, sin 1 denotes the value of the trigonometric function sine for 1 radian (using a calculator, we get sin $1 \approx 0.84147$). If we mean degrees, we will say so: $\sin 1° \approx 0.01745$.

By definition,
$$360° = 2\pi \text{ rad}$$
In words, one full revolution equals 360 degrees or 2π radians. Dividing by 360, we get
$$1° = \frac{2\pi}{360} \text{ rad} = \frac{\pi}{180} \text{ rad}$$
Thus, to convert from degrees to radians, we multiply by $\pi/180$. Conversely,
$$1 \text{ rad} = \frac{180°}{\pi}$$
says that in order to convert radians into degrees, we multiply by $180°/\pi$.

Example 2.3.1 Converting Degrees to Radians

To find 60° in radians, we convert
$$60° = 60° \times \frac{\pi \text{ rad}}{180°} = \frac{\pi}{3} \text{ rad}$$
Likewise, $90° = \pi/2$ rad, $135° = 3\pi/4$ rad, etc.

Example 2.3.2 Converting Radians to Degrees

Similarly, to find 2 radians in degrees, we convert
$$2 \text{ rad} = 2 \text{ rad} \left(\frac{180°}{\pi \text{ rad}}\right) \approx 114.59°$$
Similarly, we compute that $\pi/6$ rad $= 30°$, $5\pi/4$ rad $= 225°$, etc.

It is fairly common practice to omit the word "radians" (but not "degrees"); i.e., if no unit is specified, we mean radians. However, for many calculators, the default unit is degrees. So we need to be careful in evaluating expressions that involve angles.

For an acute angle, the trigonometric ratios are defined as ratios of the lengths of the sides in a right triangle (see Figure 2.3.33):

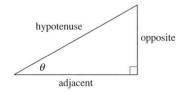

FIGURE 2.3.33
Trigonometric ratios for acute angle θ

$$\sin \theta = \frac{\text{opposite}}{\text{hypotenuse}} \qquad \csc \theta = \frac{1}{\sin \theta} = \frac{\text{hypotenuse}}{\text{opposite}}$$

$$\cos \theta = \frac{\text{adjacent}}{\text{hypotenuse}} \qquad \sec \theta = \frac{1}{\cos \theta} = \frac{\text{hypotenuse}}{\text{adjacent}}$$

$$\tan \theta = \frac{\sin \theta}{\cos \theta} = \frac{\text{opposite}}{\text{adjacent}} \qquad \cot \theta = \frac{1}{\tan \theta} = \frac{\text{adjacent}}{\text{opposite}}$$

Example 2.3.3 Values of Trigonometric Ratios for $\pi/6, \pi/4$, and $\pi/3$

The values of the trigonometric ratios for $\pi/6, \pi/4$, and $\pi/3$ radians can be calculated from the two triangles shown in Figure 2.3.34. (Recall that $\pi/6$ rad $= 30°$, $\pi/4$ rad $= 45°$, and $\pi/3$ rad $= 60°$.)

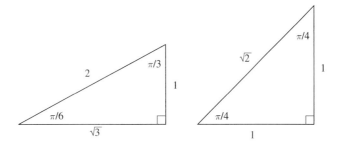

FIGURE 2.3.34
Calculating trigonometric ratios

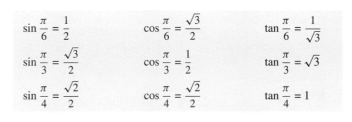

The remaining trigonometric ratios are the reciprocals of the ones given in the table. ▲

To calculate the sine, cosine, and other trigonometric values for obtuse or negative angles, the above definition does not apply, and we proceed in the following way.

Let L be the terminal arm of an angle θ; see Figure 2.3.35. Choose a point P anywhere on L (as long as it is not the origin), and denote its coordinates by (x, y). Let r be the distance between P and the origin; thus,

$$r = \sqrt{x^2 + y^2} \geq 0$$

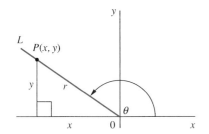

FIGURE 2.3.35
Trigonometric ratios for an obtuse angle

We define

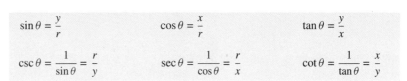

Keep in mind that x and y are coordinates of a point and thus can be positive or negative.

The ratios $\sin \theta$ and $\cos \theta$ are always defined, since r is never zero. We compute

$$\sin^2 \theta + \cos^2 \theta = \frac{y^2}{r^2} + \frac{x^2}{r^2} = \frac{x^2 + y^2}{r^2} = 1$$

since $x^2 + y^2 = r^2$. Thus, we obtain the **basic trigonometric identity**

$$\sin^2 \theta + \cos^2 \theta = 1$$

The ratios $\tan \theta$ and $\sec \theta$ are not defined when $x = 0$. In that case, P lies on the positive or negative y-axis, i.e., the angle is $\theta = \pi/2, 3\pi/2, 5\pi/2, -\pi/2$, etc. So $\tan \theta$ and $\sec \theta$ are not defined when $\theta = \pi/2 + \pi k$, where k is an integer.

The ratios $\cot \theta$ and $\csc \theta$ are not defined when $y = 0$, i.e., when P is on the x-axis. Thus, they are not defined when $\theta = 0, \pi, 2\pi, -\pi, \ldots = \pi k$.

Note that for an acute angle (i.e., when P is in the first quadrant), the two definitions (the one for acute angles and the one for general angles in the box above) agree.

Example 2.3.4 Trigonometric Ratios for $\theta = 3\pi/4$

The line that makes an angle of $\theta = 3\pi/4$ with respect to the x-axis has slope -1. We choose the point $(-1, 1)$ as P; see Figure 2.3.36. Thus, $r = \sqrt{2}$ and it follows that

$$\sin\frac{3\pi}{4} = \frac{y}{r} = \frac{1}{\sqrt{2}} \qquad \cos\frac{3\pi}{4} = \frac{x}{r} = -\frac{1}{\sqrt{2}} \qquad \tan\frac{3\pi}{4} = \frac{y}{x} = -1$$

It is also possible (and useful!) to use the unit circle to define trigonometric ratios. Let P be the point of intersection of a circle of radius 1 and the line whose angle (positive or negative) with respect to the x-axis is θ; see Figure 2.3.37. By definition, the coordinates of P are $(\cos\theta, \sin\theta)$.

From Figure 2.3.37, using the Pythagorean Theorem, we recover the basic trigonometric identity

$$\sin^2\theta + \cos^2\theta = 1$$

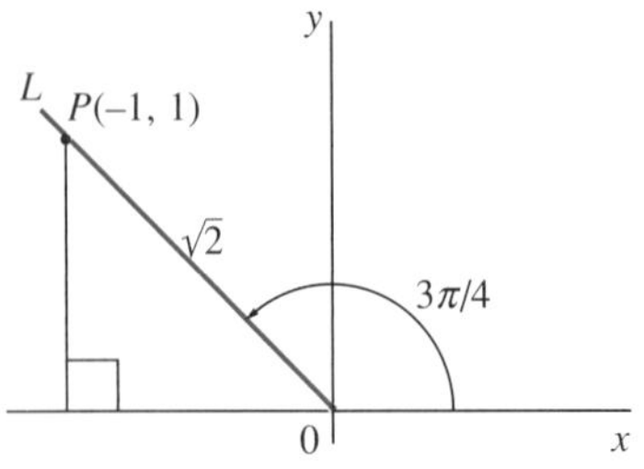

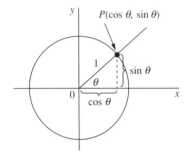

FIGURE 2.3.36
Trigonometric ratios for $\theta = 3\pi/4$

Example 2.3.5 Trigonometric Ratios for $\theta = \pi/2$ and $\theta = \pi$

We use the unit circle definitions. The point of intersection of the angle $\theta = \pi/2$ and the unit circle is $(0, 1)$. Thus,

$$\sin\frac{\pi}{2} = 1 \qquad \cos\frac{\pi}{2} = 0 \qquad \tan\frac{\pi}{2} \text{ is not defined}$$

Likewise, the point of intersection of the angle $\theta = \pi$ and the unit circle is $(-1, 0)$. We conclude that

$$\sin\pi = 0 \qquad \cos\pi = -1 \qquad \tan\pi = 0$$

The values of the functions $\sin\theta$ and $\cos\theta$ for representative values of θ are given in Table 2.3.1.

FIGURE 2.3.37
Unit circle definition of $\sin\theta$ and $\cos\theta$

Table 2.3.1

Radians	Degrees	$\cos\theta$	$\sin\theta$	Radians	Degrees	$\cos\theta$	$\sin\theta$
0	$0°$	1	0	π	$180°$	-1	0
$\frac{\pi}{6}$	$30°$	$\frac{\sqrt{3}}{2}$	$\frac{1}{2}$	$\frac{7\pi}{6}$	$210°$	$-\frac{\sqrt{3}}{2}$	$-\frac{1}{2}$
$\frac{\pi}{4}$	$45°$	$\frac{\sqrt{2}}{2}$	$\frac{\sqrt{2}}{2}$	$\frac{5\pi}{4}$	$225°$	$-\frac{\sqrt{2}}{2}$	$-\frac{\sqrt{2}}{2}$
$\frac{\pi}{3}$	$60°$	$\frac{1}{2}$	$\frac{\sqrt{3}}{2}$	$\frac{4\pi}{3}$	$240°$	$-\frac{1}{2}$	$-\frac{\sqrt{3}}{2}$
$\frac{\pi}{2}$	$90°$	0	1	$\frac{3\pi}{2}$	$270°$	0	-1
$\frac{2\pi}{3}$	$120°$	$-\frac{1}{2}$	$\frac{\sqrt{3}}{2}$	$\frac{5\pi}{3}$	$300°$	$\frac{1}{2}$	$-\frac{\sqrt{3}}{2}$
$\frac{3\pi}{4}$	$135°$	$-\frac{\sqrt{2}}{2}$	$\frac{\sqrt{2}}{2}$	$\frac{7\pi}{4}$	$315°$	$\frac{\sqrt{2}}{2}$	$-\frac{\sqrt{2}}{2}$
$\frac{5\pi}{6}$	$150°$	$-\frac{\sqrt{3}}{2}$	$\frac{1}{2}$	$\frac{11\pi}{6}$	$330°$	$\frac{\sqrt{3}}{2}$	$-\frac{1}{2}$
π	$180°$	-1	0	2π	$360°$	1	0

Trigonometric Functions

Let x denote an angle. Using the general method of defining trigonometric ratios, we can compute the values of the functions $y = \sin x$ and $y = \cos x$ for all real numbers x (keep in mind that x represents an angle in radians).

2.3 Trigonometric and Inverse Trigonometric Functions

Since the angles x and $x + 2\pi$ are the same (think of an angle and what it looks like one full revolution later), it follows that

$$\sin(x + 2\pi) = \sin x \quad \text{and} \quad \cos(x + 2\pi) = \cos x$$

These formulas state that the values of sine and cosine repeat every 2π radians. In other words, $\sin x$ and $\cos x$ are **periodic** with period 2π. (By definition, the period is the length of the *smallest* interval over which we can define a function (or draw a graph), so that, by repetition, that part generates the function in its entire domain. In the case of $\sin x$ and $\cos x$, that interval is of length 2π.)

Figure 2.3.38 shows the graph of $y = \sin x$, obtained by plotting points.

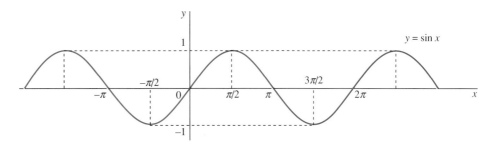

FIGURE 2.3.38
Graph of $y = \sin x$

The part of the graph of $y = \sin x$ over the interval $[0, 2\pi]$ is called the **main period.** Note that $\sin x = 0$ when $x = 0, \pi, 2\pi, -\pi$, i.e., when x is an integer multiple of π.

In Figure 2.3.39 we show the graph of $y = \cos x$.

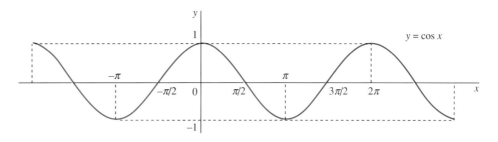

FIGURE 2.3.39
Graph of $y = \cos x$

Note that $\cos x = 0$ when $x = \pi/2, 3\pi/2, 5\pi/2, -\pi/2$, etc. In words, $\cos x = 0$ at $\pi/2$ and all points that are a multiple of π away from it. The part of the graph of $y = \cos x$ over the interval $[0, 2\pi]$ is its main period.

Remember that

$$\sin x = 0 \text{ if and only if } x = \pi k$$
$$\cos x = 0 \text{ if and only if } x = \frac{\pi}{2} + \pi k$$

(In the context of periodicity of trigonometric functions, the symbol k denotes an integer.)

The domain of both $y = \sin x$ and $y = \cos x$ is the set of all real numbers. Their range is $[-1, 1]$. We frequently write the range in the form

$$-1 \leq \sin x \leq 1 \quad \text{and} \quad -1 \leq \cos x \leq 1$$

The graph of $y = \cos x$ is a mirror image of itself when reflected across the y-axis; thus

$$\cos(-x) = \cos x$$

We say that $y = \cos x$ is an **even function.** The graph of $y = \sin x$ satisfies

$$\sin(-x) = -\sin x$$

(see Figure 2.3.40); in words, it is symmetric with respect to the origin. We say that $y = \sin x$ is an **odd function.**

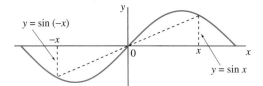

FIGURE 2.3.40

$\sin x$ is an odd function

Note that the graphs of $y = \sin x$ and $y = \cos x$ have the same shape, but are shifted from each other by $\pi/2$: shifting $y = \cos x$ to the right $\pi/2$ units gives $y = \sin x$; shifting $y = \sin x$ to the left $\pi/2$ units gives $y = \cos x$; thus

$$\sin x = \cos(x - \pi/2) \quad \text{and} \quad \cos x = \sin(x + \pi/2)$$

The function $y = \tan x = \sin x / \cos x$ is not defined when $\cos x = 0$, i.e., when $x = \pi/2 + \pi k$. Its range consists of all real numbers.

The graph of $y = \tan x$ is drawn in Figure 2.3.41. From the definition, we see that $\tan x = 0$ whenever $\sin x = 0$; thus, the zeros of $y = \tan x$ are πk, where k is an integer.

The function $y = \tan x$ is periodic with period π. The part of the graph over the interval $(-\pi/2, \pi/2)$ is defined to be its main period.

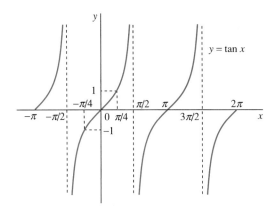

FIGURE 2.3.41

Graph of $y = \tan x$

Example 2.3.6 Relationship between $\tan x$ and $\sec x$

Recall that $\sec x = 1/\cos x$. We prove that

$$\sec^2 x = 1 + \tan^2 x$$

for $x \neq \dfrac{\pi}{2} + \pi k$, where k is an integer.

Start with

$$\sin^2 x + \cos^2 x = 1$$

and divide both sides by $\cos^2 x$:

$$\frac{\sin^2 x}{\cos^2 x} + \frac{\cos^2 x}{\cos^2 x} = \frac{1}{\cos^2 x}$$

$$\tan^2 x + 1 = \sec^2 x$$

In the above calculation we assumed that $\cos x \neq 0$ (so that we did not divide by zero). Thus, the identity is true when $\cos x \neq 0$, i.e., when $x \neq \pi/2 + \pi k$.

Note that the domain for both $y = \tan x$ and $y = \sec x$ consists of all real numbers x such that $x \neq \pi/2 + \pi k$.

Example 2.3.7 Trigonometric Identities

We mention a few of many identities involving trigonometric functions (see also Exercises 27–32 and 33–39). In Exercise 40 we prove that for any two numbers x and y

$$\cos(x - y) = \cos x \cos y + \sin x \sin y$$

Replacing y by $-y$,

$$\cos(x - (-y)) = \cos x \cos(-y) + \sin x \sin(-y)$$

and then using the symmetry formulas $\sin(-y) = -\sin y$ and $\cos(-y) = \cos y$, we obtain the **angle addition formula**

$$\cos(x + y) = \cos x \cos y - \sin x \sin y$$

The **double-angle formula** now follows by substituting $y = x$:

$$\cos 2x = \cos^2 x - \sin^2 x$$

In Exercise 40 we derive the **angle addition formula**

$$\sin(x + y) = \sin x \cos y + \cos x \sin y$$

By replacing y by x, we obtain the **double-angle formula**

$$\sin 2x = 2 \sin x \cos x$$

Next, we solve equations involving trigonometric functions (which we will use later, for instance in finding critical points). Our strategy is the following: first—if needed—we reduce the equation to a single trigonometric function (sin, cos, or tan); then we find all solutions in the main period of that function; finally, knowing the period, we list all solutions.

To solve the equation in the main period, we reason by using the symmetry of the graph and the values of the function in the first quadrant.

Example 2.3.8 Solving a Trigonometric Equation I

Find all x such that $\sin x = 1/2$.

The graph in Figure 2.3.42 shows that this equation has two solutions in the main period $[0, 2\pi]$ and that they lie in the first and second quadrants. From memory, or by looking at Table 2.3.1, we get that $\sin(\pi/6) = 1/2$. Due to the symmetry of the sine graph, the second solution is $\pi/6$ units to the left of π, i.e., $\pi - \pi/6 = 5\pi/6$.

It follows that all solutions of the equation are $x = \pi/6 + 2\pi k$ and $x = 5\pi/6 + 2\pi k$, where k is an integer (recall that the period of $\sin x$ is 2π).

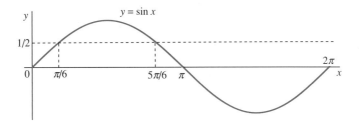

FIGURE 2.3.42
Solving the equation $\sin x = 1/2$

Example 2.3.9 Solving a Trigonometric Equation II

Solve the equation $\sin 2x + \sin x = 0$.

Using the double-angle formula for $\sin x$, we rewrite the equation as

$$2 \sin x \cos x + \sin x = 0$$
$$\sin x (2 \cos x + 1) = 0$$

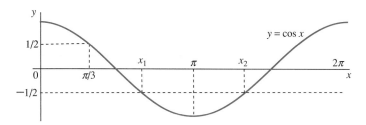

FIGURE 2.3.43
Solving the equation $\cos x = -1/2$

Thus, $\sin x = 0$ or $2 \cos x + 1 = 0$. The first equation is straightforward: using the graph in Figure 2.3.38 we find that $x = \pi k$, where k is an integer.

Next, write $2 \cos x + 1 = 0$ as $\cos x = -1/2$. Looking at the graph in Figure 2.3.43, we see that there are two solutions of this equation in $[0, 2\pi]$, one in the second quadrant and one in the third quadrant. To relate the solutions to the first quadrant, we recall that $\cos(\pi/3) = 1/2$.

By the symmetry of the graph, the distance from π to x_1 is the same as the distance from 0 to $\pi/3$. Thus, $x_1 = \pi - \pi/3 = 2\pi/3$. The distance from π to x_2 is $\pi/3$ as well, and thus $x_2 = \pi + \pi/3 = 4\pi/3$. We conclude that all solutions of the given equation are $x = \pi k$, $x = 2\pi/3 + 2\pi k$, and $x = 4\pi/3 + 2\pi k$, where k is an integer. ▲

Since $\tan x$ is an increasing function, there is only one solution of the equation $\tan x = -1$ in its main period $(-\pi/2, \pi/2)$. Recalling that $\tan(\pi/4) = 1$ and the symmetry of the graph (Figure 2.3.41), we conclude that $\tan(-\pi/4) = -1$. Thus, all solutions of $\tan x = -1$ are given by $x = -\pi/4 + \pi k$, where k is an integer.

To solve $\sin x = -0.75$ we argue in the same way. Using a calculator, we find that $\arcsin(0.75) \approx 0.848$ (radians). Thus (Figure 2.3.44), $x_1 \approx \pi + 0.848 \approx 3.990$ and $x_2 \approx 2\pi - 0.848 \approx 5.435$ and the solutions are $x \approx 3.990 + 2\pi k$ and $x \approx 5.435 + 2\pi k$, where k is an integer. (arcsin is the inverse function of sin. More about it later in this section.)

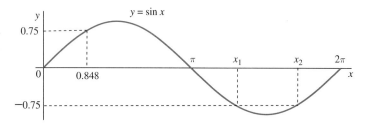

FIGURE 2.3.44
Solving the equation $\sin x = -0.75$

Describing Oscillations

A measurement is said to **oscillate** as a function of time if the values vary regularly between high and low values. Oscillations that are shaped like the graphs of the sine and cosine functions are called **sinusoidal.** Four parameters are needed to describe an oscillation with the sine or cosine function: the **average,** the **amplitude,** the **period,** and the **phase** (Figure 2.3.45).

- The **average** lies halfway between the minimum and maximum values.
- The **amplitude** is the difference between the maximum and the average (or the average and the minimum).
- The **period** is the time between successive peaks.
- The **phase** is the amount by which the graph of $y = \sin t$ or $y = \cos t$ needs to be moved left (phase is negative) or right (phase is positive).

2.3 Trigonometric and Inverse Trigonometric Functions

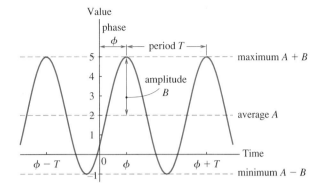

FIGURE 2.3.45

The four parameters that describe a sinusoidal oscillation based on the cosine function

In the table below we illustrate the parameters of several oscillations:

Oscillation	Minimum	Maximum	Average	Amplitude	Period	Phase
$y = 2\sin 4t$	-2	2	0	2	$\pi/2$	0
$y = 3\cos t + 5$	2	8	5	3	2π	0
$y = \cos(t - \pi/4)$	-1	1	0	1	2π	$\pi/4$
$y = \sin(2t + \pi)$	-1	1	0	1	π	$-\pi/2$
$y = \cos(t/3) - 2$	-3	-1	-2	1	6π	0

Let us verify a few entries. The function $y = \sin 4t$ is obtained by compressing $y = \sin t$ *horizontally* by a factor of 4; thus, the period of $y = \sin 4t$ is one quarter the length of the period of $y = \sin t$. A similar argument shows that, in general

the period of $y = \sin at$ and $y = \cos at$ is $2\pi/a$

To obtain $y = 3\cos t$, we stretch the graph of $y = \cos t$ *vertically* by a factor of 3; thus, the range of $y = 3\cos t$ is $[-3, 3]$. Adding 5 to the function shifts the range up by 5 units, so we conclude that the range of $y = 3\cos t + 5$ is $[2, 8]$. The minimum of $y = 3\cos t + 5$ is 2, and the maximum is 8.

Rewriting $y = \sin(2t + \pi) = \sin(2(t + \pi/2))$, we see that $y = \sin(2t + \pi)$ is obtained by first compressing the graph of $y = \sin t$ horizontally by a factor of 2 (thus getting $y = \sin 2t$), and then (since t needs to be replaced by $t + \pi/2$) shifting the resulting graph $\pi/2$ units to the left.

Note that the phase can be determined only if the internal term is in factored form, such as in the case of $y = \sin(2t + \pi) = \sin(2(t + \pi/2))$ that we just finished.

As another example, consider the function

$$f(t) = \cos(3t - 4\pi)$$

Writing $f(t)$ in the form

$$f(t) = \cos\left(3\left(t - \frac{4\pi}{3}\right)\right)$$

we see that the shift is $4\pi/3$ and the period is $2\pi/3$. To graph $f(t)$, we first compress the graph of $y = \cos t$ horizontally by a factor of 3 (thus getting $y = \cos 3t$), and then shift right $4\pi/3$ units (to reflect the fact that t was replaced by $t - 4\pi/3$).

Example 2.3.10   Building an Oscillation by Shifting and Scaling the Cosine Function

Suppose we wish to build an oscillation from the cosine function with an amplitude of 2, an average of 3, a period of 4, and a phase of 1. We construct the formula in steps.

1. To increase the amplitude by a factor of 2, we scale vertically by **multiplying** the cosine by 2 (Figure 2.3.46a). The function is now

$$y = 2\cos t$$

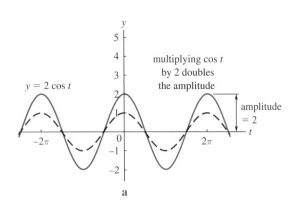
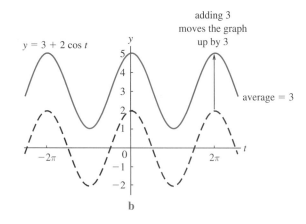
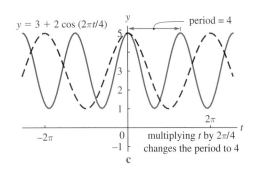
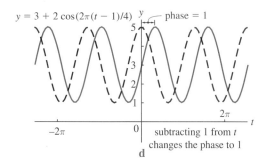

FIGURE 2.3.46
Building a function with different average, amplitude, period, and phase

2. To increase the **average** from 0 to 3, we vertically shift the function by **adding** 3 to the function (Figure 2.3.46b), making

$$y = 3 + 2\cos t$$

3. Next, we wish to decrease the period from 2π to 4. We do this by scaling horizontally by a factor of $\frac{2\pi}{4}$, or by **multiplying** the t inside the cosine by $\frac{2\pi}{4}$ (Figure 2.3.46c). Our function is now

$$y = 3 + 2\cos\left(\frac{2\pi}{4}t\right)$$

4. Finally, we shift the curve horizontally to the right so that the first peak is at 1 instead of 0. We do this by **subtracting** 1 from t (Figure 2.3.46d), arriving at the final answer of

$$y = 3 + 2\cos\left(\frac{2\pi}{4}(t - 1)\right)$$

In general, a sinusoidal oscillation with amplitude B, average A, period T, and phase ϕ (phi) can be described as a function of time, t, with the formula

$$f(t) = A + B\cos\left(\frac{2\pi}{T}(t - \phi)\right) \quad (2.3.1)$$

This function has a maximum at $t = \phi$ and a minimum at $t = \phi + \frac{T}{2}$ and takes on its average value at $t = \phi + \frac{T}{4}$ and $t = \phi + \frac{3T}{4}$. Thereafter, it repeats every T units (Figure 2.3.47).

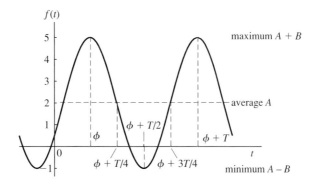

FIGURE 2.3.47

The guideposts for plotting
$f(t) = A + B \cos\left(\frac{2\pi}{T}(t - \phi)\right)$

Example 2.3.11 Plotting a Sinusoidal Function from Its Equation

Suppose we wish to plot

$$f(t) = 2 + 0.4 \cos\left(\frac{2\pi}{10}(t - 7)\right)$$

The amplitude is 0.4, the average is 2, the period is 10, and the phase is 7 (Figure 2.3.48). The maximum is the sum of the average and the amplitude, or $2 + 0.4 = 2.4$, and the minimum is the average minus the amplitude, or $2 - 0.4 = 1.6$. The first positive maximum occurs at the phase, or at $t = 7$. The average occurs 1/4 and 3/4 of the way through each cycle, or at $t = 7 + \frac{1}{4}(10) = 9.5$ and $t = 7 + \frac{3}{4}(10) = 14.5$. The minimum occurs halfway through the first period at $t = 7 + \frac{1}{2}(10) = 12$. The cycle repeats at $t = 17, 27$, and so forth.

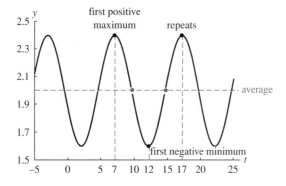

FIGURE 2.3.48

Graphing a sinusoidal oscillation from its equation

Example 2.3.12 The Daily and Monthly Temperature Cycles

Women have two cycles affecting body temperature: a daily and a monthly rhythm. The key facts about these two cycles are given in the following table for a particular woman:

	Minimum	Maximum	Average	Time of Maximum	Period
Daily cycle	36.5	37.1	36.8	2:00 P.M.	24 h
Monthly cycle	36.6	37.0	36.8	Day 16	28 days

Assuming that these cycles are sinusoidal, we can use this information to describe these cycles with the cosine function.

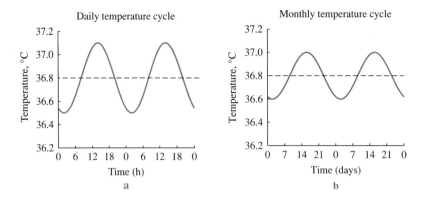

FIGURE 2.3.49

The daily and monthly temperature cycles

The amplitude of a cycle is

$$\text{amplitude} = \text{maximum} - \text{average}$$

For the daily cycle, the amplitude is

$$\text{daily cycle amplitude} = 37.1 - 36.8 = 0.3$$

For the monthly cycle, the amplitude is

$$\text{monthly cycle amplitude} = 37.0 - 36.8 = 0.2$$

The phase depends on the time chosen as the starting time. We define the daily cycle to begin at midnight and the monthly cycle to begin at menstruation. The maximum of the daily cycle occurs 14 h after the start, and that of the monthly cycle 16 days after the start. The oscillations can be described by the fundamental formula (Equation 2.3.1). For the daily cycle, with t measured in hours, the formula $P_d(t)$ is

$$P_d(t) = 36.8 + 0.3 \cos\left(\frac{2\pi(t-14)}{24}\right)$$

(Figure 2.3.49a). For the monthly cycle, with t measured in days, the formula $P_m(t)$ is

$$P_m(t) = 36.8 + 0.2 \cos\left(\frac{2\pi(t-16)}{28}\right)$$

(Figure 2.3.49b).

Inverse Trigonometric Functions

Although there are six trigonometric functions, we will construct inverses of only two, namely $y = \sin x$ and $y = \tan x$, since they appear most often in theory and applications. The knowledge and experience we will gain in doing so will help us construct, if needed, the inverse functions of the remaining four trigonometric functions.

Recall that the domain of $y = \sin x$ consists of all real numbers, and its range is $[-1, 1]$. Looking at its graph in Figure 2.3.50, we see that $y = \sin x$ does not pass the

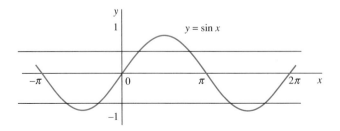

FIGURE 2.3.50

The graph of $y = \sin x$ does not pass the horizontal line test

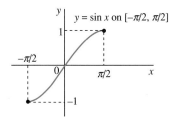

FIGURE 2.3.51

$y = \sin x$ on $[-\pi/2, \pi/2]$ passes the horizontal line test

horizontal line test: as a matter of fact, any horizontal line that crosses the graph crosses it infinitely many times. Thus, the function $y = \sin x$, *defined on the set of all real numbers,* does not have an inverse function.

For the purpose of defining the inverse function, we restrict the domain of $y = \sin x$ to $[-\pi/2, \pi/2]$; see Figure 2.3.51. Note that this graph passes the horizontal line test. However, extending the graph to the left of $-\pi/2$ or to the right of $\pi/2$ violates the test.

Thus, the interval $[-\pi/2, \pi/2]$ is the largest interval on which the graph of $y = \sin x$ satisfies the horizontal line test. The length of the interval $[-\pi/2, \pi/2]$ is π. Of course, we can find other intervals on which $y = \sin x$ satisfies the horizontal line test, but none is longer than π. That is what we mean by *the largest interval* in the sentence above.

So the function $y = \sin x$ for x in $[-\pi/2, \pi/2]$ has an inverse function. We define $y = \arcsin x$ to be the inverse function of $y = \sin x$. Quite often, the generic inverse function symbol $y = \sin^{-1} x$ is used. Since this might lead to confusion, we will use $y = \arcsin x$.

Warning: The meaning of $y = \sin^{-1} x$ depends on the context. If the context is inverse functions, then $y = \sin^{-1} x$ is the same as $y = \arcsin x$. But if it is not, $y = \sin^{-1} x$ could represent the reciprocal of $y = \sin x$, which is $y = \csc x$. It is very important to know the context, since $y = \arcsin x$ and $y = \csc x$ *are not the same function!*

In Figure 2.3.52 we show the domains and ranges of $y = \sin x$ and $y = \arcsin x$. We see that the domain of $y = \arcsin x$ is $[-1, 1]$ and its range is $[-\pi/2, \pi/2]$.

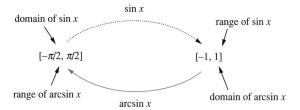

FIGURE 2.3.52

The function $y = \sin x$ and its inverse, $y = \arcsin x$

Example 2.3.13 Calculating arcsin

What is $\arcsin 1/2$? We need to find the angle *in the interval* $[-\pi/2, \pi/2]$ whose sine is $1/2$; from memory, we recall that $\sin \pi/6 = 1/2$; thus, $\arcsin 1/2 = \pi/6$.

Likewise, to find $\arcsin 0$ means to find the angle with sine 0 that lies in the interval $[-\pi/2, \pi/2]$. There are many angles whose sine is zero: $0, \pi, 2\pi, -\pi$, etc.—but there is *only one,* namely 0, that lies in the range $[-\pi/2, \pi/2]$ of $y = \arcsin x$. Thus, $\arcsin 0 = 0$.

To find $\arcsin 2.3$ we have to identify an angle whose sine is 2.3. Since the sine of an angle is always between -1 and 1, we conclude that $\arcsin 2.3$ is not defined. ▲

To review:

$y = \arcsin x$ if and only if $\sin y = x$ and y is in $[-\pi/2, \pi/2]$.

Recall that a function and its inverse cancel each other. Thus,

$$\arcsin(\sin x) = x \text{ if } x \text{ is in } [-\pi/2, \pi/2]$$
$$\sin(\arcsin x) = x \text{ if } x \text{ is in } [-1, 1]$$

To obtain the graph of $y = \arcsin x$ we reflect the graph of $y = \sin x$ with respect to the diagonal $y = x$; see Figure 2.3.53.

We now retrace our steps to construct the inverse function to $y = \tan x$. Since $y = \tan x$ in its domain does not satisfy the horizontal line test (Figure 2.3.54a), we restrict the domain to the interval $(-\pi/2, \pi/2)$; see Figure 2.3.54b. The range still consists of all real numbers.

We define $y = \arctan x$ to be the inverse function to $y = \tan x$ on $(-\pi/2, \pi/2)$. Quite often, the symbol $y = \tan^{-1} x$ is used instead of $y = \arctan x$. The warning that

FIGURE 2.3.53

The graphs of $y = \sin x$ and $y = \arcsin x$

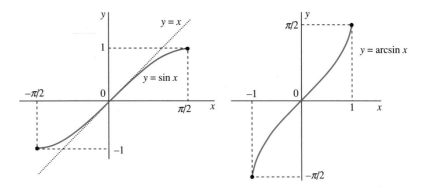

FIGURE 2.3.54

The graph of $y = \tan x$ and its restriction to $(-\pi/2, \pi/2)$

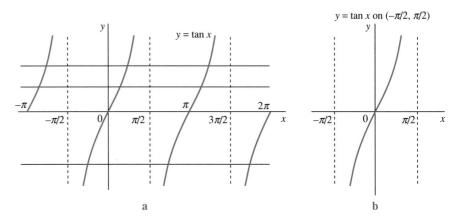

a

b

FIGURE 2.3.55

The function $y = \tan x$ and its inverse, $y = \arctan x$

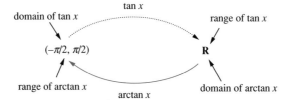

we stated earlier applies here as well: the meaning of $y = \tan^{-1} x$ depends on the context. In the case of inverse functions, $y = \tan^{-1} x$ is the same as $y = \arctan x$. Otherwise, $y = \tan^{-1} x$ could denote the reciprocal of $y = \tan x$, which is $y = \cot x$. Keep in mind that $y = \arctan x$ and $y = \cot x$ are not the same function!

The diagram in Figure 2.3.55 shows the domains and ranges of $y = \tan x$ and $y = \arctan x$. In words, the function $y = \arctan x$ takes a real number x and returns the angle y in $(-\pi/2, \pi/2)$ whose tangent is x.

Example 2.3.14 Calculating arctan

What is arctan 1? To calculate arctan 1, we need to find the angle *in the interval* $(-\pi/2, \pi/2)$ whose tangent is 1. From memory, we know that $\tan \pi/4 = 1$; thus, $\arctan 1 = \pi/4$.

As another example, we calculate arctan 0. This means that we need to find the angle with tangent 0 that lies in $(-\pi/2, \pi/2)$. There are many angles whose tangent is zero: $0, \pi, 2\pi, -\pi, -2\pi$, etc. But there is *only one,* namely 0, that belongs to the range $(-\pi/2, \pi/2)$ of arctan x; thus, $\arctan 0 = 0$.

Reasoning in the same way, we find that $\arctan(-1/\sqrt{3}) = -\pi/6$. Using a calculator, we find that $\tan 0.57 \approx 0.640969$; therefore, $\arctan 0.640969 \approx 0.57$. (Keep in mind that angles are in radians. So 0.57 represents 0.57 radians.)

2.3 Trigonometric and Inverse Trigonometric Functions

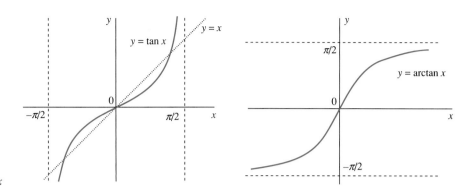

FIGURE 2.3.56

The graphs of $y = \tan x$ and $y = \arctan x$

To review:

$y = \arctan x$ if and only if $\tan y = x$ and y is in $(-\pi/2, \pi/2)$

> $\arctan(\tan x) = x$ if x is in $(-\pi/2, \pi/2)$
> $\tan(\arctan x) = x$ for all real numbers x

To obtain the graph of $y = \arctan x$ we reflect the graph of $y = \tan x$ with respect to the line $y = x$; see Figure 2.3.56.

Example 2.3.15 Analyzing Blood Stains

A DNA analysis of a blood sample is often used to identify a person involved in a crime. However, the blood can reveal a lot more—a careful analysis of blood splatters (usually performed by forensic specialists) provides useful information about the events that led to their formation.

As it moves through the air (its motion caused by bleeding, or resulting from a blow or from a bullet wound), a blood drop assumes the approximate shape of a sphere. (That it is roughly spherical rather than teardrop shaped is due to the strong surface tension of the blood.) If a blood drop drips, i.e., is acted upon by gravity only, it will form a nearly disk-like stain on the floor (Figure 2.3.57a). However, if a blood drop leaves the source with an initial velocity (caused by a bullet, for example), it will fly along a parabola and hit the surface of the floor at an angle, called the *impact angle* (Figure 2.3.57b). Experiments show that the smaller the impact angle, the more elongated the stain. For instance, the angle of impact in Figure 2.3.57c was about 70 degrees, and in Figure 2.3.57d it was about 20 degrees.

Forensic specialists work backwards—by looking at the blood stains on the floor they try to determine the location of the source (called the *area of convergence*). All the information they need is extracted from the fact that the elongated splatters look roughly like ellipses (elongated disks).

In Figure 2.3.58a the ellipse that best fits the splatter from Figure 2.3.57d has been inscribed. The direction of the largest elongation (Figure 2.3.58b) points toward

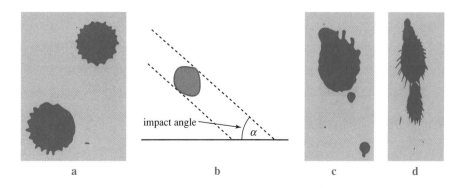

FIGURE 2.3.57

Various shapes of blood stains and impact angle

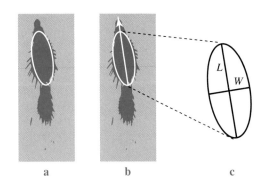

FIGURE 2.3.58

Analyzing a blood splatter

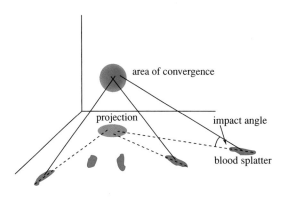

FIGURE 2.3.59

Angles of impact lead to the area of convergence

the projection of the area of convergence onto the floor (see Figure 2.3.59). The impact angle, α, is calculated from

$$\alpha = \arcsin \frac{W}{L}$$

where W is the width and L is the length of the ellipse (Figure 2.3.58c). For instance, if $W = 2.8$ mm and $L = 7.9$ mm, then

$$\alpha = \arcsin \frac{2.8}{7.9} \approx \arcsin 0.35 \approx 0.36 \text{ rad}$$

or $\alpha \approx 20.8$ degrees.

Following the angles of impact from several splatters (all calculated in the way just shown) leads to the area of convergence; see Figure 2.3.59.

Note The derivation of the formula $\alpha = \arcsin(W/L)$ is not straightforward—it requires a knowledge of math, biology, physics, and chemistry. The Royal Canadian Mounted Police offers programs in Bloodstain Pattern Analysis (see http://www.rcmp-grc.gc.ca/fsis-ssji/fis-sij/bpa-ats-eng.htm). As well, the Ontario Police College offers a one-year program in bloodstain analysis for its law-enforcement officers.

Example 2.3.16 Model for World Population

One of many models used to analyze human population growth is given by

$$P(t) = 4.42857 \left(\frac{\pi}{2} - \arctan \frac{2007 - t}{42} \right)$$

where t represents a calendar year and $P(t)$ is the population in billions [Source: S. P. Kapitza, The Phenomenological Theory of World Population Growth, *Physics-Uspekhi* 39(1): 57–71, 1996]. In this case, the formula works both ways, by modelling past population as well as trying to predict future growth. We illustrate the formula with a few values in Table 2.3.2.

Table 2.3.2

Year, t	Predicted Population, P(t), in Billions
1700	0.602
1850	1.158
2000	6.225
2009	7.167
2020	8.286
2050	10.487

S. P. Kapitza, "The Phenomenological Theory of World Population Growth," *Physics-Uspekhi*, Vol. 39 (1), Pg. 57-51, 1996.

Summary We investigated **trigonometric functions**, discussing a variety of important properties that they possess. **Sinusoidal** oscillations can be described with the **sine** or the **cosine** function. Four factors change the shape of the graph: the **average** (the middle value), the **amplitude** (the distance from the middle to the minimum or maximum), the **period** (the time between successive maxima), and the **phase** (the amount by which the graph is shifted). Functions with these parameters can be created by **shifting** and **scaling** the function vertically and horizontally. Near the end of the section we constructed inverse functions for the sine and tangent functions.

2.3 Exercises

Mathematical Techniques

1–6 ■ Convert each angle from radians to degrees.

1. $\dfrac{5\pi}{3}$
2. $-\dfrac{\pi}{4}$
3. 33π
4. $-\dfrac{7\pi}{2}$
5. $\dfrac{5\pi}{8}$
6. 4

7–11 ■ Convert each angle from degrees to radians.

7. $270°$
8. $-60°$
9. $36°$
10. $330°$
11. $49°$

12–17 ■ Find exact values (or say they are not defined) of $\sin\theta$, $\cos\theta$, $\tan\theta$, and $\sec\theta$ for each angle in radians.

12. $\theta = \dfrac{3\pi}{2}$
13. $\theta = \dfrac{5\pi}{4}$
14. $\theta = 7\pi$
15. $\theta = \dfrac{5\pi}{6}$
16. $\theta = \dfrac{7\pi}{4}$
17. $\theta = -\dfrac{3\pi}{4}$

18–21 ■ For the given trigonometric ratio, find the remaining five trigonometric ratios.

18. $\cos\theta = \dfrac{4}{5}, 0 \leq \theta \leq \pi/2$
19. $\tan\theta = 3, 0 \leq \theta \leq \pi/2$
20. $\sin\theta = \dfrac{1}{5}, 0 \leq \theta \leq \pi/2$
21. $\cos\theta = \dfrac{4}{5}, 3\pi/2 \leq \theta \leq 2\pi$

22–26 ■ In the same coordinate system, graph each pair of functions. Choose the interval on the x-axis to show at least one full period of each function.

22. $y = \cos x$ and $y = \cos 4x$
23. $y = \sin x$ and $y = \sin(x/3)$
24. $y = \tan x$ and $y = -2\tan x$
25. $y = \cos x$ and $y = 1 + \cos(x - \pi)$
26. $y = \sin 2x$ and $y = \sin(2x + 3)$

27–32 ■ The following are some of the most important trigonometric identities. Check each of them at the points (a) $\theta = 0$, (b) $\theta = \pi/4$, (c) $\theta = \pi/2$, (d) $\theta = \pi$.

27. $\cos\frac{\theta}{2} = \sqrt{\frac{1+\cos\theta}{2}}$ for $0 \le \theta \le \pi$ (using the positive square root). Check only at points (a), (c), and (d).
28. $\sin^2\theta + \cos^2\theta = 1$
29. $\cos(\theta - \pi) = -\cos\theta$
30. $\cos\left(\theta - \frac{\pi}{2}\right) = \sin\theta$
31. $\cos 2\theta = \cos^2\theta - \sin^2\theta$
32. $\sin 2\theta = 2\sin\theta\cos\theta$

▼ 33–39 ■ Prove each trigonometric identity using formulas mentioned in this section.

33. $\sin(x - y) = \sin x \cos y - \cos x \sin y$
34. $(\sin x + \cos x)^2 = 1 + \sin 2x$
35. $\tan^2 x - \sin^2 x = \tan^2 x \sin^2 x$
36. $\sin^2 x - \sin^2 y = \sin(x+y)\sin(x-y)$
37. $\tan 2x = \frac{2\tan x}{1 - \tan^2 x}$
38. $\frac{1 - \sin x}{\cos x} = \frac{\cos x}{1 + \sin x}$
39. $\sin x \sin 2x + \cos x \cos 2x = \cos x$

40. Label the lengths of the sides in a triangle by a, b, and c, and let γ be the angle opposite the side of length c. Recall the law of cosines: $c^2 = a^2 + b^2 - 2ab\cos\gamma$.
    a. Write the law of cosines formula for the triangle OAB in the figure (A and B lie on a circle of radius 1).
    b. Find the coordinates of the points A and B (use the unit circle definition), and then use the distance formula to find c^2.
    c. Combine the expressions for c^2 from (a) and (b) and simplify to obtain $\cos(x - y) = \cos x \cos y + \sin x \sin y$.
    d. Write $\sin(x + y) = \cos(x + y - \pi/2)$ (why is this true?) and then expand $\cos(x + y - \pi/2) = \cos(x + (y - \pi/2))$ using the angle sum formula. Simplify to obtain $\sin(x + y) = \sin x \cos y + \cos x \sin y$.

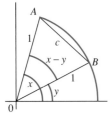

▼ 41–44 ■ The following are alternative ways to write formulas for sinusoidal oscillations. Convert them to the standard form (Equation 2.3.1) and sketch a graph.

41. $r(t) = 5(2 + 1\cos(2\pi t))$
42. $g(t) = 2 + 1\sin t$. Use Exercise 30 to change sine into cosine.
43. $f(t) = 2 - 1\cos t$. Use Exercise 29 to get rid of the negative amplitude.
44. $h(t) = 2 + 1\cos(2\pi t - 3)$

▼ 45–49 ■ Find *all* values x in $[0, 2\pi]$ that satisfy each equation.

45. $\sin x = \frac{1}{2}$
46. $\tan x = -1$
47. $\sec x = 2$
48. $2\sin^2 x = 1$
49. $\sin x = \cos x$

▼ 50–53 ■ Solve each equation.

50. $\tan x = \sin x$
51. $\sin 2x = \sin x$
52. $\tan(2x - 5) - \sqrt{3} = 0$
53. $2\cos^2 x - 5\cos x - 3 = 0$

▼ 54–59 ■ Find the exact value of each expression.

54. $\arctan(-1)$
55. $\arcsin(-1/2)$
56. $\arcsin(\sqrt{3}/2)$
57. $\arctan\sqrt{3}$
58. $\arcsin(1/\sqrt{2})$
59. $\arcsin(-1/\sqrt{2})$

▼ 60–63 ■ In the same coordinate system, graph each pair of functions. Indicate the domains of the functions in your graphs.

60. $y = \arcsin x$ and $y = \arcsin(x - 2)$
61. $y = \arcsin x$ and $y = -\arcsin x + 4$
62. $y = \arctan x$ and $y = \arctan 2x$
63. $y = \arctan x$ and $y = -3\arctan x$

Applications

▼ 64–67 ■ Estimate the average, minimum, maximum, amplitude, period, and phase from the graph of each oscillation, assuming it is a cosine graph.

64.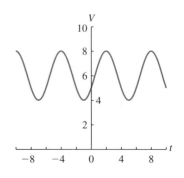

65.
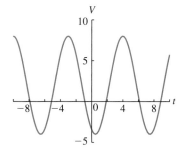

66.
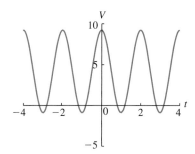

67.
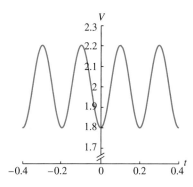

▰ 68–71 ▪ Graph each function. Give the average, maximum, minimum, amplitude, period, and phase of each and mark them on your graph.

68. $f(x) = 3 + 4\cos\left(2\pi\left(\dfrac{x-1}{5}\right)\right)$

69. $g(t) = 4 + 3\cos(2\pi(t-5))$

70. $h(z) = 1 + 5\cos\left(2\pi\left(\dfrac{z-3}{4}\right)\right)$

71. $W(y) = -2 + 3\cos\left(2\pi\left(\dfrac{y+0.1}{0.2}\right)\right)$

▰ 72–75 ▪ Describe each event as an oscillation using formula (2.3.1).

72. A population of red foxes changes periodically with a cycle of 12 years. The maximum of 580 foxes is reached 8 years into the cycle, and the minimum number of foxes is 220.

73. A population of fish oscillates between a high of 3.8 million and a low of 1.8 million, with the cycle lasting six months. The maximum is reached one month into the cycle.

74. The membrane potential in a certain type of neuron jumps from a high of −40 mV (millivolts) to a low of −80 mV 22 times per second. Assume that the highs occur at the start of each cycle.

75. The aortic pressure changes from a maximum of 120 mmHg (systolic pressure) to a minimum of 80 mmHg (diastolic) pressure during each heartbeat. Assume that there are 72 heartbeats per minute and that the maximum occurs at the start of each cycle.

76. Describe the oscillation given by the formula $y = 2e^{\sin 3t}$.

77. Describe the behaviour of the population whose size is modelled by $p(t) = p(0)e^{1.4\cos 2t}$.

▰ 78–83 ▪ Oscillations are often combined with growth or decay. Plot a graph of each function, and describe in words what you see. Make up a biological process that might have produced the result.

78. $f(t) = 1 + t + \cos(2\pi t)$ for $0 < t < 4$, where t is measured in days.

79. $h(t) = t + 0.2\sin(2\pi t)$ for $0 < t < 4$, where t is measured in days.

80. $g(t) = e^t \cos(2\pi t)$ for $0 < t < 3$, where t is measured in years.

81. $W(t) = e^{-t}\cos(2\pi t)$ for $0 < t < 3$, where t is measured in years.

82. $H(t) = \cos(e^t)$ for $0 < t < 3$, where t is measured in years.

83. $b(t) = \cos(e^{-t})$ for $0 < t < 3$, where t is measured in years.

Computer Exercises

▰ 84–87 ▪ Sleepiness has two cycles, a circadian rhythm with a period of approximately 24 h and an ultradian rhythm with a period of approximately 4 h. Both have phase 0 and average 0, but the amplitude of the circadian rhythm is 1 sleepiness unit, and that of the ultradian is 0.4 sleepiness unit.

84. Find the formula and sketch the graph of sleepiness over the course of a day due to the circadian rhythm.

85. Find the formula and sketch the graph of sleepiness over the course of a day due to the ultradian rhythm.

86. Sketch the graph of the two cycles combined.

87. At what time of day are you sleepiest? At what time of day are you least sleepy?

88. Consider the following functions:

$$f_1(x) = \cos\left(x - \dfrac{\pi}{2}\right)$$

$$f_3(x) = \dfrac{\cos\left(3x - \dfrac{\pi}{2}\right)}{3}$$

$$f_5(x) = \dfrac{\cos\left(5x - \dfrac{\pi}{2}\right)}{5}$$

$$f_7(x) = \dfrac{\cos\left(7x - \dfrac{\pi}{2}\right)}{7}$$

a. Plot them all on one coordinate system.

b. Plot the sum $f_1(x) + f_3(x)$.

c. Plot the sum $f_1(x) + f_3(x) + f_5(x)$.

d. Plot the sum $f_1(x) + f_3(x) + f_5(x) + f_7(x)$.

e. What does this sum look like?

f. Try to guess the pattern, and add on $f_9(x)$ and $f_{11}(x)$. This is an example of a **Fourier series,** a sum of cosine functions that add up to a **square wave** that jumps between values of -0.8 and 0.8.

89. Plot the function $f(x) = \cos(2\pi \cdot 440x) + \cos(2\pi \cdot 441x)$. Describe the result. If these were sounds, what might you hear?

Chapter Summary: Key Terms and Concepts

Define or explain the meaning of each term and concept.

Elementary models: linear function, slope, line, point-slope form; proportional and inversely proportional relationships; power function, allometric relation

Exponential functions: laws of exponents, natural exponential function; natural logarithm function, laws of logarithms; exponential growth, doubling time; exponential decay, half-life; semilog graph, log-log graph

Trigonometric functions: angle, radians, trigonometric ratios; trigonometric function, periodic function, main period; oscillation, average, amplitude, period, phase; inverse trigonometric functions, cancellation formulas

Concept Check: True/False Quiz

Determine whether each statement is true or false. Give a reason for your choice.

1. If A is inversely proportional to B, then B is inversely proportional to A.

2. $\sin(x + \pi) = -\sin x$.

3. A quantity that doubles every three months increases six-fold in six months.

4. The semilog graph of $M(t) = M(0)e^{2t}$ is a line of slope 2.

5. For the function $y = 3x - 2$, the ratio of output, y, to input, x, is constant.

6. The period of $g(t) = \cos(\pi t)$ is π.

7. If 75% of a drug is absorbed in three hours, then its half-life is two hours.

8. If A is proportional to the square root of B, then quadrupling B will double A.

9. The amplitude of $f(t) = 3 + 0.4\cos(2t - 1)$ is 0.4.

10. From $\sin(\pi/6) = 1/2$ we conclude that $\arcsin(1/2) = \pi/6$.

Supplementary Problems

1. Suppose you have a culture of bacteria, where the density of each bacterium is 2 g/cm³.

   a. If each bacterium is 5 μm × 5 μm × 20 μm in size, find the number of bacteria if their total mass is 30 g. Recall that 1 μm = 10^{-6} m.

   b. Suppose you learn that the sizes of bacteria range from 4 μm × 5 μm × 15 μm to 5 μm × 6 μm × 25 μm. What is the range of the possible number of bacteria making up the total mass of 30 g?

2. Suppose the number of bacteria in a culture is a linear function of time.

   a. If there are 2×10^8 bacteria in your lab at 5 P.M. on Tuesday, and 5×10^8 bacteria the next morning at 9 A.M., find the equation of the line describing the number of bacteria in your culture as a function of time.

   b. At what time will your culture have 1.1×10^9 bacteria?

   c. The lab across the hall also has a bacterial culture where the number of bacteria is a linear function of time. If they have 2×10^8 bacteria at 5 P.M. on Tuesday, and 3.4×10^8 bacteria the next morning at 9 A.M., when will your culture have twice as many bacteria as theirs?

3. Consider the functions $f(x) = e^{-2x}$ and $g(x) = x^3 + 1$.

   a. Find the inverses of f and g, and use them to find when $f(x) = 2$ and when $g(x) = 2$.

   b. Find $f \circ g$ and $g \circ f$ and evaluate each at $x = 2$.

   c. Find the inverse of $g \circ f$. What is the domain of this function?

4. The number of bacteria (in millions) in a lab are given in the table:

Time, t (h)	Number, b_t
0	1.5
1	3
2	4.5
3	5
4	7.5
5	9

   a. Graph these points.

   b. Find the line connecting them and the time, t, at which the value does not lie on the line.

   c. Find the equation of the line, and use it to find what the value at t would have to be to lie on the line.

   d. How many bacteria would you expect at time $t = 7$ h?

5. Convert each angle from degrees to radians, and find the sine and cosine of each.

   a. $\theta = 60°$

   b. $\theta = -60°$

   c. $\theta = 110°$

   d. $\theta = -190°$

   e. $\theta = 1160°$

6. Suppose the body temperature, H, of a bird follows the equation

   $$H = 38 + 3\cos\left(\frac{2\pi(t - 0.4)}{1.2}\right)$$

   where t is measured in days and H is measured in degrees Celsius.

   a. Sketch a graph of the body temperature of this bird.

   b. Write the equation if the period changes to 1.1 days. Sketch a graph.

   c. Write the equation if the amplitude increases to 3.5°C. Sketch a graph.

   d. Write the equation if the average decreases to 37.5°C. Sketch a graph.

7. Graph each function on the interval [0, 20] and describe its behaviour in words. In particular, state the domain, describe what kind of oscillation it represents (what line does it oscillate about?), identify the region in the xy-plane that contains the graph, and find the points of intersection of the graph and the line $y = x$.

   a. $f(x) = x + \sin x$

   b. $f(x) = x + \sin 2x$

   c. $f(x) = x + \sin 4x$

8. Sketch the graph of the function $f(t) = M(1 - e^{-kt})$, where k and M are positive constants.

9. a. Explain how to obtain the graphs of $|f(x)|$ and $f(|x|)$ from the graph of $f(x)$.

   b. Sketch the graphs of the functions $y = \sin|x|$ and $y = |\sin x|$.

10. A culture of bacteria has mass 3×10^{-3} g and consists of spherical cells of mass 2×10^{-10} g and density 1.5 g/cm^3.

    a. How many bacteria are in the culture?

    b. What is the radius of each bacterium?

    c. If the bacteria were mashed into mush, how much volume would they take up?

11. A person develops a small liver tumour. It grows according to

    $$S(t) = S(0)e^{\alpha t}$$

    where $S(0) = 1$ g and $\alpha = 0.1$/day. At time $t = 30$ days, the tumour is detected and treatment begins. The size of the tumour then decreases linearly with slope -0.4 g/day.

    a. Write the equation for tumour size at $t = 30$.

    b. Sketch a graph of the size of the tumour over time.

    c. When will the tumour disappear completely?

12. a. Given below is the graph of a function $y = f(x)$. Sketch the graph of the function $y = 1/f(x)$.

    b. Using what you have discovered, sketch the graph of $y = \sec x$.

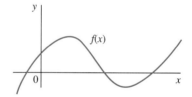

13. The population of fish in a limited ecosystem is given by

    $$P(t) = \frac{20{,}000}{400 + 600e^{-t}}$$

    where t is time in years, $0 \le t \le 10$.

    a. By constructing a table of values, sketch the graph of $P(t)$ without using graphing technology.

    b. What is the initial population, $P(0)$? Describe what happens to $P(t)$ as t increases.

14. Suppose a person's head diameter, D, and height, H, grow according to

    $$D(t) = 10e^{0.03t}$$
    $$H(t) = 50e^{0.09t}$$

    during the first 15 years of life.

    a. Find D and H at $t = 0$, $t = 7.5$, and $t = 15$.

    b. Sketch graphs of these two measurements as functions of time.

c. Sketch semilog graphs of these two measurements as functions of time.

d. Find the doubling time of each measurement.

15. On another planet, people have three hands and like to compute tripling times instead of doubling times.

    a. Suppose a population follows the equation $b(t) = 3 \times 10^3 e^{0.333t}$, where t is measured in hours. Find the tripling time.

    b. Suppose a population has a tripling time of 33 h. Find the equation for population size, $b(t)$, if $b(0) = 3 \times 10^3$.

16. Suppose vehicles are moving at 72 km/h. Each car carries an average of 1.5 people, and all are carefully keeping a 2-s following distance (getting no closer than the distance a car travels in 2 s) on a three-lane highway.

    a. How far is it between vehicles?

    b. How many vehicles per kilometre are there?

    c. How many people will pass a given point in an hour?

    d. If commuter number oscillates between this maximum (at 8:00 A.M.) and a minimum that is one-third as large (at 8:00 P.M.) on a 24-h cycle, give a formula for the number of people passing the given point as a function of time of day.

Chapter 3

Discrete-Time Dynamical Systems

In this chapter, we introduce the concept of a **discrete-time dynamical system,** which is a major mathematical object that is used in applications across a wide range of disciplines. A discrete-time dynamical system consists of an **initial value** (also called a starting value or an initial condition) and a **rule** (sometimes called a dynamic rule) that transforms the system from the present state to a state one step ahead.

For instance, the initial value "the present population is 500" and the dynamic rule "the population doubles every three years" constitute a discrete-time dynamical system. Based on these data, we calculate the **solution** of the system, which tells us how the system evolves with time.

Using calculus, we study the **updating function,** which is an algebraic expression (formula) of the dynamic rule. By studying properties of the updating function, we learn about the corresponding dynamical system. For instance, we investigate **equilibrium points,** which are the points left unchanged by the system. As well, to further investigate a dynamical system, we develop a set of algebraic rules and geometric tools (such as **cobwebbing**).

We study a number of **applications,** including unlimited and limited population growth, exponential decay, the dynamics of the consumption of coffee and alcohol, gas exchange in the lungs, and the dynamics of competing populations. In Chapter 4 we develop theoretical concepts (limits, continuity, and derivatives) that will allow us to further analyze these applications. That is why Chapter 4 will feel a bit dry (we will see fewer applications than in this chapter)—but it will be worth it.

3.1 Introduction to Discrete-Time Dynamical Systems

Suppose we collect data on how much several bacterial cultures grow in one hour or on how much trees grow in one year. How can we predict what will happen in the long run? In this section, we begin addressing these dynamical problems, which form the theme of this chapter and indeed a good part of this book. We follow the basic steps of applied mathematics: **quantifying the basic measurement** and describing the **dynamical rule.** We will learn how to summarize the rule with a **discrete-time dynamical system** or an **updating function** that describes change. From the discrete-time dynamical system and a starting point, called an **initial condition,** we will compute a **solution** that gives the values of the measurement as a function of time.

Discrete-Time Dynamical Systems and Updating Functions

A discrete-time dynamical system describes the relation between a quantity measured at the beginning and the end of an experiment or a given time interval. If the measurement is represented by the variable m, we will use the notation m_t to denote the measurement at the beginning of the experiment and m_{t+1} to denote the measurement at the end of the experiment (Figure 3.1.1). Think of t as the current time and of $t + 1$ as the time one step (one unit of time) into the future. The relation between the

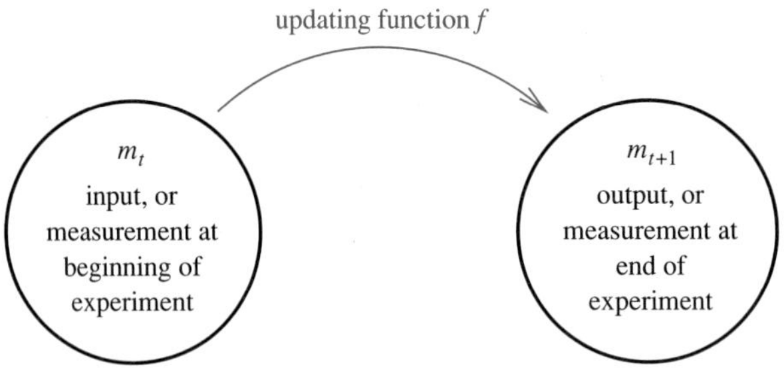

FIGURE 3.1.1

Notation for a discrete-time dynamical system

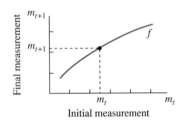

FIGURE 3.1.2

Updating function

initial measurement, m_t, and the final measurement, m_{t+1}, is given by the **discrete-time dynamical system**

$$m_{t+1} = f(m_t) \tag{3.1.1}$$

Note that this is the usual functional notation $y = f(x)$, where m_t represents the independent variable and m_{t+1} is the corresponding value of the function f. The function f is called the **updating function**. To graph f, we use a coordinate system with axes labelled m_t and m_{t+1} (see Figure 3.1.2; keep in mind that m_t is like x and m_{t+1} is like y). Thus, the updating function f accepts the initial value (input), m_t, and returns the final value (output), m_{t+1}. Formula 3.1.1 is often referred to as a **recursion** or a **recursive relation**.

Let us consider a few examples.

Example 3.1.1 A Discrete-Time Dynamical System for a Bacterial Population

Recall the data introduced in Example 1.3.2. Several bacterial cultures with different initial population sizes are grown in controlled conditions for one hour and then carefully measured:

Colony	Initial Population, b_t	Final Population, b_{t+1}
1	0.47	0.94
2	3.3	6.6
3	0.73	1.46
4	2.8	5.6
5	1.5	3.0
6	0.62	1.24

Note that we have replaced b_i (the initial population) with b_t (the population at time t) and b_f (the final population) with b_{t+1} (the population at time $t + 1$). In this case, the number 1 in $t + 1$ represents one hour (so $t + 1$ is one hour later than t).

In each colony, the population doubled in size. We can describe this with the discrete-time dynamical system

$$b_{t+1} = 2b_t$$

The updating function, f, is given by

$$f(b_t) = 2b_t$$

As we have seen, the graph of the updating function plots the initial measurement, b_t, on the horizontal axis and the final measurement, b_{t+1}, on the vertical axis (Figure 3.1.3).

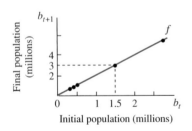

FIGURE 3.1.3

Graph of the updating function for a bacterial population

Example 3.1.2 A Discrete-Time Dynamical System for Tree Growth

Suppose that we measure the height of a tree in one year and then again in the following year. Denoting the initial height by h_t and the final height by h_{t+1}, we find that

$$h_{t+1} = h_t + 0.8$$

metres. This time, the time unit 1 represents one year. The updating function is

$$f(h_t) = h_t + 0.8$$

(see Figure 3.1.4). For example, the above dynamical system predicts that a tree whose present height is $h_t = 12.2$ m will grow to the height of

$$h_{t+1} = f(12.2) = 12.2 + 0.8 = 13 \text{ m}$$

in one year.

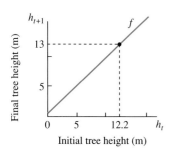

FIGURE 3.1.4
Updating function for tree growth

Example 3.1.3 Discrete-Time Dynamical System for Mites

Assume that a lizard in a pet store is infested by mites. The final number of mites, x_{t+1}, is related to the initial number of mites, x_t, by the formula

$$x_{t+1} = 2x_t + 30$$

where t represents time in weeks.

This formula represents the discrete-time dynamical system for the population of mites. The updating function is

$$f(x_t) = 2x_t + 30$$

The discrete-time dynamical systems for bacterial populations, tree height, and mite number were all derived from data. Often, dynamical rules can instead be derived directly from the principles governing a system.

Example 3.1.4 Dynamics of Absorption of Pain Medication

A patient is on methadone, a medication used to relieve chronic, severe pain (for instance after certain types of surgery). It is known that every day, the patient's body absorbs half of the methadone. In order to maintain an appropriate level of the drug, a new dosage containing 1 unit of methadone is administered at the end of each day.

By M_t we denote the amount of methadone in the patient's body at time t. Using the language of dynamical systems, we say that M_t is the initial amount. How does the final amount, M_{t+1} (unit of time is one day), depend on M_t?

Due to absorption, M_t is reduced to half, i.e., to $0.5M_t$, within a day (see Figure 3.1.5). Administering a new dosage will increase that amount by 1, and thus

$$M_{t+1} = 0.5M_t + 1$$

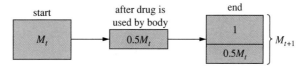

FIGURE 3.1.5
The dynamics of pain medication

So, if the patient's body contains $M_t = 3$ units of methadone at the start of the day, the amount at the end of the day is

$$M_{t+1} = 0.5(3) + 1 = 2.5$$

If $M_t = 1$, then

$$M_{t+1} = 0.5(1) + 1 = 1.5$$

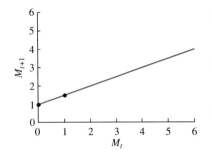

FIGURE 3.1.6

A graph of the updating function for pain medication

and so on. The updating function

$$f(M_t) = 0.5M_t + 1$$

is a line of slope 0.5 and vertical intercept 1 (Figure 3.1.6).

Solutions

A discrete-time dynamical system describes some quantity at the end of an experiment as a function of that same quantity at the beginning. What if we were to continue the experiment? A bacterial population growing according to $b_{t+1} = 2b_t$ would double again and again. A tree growing according to $h_{t+1} = h_t + 0.8$ would add more and more metres to its height. An infested lizard would become even more heavily infested.

To describe a situation where a dynamical process is repeated many times, we let m_0 represent the measurement at the beginning, m_1 the measurement after one time step (i.e., one unit of time), m_2 the measurement after two time steps, and so forth (Figure 3.1.7). In general, we define

m_t = measurement t units of time after the beginning of the experiment

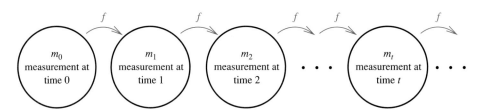

FIGURE 3.1.7

The repeated action of an updating function

Our goal is to find the values of m_t for all values of t. Before we can do so, however, we must know where we are starting. The starting value, m_0, is called the **initial condition.**

Definition 3.1.1 The sequence of values of m_t for $t = 0, 1, 2, \ldots$ is the **solution** of the discrete-time dynamical system $m_{t+1} = f(m_t)$ starting from the **initial condition** m_0.

We say that $m_0, m_1, m_2, m_3, \ldots$ is a **recursively defined sequence,** or a **recursion,** starting from m_0.

The graph of a solution is a discrete set of points with the time, t, on the horizontal axis and the measurement, m_t, on the vertical axis. The point with coordinates $(0, m_0)$ represents the initial condition. The point with coordinates $(1, m_1)$ describes the measurement at $t = 1$, and so forth (Figure 3.1.8). The point $(100, m_{100})$ represents the measurement 100 units of time after the start of the experiment.

How do we calculate m_{100}?

To get m_{100}, we need m_{99}, since $m_{100} = f(m_{99})$. To get m_{99}, we need m_{98}, and so forth. Thus, starting with m_0, we need to apply the updating function f 100 times to get m_{100}. In general, to calculate m_t from m_0 we need to apply f a total of t times. Is there a faster way to do this?

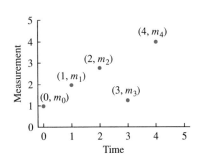

FIGURE 3.1.8

The graph of a solution

The answer is yes, in some simple cases, as we will witness in a moment—we will build a formula that will allow us to compute m_{100} directly from m_0, without going through all the intermediate steps. For general dynamical systems, however, such formulas are either hard to find, or they do not exist at all.

Example 3.1.5 A Solution of the Bacterial Discrete-Time Dynamical System from Example 3.1.1

Suppose we begin with one million bacteria, which corresponds to an initial condition of $b_0 = 1$ (with bacterial population measured in millions). If the bacteria obey the discrete-time dynamical system $b_{t+1} = 2b_t$, then

$$b_1 = 2b_0 = 2 \cdot 1 = 2$$
$$b_2 = 2b_1 = 2 \cdot 2 = 4$$
$$b_3 = 2b_2 = 2 \cdot 4 = 8$$

Examining these results, we notice that

$$b_1 = 2 \cdot 1$$
$$b_2 = 2^2 \cdot 1$$
$$b_3 = 2^3 \cdot 1$$

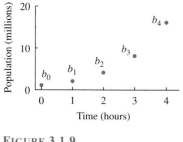

FIGURE 3.1.9

A solution: bacterial population size as a function of time

After three hours, the population has doubled three times and is $2^3 = 8$ times the original population. We graph the solution by plotting the time, t, on the horizontal axis and the number of bacteria after t hours (b_t) on the vertical axis (Figure 3.1.9). The graph consists only of a discrete set of points describing the hourly measurements—hence the name *discrete-time dynamical system*. Sometimes, we will connect the points in a solution with line segments to make the pattern easier to see.

After t hours, the population will have doubled t times and will have reached the size

$$b_t = 2^t \cdot 1 \tag{3.1.2}$$

This formula describes the solution of the discrete-time dynamical system with initial condition $b_0 = 1$. It predicts the population after t hours of reproduction for any value of t. For example, we can compute

$$b_8 = 2^8 \cdot 1 = 256$$

without ever computing b_1, b_2, and other intermediate values.

Now suppose that we start the system $b_{t+1} = 2b_t$ with a different initial condition of $b_0 = 0.3$. We can find subsequent values by repeatedly applying the discrete-time dynamical system as follows:

$$b_1 = 2b_0 = 2 \cdot 0.3$$
$$b_2 = 2b_1 = 2(2 \cdot 0.3) = 2^2 \cdot 0.3$$
$$b_3 = 2b_2 = 2(2^2 \cdot 0.3) = 2^3 \cdot 0.3$$

and so on. After t hours, the population has doubled t times and reached the size

$$b_t = 2^t \cdot 0.3 \text{ million}$$

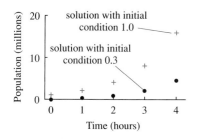

FIGURE 3.1.10

Solutions starting from two different initial conditions

This solution is **different** from the one corresponding to the initial condition $b_0 = 1$ (Figure 3.1.10). Although the two solutions get farther and farther apart, the ratio always remains the same (see Exercise 56).

The results we obtained in Example 3.1.5 illustrate this important general model:

Basic Exponential Discrete-Time Dynamical System

If $b_{t+1} = rb_t$ with initial condition b_0, then $b_t = b_0 r^t$.

In other words, if a population (or whatever else we are studying) changes according to

$$b_{t+1} = rb_t$$

then the population b_t at time t is equal to the initial population, b_0, multiplied by the rate, r, to the power of t.

How can we figure this out without much calculation? Assume that for the sake of argument, the time is measured in hours.

118   **Chapter 3**   Discrete-Time Dynamical Systems

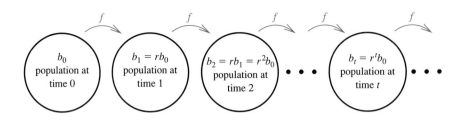

FIGURE 3.1.11
Bacterial population growth

The formula $b_{t+1} = rb_t$ states that during each hour, the population is multiplied by the same factor, r. Thus, if the initial population is b_0, the population one hour later will be $b_1 = rb_0$. The population two hours later will be r times the population one hour later:

$$b_2 = rb_1 = r(rb_0) = r^2 b_0$$

Likewise,

$$b_3 = rb_2 = r(r^2 b_0) = r^3 b_0$$

and so on (see Figure 3.1.11).

Example 3.1.6   Two Solutions of the Tree Height Discrete-Time Dynamical System

In Example 3.1.2 we studied the tree height discrete-time dynamical system

$$h_{t+1} = h_t + 0.8$$

Suppose the tree begins with a height of $h_0 = 10$ m. Then

$$h_1 = h_0 + 0.8 = 10.8 \text{ m}$$
$$h_2 = h_1 + 0.8 = 11.6 \text{ m}$$
$$h_3 = h_2 + 0.8 = 12.4 \text{ m}$$

and so on. Each year, the height of the tree increases by 0.8 m. After three years, the height is 2.4 m greater than the original height. After t years the tree has added 0.8 m to its height t times, meaning that the height will have increased by a total of $0.8t$ m. Therefore, the solution is

$$h_t = 10 + 0.8t \text{ m}$$

This formula predicts the height (in theory) after t years of growth for any t. We can compute, for instance,

$$h_8 = 10 + 0.8(8) = 16.4 \text{ m}$$

without computing h_1, h_2, and other intermediate values (Figure 3.1.12).

If the tree began at the smaller size of 2 m, the size for the first few years would be

$$h_1 = h_0 + 0.8 = 2.8 \text{ m}$$
$$h_2 = h_1 + 0.8 = 3.6 \text{ m}$$
$$h_3 = h_2 + 0.8 = 4.4 \text{ m}$$

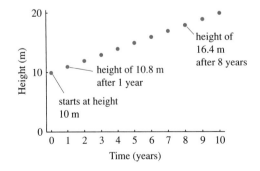

FIGURE 3.1.12
A solution: tree height as a function of time

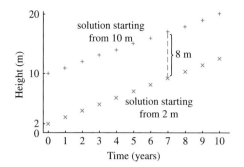

FIGURE 3.1.13

Two solutions for tree height as a function of time

and so on. Again, the tree adds $0.8t$ m of height in t years, so the height is

$$h_t = 2 + 0.8t \text{ m}$$

The solution with this smaller initial condition is always exactly 8 m less than the solution found before (Figure 3.1.13).

The solutions we obtained in Example 3.1.6 are worth remembering:

Basic Additive Discrete-Time Dynamical System

If $h_{t+1} = h_t + a$ with initial condition h_0, then $h_t = h_0 + at$.

In the above, we assume that a is a fixed real number.

This formula is easy to verify: substituting $t = 0$ into $h_{t+1} = h_t + a$, we get

$$h_1 = h_0 + a$$

Similarly,

$$h_2 = h_1 + a = (h_0 + a) + a = h_0 + 2a$$
$$h_3 = h_2 + a = (h_0 + 2a) + a = h_0 + 3a$$

and so on. In other words, the solution of the dynamical system $h_{t+1} = h_t + a$ forms an arithmetic sequence (or arithmetic progression).

We will next investigate some cases where computing the solution step by step is straightforward but finding a formula for the solution is tricky. Remarkably, there are simple discrete-time dynamical systems for which it is *impossible* to write a formula for a solution. We will meet such systems in Chapter 6.

Example 3.1.7 Finding a Solution of the Pain Medication Discrete-Time Dynamical System

Consider the discrete-time dynamical system for the pain medication (Example 3.1.4) given by

$$M_{t+1} = 0.5M_t + 1$$

Suppose we begin from an initial condition of $M_0 = 5$. Then

$$M_1 = 0.5 \cdot 5 + 1 = 3.5$$
$$M_2 = 0.5 \cdot 3.5 + 1 = 2.75$$
$$M_3 = 0.5 \cdot 2.75 + 1 = 2.375$$
$$M_4 = 0.5 \cdot 2.375 + 1 = 2.1875$$

The values are getting closer and closer to 2 (Figure 3.1.14). More careful examination indicates that the results move exactly *halfway* toward 2 each step. In particular, we find that the difference between the value of M_t and 2 is

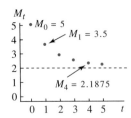

FIGURE 3.1.14

Amount of medication as a function of time

$$M_0 - 2 = 5 - 2 = 3$$
$$M_1 - 2 = 3.5 - 2 = 1.5 = 0.5 \cdot 3$$
$$M_2 - 2 = 2.75 - 2 = 0.75 = 0.25 \cdot 3 = 0.5^2 \cdot 3$$
$$M_3 - 2 = 2.375 - 2 = 0.375 = 0.125 \cdot 3 = 0.5^3 \cdot 3$$

Solving for $M_0, M_1, M_2, \ldots$, we obtain

$$M_0 = 2 + 3$$
$$M_1 = 2 + 0.5 \cdot 3$$
$$M_2 = 2 + 0.5^2 \cdot 3$$
$$M_3 = 2 + 0.5^3 \cdot 3$$

and notice that

$$M_t = 2 + 0.5^t \cdot 3$$

Obviously, finding patterns in this way and translating them into formulas can be tricky. It is much more important to *describe* the behaviour of solutions with a graph or in words. In this case, our calculations suggest that the solution moves closer and closer to 2. In Section 3.2, we will develop a powerful graphical method to deduce this pattern with a minimum of calculation. ▲

The method we just used is neither intuitive nor easy to do. As well, it relies on the assumption that the numbers in the sequence $M_1, M_2, M_3, \ldots$ are getting closer and closer to 2. (Is that a sound assumption; i.e., does calculating only the numbers M_1 through M_4 suffice to state that the sequence $M_1, M_2, M_3, \ldots$ approaches 2? How do we know that it does not approach 2.1, or 2.05?)

We now present an alternative calculation. Assume that the solution of

$$M_{t+1} = 0.5 M_t + 1$$

with initial condition $M_0 = 5$ is of the form

$$M_t = a(0.5^t) + b$$

where a and b are real numbers whose values we now determine.

Substituting $t = 0$, we get

$$M_0 = a(0.5^0) + b$$

i.e., $5 = a + b$. Likewise, substituting $t = 1$ yields

$$M_1 = a(0.5^1) + b$$

and so $3.5 = 0.5a + b$.

In this way, we obtain a system of two equations with unknowns a and b:

$$a + b = 5$$
$$0.5a + b = 3.5$$

Subtracting the second equation from the first, we get

$$0.5a = 1.5$$

and $a = 3$. From either equation we obtain that $b = 2$. Therefore,

$$M_t = 3 \cdot 0.5^t + 2$$

To finish, we show that $M_t = 3 \cdot 0.5^t + 2$ indeed solves the given dynamical system. From

$$M_{t+1} = 3 \cdot 0.5^{t+1} + 2$$

and
$$0.5M_t + 1 = 0.5(3 \cdot 0.5^t + 2) + 1 = 3 \cdot 0.5^{t+1} + 1 + 1 = 3 \cdot 0.5^{t+1} + 2$$
we conclude that M_t satisfies
$$M_{t+1} = 0.5M_t + 1$$
The initial condition is satisfied as well, since
$$M_0 = 3 \cdot 0.5^0 + 2 = 5$$

Note The initial dose, M_0, of the medication administered is referred to as the *loading dose*. Because it is supposed to start acting as soon as possible (say, after a surgery), it is fairly high—in our case it measures five units. Subsequently, in order to keep the medication at a certain level, the patient is given a lower *maintenance dose*. In our case, the maintenance dose measures one unit.

Example 3.1.8 A Solution of the Pain Medication Discrete-Time Dynamical System from Example 3.1.4 with Different Initial Condition

If we begin with an initial amount of $M_0 = 1$, then
$$M_1 = 0.5 \cdot 1 + 1 = 1.5$$
$$M_2 = 0.5 \cdot 1.5 + 1 = 1.75$$
$$M_3 = 0.5 \cdot 1.75 + 1 = 1.875$$
$$M_4 = 0.5 \cdot 1.875 + 1 = 1.9375$$

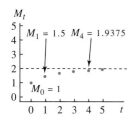

FIGURE 3.1.15

Amount of medication as a function of time

(See Figure 3.1.15.) Unlike graphs of bacterial populations (Example 3.1.5) and tree size (Example 3.1.6), the graphs of solutions starting from different initial conditions look different (compare Figures 3.1.14 and 3.1.15).

Using the technique we just introduced, we obtain the solution
$$M_t = -0.5^t + 2$$
(see also Exercise 27).

In Section 4.2, we will use the fundamental idea of the **limit** to study more carefully what it means for the sequence of values that define a solution to get closer and closer to some number (such as 2 in Examples 3.1.7 and 3.1.8).

Example 3.1.9 Special Solution of the Pain Medication Discrete-Time Dynamical System

Once again, we consider the system from Example 3.1.4,
$$M_{t+1} = 0.5M_t + 1$$
but this time with initial condition $M_0 = 2$. From
$$M_1 = 0.5M_0 + 1 = 0.5(2) + 1 = 2$$
we see that the output (M_1) is the same as the input (M_0). Likewise,
$$M_2 = 0.5M_1 + 1 = 0.5(2) + 1 = 2$$
$$M_3 = 0.5M_2 + 1 = 0.5(2) + 1 = 2$$
and so $M_t = 2$ for all t.

In other words, the value $M_0 = 2$ is not changed by the dynamical system (clearly, $M_0 = 5$ and $M_0 = 1$ from Examples 3.1.7 and 3.1.8 do not share this property).

Values that remain unchanged when subjected to a dynamical system are called **equilibria**. As they are important for our understanding of how the system works, we study them extensively in the following section.

Example 3.1.10 A Solution of the Mite Population Discrete-Time Dynamical System

Recall the discrete-time dynamical system (Example 3.1.3)

$$x_{t+1} = 2x_t + 30$$

for mites. If we start the lizard off with $x_0 = 10$ mites, we compute

$$x_1 = 2x_0 + 30 = 50$$
$$x_2 = 2x_1 + 30 = 130$$
$$x_3 = 2x_2 + 30 = 290$$

The pattern is not at all obvious in this case. There is a pattern, however, which it is a good challenge to find (Exercise 40).

Manipulating Updating Functions

All of the operations that can be applied to ordinary functions can be applied to updating functions, but with special interpretations. We will study **composition** of an updating function with itself, find the **inverse** of an updating function, and **convert the units** or **translate the dimensions** of a discrete-time dynamical system.

Composition Consider the discrete-time dynamical system

$$m_{t+1} = f(m_t)$$

with updating function f. What does the composition $f \circ f$ mean? The updating function **updates** the measurement by one time step. Then

$$(f \circ f)(m_t) = f(f(m_t)) \quad \text{definition of composition}$$
$$= f(m_{t+1}) \quad \text{definition of updating function}$$
$$= m_{t+2} \quad \text{updating function applied to } m_{t+1}$$

Therefore,

$$(f \circ f)(m_t) = m_{t+2}$$

The composition of an updating function with itself corresponds to a two-step updating function (Figure 3.1.16).

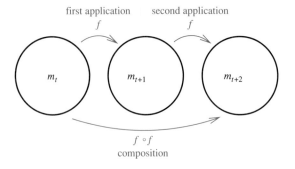

FIGURE 3.1.16

Composition of an updating function with itself

Example 3.1.11 Composition of the Bacterial Population Updating Function with Itself

Recall that in Example 3.1.1, the bacterial updating function is $f(b_t) = 2b_t$. The function $f \circ f$ takes the population size at time t as input and returns the population size two hours later, at time $t + 2$, as output. We can compute $f \circ f$ with the steps

$$(f \circ f)(b_t) = f(f(b_t)) = f(2b_t) = 2(2b_t) = 4b_t$$

After two hours, the population is four times as large, having doubled twice.

3.1 Introduction to Discrete-Time Dynamical Systems

Example 3.1.12 Composition of the Mite Population Updating Function with Itself

The composition of the mite population updating function $f(x_t) = 2x_t + 30$ with itself (see Example 3.1.3) gives

$$(f \circ f)(x_t) = f(f(x_t)) = f(2x_t + 30) = 2(2x_t + 30) + 30 = 4x_t + 90$$

Suppose we started with $x_t = 10$ mites. After one week, we would find $f(10) = 2 \cdot 10 + 30 = 50$ mites. After the second week, we would find $f(50) = 2 \cdot 50 + 30 = 130$ mites. Using the composition of the updating function with itself, we can compute the number of mites after two weeks, skipping over the intermediate value of 50 mites after one week, finding

$$(f \circ f)(10) = 4 \cdot 10 + 90 = 130$$

Thus, composing the updating function with itself produces a function that allows us to jump two time units into the future. Beyond this obvious reason, why is this useful? Consider the following example.

Example 3.1.13 Oscillating Dynamical System

Consider the dynamical system

$$x_{t+1} = 3.35 x_t (1 - x_t)$$

with initial condition $x_0 = 0.5$. Using a calculator, we compute

$$x_1 = 3.35(0.5)(1 - 0.5) = 0.8375$$
$$x_2 = 3.35(0.8375)(1 - 0.8375) = 0.4559$$
$$x_3 = 3.35(0.4559)(1 - 0.4559) = 0.8310$$
$$x_4 = 3.35(0.8310)(1 - 0.8310) = 0.4705$$

and so on: $x_5 = 0.8346$, $x_6 = 0.4625$, $x_7 = 0.8328$, $x_8 = 0.4665$, etc.

Starting with 0.5 and selecting every other term, we obtain the sequence

$$0.5, 0.4559, 0.4705, 0.4625, 0.4665, \ldots$$

which seems to be approaching 0.46. Staring with $x_1 = 0.8375$ and selecting every other term again, we arrive at the sequence

$$0.8375, 0.8310, 0.8346, 0.8328, \ldots$$

which seems to be approaching 0.83.

We have discovered an important feature of the system: it oscillates (i.e., jumps between lower and higher values), with high values approaching a certain number, and low values approaching a certain number; see Figure 3.1.17.

In nature, a system with this property (with appropriate units) could describe seasonal lows and highs in populations of certain species (such as fish).

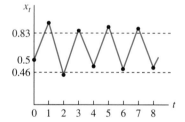

FIGURE 3.1.17

Oscillations in the dynamical system (vertical axis not to scale)

Inverses Consider again the general discrete-time dynamical system

$$m_{t+1} = f(m_t)$$

with updating function f. What does the inverse, f^{-1}, mean? The updating function updates the measurement by one time step, i.e., goes one time step into the future. Thus, the inverse function will go one step into the past,

$$f^{-1}(m_{t+1}) = m_t$$

So, the inverse of an updating function corresponds to an "updating" function that goes **backwards** in time (Figure 3.1.18).

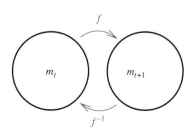

FIGURE 3.1.18

Inverse of an updating function

Example 3.1.14 Inverse of the Bacterial Population Updating Function

The bacterial population updating function is $f(b_t) = 2b_t$. We find the inverse by writing the discrete-time dynamical system

$$b_{t+1} = 2b_t$$

and solving for the input variable b_t. Dividing both sides by 2 gives

$$b_t = \frac{b_{t+1}}{2}$$

The inverse function is

$$f^{-1}(b_{t+1}) = \frac{b_{t+1}}{2}$$

If **multiplying** by 2 describes how the population changes forward in time, **dividing** by 2 describes how it changes backwards in time.

For example, if $b_t = 3$, then

$$b_{t+1} = f(b_t) = 2b_t = 2(3) = 6$$

If we go backwards from $b_{t+1} = 6$ using the inverse of the updating function, we find

$$b_t = f^{-1}(b_{t+1}) = f^{-1}(6) = \frac{6}{2} = 3$$

exactly where we started.

Example 3.1.15 Inverse of the Mite Population Updating Function

To find the inverse of the mite population updating function $f(x_t) = 2x_t + 30$, we solve $x_{t+1} = 2x_t + 30$ for x_t:

$$x_{t+1} = 2x_t + 30$$
$$2x_t = x_{t+1} - 30$$
$$x_t = \frac{x_{t+1} - 30}{2}$$
$$x_t = 0.5x_{t+1} - 15$$

Therefore,

$$x_t = f^{-1}(x_{t+1}) = 0.5x_{t+1} - 15$$

Suppose we started with $x_t = 10$ mites. After one week, we would find

$$h(10) = 2 \cdot 10 + 30 = 50$$

mites. Applying the inverse, we find

$$h^{-1}(50) = 0.5 \cdot 50 - 15 = 10$$

mites. The inverse function takes us back to where we started.

One reason that inverse functions are important in dynamical systems is that they allow us to go into the past. Here is an example, related to the system $b_{t+1} = 2b_t$ that we studied in Example 3.1.14.

Assume that there are 4000 bacteria in a culture, i.e., $b_t = 4000$. One hour later, the population will double, i.e., $b_{t+1} = 8000$. Now, let us go back, using the inverse function

$$b_t = f^{-1}(b_{t+1}) = \frac{b_{t+1}}{2}$$

So, if $b_{t+1} = 8000$, then

$$b_t = \frac{b_{t+1}}{2} = \frac{8000}{2} = 4000$$

which is the number of bacteria at the beginning of the experiment. Not a big deal, we knew that already. But even before we started our experiment, the bacterial culture was growing (how did it reach its size of 4000 otherwise?). So, assuming that the dynamics of growth did not change, we can compute that one hour before our experiment started, the bacterial count was

$$b_{t-1} = \frac{b_t}{2} = \frac{4000}{2} = 2000$$

One hour before that (i.e., two hours before the start of our experiment), the bacterial count was

$$b_{t-2} = \frac{b_{t-1}}{2} = \frac{2000}{2} = 1000$$

Thus—given a dynamical system—we can run it backwards by using the inverse of the updating function in order to recover its history (if it makes sense).

Units and Dimensions

The updating function $f(b_t) = 2b_t$ accepts as input positive numbers with units of bacteria. If we measure this quantity in different units (say, thousands), we must convert the updating function itself into the new units. If we measure a different quantity, such as total mass or volume, we can translate the updating function into different dimensions.

Example 3.1.16 Describing the Dynamics of Tree Height in Centimetres

Suppose we wish to study tree height (Example 3.1.2) in units of centimetres rather than metres. In metres, the discrete-time dynamical system is

$$h_{t+1} = f(h_t) = h_t + 0.8$$

First, we define a new variable to represent the measurement in the new units. Let H_t be tree height measured in centimetres. Then $H_t = 100 h_t$, because there are 100 centimetres in a metre. We wish to find a discrete-time dynamical system that gives a formula for H_{t+1} in terms of H_t (Figure 3.1.19):

$$\begin{aligned} H_{t+1} &= 100 h_{t+1} \\ &= 100(h_t + 0.8) \\ &= 100 h_t + 80 \\ &= H_t + 80 \end{aligned}$$

The discrete-time dynamical system in the new units corresponds to adding 80 centimetres to the height, which is equivalent to adding 0.8 metres. Although the underlying process is the same, the discrete-time dynamical system and the corresponding updating function are different, just as the numerical values of measurements are different in different units.

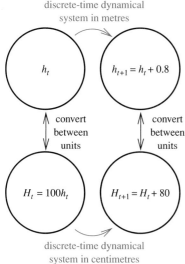

FIGURE 3.1.19

Finding the discrete-time dynamical system for trees in centimetres

Example 3.1.17 Describing the Dynamics of Bacterial Mass

Suppose we wish to study the bacterial population in terms of mass rather than number. Assume that $b_{t+1} = 2b_t$ represents the dynamics of growth (b_t represents the number of bacteria at time t). At time t, the mass, denoted by m_t, is

$$m_t = \mu b_t$$

where μ is the mass per bacterium. The updated mass, m_{t+1}, is

$$\begin{aligned} m_{t+1} &= \mu b_{t+1} \\ &= \mu \cdot 2 b_t \\ &= 2 \cdot \mu b_t \\ &= 2 m_t \end{aligned}$$

Chapter 3 Discrete-Time Dynamical Systems

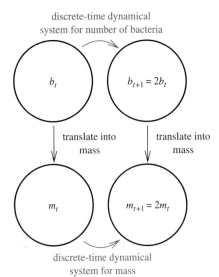

FIGURE 3.1.20
Finding the discrete-time dynamical system for bacteria in terms of mass

This new discrete-time dynamical system doubles its input just as the original discrete-time dynamical system did, but it takes mass as its input rather than numbers of bacteria (Figure 3.1.20).

Summary Starting from data or an understanding of a biological process, we can derive a **discrete-time dynamical system** (also called **recursion**), the **dynamical rule** that tells how a measurement changes from one time step to the next. The **updating function** describes the relation between measurements at times t and $t + 1$. The **composition** of the updating function with itself produces a two-step discrete-time dynamical system, and the **inverse** of the updating function produces a backward discrete-time dynamical system. Like all biological relations, a discrete-time dynamical system can be described in different units and dimensions. Repeated application of a discrete-time dynamical system starting from an **initial condition** generates a **solution,** the value of the measurement as a function of time. With the proper combination of diligence, cleverness, and luck, it is sometimes possible to find a formula for the solution.

3.1 Exercises

Mathematical Techniques

1–4 ▪ Write the updating function associated with each discrete-time dynamical system, and evaluate it at the given arguments. Which are linear?

1. $p_{t+1} = p_t - 2$; evaluate at $p_t = 5$, $p_t = 10$, and $p_t = 15$.

2. $m_{t+1} = \dfrac{m_t^2}{m_t + 2}$; evaluate at $m_t = 0$, $m_t = 8$, and $m_t = 20$.

3. $x_{t+1} = x_t^2 + 2$; evaluate at $x_t = 0$, $x_t = 2$, and $x_t = 4$.

4. $Q_{t+1} = \dfrac{1}{Q_t + 1}$; evaluate at $Q_t = 0$, $Q_t = 1$, and $Q_t = 2$.

5–8 ▪ Compose with itself the updating function associated with each discrete-time dynamical system. Find the two-step discrete-time dynamical system. Check that the result of applying the original discrete-time dynamical system to the given initial condition twice matches the result of applying the new discrete-time dynamical system to the given initial condition once.

5. Volume follows $v_{t+1} = 1.5v_t$, starting from $v_0 = 1220 \,\mu m^3$.

6. Length obeys $l_{t+1} = l_t - 1.7$, starting from $l_0 = 13.1$ cm.

7. Population size follows $n_{t+1} = 0.5n_t$, starting from $n_0 = 1200$.

8. Medication concentration obeys $M_{t+1} = 0.75M_t + 2$, starting from the initial condition $M_0 = 16$.

9–14 ▪ Find the backward discrete-time dynamical system associated with each discrete-time dynamical system. Use it to find the value at the previous time.

9. $v_{t+1} = 1.5v_t$. Find v_0 if $v_1 = 1220 \,\mu m^3$.

10. $l_{t+1} = l_t - 1.7$. Find l_0 if $l_1 = 13.1$ cm.

11. $n_{t+1} = 0.5n_t$. Find n_0 if $n_1 = 1200$.

12. $M_{t+1} = 0.75M_t + 2$. Find M_0 if $M_1 = 16$.

13. $p_{t+1} = \dfrac{4}{p_t^3}$. Find p_0 if $p_1 = 32$.

14. $m_{t+1} = \dfrac{m_t}{m_t + 2}$. Find m_0 if $m_1 = 1/5$.

15–16 ■ Find the composition of each mathematically elegant updating function with itself, and find the inverse function.

15. The updating function $f(x) = \dfrac{x}{1+x}$. Remember to put things over a common denominator to simplify the composition.

16. The updating function $h(x) = \dfrac{x}{x-1}$. Remember to put things over a common denominator to simplify the composition.

17–20 ■ Find and graph the first five values of each discrete-time dynamical system, starting from the given initial condition. Compare the graph of the solution with the graph of the updating function.

17. $v_{t+1} = 1.5 v_t$, starting from $v_0 = 1220 \, \mu\text{m}^3$.

18. $l_{t+1} = l_t - 1.7$, starting from $l_0 = 13.1$ cm.

19. $n_{t+1} = 0.5 n_t$, starting from $n_0 = 1200$.

20. $M_{t+1} = 0.75 M_t + 2$, starting from the initial condition $M_0 = 16$.

21–24 ■ Using a formula for the solution, you can project far into the future without computing all the intermediate values. Find the following, and indicate whether the results are reasonable.

21. From the solution found in Exercise 17, find the volume at $t = 20$.

22. From the solution found in Exercise 18, find the length at $t = 20$.

23. From the solution found in Exercise 19, find the number at $t = 20$.

24. From the solution found in Exercise 20, find the concentration at $t = 20$.

25–27 ■ Find a formula for the solution of each dynamical system.

25. $b_{t+1} = 0.8 b_t - 2, b_0 = 10$.

26. $x_{t+1} = 3 x_t + 0.4, x_0 = 3$.

27. $a_{t+1} = b a_t + c, a_0 = 1$ (b and c are nonzero constants).

28–31 ■ Experiment with the following mathematically elegant updating functions and try to find the solution.

28. Consider the updating function
$$f(x) = \dfrac{x}{1+x}$$
Starting from an initial condition of $x_0 = 1$, compute x_1, x_2, x_3, and x_4, and try to spot the pattern.

29. Use the updating function in Exercise 28, but start from the initial condition $x_0 = 2$.

30. Consider the updating function
$$g(x) = 4 - x$$
Start from an initial condition of $x_0 = 1$, and try to spot the pattern. Experiment with a couple of other initial conditions. How would you describe your results in words?

31. Consider the updating function
$$h(x) = \dfrac{x}{x-1}$$
from Exercise 16. Start from an initial condition of $x_0 = 3$, and try to spot the pattern. Experiment with a couple of other initial conditions. How would you describe your results in words?

Applications

32–35 ■ Consider the following actions. Which of them commute (produce the same answer when done in either order)?

32. A population doubles in size; 10 individuals are removed from a population. Try starting with 100 individuals, and then try to figure out what happens in general.

33. A population doubles in size; population size is divided by 4. Try starting with 100 individuals, and then try to figure out what happens in general.

34. An organism grows by 2 cm; an organism shrinks by 1 cm.

35. A person loses half her money; a person gains $10.

36–39 ■ Use the formula for the solution to find the following, and indicate whether the results are reasonable.

36. Using the solution for tree height, $h_t = 10 + 0.8 t$ (Example 3.1.6), find the tree height after 20 years.

37. Using the solution for tree height, $h_t = 10 + 0.8 t$ (Example 3.1.6), find the tree height after 100 years.

38. Using the solution for bacterial population number, $b_t = 2^t \cdot 1$ (Equation 3.1.2), find the bacterial population after 20 hours. If an individual bacterium weighs about 10^{-12} g, how much will the whole population weigh?

39. Using the solution for bacterial population number, $b_t = 2^t \cdot 1$ (Equation 3.1.2), find the bacterial population after 40 hours.

40–41 ■ Try to find a formula for the solution of the given discrete-time dynamical system.

40. Find the pattern in the number of mites on a lizard, starting with $x_0 = 10$ and following the discrete-time dynamical system $x_{t+1} = 2 x_t + 30$. (*Hint:* Add 30 to the number of mites.)

41. Try to find the pattern in the number of mites on a lizard, starting with $x_0 = 10$ and following the discrete-time dynamical system $x_{t+1} = 2 x_t + 20$.

42–45 ■ The following tables display data from four experiments:

1. Cell volume after 10 minutes in a watery bath
2. Fish mass after one week in a chilly tank
3. Gnat population size after three days without food
4. Yield (in bushels) of several varieties of soybeans before and one month after fertilization

For each, graph the new value as a function of the initial value, find a simple discrete-time dynamical system, and determine the missing value in the table.

42.

Cell Volume (μ m³)	
Initial, v_t	Final, v_{t+1}
1220	1830
1860	2790
1080	1620
1640	2460
1540	2310
1420	??

43.

Fish Mass (g)	
Initial, m_t	Final, m_{t+1}
13.1	11.4
18.2	16.5
17.3	15.6
16.0	14.3
20.5	18.8
1.5	??

44.

Gnat Population	
Initial, n_t	Final, n_{t+1}
$1.2 \cdot 10^3$	$6.0 \cdot 10^2$
$2.4 \cdot 10^3$	$1.2 \cdot 10^3$
$1.6 \cdot 10^3$	$8.0 \cdot 10^2$
$2.0 \cdot 10^3$	$1.0 \cdot 10^3$
$1.4 \cdot 10^3$	$7.0 \cdot 10^2$
$8.0 \cdot 10^2$	??

45.

Soybean Yield per Acre (bushels)	
Initial, Y_t	Final, Y_{t+1}
100	210
50	110
200	410
75	160
95	200
250	??

46–49 ▪ Recall the data used for Section 1.4, Exercises 60–63.

Age, a (days)	Length, L (cm)	Tail Length, T (cm)	Mass, M (g)
0.5	1.5	1.0	1.5
1	3.0	0.9	3.0
1.5	4.5	0.8	6.0
2	6.0	0.7	12.0
2.5	7.5	0.6	24.0
3	9.0	0.5	48.0

These data define several discrete-time dynamical systems. For example, between the first measurement (on day 0.5) and the second (on day 1), the length increases by 1.5 cm. Between the second measurement (on day 1) and the third (on day 1.5), the length again increases by 1.5 cm.

46. Graph the length at the second measurement as a function of length at the first, the length at the third measurement as a function of length at the second, and so on. Find the discrete-time dynamical system that reproduces the results.

47. Find and graph the discrete-time dynamical system for tail length.

48. Find and graph the discrete-time dynamical system for mass.

49. Find and graph the discrete-time dynamical system for age.

50–51 ▪ Consider the discrete-time dynamical system $b_{t+1} = 2b_t$ for a bacterial population (Example 3.1.1).

50. Write a discrete-time dynamical system for the total volume of bacteria (suppose each bacterium takes up 10^4 μm³).

51. Write a discrete-time dynamical system for the total area taken up by the bacteria (suppose the thickness is 20 μm).

52–53 ▪ Recall the equation $r_{t+1} = r_t + 0.8$ for tree height.

52. Write a discrete-time dynamical system for the total volume of a cylindrical tree.

53. Write a discrete-time dynamical system for the total volume of a spherical tree given that $r_{t+1} = r_t + 0.8$ (this is kind of tricky).

54–55 ▪ Consider the following data describing the levels of a medication in the blood of two patients over the course of several days (measured in milligrams per litre):

Day	Medication Level in Patient 1 (mg/L)	Medication Level in Patient 2 (mg/L)
0	20	0
1	16	2
2	13	3.2
3	10.75	3.92

54. Graph three points on the updating function for the first patient. Find a linear discrete-time dynamical system for the first patient.

55. Graph three points on the updating function for the second patient, and find a linear discrete-time dynamical system.

56–57 ■ For the following discrete-time dynamical systems, compute solutions starting from each of the given initial conditions. Then find the difference between the solutions as a function of time, and the ratio of the solutions as a function of time. In which cases is the difference constant, and in which cases is the ratio constant? Can you explain why?

56. Two bacterial populations follow the discrete-time dynamical system $b_{t+1} = 2b_t$, but the first starts with initial condition $b_0 = 1 \times 10^6$ and the second starts with initial condition $b_0 = 3 \times 10^5$ (in millions of bacteria).

57. Two trees follow the discrete-time dynamical system $h_{t+1} = h_t + 0.8$, but the first starts with initial condition $h_0 = 10$ m and the second starts with initial condition $h_0 = 2$ m.

58–61 ■ Derive and analyze discrete-time dynamical systems that describe the following contrasting situations.

58. A population of bacteria doubles every hour, but 1×10^6 individuals are removed after reproduction to be converted into valuable biological by-products. The population begins with $b_0 = 3 \times 10^6$ bacteria.

    a. Find the population after one, two, and three hours.
    b. How many bacteria were harvested?
    c. Write the discrete-time dynamical system.
    d. Suppose you waited to harvest bacteria until the end of three hours. How many could you remove and still match the population b_3 found in part (a)? Where did all the extra bacteria come from?

59. Suppose that a population of bacteria doubles every hour but that 1×10^6 individuals are removed *before* reproduction to be converted into valuable biological by-products. Suppose the population begins with $b_0 = 3 \times 10^6$ bacteria.

    a. Find the population after one, two, and three hours.
    b. Write the discrete-time dynamical system.
    c. How does the population compare with that in the previous problem? Why is it doing worse?

60. Suppose the fraction of individuals with some superior gene increases by 10% each generation.

    a. Write the discrete-time dynamical system for the fraction of organisms with the gene (denote the fraction at time t by f_t and figure out the formula for f_{t+1}).
    b. Write the solution, starting from an initial condition of $f_0 = 0.0001$.
    c. Will the fraction reach 1.0? Does the discrete-time dynamical system make sense for all values of f_t?

61. The Weber-Fechner law describes how human beings perceive differences. Suppose, for example, that a person first hears a tone with a frequency of 400 hertz (cycles per second). He is then tested with higher tones until he can hear the difference. The ratio between these values describes how well this person can hear differences.

    a. Suppose the next tone he can distinguish has a frequency of 404 hertz. What is the ratio?
    b. According to the Weber-Fechner law, the next higher tone will be greater than 404 by the same ratio. Find this tone.
    c. Write the discrete-time dynamical system for this person. Find the fifth tone he can distinguish.
    d. Suppose the experiment is repeated on a musician, and she manages to distinguish 400.5 hertz from 400 hertz. What is the fifth tone she can distinguish?

62–63 ■ The total mass of a population of bacteria will change if the number of bacteria changes, if the mass per bacterium changes, or if both of these variables change. Try to derive a discrete-time dynamical system for the total mass in the following situations.

62. The number of bacteria doubles each hour, and the mass of each bacterium triples during the same time.

63. The number of bacteria doubles each hour, and the mass of each bacterium increases by 10^{-9} g. What seems to go wrong with this calculation? Can you explain why?

3.2 Analysis of Discrete-Time Dynamical Systems

In the previous section, we defined discrete-time dynamical systems that describe what happens during a single time step, and their solution as the sequence of values taken on over many time steps. Finding a formula for the solution is often difficult or impossible. Nonetheless, we can usually deduce the behaviour of the solution with simpler methods. This section introduces two such methods. **Cobwebbing** is a graphical technique that makes it possible to sketch solutions without calculations. Algebraically, we will learn how to solve for **equilibria,** the values for which the discrete-time dynamical system remains unchanged.

Cobwebbing: A Graphical Solution Technique

Consider the general discrete-time dynamical system

$$m_{t+1} = f(m_t)$$

with updating function f shown in Figure 3.2.21. We now describe how, using the diagonal (i.e., the line $m_{t+1} = m_t$), we can determine the behaviour of solutions graphically. The technique is called **cobwebbing.**

Suppose we are given some initial condition m_0. To find m_1, we apply the updating function

$$m_1 = f(m_0)$$

Graphically, m_1 is the vertical coordinate of the point on the graph of the updating function directly above m_0 (Figure 3.2.22a). Similarly, m_2 is the vertical coordinate of the point on the graph of the updating function directly above m_1, and so on.

The missing step is moving m_1 from the vertical axis onto the horizontal axis. The trick is to **reflect** it off the diagonal line that has equation $m_{t+1} = m_t$. Move the point (m_0, m_1) horizontally until it intersects the diagonal. Moving a point horizontally does not change the vertical coordinate, and therefore the intersection with the diagonal is the point (m_1, m_1) (Figure 3.2.22b). The point $(m_1, 0)$ lies directly below (Figure 3.2.22c).

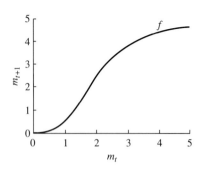

FIGURE 3.2.21

Graph of the updating function

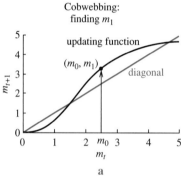

a

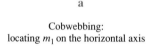

b

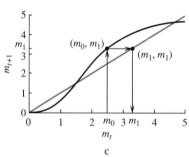

c

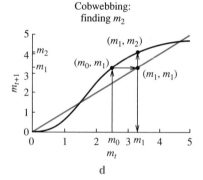

d

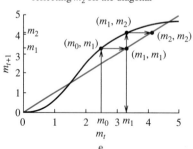

e

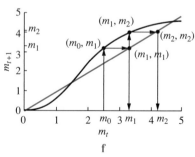

f

FIGURE 3.2.22

Cobwebbing: The first steps

3.2 Analysis of Discrete-Time Dynamical Systems 131

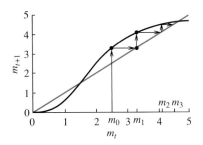

FIGURE 3.2.23
Cobwebbing

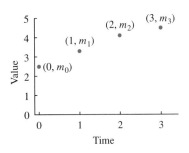

FIGURE 3.2.24
The solution derived from a cobweb diagram

What have we done? Starting from the initial value m_0, plotted on the horizontal axis, we used the updating function to find m_1 on the vertical axis and the reflecting trick to project m_1 onto the horizontal axis. Now we repeat this process. Move m_1 vertically to the graph of the updating function (Figure 3.2.22d), thus getting the point (m_1, m_2). Next, we move this point horizontally until we reach the diagonal (Figure 3.2.22e), and then vertically until we reach the horizontal axis. This way, we have identified the point m_2 (Figure 3.2.22f). Continuing in the same way, we obtain m_3, m_4, and so on. Because the lines reaching all the way to the horizontal axis are unnecessary, they are generally omitted to make the diagram cleaner (Figure 3.2.23).

Having found m_1, m_2, and m_3 on our cobwebbing graph, we can sketch the graph of the solution that shows the measurement as a function of time. In Figure 3.2.22, we began at $m_0 = 2.5$. This is plotted as the point $(0, m_0) = (0, 2.5)$ in the solution (Figure 3.2.24). The value m_1 is approximately 3.2 and is plotted as the point $(1, m_1)$ in the solution. The values of m_2 and m_3 increase more slowly and are shown accordingly on the graph.

Without plugging numbers into the updating function, using the graph instead, we figured out the behaviour of a solution starting from a given initial condition.

Similarly, we can find how the given system would behave over time if we started from a different initial condition, $m_0 = 1.2$ (Figure 3.2.25a). In this case, the diagonal lies above the graph of the updating function, so reflecting off the diagonal moves points to the left. Therefore, the solution decreases, as shown in Figure 3.2.25b.

The steps for cobwebbing are summarized in the following algorithm.

FIGURE 3.2.25
Cobweb and solution with a different initial condition

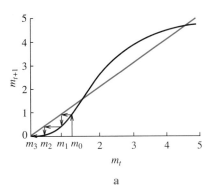
a

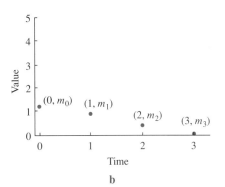
b

▶▶ **Algorithm 3.2.1** Using Cobwebbing to Find the Solution of $m_{t+1} = f(m_t)$ with Initial Condition m_0

1. Graph the updating function and the diagonal.
2. Starting from the initial condition on the horizontal axis, go vertically to the updating function and over to the diagonal.
3. Repeat going vertically to the updating function and over to the diagonal for as many steps as needed to find the pattern.
4. Sketch the solution at times 0, 1, 2, and so forth.

Example 3.2.1 Cobwebbing and Solution of the Pain Medication Model

Consider the discrete-time dynamical system for the pain medication (Example 3.1.4),

$$M_{t+1} = 0.5M_t + 1$$

The updating function, $M_{t+1} = 0.5M_t + 1$, is a line with slope 0.5 and intercept 1, and so it is less steep than the diagonal, $M_{t+1} = M_t$. If we begin at $M_0 = 5$, the cobweb and

FIGURE 3.2.26

Cobweb and solution of the medication model: $M_0 = 5$

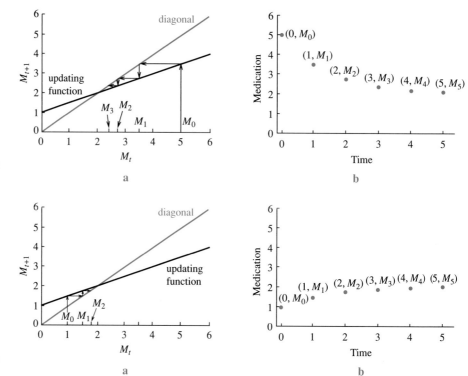

FIGURE 3.2.27

Cobweb and solution of the medication model: $M_0 = 1$

solution decrease more and more slowly over time (Figure 3.2.26). If we begin at $M_0 = 1$ instead, the cobweb and solution increase (at a slower and slower pace) over time (Figure 3.2.27).

Equilibria: Geometric Approach

The points where the graph of the updating function intersects the diagonal play a special role in cobweb diagrams. These points also play an essential role in understanding the behaviour of discrete-time dynamical systems.

Consider the discrete-time dynamical systems plotted in Figure 3.2.28. The first describes a population of plants at time t (denoted by P_t) and the second a population of birds at time t (denoted by B_t). Each graph includes the diagonal line used in cobwebbing.

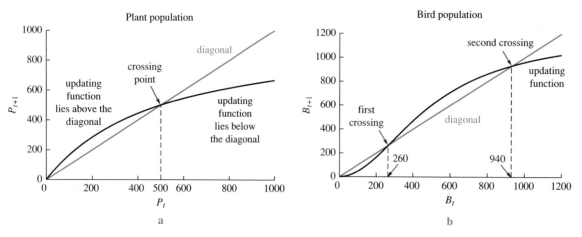

FIGURE 3.2.28

Dynamics of two populations

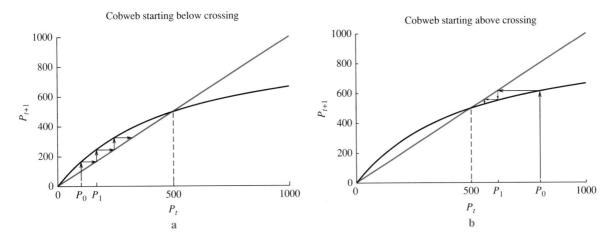

FIGURE 3.2.29

Behaviour of plant population with two different initial conditions

If we begin cobwebbing from an initial condition where the graph of the updating function lies *above* the diagonal, the population increases (Figure 3.2.29a). In contrast, if we begin cobwebbing from an initial condition where the graph of the updating function lies *below* the diagonal, the population decreases (Figure 3.2.29b). The plant population will thus increase if the initial condition lies below the crossing point, but it will decrease if it lies above.

Similarly, the updating function for the bird population lies below the diagonal for initial conditions less than the first crossing, and the population decreases (Figure 3.2.30a). The updating function is above the diagonal for initial conditions between the crossings, and the population increases (Figure 3.2.30b). Finally, the updating function is again below the diagonal for initial conditions greater than the second crossing, and the population decreases (Figure 3.2.30c).

What happens at points where the updating function crosses the diagonal, such as P_0 in Figure 3.2.31a? If we start cobwebbing from P_0, nothing much happens. The cobweb goes up to the crossing point and gets stuck there (Figure 3.2.31a). Thus, the population neither increases nor decreases, but rather remains the same (Figure 3.2.31b). The point P_0 is an example of an **equilibrium point,** which we now define.

Definition 3.2.1 A point m^* is called an equilibrium of the discrete-time dynamical system

$$m_{t+1} = f(m_t)$$

if $f(m^*) = m^*$.

This definition says that the discrete-time dynamical system leaves m^* unchanged. These points can be found graphically by looking for intersections of the graph of the updating function with the diagonal line.

When there is more than one equilibrium, they are called **equilibria.** The plant population has two equilibria, one at $P^* = 0$ and the other at $P^* = 500$ (see Figure 3.2.28a). The bird population has three equilibria, located at $B^* = 0$, $B^* = 260$, and $B^* = 940$ (see Figure 3.2.28b).

Why does the graphical method of finding equilibria work? The diagonal has the equation

$$m_{t+1} = m_t$$

and can be thought of as a discrete-time dynamical system that leaves *all* inputs unchanged, i.e., always returns an output equal to its input. The intersections of the graph of the updating function with the diagonal are the points that the updating function leaves unchanged. These are the equilibria.

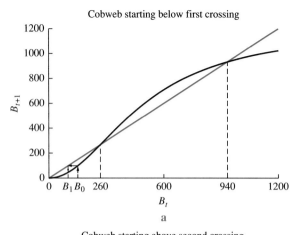
a

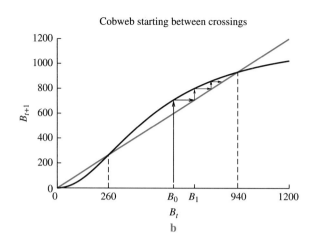
b

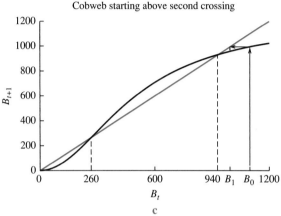
c

FIGURE 3.2.30
Behaviour of bird population starting from three different initial conditions

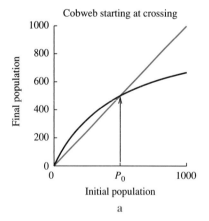
a

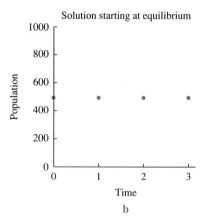
b

FIGURE 3.2.31
Behaviour of plant population starting from an equilibrium

Note that the equilibria exhibit different behaviour when we cobweb starting at *nearby* points. In Figure 3.2.29a we see that the solution starting near $P^* = 0$ *moves away* from it. However, the solutions that start near $P^* = 500$ (both to the left and to the right of it) *move toward* the equilibrium (look at Figures 3.2.29a and 3.2.29b and imagine that the starting values P_0 are closer to $P^* = 500$).

Definition 3.2.2 An equilibrium m^* is called **stable** if the solutions that start near m^* stay near or approach m^*. If the solutions that start near m^* move away from it, then m^* is an **unstable** equilibrium.

Figure 3.2.30a shows that the solution that starts at the point B_0 near the equilibrium $B^* = 0$ moves closer to it. Thus, $B^* = 0$ is a stable equilibrium. The same figure shows that the solution that starts near $B^* = 260$ and to the left of it moves away from it. Figure 3.2.30b suggests that the solution starting near $B^* = 260$ and to the right of it moves away from it. Thus, $B^* = 260$ is an unstable equilibrium. Finally, from Figure 3.2.30b and 3.2.30c we conclude that the equilibrium $B^* = 940$ is stable.

In Example 3.2.7 we investigate the case of a stable equilibrium for which a solution that starts near it remains there (but does not approach it). Further examples of stable and unstable equilibria appear in Section 3.4. We will discuss the concept of stability in more depth in Sections 6.7 and 6.8.

Equilibria: Algebraic Approach

In Example 3.1.9 we discovered an input, $M_0 = 2$, with a special property: the given dynamical system did not change its value, i.e., the output was equal to 2 as well. In other words, the value $M_0 = 2$ was left unchanged by the dynamical system (so it is an equilibrium point!). We now explore this phenomenon in detail.

First of all—how do we find equilibria algebraically? The answer is given in the following algorithm.

▶▶ **Algorithm 3.2.2** Solving for Equilibria of the Discrete-Time Dynamical System $m_{t+1} = f(m_t)$

1. Denoting the equilibrium point by m^*, write $m^* = f(m^*)$.

2. Solve the equation for m^*.

3. Think about the results. Do they make sense in the context of the given system?

Example 3.2.2 The Equilibrium of the Pain Medication Discrete-Time Dynamical System

Recall the discrete-time dynamical system for pain medication,

$$M_{t+1} = 0.5 M_t + 1$$

which we studied in Example 3.1.4. Let M^* stand for an equilibrium. The equation for equilibrium says that M^* is unchanged by the discrete-time dynamical system, or

$$M^* = 0.5 M^* + 1$$

We now solve this linear equation:

$$M^* = 0.5 M^* + 1$$
$$0.5 M^* = 1$$
$$M^* = \frac{1}{0.5} = 2$$

The equilibrium value is 2. We can check this by plugging $M_t = 2$ into the discrete-time dynamical system, finding that

$$M_{t+1} = 0.5 \cdot 2 + 1 = 2$$

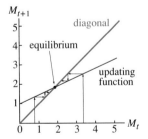

FIGURE 3.2.32
Equilibrium of the pain medication discrete-time dynamical system

Thus, the amount of 2 units of methadone is indeed unchanged over a course of days. Figure 3.2.32 shows that solutions that start near $M^* = 2$ move closer to it. Thus, $M^* = 2$ is a stable equilibrium.

Example 3.2.3 The Equilibrium of the Bacterial Discrete-Time Dynamical System

To find the equilibria for the bacterial population discrete-time dynamical system

$$b_{t+1} = 2 b_t$$

(Example 3.1.1), we write the equation for equilibria,

$$b^* = 2 b^*$$

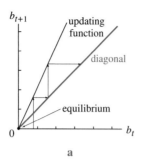
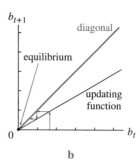

FIGURE 3.2.33

Equilibrium of the bacterial discrete-time dynamical system

from which we get $b^* = 0$. So there is one solution for equilibrium, $b^* = 0$, which makes sense: the population doubles every hour, and the only number that remains the same after doubling is zero. This conclusion is consistent with Figure 3.2.33a—the diagonal and the updating function intersect at the origin. Cobwebbing shows that the solutions that start near $b^* = 0$ move away from it (and so $b^* = 0$ is an unstable equilibrium). This means that even if there is a small number of bacteria to start with, the population will grow.

Now consider the dynamical system

$$b_{t+1} = 0.6b_t$$

From $b^* = 0.6b^*$ we compute the equilibrium $b^* = 0$. But this time, it is a stable equilibrium, as shown using cobwebbing in Figure 3.2.33b. Thus, a population that starts with a small number of individuals will decrease in size and become extinct. ▲

Example 3.2.4   A Discrete-Time Dynamical System with No Equilibrium

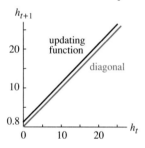

FIGURE 3.2.34

Discrete-time dynamical system for tree growth

The updating function for a growing tree (Example 3.1.2) following the discrete-time dynamical system

$$h_{t+1} = h_t + 0.8$$

has a graph that is parallel to the diagonal (Figure 3.2.34). To solve for the equilibria, we need to solve the equation

$$h^* = h^* + 0.8$$

Subtracting h^* from both sides, we get $0 = 0.8$; i.e., the given equation has no solution. The graph of the updating function and the graph of the diagonal do not intersect because they are parallel lines. Sounds logical—something that grows 0.8 m per year cannot remain unchanged. ▲

Example 3.2.5   Biologically Unrealistic Equilibrium

The graph of the updating function associated with a mite population (Example 3.1.3) that follows the discrete-time dynamical system

$$x_{t+1} = 2x_t + 30$$

lies above the diagonal for all values of x_t (Figure 3.2.35). To solve for the equilibria, we start with

$$x^* = 2x^* + 30$$

Solving for x^* gives $x^* = -30$. If we check by substituting $x_t = -30$ into the discrete-time dynamical system, we find

$$x_{t+1} = 2 \cdot (-30) + 30 = -30$$

which is indeed equal to x_t.

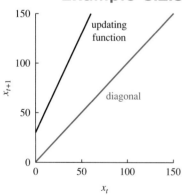

FIGURE 3.2.35

The discrete-time dynamical system for mites

Although there is a **mathematical** equilibrium, there is no **biological** equilibrium, since x_t cannot be negative. If we extend the graph to include biologically meaningless negative values, we see that the graph of the updating function does intersect the diagonal (Figure 3.2.36). ▲

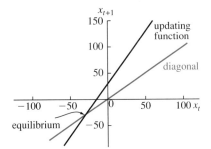

FIGURE 3.2.36
Extending the discrete-time dynamical system for mites to include a negative domain

Example 3.2.6 Population of Codfish in Coastal Regions of Eastern Canada

Among its many objectives and activities, Canada's Atlantic Zone Monitoring Program collects data on the population of fish. The station on the eastern coast of Newfoundland monitors the Southeast Grand Banks section of the Atlantic Ocean.

The population of fish in an ocean (or lake, or river) can vary widely as it reacts to changes in the number of predators, the availability of food, the water temperature, or pollution. When all of these factors stabilize, the fish population tends to stabilize as well. A simplified model for the population of codfish in one coastal subsection of the Southeast Grand Banks is given by

$$n_{t+1} = -0.6n_t + 5.3$$

where n_t is the number of codfish in millions and t is time.

Suppose that $n_0 = 1$. We calculate

$$n_1 = -0.6n_0 + 5.3 = -0.6(1) + 5.3 = 4.7$$
$$n_2 = -0.6n_1 + 5.3 = -0.6(4.7) + 5.3 = 2.48$$

and, similarly,

$$n_3 = 3.81$$
$$n_4 = 3.01$$
$$n_5 = 3.49$$
$$n_6 = 3.20$$
$$n_7 = 3.38$$

and so on.

As we can see, the model does simulate fluctuations: first, they are large (from 1 million to 4.7 million, then back to 2.48 million), but, with time, they tend to get smaller and smaller, and the cod population seems to be stabilizing.

Next, we solve for the equilibrium

$$n^* = -0.6n^* + 5.3$$
$$1.6n^* = 5.3$$
$$n^* = \frac{5.3}{1.6}$$
$$n^* \approx 3.31$$

Looking at the numerical solutions, it seems that the codfish population approaches an equilibrium value of 3.31 million. To confirm our hypothesis, and to show that n^* is a stable equilibrium, we sketch the cobweb diagram (Figure 3.2.37).

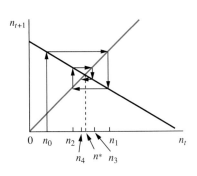

FIGURE 3.2.37
Cobwebbing codfish population

Example 3.2.7 — A System for Which Solutions That Start Near the Equilibrium Stay Near the Equilibrium

Consider the dynamical system $n_{t+1} = -n_t + 6$. From $n^* = -n^* + 6$ we find $2n^* = 6$, and so $n^* = 3$ is an equilibrium. If we start at $n_0 = 2.5$, then

$$n_1 = -n_0 + 6 = -2.5 + 6 = 3.5$$
$$n_2 = -n_1 + 6 = -3.5 + 6 = 2.5 = n_0$$
$$n_3 = -n_2 + 6 = -2.5 + 6 = 3.5 = n_1$$

and so on. Picking any other value near n^* will produce the same effect: the values oscillate back and forth across the equilibrium (see Figure 3.2.38). Thus, $n^* = 3$ is a stable equilibrium.

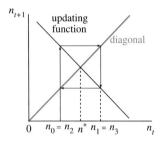

FIGURE 3.2.38
Cobwebbing the system $n_{t+1} = -n_t + 6$

Next, we consider two discrete-time dynamical systems that involve parameters.

Example 3.2.8 — Equilibria of the Pain Medication Model with a Dosage Parameter

Consider the medication discrete-time dynamical system with the parameter S,

$$M_{t+1} = 0.5 M_t + S$$

where S represents the daily dosage. Using the algorithm for finding equilibria, we get

$$M^* = 0.5 M^* + S$$
$$0.5 M^* = S$$
$$M^* = 2S$$

So, the equilibrium value is proportional to S, the daily dosage. Recall that in our original model (Example 3.2.2) S was equal to 1. The equilibrium was $M^* = 2$, which matches the above calculation.

Example 3.2.9 — Equilibria of the Pain Medication Model with Absorption

Consider the medication discrete-time dynamical system with parameter α,

$$M_{t+1} = (1 - \alpha) M_t + 1$$

where the parameter α represents the fraction of existing medication absorbed by the body during a given day. For example, if $\alpha = 0.1$, 10% of the medication is absorbed by the body and 90% remains. To find the equilibrium, we solve

$$M^* = (1 - \alpha) M^* + 1$$
$$M^* - (1 - \alpha) M^* = 1$$
$$\alpha M^* = 1$$
$$M^* = \frac{1}{\alpha}$$

The equilibrium value is inversely proportional to α and is therefore smaller when the fraction absorbed is larger. When $\alpha = 0.1$, i.e., when the body absorbs 10% of the medication each day, the equilibrium is

$$M^* = \frac{1}{0.1} = 10$$

In contrast, if the body absorbs 50% of the medication each day, leading to a larger value of $\alpha = 0.5$, then

$$M^* = \frac{1}{0.5} = 2$$

Thus, the body that absorbs more reaches a lower equilibrium.

Summary We have developed a graphical technique called **cobwebbing** to estimate the solutions of discrete-time dynamical systems. By examining the diagrams used for cobwebbing, we found that intersections of the graph of the updating function with the diagonal line play a special role. These **equilibria** are points that are unchanged by the discrete-time dynamical system. Using cobwebbing, we can decide whether an equilibrium is **stable** or **unstable**. In some cases we can solve for equilibria in general, without substituting numerical values for the parameters. Solving the equations in this way can help clarify the underlying biological processes.

3.2 Exercises

Mathematical Techniques

1–2 ■ The following steps are used to build a cobweb diagram. Follow them for the given discrete-time dynamical systems based on bacterial populations.

a. Graph the updating function.
b. Use your graph of the updating function to find the point (b_0, b_1).
c. Reflect it off the diagonal to find the point (b_1, b_1).
d. Use the graph of the updating function to find (b_1, b_2).
e. Reflect off the diagonal to find the point (b_2, b_2).
f. Use the graph of the updating function to find (b_2, b_3).
g. Sketch the solution as a function of time.

1. The discrete-time dynamical system $b_{t+1} = 2b_t$ with $b_0 = 1$.
2. The discrete-time dynamical system $n_{t+1} = 0.5n_t$ with $n_0 = 1$.

3–6 ■ Cobweb each discrete-time dynamical system for three steps, starting from the given initial condition. Find a formula for the solution and compare your geometric answer with the algebraic answer.

3. $v_{t+1} = 1.5v_t$, starting from $v_0 = 1220\ \mu\text{m}^3$.
4. $l_{t+1} = l_t - 1.7$, starting from $l_0 = 13.1$ cm.
5. $n_{t+1} = 0.5n_t$, starting from $n_0 = 1200$.
6. $M_{t+1} = 0.75M_t + 2$, starting from $M_0 = 16$ mg/L (to find the algebraic solution use either the strategy employed in Example 3.1.7 or the procedure explained in the text following the example).

7–12 ■ Graph the updating functions associated with the following discrete-time dynamical systems, and cobweb for four steps, starting from the given initial condition.

7. $x_{t+1} = 2x_t - 1$, starting from $x_0 = 2$.
8. $z_{t+1} = 0.9z_t + 1$, starting from $z_0 = 3$.
9. $w_{t+1} = -0.5w_t + 3$, starting from $w_0 = 0$.
10. $x_{t+1} = 4 - x_t$, starting from $x_0 = 1$.
11. $x_{t+1} = \dfrac{x_t}{1 + x_t}$, starting from $x_0 = 1$.
12. $x_{t+1} = \dfrac{x_t}{x_t - 1}$, starting from $x_0 = 3$. Graph for $x_t > 1$.

13–16 ■ Find the equilibria of each discrete-time dynamical system from the graph of its updating function. Label the coordinates of the equilibria. Determine whether each equilibrium is stable or unstable. If you are having difficulties cobwebbing for stability, explain why.

13.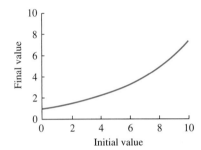

14.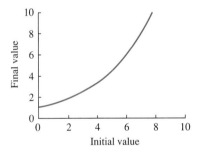

15.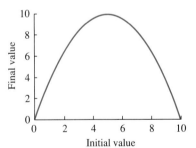

16.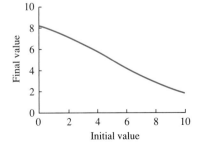

17–18 ■ Sketch graphs of the following updating functions over the given range, and mark the equilibria. Find the equilibria algebraically if possible.

17. $f(x) = x^2$ for $0 \leq x \leq 2$

18. $g(y) = y^2 - 1$ for $0 \leq y \leq 2$

19–22 ■ Graph each discrete-time dynamical system. Solve for the equilibria algebraically, and identify equilibria and the regions where the updating function lies above the diagonal on your graph.

19. $c_{t+1} = 0.5c_t + 8.0$, for $0 \leq c_t \leq 30$

20. $b_{t+1} = 3b_t$, for $0 \leq b_t \leq 10$

21. $b_{t+1} = 0.3b_t$, for $0 \leq b_t \leq 10$

22. $b_{t+1} = 2b_t - 5$, for $0 \leq b_t \leq 10$

23–30 ■ Find the equilibria of each discrete-time dynamical system. Use cobwebbing to check each equilibrium for stability.

23. $v_{t+1} = 1.5v_t$

24. $l_{t+1} = l_t - 1.7$

25. $x_{t+1} = 2x_t - 1$

26. $z_{t+1} = 0.9z_t + 1$

27. $w_{t+1} = -0.5w_t + 3$

28. $x_{t+1} = 4 - x_t$. Describe what happens when you cobweb starting at a point near the equilibrium.

29. $x_{t+1} = \dfrac{x_t}{1 + x_t}$. Use a graphing device to obtain an accurate plot of the updating function.

30. $x_{t+1} = \dfrac{x_t}{x_t - 1}$. Use a graphing device to obtain an accurate plot of the updating function.

31–34 ■ Find the equilibria of each discrete-time dynamical system with parameter. Identify values of the parameter for which there is no equilibrium, for which the equilibrium is negative, and for which there is more than one equilibrium.

31. $w_{t+1} = aw_t + 3$

32. $x_{t+1} = b - x_t$

33. $x_{t+1} = \dfrac{ax_t}{1 + x_t}$

34. $x_{t+1} = \dfrac{x_t}{x_t - K}$

Applications

35–40 ■ Cobweb the following discrete-time dynamical system for four steps, starting from the given initial condition.

35. An alternative tree growth discrete-time dynamical system with form $h_{t+1} = h_t + 5$ with initial condition $h_0 = 10$.

36. The mite population discrete-time dynamical system (Example 3.2.5) $x_{t+1} = 2x_t + 30$ with initial condition $x_0 = 0$.

37. The model for the cell volume, $v_{t+1} = 1.5v_t$, studied in Section 3.1, Exercise 42, starting from an initial volume of 1420.

38. The model for the fish mass, $m_{t+1} = m_t - 1.7$, studied in Section 3.1, Exercise 43, starting from an initial mass of 13.1.

39. The model for the gnat population size, $n_{t+1} = 0.5n_t$, studied in Section 3.1, Exercise 44, starting from an initial population of 800.

40. The model for the yield of soybeans, $y_{t+1} = 2y_t + 10$, studied in Section 3.1, Exercise 45, starting from an initial yield of 20.

41–42 ■ Reconsider the data describing the levels of a medication in the blood of two patients over the course of several days (measured in milligrams per litre), used in Section 3.1, Exercises 54 and 55.

Day	Medication Level in Patient 1 (mg/L)	Medication Level in Patient 2 (mg/L)
0	20	0
1	16	2
2	13	3.2
3	10.75	3.92

41. For the first patient, graph the updating function, and cobweb starting from the initial condition on day 0. Find the equilibrium.

42. For the second patient, graph the updating function, and cobweb starting from the initial condition on day 0. Find the equilibrium.

43–44 ■ Cobweb and find the equilibrium of each discrete-time dynamical system.

43. Consider a bacterial population that doubles every hour, but 10^6 individuals are removed after reproduction (Section 3.1, Exercise 58). Cobweb starting from $b_0 = 3 \times 10^6$ bacteria. Is the result consistent with the result of Exercise 58?

44. Consider a bacterial population that doubles every hour, but 10^6 individuals are removed before reproduction (Section 3.1, Exercise 59). Cobweb starting from $b_0 = 3 \times 10^6$ bacteria. Is the result consistent with the result of Exercise 59?

45–46 ■ Consider the following general models for bacterial populations with harvest.

45. Consider a bacterial population that doubles every hour, but h individuals are removed after reproduction. Find the equilibrium. Does it make sense?

46. Consider a bacterial population that increases by a factor of r every hour, but 1×10^6 individuals are removed after reproduction. Find the equilibrium. What values of r produce a positive equilibrium?

Computer Exercises

47. a. Use a computer (it may have a special feature for this) to find and graph the first 10 points on the solutions of each discrete-time dynamical system. The first two describe populations with reproduction and immigration

of 100 individuals per generation, and the last two describe populations that have 100 individuals harvested or removed each generation.

i. $b_{t+1} = 0.5b_t + 100$, starting from $b_0 = 100$.

ii. $b_{t+1} = 1.5b_t + 100$, starting from $b_0 = 100$.

iii. $b_{t+1} = 1.5b_t - 100$, starting from $b_0 = 201$.

iv. $b_{t+1} = 1.5b_t - 100$, starting from $b_0 = 199$.

b. What happens if you run the last one (part iv) for 15 steps? What is wrong with the model?

48. Compose the medication discrete-time dynamical system $M_{t+1} = 0.5M_t + 1$ with itself 10 times. Plot the resulting function. Use this composition to find the amount of medication after 10 days, starting from amounts of 1, 5, and 18 units. If the goal is to reach a stable concentration of 2 units, do you think this is a good therapy?

3.3 Modelling with Discrete-Time Dynamical Systems

In this section we investigate several discrete-time dynamical systems that describe the **consumption of drugs** (caffeine and alcohol) and model a variety of ways a **population** can change (unlimited growth, limited growth, decline in growth).

The common theme is the **per capita production,** which arises in various forms (e.g., constant functions, nonlinear functions). Algebraically, some models are fairly straightforward (consumption of caffeine, for instance), and some (alcohol consumption or limited population growth) are more involved. Although the models we discuss here do not attempt to capture all of the complexities of the situations they investigate, nevertheless they provide us with good and useful insights.

Absorption of Caffeine

Evidence suggests that, on average, our body eliminates caffeine at a constant rate of about 13% per hour. That means that one hour after quickly finishing a small Second Cup coffee (assuming our body was clean of caffeine before that, so that now it contains about 240 mg of caffeine), the amount of caffeine present in our body will be

$$240 \text{ mg} - 0.13(240 \text{ mg}) = 208.80 \text{ mg}$$

Of course, we could have said that 87% of the caffeine is still present in our body and calculated the amount as $0.87(240 \text{ mg}) = 208.80$ mg.

Let us build the model for the elimination and consumption of caffeine.

By c_t we denote the amount (in milligrams) of caffeine at time t (in hours). We need to find a formula for c_{t+1}, i.e., for the amount of caffeine one hour later. Due to elimination, the original amount of caffeine will decrease by 13%, i.e., will fall to

$$c_t - 0.13c_t = 0.87c_t$$

Assuming that at the end of the same time interval we consume d extra milligrams of caffeine, we get that

$$c_{t+1} = 0.87c_t + d$$

where $d \geq 0$. This dynamical system is an example of a substance absorption (elimination) and replacement (consumption) model. Of course, when $d = 0$, there is no replacement as no new caffeine is introduced into the body.

Example 3.3.1 Dynamics of Caffeine Absorption: Calculations

If we have two double espressos (at 200 mg of caffeine each) and a Red Bull (80 mg of caffeine) at 10 P.M., will we be able to fall asleep by midnight?

We consider the caffeine absorption and replacement dynamical system

$$c_{t+1} = 0.87c_t + d$$

where $c_0 = 200$ mg $+ 200$ mg $+ 80$ mg $= 480$ mg. Since there is no replacement (we assume that after the two espressos and the Red Bull we ingest no more caffeine), we set $d = 0$. Note that the initial condition assumes that there was no caffeine present in our body before we consumed the caffeine described here.

The solution of

$$c_{t+1} = 0.87 c_t$$

with initial condition $c_0 = 480$ mg is given by

$$c_t = 0.87^t \cdot 480$$

where t is time in hours measured from 10 P.M. We conclude that the amount of caffeine still present in our body at midnight will be

$$c_2 = 0.87^2 \cdot 480 \text{ mg} \approx 363.31 \text{ mg}$$

So there is still a fairly significant amount of caffeine left in our body.

How do we answer our question? Can we fall asleep with 363.31 mg of caffeine circulating through our body?

It is known that the same level of caffeine affects people in significantly different ways, so it is not possible to define a threshold, that is, to specify the amount of caffeine that, if reached or exceeded, would prevent one from falling asleep.

Let us try to provide a better answer. First, we calculate the half-life of caffeine. From

$$c_t = 0.87^t \cdot 480$$

we get

$$240 = 0.87^t \cdot 480$$
$$0.87^t = 0.5$$
$$\ln 0.87^t = \ln 0.5$$
$$t \ln 0.87 = \ln 0.5$$
$$t = \frac{\ln 0.5}{\ln 0.87}$$
$$t \approx 4.98$$

So the half-life of caffeine is about 5 h.

Reading advice on healthy sleeping from various sources, we find suggestions that it is a good idea to stop drinking coffee (ingesting caffeine) at around 2 P.M. What is a possible rationale behind this?

We just calculated that every 5 h the amount of caffeine in our body halves. So, after two half-life intervals—which end at midnight, the time we might want to go to bed—a quarter of the original amount of caffeine will be present. Thus, even a relatively high amount of caffeine taken by 2 P.M. will decay to a relatively small amount by midnight.

According to Health Canada guidelines, the recommended maximum daily caffeine intake for healthy adults is 400 mg. So if we have those two double espressos and the Red Bull at 10 P.M., the amount of caffeine present at midnight (363.31 mg) is quite close to the suggested maximum daily intake. So, the best answer to our question is—do not do it!

Elementary Population Models

In previous sections we studied the case where the bacterial population doubles every hour—that is, each bacterium divides once and both daughter bacteria survive.

In a more realistic scenario, we assume that only a fraction of the daughters survive. For example, if each offspring has a 75% chance of survival, then there will be, on average, 1.5 surviving offspring per parent. In this case, the dynamical system for the bacteria population can be written as

$$b_{t+1} = 1.5 b_t$$

Likewise, the system

$$b_{t+1} = 1.2 b_t$$

describes the situation where each of the two daughters has a 60% chance of survival, and thus each parent produces, on average, 1.2 bacteria.

Motivated by this, we consider the general system

$$b_{t+1} = r b_t$$

where r represents the number of new bacteria produced *per bacterium*. The constant r is called the **per capita production.**

This important equation of population biology states that the population at time $t+1$ is equal to the per capita production (the number of new bacteria per old bacterium) multiplied by the population at time t, or

new population = per capita production × old population

Recall that the solution of $b_{t+1} = r b_t$ is given by

population at time t = initial population × rate r to the power of t

i.e., $b_t = b_0 r^t$, where b_0 is the initial condition.

Example 3.3.2 Discrete-Time Dynamical System for Bacteria

Recall that we studied the case

$$b_{t+1} = 2 b_t$$

in detail in Example 3.1.5. This is the case where each daughter survives with the chance of $r/2 = 2/2 = 100\%$, i.e., each parent is replaced by two daughters, and so the population doubles each hour. For example, if $b_0 = 1000$, then $b_1 = 2000$ (Figure 3.3.39a).

Consider the case where $r = 1.5$ (recall that this means that each of the two offspring has an $r/2 = 1.5/2 = 75\%$ chance of survival, or, put differently, three of four offspring will survive). The discrete-time system is

$$b_{t+1} = 1.5 b_t$$

If $b_0 = 1000$, then $b_1 = 1500$, so the population increases by 50% (Figure 3.3.39b) in each generation.

The case when $r = 1$,

$$b_{t+1} = b_t$$

describes the population in which each daughter survives with a chance of $r/2 = 1/2 = 50\%$; stated differently, one in two offspring will survive. This means that the surviving offspring replaces the parent, and thus there is no change in the numbers. Indeed, if $b_0 = 1000$, then $b_1 = 1000$ as well (Figure 3.3.39c).

The system

$$b_{t+1} = 0.5 b_t$$

represents a population with high mortality: the chance of a daughter surviving is $r/2 = 0.5/2 = 25\%$; i.e., one in four offspring will survive. In this case, starting with $b_0 = 1000$ bacteria, the population will fall to $b_1 = 500$ bacteria in one hour. Thus, every hour, the population decreases by 50% (Figure 3.3.39d).

Numerical values for the four different dynamics (all with the same initial condition) are given in Table 3.3.1.

In the first two columns, $r > 1$ and the population increases every hour (at different rates). When $r = 1$, the population remains the same hour after hour. In the final

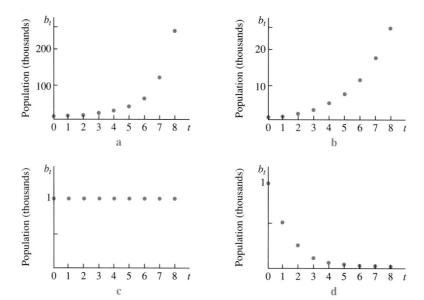

FIGURE 3.3.39

Growing, unchanged, and declining bacterial populations

Table 3.3.1

	Population (in Thousands)			
t	r = 2	r = 1.5	r = 1	r = 0.5
0	1	1	1	1
1	2	1.5	1	0.5
2	4	2.25	1	0.25
3	8	3.375	1	0.125
4	16	5.0625	1	0.0625
5	32	7.59375	1	0.03125
6	64	11.39063	1	0.01563
7	128	17.08594	1	0.00781
8	256	25.62891	1	0.00391

Table 3.3.2

Value of r	Behaviour of Population
r > 1	population increases
r = 1	population remains constant
r < 1	population decreases

column, when $r < 1$, the population decreases. We summarize these observations in Table 3.3.2.

Our model, $b_{t+1} = rb_t$, assumes that the per capita production is constant; i.e., it assumes that each bacterium produces r new bacteria (on average) *no matter what the total population is* (as shown in Figure 3.3.40).

Consider the case $r = 1.5$. Regardless of the current population of bacteria (b_t could be 100, or 5000, or 1,000,000), each bacterium will produce, on average, 1.5 offspring. Is this realistic?

Resources (space, food) are always limited, and, as the number of bacteria grows, so does competition for resources. This will, in turn, force the population to reduce its production of offspring. So if we wish to build a model for the population that reflects

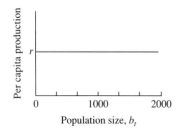

FIGURE 3.3.40

Constant per capita production

Table 3.3.3

Population, b_t	Per Capita Production, $p(b_t)$
10	1.980
100	1.818
500	1.333
1000	1.000
1500	0.800
2000	0.667

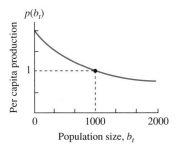

FIGURE 3.3.41

Decreasing per capita production

the real-life situation more faithfully, we have to make sure that it accounts for the reduced per capita production caused by increases in population. In other words, we need to replace the constant r with an appropriately chosen decreasing function.

Researchers in population dynamics use all kinds of functions for this purpose. We will discuss one of them in the context of logistic growth in Chapter 6. Here, we consider the function

$$p(b_t) = \frac{2}{1 + 0.001 b_t}$$

To get a better feel for p, we build a table of values (Table 3.3.3) and sketch its graph (Figure 3.3.41).

This is what we are looking for—larger populations have smaller per capita production. So, using

new population = per capita production × old population

we obtain

$$b_{t+1} = \left(\frac{2}{1 + 0.001 b_t}\right) b_t$$

where b_0, as usual, denotes the initial population. This dynamical system is an example of a model for **limited population.** Through a sequence of examples, we now investigate this model.

Example 3.3.3 Model for Limited Population: Sample Calculations

Assume that we start with $b_0 = 300$. Then

$$b_1 = \left(\frac{2}{1 + 0.001 b_0}\right) b_0 = \left(\frac{2}{1 + 0.001(300)}\right)(300) \approx 461.54$$

$$b_2 = \left(\frac{2}{1 + 0.001 b_1}\right) b_1 \approx \left(\frac{2}{1 + 0.001(461.54)}\right)(461.54) \approx 631.58$$

and so on:

$$b_3 \approx 774.19$$
$$b_4 \approx 872.73$$
$$b_5 \approx 932.04$$
$$b_6 \approx 964.82$$
$$b_7 \approx 982.10$$
$$b_8 \approx 990.97$$
$$b_9 \approx 995.46$$

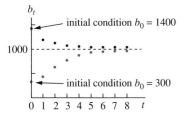

FIGURE 3.3.42

Limited population with two different initial conditions

Thus, the population increases, but—unlike in the case of exponential growth we studied before—there seems to be a limit to the growth (looking at the numbers, we see that this limit could be 1000; see Figure 3.3.42).

Chapter 3 Discrete-Time Dynamical Systems

Starting with initial condition $b_0 = 1400$, we get

$$b_1 = \left(\frac{2}{1+0.001b_0}\right) b_0 = \left(\frac{2}{1+0.001(1400)}\right)(1400) \approx 1166.67$$

$$b_2 = \left(\frac{2}{1+0.001b_1}\right) b_1 \approx \left(\frac{2}{1+0.001(1166.67)}\right)(1166.67) \approx 1076.92$$

and

$$b_3 \approx 1037.04$$
$$b_4 \approx 1018.18$$
$$b_5 \approx 1009.01$$
$$b_6 \approx 1004.48$$
$$b_7 \approx 1002.24$$

In this case the population decreases, and again, it seems to be approaching 1000; see Figure 3.3.42.

Example 3.3.4 Model for Limited Population: Equilibria

In order to calculate the equilibria for the discrete-time dynamical system

$$b_{t+1} = \left(\frac{2}{1+0.001b_t}\right) b_t$$

we solve the equation

$$b^* = \left(\frac{2}{1+0.001b^*}\right) b^*$$

for b^*. We bring both terms to the same side and factor:

$$b^* - \left(\frac{2}{1+0.001b^*}\right) b^* = 0$$

$$b^* \left(1 - \frac{2}{1+0.001b^*}\right) = 0$$

Thus, $b^* = 0$ or

$$1 - \frac{2}{1+0.001b^*} = 0$$

$$\frac{2}{1+0.001b^*} = 1$$

$$1 + 0.001b^* = 2$$

after cross-multiplying. It follows that

$$0.001b^* = 1$$

and so

$$b^* = \frac{1}{0.001} = 1000$$

We conclude that there are two equilibria: $b^* = 0$ represents extinction, and $b^* = 1000$ represents the population whose per capita production is

$$p(1000) = \frac{2}{1+0.001(1000)} = 1$$

In other words, each bacterium produces one new bacterium, so the population just breaks even (remains constant at 1000).

Example 3.3.5 Model for Limited Population: Cobwebbing

To fully understand our model, we investigate it geometrically. Using a graphing device, we produce the graph of the updating function

$$f(b_t) = \left(\frac{2}{1 + 0.001 b_t}\right) b_t$$

in Figure 3.3.43. Cobwebbing confirms our earlier conjectures: starting with $b_0 = 300$, the solution increases and approaches 1000 (Figure 3.3.43a). The solution with initial condition $b_0 = 1400$ decreases and approaches 1000 as well (Figure 3.3.43b). Note that $b_0 = 0$ is an unstable equilibrium, whereas $b_0 = 1000$ is stable.

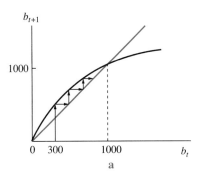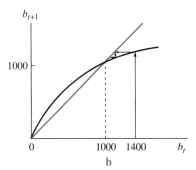

FIGURE 3.3.43

Cobwebbing the limited population model

Dynamics of Alcohol Use

Using our knowledge of dynamical systems, we now build a model for the consumption and elimination of alcohol, so that we can understand better how it works and how it affects people who consume it.

To construct our model, we need to understand how the human body processes alcohol. In the case of caffeine, the body tends to eliminate a *fixed proportion* (about 13% per hour), no matter what the amount. Thus, if we have 200 mg of caffeine in our body, 26 mg will be eliminated within an hour. If we have twice as much caffeine (400 mg), then $0.13(400 \text{ mg}) = 52$ mg (i.e., twice as much) will be eliminated over the next hour.

The elimination of alcohol is different: neither the amount that breaks down per hour nor the proportion is constant. It has been determined that the amount of alcohol that is broken down (eliminated) by the liver depends on the amount of alcohol present in the body. The larger the amount, the smaller the proportion of alcohol that is eliminated.

For the model and for subsequent calculations, we have to define a unit of alcohol. By a commonly accepted definition, **one drink** contains 14 g of alcohol, which is equivalent to 1.5 oz (44 mL) of 80-proof alcohol (vodka, rum, tequila, etc.), 5 oz (144 mL) of white wine, or 12 oz (355 mL) of beer.

By a_t we denote the amount of alcohol (in grams) at time t. Let $r(a_t)$ be the rate of elimination of alcohol (percent per hour) at the moment when the amount of alcohol in the body is a_t. Consider the function (derived from experimental data)

$$r(a_t) = \frac{10.1}{4.2 + a_t}$$

where $a_t \geq 5.9$ g. For values of a_t smaller than 5.9 g, the rate $r(a_t)$ is bigger than 1 (more than 100%) per hour, which does not make sense. Note that $r(a_t)$ is very similar to the per capita production function we studied in the limited growth population model.

To understand $r(a_t)$ better, we calculate a few values:

Amount of Alcohol, a_t, in Grams	Rate of Elimination, $r(a_t)$
6	0.9902
7	0.9012
14	0.5550
28	0.3137
42	0.2186
70	0.1361

Looking at the values, we see that we managed to do what we needed: smaller amounts of alcohol correspond to larger elimination rates, and vice versa. More precisely, if the body contains 14 g of alcohol (equivalent to one drink), it will be able to eliminate 55.5% of it within one hour (of course assuming that no new alcohol is introduced into the body). However, if someone has three rapid drinks (i.e., consumes 42 g of alcohol), then only 21.86% of the 42 g of alcohol will be eliminated within one hour.

Note that 55.5% of 14 g is 7.8 g; as well, 31.37% of 28 g is 8.8 g and 21.86% of 42 g is 9.2 g. So—as a ball-park, for a reasonable range of a_t—the net elimination is about 8 g to 9 g of alcohol per hour.

Now we are ready to build the model. If a_t is the current amount of alcohol in the body, what will a_{t+1} be?

The amount of alcohol eliminated is equal to

$$\text{amount of alcohol} \times \text{elimination rate}$$

i.e.,

$$a_t r(a_t) = a_t \left(\frac{10.1}{4.2 + a_t} \right) = \frac{10.1 a_t}{4.2 + a_t}$$

So the amount of alcohol still present in the body is

$$\text{amount at time } t - \text{eliminated amount}$$

i.e.,

$$a_t - \frac{10.1 a_t}{4.2 + a_t}$$

So, because a_{t+1} is

$$\text{amount still present} + \text{(possibly) new amount}$$

we get

$$a_{t+1} = a_t - \frac{10.1 a_t}{4.2 + a_t} + d$$

where d is the amount of alcohol consumed between times t and $t + 1$. (Actually, to be precise, d is the amount of alcohol consumed right at the end of the hour—since we did not incorporate it into the part that the body eliminates in the one-hour interval.)

This formula represents the dynamics of the elimination and consumption of alcohol. In the following examples we examine a few scenarios.

Example 3.3.6 Dynamics of Alcohol: Half a Drink per Hour

Assume that a student has two rapid drinks and then decides to consume half a drink every hour. What will the long-term effects be?

The initial condition (two rapid drinks) is $a_0 = 28$ g, and $d = 7$ g (or half a drink every hour). The corresponding discrete-time dynamical system is

$$a_{t+1} = a_t - \frac{10.1 a_t}{4.2 + a_t} + 7$$

where t is time in hours.

In a moment we will calculate a solution. But first, let us try to figure out a_1 without the above formula. The initial amount of alcohol in the body is 28 g. Looking at the table, we see that, with that amount of alcohol present, the elimination rate is 31.37% per hour. So, $0.3137(28) \approx 8.7836$ g of alcohol will be eliminated in an hour, and the amount still left in the body is $28 \text{ g} - 8.7836 \text{ g} \approx 19.2164$ g. Adding half a drink (7 g), we obtain that, at the end of the hour, the body contains 26.2164 g of alcohol.

To confirm this, and to calculate the solution, we use the model:

$$a_1 = a_0 - \frac{10.1 a_0}{4.2 + a_0} + 7 = 28 - \frac{10.1(28)}{4.2 + 28} + 7 \approx 26.2174$$

The difference between this value and the one we obtained earlier is due to round-off error. Similarly, we get

$$a_2 = a_1 - \frac{10.1 a_1}{4.2 + a_1} + 7 \approx 26.2174 - \frac{10.1(26.2174)}{4.2 + 26.2174} + 7 \approx 24.5120$$

and

$$a_3 \approx 22.8894$$
$$a_4 \approx 21.3553$$
$$a_5 \approx 19.9152$$
$$a_6 \approx 18.5743$$

So, the total amount of alcohol decreases, in spite of the fact that the student consumes half a drink every hour. How is this possible?

Since there is a relatively small amount of alcohol present, the body manages to eliminate all of the new alcohol (the half-drink every hour), as well as some of the alcohol left from the initial consumption.

Next, we calculate the equilibrium, and then, using cobwebbing, show that the above solution will approach it.

We start from

$$a^* = a^* - \frac{10.1 a^*}{4.2 + a^*} + 7$$

subtract a^* from both sides, and rearrange terms:

$$\frac{10.1 a^*}{4.2 + a^*} = 7$$

Now, cross-multiply and solve:

$$10.1 a^* = 29.4 + 7 a^*$$
$$3.1 a^* = 29.4$$
$$a^* = \frac{29.4}{3.1} \approx 9.5$$

Cobwebbing (see Figure 3.3.44) shows that the above solution decreases toward the equilibrium of approximately 9.5 g.

As well—making use of the cobweb diagram again—we see that if the student starts with a small amount of alcohol in the body (less than 9.5 g), or no alcohol at all, the routine of consuming half a drink every hour will increase the alcohol content, but only to the limit of 9.5 g. (Thus, $a^* \approx 9.5$ is a stable equilibrium.)

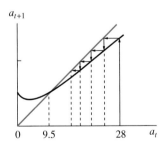

FIGURE 3.3.44
Cobwebbing alcohol consumption model (horizontal axis not to scale)

Example 3.3.7 Dynamics of Alcohol: One Drink per Hour

The system

$$a_{t+1} = a_t - \frac{10.1 a_t}{4.2 + a_t} + 14$$

with $a_0 = 14$ describes the scenario where a person starts with one drink, and then keeps consuming one drink per hour.

Let us try to find an equilibrium. We start with

$$a^* = a^* - \frac{10.1a^*}{4.2 + a^*} + 14$$

then subtract a^* from both sides and rearrange terms:

$$\frac{10.1a^*}{4.2 + a^*} = 14$$

Next, cross-multiply and solve:

$$10.1a^* = 58.8 + 14a^*$$
$$-3.9a^* = 58.8$$
$$a^* \approx -15.1$$

Although mathematically correct, the equilibrium of -15.1 does not make sense in the context of this application.

Let us calculate a solution: $a_0 = 14$ and so

$$a_1 = a_0 - \frac{10.1a_0}{4.2 + a_0} + 14 = 14 - \frac{10.1(14)}{4.2 + 14} + 14 \approx 20.2308$$

Similarly,

$$a_2 \approx 25.8671$$
$$a_3 \approx 31.1779$$
$$a_4 \approx 36.2770$$
$$a_5 \approx 41.2250$$

So the amount of alcohol will increase: after five hours, the total amount in the body will be almost equal to the consumption of three drinks (42 g).

Cobwebbing (see Figure 3.3.45) confirms that the consumption of one drink per hour will keep increasing the amount of alcohol in the body.

We have seen that continuous consumption of half a drink every hour will make the alcohol content in the body stabilize at a low amount (9.5 g). However, the one-drink-per-hour routine will keep increasing the alcohol level. ▲

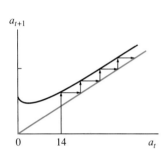

FIGURE 3.3.45

Cobwebbing alcohol consumption model

Example 3.3.8 Dynamics of Alcohol: Pure Elimination

Suppose that a person decides to stop drinking at the moment when the alcohol in their body reaches 60 g. How long will it take to eliminate almost all of the alcohol (so that there is less than 1 g left)?

The initial condition is $a_0 = 60$ g, and $d = 0$ g (since no new alcohol is consumed). The corresponding dynamical system is

$$a_{t+1} = a_t - \frac{10.1a_t}{4.2 + a_t}$$

Starting with $a_0 = 60$, we compute

$$a_1 = a_0 - \frac{10.1a_0}{4.2 + a_0} = 60 - \frac{10.1(60)}{4.2 + 60} \approx 50.56$$
$$a_2 \approx 41.23$$
$$a_3 \approx 32.07$$
$$a_4 \approx 23.14$$
$$a_5 \approx 14.59$$
$$a_6 \approx 6.75$$
$$a_7 \approx 0.52$$

So, after seven hours, there will be approximately 0.52 g of alcohol left unabsorbed in the body. Note that $a_8 = -0.59$, which does not make sense.

Summary In this section, we investigated several models: **unlimited population growth** and decay, **limited population growth,** and **dynamics of consumption of two drugs:** caffeine and alcohol. Although ideal and generalized in many ways, the models nevertheless provide us with useful insights into the real-life situations they describe.

3.3 Exercises

Mathematical Techniques

1–3 ▪ Find each population.

1. If $b_{t+1} = 1.3b_t$ and $b_0 = 12{,}000$, find b_8.
2. If $b_{t+1} = 1.01b_t$ and $b_0 = 100$, find b_{500}.
3. If $b_{t+1} = 0.99b_t$ and $b_0 = 100$, find b_{500}.

4–5 ▪ Assume that the population changes according to $b_{t+1} = rb_t$. Find the per capita production, r, and the initial population, b_0, given the following measurements.

4. $b_1 = 6$ and $b_3 = 600$.
5. $b_2 = 2000$ and $b_3 = 3000$.

6–8 ▪ Find the per capita production for an exponentially changing population.

6. The population increased from 100 to 400 in 4 months (in units of new members per member per month).
7. The population increased from 1200 to 400,000 in 25 years (in units of new members per member per year).
8. The population decreased from one million to one thousand in one year (in units of new members per member per year).
9. Consider a population that changes according to $p_{t+1} = rp_t(1 - p_t)$, where $r > 1$ is a constant. Identify the per capita production, and describe its graph in words. Find the equilibrium population.
10. Population A increased from 125,000 to 313,000 in two months. Population B increased from 2500 to 9000 in three months. Assuming that both populations change exponentially, which one has the larger per capita production?

11–14 ▪ Consider the model for limited population,
$$b_{t+1} = \frac{12}{1 + 0.001 b_t} b_t$$

11. Let $b_0 = 1000$. Using a calculator or a computer, find b_1, b_2, b_3, b_4, and b_5. What value do these numbers approach?
12. Let $b_0 = 20{,}000$. Using a calculator or a computer, find b_1, b_2, b_3, b_4, and b_5. What value do these numbers approach?
13. Calculate the equilibria and explain their meaning.
14. Using cobwebbing, show that the population that starts at $b_0 = 12{,}000$ will decrease toward one of the equilibrium points calculated in Exercise 13.

15–18 ▪ Consider the alcohol consumption model
$$a_{t+1} = a_t - \frac{10.1 a_t}{4.2 + a_t} + d$$
where d is the constant amount of alcohol that is consumed every hour.

15. If $a_0 = 42$ g and $d = 7$ g, find a_5 and give an interpretation.
16. If $a_0 = 0$ g and $d = 28$ g, find a_5 and give an interpretation.
17. If $a_0 = 28$ g and $d = 14$ g, find a_5 and give an interpretation.
18. If there are 50 g of alcohol in the body and no new alcohol is consumed, how long will it take for the alcohol level to fall below 1 g?
19. Find all equilibria of the population whose per capita production is given by $\frac{rp_t}{1 + p_t^2}$, where $r > 0$.

Applications

20–23 ▪ Find the solution of each discrete-time dynamical system, express it in exponential notation, and solve for when the given time reaches the given target. Sketch the graph of the solution.

20. A population follows the discrete-time dynamical system $b_{t+1} = rb_t$ with $r = 1.5$ and $b_0 = 10^6$. When will the population reach 10^7?
21. A population follows the discrete-time dynamical system $b_{t+1} = rb_t$ with $r = 1.2$ and $b_0 = 10^6$. When will the population reach 10^7?
22. A population follows the discrete-time dynamical system $b_{t+1} = rb_t$ with $r = 1.01$ and $b_0 = 10^6$. When will the population reach 10^7?
23. A population follows the discrete-time dynamical system $b_{t+1} = rb_t$ with $r = 0.9$ and $b_0 = 10{,}000$. When will the population reach 1,000?
24. Consider the dynamical system (Beverton-Holt recruitment curve)
$$n_{t+1} = \frac{rn_t}{1 + (r-1)n_t/k}$$
where $k > 0, r > 1$, and n_t represents some population.

a. Find all meaningful equilibria.

b. Let $k = 90$ and $r = 10$, and assume that $n_0 = 30$. Compute the first four values of n_t and plot them on a graph. Next, assume that $n_0 = 120$. Compute the first four values of n_t and plot them on the same graph. Based on your numerical experiment, give a possible interpretation of the constant k.

25. We modify the dynamical system $b_{t+1} = rb_t$ by replacing the constant per capita production rate with a decreasing function different from the one we studied in Examples 3.3.3–3.3.5. To be specific, we assume that the ratio b_{t+1}/b_t starts (i.e., when $t = 0$) at some value $r > 1$ and decreases as the reciprocal of a linear function until it reaches 1 when $t = N$.

    a. Write (for convenience, consider the reciprocal)
    $$\frac{b_t}{b_{t+1}} = \alpha b_t + \beta$$
    Using the fact that $b_0/b_1 = 1/r$ and $b_N/b_{N+1} = 1$, find α and β.

    b. Let $K = b_N = b_{N+1}$. Show that
    $$b_{t+1} = \frac{r(K - b_0)b_t}{K - rb_0 + (r - 1)b_t}$$
    This model is one version of the Beverton-Holt discrete-time population model.

26. Another variant of the Beverton-Holt population model is given by
    $$b_{t+1} = \frac{rb_t}{1 + \frac{1}{M}b_t}$$
    where $r > 1$ and M is a positive constant.

    a. Show that the solution of the system is given by
    $$b_t = \frac{(r-1)Mb_0}{b_0 + ((r-1)M - b_0)r^{-t}}$$

    b. We have not formally introduced limits. However, try to justify the fact that, as t increases, b_t approaches $(r-1)M$. The quantity $(r-1)M$ is called the *carrying capacity* of the environment.

27–29 ■ If the number of individuals falls below a certain critical number, m, the population faces extinction (since it is no longer capable of avoiding the damaging effects of inbreeding and is unable to cope with changes in the environment). Calculate the time when the following populations will fall to the level that will threaten their survival. Assume that the per capita production rate is constant in each case.

27. The population of black rhinoceroses in Africa was 2500 in 1993 and 2410 in 2004. It has been estimated that $m = 500$.

28. In 1990, there were about 5000 southern mountain caribou in British Columbia. In 2009, only about 1900 remained. The critical number is $m = 500$.

29. In 2005, there were about 25,000 Beluga whales in Western Hudson Bay. (This estimate is based on aerial surveys conducted by Fisheries and Oceans Canada. The surveys could not detect submerged whales, so the actual number of whales could be significantly higher.) Facing losses due to hunting and pollution, the population is believed to be decreasing, although its exact size is hard to estimate. Suppose that a more recent estimate (year 2010) is 24,000 whales. Take the minimum survival to be $m = 200$.

30–33 ■ Assume that the dynamics of caffeine absorption is given by $c_{t+1} = 0.87c_t + d$, where t is time in hours and d is the amount of caffeine taken every hour.

30. If the initial amount of caffeine is $c_0 = 1000$ mg (equivalent to five to six double espressos!) and no new caffeine is consumed, estimate the time needed for 90% of the caffeine to be eliminated from the body (i.e., 10% left).

31. If the initial amount of caffeine is $c_0 = 600$ mg and no new caffeine is consumed, estimate the time needed for 50% of the caffeine to be eliminated from the body.

32. If the initial amount of caffeine is $c_0 = 40$ mg and every hour we consume a small coffee (60 mg of caffeine), will the total amount of caffeine in our body increase or decrease? Find the equilibrium amount of caffeine.

33. Let $c_0 = 100$ mg and $d = 120$ mg. Find the equilibrium. How long will it take for the caffeine to reach 90% of the equilibrium?

34. *Consider* the alcohol consumption model
    $$a_{t+1} = a_t - \frac{10.1a_t}{4.2 + a_t} + d$$
    where d is the constant amount that is consumed every hour. For which values of d is there an equilibrium? For which values is there no equilibrium?

Computer Exercises

35. Population A follows the dynamics $a_{t+1} = 1.3a_t$ with $a_0 = 100$. Population B can be described by $b_{t+1} = 1.1b_t$ with $b_0 = 1000$.

    a. Which population is larger when $t = 10$?
    b. Which population reaches 3000 first?
    c. When does population A become larger than population B?
    d. When does population A reach 1 million?
    e. What is the size of population A when population B reaches 1 million?

36. Consider the model for limited population,
    $$b_{t+1} = \frac{10.4}{1 + 0.0005b_t}b_t$$
    where t is time in years.

    a. What is the positive equilibrium value?
    b. Let $b_0 = 2000$. How many years will it take for the population to reach 95% of the equilibrium value?
    c. Let $b_0 = 2000$. How many years will it take for the population to come within 1% of the equilibrium value?
    d. Let $b_0 = 100,000$. Comment on the rate at which the population will approach the equilibrium. In reality, what could possibly cause this kind of change?

3.4 Nonlinear Dynamics Model of Selection

Some discrete-time dynamical systems we have studied so far (bacterial populations, caffeine, tree height, and mite populations) are said to be **linear** because the updating function is linear. We now derive a model of two competing bacterial populations that leads naturally to a discrete-time dynamical system that is not linear. **Nonlinear dynamical systems** can have much more complicated behaviour than linear systems, as we witnessed working with the limited growth model. For example, they may have more than one equilibrium. By comparing the two equilibria in this model of selection, we will catch another glimpse of an important theme of this book, the **stability** of equilibria.

A Model of Selection

Our original model of bacterial growth followed the population of a single type of bacterium, denoted by b_t at time t. Suppose that a mutant type with population m_t appears and begins competing. If the original type (or **wild type**) has a per capita production of 1.5 and the mutant type has a per capita production of 2 (Figure 3.4.46), the two populations will follow the discrete-time dynamical systems

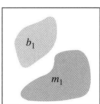
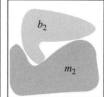

FIGURE 3.4.46
An invasion by mutant bacteria

$$b_{t+1} = 1.5b_t \quad \text{discrete-time dynamical system for wild type}$$
$$m_{t+1} = 2m_t \quad \text{discrete-time dynamical system for mutants}$$
(3.4.1)

The per capita production of the mutant type is greater than that of the wild type, perhaps because it is better able to survive. Over time, we would expect the population to include a larger and larger proportion of mutant bacteria. The establishment of this mutant is an example of **selection**. Selection occurs when the frequency of a genetic type changes over time.

Imagine observing this mixed population for many hours. Counting all of the bacteria each hour would be impossible. Nonetheless, we could track the mutant invasion by taking a sample and measuring the fraction of the mutant type by counting or using a specific stain. If this fraction became larger and larger, we would know that the mutant type was taking over.

How can we model the dynamics of the fraction? The vital first step is to define a new variable. In this case, we set p_t to be the fraction (or the percent) of mutants at time t. Then

$$\begin{aligned} p_t &= \frac{\text{number of mutants}}{\text{total number of bacteria}} \\ &= \frac{\text{number of mutants}}{\text{number of mutants} + \text{number of wild type}} \\ &= \frac{m_t}{m_t + b_t} \end{aligned}$$
(3.4.2)

In other words, we think of p_t as the percent of mutants relative to the total population of bacteria.

What is the fraction (percent) of the wild type? It is the number of wild type divided by the total number of bacteria, or

$$\text{fraction (percent) of wild type} = \frac{\text{number of wild type}}{\text{total number of bacteria}}$$

$$= \frac{\text{number of wild type}}{\text{number of mutants} + \text{number of wild type}}$$

$$= \frac{b_t}{m_t + b_t}$$

Note that

$$\text{fraction of mutants} + \text{fraction of wild type} = \frac{m_t}{m_t + b_t} + \frac{b_t}{m_t + b_t} = 1$$

as it should—because all bacteria are of one of these two types, the fractions must add up to 1 (i.e., 100%). Thus,

$$\text{fraction of mutants} = p_t = \frac{m_t}{m_t + b_t}$$

$$\text{fraction of wild type} = 1 - p_t = \frac{b_t}{m_t + b_t} \quad (3.4.3)$$

Example 3.4.1 Finding the Fractions of Mutants and Wild Type

Assume that $b_t = 3 \cdot 10^6$ and $m_t = 2 \cdot 10^5$. The total number of bacteria is $b_t + m_t = 3.2 \cdot 10^6$, and the fraction of the mutant type is

$$p_t = \frac{m_t}{m_t + b_t} = \frac{2 \cdot 10^5}{3.2 \cdot 10^6} = 0.0625.$$

Thus, mutant bacteria represent 6.25% of the total population. The fraction of the wild type is

$$\frac{b_t}{m_t + b_t} = \frac{3 \cdot 10^6}{3.2 \cdot 10^6} = 0.9375.$$

We conclude that the wild bacteria represent 93.75% of the total population of bacteria. Clearly,

$$\text{fraction of mutants} + \text{fraction of wild type} = 0.0625 + 0.9375 = 1$$

Our goal is to express p_{t+1} in terms of p_t, i.e., to find a formula that relates the fraction of the mutant type one hour later to the present fraction of the mutant type. We consider an example first.

Example 3.4.2 Finding an Updated Fraction

Recall that, with $b_t = 3 \cdot 10^6$ and $m_t = 2 \cdot 10^5$ as in Example 3.4.1, the fraction of mutants is $p_t = 0.0625$.

The updated populations are

$$m_{t+1} = 2m_t = 4 \cdot 10^5$$
$$b_{t+1} = 1.5 b_t = 4.5 \cdot 10^6$$

The updated fraction of the mutant type, p_{t+1}, is

$$p_{t+1} = \frac{4 \cdot 10^5}{4 \cdot 10^5 + 4.5 \cdot 10^6} \approx 0.0816$$

As expected, the fraction has increased. We might expect that the fraction of mutants would increase by a factor equal to the ratio $\frac{2}{1.5} \approx 1.333$ of the per capita productions of the two types. In fact,

$$\frac{p_{t+1}}{p_t} = \frac{0.0816}{0.0625} \approx 1.3056$$

which is slightly less. We will soon see why the mutant increases more slowly than we might at first expect.

We follow these same steps to find the discrete-time dynamical system for p_t. By definition, the fraction of mutants is

$$p_{t+1} = \frac{m_{t+1}}{m_{t+1} + b_{t+1}}$$

Using the discrete-time dynamical systems for the two types (Equation 3.4.1), we find

$$p_{t+1} = \frac{2m_t}{2m_t + 1.5b_t} \tag{3.4.4}$$

Although mathematically correct, this is not a satisfactory discrete-time dynamical system. We have supposed that the actual values of m_t and b_t are impossible to measure. The discrete-time dynamical system must give the new fraction, p_{t+1}, in terms of the old fraction, p_t, which we can measure by sampling.

We can do this by using an algebraic trick: dividing the numerator and the denominator by the same thing. Because the definition of p_t has the total population $m_t + b_t$ in the denominator, we divide it into the numerator and denominator, finding

$$p_{t+1} = \frac{2\frac{m_t}{m_t + b_t}}{2\frac{m_t}{m_t + b_t} + 1.5\frac{b_t}{m_t + b_t}}$$

We can simplify by substituting

$$p_t = \frac{m_t}{m_t + b_t}$$

(Equation 3.4.2) and

$$1 - p_t = \frac{b_t}{m_t + b_t}$$

(Equation 3.4.3), thus finding

$$p_{t+1} = \frac{2p_t}{2p_t + 1.5(1 - p_t)} \tag{3.4.5}$$

This is the discrete-time dynamical system we sought, giving a formula for the fraction at time $t + 1$ in terms of the fraction at time t.

Example 3.4.3 Using the Discrete-Time Dynamical System to Find the Updated Fraction

If $p_t = 0.0625$, as in Example 3.4.1, the discrete-time dynamical system tells us that

$$p_{t+1} = \frac{2 \cdot 0.0625}{2 \cdot 0.0625 + 1.5(1 - 0.0625)} \approx 0.0816$$

This matches the answer we found before but is based only on *measurable quantities*.

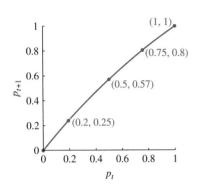

FIGURE 3.4.47

Graph of updating function from the selection model

This calculation illustrates one of the great strengths of mathematical modelling. Our *derivation* of this measurable discrete-time dynamical system used the values m_t and b_t, which are impossible to measure. But just because things cannot be measured in practice does not mean they cannot be measured in principle. These values do exist, and they can be worked with mathematically. One can think of mathematical models as a way to see the invisible.

The discrete-time dynamical system for the fraction (Equation 3.4.5) is not linear because it involves division. The graph of the function is curved (Figure 3.4.47). We drew it by plotting points: taking the values for the fraction p_t (which must lie between 0 and 1), we obtained the points (0, 0), (0.2, 0.25), (0.5, 0.57), (0.75, 0.8), and (1, 1) on the graph.

To find the equilibria, we simplify Equation 3.4.5 first:

$$p_{t+1} = \frac{2p_t}{2p_t + 1.5(1 - p_t)} = \frac{2p_t}{2p_t + 1.5 - 1.5p_t} = \frac{2p_t}{0.5p_t + 1.5}$$

Now from

$$p^* = \frac{2p^*}{0.5p^* + 1.5}$$

we get

$$p^* - \frac{2p^*}{0.5p^* + 1.5} = 0$$

$$p^*\left(1 - \frac{2}{0.5p^* + 1.5}\right) = 0$$

Therefore, $p^* = 0$ or

$$\frac{2}{0.5p^* + 1.5} = 1$$

$$0.5p^* + 1.5 = 2$$

$$0.5p^* = 0.5$$

and so $p^* = 1$. These equilibria correspond to extinction of the mutant (at $p^* = 0$) and extinction of the wild type (at $p^* = 1$).

The Discrete-Time Dynamical System and Equilibria

We can gain a better understanding of this process by studying the general case. Suppose that the mutant type has per capita production s and the wild type has per capita production r (Figure 3.4.48). The populations follow

$$\begin{aligned} m_{t+1} &= sm_t \\ b_{t+1} &= rb_t \end{aligned} \tag{3.4.6}$$

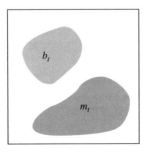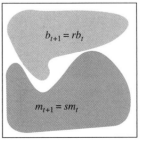

FIGURE 3.4.48

The general case

We can follow the steps above to derive the discrete-time dynamical system for the fraction:

$$p_{t+1} = \frac{m_{t+1}}{m_{t+1} + b_{t+1}}$$

$$= \frac{sm_t}{sm_t + rb_t}$$

$$= \frac{s\frac{m_t}{m_t + b_t}}{s\frac{m_t}{m_t + b_t} + r\frac{b_t}{m_t + b_t}}$$

$$= \frac{sp_t}{sp_t + r(1 - p_t)}$$

This gives the general form

$$p_{t+1} = \frac{sp_t}{sp_t + r(1 - p_t)} \tag{3.4.7}$$

Example 3.4.4 Substituting Parameters into the General Discrete-Time Dynamical System

The derivation in the previous subsection considered the case $s = 2$ and $r = 1.5$. Substituting these parameter values into the general form for bacterial selection gives

$$p_{t+1} = \frac{2p_t}{2p_t + 1.5(1 - p_t)}$$

matching what we found before.

The updating function for the general model (3.4.7) is given by

$$f(p_t) = \frac{sp_t}{sp_t + r(1 - p_t)} = \frac{sp_t}{r + (s - r)p_t}$$

Consider the case $s > r$ (i.e., the mutants have higher per capita production than the wild type). Note that $f(0) = 0$ and $f(1) = 1$; see Figure 3.4.49a. Because (keep in mind that $p_t \leq 1$)

$$r + (s - r)p_t \leq r + (s - r) = s$$

we conclude that

$$f(p_t) = \frac{sp_t}{r + (s - r)p_t} \geq \frac{sp_t}{s} = p_t$$

Consequently, the graph of the updating function lies above the diagonal except at the intersection points, $p_t = 0$ and $p_t = 1$, as shown in Figure 3.4.49 (the plot was obtained using a graphing device). This means that any value of p_t between 0 and 1 will be increased by the discrete-time dynamical system, consistent with the higher per capita production of the mutants. The cobwebbing moves up, indicating this increase (Figure 3.4.49b).

Note What we just said is true because the updating function has positive slope. If the updating function lies above the diagonal but has negative slope, then the solution is not increasing (see Example 3.2.6).

What happens if the per capita production of the wild type exceeds that of the mutants? With $r = 2.1$ and $s = 1.7$, the dynamical system is

$$p_{t+1} = \frac{1.7p_t}{1.7p_t + 2.1(1 - p_t)}$$

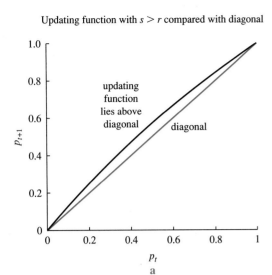
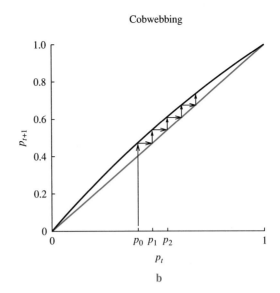

FIGURE 3.4.49

Dynamics when mutants reproduce more quickly

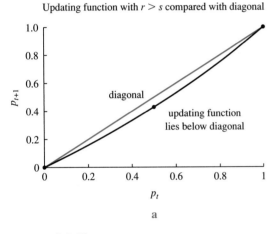
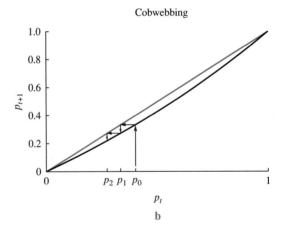

FIGURE 3.4.50

Dynamics when the wild type reproduces more quickly

The points $(0, 0)$, $(0.5, 0.45)$, and $(1, 1)$ lie on the graph, which itself lies below the diagonal (Figure 3.4.50a). Values of p_t between 0 and 1 are decreased by the discrete-time dynamical system, as shown by the decreasing cobweb (Figure 3.4.50b). This is consistent with the lower production of the mutants.

Finally, what happens if the two types have equal per capita production? If $r = s$, the discrete-time dynamical system simplifies to

$$p_{t+1} = \frac{sp_t}{sp_t + s(1 - p_t)}$$
$$= \frac{sp_t}{sp_t + s - sp_t}$$
$$= \frac{sp_t}{s}$$
$$= p_t$$

In this case, the discrete-time dynamical system leaves all values unchanged. When both types reproduce equally well, the fraction of the mutant neither increases nor decreases, and every value of p_t is an equilibrium. This makes biological sense; there is no selection in this case. This does not say that the *total number* of bacteria is unchanged; if $r = s = 2$, the total number will double each hour. The *fraction* of mutants remains the same.

We now solve the general system

$$p_{t+1} = \frac{sp_t}{sp_t + r(1 - p_t)}$$

for the equilibria. Start from

$$p^* = \frac{sp^*}{sp^* + r(1 - p^*)}$$

move all terms to the left, expand, and factor:

$$p^* - \frac{sp^*}{sp^* + r(1 - p^*)} = 0$$

$$p^*\left(1 - \frac{s}{sp^* + r(1 - p^*)}\right) = 0$$

Thus, $p^* = 0$ or

$$\frac{s}{sp^* + r(1 - p^*)} = 1$$

$$sp^* + r(1 - p^*) = s$$

Again, move all terms to the left and factor:

$$sp^* + r(1 - p^*) - s = 0$$

$$sp^* + r - rp^* - s = 0$$

$$p^*(s - r) - (s - r) = 0$$

$$(s - r)(p^* - 1) = 0$$

We conclude that either $s = r$ or $p^* = 1$.

What do these three equilibria mean? If $s = r$, the discrete-time dynamical system leaves all values unchanged, and every value of p_t is an equilibrium. Otherwise, the equilibria are $p^* = 0$ and $p^* = 1$. When $p^* = 0$, the population consists entirely of the wild type. Because our model includes no mutation or immigration, there is nowhere for the mutant type to arise. Similarly, when $p^* = 1$ the population consists entirely of the mutant type, and the wild type will never arise. These equilibria correspond to the extinction equilibrium for a population of one type of bacteria: at $p^* = 0$ the mutants are extinct, and at $p^* = 1$ the wild type are extinct.

Figure 3.4.51 shows many steps of cobwebbing with $s = 2$ and $r = 1.5$, starting near the equilibrium $p^* = 0$. The solution moves slowly away from 0, moves swiftly through the halfway point at $p_t = 0.5$, and then slowly approaches the other equilibrium at $p_t = 1$.

If we started *exactly* at $p_0 = 0$, the solution would remain at $p_t = 0$ for all times t. Similarly, if we started *exactly* at $p_0 = 1$, the solution would remain at $p_t = 1$ for all times t. The two equilibria behave quite differently, however, if our starting point is nearby. A solution starting *near* $p^* = 0$ moves steadily *away from* the equilibrium (Figure 3.4.51). A solution starting *near* $p^* = 1$ moves *toward* the equilibrium (Figure 3.4.52). Thus, $p^* = 0$ is an unstable equilibrium and $p^* = 1$ is a stable equilibrium.

We will derive powerful methods to analyze discrete-time dynamical systems and determine whether their equilibria are stable or unstable. Because these techniques require the **derivative,** a central idea from calculus, we must first study the foundational notions of limits and rate of change.

Summary   As an example of a **nonlinear dynamical system,** a discrete-time dynamical system with a curved graph, we derived the equation for the fraction of mutants invading a

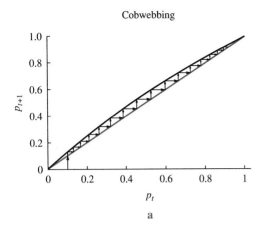
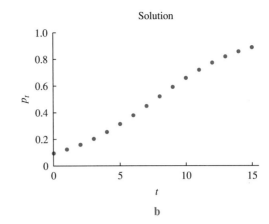

FIGURE 3.4.51
Solution of the selection model starting near $p_0 = 0$

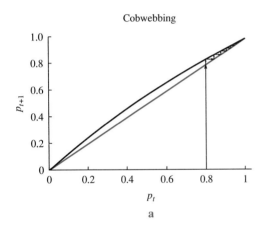

FIGURE 3.4.52
Solution of the selection model starting near $p_0 = 1$

population of wild type bacteria. This dynamical system, unlike the linear ones studied hitherto, has two equilibria. One of these equilibria is **unstable**; solutions starting nearby move farther and farther away. The other is **stable**; solutions starting nearby move closer and closer to the equilibrium.

3.4 Exercises

Mathematical Techniques

▼1–4 ▪ A population consists of 200 red birds and 800 blue birds. Find the fraction of red birds and blue birds after the following. Check that the fractions add up to 1.

1. The population of red birds doubles and the population of blue birds remains the same.

2. The population of blue birds doubles and the population of red birds remains the same.

3. The population of red birds is multiplied by a factor of r and the population of blue birds remains the same.

4. The population of blue birds is multiplied by a factor of s and the population of red birds remains the same.

▼5–6 ▪ Sketch graphs of the following functions.

5. $f(x) = \dfrac{x}{x+1}$ for $0 \leq x \leq 2$.

6. $g(x) = \dfrac{3x}{2x+1}$ for $0 \leq x \leq 2$.

▼7–10 ▪ Using the discrete-time dynamical system and the derivation of Equation 3.4.7, find p_t, m_{t+1}, b_{t+1}, and p_{t+1} in each situation.

7. $s = 1.2$, $r = 2$, $m_t = 1.2 \cdot 10^5$, $b_t = 3.5 \cdot 10^6$

8. $s = 1.2$, $r = 2$, $m_t = 1.2 \cdot 10^5$, $b_t = 1.5 \cdot 10^6$

9. $s = 0.3$, $r = 0.5$, $m_t = 1.2 \cdot 10^5$, $b_t = 3.5 \cdot 10^6$

10. $s = 1.8$, $r = 1.8$, $m_t = 1.2 \cdot 10^5$, $b_t = 3.5 \cdot 10^6$

11–12 ■ Solve for the equilibria of each discrete-time dynamical system.

11. $p_{t+1} = \dfrac{p_t}{p_t + 2(1 - p_t)}$

12. $p_{t+1} = \dfrac{4p_t}{4p_t + 0.5(1 - p_t)}$

13–14 ■ Find all non-negative equilibria of the following mathematically elegant discrete-time dynamical systems.

13. $x_{t+1} = \dfrac{x_t}{1 + ax_t}$, where a is a positive parameter. What happens to this system if $a = 0$?

14. $x_{t+1} = \dfrac{x_t}{a + x_t}$, where a is a positive parameter. What happens to this system if $a = 0$?

15–18 ■ Identify stable and unstable equilibria on the following graphs of updating functions.

15.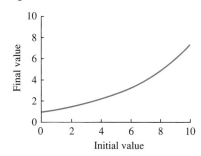

16.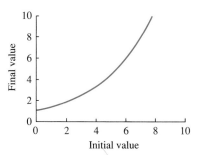

17.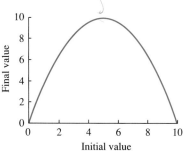

18.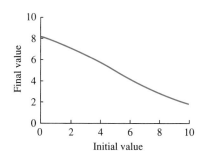

Applications

19–22 ■ Find and graph the updating function for each case of the selection model (Equation 3.4.7). Cobweb starting from $p_0 = 0.1$ and $p_0 = 0.9$. Which equilibria seem to be stable in each case?

19. $s = 1.2, r = 2$

20. $s = 1.8, r = 0.8$

21. $s = 0.3, r = 0.5$. Compare with the result of Exercise 19.

22. $s = 1.8, r = 1.8$

23–24 ■ For each discrete-time dynamical system, indicate which of the equilibria are stable and which are unstable.

23.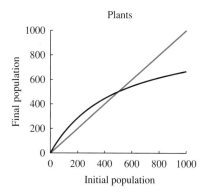

24.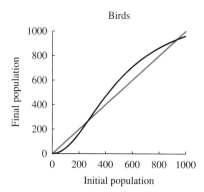

25–28 ■ This section ignores the important evolutionary force of mutation. This series of problems builds models that consider mutation without reproduction. Suppose that 20% of wild type bacteria transform into mutants and that 10% of mutants transform back into wild type (revert). In each case, find the following.

a. The number of wild type bacteria that mutate and the number of mutants that revert.

b. The number of wild type bacteria and the number of mutants after mutation and reversion.

c. The total number of bacteria before and after mutation. Why is it the same?

d. The fraction of mutants before and after mutation.

25. Begin with $1 \cdot 10^6$ wild type and $1 \cdot 10^5$ mutants.

26. Begin with $1 \cdot 10^5$ wild type and $1 \cdot 10^6$ mutants.

27. Begin with b_t wild type and m_t mutants. Find the discrete-time dynamical system for the fraction, p_t, of mutants (divide m_{t+1} by $b_{t+1} + m_{t+1}$ to find p_{t+1} and use the fact that $b_{t+1} + m_{t+1} = b_t + m_t$). Find the equilibrium fraction of mutants. Cobweb starting from the initial condition in Exercise 25. Is the equilibrium stable?

28. Begin with b_t wild type and m_t mutants, but suppose that a fraction 0.1 mutate and a fraction 0.2 revert. Find the discrete-time dynamical system and the equilibrium fraction of mutants.

29–32 ■ This series of problems combines mutation with selection. In one simple scenario, mutations occur in only one direction (wild type turn into mutants but not vice versa), but wild type and mutants have different per capita production. Suppose that a fraction 0.1 of wild type mutate each generation but that each wild type individual produces 2 offspring while each mutant produces only 1.5 offspring. In each case, find the following.

a. The number of wild type bacteria that mutate.

b. The number of wild type bacteria and the number of mutants after mutation.

c. The number of wild type bacteria and the number of mutants after reproduction.

d. The total number of bacteria after mutation and reproduction.

e. The fraction of mutants after mutation and reproduction.

29. Begin with $1 \cdot 10^6$ wild type and $1 \cdot 10^5$ mutants.

30. Begin with $1 \cdot 10^5$ wild type and $1 \cdot 10^6$ mutants.

31. Begin with b_t wild type and m_t mutants. Find the discrete-time dynamical system for the fraction, p_t, of mutants. Find the equilibrium fraction of mutants. Cobweb starting from the initial condition in Exercise 29. Is the equilibrium stable?

32. Begin with b_t wild type and m_t mutants, but suppose that a fraction 0.2 mutate and that the per capita production of mutants is 1. Find the discrete-time dynamical system and the equilibrium fraction of mutants.

33–36 ■ The model of selection studied in this section is similar in many ways to a model of migration. Suppose two adjacent islands have populations of butterflies, with x_t on the first island and y_t on the second. Each year, 20% of the butterflies from the first island fly to the second and 30% of the butterflies from the second island fly to the first.

33. Suppose there are 100 butterflies on each island at time $t = 0$. How many are on each island at $t = 1$? At $t = 2$?

34. Suppose there are 200 butterflies on the first island and none on the second at time $t = 0$. How many are on each island at $t = 1$? At $t = 2$?

35. Find equations for x_{t+1} and y_{t+1} in terms of x_t and y_t.

36. Divide both sides of the discrete-time dynamical system for x_t by $x_{t+1} + y_{t+1}$ to find a discrete-time dynamical system for the fraction p_t on the first island. What is the equilibrium fraction?

37–38 ■ The following two problems extend the migration models to include some reproduction. Each year, 20% of the butterflies from the first island fly to the second and 30% of the butterflies from the second island fly to the first. Again, x_t represents the number of butterflies on the first island, y_t represents the number of butterflies on the second island, and p_t represents the fraction of butterflies on the first island. In each case:

a. Start with 100 butterflies on each island and find the number after migration and after reproduction.

b. Find equations for x_{t+1} and y_{t+1} in terms of x_t and y_t.

c. Find the discrete-time dynamical system for p_{t+1} in terms of p_t.

d. Find the equilibrium, p^*.

e. Sketch a graph and cobweb from a reasonable initial condition.

37. Each butterfly that begins the year on the first island produces one additional butterfly after migration (whether it finds itself on the first or the second island). Those that begin the year on the second island do not reproduce. No butterflies die.

38. Now suppose that the butterflies that do not migrate reproduce (making one additional butterfly each) and those that do migrate fail to reproduce from exhaustion. No butterflies die.

39–42 ■ The model describing the dynamics of the pain medication in the bloodstream,

$$M_{t+1} = 0.5M_t + 1$$

becomes nonlinear if the fraction of medication used is a function of the concentration. In the basic model, half is used no matter how much there is. More generally,

new concentration = old concentration
− fraction used
× old concentration
+ supplement

Suppose that the fraction used is a *decreasing function* of the concentration.

39. Suppose that

$$\text{fraction used} = \frac{0.5}{1 + 0.1M_t}$$

Write the discrete-time dynamical system and solve for the equilibrium. Why is the equilibrium larger than the value of $M^* = 2$ that we found for the basic model?

40. Suppose that

$$\text{fraction used} = \frac{0.5}{1 + 0.4M_t}$$

Write the discrete-time dynamical system and solve for the equilibrium. Why is the equilibrium larger than the value of $M^* = 2$ that we found for the basic model?

41. Suppose that
$$\text{fraction used} = \frac{\beta}{1 + 0.1 M_t}$$
for some parameter $\beta \leq 1$. Write the discrete-time dynamical system and solve for the equilibrium. Sketch a graph of the equilibrium as a function of β. Cobweb starting from $M_0 = 1$ in the cases $\beta = 0.05$ and $\beta = 0.5$.

42. Suppose that
$$\text{fraction used} = \frac{0.5}{1 + \alpha M_t}$$
for some parameter α. Write the discrete-time dynamical system and solve for the equilibrium. Sketch a graph of the equilibrium as a function of α. What happens when $\alpha > 0.5$? Can you explain this in biological terms? Cobweb starting from $M_0 = 1$ in the cases $\alpha = 0.1$ and $\alpha = 1$.

43–46 ▪ Our models of bacterial population growth neglect the fact that bacteria produce fewer offspring in large populations. The following problems introduce two important models of this process, having the form
$$b_{t+1} = r(b_t) b_t$$
where the per capita production, r, is a function of the population size, b_t. In each case:

a. Graph the per capita production as a function of population size.
b. Write the discrete-time dynamical system and graph the updating function.
c. Find the equilibria.
d. Cobweb and indicate whether the equilibrium seems to be stable.

43. One widely used nonlinear model of competition is the **logistic** model, where per capita production is a linearly decreasing function of population size. Suppose that the per capita production is $r(b) = 2\left(1 - \frac{b}{1 \cdot 10^6}\right)$.

44. In an alternative model, the per capita production decreases as the reciprocal of a linear function. Suppose that the per capita production is $r(b) = \dfrac{2}{1 + \frac{b}{1 \cdot 10^6}}$.

45. In another alternative model, called the Ricker model, the per capita production decreases exponentially. Suppose that per capita production is $r(b) = 2e^{-\frac{b}{1 \cdot 10^6}}$.

46. In a model with an **Allee effect,** organisms reproduce poorly when the population is small. In one case, per capita production follows $r(b) = \dfrac{4b}{1 + b^2}$.

Computer Exercises

47. Consider the discrete-time dynamical system
$$x_{t+1} = r x_t (1 - x_t)$$

similar to the form in Exercise 43. Plot the updating function and have a computer find solutions for 50 steps starting from $x_0 = 0.3$ for the following values of r:

a. Some value of r between 0 and 1. What is the only equilibrium?
b. Some value of r between 1 and 2. Where are the equilibria? Which one seems to be stable?
c. Some value of r between 2 and 3. Where are the equilibria? Which one seems to be stable?
d. Try several values of r between 3 and 4. What is happening to the solution? Is there any stable equilibrium?
e. The solution is **chaotic** when $r = 4$. One property of chaos is **sensitive dependence on initial conditions.** Compare a solution starting from $x_0 = 0.3$ with one starting at $x_0 = 0.30001$. Even though they start off very close, they soon separate and become completely different. Why might this be a problem for a scientific experiment?

48. Consider the equation describing the dynamics of selection (Equation 3.4.7),
$$p_{t+1} = \frac{s p_t}{s p_t + r(1 - p_t)}$$
Suppose you have two cultures, 1 and 2. In 1, the mutant does better than the wild type, and in 2 the wild type does better. In particular, suppose that $s = 2$ and $r = 0.3$ in culture 1 and that $s = 0.6$ and $r = 2$ in culture 2. Define discrete-time dynamical systems f_1 and f_2 to describe the dynamics in the two cultures.

a. Graph the functions f_1 and f_2 along with the identity function. Find the first five values of solutions, starting from $p_0 = 0.02$ and $p_0 = 0.98$ in each culture. Explain what each solution is doing and why.
b. Suppose you change the experiment. Begin by taking a population with a fraction p_0 of mutants. Split this population in half, and place one half in culture 1 and the other half in culture 2. Let the bacteria reproduce once in each culture, and then mix them together. Split the mix in half and repeat the process. The updating function is
$$f(p) = \frac{f_1(p) + f_2(p)}{2}$$
Can you derive this? Plot this updating function along with the diagonal. Have a computer find the equilibria and label them on your graph. Do they make sense?
c. Use cobwebbing to figure out which equilibria are stable.
d. Find one solution starting from $p_0 = 0.001$ and another starting from $p_0 = 0.999$. Are these results consistent with the stability of the equilibria? Explain why the solutions do what they do. Why don't they move toward $p = 0.5$?

3.5 A Model of Gas Exchange in the Lung

The exchange of materials between an organism and its environment is one of the most fundamental biological processes. By following the amount of chemical step by step through the breathing process, we can derive a discrete-time dynamical system that models this process for a simplified lung. This discrete-time dynamical system describes how the outside air mixes with the internal air, taking the form of a **weighted average.** This model provides a framework we can use to study more complicated biological processes such as the absorption or release of a chemical.

A Model of the Lungs

Consider a simplified breathing process. Suppose a lung has a volume of 3 L (litres) when full. With each breath, 0.6 L of the air is exhaled and replaced by 0.6 L of outside (or ambient) air. After exhaling, the volume of the lung is 2.4 L, and it returns to 3 L after inhaling (Figure 3.5.53).

Suppose further that the lung contains a particular chemical with concentration 2 millimoles per litre before exhaling. (A mole is a chemical unit indicating 6.02×10^{23} molecules, and a millimole (mmol) represents 6.02×10^{20} molecules.) The ambient air has a chemical concentration of 5 mmol/L. What is the chemical concentration after one breath?

We must track three quantities through these steps: the volume (Figure 3.5.53), the total amount of chemical, and the chemical concentration (Figure 3.5.54). To find the total amount from the concentration, we use the fundamental relation

$$\text{total amount} = \text{concentration} \times \text{volume}$$

Conversely, to find the concentration from the total amount, we rearrange the fundamental relation to

$$\text{concentration} = \frac{\text{total amount}}{\text{volume}}$$

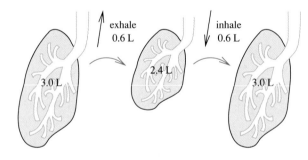

Figure 3.5.53

Gas exchange in the lung: the volume

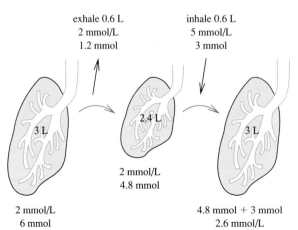

Figure 3.5.54

Gas exchange in the lung: the concentration

3.5 A Model of Gas Exchange in the Lung

One basic biological assumption underlies our reasoning: the air breathed out has a concentration equal to that of the whole lung. This means that the air in the lung is completely mixed at each breath, which is not exactly true.

Assuming also that neither air nor chemical is produced or used while breathing, we can track the process step by step:

Step	Volume (L)	Total Chemical (mmol)	Concentration (mmol/L)	What We Did
Air in lung before breath	3	6	2	Multiplied volume of lung (3) by concentration (2) to get 6.
Air exhaled	0.6	1.2	2	Multiplied volume exhaled (0.6) by concentration (2) to get 1.2.
Air in lung after exhalation	2.4	4.8	2	Multiplied volume remaining (2.4) by concentration (2) to get 4.8.
Air inhaled	0.6	3	5	Multiplied volume inhaled (0.6) by ambient concentration (5) to get 3.
Air in lung after breath	3	7.8	2.6	Found total by adding 4.8 + 3 = 7.8 and divided by volume (3) to get 2.6.

Let us create a discrete-time dynamical system that will describe the breathing process. From the analysis we just finished, we realize that the original concentration of 2 mmol/L is updated to 2.6 mmol/L after a breath. To write the discrete-time dynamical system, we must figure out the concentration after a breath, c_{t+1}, as a function of the concentration before the breath, c_t. We follow the same steps but replace 2 with c_t (Figure 3.5.55):

Step	Volume (L)	Total Chemical (mmol)	Concentration (mmol/L)	What We Did
Air in lung before breath	3	$3c_t$	c_t	Multiplied volume of lung (3) by concentration (c_t) to get $3c_t$.
Air exhaled	0.6	$0.6c_t$	c_t	Multiplied volume exhaled (0.6) by concentration (c_t) to get $0.6c_t$.
Air in lung after exhalation	2.4	$2.4c_t$	c_t	Multiplied volume remaining (2.4) by concentration (c_t) to get $2.4c_t$.
Air inhaled	0.6	3	5	Multiplied volume inhaled (0.6) by ambient concentration (5) to get 7.5.
Air in lung after breath	3	$3 + 2.4c_t$	$1 + 0.8c_t$	Found total by adding $3 + 2.4c_t$ and divided by volume (3) to get $1 + 0.8c_t$.

The discrete-time dynamical system is therefore

$$c_{t+1} = 1 + 0.8c_t$$

Checking, we find that an input of $c_t = 2$ gives

$$c_{t+1} = 1 + 0.8 \cdot 2 = 2.6$$

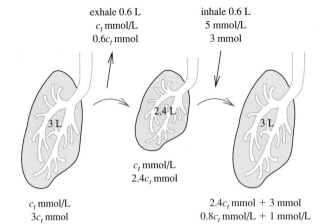

FIGURE 3.5.55

Gas exchange in the lung: finding the discrete-time dynamical system

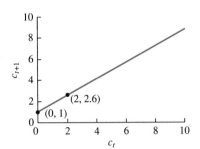

FIGURE 3.5.56

Updating function for the lung model

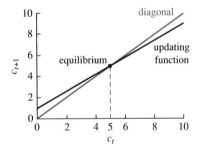

FIGURE 3.5.57

Equilibrium of the lung discrete-time dynamical system

as found above. The graph of the updating function is a line with vertical-intercept 1 and slope 0.8 (Figure 3.5.56).

We can solve for equilibria and use cobwebbing to better understand this discrete-time dynamical system. Let the variable c^* stand for an equilibrium. The equation for equilibrium says that an input of c^* is unchanged by the discrete-time dynamical system, or

$$c^* = 1 + 0.8c^*$$

The solutions of this equation are equilibria (Figure 3.5.57). To solve, we write

$$c^* = 1 + 0.8c^*$$
$$c^* - 0.8c^* = 1$$
$$0.2c^* = 1$$
$$c^* = \frac{1}{0.2} = 5$$

The equilibrium value is 5 mmol/L. We can check this by plugging $c_t = 5$ into the discrete-time dynamical system, finding

$$c_{t+1} = 1 + 0.8 \cdot 5 = 5$$

A concentration of 5 is indeed unchanged by the breathing process.

We can use cobwebbing to check whether solutions move toward or away from this equilibrium. Recall that cobwebbing is a graphical procedure for finding approximate solutions. Both the cobweb starting from $c_0 = 10$ (Figure 3.5.58) and the one starting from $c_0 = 0$ (Figure 3.5.59) produce solutions that approach the equilibrium at $c^* = 5$. Thus, the breathing process forces the concentration to stabilize: no matter what its initial value, over time, the concentration will approach 5 mmol/L (which is the concentration of the ambient air that is breathed in). In other words, the equilibrium, $c^* = 5$, is stable.

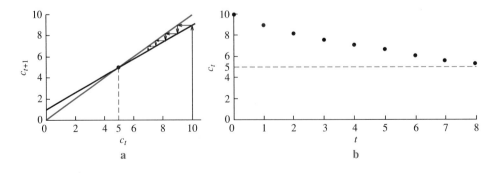

FIGURE 3.5.58

Cobwebbing the lung discrete-time dynamical system: $c_0 = 10$

3.5 A Model of Gas Exchange in the Lung

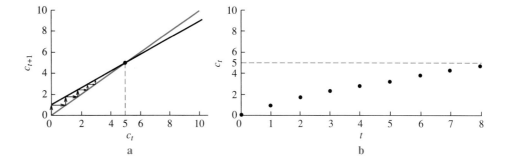

FIGURE 3.5.59
Cobwebbing the lung discrete-time dynamical system: $c_0 = 0$

The Lung System in General

In the previous subsection, we assumed that the lung had a volume of 3 L, that 0.6 L of air was exhaled and inhaled, and that the ambient concentration of chemical was 5 mmol/L. Suppose, more generally, that the lung has a volume of V litres, that W litres of air is exhaled and inhaled at each breath, and that the ambient concentration of chemical is γ (mmol/L). Our goal is to find the discrete-time dynamical system giving c_{t+1} as a function of c_t, which we can do by again following the breathing process step by step (Figure 3.5.60):

Step	Volume (L)	Total Chemical (mmol)	Concentration (mmol/L)	What We Did
Air in lung before breath	V	$c_t V$	c_t	Multiplied volume of lung (V) by concentration (c_t) to get $c_t V$.
Air exhaled	W	$c_t W$	c_t	Multiplied volume exhaled (W) by concentration (c_t) to get $c_t W$.
Air in lung after exhalation	$V - W$	$c_t(V - W)$	c_t	Multiplied volume remaining ($V - W$) by concentration (c_t) to get $c_t(V - W)$.
Air inhaled	W	γW	γ	Multiplied volume inhaled (W) by ambient concentration (γ) to get γW.
Air in lung after breath	V	$c_t(V - W) + \gamma W$	$\dfrac{c_t(V - W) + \gamma W}{V}$	Found total by adding $c_t(V - W)$ to γW and then divided by volume (V).

The new concentration appears at the end of the last line of the table, giving the discrete-time dynamical system

$$c_{t+1} = \frac{c_t(V - W) + \gamma W}{V}$$

This equation can be simplified by multiplying out the first term and dividing out the V:

$$c_{t+1} = \frac{c_t(V - W) + \gamma W}{V}$$
$$= \frac{c_t V - c_t W + \gamma W}{V}$$

168 Chapter 3 Discrete-Time Dynamical Systems

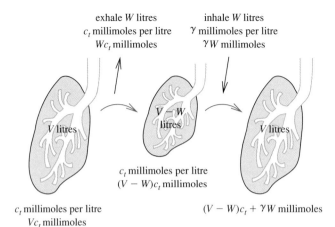

FIGURE 3.5.60
Gas exchange in the lung: general case

$$= c_t - c_t \frac{W}{V} + \gamma \frac{W}{V}$$

The two values W (volume of exhaled air) and V (volume of the lung) appear only as the ratio $\frac{W}{V}$, which is the fraction of the total volume exchanged at each breath. For example, when $W = 0.6$ L and $V = 3$ L, then $\frac{W}{V} = 0.2$, which means that 20% of air is exhaled with each breath. Defining a new parameter

$$q = \frac{W}{V} = \text{fraction of air exchanged}$$

we write the discrete-time dynamical system as

$$c_{t+1} = c_t - c_t q + \gamma q$$

or, after combining terms with c_t, as **the general lung discrete-time dynamical system,**

$$c_{t+1} = (1 - q)c_t + q\gamma \qquad (3.5.1)$$

Example 3.5.1 Finding the Discrete-Time Dynamical System with Specific Parameter Values

In the original example, $W = 0.6$ and $V = 3$, giving $q = \frac{W}{V} = 0.2$. Using $\gamma = 5$, the general equation matches our original discrete-time dynamical system because

$$c_{t+1} = (1 - 0.2)c_t + 0.2 \cdot 5 = 0.8c_t + 1$$

Let us think a bit more about the meaning of the parameter q in the general lung dynamical system

$$c_{t+1} = (1 - q)c_t + q\gamma$$

After a breath, the air in the lung is a mix of old air and ambient air (Figure 3.5.61). The fraction $1 - q$ is old air that remains in the lung, and the remaining fraction, q, is ambient air. If $q = 0.5$, half of the air in the lung after a breath came from outside, and from

$$c_{t+1} = \frac{1}{2}c_t + \frac{1}{2}\gamma$$

we conclude that c_{t+1} is the average of the previous concentration and the ambient concentration.

If q is small, little of the internal air is replaced with ambient air, and c_{t+1} is close to c_t. If q is near 1, most of the internal air is replaced with ambient air. The air in the lung then resembles ambient air, and c_{t+1} is close to the ambient concentration, γ.

The right-hand side of the lung equation is an example of a **weighted average.**

3.5 A Model of Gas Exchange in the Lung

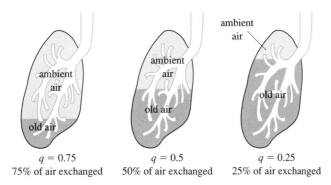

FIGURE 3.5.61
Effects of different values of q

Definition 3.5.1   A weighted average of two values, x and y (which places the weight q on x and the weight $1 - q$ on y), is the sum of the form $qx + (1 - q)y$ for some value of q between 0 and 1.

When $q = 1/2$, the weighted average is the ordinary average. The concentration in the lung after breathing is a weighted average: the fraction $1 - q$ of air is left over from the previous breath, and the fraction q is ambient air.

Example 3.5.2   Weighted Average

Suppose $x = 2$ and $y = 5$. Then the weighted average that places a weight $q = 0.8$ on x and a weight $1 - q = 0.2$ on y is

$$qx + (1 - q)y = 0.8 \cdot 2 + 0.2 \cdot 5 = 2.6$$

Less weight is placed on y, and the weighted average is closer to x.

Example 3.5.3   Contrasting Weighted Average

Suppose $x = 2$ and $y = 5$, as in Example 3.5.2. The weighted average that places a weight $q = 0.2$ on x and a weight $1 - q = 0.8$ on y is

$$qx + (1 - q)y = 0.2 \cdot 2 + 0.8 \cdot 5 = 4.4$$

More weight is placed on y, and the weighted average is closer to y.

Example 3.5.4   Ordinary Average

Suppose $x = 2$ and $y = 5$, as in Examples 3.5.2 and 3.5.3. The ordinary average places equal weights $q = 0.5$ on x and $1 - q = 0.5$ on y, and is equal to

$$qx + (1 - q)y = 0.5 \cdot 2 + 0.5 \cdot 5 = 3.5$$

This value is exactly halfway between x and y.

Example 3.5.5   Weighted Average Applied to Liquids

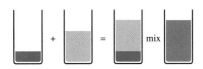

FIGURE 3.5.62
Mixing liquids as a weighted average

Suppose 1 L of liquid with a concentration of 10 mmol/L of salt is mixed with 3 L of liquid with a concentration of 5 mmol/L of salt (Figure 3.5.62). What is the concentration of the resulting mixture? We can think of this as a weighted average. The 4 L of the mixture contains 1 L of the high-salt solution (or a fraction of 0.25) and 3 L of the low-salt solution (or a fraction of 0.75). The resulting concentration is the weighted average

$$0.25 \cdot 10 \, \frac{\text{mmol}}{\text{L}} + 0.75 \cdot 5 \, \frac{\text{mmol}}{\text{L}} = 6.25 \, \frac{\text{mmol}}{\text{L}}$$

We could work this out explicitly by computing the total amount of salt and the total volume. There are 10 mmol of salt from the first solution and 15 mmol from the second (multiplying the concentration of 5 mmol/L by the volume of 3 L), for a total of 25 mmol in 4 L. The concentration is

$$\frac{25 \text{ mmol}}{4 \text{ L}} = 6.25 \text{ mmol/L}$$

The weighted average provides perhaps a simpler way to find this answer. ▲

Example 3.5.6 Weighted Average with More Than Two Components

Weighted averages also work when more than two solutions are mixed. Suppose 1 L of liquid with a concentration of 10 mmol/L of salt is mixed with 3 L of liquid with a concentration of 5 mmol/L of salt and 1 L of liquid with a concentration of 2 mmol/L of salt. What is the concentration of the resulting mixture? In this case, the 5-L mixture is composed of 20% (or 0.20) of the high-salt solution, 60% (or 0.60) of the medium-salt solution, and 20% (or 0.20) of the low-salt solution. The resulting concentration is the weighted average

$$0.20 \cdot 10 \text{ mmol/L} + 0.60 \cdot 5 \text{ mmol/L} + 0.20 \cdot 2 \text{ mmol/L} = 5.4 \text{ mmol/L}$$

If we work this out explicitly, there is a total of 10 mmol from the first solution, 15 mmol from the second, and 2 mmol from the last, for a total of 27 mmol in the 5-L mixture. The concentration is

$$\frac{27 \text{ mmol}}{5 \text{ L}} = 5.4 \text{ mmol/L}$$

▲

The Equilibrium of the Lung Discrete-Time Dynamical System The general discrete-time dynamical system for the lung model is

$$c_{t+1} = (1-q)c_t + q\gamma$$

Following the steps for finding equilibria gives

$$\begin{aligned} c^* &= (1-q)c^* + q\gamma \\ c^* &= c^* - qc^* + q\gamma \\ qc^* - q\gamma &= 0 \\ q(c^* - \gamma) &= 0 \\ q = 0 \text{ or } c^* &= \gamma \end{aligned}$$

What do these results mean? The first case, $q = 0$, occurs when no air is exchanged. Because there is no expression for c^* in this case, *any* value of c_t is an equilibrium. This make sense because a lung that is exchanging no air is, in fact, at equilibrium.

The second case is more interesting. It says that the equilibrium value of the concentration is equal to the ambient concentration. Exchanging air with the outside world has no effect when the inside and the outside match. Doing the calculation in general explains why the equilibrium of 5 mmol/L found earlier in this section must match the ambient concentration of 5 mmol/L.

Lung Dynamics with Absorption

Our model of chemical dynamics in the lung ignored any absorption of the chemical by the body. We can now consider the dynamics of oxygen, which is in fact absorbed by blood. How will this change the discrete-time dynamical system and the resulting solution and equilibrium?

We can use the weighted average to derive the discrete-time dynamical system describing absorption. Suppose that a fraction, q, of air is exchanged each breath, that ambient air has a concentration of γ, and that a fraction, α, of oxygen is absorbed before breathing out (Figure 3.5.63). After absorption, the concentration in the lung is

3.5 A Model of Gas Exchange in the Lung

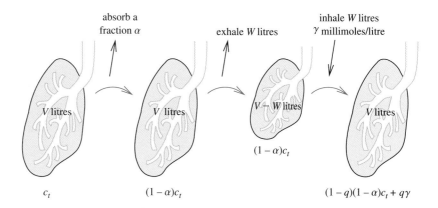

FIGURE 3.5.63
Dynamics of a lung with absorption

$(1 - \alpha)c_t$. Mixing produces a weighted average with a fraction $1 - q$ of this old air and a fraction q of ambient air, giving the discrete-time dynamical system

$$c_{t+1} = (1 - q)(1 - \alpha)c_t + q\gamma$$

If $\alpha = 0$, this reduces to the original model of a lung without absorption.

Example 3.5.7 Absorption of Oxygen by the Lung

Consider again a lung that has a volume of 3 L and that replaces 0.6 L at each breath with ambient air (as in Figure 3.5.54). Suppose now that we are tracking oxygen with an ambient concentration of 21%. Assume that 30% of the oxygen in the lung is absorbed at each breath. We then have

$$q = 0.2$$
$$\alpha = 0.3$$
$$\gamma = 0.21$$

The discrete-time dynamical system is then

$$c_{t+1} = 0.8 \cdot 0.7c_t + 0.2 \cdot 0.21 = 0.56c_t + 0.042$$

The equilibrium concentration in the lung solves

$$c^* = 0.56c^* + 0.042$$
$$0.44c^* = 0.042$$
$$c^* \approx 0.095$$

The equilibrium concentration of oxygen in the lung, which is equal to the concentration of oxygen in the air breathed out, is about 9.5%, or less than half the ambient concentration.

As a consequence of absorption, the equilibrium concentration will be lower than the ambient concentration. By solving for the equilibrium of the system in general, we can investigate how the equilibrium depends on the fraction absorbed. To find the equilibrium, we solve

$$c^* = (1 - q)(1 - \alpha)c^* + q\gamma$$
$$c^* - (1 - q)(1 - \alpha)c^* = q\gamma$$
$$c^*(1 - (1 - q)(1 - \alpha)) = q\gamma$$
$$c^* = \frac{q\gamma}{1 - (1 - q)(1 - \alpha)}$$

As a check, if we substitute $\alpha = 0$, we find

$$c^* = \frac{q\gamma}{1 - (1 - q)} = \frac{q\gamma}{q} = \gamma$$

matching the equilibrium in the case of no absorption.

Example 3.5.8 The Equilibrium Concentration of Oxygen as a Function of α

With the parameter values $q = 0.2$ and $\gamma = 0.21$, we find

$$c^* = \frac{0.2 \cdot 0.21}{1 - 0.8(1 - \alpha)} = \frac{0.042}{1 - 0.8(1 - \alpha)}$$

By substituting values of α ranging from $\alpha = 0$ to $\alpha = 1$, we can plot the equilibrium concentration as a function of absorption (Figure 3.5.64).

Example 3.5.9 Finding α from the Equilibrium Concentration of Oxygen

Suppose that the actual oxygen concentration in exhaled air is approximately 15%. What fraction of oxygen is in fact absorbed? We can find this by solving for the value of α that produces $c^* = 0.15$:

$$0.15 = \frac{0.042}{1 - 0.8(1 - \alpha)}$$
$$0.15(1 - 0.8(1 - \alpha)) = 0.042$$
$$0.15(0.2 + 0.8\alpha) = 0.042$$
$$0.2 + 0.8\alpha = \frac{0.042}{0.15} = 0.28$$
$$0.8\alpha = 0.08$$
$$\alpha = 0.1$$

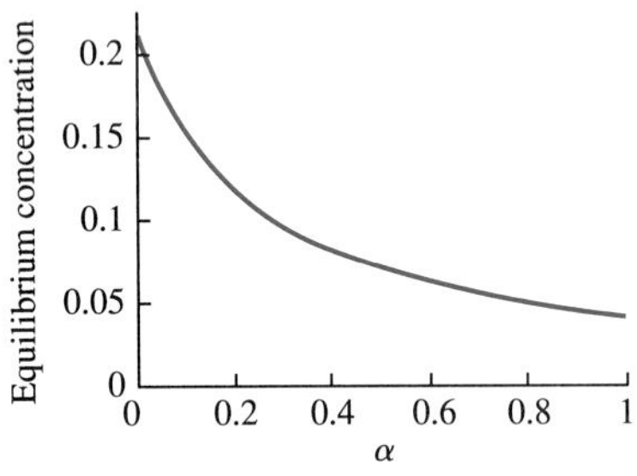

FIGURE 3.5.64
Equilibrium as a function of α

Rather surprisingly, the lung absorbs only about 10% of the available oxygen, leading to exhaled air that has nearly 30% less oxygen than ambient air.

Summary This section develops a mathematical model of chemical concentration in the lung. Starting from an understanding of how a lung exchanges air, we derived a discrete-time dynamical system for the chemical concentration. The discrete-time dynamical system can be described as a **weighted average** of the internal concentration and the **ambient concentration.** The equilibrium is equal to the ambient concentration, and cobwebbing diagrams indicate that solutions approach this equilibrium. Including absorption produces a slightly more complicated model, with an equilibrium that is less than the ambient concentration. We used this model to investigate the dynamics of oxygen in the lung.

3.5 Exercises

Mathematical Techniques

1–2 ■ Use the idea of the weighted average to find the following.

1. 1 L of water at 30°C is mixed with 2 L of water at 100°C. What is the temperature of the resulting mixture?

2. In a class of 52 students, 20 scored 50 on a test, 18 scored 75, and the rest scored 100. What was the average score?

3–6 ■ Express the following weighted averages in terms of the given variables.

3. 1 L of water at temperature T_1 is mixed with 2 L of water at temperature T_2. What is the temperature of the resulting mixture? Set $T_1 = 30$ and $T_2 = 100$, and compare with the result of Exercise 1.

4. V_1 litres of water at 30°C is mixed with V_2 litres of water at 100°C. What is the temperature of the resulting mixture? Set $V_1 = 1$ and $V_2 = 2$, and compare with the result of Exercise 1.

5. V_1 litres of water at temperature T_1 is mixed with V_2 litres of water at temperature T_2. What is the temperature of the resulting mixture?

6. V_1 litres of water at temperature T_1 is mixed with V_2 litres of water at temperature T_2 and V_3 litres of water at temperature T_3. What is the temperature of the resulting mixture?

7–8 ■ The following are similar to examples of weighted averages with absorption.

7. 1 L of water at 30°C is to be mixed with 2 L of water at 100°C, as in Exercise 1. Before mixing, however, the temperature of each moves halfway to 0°C (so the 30°C water cools to 15°C). What is the temperature of the resulting mixture? Is this half the temperature of the result in Exercise 1?

8. In a class of 52 students, 20 scored 50 on a test, 18 scored 75, and the rest scored 100. The professor suspects cheating, however, and deducts 10 from each score. What is the average score after the deduction? Is it exactly 10 less than the average found in Exercise 2?

Applications

9–12 ■ Suppose that the volume of the lung is V litres, the amount breathed in and out is W litres, and the ambient concentration is γ millimoles per litre. For each of the given sets of parameter values and the given initial condition, find the following:

a. The amount of chemical in the lung before breathing.
b. The amount of chemical breathed out.
c. The amount of chemical in the lung after breathing out.
d. The amount of chemical breathed in.
e. The amount of chemical in the lung after breathing in.
f. The concentration of chemical in the lung after breathing in.
g. A comparison of this result with the result of using the general lung discrete-time dynamical system (Equation 3.5.1). Remember that $q = W/V$.

9. $V = 2$ L, $W = 0.5$ L, $\gamma = 5$ mmol/L, $c_0 = 1$ mmol/L
10. $V = 1$ L, $W = 0.1$ L, $\gamma = 8$ mmol/L, $c_0 = 4$ mmol/L
11. $V = 1$ L, $W = 0.9$ L, $\gamma = 5$ mmol/L, $c_0 = 9$ mmol/L
12. $V = 10$ L, $W = 0.2$ L, $\gamma = 1$ mmol/L, $c_0 = 9$ mmol/L

13–16 ■ Find and graph the updating function in the following cases. Cobweb for three steps, starting from the points indicated in the earlier problems. Sketch the solutions.

13. The situation in Exercise 9.
14. The situation in Exercise 10.
15. The situation in Exercise 11.
16. The situation in Exercise 12.

17–20 ■ Find the lung discrete-time dynamical system with the following parameter values, and compute the equilibrium. Check that it matches the formula $c^* = \gamma$.

17. $V = 2$ L, $W = 0.5$ L, $\gamma = 5$ mmol/L, $c_0 = 1$ mmol/L.
18. $V = 1$ L, $W = 0.1$ L, $\gamma = 8$ mmol/L, $c_0 = 4$ mmol/L.
19. $V = 1$ L, $W = 0.9$ L, $\gamma = 5$ mmol/L, $c_0 = 9$ mmol/L.
20. $V = 10$ L, $W = 0.2$ L, $\gamma = 1$ mmol/L, $c_0 = 9$ mmol/L.

21–22 ■ The following problems investigate what happens if the breathing rate changes in the models of absorption examined in "Lung Dynamics with Absorption," Section 3.5. Use an external concentration of $\gamma = 0.21$, as before.

21. Find the equilibrium oxygen concentration if the fraction of air exchanged is $q = 0.4$ and the fraction absorbed is $\alpha = 0.1$. Can you explain why the concentration becomes higher even though the person is breathing more?

22. Find the equilibrium oxygen concentration if the fraction of air exchanged decreases to $q = 0.1$ and the fraction absorbed decreases to $\alpha = 0.05$. Think of this as a person gasping for breath. Why is the concentration nearly the same as the value found in Example 3.5.9? Does this mean that gasping for breath is OK?

23–26 ■ The following problems investigate absorption that is not proportional to the concentration in the lung, as in "Lung Dynamics with Absorption," Section 3.5. Assume an external concentration of $\gamma = 0.21$ and $q = 0.2$.

23. Suppose that the oxygen concentration is reduced by 2% at each breath. Find the discrete-time dynamical system and the equilibrium. Are there values of c_t for which the system does not make sense?

24. Suppose that the oxygen concentration is reduced by 3% at each breath. Find the discrete-time dynamical system and the equilibrium. Are there values of c_t for which the system does not make sense?

25. Suppose that the amount absorbed is $0.2(c_t - 0.05)$ if $c_t \geq 0.05$. This models a case where the only oxygen available is that in excess of the concentration in the blood, which corresponds roughly to 5%.

26. Consider a case like Exercise 25, but suppose that the amount absorbed is $0.1(c_t - 0.05)$ if $c_t \geq 0.05$. Why is the concentration different from that found in Example 3.5.9?

27–28 ■ On the basis of the problems investigating absorption that is not proportional to the concentration in the lung (Exercises 23–26a), find the value of the parameter that produces an exhaled concentration of exactly 0.15. Assume $\gamma = 0.21$ and $q = 0.2$.

27. Suppose that the concentration is reduced by an amount A (generalizing the case in Exercises 23 and 24). Does the amount of oxygen absorbed match that found in Example 3.5.9?

28. Suppose that the amount absorbed is $\alpha(c_t - 0.05)$ (generalizing the case where only available oxygen is absorbed in Exercises 25 and 26a). Does the amount of oxygen absorbed match that found in Example 3.5.9?

29–30 ■ The following problems investigate production of carbon dioxide by the lung. Suppose that the concentration increases by an amount S before the air is exchanged. Assume an external concentration of carbon dioxide of $\gamma = 0.0004$ and $q = 0.2$.

29. Suppose $S = 0.001$. Write the discrete-time dynamical system and find its equilibrium. Compare the equilibrium with the external concentration.

30. The actual concentration of carbon dioxide in exhaled air is about 0.04, or 100 times the external concentration. Find the value of S that gives this as the equilibrium.

31–32 ■ A bacterial population that has per capita production $r < 1$ but is supplemented each generation follows a discrete-time dynamical system much like that of the lung. Use the following steps to build the discrete-time dynamical system in the two given cases.

a. Starting from $3 \cdot 10^6$ bacteria, find the number after reproduction.
b. Find the number after the new bacteria are added.
c. Find the discrete-time dynamical system.

31. A population of bacteria has per capita production $r = 0.6$, and $1 \cdot 10^6$ bacteria are added each generation.

32. A population of bacteria has per capita production $r = 0.2$, and $5 \cdot 10^6$ bacteria are added each generation.

33–35 ▪ Find the equilibrium population of bacteria in the following cases with supplementation.

33. A population of bacteria has per capita production $r = 0.6$, and $1 \cdot 10^6$ bacteria are added each generation (as in Exercise 31).

34. A population of bacteria has per capita production $r = 0.2$, and $5 \cdot 10^6$ bacteria are added each generation (as in Exercise 32a).

35. A population of bacteria has per capita production $r < 1$, and $1 \cdot 10^6$ bacteria are added each generation. What happens to the equilibrium if $r = 0$? What happens if r is close to 1? Do these results make biological sense?

36–39 ▪ Lakes receive water from streams each year and lose water to outflowing streams and evaporation. The following values are based on the Great Salt Lake in Utah. The lake receives 3×10^6 m³ of water per year with salinity of 1 part per thousand (concentration 0.001). The lake contains $3.3 \cdot 10^7$ m³ of water and starts with no salinity. Assume that the water that flows out is well mixed, having a concentration equal to that of the entire lake. Compute the discrete-time dynamical system by finding (a) the total salt before the inflow; (b) total water; (c) total salt and salt concentration after inflow; and (d) total water, total salt, and salt concentration after outflow or evaporation,

36. There is no evaporation, and $3 \cdot 10^6$ m³ of water flows out each year.

37. $1.5 \cdot 10^6$ m³ of water flows out each year, and $1.5 \cdot 10^6$ m³ evaporates. No salt is lost through evaporation.

38. A total of $3 \cdot 10^6$ m³ of water evaporates, and there is no outflow.

39. Assume instead that $2 \cdot 10^6$ m³ of water evaporates and that there is no outflow. The volume of this lake is increasing.

40–43 ▪ Find the equilibrium concentration of salt in a lake in the following cases. Describe the result in words by comparing the equilibrium salt level with the salt level of the water flowing in.

40. The situation described in Exercise 36.
41. The situation described in Exercise 37.
42. The situation described in Exercise 38.
43. The situation described in Exercise 39a.

44–45 ▪ A lab is growing and harvesting a culture of valuable bacteria described by the discrete-time dynamical system

$$b_{t+1} = rb_t - h$$

The bacteria have per capita production r, and h are harvested each generation.

44. Suppose that $r = 1.5$ and $h = 1 \cdot 10^6$ bacteria. Sketch the updating function, and find the equilibrium both algebraically and graphically.

45. Without setting r and h to particular values, find the equilibrium algebraically. Does the equilibrium get larger when h gets larger? Does it get larger when r gets larger? If the answers seem odd (as they should), look at a cobweb diagram to try to figure out why.

Computer Exercises

46. Use a computer to graph each trigonometric discrete-time dynamical system.

   a. $x_{t+1} = \cos x_t$
   b. $y_{t+1} = \sin y_t$
   c. $z_{t+1} = \sin z_t + \cos z_t$

   Find the equilibria of each, and produce cobweb diagrams starting from three different initial conditions. Do the diagrams help make sense of the solutions?

47. Consider the discrete-time dynamical system

$$x_{t+1} = e^{ax_t}$$

   for the following values of the parameter a. Use a computer to graph the function and the diagonal to look for equilibria. Cobweb starting from $x_0 = 1$ in each case.

   a. $a = 0.3$
   b. $a = 0.4$
   c. $a = 1/e$

Chapter Summary: Key Terms and Concepts

Define or explain the meaning of each term and concept.
Discrete-time dynamical system: updating function, initial condition; solution; basic exponential discrete-time dynamical system, basic additive discrete-time dynamical system, inverse (backward) dynamical system; equilibrium, stable and unstable equilibria; cobwebbing
Applications: per capita production, limited population model, model of selection; weighted average

Concept Check: True/False Quiz

Determine whether each statement is true or false. Give a reason for your choice.

1. If m^* is an equilibrium of the system $m_{t+1} = f(m_t)$, then $1/m^*$ is an equilibrium of the (backward in time) system $m_t = f^{-1}(m_{t+1})$.

2. The updating function of the system $m_{t+1} = \dfrac{2m_t - 3}{m_t^2 + 1}$ is $f(x) = x - \dfrac{2x - 3}{x^2 + 1}$.

3. Assume that $H_{t+1} = 2H_t + 3$, where H_t and H_{t+1} are in metres. In units of millimetres, this dynamical system can be expressed as $h_{t+1} = 2000h_t + 3000$.

4. In the population model $b_{t+1} = 4(b_t - 2b_t^2)$, the per capita production rate is $4(1 - 2b_t)$.

5. $m^* = 2$ is an equilibrium of the dynamical system $m_{t+1} = e^{m_t - 2} + 1$.

6. The system $b_{t+1} = 3b_t + 120$ for the number of bacteria has no biologically significant equilibria.

7. The equilibrium $m^* = 0$ of the dynamical system $m_{t+1} = 1.02m_t$ is stable.

8. If $m_{t+1} = 0.6m_t$ and $m_0 = 0.2$, then $m_{t+1} = 0.2 \cdot 0.6^t$.

9. The dynamical system $b_{t+1} = b_t^3 - 2b_t^2 + 6$ has four equilibrium points.

10. 1 L of liquid with a concentration of 20 g/L of salt is mixed with 4 L of liquid with a concentration of 10 g/L. The resulting mixture has a concentration of 12 g/L.

Supplementary Problems

1. A lab has a culture of a new kind of bacteria where each individual takes two hours to split into three bacteria. Suppose that these bacteria never die and that all offspring are OK.

    a. Write an updating function describing this system.

    b. Suppose there are $2 \cdot 10^7$ bacteria at 9 A.M. How many will there be at 5 P.M.?

    c. Write an equation for how many bacteria there are as a function of how long the culture has been running.

    d. When will this population reach 10^9?

2. The number of bacteria in another lab follows the discrete-time dynamical system

    $$b_{t+1} = \begin{cases} 2b_t & b_t \leq 1 \\ -0.5(b_t - 1) + 2 & b_t > 1 \end{cases}$$

    where t is measured in hours and b_t in millions of bacteria.

    a. Graph the updating function. For what values of b_t does it make sense?

    b. Find the equilibrium.

    c. Cobweb starting from $b_0 = 0.4$ million bacteria. What do you think happens to this population?

3. Find all equilibria for the population, p_t, whose per capita production is given by $2p_t/(1 + p_t^2)$.

4. Find the solution of the dynamical system

    $$M_{t+1} = 0.5M_t + 1$$

    with initial condition $M_0 = 1$. Hint: Assume that $M_t = a(0.5^t) + b$ and use the technique that we introduced after Example 3.1.7 to find a and b. Verify that your solution is correct.

5. Consider the medication discrete-time dynamical system with both parameters from Examples 3.2.8 and 3.2.9:

    $$M_{t+1} = (1 - \alpha)M_t + S$$

    Find the equilibrium, M^*. Discuss possible ways of changing the parameters α and S to make the equilibrium increase.

6. What is the concentration of a mixture of 2 mL of water with a salt concentration of 0.85 mol/L and 5 mL of water with a salt concentration of 0.7 mol/L?

7. As in Exercise 6, 2 mL of water with a salt concentration of 0.85 mol/L is to be mixed with 5 mL of water with a salt concentration of 0.7 mol/L. Before mixing, however, evaporation leads the concentration of each component to double. What is the concentration of the mixture? Is it exactly twice the concentration found in Exercise 6?

8. A population, b_t, of bacteria has per capita production $r = 0.05b_t$, and a fixed number $B > 0$ of bacteria are added each generation. Discuss how the equilibrium of the population depends on B.

9. Identify the values (if they exist) of the parameter r for which the system

    $$b_{t+1} = \dfrac{rb_t}{1 + 3b_t}$$

    has no equilibria, for which it has one equilibrium, and for which there is more than one equilibrium.

10. Find all equilibrium points of the system

    $$n_{t+1} = \dfrac{2n_t}{1 + 3n_t/100}$$

11. The butterflies on a particular island are not doing very well. Each autumn, every butterfly produces on average 1.2 eggs and then dies. Half of these eggs survive the winter and

produce new butterflies by late summer. At this time, 1000 butterflies arrive from the mainland to escape overcrowding.

a. Write a discrete-time dynamical system for the population on this island.

b. Graph the updating function and cobweb starting from 1000.

c. Find the equilibrium number of butterflies.

12. Two similar objects are left to cool for one hour. One starts at 80°C and cools to 70°C, and the other starts at 60°C and cools to 55°C. Suppose the discrete-time dynamical system for cooling objects is linear.

a. Find the discrete-time dynamical system. Find the temperature of the first object after two hours. Find the temperature after one hour of an object starting at 20°C.

b. Graph the updating function, and cobweb starting from 80°C.

c. Find the equilibrium. Explain what the equilibrium means.

13. A culture of bacteria increases in area by 10% each hour. Suppose the area is 2 cm² at 2:00 P.M.

a. What will the area be at 5:00 P.M.?

b. Write the relevant discrete-time dynamical system, and cobweb starting from 2.

c. What was the area at 1:00 P.M.?

d. If all bacteria are the same size and each adult produces two offspring each hour, what fraction of offspring must survive?

e. If the culture medium is only 10 cm² in size, when will it be full?

14. Candidates Dewey and Howe are competing for fickle voters. Exactly 100,000 people are registered to vote in the election, and each will vote for one of these two candidates. Each week, some voters switch their allegiance. Twenty percent of Dewey's supporters switch to Howe each week. Howe's supporters are more likely to switch when Dewey is doing well: the fraction switching from Howe to Dewey is proportional to Dewey's percentage of the vote—none switch if Dewey commands 0% of the vote, and 50% switch if Dewey commands 100% of the vote. Suppose Howe starts with 90% of the vote.

a. Find the number of votes Dewey and Howe have after a week.

b. Find Dewey's percentage after a week.

c. Find the discrete-time dynamical system describing Dewey's percentage.

d. Graph the updating function and find the equilibrium or equilibria.

e. Who will win the election?

15. An organism is breathing a chemical that modifies the depth of its breaths. In particular, suppose that the fraction, q, of air exchanged is given by

$$q = \frac{c_t}{c_t + \gamma}$$

where γ is the ambient concentration and c_t is the concentration in the lung. After a breath, a fraction q of the air came from outside, and a fraction $1 - q$ remained from inside. Suppose $\gamma = 0.5$ mol/L.

a. Describe the breathing of this organism.

b. Find the discrete-time dynamical system for the concentration in the lung.

c. Find the equilibrium or equilibria.

16. Lint is building up in a dryer. With each use, the old amount of lint, x_t, is divided by $1 + x_t$, and 0.5 linton (the units of lint) is added.

a. Find the discrete-time dynamical system and graph the updating function.

b. Cobweb starting from $x_0 = 0$. Graph the associated solution.

c. Find the equilibrium or equilibria.

17. Suppose people in a bank are waiting in two separate lines. Each minute several things happen: some people are served, some people join the lines, and some people switch lines. In particular, suppose that 1/10 of the people in the first line are served, and 3/10 of the people in the second line are served. Suppose that the number of people who join each line is equal to 1/10 of the total number of people in both lines and that 1/10 of the people in each line switch to the other.

a. Suppose there are 100 people in each line at the beginning of a minute. Find how many people are in each line at the end of the minute.

b. Write a discrete-time dynamical system for the number of people in the first line and another discrete-time dynamical system for the number of people in the second.

c. Write a discrete-time dynamical system for the fraction of people in the first line.

18. A gambler faces off against a small casino. She begins with $1000, and the casino starts with $11,000. In each round, the gambler loses 10% of her current funds to the casino, and the casino loses 2% of its current funds to the gambler.

a. Find the amount of money each has after one round.

b. Find a discrete-time dynamical system for the amount of money the gambler has and another for the amount of money the casino has.

c. Find the discrete-time dynamical system for the fraction, p, of money the gambler has.

d. Find the equilibrium fraction of the money held by the gambler.

e. Using the fact that the total amount of money is constant, find the equilibrium amount of money held by the gambler.

19. Let V represent the volume of a lung and c the concentration of some chemical inside. Suppose the internal surface area is proportional to the volume, and a lung with volume 400 cm^3 has a surface area of 100 cm^2. The lung absorbs the chemical at a rate per unit surface area of

$$R = \alpha \left(\frac{c}{4 \times 10^{-2} + c} \right)$$

Time is measured in seconds, surface area in square centimetres, and volume in cubic centimetres. The parameter α takes on the value 6 in the appropriate units.

a. Find surface area as a function of volume. Make sure your dimensions make sense.

b. What are the units of R? What must the units of α be?

c. Suppose that $c = 1 \cdot 10^{-2}$ and $V = 400$. Find the total amount of chemical absorbed.

d. Suppose that $c = 1 \cdot 10^{-2}$. Find the total chemical absorbed as a function of V.

20. An Alberta millionaire (with $1,000,001 in assets in 2014) got rich by clever investments. She managed to earn 10% interest per year for the last 20 years, and she plans to do the same in the future.

a. How much did she have in 1994?

b. When will she have $5,000,001?

c. Write the discrete-time dynamical system and graph the updating function.

d. Write and graph the solution.

21. A major university hires a famous Saskatchewan millionaire to manage its endowment. The millionaire decides to follow this plan each year:

- Spend 25% of all funds above $10 million on university operations.
- Invest the remainder at 10% interest.
- Collect $5 million in donations from wealthy alumni.

a. Suppose the endowment has $340 million to start. How much will it have after spending on university operations? After collecting interest on the remainder? After the donations roll in?

b. Find the discrete-time dynamical system.

c. Graph the updating function, and cobweb starting from $340 million.

22. Another major university hires a famous Ontario millionaire to manage its endowment. This millionaire starts with $340 million, brings back $35.5 million the next year, and claims to be able to guarantee a linear increase in funds thereafter.

a. How much money will this university have after eight years?

b. Graph the endowment as a function of time.

c. Write the discrete-time dynamical system, graph, and cobweb starting from $340 million.

d. Which university do you think will do better in the long run? Which millionaire would you hire?

23. Suppose traffic volume on a particular road has been as shown in the following table:

Year	Vehicles
1980	40,000
1990	60,000
2000	90,000
2010	135,000

a. Sketch a graph of traffic over time.

b. Find the discrete-time dynamical system that describes this traffic.

c. What was the traffic volume in 1970?

d. Give a formula for the predicted traffic in the year 2060.

e. Find the half-life or the doubling time (whichever is appropriate) of the traffic volume.

24. In order to improve both the economy and the quality of life, policies are designed to encourage growth and decrease traffic flow. In particular, the number of cars is encouraged to increase by a factor of 1.6 over each 10-year period, but the commuters from 10,000 cars are to choose to ride comfortable new trains instead of driving.

a. If there were 40,000 people commuting by car in 1980, how many would there have been in 1990?

b. Find the discrete-time dynamical system describing the number of people commuting by car.

c. Find the equilibrium.

d. Graph the updating function, and cobweb starting from an initial number of 40,000.

e. In the long run, will there be more or less traffic with this policy than with the policy that led to the data in the previous problem? Why?

Project

1. Combine the model of selection from the chapter with the models of mutation and reversion. Assume that wild type have per capita production r, mutants have per capita production s, a fraction μ of the offspring of the wild type mutate into the mutant type, and a fraction ν of the offspring of the mutant type revert. First set $b_t = 4 \cdot 10^6$, $m_t = 2 \cdot 10^5$, $\mu = 0.2$, $\nu = 0.1$, $r = 1.5$, and $s = 2$.

a. How many of the wild type individuals will there be after production and before mutation? How many of these will mutate? How many will remain the wild type?

b. How many mutant individuals will there be after production and before mutation? How many of these will revert? How many will not?

c. Find the total number of wild type after reproduction, mutation, and reversion.

d. Find the total number of mutants after reproduction, mutation, and reversion.

e. Find the total number of bacteria after reproduction, mutation, and reversion. Why is it different from the initial number?

f. Find the fraction of mutants to begin with and the fraction after reproduction, mutation, and reversion.

Now, treat b_t and m_t as variables. The following steps will help you find the updating function.

g. Use the above steps to find m_{t+1} in terms of b_t and m_t.

h. Find the total number of bacteria, $b_{t+1} + m_{t+1}$, in terms of b_t and m_t.

i. Divide your equation for m_{t+1} by $b_{t+1} + m_{t+1}$ to find an expression for p_{t+1} in terms of b_t and m_t.

j. Divide the numerator and denominator by $m_t + b_t$ as in the derivation of Equation 3.4.5, and write the updating function in terms of p_t.

k. Find the equilibrium of this updating function.

Finally, we can do this in general by treating r, s, μ, and ν as parameters. Do exactly the above steps to find the updating function. After doing so, study the following special cases. In each case, explain your answer.

l. $r = s$ (no selection) and mutation in only one direction ($\mu = 0$ and $\nu > 0$). Find the updating function and the equilibrium. What does this mean?

m. $r = s$ and $\mu = \nu$. Find the updating function and the equilibrium. What does this mean?

n. $r > s$, $\nu = 0$, and $\mu > 0$. This means that the wild type has a reproductive advantage but keeps mutating. Find the updating function and the equilibrium. The result is called mutation-selection balance. Can you guess why?

Chapter 4

Limits, Continuity, and Derivatives

Discrete-time dynamical systems describe biological change when measurements are made at discrete intervals. We now develop methods to describe a quantity that changes **continuously.** This description requires the two central ideas of differential calculus: the **limit** and the **instantaneous rate of change,** or the **derivative.** We will find a **geometric** interpretation of the derivative as the **slope of the tangent line,** which will help us to graph and analyze complicated functions.

We start by developing the concept of **average rate of change,** whose geometric interpretation is the slope of a **secant line.** We define the **limit** of a function and develop its basic properties. The limit is the most important concept in calculus, as all of its fundamental ideas—**continuity, derivative,** and **integral** (which we discuss in Chapter 7)—are based on it. We spend some time learning how to calculate limits numerically and algebraically. With the help of limits we can identify certain features of the graphs of functions, such as **vertical** and **horizontal asymptotes.** At the end of the chapter we discuss **continuity** and **derivatives** and **differentiability.** Algebraic aspects of derivatives are postponed until the next chapter.

4.1 Investigating Change

Discrete-time dynamical systems are a powerful tool for describing the dynamics of biological systems when change can be accurately described by measurements made at **discrete times.** In order for us to understand other systems fully, however, measurements must be made at all times, or **continuously.** We have only to think of the growth of a plant or the motion of an animal to realize that some change is best described by a continuous set of measurements.

The Average Rate of Change

Suppose we measure a bacterial population continuously and find that the population size, $b(t)$ (in millions), follows the equation

$$b(t) = 2^t$$

where t is measured in hours (Figure 4.1.1). How can we best describe the growth of this population?

More precisely, we would like to answer the following question: how does the population change at a particular moment? We know that when $t = 1$, the population count is 2 million. But how fast is it changing at that moment?

If we check the population size every hour, we find the data shown in the first two columns of Table 4.1.1, as well as in Figure 4.1.2.

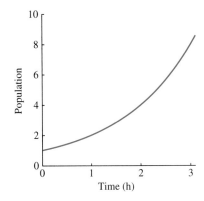

FIGURE 4.1.1

A bacterial population measured continuously

Table 4.1.1

t (hours)	$b(t)$ (millions)	Change $\Delta b = b(t) - b(t-1)$	Average Rate of Change $\dfrac{\Delta b}{\Delta t} = \dfrac{b(t) - b(t-1)}{1}$
0	1		
1	2	1	1
2	4	2	2
3	8	4	4
4	16	8	8

One thing we can do is to find out how the population count changes between two consecutive measurements. In other words, we can calculate the **change in function**

$$\Delta b = \text{population now minus population one hour ago}$$

i.e.,

$$\Delta b = b(t) - b(t-1)$$

(recall that Δ (the Greek capital letter delta) means "change in").

For example, the bacterial population increased by

$$\Delta b = b(2) - b(1) = 4 - 2 = 2$$

million between $t = 1$ and $t = 2$. The changes, Δb, have been recorded in the third column of Table 4.1.1; the change in population is not defined when $t = 0$ because there was no previous measurement.

The change in function, Δb, by itself, does not give us much. If we are told that some quantity $b(t)$ has changed by 2 units (and nothing else), we do not know *how* it changed—was it fast, or slow, or at the same pace, or did the pace vary?

Clearly, we need something better than just Δb. To discover what it is, consider the following question: There are two quantities, A and B, and we know that A changed by three units in two hours and B changed by five units in three hours. Which quantity changed faster?

Since the quantity A changed by three units in two hours, it changed, *on average*, by $3/2 = 1.5$ units per hour. The quantity B changed *on average* by $5/3 \approx 1.67$ units per hour. Now that both quantities have been described using the same measure—the *average rate of change*—we compare them, and say that, *on average*, quantity B changed more quickly than quantity A.

In general, we define the **average rate of change** of a function $f(t)$ as

$$\text{average rate of change} = \frac{\text{change in function}}{\text{change in variable}} = \frac{\Delta f}{\Delta t}$$

Thus, the **average rate of change of $f(t)$ from $t = t_1$ to $t = t_2$** (or the **average rate of change of $f(t)$ over the interval $[t_1, t_2]$**) is given by the quotient

$$\frac{\Delta f}{\Delta t} = \frac{f(t_2) - f(t_1)}{t_2 - t_1}$$

For example, the average rate of change of the bacterial population $b(t)$ between $t = 1$ and $t = 2$ is

$$\frac{\Delta b}{\Delta t} = \frac{b(2) - b(1)}{2 - 1} = \frac{4 - 2}{1} = 2$$

in units of millions of bacteria per hour. In the right-most column in Table 4.1.1, we have calculated the average rates of change for the data given.

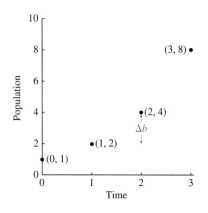

FIGURE 4.1.2

Hourly measurement of a bacterial population

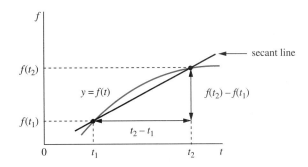

FIGURE 4.1.3

Calculating the slope of the secant line

As the quotient of the change in the function over the change in the variable, the average rate of change has a very useful geometric interpretation—it represents the slope of a line.

By definition, a **secant line** is a line that connects two points on the graph of a function. In Figure 4.1.3 we drew the points $(t_1, f(t_1))$ and $(t_2, f(t_2))$ on the graph of $f(t)$ and connected them with a secant line. (Quite often we single out one of the two points as a **base point** to focus on more detailed study.) The slope of the secant line,

$$\frac{f(t_2) - f(t_1)}{t_2 - t_1}$$

is exactly the same as the average rate of change of $f(t)$ over $[t_1, t_2]$.

Going back to our original question—now we know that $b(t)$ changes on average by two million bacteria per hour between $t = 1$ and $t = 2$. What does this tell us about the change in $b(t)$ when $t = 1$?

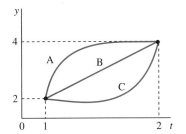

FIGURE 4.1.4

Quantities A, B, and C have the same average rate of change

Nothing much (yet), as the following example shows. The three quantities, A, B, and C, whose graphs are given in Figure 4.1.4 all change by two units between $t = 1$ and $t = 2$. Quantity A shows a rapid initial increase, and then slows down. Quantity B seems to be increasing at the same rate throughout. Quantity C starts by slowly decreasing, and then later increases quickly. As we can see, the three quantities change in significantly different ways and yet have the same average rate of change.

The problem is that the time interval $[1, 2]$ that we used to compute the average rates of change is too large to capture the differences in the ways in which the three quantities change.

So, if we wish to describe how the population $b(t)$ changes when $t = 1$, we need to get closer to $t = 1$. In other words, we need to calculate average rates of change over shorter time intervals.

Example 4.1.1 Average Rate of Change and Secant Line

Find the average rate of change of the bacterial population $b(t) = 2^t$ from $t = 1$ to $t = 1.5$ and interpret it geometrically.

The average rate of change is

$$\frac{\Delta b}{\Delta t} = \frac{b(1.5) - b(1)}{1.5 - 1} = \frac{2^{1.5} - 2^1}{0.5} = \frac{2.8284 - 2}{0.5} \approx 1.6569$$

in units of millions of bacteria per hour.

The average rate of change represents the slope of the secant line joining the points $(1, 2)$ and $(1.5, 2.8284)$; see Figure 4.1.5. Using the point-slope form

$$y - y_0 = m(x - x_0)$$

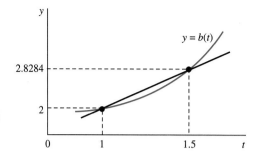

FIGURE 4.1.5

The secant line connecting $t = 1$ and $t = 1.5$

where m is the slope and (x_0, y_0) is the base point, we compute the equation of the secant line to be

$$y - 2 = 1.6569(t - 1)$$
$$y = 1.6569t + 0.3431$$

Although $t = 1.5$ is still far from $t = 1$ (and so does not help us assess the change in $b(t)$ when $t = 1$), Example 4.1.1 illustrates the beginning of the process that will provide us with an answer. We now fully describe this process.

Instantaneous Rate of Change

To improve our estimate for the average rate of change of $b(t) = 2^t$ at $t = 1$, we keep moving closer to $t = 1$. As the time, Δt, between measurements becomes smaller, the average rate of change between times $t = 1$ and $t = 1 + \Delta t$ (note that Δt could be positive or negative) becomes a better and better description of what is happening exactly at $t = 1$. We visualize this process by drawing secant lines with the value Δt getting smaller and smaller; see Figure 4.1.6.

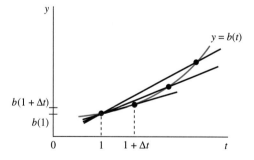

FIGURE 4.1.6

Secant lines with small values of Δt

The average rates of change over smaller and smaller intervals based at $t = 1$ are recorded in Table 4.1.2.

Table 4.1.2

Δt	Interval	Average Rate of Change	Δt	Interval	Average Rate of Change
−0.2	[0.8, 1]	1.2945	0.2	[1, 1.2]	1.4870
−0.1	[0.9, 1]	1.3393	0.1	[1, 1.1]	1.4355
−0.01	[0.99, 1]	1.3815	0.01	[1, 1.01]	1.3911
−0.001	[0.999, 1]	1.3858	0.001	[1, 1.001]	1.3868
−0.0001	[0.9999, 1]	1.3862	0.0001	[1, 1.0001]	1.3863

As Δt becomes smaller, there is less and less time for anything to happen, and the **change** in population, Δb, becomes small. However, the **average rate of change,** $\Delta b/\Delta t$, does not become small because the change takes place over shorter and shorter times (i.e., we are dividing two small numbers).

In fact, the average rate of change approaches a value of about 1.3863. We would like to define this value as the **instantaneous rate of change,** the rate of change *exactly* at $t = 1$ (which—finally—would answer the question we asked at the beginning of this section).

In order to define the instantaneous rate of change, we need to formalize this process of making Δt smaller and smaller (i.e., making "Δt approach 0") and computing average rates of change over intervals whose length, Δt, is shrinking. Said differently, we need to calculate the slopes of secant lines

$$\frac{\Delta b}{\Delta t} = \frac{b(1 + \Delta t) - b(1)}{\Delta t}$$

as the point $(1 + \Delta t, b(1 + \Delta t))$ slides along the graph of $y = b(t)$ toward the base point $(1, b(1)) = (1, 2)$; see Figure 4.1.6. As well, in Figure 4.1.7 we show secant lines with base point $(1, 2)$ for $\Delta t = 0.5$, $\Delta t = 0.3$, and $\Delta t = 0.1$.

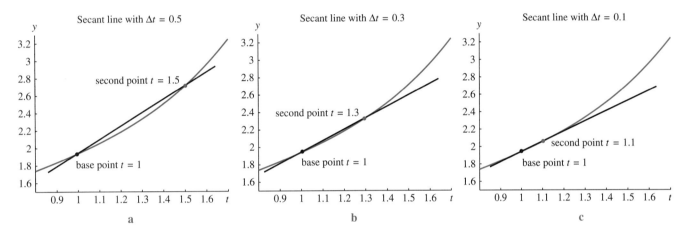

FIGURE 4.1.7

Secant lines

We often use the notation $\Delta t \to 0$ to say that the quantity Δt becomes smaller and smaller.

This formal process of obtaining the instantaneous rate of change relies on the concept of the **limit,** which we will define in the next section and explore throughout this chapter. It will turn out that the instantaneous rate of change is the limit of the average rates of change as Δt approaches 0.

The line that is the limiting position of the secant lines as Δt approaches 0 is called the **tangent line.** Thus, the slope of the tangent line is the limit of the slopes of the secant lines as $\Delta t \to 0$.

Back to our example: assuming that the limit of the slope of the secant lines (with base point at $(1, 2)$) is equal to 1.3863 (the value we deduced from Table 4.1.2), we calculate the equation of the tangent line:

$$y - 2 = 1.3863(x - 1)$$

$$y = 1.3863x + 0.6137$$

In Figure 4.1.8 we illustrate an important property of the tangent line: as we zoom in on the graph of the function near the base point (i.e., near the point of tangency), the curve looks more and more like its tangent. (Not all functions share this property—we will discuss this in Section 4.5.)

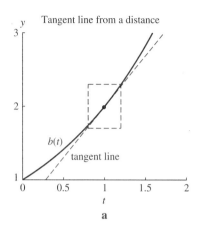
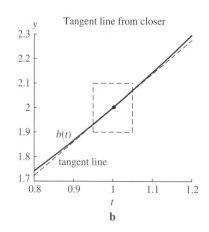
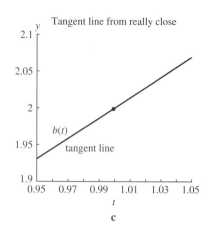

FIGURE 4.1.8

Zooming in on the curve and its tangent

The most familiar instantaneous rate of change is speed, i.e., the rate of change of position as measured by a speedometer. The speed is estimated by measuring the position at one time and again at a later time, and then dividing the distance moved by the time elapsed:

$$\text{average speed} = \text{average rate of change} = \frac{\text{change in position}}{\text{change in time}}$$

Again, this value tells only what happened on **average** during the interval. To find the speed *at a particular moment*, we use the idea of the instantaneous rate of change.

Example 4.1.2 Average Speed and Instantaneous Speed

Assume that the position of a car (i.e., the distance in metres from the start) is given by the function $s(t) = 0.8t^2 + 2t$, where the time, t, is in seconds. Find its speed when $t = 4$.

Our strategy is to approximate the speed at $t = 4$ (i.e., the instantaneous speed) by average speeds over shorter and shorter time intervals. For instance, the average speed from $t = 4$ to $t = 4.2$ is

$$\frac{\Delta s}{\Delta t} = \frac{s(4.2) - s(4)}{4.2 - 4} = \frac{22.512 - 20.8}{0.2} = \frac{1.712}{0.2} = 8.56 \text{ m/s}$$

The average speeds calculated over shrinking time intervals (see Table 4.1.3) seem to be approaching 8.4 m/s. We conclude that the instantaneous speed of the car when $t = 4$ is 8.4 m/s.

Table 4.1.3

Δt	Interval	Average speed (m/s)
0.1	[4, 4.1]	8.48000
0.05	[4, 4.05]	8.44000
0.01	[4, 4.01]	8.40800
0.001	[4, 4.001]	8.40080
0.0001	[4, 4.0001]	8.40010
0.00001	[4, 4.00001]	8.40000

Example 4.1.3 Finding a Tangent Line

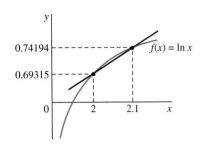

FIGURE 4.1.9

Secant line to $y = \ln x$ (units on x-axis not to scale)

Find the equation of the tangent line to the graph of $f(x) = \ln x$ at $x = 2$.

To start, we calculate the slopes of the secant lines,

$$\text{slope} = \frac{\text{change in } f}{\text{change in } x} = \frac{\Delta f}{\Delta x} = \frac{f(2 + \Delta x) - f(2)}{\Delta x} = \frac{\ln(2 + \Delta x) - \ln 2}{\Delta x}$$

for smaller and smaller values of Δx, both positive and negative (i.e., we let Δx approach 0); see Table 4.1.4 and Figure 4.1.9, where we illustrate the case $\Delta x = 0.1$.

The values $\Delta f/\Delta x$ in the table suggest that the slope of the tangent is 0.5.

Table 4.1.4

Δx	Slope	Δx	Slope
−1	0.69315	1	0.40547
−0.5	0.57536	0.5	0.44629
−0.1	0.51293	0.1	0.48790
−0.01	0.50125	0.01	0.49875
−0.001	0.50013	0.001	0.49988

We now compute the equation of the tangent line (the base point is $(2, \ln 2)$):

$$y - \ln 2 = 0.5(x - 2)$$

$$y = 0.5x + \ln 2 - 1$$

We have to be careful when calculating expressions where Δx is supposed to approach zero. Since the average rate of change is

$$\frac{\Delta f}{\Delta x} = \frac{f(x + \Delta x) - f(x)}{\Delta x}$$

we cannot allow its denominator, Δx, to be zero. Thus, the phrase "Δx approaches zero" needs to include the fact that $\Delta x \neq 0$.

To conclude, we find the instantaneous rate of change by computing the average rate of change with smaller and smaller values of Δx, but without ever reaching $\Delta x = 0$. Graphically, this corresponds to moving the second point closer and closer to the base point and seeing that the secant line gets closer and closer to the tangent line.

Summary To describe how a function changes, we defined the **average rate of change,** and the **instantaneous rate of change.** The average rate of change represents the **slope of the secant line** connecting two data points. By calculating the average rates of change over shorter and shorter intervals, we obtain an estimate for the instantaneous rate of change. Geometrically, we have estimated the **slope of the tangent line.**

4.1 Exercises

Mathematical Techniques

1–4 ■ For each function, find the average rate of change between the given base point, t_0, and times $t_0 + \Delta t$ for the following four values of Δt: $\Delta t = 1$, $\Delta t = 0.5$, $\Delta t = 0.1$, and $\Delta t = 0.01$.

1. $f(t) = 2 + 3t$ with base point $t_0 = 1$.
2. $h(t) = t^2 + 1$ with base point $t_0 = 0$.
3. $G(t) = e^{2t}$ with base point $t_0 = 0$.
4. $G(t) = e^{-t}$ with base point $t_0 = 0$.

5–8 ■ For each function, find the equation of the secant line connecting the given base point, t_0, and times $t_0 + \Delta t$ for $\Delta t = 1$,

$\Delta t = 0.5$, $\Delta t = 0.1$, and $\Delta t = 0.01$. Sketch the function and each of the secant lines.

5. $f(t) = 2 + 3t$ with base point $t_0 = 1$.
6. $h(t) = t^2 + 1$ with base point $t_0 = 0$.
7. $G(t) = e^{2t}$ with base point $t_0 = 0$.
8. $G(t) = e^{-t}$ with base point $t_0 = 0$.

9–12 ■ Using the results in Exercises 1–4, take a guess at the limit of the slopes of the secants, and find the slope and equation of the tangent line.

9. $f(t) = 2 + 3t$ with base point $t_0 = 1$. Call the tangent line function $L(t)$ (based on Exercise 1).
10. $h(t) = t^2 + 1$ with base point $t_0 = 0$. Call the tangent line function $L(t)$ (based on Exercise 2).
11. $G(t) = e^{2t}$ with base point $t_0 = 0$. Call the tangent line function $L(t)$ (based on Exercise 3).
12. $G(t) = e^{-t}$ with base point $t_0 = 0$. Call the tangent line function $L(t)$ (based on Exercise 4).
13. Consider the function $f(x) = \arctan x$. Using $x_0 = 0$ as the base point, find the slopes of the secant lines with the following values of Δx: 0.5, 0.1, 0.01, 0.005, −0.5, −0.1, and −0.005. Guess the value of the slope of the tangent line. Use your guess to find the equation of the tangent line at $x_0 = 0$.
14. Consider the function $f(x) = \sqrt[3]{x}$. Using $x_0 = 0$ as the base point, find the slopes of the secant lines with the following values of Δx: 0.1, 0.01, 0.001, 10^{-4}, 10^{-5}, and 10^{-10}. Explain what is happening. Sketch the graph of the average rate of change of $f(x)$ near $x_0 = 0$.
15. The function $m(t)$ represents the number of monkeys at time t (in months); assume that $t = 1$ represents January. What does the quantity $m(6) - m(2)$ represent? What does the quantity $(m(6) - m(2))/4$ represent? What does the instantaneous rate of change when $t = 6$ represent? What are its units?
16. The function $c(t)$ represents the concentration of a pollutant in the air (in grams per cubic metre) on day t; assume that $t = 0$ represents Sunday. What does the quantity $c(5) - c(3)$ represent? What does the quantity $(c(5) - c(3))/2$ represent? What does the instantaneous rate of change when $t = 5$ represent? What are its units?

Applications

17–18 ■ For each equation for population size, find the following and illustrate on a graph.

a. The population at times 0, 1, and 2.
b. The average rate of change between times 0 and 1.
c. The average rate of change between times 1 and 2.

17. A population of bacteria described by the formula $b(t) = 1.5^t$, where the time, t, is measured in hours.
18. A population of bacteria described by the formula $b(t) = 1.2^t$, where the time, t, is measured in hours.

19–22 ■ a. For each equation for population size, find the following.

  i. The average rate of change between times 0 and 1.
  ii. The average rate of change between times 0 and 0.1.
  iii. The average rate of change between times 0 and 0.01.
  iv. The average rate of change between times 0 and 0.001.

b. What do you think the limit is?
c. Graph the tangent line.

19. A population following $b(t) = 1.5^t$.
20. A population following $b(t) = 2^t$.
21. A population following $h(t) = 5t^2$.
22. A bacterial population following $b(t) = (1 + 2t)^3$.

23–26 ■ For each bacterial population, find the average rate of change during the first hour and during the first and second half-hours. Graph the data and the secant lines associated with the half-hour average rates of change. Which populations change more rapidly during the first half-hour?

23. $b(t) = 3(2^t)$
24. $b(t) = e^{0.5t}$
25. $b(t) = 2e^{-0.5t}$
26. $b(t) = 3(0.5^t)$

27–28 ■ Consider the following data about a tree.

Age	Height (m)	Mass (tonnes)
0	10.11	30
1	11.18	39.1
2	12.40	50.6
3	13.74	65.8
4	15.01	85.9
5	16.61	111.6
6	18.27	144.2
7	20.17	187.7
8	22.01	244.1
9	24.45	319.2
10	26.85	414.2

For each of the measurements:

a. Estimate the rate of change at each age.
b. Graph the rate of change as a function of age.
c. Find and graph the rate of change divided by the value as a function of age.

27. The height.
28. The mass.

Computer Exercises

29. Consider the function
$$f(x)=\sqrt{1-x^2}$$
defined for $-1 \leq x \leq 1$. This is the equation for a semicircle. The tangent line at the base point $(\sqrt{2}/2, \sqrt{2}/2)$ has slope -1. Graph this tangent line. Now zoom in on the base point. Does the circle look more and more like the tangent line? How far do you need to go before the circle looks flat? Would a tiny insect be able to tell that his world was curved?

30. Suppose a bacterial population oscillates with the formula
$$b(t) = 2 + \cos t$$

   a. Graph this function.
   b. Find and graph the function that gives the rate of change between times t and $t+1$ as a function of t for $0 \leq t \leq 10$.
   c. Find and graph the function that gives the rate of change between times t and $t+0.1$ as a function of t for $0 \leq t \leq 10$.
   d. Try the same with smaller values of Δt. Do you have any idea what the limit might be?

31. Repeat the steps in Exercise 30 for a bacterial population that follows the formula
$$b(t) = e^{-0.1t}(2 + \cos t)$$

4.2 Limit of a Function

The derivative, the mathematical version of the instantaneous rate of change and the slope of a curve, includes a **limit** in its definition. We will now study the mathematical and scientific basis of this fundamental idea. By understanding the useful properties of limits, we will be able to calculate limits of many functions. We approach limits from the **numeric** (by analyzing table of values), **geometric** (by analyzing graphs), and **algebraic** (by developing and using limit laws) viewpoints.

Limit

To start, we look at an example. We are interested in the behaviour of the function
$$f(x) = \frac{x^3 - 1}{x - 1}$$
for values of x that are close to 1.

Note that $f(x)$ is not defined when $x = 1$. If $x \neq 1$, then
$$f(x) = \frac{x^3 - 1}{x - 1} = \frac{(x-1)(x^2 + x + 1)}{x - 1} = x^2 + x + 1$$

Thus, the graph of $f(x)$ is the parabola $y = x^2 + x + 1$ from which the point $(1, 3)$ has been removed (Figure 4.2.10a).

Examining the values of $f(x)$ in Table 4.2.1, we notice that as x gets closer to 1 (either from the left or from the right side), the values of $f(x)$ seem to get closer to 3.

Table 4.2.1

x	f(x)	x	f(x)
0.5	1.750000	1.2	3.640000
0.8	2.440000	1.1	3.310000
0.9	2.710000	1.05	3.152500
0.99	2.970100	1.01	3.030100
0.999	2.997001	1.001	3.003001

As a matter of fact, we can make the values of $f(x)$ as close to 3 as desired by taking x close enough to 1. For instance, to make $f(x)$ within 0.05 of 3, i.e.,
$$2.95 < f(x) < 3.05$$
it is enough to take values of x in the interval $(0.99, 1.01)$, as we can check by looking at Table 4.2.1; see Figure 4.2.10b.

FIGURE 4.2.10

Limit of a function

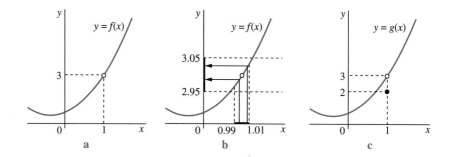

Can we make $f(x)$ fall within 0.00005 of 3, i.e., make it satisfy

$$2.99995 < f(x) < 3.00005$$

We continue building the table of values (Table 4.2.2).

Table 4.2.2

x	f(x)	x	f(x)
0.9999	2.999700	1.0001	3.000300
0.99999	2.999970	1.00001	3.000030
0.999999	2.999997	1.000001	3.000003

If we take x in (0.99999, 1.00001), the corresponding values for $f(x)$ will satisfy

$$2.99997 < f(x) < 3.00003$$

which is a bit better than needed.

The measure of closeness (first we took 0.05 and then 0.00005) is sometimes called the **tolerance** (or **prescribed precision**). It appears that we can make $f(x)$ fall within any given tolerance from 3, as long as we force x to be close enough to 1. This is what is meant by saying that the **limit of $f(x)$ is 3 as x approaches 1.** We write this fact as

$$\lim_{x \to 1} f(x) = 3$$

Before stating the definition of the limit, we give a precise meaning to the word *near*, which is used in various contexts in math, including limits and continuity. We say that a number x is **near** some number a if x lies in an open interval that contains a, and possibly $x \neq a$. In a moment we will see that the part "and possibly $x \neq a$" is related to the definition of the limit. The phrase "$f(x)$ **is defined near** a" means that $f(x)$ is defined for the values of x near a; i.e., $f(x)$ is defined for all x in some open interval that contains a, except possibly at a.

Definition 4.2.1 Assume that $f(x)$ is defined near a. We say that the **limit of $f(x)$ as x approaches a is equal to L,** and write

$$\lim_{x \to a} f(x) = L$$

if we can make the values of $f(x)$ as close to L as desired by taking x close enough to a, but not equal to a.

By "taking x close enough to a" we mean that we can take any x in some small interval that contains a, and $x \neq a$; see Examples 4.2.1 and 4.2.2 to see how that interval is found.

In other words, $\lim_{x \to a} f(x) = L$ means that the values $f(x)$ get closer and closer to L as the numbers x get closer and closer to a (from either side), but $x \neq a$.

Note that the behaviour of $f(x)$ *at* a is irrelevant for the limit. Going back to our introductory example, define

$$g(x) = \begin{cases} \dfrac{x^3 - 1}{x - 1} & \text{if } x \neq 1 \\ 2 & \text{if } x = 1 \end{cases}$$

i.e., $g(x)$ is the same as $f(x)$ for $x \neq 1$; $f(x)$ is not defined at $x = 1$, whereas $g(x)$ is defined, and is equal to 2. Nevertheless,

$$\lim_{x \to 1} g(x) = \lim_{x \to 1} f(x) = 3$$

(compare $f(x)$ in Figure 4.2.10a and $g(x)$ in Figure 4.2.10c). Replacing 2 in the definition of $g(x)$ with any other number does not change the value of the limit of $g(x)$ as x approaches 1.

To further illustrate this important point, we note that all three functions in Figure 4.2.11 satisfy $\lim_{x \to a} f(x) = L$.

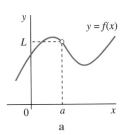
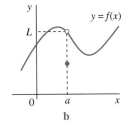
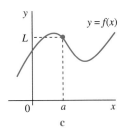

a b c

FIGURE 4.2.11

All three functions have the same limit as x approaches a

For the function in Figure 4.2.11a, the value $f(a)$ is not defined. In Figure 4.2.11b, $f(a)$ is defined, but it is not equal to L. In Figure 4.2.11c, $f(a)$ is defined and is equal to L.

As we can see, the limit (alone) is not able to describe significant differences in the relationship between the behaviour of the three functions *near* a and *at* a. The concept of continuity (which we discuss in a forthcoming section) will help us accomplish this.

Example 4.2.1 How Close Must the Input Be?

Consider the function

$$f(x) = \frac{4x + 2x^2}{x}$$

which is not defined at $x = 0$. For any $x \neq 0$, $f(x) = 4 + 2x$ because we can divide the numerator by x. We will soon show that

$$\lim_{x \to 0} f(x) = 4$$

The idea of the limit says that we can pick the values of x close enough to zero (but not equal to zero) to guarantee that $f(x)$ gets as close to 4 as desired. For example, if we wish $f(x)$ to be within 0.1 of the limit, we require

$$3.9 < f(x) < 4.1$$
$$3.9 < 4 + 2x < 4.1$$
$$-0.1 < 2x < 0.1$$
$$-0.05 < x < 0.05$$

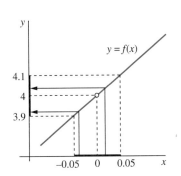

FIGURE 4.2.12

Finding how close the input must be

Thus, if the inputs x lie inside the interval $(-0.05, 0.05)$, and $x \neq 0$, then the corresponding outputs will be in $(3.9, 4.1)$; i.e., $f(x)$ will be close to 4, within the required tolerance of 0.1 (Figure 4.2.12).

If we wish $f(x)$ to be within 0.01 of the limit, we require

$$3.99 < f(x) < 4.01$$
$$3.99 < 4 + 2x < 4.01$$
$$-0.01 < 2x < 0.01$$
$$-0.005 < x < 0.005$$

and $x \neq 0$. Getting closer to the limit requires inputs that are closer to 0. Because it appears that we can do this for *any* prescribed precision, we conclude that

$$\lim_{x \to 0} \frac{4x + 2x^2}{x} = 4$$

Example 4.2.2   Administering a Drug

At time $t = 0$, a patient is given an initial dose of 15 units of morphine (a pain reliever). As the patient's body absorbs the drug, its amount, $M(t)$, decreases. It has been determined that at time t there are still

$$M(t) = \frac{36}{2.4 + 1.7t}$$

units of morphine left (t is time in hours). Knowing that the patient needs a new dose of morphine when $M(t)$ falls to five units, find the time when the new dose needs to be administered.

From

$$M(t) = \frac{36}{2.4 + 1.7t} = 5$$

by cross-multiplying, we get

$$12 + 8.5t = 36$$
$$t = \frac{24}{8.5} \approx 2.82$$

So the new dose needs to be administered 2.82 hours after the initial dose.

Note that while this is a perfectly sound mathematical calculation, it is not at all realistic. Imagine hospital staff keeping track of each patient and administering drugs at an *exact* time (such as the one calculated above).

A realistic situation is given by answering the following question: the patient needs a new dose when the level of morphine, $M(t)$, comes to within 0.4 of 5 units. When should the patient be given a new dose?

Note that, now, this is related to a limit question (requesting precision). This time, we find t from the requirement that

$$5 - 0.4 < M(t) < 5 + 0.4$$

i.e.,

$$4.6 < \frac{36}{2.4 + 1.7t} < 5.4$$

Taking reciprocals and multiplying by 36, we get

$$\frac{1}{4.6} > \frac{2.4 + 1.7t}{36} > \frac{1}{5.4}$$

$$7.8 > 2.4 + 1.7t > 6.7$$

$$5.4 > 1.7t > 4.3$$

$$3.2 > t > 2.5$$

Thus, a new dose has to be administered 2.5 hours to 3.2 hours after the initial dose.

In the previous section, we studied average rates of change

$$\frac{\Delta b}{\Delta t} = \frac{b(1 + \Delta t) - b(1)}{\Delta t} = \frac{2^{1+\Delta t} - 2}{\Delta t}$$

for the function $b(t) = 2^t$ at the base point $t = 1$. To calculate the instantaneous rate of change, we need to find

$$\lim_{\Delta t \to 0} \frac{\Delta b}{\Delta t} = \lim_{\Delta t \to 0} \frac{2^{1+\Delta t} - 2}{\Delta t}$$

From the table of values (Table 4.1.2) we saw that as Δt becomes smaller and smaller, the average rate of change gets closer and closer to a value near 1.386. We could not take the smallest possible value, $\Delta t = 0$, because that would have led to division of zero by zero.

We can think of the average rate of change

$$\frac{\Delta b}{\Delta t} = \frac{2^{1+\Delta t} - 2}{\Delta t}$$

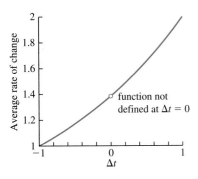

FIGURE 4.2.13
The limit of the average rate of change

as a **function** of Δt that is defined at all points except $\Delta t = 0$ (Figure 4.2.13). We guessed that the limit is 1.386 because the values get closer and closer to 1.386.

We could check by taking even smaller values of Δt, as shown in Table 4.2.3.

Table 4.2.3

Δt	Interval	$\frac{\Delta b}{\Delta t} = \frac{2^{1+\Delta t} - 2}{\Delta t}$
0.0001	[1, 1.0001]	1.3863424
0.00001	[1, 1.00001]	1.3862992
0.000001	[1, 1.000001]	1.3862948

So the values of $\Delta b / \Delta t$ get closer and closer to (rounded to five decimal places) 1.38629.

But when is "closer and closer" close enough, so that we can be fairly confident that we have found the true value of the limit or that we are "close" to its true value?

This mathematical question can be understood by thinking of it *scientifically*. What does it mean scientifically for two quantities to be close to one another?

> Two quantities are close when they are too similar to distinguish with a precise measuring device.

Consider the measurement of temperature. With a crude measuring device, such as waving your hand around in the air, it might be difficult to distinguish between temperatures of 20°C and 21°C. Without a more precise device, the two temperatures are effectively the same. With an ordinary thermometer, we can distinguish between 20°C and 21°C, but we may not be able to distinguish between 20°C and 20.1°C. Two temperatures are *exactly* equal if no thermometer, no matter how precise, can distinguish between them.

We can translate this scientific idea into the mathematical idea of a limit. The average rate of change is a function of Δt (Figure 4.2.13). If we can measure the rate of change with an accuracy of only 0.2, we do not need a very small value of Δt to be within 0.2 of the limit (Figure 4.2.14a). For a more precise measurement of the average rate of change with an accuracy of 0.01, we need a much smaller Δt to be within measurement accuracy of the limit (Figure 4.2.14b). No matter how precisely we want to measure the rate of change, we can always pick a sufficiently small Δt to be within measurement accuracy of the limit.

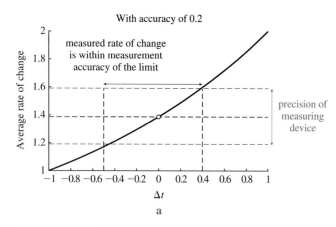
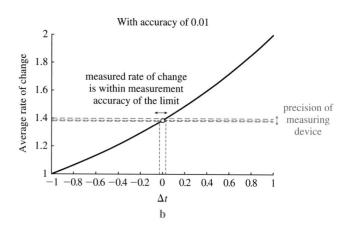

FIGURE 4.2.14

An experimental output approaching a limit

Calculating Limits

How do we calculate

$$\lim_{x \to a} f(x)$$

for a given function, $f(x)$, and the given value of a?

We have already investigated several limits using a **numeric approach.** Note that calculating the table of values of $f(x)$ for the values of x that are getting closer and closer to a is not a proof—at best, it is an indication of what the limit *could be equal to* (however, even as such, it could be quite useful). Things could go wrong, as the following example shows.

Example 4.2.3 Numeric Calculation Suggests an Incorrect Value for the Limit

Let

$$f(x) = \frac{\sqrt{x^4 + 16} - 4}{x^4}$$

To find $\lim_{x \to 0} f(x)$, we construct a table of values (Table 4.2.4) using a calculator capable of handling 10 decimal places.

Table 4.2.4

x	f(x)	x	f(x)
0.1	0.125000	−0.1	0.125000
0.01	0.100000	−0.01	0.100000
0.001	0.000000	−0.001	0.000000
0.0001	0.000000	−0.0001	0.000000

The conclusion we might draw from these values, namely that the limit is 0, is incorrect (in Example 4.2.15 we will show that the true value of the limit is $1/8 = 0.125$). What went wrong? The fourth power of a small number is a very small number, and the calculator (due to the limited number of decimal places it operates with) thinks it is zero! To illustrate this: take $x = 0.001$; then $x^4 = 0.000000000001$, but for the calculator we are using this number is indistinguishable from zero! Thus, the expression under the square root is 16, which makes the numerator—and thus all of $f(x)$—equal to zero (this explains the zero entries in Table 4.2.4).

4.2 Limit of a Function

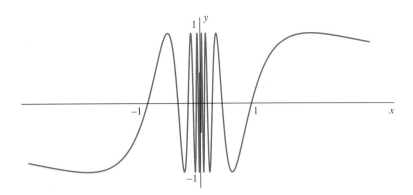

FIGURE 4.2.15
The graph of $f(x) = \sin(\pi/x)$

Although we cannot use it to prove that a function has a limit, the numeric approach (i.e., looking at the values of a function) can help us show that a limit does not exist. Consider the following example.

Example 4.2.4 Using Numeric Calculation to Investigate a Limit

Show that the values of the function $f(x) = \sin(\pi/x)$ do not approach any real number as x approaches zero.

A computer-generated graph of $f(x)$ is shown in Figure 4.2.15. Setting $\sin(\pi/x) = 0$, we get $\pi/x = \pi k$, $1/x = k$, and $x = 1/k$ (k is a nonzero integer). Thus, $\sin(\pi/x) = 0$ for infinitely many values of x that approach 0. From $\sin(\pi/x) = 1$ we get

$$\frac{\pi}{x} = \frac{\pi}{2} + 2\pi k = \frac{\pi + 4k\pi}{2}$$

$$\frac{1}{x} = \frac{1 + 4k}{2}$$

$$x = \frac{2}{4k + 1}$$

where k is an integer. So, the function $f(x) = \sin(\pi/x)$ takes on the value 1 for infinitely many values of x that approach zero. In Exercise 68 we show that the same is true not just for 0 and 1, but for any number between -1 and 1. The graph of $f(x) = \sin(\pi/x)$ keeps oscillating between -1 and 1 as x approaches zero, and does not approach any particular number. We say that the limit of $\sin(\pi/x)$ as x approaches zero does not exist.

In many situations, we can deduce the value of a limit **geometrically**, looking at the graph of a function.

Example 4.2.5 Finding Limits from the Graph

Consider the function whose graph is given in Figure 4.2.16.

We see that as x approaches 2, the values of $f(x)$ get closer and closer to 4; thus

$$\lim_{x \to 2} f(x) = 4$$

But what is the value of the limit of $f(x)$ as x approaches 5?

To describe what happens near 5, we have to be more specific about saying *how x approaches 5*. We see that $f(x)$ approaches 1 as x approaches 5, but only if x is larger than 5, i.e., if it approaches 5 from the right. We write

$$\lim_{x \to 5^+} f(x) = 1$$

and say that the **right-hand limit** of $f(x)$ as x approaches 5 is 1. We keep in mind that the symbol $x \to 5^+$ means $x \to 5$ *and* $x > 5$.

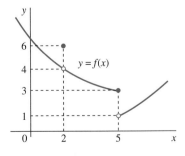

FIGURE 4.2.16
Finding limits from a graph

Likewise, the values of $f(x)$ approach 3 as $x \to 5^-$, i.e., as x approaches 5 and is smaller than 5. We say that the **left-hand limit** of $f(x)$ as x approaches 5 is 3, and write

$$\lim_{x \to 5^-} f(x) = 3$$

Thus, $x \to 5^-$ means $x \to 5$ *and* $x < 5$.

In situations such as this, where the left-hand and right-hand limits are not equal, we say that the function $f(x)$ **does not have a limit** as x approaches 5, or that the limit

$$\lim_{x \to 5} f(x)$$

does not exist.

Next, we formally define the right-hand limit. As was the case with Definition 4.2.1, we need to make sure that the function involved is defined so that the limit makes sense. So, we assume that $f(x)$ is defined for all x near a, such that $x > a$.

Definition 4.2.2 We say that the **right-hand limit** of $f(x)$ as x approaches a (or the **limit of $f(x)$ as x approaches a from the right**) is equal to L, and write

$$\lim_{x \to a^+} f(x) = L$$

if we can make the values of $f(x)$ as close to L as desired by taking any x close enough to a *and* larger than a.

Replacing "larger than" by "smaller than" in the last line of the definition, and assuming that $f(x)$ is defined for all x near a such that $x < a$, we obtain the **left-hand limit** (or, **limit from the left**)

$$\lim_{x \to a^-} f(x) = L$$

Left-hand and right-hand limits are called **one-sided limits.**

From Definitions 4.2.1 and 4.2.2 we conclude that

$$\lim_{x \to a} f(x) = L \text{ if and only if } \lim_{x \to a^+} f(x) = L \text{ and } \lim_{x \to a^-} f(x) = L$$

Thus, if the one-sided limits are equal to the same number L, then the limit of the function exists and is equal to L. If

$$\lim_{x \to a^+} f(x) \neq \lim_{x \to a^-} f(x)$$

then we say that

$$\lim_{x \to a} f(x)$$

does not exist.

Keep in mind that the phrase "$\lim_{x \to a} f(x)$ exists" means that $\lim_{x \to a} f(x)$ is a real number.

Example 4.2.6 One-Sided Limits of the Function $y = \sqrt{x}$

Recalling the graph of $y = \sqrt{x}$ that we drew in Section 1.3, we see that $\lim_{x \to 0^+} \sqrt{x} = 0$. However, $\lim_{x \to 0^-} \sqrt{x}$ does not exist, since the function $y = \sqrt{x}$ is not defined for negative numbers.

Consequently, we say that the two-sided limit $\lim_{x \to 0} \sqrt{x}$ does not exist.

Note that for the function $f(x)$ shown in Figure 4.2.16,

$$\lim_{x \to 2^-} f(x) = 4 \quad \text{and} \quad \lim_{x \to 2^+} f(x) = 4$$

Thus, we say that $\lim_{x \to 2} f(x)$ exists and is equal to 4. We write $\lim_{x \to 2} f(x) = 4$.

Example 4.2.7 Calculation of a Limit

Compute $\lim_{x \to 0} f(x)$, where $f(x) = |x|/x$.

Using the definition of absolute value, we write

$$f(x) = \frac{|x|}{x} = \begin{cases} x/x & \text{if } x > 0 \\ -x/x & \text{if } x < 0 \end{cases} = \begin{cases} 1 & \text{if } x > 0 \\ -1 & \text{if } x < 0 \end{cases}$$

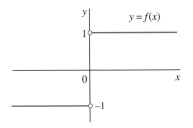

FIGURE 4.2.17
Graph of $f(x) = |x|/x$

(see Figure 4.2.17). Note that $x = 0$ is not in the domain of $f(x)$.

The function $f(x)$ is a constant function $f(x) = 1$ if $x > 0$. Thus,

$$\lim_{x \to 0^+} f(x) = 1$$

Likewise, as x approaches 0 from the left, the values of $f(x)$ approach (are actually equal to) -1, i.e.,

$$\lim_{x \to 0^-} f(x) = -1$$

Since the one-sided limits are not the same, we conclude that $\lim_{x \to 0} f(x)$ does not exist.

Calculating Limits Algebraically—Limit Laws

Limits have many properties that simplify their computation. The basic idea is an important one; we first find the limits of some simple functions and then build limits of more complicated functions by combining these simple pieces with rules that tell how limits add, subtract, multiply, and divide.

The simple functions we begin with are **constant functions,** with formula

$$f(x) = c$$

where c is any real number, and the **identity function** with formula

$$f(x) = x$$

From the graphs of $f(x) = c$ (Figure 4.2.18) and $f(x) = x$ (Figure 4.2.19) we see that for any value of the constant c and for any real number a,

$$\lim_{x \to a} c = c \quad \text{and} \quad \lim_{x \to a} x = a$$

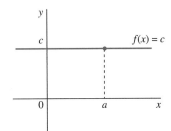

FIGURE 4.2.18
Limit of a constant function

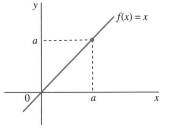

FIGURE 4.2.19
Limit of the identity function

To calculate limits of more complicated functions, we use the following **limit laws.**

Theorem 4.2.1 **Properties of Limits**

Suppose $f(x)$ and $g(x)$ are functions such that $\lim_{x \to a} f(x)$ and $\lim_{x \to a} g(x)$ exist. Then:

a. The limit of the sum is the sum of the limits, or
$$\lim_{x \to a}[f(x) + g(x)] = \lim_{x \to a} f(x) + \lim_{x \to a} g(x)$$

b. The limit of the difference is the difference of the limits, or
$$\lim_{x \to a}[f(x) - g(x)] = \lim_{x \to a} f(x) - \lim_{x \to a} g(x)$$

c. The limit of the product of a constant, c, and a function is the product of the constant and the limit of the function, or
$$\lim_{x \to a}[cf(x)] = c \cdot \lim_{x \to a} f(x)$$

d. The limit of the product is the product of the limits, or
$$\lim_{x \to a}[f(x)g(x)] = \left[\lim_{x \to a} f(x)\right] \cdot \left[\lim_{x \to a} g(x)\right]$$

e. Suppose $\lim_{x \to a} g(x) \neq 0$. Then the limit of the quotient is the quotient of the limits, or
$$\lim_{x \to a}\left[\frac{f(x)}{g(x)}\right] = \frac{\lim_{x \to a} f(x)}{\lim_{x \to a} g(x)}$$

Note that all laws and properties that we stated in the theorem for (two-sided) limits hold for one-sided limits as well.

Example 4.2.8 Finding the Limit of a Polynomial

If we use law (d) with $f(x) = g(x) = x$, we get that
$$\lim_{x \to a} x^2 = \lim_{x \to a}(x \cdot x) = \left(\lim_{x \to a} x\right)\left(\lim_{x \to a} x\right) = a \cdot a = a^2$$

Likewise,
$$\lim_{x \to a} x^n = a^n$$

for any positive integer n. For instance,
$$\lim_{x \to -4} x^3 = (-4)^3 = -64 \quad \text{and} \quad \lim_{x \to 2} x^{10} = 2^{10} = 1024$$

Using what we just discovered together with the limit laws, we can calculate the limit of a polynomial:

$$\lim_{x \to -1}(x^3 - 2x^2 + 4) = \lim_{x \to -1} x^3 - 2 \lim_{x \to -1} x^2 + \lim_{x \to -1} 4$$
$$= (-1)^3 - 2(-1)^2 + 4 = 1$$

Analyzing the limits we calculated in Example 4.2.8, we see that the answer could have been obtained by direct substitution—i.e., by replacing x in the function with the given value, a. Clearly, we can calculate the limit of any polynomial in the same way. As well, using limit law (e), we conclude that direct substitution applies to all rational functions as well.

Theorem 4.2.2 Direct Substitution Rule

If $f(x)$ is a polynomial or a rational function (in which case a must be in the domain of $f(x)$), then

$$\lim_{x \to a} f(x) = f(a)$$

Example 4.2.9 The Limit of a Rational Function

We can compute the limit

$$\lim_{x \to 5} \frac{2x^2 + 3x}{2x + 1} = \frac{2 \cdot 5^2 + 3 \cdot 5}{2 \cdot 5 + 1} = \frac{65}{11} \approx 5.91$$

by plugging in $x = 5$ because the denominator is not equal to 0 at $x = 5$.

Example 4.2.10 Finding the Limit of the Average Rate of Change Algebraically

Consider finding the instantaneous rate of change of the distance travelled by a falling rock, which follows the quadratic function (approximately the value on Earth)

$$y(t) = 5t^2$$

where t is measured in seconds and y is measured in metres fallen (Figure 4.2.20). We wish to find the rate of change at $t = 1$. The change in distance, Δy, between times 1 and $1 + \Delta t$ is

$$\Delta y = y(1 + \Delta t) - y(1)$$
$$= 5(1 + \Delta t)^2 - 5$$
$$= 5\left(1 + 2\Delta t + \Delta t^2\right) - 5$$
$$= 10\Delta t + 5\Delta t^2$$

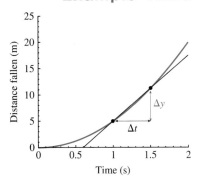

FIGURE 4.2.20
The distance travelled by a falling rock and the average rate of change

Thus, the average rate of change (i.e., the average speed) over the interval $t = 1$ to $t = 1 + \Delta t$ is

$$\frac{\Delta y}{\Delta t} = \frac{10\Delta t + 5\Delta t^2}{\Delta t} = \frac{\Delta t(10 + 5\Delta t)}{\Delta t}$$

As long as $\Delta t \neq 0$, we can divide out the Δt, so that

$$\frac{\Delta y}{\Delta t} = 10 + 5\Delta t$$

(Figure 4.2.21). The instantaneous rate of change is

$$\lim_{\Delta t \to 0} \frac{\Delta y}{\Delta t} = \lim_{\Delta t \to 0} (10 + 5\Delta t) = 10$$

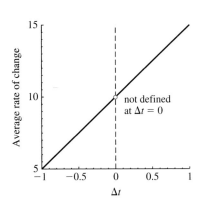

FIGURE 4.2.21
The average rate of change of distance for the falling rock

Thus, the speed of the rock one second after it is dropped is 10 m/s.

By repeated use of law (d) we can prove that

$$\lim_{x \to a} [f(x)]^n = \left[\lim_{x \to a} f(x)\right]^n$$

So, if we can calculate $\lim_{x \to a} f(x)$ by direct substitution, then

$$\lim_{x \to a} [f(x)]^n = [f(a)]^n$$

Furthermore, it can be proven that the direct substitution rule applies to root functions

$$\lim_{x \to a} \sqrt[n]{x} = \sqrt[n]{a}$$

as long as a is in the domain of the function $y = \sqrt[n]{x}$.

Through a sequence of results (that we do not prove here) we arrive at the following general fact:

Direct Substitution Rule

If $f(x)$ is an algebraic, exponential, logarithmic, trigonometric, or inverse trigonometric function, then

$$\lim_{x \to a} f(x) = f(a)$$

for all a in the domain of $f(x)$.

Recall that an algebraic function is obtained from polynomials using composition, elementary algebraic operations, and power and root functions.

Example 4.2.11 Direct Substitution and When It Fails to Work

Consider the function $f(x) = \frac{x^3 - 4x}{x - 2}$.

We compute

$$\lim_{x \to -1} f(x) = \lim_{x \to -1} \frac{x^3 - 4x}{x - 2} = \frac{(-1)^3 - 4(-1)}{(-1) - 2} = -1$$

and

$$\lim_{x \to 0} f(x) = \lim_{x \to 0} \frac{x^3 - 4x}{x - 2} = \frac{(0)^3 - 4(0)}{0 - 2} = 0$$

However, we cannot calculate

$$\lim_{x \to 2} f(x)$$

by direct substitution since $a = 2$ is not in the domain of $f(x)$. The trick is to simplify before taking the limit. We start by factoring

$$\lim_{x \to 2} \frac{x^3 - 4x}{x - 2} = \lim_{x \to 2} \frac{x(x-2)(x+2)}{x - 2}$$
$$= \lim_{x \to 2} [x(x+2)]$$
$$= (2)(2+2) = 8$$

Note that after cancelling the fraction, we did evaluate the limit by direct substitution. One more comment—we were allowed to cancel the fraction by $x - 2$, since $x \to 2$ implies that $x \neq 2$, and thus $x - 2$ is not zero!

Example 4.2.12 Calculation of a Limit

Find

$$\lim_{h \to 0} \frac{(2+h)^2 - 4}{h}$$

(We will study limits of this kind extensively in the context of derivatives.)

Since $h = 0$ makes the denominator equal to zero, we cannot use the direct substitution rule right away. Instead, we start by simplifying:

$$\lim_{h \to 0} \frac{(2+h)^2 - 4}{h} = \lim_{h \to 0} \frac{4 + 4h + h^2 - 4}{h}$$

$$= \lim_{h \to 0} \frac{h(4+h)}{h}$$
$$= \lim_{h \to 0} (4+h) = 4$$

Note that $h \to 0$ implies that $h \neq 0$ (that is part of the definition of $h \to 0$!); that's why we were able to cancel the h in the fraction. In the last line we calculated the limit by direct substitution.

Example 4.2.13   Calculation of One-Sided Limits

Find $\lim_{x \to 3} f(x)$ if

$$f(x) = \begin{cases} \dfrac{1}{x} + 2 & \text{if } x \geq 3 \\ \sqrt{x^2+16} - \dfrac{8}{3} & \text{if } x < 3 \end{cases}$$

Since the definition of $f(x)$ changes at 3, we need to use one-sided limits.
We compute

$$\lim_{x \to 3^+} f(x) = \lim_{x \to 3^+} \left(\frac{1}{x} + 2\right) = \frac{1}{3} + 2 = \frac{7}{3}$$

and

$$\lim_{x \to 3^-} f(x) = \lim_{x \to 3^-} \left(\sqrt{x^2+16} - \frac{8}{3}\right) = \sqrt{9+16} - \frac{8}{3} = \frac{7}{3}.$$

The one-sided limits exist and are equal, and we conclude that

$$\lim_{x \to 3} f(x) = \frac{7}{3}$$

Example 4.2.14   Limits of Transcendental Functions

To find

$$\lim_{x \to 2} \ln(x^3 + 4xe^x)$$

we use direct substitution, since $x = 2$ is in the domain of the function:

$$\lim_{x \to 2} \ln(x^3 + 4xe^x) = \ln((2)^3 + 4(2)e^2) = \ln(8 + 8e^2) \approx 4.206$$

Likewise

$$\lim_{x \to 0} \frac{\sin x + 3}{\cos x} = \frac{\sin 0 + 3}{\cos 0} = \frac{0+3}{1} = 3$$

and

$$\lim_{x \to 10} (2x \arctan x - 4) = 2(10) \arctan 10 - 4$$
$$= 20 \arctan 10 - 4 \approx 25.423$$

Example 4.2.15   Calculation of the Limit from Example 4.2.3

To compute

$$\lim_{x \to 0} \frac{\sqrt{x^4+16} - 4}{x^4}$$

we cannot use direct substitution because the denominator is zero when $x=0$. We simplify first:

$$\lim_{x \to 0} \frac{\sqrt{x^4+16} - 4}{x^4} = \lim_{x \to 0} \left(\frac{\sqrt{x^4+16} - 4}{x^4} \cdot \frac{\sqrt{x^4+16} + 4}{\sqrt{x^4+16} + 4}\right)$$

$$= \lim_{x \to 0} \frac{x^4 + 16 - 16}{x^4(\sqrt{x^4 + 16} + 4)}$$

$$= \lim_{x \to 0} \frac{1}{\sqrt{x^4 + 16} + 4}$$

$$= \frac{1}{\sqrt{16} + 4} = \frac{1}{8}$$

Example 4.2.16 Calculation of a Limit Involving Absolute Value

Find
$$\lim_{x \to 5} \frac{2x - 10}{|x - 5|}$$

To start, we recall what absolute value means:

$$|x - 5| = \begin{cases} x - 5 & \text{if } x - 5 \geq 0 \\ -(x - 5) & \text{if } x - 5 < 0 \end{cases} = \begin{cases} x - 5 & \text{if } x \geq 5 \\ -(x - 5) & \text{if } x < 5 \end{cases}$$

Since the denominator is zero when $x = 5$, we cannot use the direct substitution. The fact that the definition of the function changes at $x = 5$ suggests that we compute the one-sided limits (as in Example 4.2.13). Thus (recall that $x \to 5^+$ implies that $x > 5$),

$$\lim_{x \to 5^+} \frac{2x - 10}{|x - 5|} = \lim_{x \to 5^+} \frac{2x - 10}{x - 5} = \lim_{x \to 5^+} \frac{2(x - 5)}{x - 5} = \lim_{x \to 5^+} 2 = 2$$

and

$$\lim_{x \to 5^-} \frac{2x - 10}{|x - 5|} = \lim_{x \to 5^-} \frac{2x - 10}{-(x - 5)} = \lim_{x \to 5^-} \frac{2(x - 5)}{-(x - 5)} = \lim_{x \to 5^-} (-2) = -2$$

We conclude that $\lim_{x \to 5} \frac{2x - 10}{|x - 5|}$ does not exist.

When direct substitution or other methods fail, we might be able to use the following result.

Theorem 4.2.3 The Squeeze Theorem (or the Sandwich Theorem)
Suppose that $m(x) \leq f(x) \leq M(x)$ for all x near a. If

$$\lim_{x \to a} m(x) = \lim_{x \to a} M(x) = L$$

then

$$\lim_{x \to a} f(x) = L$$

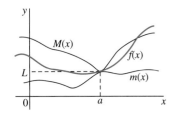

FIGURE 4.2.22
The Squeeze Theorem

Recall that "x near a" includes the possibility that $x \neq a$. In words—if $f(x)$ is squeezed between two functions $m(x)$ and $M(x)$ near a, and if $m(x)$ and $M(x)$ approach the same value L as $x \to a$, then, as $x \to a$, $f(x)$ approaches L as well; see Figure 4.2.22.

As an illustration, suppose that we know that $2x \ln x + 5 \leq f(x) \leq 7x^2 - 2$ for all x near 1. Then from

$$\lim_{x \to 1} (2x \ln x + 5) = 5 \quad \text{and} \quad \lim_{x \to 1} (7x^2 - 2) = 5$$

with the help of the Squeeze Theorem we conclude that $\lim_{x \to 1} f(x) = 5$.

Example 4.2.17 Using the Squeeze Theorem

Show that
$$\lim_{x \to 0} x^2 \sin(\pi/x) = 0$$

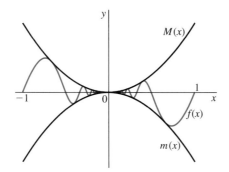

FIGURE 4.2.23
Visualizing $-x^2 \leq x^2 \sin(\pi/x) \leq x^2$

We cannot use Theorem 4.2.1 part (d) to write

$$\lim_{x \to 0} x^2 \sin(\pi/x) = \lim_{x \to 0} x^2 \cdot \lim_{x \to 0} \sin(\pi/x)$$

because $\lim_{x \to 0} \sin(\pi/x)$ does not exist (see Example 4.2.4). To use the Squeeze Theorem we need an inequality, so we start with

$$-1 \leq \sin(\pi/x) \leq 1$$

which is true because the sine of any real number is between -1 and 1. Multiplying all sides by x^2 (and keeping the inequalities since $x^2 \geq 0$), we obtain

$$-x^2 \leq x^2 \sin(\pi/x) \leq x^2$$

Thus, the function $f(x) = x^2 \sin(\pi/x)$ is squeezed between $m(x) = -x^2$ and $M(x) = x^2$; see Figure 4.2.23. Since

$$\lim_{x \to 0} m(x) = \lim_{x \to 0} (-x^2) = 0 \quad \text{and} \quad \lim_{x \to 0} M(x) = \lim_{x \to 0} x^2 = 0$$

the Squeeze Theorem implies that $\lim_{x \to 0} x^2 \sin(\pi/x) = 0$.

Summary We developed the concept of a **limit** and defined **one-sided limits.** After investigating limits **numerically** (using a table of values), we learned how to find limits from **graphs,** as well as **algebraically,** using the **limit laws.** In cases when the limit laws cannot be used, the **Squeeze Theorem** might provide an answer.

4.2 Exercises

Mathematical Techniques

1. Explain in your own words what is meant by saying that, for some function $f(x)$, $\lim_{x \to 0^-} f(x) = -2$ and $\lim_{x \to 0^+} f(x) = 4$.

2. Draw a possible graph of a function $f(x)$ that satisfies $\lim_{x \to 1^-} f(x) = 3$, $\lim_{x \to 1^+} f(x) = 4$, and $f(1) = 5$.

3. Draw a possible graph of a function $f(x)$ that satisfies $\lim_{x \to 3^-} f(x) = 2$, $\lim_{x \to 3^+} f(x) = 2$, and $f(3) = -2$.

4–7 ▪ Given below each question is the graph of a function $f(x)$. Find the value of each limit (if the limit does not exist, explain why).

4. a. $\lim_{x \to 3^-} f(x)$ b. $\lim_{x \to 3^+} f(x)$ c. $\lim_{x \to 3} f(x)$
    d. $\lim_{x \to 5^-} f(x)$ e. $\lim_{x \to 5^+} f(x)$ f. $\lim_{x \to 5} f(x)$
    g. $f(3)$ h. $f(5)$

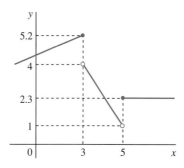

5. a. $\lim_{x \to 1^-} f(x)$ b. $\lim_{x \to 1^+} f(x)$ c. $\lim_{x \to 1} f(x)$
    d. $\lim_{x \to 4^-} f(x)$ e. $\lim_{x \to 4^+} f(x)$ f. $\lim_{x \to 4} f(x)$
    g. $f(1)$ h. $f(4)$

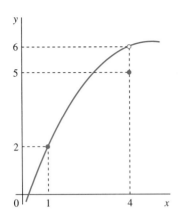

6. a. $\lim_{x \to 3^-} f(x)$ b. $\lim_{x \to 3^+} f(x)$ c. $\lim_{x \to 3} f(x)$
   d. $\lim_{x \to 2^-} f(x)$ e. $\lim_{x \to 2^+} f(x)$ f. $\lim_{x \to 2} f(x)$
   g. $f(3)$ h. $f(2)$

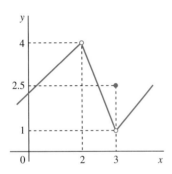

7. a. $\lim_{x \to 3^-} f(x)$ b. $\lim_{x \to 3^+} f(x)$ c. $\lim_{x \to 3} f(x)$
   d. $\lim_{x \to -1.3^-} f(x)$ e. $\lim_{x \to -1.3^+} f(x)$ f. $\lim_{x \to -1.3} f(x)$
   g. $f(3)$ h. $f(-1.3)$

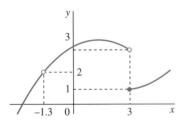

8. Sketch a possible graph of a function $f(x)$ that satisfies all of the following conditions: $\lim_{x \to 0^-} f(x) = 2$, $\lim_{x \to 0^+} f(x) = -1$, $f(0)$ is not defined, $f(2) = -1$, and $\lim_{x \to 2} f(x) = 1$.

9. Sketch a possible graph of a function $f(x)$ that satisfies all of the following conditions: $\lim_{x \to 0^-} f(x) = 2$, $\lim_{x \to 0^+} f(x) = 2$, $f(0) = 1$, $f(1)$ is undefined, $f(2) = 4$, and $\lim_{x \to 2} f(x)$ does not exist.

10–13 ▪ Sketch the graph of each function and determine the left-hand limit, the right-hand limit, and the (two-sided) limit at the given number a.

10. $f(x) = \begin{cases} -1 & \text{if } x \leq 2 \\ 3 - x & \text{if } x > 2 \end{cases}$ $a = 2$

11. $f(x) = \begin{cases} \sin x & \text{if } x \leq \pi \\ 0 & \text{if } x > \pi \end{cases}$ $a = \pi$

12. $g(x) = \begin{cases} e^x & \text{if } x < 1 \\ 0 & \text{if } x = 1 \\ e^{-x} & \text{if } x > 1 \end{cases}$ $a = 1$

13. $g(x) = \begin{cases} |x| & \text{if } x < 1 \\ |x - 2| & \text{if } x > 1 \end{cases}$ $a = 1$

14–17 ▪ Using a graphing device, sketch the graph of the function to verify each estimate. If needed, plot the graph over smaller and smaller intervals.

14. $\lim_{x \to 0} (1 + x)^{1/x} \approx e$

15. $\lim_{x \to 0} \frac{\sin x}{x} \approx 1$ (x is in radians)

16. $\lim_{x \to 0} \frac{1 - \cos x}{x} \approx 0$ (x is in radians)

17. $\lim_{x \to 0^+} x^x \approx 1$ (the function is defined for $x > 0$)

18–21 ▪ Using a graphing device, estimate each limit. If needed, plot the graph over smaller and smaller intervals.

18. $\lim_{x \to 0^+} x \ln x$

19. $\lim_{x \to 0} \frac{e^{2x} - 1}{x}$

20. $\lim_{x \to 1^-} \frac{\ln(1 - x)}{x}$

21. $\lim_{x \to 1^+} \sqrt{\ln x}$

22–25 ▪ Using the results from Exercises 14–17, find the combined limits using Theorem 4.2.1. Indicate how the new function was built.

22. $\lim_{x \to 0} 5(1 + x)^{1/x}$.

23. $\lim_{x \to 0} 3 \left(\frac{\sin x}{x} \right) + 4$

24. $\lim_{x \to 0} (1 + x)^{1/x} \left(\frac{1 - \cos x}{x} \right)$

25. $\lim_{x \to 0^+} \frac{(1 + x)^{1/x}}{x^x}$

26–29 ■ The given functions all have limits of 0 as $x \to 0^+$. For each function, find how close the input must be to 0 for the output to be (a) within 0.1 of 0, and (b) within 0.01 of 0. Sketch a graph of each function for $x < 1$, and indicate which functions approach 0 quickly and which approach 0 slowly.

26. $f_1(x) = \sqrt{x}$

27. $f_2(x) = x$

28. $f_3(x) = x^2$

29. $f_4(x) = x^4$

30–55 ■ Find the limit, or explain why it does not exist.

30. $\lim_{x \to 0} \left(x^3 - 2x + \dfrac{1}{x-1} \right)$

31. $\lim_{x \to \pi} (\sin x + \tan^2 x - 3)$

32. $\lim_{x \to 2} \dfrac{x-1}{x^2-1}$

33. $\lim_{x \to 1} \dfrac{x-1}{x^2-1}$

34. $\lim_{x \to 0} \dfrac{x^2 + 5x + 4}{x+4}$

35. $\lim_{x \to -4^+} \dfrac{x^2 + 5x + 4}{x+4}$

36. $\lim_{x \to 0} \dfrac{e^x - 1}{2}$

37. $\lim_{x \to 3} \dfrac{x^3 - 27}{2x - 6}$

38. $\lim_{x \to 0} \dfrac{x^2 - 3x}{x^3 - 9x}$

39. $\lim_{x \to 0^-} \sqrt{x}$

40. $\lim_{x \to 5^-} \sqrt{5-x}$

41. $\lim_{x \to 0} \dfrac{\sqrt{x}}{x}$

42. $\lim_{h \to 0} \dfrac{(4+h)^2 - 16}{h}$

43. $\lim_{h \to 0} \dfrac{\sqrt{1-h} - 1}{h}$

44. $\lim_{h \to 0} \dfrac{(2+h)^3 - 8}{h}$

45. $\lim_{x \to 0} \dfrac{|x|}{x}$

46. $\lim_{x \to 1} \dfrac{1 - |x|}{x - 1}$

47. $\lim_{x \to 1} \dfrac{1 - \sqrt{1-x}}{x}$

48. $\lim_{x \to 4} \dfrac{\sqrt{x} - 2}{4 - x}$

49. $\lim_{x \to 2} \dfrac{\frac{1}{x} - \frac{1}{2}}{x - 2}$

50. $\lim_{x \to \pi} \ln[(x^3 + 3x + \cos x)]$

51. $\lim_{x \to 0} \left(\arcsin \dfrac{x-2}{x+2} \right)$

52. $\lim_{x \to -1/2} \sqrt{\arcsin x + 12}$

53. $\lim_{x \to 1^+} (\arcsin x)$

54. $\lim_{x \to 1} [\arctan(x-2)]$

55. $\lim_{x \to 0} \left(\dfrac{\arctan x - x}{1 + x} \right)$

56–61 ■ Using the fact that $\lim_{x \to 0} \dfrac{\sin x}{x} = 1$ (x is in radians), compute each limit. (In Exercise 15 we estimated $\lim_{x \to 0} \dfrac{\sin x}{x} \approx 1$; in Section 5.4 we will prove that the limit is actually equal to 1.)

56. $\lim_{x \to 0} \dfrac{\sin 5x}{x}$

57. $\lim_{x \to 0} \dfrac{1 - \cos x}{x}$ (x is is radians)

58. $\lim_{x \to 0} \dfrac{\sin 5x}{\sin 3x}$

59. $\lim_{x \to 0} \dfrac{\sin(x^2)}{2x}$

60. $\lim_{x \to 0} \dfrac{\sin 2x}{\tan 3x}$

61. $\lim_{x \to 0^+} \dfrac{\sin x}{1 - \cos x}$

62–67 ■ Find the average rate of change of each function as a function of Δx, and find the limit as $\Delta x \to 0$. Graph the function and indicate the rate of change on your graph.

62. $f(x) = 5x + 7$ near $x = 0$.

63. $f(x) = 5x + 7$ near $x = 1$.

64. $f(x) = 5x^2$ near $x = 0$.

65. $f(x) = 5x^2$ near $x = 1$.

66. $f(x) = 5x^2 + 7x + 3$ near $x = 1$.

67. $f(x) = 5x^2 + 7x + 3$ near $x = 2$.

68. Consider the function $f(x) = \sin(\pi/x)$.

   a. Find all x for which $f(x) = -1$.

   b. Show that $f(x) = 1/2$ for infinitely many values of x that approach zero.

   c. Let a be any number in $[-1, 1]$. Show that $f(x) = a$ for infinitely many values of x that approach zero.

69. Show that the limit of $f(x) = \cos(1/x)$ as x approaches 0 does not exist. (Hint: Look at Example 4.2.4 and Exercise 68.)

70. Assume that you know that $e^{-2x} \leq f(x) \leq 2e^{-x}$ and you need to find $\lim_{x \to 0} f(x)$. Explain why the Squeeze Theorem does not apply. Nevertheless, what can you say about $\lim_{x \to 0} f(x)$?

71. Assume that you know that $\sin x \leq f(x) \leq \tan x$ and you need to find $\lim_{x \to \pi/4} f(x)$. Explain why the Squeeze Theorem does not apply. Nevertheless, what can you say about $\lim_{x \to \pi/4} f(x)$?

72. Using the Squeeze Theorem, show that $\lim_{x \to 0} x^4 \cos(3/x) = 0$.

73. Using the Squeeze Theorem, show that $\lim_{x \to 0} 3x^2 \sin(1/x) = 0$.

Applications

74–75 ▪ Suppose we are interested in measuring the properties of a substance at a temperature of absolute zero (which is 0 kelvins). However, we cannot measure these properties directly because it is impossible to reach absolute 0. Instead, properties are measured for small values of the temperature T, measured in kelvins (K). For each of the following, find

a. The limit as $T \to 0^+$.

b. How close we would be to the limit if we measured the property at $2°$K.

c. How close we would be to the limit if we measured the property at $1°$K.

d. About how cold the temperature would have to be for the property to be within 1% of its limit.

74. The volume, $V(T)$ (in cubic centimetres), follows $V(T) = 1 + T^2$.

75. The hardness, $H(T)$, follows $H(T) = \dfrac{10}{1+T}$.

76–79 ▪ For each population, the instantaneous rate of change of the population size at $t = 0$ is exactly one million bacteria per hour. If you computed the average rate of change between $t = 0$ and $t = \Delta t$, how small would Δt have to be before your value was within 1% of the instantaneous rate of change?

76. $b(t) = t + t^2$

77. $b(t) = t + 0.1t^2$

78. $b(t) = e^t$ (this cannot be solved algebraically)

79. $b(t) = \sin t$ (this cannot be solved algebraically)

80–83 ▪ Scientifically, two quantities are close if an accurate measuring device is required to detect the difference. In real life, accuracy costs money. How much would it cost to measure the differences in the following circumstances?

80. A piano tuner is trying to get the note A on a piano to have a frequency of exactly 440 hertz (Hz), or cycles per second. An electronic tuner capable of detecting a difference of x cycles per second costs $\dfrac{5}{x}$ dollars.

   a. How much would it cost to make sure the note was within 1 Hz of 440 Hz?

   b. How much would it cost to make sure the note was within 0.1 Hz of 440 Hz?

   c. How much would it cost to make sure the note was within 0.01 Hz of 440 Hz?

81. The army is developing satellite-based targeting systems. A system that can send a missile within y metres of its target costs $\dfrac{1}{y^2}$ million dollars.

   a. How much would it cost to hit within 10 m?

   b. How much would it cost to hit within 1 m?

   c. How much would it cost to hit within 1 cm?

82. Suppose a body has temperature B and is cooling toward room temperature of $20°$C according to the function $B(t) = 20 + 17e^{-t}$, where t is measured in hours. A $10 thermocouple can detect a difference of $0.1°$C, a $100 thermocouple can detect a difference of $0.01°$C, and so forth.

   a. How much would it cost to detect the difference after one hour?

   b. How much would it cost to detect the difference after five hours?

   c. How much would it cost to detect the difference after 10 hours?

83. Some dangerously radioactive and toxic radium was dumped in the desert in 1970. It has a half-life of 50 years, and the initial level of radioactivity was $r = 10$ rads. Nobody remembers where it was. How much will it cost to find it in the following years if detecting radioactivity at level r costs $\dfrac{5}{r}$ thousand dollars?

   a. How much will it cost to find in the year 2020?

   b. How much will it cost to find in the year 2070?

   c. How much will it cost to find in the year 2150?

Computer Exercises

84–85 ▪ In each case, use a plotting device to investigate the limit of $f(x)$ as $x \to 0$.

84. $f(x) = \sin(1/x)$. Describe what you see as you plot the graph of $f(x)$ over $[-0.1, 0.1]$, $[-0.01, 0.01]$, and $[-0.001, 0.001]$. Is it possible to determine the limit? Explain what the graphs look like.

85. $f(x) = \left(\dfrac{1}{x^2} - \dfrac{2}{x} + 1\right) - \left(\dfrac{1}{x} - 1\right)^2$. Plot the graph of $f(x)$ on the intervals $[0.001, 0.01]$ and $[0.0001, 0.001]$. Is it possible to determine the right-hand limit? Compute the right-hand limit of $f(x)$ by hand.

4.3 Infinite Limits and Limits at Infinity

Continuing our study of limits, we discuss **infinite limits** and **limits at infinity**. We explain what it means for a function to **go to infinity** and use one-sided limits to algebraically describe this kind of behaviour. Models in biology and elsewhere sometimes require that we identify the **long-term behaviour** of the functions involved. In other words, we are asked to identify what happens to a function if its variable increases beyond any bounds. Next, we learn how to **compare the rates** at which functions approach infinity and zero (as their variable approaches infinity), and, as an application, examine a variety of absorption rates. Finally, we discuss **limits of sequences**, due to their relevance to discrete-time dynamical systems.

Infinite Limits

What could the statement "a function gets infinitely large" possibly mean? As with ordinary limits, let us think about measuring some quantity. Of course, it is impossible to measure infinity. Instead, we encounter the values that exceed the capacity of our measuring device. For instance, a room thermometer might explode if placed too close to a source of high heat (say, inside a furnace). A value can be thought of as being infinitely large if it exceeds the capacity of *every* possible measuring device.

Example 4.3.1 A Function with an Infinitely Large Limit

Consider the function

$$f(x) = \frac{1}{x^2}$$

(see Figure 4.3.24 and Table 4.3.1). If we plug in values of x that are closer and closer to 0, we see that $f(x)$ becomes larger and larger.

Table 4.3.1

x	$f(x)$
± 1	1
± 0.1	100
± 0.01	10,000
± 0.001	1,000,000
$\pm 10^{-4}$	10^8

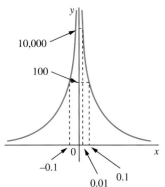

FIGURE 4.3.24
A function with a limit of positive infinity (axes not to scale)

What does "larger and larger" mean?
If we wish to make $f(x)$ larger than, say, 1000, we can take (look at Table 4.3.1) $x = 0.01$. As a matter of fact, if we take any x in the interval $(-0.01, 0.01)$ such that $x \neq 0$, then

$$x^2 < 0.0001$$

and therefore

$$\frac{1}{x^2} > \frac{1}{0.0001} = 10,000 > 1000$$

(so we made the function $f(x)$ larger than 1000).

Similarly, if we wish to make $f(x)$ larger than one million (10^6), we can take $x = 10^{-4}$. In that case,

$$f(x) = \frac{1}{(10^{-4})^2} = \frac{1}{10^{-8}} = 10^8 > 10^6$$

Again, we can take the whole interval $(-10^{-4}, 10^{-4})$; for every nonzero x in it, the function satisfies $f(x) > 10^6$.

It appears that we can make $1/x^2$ as large as we wish by choosing x to be close enough to 0 (but not equal to 0). By "x close enough to 0" we mean any x in a suitably chosen interval containing 0, such as the intervals $(-0.01, 0.01)$ and $(-10^{-4}, 10^{-4})$ in this example. To describe this conclusion in terms of limits, we write

$$\lim_{x \to 0} \frac{1}{x^2} = \infty$$

It is important to remember that ∞ *is not a real number!* It is a symbol that describes, for instance, what happens to $1/x^2$ as x gets closer and closer to 0. In general,

$$\lim_{x \to a} f(x) = \infty$$

describes the fact that the values of the function $f(x)$ increase beyond any bounds as x approaches a.

Similarly, a limit is equal to **negative infinity,** which is written $-\infty$, when the output becomes smaller than every possible value. We write

$$\lim_{x \to a} f(x) = -\infty$$

Example 4.3.2   A Function with a Limit of Negative Infinity

The function

$$g(x) = -\frac{1}{x^2}$$

has a limit of negative infinity as x approaches 0 (Figure 4.3.25 and Table 4.3.2). We write

$$\lim_{x \to 0} \left(-\frac{1}{x^2} \right) = -\infty$$

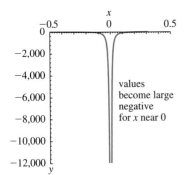

Figure 4.3.25

A function with a limit of negative infinity

Table 4.3.2

x	g (x)
±1	−1
±0.1	−100
±0.01	−10,000
±0.001	−1,000,000
±10^{-4}	−10^8

to say that the values of $g(x)$ can be made as large negative as desired by taking x close enough to 0 (but not equal to 0).

Definition 4.3.1   Assume that $f(x)$ is defined on some interval around a, except possibly at a. We say that the **limit of $f(x)$ as x approaches a is ∞** and write

$$\lim_{x \to a} f(x) = \infty$$

if we can make the values of $f(x)$ as large as desired by taking x close enough (but not equal) to a.

Similarly,

$$\lim_{x \to a} f(x) = -\infty$$

means we can make the values of $f(x)$ as small (i.e., as large negative) as desired by taking x close enough (but not equal) to a.

Once again—keep in mind that "x close enough to a" means that x is in a suitably chosen open interval that contains a, and $x \neq a$.

Remember that ∞ and $-\infty$ are *not real numbers*—their precise meaning is given in Definition 4.3.1.

Definition 4.3.1 extends to one-sided limits. For instance, to define

$$\lim_{x \to a^-} f(x) = \infty$$

all we need to do is to replace "x close enough to a" by "x close enough to a and less than a." We define

$$\lim_{x \to a^-} f(x) = -\infty, \quad \lim_{x \to a^+} f(x) = \infty, \quad \text{and} \quad \lim_{x \to a^+} f(x) = -\infty$$

in a similar fashion.

Example 4.3.3 A Function with Different, but Infinite, Left-Hand and Right-Hand Limits

Consider the function

$$f(x) = \frac{1}{x}$$

(Figure 4.3.26). Suppose we wish to find the limit as x approaches 0. This function increases to positive infinity for $x > 0$ and to negative infinity for $x < 0$:

x	f(x)
1	1
0.1	10
0.01	100
0.001	1000
−1	−1
−0.1	−10
−0.01	−100
−0.001	−1000

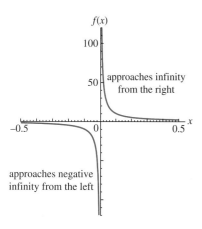

FIGURE 4.3.26

A function with a left-hand limit of $-\infty$ and a right-hand limit of ∞

The left-hand and right-hand limits help us to separate out these two types of behaviour. As we move toward 0 from the left (negative x), the values of the function get smaller and smaller, falling below any (negative) number we pick. As we move toward 0 from the right (positive x), the values of the function get larger and larger, surpassing any (positive) number we pick. In mathematical notation,

$$\lim_{x \to 0^-} \frac{1}{x} = -\infty \quad \text{and} \quad \lim_{x \to 0^+} \frac{1}{x} = \infty$$

Example 4.3.4 Infinite Limits

Find the limit of $f(x) = \frac{4}{x-5}$ as $x \to 5^-$ and as $x \to 5^+$.

As x approaches 5 from the right, the denominator becomes a smaller and smaller *positive number* (since $x > 5$), which makes the fraction become larger and larger positive. Thus, we write

$$\lim_{x \to 5^+} \frac{4}{x-5} = +\infty$$

See Figure 4.3.27.

As x approaches 5 so that $x < 5$, the denominator, $x - 5$, becomes a smaller and smaller *negative number*, which makes the fraction become larger and larger negative. Therefore,

$$\lim_{x \to 5^-} \frac{4}{x-5} = -\infty$$

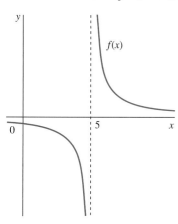

FIGURE 4.3.27

Infinite limits

Definition 4.3.2 The line $x = a$ is called a **vertical asymptote** of the graph of $y = f(x)$ if any of the following hold:

$$\lim_{x \to a} f(x) = \infty \quad \text{or} \quad \lim_{x \to a^-} f(x) = \infty \quad \text{or} \quad \lim_{x \to a^+} f(x) = \infty \quad \text{or}$$

$$\lim_{x \to a} f(x) = -\infty \quad \text{or} \quad \lim_{x \to a^-} f(x) = -\infty \quad \text{or} \quad \lim_{x \to a^+} f(x) = -\infty$$

So far in this section, we have shown that $x = 0$ is a vertical asymptote of the graphs of the functions $y = \frac{1}{x^2}$, $y = -\frac{1}{x^2}$, and $y = \frac{1}{x}$. As well, in Example 4.3.4 we proved that $x = 5$ is a vertical asymptote of the graph of $f(x) = \frac{4}{x-5}$.

In Figure 4.3.28 we show several functions that have a vertical asymptote at a.

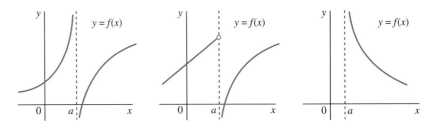

FIGURE 4.3.28

Functions with vertical asymptote at a

Looking at the graph of $y = \tan x$ (Figure 4.3.29), we see that

$$\lim_{x \to (\pi/2)^-} \tan x = \infty \quad \text{and} \quad \lim_{x \to (\pi/2)^+} \tan x = -\infty$$

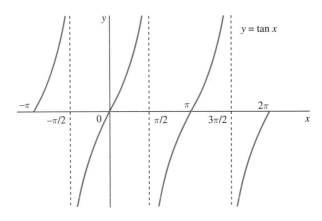

FIGURE 4.3.29

Vertical asymptotes of the graph of $y = \tan x$

Thus, $x = \pi/2$ is a vertical asymptote of the graph of $y = \tan x$. Because $y = \tan x$ is periodic with period π, we conclude that $x = \pi/2 + \pi k$ (where k is an integer) are its vertical asymptotes.

Example 4.3.5 The Limit of $\ln x$ as $x \to 0^+$

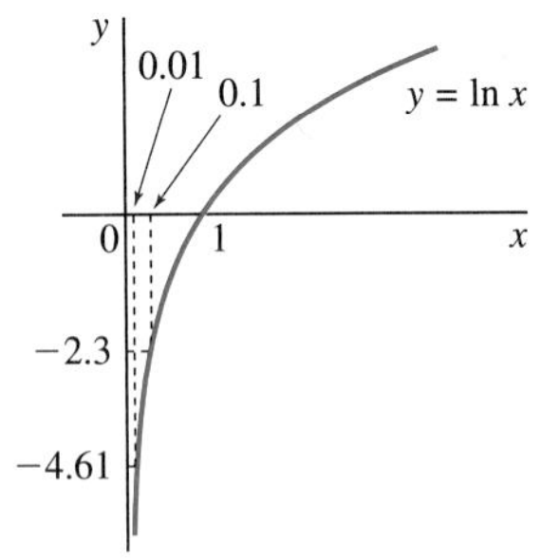

During our study of the natural logarithm, we indicated that $\ln x$ is not defined for 0 or negative values of x and that the graph of $\ln x$ rises from negative infinity near $x = 0$ (Figure 4.3.30):

FIGURE 4.3.30

The natural logarithm approaching negative infinity

x	ln x
1	0
0.1	−2.30
0.01	−4.61
0.001	−6.91
10^{-6}	−13.82
10^{-9}	−20.72

The value of ln x becomes more and more negative, although very slowly. We can now describe this in terms of the limit as

$$\lim_{x \to 0^+} \ln x = -\infty$$

Example 4.3.6 How Close Must the Input Be?

Consider the function $y = \ln x$. If the limit as $x \to 0^+$ is indeed negative infinity, we should be able to choose inputs x so small that the output is less than -10, or -100, or any large negative number. When is $\ln x < -10$? Solving yields

$$\ln x < -10$$
$$x < e^{-10} = 4.5 \times 10^{-5}$$

Graphically, the inputs on the horizontal axis must lie close enough to zero for the output to be below the horizontal line at $y = -10$ (Figure 4.3.31). A very small input is required to get an output even this small. To make $\ln x < -100$, we need $x < e^{-100} \approx 3.7 \cdot 10^{-44}$. If the units on the x-axis were metres, then x would have to be many orders of magnitude closer to zero than the distance between a nucleus of an atom and its electrons. We will later think of this as meaning that $\ln x$ approaches negative infinity "slowly."

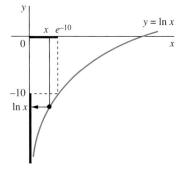

FIGURE 4.3.31
Finding how small the input must be

Nothing in nature is infinite, so if a function $f(t)$ represents some biological quantity that changes with time, it is very unlikely that it will approach ∞ or $-\infty$ as t approaches some value t_0. Thus, we do not often meet vertical asymptotes in biological applications (we *do* meet them in more advanced models). However, if we build a model for some quantity $f(t)$, and it turns out that $f(t)$ has a vertical asymptote as $t \to t_0$ (sometimes we say that "$f(t)$ blows up in finite time"), then we can conclude that the model has a limitation; i.e., beyond certain times it no longer accurately represents the biological quantity that is modelled.

Limits at Infinity

So far, we have studied limits of functions where x approaches some real number a. For instance, the formula

$$\lim_{x \to a} f(x) = \infty$$

characterizes a function whose values grow beyond any bounds as x approaches a. Instead of using lim, we can write $f(x) \to \infty$ as $x \to a$.

Now, we investigate what happens to a function $f(x)$ at its "ends," i.e., as x becomes larger and larger, or smaller and smaller (meaning larger and larger negative). In analogy to $f(x) \to \infty$, when we write

$$x \to \infty$$

we mean that the values of x can be made arbitrarily large. Likewise, the symbol

$$x \to -\infty$$

says that x can be made as small (as large negative) as desired.

We look at an example first.

Consider the function $f(x) = \dfrac{2x^2}{x^2 + 3}$. In Table 4.3.3 we recorded the values $f(x)$ as x becomes large positive and large negative, and in Figure 4.3.32 we show the graph of $f(x)$.

We see that as x grows larger and larger, the values of $f(x)$ get closer and closer to 2. To make $f(x)$ within 0.01 of 2, i.e., to make $f(x)$ satisfy

$$1.99 < f(x) < 2.01$$

Table 4.3.3

x	$f(x) = \dfrac{2x^2}{x^2+3}$
±10	1.941748
±50	1.997603
±100	1.999400
±200	1.999850
±1000	1.999994

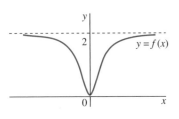

FIGURE 4.3.32
Graph of $f(x) = \dfrac{2x^2}{x^2+3}$

we need to take $x > 50$ (look at Table 4.3.3). To make $f(x)$ even closer, say, within 0.0001 of 2, i.e., make it satisfy

$$1.9999 < f(x) < 2.0001$$

we take $x > 1000$.

Thus, it appears that we can make $f(x)$ as close to 2 as desired by taking x large enough. Using limits, we write what we just concluded as

$$\lim_{x \to \infty} \frac{2x^2}{x^2+3} = 2$$

Arguing in a similar way, we show that

$$\lim_{x \to -\infty} \frac{2x^2}{x^2+3} = 2$$

Definition 4.3.3 Assume that $f(x)$ is defined for all $x > a$, where a is some real number. We say that the **limit of $f(x)$ as x approaches ∞ is L,** and write

$$\lim_{x \to \infty} f(x) = L$$

if we can make the values of $f(x)$ as close to L as desired by taking x large enough.

Similarly, assume that $f(x)$ is defined for $x < a$, where a is a real number. Then

$$\lim_{x \to -\infty} f(x) = L$$

means that the values of $f(x)$ can be made as close to L as desired by taking x small (i.e., large negative) enough.

The phrase "by taking x large enough" means that we can take any x larger than some number N (for instance, in the example we just discussed, to make $1.9999 < f(x) < 2.0001$ we took $x > 1000$; so $N = 1000$). Likewise, "by taking x small enough" means that we can take any x smaller than some number N (see Examples 4.3.7 and 4.3.8).

Sometimes, instead of

$$\lim_{x \to \infty} f(x) = L$$

we write

$$f(x) \to L \text{ as } x \to \infty$$

To say that a function approaches a limit L as x approaches infinity means that its values get closer and closer to L as x becomes large (Figure 4.3.33). In other words, a function f approaches the limit L as x approaches infinity if the measurement eventually becomes indistinguishable from the limit.

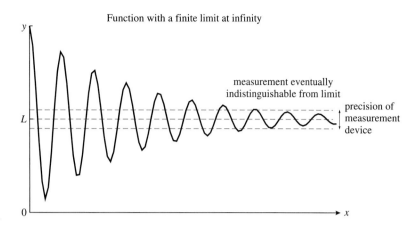

FIGURE 4.3.33
A function with a finite limit at infinity

Example 4.3.7 Limit of the Exponential Function

Looking at the graph of $f(x) = e^{-x}$ (Figure 4.3.34), we claim that

$$\lim_{x \to \infty} e^{-x} = 0$$

This means that we should be able to make the values of $f(x) = e^{-x}$ as small (i.e., as close to zero) as desired by picking x large enough.

How can we make e^{-x} smaller than, say, 0.001? From

$$e^{-x} < 0.001$$

applying ln to both sides, we get

$$-x < \ln(0.001) \approx -6.908$$

$$x > 6.908$$

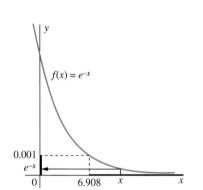

FIGURE 4.3.34
Graph of $f(x) = e^{-x}$

Thus, if $x > 6.908$, then e^{-x} falls within 0.001 of 0 (Figure 4.3.34).

Likewise, to make e^{-x} fall within 10^{-10} of 0, we need values of x that satisfy

$$e^{-x} < 10^{-10}$$
$$-x < \ln(10^{-10})$$
$$-x < -10 \ln 10 \approx -23.026$$
$$x > 23.026$$

Clearly, we can repeat the above for any precision we pick.

In this example we used the fact that the graph of $f(x) = e^{-x}$ is decreasing for all x, as suggested in Figure 4.3.34. That this is indeed so we will be able to prove later, using derivatives.

In a similar way we show that

$$\lim_{x \to \infty} e^{-ax} = 0$$

for any positive constant a.

Example 4.3.8 Limits of the Reciprocal Function

Let $f(x) = \dfrac{1}{x}$. As x grows larger and larger positive, the values of $1/x$ get smaller and smaller positive. Thus,

$$\lim_{x \to \infty} \frac{1}{x} = 0$$

(see Figure 4.3.35). Likewise, as x grows larger and larger negative, the values of $1/x$

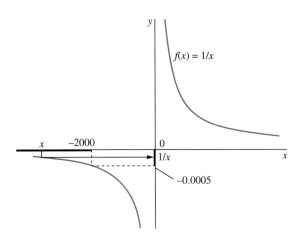

FIGURE 4.3.35
Limits at infinity of $f(x) = \dfrac{1}{x}$

get smaller and smaller negative, which means that
$$\lim_{x \to -\infty} \frac{1}{x} = 0$$
We check the latter limit by picking a tolerance—say we wish to force $1/x$ to be in the interval
$$-0.0005 < \frac{1}{x} < 0$$
—and finding appropriate values for x.
From
$$-0.0005 < \frac{1}{x}$$
we get, after multiplying by x (keep in mind that x is negative!)
$$-0.0005x > 1$$
$$x < \frac{1}{-0.0005} = -2000$$
Thus, if $x < -2000$, then $1/x$ falls within the chosen tolerance from 0.

It can be proven that all the limit laws that we stated in the previous section hold for limits as $x \to \infty$ and $x \to -\infty$. As well, one can prove that, for $r > 0$,
$$\lim_{x \to \infty} \frac{1}{x^r} = 0$$
If x^r is defined for negative values of x, and $r > 0$, then
$$\lim_{x \to -\infty} \frac{1}{x^r} = 0$$
(For instance, x^r is defined for all negative x if r is an even positive integer.)

Definition 4.3.4 The line $y = L$ is called a **horizontal asymptote** of the graph of $y = f(x)$ if either
$$\lim_{x \to \infty} f(x) = L \quad \text{or} \quad \lim_{x \to -\infty} f(x) = L$$

In the introduction to this subsection, we proved that $\dfrac{2x^2}{x^2 + 3} \to 2$ as $x \to \infty$ and as $x \to -\infty$. Thus, $y = 2$ is a horizontal asymptote of the graph of $f(x) = \dfrac{2x^2}{x^2 + 3}$ (Figure 4.3.32).

From Example 4.3.7 we conclude that $y = 0$ is a horizontal asymptote of $f(x) = e^{-x}$.

As well, $y = 0$ is a horizontal asymptote of the graph of $f(x) = 1/x$ (Example 4.3.8). The line $y = c$ is a horizontal asymptote of the constant function $f(x) = c$.

From the graph of $f(x) = \arctan x$ (Figure 4.3.36) we deduce that

$$\lim_{x \to \infty} \arctan x = \frac{\pi}{2} \quad \text{and} \quad \lim_{x \to -\infty} \arctan x = -\frac{\pi}{2}$$

i.e., the graph of $y = \arctan x$ has two horizontal asymptotes: $y = \pi/2$ and $y = -\pi/2$.

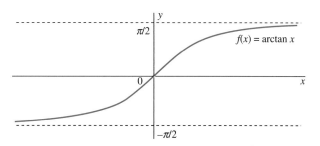

FIGURE 4.3.36

Asymptotes of $y = \arctan x$

Example 4.3.9 Calculating the Limit of a Rational Function

Find

$$\lim_{x \to \infty} \frac{2x^3 - 7x + 100}{5x^3 + 3x^2}$$

The trick is to divide both the numerator and the denominator by x^3. We get

$$\lim_{x \to \infty} \frac{2x^3 - 7x + 100}{5x^3 + 3x^2} = \lim_{x \to \infty} \frac{2 - \frac{7}{x^2} + \frac{100}{x^3}}{5 + \frac{3}{x}}$$

$$= \frac{\lim_{x \to \infty} 2 - \lim_{x \to \infty} \frac{7}{x^2} + \lim_{x \to \infty} \frac{100}{x^3}}{\lim_{x \to \infty} 5 + \lim_{x \to \infty} \frac{3}{x}}$$

$$= \frac{2 - 0 + 0}{5 + 0} = \frac{2}{5}$$

Thus, $y = 2/5$ is a horizontal asymptote of the given function.

It is not a coincidence that the answer, 2/5, is the ratio of the coefficients of x^3 in the original fraction. Note that the answer would not change if we replaced $-7x$ in the numerator by $+7000x$ or $3x^2$ in the denominator by $3{,}000{,}000x^2$. As well, if we added an x^2 term on the top, the answer would still be 2/5.

Thus, an important conclusion (that we will justify soon) is that the behaviour of a rational function at ∞ (or at $-\infty$) is determined by the highest powers of x in the numerator and the denominator. No other terms matter!

For example:

$$\lim_{x \to \infty} \frac{x^4 - 2x^3 + 4x - 14}{x^3 - 5x + 6x^4} = \lim_{x \to \infty} \frac{x^4}{6x^4} = \lim_{x \to \infty} \frac{1}{6} = \frac{1}{6}$$

If a quantity changes with time, the horizontal asymptote (if it exists) describes its long-term behaviour. Consider the following examples.

Example 4.3.10 Fish Growth Model

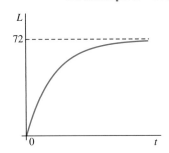

FIGURE 4.3.37
Length of a walleye

The function
$$L(t) = 72\left(1 - e^{-0.09t}\right)$$

models the growth of walleye, where $L(t)$ is the length in centimetres, and the time, t, is in years. (Also known as pickerel, walleye is a freshwater fish native to most of Canada; see Example 7.5.8 for more detail.) Taking the limit as $t \to \infty$, we obtain

$$\lim_{x \to \infty} L(t) = \lim_{x \to \infty} 72\left(1 - e^{-0.09t}\right) = 72$$

since $e^{-0.09t} \to 0$ as $t \to \infty$. Thus, the model predicts that a walleye will keep growing, its length approaching 72 cm; see Figure 4.3.37. The calculation above shows that $L = 72$ is a horizontal asymptote of the graph of $L(t)$.

Example 4.3.11 Logistic Population Model

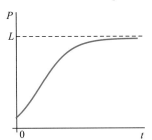

FIGURE 4.3.38
Logistic curve

An important model (called the **logistic model**) describes a population change (could be human population, or bacteria, animals, etc.) using the formula

$$P(t) = \frac{L}{1 + Ce^{-kt}}$$

where $P(t)$ is the population at time t and L, C, and k are positive constants. Figure 4.3.38 shows the graph of $P(t)$ with $L = 200$, $C = 5$, and $k = 0.2$. From

$$\lim_{x \to \infty} P(t) = \lim_{x \to \infty} \frac{L}{1 + Ce^{-kt}} = L$$

(since $e^{-kt} \to 0$ as $t \to \infty$) we conclude that, in the long term, the population size approaches L. This value, which gives the **carrying capacity** of the ecosystem supporting the population, is an important biological quantity. The graph of $P(t)$ is called a **logistic curve.**

The calculation of the limit proves that the logistic curve has a horizontal asymptote of value L. Quite often, we say that $P(t)$ **plateaus** at L. (For more detail about the logistic model, see Example 8.4.5.)

Example 4.3.12 Limit of an Exponential Function

Calculate $\lim_{x \to \infty} e^{-1/x^3}$.

As $x \to \infty$, $-\frac{1}{x^3} \to 0$. Thus, the numbers in the exponent of e approach zero, and we conclude that $e^{-1/x^3} \to e^0 = 1$.

What we said in the previous example is correct, and—intuitively—sounds right. That we are allowed to use this argument is a consequence of the fact that the exponential function is continuous (we will discuss this fully in the next section).

Example 4.3.13 Limits at Infinity of Trigonometric Functions

Looking at the graph of $f(x) = \sin x$, we see that the values oscillate between -1 and 1; since they do not approach any particular real number, we conclude that

$$\lim_{x \to \infty} \sin x$$

does not exist. Likewise, the limit of $\sin x$ as $x \to -\infty$ does not exist, and the same is true for $\cos x$.

All trigonometric functions are periodic—thus, they keep oscillating, rather than approaching a particular real number. So, if $f(x)$ is any of the six trigonometric functions, then

$$\lim_{x \to \infty} f(x) \quad \text{and} \quad \lim_{x \to -\infty} f(x) \quad \text{do not exist}$$

Example 4.3.14 Limit Involving Exponential and Trigonometric Functions

Find
$$\lim_{x \to \infty} e^{-x} \sin x$$

Because the limit of $\sin x$ as $x \to \infty$ does not exist, we cannot use the "limit of the product is the product of the limits" law. Instead, we use the Squeeze Theorem, in much the same way as in Example 4.2.17. We start with
$$-1 \leq \sin x \leq 1$$
and multiply all sides by e^{-x} (which is positive) to get
$$-e^{-x} \leq e^{-x} \sin x \leq e^{-x}$$

Because $e^{-x} \to 0$ as $x \to \infty$, the Squeeze Theorem implies that
$$\lim_{x \to \infty} e^{-x} \sin x = 0$$

See Figure 4.3.39 for the graph of $f(x) = e^{-x} \sin x$ and the two functions $y = e^{-x}$ and $y = -e^{-x}$ that bound it. ▲

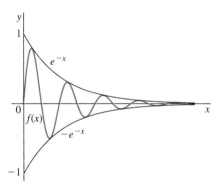

FIGURE 4.3.39

The function $y = e^{-x} \sin x$ squeezed between $y = -e^{-x}$ and $y = e^{-x}$

"The limit exists," "the limit does not exist"—what do these phrases mean? In short, we use the following convention:

- "The limit exists" means that the limit is a unique real number.
- "The limit does not exist" includes all other cases.

Clearly, the category "does not exist" is very broad, and we try to provide additional information—which is useful in many situations, such as analyzing and drawing graphs, or comparing features of different functions. See Table 4.3.4.

Infinite Limits at Infinity

So far, we have investigated functions whose limit at infinity is a real number and functions that do not have a limit at infinity. There is one more possibility left, namely functions that satisfy

$$\lim_{x \to \infty} f(x) = \infty \quad \text{or} \quad \lim_{x \to \infty} f(x) = -\infty \quad \text{or} \quad \lim_{x \to -\infty} f(x) = \infty \quad \text{or} \quad \lim_{x \to -\infty} f(x) = -\infty$$

A function approaches infinity as x approaches infinity if its values grow larger than any number as x becomes arbitrarily large (Figure 4.3.40a). In other words, the function f approaches infinity as x approaches infinity if the output eventually overflows any given measurement device. Similarly, a function approaches negative infinity as x approaches infinity if its values can be made as small (i.e., as large negative) as desired by letting x assume arbitrarily large values (Figure 4.3.40b).

Table 4.3.4

Reference	Limit Calculated	Answer (with Additional Information)
Example 4.2.11	$\lim_{x \to 2} \dfrac{x^3 - 4x}{x - 2} = 8$	The limit exists and is equal to 8.
Example 4.3.8	$\lim_{x \to \infty} \dfrac{1}{x} = 0$	The limit exists and is equal to 0.
Example 4.2.6	$\lim_{x \to 0^-} \sqrt{x}$	The limit does not exist because the function is not defined.
Example 4.2.7	$\lim_{x \to 0} \dfrac{|x|}{x}$	The limit does not exist because the left-hand limit is -1 and the right-hand limit is 1.
Example 4.3.1	$\lim_{x \to 0} \dfrac{1}{x^2}$	The limit does not exist; both one-sided limits are equal to $+\infty$.
Example 4.3.3	$\lim_{x \to 0} \dfrac{1}{x}$	The limit does not exist; the left-hand limit is $-\infty$ and the right-hand limit is $+\infty$.
Example 4.3.13	$\lim_{x \to \infty} \sin x$	The limit does not exist; the function oscillates, assuming all values between -1 and 1.

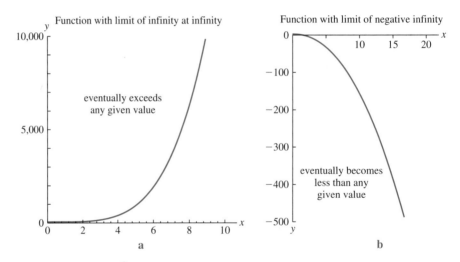

FIGURE 4.3.40

Functions with infinite limits at infinity

Consider $f(x) = x^2$. Clearly, as x becomes larger and larger, so does its square, x^2. As a matter of fact, we can make x^2 larger than any number we choose by taking x large enough. For instance, to make $x^2 > 1{,}000{,}000$, we take

$$x > \sqrt{1{,}000{,}000} = 1000$$

In general, to make $x^2 > M$, where M is any positive number, we take $x > \sqrt{M}$. Thus,

$$\lim_{x \to \infty} x^2 = \infty$$

Since we can make $x^2 > M$ by taking $x < -\sqrt{M}$, it follows that

$$\lim_{x \to -\infty} x^2 = \infty$$

By examining tables of values, or graphs, we conclude that

$$\lim_{x \to \infty} x^n = \infty \text{ if } n \text{ is a positive integer}$$

and

$$\lim_{x \to -\infty} x^n = \infty \text{ if } n \text{ is an even positive integer}$$

$$\lim_{x \to -\infty} x^n = -\infty \text{ if } n \text{ is an odd positive integer}$$

Example 4.3.15 Limit at Infinity of a Polynomial

In Table 4.3.5 we compare the values of the polynomial $f(x) = 2x^3 - 3x^2 + 4x + 3$ with the values of $2x^3$ as x grows large.

Table 4.3.5

x	$2x^3$	$2x^3 - 3x^2 + 4x + 3$	$\dfrac{2x^3 - 3x^2 + 4x + 3}{2x^3}$
10	2,000	1,743	0.81750
50	250,000	242,703	0.970812
100	2,000,000	1,970,403	0.985202
500	250,000,000	249,252,003	0.997008
1000	2,000,000,000	1,997,004,003	0.998502
5000	250,000,000,000	249,925,020,003	0.999700

As x increases, the ratio of $f(x) = 2x^3 - 3x^2 + 4x + 3$ to $2x^3$ approaches 1. Thus, the behaviour of the two polynomials at ∞ is the same, and we write

$$\lim_{x \to \infty} (2x^3 - 3x^2 + 4x + 3) = \lim_{x \to \infty} 2x^3$$

In other words, the behaviour of the given polynomial as $x \to \infty$ is determined by its term of highest degree, $2x^3$. Similar analysis would show that the same is true in the case when $x \to -\infty$. ▲

We now show that what appeared to be true in Example 4.3.15 is indeed true. Consider a polynomial

$$f(x) = a_n x^n + a_{n-1} x^{n-1} + \cdots + a_1 x + a_0$$

of degree n (that is, $a_n \neq 0$). Then

$$\lim_{x \to \infty} f(x) = \lim_{x \to \infty} (a_n x^n + a_{n-1} x^{n-1} + \cdots + a_1 x + a_0)$$

$$= \lim_{x \to \infty} a_n x^n \left(1 + \frac{a_{n-1}}{a_n} \frac{1}{x} + \cdots + \frac{a_1}{a_n} \frac{1}{x^{n-1}} + \frac{a_0}{a_n} \frac{1}{x^n}\right)$$

Since

$$\lim_{x \to \infty} \left(1 + \frac{a_{n-1}}{a_n} \frac{1}{x} + \cdots + \frac{a_1}{a_n} \frac{1}{x^{n-1}} + \frac{a_0}{a_n} \frac{1}{x^n}\right) = 1 + 0 + \cdots + 0 + 0 = 1$$

we conclude that

$$\lim_{x \to \infty} (a_n x^n + a_{n-1} x^{n-1} + \cdots + a_1 x + a_0) = \lim_{x \to \infty} a_n x^n$$

> The limit of a polynomial as $x \to \infty$ or $x \to -\infty$ is determined by its term of highest degree.

Note that in the above we can replace $x \to \infty$ by $x \to -\infty$ and obtain the same result. We have arrived at an important result:
For example,

$$\lim_{x \to \infty} (33 - 2.4x + 5x^3 - x^4) = \lim_{x \to \infty} (-x^4) = -\infty$$

and

$$\lim_{x \to -\infty} (5x^3 - x^2 + 11x + 4) = \lim_{x \to -\infty} (5x^3) = -\infty$$

So, what we claimed after examining the rational function in Example 4.3.9 was true. With this in mind, we can now calculate the limit in Example 4.3.9 fairly quickly:

$$\lim_{x\to\infty} \frac{2x^3 - 7x + 100}{5x^3 + 3x^2} = \lim_{x\to\infty} \frac{2x^3}{5x^3} = \lim_{x\to\infty} \frac{2}{5} = \frac{2}{5}$$

Likewise,

$$\lim_{x\to-\infty} \frac{2x^4 - 3x^2 + 11}{7x^2 - 3x + 4} = \lim_{x\to-\infty} \frac{2x^4}{7x^2} = \frac{2}{7} \lim_{x\to-\infty} x^2 = \infty$$

In the last step, we used the fact that a nonzero constant times a quantity that keeps increasing beyond any bounds produces another quantity that keeps increasing beyond any bounds; sometimes, we write this as

$$\frac{2}{7} \cdot \infty = \infty$$

The following limit laws for infinite limits are intuitively clear (we can argue as above). We present them here without formal proof.

Theorem 4.3.1 **Limit Laws for Infinite Limits**

(a) Assume that $\lim_{x\to\infty} f(x) = \infty$ and $\lim_{x\to\infty} g(x) = \infty$, or $\lim_{x\to\infty} g(x) = L$, where L is a real number. Then

$$\lim_{x\to\infty} (f(x) + g(x)) = \infty$$

(b) Assume that $\lim_{x\to\infty} f(x) = -\infty$ and $\lim_{x\to\infty} g(x) = -\infty$, or $\lim_{x\to\infty} g(x) = L$, where L is a real number. Then

$$\lim_{x\to\infty} (f(x) + g(x)) = -\infty$$

(c) Assume that $\lim_{x\to\infty} f(x) = \infty$, and let c be a nonzero real number. If $c > 0$, then

$$\lim_{x\to\infty} cf(x) = \infty$$

and if $c < 0$, then

$$\lim_{x\to\infty} cf(x) = -\infty$$

(d) Assume that $\lim_{x\to\infty} f(x) = \infty$ and $\lim_{x\to\infty} g(x) = \infty$. Then

$$\lim_{x\to\infty} (f(x)g(x)) = \infty$$

The signs obey the usual convention: if one of the limits of $f(x)$ or $g(x)$ is $-\infty$ and the other is ∞, then the limit of the product is $-\infty$. If both limits are $-\infty$, then the limit of the product is ∞.

(e) Assume that $\lim_{x\to\infty} f(x) = \infty$ or $\lim_{x\to\infty} f(x) = -\infty$. For any real number c,

$$\lim_{x\to\infty} \frac{c}{f(x)} = 0$$

All the laws stated in the theorem hold if $x \to \infty$ is replaced by $x \to -\infty$. Sometimes, we abbreviate these laws in calculations and write

$$\infty + 3.5 = \infty \quad \text{or} \quad \infty - 100 = \infty \quad \text{or} \quad -\infty + 67 = -\infty \quad \text{or} \quad \infty + \infty = \infty$$

$$\text{or} \quad \infty \cdot \infty = \infty \quad \text{or} \quad (-4) \cdot \infty = -\infty \quad \text{or} \quad \infty \cdot (-\infty) = -\infty, \text{ etc.}$$

Note that the expressions

$$\infty - \infty, \frac{\infty}{\infty}, \text{ and } 0 \cdot \infty$$

are not mentioned in Theorem 4.3.1. They are called **indeterminate forms** and their behaviour is somewhat more complicated. We will study limits that involve these expressions in Chapter 6.

Comparing Functions at Infinity

Functions Approaching ∞ at ∞ Many of the fundamental functions of biology approach infinity at infinity, including the natural logarithm function $f(x) = \ln x$, the power function $f(x) = x^n$ with positive n, and the exponential function $f(x) = e^x$.

To reason about more complicated functions we must *compare* the behaviours of these basic functions. The graph of $f(x) = e^x$ increases very quickly, whereas that of $f(x) = \ln x$ increases slowly. Is there a precise way in which the exponential function increases "more quickly" than the logarithmic function? What does it mean to approach infinity "more quickly" or "more slowly"? These relations are summarized in the following definition.

Definition 4.3.5 Suppose

$$\lim_{x \to \infty} f(x) = \infty$$

$$\lim_{x \to \infty} g(x) = \infty$$

1. The function $f(x)$ approaches infinity **more quickly** than $g(x)$ as x approaches infinity if

$$\lim_{x \to \infty} \frac{f(x)}{g(x)} = \infty$$

2. The function $f(x)$ approaches infinity **more slowly** than $g(x)$ if

$$\lim_{x \to \infty} \frac{f(x)}{g(x)} = 0$$

3. $f(x)$ and $g(x)$ approach infinity at the **same rate** if

$$\lim_{x \to \infty} \frac{f(x)}{g(x)} = L$$

where L is any real number other than 0.

When $f(x)$ approaches infinity more quickly than $g(x)$, $f(x)$ gets farther and farther ahead of $g(x)$. For instance, compare $f(x) = x^3$ and $g(x) = \sqrt{x}$ in Table 4.3.6: when $x = 10$, x^3 is over 300 times as large as $\sqrt{x}$; when $x = 100$, x^3 is 100 thousand times as large as $\sqrt{x}$; when $x = 1000$, x^3 is more than 30 million times as large as $\sqrt{x}$.

Table 4.3.6

x	$\ln x$	$\sqrt{x}$	x^2	x^3	$e^{0.1x}$	e^x
10	2.30	3.16	100	1000	2.72	22,026.47
50	3.91	7.07	2500	$1.25 \cdot 10^5$	148.41	$5.18 \cdot 10^{21}$
100	4.61	10	10^4	10^6	22,026.47	$2.69 \cdot 10^{43}$
500	6.21	22.36	$2.5 \cdot 10^5$	$1.25 \cdot 10^8$	$5.18 \cdot 10^{21}$	$1.40 \cdot 10^{217}$
1000	6.91	31.62	10^6	10^9	$2.69 \cdot 10^{43}$	$1.97 \cdot 10^{434}$

When $f(x)$ approaches infinity more slowly, $f(x)$ falls farther and farther behind $g(x)$. For instance, we look at the value of the fraction

$$\frac{f(x)}{g(x)} = \frac{\ln x}{x^2}$$

When $x = 10$, then $\ln 10/10^2 \approx 0.023026$; if we take $x = 100$, then $\ln 100/100^2 \approx 0.000461$; when $x = 1000$, then $\ln 1000/1000^2 \approx 0.000007$. This calculation suggests that the limit of $f(x)/g(x)$ is zero as $x \to \infty$.

When the two functions approach infinity at the same rate, neither gets ahead or falls behind; see Table 4.3.5, where we compared $f(x) = 2x^3 - 3x^2 + 4x + 3$ and $g(x) = 2x^3$. In that case, we use the last part of Definition 4.3.5 with the value $L = 1$. *More quickly, more slowly,* and *at the same rate* act like *greater than, less than,* and *equal to* for numbers. They provide a way to compare the "sizes" of functions.

In Table 4.3.6 we compare the behaviour of several functions as $x \to \infty$.

We see striking differences in the rates at which these functions increase. For instance, when $x = 1000$, the natural logarithm function is not yet equal to 7, whereas x^2 reaches one million, and x^3 is equal to one billion. Exponential functions far surpass these values: e^{500} is far beyond any number that could possibly have some physical meaning for us (or for the whole universe).

As well, note that the relationships change: $x^3 > e^{0.1x}$ when $x = 100$, but $e^{0.1x} > x^3$ for $x = 500$.

The basic functions are shown in increasing order in Table 4.3.7 and Figure 4.3.41. The constant a in front of each function can be any positive number and does not change the order of the functions. Any power function, however small the power n, beats the logarithm. Any exponential function with a positive parameter β in the exponent beats any power function.

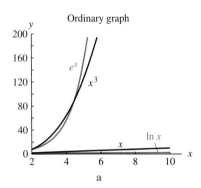

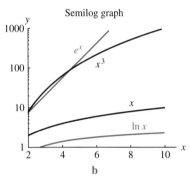

FIGURE 4.3.41
The behaviour of the basic functions that approach infinity

Table 4.3.7   The Basic Functions in Increasing Order of Speed

Function	Comments
$a \ln x$	Goes to infinity slowly
ax^n with $n > 0$	Approaches infinity more quickly for larger n
$ae^{\beta x}$ with $\beta > 0$	Approaches infinity more quickly for larger β

In Chapter 6, we will use Definition 4.3.5 and L'Hôpital's rule to show that what we claim in Table 4.3.7 is true. For instance, we will be able to show that

$$\lim_{x \to \infty} \frac{e^x}{x^3} = \infty$$

or, in general,

$$\lim_{x \to \infty} \frac{ae^{\beta x}}{cx^n} = \infty$$

when $a > 0$, $c > 0$, and $\beta > 0$. As well, using the concept of **leading behaviour** that we develop in Section 6.4, we will be able to compare more complicated functions.

Example 4.3.16   Ordering a Set of Functions

To order the functions

$$y = 0.1e^{2x}, \ y = 4.5 \ln x, \ y = 23.2x^{0.5}, \ y = 10.1e^{0.2x}, \ y = 0.03x^4$$

in the limit as $x \to \infty$ in increasing order, first spot functions of the three types: logarithmic, power, and exponential. There is only one logarithmic function, which is therefore the slowest. There are two power functions, with $23.2x^{0.5}$ having the smaller power and $0.03x^4$ having the larger. There are two exponential functions, with $y = 10.1e^{0.2x}$ having the smaller parameter (0.2) inside the exponent and $y = 0.1e^{2x}$ having the larger parameter (2) inside the exponent. In increasing order, these functions are $y = 4.5 \ln x, \ y = 23.2x^{0.5}, \ y = 0.03x^4, \ y = 10.1e^{0.2x}, \ y = 0.1e^{2x}$. The constants in front do not affect the ordering. ▲

Two remarks concerning our discussion: First, comparing complicated functions (compositions of power, logarithmic, and exponential functions) might require special techniques (such as L'Hôpital's rule, which we discuss in Chapter 6). Second, we need to keep in mind that the comparisons we have done hold for *large values* of x, i.e., as x approaches ∞. Consider the following example.

Example 4.3.17   Functions That Take a Long Time to Get into Order

The power function $y = x^2$ does eventually grow faster than $y = 1000x$ because it has a larger power (Figure 4.3.42). However, $x^2 < 1000x$ for $x < 1000$. If x cannot realistically take on values greater than 1000, the comparison in Table 4.3.7 is not relevant.

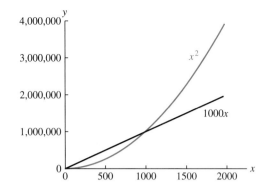

FIGURE 4.3.42

A faster function eventually overtaking a slower function

When a comparison includes a large or a small parameter (1000 in this case), we must first check whether the faster function becomes larger for biologically reasonable values of the argument.

Functions Approaching 0 at ∞ We use a similar approach to compare the rate at which functions approach a limit of zero.

Definition 4.3.6 Suppose that

$$\lim_{x \to \infty} f(x) = 0$$
$$\lim_{x \to \infty} g(x) = 0$$

1. The function $f(x)$ approaches zero **more quickly** than $g(x)$ as x approaches infinity if

$$\lim_{x \to \infty} \frac{f(x)}{g(x)} = 0$$

2. The function $f(x)$ approaches zero **more slowly** than $g(x)$ as x approaches infinity if

$$\lim_{x \to \infty} \frac{f(x)}{g(x)} = \pm\infty$$

3. $f(x)$ approaches zero at the same rate as $g(x)$ if

$$\lim_{x \to \infty} \frac{f(x)}{g(x)} = L$$

where L is any real number other than zero.

Be careful not to confuse this with the definition for functions approaching infinity (Definition 4.3.5). The function $f(x)$ approaches zero more quickly if it becomes *small* more quickly than $g(x)$. For instance, compare $f(x) = e^{-x}$ and $g(x) = x^{-2}$ by looking at the values in Table 4.3.9: when $x = 10$, the value of e^{-x}/x^{-2} is $e^{-10}/10^{-2} \approx 0.0045$; when $x = 50$, then $e^{-50}/50^{-2} \approx 4.82 \cdot 10^{-19}$, and when $x = 100$, then $e^{-100}/100^{-2} \approx 3.72 \cdot 10^{-40}$.

The basic examples are reciprocals of the functions in Table 4.3.7 (Table 4.3.8 and Figure 4.3.43). If $f(x)$ approaches *infinity* quickly, the reciprocal $\frac{1}{f(x)}$ approaches

Table 4.3.8 The Basic Functions Approaching Zero from Slowest to Fastest

Function	Comments
ax^{-n} with $n > 0$	Approaches zero more quickly for larger n
$ae^{-\beta x}$ with $\beta > 0$	Approaches zero more quickly for larger β
$ae^{-\beta x^2}$ with $\beta > 0$	Approaches zero really quickly

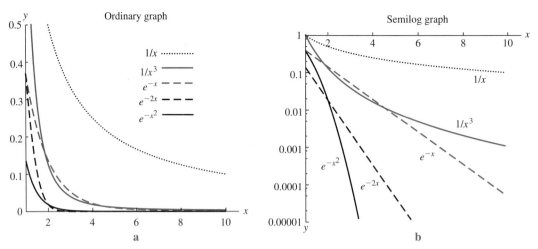

FIGURE 4.3.43

The behaviour of the basic functions that approach zero

Table 4.3.9

x	$x^{-1/2}$	x^{-1}	x^{-2}	e^{-x}	e^{-2x}	e^{-x^2}
10	0.316	0.1	0.01	$4.5 \cdot 10^{-5}$	$2.1 \cdot 10^{-9}$	$3.7 \cdot 10^{-44}$
50	0.141	0.02	0.0004	$1.9 \cdot 10^{-22}$	$3.7 \cdot 10^{-44}$	$1.8 \cdot 10^{-1086}$
100	0.1	0.01	0.0001	$3.7 \cdot 10^{-44}$	$1.4 \cdot 10^{-87}$	$1.1 \cdot 10^{-4343}$
500	0.045	0.002	$4 \cdot 10^{-6}$	$7.1 \cdot 10^{-218}$	$5.1 \cdot 10^{-435}$	$2.4 \cdot 10^{-108,574}$

0 quickly. Because these functions are hard to distinguish in a usual coordinate system (Figure 4.3.43a), we draw a semilog graph (Figure 4.3.43b). Again, the positive constant a does not change the ordering of the functions.

In Table 4.3.9 we compare several functions that approach zero as x approaches ∞, illustrating the claim we make in Table 4.3.8. As in the case of functions approaching ∞, we see a huge difference in the rates at which these functions approach 0.

Example 4.3.18 Ordering Functions That Approach Zero

To order the functions

$$y = 0.1e^{-2x}, \ y = 23.2x^{-0.5}, \ y = 10.1e^{-0.2x}, \ y = 0.03x^{-4}$$

from the one that approaches zero most quickly to the one that approaches zero most slowly, first identify the functions as exponential and power functions. The fastest is the exponential function with the more negative parameter, $y = 0.1e^{-2x}$, followed by the exponential function with the less negative parameter, $y = 10.1e^{-0.2x}$, then the power function with the more negative power, $y = 0.03x^{-4}$ and finally the power function with the less negative power, $y = 23.2x^{-0.5}$. From fastest to slowest approaching zero, they are $y = 0.1e^{-2x}, y = 10.1e^{-0.2x}, y = 0.03x^{-4}, y = 23.2x^{-0.5}$.

Application to Absorption Functions

Suppose the function

$$p(c) = 0.5 \left(1 - e^{-0.6c}\right)$$

describes the fraction of chemical absorbed by a lung during each breath. The total amount absorbed with each breath is

$$\alpha(c) = p(c)cV = 0.5 \left(1 - e^{-0.6c}\right) cV$$

4.3 Infinite Limits and Limits at Infinity

Table 4.3.10 Absorption Functions

$\alpha(c)$	Description	Figure
Ac	Linear absorption	4.3.44a
$\dfrac{Ac}{k+c}$	Saturated absorption	4.3.44b
$\dfrac{Ac^2}{k+c^2}$	Saturated absorption with threshold	4.3.44c
$Ace^{-\beta c}$	Saturated absorption with overcompensation I	4.3.44d
$\dfrac{Ac}{k+c^2}$	Saturated absorption with overcompensation II	4.3.44e
$Ac(1+kc)$	Enhanced absorption	4.3.44f

the product of the fraction absorbed; the concentration, c; and the volume, V. What does the graph of the function $p(c)$ look like? What does the graph of the total amount absorbed look like?

The amount of a chemical or resource used as a function of the amount available is important throughout biology. For consumers, like predators, this relation is called the **functional response.** In chemical reactions, this relation is often described by **Michaelis-Menten** or **Monod** reaction kinetics.

Several possible absorption functions are given in Table 4.3.10 (see Figure 4.3.44 for graphs). Each describes the total amount of chemical absorbed as a function of the concentration, c. The parameter A is a measure of efficiency, with small values producing low absorption and large values producing high absorption. The parameters k and β describe the shape of the function.

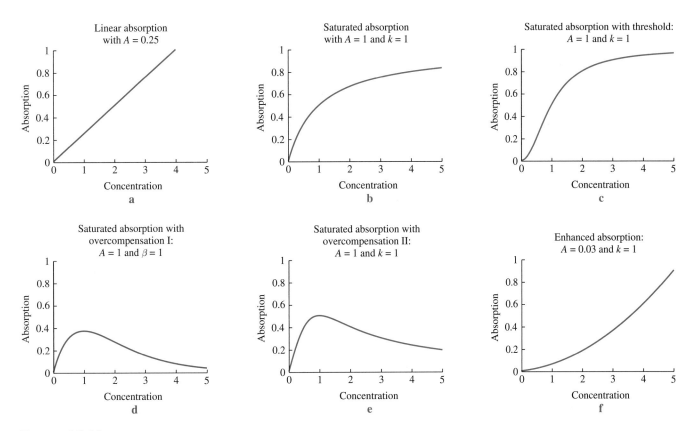

FIGURE 4.3.44

Various absorption functions

Table 4.3.11 Analyzing Absorption Functions

$\alpha(c)$	Numerator	Denominator	Behaviour at Infinity
Ac	Linear	None	Approaches infinity
$\dfrac{Ac}{k+c}$	Linear	Linear	Approaches nonzero constant
$\dfrac{Ac^2}{k+c^2}$	Quadratic	Quadratic	Approaches nonzero constant
$Ace^{-\beta c}$	Linear	Exponential	Approaches zero
$\dfrac{Ac}{k+c^2}$	Linear	Quadratic	Approaches zero
$Ac(1+kc)$	Quadratic	None	Approaches infinity

In each case, absorption is zero when $c = 0$. The behaviour of the absorption function for large values of c describes absorption at high concentrations.

How can we use the facts about the basic functions to understand the absorption functions in Table 4.3.10? The results are summarized in Table 4.3.11. We compare the numerator and denominator of the absorption function as functions of the concentration, c. If the numerator grows more quickly than the denominator, absorption grows without bound as c gets large. If the numerator and denominator grow at the same rate, absorption approaches a constant as c gets large (Figure 4.3.44b and c). If the denominator grows faster than the numerator, absorption approaches zero as c gets large (Figure 4.3.44d and e).

With linear absorption (Figure 4.3.44a), absorption grows without bound (there is no denominator to balance the numerator). Because there are almost always limits to absorption, the saturated absorption functions (Figure 4.3.44b and c) provide more reasonable models.

The saturated absorption form represented in Figure 4.3.44b, the ratio of linear functions, is among the most important in biology and is known as the **Michaelis-Menten** or **Monod** equation. In Chapter 6, we will formalize the calculation of the behaviour at infinity with the **method of leading behaviour.**

In the forms shown in Figure 4.3.44d and e, the denominator grows faster than the numerator (both exponential and quadratic functions grow faster than linear functions). These functions begin at 0, increase to a maximum, and eventually decrease again to 0. This behaviour is called **overcompensation** because absorption decreases when the concentration is too large.

Limits of Sequences

Next, we study the limit of a solution of a discrete-time dynamical system. The solution of the bacterial discrete-time dynamical system $b_{t+1} = rb_t$ is

$$b_t = b_0 r^t = b_0 e^{\ln(r^t)} = b_0 e^{(\ln r)t}$$

Keep in mind that the populations b_t are defined only at positive integer values of t. The solution

$$b_0, b_1, b_2, b_3, \ldots$$

is an example of a **sequence** of real numbers.

To find the limit, define the **associated function** $b(t)$ as the exponential function

$$b(t) = b_0 e^{(\ln r)t}$$

defined for all values of t. This function fills in the gaps in the sequence (Figure 4.3.45). If the associated function has a limit, whether it is zero, infinity, or some other value, the sequence will share that limit. In this case, the associated function is an exponential function. If $r > 1$, the parameter in the exponent is positive, and the limit of the function—and therefore of the sequence—is infinity. If $r < 1$, the parameter in the

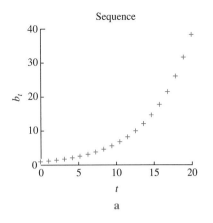
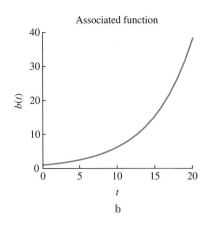

FIGURE 4.3.45

A sequence and its associated function

exponent is negative, and the limit of the function—and therefore of the sequence—is zero. If $r = 1$, the function has the constant value b_0, and the function and the sequence share the limit b_0.

This definition has an important connection with the idea of a **stable equilibrium.** If the sequence of points that represents a solution approaches a particular value as a limit, that limit is a stable equilibrium (Figure 4.3.46).

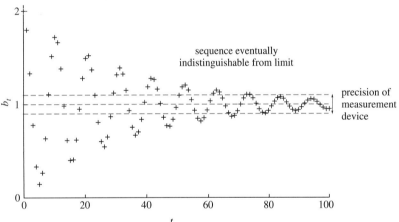

FIGURE 4.3.46

The limit of a sequence at infinity

Example 4.3.19 The Limit of a Solution of the Pain Medication Discrete-Time Dynamical System

In Example 3.1.7, we found the solution of the discrete-time dynamical system

$$M_{t+1} = 0.5 M_t + 1$$

describing the amount of pain medication given to a patient after surgery to be

$$M_{t+1} = 3 \cdot 0.5^t + 2$$

with initial condition $M_0 = 5$. The associated function $M(t)$ can be written as

$$M(t) = 3 \cdot 0.5^t + 2 = 3 e^{(\ln 0.5) t} + 2 = 3 e^{-0.693 t} + 2$$

where $M(0) = 5$.

The first term is an exponential function with a negative coefficient in the power, implying that this term approaches 0 as the argument t approaches infinity. The whole associated function then approaches the value 2. The sequence of points that represents the solution also approaches 2.

Example 4.3.20 A Case Where the Sequence and Associated Function Behave Differently

If the associated function has a limit, the sequence shares that limit. The sequence, however, may have a limit even when the associated function does not. Consider the sequence

$$a_t = \sin(2\pi t)$$

The associated function, $a(t) = \sin(2\pi t)$, has no limit because it oscillates forever. The sequence, on the other hand, takes on only the value zero and has the limit zero (Figure 4.3.47).

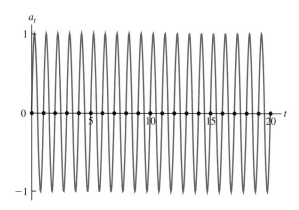

FIGURE 4.3.47

A sequence with a different limit from its associated function

Summary We expand the concept of the limit to include infinity. A function is said to **approach infinity** if it can be made larger than any real number by appropriately choosing the values of its variable. We use **limits at infinity** to determine the behaviour of a function as its variable becomes a larger and larger positive or a larger and larger negative number. One function approaches infinity **more quickly** than another if the limit of their ratio is infinity. Exponential functions approach infinity more quickly than power functions, which in turn approach infinity more quickly than logarithmic functions. Conversely, one function approaches zero more quickly than another if the limit of their ratio is zero. Exponential functions with negative parameters approach zero more quickly than power functions with negative powers. Using limits at infinity, we analyzed the limits of **sequences,** lists of numbers generated as solutions of discrete-time dynamical systems, by studying the **associated function.**

4.3 Exercises

Mathematical Techniques

1. Explain in your own words what is meant by saying that $\lim\limits_{x \to 2^-} f(x) = -\infty$ and $\lim\limits_{x \to 0^+} f(x) = \infty$.

2. Explain in your own words the meaning of the equations $\lim\limits_{x \to \infty} f(x) = 3$ and $\lim\limits_{x \to -\infty} f(x) = 0$.

3. Find a formula for a function $f(x)$ that satisfies $\lim\limits_{x \to 1^-} f(x) = \infty$ and $\lim\limits_{x \to 1^+} f(x) = -\infty$.

4. Find a formula for a function $f(x)$ that satisfies $\lim\limits_{x \to 1^-} f(x) = \infty$ and $\lim\limits_{x \to 1^+} f(x) = \infty$.

5. Find a formula for a function $f(x)$ that satisfies $\lim\limits_{x \to 0^-} f(x) = -\infty$ and $\lim\limits_{x \to 1} f(x) = \infty$.

6. Given below is the graph of the function $f(x) = \sec x$. Using one-sided limits, describe its behaviour at $-\pi/2$, $\pi/2$, and $3\pi/2$.

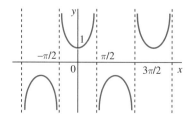

7. Given below is the graph of the function $f(x) = \cot x$. Using one-sided limits, describe its behaviour at $-\pi$, π, and 2π.

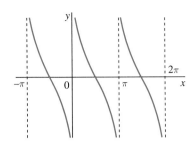

8–19 ■ Determine whether the limit is ∞ or $-\infty$.

8. $\lim_{x \to 2^+} \dfrac{5}{x-2}$

9. $\lim_{x \to 2^-} \dfrac{5}{x-2}$

10. $\lim_{x \to 2} \dfrac{5}{(x-2)^4}$

11. $\lim_{x \to 1^-} \dfrac{x}{x^2-1}$

12. $\lim_{x \to 1^+} \dfrac{x}{x^2-1}$

13. $\lim_{x \to 0} \dfrac{x^4+2}{|x|}$

14. $\lim_{x \to -7^+} \sqrt{\dfrac{1}{x+7}}$

15. $\lim_{x \to \pi} (\tan x + \cot x)$

16. $\lim_{x \to \pi/2^+} \sec x$

17. $\lim_{x \to \pi/2^-} \sec x$

18. $\lim_{x \to 6^+} \ln(x-6)$

19. $\lim_{x \to 0^+} [\ln(x^2 + e^x - 1)]$

20–25 ■ Find the vertical asymptotes (or else say there are none) of the graph of each function.

20. $f(x) = \dfrac{x-3}{x^2-1}$

21. $g(x) = \dfrac{x^2-1}{x^3-x}$

22. $f(x) = \dfrac{x^2-1}{x^2+1}$

23. $h(x) = \dfrac{x^2-7x+10}{x^2-4}$

24. $g(x) = \dfrac{1}{x} - \dfrac{1}{|x|}$

25. $h(x) = e^{-1/x^2}$

26. Sketch the graph of the function $f(x) = \ln |x|$ and find its vertical asymptote(s). Using limits, describe the behaviour of $f(x) = \ln |x|$ at its asymptote(s).

27–30 ■ The given functions all have limits of ∞ as $x \to 0^+$. For each function, find how close the input must be to 0 for the output to be (a) greater than 10, and (b) greater than 100. Sketch a graph of each function for $x < 1$, and indicate which functions approach infinity quickly and which approach infinity slowly.

27. $g_1(x) = \dfrac{1}{\sqrt{x}}$

28. $g_2(x) = \dfrac{1}{x}$

29. $g_3(x) = \dfrac{1}{x^2}$

30. $g_4(x) = \dfrac{1}{x^4}$

31–57 ■ Compute each limit.

31. $\lim_{x \to -\infty} x^{-5}$

32. $\lim_{x \to \infty} x^{-5}$

33. $\lim_{x \to \infty} x^{-0.5}$

34. $\lim_{x \to -\infty} (2x^{-4} - 3)$

35. $\lim_{x \to -\infty} (-x^{-3})$

36. $\lim_{x \to \infty} 0.7^x$

37. $\lim_{x \to \infty} 0.7^{-x}$

38. $\lim_{x \to -\infty} (2.2^x - 4)$

39. $\lim_{x \to -\infty} \sqrt{1-x}$

40. $\lim_{x \to \infty} \dfrac{1}{\sqrt{x-100}}$

41. $\lim_{x \to \infty} \sqrt{\dfrac{1}{x^2+4}}$

42. $\lim_{x \to \infty} \dfrac{x^3-6x+4}{3-4x^3}$

43. $\lim_{x \to -\infty} \dfrac{x-6x^4+4x^3}{3x^4-x^3}$

44. $\lim_{x \to \infty} \dfrac{2x^2+x-3}{x^3+23x+1}$

45. $\lim_{x \to \infty} \dfrac{(x-1)(x-2)(x-3)(x-4)}{2x^4+5}$

46. $\lim_{x \to \infty} \dfrac{x^5-x+2}{x^4+2x^2-7}$

47. $\lim_{x \to \infty} \sqrt{x + \dfrac{1}{x}}$

48. $\lim_{x \to \infty} (2x^2 - 45x + 2)$

49. $\lim_{x \to -\infty} (-5x^3 + 14x^2 - 100)$

50. $\lim_{x \to -\infty} \ln(3-x^3)$

51. $\lim_{x \to \infty} [\ln x - \ln(x-1)]$

52. $\lim_{x \to \infty} [\arctan(3x+5)]$

53. $\lim_{x \to -\infty} [\arctan(1-3x^4)]$

54. $\lim_{x \to -\infty} e^{1/x}$

55. $\lim_{x \to \infty} e^{-x^2}$

56. $\lim_{x \to \infty} \dfrac{e^x - 2e^{-x}}{3e^x + e^{-x}}$

57. $\lim_{x \to -\infty} \dfrac{e^x - 2e^{-x}}{3e^x + e^{-x}}$

58–65 ■ Find all horizontal asymptotes of the graph of each function.

58. $f(x) = \dfrac{x}{3x^4 - 1}$

59. $h(x) = \dfrac{(x-1)(x+4)}{x^2 + 2}$

60. $g(x) = \ln(4 + 3e^{-x})$

61. $f(x) = \arctan(\ln x)$

62. $f(x) = \arctan(e^{-x})$

63. $f(x) = \dfrac{e^x}{1 - e^x}$

64. $f(x) = \sqrt{1 - \dfrac{2}{x}}$

65. $f(x) = \sin(1/x)$

66. Find $\lim_{x \to \infty} e^{-x^2} \sin(3x)$ using the Squeeze Theorem.

67. Find $\lim_{x \to \infty} e^{-7x} \cos^2 x$ using the Squeeze Theorem.

68–73 ■ For each pair of functions, say which approaches ∞ more quickly as x approaches infinity. Explain which rule you used to compare each pair. Compute the value of each function at $x = 1$, $x = 10$, and $x = 100$. How do these compare with the order of the functions in the limit? If they are different, how large would x have to be for the values to match the order in the limit?

68. $y = x^2$ and $y = e^{2x}$

69. $y = x^3$ and $y = 1000x$

70. $y = x^{3.5}$ and $y = 0.1x^{10}$

71. $y = 5e^x$ and $y = e^{5x}$

72. $y = 0.1x^{0.5}$ and $y = 30 \ln x$

73. $y = 10x^{0.1}$ and $y = x^{0.5}$

74. By drawing their graphs in a semilog plot on the interval [0, 100], compare the functions $y = e^{2.2x}$, $y = e^{0.4x}$, $y = e^x$, and $y = x^5$ in terms of how fast they approach infinity as $x \to \infty$.

75. By drawing their graphs in a log-log plot, compare the functions $y = 0.1x^7$, $y = x^6$, $y = 100x^5$, and $y = 0.01x^8$ in terms of how fast they approach infinity as $x \to \infty$.

76–81 ■ For each pair of functions, say which approaches zero more quickly as x approaches infinity. Explain which rule you used to compare each pair. Compute the value of each function at $x = 1$, $x = 10$, and $x = 100$. How do these compare with the order of the functions in the limit? If they are different, how large would x have to be for the values to match the order in the limit?

76. $y = e^{-2x}$ and $y = x^{-10}$

77. $y = 10e^{-x}$ and $y = 0.1e^{-0.2x}$

78. $y = 1000/x$ and $y = x^{-3.5}$

79. $y = x^{-0.1}$ and $y = 25x^{-0.2}$

80. $y = x^{-2}$ and $y = 30/\ln x$

81. $y = 1/\ln x$ and $y = 30x^{-0.1}$

82. By drawing their graphs in a semilog plot on the interval [0, 100], compare the functions $y = e^{-1.2x}$, $y = x^{-2}$, $y = e^{-x}$, and $y = e^{-0.3x}$ in terms of how fast they approach zero as $x \to \infty$.

83. By drawing their graphs in a log-log plot, compare the functions $y = 100x^{-3}$, $y = 10x^{-2}$, $y = x^{-1}$, and $y = 1000x^{-4}$ in terms of how fast they approach zero as $x \to \infty$.

84–87 ■ The following are possible absorption functions. What happens to each as c approaches infinity? Assume that all parameters take on positive values.

84. $y = \dfrac{\beta c^2}{1 + e^c}$

85. $y = \dfrac{Ac}{\ln(1 + c)}$

86. $y = \dfrac{\gamma(e^c - 1)}{e^{2c}}$

87. $y = \dfrac{c^2}{1 + 10c}$

88–89 ■ Recall that if the associated function has a limit, the sequence shares the limit. However, the sequence might have a limit even when the associated function does not.

a. Find the limit of the sequence.

b. Explain why the associated function does not have a limit.

88. $a_t = \cos 2\pi t$

89. $a_t = t \sin \pi t$

90–93 ■ Read Definition 4.3.3. By looking at Table 4.3.9 (or by calculating appropriate values if needed), answer each question.

90. Find x large enough so that $x^{-2} < 0.005$; i.e., find a number N such that for every $x > N$, $x^{-2} < 0.005$.

91. Find x large enough so that $x^{-2} < 10^{-7}$; i.e., find a number N such that for every $x > N$, $x^{-2} < 10^{-7}$.

92. Find x large enough so that $e^{-x} < 10^{-15}$.

93. Find x large enough so that $x^{-1/2} < 0.01$.

Applications

94–97 ■ A bacterial population that obeys the discrete-time dynamical system $b_{t+1} = rb_t$ with initial condition b_0 has solution $b_t = b_0 r^t$. For the following values of r and b_0, state which populations increase to infinity and which decrease to zero. For those increasing to infinity, find the time when the population will reach 10^{10}. For those decreasing to zero, find the time when the population will reach 10^3.

94. $b_0 = 10^8$ and $r = 1.1$

95. $b_0 = 10^8$ and $r = 1.5$

96. $b_0 = 10^8$ and $r = 0.5$

97. $b_0 = 10^8$ and $r = 0.9$

98–99 ■ In the polymerase chain reaction (PCR) used to amplify DNA, some sequences of DNA that are produced are too long and others are the right length. Denote the number of overly long pieces after t generations of the process by l_t and the number of pieces of the right length by r_t. The dynamics follow approximately

$$l_{t+1} = l_t + 2$$
$$r_{t+1} = 2r_t$$

because two new overly long pieces are produced each step while the number of good pieces doubles. Suppose that $l_0 = 0$ and $r_0 = 2$.

98. Find expressions for l_t and r_t, and compute the fraction of pieces that are too long after 1, 5, 10, and 20 generations of the process.

99. Find the ratio of the number of pieces that are too long to the total number of pieces as a function of time. What is the limit? How long would you have to wait to make sure that less than one in a million pieces are too long? (This can't be solved exactly, just plug in some numbers.)

100–103 ■ Consider the discrete-time dynamical system for medication given by $M_{t+1} = 0.5M_t + 1$ with $M_0 = 3$.

100. Find the equilibrium.

101. The solution is $M_t = 2 + 0.5^t \cdot 3$. Find the limit as $t \to \infty$.

102. How long will the solution take to be within 1% of the equilibrium?

103. What are two ways to show that this equilibrium is stable?

104–107 ■ Try to think of a biological mechanism that could produce the following relations between the amount absorbed and the concentration of chemical.

104. Saturated absorption, but with an infinite limit as the concentration approaches infinity.

105. Saturated absorption with a finite limit as the concentration approaches infinity.

106. Overcompensation.

107. Enhanced absorption.

108–109 ■ In real life, the (monetary) cost is—in many cases—an integral part of developing and conducting experiments. How much would it cost to measure the value under each circumstance?

108. We are interested in measuring the pressure at different depths below the surface of the ocean. Pressure increases by approximately 1 atmosphere (atm) for every 10 m of depth below the surface (for example, at a depth of 20 m, there are approximately 3 atm of pressure: 2 atm due to the ocean and 1 atm due to the atmosphere itself). Measuring a pressure of x atmospheres without crushing the device costs x^2 dollars.

   a. How much would it cost to measure the pressure 100 m down?

   b. How much would it cost to measure the pressure 1000 m down?

   c. How much would it cost to measure the pressure 5000 m down?

109. Solar scientists want to measure the temperature inside the Sun by sending in probes. Imagine that temperature increases by 1 million degrees Celsius for every 10,000 km below the surface. A probe that can handle a temperature of x million degrees Celsius costs x^3 million dollars.

   a. How much would it cost to measure the temperature 10,000 km down?

   b. How much would it cost to measure the temperature 100,000 km down?

   c. How much would it cost to measure the temperature 200,000 km down?

Computer Exercises

110. Use a computer to find out how large x must be before the faster function finally overtakes the slower function.

   a. $y = e^{0.1x}$ catches up with $y = x^3$.

   b. $y = 0.1e^x$ catches up with $y = x^3$.

   c. $y = 0.1e^{0.1x}$ catches up with $y = x^3$.

   d. $y = 0.1x$ catches up with $y = \ln x$.

   e. $y = x^{0.1}$ catches up with $y = \ln x$.

4.4 Continuity

By describing mathematically what it means for a function to have no "gaps" or "jumps," we arrive at the definition of a **continuous function.** A function is continuous at a number a if its behaviour *near a* (expressed in terms of the limit) matches its behaviour *at a*. Scientifically, continuous functions represent relations between measurements where a small change in the input produces a small change in the output. In the forthcoming chapters we will see how numerous phenomena in biology and elsewhere can be modelled using continuous functions. Functions that are not continuous do occur in biological systems and can generate novel situations, such as **hysteresis.**

Continuous Functions

We can find
$$\lim_{x \to 0}(4x^2 + 9x + 1)$$
in several ways. We can plug in values of x that are close to 0 and see that the values approach 1. Alternatively, we could use the limit laws or direct substitution. In Section 4.2 we discovered that *in some cases* we can calculate the limit of a function as x approaches a by calculating the value of the function at a (using what we called the direct substitution rule). When exactly does this work?

It works when the function is **continuous.** Our intuitive notion of "continuous" is that the graph is "connected," without jumps or gaps. In Figure 4.4.48a we have drawn a graph of a continuous function. Figure 4.4.48b shows what a discontinuous function might look like ("discontinuous" is a synonym for "not continuous").

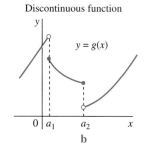

FIGURE 4.4.48

Continuous and discontinuous functions

The mathematical definition captures this intuitive idea.

Definition 4.4.1   A function f is **continuous** at a number a if
$$\lim_{x \to a} f(x) = f(a)$$
Otherwise, we say the function is **discontinuous** at a.

Note that the requirement
$$\lim_{x \to a} f(x) = f(a)$$
means that

(i) $\lim_{x \to a} f(x)$ is a real number.

(ii) $f(x)$ is defined at a (i.e., a is in the domain of $f(x)$).

(iii) the two real numbers $\lim_{x \to a} f(x)$ and $f(a)$ are equal.

To enhance our understanding of this definition, in Figure 4.4.49 we show several graphs of functions that are not continuous at a number a.

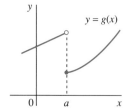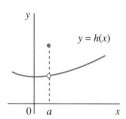

FIGURE 4.4.49

Functions that are not continuous at a

The function f is not continuous at a because a is not in its domain (so (ii) fails to hold; note that (i) holds!). The function g is not continuous at a: the one-sided limits are not equal, and thus g does not have a limit as $x \to a$ (so (i) does not hold; for the

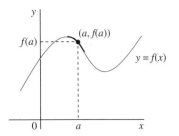

FIGURE 4.4.50
Intuitive understanding of continuity

record, (ii) does hold). The function h is defined at a and has a limit as $x \to a$; however,

$$\lim_{x \to a} h(x) \neq h(a)$$

i.e., requirement (iii) is not satisfied.

As mentioned, the concept of continuity of a function is close to our intuitive notion of the word "continuous." A continuous function cannot have holes in its graph (such as function f in Figure 4.4.49), jumps (such as functions g and h in Figure 4.4.49), or vertical asymptotes (such as $f(x) = 1/x$ at $x = 0$). In other words, a function is continuous at a if we can draw its graph *near a* (i.e., from one side of the point $(a, f(a))$ to the other side; see Figure 4.4.50) without lifting our pencil.

Note Air temperature over time is an example of a continuous function. It cannot have gaps—for instance, if the temperature at 9 A.M. was measured to be 15°C, and at 11 A.M. it is 21°C, then sometime between 9 A.M. and 11 A.M. it had to be exactly 16.7°C, or 20°C (i.e., the temperature has to assume all intermediate values; later, in Section 6.3 we will generalize this observation in the statement of the Intermediate Value Theorem).

Trajectories of various motions (particles, planets revolving around stars, or cats running) are examples of continuous curves. Sudden jumps within electric circuits and hysteresis (which we discuss at the end of this section) are examples of discontinuous functions.

Note One reason that we like continuous functions is that they are predictable. Assume that we are watching a bird flying and then close our eyes for a second. When we open our eyes, we know that the bird will be somewhere around the location where we last saw it. Contrary to this situation, a discontinuous function might not be predictable—for instance, it might jump suddenly from one value to another that is 10 times as large. Or—suppose that you are driving a car, turn around a corner, and see that a traffic light is green. You cannot know if two seconds later it will still be green, or if it will change to red.

The requirement for continuity,

$$\lim_{x \to a} f(x) = f(a)$$

says that the values of $f(x)$ get closer and closer to $f(a)$ as x gets closer and closer to a. Actually, we can *make the values* of $f(x)$ as close as we wish to $f(a)$ by picking x close enough to a. Thus, small changes in the variable (input) produce small changes in the values (output) of a continuous function. This is not true for discontinuous functions.

Example 4.4.1 Using the Definition to Check the Continuity of a Function

Show that the function

$$f(x) = \begin{cases} \dfrac{x^2 + 2x - 3}{x - 1} & \text{if } x \neq 1 \\ 4 & \text{if } x = 1 \end{cases}$$

is continuous at $x = 1$.

According to Definition 4.4.1, we need to prove that

$$\lim_{x \to 1} f(x) = f(1)$$

Since

$$\lim_{x \to 1} f(x) = \lim_{x \to 1} \frac{x^2 + 2x - 3}{x - 1} = \lim_{x \to 1} \frac{(x - 1)(x + 3)}{x - 1} = \lim_{x \to 1} (x + 3) = 4$$

and $f(1) = 4$, we conclude that $f(x)$ is indeed continuous at $x = 1$.

Example 4.4.2 Using the Definition of Continuity

Consider the function defined piecewise by

$$f(x) = \begin{cases} 3 & \text{if } x \leq -1 \\ 2x^2 + 1 & \text{if } -1 < x < 2 \\ \dfrac{9}{2x+1} & \text{if } x \geq 2 \end{cases}$$

Determine whether or not $f(x)$ is continuous at $x = -1$ and $x = 2$.

We check continuity at $x = -1$ first. Since the definition of $f(x)$ changes at -1, we need to compute one-sided limits. From

$$\lim_{x \to -1^-} f(x) = \lim_{x \to -1^-} 3 = 3 \quad \text{and} \quad \lim_{x \to -1^+} f(x) = \lim_{x \to -1^+} (2x^2 + 1) = 3$$

it follows that $\lim_{x \to -1} f(x) = 3$. Since $f(-1) = 3$ (calculated using the top line in the definition of $f(x)$), we conclude that $f(x)$ is continuous at $x = -1$.

Using the same approach, we examine $x = 2$. This time

$$\lim_{x \to 2^-} f(x) = \lim_{x \to 2^-} (2x^2 + 1) = 9 \quad \text{and} \quad \lim_{x \to 2^+} f(x) = \lim_{x \to 2^+} \frac{9}{2x+1} = \frac{9}{5}$$

and so $\lim_{x \to 2} f(x)$ does not exist. Consequently, $f(x)$ is not continuous at $x = 2$. See Figure 4.4.51.

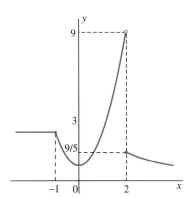

FIGURE 4.4.51
Graph of $f(x)$ from Example 4.4.2

Example 4.4.3 Discontinuous Functions in Modelling of Cancer Treatments

Chemotherapy and surgery are two frequently used treatments for a wide spectrum of cancers. How should they be used—chemotherapy first and then surgery, or the other way around? In order to answer this question (which is far from trivial) for the case of ovarian cancer, researchers have built mathematical models that track the number of cancer cells as both sequences of treatments (chemotherapy then surgery, surgery then chemotherapy) are applied. (See M. Kohandel et al. Mathematical modeling of ovarian cancer treatments: Sequencing of surgery and chemotherapy. *Journal of Theoretical Biology* 242: 62–68, 2006.)

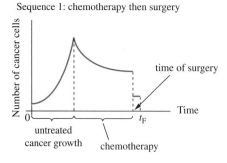

FIGURE 4.4.52
Number of cancer cells and reactions to treatments
M. Kohandel et al, "Mathematical Modeling of Ovarian Cancer Treatments: Sequencing of Surgery and Chemotherapy," *Journal of Theoretical Biology*, Vol. 242, Issue 1, Pg. 62-68, 2006.

Figure 4.4.52 (adapted from the graphs in the quoted reference) shows how the number of cancer cells changes depending on the treatment. Note that both functions have a discontinuity at the moment of surgery, when a number of cancer cells are removed. The size of the jump is highly relevant to the decision about the sequence of treatments that is applied. The effects of the two sequences are assessed at time t_F.

In Figure 4.4.49 we showed examples of functions that are not continuous. We now discuss these and related cases in more depth.

Some discontinuous functions have jumps, such as the **signum** function, defined by

$$\operatorname{sgn} x = \begin{cases} 1 & \text{if } x > 0 \\ 0 & \text{if } x = 0 \\ -1 & \text{if } x < 0 \end{cases}$$

In words, $\operatorname{sgn} x$ returns 1 if the number x is positive, -1 if it is negative, and 0 if it is zero. We show its graph in Figure 4.4.53.

At $x = 0$, the value of the function is 0, the left-hand limit is -1, and the right-hand limit is 1. Because the left-hand limit and the right-hand limit do not match, the limit does not exist. Furthermore, neither of these limits matches the value of the function. Thus, $\operatorname{sgn} x$ is not continuous at zero.

Functions with limits of infinity or negative infinity as x approaches a real number cannot be graphed without a jump (however, these functions can be continuous at all points in their domain; $f(x) = 1/x^2$, for example). In Figure 4.4.54a, the limit is infinity from both the left and the right. However, the function cannot be evaluated at $x = 0$; it is said to have an infinite discontinuity at zero. In Figure 4.4.54b, the limit is negative infinity from the left and positive infinity from the right. This function has an infinitely large jump.

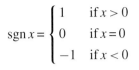

FIGURE 4.4.53

The function $y = \operatorname{sgn} x$

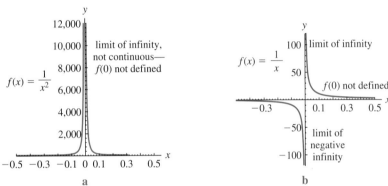

FIGURE 4.4.54

Two functions that are not defined at $x = 0$

Continuous functions are useful because we can find limits by direct substitution. How can we recognize a continuous function? Like limits, continuous functions can be combined in many useful ways, including addition, multiplication, division, and composition. Mathematically, we build a collection of continuous functions on the results of three theorems, one giving the basic continuous functions and the other two saying how they can be combined to form continuous functions.

Continuity is defined at *numbers* (sometimes referred to as points). A function can be continuous at some numbers and discontinuous at others. The function in Figure 4.4.48b is discontinuous *at a_1 and at a_2* and continuous *at all other numbers*. The functions in Figure 4.4.49 are not continuous *at a*. The function $f(x) = \operatorname{sgn} x$ is discontinuous *at $x = 0$* only.

When we say that a function is continuous without specifying a particular number, we mean that the function is continuous at *all* numbers in its domain.

Definition 4.4.2 A function is **continuous on an interval** if it is continuous at every number in the interval.

Thus, a function f is continuous on an interval (c, d) if

$$\lim_{x \to a} f(x) = f(a)$$

for *all* numbers a in (c, d). A function f is continuous on a closed interval $[c, d]$ if it is continuous at every number (point) inside the interval (i.e., if it is continuous on (c, d)), and at the endpoints c and d.

How do we define continuity at the endpoints? We cannot approach endpoints from both sides, since f is not defined to the left of c or to the right of d (see

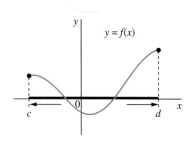

FIGURE 4.4.55
Continuity at the endpoints of an interval

Figure 4.4.55). So we modify the definition of continuity to account for the fact that x has to approach the endpoints *from within the interval*.

Thus, f is **continuous at the endpoint c** of the interval $[c, d]$ if

$$\lim_{x \to c^+} f(x) = f(c)$$

(we say that f is **continuous from the right at c**); it is **continuous at the endpoint d** of the interval $[c, d]$ if

$$\lim_{x \to d^-} f(x) = f(d)$$

(i.e., f is **continuous from the left at d**). See Figure 4.4.55.

Reading the next theorem, keep in mind that *continuous* (without specifying the number) means *continuous at all numbers in the domain*.

Theorem 4.4.1 The Basic Continuous Functions

a. The constant function $f(x) = c$ is continuous.
b. The identity function $f(x) = x$ is continuous.
c. The exponential function $f(x) = e^x$ is continuous.
d. The logarithmic function $f(x) = \ln x$ is continuous for $x > 0$.
e. The absolute value function $f(x) = |x|$ is continuous.
f. The trigonometric functions $f(x) = \sin x$ and $f(x) = \cos x$ are continuous.

Theorem 4.4.2 Combining Continuous Functions

Assume that the functions $f(x)$ and $g(x)$ are continuous at a. Then

a. The sum, $f(x) + g(x)$, and the difference, $f(x) - g(x)$, are continuous at a.
b. The products $cf(x)$ (where c is a real number) and $f(x)g(x)$ are continuous at a.
c. The quotient $\dfrac{f(x)}{g(x)}$ is continuous at a if $g(a) \neq 0$.

In reading the assumptions in the following theorem, keep in mind the definition of the composition of two functions:

$$(f \circ g)(a) = f(g(a))$$

Theorem 4.4.3 Continuity of Composition of Functions

If $g(x)$ is continuous at a and $f(x)$ is continuous at $g(a)$, then the composition $(f \circ g)(x) = f(g(x))$ is continuous at a.

We will not prove these theorems. Instead, we show how we can use them to build a library of continuous functions.

Theorem 4.4.1b tells us that $f(x) = x$ is continuous at every number a; by repeatedly using Theorem 4.4.2b we conclude that $f(x) = x^2$, $f(x) = x^3$, and, in general, $f(x) = x^n$ (n is a positive integer) are continuous at a.

Since multiplication by a constant (Theorem 4.4.2b) and addition and subtraction (Theorem 4.4.2a) preserve the continuity of the functions involved, it follows that a polynomial is continuous at every real number a. From Theorem 4.4.2c we conclude that a rational function is continuous at all numbers a in its domain (i.e., the quotient of two polynomials is continuous as long as we do not divide by zero).

Because $f(x) = \sin x$ and $f(x) = \cos x$ are continuous for all real numbers, we conclude that $f(x) = \tan x = \sin x / \cos x$ is continuous at all numbers a for which $\cos a \neq 0$.

Note that the inverse function of a continuous function is continuous. The graph of f^{-1} is the mirror image of the graph of f, so if f has no jumps or holes, then f^{-1} will not have them either. Thus, $f(x) = \ln x$ is continuous at all a for which it is defined, i.e., for $a > 0$. Likewise, $f(x) = \arcsin x$ and $f(x) = \arctan x$ are continuous in their domains.

The following theorem summarizes these results. As well, it adds a few more functions to the list of continuous functions.

Theorem 4.4.4 **Continuous Functions**

The following functions are continuous at all numbers in their domain:

a. polynomials

b. rational functions

c. roots

d. algebraic functions

e. absolute value function

f. exponential and logarithmic functions

g. trigonometric and inverse trigonometric functions

Next we quote a useful result that (besides its theoretical importance) will help us evaluate limits.

Theorem 4.4.5 **Interchanging a Limit and a Continuous Function**

Assume that a function g satisfies

$$\lim_{x \to a} g(x) = b$$

and that a function f is continuous at b. Then

$$\lim_{x \to a} f(g(x)) = f(b)$$

This is intuitively clear: by assumption on g, if x is close to a, then $g(x)$ is close to b. But f is continuous at b, so $f(g(x))$ must be close to $f(b)$.

Substituting the assumption $\lim_{x \to a} g(x) = b$ into the right side of the formula in the theorem, we get

$$\lim_{x \to a} f(g(x)) = f\left(\lim_{x \to a} g(x)\right)$$

In words, a limit and a continuous function can be interchanged, as the following example shows.

Example 4.4.4 Calculating a Limit Using Theorem 4.4.5

Compute

$$\lim_{x \to 1} \left[\sin\left(x^3 - \frac{2x}{1 + 3x^2}\right)\right]$$

Since $y = \sin x$ is a continuous function, we write

$$\lim_{x \to 1} \left[\sin\left(x^3 - \frac{2x}{1 + 3x^2}\right)\right] = \sin\left(\lim_{x \to 1}\left(x^3 - \frac{2x}{1 + 3x^2}\right)\right)$$

$$= \sin\left(1^3 - \frac{2(1)}{1 + 3(1)^2}\right)$$

$$= \sin\frac{1}{2} \approx 0.479$$

(recall that we work in radians).

In Example 4.3.12 in the previous section we showed that

$$\lim_{x \to \infty} e^{-1/x^3} = 1$$

The exponential function $f(x) = e^{-1/x^3}$ is continuous, and therefore we can use Theorem 4.4.5:

$$\lim_{x \to \infty} e^{-1/x^3} = e^{\lim_{x \to \infty}(-1/x^3)} = e^0 = 1$$

since

$$\lim_{x \to \infty}\left(-\frac{1}{x^3}\right) = 0$$

Next, we illustrate the concepts covered in this section. Theorems 4.4.1 to 4.4.5 help us identify continuous functions and tell us how to work with them. To recognize numbers where a function is not continuous we use the summary given in Table 4.4.1.

Table 4.4.1

Points Where a Function May Be Undefined or Discontinuous
Division by 0
Logs of 0
Points where the definition of a function changes

Example 4.4.5 Recognizing a Discontinuous Function

Consider the function $f(x) = \tan x$. Once we recall that

$$\tan x = \frac{\sin x}{\cos x}$$

we note that there is division by zero when $\cos x = 0$, i.e., when $x = -\pi/2$, $x = \pi/2$, and so forth (in general $x = \pi/2 + k\pi$, where k is an integer). These are the points where $f(x) = \tan x$ is not defined; thus, it not continuous there either (see Figure 4.4.56).

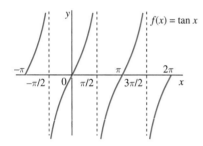

FIGURE 4.4.56
Graph of the function $f(x) = \tan x$

Example 4.4.6 Evaluating the Limit of a Continuous Function I

We can find

$$\lim_{x \to 0} f(x)$$

where

$$f(x) = e^{x^2 + 3}$$

by evaluating $f(0)$ if f is continuous at $x = 0$. This function is built up as a **composition** of the polynomial function $x^2 + 3$ and the exponential function. Both pieces are continuous (Theorem 4.4.1). The composition of these pieces is also continuous (Theorem 4.4.3). Therefore,

$$\lim_{x \to 0} e^{x^2 + 3} = e^{0^2 + 3} = e^3 \approx 20.09$$

Example 4.4.7 Evaluating the Limit of a Continuous Function II

To find
$$\lim_{x \to 0} \frac{e^{x^2+3}}{2 + x + \ln(x+1)}$$
we must take care in taking apart the function. The numerator is the function $f(x) = e^{x^2+3}$ from the previous example, which is continuous at all numbers x. The denominator, $h(x) = 2 + x + \ln(x+1)$, is the sum of the linear function $y = 2 + x$, which is continuous for all numbers, and the natural log $y = \ln(x+1)$, which is continuous for $x > -1$. The quotient will then be continuous at 0 as long as $h(0) \neq 0$. Since
$$h(0) = 2 + 0 + \ln(0 + 1) = 2 + 0 + 0 = 2 \neq 0$$
$h(x)$ is continuous and
$$\lim_{x \to 0} \frac{e^{x^2+3}}{2 + x + \ln(x+1)} = \frac{e^{0^2+3}}{2} = \frac{e^3}{2} \approx 10.04$$

Example 4.4.8 Recognizing Division by Zero

Division by 0 can be disguised. Suppose we are asked to find
$$\lim_{x \to 1^-} F(x) = \lim_{x \to 1^-} (1 - x)^{-3}$$
Note that we can write
$$F(x) = \frac{1}{(1-x)^3}$$
Because the denominator is equal to 0 at $x = 1$, this function is not continuous at $x = 1$ (because $F(1)$ is not defined).

For the record,
$$\lim_{x \to 1^-} F(x) = \lim_{x \to 1^-} \frac{1}{(1-x)^3} = +\infty$$
since $1 - x > 0$ as $x \to 1^-$.

Example 4.4.9 A Discontinuous Function That Is Defined Piecewise

Consider the function defined piecewise
$$f(x) = \begin{cases} 3x + \dfrac{4}{x} & \text{if } x > 2 \\ 2x + 1 & \text{if } x \leq 2 \end{cases}$$

The function in the top line is the sum of the polynomial $y = 3x$ (which is continuous for all x, and hence for $x > 2$) and the rational function $y = \dfrac{4}{x}$ (which is continuous when $x \neq 0$, and thus continuous when $x > 2$). Thus, if $x > 2$, then
$$f(x) = 3x + \frac{4}{x}$$
is continuous. When $x < 2$, then
$$f(x) = 2x + 1$$
is continuous, because it is a polynomial.

From what we just said we conclude that $f(x)$ is continuous at all $x \neq 2$.

Now we need to check the continuity at the number $x = 2$ where the definition of $f(x)$ changes. To do this, we need to see whether or not
$$\lim_{x \to 2} f(x) = f(2)$$

holds. Computing the one-sided limits, we obtain

$$\lim_{x \to 2^+} f(x) = \lim_{x \to 2^+} \left(3x + \frac{4}{x}\right) = 3(2) + \frac{4}{2} = 8$$

and

$$\lim_{x \to 2^-} f(x) = \lim_{x \to 2^-} (2x + 1) = 2(2) + 1 = 5$$

The two limits are not equal. Therefore

$$\lim_{x \to 2} f(x)$$

does not exist, and $f(x)$ is not continuous at $x = 2$.

Example 4.4.10 A Continuous Function That Is Defined Piecewise

Find all x where the function

$$f(x) = \begin{cases} 3e^{-8x} - 2 & \text{if } x < 0 \\ 1 & \text{if } x = 0 \\ \ln(x+1) + 1 & \text{if } x > 0 \end{cases}$$

is not continuous.

As in the previous example, we examine the pieces first. The function e^{-8x} is continuous at all x, as a composition of an exponential function and a polynomial. Multiplication by a constant and subtraction of a constant preserve the continuity; hence $y = 3e^{-8x} - 2$ is continuous for all x (and thus continuous at $x < 0$, where we need it).

The fact that $x > 0$ guarantees that $\ln(x + 1)$ is defined—and hence continuous—at $x > 0$. Adding the constant to it keeps it continuous; i.e., $y = \ln(x+1) + 1$ is continuous at $x > 0$.

For the continuity at $x = 0$, we check whether or not

$$\lim_{x \to 0} f(x) = f(0)$$

is true. Because the definition of $f(x)$ changes at $x = 0$, we need to use one-sided limits:

$$\lim_{x \to 0^+} f(x) = \lim_{x \to 0^+} (\ln(x+1) + 1) = \ln 1 + 1 = 1$$

and

$$\lim_{x \to 0^-} f(x) = \lim_{x \to 0^-} (3e^{-8x} - 2) = 3e^0 - 2 = 1$$

So,

$$\lim_{x \to 0} f(x) = 1$$

and since $f(0) = 1$, it follows that $f(x)$ is continuous at $x = 0$. Thus, it turns out that $f(x)$ is continuous at all real numbers.

Input and Output Precision

In applied mathematics, functions describe relationships between measurements. Continuous functions represent a special sort of relationship between measurements where a small change in the input produces only a small change in the output.

Example 4.4.11 Input and Output Tolerances for a Continuous Function I

Recall the discrete-time dynamical system

$$b_{t+1} = 2b_t$$

describing a bacterial population. Note that the updating function, $f(b_t) = 2b_t$, is continuous. Suppose we want a population of $2 \cdot 10^6$ at time $t = 1$. To hit $2 \cdot 10^6$ exactly, we require $b_0 = 10^6$. How close must b_0 be to 10^6 for b_1 to be within $0.1 \cdot 10^6$ of our target (Figure 4.4.57)? We require

$$1.9 \cdot 10^6 \leq b_1 \leq 2.1 \cdot 10^6$$
$$1.9 \cdot 10^6 \leq 2b_0 \leq 2.1 \cdot 10^6$$
$$0.95 \cdot 10^6 \leq b_0 \leq 1.05 \cdot 10^6$$

As long as we can guarantee that b_0 is within $0.05 \cdot 10^6$ of 10^6, the output is within our tolerance. A small error in the input produces a small (but somewhat larger) error in the output. In other words, the output tolerance of $0.1 \cdot 10^6$ translates into an input tolerance of $0.05 \cdot 10^6$.

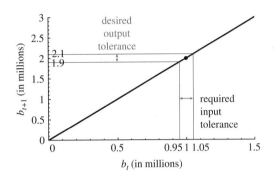

FIGURE 4.4.57
Tolerances and continuous functions

Example 4.4.12 Input and Output Tolerances for a Continuous Function II

Consider the discrete-time dynamical system

$$M_{t+1} = 0.5M_t + S$$

for the amount of pain medication in the blood (Example 3.1.4), modified so that the dosage is a variable, S, rather than the constant, 1. Suppose the concentration on day t is $M_t = 1$ and we wish to hit $M_{t+1} = 2$. What value of S should be used? The value of M_{t+1} is

$$M_{t+1} = 0.5 \cdot 1 + S = 0.5 + S$$

Think of M_{t+1} as a function of S. To hit $M_{t+1} = 2$ exactly, we require $S = 1.5$. Nothing in biology is exact, and perhaps we need only be within 0.2 of the target. Because M_{t+1} is a continuous function of S, there will be a tolerance interval around $S = 1.5$. For M_{t+1} to be between 1.8 and 2.2,

$$1.8 \leq 0.5 + S \leq 2.2$$
$$1.3 \leq S \leq 1.7$$

(Figure 4.4.58). If the input, S, is within 0.2 of 1.5, the output, M_{t+1}, will be within the desired tolerance of 2.

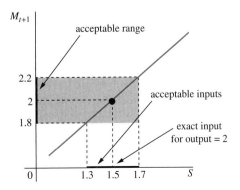

FIGURE 4.4.58
Tolerance and continuous function

Example 4.4.13 Failure to Find the Input Tolerance of a Discontinuous Function

In contrast, if the updating function is discontinuous, a tiny change in the input can produce a huge change in the output. Suppose we wish to have a voltage $V_1 = 5$ from the discrete-time dynamical system

$$V_{t+1} = \begin{cases} 3V_t & \text{if } V_t > 2 \\ 2V_t + 1 & \text{if } V_t \leq 2 \end{cases}$$

We could hit $V_1 = 5$ exactly with $V_0 = 2$. But if the input is even a tiny bit too high, the output will be different. If $V_0 = 2.001$, then $V_1 = 6.003$ (Figure 4.4.59). If the output tolerance was 0.1, meaning that values between 4.9 and 5.1 were satisfactory, there is no input tolerance at all.

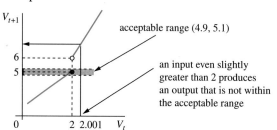

FIGURE 4.4.59
A discontinuous function with no input tolerance

Relations that are described by functions that are not continuous are difficult to deal with experimentally. Because no measurement can be controlled exactly, we hope that small errors in one measurement will not result in large errors in another. Systems with threshold behaviour, or that display hysteresis (as in the next section), can include discontinuous relations and must be treated with caution.

Hysteresis

Different left-hand and right-hand limits arise in biological systems with **hysteresis** (Figure 4.4.60). In the experiment shown, a neuron is stimulated with different levels of electric current. Low levels of current generally have no effect, whereas high levels tend to induce an oscillation. If the current is started at a low level and is steadily increased, the neuron switches from no oscillation (designated as period 0 on the graph) to an oscillation with period 2 when the input current crosses 3. If the current is instead started at a high value and steadily decreased, the neuron switches from an oscillation of period 1 to no oscillation when the input current crosses 2. In both cases, the jump indicates that the left-hand and right-hand limits do not match.

In general, hysteresis refers to a specific behaviour of a system as a response to a stimulus. The system does not react to the stimulus until the stimulus reaches a certain level (called threshold; it is said that the reaction lags behind the stimulus). However, when the stimulus falls below the threshold, the reaction continues (usually until some other stimulus forces it to stop).

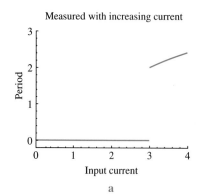
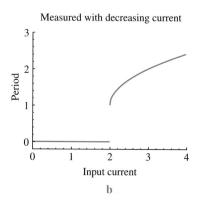

FIGURE 4.4.60
Hysteresis

4.4 Exercises

Mathematical Techniques

1. Write a formula expressing the fact that a function $f(x)$ is continuous at $x = -2$.

2. If $\lim_{x \to 3} f(x) = 6$, is it possible to conclude that $f(x)$ is continuous at $x = 3$? Explain.

3. Sketch the graph of a function that is continuous on the interval $[0, 2]$, except at $x = 1$.

4. Find a formula for a function that is continuous on the interval $[-3, 2]$, except at $x = 0$.

5. Find a formula for a function that is continuous at all real numbers except $x = -4$ and $x = -2$.

6. Find a formula for a function $g(x)$ that makes the function $\sin(g(x))$ discontinuous at $x = \pi$.

7. If $f(x)$ is continuous for all real numbers, where is $1/f(x)$ not continuous?

8. As we know, the sum of two continuous functions is continuous. What about the sum of two discontinuous functions? Find an example (sketch graphs, or find formulas) of functions $f(x)$ and $g(x)$ that are *not continuous* at a, but their sum, $f(x) + g(x)$, *is continuous* at a.

9. Show by example that the product of two discontinuous functions can be a continuous function.

10–19 ▪ Describe how each function is built out of basic continuous functions. Identify points where they are not continuous.

10. $l(t) = 5t + 6$
11. $p(x) = x^5 + 6x^3 + 7$
12. $f(x) = \dfrac{e^x}{x+1}$
13. $h(y) = y^2 \ln(y - 1)$ for $y > 1$
14. $g(z) = \dfrac{\ln(z-1)}{z^2}$ for $z > 1$
15. $F(t) = \cos(e^{t^2})$
16. $f(x) = \dfrac{\sin x}{3 - 2x}$
17. $a(t) = t^2$ if $t > 0$ and 0 if $t \leq 0$
18. $r(w) = (1 - w)^{-4}$
19. $q(z) = (1 + z^2)^{-2}$

20–29 ▪ Using the given functions, find the limits by direct substitution (if possible). Otherwise, indicate whether the limit is infinity or negative infinity. Compute the value of the function 0.1 and 0.01 above and below the limiting value of the argument as a test of whether your answer is correct.

20. $\lim_{t \to 5} l(t)$ where $l(t) = 5t + 6$.
21. $\lim_{x \to 2} p(x)$ where $p(x) = x^5 + 6x^3 + 7$.
22. $\lim_{x \to 0} f(x)$ where $f(x) = \dfrac{e^x}{x+1}$.
23. $\lim_{y \to 1^+} h(y)$ where $h(y) = y^2 \ln(y - 1)$ for $y > 1$.
24. $\lim_{z \to 2} g(z)$ where $g(z) = \dfrac{\ln(z-1)}{z^2}$ for $z > 1$.
25. $\lim_{y \to 2} h(y)$ where $h(y) = y^2 \ln(y - 1)$ for $y > 1$.
26. $\lim_{z \to 1^+} g(z)$ where $g(z) = \dfrac{\ln(z-1)}{z^2}$ for $z > 1$.
27. $\lim_{t \to 2} F(t)$ where $F(t) = \cos(e^{t^2})$.
28. $\lim_{w \to 1} r(w)$ where $r(w) = (1 - w)^{-4}$.
29. $\lim_{t \to 0} a(t)$ where $a(t) = t^2$ if $t > 0$ and 0 if $t \leq 0$.

30–33 ▪ For each function, find the input tolerance necessary to achieve the given output tolerance.

30. How close must the input be to $x = 0$ for $f(x) = x + 2$ to be within 0.1 of 2?

31. How close must the input be to $x=1$ for $f(x)=2x+1$ to be within 0.1 of 3?

32. How close must the input be to $x=1$ for $f(x)=x^2$ to be within 0.1 of 1?

33. How close must the input be to $x=2$ for $f(x)=5x^2$ to be within 0.1 of 20?

34–37 ▪ Sketch the graph of each function and identify all points where it is not continuous.

34. $f(x) = \begin{cases} -1 & \text{if } x \leq 0 \\ x-1 & \text{if } 0 < x < 3 \\ 3-x & \text{if } x \geq 3 \end{cases}$

35. $f(x) = \begin{cases} e^{2x} & \text{if } x \leq 0 \\ e^{-2x} & \text{if } x > 0 \end{cases}$

36. $f(x) = \begin{cases} \frac{x^2-4}{x-2} & \text{if } x \neq 2 \\ 0 & \text{if } x = 2 \end{cases}$

37. $f(x) = \begin{cases} x^2 & \text{if } x \leq 4 \\ \frac{1}{x-3} & \text{if } x > 4 \end{cases}$

38–43 ▪ Find all points (if any) where the function is not continuous.

38. $f(x) = \arctan \sqrt{x}$

39. $g(x) = \sqrt{\arctan x}$

40. $f(x) = \frac{1}{\sin x}$

41. $f(x) = \begin{cases} \frac{x-1}{x-2} & \text{if } x > 2 \\ \frac{1}{x-2} & \text{if } x \leq 2 \end{cases}$

42. $g(x) = \begin{cases} \sqrt{1+\frac{1}{x}} & \text{if } x > 0 \\ x & \text{if } x \leq 0 \end{cases}$

43. $f(x) = \begin{cases} x & \text{if } x \neq 2 \text{ and } x \neq 4 \\ 2 & \text{if } x = 2 \\ -1 & \text{if } x = 4 \end{cases}$

44–45 ▪ Consider the Heaviside function, defined by

$$H(x) = \begin{cases} 0 & \text{if } x < 0 \\ 1 & \text{if } x \geq 0 \end{cases}$$

44. How close must the input be to $x=1$ for $H(x)$ to be within 0.1 of 1?

45. How close must the input be to $x=0$ for $H(x)$ to be within 0.1 of 0?

46–47 ▪ We can build different continuous approximations of signum (the function giving the sign of a number) as follows. For each case,

a. Graph the continuous function.

b. Find the formula.

c. Indicate how close the input would have to be to 0 for the output to be within 0.1 of 0.

46. A continuous function that is -1 for $x \leq -0.1$, 1 for $x \geq 0.1$, and linear for $-0.1 < x < 0.1$.

47. A continuous function that is -1 for $x \leq -0.01$, 1 for $x \geq 0.01$, and linear for $-0.01 < x < 0.01$.

Applications

48–51 ▪ Find the accuracy of input necessary to achieve the desired output accuracy.

48. Suppose the mass of an object as a function of volume is given by $M = \rho V$. If $\rho = 2$ g/cm^3, how close must V be to 2.5 cm^3 for M to be within 0.2 g of 5 g?

49. The area of a disk as a function of its radius is given by $A = \pi r^2$. How close must r be to 2 cm to guarantee an area within 0.5 cm^2 of 4π?

50. The flow rate, F, through a vessel is proportional to the fourth power of the radius, or

$$F(r) = ar^4$$

Suppose $a = 1$ (units of a are 1/cm·s). How close must r be to 1 cm to guarantee a flow within 5% of 1 cm^3/s?

51. Consider an organism growing according to $S(t) = S(0)e^{\alpha t}$. Suppose $\alpha = 0.001/$s and $S(0) = 1$ mm. At time 1000 s, $S(t) = 2.71828$ mm. How close must t be to 1000 s to guarantee a size within 0.1 mm of 2.71828 mm?

52–55 ▪ Suppose a population of bacteria follows the discrete-time dynamical system

$$b_{t+1} = 2b_t$$

and we wish to have a population within 10^8 of 10^9 at $t = 10$.

52. What values of b_9 produce a result within the desired tolerance? What is the input tolerance?

53. What values of b_5 produce a result within the desired tolerance? What is the input tolerance? Why is it harder to hit the target from here?

54. What values of b_0 produce a result within the desired tolerance? What is the input tolerance?

55. How would your answers differ if the discrete-time dynamical system were $b_{t+1} = 5b_t$? Would the tolerances be larger or smaller? Why?

56–59 ▪ Suppose the amount of toxin in a culture declines according to $T_{t+1} = 0.5T_t$, and we wish to have a concentration within 0.02 of 0.5 g/L at $t = 10$.

56. What values of T_9 produce a result within the desired tolerance? What is the input tolerance?

57. What values of T_5 produce a result within the desired tolerance? What is the input tolerance?

58. What values of T_0 produce a result within the desired tolerance? What is the input tolerance?

59. How would your answers differ if the discrete-time dynamical system were $T_{t+1} = 0.1T_t$? Would the tolerances be larger or smaller? Why?

60–61 ■ The following questions are based on examples of hysteresis involving children.

60. A child outside is swinging on a swing that makes a horrible screeching noise. Starting from when the swing is farthest back, the pitch of the screeching noise increases as it swings forward and then decreases as it swings back.

    a. Draw a graph of the pitch as a function of position without hysteresis.
    b. Draw a graph with hysteresis. Which graph seems more likely?
    c. Imagine what each noise sounds like. Which is more irritating?

61. Little Billy walks due east to school but must cross from the south side to the north side of the street. Because he is a very careful child, he crosses quickly at the first possible opportunity.

    a. Graph little Billy's latitude as a function of distance from home on the way to school.
    b. Graph little Billy's latitude as a function of distance from home on the way home.
    c. Is this an example of hysteresis?

Computer Exercise

62. Graph the function $f(x) = \sin\left(\dfrac{1}{x}\right)$. What happens near $x = 0$?

4.5 Derivatives and Differentiability

We are now ready to compute one particularly important limit—the limit of the average rates of change. When it exists, this limit is the **instantaneous rate of change,** or the **derivative,** and the function involved is said to be **differentiable.** We discover that there are functions that do not have a derivative. We then use the definition to compute the derivatives of several functions. Along the way, we will see that the derivative, or the slope of the graph, can help us reason more effectively about the measurements depicted by the graph. We can tell whether a function is increasing or decreasing simply by checking whether the derivative is positive or negative.

Differentiable Functions

At the beginning of this chapter, we studied several phenomena that have the same underlying structure. Given a function $f(x)$, we defined the average rate of change over the interval $[x, x + \Delta x]$ as

$$\frac{\Delta f}{\Delta x} = \frac{f(x + \Delta x) - f(x)}{\Delta x}$$

Recall that the symbol Δ denotes change:

$$\Delta f = f(x + \Delta x) - f(x)$$

Δf is the change in the function corresponding to the change, Δx, in the variable x.

Depending on the context, we use different symbols. For instance, the average rate of change of a function $b(t)$ is

$$\frac{\Delta b}{\Delta t} = \frac{b(t + \Delta t) - b(t)}{\Delta t}$$

The slope of a secant line and the average velocity are among the most important examples of average rates of change.

We will continue using various symbols to reflect the meaning of the functions and variables involved in applications that we study. However—as somewhat common practice suggests—in abstract situations (i.e., when functions and variables have no meaning attached to them) we will often use h to represent the change in the variable x (thus, h has the same meaning as Δx and Δt).

With this in mind, we now define the derivative.

Definition 4.5.1 The **derivative $f'(x)$ of a function $f(x)$ at a number x** is given by

$$f'(x) = \lim_{h \to 0} \frac{f(x+h) - f(x)}{h}$$

provided that the limit exists (i.e., that it is a real number).

The average rate of change (the slope of the secant)

$$\frac{f(x+h) - f(x)}{h}$$

is also called the **difference quotient.** Thus, the derivative is the limit of the difference quotients as $h \to 0$.

Using Δx instead of h, we write

$$f'(x) = \lim_{\Delta x \to 0} \frac{f(x + \Delta x) - f(x)}{\Delta x}$$

Besides $f'(x)$, commonly used symbols for the derivative are

$$\frac{df}{dx}, \quad \frac{d}{dx}f(x), \quad y', \quad \frac{dy}{dx}$$

It is important that we familiarize ourselves with this variety in notation as (unfortunately) we find all of the above (and a few more!) in books on applied mathematics and journal articles in biology, medicine, and other disciplines.

Recalling what we said in the opening section of this chapter, we interpret the derivative as the **instantaneous rate of change** of a function, the **speed** of motion, or the **slope of a tangent,** depending on the context.

Example 4.5.1 Calculating the Derivative at a Number

Find $f'(2)$ for $f(x) = \dfrac{1}{x}$.

Using Definition 4.5.1 with $x = 2$, we get

$$f'(2) = \lim_{h \to 0} \frac{f(2+h) - f(2)}{h}$$

$$= \lim_{h \to 0} \frac{\frac{1}{2+h} - \frac{1}{2}}{h}$$

$$= \lim_{h \to 0} \frac{\frac{2 - (2+h)}{2(2+h)}}{h}$$

$$= \lim_{h \to 0} \frac{-h}{2(2+h)} \cdot \frac{1}{h}$$

$$= \lim_{h \to 0} \frac{-1}{2(2+h)} = -\frac{1}{4}$$

Thus, $-1/4$ is the slope of the tangent to the graph of the function $f(x) = 1/x$ at $x = 2$; see Figure 4.5.61.

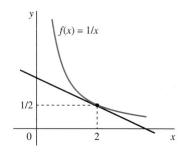

FIGURE 4.5.61

Tangent to the function $f(x) = 1/x$ at $x = 2$

Note that the derivative $f'(x)$ is defined at a particular number x and is, therefore, a real number. However, if we let x vary, then $f'(x)$ becomes a function, as our next example illustrates.

Example 4.5.2 Calculating the Derivative Function

Find $f'(x)$ for $f(x) = \dfrac{1}{x}$.

This time, the value of x is not specified. Using Definition 4.5.1, we get

$$f'(x) = \lim_{h \to 0} \frac{f(x+h) - f(x)}{h}$$

$$= \lim_{h \to 0} \frac{\frac{1}{x+h} - \frac{1}{x}}{h}$$

$$= \lim_{h \to 0} \frac{\frac{x-(x+h)}{x(x+h)}}{h}$$

$$= \lim_{h \to 0} \frac{-h}{x(x+h)} \frac{1}{h}$$

$$= \lim_{h \to 0} \frac{-1}{x(x+h)} = -\frac{1}{x^2}$$

So the derivative of the function $f(x) = \frac{1}{x}$ is the function $f'(x) = -\frac{1}{x^2}$.

Before we start calculating derivatives in a systematic way, we spend bit more time reflecting on the concept of the derivative.

Definition 4.5.2 We say that a function $f(x)$ is **differentiable at a number** x if it has a derivative at x, i.e., if $f'(x)$ is a real number. Furthermore, $f(x)$ is **differentiable on an open interval** (c, d) if it is differentiable at every number x in (c, d).

Thus, for a function $f(x)$, the phrases "has a derivative at x" and "is differentiable at x" have the same meaning. But why do we need two definitions for the same thing?

The functions we study in this book are special in the sense that they depend on *one variable only*. In general, for functions of two or more variables, the concepts of "having a derivative" and "being differentiable" *are not the same.*

To understand derivatives (differentiability) a bit better, we explore cases of functions that fail to have a derivative (i.e., that are not differentiable).

Example 4.5.3 The Derivative at a Corner Does Not Exist

Recall that the absolute value function

$$f(x) = |x| = \begin{cases} x & \text{if } x \geq 0 \\ -x & \text{if } x < 0 \end{cases}$$

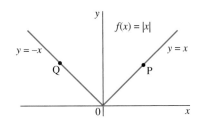

FIGURE 4.5.62

The function $f(x) = |x|$ is not differentiable at $x = 0$

is continuous for all real numbers x. As we know (see Figure 4.5.62), its graph has a corner at $x = 0$. Let us try to find $f'(0)$.

Using Definition 4.5.1, we compute

$$f'(0) = \lim_{h \to 0} \frac{f(0+h) - f(0)}{h} = \lim_{h \to 0} \frac{|h| - |0|}{h} = \lim_{h \to 0} \frac{|h|}{h}$$

To proceed, we must use one-sided limits

$$\lim_{h \to 0^+} \frac{|h|}{h} = \lim_{h \to 0^+} \frac{h}{h} = \lim_{h \to 0^+} 1 = 1$$

and

$$\lim_{h \to 0^-} \frac{|h|}{h} = \lim_{h \to 0^-} \frac{-h}{h} = \lim_{h \to 0^-} (-1) = -1$$

We conclude that

$$\lim_{h \to 0} \frac{|h|}{h}$$

does not exist, and, therefore, $f'(0)$ does not exist. In words, the derivative of $f(x) = |x|$ at $x = 0$ is not defined.

Let us visualize this: can we draw the tangent to $f(x) = |x|$ at $x = 0$? If so, what would its slope be? Consider the branch $y = x$ of the graph of $f(x) = |x|$. The secant line connecting any point P on $y = x$ to the origin coincides with $y = x$, i.e., its slope is equal to 1 (see Figure 4.5.62). Thus, as P approaches the origin, all slopes are 1, and if there were a tangent at $x=0$, its slope would have to be 1 (note that this corresponds to the calculation of the right-hand limit above).

Now consider the secant lines connecting the origin and points Q on the branch $y=-x$ of the graph of $f(x)=|x|$. Since all those secant lines have slope -1, we conclude that if the tangent existed, its slope would have to be -1 (see the left-hand limit calculation above). Since no line can have a slope of 1 and -1 at the same time, we conclude that there can be no tangent line at the origin, i.e., for $x=0$.

The shape of the graph of $f(x)=|x|$ at $x=0$ is called a **corner.** In general, a corner is formed by two not necessarily straight curves; see Figure 4.5.63a. A special type of corner is called a **cusp.** A cusp is formed by two curves that have vertical tangent lines at the point where they meet; see Figure 4.5.63b.

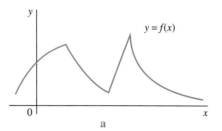
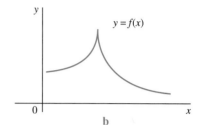

FIGURE 4.5.63

A function with three corners and a function with a cusp

Thus, a function is not differentiable at those numbers x where its graph has a corner or a cusp.

Example 4.5.4 The Derivative at a Discontinuity Does Not Exist

The function
$$f(x) = \begin{cases} 2x & \text{if } x > 3 \\ 4 & \text{if } x \leq 3 \end{cases}$$
is not continuous at $x=3$; see Figure 4.5.64.

Using Definition 4.5.1, we try to calculate the derivative of $f(x)$ at $x=3$:
$$f'(3) = \lim_{h \to 0} \frac{f(3+h) - f(3)}{h} = \lim_{h \to 0} \frac{f(3+h) - 4}{h}$$

Because $f(x)$ changes definition at $x=3$, we need to examine one-sided limits. Thus
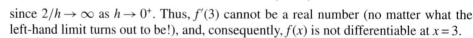
$$= \lim_{h \to 0^+} \frac{2 + 2h}{h}$$
$$= \lim_{h \to 0^+} \left(\frac{2}{h} + 2\right) = \infty$$

since $2/h \to \infty$ as $h \to 0^+$. Thus, $f'(3)$ cannot be a real number (no matter what the left-hand limit turns out to be!), and, consequently, $f(x)$ is not differentiable at $x=3$.

Alternatively, look at the slopes of the secant lines connecting the base point $(3, 4)$ with points P on the branch $y=2x$; see Figure 4.5.64. As P slides along the line $y=2x$ toward the point $(3, 6)$, the slopes of the secant lines keep increasing, their limit being $+\infty$.

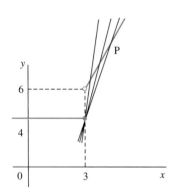

FIGURE 4.5.64

The function $f(x)$ cannot have a tangent at $x=3$

The conclusion of Example 4.5.4 holds in general: if a function is not continuous at x, then it is not differentiable there either.

In Section 6.3, we will rephrase this conclusion in the form of a theorem that states that a function that is differentiable at x must be continuous there as well.

Example 4.5.5 Vertical Tangent

Compute $f'(0)$ for $f(x) = \sqrt[3]{x} = x^{1/3}$.

By definition,

$$f'(0) = \lim_{h \to 0} \frac{f(0+h) - f(0)}{h}$$
$$= \lim_{h \to 0} \frac{h^{1/3} - 0}{h}$$
$$= \lim_{h \to 0} \frac{1}{h^{2/3}} = \infty$$

Because the limit is infinite, we have to say that $f(x) = x^{1/3}$ does not have a derivative at $x = 0$. Looking at the graph (Figure 4.5.65), we see the reason: as P slides along the curve toward the origin, the slopes of the secant lines keep increasing and approach ∞. Thus, the tangent line at $x = 0$ is vertical (and so its slope is not a real number). ▲

To summarize:

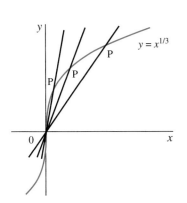

FIGURE 4.5.65

The function $f(x) = x^{1/3}$ has a vertical tangent at $x = 0$

The Three Ways a Function Can Fail to Be Differentiable
The graph has a discontinuity.
The graph has a corner or a cusp.
The graph has a vertical tangent.

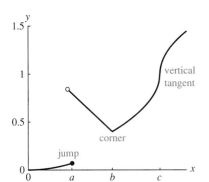

FIGURE 4.5.66

The graph of a function that is not differentiable at $x = a$, $x = b$, and $x = c$

If Figure 4.5.66 represents the graph of position against time, the derivative is the velocity. At a point of discontinuity (such as point *a*) the object jumped instantly from one place to another. The idea of speed or instantaneous rate of change makes no sense. At a corner (such as point *b*), the object has instantly changed direction and has no well-defined velocity. At a point with a vertical tangent line (such as point *c*), the tangent line has infinite slope, meaning that the object has infinite speed at this time.

Recall that if a function has a well-defined tangent line with finite slope at some point, the function is said to be **differentiable** at that point. If we say that a function is differentiable without mentioning a particular point, the function is differentiable on its entire domain. In Example 4.5.2 we proved that the function $f(x) = 1/x$ is differentiable at all $x \neq 0$. In a moment we will show that a linear function $y = mx + b$ is differentiable at all real numbers. As well, in Example 4.5.6 we will prove that a quadratic function $y = ax^2 + bx + c$ is differentiable for all x. Once we learn the rules for calculating derivatives and figure out differentiation formulas for various functions in Chapter 5, our list of differentiable functions will expand.

The function shown in Figure 4.5.66 is differentiable everywhere except at *a*, *b* and *c*. Points where a function fails to have a derivative represent one type of **critical point.**

Graphs and Derivatives

We now go back to the derivative and its interpretation as the slope of a tangent line.

The graph of the linear function $f(x) = mx + b$ is a line of slope m. The secant line connecting any two points on a line coincides with the line. Thus, in the formula

$$\text{tangent line} = \text{limit of secant lines}$$

we are computing the limit of the constant expression (all secant lines are the line $f(x) = mx + b$). We conclude that the tangent line to a given line is the line itself (note that we have already used this argument in Example 4.5.3), and thus

$$f'(x) = \text{slope of the tangent line} = \text{slope of the given line} = m$$

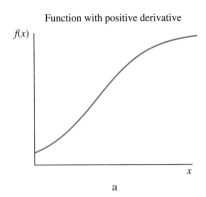
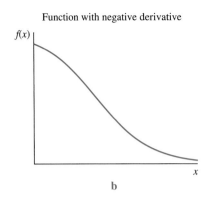
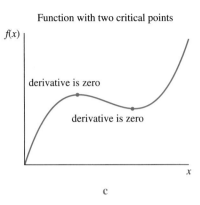

FIGURE 4.5.67

Functions with positive and negative derivatives

We can confirm this reasoning algebraically:

$$f'(x) = \lim_{h \to 0} \frac{f(x+h) - f(x)}{h}$$

$$= \lim_{h \to 0} \frac{m(x+h) + b - (mx + b)}{h}$$

$$= \lim_{h \to 0} \frac{mh}{h}$$

$$= \lim_{h \to 0} m = m$$

Lines can have positive, negative, or zero slope, corresponding to increasing, decreasing, and constant functions. Because the derivative is equal to the slope, this same correspondence holds in general. If the derivative is positive, the rate of change of the function is positive and the function is **increasing** (Figure 4.5.67a). If the derivative is negative, the function is **decreasing** (Figure 4.5.67b). If the derivative is zero at some point, the function is neither increasing nor decreasing (Figure 4.5.67c) there.

Points where the derivative is zero are the other type of **critical point** and will prove extremely useful for finding maxima and minima of functions (Chapter 6). The following definition combines the two types of critical points we identified in this section.

Definition 4.5.3   A number x is called a **critical number** (or a **critical point**) of a function $f(x)$ if it is in the domain of $f(x)$ and $f'(x) = 0$ or the derivative is not defined at x.

The geometric interpretation is that at a critical number x, the graph of f has

- a horizontal tangent ($f'(x) = 0$), or
- a corner (cusp), a discontinuity, or a vertical tangent ($f'(x)$ is not defined).

The derivative encodes more information than simply whether the function is increasing or decreasing. We know that lines with large slopes are steep. Large values of the derivative are thus associated with points where the graph of the function is steep.

Example 4.5.6   Analyzing the Graph of the Parabola $f(x) = ax^2 + bx + c$

Using the definition, compute the derivative of the function $f(x) = ax^2 + bx + c$, where $a \neq 0$. Identify the intervals where $f(x) = ax^2 + bx + c$ is increasing and where it is decreasing.

We compute

$$f'(x) = \lim_{h \to 0} \frac{f(x+h) - f(x)}{h}$$

$$= \lim_{h \to 0} \frac{a(x+h)^2 + b(x+h) + c - (ax^2 + bx + c)}{h}$$

$$= \lim_{h \to 0} \frac{ax^2 + 2axh + ah^2 + bx + bh + c - ax^2 - bx - c}{h}$$

$$= \lim_{h \to 0} \frac{2axh + ah^2 + bh}{h}$$

$$= \lim_{h \to 0} \frac{h(2ax + ah + b)}{h}$$

$$= \lim_{h \to 0} (2ax + ah + b) = 2ax + b$$

Therefore, if $f(x) = ax^2 + bx + c$, then $f'(x) = 2ax + b$.

From

$$f'(x) = 2ax + b = 0$$

we obtain the (only) critical point

$$x = -\frac{b}{2a}$$

We know that the graph of $f(x)$ is a parabola that opens upward when $a > 0$ and downward when $a < 0$.

Assume that $a > 0$. In Figure 4.5.68 we show the parabola and its derivative function (which is a line of slope $2a$). Note that the critical point $x = -b/2a$ is the x-intercept of the graph of $f'(x)$.

Looking at Figure 4.5.68b, we see that the graph of $f'(x)$ lies below the x-axis (i.e., $f'(x) < 0$) if $x < -b/2a$; thus, $f(x)$ is decreasing if $x < -b/2a$. When $x > -b/2a$, the derivative, $f'(x)$, is positive and so $f(x)$ is increasing.

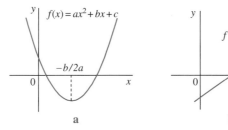

FIGURE 4.5.68
The function $f(x) = ax^2 + bx + c$ and its derivative

The case $a < 0$ is argued analogously (see Exercise 7).

Reading the formula

$$(ax^2 + bx + c)' = 2ax + b$$

with $a = 1$, $b = 0$, and $c = 0$, we obtain the derivative formula

$$(x^2)' = 2x$$

Example 4.5.7 Showing That a Function Is Decreasing

Show that the function $f(x) = x^{-1/2}$ is decreasing for all $x > 0$.

We are going to show that the derivative of $f(x)$ is negative for all $x > 0$. Writing

$$f(x) = x^{-1/2} = \frac{1}{\sqrt{x}}$$

we compute

$$f'(x) = \lim_{h \to 0} \frac{f(x+h) - f(x)}{h}$$

$$= \lim_{h \to 0} \frac{\frac{1}{\sqrt{x+h}} - \frac{1}{\sqrt{x}}}{h}$$

$$= \lim_{h \to 0} \left(\frac{1}{h} \frac{\sqrt{x} - \sqrt{x+h}}{\sqrt{x}\sqrt{x+h}} \right)$$

$$= \lim_{h \to 0} \left(\frac{\sqrt{x} - \sqrt{x+h}}{h\sqrt{x}\sqrt{x+h}} \cdot \frac{\sqrt{x} + \sqrt{x+h}}{\sqrt{x} + \sqrt{x+h}} \right)$$

$$= \lim_{h \to 0} \frac{x - (x+h)}{h\sqrt{x}\sqrt{x+h}(\sqrt{x} + \sqrt{x+h})}$$

$$= \lim_{h \to 0} \frac{-1}{\sqrt{x}\sqrt{x+h}(\sqrt{x} + \sqrt{x+h})}$$

$$= \frac{-1}{x(2\sqrt{x})} = -\frac{1}{2x^{3/2}}$$

From $x^{3/2} = \sqrt{x^3}$, knowing that the square root is always positive or zero, we conclude that $x^{3/2} > 0$ for all $x > 0$. Thus, $f'(x) < 0$ and $f(x)$ is decreasing for all $x > 0$.

Example 4.5.8 Identifying Regions with Positive and Negative Rates of Change

Suppose the volume of a cell is given graphically (Figure 4.5.69a). The volume is increasing when the graph is increasing, or between times 0 and about 0.6, and again between times 1.5 and about 2.6. It is decreasing between times 0.6 and 1.5, and again after time 2.6. The graph of the derivative is positive when the function is increasing and negative when the function is decreasing. Furthermore, as Figure 4.5.69b shows, the derivative takes on its largest positive value when the function is increasing most quickly, at about time 2, and its largest negative value when the function is decreasing most quickly, at about time 1. (Note that it is not possible to determine (from the graph of f) the exact location of the points where the derivative of f has the largest positive and the largest negative values.)

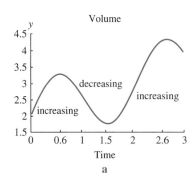
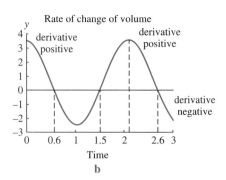

FIGURE 4.5.69
The changing volume of a cell

The following table summarizes the relationship between a function and its derivative:

f Is Increasing on an Interval (c, d)
The derivative, f', is positive on (c, d).
The rate of change of f is positive for all x in (c, d).
The slope of the tangent is positive for all x in (c, d).

4.5 Derivatives and Differentiability

f Is Decreasing on an Interval (c, d)

The derivative, f', is negative on (c, d).

The rate of change of f is negative for all x in (c, d).

The slope of the tangent is negative for all x in (c, d).

x Is a Critical Number (critical point) of f

if x is in the domain of f and

$f'(x) = 0$; i.e., the tangent at x is horizontal

or

$f'(x)$ is not defined; i.e., f is not differentiable at x.

Example 4.5.9 Graphing the Derivative of a Function

The graph of a function $f(x)$ is shown in Figure 4.5.70a. Sketch the graph of its derivative, $f'(x)$.

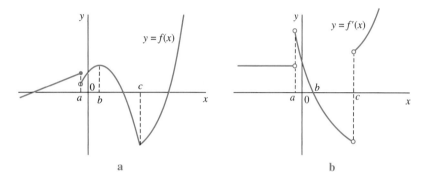

FIGURE 4.5.70

Graphs of a function and its derivative

The graph of $f(x)$ has a discontinuity at a and a corner at c. Consequently, $f'(a)$ and $f'(c)$ are not defined. When $x < a$, the graph of $f(x)$ is a straight line with constant (positive) slope, and so $f'(x)$ is a constant function—its graph is a horizontal line lying above the x-axis (the value, i.e., its y-intercept, cannot be determined from the graph, since the units on the coordinate axes are not given).

Between a and b the function $f(x)$ is increasing, i.e., $f'(x) > 0$. Moving from a toward b, we see that the slopes are decreasing. Thus, on (a, b), the graph of $f'(x)$ is a positive decreasing function. At b, $f(x)$ has a horizontal tangent, i.e., $f'(b) = 0$. Between b and c the function $f(x)$ is decreasing, and its slopes are becoming smaller and smaller (more and more negative). It follows that the graph of $f'(x)$ between b and c lies below the x-axis and is decreasing. See Figure 4.5.70b. (Note: If $f(x)$ were part of a quadratic parabola, then the graph of $f'(x)$ would have been a line; since we do not know what the graph of $f(x)$ is, we drew an arbitrary decreasing function on (a, c) crossing the x-axis at b; that's all we can say with certainty.)

On (c, ∞), the graph of $f(x)$ is increasing (thus, $f'(x) > 0$), and the slopes are increasing as well (thus $f'(x)$ is increasing). Again—since we have no additional information, we show a generic positive and increasing graph for $f'(x)$ on (c, ∞).

The graph of $f'(x)$ is drawn in Figure 4.5.70b. Note that a, b, and c are critical points of $f(x)$.

In Chapter 6, we resume our analysis of functions using derivatives by investigating extreme values (minimum and maximum) and by drawing graphs.

We now summarize the notation for the derivative and introduce a related concept. If $y = f(t) = 5t^2 - 6t + 4$, we write

$$f'(t) = 10t - 6 \quad \text{or} \quad y' = 10t - 6 \quad \text{(prime notation)}$$

$$\frac{d}{dt}(5t^2 - 6t + 4) = 10t - 6 \quad \text{(differential notation)}$$

or

$$\frac{d}{dt}f(t) = 10t - 6 \quad \text{or} \quad \frac{dy}{dt} = 10t - 6 \quad \text{(differential notation)}$$

The equation

$$\frac{dy}{dt} = 10t - 6$$

is an example of a **differential equation,** giving a formula for the rate of change as a function of time. We will study differential equations in Chapter 8.

Approximating Derivatives

So far we have learned how to find the derivative of a function given by a formula or as a graph. What can we do if a function is given by a table of values?

Consider the function $f(x)$ whose values are given in Table 4.5.1. What can we say about $f'(3)$? Clearly, we cannot compute the limit

$$f'(3) = \lim_{h \to 0} \frac{f(3 + h) - f(3)}{h}$$

because we do not know the values $f(3 + h)$ of the function f near 3. So the best we can do is to **approximate the derivative** by calculating an appropriate average rate of change. To do so, we pick the points closest to the base point $x = 3$. Looking at Table 4.5.1, we see that $x = 4$ is the nearest point to the right of $x = 3$, and $f(4) = 7.7$. Thus

$$f'(3) \approx \frac{f(4) - f(3)}{4 - 3} = \frac{7.7 - 8.3}{1} = -0.6$$

We say that we approximated $f'(3)$ using the **forward difference quotient** (*forward* refers to the fact that the second point we need for the average rate of change is larger than the base point). We obtain another approximation by calculating the **backward difference quotient;** using the fact that $f(2) = 11.1$, we obtain

$$f'(3) \approx \frac{f(2) - f(3)}{2 - 3} = \frac{11.1 - 8.3}{-1} = -2.8$$

By calculating the average of the two estimates, we obtain another approximation, called the **central difference quotient:**

$$f'(3) \approx \frac{-0.6 + (-2.8)}{2} = -1.7$$

Table 4.5.1

x	f(x)
1	12.4
2	11.1
3	8.3
4	7.7
5	8.1

Keep in mind that the three values we obtained are *approximations* of $f'(3)$. How good they are (i.e., how close they are to the true value of $f'(3)$) depends on the way

$f(x)$ behaves in the interval $[2, 4]$. If f does not vary widely within the interval, the answers we obtained could be reasonable approximations. If, however, $f(2.5) = 30$ or $f(3.3) = -40$, our estimates are worthless.

Example 4.5.10   Approximating the Rate of Change of a Population

Table 4.5.2 shows the population, $P(t)$, of Nova Scotia, based on Government of Canada census data. Approximate $P'(2006)$ and interpret your answer.

Calculating the forward difference quotient, we obtain

$$P'(2006) \approx \frac{P(2011) - P(2006)}{2011 - 2006} = \frac{921{,}727 - 913{,}462}{5} = 1653$$

In 2006 (based on future data), the population of Nova Scotia was changing (increasing) at a rate of approximately 1653 people per year. The backward difference quotient is

$$P'(2006) \approx \frac{P(2001) - P(2006)}{2001 - 2006} = \frac{908{,}007 - 913{,}462}{-5} = 1091$$

So, based on historic data (census of 2001), the population in 2006 was changing at a rate of approximately 1091 people per year. On average (i.e., computing the central difference), we find that the rate of increase in population in 2006 was approximately $(1653 + 1091)/2 = 1372$ people per year. Since there have been no large fluctuations in the population of Nova Scotia between 2001 and 2011, our approximations can be assumed to be reasonable.

Table 4.5.2

Year, t	Population, $P(t)$
1996	909,282
2001	908,007
2006	913,462
2011	921,727

Statistics Canada. 2011. *Population and Dwelling Counts, for Canada, Provinces and Territories, 2011 and 2006 Censuses*, "Population and dwelling count highlight tables, 2011 Census." Statistics Canada Catalogue no. 98-310-XWE2011002. Ottawa, Ontario. http://www12.statcan.gc.ca/census-recensement/2011/dp-pd/hlt-fst/pd-pl/Table-Tableau.cfm?LANG=Eng&T=101&S=50&O=A; Statistics Canada. 2007. *Population and Dwelling Counts, for Canada, Provinces and Territories, 2006 and 2001 Censuses, 100% Data* (table). "Population and dwelling count highlight tables, 2006 Census." "2006 Census: Release topics." Census. Statistics Canada Catalogue no. 97-550-XWE2006002. Ottawa, Ontario. March 13. http://www12.statcan.ca/english/census06/data/popdwell/Table.cfm?T=101; Statistics Canada. 2001. *Population and Dwelling Counts, for Canada, Provinces and Territories, 2001 and 1996 Censuses, 100% Data* (table). "Population and dwelling count highlight tables, 2001 Census." "2001 Census: Release topics." Census. Statistics Canada Catalogue no. 92-377-XIE02001. Ottawa, Ontario. http://www12.statcan.ca/english/census01/products/standard/popdwell/Table-PR.cfm.

Summary   The derivative of a function is defined as the limit of the average rates of change. If that limit is a real number, we say that the function is **differentiable.** A function might fail to have a derivative at a particular point in three ways: if the graph has a jump, has a corner or a cusp, or has a vertical tangent. The graphs of linear and quadratic functions have none of these problems, and we can compute their derivatives directly from the definition. The derivative of a linear function is equal to its slope. The derivative is positive when the function is increasing and negative when the function is decreasing. **Critical points** are the points in the domain of a function where the derivative is either zero or is not defined. Using various **difference quotients (forward, backward,** and **central)**, we obtain approximations of the derivative.

4.5 Exercises

Mathematical Techniques

1. Write a formula involving a limit expressing the fact that $f'(-2) = 4$.

2. Explain why $f(x) = mx + b$ is differentiable for all real numbers x.

3. Explain why $f(x) = ax^2 + bx + c$ is differentiable for all real numbers x.

4. Find a formula for a function that satisfies $f'(3) = 1$.

5. Sketch the graph of a continuous function that satisfies $f'(a) = 0$ at three different numbers a in its domain.

6. Sketch the graph of a continuous increasing function that satisfies $f'(a) = 0$ at three different numbers a in its domain.

7. Let $f(x) = ax^2 + bx + c$ with $a < 0$. Show that $f(x)$ is increasing when $x < -b/2a$ and decreasing when $x > -b/2a$.

8–9 ■ On each graph, identify

a. Points where the function is not continuous.

b. Points where the function is not differentiable (and say why).

c. Points where the derivative is zero.

8.

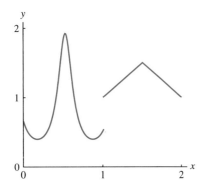

9.

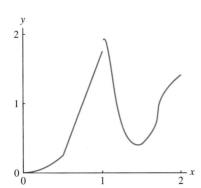

10–14 ■ Sketch the graph of a *continuous* function that satisfies each set of conditions.

10. $f(2) = 4, f'(2) = 1, f'(4) = 0$, and $f'(5) = -1$.

11. $f(0) = 1, f'(0) = -1, f(2) = -2$, and $f'(2) = 4$.

12. $f(-4) = 2, f'(0) = 0, f(2) = -2$, and $f'(x) > 0$ for $x > 2$.

13. $f(1) = 2, f'(1) = 0$, and $f'(x) > 0$ at all $x \neq 1$.

14. $f(0) = 0, f'(1)$ does not exist, and $f'(x) < 0$ at all $x \neq 1$.

15–21 ■ Sketch the graph of the function $f(x)$. In the same coordinate system, sketch the graph of its derivative, $f'(x)$.

15. $f(x) = \begin{cases} 3 & \text{if } x \leq 0 \\ 4 & \text{if } x > 0 \end{cases}$

16. $f(x) = \begin{cases} x + 4 & \text{if } x \leq 2 \\ 10 - 2x & \text{if } x > 2 \end{cases}$

17. $f(x) = \begin{cases} x^2 & \text{if } x \geq 0 \\ -x^2 & \text{if } x < 0 \end{cases}$

18. $f(x) = 10^x$

19. $f(x) = \cos x$

20. $f(x) = \ln x$

21. $f(x) = \arctan x$

22–28 ■ Identify the limit expressions below as the derivative $f'(a)$ (i.e., identify the function, $f(x)$, and the number, a, where the derivative is calculated).

22. $\lim\limits_{h \to 0} \dfrac{\sqrt{4+h} - 2}{h}$

23. $\lim\limits_{\Delta x \to 0} \dfrac{3(1 + \Delta x)^4 - 3}{\Delta x}$

24. $\lim\limits_{h \to 0} \dfrac{(3+h)^3 - 27}{h}$

25. $\lim\limits_{\Delta x \to 0} \dfrac{(10 + \Delta x)^2 - 100}{\Delta x}$

26. $\lim\limits_{\Delta x \to 0} \dfrac{\sin(\pi + \Delta x)}{\Delta x}$

27. $\lim\limits_{h \to 0} \dfrac{e^h - 1}{h}$

28. $\lim\limits_{h \to 0} \dfrac{\frac{1}{5+h} - \frac{1}{5}}{h}$

29–32 ■ Find the derivative of each function. Write your answers in both differential and prime notation. Which functions are increasing and which are decreasing?

29. $M(x) = 0.5x + 2$

30. $L(t) = 2t + 30$

31. $g(y) = -3y + 5$

32. $Q(z) = -3.5 \cdot 10^8$

33–34 ■ For each quadratic function, find the slope of the secant line connecting $x = 1$ and $x = 1 + \Delta x$, and the slope of the tangent line at $x = 1$ by taking the limit.

33. $f(x) = 4 - x^2$

34. $g(x) = x + 2x^2$

35–36 ▪ For each quadratic function, find the slope of the secant line connecting x and $x + \Delta x$, and the slope of the tangent line as a function of x. Write your result in both differential and prime notation.

35. $f(x) = 4 - x^2$ (based on Exercise 33).

36. $g(x) = x + 2x^2$ (based on Exercise 34).

37–38 ▪ For each quadratic function, graph the function and the derivative. Identify critical points, points where the function is increasing, and points where the function is decreasing.

37. $f(x) = 4 - x^2$ (based on Exercise 35).

38. $g(x) = x + 2x^2$ (based on Exercise 36).

39–43 ▪ Using Definition 4.5.1, compute the derivative of each function.

39. $f(x) = x^3 + x^2 + x$

40. $f(x) = 2x + \sqrt{x}$

41. $f(x) = \dfrac{x+1}{x-1}$

42. $f(x) = \dfrac{1}{\sqrt{x+1}}$

43. $f(x) = \dfrac{4}{x^2}$

44–47 ▪ Find the equation of the line tangent to the graph of the function $f(x)$ at the given point.

44. $f(x) = 3 - x^2, x = -2$

45. $f(x) = \dfrac{1}{x}, (4, 1/4)$

46. $f(x) = x - 2\sqrt{x}, x = 1$

47. $f(x) = x^3 - 1, (0, -1)$

48–49 ▪ On each figure, label the following points and sketch the derivative.

a. One point where the derivative is positive.
b. One point where the derivative is negative.
c. The point with maximum derivative.
d. The point with minimum (most negative) derivative.
e. Points with derivative of zero (critical points).

48.
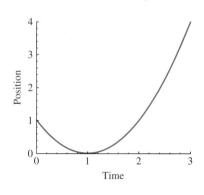

49.
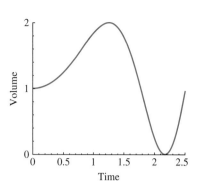

50–53 ▪ On each figure, identify which of the curves is the graph of the derivative of the other.

50.
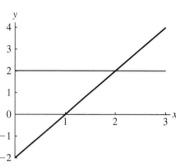

51.
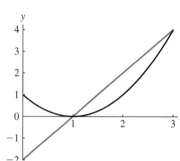

52.
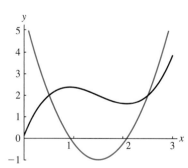

53.
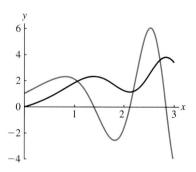

54–57 ▪ The following functions all fail to be differentiable at $x = 0$. In each case, graph the function, see what happens when you try to compute the derivative as the limit of the slopes of secant lines, and say something about the tangent line.

54. The function $f(x) = |x| + 3$.

55. The square root function $f(x) = \sqrt{x}$. (Because this function is defined only for $x \geq 0$, you can only use $\Delta x > 0$.)

56. The Heaviside function (Section 4.4, Exercises 44 and 45), defined by

$$H(x) = \begin{cases} 0 & \text{if } x < 0 \\ 1 & \text{if } x \geq 0 \end{cases}$$

57. The signum function, defined by

$$\text{sgn } x = \begin{cases} -1 & \text{if } x < 0 \\ 0 & \text{if } x = 0 \\ 1 & \text{if } x > 0 \end{cases}$$

58. Find forward, backward, and central difference quotient approximations of the value $f'(2)$ for the function $f(x)$ given by Table 4.5.1.

59. Approximate of the rate of change $P'(2001)$ of the population of Nova Scotia given in Table 4.5.2 using forward, backward, and central difference quotients.

60–63 ▪ Consider the function $f(x)$ whose values are given in the table. Find forward, backward, and central difference quotients at the indicated base point.

x	f(x)	x	f(x)
−3	16.2	1	6.4
−2	12.8	2	3.1
−1	8.2	3	4.2
0	8.8	4	6.7

60. Base point $x = -2$.
61. Base point $x = 0$.
62. Base point $x = 2$.
63. Base point $x = 3$.

Applications

64–67 ▪ The following graphs show the temperature of different solutions with chemical reactions as functions of time. Graph the rate of change of temperature in each case, and indicate when the solution is warming up and when it is cooling down.

64.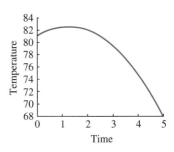

65.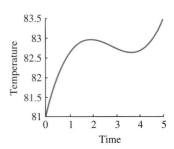

66.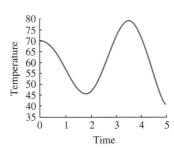

67.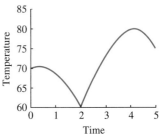

68–71 ▪ A bear sets off in pursuit of a hiker. Graph the position of the bear and that of the hiker as functions of time from the following descriptions.

68. Both move at constant speed, but the bear is faster and eventually catches the hiker.

69. Both increase speed until the bear catches the hiker.

70. The bear increases speed, and the hiker steadily slows down until the bear catches the hiker.

71. The bear runs at constant speed, and the hiker steadily runs faster until the bear gives up and stops. The hiker slows down and stops soon after that.

72–75 ▪ An object dropped from a height of 100 m has distance above the ground of

$$M(t) = 100 - \frac{a}{2}t^2$$

where a is the acceleration of gravity. For each of the following celestial bodies and with the given acceleration, find the time when the object hits the ground and the speed of the object at that time.

72. On Earth, where $a = 9.78$ m/s^2.

73. On the Moon, where $a = 1.62$ m/s^2.

74. On Jupiter, where $a = 22.88$ m/s^2.

75. On Mars's moon Deimos, where $a = 2.15 \cdot 10^{-3}$ m/s^2.

Computer Exercises

76. a. On a graph of the function
$$y(t) = 10t^{2.5}$$
have a computer plot the following and compute the slope (use base point $t = 1$).
   i. The secant line with $\Delta t = 1$.
   ii. The secant line with $\Delta t = -1$.
   iii. The secant line with $\Delta t = 0.1$.
   iv. The secant line with $\Delta t = -0.1$.
   b. Use smaller and smaller values of Δt and try to estimate the slope of the tangent.

77. The functions we have seen that are not differentiable fail at only a few points. Surprisingly, it is possible to find continuous curves that have a corner at every point. These curves are called fractals. The following construction creates a fractal called Koch's snowflake.

   a. Draw an equilateral triangle.
   b. Take the middle third of each side and expand it as shown.
   c. Do the same for the middle third of each straight piece.
   d. Repeat the process for as long as you can.

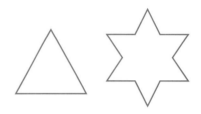

If this process is continued for an infinite number of steps, the resulting curve has a corner at every point.

Chapter Summary: Key Terms and Concepts

Define or explain the meaning of each term and concept.
Quantifying and visualizing change: average rate of change, instantaneous rate of change; secant line, tangent line
Limits: definition of the limit, right-hand limit, left-hand limit; limit exists, limit does not exist; limit laws, direct substitution rule, Squeeze Theorem; limit is ∞, limit is $-\infty$; vertical asymptote, horizontal asymptote; function approaches infinity more quickly than another function as $x \to \infty$, function approaches zero more quickly than another function as $x \to \infty$
Continuity: definition of continuity at a point and on an interval; algebraic properties of continuous functions, interchanging a limit and a continuous function; discontinuous function; hysteresis
Derivative: difference quotient, definition of the derivative; derivative as a real number, derivative as a function; differentiable function, corner, vertical tangent; critical number (critical point); forward difference quotient, backward difference quotient, central difference quotient

Concept Check: True/False Quiz

Determine whether each statement is true or false. Give a reason for your choice.

1. If $\lim_{x \to 5} f(x) = 4$, then $f(x)$ is continuous at $x = 5$ and $f(5) = 4$.

2. The average rate of change of $y = \sin x$ on $[0, \pi/2]$ is $\pi/2$.

3. If $f(x)$ is continuous at $x = -1$, then $g(x) = e^{f(x)}$ is continuous at $x = -1$.

4. The instantaneous rate of change of $f(x) = x^2 + x$ at $x = 10$ is 21.

5. If $f(x) > g(x)$ for all $x > 0$, then $\lim_{x \to \infty} f(x) > \lim_{x \to \infty} g(x)$.

6. The function $g(x) = |x - 6|$ is differentiable at $x = 5$.

7. If $f(x)$ is continuous at $x = 2$, then $\lim_{x \to 2} f(x) = f(2)$.

8. If the graph of $f(x)$ is a line, then the graph of $f'(x)$ is a horizontal line.

9. If the domain of $f(x)$ consists of all real numbers, then $f(x)$ has no vertical asymptotes.

10. The graph of $f(x) = \arctan x^3$ has two horizontal asymptotes.

Supplementary Problems

1–6 ▪ Find the limit of each function, or explain why it does not exist.

1. $\lim_{x \to 0} (1 + x^2)$

2. $\lim_{x \to -2} (1 + x^2)$

3. $\lim_{x \to 1} \dfrac{1}{x^2 - 1}$

4. $\lim_{x \to 0} \dfrac{e^x}{1 + e^{2x}}$

5. $\lim_{x \to 1^+} \dfrac{1}{x - 1}$

6. $\lim_{x \to 1^-} \dfrac{1}{x - 1}$

7–13 ▪ Find each limit, or explain why it does not exist.

7. $\lim_{x \to \pi} [\sec(2x)]$

8. $\lim_{x \to 0^-} \left(\dfrac{1}{x} - \dfrac{1}{|x|} \right)$

9. $\lim_{x \to 0^+} \left(\dfrac{1}{x} - \dfrac{1}{|x|} \right)$

10. $\lim_{x \to \infty} \dfrac{e^x - 2e^{-x}}{3e^x + e^{-x}}$

11. $\lim_{x \to -\infty} \dfrac{e^x - 2e^{-x}}{3e^x + e^{-x}}$

12. $\lim_{x \to 1} \dfrac{1}{x^4 - 1}$

13. $\lim_{x \to 2} \dfrac{1}{x^4 - 1}$

14. Calculate $\lim_{x \to 0} \left(e^x - 1 - x - \dfrac{1}{2}x^2 - \dfrac{1}{6}x^3 \right)$. Based on your answer, what can you conclude about the functions $f(x) = e^x$ and $g(x) = 1 + x + \dfrac{1}{2}x^2 + \dfrac{1}{6}x^3$?

15. Calculate $\lim_{x \to 0} \left(\sin x - x + \dfrac{1}{6}x^3 \right)$. Based on your answer, what can you conclude about the functions $f(x) = \sin x$ and $g(x) = x - \dfrac{1}{6}x^3$?

16–25 ▪ Find all numbers x where each function is continuous.

16. $f(x) = \dfrac{x - 1}{x^2 - 1}$

17. $g(x) = \dfrac{x^2 - 1}{x - 1}$

18. $f(x) = e^{-3/x}$

19. $g(x) = \sin \dfrac{1}{x}$

20. $g(x) = \dfrac{x}{\ln(x - 2)}$

21. $g(x) = \ln(x^2 - 3)$

22. $f(x) = \arcsin(2x - 12)$

23. $y = \arctan \dfrac{x}{x - 1}$

24. $g(x) = \sqrt{\sin x}$

25. $g(x) = \sqrt{\tan x}$

26. Find a formula for a function that is continuous at all real numbers except $x = 0, x = 1$, and $x = 2$.

27. If $f(x)$ is continuous for all real numbers, where is $\ln f(x) + \sqrt{f(x)}$ not continuous?

28. Sketch the graph of a continuous function that is not differentiable at $x = 0, x = 1$, and $x = 2$.

29–33 ▪ Using Definition 4.5.1, compute the derivative of each function.

29. $f(x) = \dfrac{4x + 1}{x - 6}$

30. $f(x) = x^5$

31. $f(x) = \sqrt{x - 1} + \sqrt{x + 1}$

32. $f(x) = -\dfrac{2}{x^2}$

33. $f(x) = \pi x - e$

34. Find an example of a function $f(x)$ that is not differentiable at $x = 0$ but whose square, $f^2(x)$, is differentiable there.

35–36 ▪ Find the equation of the line tangent to the graph of the function $f(x)$ at the given point.

35. $f(x) = \sqrt{3 + 4x}, x = 0$

36. $f(x) = \dfrac{x}{x - 2}, x = 4$

37. Sketch the graph of the function $f(x) = \arcsin x$. In the same coordinate system, draw the graph of its derivative.

38. A population of mosquitoes has size
$$N(t) = 1000 + 10t^2$$
where t is measured in years.

  a. What units should follow the 1000 and the 10?

  b. Graph the population between $t = 0$ and $t = 10$.

  c. Find the population after seven years and eight years. Find the approximate growth rate between these two times. Where does this approximate growth rate appear on your graph?

  d. Find and graph the derivative, $N'(t)$.

  e. Find the per capita rate of growth. Is it increasing?

39. Suppose the volume of a cell is described by the function $V(t) = 2 - 2t + t^2$, where t is measured in minutes and V is measured in thousands of cubic microns.

  a. Graph the secant line from time $t = 2$ to $t = 2.5$.

  b. Find the equation of this line.

c. Find the value of the derivative at $t = 2$, and express it in both differential and prime notation. Don't forget the units.

d. What is happening to the cell volume at time $t = 2$?

e. Graph the derivative of V as a function of time.

40. Suppose a machine is invented to measure the amount of knowledge in a student's head in units called "factoids." One student is measured at $F(t) = t^3 - 6t^2 + 9t$ factoids at time t, where t is measured in weeks.

    a. Find the rate at which the student is gaining (or losing) knowledge as a function of time (be sure to give the units).

    b. During what time between $t = 0$ and $t = 11$ is the student losing knowledge?

    c. Sketch a graph of the function $F(t)$.

41. The following measurements are made of a plant's height in centimetres and its rate of growth.

Day	Height	Growth Rate
1	8	5.2
2	15	9.4
3	28	17.8
4	53	34.6

    a. Graph these data. Make sure to give units.

    b. Write the equation of a secant line connecting two of these data points and use it to guess the height on day 5. Why did you pick the points you did?

    c. Write the equation of a tangent line and use it to guess the height on day 5.

d. Which guess do you think is better?

e. How do you think the growth rate might have been measured?

42. Suppose the fraction of mutants in a population follows the discrete-time dynamical system

$$p_{t+1} = \frac{2p_t}{2p_t + 1.5(1 - p_t)}$$

Suppose you wish the fraction at time $t = 1$ to be close to $p_1 = 0.4$.

a. What would p_0 have to be to hit 0.4 exactly?

b. How close would p_0 have to be to this value to produce p_1 within 0.1 of the target?

c. What happens to the input tolerance as the output tolerance becomes smaller (as p_1 is required to be closer and closer to 0.4)?

43. Suppose a system exhibits hysteresis. As the temperature, T, increases, the voltage response follows

$$V_i(T) = \begin{cases} T & \text{if } T \leq 50 \\ 100 & \text{if } 50 < T \leq 100 \end{cases}$$

As the temperature decreases, however, the voltage response follows

$$V_d(T) = \begin{cases} T & \text{if } T \leq 20 \\ 100 & \text{if } 20 < T \leq 100 \end{cases}$$

a. Graph these functions.

b. Find the left- and right-hand limits of V_i and V_d as T approaches 50.

c. Find the left- and right-hand limits of V_i and V_d as T approaches 20.

Project

1. Periodic hematopoiesis is a disease characterized by large oscillations in the red blood cell count. Red blood cells are generated by a feedback mechanism that approximately obeys the following equation:

$$x_{t+1} = \frac{\tau}{1 + \gamma\tau} F(x_t) + \frac{1}{1 + \gamma\tau} x_t$$

The function $F(x_t)$ describes production as a function of number of cells and takes the form of a Hill function:

$$F(x) = F_0 \frac{\theta^n}{\theta^n + x^n}$$

The terms are described as follows:

Parameter	Meaning	Normal Value
x_t	Number of cells at time t (in billions)	About 330
τ	Time for cell development	5.7 days
γ	Fraction of cells that die each day	0.0231
F_0	Maximum production of cells	76.2 billion
θ	Value where cell production is halved	247 billion
n	Shape parameter of Hill function	7.6

a. Graph and explain $F(x)$ with these parameter values. Does this sort of feedback system make sense?

b. Use a computer to find the equilibrium with normal parameters.

c. Graph and cobweb the updating function. If you start near the equilibrium, do values remain nearby? What would a solution look like?

d. Certain autoimmune diseases increase γ, the death rate of cells. Study what happens to the equilibrium and the solution as γ increases. Explain your results in biological terms.

e. Explore what would happen if the value of n were decreased. Does the equilibrium become more sensitive to small changes in γ?

Mackey, M. C., Mathematical models of hematopoietic cell replication and control. Pages 149–178 in H. G. Othmer, F. R. Adler, M. A. Lewis, and J. C. Dallon, editors, *Case Studies in Mathematical Modeling*. Prentice-Hall, Upper Saddle River, NJ (1997).

Chapter 5

Working with Derivatives

In the previous chapter we used the definition of the derivative and the limit to calculate derivatives of some functions. Derivatives of complicated functions can be computed easily if we break them down into simpler components. If we have one set of rules for computing the derivatives of these components and another set for putting these derivatives together, we will be able to find the derivative of almost any function. The sections in this chapter are dedicated to developing and practising these rules. We will learn how to calculate the derivatives of **power, exponential, logarithmic, trigonometric,** and **inverse trigonometric functions.** We will then combine them using the **sum rule,** the **product rule,** the **quotient rule,** and the **chain rule.**

As well, we will use derivatives to gain information about the behaviour of a function. For instance, using the second derivative, we can describe the curvature of a graph by measuring its **concavity.** By using second- and higher-order derivatives, we are able to build a powerful tool—**Taylor polynomials**—that allows us to simplify calculations that would otherwise be difficult, or perhaps even impossible, to do.

5.1 Derivatives of Powers, Polynomials, and Exponential Functions

This section presents three rules for putting derivatives together. The **power rule** gives a formula for the derivative of a power function. The **sum rule** states that the derivative of a sum is the sum of the derivatives. The **constant product rule** states that multiplying a function by a constant multiplies the derivative by the same constant. With these tools, we can compute the derivative of any **polynomial.** Near the end of the section we learn how to differentiate the **natural exponential function.**

Derivatives of Constant and Power Functions

The graph of the constant function

$$f(x) = c$$

is a horizontal line. Its slope is zero everywhere, and so

$$f'(x) = 0$$

for all real numbers x. Alternatively, we use the definition

$$f'(x) = \lim_{h \to 0} \frac{f(x+h) - f(x)}{h} = \lim_{h \to 0} \frac{c - c}{h} = \lim_{h \to 0} 0 = 0$$

Next, consider $f(x) = x$. We compute

$$f'(x) = \lim_{h \to 0} \frac{f(x+h) - f(x)}{h} = \lim_{h \to 0} \frac{x + h - x}{h} = \lim_{h \to 0} \frac{h}{h} = \lim_{h \to 0} 1 = 1$$

Geometrically, this result is clear: the graph of $f(x) = x$ is a line of slope 1.

In Section 4.5 we showed that

$$\frac{d}{dx}(x^2) = 2x$$

To find the derivative of any positive integer power function $f(x) = x^n$, we write

$$f'(x) = \lim_{h \to 0} \frac{f(x+h) - f(x)}{h} = \lim_{h \to 0} \frac{(x+h)^n - x^n}{h}$$

We need to simplify the fraction—and to do so, we need to expand $(x+h)^n$. How do we do that? If n is small, we recall the following formulas (if not—we multiply $(x+h)$ by itself as many times as needed):

$$(x+h)^2 = x^2 + 2xh + h^2$$
$$(x+h)^3 = x^3 + 3x^2h + 3xh^2 + h^3$$
$$(x+h)^4 = x^4 + 4x^3h + 6x^2h^2 + 4xh^3 + h^4$$

To calculate the derivative of x^4, we proceed as follows:

$$\frac{d}{dx}(x^4) = \lim_{h \to 0} \frac{(x+h)^4 - x^4}{h}$$
$$= \lim_{h \to 0} \frac{(x^4 + 4x^3h + 6x^2h^2 + 4xh^3 + h^4) - x^4}{h}$$
$$= \lim_{h \to 0} \frac{4x^3h + 6x^2h^2 + 4xh^3 + h^4}{h}$$
$$= \lim_{h \to 0} (4x^3 + 6x^2h + 4xh^2 + h^3)$$
$$= 4x^3$$

Note Look at the expression

$$(x+h)^4 = x^4 + 4x^3h + 6x^2h^2 + 4xh^3 + h^4$$

The only term that appears in the derivative is $4x^3$. The first term (x^4) cancelled when we subtracted x^4, and the remaining terms ($6x^2h^2 + 4xh^3 + h^4$) are gone because, after dividing by h, they still contain h (and thus approach zero as $h \to 0$).

To generalize this calculation, we need to know the formula for $(x+h)^n$, which is provided by the binomial theorem.

Theorem 5.1.1 Binomial Theorem

Let n be a positive integer. Then for any two real numbers x and h,

$$(x+h)^n = x^n + nx^{n-1}h + \frac{n(n-1)}{2}x^{n-2}h^2 + \cdots + nxh^{n-1} + h^n$$

As in the case of the derivative of x^4 above—the derivative of x^n will come from the second term of the expansion given in the theorem ($nx^{n-1}h$) after it is divided by h. Thus,

$$\frac{d}{dx}(x^n) = nx^{n-1}$$

Therefore, we only need to know the first two coefficients in the binomial theorem expansion!

For completeness (and as practice), we provide the details:

$$\frac{d}{dx}(x^n) = \lim_{h \to 0} \frac{(x+h)^n - x^n}{h}$$
$$= \lim_{h \to 0} \frac{\left(x^n + nx^{n-1}h + \frac{n(n-1)}{2}x^{n-2}h^2 + \cdots + nxh^{n-1} + h^n\right) - x^n}{h}$$
$$= \lim_{h \to 0} \frac{nx^{n-1}h + \frac{n(n-1)}{2}x^{n-2}h^2 + \cdots + nxh^{n-1} + h^n}{h}$$
$$= \lim_{h \to 0} \left(nx^{n-1} + \frac{n(n-1)}{2}x^{n-2}h + \cdots + nxh^{n-2} + h^{n-1}\right)$$
$$= nx^{n-1}$$

An alternative proof of the derivative formula for $f(x) = x^n$, which uses mathematical induction, is discussed in Exercises 54–57 in Section 5.2.

Theorem 5.1.2 **The Power Rule for Derivatives: Positive Integer Powers**

Assume that n is a positive integer. Then

$$(x^n)' = nx^{n-1} \quad \text{prime notation}$$

$$\frac{d}{dx}(x^n) = nx^{n-1} \quad \text{differential notation}$$

Thus, if $f(x) = x^{24}$, then $f'(x) = 24x^{23}$. As well, $(x^9)' = 9x^8$. Using differential notation,

$$\frac{d}{dt}(t^8) = 8t^7 \quad \text{and} \quad \frac{d}{dt}(t) = 1t^{1-1} = t^0 = 1$$

Example 5.1.1 The Power Rule in Action

With $n = 1$, the graph of the power function $y = x$ is a line of (constant) slope equal to one; the graph of its derivative, $y' = 1$, is a horizontal line (Figure 5.1.1a). With $n = 2$, the graph of the power function $y = x^2$ is a parabola; its derivative, $y' = 2x$, is a line of slope two (Figure 5.1.1b). The function changes from decreasing to increasing at $x = 0$, indicated by the derivative crossing from negative to positive values. With $n = 3$, the power function $y = x^3$ is always increasing but has a critical point at $x = 0$; indeed, the derivative $y' = 3x^2$ is zero when $x = 0$ (Figure 5.1.1c).

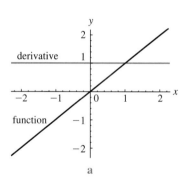
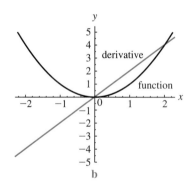
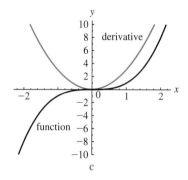

FIGURE 5.1.1
The first few power functions and their derivatives

In Section 4.5 we showed that

$$\frac{d}{dx}\left(\frac{1}{x}\right) = -\frac{1}{x^2}$$

(see Figure 5.1.2). The derivative is always negative, indicating that the function is decreasing (except at $x = 0$ where it is not defined). Near $x = 0$, the slope approaches negative infinity.

Note that we can write the differentiation formula for $1/x$ as

$$\frac{d}{dx}(x^{-1}) = -x^{-2}$$

Thus, the formula of Theorem 5.1.2 applies to $n = -1$ as well. As a matter of fact, it applies to *all real numbers n*.

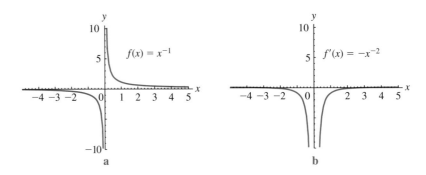

FIGURE 5.1.2

Power function with a negative power and its derivative

Theorem 5.1.3 The Power Rule for Derivatives: General Case

Let p be any real number, and assume that $f(x) = x^p$ is defined. Then

$$\frac{d}{dx}(x^p) = px^{p-1}$$

The proof suggested in Exercises 58–59 in Section 5.3 uses the rules for finding derivatives of exponential functions and functional compositions.

Example 5.1.2 The Power Rule Applied to the Square Root

The power rule can be used to find the derivative of $\sqrt{x} = x^{1/2}$:

$$\frac{d}{dx}(x^{1/2}) = \frac{1}{2}x^{1/2-1} = \frac{1}{2}x^{-1/2} = \frac{1}{2x^{1/2}} = \frac{1}{2\sqrt{x}}$$

As x approaches zero, the slope becomes infinite. For large x, the slope becomes small (Figure 5.1.3).

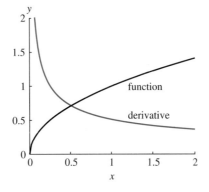

FIGURE 5.1.3

The square root function and its derivative

Example 5.1.3 The Power Rule

The derivatives of several functions, computed using the power rule, are given in the table below:

Function	Derivative
$\dfrac{1}{x^6} = x^{-6}$	$(-6)x^{-7} = -\dfrac{6}{x^7}$
$\dfrac{1}{\sqrt{x}} = x^{-1/2}$	$-\dfrac{1}{2}x^{-1/2-1} = -\dfrac{1}{2x^{3/2}}$
$x^{0.43}$	$0.43x^{0.43-1} = 0.43x^{-0.57}$
x^π	$\pi x^{\pi-1}$

The Sum (Difference) Rule for Derivatives

Next, consider two differentiable functions, $f(x)$ and $g(x)$, and let $s(x) = f(x) + g(x)$. What is the derivative of $s(x)$? Let's use the definition to see what we get:

$$\begin{aligned}
s'(x) &= \lim_{h \to 0} \frac{s(x+h) - s(x)}{h} \\
&= \lim_{h \to 0} \frac{[f(x+h) + g(x+h)] - [f(x) + g(x)]}{h} \\
&= \lim_{h \to 0} \left(\frac{f(x+h) - f(x)}{h} + \frac{g(x+h) - g(x)}{h} \right) \\
&= \lim_{h \to 0} \frac{f(x+h) - f(x)}{h} + \lim_{h \to 0} \frac{g(x+h) - g(x)}{h} \\
&= f'(x) + g'(x)
\end{aligned}$$

Thus, we have proven the **sum rule** for derivatives:

$$\frac{d}{dx}(f(x) + g(x)) = \frac{d}{dx} f(x) + \frac{d}{dx} g(x)$$

In the same way we prove the **difference rule:**

$$\frac{d}{dx}(f(x) - g(x)) = \frac{d}{dx} f(x) - \frac{d}{dx} g(x)$$

The sum and difference rules can be extended to any number of terms. For instance,

$$(f + g + h)' = [(f + g) + h]' = (f + g)' + h' = f' + g' + h'$$

Example 5.1.4 Using the Sum (Difference) Rule

The derivative of the function

$$f(x) = x^{-2} + x^{4.3} + 3 - x^{-1/2}$$

is computed to be

$$\begin{aligned}
f'(x) &= (x^{-2})' + (x^{4.3})' + (3)' - (x^{-1/2})' \\
&= -2x^{-2-1} + 4.3 x^{4.3-1} + 0 - \left(-\frac{1}{2} x^{-1/2-1}\right) \\
&= -2x^{-3} + 4.3 x^{3.3} + \frac{1}{2} x^{-3/2}
\end{aligned}$$

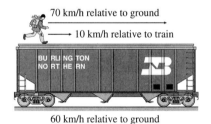

FIGURE 5.1.4
Velocities add

Why does the sum rule work? In one interpretation, the derivative of the position of a moving object is its velocity. Suppose a train has a speed of 60 km/h and a person is running forward on the train at a speed of 10 km/h. Relative to the ground, the person is moving at 60 km/h + 10 km/h = 70 km/h, the sum of the two velocities (Figure 5.1.4). Similarly, if the person is walking backwards on a train at a speed of 5 km/h, her velocity relative to the ground is 60 km/h − 5 km/h = 55 km/h. Velocities add (and subtract). Derivatives, the mathematical version of velocity, also add (and subtract).

Derivatives of Polynomials

Recall that a polynomial is a function built from power functions with non-negative integer powers that are multiplied by constants and added or subtracted. For example, the function

$$p(x) = 2x^3 - 7x^2 + 7x + 1$$

is a polynomial. Using the power rule, the sum (difference) rule, and a new rule, the **constant multiple** (or **constant product**) **rule,** we can find the derivative of any polynomial.

Theorem 5.1.4 The Constant Multiple Rule for Derivatives

Assume that $f(x)$ is a differentiable function. For any real number c,

$$\frac{d}{dx}(cf(x)) = c\frac{d}{dx}(f(x))$$

The constant multiple rule says that constants can be factored out of derivatives. To see why this is true, we consider an example—we compare the derivatives of $f(x)$ and $2f(x)$; see Figure 5.1.5.

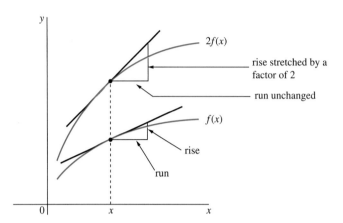

FIGURE 5.1.5

The constant multiple rule

Recall that the derivative is the slope of the tangent. What is the relationship between the slope of $2f(x)$ at x and the slope of $f(x)$ at x? The graph of $2f(x)$ is obtained by expanding (stretching) the graph of $f(x)$ vertically by a factor of 2. Since the expansion is vertical, the rise is doubled, whereas the run remains the same. Thus, the slope of $2f(x)$ is twice the slope of $f(x)$; i.e.,

$$(2f(x))' = 2f'(x)$$

To prove the constant multiple rule in general, we use the definition of the derivative and the properties of limits. Let $g(x) = cf(x)$. Then

$$g'(x) = \lim_{h \to 0} \frac{g(x+h) - g(x)}{h} = \lim_{h \to 0} \frac{cf(x+h) - cf(x)}{h} = c\lim_{h \to 0} \frac{f(x+h) - f(x)}{h} = cf'(x)$$

Example 5.1.5 Comparing Surface Area and Volume

In Example 2.1.14 in Section 2.1 we derived the relationship

$$S = aV^{2/3}$$

between the surface area, S, of a sphere and its volume, V. The value of the constant is $a = (4\pi)^{1/3}3^{2/3} \approx 4.84$. If we imagine changing the volume of a spherical object without changing its shape, the surface area will change according to

$$S' = a(V^{2/3})' = a\frac{2}{3}V^{-1/3} = \frac{2a}{3}V^{-1/3}$$

Thus, the surface area will increase as the volume increases (since its derivative is positive), but at a *decreasing rate* (Figure 5.1.6). In words, as an animal grows larger, the ratio of its surface area to its volume decreases.

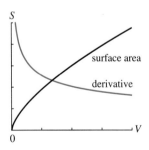

FIGURE 5.1.6

Surface area and its derivative

Example 5.1.6 Finding the Derivative of a Polynomial

Using the rules that we have seen so far, we can now compute the derivative of the polynomial

$$p(x) = 2x^3 - 7x^2 + 7x + 1$$

by breaking it into component parts:

$$\frac{d}{dx}(p(x)) = \frac{d}{dx}(2x^3) - \frac{d}{dx}(7x^2) + \frac{d}{dx}(7x) + \frac{d}{dx}(1)$$
$$= 2\frac{d}{dx}(x^3) - 7\frac{d}{dx}(x^2) + 7\frac{d}{dx}(x) + 0$$
$$= 2 \cdot 3x^2 - 7 \cdot 2x + 7 \cdot 1$$
$$= 6x^2 - 14x + 7$$

Note that the derivative of a polynomial is another polynomial. Because the power rule reduces the power by one, the degree of the derivative is always one less than the degree of the original polynomial.

Example 5.1.7 Using the Derivative to Sketch the Graph of a Function

Determine the intervals where the function

$$f(x) = 4x + \frac{10}{x^2}$$

is increasing and where it is decreasing. Use that information to draw a rough sketch of the graph of $f(x)$ for $x > 0$.

Write $f(x) = 4x + 10x^{-2}$ and differentiate:

$$f'(x) = 4 + 10(-2)x^{-3} = 4 - \frac{20}{x^3}$$

We compute

$$f'(x) = 4 - \frac{20}{x^3} = 0$$
$$\frac{20}{x^3} = 4$$
$$x^3 = 5$$
$$x = \sqrt[3]{5} \approx 1.7$$

Thus, $f(x)$ has a critical point at $x = \sqrt[3]{5}$. If $0 < x < \sqrt[3]{5}$, then $f'(x) < 0$ and $f(x)$ is decreasing. If $x > \sqrt[3]{5}$, then $f'(x) > 0$ and $f(x)$ is increasing.

Note that $f(x)$ is positive for $x > 0$, as the sum of two positive functions. As x approaches zero, $f(x)$ approaches infinity. As well, $f(x)$ approaches infinity as x approaches infinity. We have sketched the graph of $f(x)$ in Figure 5.1.7.

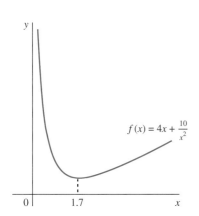

Figure 5.1.7

Graph of $f(x) = 4x + 10x^{-2}$

Example 5.1.8 Describing Bird Reproduction

Consider the following model of offspring production. Take N to be the number of eggs a bird lays. Suppose that each chick survives with probability

$$P(N) = 1 - 0.1N$$

where $1 \leq N \leq 10$ (Figure 5.1.10a). This formula models the fact that the more offspring a parent has, the smaller chance each offspring has of surviving. For instance, if a bird lays $N = 1$ egg, the chick has a 90% chance of surviving. If the bird lays $N = 5$ eggs, each chick has only a 50% chance of surviving. If the bird lays $N = 10$ eggs, none of the chicks survive. The total number of offspring that are likely to survive, $S(N)$, is the product of the number of eggs, N, and the probability of survival, $P(N)$, or

$$S(N) = N \cdot P(N) = N(1 - 0.1N) = N - 0.1N^2$$

(Figure 5.1.10b). For example, when $N = 1$, then

$$S(1) = 1 - 0.1(1)^2 = 0.9$$

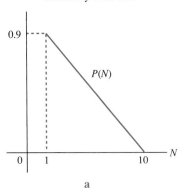
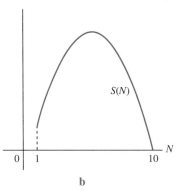
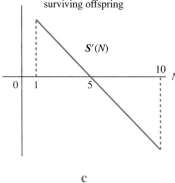

FIGURE 5.1.10

A model of offspring survival

offspring survive on average. When $N = 5$, then

$$S(5) = 5 - 0.1(5)^2 = 2.5$$

offspring survive on average.

Note that $S(N)$ is a polynomial. Its derivative is

$$S'(N) = (N - 0.1N^2)' = 1 - 0.1(2N) = 1 - 0.2N$$

At $N = 1$ the derivative is positive (Figure 5.1.10c), meaning that a higher value of N (i.e., more eggs) will likely produce more surviving offspring. At $N = 5$, the derivative is zero. For $N > 5$ the derivative is negative. This means that the number of surviving offspring is likely to *decrease* when the number of eggs becomes too large.

The Exponential Function

Suppose we wish to find the derivative of the function $b(t) = 2^t$. The best way to find the derivative of an unfamiliar function is to return to the definition

$$\frac{db}{dt} = \lim_{h \to 0} \frac{b(t+h) - b(t)}{h}$$

The derivative of the function 2^t is

$$\begin{aligned}
b'(t) &= \lim_{h \to 0} \frac{2^{t+h} - 2^t}{h} && \text{definition of the derivative} \\
&= \lim_{h \to 0} \frac{2^t 2^h - 2^t}{h} && \text{law of exponents} \\
&= \lim_{h \to 0} \frac{2^t (2^h - 1)}{h} && \text{factor out } 2^t \\
&= 2^t \lim_{h \to 0} \frac{2^h - 1}{h} && \text{pull the constant out of the limit}
\end{aligned}$$

The quantity 2^t acts as a constant because the limit depends on h, not on t. Thus, the derivative $b'(t)$ is the product of two factors, the function $b(t) = 2^t$ and the *number*

$$\lim_{h \to 0} \frac{2^h - 1}{h}$$

In other words, the rate of change of the exponential function $b(t) = 2^t$ is proportional to the function itself. In Section 5.3 we will find the exact value of

$$\lim_{h \to 0} \frac{2^h - 1}{h}$$

Here, we find an approximation of it numerically; see Table 5.1.1.

Table 5.1.1

h	$\dfrac{2^h - 1}{h}$	h	$\dfrac{2^h - 1}{h}$
0.01	0.69556	−0.01	0.69075
0.001	0.69339	−0.001	0.69291
0.0001	0.69317	−0.0001	0.69312
0.00005	0.69316	−0.00005	0.69314

Thus, taking the average of the bottom two values in the table,

$$\lim_{h \to 0} \frac{2^h - 1}{h} \approx 0.69315$$

and so

$$b'(t) \approx 2^t (0.69315)$$

i.e.,

$$(2^t)' \approx 0.69315 \cdot 2^t$$

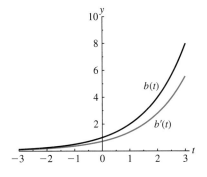

FIGURE 5.1.8

The function $b(t) = 2^t$ and its derivative

The derivative, $b'(t)$, of the function $b(t)$ is approximately 0.69315 times the function itself (Figure 5.1.8). The graph of the derivative looks like the graph of the function but is scaled down by a constant factor.

What happens when we try to find the derivative of the natural exponential function $f(x) = e^x$? We can follow the same steps to compute

$$\begin{aligned}
\frac{d(e^x)}{dx} &= \lim_{h \to 0} \frac{e^{x+h} - e^x}{h} && \text{definition of the derivative} \\
&= \lim_{h \to 0} \frac{e^x e^h - e^x}{h} && \text{law of exponents} \\
&= \lim_{h \to 0} \frac{e^x (e^h - 1)}{h} && \text{factor out } e^x \\
&= e^x \lim_{h \to 0} \frac{e^h - 1}{h} && \text{pull the constant out of the limit}
\end{aligned}$$

Thus, we are left with

$$\lim_{h \to 0} \frac{e^h - 1}{h}$$

As before, we investigate the limit numerically (Table 5.1.2).

Table 5.1.2

h	$\dfrac{e^h - 1}{h}$	h	$\dfrac{e^h - 1}{h}$
0.1	1.05171	−0.1	0.95163
0.01	1.00502	−0.01	0.99502
0.001	1.00050	−0.001	0.99950
0.0001	1.00005	−0.0001	0.99995

The limit seems to be 1. In fact, this is how we define the number e.

Definition 5.1.1 The number e is the number for which

$$\lim_{h \to 0} \frac{e^h - 1}{h} = 1$$

270 Chapter 5 Working with Derivatives

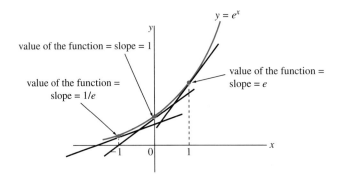

FIGURE 5.1.9

The function $y = e^x$ and its derivative

Note It can be shown that e is an irrational number, $e = 2.7182818284590\ldots$.

Finishing the calculation of the limit, we obtain

$$\frac{d(e^x)}{dx} = e^x \lim_{h \to 0} \frac{e^h - 1}{h} = e^x \cdot 1 = e^x$$

The exponential function is its own derivative. At every point on the graph, the slope of the curve is equal to the height of the curve (Figure 5.1.9).

Example 5.1.9 A Related Function That Is Its Own Derivative

Are there any other functions with this remarkable property? The function $g(x) = 2e^x$ has the derivative

$$(2e^x)' = 2(e^x)' = 2e^x$$

or, in differential notation,

$$\frac{d(2e^x)}{dx} = 2e^x$$

In general, any constant multiple of the exponential function is its own derivative, and these are the **only** functions with this property (we will prove this in Chapter 8). Therefore, functions of the form $f(t) = Ke^t$ (where K is a real number) satisfy the **differential equation**

$$\frac{df(t)}{dt} = f(t)$$

which says that the rate of change of a function is equal to the function itself.

Differential equations are very important in mathematics and its applications. We will study them in detail in Chapters 7 and 8.

Example 5.1.10 Tangent Line to the Graph of an Exponential Function

We calculate the equation of the line that is tangent to the graph of

$$f(x) = 5e^x$$

at the point where $x = 1$.

Since $f'(x) = 5e^x$, it follows that the slope of the tangent is

$$m = f'(1) = 5e$$

The point of tangency is $(1, f(1)) = (1, 5e)$, and so the equation of the tangent line (in point-slope form) is

$$y - 5e = 5e(x - 1)$$

i.e., $y = 5ex$. Note that the tangent line goes through the origin.

Example 5.1.11 Differentiating Exponential and Power Functions

Find the derivative of the function

$$f(x) = 8ex - x^e + e^{x+4} - e^e$$

We differentiate each term separately. The first term, $y = 8ex$, is the product of the constant $8e$ and the function $y = x$. Thus, by the constant multiple rule, $y' = 8e \cdot 1 = 8e$. The term $y = x^e$ is a power of x; the power rule (Theorem 5.1.3) gives $y' = ex^{e-1}$. By writing $y = e^{x+4} = e^x e^4$, we realize that y is the product of the constant e^4 and the exponential function e^x. Keeping the constant and differentiating the function e^x, we obtain $y' = e^x e^4 = e^{x+4}$. The last term has no x. It is a constant function, and its derivative is zero. We are done:

$$f'(x) = 8e - ex^{e-1} + e^{x+4}$$

In the following table we summarize the differentiation formulas that we have discussed in this section:

Function or Rule	Derivative
constant function	$(c)' = 0$
power rule	$(x^p)' = px^{p-1}$
sum/difference rule	$(f(x) \pm g(x))' = f'(x) \pm g'(x)$
constant multiple (product) rule	$(cf(x))' = cf'(x)$
natural exponential function	$(e^x)' = e^x$

Summary By combining three rules, the **power rule**, the **sum (difference) rule**, and the **constant multiple rule**, we can find the derivative of any polynomial. The power rule gives the formula for the derivative of the power function, $f(x) = x^p$, for any real number p. The sum rule states that the derivative of the sum is the sum of the derivatives, corresponding to the physical fact that velocities add. The constant multiple rule says that we can factor the constant out of the derivative. The derivative of the natural exponential function is equal to itself.

5.1 Exercises

Mathematical Techniques

1–8 ■ Find the derivative of each power function.

1. $y = x^5$
2. $y = x^{-5}$
3. $y = x^{0.2}$
4. $y = x^{-0.2}$
5. $y = x^e$
6. $y = x^{-e}$
7. $y = x^{1/e}$
8. $y = x^{-1/e}$

9–12 ■ Find the derivative of each polynomial function. Indicate where you used the sum (difference), constant multiple, and power rules.

9. $f(x) = 3x^2 + 3x + 1$
10. $s(x) = 1 - x + x^2 - x^3 + x^4$
11. $g(z) = 3z^3 + 2z^2$
12. $p(x) = 1 + x + \dfrac{x^2}{2} + \dfrac{x^3}{6} + \dfrac{x^4}{24}$

13–16 ■ Use the binomial theorem to compute the following.

13. $y = (x + 1)^3$
14. $y = (x + 2)^3$
15. $y = (2x + 1)^3$
16. $y = (x + 1)^4$

17–42 ■ Find the derivative of each function.

17. $f(x) = x^3 - 12x + 4$
18. $f(v) = 3v^6 - v - 27$
19. $g(x) = \pi x^4 + 2$
20. $h(x) = e^4 + \pi$
21. $h(x) = \dfrac{x}{4} + 3e^x - 2$
22. $f(z) = \dfrac{3e^z}{11} - \dfrac{e}{3}$
23. $f(x) = 3x^{1.5} - x^{-3.2}$
24. $q(x) = \dfrac{12}{x^2} - \dfrac{3}{\sqrt{x}}$
25. $m(t) = t^{-3/4} - 2.16t^{1/2}$

26. $g(x) = \sqrt[3]{x} - \dfrac{6}{\sqrt[5]{x}}$

27. $A(r) = \pi r^2$

28. $V(r) = \dfrac{4}{3}\pi r^3$

29. $f(x) = x(2x+1)$

30. $g(t) = \sqrt{t}(t^2 - t - 1)$

31. $g(t) = \dfrac{\sqrt{t} - 5t^2 + 1}{t}$

32. $f(x) = \left(x + x^{-2}\right)^2$

33. $f(x) = (3x-2)^4$

34. $f(x) = (Ax + B)^2$

35. $f(x) = (mx + nx^2)^3$

36. $f(t) = 2^{-4} + t^{-4} - 4e$

37. $u(x) = \dfrac{a}{x} + \dfrac{b}{x^2} + \dfrac{c}{\sqrt{x}}$

38. $u(x) = \dfrac{1}{ax} + \dfrac{6}{5}x^2$

39. $f(z) = 3e^{z+12} - 1$

40. $f(u) = 0.4 - 1.53 e^{u-1}$

41. $f(x) = \dfrac{9e^x - 2\sqrt[3]{x} + 4}{10}$

42. $u(x) = \dfrac{3}{\sqrt{x}} - \dfrac{1}{2}(x^2 - x^{-3})$

43–46 ▪ Find the equation of the tangent line to each curve at the given base point.

43. $f(x) = \sqrt[3]{x},\ x = 8$

44. $f(x) = x^{-1},\ x = -1$

45. $f(x) = 7x^3 - 10x^2 - 3x + 2,\ x = 1$

46. $f(x) = \dfrac{\sqrt{x}}{2} - \dfrac{3}{\sqrt{x}},\ x = 1$

47. Find all points where the graph of $y = 3e^x - x + 7$ has a horizontal tangent.

48. Find all points where the graph of $y = x^3 - 5x + 14$ has a tangent of slope 2.

49–50 ▪ Imitating the procedure that was used to find the derivative of $b(t) = 2^t$, find the (approximate) derivative of each function.

49. $f(x) = 5^x$

50. $f(x) = 0.3^x$

51–54 ▪ Use the derivative to sketch the graph of each function.

51. $f(x) = 1 - 2x + x^2$ for $0 \le x \le 2$.

52. $g(x) = 4x - x^2$ for $0 \le x \le 5$.

53. $h(x) = x^3 - 3x$ for $0 \le x \le 2$.

54. $F(x) = x + \dfrac{1}{x}$ for $0 < x \le 2$.

55–60 ▪ Try to guess functions that have the following as their derivatives.

55. $y = 2$

56. $y = 2x$

57. $y = 15x^{14}$

58. $y = x^{14}$

59. $y = -x^{-2}$

60. $y = 3x^{-4}$

Applications

61. In the early phase of the epidemic, the number of AIDS cases in the United States grew approximately according to a cubic equation, $A(t) = 175t^3$, where t is measured in years since the beginning of the epidemic in 1980. Find and interpret the derivative.

62. Let $F(r) = 1.5r^4$, where F represents the flow in cubic centimetres per second through a pipe of radius r. If $r = 1$, how much will a small increase in radius change the flow (try it with $\Delta r = 0.1$)? If $r = 2$, how will a small increase in radius change the flow?

63. The area of a circle as a function of its radius is given by $A(r) = \pi r^2$, with area measured in square centimetres and radius measured in centimetres. Find the derivative of area with respect to radius. On a geometric diagram, illustrate the area corresponding to $\Delta A = A(r + \Delta r) - A(r)$. What is a geometric interpretation of the derivative? Do the units make sense?

64. The volume of a sphere as a function of its radius is given by $V(r) = \dfrac{4}{3}\pi r^3$. Find the derivative of volume with respect to radius. On a geometric diagram, illustrate the volume corresponding to $\Delta V = V(r + \Delta r) - V(r)$. What is a geometric interpretation of this derivative? Do the units make sense?

65–66 ▪ One car is towing another using a rigid 15-m pole. Sketch the positions and speeds of the two cars as functions of time in the following circumstances.

65. The car starts from a stop, slowly speeds up, cruises for a while, and then abruptly stops.

66. The car starts from a stop, goes slowly in reverse for a short time, stops, goes forward slowly, and then goes forward more quickly.

67–70 ▪ A passenger is travelling on a luxury train that is moving west at 120 km/h.

67. The passenger starts running east at 15 km/h. What is her velocity relative to the ground?

68. While running, the passenger flips a dinner roll over her shoulder (west) at 40 km/h. What is the velocity of the roll relative to the train? What is the velocity of the roll relative to the ground?

69. A roll weevil jumps east off the roll at 10 km/h. What is the velocity of the roll weevil relative to the passenger? What is the velocity of the roll weevil relative to the train? What is the velocity of the roll weevil relative to the ground?

70. A roll weevil flea jumps west off the roll weevil at 20 km/h. What is the velocity of the roll weevil flea relative to the roll? What is the velocity of the roll weevil flea relative to the passenger? What is the velocity of the roll weevil flea relative to the train? What is the velocity of the roll weevil flea relative to the ground?

▼ 71–72 ▪ Each measurement is the sum of two components.

a. Find the formula for the sum.
b. Find the derivative of each component. What are the units?
c. Find the derivative of the sum, and check that the sum rule worked.
d. Describe what is happening.
e. Sketch the graph of each component and the total as functions of time.

71. A population of bacteria consists of two types, a and b. The first follows $a(t) = 1 + t^2$, and the second follows $b(t) = 1 - 2t + t^2$, where populations are measured in millions and time is measured in hours. The total population is $P(t) = a(t) + b(t)$.

72. The above-ground volume (stem and leaves) of a plant is $V_a(t) = 3t + 20 + \frac{t^2}{2}$, and the below-ground volume (roots) is $V_b(t) = -t + 40$, where t is measured in days and volumes are measured in cubic centimetres. The total volume is $V(t) = V_a(t) + V_b(t)$.

▼ 73–76 ▪ An object tossed upward at 10 m/s from a height of 100 m has distance above the ground of

$$M(t) = 100 + 10t - \frac{a}{2}t^2$$

where a is the acceleration due to gravity. For each celestial body, with the given acceleration, find the time when the object reaches a critical point, how high it gets, the time when it hits the ground, and the speed of the object at that time. Sketch the position of the object as a function of time.

73. On Earth, where $a = 9.78$ m/s^2.
74. On the Moon, where $a = 1.62$ m/s^2.
75. On Jupiter, where $a = 22.88$ m/s^2.
76. On Mars's moon Deimos, where $a = 2.15 \cdot 10^{-3}$ m/s^2.

▼ 77–80 ▪ Try different power functions of the form $s(t) = t^n$ to guess a solution for each of the following differential equations describing the size of an organism, measured in kilograms. Find the size at $t = 1$ and $t = 2$. Can you explain why some grow so much faster than others?

77. $\frac{dS}{dt} = 6t$

78. $\frac{dS}{dt} = 6t^2$

79. $\frac{dS}{dt} = 6\frac{S(t)}{t}$

80. $\frac{dS}{dt} = 12\frac{S(t)}{t}$

Computer Exercise

81. Consider the polynomials

$$p_0(x) = 1$$
$$p_1(x) = 1 + x$$
$$p_2(x) = 1 + x + \frac{x^2}{2}$$
$$p_3(x) = 1 + x + \frac{x^2}{2} + \frac{x^3}{6}$$

To find the fourth polynomial in the series, add a term equal to the last term of $p_3(x)$ multiplied by $\frac{x}{4}$. To find the fifth polynomial in the series, add a term equal to the last term of $p_4(x)$ multiplied by $\frac{x}{5}$, and so forth.

a. Find the first six polynomials in this series.
b. Plot them all for $-3 \leq x \leq 3$.
c. Take the derivative of $p_6(x)$. What is it equal to? Compare the graph of $p_6(x)$ with the graph of its derivative.
d. Can you guess what function these polynomials are approaching?

5.2 Derivatives of Products and Quotients

In this section, we learn how to compute the derivatives of products and quotients of known functions by deriving the **product rule** and the **quotient rule**. These two rules enable us to study the derivatives of **rational functions**, functions that are the quotients of two polynomials.

The Product Rule

Tempting as it might sound, the derivative of the product is *not* the product of the derivatives (see Exercises 46–47). Finding the derivative of a product is a bit more complicated. But as with the derivative of the sum, there is a useful geometric argument to help us derive and understand the formula.

Suppose the density, $\rho(t)$ (Greek letter rho), and the volume, $V(t)$, of an object are differentiable functions of time. The mass, $M(t)$, the product of the density and the volume, is also a function of time and has the formula

$$M(t) = \rho(t)V(t)$$

What is the derivative of the mass? For example, if both the density and the volume are increasing, the mass is also increasing. But if the density increases while the volume decreases, we cannot tell whether the mass increases without more information.

$M(t)$ can be represented geometrically as the area of a rectangle with base $\rho(t)$ and height $V(t)$ (Figure 5.2.11). Between times t and $t + \Delta t$, the density changes from $\rho(t)$ to $\rho(t) + \Delta\rho$ and the volume changes from $V(t)$ to $V(t) + \Delta V$. The mass at time $t + \Delta t$ is then

$$M(t + \Delta t) = (\rho(t) + \Delta\rho)(V(t) + \Delta V)$$
$$= \rho(t)V(t) + \rho(t)\Delta V + V(t)\Delta\rho + \Delta\rho\Delta V$$
$$= M(t) + \rho(t)\Delta V + V(t)\Delta\rho + \Delta\rho\Delta V$$

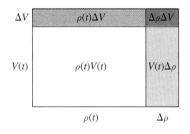

FIGURE 5.2.11

The product rule

The change in mass, ΔM, is

$$\Delta M = M(t + \Delta t) - M(t)$$
$$= \rho(t)\Delta V + V(t)\Delta\rho + \Delta\rho\Delta V$$

Each term in ΔM corresponds to one of the shaded regions in Figure 5.2.11.

How does this help us compute the derivative of the product? To find the derivative of M with respect to t, we divide ΔM by Δt and take the limit. ΔM divided by Δt is

$$\frac{\Delta M}{\Delta t} = \frac{\rho(t)\Delta V + V(t)\Delta\rho + \Delta\rho\Delta V}{\Delta t}$$
$$= \rho(t)\frac{\Delta V}{\Delta t} + V(t)\frac{\Delta\rho}{\Delta t} + \frac{\Delta\rho\Delta V}{\Delta t}$$

Taking the limit, we get

$$\lim_{\Delta t \to 0} \frac{\Delta M}{\Delta t} = \rho(t) \lim_{\Delta t \to 0} \frac{\Delta V}{\Delta t} + V(t) \lim_{\Delta t \to 0} \frac{\Delta\rho}{\Delta t} + \lim_{\Delta t \to 0} \frac{\Delta\rho\Delta V}{\Delta t}$$

where we used the properties of limits to break up the sum and take constants outside the limits. The first two terms are

$$\rho(t) \lim_{\Delta t \to 0} \frac{\Delta V}{\Delta t} = \rho(t) \lim_{\Delta t \to 0} \frac{V(t + \Delta t) - V(t)}{\Delta t} = \rho(t)V'(t)$$

$$V(t) \lim_{\Delta t \to 0} \frac{\Delta\rho}{\Delta t} = V(t) \lim_{\Delta t \to 0} \frac{\rho(t + \Delta t) - \rho(t)}{\Delta t} = V(t)\rho'(t)$$

by the definition of the derivative (here we need the fact that $V(t)$ and $\rho(t)$ are differentiable). The final term can be analyzed as the product

$$\lim_{\Delta t \to 0} \frac{\Delta\rho\Delta V}{\Delta t} = \lim_{\Delta t \to 0} \Delta\rho \lim_{\Delta t \to 0} \frac{\Delta V}{\Delta t}$$

However,

$$\lim_{\Delta t \to 0} \Delta\rho = \lim_{\Delta t \to 0} (\rho(t + \Delta t) - \rho(t)) = 0$$

by direct substitution, because $\rho(t)$ is a continuous function. (Since $\rho(t)$ is differentiable, it is continuous as well; we mentioned that differentiable implies continuous in Section 4.5; we will formally state it in Section 6.3.) Thus,

$$\lim_{\Delta t \to 0} \frac{\Delta\rho\Delta V}{\Delta t} = 0$$

and we conclude that
$$M'(t) = \lim_{\Delta t \to 0} \frac{\Delta M}{\Delta t} = \rho(t)V'(t) + V(t)\rho'(t)$$
We summarize this calculation in the following theorem.

Theorem 5.2.1 **The Product Rule for Derivatives**

Suppose that
$$p(x) = f(x)g(x)$$
where f and g are differentiable. Then
$$p'(x) = f(x)g'(x) + g(x)f'(x) \quad \text{prime notation}$$
$$\frac{dp}{dx} = f(x)\frac{dg}{dx} + g(x)\frac{df}{dx} \quad \text{differential notation}$$

If c is a constant and $p(x) = cf(x)$, then, by the product rule,
$$p'(x) = (c \cdot f(x))'$$
$$= c \cdot f'(x) + f(x) \cdot c'$$
$$= c \cdot f'(x)$$

since $c' = 0$. Thus, we have obtained the **constant multiple rule**
$$(c \cdot f(x))' = c \cdot f'(x)$$

(which we have already discussed in the previous section). In words, multiplying a function by a constant multiplies the derivative by that same constant.

Example 5.2.1 Using the Product Rule

To find the derivative of
$$p(x) = (x^2 - 4)(3x + 1)$$
think of $p(x)$ as the product of $f(x) = x^2 - 4$ and $g(x) = 3x + 1$. Both $f(x)$ and $g(x)$ are polynomials (hence differentiable), and $f'(x) = 2x$ and $g'(x) = 3$. Then, by the product rule,
$$p'(x) = (x^2 - 4) \cdot 3 + (3x + 1) \cdot 2x = 9x^2 + 2x - 12$$
To check, we multiply the function out, finding
$$p(x) = 3x^3 + x^2 - 12x - 4$$
Using the differentiation rules from the previous section gives
$$p'(x) = 3(3x^2) + 2x - 12 = 9x^2 + 2x - 12$$
which matches the result found with the product rule.

Example 5.2.2 Using the Product Rule

Compute the derivative of
$$f(x) = \sqrt{x}\left(3x^4 - 2 + \frac{1}{x}\right)$$
Write the function as
$$f(x) = x^{1/2}(3x^4 - 2 + x^{-1})$$
and apply the product rule:
$$f'(x) = x^{1/2}(3x^4 - 2 + x^{-1})' + (3x^4 - 2 + x^{-1})(x^{1/2})'$$
$$= x^{1/2}(12x^3 - x^{-2}) + (3x^4 - 2 + x^{-1})\tfrac{1}{2}x^{-1/2}$$

$$= 12x^{7/2} - x^{-3/2} + \tfrac{3}{2}x^{7/2} - 2x^{-1/2} + \tfrac{1}{2}x^{-3/2}$$
$$= \frac{27}{2}x^{7/2} - \frac{1}{2}x^{-3/2} - 2x^{-1/2}$$

Example 5.2.3 Applying the Product Rule

Suppose the volume, V, of a plant is increasing according to

$$V(t) = 100 + 12t$$

where t is measured in days and V is measured in cubic centimetres, and that the density is decreasing according to

$$\rho(t) = 0.8 - 0.05t$$

where ρ is measured in grams per cubic centimetre. Is the mass of the plant increasing or decreasing? We can use the product rule to find out:

$$M'(t) = \rho(t)V'(t) + V(t)\rho'(t)$$
$$= (0.8 - 0.05t)12 + (100 + 12t)(-0.05)$$
$$= 9.6 - 0.6t - 5 - 0.6t$$
$$= 4.6 - 1.2t$$

At $t=0$, the derivative is positive and the mass is increasing. The same is true when $t=1, t=2$, and $t=3$. However, by $t=4$ the decrease in density has overwhelmed the increase in volume, and the mass is decreasing (Figures 5.2.12 and 5.2.13).

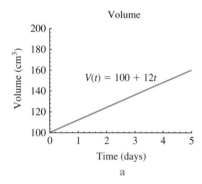
a

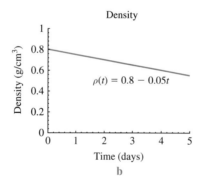
b

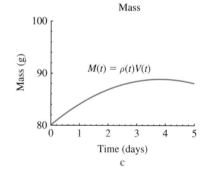
c

FIGURE 5.2.12
The volume, density, and mass of a plant

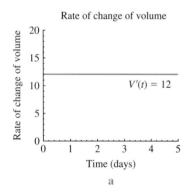
a

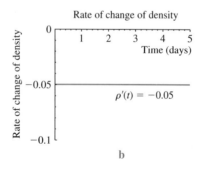
b

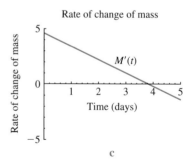
c

FIGURE 5.2.13
The rates of change of volume, density, and mass of a plant

Example 5.2.4   Finding a Tangent Line

Find the equation of the line tangent to the graph of $f(x) = x^2 e^x$ at the point $(1, e)$.

To find the slope of the tangent, we differentiate:

$$f'(x) = x^2 e^x + e^x(2x) = x^2 e^x + 2xe^x$$

Thus, $f'(1) = e + 2e = 3e$, and the equation of the tangent line is

$$y - e = 3e(x - 1)$$

i.e., $y = 3ex - 2e$.

The Quotient Rule

We are interested in finding the derivative of a function $w(x)$ defined as the quotient of $f(x)$ and $g(x)$:

$$w(x) = \frac{f(x)}{g(x)}$$

With a bit of a trick, we can find $w'(x)$ from the product rule. First, multiply both sides by $g(x)$

$$g(x)w(x) = f(x)$$

Now take the derivative of both sides using the product rule,

$$g(x)w'(x) + g'(x)w(x) = f'(x)$$

This can be thought of as an equation for $w'(x)$, the derivative of the quotient, and can be solved for $w'(x)$ as follows:

$$g(x)w'(x) = f'(x) - g'(x)w(x)$$
$$w'(x) = \frac{f'(x) - g'(x)w(x)}{g(x)}$$

To write the answer entirely in terms of the component functions $f(x)$ and $g(x)$, substitute in $w(x) = f(x)/g(x)$:

$$w'(x) = \frac{f'(x) - g'(x)\frac{f(x)}{g(x)}}{g(x)}$$

Multiplying both the numerator and the denominator by $g(x)$, we get

$$w'(x) = \frac{g(x)f'(x) - f(x)g'(x)}{[g(x)]^2}$$

We summarize this result in the following theorem.

Theorem 5.2.2   The Quotient Rule for Derivatives

Suppose

$$w(x) = \frac{f(x)}{g(x)}$$

where $f(x)$ and $g(x)$ are differentiable functions and $g(x) \neq 0$. Then

$$w'(x) = \frac{g(x)f'(x) - f(x)g'(x)}{[g(x)]^2}$$

Example 5.2.5   Applying the Quotient Rule

We can use the derivative to check whether the ratio of polynomials

$$w(x) = \frac{x^3 + 2x}{1 + x^2}$$

is an increasing function. We write

$$w(x) = \frac{f(x)}{g(x)}$$

where $f(x) = x^3 + 2x$ and $g(x) = 1 + x^2$. Applying the quotient rule, we obtain

$$w'(x) = \frac{g(x)f'(x) - f(x)g'(x)}{[g(x)]^2}$$

$$= \frac{(1+x^2)(3x^2+2) - (x^3+2x)2x}{(1+x^2)^2}$$

This can be simplified as

$$w'(x) = \frac{(3x^4 + 5x^2 + 2) - (2x^4 + 4x^2)}{(1+x^2)^2} = \frac{x^4 + x^2 + 2}{(1+x^2)^2}$$

All terms in this expression are positive, meaning that the derivative is positive and this function is increasing for all x (Figure 5.2.14). This would be difficult to show simply by plotting points.

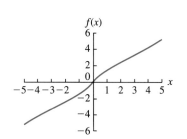

Figure 5.2.14

The increasing rational function $w(x) = \dfrac{x^3 + 2x}{1 + x^2}$

Example 5.2.6 Finding the Derivative of e^{-x}

To find the derivative of

$$f(x) = e^{-x} = \frac{1}{e^x}$$

we use the quotient rule

$$f'(x) = \frac{(e^x)(1)' - (1)(e^x)'}{(e^x)^2} = \frac{-e^x}{(e^x)^2} = -\frac{1}{e^x} = -e^{-x}$$

Thus, $(e^{-x})' = -e^{-x}$, i.e., the derivative of e^{-x} is the negative of itself. In other words, $f(x) = e^{-x}$ is a solution of the differential equation

$$\frac{df(x)}{dx} = -f(x)$$

The chain rule (which we study in the next section) will make calculating the derivative of e^{-x} much quicker.

Example 5.2.7 Checking That a Function Is Increasing

Show that the function

$$f(x) = \frac{e^x - e^{-x}}{e^x + e^{-x}}$$

is increasing for all x.

Using the quotient rule and the result of Example 5.2.6, we obtain

$$f'(x) = \frac{(e^x + e^{-x})(e^x - e^{-x})' - (e^x - e^{-x})(e^x + e^{-x})'}{(e^x + e^{-x})^2}$$

$$= \frac{(e^x + e^{-x})(e^x + e^{-x}) - (e^x - e^{-x})(e^x - e^{-x})}{(e^x + e^{-x})^2}$$

$$= \frac{(e^{2x} + 2e^x e^{-x} + e^{-2x}) - (e^{2x} - 2e^x e^{-x} + e^{-2x})}{(e^x + e^{-x})^2}$$

$$= \frac{4}{(e^x + e^{-x})^2}$$

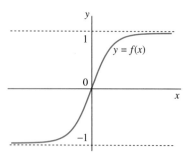

Figure 5.2.15

Graph of $f(x)$ from Example 5.2.7

Clearly, $f'(x) > 0$ for all x, and so $f(x)$ is an increasing function. See Figure 5.2.15.

An important family of functions used to describe biological phenomena (such as the properties of oxygen binding to hemoglobin, or certain activities of enzymes) is the set of **Hill functions,** with the form

$$h(x) = \frac{x^n}{1+x^n} \qquad (5.2.1)$$

where $x \geq 0$ and n can be any positive number. We will compute the derivatives of these functions with $n = 1$ and $n = 2$.

Example 5.2.8 Computing the Derivatives of Hill Functions

With $n = 1$, we get

$$h(x) = \frac{x}{1+x}$$

and so

$$h'(x) = \frac{(1+x)(x)' - (x)(1+x)'}{(1+x)^2}$$
$$= \frac{(1+x) - x}{(1+x)^2}$$
$$= \frac{1}{(1+x)^2}$$

Note that $h'(x)$ is always positive and takes on the value $h'(0) = 1$ at $x = 0$. When $n = 2$, we get

$$h(x) = \frac{x^2}{1+x^2}$$

and thus

$$\frac{dh}{dx} = \frac{(1+x^2)\frac{d(x^2)}{dx} - x^2\frac{d(1+x^2)}{dx}}{(1+x^2)^2}$$
$$= \frac{(1+x^2)2x - x^2 \cdot 2x}{(1+x^2)^2}$$
$$= \frac{2x}{(1+x^2)^2}$$

This derivative is positive when $x > 0$, but it takes on the value $h'(0) = 0$ at $x = 0$. Both Hill functions are increasing, but with different shapes (Figure 5.2.16). In Section 6.5 we will analyze Hill functions in more detail.

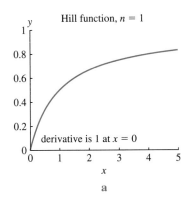
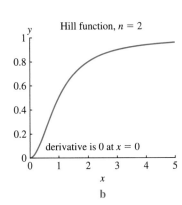

FIGURE 5.2.16
Two Hill functions

We summarize the product rule, the constant multiple rule, and the quotient rule in the following tables:

Prime Notation

Rule	Formula
product rule	$(f(x)g(x))' = f(x)g'(x) + g(x)f'(x)$
constant multiple rule	$(cf(x))' = cf'(x)$
quotient rule	$\left(\dfrac{f(x)}{g(x)}\right)' = \dfrac{g(x)f'(x) - f(x)g'(x)}{[g(x)]^2}$

Differential Notation

Rule	Formula
product rule	$\dfrac{d(fg)}{dx} = f(x)\dfrac{dg}{dx} + g(x)\dfrac{df}{dx}$
constant multiple rule	$\dfrac{d(cf)}{dx} = c\dfrac{df}{dx}$
quotient rule	$\dfrac{d}{dx}\left(\dfrac{f}{g}\right) = \dfrac{g(x)\dfrac{df}{dx} - f(x)\dfrac{dg}{dx}}{[g(x)]^2}$

Summary We developed rules for differentiating products and quotients. With these rules, we can find derivatives of **rational functions** (the ratios of polynomials), such as the **Hill functions.**

5.2 Exercises

Mathematical Techniques

1–18 ■ Find the derivative of each function using the product rule.

1. $f(x) = (2x + 3)(-3x + 2)$
2. $g(z) = (5z - 3)(z + 2)$
3. $r(y) = (5y - 3)(y^2 - 1)$
4. $s(t) = (t^2 + 2)(3t^2 - 1)$
5. $f(x) = (x^3 - x)e^x$
6. $g(t) = (e^t - 6t)(t^{-1} + 12)$
7. $h(x) = x^{-2}e^{-x}$
8. $f(x) = (a + be^x)(c + de^{-x})$
9. $g(t) = \sqrt[3]{t^2}\left(7t - \sqrt[3]{t}\right)$
10. $g(x) = ax^{3/4}e^{x-2}$
11. $u(z) = \left(A + Bz^{-4}\right)\left(A - Bz^{-1}\right)$
12. $g(v) = \sqrt{v}\left(6v + e^{v+9}\right)$
13. $f(x) = \dfrac{3}{5}\sqrt{x}(x - 1)^2$
14. $h(x) = \dfrac{7xe^x}{3} + \dfrac{1}{3x}$
15. $h(x) = (x + 2)(2x + 3)(-3x + 2)$ (apply the product rule twice)
16. $F(w) = (w - 1)(2w - 1)(3w - 1)$ (apply the product rule twice)
17. $f(x) = 0.5x^3(x^4 - 1)e^x$
18. $f(x) = \dfrac{2\sqrt{x}(x - 1)e^{-x}}{9}$

19–32 ■ Find the derivative of each function using the quotient rule.

19. $f(x) = \dfrac{1 + x}{2 + x}$
20. $f(x) = \dfrac{x^2}{1 + 2x^3}$
21. $g(z) = \dfrac{1 + z^2}{1 + 2z^3}$
22. $h(z) = \dfrac{1 + 2z^3}{1 + z^2}$
23. $F(x) = \dfrac{1 + x}{(2 + x)(3 + x)}$ (apply the product rule to the denominator)
24. $G(x) = \dfrac{(1 + x)(2 + x)}{3 + x}$ (apply the product rule to the numerator)
25. $f(x) = \dfrac{e^x}{7x}$
26. $u(x) = \dfrac{e^{-x}}{3x^3}$
27. $g(t) = \dfrac{e^t - 2t}{e^t + t}$

28. $g(x) = \dfrac{2}{e^x + e^{-x}}$

29. $g(x) = \dfrac{a + be^x}{c + de^x}$

30. $f(t) = \dfrac{a}{b + ce^{-t}}$

31. $h(x) = \dfrac{e^{x+5} + e^{-x-5}}{2}$

32. $g(x) = \dfrac{xe^{-x}}{4} + \dfrac{4}{x^2 + 1}$

33–43 ▪ Compute each derivative.

33. $f'(1)$ if $f(x) = (x^4 - 11x + 2)(2 - x)$.

34. $g'(4)$ if $g(x) = x^{3/2}(x^2 - 3x)$.

35. $f'(0)$ if $f(x) = \pi(x^{12} + x^7 - 2)$.

36. $f'(1)$ if $f(x) = (33 + ex)(e - x)$.

37. $f'(1)$ if $f(x) = \dfrac{x^4 - 11x + 2}{2x - 1}$.

38. $g'(0)$ if $g(x) = \dfrac{2x - 4}{x^4 + 1}$.

39. $f'(1)$ if $f(x) = \dfrac{12}{x^2 + x + 1}$.

40. $f'(0)$ if $f(x) = \dfrac{1 + x^2}{e^x}$.

41. $f'(1)$ if $f(t) = \dfrac{3e^t}{t^2 + t + 1}$.

42. $g'(0)$ if $g(x) = \dfrac{x + e^x}{e^x - x}$.

43. $g'(1)$ if $g(x) = 2xe^x + e - 1$.

44–45 ▪ For each function, use base points $x_0 = 1$ and $\Delta x = 0.1$ to compute Δf and Δg. Find $\Delta(fg)$ (the change in the product) by computing $f(x_0 + \Delta x)g(x_0 + \Delta x) - f(x_0)g(x_0)$. Check that $\Delta(fg) = g(x_0)\Delta f + f(x_0)\Delta g + \Delta f \Delta g$. Try the same with $\Delta x = 0.01$ and see whether the term $\Delta f \Delta g$ becomes very small.

44. $f(x) = 2x + 3$ and $g(x) = -3x + 2$.

45. $f(x) = x^2 + 2$ and $g(x) = 3x^2 - 1$.

46–47 ▪ Suppose $p(x) = f(x)g(x)$. Test out the *incorrect* formula $p'(x) = f'(x)g'(x)$ on the following functions.

46. $f(x) = x$ and $g(x) = x^2$.

47. $f(x) = 1$ and $g(x) = x^3$.

48–51 ▪ Use the first derivative to sketch the graph of each function on the given domain. Identify regions where the function is increasing and regions where it is decreasing.

48. $f(x) = (1 - x)e^x$ for $-2 \le x \le 1$.

49. $g(x) = (2 - x)e^x$ for $-2 \le x \le 1$.

50. $G(z) = \dfrac{e^z}{z^2}$ for $1 \le z \le 3$.

51. $F(z) = \dfrac{z^3}{e^z}$ for $0 \le z \le 5$.

52–53 ▪ Suppose that $f(x)$ is a positive increasing function defined for all x.

52. Use the product rule to show that $f(x)^2$ is also increasing.

53. Use the quotient rule to show that $\dfrac{1}{f(x)}$ is decreasing.

54–57 ▪ For positive integer powers, it is possible to derive the power rule with **mathematical induction.** The idea is to show that a formula is true for $n = 1$ and then to show that whenever it is true for some particular n, it must also be true for $n + 1$.

54. Check that the power rule is true for $n = 1$.

55. Use the product rule on $x^2 = x \cdot x$ to check the power rule for $n = 2$ using only the power rule with $n = 1$.

56. Use the product rule on $x^3 = x^2 \cdot x$ to check the power rule for $n = 3$ using only the power rule with $n = 1$ and $n = 2$.

57. Assuming that the power rule is true for n, find $\dfrac{d(x^{n+1})}{dx}$ using the product rule, and check that it too satisfies the power rule.

Applications

58–61 ▪ The total mass of the population is the product of the number of individuals and the mass of each individual. In each case, time is measured in years and mass is measured in kilograms.

a. Find the total mass as a function of time.

b. Compute the derivative.

c. Find the population, the mass of each individual, and the total mass at the time when the derivative is equal to zero.

d. Sketch the graph of the total mass over the next 100 years.

58. The population, $P(t)$, is $P(t) = 2 \cdot 10^6 + 2 \cdot 10^4 t$ and the mass per person, $W(t)$, is $W(t) = 80 - 0.5t$.

59. The population, $P(t)$, is $P(t) = 2 \cdot 10^6 - 2 \cdot 10^4 t$ and the mass per person, $W(t)$, is $W(t) = 80 + 0.5t$.

60. The population, $P(t)$, is $P(t) = 2 \cdot 10^6 + 1000t^2$ and the mass per person, $W(t)$, is $W(t) = 80 - 0.5t$.

61. The population, $P(t)$, is $P(t) = 2 \cdot 10^6 + 2 \cdot 10^4 t$ and the mass per person, $W(t)$, is $W(t) = 80 - 0.005t^2$.

62–63 ▪ In each situation (extending Section 5.1, Exercise 72), the mass is the product of the density and the volume. In each case, time is measured in days and density is measured in grams per cubic centimetre.

a. Find the mass as a function of time.

b. Compute the derivative.

c. Sketch the graph of the mass over the first 30 days.

62. The above-ground volume is $V_a(t) = 3t + 20$ and the above-ground density is $\rho_a(t) = 1.2 - 0.01t$.

63. The below-ground volume is $V_b(t) = -1t + 40$ and the below-ground density is $\rho_b(t) = 1.8 + 0.02t$.

64–67 ▪ Suppose that the fraction of chicks that survive, $P(N)$, as a function of the number, N, of eggs laid is given by the following forms (variants of the model studied in Example 5.1.8). The total number of offspring that survive is $S(N) = N \cdot P(N)$. Find the expected number of surviving offspring when the bird lays 1, 5, or 10 eggs. Find $S'(N)$. Sketch the graph of $S(N)$. What do you think is the best strategy for each bird?

64. $P(N) = 1 - 0.08N$
65. $P(N) = 1 - 0.16N$. What seems strange if $N = 10$?
66. $P(N) = \dfrac{1}{1 + 0.5N}$
67. $P(N) = \dfrac{1}{1 + 0.1N^2}$

68–69 ▪ Find the derivative of the updating function from Equation 3.4.7, $f(p) = \dfrac{sp}{sp + r(1-p)}$, with the following values of the parameters s and r.

68. $s = 1.2, r = 2$
69. $s = 1.8, r = 0.8$

70–71 ▪ Suppose that the mass, $M(t)$, of an insect (in grams) and the volume, $V(t)$ (in cubic centimetres), are known functions of time (in days).

a. Find the density, $\rho(t)$, as a function of time.
b. Find the derivative of the density.
c. At what times is the density increasing?
d. Sketch the graph of the density over the first five days.

70. $M(t) = 1 + t^2$ and $V(t) = 1 + t$
71. $M(t) = 1 + t^2$ and $V(t) = 1 + 2t$

72–73 ▪ In a discrete-time dynamical system describing the growth of a population in the absence of immigration and emigration, the final population is the product of the initial population and the per capita production. Represent the initial population by b_t. In each case, find the final population as a function, $f(b_t)$, of the initial population; find the derivative; and sketch the function.

72. Per capita production is $2\left(1 - \dfrac{b_t}{1000}\right)$. Sketch $f(b_t)$ for $0 \le b_t \le 1000$.
73. Per capita production is $\dfrac{2}{1 + \dfrac{b_t}{1000}}$. Sketch $f(b_t)$ for $0 \le b_t \le 2000$.

74–75 ▪ The following steps should help you to figure out what happens to the Hill function, $h_n(x) = \dfrac{x^n}{1 + x^n}$, for large values of n.

a. Compute the value of the function at $x = 0$, $x = 1$, and $x = 2$.
b. Compute the derivative and evaluate at $x = 0$, $x = 1$, and $x = 2$.
c. Sketch the graph of $h_n(x)$ and $h'_n(x)$.
d. $h_n(x)$ can be thought of as representing a response to a stimulus of strength x. Would the response work as a good filter, giving a small output for inputs less than 1 and a large output for inputs greater than 1?

74. With $n = 3$.
75. With $n = 10$.

76–79 ▪ Suppose a population of bacteria grow according to $P(t) = 10e^t$. Find the first derivative to graph the total mass when the mass per individual, $m(t)$, has the following forms. Is the total mass ever greater than it is at $t = 0$? When does the total mass reach zero?

76. $m(t) = 1 - \dfrac{t}{2}$
77. $m(t) = 1 - t$
78. $m(t) = 1 - t^2$
79. $m(t) = 1 - \dfrac{t^2}{4}$

80–82 ▪ The following are differential equations that could describe a bacterial population. For each, describe what the equation says and check that the given solution works. Indicate whether the solution is an increasing or a decreasing function.

80. $\dfrac{db}{dt} = e^t$ has solution $b(t) = e^t$.
81. $\dfrac{db}{dt} = b(t)$ has solution $b(t) = e^t$.
82. $\dfrac{db}{dt} = -b(t)$ has solution $b(t) = e^{-t}$.

Computer Exercises

83. Consider the functions
$$g_n(x) = 1 + x + x^2 + \cdots + x^n$$
for various values of n. We will compare these functions with
$$g(x) = \dfrac{1}{1-x}$$

a. Plot the functions $g_1(x)$, $g_3(x)$, $g_5(x)$, and $g(x)$ on the intervals $0 \le x \le 0.5$ and $0 \le x \le 0.9$.
b. Take the derivative of $g(x)$. Can you see how the derivative is related to $g(x)$ itself? In other words, what function could you apply to $g(x)$ to get $g'(x)$?
c. Apply the function found in part (b) to $g_1(x)$, $g_3(x)$, and $g_5(x)$. Can you see why these functions are good approximations to $g(x)$? Can you see why these approximations are best for small values of x? What happens to the approximations for x near 1?

84. Consider the function
$$r(x) = \dfrac{f(x)}{g(x)} = \dfrac{1 + x}{2 + x^2 + x^3}$$

a. Make one graph of $f(x)$ and $g(x)$ for $0 \le x \le 1$, and another of $r(x)$. Could you have guessed the shape of $r(x)$ from looking at the graphs of $f(x)$ and $g(x)$?
b. What happens at the critical point of $r(x)$?
c. Find the exact location of the critical point $x = x_c$.
d. Compare $\dfrac{f'(x_c)}{f(x_c)}$ with $\dfrac{g'(x_c)}{g(x_c)}$. Why are they equal?

5.3 The Chain Rule and the Derivatives of Logarithmic Functions

Composition is an important way to combine functions. If we know the derivatives of the component parts, the **chain rule** gives the formula for the derivative of the composition. Using the chain rule (along with the sum, product, power, and quotient rules and the derivatives of special functions), we can find the derivative of *any* function that can be built from polynomial, exponential, and trigonometric functions. We can also apply the chain rule to compute the derivatives of **inverse functions,** such as logarithmic functions.

The Derivative of a Composite Function

Suppose we have to differentiate the function

$$F(x) = e^{-x^2}$$

None of the differentiation rules we have seen so far apply: $F(x)$ is neither a product nor a quotient. Instead, it is a composition: if we define $f(u) = e^u$ and $g(x) = -x^2$, then $F = f \circ g$, i.e.,

$$F(x) = f(g(x)) = f(-x^2) = e^{-x^2}$$

Likewise, the function

$$H(x) = \frac{1}{1 + x^5}$$

is a composition. Take $f(u) = \frac{1}{u}$ and $g(x) = 1 + x^5$. Then

$$H(x) = f(g(x)) = f(1 + x^5) = \frac{1}{1 + x^5}$$

When we write $F(x) = (f \circ g)(x) = f(g(x))$, we refer to g as the **inner function** and to f as the **outer function.** In order to calculate the value of the composition $f(g(x))$ at a given value of x, we evaluate the inner function first, and then we apply the outer function to the value of the inner function.

Consider an example: for $F(x) = e^{-x^2}$, the inner function is $g(x) = -x^2$ and the outer function is $f(u) = e^u$. To calculate $F(0.6)$ we start with the inner function

$$g(0.6) = -(0.6)^2 = -0.36$$

and then exponentiate -0.36 (i.e., apply the outer function to -0.36) to get

$$F(0.6) = e^{-0.36} \approx 0.6977$$

(see Figure 5.3.17). We often do not formally break down the composition into component functions; instead, we write

$$F(0.6) = e^{-(0.6)^2} = e^{-0.36} \approx 0.6977$$

Now we turn to differentiating the composition. The chain rule—which we state next—will tell us that the derivative of the composition is calculated as the product of the derivatives of its component functions.

FIGURE 5.3.17
Calculating the composition
$F(0.6) = f(g(0.6))$

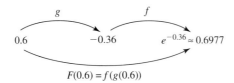

Theorem 5.3.3 The Chain Rule for Derivatives

Let
$$F(x) = (f \circ g)(x) = f(g(x))$$
and assume that g is differentiable at x and f is differentiable at $g(x)$. Then F is differentiable at x and
$$F'(x) = f'(g(x))g'(x)$$

Thus, to compute the derivative of the composition, we differentiate the outer function and evaluate it at the inner function and then multiply by the derivative of the inner function.

Writing $F = f(u)$ and $u = g(x)$, we express the chain rule in differential notation as
$$\frac{dF}{dx} = \frac{df}{du}\frac{du}{dx}$$

Why does the chain rule work, and how do we prove it? We give an example suggesting that it should be as claimed in the theorem. A complete proof is presented at the end of this section.

As we dive into a lake, or sea, we feel the increase in water pressure. From physics, we know that the pressure at a water depth of d metres is given by $p(d) = 1 + 0.0968d$ atmospheres (units of pressure). Thus, the *rate of change* of pressure with respect to depth is 0.0968 atm/m.

Now assume that we dive vertically with a speed of two metres per second. For every metre we dive, we will experience an increase in pressure of 0.0968 atm. Since every second we dive for two metres, we will experience an increase in pressure of $2 \cdot 0.0968 = 0.1936$ atm/s.

So, in our case, pressure depends on depth, and depth depends on time (composition!). We have just seen that

rate of change of pressure with respect to time (0.1936) =

rate of change of pressure with respect to depth (0.0968)

times

rate of change of depth with respect to time (2)

which is exactly what the chain rule claims to be true.

Example 5.3.1 Finding the Derivative of $F(x) = e^{-x^2}$

We know that $F(x) = f(g(x))$, where $f(u) = e^u$ and $g(x) = -x^2$. Since $f'(u) = e^u$ and $g'(x) = -2x$, we get
$$F'(x) = f'(g(x))g'(x) = e^{-x^2}(-2x) = -2xe^{-x^2}$$

Alternatively, using differential notation, we write
$$\frac{dF}{dx} = \frac{df}{du}\frac{du}{dx} = e^u(-2x) = -2xe^{-x^2}$$

Note that $F'(x)$ is positive if $x < 0$ and negative if $x > 0$; see Figure 5.3.18.

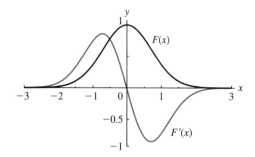

FIGURE 5.3.18

The function $F(x) = e^{-x^2}$ and its derivative

5.3 The Chain Rule and the Derivatives of Logarithmic Functions

The steps in calculating the chain rule are summarized in the following algorithm.

▶▶ **Algorithm 5.3.1** Using the Chain Rule

1. Write the function as a composition.
2. Take the derivatives of the component pieces.
3. Multiply the derivatives together.
4. Put everything in terms of the original variable.

Let us consider another example.

Example 5.3.2 Finding the Derivative of $H(x) = \dfrac{1}{1+x^5}$

Write $H(x) = f(g(x))$, where $f(u) = \dfrac{1}{u}$ is the outer function and $g(x) = 1 + x^5$ is the inner function. Since

$$f'(u) = -\frac{1}{u^2} \quad \text{and} \quad g'(x) = 5x^4$$

it follows that

$$H'(x) = f'(g(x))g'(x) = -\frac{1}{(1+x^5)^2}(5x^4) = -\frac{5x^4}{(1+x^5)^2}$$

Alternatively, using differential notation, we write

$$\frac{dH}{dx} = \frac{df}{du}\frac{du}{dx} = -\frac{1}{u^2}(5x^4) = -\frac{5x^4}{(1+x^5)^2}$$

With practice, we realize that it is not necessary to break down the function into component pieces. For instance, we compute

$$(e^{x+12})' = e^{x+12}(x+12)' = e^{x+12}(1) = e^{x+12}$$
$$(e^{-x})' = e^{-x}(-x)' = e^{-x}(-1) = -e^{-x}$$

and, for any constant a,

$$(e^{ax})' = e^{ax}(ax)' = e^{ax}(a) = ae^{ax}$$

Example 5.3.3 Analyzing Drug Concentration

The concentration of a drug introduced into the blood at time $t = 0$ is given by

$$c(t) = 2.3e^{-0.2t} - 1.65e^{-0.4t} \text{ grams per litre}$$

where t is time in hours. During which interval is the concentration increasing? Decreasing? Identify its largest value. What is the long-term behaviour of $c(t)$?

Setting the derivative of $c(t)$ equal to zero,

$$c'(t) = 2.3(-0.2)e^{-0.2t} - 1.65(-0.4)e^{-0.4t} = 0$$

we obtain

$$-0.46e^{-0.2t} + 0.66e^{-0.4t} = 0$$
$$e^{-0.4t}\left(-0.46e^{0.2t} + 0.66\right) = 0$$
$$-0.46e^{0.2t} + 0.66 = 0$$

Note that the first factor is nonzero. We obtain

$$e^{0.2t} = \frac{0.66}{0.46}$$
$$t = \frac{1}{0.2}\ln(0.66/0.46) = 5\ln(33/23) \approx 1.805$$

FIGURE 5.3.19

The concentration function, $c(t)$

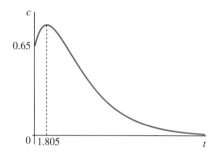

Thus, $t \approx 1.805$ is the only critical point. Note that in the expression

$$c'(t) = e^{-0.4t}\left(-0.46e^{0.2t} + 0.66\right)$$

the first factor is positive. Consequently, when $t < 5\ln(33/23)$, the derivative $c'(t)$ is positive, and the concentration increases from its initial value, $c(0) = 0.65$.

For $t > 5\ln(33/23)$ the concentration decreases, and thus $c(t)$ reaches its largest value at $t = 5\ln(33/23) \approx 1.805$. The largest concentration is approximately equal to $c(1.805) \approx 0.802$ g/L. Because both exponential terms in $c(t)$ approach zero as $t \to \infty$, we conclude that in the long term the concentration decreases to zero. See Figure 5.3.19.

Example 5.3.4 Cell Survival Fraction

In radiotherapy treatment, the function

$$S(D) = e^{-\alpha D - \beta D^2}$$

(where $\alpha, \beta > 0$) models the fraction (percent) of cancer cells surviving a treatment with a radiation dose D, measured in grays (Gy). Find and interpret $S'(D)$. What are its units?

Using the chain rule, we find

$$S'(D) = e^{-\alpha D - \beta D^2}\left(-\alpha D - \beta D^2\right)' = -e^{-\alpha D - \beta D^2}(\alpha + 2\beta D)$$

The quantity $S'(D)$ represents the rate of change in the percent of cells surviving the treatment with respect to the radiation dose. Since $S'(D) < 0$ for $D > 0$, we conclude that the survival rate decreases as the dose increases. The rate of change $S'(D)$ is measured in percent per gray.

Example 5.3.5 Using the Chain Rule

a. Find $F'(x)$ if $F(x) = (x^3 + 2x^2 - 7)^{12}$.

The outer function is a power and the inner function is a polynomial. Since

$$F'(x) = \text{(derivative of the outer function evaluated at the inner function)}$$
$$\cdot \text{(derivative of the inner function)}$$

we obtain

$$F'(x) = 12(x^3 + 2x^2 - 7)^{11} \cdot (x^3 + 2x^2 - 7)'$$
$$= 12(x^3 + 2x^2 - 7)^{11}(3x^2 + 4x)$$

b. Let $H(x) = e^{\sqrt{x}}$. Then

$$H'(x) = \text{(derivative of the exponential function evaluated at } \sqrt{x}) \cdot \text{(derivative of } \sqrt{x})$$
$$= e^{\sqrt{x}}(\sqrt{x})' = e^{\sqrt{x}}\frac{1}{2\sqrt{x}} = \frac{e^{\sqrt{x}}}{2\sqrt{x}}$$

Recall that

$$(\sqrt{x})' = (x^{1/2})' = \frac{1}{2}x^{-1/2} = \frac{1}{2\sqrt{x}}$$

c. Find $f'(x)$ if $f(x) = \sqrt{xe^x + 7x^3 - 1}$.

Writing $f(x) = (xe^x + 7x^3 - 1)^{1/2}$, we find

$$f'(x) = \tfrac{1}{2}(xe^x + 7x^3 - 1)^{-1/2}(xe^x + 7x^3 - 1)'$$
$$= \tfrac{1}{2}(xe^x + 7x^3 - 1)^{-1/2}(e^x + xe^x + 21x^2)$$
$$= \frac{e^x + xe^x + 21x^2}{2\sqrt{xe^x + 7x^3 - 1}}$$

Note that we used the product rule to find $(xe^x)'$.

Using the chain rule, we can compute the derivative of the exponential function $f(x) = a^x$.

Recall that in the previous section we arrived at the formula

$$(2^t)' = 0.69315 \cdot 2^t$$

by numerically investigating the limit that was involved. Let's now compute the derivative exactly.

Since $A = e^{\ln A}$ for $A > 0$, we write

$$2^t = e^{\ln 2^t} = e^{t \cdot \ln 2} = e^{(\ln 2)t}$$

By the chain rule,

$$(2^t)' = (e^{(\ln 2)t})' = e^{(\ln 2)t}((\ln 2)t)' = e^{(\ln 2)t} \ln 2$$

Since $e^{(\ln 2)t} = 2^t$, we obtain

$$(2^t)' = 2^t \ln 2$$

So the number 0.69315 that we arrived at numerically is an approximation of $\ln 2$.

In general, to differentiate $f(x) = a^x$, we start with

$$f(x) = a^x = e^{\ln a^x} = e^{x \cdot \ln a} = e^{(\ln a)x}$$

and then use the chain rule to get

$$f'(x) = e^{(\ln a)x} \ln a = a^x \ln a$$

Thus,

$$(a^x)' = a^x \ln a$$

Example 5.3.6 Differentiating the General Exponential Function

According to the formula that we just derived,

$$(13^x)' = 13^x \ln 13 \approx 2.5649 \cdot 13^x$$

and

$$(0.7^x)' = 0.7^x \ln 0.7 \approx -0.3567 \cdot 0.7^x$$

To compute the derivative of

$$F(x) = 13^{x^2}$$

we write $F(x) = f(g(x))$, where $f(u) = 13^u$ and $g(x) = x^2$. Thus

$$\frac{dF}{dx} = \frac{df}{du}\frac{du}{dx} = (13^u \ln 13)(2x) = 13^{x^2} \ln 13 \cdot 2x$$

Example 5.3.7 Differentiating a Function That Is Not Given Explicitly I

Define the function $H(x) = e^{p(x)} + (p(x))^{2.7}$, where $p(x)$ is a differentiable function. Find $H'(x)$.

To differentiate $e^{p(x)}$, we write it as the composition $f(g(x))$, where $f(u) = e^u$ and $g(x) = p(x)$. It follows that

$$(e^{p(x)})' = f'(g(x))g'(x) = e^{p(x)} p'(x)$$

Now consider $(p(x))^{2.7}$. The power is the outer function, and $p(x)$ is the inner function. Thus,
$$[(p(x))^{2.7}]' = 2.7(p(x))^{1.7}p'(x)$$
Combining the two derivatives, we write
$$H'(x) = e^{p(x)}p'(x) + 2.7(p(x))^{1.7}p'(x) \qquad \blacktriangle$$

Example 5.3.8 Differentiating a Function That Is Not Given Explicitly II

Assume that $f(x)$ is a differentiable function. To differentiate $(3f(x) - 5)^7$, we use the chain rule:
$$\begin{aligned}((3f(x) - 5)^7)' &= 7(3f(x) - 5)^6 (3f(x) - 5)' \\ &= 7(3f(x) - 5)^6 (3f'(x)) \\ &= 21(3f(x) - 5)^6 f'(x)\end{aligned}$$

Likewise,
$$\begin{aligned}\left(e^{3f(x)-5}\right)' &= e^{3f(x)-5}(3f(x) - 5)' \\ &= e^{3f(x)-5}(3f'(x)) \\ &= 3e^{3f(x)-5}f'(x)\end{aligned}$$

and
$$\begin{aligned}(f(2^x - 7))' &= f'(2^x - 7)(2^x - 7)' \\ &= f'(2^x - 7)(2^x \ln 2) \\ &= 2^x(\ln 2)f'(2^x - 7)\end{aligned} \qquad \blacktriangle$$

Derivatives of Logarithmic Functions

We now use the differentiation rules to calculate the derivatives of logarithmic functions.

Start with the identity $e^{\ln x} = x$ and differentiate both sides:
$$\left(e^{\ln x}\right)' = (x)'$$

By the chain rule,
$$\left(e^{\ln x}\right)' = e^{\ln x}(\ln x)' = x(\ln x)'$$

The derivative of the right side of the first equation above is $(x)' = 1$, and therefore
$$x(\ln x)' = 1$$
$$(\ln x)' = \frac{1}{x}$$

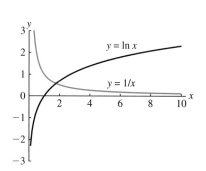

FIGURE 5.3.20

The natural logarithm and its derivative

Because the function $\ln x$ is defined only for $x > 0$, its derivative is also defined only on this domain. See Figure 5.3.20.

Recall the conversion formula for logarithms
$$\log_a x = \frac{\ln x}{\ln a} = \frac{1}{\ln a} \ln x$$

By the constant product rule,
$$(\log_a x)' = \frac{1}{\ln a}(\ln x)' = \frac{1}{\ln a}\frac{1}{x} = \frac{1}{x \ln a}$$

Thus, the derivative of $\log_a x$ is proportional to $1/x$.

Example 5.3.9 The Derivative of $x \ln x$

The derivative of the function $g(x) = x \ln x$ is
$$\frac{dg}{dx} = x\frac{d(\ln x)}{dx} + \ln x \frac{dx}{dx}$$

$$= x \cdot \frac{1}{x} + \ln x \cdot 1$$
$$= 1 + \ln x$$

Example 5.3.10 Finding the Derivative Using the Laws of Logarithms

We can find the derivative of the function $f(x) = \ln 2x$ using the fact that $\ln 2x = \ln 2 + \ln x$.

We compute
$$(\ln 2x)' = (\ln 2 + \ln x)'$$
$$= 0 + \frac{1}{x} = \frac{1}{x}$$

Because
$$\ln 2x = \ln 2 + \ln x$$

we see that the graph of $f(x) = \ln 2x$ is obtained by translating the graph of $f(x) = \ln x$ vertically by $\ln 2 \approx 0.69315$ units. Therefore, the two derivatives $(\ln 2x)'$ and $(\ln x)'$ have to be equal (Figure 5.3.21).

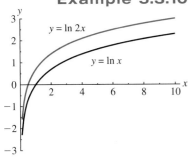

FIGURE 5.3.21

The functions $f(x) = \ln x$ and $f(x) = \ln 2x$ have the same derivative

Example 5.3.11 Finding the Derivative Using the Laws of Logarithms

Let
$$f(x) = \ln \frac{e^x \sqrt{x}}{14}$$

Find $f'(1)$.

As suggested by the previous example, we simplify using the laws of logarithms before calculating the derivative. We get
$$f(x) = \ln \frac{e^x \sqrt{x}}{14}$$
$$= \ln(e^x \sqrt{x}) - \ln 14$$
$$= \ln e^x + \ln(\sqrt{x}) - \ln 14$$
$$= x + \frac{1}{2} \ln x - \ln 14$$

since $\ln e^x = x$ and
$$\ln(\sqrt{x}) = \ln(x^{1/2}) = \frac{1}{2} \ln x$$

Thus, the derivative is
$$f'(x) = 1 + \frac{1}{2} \frac{1}{x} = 1 + \frac{1}{2x}$$

and so $f'(1) = 3/2$.

Keep in mind that the derivative represents the rate of change. Thus, the result $f'(1) = 3/2$ from the previous example means that when $x = 1$, the function $f(x)$ increases (the derivative is positive) at a rate of approximately $3/2 = 1.5$ units per unit change in x.

Example 5.3.12 Differentiating Logarithmic Functions

a. Find $f'(x)$ if $f(x) = \ln(3 + 2x - x^2)$.

The outer function is a logarithm and the inner function is a polynomial. Since
$$f'(x) = \text{(derivative of the outer function evaluated at the inner function)}$$
$$\cdot \text{(derivative of the inner function)}$$

we obtain
$$f'(x) = \frac{1}{3 + 2x - x^2} \cdot (3 + 2x - x^2)'$$

$$= \frac{1}{3+2x-x^2}(2-2x)$$
$$= \frac{2-2x}{3+2x-x^2}$$

b. If $f(x) = \log_{10}(4x - e^x)$, then
$$f'(x) = \frac{1}{(4x-e^x)\ln 10} \cdot (4x-e^x)'$$
$$= \frac{1}{(4x-e^x)\ln 10}(4-e^x)$$
$$= \frac{4-e^x}{(4x-e^x)\ln 10}$$

c. Find $g'(x)$ if $g(x) = 0.7^{\ln x}$.

Recall that $(0.7^x)' = 0.7^x \ln 0.7$. By the chain rule,
$$\left(0.7^{\ln x}\right)' = 0.7^{\ln x} \ln 0.7 (\ln x)' = 0.7^{\ln x} \ln 0.7 \left(\frac{1}{x}\right) = \frac{0.7^{\ln x} \ln 0.7}{x}$$ △

Example 5.3.13 Using the Chain Rule on a Multiple Composition

Consider the function
$$F(x) = \ln((1-x^2)^4 + 7)$$

By the chain rule,
$$F'(x) = \frac{1}{(1-x^2)^4 + 7}((1-x^2)^4 + 7)'$$

Using the chain rule to compute the derivative in the second factor, we get
$$((1-x^2)^4 + 7)' = 4(1-x^2)^3(1-x^2)' + 0$$
$$= 4(1-x^2)^3(-2x)$$
$$= -8x(1-x^2)^3$$

Thus,
$$F'(x) = \frac{1}{(1-x^2)^4 + 7}(-8x(1-x^2)^3) = -\frac{8x(1-x^2)^3}{(1-x^2)^4 + 7}$$ △

Exponential Measurements and Relative Rate of Change

Exponential measurements are generally written in the form
$$P(t) = P(0)e^{\alpha t}$$

The constant $P(0)$ represents the value of the measurement at $t = 0$. The parameter α determines whether and how fast the measurement is changing as a function of time. If α is negative, the measurement is decreasing, and it decreases more quickly the more negative α is. If α is positive, the measurement is increasing, and it increases more quickly the larger α is.

We can find the derivative of $P(t)$ with the chain rule.
$$P'(t) = P(0)(e^{\alpha t})' = P(0)e^{\alpha t}\alpha$$

Rearranging the terms, we get
$$P'(t) = \alpha P(0)e^{\alpha t} = \alpha P(t)$$

Thus, if the measurement $P(t)$ is given by the exponential function, then the rate of change of $P(t)$ is proportional to itself.

In Chapter 8 we will show that the statement is true the other way around—namely, if the rate of change of some measurement $P(t)$ is proportional to $P(t)$, then

$$P(t) = Ae^{\alpha t}$$

for some constants A and α.

What are the meanings of A and α? Since

$$P(0) = Ae^{\alpha \cdot 0} = A$$

it follows that A represents the initial measurement (i.e., the measurement at time $t = 0$). Dividing both sides of the equation

$$P'(t) = \alpha P(t)$$

by $P(t)$, we get

$$\frac{P'(t)}{P(t)} = \alpha$$

So α represents the **relative rate of change** of $P(t)$, that is, the rate of change of $P(t)$ per unit of $P(t)$.

Note that, by the chain rule,

$$(\ln P(t))' = \frac{1}{P(t)} P'(t) = \frac{P'(t)}{P(t)}$$

Thus, the relative rate of change of a function is the derivative of the natural logarithm of the function.

Example 5.3.14 Rate of Change and Relative Rate of Change

Assume that $P(t) = e^t + 3t + 100$ and $Q(t) = e^t + 3t + 1000$ represent populations of monkeys living on two different islands, Island P and Island Q, where t is the time in months.

Pick a time, say, $t = 2$. From $P'(t) = e^t + 3$ and $Q'(t) = e^t + 3$ we conclude that

$$P'(2) = Q'(2) = e^2 + 3 \approx 10.4$$

In words, the **rate of change** of both populations is the same and is approximately equal to 10.4 monkeys per month. The actual populations at $t = 2$ are calculated to be

$$P(2) = e^2 + 3(2) + 100 \approx 113.4$$

and

$$Q(2) = e^2 + 3(2) + 1000 \approx 1013.4$$

i.e., 113 and 1013 monkeys, respectively.

Clearly, the rate of change does not tell the whole story. Although the population of monkeys on Island Q is about *nine times as large as* the population of monkeys on Island P, *the rate (monkeys per month) at which the two populations increase is the same*. How do we quantify the difference in the behaviour of the two populations, i.e., quantify the fact that the monkeys on Island P are much more prolific than the monkeys on Island Q?

To answer this question, for each population, we relate the rate of change to the total number, i.e., we look at the **relative rates of change**

$$\frac{P'(t)}{P(t)} \quad \text{and} \quad \frac{Q'(t)}{Q(t)}$$

Now the units are monkeys per month (that's the numerator) per monkey. The relative rate of change of the population on Island P at time $t = 2$ is

$$\frac{P'(2)}{P(2)} = \frac{10.4}{113.4} \approx 0.09$$

In words, for each monkey on Island P there will be 0.09 new monkeys per month. Or (multiplying the numbers by 100), within each group of 100 monkeys on island P there will be an increase of 9 monkeys per month. On the other hand,

$$\frac{Q'(2)}{Q(2)} = \frac{10.4}{1013.4} \approx 0.01$$

So, the relative rate of change of the monkey population on island Q is 0.01 monkeys per monkey per month (or, within each group of 100 monkeys there will be an increase of 1 monkey per month). Thus, the relative rate of increase in population on Island Q is about a ninth of the relative rate of increase in population on Island P. ▲

Example 5.3.15 Relative Rate of Change of Population

Assume that the function $P(t)$ represents the population of elephants in Serengeti National Park. One unit of time, t, represents 10 years. It has been determined that

$$\frac{P'(t)}{P(t)} = 0.1$$

The fact that the relative rate of change is 0.1 means that the population of elephants increases at a rate of 0.1 elephants *per elephant per 10 years*.
From the discussion preceding Example 5.3.14,

$$\frac{P'(t)}{P(t)} = 0.1$$

implies that

$$P(t) = P(0)e^{0.1t}$$

where $P(0)$ is the initial population of elephants.
Thus, 10 years later, the number of elephants will be

$$P(1) = P(0)e^{0.1(1)} = 1.105P(0)$$

Twenty years later, it will be

$$P(2) = P(0)e^{0.1(2)} = 1.221P(0)$$

and so forth. The population will increase exponentially; see Figure 5.3.22. ▲

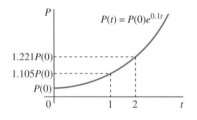

Figure 5.3.22
Exponentially increasing population of elephants

Proof of the Chain Rule

To conclude this section, we present a full proof of the chain rule. We consider the composition

$$F(x) = (f \circ g)(x) = f(g(x))$$

under the assumption that g is differentiable at x and f is differentiable at $g(x)$. The claim is that F is differentiable at x and

$$F'(x) = f'(g(x))g'(x)$$

Recall that

$$g'(x) = \lim_{\Delta x \to 0} \frac{g(x + \Delta x) - g(x)}{\Delta x} = \lim_{\Delta x \to 0} \frac{\Delta g}{\Delta x}$$

i.e., $g'(x)$ is the limit of $\frac{\Delta g}{\Delta x}$ as $\Delta x \to 0$. Let us compare how much the two quantities differ by defining

$$\varepsilon = \frac{\Delta g}{\Delta x} - g'(x).$$

Then
$$\lim_{\Delta x \to 0} \varepsilon = \lim_{\Delta x \to 0}\left(\frac{\Delta g}{\Delta x} - g'(x)\right) = \lim_{\Delta x \to 0}\frac{\Delta g}{\Delta x} - g'(x) = g'(x) - g'(x) = 0$$

Now we rearrange the terms in
$$\varepsilon = \frac{\Delta g}{\Delta x} - g'(x)$$
to obtain
$$\frac{\Delta g}{\Delta x} = g'(x) + \varepsilon, \quad \text{i.e.,} \quad \Delta g = (g'(x) + \varepsilon)\Delta x$$

This is important, so let's repeat: if g is differentiable at x, then we can write
$$\Delta g = (g'(x) + \varepsilon)\Delta x$$
where $\varepsilon \to 0$ as $\Delta x \to 0$. Applying the same argument to the function f differentiable at u, we get
$$\Delta f = (f'(u) + \eta)\Delta u$$
where $\eta \to 0$ as $\Delta u \to 0$.

Now consider $F(x) = f(g(x))$, i.e., $F(x) = f(u)$, where $u = g(x)$. Using the expression for Δf we just derived, we write (recalling that since $u = g(x)$, we get $\Delta u = \Delta g$)
$$\Delta F = \Delta f = (f'(u) + \eta)\Delta u = (f'(u) + \eta)\Delta g$$
$$= (f'(u) + \eta)(g'(x) + \varepsilon)\Delta x$$

Dividing by Δx and substituting $g(x)$ for u, we obtain
$$\frac{\Delta F}{\Delta x} = (f'(g(x)) + \eta)(g'(x) + \varepsilon)$$
and
$$F'(x) = \lim_{\Delta x \to 0}\frac{\Delta F}{\Delta x} = \lim_{\Delta x \to 0}(f'(g(x)) + \eta)(g'(x) + \varepsilon)$$

We know that $\varepsilon \to 0$ since $\Delta x \to 0$. As well, from
$$\Delta u = \Delta g = (g'(x) + \varepsilon)\Delta x$$
we conclude that $\Delta u \to 0$ as well. Thus, $\eta \to 0$ and
$$F'(x) = \lim_{\Delta x \to 0}(f'(g(x)) + \eta)(g'(x) + \varepsilon) = f'(g(x))g'(x)$$

In this section we have added the following to our list of differentiation formulas:

Function or Rule	Derivative
general exponential function	$(a^x)' = a^x \ln a$
natural logarithm function	$(\ln x)' = \dfrac{1}{x}$
general logarithm function	$(\log_a x)' = \dfrac{1}{x \ln a}$
chain rule (prime notation)	$[f(g(x))]' = f'(g(x))g'(x)$
chain rule (differential notation)	If $F = f(u)$ and $u = g(x)$, then $\dfrac{dF}{dx} = \dfrac{df}{du}\dfrac{dg}{dx}$.

Summary The **chain rule** states that the derivative of the composition of two functions is computed by multiplying the derivative of the outer function (evaluated at the inner function) by the derivative of the inner function. Using the chain rule, we derived the formulas for the derivatives of the general exponential and logarithmic functions. To compare the rates of change for different measurements, we computed their **relative rates of change.**

5.3 Exercises

Mathematical Techniques

1–5 ■ Write each function in the form $F(x)=f(g(x))$; i.e., identify the inner function g and the outer function f. Then use the chain rule to compute $F'(x)$.

1. $F(x) = \sqrt{e^{x+1}}$
2. $F(x) = e^{\sqrt{x}}$
3. $F(x) = \left(x - 4 - \dfrac{2}{x}\right)^4$
4. $F(x) = \ln(x + \ln x)$
5. $F(x) = \dfrac{1}{4e^x + 3}$

6–14 ■ Compute the derivative of each function.

6. $f(x) = 5^6 + e^4 + \ln 3$
7. $f(x) = x + 4\ln x$
8. $h(x) = 2x^2 - 2\ln x$
9. $g(z) = (z+4)\ln z$
10. $f(x) = x^2 \ln x$
11. $F(w) = e^w \ln w$
12. $F(y) = \dfrac{\ln y}{e^y}$
13. $s(x) = \ln x^2$
14. $r(x) = \ln(x^2 e^x)$

15–16 ■ Use the first derivative to sketch the graph of each function on the given domain. Identify the intervals where the function is increasing and the intervals where it is decreasing.

15. $M(x) = (x+2)\ln x$ for $1 \leq x \leq 3$.
16. $L(x) = \dfrac{x}{2} - \ln x$ for $1 \leq x \leq 3$.

17–41 ■ Compute each derivative using the chain rule.

17. $g(x) = (1 + 3x)^2$
18. $h(x) = (1 + 2x)^3$
19. $f_1(t) = (1 + 3t)^{30}$
20. $f_2(t) = (1 + 2t^2)^{15}$
21. $r(x) = \dfrac{(1+3x)^2}{(1+2x)^3}$
22. $p(z) = (1 + 3z)^2 (1 + 2z)^3$
23. $f(x) = \sqrt{1 - x - 2x^3}$
24. $h(x) = \dfrac{12}{\sqrt[4]{x^2 + x + 1}}$
25. $f(x) = \sqrt{2 - \dfrac{x}{x - 2}}$
26. $f(x) = (e^x - 2)^{-3}$
27. $F(z) = \left(1 + \dfrac{2}{1+z}\right)^3$
28. $G(w) = \left[\left(2 - \dfrac{3}{1-w}\right)^2 + 1\right]^2$
29. $f(x) = e^{-3x}$
30. $h(x) = 2^x 3^x$
31. $g(y) = \ln(1 + y)$
32. $A(z) = \ln(1 + e^z)$
33. $f(x) = \log_{10} x^3$
34. $f(x) = \ln x^4 + \ln^4 x$
35. $f(x) = (2 + \log_{10} x)^2$
36. $f(x) = \log_5(\ln x)$
37. $f(x) = \sqrt{\ln x + x - 4}$
38. $f(x) = \ln(\ln x) + \ln(\ln 6)$
39. $G(x) = 8e^{x^2}$
40. $s(w) = 4.2\sqrt{1 + e^w}$
41. $L(x) = \ln \sqrt{\ln x}$

42. Find the equation of the line that is tangent to the graph of the function $f(x) = e^{-x^2}$ at the point where $x = 1$.

43. Find the equation of the line that is tangent to the graph of the function $f(x) = \ln(x^2 + 1)$ at the point where $x = 0$.

44. Find all points where the graph of the function $f(x) = e^{x^2 + x - 1}$ has a horizontal tangent line.

45. Let $F(x) = f(g(x))$. If $g'(6) = -2, f'(4) = 3,$ and $g(6) = 4$, find $F'(6)$.

46–49 ■ Rewrite each function using the identity $f(x) = e^{\ln f(x)}$ and then calculate its derivative.

46. $f(x) = x^x$
47. $f(x) = (\ln x)^x$
48. $f(x) = x^{e^x}$
49. $f(x) = (1 + x)^{2+x}$

50–57 ■ Compute the derivative of each function in the two ways given.

50. $F(x) = \dfrac{1}{1 + e^x}$, first using the quotient rule and then using the chain rule.

51. $H(y) = \dfrac{1}{1 + y^3}$, first using the quotient rule and then using the chain rule.

52. $g(x) = \ln 3x$, first using a law of logarithms and then using the chain rule.

53. $h(x) = \ln x^3$, first using a law of logarithms and then using the chain rule.

54. $F(x) = (1 + 2x)^2$, first by expanding the binomial and taking the derivative of the polynomial, and then with the chain rule.

55. $F(x) = (1+2x)^3$, first by expanding the binomial and taking the derivative of the polynomial, and then with the chain rule.

56. $F(x) = x^3$, first with the power rule and then by writing $F(x)$ using the exponential function and using the chain rule.

57. $F(x) = x^{-5}$, first with the power rule and then by writing $F(x)$ using the exponential function and using the chain rule.

▼ 58–59 ▪ We can use laws of exponents and the chain rule to check the power rule.

58. Write $f(x) = x^n$ using the exponential function.

59. Take the derivative and then rewrite as a power function.

▼ 60–61 ▪ The equation for the top half of the circle of radius 1 is $f(x) = \sqrt{1-x^2}$. We can find the slope of the tangent line with the chain rule, or we can find it geometrically.

60. Find the derivative of $f(x)$ with the chain rule.

61. Find the slope of the ray connecting the centre of the circle at $(0,0)$ to the point $(x, f(x))$ on the circle. Then use the fact that the tangent line to a circle is perpendicular to the ray to find the slope of the tangent line. Check that it matches the result with the chain rule.

Applications

62. The concentration of a pollutant (measured in ppm = parts per million) at a fixed location x units away from the source changes according to the formula

$$c(t) = \frac{N}{\sqrt{4\pi kt}} e^{-x^2/4kt}$$

where the positive quantities N and k characterize the pollutant. Find and interpret $c'(t)$. What are its units?

63. The concentration of a pollutant (measured in ppm = parts per million) at a location x units away from the source at a fixed time t is given by

$$c(x) = \frac{N}{\sqrt{4\pi kt}} e^{-x^2/4kt}$$

where the positive quantities N and k characterize the pollutant. Find and interpret $c'(x)$. What are its units?

▼ 64–67 ▪ The following functional compositions describe connections between measurements (as in Section 1.4, Exercises 64–67). Find the derivative of the composition using the chain rule.

64. The number of mosquitoes (M) that end up in a room is a function of how wide the window is open (W, in square centimetres) according to $M(W) = 5W + 2$. The number of bites (B) depends on the number of mosquitoes according to $B(M) = 0.5M$. Find the derivative of B as a function of W.

65. The temperature of a room (T) in degrees Celsius is a function of how wide the window is open (W) according to $T(W) = 40 - 0.2W$. How long you sleep (S, measured in hours) is a function of the temperature according to $S(T) = 14 - T/5$. Find the derivative of S as a function of W.

66. The number of viruses (V, measured in trillions) that infect a person is a function of the degree of immunosuppression (I, the fraction of the immune system that is turned off by medication) according to $V(I) = 5I^2$. The fever (F, measured in degrees Celsius) associated with an infection is a function of the number of viruses according to $F(V) = 37 + 0.4V$. Find the derivative of F as a function of I.

67. The length of an insect (L, in millimetres) is a function of the temperature during development (T, measured in degrees Celsius) according to $L(T) = 10 + T/10$. The volume of the bug (V, in cubic millimetres) is a function of the length according to $V(L) = 2L^3$. The mass (M, in milligrams) depends on volume according to $M(V) = 1.3V$. Find the derivative of M as a function of T.

▼ 68–69 ▪ The amount of carbon-14 (^{14}C) per gram left t years after the death of an organism is given by

$$Q(t) = Q_0 e^{-0.000122t}$$

where Q_0 is the amount left per gram at the time of death. Suppose $Q_0 = 6 \cdot 10^{10}$ ^{14}C atoms/g.

68. Find the derivative of $Q(t)$.

69. If ^{14}C were lost at a constant rate equal to the rate at $t = 0$, how long would it take for half of the ^{14}C to disappear? How does this compare with the half-life of ^{14}C? Why are the results different?

▼ 70–75 ▪ Find the derivatives of the following absorption functions. Compute the value of the derivative at $c = 0$.

70. $\alpha(c) = 5c$

71. $\alpha(c) = \dfrac{5c}{1+c}$

72. $\alpha(c) = \dfrac{5c^2}{1+c^2}$

73. $\alpha(c) = \dfrac{5c}{e^{2c}}$

74. $\alpha(c) = \dfrac{5c}{1+c^2}$

75. $\alpha(c) = 5c(1+c)$.

▼ 76–80 ▪ Check the given solutions to the following differential equations. Which solutions are increasing and which are decreasing?

76. $\dfrac{db}{dt} = 3b(t)$ has solution $b(t) = 100e^{3t}$ if $b(0) = 100$.

77. $\dfrac{db}{dt} = -2b(t)$ has solution $b(t) = 10e^{-2t}$ if $b(0) = 10$.

78. $\dfrac{db}{dt} = 1 + 2b(t)$ has solution $b(t) = 3e^{2t} - 0.5$ if $b(0) = 2.5$.

79. $\dfrac{db}{dt} = 10 - 2b(t)$ has solution $b(t) = 5 + 20e^{-2t}$ if $b(0) = 25$.

80. $\dfrac{db}{dt} = e^{-b(t)}$ has solution $b(t) = \ln t$ if $b(1) = 0$.

5.4 Derivatives of Trigonometric and Inverse Trigonometric Functions

The last group of special functions that are important in biology consists of trigonometric and inverse trigonometric functions. In this section, we compute the derivatives of these special functions.

Derivatives of Sine and Cosine

All trigonometric functions are periodic, and so their derivatives are periodic functions as well.

In Figure 5.4.23 we show the graph of $f(x) = \sin x$ (keep in mind that x is in radians). What does the graph of its derivative look like?

Between $x = 0$ and $x = \pi/2$ the graph of $f(x) = \sin x$ is increasing, so its derivative, $f'(x)$, is positive. As we move from $x = 0$ to $x = \pi/2$, we notice that the slopes are decreasing; i.e., $f'(x)$ must be decreasing on $(0, \pi/2)$. At $x = \pi/2$ the graph of $f(x) = \sin x$ has a horizontal tangent, and thus $f'(\pi/2) = 0$. Continuing in the same way, we realize that the graph of $f'(x)$ might look like the bottom graph in Figure 5.4.23.

It seems that $f(x) = \sin x$ has the largest slope at $x = 0$ (and then, again, at points that are a multiple of 2π away from it). From

$$f'(0) = \lim_{h \to 0} \frac{\sin(0 + h) - \sin 0}{h} = \lim_{h \to 0} \frac{\sin h}{h}$$

using a small value for h, say $h = 0.001$, we obtain the estimate

$$f'(0) \approx \frac{\sin(0.001)}{0.001} \approx \frac{0.001000}{0.001} = 1.000$$

Likewise, the smallest slope (i.e., largest negative slope) seems to occur at $x = \pi$ (and then at $x = \pi + 2k\pi$, where k is an integer). We estimate

$$f'(\pi) \approx \frac{\sin(\pi + 0.001) - \sin \pi}{0.001} \approx \frac{-0.001000 - 0}{0.001} = -1.000$$

Thus, the graph of $f'(x)$ seems to be bounded by -1 from below and by 1 from above.

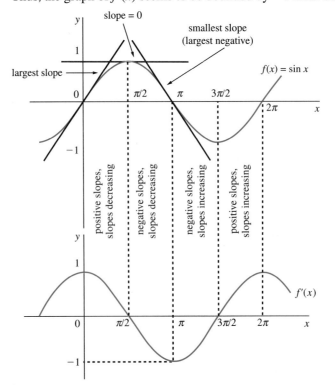

FIGURE 5.4.23

Finding the slopes of $f(x) = \sin x$

5.4 Derivatives of Trigonometric and Inverse Trigonometric Functions

Our analysis suggests that $f'(x) = (\sin x)'$ might be $\cos x$. A similar analysis would suggest that $(\cos x)' = -\sin x$.

We now prove that these formulas are indeed true.

Because the functions sine and cosine cannot be built out of polynomials and exponential functions with products, sums, and quotients, we must compute their derivatives from the definition

$$(\sin x)' = \lim_{h \to 0} \frac{\sin(x+h) - \sin x}{h}$$

$$(\cos x)' = \lim_{h \to 0} \frac{\cos(x+h) - \cos x}{h}$$

These formulas do us little good unless we can expand the expressions for $\sin(x+h)$ and $\cos(x+h)$. When finding the derivative of the exponential function, we used a law of exponents to rewrite e^{x+h} as $e^x e^h$. Trigonometric functions have similarly useful sum laws, called the **angle addition formulas:**

$$\sin(x+h) = \sin x \cos h + \cos x \sin h \qquad (5.4.1)$$

$$\cos(x+h) = \cos x \cos h - \sin x \sin h \qquad (5.4.2)$$

(see Section 2.3). We will use the sum law for sine to find the derivative:

$$\lim_{h \to 0} \frac{\sin(x+h) - \sin x}{h}$$

$$= \lim_{h \to 0} \frac{\sin x \cos h + \cos x \sin h - \sin x}{h} \qquad \text{apply the sum law for sine}$$

$$= \lim_{h \to 0} \frac{\sin x [\cos h - 1] + \cos x \sin h}{h} \qquad \text{combine terms involving } \sin x$$

$$= \lim_{h \to 0} \frac{\sin x [\cos h - 1]}{h} + \lim_{h \to 0} \frac{\cos x \sin h}{h} \qquad \text{break up the limit}$$

$$= \sin x \lim_{h \to 0} \frac{\cos h - 1}{h} + \cos x \lim_{h \to 0} \frac{\sin h}{h} \qquad \text{pull out terms without } h$$

Although this might not look much better than the limit we started with, it has an important simplification, much like that found with the exponential function. All terms involving x have come outside the limits. If we could figure out the numerical values of

$$\lim_{h \to 0} \frac{\cos h - 1}{h}$$

and

$$\lim_{h \to 0} \frac{\sin h}{h}$$

we would be finished.

In Derivation of the Key Limits later in this section, we show that

$$\lim_{h \to 0} \frac{\cos h - 1}{h} = 0$$

$$\lim_{h \to 0} \frac{\sin h}{h} = 1$$

Substituting these values into the formulas, we obtain

$$(\sin x)' = \sin x \lim_{h \to 0} \frac{\cos h - 1}{h} + \cos x \lim_{h \to 0} \frac{\sin h}{h}$$

$$= \sin x \cdot 0 + \cos x \cdot 1 = \cos x$$

Similarly, the derivative of $\cos x$ is

$$\lim_{h \to 0} \frac{\cos(x+h) - \cos x}{h}$$

$$= \lim_{h \to 0} \frac{\cos x \cos h - \sin x \sin h - \cos x}{h} \qquad \text{apply the sum law for cosine}$$

$$= \lim_{h \to 0} \frac{\cos x[\cos h - 1] - \sin x \sin h}{h} \qquad \text{combine terms involving } \cos x$$

$$= \lim_{h \to 0} \frac{\cos x[\cos h - 1]}{h} - \lim_{h \to 0} \frac{\sin x \sin h}{h} \qquad \text{break up the limit}$$

$$= \cos x \lim_{h \to 0} \frac{[\cos h - 1]}{h} - \sin x \lim_{h \to 0} \frac{\sin h}{h} \qquad \text{pull out terms without } h$$

$$= \cos x \cdot 0 - \sin x \cdot 1 \qquad \text{use the limits above}$$

$$= -\sin x$$

In summary,

$$\frac{d[\sin x]}{dx} = \cos x$$

$$\frac{d[\cos x]}{dx} = -\sin x$$

To recall where the negative sign goes, remember the graphs of sine and cosine. At $x=0$, the cosine is flat but beginning to decrease, meaning that the derivative must begin at 0 and become negative. At $x=0$, the sine is increasing, meaning that the derivative should take on a positive value at that point.

Example 5.4.1 A Derivative Involving a Trigonometric Function

We can use the rules for differentiation to find the derivatives of more complicated functions. For example, we could try to determine whether

$$f(x) = x + \sin x$$

is an increasing function. We compute the derivative

$$f'(x) = 1 + \cos x$$

Because $\cos x \geq -1$ for all x, we conclude that

$$f'(x) = 1 + \cos x \geq 0$$

and thus $f(x)$ is increasing (Figure 5.4.24).

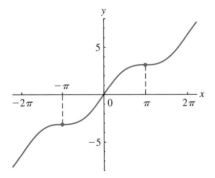

FIGURE 5.4.24
The graph of $x + \sin x$

Example 5.4.2 Derivatives Involving Trigonometric Functions

a. Let

$$F(x) = \cos(2 + x - 2x^2)$$

To find $F'(x)$, we use the chain rule. Since cosine is the outside function, we differentiate it first:

$$F'(x) = -\sin(2 + x - 2x^2) \cdot (2 + x - 2x^2)'$$
$$= -\sin(2 + x - 2x^2) \cdot (1 - 4x)$$
$$= -(1 - 4x)\sin(2 + x - 2x^2)$$

b. If $H(x) = e^{\cos x}$, then

$$H'(x) = e^{\cos x}(\cos x)' = e^{\cos x}(-\sin x) = -e^{\cos x}\sin x$$

c. To differentiate the function

$$G(x) = \sin^3 x + \sin x^3$$

we rewrite it using brackets:

$$G(x) = (\sin x)^3 + \sin(x^3)$$

To differentiate $(\sin x)^3$, we differentiate the power first:

$$((\sin x)^3)' = 3(\sin x)^2(\sin x)' = 3(\sin x)^2 \cos x$$

To differentiate $\sin(x^3)$, we differentiate sine first:

$$(\sin(x^3))' = \cos(x^3)(x^3)' = 3x^2 \cos(x^3)$$

Thus,

$$G'(x) = 3(\sin x)^2 \cos x + 3x^2 \cos(x^3)$$
$$= 3\sin^2 x \cos x + 3x^2 \cos x^3$$

We wrote sinusoidal oscillations as

$$f(t) = A + B\cos\left(\frac{2\pi}{T}(t - \phi)\right)$$

(Equation 2.3.1), where A is the average, B is the amplitude, T is the period, and ϕ is the phase.

We now find $f'(t)$ using the chain rule:

$$f'(t) = 0 + B\left(-\sin\left(\frac{2\pi}{T}(t - \phi)\right)\right) \cdot \left(\frac{2\pi}{T}(t - \phi)\right)'$$
$$= -B\sin\left(\frac{2\pi}{T}(t - \phi)\right) \cdot \left(\frac{2\pi}{T}\right)$$
$$= -\frac{2\pi B}{T}\sin\left(\frac{2\pi}{T}(t - \phi)\right)$$

Example 5.4.3 Derivatives of Oscillations

The chain rule can be applied to find the derivatives of the daily and monthly temperature cycles introduced in Section 2.3 (written in units of days):

$$P_d(t) = 36.8 + 0.3\cos[2\pi(t - 0.583)]$$

$$P_m(t) = 36.8 + 0.2\cos\left[\frac{2\pi(t - 16)}{28}\right]$$

FIGURE 5.4.25

The daily temperature cycle and its derivative

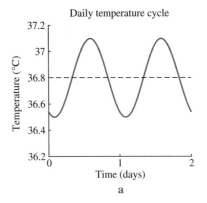
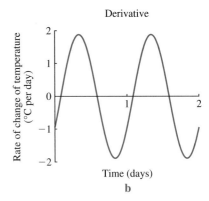

FIGURE 5.4.26

The monthly temperature cycle and its derivative

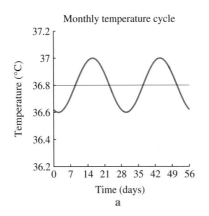
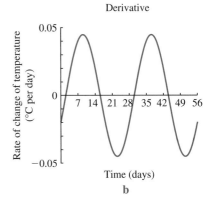

Using the rule for finding the derivative of a general trigonometric function, along with the constant multiple rule, we get

$$\frac{dP_d}{dt} = 0.3 \frac{d}{dt}\{\cos[2\pi(t - 0.583)]\}$$
$$= -0.3 \cdot 2\pi \sin[2\pi(t - 0.583)]$$
$$\frac{dP_m}{dt} = 0.2 \frac{d}{dt}\left\{\cos\left[\frac{2\pi(t - 16)}{28}\right]\right\}$$
$$= -0.2 \cdot \frac{2\pi}{28} \sin\left[\frac{2\pi(t - 16)}{28}\right]$$

The scales on the graphs of the two derivatives are very different (Figures 5.4.25 and 5.4.26). The daily oscillation has a much greater rate of change than the monthly one, nearly 2 degrees per day, with the monthly oscillation reaching only 0.05 degrees per day. Biologically, this corresponds to the fact that the daily oscillation is much faster. Mathematically, this results from the factor of 28 dividing the amplitude of the derivative of the monthly rhythm.

Other Trigonometric Functions

Using the quotient rule, we compute

$$(\tan x)' = \left(\frac{\sin x}{\cos x}\right)'$$
$$= \frac{\cos x (\sin x)' - \sin x (\cos x)'}{\cos^2 x}$$

5.4 Derivatives of Trigonometric and Inverse Trigonometric Functions

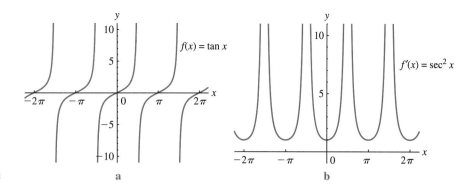

FIGURE 5.4.27
The tangent function and its derivative

$$= \frac{\cos x \cos x - \sin x(-\sin x)}{\cos^2 x}$$

$$= \frac{1}{\cos^2 x} = \sec^2 x$$

(Recall that $\sin^2 x + \cos^2 x = 1$.) Since $\sec^2 x > 0$ for all x (where it is defined), it follows that $y = \tan x$ is an increasing function (to be precise—increasing on the intervals $(-\pi/2, \pi/2), (\pi/2, 3\pi/2)$, etc.). See Figure 5.4.27.

To compute the derivative of sec x we use the chain rule:

$$(\sec x)' = ((\cos x)^{-1})'$$
$$= (-1)(\cos x)^{-2}(\cos x)'$$
$$= (-1)(\cos x)^{-2}(-\sin x)$$
$$= \frac{\sin x}{\cos^2 x}$$
$$= \frac{1}{\cos x} \cdot \frac{\sin x}{\cos x} = \sec x \tan x$$

Similarly (see Exercises 33 and 35), we prove that

$$(\csc x)' = -\cot x \csc x$$
$$(\cot x)' = -\csc^2 x$$

Example 5.4.4 Derivatives Involving Trigonometric Functions

To differentiate

$$F(x) = \sec \sqrt{x}$$

we differentiate secant first:

$$F'(x) = \sec \sqrt{x} \tan \sqrt{x} \cdot (\sqrt{x})'$$
$$= \sec \sqrt{x} \tan \sqrt{x} \cdot \frac{1}{2\sqrt{x}}$$
$$= \frac{\sec \sqrt{x} \tan \sqrt{x}}{2\sqrt{x}}$$

If $G(x) = e^{\tan x}$, then

$$G'(x) = e^{\tan x}(\tan x)' = e^{\tan x} \sec^2 x$$

Derivation of the Key Limits

Finding the derivatives of sine and cosine requires computing the following limits:

$$\lim_{h \to 0} \frac{\cos h - 1}{h}$$

$$\lim_{h \to 0} \frac{\sin h}{h}$$

Note that these limits cannot be calculated by direct substitution (one of the reasons that they are challenging) since, as $h \to 0$, both $\sin h$ and $\cos h - 1$ approach zero. Testing with a calculator gives the following table. It seems as though the first limit is 0 and the second is 1. Important—keep in mind that h is in radians!

h	$\dfrac{\cos h - 1}{h}$	$\dfrac{\sin h}{h}$
1	−0.45970	0.84147
0.1	−0.04996	0.99833
0.01	−0.00500	0.99998
0.001	−0.00050	1.00000

We can see why this occurs from a geometric diagram of sine and cosine. First, we show that

$$\lim_{h \to 0} \frac{\sin h}{h} = 1$$

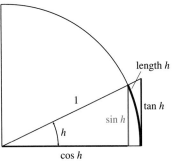

FIGURE 5.4.28

Geometric components of sine on the graph of the unit circle

In Figure 5.4.28, the length of the arc is equal to the angle in radians (which is h). By similar triangles, the length of the line segment that is tangent to the circle is $\tan h$.

The right triangle with sides $\cos h$ and $\sin h$ is contained in the sector of the circle of radius 1 with central angle h, which, in turn, is contained in the right triangle with sides 1 and $\tan h$. Comparing the three areas, we get (recall that the area of a sector of a circle of radius r and central angle h (in radians!) is $r^2 h/2$)

$$\frac{1}{2} \sin h \cos h \leq \frac{1}{2}(1)^2 h \leq \frac{1}{2} 1 \cdot \tan h$$

i.e.,

$$\sin h \cos h \leq h \leq \frac{\sin h}{\cos h}$$

Assume that $h > 0$ and h is small (so that it is in the first quadrant, where $\sin h > 0$ and $\cos h > 0$). Dividing by $\sin h$ and computing the reciprocals of all sides yields

$$\cos h \leq \frac{h}{\sin h} \leq \frac{1}{\cos h}$$

$$\frac{1}{\cos h} \geq \frac{\sin h}{h} \geq \cos h$$

Finally, computing the limit as $h \to 0$ (keeping in mind that we assumed that $h > 0$ so it's the right-hand limit), we get

$$\lim_{h \to 0^+} \frac{1}{\cos h} \geq \lim_{h \to 0^+} \frac{\sin h}{h} \geq \lim_{h \to 0^+} \cos h$$

$$1 \geq \lim_{h \to 0^+} \frac{\sin h}{h} \geq 1$$

(recall that $\lim_{h \to 0^+} \cos h = 1$ by direct substitution) and therefore, by the Squeeze Theorem,

$$\lim_{h \to 0^+} \frac{\sin h}{h} = 1$$

In a similar way we can show that the left-hand limit is equal to 1 as well. Thus, we have proven that

$$\lim_{h \to 0} \frac{\sin h}{h} = 1$$

To compute
$$\lim_{h \to 0} \frac{\cos h - 1}{h}$$
multiply both the numerator and the denominator by $\cos h + 1$:
$$\lim_{h \to 0} \frac{\cos h - 1}{h} = \lim_{h \to 0} \left(\frac{\cos h - 1}{h} \cdot \frac{\cos h + 1}{\cos h + 1} \right)$$
$$= \lim_{h \to 0} \frac{\cos^2 h - 1}{h(\cos h + 1)}$$
$$= \lim_{h \to 0} \frac{-\sin^2 h}{h(\cos h + 1)}$$

(we used the formula $\sin^2 h + \cos^2 h = 1$ to rewrite the numerator). Thus
$$\lim_{h \to 0} \frac{\cos h - 1}{h} = -\lim_{h \to 0} \left(\frac{\sin h}{h} \cdot \frac{\sin h}{\cos h + 1} \right)$$
$$= -\left(\lim_{h \to 0} \frac{\sin h}{h} \right) \left(\lim_{h \to 0} \frac{\sin h}{\cos h + 1} \right)$$
$$= -1 \cdot 0 = 0$$

Note that we used direct substitution to compute
$$\lim_{h \to 0} \frac{\sin h}{\cos h + 1} = \frac{\sin 0}{\cos 0 + 1} = \frac{0}{1} = 0$$

Derivatives of Inverse Trigonometric Functions

We use the same approach that we used to calculate the derivatives of $\ln x$ and $\log_a x$ in the previous section.

To calculate the derivative of $\arcsin x$ we start with
$$\sin(\arcsin x) = x$$
and differentiate both sides (using the chain rule on the left side)
$$\cos(\arcsin x) \cdot (\arcsin x)' = 1$$
$$(\arcsin x)' = \frac{1}{\cos(\arcsin x)}$$

Now we attempt to simplify the denominator.

Recall that from $\sin^2 A + \cos^2 A = 1$ we get that $\cos A = +\sqrt{1 - \sin^2 A}$, provided that $\cos A \geq 0$. In our calculation, we set $A = \arcsin x$ and (because $-\pi/2 \leq A \leq \pi/2$) it follows that $\cos A \geq 0$.

Thus
$$\frac{1}{\cos(\arcsin x)} = \frac{1}{\sqrt{1 - \sin^2(\arcsin x)}}$$
$$= \frac{1}{\sqrt{1 - (\sin(\arcsin x))^2}}$$
$$= \frac{1}{\sqrt{1 - x^2}}$$

and therefore
$$(\arcsin x)' = \frac{1}{\sqrt{1 - x^2}}$$

To calculate $(\arctan x)'$ we differentiate

$$\tan(\arctan x) = x$$

using the chain rule:

$$\sec^2(\arctan x) \cdot (\arctan x)' = 1$$

$$(\arctan x)' = \frac{1}{\sec^2(\arctan x)}$$

Apply the formula $\sec^2 A = 1 + \tan^2 A$ (see Section 2.3) with $A = \arctan x$:

$$(\arctan x)' = \frac{1}{\sec^2(\arctan x)}$$

$$= \frac{1}{1 + \tan^2(\arctan x)}$$

$$= \frac{1}{1 + (\tan(\arctan x))^2}$$

$$= \frac{1}{1 + x^2}$$

Since we will not need the remaining four inverse trigonometric functions, we do not derive formulas for their derivatives. (If we ever needed them, we would just have to adjust the calculations we have done for arcsin x and arctan x.)

Example 5.4.5 Derivatives Involving Inverse Trigonometric Functions

a. To compute the derivative of

$$F(x) = \arcsin(e^x)$$

we differentiate the outer function, arcsin, first:

$$F'(x) = \frac{1}{\sqrt{1 - (e^x)^2}}(e^x)'$$

$$= \frac{1}{\sqrt{1 - e^{2x}}} e^x$$

$$= \frac{e^x}{\sqrt{1 - e^{2x}}}$$

b. If

$$G(x) = (\arctan(5x - 2) + 6)^3$$

then

$$G'(x) = 3(\arctan(5x - 2) + 6)^2 (\arctan(5x - 2) + 6)'$$

$$= 3(\arctan(5x - 2) + 6)^2 \frac{1}{1 + (5x - 2)^2} \cdot 5$$

$$= \frac{15(\arctan(5x - 2) + 6)^2}{1 + (5x - 2)^2}$$

c. To compute the derivative of

$$H(x) = \ln(x^4 \arctan x)$$

we simplify first:

$$H(x) = \ln x^4 + \ln(\arctan x)$$
$$= 4 \ln x + \ln(\arctan x)$$

Thus,

$$H'(x) = 4\frac{1}{x} + \frac{1}{\arctan x} \cdot (\arctan x)'$$
$$= \frac{4}{x} + \frac{1}{(1+x^2)\arctan x}$$

Example 5.4.6 Model for World Population

In Section 2.3 we introduced the model

$$P(t) = 4.42857\left(\frac{\pi}{2} - \arctan\frac{2007-t}{42}\right)$$

where t represents a calendar year and $P(t)$ is the world population in billions.

To compute $P'(t)$ we use the chain rule:

$$P'(t) = 4.42857\frac{-1}{1+\left(\frac{2007-t}{42}\right)^2}\left(-\frac{1}{42}\right)$$
$$= \frac{0.10544}{1+\left(\frac{2007-t}{42}\right)^2}$$

For instance,

$$P'(2020) = \frac{0.10544}{1+\left(\frac{2007-2020}{42}\right)^2} \approx 0.0962$$

In words, the model predicts that in 2020, the total world population will be about 8.286 billion (Table 2.3.2) and will be increasing at a rate of about 0.0962 billion (96.2 million) people per year.

Summary To complete our collection of building blocks, we derived the derivatives of trigonometric and inverse trigonometric functions:

Function	Derivative
$\sin x$	$\cos x$
$\cos x$	$-\sin x$
$\tan x$	$\sec^2 x$
$\sec x$	$\sec x \tan x$
$\cot x$	$-\csc^2 x$
$\csc x$	$-\csc x \cot x$
$\arcsin x$	$\dfrac{1}{\sqrt{1-x^2}}$
$\arctan x$	$\dfrac{1}{1+x^2}$

5.4 Exercises

Mathematical Techniques

1–31 ■ Find the derivative of each function.

1. $f(x) = \sin x + 3\sec x - \tan x$
2. $g(x) = \cos x + \cos 1$
3. $g(x) = \sec 3x$
4. $f(x) = \sin x + \sin 2x + \sin^2 x$
5. $y = \cos(ax+b)$
6. $f(x) = x^2 \sin x$
7. $g(x) = x^2 \cos x$
8. $h(\theta) = \sin\theta \cos\theta$

9. $q(\theta) = \dfrac{\sin\theta}{1+\cos\theta}$

10. $f(x) = x^3 \cos x$

11. $h(x) = x^4 \sin(x^3)$

12. $f(x) = x\sec x + x\tan x$

13. $y = \dfrac{\cot x}{x - \pi}$

14. $f(x) = \ln(\sin x)$

15. $h(x) = \cos x^4 + \cos^4 x$

16. $g(x) = (2 + \sec x)^2$

17. $f(x) = \cos(\ln x) + \ln(\cos x)$

18. $f(x) = \sqrt{\sin x + 3}$

19. $g(x) = x^2 \sin\dfrac{1}{x}$

20. $F(z) = 3 + \cos(2z - 1)$

21. $f(x) = \ln(\sec x + \tan x)$

22. $g(u) = \dfrac{\sin u + \cos u}{\sin u - \cos u}$

23. $y = \sec\sqrt{t+1}$

24. $y = \tan(\tan(\sec x))$

25. $G(t) = 1 + 2\cos\left[\dfrac{2\pi}{5}(t-3)\right]$

26. $f(x) = e^{\cos x}$

27. $f(x) = \cos e^x$

28. $f(x) = e^{\sec x}$

29. $f(x) = 2^{3\tan x}$

30. $g(x) = \cos(3e^x - 2)$

▲ 31. $f(x) = \cot(x + 10^x)$

▼ 32–35 ▪ Use the definitions and the derivatives of $\sin x$ and $\cos x$ to check the derivatives of the other trigonometric functions.

32. $y = \tan x$

33. $y = \cot x$

34. $y = \sec x$

▲ 35. $y = \csc x$

▼ 36–37 ▪ Use the angle addition formulas (Equations 5.4.1 and 5.4.2) to find the derivative of each of the following. Compare the result with what you get with the chain rule.

36. $y = \cos 2x$. Simplify the answer in terms of $\sin 2x$.

▲ 37. $y = \sin 2x$. Simplify the answer in terms of $\cos 2x$.

38. Find the equation of the line that is tangent to the graph of the function $f(x) = 1 + 2\sin e^x$ at the point where $x = 0$.

39. Find the equation of the line that is tangent to the graph of the function $f(x) = \ln(\cos x)$ at the point where $x = \pi/4$.

40. Find all points where the graph of the function $f(x) = \sin 3x$ has a horizontal tangent line.

41. Take the derivative of $\cos(x + y)$ with respect to x, thinking of y as a constant. Simplify the answer in terms of $\sin(x + y)$.

42. Take the derivative of $\sin(x + y)$ with respect to x, thinking of y as a constant. Simplify the answer in terms of $\cos(x + y)$.

▼ 43–48 ▪ Compute the derivative of each function. Find the value of the function and the slope at 0, $\pi/2$, and π. Sketch the graph of the function on the given domain. How would you describe the behaviour?

43. $a(x) = 3x + \cos x$ for $0 \le x \le 2\pi$.

44. $b(y) = y^2 + 3\cos y$ for $0 \le y \le 2\pi$.

45. $c(z) = e^{-z}\sin z$ for $0 \le z \le 2\pi$.

46. $r(t) = t\cos t$ for $0 \le t \le 4\pi$.

47. $s(t) = e^{0.2t}\cos t$ for $0 \le t \le 40$.

▲ 48. $p(t) = e^t[1 + 0.2\cos t]$ for $0 \le t \le 40$.

▼ 49–56 ▪ Find the derivative of each function.

49. $f(x) = \arcsin 2x + 4$

50. $f(u) = \arctan(u - 1)$

51. $g(z) = z(\arctan z)^{-1}$

52. $g(x) = x^2 \arcsin^3 x$

53. $f(x) = 2\arcsin 3x + \arcsin^2 3x$

54. $y = (\arctan x^2 + 1)^3$

55. $F(t) = \dfrac{\arctan t}{t}$

▲ 56. $f(x) = \arcsin\sqrt[3]{x} + \sqrt[3]{\arcsin x}$

57. Show that $(\arccos x)' = \dfrac{-1}{\sqrt{1-x^2}}$. Hint: Imitate the derivation of the formula for the derivative of $\arcsin x$. You will need the fact that if $y = \arccos x$ then $y \in [0, \pi]$.

58. Show that $(\arcsec x)' = \dfrac{1}{x\sqrt{x^2-1}}$. You will need the fact that if $y = \arcsec x$ then $y \in [0, \pi/2) \cup [\pi, 3\pi/2)$.

59. Define $f(x) = \arcsin x + \arccos x$.

   a. Show that $f'(x) = 0$ for all x in $[-1, 1]$.

   b. From part (a), conclude that $\arcsin x + \arccos x = \pi/2$ for all x in $[-1, 1]$.

Applications

▼ 60–63 ▪ Find the derivative of each function. Sketch the graph and check that your derivative has the correct sign when the argument is equal to 0.

60. $f(x) = 3 + 4\cos\left(2\pi\dfrac{x-1}{5}\right)$

61. $g(t) = 4 + 3\cos[2\pi(t-5)]$

62. $h(z) = 1 + 5\cos\left(2\pi\dfrac{z-3}{4}\right)$

▲ 63. $W(y) = -2 + 3\cos\left(2\pi\dfrac{y+0.1}{0.2}\right)$

64–65 ■ Consider the combination of the temperature cycles (Section 2.3)

$$P_d(t) = 36.8 + 0.3 \cos[2\pi(t - 0.583)]$$

$$P_m(t) = 36.8 + 0.2 \cos\left[\frac{2\pi(t - 16)}{28}\right]$$

generating

$$P_t(t) = 36.8 + 0.2 \cos\left[\frac{2\pi(t - 16)}{28}\right] + 0.3 \cos[2\pi(t - 0.583)]$$

64. Find the derivative of P_t.

65. Sketch the graph of the derivative over a month. If you measured only the derivative, which oscillation would be easier to see?

66–71 ■ Suppose a bacterial population follows the equation

$$\frac{db}{dt} = 0.1[1 + \cos(0.8t)]b(t)$$

66. Describe what is happening.

67. Show that $b(t) = e^{0.1t + 0.125 \sin(0.8t)}$ is a solution.

68. Graph this solution for $0 \le t \le 20$.

69. Consider the more general equation

$$\frac{db}{dt} = 0.1[1 + A \cos(0.8t)]b(t)$$

where A is a constant. How does the constant A affect the population (compare small and large values)?

70. Show that $b(t) = e^{0.1t + 0.125A \sin(0.8t)}$ is a solution.

71. Graph the solution in Exercise 70 for $A = 0$, $A = 0.5$, and $A = 2$.

72–73 ■ Use trigonometric derivatives to study the following.

72. In London, U.K., the number of hours of daylight follows roughly

$$L(t) = 12 - 4.5 \cos t$$

where t represents time measured in units, 2π corresponds to one year, and the shortest day is December 21. A plant puts out leaves in the spring in response to the *change* in day length.

  a. Find the value of t that produces the longest day. What day is this? How many hours of daylight does it have?

  b. Find the smallest value of t that produces a day of average length. What day is this? How many hours of daylight does it have?

  c. Find the rate of change of day length. When is this zero?

  d. At what time of year would it be easiest for the plant to detect changes in day length?

73. Blood flow is pulsatile. Suppose the blood flow along the artery of a whale is given by

$$F(t) = 212 \cos\left(\frac{2\pi t}{10}\right)$$

where F is measured in litres per second and t is measured in seconds.

  a. Find the average flow and the amplitude.

  b. Find the period of this flow. How many heartbeats does this whale have per minute?

  c. When is the flow zero? What is happening at these times?

  d. Find the rate of change of the flow. What does it mean when this is zero? What is the flow at these times?

Computer Exercise

74. Find the derivatives of the following functions. Where are the critical points? What happens as more and more cosines are piled up? Explain this in terms of the discrete-time dynamical system

$$x_{t+1} = \cos(x_t)$$

  a. $\cos[\cos x]$
  b. $\cos\{\cos[\cos x]\}$
  c. $\cos(\cos\{\cos[\cos x]\})$
  d. $\cos[\cos(\cos\{\cos[\cos x]\})]$

5.5 Implicit Differentiation, Logarithmic Differentiation, and Related Rates

In this section we develop useful techniques for calculating derivatives of functions. In order to differentiate a function defined by an equation, we use **implicit differentiation.** As a special case, **logarithmic differentiation** helps us differentiate functions whose derivatives cannot be found using the rules we have covered so far. The **related rates** technique enables us to discover relationships between the rates of change of various quantities.

Implicit Differentiation

All functions we have worked with so far have the same form—the dependent variable is expressed *explicitly* in terms of the independent variable, for instance, $f(x) = 24x^3$, $y = \ln(3x - 7)$, and $g(t) = \sin t - 3\cos 4t$. This is not the only way to describe functions—sometimes the relationship between the variables can be expressed *implicitly*, in the form of an equation, as in $x^2 + y^2 = 4$, $y^2 e^y = 2x^3$, and $x^3 + y^4 = y \sin x - 7$. (Unless stated otherwise, we assume that y is the dependent variable and x is the independent variable.)

In some cases, we can solve the given equation, thus expressing the function in explicit form. For example, the equation

$$3x + 4y - 5 = 0$$

(which defines the function $y = f(x)$ implicitly) can be written explicitly as

$$y = -\frac{3}{4}x + \frac{5}{4} \quad \text{or} \quad f(x) = -\frac{3}{4}x + \frac{5}{4}$$

The equation $x^2 + y^2 = 4$ (whose graph is the circle of radius two centred at the origin) determines the two functions

$$y = \sqrt{4 - x^2}$$

(upper semicircle) and

$$y = -\sqrt{4 - x^2}$$

(lower semicircle); see Figure 5.5.29.

In general, solving an equation can be difficult or impossible. The equation

$$x^3 + y^4 = y \sin x - 7$$

can be solved for y using the formula for the solution of a fourth-degree equation, but it results in a fairly complicated formula. It is not possible to solve the equation $y^2 e^y = 2x^3$ for y.

How do we differentiate functions that are given implicitly? It turns out that we do not have to solve the given equation for y in order to differentiate. Instead, we differentiate the equation as is and use the chain rule every time we have to differentiate the term that contains the dependent variable (as we have already done in Examples 5.3.7 and 5.3.8). This form of differentiation is referred to as **implicit differentiation**.

For example, differentiating the equation $x^2 + y^2 = 4$, we get

$$\frac{d}{dx}(x^2) + \frac{d}{dx}(y^2) = \frac{d}{dx}(4)$$

$$2x + 2y\frac{dy}{dx} = 0$$

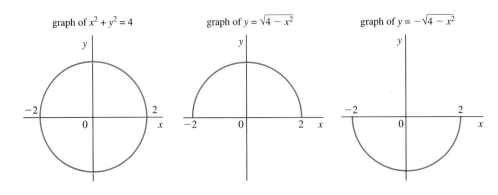

Figure 5.5.29

Equation $x^2 + y^2 = 4$ and functions $y = \pm\sqrt{4 - x^2}$

To finish the calculation, we solve for the derivative:
$$\frac{dy}{dx} = -\frac{2x}{2y} = -\frac{x}{y}$$

Now we can work with it as with any derivative. For example, the slope of the tangent line to the circle $x^2 + y^2 = 4$ at the point $(\sqrt{3}, 1)$ is
$$\frac{dy}{dx}(\sqrt{3}, 1) = -\frac{\sqrt{3}}{1} = -\sqrt{3}$$

and its equation is
$$y - 1 = -\sqrt{3}(x - \sqrt{3})$$
$$y = -\sqrt{3}x + 4$$

Example 5.5.1 Implicit Differentiation I

Find y' if $y^2 e^y = 2x^3$.

We differentiate the given equation using the product rule:
$$(y^2)' e^y + y^2 (e^y)' = (2x^3)'$$
$$2yy' e^y + y^2 e^y y' = 6x^2$$

Note that $(y^2)' = 2y \cdot y'$ and $(e^y)' = e^y \cdot y'$ by the chain rule. Now we solve for y':
$$y'(2ye^y + y^2 e^y) = 6x^2$$
$$y' = \frac{6x^2}{2ye^y + y^2 e^y} = \frac{6x^2}{ye^y(2 + y)}$$

Example 5.5.2 Implicit Differentiation II

Find dy/dx if $x^3 + y^4 = y \sin x - 7$.

Differentiating both sides (again, using the chain rule every time we differentiate a term containing y) and then solving for dy/dx, we get
$$\frac{d}{dx}(x^3) + \frac{d}{dx}(y^4) = \frac{d}{dx}(y \sin x - 7)$$
$$3x^2 + 4y^3 \frac{dy}{dx} = \frac{d}{dx}(y) \sin x + y \frac{d}{dx}(\sin x) - 0$$
$$3x^2 + 4y^3 \frac{dy}{dx} = \frac{dy}{dx} \sin x + y \cos x$$
$$4y^3 \frac{dy}{dx} - \sin x \frac{dy}{dx} = y \cos x - 3x^2$$
$$\frac{dy}{dx} = \frac{y \cos x - 3x^2}{4y^3 - \sin x}$$

Example 5.5.3 Investigating Population Change

A population $P(t)$ has been determined to satisfy the equation
$$2.8^{P(t)} + 3P(t) - 7t = 4t^3$$

where t represents time. Show that $P(t)$ is increasing for all time.

We show that the derivative of $P(t)$ is positive. Using implicit differentiation, we get
$$2.8^{P(t)} \ln 2.8 \cdot P'(t) + 3P'(t) - 7 = 12t^2$$

and so
$$P'(t) = \frac{12t^2 + 7}{2.8^{P(t)} \ln 2.8 + 3}$$

The numerator, $12t^2 + 7$, is positive for all t. Since $\ln 2.8 > 0$ and $2.8^{P(t)} > 0$ for all $P(t)$, the denominator is positive as well, and therefore $P'(t) > 0$ for all t.

Using implicit differentiation, we can compute derivatives of inverse functions. Consider the following two examples.

Example 5.5.4 Computing Derivatives of $\ln x$ and $\arctan x$

By writing $y = \ln x$ as $e^y = x$, find y'. Repeat the same strategy to find the derivative of $y = \arctan x$.

Differentiating $e^y = x$ implicitly, we obtain

$$e^y y' = 1$$
$$y' = \frac{1}{e^y} = \frac{1}{x}$$

In the last step we used the fact that $e^y = x$. Thus, $(\ln x)' = 1/x$.

Likewise, we write $y = \arctan x$ as $\tan y = x$, differentiate implicitly, and then express the derivative in terms of x:

$$\sec^2 y \, y' = 1$$
$$y' = \frac{1}{\sec^2 y} = \frac{1}{1 + \tan^2 y} = \frac{1}{1 + x^2}$$

Note that we used the identity $1 + \tan^2 y = \sec^2 y$ (see Section 2.3).

Example 5.5.5 Computing the Derivative of an Inverse Function

The equation $3x^2 e^y + 2x = x^3 - y$ defines $y = f(x)$ implicitly as a function of x. Find the derivative of its *inverse* function $x = f^{-1}(y)$.

Think of the given equation not as defining $y = f(x)$ implicitly as a function of x, but as defining $x = g(y)$ implicitly as a function of y; this means that g is the inverse function of f. By implicitly differentiating $3x^2 e^y + 2x = x^3 - y$, we obtain

$$6xx'e^y + 3x^2 e^y + 2x' = 3x^2 x' - 1$$
$$x'\left(6xe^y + 2 - 3x^2\right) = -1 - 3x^2 e^y$$
$$x' = \frac{-1 - 3x^2 e^y}{6xe^y + 2 - 3x^2}$$

Logarithmic Differentiation

How do we find the derivative of $y = x^x$? We cannot apply the power rule, $(x^n)' = nx^{n-1}$, because the exponent needs to be constant. Nor can we use the formula for the derivative of the exponential function, $(a^x)' = a^x \ln a$, since the base in x^x is not constant.

We cannot use the formulas—so first we simplify $y = x^x$ by taking logarithms

$$\ln y = \ln x^x = x \ln x$$

and then use implicit differentiation (with the product rule on the right side):

$$\frac{1}{y} y' = 1 \cdot \ln x + x \cdot \frac{1}{x}$$
$$y' = y (\ln x + 1)$$
$$y' = x^x (\ln x + 1)$$

Since we know that $y = x^x$, in the last step we replaced y by x^x (that's something we cannot always do with implicitly defined functions). This method of calculating the derivative of a function is called **logarithmic differentiation.**

Now we know how to differentiate all functions that involve expressions of the form A^B; for instance, the derivative of

$$f(x) = 3^4 + x^4 + 3^x + x^x$$

is
$$f'(x) = 0 + 4x^3 + 3^x \ln 3 + x^x (\ln x + 1) = 4x^3 + 3^x \ln 3 + x^x (\ln x + 1)$$

Example 5.5.6 Logarithmic Differentiation I

Find the derivative of the function $y = (x + e^x)^{3x^2 - 1}$.

Taking the logarithms of both sides and simplifying, we get
$$\ln y = \ln(x + e^x)^{3x^2 - 1} = (3x^2 - 1) \ln(x + e^x)$$

We proceed by implicit differentiation (with the product and chain rules on the right side):
$$\frac{1}{y} y' = 6x \ln(x + e^x) + (3x^2 - 1) \frac{1}{x + e^x} (1 + e^x)$$
$$y' = y \left(6x \ln(x + e^x) + \frac{(3x^2 - 1)(1 + e^x)}{x + e^x} \right)$$
$$y' = (x + e^x)^{3x^2 - 1} \left(6x \ln(x + e^x) + \frac{(3x^2 - 1)(1 + e^x)}{x + e^x} \right)$$

Whenever the function we need to differentiate contains products, quotients, or powers, we can use logarithmic differentiation, as the following example illustrates.

Example 5.5.7 Logarithmic Differentiation II

Find the derivative of the function $y = \dfrac{5^{3x} \sqrt{x + 4}}{(x^2 - 1)^3}$.

We simplify using the laws of logarithms:
$$y = \frac{5^{3x} \sqrt{x + 4}}{(x^2 - 1)^3}$$
$$\ln y = \ln \left(5^{3x} \sqrt{x + 4} \right) - \ln \left((x^2 - 1)^3 \right)$$
$$\ln y = \ln 5^{3x} + \ln \sqrt{x + 4} - 3 \ln(x^2 - 1)$$
$$\ln y = 3x \ln 5 + \frac{1}{2} \ln(x + 4) - 3 \ln(x^2 - 1)$$

Next, we differentiate implicitly:
$$\frac{1}{y} y' = 3 \ln 5 + \frac{1}{2} \frac{1}{x + 4} - 3 \frac{1}{x^2 - 1} 2x$$

and solve for y':
$$y' = y \left(3 \ln 5 + \frac{1}{2(x + 4)} - \frac{6x}{x^2 - 1} \right)$$
$$y' = \frac{5^{3x} \sqrt{x + 4}}{(x^2 - 1)^3} \left(3 \ln 5 + \frac{1}{2(x + 4)} - \frac{6x}{x^2 - 1} \right)$$

Logarithmic differentiation is introduced for convenience—we can do without it by rewriting the function using $A = e^{\ln A}$ and with the usual differentiation rules. For instance, to find the derivative of $y = x^x$, we write
$$y = x^x = e^{\ln x^x} = e^{x \ln x}$$

and then use the chain rule
$$y' = e^{x \ln x} (x \ln x)' = e^{x \ln x} \left(\ln x + x \frac{1}{x} \right)$$
$$y' = e^{x \ln x} (\ln x + 1)$$

$$y' = x^x (\ln x + 1)$$

Related Rates

To study how the buildup of the fluid inside the skull (hydrocephalus) affects the pressure within and around a growing tumour, researchers model the tumour as a sphere. As the tumour grows, both its volume and radius change, and so in the formula for the volume of a sphere,

$$V = \frac{4}{3}\pi r^3 \qquad (5.5.1)$$

both V and r are functions of time, t. (To keep notation simple, we use V and r instead of $V(t)$ and $r(t)$.) By calculating the derivative of Equation 5.5.1 with respect to t we will obtain the relation between the rate of increase of the volume and the rate of increase of the radius ("related rates").

In this context, using the prime notation V' for the derivative is ambiguous (is it the derivative of V with respect to r or with respect to t?) and so we use differential notation. Implicitly differentiating Equation 5.5.1 by using the chain rule on the right side, we obtain

$$\frac{dV}{dt} = \frac{4}{3}\pi(3r^2)\frac{dr}{dt} = 4\pi r^2 \frac{dr}{dt}$$

Assume that we have determined (say, by taking consecutive MRI scans) that when the radius of the tumour reaches $r = 0.4$ cm, it grows at a rate of $dr/dt = 0.005$ cm/month. Then

$$\frac{dV}{dt} = 4\pi(0.4)^2(0.005) \approx 0.01 \text{ cm}^3/\text{month}$$

Assuming that the density of the tumour is $\rho = 2.05 \cdot 10^9$ cells/cm^3, we compute that the tumour grows at a rate of

$$\frac{dV}{dt}\rho \approx 0.01 \, (2.05 \cdot 10^9) \approx 2 \cdot 10^7$$

(i.e., about 20 million) cells/month.

Now we repeat what we have just done in more general terms. Sometimes, what we know is an equation (such as Equation 5.5.1) defining a relation between quantities that depend on time (in rare cases it is some other variable). By using implicit differentiation with respect to time, we obtain the relationship between the rates of change of these quantities—hence **related rates.** To illustrate this idea further, we work on several related rates problems. We start with two classical problems and then discuss several applications. (More related rates questions can be found among the exercises at the end of this section.)

Example 5.5.8 Ripples on the Surface of a Lake

After a rock is dropped into a lake, circular ripples start moving away from the point on the surface where the rock hit it. How fast is the area of a circular ripple increasing at the moment when its radius is 2.4 m and is increasing at a rate of 0.3 m/s?

The area, A, inside the circular ripple of radius r is $A = \pi r^2$. Both A and r increase with time, and therefore the implicit differentiation gives

$$\frac{dA}{dt} = \pi(2r)\frac{dr}{dt} = 2\pi r \frac{dr}{dt}$$

We know that when $r = 2.4$ m, the radius is increasing at a rate of 0.3 m/s; thus, $dr/dt = 0.3$. Substituting into the formula for dA/dt, we obtain

$$\frac{dA}{dt} = 2\pi(2.4)(0.3) = 1.44\pi \approx 4.52$$

Thus, the area of the ripple increases by approximately 4.52 m²/s.

Example 5.5.9 Rectangle with Changing Sides

Imagine a rectangle whose length and width change in the following way: at the moment when the length reaches 4 m, it is decreasing at a rate of 0.3 m/s; at the same moment, the width of the rectangle reaches 3 m and is increasing at a rate of 0.2 m/s. Determine whether the area of the rectangle is increasing or decreasing at that moment.

Denote by l the length and by w the width of the rectangle. In the formula $A = lw$ for its area, all three quantities change with time. Using implicit differentiation and the product rule on the right side, we find

$$\frac{dA}{dt} = \frac{dl}{dt}w + l\frac{dw}{dt}$$

It is given that $l = 4$ m, $dl/dt = -0.3$ m/s, $w = 3$ m, and $dw/dt = 0.2$ m/s. Thus,

$$\frac{dA}{dt} = (-0.3)(3) + (4)(0.2) = -0.1$$

Thus, at the moment when $l = 4$ m and $w = 3$ m, the area of the rectangle is decreasing at a rate of 0.1 m²/s.

Example 5.5.10 Relation between Skull Length and Spine Length in Dinosaurs

The equation (introduced in Section 1.1)

$$\text{Sk} = 0.49\,\text{Sp}^{0.84} \tag{5.5.2}$$

relates the skull length, Sk, of a larger dinosaur to its spine length, Sp (both measured in metres). Find the rate at which the skull length is growing if it is known that when the spine length is 4.2 m, it is growing at a rate of 0.3 m/year.

Viewing both Sk and Sp as functions of time t, and implicitly differentiating Equation 5.5.2, we obtain

$$\frac{d\text{Sk}}{dt} = 0.49\,(0.84)\text{Sp}^{0.84-1}\frac{d\text{Sp}}{dt} = 0.4116\,\text{Sp}^{-0.16}\frac{d\text{Sp}}{dt}$$

When Sp = 4.2 m and $d\text{Sp}/dt = 0.3$ m/year,

$$\frac{d\text{Sk}}{dt} = 0.4116\,(4.2)^{-0.16}(0.3) \approx 0.0981$$

Thus, at the moment when the spine of a dinosaur reaches a length of 4.2 m, its skull is increasing in length by approximately 0.0981 m per year.

Note In the previous example we derived the relation (now we use prime notation for the derivatives with respect to time because we need to save fractions for something else)

$$\text{Sk}' = 0.4116\,\text{Sp}^{-0.16}\,\text{Sp}'$$

Dividing this relation by $\text{Sk} = 0.49\,\text{Sp}^{0.84}$, we obtain

$$\frac{\text{Sk}'}{\text{Sk}} = \frac{0.4116\,\text{Sp}^{-0.16}\,\text{Sp}'}{0.49\,\text{Sp}^{0.84}}$$

$$\frac{\text{Sk}'}{\text{Sk}} = 0.84\frac{\text{Sp}'}{\text{Sp}}$$

Comparing the relative rates of increase, we conclude that the skull grows slower than the spine.

Example 5.5.11 Spread of a Pollutant

The concentration of a pollutant (expressed as mass per unit volume, so its units could be grams per cubic metre) at a location x metres away from the source is given by

$$c(x) = 0.28\, e^{-0.25x^2}$$

(this is a special solution of the diffusion equation). An observer is located 5 m from the source. How does the concentration change as she runs away from the source at a speed of 4.8 m/s?

Differentiating the given equation with respect to time, we obtain

$$\frac{dc}{dt} = 0.28\, e^{-0.25x^2}(-0.25)(2x)\frac{dx}{dt} = -0.14\, x\, e^{-0.25x^2}\frac{dx}{dt}$$

It is given that when $x = 5$ m, the rate of change of x (i.e., the speed) is $dx/dt = 4.8$ m/s. Thus,

$$\frac{dc}{dt} = -0.14\,(5)\, e^{-0.25\,(5)^2}(4.8) = -3.36\, e^{-6.25} \approx -0.00648$$

Thus, the concentration is decreasing (the derivative dc/dt is negative!) at a rate of approximately 0.00648 g/m³ per second.

Example 5.5.12 Metabolic Rate

Metabolic rate is defined as the amount of energy used by an organism in a given time interval. For mammals, it can be modelled by the formula

$$E = 72\, M^{0.75}$$

where M is mass in kilograms and the metabolic rate, E, is measured in kilocalories per day. Although most researchers agree that E is proportional to $M^{0.75}$, there is no agreement on the value of the constant of proportionality; the major reason for this is that two different tissue types of the same mass use different amounts of energy. If we consider two different tissues of masses M_1 and M_2, then the metabolic rate is expressed as

$$E = 140\, M_1^{0.75} + 72\, M_2^{0.75}$$

Note that the formula states that the tissue of mass M_1 uses about twice the energy of the tissue of mass M_2.

Assume that when $M_1 = 5$ kg and $M_2 = 30$ kg, the tissue growth rates are 0.02 kg/day and 0.018 kg/day, respectively. How fast is the metabolic rate, E, growing?

Differentiating the formula for E, we obtain

$$\frac{dE}{dt} = 140(0.75)M_1^{-0.25}\frac{dM_1}{dt} + 72\,(0.75)\, M_2^{-0.25}\frac{dM_2}{dt}$$

$$= 105\, M_1^{-0.25}\frac{dM_1}{dt} + 54\, M_2^{-0.25}\frac{dM_2}{dt}$$

We know that for the first tissue, $M_1 = 5$ kg and $dM_1/dt = 0.02$ kg/day; for the second tissue, $M_2 = 30$ kg and $dM_2/dt = 0.018$ kg/day. Thus,

$$\frac{dE}{dt} = 105(5)^{-0.25}(0.02) + 54\,(30)^{-0.25}(0.018) \approx 1.82$$

So the change in the metabolic rate is approximately 1.82 kcal/day per day.

Summary We used **implicit differentiation** to find derivatives of functions defined by equations. To differentiate complicated expressions, we sometimes use **logarithmic differentiation**. The relation between the rates of change of two or more functions can be found using the **related rates** method.

5.5 Exercises

1–7 ■ Use implicit differentiation to calculate y'.

1. $x^3 + 2y^{10} = 4y$
2. $e^{4x}y^4 - \sqrt{y} = 3x$
3. $ye^x = xe^y$
4. $\cos x \sin y = 2 - \sin x$
5. $\ln(xy + 1) - y = 0$
6. $3^{xy^2} = 2x - y - 3$
7. $\tan y \sec x = 3x$
8. If $f(x) + e^{f(x)} = 4x$ and $f(0) = 2$, find $f'(0)$.
9. If $x^3y^3 - 3y^4 = 3$ and $y(1) = 13$, find $y'(1)$.
10. Find the equation of the line tangent to the circle given by $x^2 + y^2 = 6$ at the point in the fourth quadrant where $x = 1$.
11. Find the equation of the line tangent to the curve given by $(1 + x^2 + y^2)^2 + y = 2e^x + 3$ at the point $(0, 1)$.

12–19 ■ Find the derivative of each function using logarithmic differentiation.

12. $y = 7x^5$
13. $y = 6x^3 \, 5^{x^2}$
14. $f(x) = x^{3x-4}$
15. $g(x) = \sin x^{\tan x}$
16. $y = 2^{x^x + 1}$
17. $f(x) = \left(1 + \dfrac{1}{x}\right)^x$
18. $h(x) = \dfrac{(x+1)e^x}{x^3 \sin x}$
19. $y = \sqrt{\dfrac{x^4 3^x}{(x^2 + 5)^3}}$

20. We showed that the derivative of $y = x^x$ is $y' = x^x(\ln x + 1)$. Compute the derivative of x^x in two incorrect ways: by assuming that the base, x, is constant (i.e., by using the derivative of an exponential function) and then by assuming that the exponent, x, is constant (i.e., by using the power rule). Add up the two expressions and convince yourself that this is the correct answer for y'. See the next exercise.

21. Let $y = f(x)^{g(x)}$.
    a. Use logarithmic differentiation to find y'.
    b. Compute the derivative of y by assuming that the base, $f(x)$, is constant and then by assuming that the exponent, $g(x)$, is constant. Add up the two expressions and show that it is equal to the expression for y' you found in (a).

22–25 ■ Find the relation between the relative rates of change of the two quantities.

22. $P(t) = 4.5Q(t)$
23. $P(t) = Q(t)^{0.7}$
24. $B(t) = 3m(t)^{-2}$
25. $y(t) = 4.7p(t)^4$

Applications

26. Assume that a population of fish satisfies $7P(t) - t^2 + \ln P(t) = 3 + t$. Find $P'(t)$ and explain why the fish population is increasing for $t > 0$.

27. The metabolic rate of larger mammals is given by $E = 120.7 M^{0.74}$, where E is measured in kilocalories per day and the mass, M, is in kilograms. Find and interpret dE/dt for a 45-kg dog at the moment when it is gaining weight at a rate of 0.2 kg/day.

28. Another model for the metabolic rate of larger mammals claims that $E = 118 M^{0.76}$, where E is measured in kilocalories per day and the mass, M, is in kilograms. Find and interpret dE/dt for a 45-kg dog at the moment when it is gaining weight at a rate of 0.2 kg/day.

29. Two quantities, $P(t)$ and $Q(t)$, are in the allometric relation $P(t) = aQ(t)^b$. Show that the relative rates of change of $P(t)$ and $Q(t)$ are proportional to each other.

30. The quantities M and N, both dependent on time, t, are related by the equation $M^{2/3} + 4N^{2/3} = N$. What is the rate of change dM/dt when $M = 8$, $N = 1$, and $dN/dt = -2$?

31. The quantities A and P, both dependent on time, t, are related by the equation $\ln A - \ln(P^2 + 1) = A + P$. What is the rate of change dP/dt when $A = 4$, $P = 1$, and $dA/dt = 1$?

32. The **DuBois and DuBois formula**, $S = 0.2m^{0.425}h^{0.725}$, gives a good approximation of the body surface area of a human of height h (metres) and mass m (kilograms). Assume that a person weighs 65 kg and is 1.6 m tall. How fast is his surface area changing at the moment when he is gaining weight at a rate of 1.5 kg/month? Assume that his height is not changing.

33. Consider the formula $S = 0.2m^{0.425}h^{0.725}$ for the body surface area of a human of height h (metres) and mass m (kilograms). Assume that a person weighs 70 kg and is 1.8 m tall. How fast is her surface area changing at the moment when she is losing weight at a rate of 0.9 kg/month? Assume that her height is not changing.

34. In 2001, Environment Canada started calculating the wind chill based on the function $W = 13.12 + 0.6215T - 11.37v^{0.16} + 0.3965Tv^{0.16}$, where T denotes the air temperature (in degrees Celsius) and v is the wind speed (in kilometres per hour), measured at 10 m above the ground. Assume that $v = 35$ km/h and $T = -10°$C. How fast is the wind chill changing if the wind speed is increasing at a rate of 3 km/h per hour and the temperature is decreasing at a rate of 1°C per hour?

35. Consider the wind chill index $W = 13.12 + 0.6215T - 11.37v^{0.16} + 0.3965Tv^{0.16}$, where T is the air temperature (in degrees Celsius) and v is the wind speed (in kilometres per hour). Assume that, at the moment when $v = 35$ km/h and $T = -10°C$, the wind speed is decreasing at a rate of 2 km/h per hour and the temperature is decreasing at a rate of 4°C/h. How fast is the wind chill changing?

36. At the moment when the base of a triangle reaches 2.1 cm, it is decreasing at a rate of 0.3 cm/s; at the same moment, the height of the triangle reaches 2.9 cm and is increasing at a rate of 0.5 cm/s. Determine whether the area of the triangle is increasing or decreasing.

37. One side of a triangle is fixed at 4 m in length, and one is of a variable length of b metres. How fast is the area of the triangle changing at the moment when $b = 8$ m, $db/dt = 2$ m/h, and the angle between the two sides is $\pi/6$ and is changing at a rate of 0.1 rad/h?

38–40 ▪ A 12-m-long ladder leans against a vertical wall (see the figure below). Answer each question.

38. At the moment when the bottom of the ladder is 5 m from the wall, the ladder is being pulled away from the wall at a rate of 0.5 m/s. How fast is the top of the ladder sliding down the wall?

39. At the moment when the bottom of the ladder is 4 m from the wall, the ladder is being pushed toward the wall so that its top travels at a speed of 0.3 m/s. How fast is the bottom of the ladder travelling toward the wall?

40. Determine the rate of change of the angle between the ladder and the floor at the moment when the ladder is 3 m from the wall and is being pulled away from it at a rate of 0.1 m/s.

41. At the moment when the radius of a cylinder reaches 3.5 cm and its height is 12.4 cm, the radius is decreasing at a rate of 1.2 cm/h and the height is increasing at a rate of 0.4 cm/h. Determine whether the volume of the cylinder is increasing or decreasing, and at what rate.

42. According to Poiseuille's law, the velocity, $v(r)$, of blood flowing through a vessel (assumed to be cylinder-shaped with flexible walls) is given by $v(r) = K(R^2 - r^2)$, where R is the radius of the vessel and r is the distance from the centre of the vessel (see the figure below). The positive constant K is given by $K = P/4\mu L$, where L is the length of the vessel and P is the pressure difference at its ends.

Assume that at the moment when $R = 0.15$ mm, the vessel is contracting (for instance due to exposure to low temperatures) at a rate of 0.03 mm/min. How fast is the velocity of the blood changing? (Your answer will contain the constant K.)

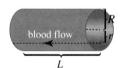

5.6 The Second Derivative, Curvature, and Concavity

A function with a positive derivative is increasing and a function with a negative derivative is decreasing. In this section, we extend the graphical interpretation of the derivative by examining the derivative of the derivative, or the **second derivative.** In particular, we will see that the second derivative tells whether the graph of a function curves upward (**concave up**) or downward (**concave down**). As done in previous sections, to enhance geometric intuition, we use the information obtained from the derivatives of a function to sketch its graph. In Section 6.5 we summarize all relevant information and draw more graphs of functions. Furthermore, just as the first derivative of position is the **velocity,** the second derivative of position is the **acceleration.**

The Second Derivative

Consider the two graphs in Figure 5.6.30. Both show functions that are increasing, but in different ways. In the first, the slope becomes steeper and steeper, indicating that the function is increasing more and more quickly. In the second, the slope becomes smaller and smaller. Although this function is increasing, it does so at a decreasing rate. These differences are clear on graphs of the derivatives of the two functions. Each derivative is positive because the functions are increasing. The derivative of the first function is increasing (Figure 5.6.31a) because the graph gets steeper and steeper. The derivative of the second function is decreasing (Figure 5.6.31b) because the graph becomes less and less steep.

5.6 The Second Derivative, Curvature, and Concavity

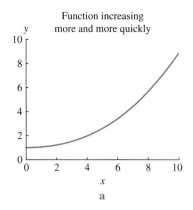

FIGURE 5.6.30
Two different increasing functions

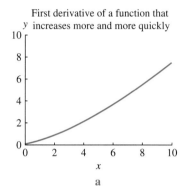

FIGURE 5.6.31
Derivatives of two increasing functions

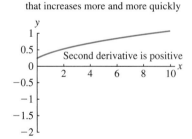

FIGURE 5.6.32
Second derivatives of two increasing functions

The derivative of any function is positive when the function is increasing and negative when the function is decreasing. We can apply this observation to the derivative itself. Because the **derivative** of the first function is increasing, the **derivative of the derivative** must be positive. Similarly, because the derivative of the second function is decreasing, the **derivative of the derivative** must be negative (Figure 5.6.32).

Definition 5.6.1 The Second Derivative

The derivative of the derivative is called the **second derivative**. We write the second derivative of f in prime notation as follows:

$$\text{the second derivative of } f = f''(x) \tag{5.6.1}$$

In differential notation,

$$\text{the second derivative of } f = \frac{d^2 f}{dx^2} \tag{5.6.2}$$

For clarity, the derivative itself is often called the **first derivative**.

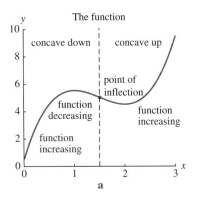 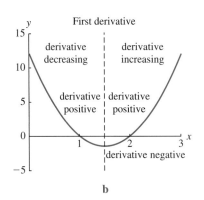 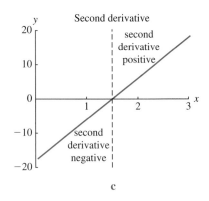

FIGURE 5.6.33

A function, its derivative, and its second derivative

In general, the result of taking n derivatives is written as follows:

$$\text{the } n\text{th derivative of } f = f^{(n)}(x) \quad \text{prime notation}$$

$$= \frac{d^n f}{dx^n} \quad \text{differential notation}$$

For a small number of derivatives, we still use ′ and write f' instead of $f^{(1)}$ and f'' instead of $f^{(2)}$. For higher-order derivatives, the $f^{(n)}$ notation is a lot more convenient.

The second derivative has a general interpretation in terms of the **curvature** of the graph. Consider the function shown in Figure 5.6.33a. There are many ways to describe this graph. First of all, the graph of the function itself is positive, meaning that the associated measurement takes on only positive values. Furthermore, the function is increasing for $x < 1$, decreasing for $1 < x < 2$, and increasing for $x > 2$. The derivative, therefore, is positive for $x < 1$, negative for $1 < x < 2$, and positive for $x > 2$ (Figure 5.6.33b).

With more careful examination, we can see that the graph breaks into two regions of **curvature**. Between $x = 0$ and $x = 1.5$, the graph curves downward (Figure 5.6.33a). This portion of the graph is said to be **concave down.** The slope of the curve is a **decreasing** function in this region (Figure 5.6.33b), implying that the **second derivative is negative** (Figure 5.6.33c). Between $x = 1.5$ and $x = 3$, the graph curves upward (Figure 5.6.33a). This portion of the graph is said to be **concave up.** In this region, the slope of the curve becomes steeper and steeper, the derivative is increasing (Figure 5.6.33b), and the **second derivative is positive** (Figure 5.6.33c).

At a point where the **second derivative** is zero, the curvature of the graph can change. If the second derivative changes sign (from positive to negative or from negative to positive), such a point is called a **point of inflection.** In Figure 5.6.33a, the graph has a point of inflection at $x = 1.5$, where the function switches from concave down to concave up and the second derivative switches from negative to positive.

To summarize, assume that $f''(x)$ is continuous:

f Is Concave Up on an Interval (c, d)
The graph of f curves (bends) upward on (c, d).
The first derivative, f', is increasing on (c, d).
The second derivative, f'', is positive on (c, d).

f Is Concave Down on an Interval (c, d)
The graph of f curves (bends) downward on (c, d).
The first derivative, f', is decreasing on (c, d).
The second derivative, f'', is negative on (c, d).

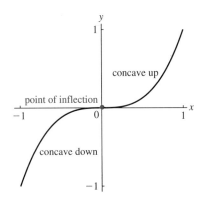

FIGURE 5.6.34

The function $f(x) = x^3$ has a point of inflection at a critical point

It is possible to define concavity for functions that are not differentiable. Since this general definition will not be needed for our purposes, we do not discuss it here.

5.6 The Second Derivative, Curvature, and Concavity

Example 5.6.1 Graphing a Power Function with the Second Derivative

Consider the power function $f(x) = x^3$ (Figure 5.6.34). Using the power rule, $f'(x) = 3x^2$ and $f''(x) = 6x$. Therefore, this function has both derivative and second derivative equal to zero at the same point, $x = 0$. Note that $f'(x) = 3x^2 > 0$ for all $x \neq 0$. Thus, $f(x)$ is increasing on $(-\infty, 0)$ and $(0, \infty)$. (Since $f(x)$ is defined at $x = 0$, we can say that $f(x)$ is increasing on $(-\infty, \infty)$.)

The second derivative, $f''(x) = 6x$, is positive when $x > 0$ and negative when $x < 0$. Thus, the graph of $f(x)$ is concave down on $(-\infty, 0)$ and concave up on $(0, \infty)$. Because the concavity changes at $x = 0$, the point $(0, 0)$ is an inflection point.

Example 5.6.2 A Point with Second Derivative Equal to Zero Does Not Have to Yield a Point of Inflection

For the power function $f(x) = x^4$ (Figure 5.6.35) we compute $f'(x) = 4x^3$ and $f''(x) = 12x^2$. Both the first and the second derivatives are zero at $x = 0$. However, the second derivative is positive for $x < 0$ and also for $x > 0$. Because the function is concave up on both sides of $x = 0$, this is not a point of inflection. See Figure 5.6.35.

Example 5.6.3 Analyzing a Function Using Derivatives

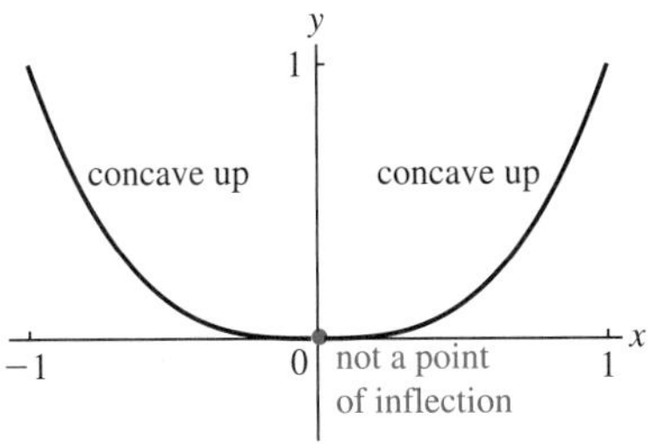

Consider the quadratic function

$$f(x) = 3x^2 - 6x + 5$$

The first and second derivatives are

$$f'(x) = 6x - 6 \text{ and } f''(x) = 6$$

Since $f''(x) > 0$ for all x, the graph of $f(x)$ is concave up for all x.

From

$$f'(x) = 6x - 6 = 6(x - 1) = 0$$

we get $x = 1$. Thus, $x = 1$ is a critical number of f.

Since $f'(x) < 0$ for $x < 1$ and $f'(x) > 0$ for $x > 1$, we conclude that f is decreasing on $(-\infty, 1)$ and increasing on $(1, \infty)$. When $x = 1$, then $f(1) = 2$; we see that $(1, 2)$ is the lowest point on the graph (Figure 5.6.36).

FIGURE 5.6.35

The function $f(x) = x^4$ does not have a point of inflection at $x = 0$

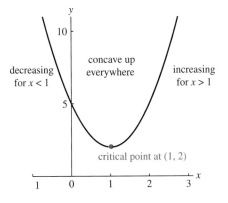

FIGURE 5.6.36

Graphing a quadratic using the first and second derivatives

Example 5.6.4 Calculating Higher-Order Derivatives I

Find a formula for $f^{(n)}(x)$ (nth derivative of $f(x)$) if $f(x) = e^{2x}$.

We compute

$$f'(x) = 2e^{2x}$$
$$f''(x) = 2(e^{2x} \cdot 2) = 2^2 e^{2x}$$
$$f^{(3)}(x) = 2^2(e^{2x} \cdot 2) = 2^3 e^{2x}$$
$$f^{(4)}(x) = 2^3(e^{2x} \cdot 2) = 2^4 e^{2x}$$

and so on. Thus,
$$f^{(n)}(x) = 2^n e^{2x}$$

Example 5.6.5 Calculating Higher-Order Derivatives II

Find a formula for $f^{(n)}(x)$ if $f(x) = \dfrac{1}{x}$.

Writing $f(x) = x^{-1}$, we calculate the first few derivatives, hoping to notice a pattern:

$$f'(x) = -x^{-2}$$
$$f''(x) = -(-2)x^{-3} = 2x^{-3}$$
$$f^{(3)}(x) = 2(-3)x^{-4} = -2 \cdot 3 x^{-4}$$
$$f^{(4)}(x) = -2 \cdot 3 \cdot (-4)x^{-5} = 2 \cdot 3 \cdot 4 x^{-5}$$

To account for the signs alternating (minus, plus, minus, plus, ...), we use $(-1)^n$. Next, the product of n and all positive integers smaller than n is denoted by $n!$ (and called **n factorial**); i.e.,

$$n! = 1 \cdot 2 \cdot 3 \cdot \cdots \cdot (n-1) \cdot n$$

In particular,

$$1! = 1$$
$$2! = 1 \cdot 2 = 2$$
$$3! = 1 \cdot 2 \cdot 3 = 6$$
$$4! = 1 \cdot 2 \cdot 3 \cdot 4 = 24$$
$$5! = 1 \cdot 2 \cdot 3 \cdot 4 \cdot 5 = 120$$

and so on.

With all this in mind, we write the required formula as

$$f^{(n)}(x) = (-1)^n n! x^{-(n+1)} = (-1)^n \frac{n!}{x^{n+1}}$$

Example 5.6.6 Calculating Higher-Order Derivatives III

Find $f^{(26)}(x)$ if $f(x) = \sin x$.

A straightforward calculation shows that

$$f'(x) = \cos x$$
$$f''(x) = -\sin x$$
$$f^{(3)}(x) = -\cos x$$
$$f^{(4)}(x) = \sin x$$
$$f^{(5)}(x) = \cos x$$

We notice that the derivatives repeat in a cycle of length 4. Thus,

$$f(x) = f^{(4)}(x) = f^{(8)}(x) = f^{(12)}(x) = \cdots = \sin x$$

It follows that $f^{(24)}(x) = \sin x$ and $f^{(26)}(x) = -\sin x$.

In Section 5.7 we will see how higher-order derivatives are used to approximate complicated functions by polynomials.

Using the Second Derivative for Graphing

When we can compute the derivative and second derivative of a function, we can use these tools to sketch graphs.

Example 5.6.7 The Second Derivative of a Linear Function

As a test, we check the second derivative of a linear function,

$$f(x) = 2x + 1$$

The first derivative is $f'(x) = 2$, a constant, implying that the second derivative is zero, which is consistent with the fact that a line is straight and has no curvature; i.e., it is neither concave up nor concave down (Figure 5.6.37).

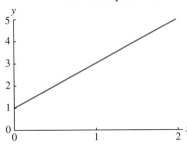

FIGURE 5.6.37
A linear function has zero curvature

The second derivative helps to categorize all **power functions.** Consider the power function

$$f(x) = x^p$$

According to the power rule, the first and second derivatives are

$$f'(x) = px^{p-1}$$
$$f''(x) = p(p-1)x^{p-2}$$

By substituting in various values of p, we can create the following table and graph the three possibilities (Figure 5.6.38). This table and the graphs apply only for $x > 0$, the values for which power functions are most often applied:

Behaviour of Power Function $f(x) = x^p$ for $x > 0$

Power	First Derivative	Second Derivative
$p > 1$	positive	positive
$0 < p < 1$	positive	negative
$p < 0$	negative	positive

Example 5.6.8 Graphing the Square Root Function

The square root function $f(x) = \sqrt{x}$ is the power function

$$f(x) = x^{1/2}$$

defined for all $x \geq 0$. When $x > 0$, the first and second derivatives are

$$f'(x) = \frac{1}{2}x^{-1/2} = \frac{1}{2\sqrt{x}} > 0$$

$$f''(x) = \frac{1}{2}\left(-\frac{1}{2}\right)x^{-3/2} = -\frac{1}{4}x^{-3/2} < 0$$

In accordance with the table, the function is increasing and its graph is concave down (Figure 5.6.38b).

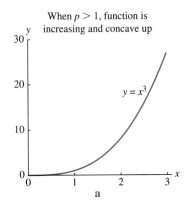
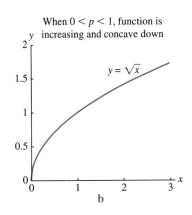
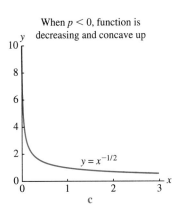

FIGURE 5.6.38
Graphs of the basic power functions

Example 5.6.9 Graphing the Reciprocal of the Square Root Function

For the function
$$f(x) = x^{-1/2} = \frac{1}{\sqrt{x}}$$
defined for $x > 0$, we compute
$$f'(x) = -\frac{1}{2}x^{-3/2} = -\frac{1}{2\sqrt{x^3}} < 0$$
$$f''(x) = -\frac{1}{2}\left(-\frac{3}{2}\right)x^{-5/2} = \frac{3}{4}x^{-5/2} = \frac{3}{4\sqrt{x^5}} > 0$$

This decreasing function is concave up (Figure 5.6.38c).

Example 5.6.10 Graphing a Cubic Polynomial

Suppose we wish to study the polynomial
$$p(x) = 2x^3 - 7x^2 + 5x + 2$$

What does the graph look like? To begin, we find the first and second derivatives
$$p'(x) = 6x^2 - 14x + 5$$
$$p''(x) = 12x - 14$$

The point of inflection occurs where $12x - 14 = 2(6x - 7) = 0$, or at $x = 7/6 \approx 1.167$, because the graph is concave down for $x < 7/6$ and concave up for $x > 7/6$. The critical numbers are the solutions of $p'(x) = 0$. Applying the quadratic formula, we find
$$x = \frac{14 \pm \sqrt{14^2 - 4(30)}}{12} = \frac{14 \pm \sqrt{76}}{12} = \frac{14 \pm \sqrt{4 \cdot 19}}{12} = \frac{14 \pm 2\sqrt{19}}{12} = \frac{7 \pm \sqrt{19}}{6}$$

Thus, the critical points are
$$\frac{7 - \sqrt{19}}{6} \approx 0.440 \quad \text{and} \quad \frac{7 + \sqrt{19}}{6} \approx 1.893$$

At these points,
$$p(0.440) \approx 3.015 \quad \text{and} \quad p(1.893) \approx -0.052$$

Figure 5.6.39 combines this information in a single graph.

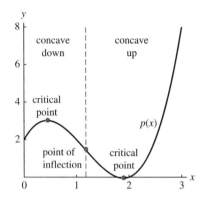

FIGURE 5.6.39

Using the first and second derivatives to graph a polynomial

The guideposts for reading and interpreting a graph based on the first and the second derivatives are summarized in the following table:

What the Graph Does	What the Derivative Does
graph increasing	derivative positive
graph decreasing	derivative negative
graph horizontal	derivative equal to zero, critical number
graph concave up	second derivative positive
graph concave down	second derivative negative
graph switches concavity	point of inflection, second derivative zero

Applications

We now discuss a few functions that appear in models in biology and elsewhere.

Example 5.6.11 Spread of an Infectious Disease

In some cases, the number of people infected with a virus, $I(t)$, is modelled by the **logistic growth function**

$$I(t) = \frac{L}{1 + Ce^{-kt}}$$

where t is time, $C = (L - I_0)/I_0$, L is a bound on the number of people who are infected, I_0 is the number of people infected initially (at time $t = 0$), and $k > 0$ is the growth rate. It is assumed that I_0 is small compared to L (actually, $I_0 < L/2$ will suffice), so that $C > 1$. (This model is studied in detail in Section 8.4.)

We are going to determine the intervals where $I(t)$ is concave up and where it is concave down. Writing $I(t) = L\left(1 + Ce^{-kt}\right)^{-1}$, we compute

$$I'(t) = L(-1)\left(1 + Ce^{-kt}\right)^{-2} Ce^{-kt}(-k) = kLCe^{-kt}\left(1 + Ce^{-kt}\right)^{-2}$$

$$I''(t) = kLC\left[e^{-kt}(-k)\left(1 + Ce^{-kt}\right)^{-2} + e^{-kt}(-2)\left(1 + Ce^{-kt}\right)^{-3} Ce^{-kt}(-k)\right]$$

$$= -k^2 LCe^{-kt}\left(1 + Ce^{-kt}\right)^{-3}\left[\left(1 + Ce^{-kt}\right) - 2Ce^{-kt}\right]$$

$$= -k^2 LCe^{-kt}\left(1 + Ce^{-kt}\right)^{-3}\left(1 - Ce^{-kt}\right)$$

The constants k, L, and C are positive; as well, both e^{-kt} and $1 + Ce^{-kt}$ are positive functions. Thus, $I'(t) > 0$ for all $t > 0$; i.e., the number of infected people is increasing.

The only way to make $I''(t) = 0$ is to set $1 - Ce^{-kt} = 0$, i.e., $Ce^{-kt} = 1$. It follows that

$$e^{-kt} = \frac{1}{C}$$

$$-kt = \ln\frac{1}{C} = -\ln C$$

$$t = \frac{-\ln C}{-k} = \frac{\ln C}{k}$$

When $t < (\ln C)/k$, then $I''(t) > 0$ and $I(t)$ is concave up. This means that, initially, the number of infected people is increasing at an increasing rate. When $t > (\ln C)/k$, the second derivative, $I''(t)$, is negative, so, although the number of infected people is increasing, it is doing so at a decreasing rate. The change in the increasing pattern occurs when $t = (\ln C)/k$, which is the point of inflection of the graph of $I(t)$. Sometimes, it is said that at that moment the infection is under control. As $t \to \infty$, the denominator in $I(t) = L/\left(1 + Ce^{-kt}\right)$ approaches 1 and so $I(t)$ approaches L. Thus, the number of infected people plateaus at L; see Figure 5.6.40.

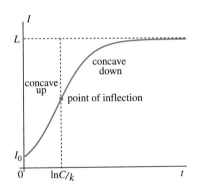

FIGURE 5.6.40
The pattern of change in $I(t)$

The gamma distributions are an important family of functions that appear in various applications in population modelling (see Ricker model in Section 6.8), probability, and elsewhere. They are defined by

$$g(x) = cx^n e^{-x}$$

where n is a positive integer, c is a constant (here we assume that $c = 1$), and $x \geq 0$.

Example 5.6.12   The Gamma Distribution with $n = 1$

With $n = 1$, the gamma distribution $g(x) = xe^{-x}$ has derivative

$$g'(x) = \frac{dx}{dx}e^{-x} + x\frac{d\left(e^{-x}\right)}{dx}$$

$$= e^{-x} - xe^{-x} = (1 - x)e^{-x}$$

This derivative is positive for $x < 1$ and negative for $x > 1$ (keeping in mind that $e^{-x} > 0$ for all x). Furthermore,

$$g''(x) = \frac{d(1-x)}{dx}e^{-x} + (1-x)\frac{d\left(e^{-x}\right)}{dx}$$

$$= -e^{-x} - (1-x)e^{-x} = (x - 2)e^{-x}$$

The second derivative is negative when $x < 2$ and positive when $x > 2$. Using the facts that $g(0) = 0$ and $g(x) > 0$ when $x > 0$, we can draw an accurate graph of this function (Figure 5.6.41).

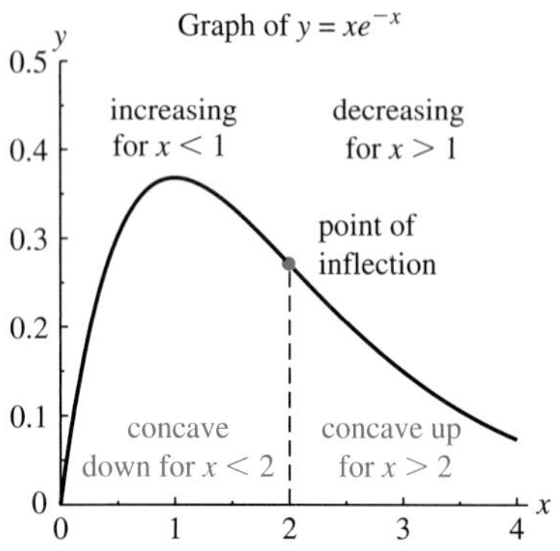

Example 5.6.13 The Gamma Distribution with $n = 2$

With $n = 2$, the gamma distribution is given by
$$g(x) = x^2 e^{-x}$$

The first derivative and the second derivative are
$$g'(x) = 2xe^{-x} + x^2 e^{-x}(-1)$$
$$= 2xe^{-x} - x^2 e^{-x}$$
$$= x(2 - x)e^{-x}$$

and
$$g''(x) = (x)'(2-x)e^{-x} + x(2-x)'e^{-x} + x(2-x)(e^{-x})'$$
$$= (2-x)e^{-x} - xe^{-x} - x(2-x)e^{-x}$$
$$= (2 - x - x - 2x + x^2)e^{-x}$$
$$= (2 - 4x + x^2)e^{-x}$$

Since we consider $x \geq 0$ only, we see that $g'(x)$ is positive when $x < 2$ and negative when $x > 2$. Thus, $g(x)$ is increasing on $(0, 2)$ and decreasing on $(2, \infty)$.

To find points of inflection, we solve
$$g''(x) = (x^2 - 4x + 2)e^{-x} = 0$$

i.e.,
$$x^2 - 4x + 2 = 0$$

Using the quadratic equation, we get
$$x = 2 - \sqrt{2} \approx 0.586 \quad \text{and} \quad x = 2 + \sqrt{2} \approx 3.414$$

To analyze concavity, we use Table 5.6.1.

Table 5.6.1

Interval	$(0, 2 - \sqrt{2})$	$(2 - \sqrt{2}, 2 + \sqrt{2})$	$(2 + \sqrt{2}, \infty)$
Second Derivative	positive	negative	positive
Function	concave up	concave down	concave up

There are two points of inflection:
$$(2 - \sqrt{2}, g(2 - \sqrt{2})) \approx (0.586, 0.191)$$

and
$$(2 + \sqrt{2}, g(2 + \sqrt{2})) \approx (3.414, 0.384)$$

See Figure 5.6.42.

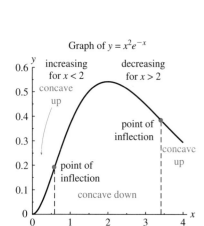

FIGURE 5.6.41
Graph of the gamma distribution with $n = 1$

FIGURE 5.6.42
Graph of $f(x) = x^2 e^{-x}$

Approximating the Second Derivative

In Section 4.5 we learned how to approximate the derivative of a function given as a table of values. Using the same idea, we now approximate the second derivative.

Example 5.6.14 Approximating the Second Derivative

Consider the function $f(x)$ whose values are given in Table 5.6.2. Using the forward difference quotients, find an approximation for $f''(3)$.

Recall that the word *forward* in *forward difference quotient* means that we use the closest point to the right of the base point, $x = 3$; thus,

$$f''(3) \approx \frac{f'(4) - f'(3)}{4 - 3}$$

To approximate $f'(4)$, we use the forward difference quotient again (this time $x = 4$ is the base point):

$$f'(4) \approx \frac{f(5) - f(4)}{5 - 4} = \frac{8.1 - 7.7}{1} = 0.4$$

Likewise,

$$f'(3) \approx \frac{f(4) - f(3)}{4 - 3} = \frac{7.7 - 8.3}{1} = -0.6$$

and therefore

$$f''(3) \approx \frac{f'(4) - f'(3)}{4 - 3} \approx \frac{0.4 - (-0.6)}{1} = 1$$

Table 5.6.2

x	f(x)
1	12.4
2	11.1
3	8.3
4	7.7
5	8.1

Our result, $f''(3) \approx 1$, is only one possible approximation. We could use any combination of difference quotients. For instance, using the backward difference quotient, we write

$$f''(3) \approx \frac{f'(2) - f'(3)}{2 - 3}$$

Using the approximation $f'(3) \approx -0.6$ from the previous example, and the forward difference quotient

$$f'(2) \approx \frac{f(3) - f(2)}{3 - 2} = \frac{8.3 - 11.1}{1} = -2.8$$

we obtain

$$f''(3) \approx \frac{f'(2) - f'(3)}{2 - 3} \approx \frac{-2.8 - (-0.6)}{1} = -2.2$$

Since we have two approximations, we average them to get $f''(3) \approx (1 - 2.2)/2 = -0.6$. Based on this result, we conclude that the graph of $f(x)$ is concave down near $x = 3$.

Acceleration

When we consider position as a function of time, the second derivative has an important physical interpretation as the **acceleration.** Suppose the position of an object is $y(t)$. The derivative, $\frac{dy}{dt}$, is the velocity, and the second derivative is the rate of change of velocity. Formally,

$$\frac{d^2y}{dt^2} = \text{acceleration} \qquad (5.6.3)$$

A positive acceleration indicates that an object is speeding up, and a negative acceleration indicates that it is slowing down.

Equation 5.6.3 is a special case of a fundamental type of differential equation studied in physics. It says that acceleration is proportional to force. In fact, we can rewrite Newton's famous law,

$$F = ma$$

where F is the force, m is the mass of the object, and a is the resulting acceleration, as

$$a = \frac{d^2y}{dt^2} = \frac{F}{m}$$

If we know the force, F, and the mass, m, we have a differential equation for the position, y.

Example 5.6.15 A Falling Rock

A rock that has fallen a distance

$$y(t) = 4.9t^2$$

in time t has second derivative, and acceleration, of

$$\frac{d^2y}{dt^2} = 9.8$$

This acceleration is positive because we are measuring how far the rock has fallen, and the downward speed is increasing.

Summary The **second derivative,** defined as the derivative of the first derivative, is positive when the graph of a function is **concave up** and negative when the graph of a function is **concave down.** A point where a function changes concavity is called a **point of inflection.** Using the first and the second derivatives, we can graph various functions. Physically, the second derivative of the position is the **acceleration.**

5.6 Exercises

Mathematical Techniques

1–4 ▪ On the figures, label

a. One critical point.

b. One point with a positive derivative.

c. One point with a negative derivative.

d. One point with a positive second derivative.

e. One point with a negative second derivative.

f. One point of inflection.

1.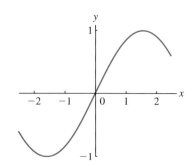

2.

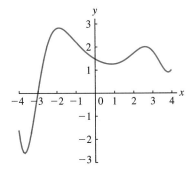

3.

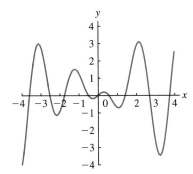

4.

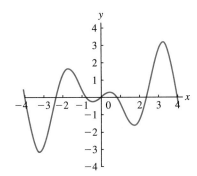

5–8 ■ Draw the graph of a function with each set of properties.

5. A function with a positive, increasing derivative.

6. A function with a positive, decreasing derivative.

7. A function with a negative, increasing (becoming less negative) derivative.

8. A function with a negative, decreasing (becoming more negative) derivative.

9–10 ■ Although there is no easy way to recognize all points with positive or negative *third* derivative, it is possible for some points (usually points of inflection).

9. On the figure for Exercise 1, find one point with negative third derivative.

10. On the figure for Exercise 2, find one point with positive third derivative.

11–24 ■ Find the first and second derivatives of the following functions.

11. $s(x) = 1 - x + x^2 - x^3 + x^4$

12. $g(z) = 3z^3 + 2z^2$

13. $h(y) = y^{10} - y^9$

14. $p(x) = 1 + x + \dfrac{x^2}{2} + \dfrac{x^3}{6} + \dfrac{x^4}{24}$

15. $F(z) = z(1+z)(2+z)$

16. $R(s) = (1+s^2)(2+s)$

17. $f(x) = \dfrac{3+x}{2x}$

18. $G(y) = \dfrac{2+y}{y^2}$

19. $f(x) = x^3 \ln x$

20. $f(x) = \arcsin x$

21. $f(x) = e^{-x^2}$

22. $f(x) = 2^{\ln x}$

23. $f(x) = \sin 2x + \sin^2 x$

24. $f(x) = \sec 4x$

25–31 ■ Recall that the notation $f^{(n)}(x)$ represents the nth derivative of a function $f(x)$. Find the indicated derivative.

25. $f^{(4)}(x)$ if $f(x) = x^4 - 3x^2 + 17x - 19$.

26. $f^{(20)}(x)$ if $f(x) = x^{19} + x - 1$.

27. $f^{(3)}(x)$ if $f(x) = e^{4x}$.

28. $f^{(4)}(x)$ if $f(x) = \dfrac{1}{x-1}$.

29. $f^{(14)}(x)$ if $f(x) = \cos x$.

30. $f^{(4)}(x)$ if $f(x) = \sqrt{x}$.

31. $f^{(5)}(x)$ if $f(x) = \ln x$.

32–35 ■ Find a formula for the derivative of each function, as indicated.

32. $f(x) = \cos x$; find $f^{(n)}(0)$.

33. $f(x) = x^5 + x^4 + x^3 + x^2 + x + 1$; find $f^{(n)}(0)$.

34. $f(x) = xe^x$; find $f^{(n)}(x)$.

35. $f(x) = xe^{-x}$; find $f^{(n)}(x)$.

36–43 ■ Find the first and second derivatives of each function and use them to sketch the graph of the function.

36. $f(x) = x^{-3}$ for $x > 0$

37. $g(z) = z + \dfrac{1}{z}$ for $z > 0$

38. $h(x) = (1-x)(2-x)(3-x)$

39. $M(t) = \dfrac{t}{1+t}$ for $t > 0$

40. $f(x) = 2x^3 + 1$ for $-5 \le x \le 5$

41. $f(x) = \dfrac{1}{x^2}$ for $0 < x \le 2$

42. $f(x) = 10x^2 - 50x$ for $-5 \le x \le 5$

43. $f(x) = x - x^2$ for $0 \le x \le 1$

44–45 ■ Identify the graph of $f(x)$ and the graphs of its first and second derivatives, $f'(x)$ and $f''(x)$.

44.

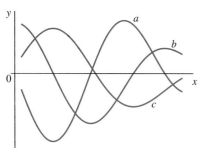

45.

46–47 ■ Find the first and second derivatives of each function (related to the gamma distribution) and sketch the graph of the function for $0 \leq x \leq 2$.

46. $G(x) = \sqrt{x} e^{-x}$

47. $G(x) = \dfrac{1}{\sqrt{x}} e^{-x}$

48–51 ■ Some higher derivatives can be found without a lot of calculation.

48. Find the tenth derivative of x^9.

49. Describe the graph of the fifth derivative of x^5.

50. Is the eighth derivative of $p(x) = 7x^8 - 8x^7 - 5x^6 + 6x^5 - 4x^3$ positive or negative?

51. Find the fifth derivative of $x(1+x)(2+x)(3+x)(4+x)$.

52–53 ■ We can approximate a function with the tangent line that matches the value of the function and its first derivative. A better approximation uses a parabola that matches the value of the function and its first and second derivatives. For each function:

a. Find the tangent line at $x = 1$.

b. Add a quadratic term to the formula of your tangent line to match the second derivative at $x = 1$.

c. Sketch the graph of the function, its tangent line, and the approximating quadratic for $0.5 < x < 1.5$.

52. $f(x) = x^{-3}$

53. $g(x) = x + \dfrac{1}{x}$

54–57 ■ Polynomials form a useful set of functions in part because the derivative of a polynomial is another polynomial. Another set of functions with this useful property is the set of **generalized polynomials,** formed as products and sums of polynomials and exponential functions. One simple group of generalized polynomials consists of the products of linear functions with the exponential function, taking the form

$$h(x) = (ax + b)e^x$$

for various values of a and b. We will call these **generalized first-order polynomials.**

54. Set $a = 1$ and $b = 1$. Find the first derivative of $h(x)$.

55. Use the results of Exercise 54 to guess the 10th derivative of $h(x)$ when $a = 1$ and $b = 1$.

56. Find a generalized first-order polynomial such that $h(1) = 0$. Where is the critical point? Where is the point of inflection?

57. Let x^* be the solution of the equation $h(x^*) = 0$. Show that the critical point of a generalized first-order polynomial is $x^* - 1$ and the point of inflection is $x^* - 2$.

58–61 ■ Consider the function $f(x)$ whose values are given in the table. Approximate each second derivative.

x	f(x)	x	f(x)
−3	16.2	1	6.4
−2	12.8	2	3.1
−1	8.2	3	4.2
0	8.8	4	6.7

58. $f''(0)$, using forward difference quotients for f'' and f'.

59. $f''(0)$, using backward difference quotients for f'' and f'.

60. $f''(-2)$, using forward difference quotients for f'' and f'.

61. $f''(1)$, using backward difference quotients for f'' and f'.

Applications

62–65 ■ The following equations give the positions as functions of time of objects tossed from towers in various exotic solar system locations. For each,

a. Find the velocity and the acceleration of the object.

b. Sketch the graph of the position for $0 \leq t \leq 3$.

c. How high was the tower? Which way was the object thrown? How does the acceleration compare with that on Earth (9.8 m/s^2)?

62. An object on Saturn that follows $p(t) = -5.2t^2 - 2t + 50$.

63. An object on the Sun that follows $p(t) = -137t^2 + 20t + 500$.

64. An object on Pluto that follows $p(t) = -0.325t^2 - 20t + 500$.

65. An object on Mercury that follows $p(t) = -1.85t^2 + 20t$.

66–69 ■ The total mass is the product of the following functions for mass and number as functions of time in years. Find the first and second derivatives and make a rough sketch of the graph for the first 100 years.

66. The population is $P(t) = 2 \cdot 10^6 + 2 \cdot 10^4 t$ and the mass in kilograms per person is $W(t) = 80 - 0.5t$.

67. The population is $P(t) = 2 \cdot 10^6 - 2 \cdot 10^4 t$ and the mass in kilograms per person is $W(t) = 80 + 0.5t$.

68. The population is $P(t) = 2 \cdot 10^6 + 1000t^2$ and the mass in kilograms per person is $W(t) = 80 - 0.5t$.

69. The population is $P(t) = 2 \cdot 10^6 + 2 \cdot 10^4 t$ and the mass in kilograms per person is $W(t) = 80 - 0.005t^2$.

70–71 ▪ The following graphs show the horizontal distance travelled by a roller coaster as a function of time. When is the roller coaster going most quickly? When is it accelerating most quickly? When is it decelerating most quickly?

70.

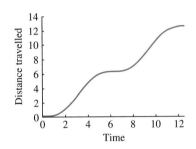

71.

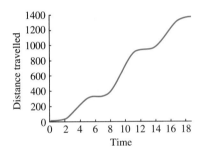

72–73 ▪ In a model of a growing population, we find the new population by multiplying the old population by the per capita production. For each case, find the second derivative of the new population as a function of the old population and sketch the graph of the population function.

72. Per capita production is $2\left(1 - \dfrac{b_t}{1000}\right)$. Consider values of b_t less than 1000.

73. Per capita production is $2b_t\left(1 - \dfrac{b_t}{1000}\right)$. Consider values of b_t less than 1000.

74–75 ▪ We can use the second derivative to study Hill functions, $h_n(x) = \dfrac{x^n}{1 + x^n}$ for $x > 0$.

74. Find the second derivative of the Hill function with $n = 1$ and describe the curvature of the graph.

75. Find the second derivative of the Hill function with $n = 2$ and describe the curvature of the graph.

76–81 ▪ The amount of food a predator eats as a function of prey density is called the **functional response**. Functional response is often broken into three categories:

▪ Type I: Linear.

▪ Type II: Increasing, concave down, finite limit.

▪ Type III: Increasing with finite limit, concave up for small prey densities, concave down for large prey densities.

76. Sketch graphs of these three types. Which of the absorption functions do they resemble? What is the optimal prey density for a predator in each case?

77. Suppose that the number of prey that escape increases linearly with the number of prey (the prey join together and fight back). Let p be the number of prey and $F(p)$ be the functional response. The number of prey captured is then $F(p) - cp$. The constant c represents how effectively the prey can fight. Write the equation for the optimal prey density (the value giving the maximum rate of prey capture) in terms of $F'(p)$.

78. Draw a graph illustrating the optimal prey density in a case with a type II functional response.

79. Draw a graph illustrating the optimal prey density in a case with a type III functional response.

80. Find the optimal prey density if $F(p) = p$. Make sure to consider separately cases with $c < 1$ and $c > 1$.

81. Find the optimal prey density if $F(p) = \dfrac{p}{1 + p}$. Make sure to consider separately cases with $c < 1$ and $c > 1$.

Computer Exercises

82. Consider again the function

$$g(x) = x^n e^{-x}$$

describing the gamma distribution.

a. Find the nonzero critical point.

b. Find the point or points of inflection. What happens when $n < 1$?

c. Graph this function for $n = 0.5$, $n = 2$, and $n = 5$. What happens for very large values of n?

83. Consider the generalized polynomial

$$R(x) = x^4 + \left(88x^3 - 76x^2 - 65x + 25\right)e^x$$

As defined in Exercises 54–57, a generalized polynomial is formed by multiplying and adding polynomials and exponential functions.

a. Find all critical points and points of inflection of $R(x)$ for $-1 \leq x \leq 1$.

b. Find the fifth derivative of $R(x)$. Does it look any simpler than $R(x)$ itself? Compare with what happens when you take many derivatives of a polynomial or of the exponential function.

84. Have a computer find all critical points and points of inflection of the function

$$P(x) = 8x^5 - 18x^4 - x^3 + 18x^2 - 7x$$

Show that these match what you see on a graph.

5.7 Approximating Functions with Polynomials

In this section, we extend the idea of the **tangent line approximation** in several ways. First, we compare the tangent line approximation with the **secant line approximation,** showing that the tangent line is the *best* linear approximation to a curve near the point of tangency but that the secant line can be more useful over larger ranges.

The tangent line matches the value and derivative of a function at a point. More accurate approximations can be found by also matching the second, third, and higher derivatives. If we use **polynomials** to match these higher derivatives, the resulting approximation generates a **Taylor polynomial.**

The Tangent and Secant Lines

Suppose we wish to approximate the exponential function $f(x) = e^x$ near $x = 0$ with a line, perhaps in order to compare a complicated dynamical system with its linear approximation. The general formula for the tangent line to the function f at base point a is obtained from the point-slope equation

$$y - f(a) = f'(a)(x - a)$$

Rather than using the generic symbol y for this linear function, we use $L(x)$ instead and write

$$L(x) = f(a) + f'(a)(x - a)$$

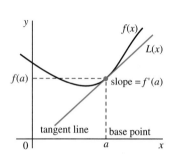

FIGURE 5.7.43

Approximating a function with the tangent line

Sometimes—for instance, in the context that we develop later in this section (Taylor polynomials)—we use the notation $T_1(x)$ instead of $L(x)$. The linear function $L(x)$ has as its graph the tangent line to $f(x)$ at a, and it matches both the value and the slope at the point of tangency (Figure 5.7.43). $L(x)$ is called the **linear approximation,** or the **tangent line approximation,** or the **linearization of $f(x)$ at $x = a$.**

Example 5.7.1   The Tangent Line to the Exponential Function

The tangent line to the graph of the exponential function

$$f(x) = e^x$$

at $x = 0$ has to match the value $f(0) = e^0 = 1$ and the slope $f'(0) = e^0 = 1$. Hence,

$$L(x) = f(0) + f'(0)(x - 0) = 1 + x$$

(Figure 5.7.44). At $x = 0$, the graphs of $f(x) = e^x$ and the tangent $L(x) = 1 + x$ have the same value (i.e., both go through the point $(0, 1)$). Near $x = 0$, the line $L(x)$ approximates the graph of $f(x) = e^x$ fairly well. For example,

$$L(0.1) = 1.1$$

which is close to the exact value $e^{0.1} \approx 1.10517$.

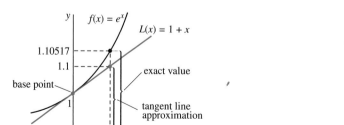

FIGURE 5.7.44

Approximating $e^{0.1}$ with the tangent line

Before computers, this sort of approximation was indispensable. As we will soon see when we study Newton's method for solving equations in Chapter 6 and Euler's method for solving differential equations in Chapter 7, the tangent line approximation remains important for dealing with more complicated problems.

Example 5.7.2 The Tangent Line to the Logarithmic Function

To estimate $\ln 0.9$ without a calculator, we note that 0.9 is close to 1. Because $\ln 1 = 0$, we expect that the estimate for $\ln 0.9$ will be close to 0 as well.

The idea is to construct the tangent to the graph of $f(x) = \ln x$ at $x = 1$ and then use it to approximate $\ln 0.9$; see Figure 5.7.45.

Let $f(x) = \ln x$. Then $f(1) = \ln 1 = 0$, and since $f'(x) = 1/x$, the slope is equal to $f'(1) = 1/1 = 1$. The tangent line is

$$L(x) = 0 + 1(x - 1) = x - 1$$

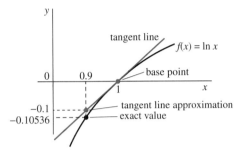

FIGURE 5.7.45

Approximating $\ln 0.9$ with the tangent line

Substituting $x = 0.9$ into $L(x)$, we find an approximate value of -0.1, close to the exact value of $\ln(0.9) \approx -0.10536$ (Figure 5.7.45).

Recall the bacterial population growing according to

$$b(t) = 2^t$$

Can we find a good linear approximation to this function for times between 0 and 1? One method is to use the tangent line at $t = 0$. First, we find the derivative

$$b'(t) = 2^t \ln 2$$

From $b(0) = 1$ and $b'(0) = \ln 2$ we compute the tangent line at $t = 0$:

$$L_0(t) = 1 + (\ln 2)(t - 0) = 1 + (\ln 2)t \approx 1 + 0.69315t$$

where the subscript 0 indicates the base point. If we are interested in approximating values near $t = 0$, the tangent line is accurate (Figure 5.7.46a).

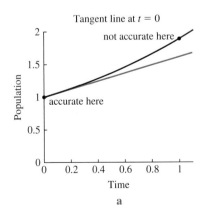
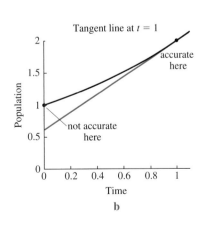
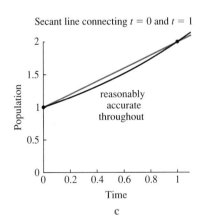

FIGURE 5.7.46

Two tangent lines and a secant line as approximations

If instead we are interested in approximating values near $t = 1$, this tangent line is quite inaccurate. So, we calculate the tangent line at $t = 1$ (we will call it $L_1(t)$). This time, $b(1) = 2$ and $b'(1) = 2\ln 2$ and thus

$$L_1(t) = 2 + (2\ln 2)(t - 1) \approx 2 + 1.38629(t - 1)$$

Near the base point $t = 1$, this tangent line provides an accurate approximation (Figure 5.7.46b).

Over the whole interval from $t = 0$ to $t = 1$, the secant line is a reasonably good approximation everywhere. The secant line has slope

$$\text{slope of secant} = \frac{\Delta b}{\Delta t} = \frac{b(1) - b(0)}{1 - 0} = 1$$

which lies between the slopes of the two tangent lines we calculated. The equation of the secant line—we will call it $S(t)$—is (we take $(0, 1)$ as the base point)

$$S(t) - 1 = 1(t - 0)$$

i.e., $S(t) = 1 + t$.

How can we quantify the accuracy of these alternative approximations? Their values near both endpoints and at the middle of the interval are given in the following table:

t	b(t)	$L_0(t)$	$L_1(t)$	S(t)
0.01	1.00696	1.00693	0.62757	1.01
0.10	1.07177	1.06931	0.75234	1.10
0.50	1.41421	1.34658	1.30686	1.50
0.90	1.86607	1.62384	1.86137	1.90
0.99	1.98618	1.68622	1.98614	1.99

Each tangent line is a good approximation near its point of tangency. The secant line is fairly close on the whole interval between $t = 0$ and $t = 1$. This is one of the two primary strengths of using the secant line, which is also called **linear interpolation**. The other is that the secant can be directly estimated from data, even when we do not know the underlying equation.

Example 5.7.3 Finding a Linear Approximation

Find the linear approximation of $f(x) = \sin x$ at $x = 0$.

From $f(0) = 0$ and $f'(0) = \cos 0 = 1$ we compute

$$L(x) = 0 + 1(x - 0) = x$$

Thus, for small values of x (in radians!) we get the approximation

$$\sin x \approx x$$

For example, if $x = 0.05$, then $L(0.05) = 0.05$ closely approximates the value $f(0.05) = \sin 0.05 \approx 0.049979$ (note that this is not the exact value either, but a very precise approximation!).

Example 5.7.4 Estimating Population Growth Using Linear Approximation

A population of rabbits in an ecosystem is modelled by

$$P'(t) = 2.2\sqrt{P(t)}$$

where $P(t)$ is the number of rabbits and t is time in months. Given that $P(5) = 900$, we will find an approximation for $P(6)$.

Recall that if

$$P'(t) = 2.2P(t)$$

This way, we obtain the approximating quadratic polynomial

$$T_2(x) = f(a) + f'(a)(x-a) + \frac{f''(a)}{2}(x-a)^2$$

The first two terms in this approximate polynomial exactly match the tangent line

$$L(x) = f(a) + f'(a)(x-a)$$

(in certain situations we will denote $L(x)$ by $T_1(x)$). The last term,

$$\frac{f''(a)}{2}(x-a)^2$$

in $T_2(x)$ is an additional correction that accounts for the concavity at a.

If the second derivative is equal to zero, this term vanishes and the best approximating quadratic is the tangent line itself.

Example 5.7.6 Approximating a Cubic with a Quadratic

Suppose we wish to approximate the function $f(x) = 1 + 2x + x^2 + 3x^3$ with a quadratic function $T_2(x)$ for the values of x that are near 1; thus,

$$T_2(x) = f(1) + f'(1)(x-1) + \frac{f''(1)}{2}(x-1)^2$$

The first and the second derivatives of f are

$$f'(x) = 2 + 2x + 9x^2$$
$$f''(x) = 2 + 18x$$

Therefore,

$$f(1) = 1 + 2 \cdot 1 + 1^2 + 3 \cdot 1^3 = 7$$
$$f'(1) = 2 + 2 \cdot 1 + 9 \cdot 1^2 = 13$$
$$f''(1) = 2 + 18 \cdot 1 = 20$$

Then

$$T_2(x) = 7 + 13(x-1) + 10(x-1)^2$$

Note that the approximating quadratic polynomial in the previous example is *not* found by ignoring the x^3 term in $f(x)$.

However, there is one case when we do obtain $T_2(x)$ from a polynomial $f(x)$ in that way (i.e., by ignoring the powers of x that are larger than two). Suppose that we wish to approximate the same polynomial

$$f(x) = 1 + 2x + x^2 + 3x^3$$

near $x = 0$. In that case,

$$T_2(x) = f(0) + f'(0)(x-0) + \frac{f''(0)}{2}(x-0)^2$$

From $f'(x) = 2 + 2x + 9x^2$ and $f''(x) = 2 + 18x$ we get $f'(0) = 2, f''(0) = 2$ and thus

$$T_2(x) = 1 + 2(x-0) + \frac{2}{2}(x-0)^2 = 1 + 2x + x^2$$

(See Exercises 28 and 29 for additional examples.)

Example 5.7.7 Approximating the Logarithmic Function with a Quadratic

In Example 5.7.2, we used the linear approximation to estimate $\ln 0.9$ by finding the tangent line to $f(x) = \ln x$ at base point $x = 1$. We can get a more accurate estimate by finding the quadratic approximation at this same base point. We have that

$$f'(x) = \frac{1}{x}$$

$$f''(x) = -\frac{1}{x^2}$$

Evaluating at $x = 1$,

$$f(1) = \ln 1 = 0$$

$$f'(1) = \frac{1}{1} = 1$$

$$f''(1) = -\frac{1}{1^2} = -1$$

Therefore, the approximating quadratic is

$$T_2(x) = 0 + 1(x-1) - \frac{1}{2}(x-1)^2 = x - 1 - \frac{1}{2}(x-1)^2$$

Substituting in $x = 0.9$, we find an approximate value of -0.105, very close to the exact value of $\ln 0.9 \approx -0.10536$. As we can see from Figure 5.7.49a, the approximation is nearly indistinguishable for x near 1. However, over a larger range, the function and the approximation diverge (Figure 5.7.49b).

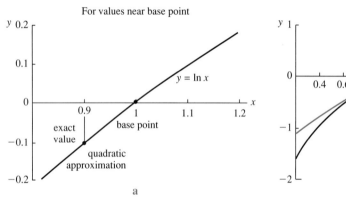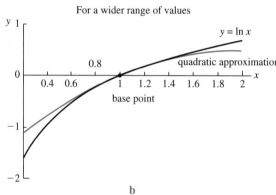

FIGURE 5.7.49

Approximating $\ln 0.9$ with a quadratic

Example 5.7.8 Finding Approximating Quadratics at Three Points

The function

$$h(x) = xe^{-x}$$

has derivatives

$$h'(x) = (1-x)e^{-x}$$
$$h''(x) = (x-2)e^{-x}$$

(see Example 5.6.12). In the same example we showed that this function has a critical point at $x = 1$ and a point of inflection at $x = 2$. Using the formula for the approximating quadratic at the three points $x = 0$, $x = 1$, and $x = 2$, we can find three different approximating quadratics that we denote by $T_{2,0}(x)$, $T_{2,1}(x)$, and $T_{2,2}(x)$, respectively:

$$T_{2,0}(x) = h(0) + h'(0)x + \frac{h''(0)}{2}x^2 = x - x^2$$

5.7 Approximating Functions with Polynomials

$$T_{2,1}(x) = h(1) + h'(1)(x-1) + \frac{h''(1)}{2}(x-1)^2 = \frac{1}{e} - \frac{1}{2e}(x-1)^2$$

$$T_{2,2}(x) = h(2) + h'(2)(x-2) + \frac{h''(2)}{2}(x-2)^2 = \frac{2}{e^2} - \frac{1}{e^2}(x-2)$$

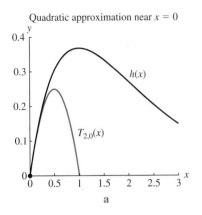
a

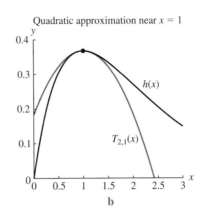
b

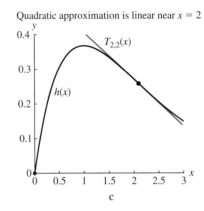
c

FIGURE 5.7.50
Three quadratic approximations to $h(x) = xe^{-x}$

(Figure 5.7.50). At the critical point, the approximating quadratic has no $x - 1$ term because $h'(1) = 0$. At the point of inflection, the quadratic term drops out, leaving us with the tangent line.

Example 5.7.9 Estimating Population Growth Using Quadratic Approximation

In Example 5.7.4 we constructed the linear approximation for the function $P(t)$ satisfying

$$P'(t) = 2.2\sqrt{P(t)}$$

such that $P(5) = 900$. Using that linear approximation, we estimated $P(6) \approx 966$. Now we use the quadratic approximation

$$T_2(t) = P(5) + P'(5)(t-5) + \frac{P''(5)}{2}(t-5)^2$$

to improve our estimate. The value $P(5) = 900$ is given, and $P'(5) = 2.2\sqrt{900} = 66$. Thus, to find $T_2(t)$, all we need is $P''(5)$.

Differentiating

$$P'(t) = 2.2\sqrt{P(t)} = 2.2 P(t)^{1/2}$$

using the chain rule, we get

$$P''(t) = 2.2 \frac{1}{2} P(t)^{-1/2} P'(t) = \frac{1.1 P'(t)}{\sqrt{P(t)}}$$

and therefore

$$P''(5) = \frac{1.1 P'(5)}{\sqrt{P(5)}} = \frac{1.1(66)}{\sqrt{900}} = 2.42$$

The quadratic approximation

$$T_2(t) = 900 + 66(t-5) + 1.21(t-5)^2$$

gives the estimate

$$T_2(6) = 900 + 66(6-5) + 1.21(6-5)^2 = 967.21$$

(i.e., 967) for the value of $P(6)$.

Taylor Polynomials ~~excluded~~

The idea of matching derivatives can be extended to the third, fourth, and higher derivatives. With each added derivative, the approximation becomes more accurate but requires a polynomial of higher *degree*. The derivation of the following formula is the same as the derivation of the quadratic approximation. The approximating polynomial is called a **Taylor polynomial** of degree n.

Definition 5.7.1 **The Taylor Polynomial of Degree n**

Suppose that the first n derivatives of a function $f(x)$ are defined at $x = a$. The **Taylor polynomial of $f(x)$ of degree n at a** (or the **nth-degree Taylor polynomial of $f(x)$ at a**) is given by

$$T_n(x) = f(a) + f'(a)(x-a) + \frac{f''(a)}{2}(x-a)^2 + \cdots$$
$$+ \frac{f^{(i)}(a)}{i!}(x-a)^i + \cdots + \frac{f^{(n)}(a)}{n!}(x-a)^n$$

We used two pieces of notation in this definition. Recall that the notation

$$f^{(i)}(x)$$

indicates the ith derivative of f. For example, we could write

$$f^{(2)}(x) = f''(x)$$

for the second derivative; $f^{(3)}(x)$ is the third derivative, and so on. Second, the terms $i!$ are called **factorials**. The value of $i!$ is the product of i and all positive integers smaller than i, or

$$i! = 1 \cdot 2 \cdot 3 \cdots (i-1) \cdot i$$

(see Example 5.6.5).

The values of factorials increase very quickly, meaning that later terms in a Taylor polynomial become ever smaller.

The Taylor polynomial of degree one is the linear approximation (hence the notation $T_1(x)$ for $L(x)$, as mentioned earlier). The Taylor polynomial of degree two is the quadratic approximation $T_2(x)$ that we discussed in the previous subsection.

Example 5.7.10 Approximating the Logarithmic Function with a Cubic

In Example 5.7.7, we used a quadratic to estimate $\ln 0.9$. To find the cubic (third-order) approximation, we must also match the third derivative. For $f(x) = \ln x$,

$$f'(x) = \frac{1}{x}$$
$$f''(x) = -\frac{1}{x^2}$$
$$f'''(x) = \frac{2}{x^3}$$

Evaluating at the point $x = 1$ yields

$$f(1) = \ln 1 = 0$$
$$f'(1) = \frac{1}{1} = 1$$
$$f''(1) = -\frac{1}{1^2} = -1$$
$$f'''(1) = \frac{2}{1^3} = 2$$

Therefore, the approximating third-degree Taylor polynomial is

$$T_3(x) = 0 + 1(x - 1) - \frac{1}{2}(x - 1)^2 + \frac{2}{6}(x - 1)^3$$

Substituting in $x = 0.9$, we find an approximate value $T_3(0.9) = -0.10533$, which is even closer to the exact value of $\ln 0.9 \approx -0.10536$.

Example 5.7.11 Taylor Polynomials for the Exponential Function

Consider again the exponential function $f(x) = e^x$. Using our new notation, we see that

$$f^{(i)}(x) = e^x$$

for all i because each derivative of the exponential function is equal to the function itself. For $a = 0$, the Taylor polynomial of degree four is

$$T_4(x) = f(0) + f'(0)(x - 0) + \frac{f''(0)}{2}(x - 0)^2 + \frac{f^{(3)}(0)}{3!}(x - 0)^3 + \frac{f^{(4)}(0)}{4!}(x - 0)^4$$

$$= f(0) + f'(0)x + \frac{f''(0)}{2}x^2 + \frac{f^{(3)}(0)}{6}x^3 + \frac{f^{(4)}(0)}{24}x^4$$

$$= 1 + x + \frac{1}{2}x^2 + \frac{1}{6}x^3 + \frac{1}{24}x^4$$

See Figure 5.7.51. We estimate that

$$e^{0.1} \approx T_4(0.1) = 1 + 0.1 + \frac{1}{2}(0.1)^2 + \frac{1}{6}(0.1)^3 + \frac{1}{24}(0.1)^4 \approx 1.1051708$$

The correct value of $e^{0.1}$ to seven decimal places is 1.1051709.

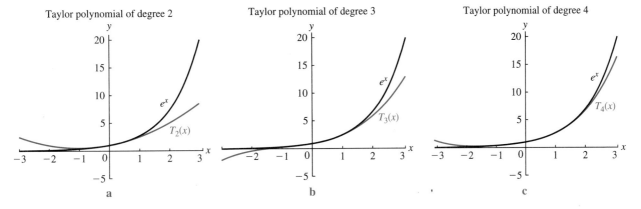

FIGURE 5.7.51
Three Taylor polynomial approximations to $f(x) = e^x$

In Section 7.5 we will see how Taylor polynomials are used to integrate functions for which standard integration methods do not work.

Summary Using lines to approximate curves is one of the central ideas in calculus. We have compared the tangent line with three other approximations. The secant line, or **linear interpolation,** has the joint virtues of using actual data and remaining fairly accurate over a broad domain. The tangent line is the **best linear approximation** near the base point. To do even better, we can use a quadratic polynomial to match both the first and second derivatives of the original function at the base point. This idea can be expanded to the **Taylor polynomial,** which is a polynomial of degree n that matches the first n derivatives of the function.

5.7 Exercises

Mathematical Techniques

1–7 ▪ Compute the required Taylor polynomial $T_n(x)$ for the function $f(x)$ near the given value $x = a$.

1. $T_1(x)$ and $T_2(x)$ for $f(x) = \sin x$ near $a = \pi/2$.
2. $T_2(x)$ and $T_4(x)$ for $f(x) = \cos x$ near $a = 0$.
3. $T_1(x)$ and $T_2(x)$ for $f(x) = \arctan x$ near $a = 0$.
4. $T_1(x)$ and $T_2(x)$ for $f(x) = \ln x$ near $a = e$.
5. $T_3(x)$ and $T_4(x)$ for $f(x) = \dfrac{1}{1-x}$ near $a = 0$.
6. $T_1(x)$ and $T_2(x)$ for $f(x) = \sqrt{x}$ near $a = 1$.
7. $T_1(x)$ and $T_2(x)$ for $f(x) = \sec x$ near $a = 0$.

8–13 ▪ Use the tangent line and secant line to estimate each value. Be sure to identify the base point a used for the tangent line approximation and the second point used for the secant line approximation. Use a calculator to compare the estimates with the exact answer.

8. 2.02^3
9. 3.03^2
10. $\sqrt{4.01}$
11. $\sqrt{6}$
12. $\sin 0.02$
13. $\cos(-0.02)$

14–19 ▪ Use the quadratic approximation to estimate each value. Compare the estimates with the exact answer.

14. 2.02^3
15. 3.03^2
16. $\sqrt{4.01}$
17. $\sqrt{6}$
18. $\sin 0.02$
19. $\cos(-0.02)$

20–23 ▪ Use the tangent line approximation to evaluate each value in two ways. First, find the tangent line to the whole function using the chain rule. Second, break the calculation into two pieces by writing the function as a composition, approximate the inner function with its tangent line, and use this value to substitute into the tangent line of the outer function. Select the closest integer as the base point where you calculate the tangent. Do your answers match?

20. $(1 + 3 \cdot 1.01)^2$
21. $\ln \sqrt{0.98}$
22. $e^{\sin 0.02}$
23. $\sin \ln((1 + 0.1)^3)$

24–27 ▪ For each of the following, find the tangent line approximation of the two values, and compare with the true value. Select the closest integer as the base point where you calculate the tangent. Indicate which approximations are too high and which are too low. From the graphs of the functions, try to explain what it is about the graph that causes this.

24. $e^{0.1}$ and $e^{-0.1}$
25. $\ln 1.1$ and $\ln 0.9$
26. 1.1^2 and 0.9^2
27. $\sqrt{1.1}$ and $\sqrt{0.9}$

28–33 ▪ Find the third-order Taylor polynomial for each function.

28. $f(x) = x^3 + 4x^2 + 3x + 1$ for x near 0.
29. $g(x) = 4x^4 + x^3 + 4x^2 + 3x + 1$ for x near 0.
30. $f(x) = 4x^2 + 3x + 1$ for x near 1.
31. $g(x) = 4x^4 + x^3 + 4x^2 + 3x + 1$ for x near 1.
32. $h(x) = \ln x$ for x near 1.
33. $h(x) = \sin x$ for x near 0.

34. The population of bacteria, $P(t)$, increases according to $P'(t) = P(t)^2$. It is known that $P(3) = 200$. Using linear and quadratic approximations, find estimates for $P(4)$.

35. The concentration of pollutants, $c(t)$, changes according to $c'(t) = 1 + 2/c(t)$. It is known that $c(9) = 1.13$. Using quadratic approximation, find an estimate for $c(10)$.

36–37 ▪ Taylor series can be used to sum some **infinite series** (sums with an infinite number of terms). Find the Taylor polynomial for each function, and use it to add up the series. Check by adding up the first terms in the series.

36. Find the Taylor polynomial of degree n for $f(x) = \dfrac{1}{1-x}$ with base point $x = 0$. Use your result to find $1 + \dfrac{1}{3} + \dfrac{1}{3^2} + \dfrac{1}{3^3} + \dfrac{1}{3^4} + \cdots$.

37. Find the Taylor polynomial of degree n for $f(x) = \ln(1+x)$ with base point $x = 0$. Use your result to find $1 - \dfrac{1}{2} + \dfrac{1}{3} - \dfrac{1}{4} + \dfrac{1}{5} - \cdots$.

Applications

38–43 ▪ Find the tangent line approximation of the following absorption functions at $c = 0$.

38. $\alpha(c) = \dfrac{5c}{1+c}$

39. $\alpha(c) = \dfrac{c}{5+c}$

40. $\alpha(c) = \dfrac{5c^2}{1+c^2}$

41. $\alpha(c) = \dfrac{5c}{e^{2c}}$

42. $\alpha(c) = \dfrac{5c}{1+c^2}$

43. $\alpha(c) = 5c(1+c)$

44–47 ▪ Consider a declining population following the formula

$$b(t) = \dfrac{1}{1+t}$$

(measured in millions). Approximate the population at each time, using (a) the tangent with base point $t = 0$, (b) the tangent with base point $t = 1$, and (c) the secant connecting times $t = 0$ and $t = 1$. Graph each of the relevant tangents and secants. Which method is best for what?

44. $t = 0.1$

45. $t = 0.5$

46. $t = 0.9$

47. $t = 1.1$

48–49 ▪ Consider the following table giving mass as a function of age.

Age, a (days)	Mass, M (g)
0.5	0.125
1	1
1.5	3.375
2	8

The data follow the equation $M(a) = a^3$. Estimate each of the following using the tangent line approximation and the secant line approximation. Which approximation is closer to the exact answer? Which method would be best if you did not know the formula for $M(a)$?

48. $M(1.25)$

49. $M(1.45)$

50–51 ▪ Consider the following table giving temperature as a function of time.

Time, t	Temperature, T (°C)
0	0.172
1	1.635
2	6.492
3	11.95
4	20.24

The data follow the equation $T(t) = t + t^2$, but there is some noise in each of the measurements, so the values are not exactly on the curve. Estimate each of the following using the tangent line approximation and the secant line approximation. Which method do you think deals best with the noise?

50. Estimate $T(1)$ using the values at $t = 0$ and $t = 2$.

51. Estimate $T(3)$ using the values at $t = 2$ and $t = 4$.

Computer Exercises

52. Find the Taylor polynomials for $f(x) = e^x$, $f(x) = \cos x$, and $f(x) = \sin x$ with base point $x = 0$ up to degree 10. Can you see the pattern? Graph the function and the Taylor polynomials T_2, T_5, and T_{10} on domains around 0 that get larger and larger. What happens to the approximation for values of x far from 0?

53. A simple equation that is impossible to solve algebraically is

$$e^x = x + 2$$

a. Graph the two sides and convince yourself there is a solution.

b. Replace e^x with its tangent line approximation at $x = 0$, and try to solve for the point where the tangent line approximation is equal to $x + 2$. This is an approximate solution. What goes wrong in this case?

c. Replace e^x with its quadratic approximation at $x = 0$, and solve for the point where the quadratic approximation is equal to $x + 2$.

d. Replace e^x with its tangent line at $x = 1$ and solve.

e. Replace e^x with its quadratic approximation at $x = 1$ and solve.

f. Find approximations of the solutions using mathematics software, and compare with your answers.

342 Chapter 5 Working with Derivatives

Chapter Summary: Key Terms and Concepts

Define or explain the meaning of each term and concept.
Derivative rules and techniques: sum (difference) rule, constant multiple (product) rule, power rule, product rule, quotient rule, chain rule; implicit differentiation, logarithmic differentiation
Derivatives of functions: constant and power functions; exponential and logarithm functions; trigonometric and inverse trigonometric functions; first derivative, second derivative, higher-order derivative
Describing change: rate of change, relative rate of change; related rates; differential equation
Derivatives and graphs: critical point; increasing function, decreasing function; concave up graph, concave down graph; inflection point
Approximations of functions: tangent line approximation, linear approximation, linearization; quadratic approximation, Taylor polynomial

Concept Check: True/False Quiz

Determine whether each statement is true or false. Give a reason for your choice.

1. Let $f(x) = 2\sqrt{g(x)}$. Then $f'(x) = g'(x)/\sqrt{g(x)}$. *true*
2. The derivative of a rational function is a rational function. *true*
3. If $f(x) = g(x)h(x)$, then $f''(x) = g''(x)h(x) + g(x)h''(x)$. *false*
4. If $y = 2^7$, then $y' = 2^7 \ln 2$. *false*
5. If $f(x) = \cos x$, then $f^{(10)}(x) = -\cos x$. *true*
6. If A is proportional to B, then the rates of change of A and B are equal. *true*
7. If A is proportional to B, then the relative rates of change of A and B are equal. *true*
8. The third-degree Taylor polynomial of e^{2x} at $x = 0$ is $1 + 2x + 2x^2 + 2x^3$. *false*
9. The tangent line approximation of $f(x) = \ln x$ at $x = 1$ gives $\ln 1.2 \approx 0.2$. *false*
10. A polynomial of degree five can have four inflection points. *false*

Supplementary Problems

1–6 ▪ Find the derivative of each function. Note any points where the derivative does not exist.

1. $F(y) = y^4 + 5y^2 - 1$
2. $a(x) = 4x^7 + 7x^4 - 28$
3. $H(c) = \dfrac{c^2}{1+2c}$
4. $h(z) = \dfrac{z}{1 + \ln z^2}$ for $z > 0$
5. For $y > 0$, $b(y) = \dfrac{1}{y^{0.75}}$.
6. For $z \geq 0$, $c(z) = \dfrac{z}{(1+z)(2+z)}$.

7–17 ▪ Find the derivative of each function.

7. $g(x) = (4 + 5x^2)^6$
8. $c(x) = \left(1 + \dfrac{2}{x}\right)^5$
9. $s(t) = \ln 2t^3$
10. $p(t) = t^2 e^{2t}$
11. $f(x) = \sin(\sin(\sin x))$
12. $f(x) = \dfrac{\sqrt{x} - 3}{2 + \sqrt{x}}$
13. $h(x) = \arcsin(\cos x)$
14. $g(x) = \arctan^2 x + \arctan x^2$
15. $f(x) = \sqrt[3]{2 + \dfrac{x}{e^{3x}}}$
16. $s(x) = e^{-3x+1} + 5 \ln 3x$
17. $g(y) = e^{3y^3 + 2y^2 + y}$

18–21 ▪ Find the derivative and other requested items for each function.

18. $f(t) = e^t \cos t$. Find one critical point.
19. $g(x) = \ln(1 + x^2)$. Find all points where $g(x)$ is decreasing.
20. $h(y) = \dfrac{1 - y}{(1 + y)^3}$. Find all values where $h(y)$ is increasing.
21. $c(z) = \dfrac{e^{2z} - 1}{z}$. What is $\lim\limits_{z \to 0} c(z)$?

22–25 ▪ Find all critical points and points of inflection and then sketch the graph of each function.

22. $f(x) = e^{-x^3}$
23. $g(x) = e^{-x^4}$
24. $h(y) = \cos(y) + \dfrac{y}{2}$
25. $F(c) = e^{-2c} + e^c$

26. Compute $f'(1)$ if $f(x) = \ln\sqrt{\dfrac{x^4 + 11x + 2}{2x - 1}}$.

27. Simplify using the laws of logarithms, and then differentiate the function $f(x) = \log_2 \dfrac{8x^4 \sin x}{e^x}$.

28. Find the equation of the line tangent to the graph of the function $f(x) = \sqrt{3 - \arctan x}$ at the point where $x = 0$.

29. Find the equation of the line tangent to the graph of $y = x^{2x-4}$ at the point where $x = 1$.

30. The figure below shows a mammogram of a tumour. The formula $V = ab(a + b)\pi/12$ can be used to obtain an approximation of the volume of the tumour from the two-dimensional image. The variable a is the largest distance across the flat image of a tumour. The variable b is the largest distance perpendicular to a. At the moment when $a = 2.4$ mm, $b = 1.1$ mm, $da/dt = 0.1$ mm/month, and $db/dt = 0.08$ mm/month, how fast is the volume increasing?

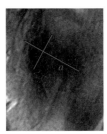

Miroslav Lovric

31. Suppose the total product in grams generated by a chemical reaction is
$$P(t) = \dfrac{t}{1+2t}$$
where t is measured in hours.

   a. Find the average rate of change between $t = 1$ and $t = 2$.
   b. Find the equation of the secant line between these times.
   c. Graph the secant line and the function $P(t)$.

   d. Sketch the graph of the tangent line at $t = 2$. Judging from your graph, is the slope of the secant larger than, smaller than, or the same as the slope of the tangent?
   e. Write the limit you would take to find the instantaneous rate of change at $t = 2$.

32–33 ■ Each of the following graphs shows how wide open, in square centimetres, a person's mouth is while asleep. Time is measured in hours after going to sleep at 10:00 P.M. In each case:

   a. Label a point where the first derivative is positive and the second derivative is negative.
   b. Label a point of inflection.
   c. Sketch the graph of the rate of change.
   d. Sketch the graph of the second derivative.

32.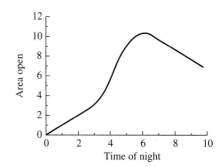

33.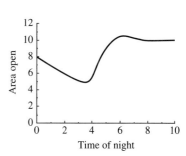

Project

1. Consider the function $f(x) = e^{-x^2}$.

   a. Find the Taylor polynomials $T_2(x)$, $T_3(x)$, $T_4(x)$, $T_5(x)$, and $T_6(x)$. (You will notice that the subscript n in $T_n(x)$ does not necessarily correspond to the degree of the polynomial.)

   b. Using a graphing calculator or a computer, sketch the graphs of the polynomials $T_2(x)$, $T_4(x)$, and $T_6(x)$ in the same coordinate system.

   c. Generate the graphs of the functions $E_2(x) = |e^{-x^2} - T_2(x)|$, $E_4(x) = |e^{-x^2} - T_4(x)|$, and $E_6(x) = |e^{-x^2} - T_6(x)|$. What do the three functions represent? What can you conclude from your graphs?

   d. Comment on the behaviour of $E_2(x)$, $E_4(x)$, and $E_6(x)$ as the values of x increase.

Chapter 6

Applications of Derivatives

We have developed techniques for calculating the derivatives of functions and interpreted the derivative as both the rate of change of a measurement and the slope of the graph of a function. In this chapter, we develop numerous applications, ranging from theoretical considerations to solving numerical problems to applications in biological systems and elsewhere.

First, we learn how to use the derivative to identify the **minima** and **maxima** of functions. The techniques developed allow us to study a wide range of **optimization** problems. In Section 6.2 we present three longer case studies in optimization, involving the strength of bones, the feeding patterns of bees, and the shape of a honeycomb. Next, we develop tools for reasoning about continuous and differentiable functions, such as the **Intermediate Value Theorem,** the **Extreme Value Theorem,** and the **Mean Value Theorem.**

Using **L'Hôpital's rule,** we are able to compare the behaviour of functions as their arguments approach zero or infinity. We develop the method of **leading behaviour,** which helps us look at complicated functions in a simple way, preserving their major features. We use the tangent line approximation to solve equations using **Newton's method.** By investigating the slope of the updating function, we derive powerful criteria for the **stability** of equilibrium points. The criteria are both geometric and algebraic in nature. Armed with our understanding of stability, we explore more complex systems, such as the **logistic dynamical system.**

6.1 Extreme Values of a Function

In this section we learn how to use the derivative to identify the **minimum** and the **maximum** values of a function. We study two important cases: the **relative extreme values** of a function on its domain, and the **absolute extreme values** of a continuous function on a closed interval.

Minima and Maxima

To start, we define extreme values.

Definition 6.1.1 We say that a function f has an **absolute** (or **global**) **maximum** at a number c if $f(x) \leq f(c)$ for all x in the domain of f. The value $f(c)$ is called the **absolute** (or **global**) **maximum value** of f.

A function f has an **absolute** (or **global**) **minimum** at a number c if $f(x) \geq f(c)$ for all x in the domain of f. The value $f(c)$ is called the **absolute** (or **global**) **minimum value** of f.

The maximum and minimum values are called the **extreme values.** To rephrase the definition, we say that $f(c)$ is an absolute maximum if f does not attain values larger than $f(c)$ anywhere in its domain (or does not attain values smaller than $f(c)$, if $f(c)$ is claimed to be the absolute minimum).

Note that a function can attain its absolute maximum or absolute minimum at many numbers. For example, $f(x) = \sin x$ attains its absolute maximum of 1 at infinitely

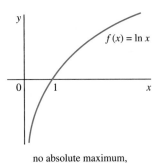
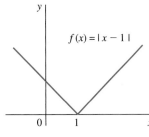
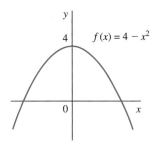

no absolute maximum, no absolute minimum

absolute minimum $f(1) = 0$, no absolute maximum

absolute maximum $f(0) = 4$, no absolute minimum

FIGURE 6.1.1

A function may or may not have absolute extreme values

many numbers $c = \frac{\pi}{2} + 2\pi k$ (k is an integer). Likewise, the minimum value of f, which is -1, occurs at all $c = \frac{3\pi}{2} + 2\pi k$.

Thus, what is unique about the absolute maximum (minimum) *is the value, and not the number(s) where it occurs.*

A function does not have to have absolute extreme values (the functions $f(x) = x$, $f(x) = x^3$, and $f(x) = \ln x$ do not have any). It might have an absolute minimum only ($f(1) = 0$ is the absolute minimum of $f(x) = |x - 1|$) or an absolute maximum only ($f(0) = 4$ is the absolute maximum of $f(x) = 4 - x^2$); see Figure 6.1.1.

If we apply Definition 6.1.1 to a constant function, $f(x) = K$, we realize that K is the absolute maximum value (attained at all numbers x) *and* the absolute minimum value (attained, again, at all numbers x).

Next, we define local extreme values.

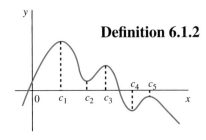

FIGURE 6.1.2

Absolute and relative extreme values

Definition 6.1.2 A function f has a **relative** (or **local**) **maximum** at a number c if $f(x) \leq f(c)$ for all x near c. The value $f(c)$ is called the **relative** (or **local**) **maximum value** of f.

A function f has a **relative** (or **local**) **minimum** at a number c if $f(x) \geq f(c)$ for all x near c. The value $f(c)$ is called the **relative** (or **local**) **minimum value** of f.

Recall that "for all x near c" means for all x in the domain of f that belong to some small open interval around c.

The function f in Figure 6.1.2 has absolute maximum value $f(c_1)$ and relative maximum values $f(c_1), f(c_3)$, and $f(c_5)$. It attains relative minimum values at c_2 and c_4 and has no absolute minimum value.

Example 6.1.1 Extreme Values of a Function Defined on a Closed Interval

Consider the function $f(x) = e^x \sin 3x$ on the interval $[-0.8, 2.2]$. Its graph is drawn in Figure 6.1.3.

The function $f(x)$ attains its absolute maximum value of approximately 2.81 at $x \approx 2.2$. The value $f(0.63) \approx 1.78$ is a relative maximum.

The value $f(-0.42) \approx -0.63$ is a relative minimum, and $f(x)$ has an absolute minimum (also a relative minimum) value of approximately -5.08 at $x \approx 1.68$.

Note A relative extreme value, $f(c)$, requires that we test $f(c)$ against the values of $f(x)$ for x in some open interval *around* c. At the endpoints of the interval we cannot do this, since any open interval that contains an endpoint cannot be fully contained within $[-0.8, 2.2]$. Therefore, a function cannot have relative extreme values at the endpoints of an interval where it is defined. That is why in the previous example we did not declare the absolute maximum value $f(2.2) \approx 2.81$ as a relative maximum value.

Having learned what the minimum and maximum are, we now start working on answering the first question that comes to mind—given a function, how do we find its extreme values, i.e., where do they occur?

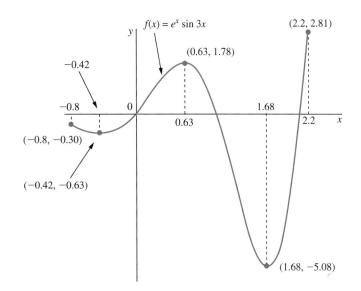

FIGURE 6.1.3

Graph of $f(x) = e^x \sin 3x$ on $[-0.8, 2.2]$

Consider the graph in Figure 6.1.4. We see that the relative minimum and maximum values occur at those numbers c where the graph has a horizontal tangent, or where the derivative does not exist (shown are the cases of a cusp and a discontinuity). Due to their importance, we give these numbers a name. (Note that we have already introduced critical numbers in Section 4.5. We repeat the definition for convenience.)

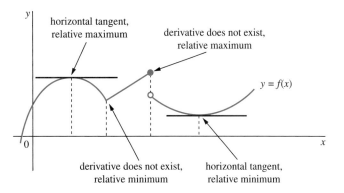

FIGURE 6.1.4

Extreme values of a function

Definition 6.1.3 A number c in the domain of a function f is called a **critical number** if $f'(c) = 0$ or if $f'(c)$ is not defined.

Often, instead of the term **critical number** we use the term **critical point,** and we say that *f* **has a critical point at** $x = c$.

Note We will use the terms *critical point* and *critical number* interchangeably, although there is a small difference. When we refer to the value of x (independent variable) *only,* the term *critical number* is more appropriate. However (assuming that x is a critical number), when we refer to the *point* $(x, f(x))$—say, when we talk about or draw graphs—we call it a critical point.

Example 6.1.2 Calculating Critical Numbers

a. To find all critical numbers of the function

$$f(x) = x^3 - 6x + 14$$

we find the derivative

$$f'(x) = 3x^2 - 6$$

Solving the equation
$$f'(x) = 3x^2 - 6 = 0$$
we get $x^2 = 2$ and $x = \pm\sqrt{2}$. Since $f'(x)$ is defined for all real numbers x, we conclude that $x = \pm\sqrt{2}$ are the only critical numbers.

b. Consider the function
$$g(x) = \sqrt[3]{x} = x^{1/3}$$
We compute
$$g'(x) = \frac{1}{3}x^{-2/3} = \frac{1}{3\sqrt[3]{x^2}}$$
The equation
$$g'(x) = \frac{1}{3\sqrt[3]{x^2}} = 0$$
has no solutions (the fraction is zero only if its numerator is zero).

However, $g'(x)$ is not defined at $x = 0$. Since $x = 0$ belongs to the domain of $g(x)$, we conclude that $x = 0$ is a critical number of $g(x)$.

c. Look at the graph of the function $f(x) = |x - 1|$ shown in Figure 6.1.1.

At $x = 1$ it has a corner, and so $f'(1)$ does not exist. If $x \neq 1$, then either $f(x) = x - 1$ (and $f'(x) = 1 \neq 0$) or $f(x) = -(x - 1) = -x + 1$ (and $f'(x) = -1 \neq 0$). Thus, $x = 1$ is the only critical number of the function $f(x)$. ▲

Our next theorem provides us with a crucial step in identifying extreme values.

Theorem 6.1.1 Fermat's Theorem

If f has a local minimum or a local maximum at a number c, and if $f'(c)$ exists, then $f'(c) = 0$.

The theorem says that in order to locate extreme values we need to look no farther than critical numbers. In other words, extreme values—if they exist—can only occur at the critical numbers.

What the theorem *does not claim* is that a critical number must yield an extreme value. The following example illustrates this point.

Example 6.1.3 Critical Numbers Might Not Yield Extreme Values

Consider the cubic function
$$f(x) = x^3$$
Solving
$$f'(x) = 3x^2 = 0$$
we get that $x = 0$ is a critical number. However, $f(x) = x^3$ is an increasing function, so it does not have a maximum or a minimum anywhere (Figure 6.1.5a).

Thus, $x = 0$ is a critical number but $f(0)$ is *not* an extreme value.

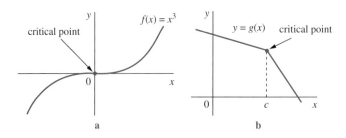

FIGURE 6.1.5

Critical numbers that do not yield extreme values

The function $g(x)$ in Figure 6.1.5b has a corner at $x = c$ (the value $g(c)$ is defined but the derivative $g'(c)$ does not exist) and so $x = c$ is a critical number. However, $g(c)$ is not an extreme value.

As we have just seen, in order to locate extreme values it does not suffice to find critical numbers. We need to have a test that will classify each critical number as a minimum, a maximum, or neither a minimum nor a maximum.

In the context of extreme values and their applications, we consider the following two important problems:

> **Extreme Values**
>
> (1) Given a function f in its domain, find its relative extreme values.
>
> (2) Given a continuous function f defined on a closed interval $[a, b]$, find its absolute extreme values.

We focus on question (1) first.

Assume that c is a critical number for a function $f(x)$. How do we decide whether or not $f(c)$ is an extreme value?

Using the function $f(x) = e^x \sin 3x$ from Example 6.1.1 (see Figure 6.1.3) to suggest the answer, we can see that $f(x)$ has a relative maximum at $c = 0.63$ because it is increasing on $(-0.42, 0.63)$ and decreasing on $(0.63, 1.68)$. In other words, $f(x)$ changes from increasing to decreasing as x passes through the critical number $c = 0.63$.

On the other hand, $f(x)$ changes from decreasing to increasing as x passes through the critical number $c = 1.68$. Thus, $f(1.68) = -5.08$ is a relative minimum.

We summarize these observations in the form of the **first derivative test.**

> **First Derivative Test**
> **Assume that f is continuous at c, where c is a critical number of f.**
>
> | | |
> |---|---|
> | If f changes from increasing to decreasing at c...
 If f' changes from positive to negative at c... | then f has a local maximum at c. |
> | If f changes from decreasing to increasing at c...
 If f' changes from negative to positive at c... | then f has a local minimum at c. |
> | If f does not change from decreasing to increasing nor from increasing to decreasing at c...
 If f' does not change sign at c... | then f has neither a local minimum nor a local maximum at c. |

The phrase "f' does not change sign" means that f' is either positive on both sides of c or negative on both sides of c.

Example 6.1.4 A Quadratic Function with a Maximum

Consider the function
$$g(x) = x(1 - x) = x - x^2$$

Its derivative is
$$g'(x) = 1 - 2x$$

The derivative $g'(x)$ is zero when $1 - 2x = 0$, i.e., when $x = 0.5$.

Since $g'(x) > 0$ for $x < 0.5$ and $g'(x) < 0$ for $x > 0.5$, the function $g(x)$ switches from increasing to decreasing (Figure 6.1.6). Thus, it has a local maximum at $x = 0.5$. The maximum value is $g(0.5) = 0.5(1 - 0.5) = 0.25$.

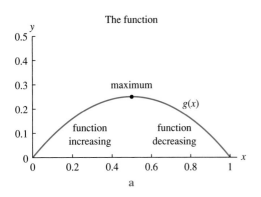
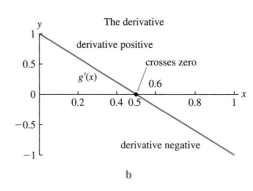

FIGURE 6.1.6

A function with a peak: $g(x) = x(1-x)$

Example 6.1.5 Finding the Maximum of a Cubic Function

Consider the function
$$f(x) = x^3 - x$$
To find the extreme values, we take the derivative
$$f'(x) = 3x^2 - 1$$
and solve the equation
$$f'(x) = 3x^2 - 1 = 0$$
$$x^2 = \frac{1}{3}$$
$$x = \pm\sqrt{\frac{1}{3}} = \pm\frac{1}{\sqrt{3}} \approx \pm 0.577$$

Since f' is defined for all x, there are no other critical points. To organize our calculations, we use the table below.

Table 6.1.1

Interval	$(-\infty, -1/\sqrt{3})$	$(-1/\sqrt{3}, 1/\sqrt{3})$	$(1/\sqrt{3}, \infty)$
First Derivative	positive	negative	positive
Function	increasing	decreasing	increasing

The function f changes from increasing to decreasing at $-1/\sqrt{3}$, and thus
$$f(-1/\sqrt{3}) \approx 0.385$$
is a local maximum. Similarly, we conclude that
$$f(1/\sqrt{3}) \approx -0.385$$
is a local minimum. See Figure 6.1.7.

Note There are other ways of keeping track of the results that we computed. For instance, we can add a bit more information to Table 6.1.1 to obtain Table 6.1.2.

Table 6.1.2

x	$x < -1/\sqrt{3}$	$x = -1/\sqrt{3}$	$-1/\sqrt{3} < x < 1/\sqrt{3}$	$x = 1/\sqrt{3}$	$1/\sqrt{3}$
$f'(x)$	+	0	−	0	+
$f(x)$	increasing	local maximum	decreasing	local minimum	increasing

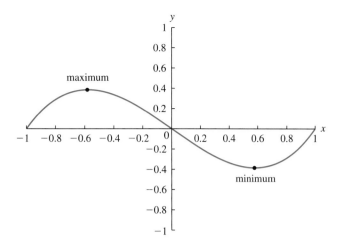

FIGURE 6.1.7

A function with a relative minimum and a relative maximum: $f(x) = x^3 - x$

Alternatively, we summarize our calculations in a diagram:

$$\begin{array}{c|c|c|c|c}
f'>0 & f'=0 & f'<0 & f'=0 & f'>0 \\
\hline
f \text{ increasing} & -1/\sqrt{3} & f \text{ decreasing} & 1/\sqrt{3} & f \text{ increasing}
\end{array} \longrightarrow x$$

Similar diagrams will be used in Chapter 8.

Example 6.1.6 Finding Extreme Values

Find the relative maximum and relative minimum values of the function

$$f(x) = x(x-5)^{1/3}$$

Note that the domain of f consists of all real numbers. Using the product rule and the chain rule, we find

$$f'(x) = (x-5)^{1/3} + x\frac{1}{3}(x-5)^{-2/3}$$

$$= (x-5)^{-2/3}\left(x - 5 + \frac{1}{3}x\right)$$

$$= \frac{1}{\sqrt[3]{(x-5)^2}}\left(\frac{4}{3}x - 5\right)$$

Requiring that $f'(x) = 0$, we get

$$\frac{4}{3}x - 5 = 0$$

$$x = \frac{15}{4}$$

So $x = 15/4$ is a critical point.

We note that f' is not defined at $x = 5$. Since $x = 5$ is in the domain of f, it is a critical point as well (vertical tangent case). As in the previous example, we analyze the first derivative in Table 6.1.3.

Table 6.1.3

Interval	$(-\infty, 15/4)$	$(15/4, 5)$	$(5, \infty)$
First Derivative	negative	positive	positive
Function	decreasing	increasing	increasing

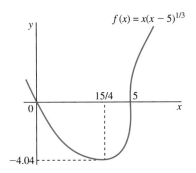

FIGURE 6.1.8
The graph of $f(x) = x(x-5)^{1/3}$

Thus, f is increasing on $(15/4, 5)$ and $(5, \infty)$ and decreasing on $(0, 15/4)$. It has a relative minimum value of $f(15/4) \approx -4.04$. Note that f does not have an extreme value at $x = 5$. See Figure 6.1.8.

An alternative to analyzing extreme values using the first derivative—in some cases—consists of using the information about concavity obtained from the second derivative. Looking at Figures 6.1.3 and 6.1.7, we see that at the points where f has a maximum its graph is concave down, and where it has a minimum its graph is concave up.

Let's show that this observation holds true in general. Assume that f has a critical number c such that $f'(c) = 0$ (what we will say does not hold for critical numbers where f' does not exist). Consider the second-degree Taylor polynomial of f at c:

$$T_2(x) = f(c) + f'(c)(x-c) + \frac{f''(c)}{2}(x-c)^2$$

$$= f(c) + \frac{f''(c)}{2}(x-c)^2$$

since $f'(c) = 0$. Near c, $f(x) \approx T_2(x)$, i.e.,

$$f(x) \approx f(c) + \frac{f''(c)}{2}(x-c)^2$$

Thus,

$$f(x) - f(c) \approx \frac{f''(c)}{2}(x-c)^2$$

If $f''(c) > 0$ (i.e., if f is concave up), then the right side above is positive, and

$$f(x) - f(c) > 0$$

In other words, $f(x) > f(c)$ for x near c, and $f(c)$ is a relative minimum.

Similarly, if f is concave down, then $f''(c) < 0$ and

$$f(x) - f(c) \approx \frac{f''(c)}{2}(x-c)^2 < 0$$

Thus $f(x) - f(c) < 0$, i.e., $f(x) < f(c)$ for x near c. Therefore, $f(c)$ is a relative maximum.

In the following table, we summarize the **second derivative test**:

Second Derivative Test
Assume that f'' is continuous near c and $f'(c) = 0$.
If $f''(c) > 0$, then f has a relative minimum at c.
If $f''(c) < 0$, then f has a relative maximum at c.
If $f''(c) = 0$, then the test provides no answer.

Example 6.1.7 What Happens When $f''(c) = 0$?

If we encounter a critical number c for which $f''(c) = 0$, we need to use other means to find out what is going on. As we will see, it could happen that $f(c)$ is a relative extreme value, or the graph of f might have a point of inflection at c.

Consider $f(x) = x^3$. Then $f'(x) = 3x^2$ (so $x = 0$ is a critical number) and $f''(x) = 6x$ (so $f''(0) = 0$). Since $f(x) = x^3$ is an increasing function, $f(0)$ is not an extreme value. In Example 5.6.1 in Section 5.6 we showed that $x = 0$ is a point of inflection.

For $f(x) = x^4$, we compute $f'(x) = 4x^3$ (so $x = 0$ is a critical number) and $f''(x) = 12x^2$ (so $f''(0) = 0$). Since f is decreasing for $x < 0$ and increasing for $x > 0$, we conclude that $f(0) = 0$ is relative minimum (see Example 5.6.2 and Figure 5.6.35).

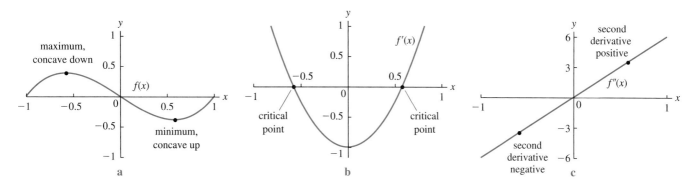

FIGURE 6.1.9
The function $f(x) = x^3 - x$, its derivative, and its second derivative

Example 6.1.8 Using the Second Derivative Test

In Example 6.1.5 we showed that $x = \pm 1/\sqrt{3}$ are critical numbers of the function $f(x) = x^3 - x$. We compute $f'(x) = 3x^2 - 1$ and $f''(x) = 6x$.

Since $f''(1/\sqrt{3}) = 6/\sqrt{3} > 0$, we see that $x = 1/\sqrt{3}$ is a relative minimum; from $f''(-1/\sqrt{3}) = -6/\sqrt{3} < 0$ we conclude that $x = -1/\sqrt{3}$ is a relative maximum (Figure 6.1.9).

Note that $f''(0) = 0$. Figure 6.1.9c shows that f'' changes sign as it passes through 0, so $(0, 0)$ is an inflection point of the graph of $f(x)$.

Example 6.1.9 Using the Second Derivative to Find Extreme Values

Consider the function $f(x) = x^2 e^{-x}$. We compute

$$f'(x) = 2xe^{-x} + x^2 e^{-x}(-1)$$
$$= (2x - x^2)e^{-x}$$

and

$$f''(x) = (2 - 2x)e^{-x} + (2x - x^2)e^{-x}(-1)$$
$$= (2 - 4x + x^2)e^{-x}$$

From

$$f'(x) = (2x - x^2)e^{-x} = x(2 - x)e^{-x} = 0$$

we conclude that $x = 0$ and $x = 2$ are critical points. (Note that there are no critical points coming from the requirement that $f'(x)$ does not exist.)

From the values of the second derivative,

$$f''(0) = 2e^{-0} = 2 > 0$$
$$f''(2) = -2e^{-2} < 0$$

we see that $f(0) = 0$ is a relative minimum and $f(2) = 4e^{-2} \approx 0.541$ is a relative maximum; see Figure 6.1.10.

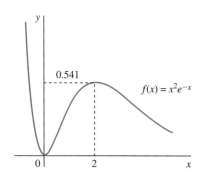

FIGURE 6.1.10
The graph of $f(x) = x^2 e^{-x}$

Example 6.1.10 The Tradeoff between Medication and Side Effects

Suppose that a patient is given a dosage, x, of some medication, and the probability of a cure is

$$P(x) = \frac{\sqrt{x}}{1+x}$$

What dosage maximizes the probability of a cure? Taking the derivative, we find

$$P'(x) = \frac{(1+x)\frac{d}{dx}(\sqrt{x}) - \sqrt{x}\frac{d}{dx}(1+x)}{(1+x)^2}$$

$$= \frac{\frac{1+x}{2\sqrt{x}} - \sqrt{x}}{(1+x)^2}$$

Multiply and divide the fraction by $2\sqrt{x}$:

$$P'(x) = \frac{1+x - 2\sqrt{x}\sqrt{x}}{(1+x)^2 2\sqrt{x}}$$

$$= \frac{1-x}{2\sqrt{x}(1+x)^2}$$

The derivative is zero when the numerator is zero, or when $1 - x = 0$, i.e., $x = 1$.

Since calculating $P''(x)$ is quite involved, we decide to use the first derivative test. Note that this is easy, as the denominator, $2\sqrt{x}(1+x)^2$, in $P'(x)$ is positive no matter which $x > 0$ we take (so the sign of $P'(x)$ is determined by the sign of its numerator).

If $x < 1$, then $1 - x > 0$, and $P(x)$ is increasing. If $x > 1$, then $1 - x < 0$, and $P(x)$ is decreasing. Thus, P has a relative maximum at $x = 1$; see Figure 6.1.11.

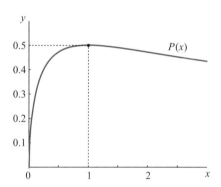

FIGURE 6.1.11

The tradeoff between treatment and side effects

Thus, the optimal dosage is $x = 1$. The fact that

$$P(1) = \frac{\sqrt{1}}{1+1} = \frac{1}{2}$$

says that the probability of a cure in this case is 50%.

Finding an optimal dosage is very important. Lower dosages might fail to cure the disease, whereas administering higher dosages could lead to side effects whose severity eclipses any possible benefits of the cure.

Example 6.1.11   Spread of a Pollutant

The concentration of a pollutant (measured in ppm = parts per million) at a fixed location x units from the source changes according to the formula

$$c(t) = \frac{N}{\sqrt{4\pi kt}} e^{-x^2/4kt}$$

where the positive quantities N and k characterize the pollutant and $t > 0$ is time. This model assumes that the pollutant is released at time $t = 0$ and then its source is shut off. When does the pollution reach its maximum value at a location x units from the source?

We rewrite $c(t)$ by separating constants from the factors involving the variable:

$$c(t) = \frac{N}{\sqrt{4\pi k}} t^{-1/2} e^{-x^2/4kt}$$

(It is common practice to leave the factor of 4 under the square root, instead of taking it out.) Using the product and the chain rules, we obtain

$$c'(t) = \frac{N}{\sqrt{4\pi k}} \left[-\frac{1}{2} t^{-3/2} e^{-x^2/4kt} + t^{-1/2} e^{-x^2/4kt} \left(-\frac{x^2}{4k}\right)(-1)t^{-2} \right]$$

$$= \frac{N}{\sqrt{4\pi k}} e^{-x^2/4kt} \left(-\frac{1}{2} t^{-3/2} + \frac{x^2}{4k} t^{-5/2} \right)$$

$$= \frac{N}{\sqrt{4\pi k}} e^{-x^2/4kt} \frac{1}{4} t^{-5/2} \left(-2t + \frac{x^2}{k} \right)$$

$$= \frac{N}{4\sqrt{4\pi k}} \frac{1}{t^{5/2}} e^{-x^2/4kt} \left(-2t + \frac{x^2}{k} \right)$$

All terms in front of the parentheses are positive, and therefore $c'(t) = 0$ when $-2t + x^2/k = 0$, i.e., when $t = x^2/(2k)$.

When $t < x^2/(2k)$, then $c'(t) > 0$ and $c(t)$ is increasing, and when $t > x^2/(2k)$, then $c'(t) < 0$ and $c(t)$ is decreasing. Thus, $c(t)$ has a local maximum at $t = x^2/(2k)$. The maximum concentration is

$$c(x^2/(2k)) = \frac{N}{\sqrt{4\pi k(x^2/(2k))}} e^{(-x^2/4k)(x^2/(2k))^{-1}}$$

$$= \frac{N}{\sqrt{2\pi x^2}} e^{(-x^2/4k)(2k/x^2)}$$

$$= \frac{N}{x\sqrt{2\pi}} e^{-1/2}$$

As the pollutant starts spreading from the source, the concentration at a location x increases, reaches its maximum when $t = x^2/(2k)$, and then starts decreasing. Note that as x increases, so does t; thus, the concentration at points farther away from the source reaches its maximum later than the concentration at locations closer to the source. As x increases, the value

$$c(x^2/(2k)) = \frac{\text{positive constant}}{x}$$

decreases. Thus, the maximum concentration is larger at locations closer to the source; see Figure 6.1.12.

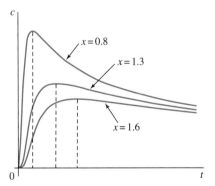

FIGURE 6.1.12

The concentration $c(t)$ with $N = 4$ and $k = 1$ at three different locations

Absolute Extreme Values

Now we turn to question (2) from page 349. To recall—we consider a continuous function f defined on a closed interval $[a, b]$. The Extreme Value Theorem (which we state in a moment) guarantees that f will have absolute extreme values. So, this situation is quite different from the context of question (1), where—as we witnessed— there are functions that do not have extreme values, or have only a maximum value or only a minimum value.

Theorem 6.1.2 Extreme Value Theorem

Assume that f is a continuous function defined on a closed interval $[a, b]$. Then f has an absolute maximum and an absolute minimum in $[a, b]$. That is, there exist numbers c_1 and c_2 in $[a, b]$ such that $f(c_1)$ is the absolute maximum and $f(c_2)$ is the absolute minimum.

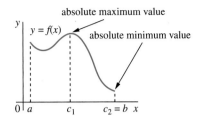

FIGURE 6.1.13
The Extreme Value Theorem

The proof of this theorem is not easy, so it is omitted. The conclusions are illustrated in Figure 6.1.13. The theorem does not guarantee that the maximum and minimum must occur strictly between a and b. Either might lie at one of the endpoints (the minimum in Figure 6.1.13 lies at the endpoint b).

The assumption on the continuity of the function is essential. If we remove it, the conclusion of the Extreme Value Theorem may no longer be true.

Neither of the two functions in Figure 6.1.14 is continuous on $[a, b]$. The absolute maximum value of f occurs at c. However, the function g does not have a maximum value. (Note that we did not say that, without continuity, the conclusion of the Extreme Value Theorem does not hold; we said "may not hold.")

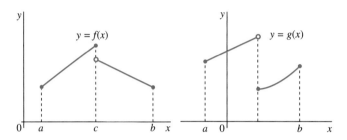

FIGURE 6.1.14
The Extreme Value Theorem does not apply

Likewise, if the interval is not closed, the theorem cannot be used. Its conclusion may or may not be true, as the functions drawn in Figure 6.1.15 demonstrate.

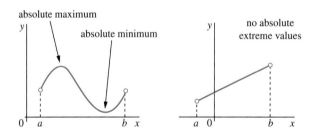

FIGURE 6.1.15
The Extreme Value Theorem does not apply

Example 6.1.12 Using the Extreme Value Theorem

Consider the function

$$f(x) = x \ln \frac{3}{1+x} = x(\ln 3 - \ln(1+x))$$

Checking whether or not $f(x)$ has extreme values requires that we calculate the derivative, $f'(x)$:

$$f'(x) = (\ln 3 - \ln(1+x)) - x\left(\frac{1}{1+x}\right) = -\frac{x}{1+x} + \ln\frac{3}{1+x}$$

and then solve the equation

$$f'(x) = 0$$

to find critical numbers. Unfortunately, this equation cannot be solved algebraically.

We now show how we can use the Extreme Value Theorem to prove that $f(x)$ does have a relative maximum.

Note that $f(0) = 0$ and $f(2) = 0$. As well, $f(x)$ is positive for $0 < x < 2$. So let's consider $f(x)$ on the interval $[0, 2]$. Since $f(x)$ is continuous, the Extreme Value Theorem

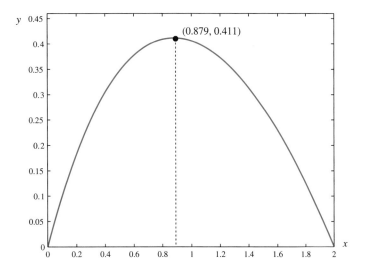

FIGURE 6.1.16

The graph of $f(x) = x(\ln 3 - \ln(1+x))$

guarantees the existence of a number c in $[0, 2]$ such that $f(c)$ is an absolute maximum. The fact that f is positive between 0 and 2 implies that $f(c) > 0$. Thus, the maximum value $f(c)$ cannot occur at the endpoints (because f is zero there), but rather *inside* the interval $[0, 2]$. We conclude that $f(c)$ is a *relative* maximum of f. ▲

Note A computer-generated graph of $f(x)$ is shown in Figure 6.1.16. Using Maple or some other equation solver, we find that the critical number is $x \approx 0.879$, and $f(0.879) \approx 0.411$ is the absolute maximum of $f(x)$ on $[0, 2]$. Later in this chapter we will learn two techniques—the bisection method (Section 6.3) and Newton's method (Section 6.6)—which will help us find numeric approximations to solutions of equations (such as the equation $f'(x) = 0$ in this example).

Let us repeat that the strength of the Extreme Value Theorem is that it guarantees the *existence* of absolute extreme values.

To find an absolute extreme value, we note that it has to occur at a critical number within the given interval (that's what Fermat's Theorem says) or at an endpoint (that's the possibility that is not captured by Fermat's Theorem):

How to Find Absolute Extreme Values
Assume that f is a continuous function defined on a closed interval $[a, b]$.

(1) Find all critical numbers (numbers where the derivative f' is equal to zero or where f is not differentiable).

(2) Compute the values of f at all critical numbers that are in $[a, b]$.

(3) Compute the values $f(a)$ and $f(b)$ of f at the endpoints of $[a, b]$.

(4) The largest value found in (2) and (3) is the absolute maximum, and the smallest value found in (2) and (3) is the absolute minimum.

Example 6.1.13 The Absolute Maximum and Minimum of a Cubic Function

To find the absolute minimum and maximum of the differentiable function $f(x) = x^3 - x$ for $-1 \leq x \leq 1$ (Figure 6.1.7), we follow the algorithm we just outlined.

Step (1): From $f'(x) = 3x^2 - 1 = 0$ we get that $x = \pm 1/\sqrt{3}$ are critical points (note that both belong to the given interval $[-1, 1]$).

Step (2): We compute

$$f(1/\sqrt{3}) \approx -0.385$$

$$f(-1/\sqrt{3}) \approx 0.385$$

Step (3): The values of f at the endpoints of the given interval are
$$f(-1) = 0$$
$$f(1) = 0$$

Step (4): Comparing the values from steps (2) and (3), we conclude that
$$f(-1/\sqrt{3}) \approx 0.385$$
is the absolute maximum value and
$$f(1/\sqrt{3}) \approx -0.385$$
is the absolute minimum value.

Example 6.1.14 Absolute Extreme Values on a Different Domain

If we extend the domain of the function $f(x) = x^3 - x$ from the previous example to the interval $[-1, 2]$, then, in our solution, we need to replace $f(1) = 0$ by $f(2) = 6$. Thus, the absolute maximum moves to $x = 2$, and the absolute minimum value is still $f(1/\sqrt{3}) \approx -0.385$; see Figure 6.1.17.

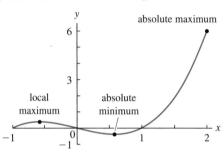

FIGURE 6.1.17

The absolute maximum of $f(x) = x^3 - x$ changes on an extended domain

Example 6.1.15 The Absolute Maximum and Minimum of a Function with a Corner

Consider the absolute value function $f(x) = |x|$ defined for $-2 \leq x \leq 3$ (Figure 6.1.18).
The only critical point is at $x = 0$ (f is not differentiable there; if $x \neq 0$ then $f'(x)$ is equal to either 1 or -1). The absolute extreme values of f must be among the values of f at the critical point
$$f(0) = 0$$
and at the endpoints
$$f(-2) = 2$$
$$f(3) = 3$$

Thus, $f(3) = 3$ is the absolute maximum value and $f(0) = 0$ is the absolute minimum value.

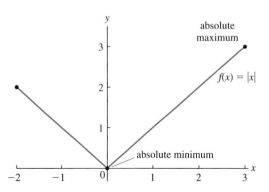

FIGURE 6.1.18

The absolute maximum and minimum of the absolute value function

Example 6.1.16 Finding Absolute Extreme Values

Find the absolute maximum and the absolute minimum of the function $f(x) = 2\sin x + \cos 2x$ on $[0, 2]$.

Solving
$$f'(x) = 2\cos x - 2\sin 2x = 2(\cos x - \sin 2x) = 0$$

with the help of the double-angle formula $\sin 2x = 2\sin x \cos x$, we obtain
$$\cos x - \sin 2x = 0$$
$$\cos x - 2\sin x \cos x = 0$$
$$\cos x(1 - 2\sin x) = 0$$

By solving $\cos x = 0$ we obtain $x = \pi/2$ (this is the only solution within $[0, 2]$). From $1 - 2\sin x = 0$ it follows that $\sin x = 1/2$ and thus there are two solutions in the main period of $\sin x$: $x = \pi/6$ and $x = 5\pi/6$ (see Example 2.3.8 in Section 2.3). However, only $x = \pi/6$ belongs to the interval $[0, 2]$. We conclude that $f(x)$ has two critical points inside $[0, 2]$: $x = \pi/6$ and $x = \pi/2$.

Finding the values of $f(x)$ at the endpoints and at the critical points inside the interval,
$$f(0) = 2\sin 0 + \cos 0 = 1$$
$$f(2) = 2\sin 2 + \cos 4 \approx 1.165$$
$$f(\pi/6) = 2\sin(\pi/6) + \cos(\pi/3) = \frac{3}{2}$$
$$f(\pi/2) = 2\sin(\pi/2) + \cos(\pi) = 1$$

we see that the absolute minimum of $f(x)$ on $[0, 2]$ is $f(0) = f(\pi/2) = 1$, and the absolute maximum is $f(\pi/6) = 3/2$. See Figure 6.1.19.

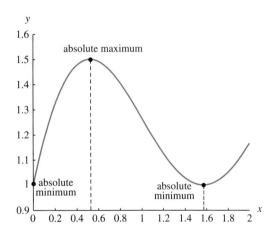

FIGURE 6.1.19
The graph of $f(x) = 2\sin x + \cos 2x$

We finish this section by discussing two more applications.

Example 6.1.17 Optimizing the Boundary of a Rectangle

Of all rectangles of area $A > 0$, identify the one with the smallest perimeter.

Denote the sides of the rectangle by x and y. It is given that $xy = A$, and we are asked to minimize the perimeter, $2x + 2y$ (Figure 6.1.20a). The requirement $xy = A$ gives $y = A/x$; this enables us to write the perimeter as a function of one variable:
$$p(x) = 2x + 2y = 2x + 2\frac{A}{x}$$

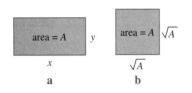

FIGURE 6.1.20

Optimizing the boundary of a rectangle

where $x > 0$. Now we minimize $p(x)$. We find

$$p'(x) = 2 - 2\frac{A}{x^2} = 0$$
$$x^2 = A$$

and so $x = \sqrt{A}$ (since x has to be positive). From $p''(x) = 4A/x^3$ it follows that

$$p''(\sqrt{A}) = 4\frac{A}{(\sqrt{A})^3} = 4\frac{A}{A^{3/2}} = \frac{4}{\sqrt{A}} > 0$$

Consequently, the second derivative test implies that

$$p(\sqrt{A}) = 2\sqrt{A} + 2\frac{A}{\sqrt{A}} = 4\sqrt{A}$$

is a relative minimum. Because $y = A/x = A/\sqrt{A} = \sqrt{A} = x$, we conclude that of all rectangles whose area is A, the square of side $\sqrt{A}$ has the smallest perimeter (Figure 6.1.20b).

Now consider not only rectangles, but *all* two-dimensional regions of area A. Which one has the smallest perimeter?

We can try to argue intuitively: of all rectangles of area A, the most optimal (in the sense of having the smallest perimeter) is the one that is the most symmetric, i.e., the square. Thus, we would expect that the most optimal shape of all regions in the plane is the one that is the most symmetric, namely the disk. It turns out that this is indeed so—but the proof is quite challenging and involves a lot more than calculus.

We can only illustrate this fact: the radius of the circle whose area is A is computed from $A = \pi r^2$ to be $r = \sqrt{A}/\sqrt{\pi}$. Its perimeter,

$$p = 2\pi r = 2\pi \frac{\sqrt{A}}{\sqrt{\pi}} = 2\sqrt{\pi}\sqrt{A} \approx 3.5449\sqrt{A}$$

is smaller than the perimeter of the square ($4\sqrt{A}$) we calculated above.

An analogous principle holds in three dimensions as well: of all regions in space with the same volume, the ball has the smallest surface area. In other words, to build a container holding a given volume with the least amount of material, we must build a sphere. That's why igloos (and domes) have the shape of a hemisphere.

Nature provides numerous illustrations: in cold weather an animal curls up, trying to approximate as closely as possible the shape of a sphere—knowing that minimizing its surface area will minimize heat loss (for the same reason we curl our fingers when it's cold), certain cells in our body are (approximately) spherical in shape, some birds build their nests in the shape of a sphere, and so on. The fact that optimization creates (or increases) symmetry is an important idea (which we cannot, unfortunately, pursue any further here).

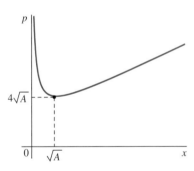

FIGURE 6.1.21

The graph of $p(x) = 2x + 2A/x$

Our analysis in Example 6.1.17 shows that the perimeter function $p(x)$ does not have a relative maximum. The graph of $p(x)$ is shown in Figure 6.1.21.

In words, we can build a rectangle with the given area A of *any* perimeter larger than $4\sqrt{A}$! For instance, the area of the rectangle with sides $10^{10}\sqrt{A}$ and $10^{-10}\sqrt{A}$ is A, and its perimeter is a bit larger than $2 \cdot 10^{10}\sqrt{A}$.

We can also reason geometrically: start with a rectangle of area A (Figure 6.1.22a), then cut off a small region (for instance a square) somewhere along its boundary and attach it someplace else (Figure 6.1.22b). This way, the perimeter increases, but the area does not change. We can repeat this process as many times as we wish (Figure 6.1.22c).

This idea explains why the shape of certain larger cells is not disk-like, but more like the shape in Figure 6.1.22e. The life-sustaining exchange of material between a cell and its surroundings occurs through its membrane. A small cell has a disk-like shape (Figure 6.1.22d) since its membrane is long enough to allow for an adequate amount of material to be exchanged. However, a large cell needs to exchange a large amount of material, and the circular shape of its membrane no longer suffices. By

FIGURE 6.1.22

Increasing the perimeter but keeping the area unchanged

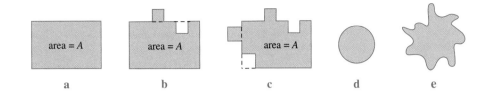

adjusting its shape (Figure 6.1.22e), the cell increases the length of its membrane, and thus increases, to the appropriate level, the amount of material exchanged with its surroundings.

Example 6.1.18 Optimizing the Work of the Heart

The human cardiovascular system is a network of blood vessels (arteries, veins, capillaries) that distribute blood to the organs and bring it back to the heart. The system operates by minimizing the energy that the heart needs to pump blood.

We consider two elements of the system: a single blood vessel and branching of a blood vessel into two vessels. The model of blood vessel that we use is a cylindrical tube of diameter d and length l; see Figure 6.1.23. Our assumption that the tube has fixed walls is a major simplification, since a blood vessel contracts and expands perpendicular to the flow of blood. Nevertheless, even with this simple model, we obtain valuable insights.

According to **Poiseuille's law**, the resistance to the flow of a fluid through a tube is proportional to the length of the tube and inversely proportional to the fourth power of the diameter:

$$R = K\frac{l}{d^4}$$

The positive constant K incorporates the properties of the fluid—in particular, its viscosity. (Viscosity is a measure of the resistance of a fluid to stress; in everyday situations, we think of water as being "thin," i.e., having low viscosity, and honey or pasta sauce as being "thick," i.e., having high viscosity.)

Minimizing the resistance to the flow of blood will optimize the work of the heart.

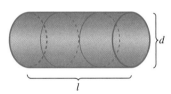

FIGURE 6.1.23

A model of a blood vessel

The resistance, R, can be increased in two ways: by making the tube longer or by making the diameter smaller. For instance, if the diameter of the tube gets reduced by 10% (in a blood vessel this happens due to a buildup of cholesterol, also called plaque), then

$$R(0.9d) = K\frac{l}{(0.9d)^4} = \frac{1}{0.9^4} \cdot K\frac{l}{d^4} \approx 1.52 \cdot K\frac{l}{d^4} = 1.52R(d)$$

i.e., the resistance increases by about 52%—this is the trouble with the narrowing of the arteries. In order to deliver the same amount of blood to the organs, the heart has to pump the blood much harder. Usually this is not possible, resulting in a smaller amount of blood delivered to various locations in the body, causing anything from very mild to life-threatening health conditions.

Now let us look at vascular branching (Figure 6.1.24): a main blood vessel of diameter d_1 branches into two smaller vessels, one of which has diameter d_2. By calculating the resistance and minimizing it, we will obtain the relationship between the branching angle and the size of the vessel.

We compute the resistance along the path from A to B to C (Figure 6.1.24). From the triangle BDC we compute

$$\sin\alpha = \frac{h}{BC} \quad \text{and} \quad \tan\alpha = \frac{h}{BD}$$

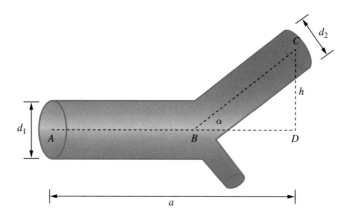

FIGURE 6.1.24

A model of vascular branching

and thus

$$BC = \frac{h}{\sin \alpha} = h \csc \alpha \quad \text{and} \quad BD = \frac{h}{\tan \alpha} = h \cot \alpha$$

Since

$$AB = a - BD = a - h \cot \alpha$$

we know the lengths that we need. The resistance from A to C is the sum of the resistances from A to B and from B to C. From Poiseuille's law, we get

$$R = K\frac{AB}{d_1^4} + K\frac{BC}{d_2^4} = K\frac{a - h\cot\alpha}{d_1^4} + K\frac{h\csc\alpha}{d_2^4}$$

We compute the branching angle, α, that will minimize the resistance. Thinking of R as a function of α, we differentiate:

$$R'(\alpha) = K\frac{0 - h(-\csc^2\alpha)}{d_1^4} + K\frac{h(-\csc\alpha\cot\alpha)}{d_2^4}$$

$$= Kh\csc\alpha\left(\frac{\csc\alpha}{d_1^4} - \frac{\cot\alpha}{d_2^4}\right)$$

Now $R'(\alpha) = 0$ implies $\csc \alpha = \frac{1}{\sin \alpha} = 0$ (no solutions for α) or

$$\frac{\csc\alpha}{d_1^4} - \frac{\cot\alpha}{d_2^4} = 0$$

$$\frac{1}{d_1^4 \sin\alpha} - \frac{\cos\alpha}{d_2^4 \sin\alpha} = 0$$

$$\frac{1}{d_1^4} - \frac{\cos\alpha}{d_2^4} = 0$$

$$\frac{\cos\alpha}{d_2^4} = \frac{1}{d_1^4}$$

$$\cos\alpha = \frac{d_2^4}{d_1^4}$$

(In the calculation above we were allowed to multiply by $\sin \alpha$ because $\alpha > 0$—otherwise there is no branching.) Thus, the resistance is minimized when the cosine of the angle is equal to the fourth power of the ratio of the diameters of the two vessels (in Exercise 60 we show that α is indeed a relative minimum of R). In Table 6.1.4 we show several branching angles for the given ratio of the smaller to the larger vessel. Thus, smaller blood vessels ideally branch off at larger angles.

Table 6.1.4

Ratio d_2/d_1	$\cos\alpha = d_2^4/d_1^4$	α (in degrees)
0.95	0.81451	35.46
0.9	0.6561	49.00
0.8	0.4096	65.82
0.7	0.2401	76.11

In an article on neural branching, we find a similar formula:

$$\cos\alpha = \frac{d_0^4 - d_2^4 - d_1^4}{2d_1^2 d_2^2}$$

where α is one of the branching angles, d_0 is the diameter of a parent neurite (the one that branches), and d_1 and d_2 are the diameters of the daughter neurites (the branches). [Source: Shea, O., et al. Biophysical constraints on neuronal branching: *Neurocomputing* 58–60: 487–495, 2004.]

Summary We have seen how to use the derivative to find **maxima** and **minima** of functions by locating **critical points** where the derivative is either zero or undefined. The **Extreme Value Theorem** guarantees that a continuous function defined on a domain that includes its endpoints must have a minimum and a maximum. To find **global** maxima and minima, we must compare values at critical points with values at the endpoints. When it is defined, the second derivative can be used to identify **local** minima and maxima. A critical point is a local minimum if the second derivative is positive and a local maximum if the second derivative is negative. If we cannot use the second derivative, we use the first derivative test. We applied the methods we learned to solve an area optimization problem and to understand how minimizing the resistance in a blood vessel determines the optimal angle for vascular branching.

6.1 Exercises

Mathematical Techniques

1. Write a formula for a function whose critical number is $x=1$.

2. Write a formula for a function whose critical numbers are $x=0$, $x=1$, and $x=2$.

3. Sketch the graph of a differentiable function that has four critical points.

4. Sketch the graph of a continuous function $f(x)$ that is increasing for all x and has two critical points.

5. Sketch the graph of a continuous function $f(x)$ that has a critical point at $x=3$ but has neither a minimum nor a maximum there. Can you find a formula for such a function?

6–21 ■ Find all critical numbers of each function.

6. $a(x) = \dfrac{x}{1+x}$
7. $f(x) = 1 + 2x - 2x^2$
8. $c(w) = w^3 - 3w$
9. $g(y) = \dfrac{y}{1+y^2}$
10. $h(z) = e^{z^2}$
11. $c(\theta) = \cos(2\pi\theta)$
12. $g(x) = x^3 - 2x^2 - 1$
13. $f(x) = x^5 - 15x$
14. $f(x) = x^{1/3} - x^{-2/3}$
15. $f(x) = \sin x$
16. $g(x) = \sin 2x$
17. $h(x) = xe^{4x}$
18. $f(x) = x^2 \ln x$
19. $g(x) = x - \cos x$
20. $f(x) = x^{1/2} - x^{3/2}$
21. $f(x) = |3x - 5|$

22–29 ■ Find the intervals on which the function $f(x)$ is increasing and on which it is decreasing. Identify all relative extreme values of $f(x)$.

22. $f(x) = x^3 - 2x^2 - 1$

23. $f(x) = x^4 - 2x^2 + 1$

24. $f(x) = \dfrac{x+1}{x^2+1}$

25. $f(x) = \dfrac{x+1}{x^2 - 2x}$

26. $f(x) = e^x + e^{-2x}$

27. $f(x) = x^2 e^{-x}$

28. $f(x) = x^2 \ln x$

29. $f(x) = \sin x + \cos x$

30–43 ■ Find the absolute minimum and the absolute maximum of each function on the interval given.

30. $a(x) = \dfrac{x}{1+x}$ for $0 \le x \le 1$.

31. $f(x) = 1 + 2x - 2x^2$ for $0 \le x \le 2$.

32. $c(w) = w^3 - 3w$ for $-2 \le w \le 2$.

33. $g(y) = \dfrac{y}{1+y^2}$ for $0 \le y \le 2$.

34. $h(z) = e^{z^2}$ for $0 \le z \le 1$.

35. $F(x) = |1 - x|$ for $0 \le x \le 3$.

36. $f(x) = x^2 + \dfrac{1}{x}$ on $[1, 2]$.

37. $g(x) = |x - 4|$ on $[0, 6]$.

38. $f(x) = \dfrac{x}{1+x^2}$ on $[0, 3]$.

39. $f(x) = \sin x$ on $[\pi/4, \pi]$.

40. $g(x) = \arctan x$ on $[-1, 1]$.

41. $f(x) = xe^{-x}$ on $[0, 4]$.

42. $f(x) = x - 2 \ln x$ on $[0.1, 10]$.

43. $f(x) = x + \sin x$ on $[-\pi, 6\pi]$.

44–49 ■ Draw the graph of a function with each set of properties.

44. A function with an absolute minimum and an absolute maximum between the endpoints.

45. A function with an absolute maximum at the left endpoint and an absolute minimum between the endpoints.

46. A differentiable function with an absolute maximum at the left endpoint, an absolute minimum at the right endpoint, and no critical points.

47. A function with an absolute maximum at the left endpoint, an absolute minimum at the right endpoint, and at least one critical point.

48. A function with an absolute minimum and an absolute maximum between the endpoints, but no point with derivative equal to zero.

49. A function that never reaches an absolute maximum.

50–55 ■ Find the second derivative at the critical points of each function. Classify the critical points as yielding local minima or local maxima. Use the second derivative to draw an accurate graph of the function for the given range.

50. $a(x) = \dfrac{x}{1+x}$ for $0 \le x \le 1$.

51. $f(x) = 1 + 2x - 2x^2$ for $0 \le x \le 2$.

52. $c(w) = w^3 - 3w$ for $-2 \le w \le 2$.

53. $g(y) = \dfrac{y}{1+y^2}$ for $0 \le y \le 2$.

54. $h(z) = e^{z^2}$ for $-1 \le z \le 1$.

55. $c(\theta) = \cos 2\pi\theta$ for $-1 \le \theta \le 1$.

56–59 ■ Suppose $f(x)$ is a positive function with a maximum at x^*. We can often find maxima and minima of other functions composed with $f(x)$. For each function $h(x) = g[f(x)]$:

a. Show that h has a critical point at x^*.

b. Compute the second derivative at this point.

c. Check whether the function has a minimum or a maximum and explain.

d. Check your result using the function $f(x) = xe^{-x}$ for $x \ge 0$, which has a maximum at $x = 1$. Sketch the graph of $f(x)$ and $h(x)$ in this case.

56. $g(f) = \dfrac{1}{f}$

57. $g(f) = 1 - f$

58. $g(f) = \ln f$

59. $g(f) = f - f^2$

60. Consider the expression

$$R'(\alpha) = Kh \csc\alpha \left(\dfrac{\csc\alpha}{d_1^4} - \dfrac{\cot\alpha}{d_2^4} \right)$$

for the resistance to the flow of blood in the vascular branching model (Example 6.1.18). Why is it reasonable to assume that $\csc\alpha > 0$? By analyzing the sign of the term in parentheses, show that the value of α given by $\cos\alpha = d_2^4/d_1^4$ yields a local minimum of R.

61. According to Poiseuille's law, by how much does the resistance increase if the diameter of the tube gets reduced by 30% (as in the case of severely clogged arteries)?

62. Find the optimal branching angle (in degrees) in the case where the diameter of the larger vessel is twice the diameter of the smaller one. To do this, use the formula given in Example 6.1.18.

Applications

63. Organic waste deposited in a lake at $t = 0$ decreases the oxygen content of the water. Suppose the oxygen content is $C(t) = t^3 - 30t^2 + 6000$ for $0 \le t \le 25$. Find the maximum and minimum oxygen content during this time.

64. The size of a population of bacteria introduced to a nutrient grows according to

$$N(t) = 5000 + \frac{30{,}000t}{100 + t^2}$$

Find the maximum size of this population for $t \geq 0$.

65. The **Shannon index** (also known as the **Shannon-Wiener index**) can be used to measure the diversity of species living in an ecosystem. In the case of two species, it is defined by the formula $H = -a \ln a - b \ln b$, where a is the percentage of species A in the ecosystem and b is the percentage of species B in the ecosystem.

   a. Using $a + b = 1$ (why is this true?), reduce H to a function of one variable.

   b. What is the maximum value of H and when does it occur? Explain what it means for the diversity of species in the given ecosystem.

66. **Simpson's diversity index** is a common way of assessing the diversity of species living in a region. (Although it is usually used for studying vegetation, the index has been applied to animals as well.) In the case of two species, it is defined by $D = 1 - a^2 - b^2$, where a is the percentage of species A in the ecosystem and b is the percentage of species B in the ecosystem. Find the maximum value of D and explain its meaning. (Hint: Use the fact that $a + b = 1$ to reduce D to a function of one variable.)

67. Find all points on the curve $y = 4/x$ that are closest to the origin.

68. Find all points on the line $y = 2x - 5$ that are closest to the point $(3, 0)$.

69. Find the point A on the x-axis that minimizes the sum of the squares of the distances from A to the origin and from A to the point $(1, 4)$.

70. Find the point A on the y-axis that minimizes the sum of the squares of the distances from A to the origin and from A to the point $(1, 4)$.

71. Find the rectangle of the largest area that can be inscribed in a circle of radius R.

72. Find the rectangle of the largest perimeter that can be inscribed in a circle of radius R.

73. Denote by α the angle between the two equal sides of an isosceles triangle. Which value of α will maximize the area of the triangle?

74. Find the rectangle of the largest area that can be inscribed in the region in the first quadrant bounded by the parabola $y = 9 - x^2$.

75. Find the rectangle of the smallest perimeter that can be inscribed in the region in the first quadrant bounded by the hyperbola $y = 1/x$.

76. Find the radius and the height of the closed cylindrical can (i.e., with a lid) that holds 20 cm^3 of liquid and has minimum surface area (that is, that requires the least amount of material to manufacture).

77. You have to construct a closed rectangular box with a square base using S square metres of material. Find the dimensions of the box that has the largest volume.

78. You have to construct a closed rectangular box with a square base of volume V cubic metres. Find the dimensions of the box that requires the least amount of material to construct.

79. A wall in a room contains two windows: one in the shape of a square and the other in the shape of a disk. If the total perimeter (i.e., the perimeter of both windows) is 6 m, what dimensions of the two windows will minimize the amount of light coming in?

80. Consider a window in the shape of a rectangle with a half-disk on top of it. If the perimeter of the window is P metres, what dimensions will maximize its area?

81–82 ▪ No calculus book is complete without optimization problems involving fences.

81. A farmer owns 1000 m of fence and wants to enclose the largest possible rectangular area. The region to be fenced has a straight canal on one side and a perpendicular and perfectly straight ancient stone wall on another. The area thus needs to be fenced on only two sides. What is the largest area she can enclose?

82. A farmer owns 1000 m of fence and wants to enclose the largest possible rectangular area. The region to be fenced has a straight canal on one side and thus needs to be fenced on only three sides. What is the largest area she can enclose?

6.2 Three Case Studies in Optimization

In this section we study the **strength of bones** and the **feeding patterns of bees,** and we try to understand why bees build their **honeycombs in the shape of a hexagon.**

Strength of Bones

Long bones (such as the bones in the legs) in mammals are hollow, filled with blood cell–producing marrow. Although lightweight, they are strong enough to support the entire body, enabling it to move in various ways (walking, running, climbing, jumping, etc.).

Adapted from: R. Alexander, Optima for Animals (Princeton: Princeton University Press) 1996.

In this example we focus on mechanical properties of bones, i.e., on their strength. To model the relatively straight part of a bone (called the diaphysis, or shaft) we use a hollow tube. Physics tells us that a hollow tube is stiffer and more resistant to forces (bending) than a solid tube of the same length and the same weight. The advantage of a hollow bone is that it is light—thus, less energy is required for any kind of movement.

Assume that a bone is a tube of radius r and that the marrow cavity has radius mr, where m is a number between 0 and 1; see Figure 6.2.25. The value $m = 0$ characterizes a solid bone, whereas values of m close to 1 describe thin-walled bones (note that m must be smaller than 1; the case $m = 1$ describes a bone that is all marrow).

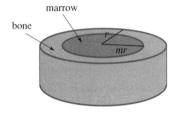

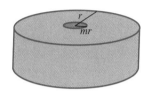

thick bone (m close to 0)

thin bone (m close to 1)

FIGURE 6.2.25
Model (cross-section) of a bone

A bone experiences a stress, also known as a bending moment, whenever the direction of the force acting on it is not in line (i.e., not parallel) to its axis of symmetry (= axis of rotation). The formula

$$r = \left(\frac{M}{K}\right)^{1/3} \left(1 - m^4\right)^{-1/3}$$

from engineering relates the radius, r, of the tube (bone) to the bending moment, M (assumed constant), that the tube is supposed to withstand. The positive constant K incorporates the strength of the material. The graph in Figure 6.2.26 shows the function $g(m) = \left(1 - m^4\right)^{-1/3}$. The radius, r, is a scaled version of $g(m)$.

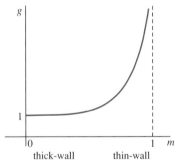

FIGURE 6.2.26
The graph of $g(m) = \left(1 - m^4\right)^{-1/3}$

As we can see from the formula for r and the graph of $g(m)$, smaller values of m (thick-walled bones) require a smaller radius r, whereas thin-walled bones (large m) require a larger radius to withstand the same momentum; see Figure 6.2.27.

Marrow is lighter than bone (we use the experimental fact stating that the density of marrow is about one-half of the density of bone). So thick-walled bones are narrow (small r), but heavy. Thin-walled bones are light, but need larger r. What is the most optimal ratio of bone to marrow? How can we make a bone that is light, but not compromise its strength?

To answer the question, we minimize the total mass, $T(m)$, i.e., the mass of the bone plus the mass of the marrow. Our calculations are based on a cross-section (this way, we work with area rather than with volume). We assume that the density, ρ, of a bone is constant.

The cross-sectional area of the bone is (look at Figure 6.2.25)

$$\pi r^2 - \pi(mr)^2 = \pi r^2 \left(1 - m^2\right)$$

and so its mass is (density times cross-sectional area)

$$\text{mass of bone} = \rho \pi r^2 \left(1 - m^2\right)$$
$$= \rho \pi \left[\left(\frac{M}{K}\right)^{1/3} \left(1 - m^4\right)^{-1/3}\right]^2 \left(1 - m^2\right)$$

thick-walled bone

thin-walled bone

FIGURE 6.2.27
These two bones are able to withstand the same momentum

$$= \rho\pi \left(\frac{M}{K}\right)^{2/3} \left(1 - m^2\right) \left(1 - m^4\right)^{-2/3}$$

The cross-sectional area of the marrow is $\pi(mr)^2$ and its mass is (the density is $\rho/2$)

$$\text{mass of marrow} = \frac{\rho}{2}\pi m^2 r^2$$

$$= \frac{\rho}{2}\pi m^2 \left[\left(\frac{M}{K}\right)^{1/3} \left(1 - m^4\right)^{-1/3}\right]^2$$

$$= \frac{1}{2}\rho\pi \left(\frac{M}{K}\right)^{2/3} m^2 \left(1 - m^4\right)^{-2/3}$$

The total mass is

$$T(m) = \rho\pi \left(\frac{M}{K}\right)^{2/3} \left(1 - m^2\right) \left(1 - m^4\right)^{-2/3} + \frac{1}{2}\rho\pi \left(\frac{M}{K}\right)^{2/3} m^2 \left(1 - m^4\right)^{-2/3}$$

$$= \rho\pi \left(\frac{M}{K}\right)^{2/3} \left(1 - m^2 + \frac{m^2}{2}\right) \left(1 - m^4\right)^{-2/3}$$

$$= \frac{1}{2}\rho\pi \left(\frac{M}{K}\right)^{2/3} \left(2 - m^2\right) \left(1 - m^4\right)^{-2/3}$$

Since ρ, K, and M are positive constants, we minimize the function

$$f(m) = \left(2 - m^2\right) \left(1 - m^4\right)^{-2/3}$$

We compute

$$f'(m) = -2m \left(1 - m^4\right)^{-2/3} + \left(2 - m^2\right) \left(-\frac{2}{3}\right) \left(1 - m^4\right)^{-5/3} \left(-4m^3\right)$$

$$= -2m \left(1 - m^4\right)^{-5/3} \left[\left(1 - m^4\right) - \frac{4}{3}\left(2 - m^2\right) m^2\right]$$

$$= -2m \left(1 - m^4\right)^{-5/3} \left[\frac{1}{3}m^4 - \frac{8}{3}m^2 + 1\right]$$

$$= -\frac{2}{3}m \left(1 - m^4\right)^{-5/3} \left(m^4 - 8m^2 + 3\right)$$

Now $f'(m) = 0$ implies that $m = 0$ or

$$m^4 - 8m^2 + 3 = 0$$

By using the quadratic formula we find $m^2 = 4 \pm \sqrt{13}$ and

$$m = \pm\sqrt{4 \pm \sqrt{13}}$$

Since m needs to be positive, we choose the plus sign for the outer square root; as well, m needs to be between 0 and 1, so that

$$m = \sqrt{4 - \sqrt{13}} \approx 0.628$$

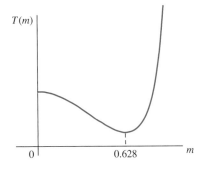

FIGURE 6.2.28

Mass of a bone as function of marrow cavity radius

(Note that the remaining critical number, $m = 0$, does not make sense in the context of this application: large bones of mammals are not solid.)

The graph of $f(m)$ is drawn in Figure 6.2.28. In Exercise 2 we show that $m = \sqrt{4 - \sqrt{13}}$ is indeed a minimum, as suggested by the graph.

Thus, of all bones that are capable of withstanding the same moment, the one whose marrow cavity radius is 62.8% of the total radius is the lightest. In Table 6.2.1 we show values of m for two different bones in several animals (source: McNeill Alexander, R. *Optima for Animals*. Princeton University Press, Princeton, NJ, 1996). Some are close to the value m that we obtained, but some are not (one possible reason for the discrepancy could lie in the fact that we assumed that the bone is a straight tube of uniform thickness).

Table 6.2.1

Animal	m for Femur (thigh)	m for Humerus (forelimb)
hare	0.57	0.55
fox	0.63	0.59
lion	0.56	0.42
camel	0.62	0.66
buffalo	0.54	0.51

Maximizing the Rate of Food Intake

Consider a honey bee going from flower to flower collecting nectar. After she finds a flower, she sucks up nectar at a slower and slower rate as the flower is depleted. However, she does not want to leave too soon because she must fly some distance to find the next flower. When should she give up and leave? If she stays a long time at each flower, she will get most of the nectar from that flower but will visit few flowers. If she stays a short time at each flower, she gets to skim the best nectar off the top but will spend most of her time flying to new flowers (Figure 6.2.29).

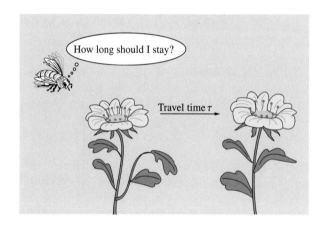

FIGURE 6.2.29
The bee's maximization problem

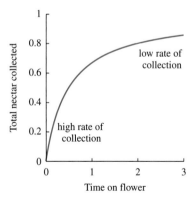

FIGURE 6.2.30
Nectar collected as a function of time

First, we must formulate this as a *maximization problem*. What is the bee trying to achieve? Her job is to bring back as much nectar as possible over the course of a day. To do so, she should maximize the *rate per visit,* which includes the travel time, τ, between flowers, measured in minutes. Suppose $F(t)$ is the amount of nectar collected by a bee that stays on a flower for time t (Figure 6.2.30 shows a likely shape of the graph of $F(t)$). The rate, $R(t)$, must take travel time into account:

$$\text{rate at which nectar is collected, } R(t) = \frac{\text{nectar per visit}}{\text{total time per visit}}$$

$$= \frac{\text{nectar per visit}}{\text{time on flower + travel time}}$$

$$= \frac{F(t)}{t + \tau}$$

Example 6.2.1 Computing the Maximum with a Particular Parameter Value

As a particular case, assume that the travel time, τ, is equal to 1 min and that

$$F(t) = \frac{t}{t + 0.5}$$

Then

$$R(t) = \frac{F(t)}{t+1} = \frac{t}{(t+0.5)(t+1)}$$

The derivative can be found with the quotient rule and some algebra to be

$$\frac{dR}{dt} = \frac{(t+0.5)(t+1) - t[(t+0.5) + (t+1)]}{[(t+0.5)(t+1)]^2}$$

$$= \frac{0.5 - t^2}{[(t+0.5)(t+1)]^2}$$

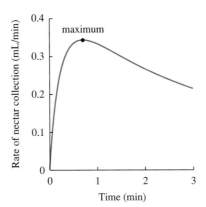

FIGURE 6.2.31

The average rate of return for the bee

This derivative is zero when the numerator is zero, or when

$$0.5 - t^2 = 0$$

which has a positive solution $t = \sqrt{0.5} \approx 0.707$. The derivative, dR/dt, is positive when $0 \le t < \sqrt{0.5}$ and negative for $t > \sqrt{0.5}$, implying that the value $t = \sqrt{0.5}$ gives a maximum (Figure 6.2.31). The coordinates of the maximum are $(\sqrt{0.5}, R(\sqrt{0.5})) \approx (0.707, 0.343)$. △

By looking at the problem in general, without substituting in a specific functional form for $F(t)$ or a value of τ, we can find a graphical method to solve the bee's problem. We differentiate $R(t)$ with the quotient rule, finding

$$\frac{dR}{dt} = \frac{(t+\tau)F'(t) - F(t)}{(t+\tau)^2}$$

Because the denominator is positive, the critical points occur where the numerator is zero, or

$$(t+\tau)F'(t) = F(t)$$

or

$$F'(t) = \frac{F(t)}{t+\tau} = R(t)$$

The solution of this equation is a local maximum as long as $F(t)$ is concave down (Exercise 12). This fundamental equation says that the bee should leave when the derivative of F, the instantaneous rate of nectar collection, is equal to the average rate. This **Marginal Value Property** (also referred to as the **Marginal Value Theorem**) is a powerful tool in ecology and elsewhere. The idea is simple: leave when you can do better elsewhere. More precisely, the optimum time that an animal collecting food remains in a certain small area (usually called a patch) is reached when the rate of food collection from that patch falls to a certain level (which is assumed to be the same for all patches). In our case, that level is the average rate of nectar collection, $F(t)/(t + \tau)$. The Marginal Value Theorem models the food collection situation accurately, in that it predicts that an animal

- remains a shorter time in a patch with less food than in a patch with more food,
- moves from one patch to another more quickly if the food-rich patches are close to each other, and
- moves more quickly from patch to patch if there is generally more food available.

Graphically, the slope of the food collection curve at the critical point is equal to the slope of the line connecting that point with a point at negative τ (Figure 6.2.32a), because that line has a "rise" of $F(t)$ and a "run" of $t + \tau$. The slope of the line is then

$$\frac{F(t)}{t + \tau} = R(t)$$

From the graph, we can see that the optimal time to remain becomes shorter when the travel time between flowers is shorter (Figure 6.2.32b).

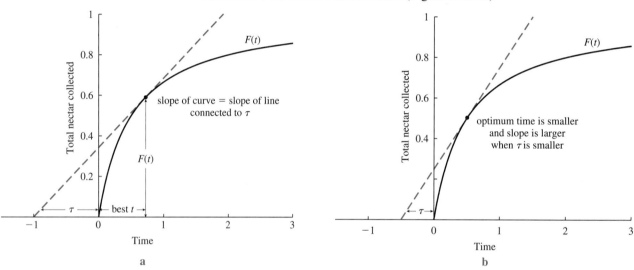

FIGURE 6.2.32
The Marginal Value Theorem: graphical method

Example 6.2.2 Using the Marginal Value Theorem to Solve a Maximization Problem

Consider the situation in Example 6.2.1 with the nectar collected as a function of time equal to

$$F(t) = \frac{t}{t + 0.5}$$

and a travel time of $\tau = 1$. In Example 6.2.1 we found that

$$R(t) = \frac{F(t)}{t + 1} = \frac{t}{(t + 0.5)(t + 1)}$$

by writing down the formula for $R(t)$ and differentiating. Alternatively, we can find the maximum with the Marginal Value Property by solving for the time when $F'(t) = R(t)$. The derivative of $F(t)$ can be found with the quotient rule as

$$F'(t) = \frac{(t + 0.5)\frac{dt}{dt} - t\frac{d(t + 0.5)}{dt}}{(t + 0.5)^2}$$

$$= \frac{(t+0.5)-t}{(t+0.5)^2}$$

$$= \frac{0.5}{(t+0.5)^2}$$

To find the maximum, we solve $F'(t) = R(t)$, or

$$\frac{0.5}{(t+0.5)^2} = \frac{t}{(t+0.5)(t+1)}$$

$$\frac{0.5}{t+0.5} = \frac{t}{t+1}$$

$$0.5(t+1) = t(t+0.5)$$

$$0.5t + 0.5 = t^2 + 0.5t$$

$$0.5 = t^2$$

$$t \approx 0.707$$

This matches the result found by differentiating $R(t)$ directly.

Example 6.2.3 The Effect of Increasing Travel Time

Suppose that the nectar collected as a function of time follows

$$F(t) = \frac{t}{t+0.5}$$

as in Example 6.2.2, but that the travel time is $\tau = 10$. Then

$$R(t) = \frac{F(t)}{t+10} = \frac{t}{(t+0.5)(t+10)}$$

The Marginal Value Property states that the optimal time for the bee to depart is when $F'(t) = R(t)$. We found the derivative of $F(t)$ in Example 6.2.2, so we can find the maximum by solving

$$\frac{0.5}{(t+0.5)^2} = \frac{t}{(t+0.5)(t+10)} \quad \text{the equation } F'(t) = R(t)$$

$$\frac{0.5}{t+0.5} = \frac{t}{t+10}$$

$$0.5(t+10) = t(t+0.5)$$

$$0.5t + 5 = t^2 + 0.5t$$

$$5 = t^2$$

$$t \approx 2.236$$

As expected, the optimal time to remain on the flower becomes longer when travel time becomes longer. Nonetheless, the bee still spends a smaller *fraction* of its time on flowers in this case. With $\tau = 10$, the fraction of time on flowers is

$$\text{fraction of time on flowers} = \frac{\text{time on flowers}}{\text{time on flowers} + \text{travel time}}$$

$$= \frac{2.236}{2.236 + 10} \approx 0.18$$

This bee is predicted to spend 18% of its time on flowers. The bee in Example 6.2.2, in contrast, is predicted to spend

$$\text{fraction of time on flowers} = \frac{0.707}{0.707 + 1} \approx 0.41$$

or 41% of its time on flowers.

Optimizing Area and the Shape of a Honeycomb

We start with a common optimization situation to introduce our reasoning about the structure of a honeycomb.

Consider the following problem: with a piece of a rope of length 12 metres we have to enclose a rectangular region of maximum area. In Figure 6.2.33a we show one case: the sides are of length 2 and 4 (so the perimeter is 12, as it is supposed to be), and the area is $2 \cdot 4 = 8$.

Now take away a bit from the longer side and add it to the shorter one (so that we do not change the perimeter); for example, consider the rectangle with sides 2.4 and 3.6, shown in Figure 6.2.33b. Its area is $2.4 \cdot 3.6 = 8.64$, so we did better! To improve further, we do the same: take away, say, 0.3 from the longer side and add to the shorter, so that the sides are 2.7 and 3.3; see Figure 6.2.33c. This time, the area is $2.7 \cdot 3.3 = 8.91$. So, it seems that, as long as one side of the rectangle is longer than the other, we can make the area larger by taking a bit away from the longer side and giving it to the shorter side. Continuing in this way, we would arrive at the rectangle with sides 3 and 3 (i.e., the square), whose area is 9. And that seems to be our answer! In words, of all rectangles with a fixed perimeter, the square has the largest area.

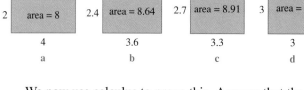

FIGURE 6.2.33
Optimizing area with a given perimeter

We now use calculus to prove this. Assume that the perimeter of a rectangle is P, and we need to maximize its area. Label one of its sides by x; then the other side (call it y) is obtained from the requirement that the perimeter is P (Figure 6.2.34). So from $2x + 2y = P$ we obtain

$$y = \frac{P}{2} - x$$

The area of the rectangle is

$$A(x) = xy = x\left(\frac{P}{2} - x\right) = \frac{P}{2}x - x^2$$

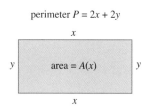

FIGURE 6.2.34
Optimizing area with a given perimeter

where $x > 0$ (and, of course, limited by the requirement on the perimeter). From

$$A'(x) = \frac{P}{2} - 2x = 0$$

we obtain the critical point $x = P/4$. Note that the graph of $A(x)$ is a parabola, so $x = P/4$ is the x-coordinate of its vertex. Since the parabola opens downward, $x = P/4$ gives a maximum, and the maximum area is

$$A\left(\frac{P}{4}\right) = \frac{P}{2}\frac{P}{4} - \left(\frac{P}{4}\right)^2 = \frac{P^2}{16}$$

So the optimal rectangle (in the context we discussed) is the square of side length $P/4$.

Now look at regular polygons (i.e., all sides and all angles are equal) with the same perimeter, P. Which one has the largest area?

6.2 Three Case Studies in Optimization

Consider a regular polygon with n sides, and label its side length by a; Figure 6.2.35a. Then $P = na$, and the length of each side is $a = P/n$.

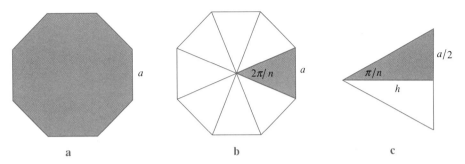

FIGURE 6.2.35
Regular polygon

We divide the polygon into n isosceles triangles (Figure 6.2.35b) and calculate the area of one of them (Figure 6.2.35c). From

$$\tan(\pi/n) = \frac{a/2}{h}$$

we find

$$h = \frac{a}{2} \cot(\pi/n)$$

and therefore the area of a triangle is

$$\frac{1}{2} \cdot a \cdot \frac{a}{2} \cot(\pi/n) = \frac{a^2}{4} \cot(\pi/n) = \frac{(P/n)^2}{4} \cot(\pi/n) = \frac{P^2}{4n^2} \cot(\pi/n)$$

The area of the polygon is

$$A(n) = n \frac{P^2}{4n^2} \cot(\pi/n) = \frac{P^2}{4} \cdot \frac{1}{n} \cot(\pi/n)$$

We are asked to find the value of n that maximizes $A(n)$.

To use calculus, we analyze the function

$$A(x) = \frac{P^2}{4} \cdot \frac{\cot(\pi/x)}{x}$$

for $x \geq 3$. Differentiate:

$$A'(x) = \frac{P^2}{4} \cdot \frac{-\csc^2(\pi/x)\left(-\frac{\pi}{x^2}\right)x - \cot(\pi/x)}{x^2}$$

$$= \frac{P^2}{4} \cdot \frac{\pi \csc^2(\pi/x) - x \cot(\pi/x)}{x^3}$$

In Exercise 23 we show that $A'(x) > 0$ for $x \geq 3$. In other words, $A(x)$ has no critical points—it is an increasing function for all $x \geq 3$. Using a graphing calculator or math software, we obtain the values in Table 6.2.2 and the graph in Figure 6.2.36.

Table 6.2.2

Number of Sides, n	Area Is $P^2/4$ Times
4	0.25
10	0.30777
50	0.31789
100	0.31821

Going back to the polygons, the answer is—the more sides, the larger the area! Thus, it seems plausible to conclude that of all polygons with a given perimeter, the polygon with infinitely many sides (i.e., the circle) encloses the largest area.

374 Chapter 6 Applications of Derivatives

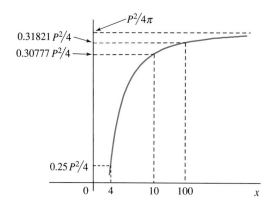

FIGURE 6.2.36
Graph of $A(x)$ (axes not to scale)

In Exercise 24 we show that the limit of the numbers in Table 6.2.2 is
$$\lim_{x \to \infty} \frac{\cot(\pi/x)}{x} = \frac{1}{\pi} \approx 0.31831$$
which gives a maximum area of
$$A(x) = \frac{P^2}{4} \cdot \frac{1}{\pi} = \frac{P^2}{4\pi}$$
The radius of the circle whose perimeter is P is equal to $r = P/2\pi$. The area is
$$A = \pi \left(\frac{P}{2\pi}\right)^2 = \frac{P^2}{4\pi}$$
exactly as above!

Bees are clever and want to use the least amount of material to build their honeycomb. We have just shown that the circle is the most optimal shape. That is true, if we want to build one cell. But a honeycomb has many cells, and therein lies another problem. It is not possible to arrange circles so that there are no gaps; see Figure 6.2.37.

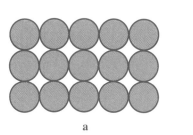
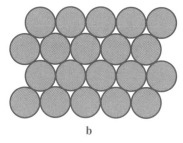

a b

FIGURE 6.2.37
Arranging circles

Long ago, mathematicians proved that there are only three regular polygons that one can use to cover a plane with no gaps and no overlaps: triangles, squares, and hexagons. In other words, to tile a floor *with one type of tile only,* we have three choices: equilateral triangle, square, or hexagon. See Figure 6.2.38.

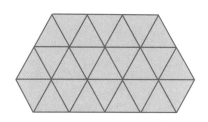
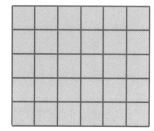
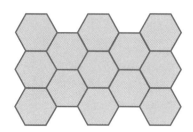

FIGURE 6.2.38
Covering a plane with no gaps and no overlaps

FIGURE 6.2.39
Honeycomb with hexagonal cells
StudioSmart/Shutterstock

Now we put everything together: circles are not good for cell shape because they generate waste (of space as well as of wax needed to build them). So the bees select the shape (among the triangle, square, and hexagon—no gaps, hence no waste) that's closest to a circle—hence the hexagon! (See Figure 6.2.39.)

Summary We modelled a bone as a tube and used optimization techniques to find the optimal (in the sense of maximum strength) ratio between the thickness of the bone wall and the marrow cavity. Next, we found the optimal length of time a bee should spend harvesting nectar from a flower. The solution is an example of the **Marginal Value Property,** which states that the best time to leave occurs when the rate of collecting resources falls below the average rate. In order to minimize the amount of material and avoid gaps, bees build their honeycombs with hexagonal cross-sections.

6.2 Exercises

1–4 ■ Consider the case study on the strength of bone.

1. The formula $r = (M/K)^{1/3} (1 - m^4)^{-1/3}$ shows how the radius of the bone relates to the amount of marrow inside the bone. Compute $r'(0.1)$, $r'(0.5)$, and $r'(0.9)$, and interpret your answers.

2. By using the first derivative test, show that the number $m = \sqrt{4 - \sqrt{13}}$ is a relative minimum of the function $f(m) = (2 - m^2)(1 - m^4)^{-2/3}$.

3. We made the assumption that the density of marrow is about one-half the density of bone. Assuming that the density of marrow is 55% of the density of bone, minimize the total mass. Comment on how the minimum obtained this time compares to the values in Table 6.2.1.

4. We made the assumption that the density of marrow is about one-half the density of bone. Assuming that the density of marrow is 45% of the density of bone, minimize the total mass. Comment on how the minimum obtained this time compares to the values in Table 6.2.1.

5–6 ■ Consider the bee confronted by the problem described in the subsection Maximizing the Rate of Food Intake.

5. Find the optimal strategy with the following travel times, τ, and illustrate the graphical method of solution. For each value of τ, find the equation of the tangent line at the optimal t and show that it goes through the point $(-\tau, 0)$.

   a. $\tau = 2$
   b. $\tau = 0.5$
   c. $\tau = 0.1$

6. Find the solution in general (without substituting a value for τ). What is the limit as τ approaches zero? Does this answer make sense?

7–8 ■ Suppose that the total food collected by a bee follows

$$F(t) = \frac{t}{c + t}$$

where c is some parameter.

7. If $\tau = 1$, find the optimal departure time in the following circumstances.

   a. $c = 2$
   b. $c = 1$
   c. $c = 0.1$

8. Find the solution in general (without substituting a value for c). What does the parameter c mean biologically (think about how long it takes the bee to collect half the nectar)? Explain why the bee leaves sooner when c is smaller.

9–10 ■ Mathematical models can help us to estimate values that are difficult to measure. Consider again a bee sucking nectar from a flower, with

$$F(t) = \frac{t}{0.5 + t}$$

9. If the bee remains for a length of time, t, on the flower, estimate the travel time, τ, assuming that the bee understands the Marginal Value Theorem, for the following values of t:

a. $t = 1$
b. $t = 0.1$
c. $t = 4$

10. Find the solution in general (without substituting a value for t).

11–12 ▪ We never showed that the value found in computing the optimal t with the Marginal Value Theorem is in fact a maximum. For each of the following forms for the function $F(t)$, find the second derivative of $R(t)$ at the point where $F'(t) = \dfrac{F(t)}{t+\tau}$, and check whether the solution is a maximum.

11. Suppose $F(t) = \dfrac{t}{1+t}$ and travel time is $\tau = 1$.

12. Suppose $F(t)$ is any function with $F''(t) < 0$ and travel time is $\tau = 1$.

13–16 ▪ Animals must survive predation in addition to maximizing their rate of food intake. One theory assumes that they try to maximize the ratio of food collected relative to predation risk. Suppose that different flowers with nectar of quality n attract $P(n)$ predators. For example, flowers with higher-quality nectar (large values of n) might attract more predators (large value of $P(n)$). Bees must decide which flowers to use. For each of the following forms of $P(n)$, find the function $R(n) = n/P(n)$ that the bees are trying to maximize, and find the optimal n.

13. Suppose that $P(n) = 1 + n^2$. Find the optimal n for the bees.

14. Suppose that $P(n) = 1 + n$. Find the optimal n for the bees and draw a graph like that for the Marginal Value Theorem. Does this make sense? Why is the result so different?

15. Find the condition for the maximum for a general function $P(n)$ by solving for $P'(n)$. Use this condition to find the optimal n for the cases $P(n) = 1 + n^2$ and $P(n) = 1 + n$.

16. Find a graphical interpretation of the condition in the previous problem, and test it on $P(n) = 1 + n^2$ and $P(n) = 1 + n$.

17–18 ▪ Find the maximum harvest from a population following the discrete-time dynamical system

$$N_{t+1} = rN_t(1 - N_t) - hN_t$$

for the given values of r.

a. Find the equilibrium population as a function of h. What is the largest h consistent with a positive equilibrium?
b. Find the equilibrium harvest as a function of h.
c. Find the harvesting effort that maximizes harvest.
d. Find the maximum harvest.

17. $r = 2$

18. $r = 1.5$

19–20 ▪ Find the conditions for stability of the equilibrium of

$$N_{t+1} = rN_t(1 - N_t) - hN_t$$

for each value of r. Show that the equilibrium N^* is stable when h is set to the value that maximizes the long-term harvest. Graph the updating function and cobweb.

19. $r = 2.5$

20. $r = 1.5$, as in Exercise 18.

21–22 ▪ Calculate the maximum long-term harvest for an alternative model of competition that obeys the discrete-time dynamical system

$$N_{t+1} = \dfrac{rN_t}{1 + kN_t} - hN_t$$

Try the following steps for the given values of the parameters r and k.

a. Find the equilibrium as a function of h.
b. What is the largest value of h consistent with a positive equilibrium?
c. Find the harvest level giving the maximum long-term harvest.
d. Sketch the graph of $P(h)$ and compute the value at the maximum.

21. With $r = 2.5$ and $k = 1$.

22. With $r = 1.5$ and $k = 1$.

23. This exercise relates to the calculation of the optimal area for a honeycomb. We show that the numerator, $\pi \csc^2(\pi/x) - x\cot(\pi/x)$, in the expression for $A'(x)$ is positive when $x > 0$.

a. Consider the function $f(x) = x - \sin x$. Show that $f(x)$ is increasing for all $x > 0$, and conclude that $\sin x < x$ for all $x > 0$.
b. Show that $\pi \csc^2(\pi/x) - x\cot(\pi/x) > 0$ is equivalent to

$$\pi - x\sin(\pi/x)\cos(\pi/x) > 0$$

c. Using the double-angle formula, show that the inequality in (b) is further equivalent to

$$\sin(2\pi/x) < \dfrac{2\pi}{x}$$

d. Use (a) to finish the proof.

24. Using the fact that $\lim\limits_{h \to 0} \dfrac{\sin h}{h} = 1$ or L'Hôpital's rule (which will formally be introduced later in this chapter), show that

$$\lim_{x \to \infty} x\sin(\pi/x) = \pi$$

and use it to prove the formula

$$\lim_{x \to \infty} \dfrac{\cot(\pi/x)}{x} = \dfrac{1}{\pi}$$

which was used in the honeycomb case study.

6.3 Reasoning about Functions: Continuity and Differentiability

Continuous and differentiable functions have many useful properties that can be employed to reason about the biological processes they describe without doing a great deal of algebra. In particular, we can make deductions about the solutions of equations, the existence of maxima and minima, or the values of the derivative. We use the **Intermediate Value Theorem** to show, without solving any equations, that a discrete-time dynamical system has an equilibrium; we have already used the **Extreme Value Theorem** to show, without computing any derivatives, that a function has a maximum; the **Mean Value Theorem** helps us find the value of a derivative without taking any limits.

Continuous Functions: The Intermediate Value Theorem

Consider the dynamical system model (i.e., a recursive relation)

$$p_{t+1} = 3.2 p_t e^{-p_t} + 6.5$$

where p_t represents population at time t. Models such as this one are used to study reproduction in fisheries (we will briefly discuss a somewhat simpler model—called the Ricker model—in Section 6.8). As with any dynamical system, it is important to know if this model has one or more equilibrium points.

Recall that, in order to find equilibrium points, we have to solve the equation

$$p^* = 3.2 p^* e^{-p^*} + 6.5$$

for p^*. The problem we face here is that we cannot solve this equation algebraically (i.e., there is no formula for it).

We rewrite the equation in the form

$$f(x) = 0$$

where

$$f(x) = 3.2 x e^{-x} + 6.5 - x$$

Thus, the solutions of the equation $f(x) = 0$ (if any) are equilibrium points of the given dynamical system. Note that f is a continuous function, and that—geometrically—we are looking for its x-intercepts.

To get a feeling for f, using a calculator we compute its value for several values of x:

$$f(0) = 6.5$$
$$f(1) \approx 6.67721$$
$$f(3) \approx 3.97796$$
$$f(5) \approx 1.60781$$
$$f(7) \approx -0.47957$$

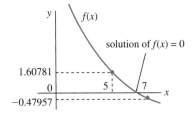

FIGURE 6.3.40

The graph of $f(x) = 3.2xe^{-x} + 6.5 - x$ must cross the x-axis

Think of the graph of f. At $x = 0$, $x = 1$, $x = 3$, and $x = 5$ it lies above the x-axis, and at $x = 7$ it is below the x-axis. Thus, if the graph of f has no gaps and no holes (and it does not have them, since f is continuous), it seems logical to conclude that it must cross the x-axis somewhere between $x = 5$ and $x = 7$; see Figure 6.3.40. Thus, the equation $f(x) = 0$ has a solution, and it lies between $x = 5$ and $x = 7$.

That our argument is mathematically sound is due to the theorem that we will state in a moment. Thus, if a *continuous* function assumes a positive value somewhere and a negative value somewhere else, then there must be a number where the function is zero.

Here is another example: if the water level in a lake rises from 2.4 m on Monday to 6.7 m on Thursday, then some time between Monday and Thursday the water level had to reach 5.5 m at least once (it could happen that the water level oscillated and reached 5.5 m several times).

Theorem 6.3.1 Intermediate Value Theorem

If $f(x)$ is a continuous function defined on a closed interval $[a, b]$ and N is a number between $f(a)$ and $f(b)$, then there is a number c in $[a, b]$ such that $f(c) = N$.

Although the statement of the theorem looks fairly obvious, the proof is not. The proof requires facts and techniques that are beyond what can be done at this level, so it is usually postponed until upper-year analysis textbooks. The idea of the Intermediate Value Theorem is shown in Figure 6.3.41. The theorem guarantees that there is *at least* one crossing point, but there may be more.

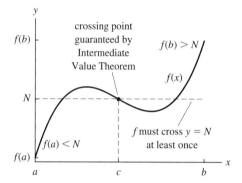

FIGURE 6.3.41

The Intermediate Value Theorem

Example 6.3.1 The Intermediate Value Theorem Applied to Height

Physically, the Intermediate Value Theorem says that if you grew from 2 ft to 6 ft in height, you must have been exactly 4 ft tall at some time (Figure 6.3.42). How does this fit into the Intermediate Value Theorem?

Denote by $H(t)$ the height (in feet); the variable t represents age in years. $H(t)$ is assumed to be continuous (although we cannot prove it mathematically, it sounds reasonable!). The interval over which $H(t)$ is defined is, say, $[0, 15]$. The value $N = 4$ that we want H to assume is between the values of H at the endpoints, i.e., between $H(0) = 2$ and $H(15) = 6$. So all assumptions of the theorem are satisfied. The theorem claims that there is a number c between 0 and 15 such that $H(c) = N = 4$.

In words, at a certain age c your height was exactly 4 ft.

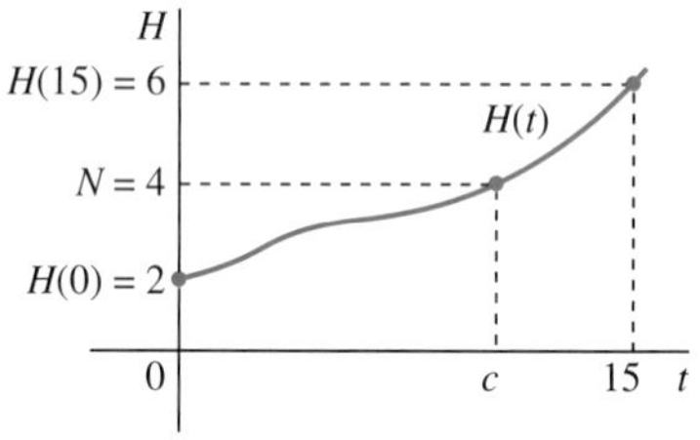

FIGURE 6.3.42

The Intermediate Value Theorem applied to height

Example 6.3.2 Application of the Intermediate Value Theorem

If you accelerate from 0 to 60 km/h in 10 seconds, you must have been going exactly 31.4 km/h at some time. In this case, the function $f(t)$ is the speed in kilometres per hour at time t, where t is measured in seconds. It sounds reasonable to assume that

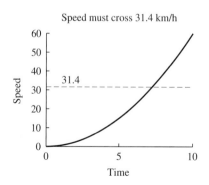

FIGURE 6.3.43

The Intermediate Value Theorem applied to speed

$f(t)$ is continuous. The domain of f is the interval $[0, 10]$; see Figure 6.3.43. The value $N = 31.4$ that we want f to assume lies between $f(0) = 0$ and $f(10) = 60$.

The Intermediate Value Theorem implies that at some time c between 0 seconds and 10 seconds, $f(c) = 31.4$. That is, the speed, f, assumed the exact value of 31.4 km/h.

Example 6.3.3 The Intermediate Value Theorem Applied to Solving an Equation

Suppose we wish to show that there is a value of $x > 0$ that solves the equation

$$e^x = 5x + 10$$

Rewrite the equation as

$$e^x - 5x - 10 = 0$$

and define the function

$$f(x) = e^x - 5x - 10$$

Any number x that is a solution of the equation $e^x = 5x + 10$ is also a solution of $f(x) = 0$ (and vice versa). Geometrically, the solutions of the equation $e^x = 5x + 10$ are the x-coordinates of the points where the graphs of $y = e^x$ and $y = 5x + 10$ intersect; see Figure 6.3.44a. These x-coordinates are exactly the same as the x-intercepts of the graph of $f(x)$, i.e., the solutions of $f(x) = 0$; see Figure 6.3.44b.

Note that f is a continuous function for all numbers x (and thus it will be continuous on any interval that we might wish to use).

Our strategy is to find an interval $[a, b]$ so that f is positive at one end and negative at the other. In this way, $N = 0$ (the value that we want f to assume) will be between the values $f(a)$ and $f(b)$, and we will be able to use the theorem.

Checking $x = 0$ gives $f(0) = -9$. Now we try various numbers x until we find one for which the function is positive:

x	f(x)
1	$e^1 - 5 - 10 \approx -12.28 < 0$
2	$e^2 - 10 - 10 \approx -12.61 < 0$
3	$e^3 - 15 - 10 \approx -4.91 < 0$
4	$e^4 - 20 - 10 \approx 24.60 > 0$

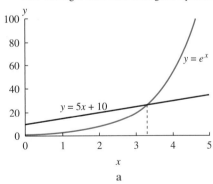
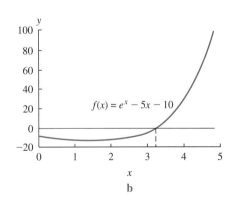

FIGURE 6.3.44
The Intermediate Value Theorem applied to solving an equation

So, $f(3) < 0$ and $f(4) > 0$. Applying the Intermediate Value Theorem to the function f defined on $[3, 4]$, we conclude that there is a number c in $[3, 4]$ such that $f(c) = 0$; i.e., c is the solution of the given equation.

Note that we had other choices for the interval: we could have taken $[2, 4]$, or $[0, 4]$, or (knowing that $f(5) = e^5 - 25 - 10 \approx 113.41 > 0$) we could have used $[3, 5]$.

Example 6.3.4 Caution Needed When Using the Intermediate Value Theorem

As with any theorem, we need to be careful to use the Intermediate Value Theorem only if *all assumptions are satisfied.*

Consider the function $f(x) = \frac{1}{x}$ on the interval $[-1, 2]$. The number $N = 0$ is between $f(-1) = -1$ and $f(2) = 1/2$. However, there is no number c in $[-1, 2]$ such that

$$f(c) = \frac{1}{c} = 0$$

(in other words, the function $f(x) = \frac{1}{x}$ does not assume the value zero). This does not contradict the Intermediate Value Theorem, since f is not continuous on $[-1, 2]$, and thus—because an assumption fails to hold—the theorem does not apply.

Keep in mind that the statement of the theorem can still be true, even though an assumption is not satisfied. For instance, the function g shown in Figure 6.3.45 is not continuous on $[a, b]$, but it does assume all values between $g(a)$ and $g(b)$.

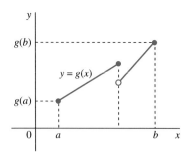

Figure 6.3.45

Function g satisfies the conclusions of the Intermediate Value Theorem

In Example 6.3.3 we showed that the equation

$$f(x) = e^x - 5x - 10 = 0$$

has a solution that lies between 3 and 4. Can we do better? Can we give a better estimate for the interval where the solution lies?

The idea is to repeat what we did in the example, in a systematic way. Take the midpoint, $x = 3.5$, of the interval $[3, 4]$ and compute

$$f(3.5) = e^{3.5} - 5(3.5) - 10 \approx 5.61545$$

Earlier, we computed $f(3) \approx -4.91447$ and $f(4) \approx 24.59815$. By using the midpoint, we now divide the interval $[3, 4]$ into two subintervals, $[3, 3.5]$ and $[3.5, 4]$. Which of the two contains the solution?

From the Intermediate Value Theorem, we know that the interval at whose ends the function has opposite signs is the one that contains the solution. Since $f(3) < 0$, $f(3.5) > 0$, and $f(4) > 0$, we see that the solution must lie in $[3, 3.5]$.

Now all we need to do is to repeat this process—we calculate the value of f at the midpoint, 3.25, of the interval $[3, 3.5]$ and pick the subinterval of $[3, 3.5]$ at whose ends f has opposite signs. We organize our calculations in Table 6.3.1. (Recall that the midpoint of an interval $[a, b]$ is $(a + b)/2$.)

Table 6.3.1

Midpoint	Value of f	Interval Containing the Solution
	$f(3) \approx -4.91447 < 0$	$[3, 4]$
	$f(4) \approx 24.59815 > 0$	
3.5	$f(3.5) \approx 5.61545 > 0$	$[3, 3.5]$
3.25	$f(3.25) \approx -0.45966 < 0$	$[3.25, 3.5]$
3.375	$f(3.375) \approx 2.34928 > 0$	$[3.25, 3.375]$
3.3125	$f(3.3125) \approx 0.89117 > 0$	$[3.25, 3.3125]$
3.28125	$f(3.28125) \approx 0.20276 > 0$	$[3.25, 3.28125]$
3.265625	$f(3.265625) \approx -0.13165 < 0$	$[3.265625, 3.28125]$

Thus, we have narrowed down the interval that contains the solution to $[3.265625, 3.28125]$. A very good approximation of the solution (obtained using other means, or by repeating the work in the table for many more steps) is $x \approx 3.271812$.

As we can see, in every step of the algorithm that we used in the previous example, we halve the length of the interval. By making the interval shrink, we obtain better

6.3 Reasoning about Functions: Continuity and Differentiability

and better approximations of the solution. This algorithm is called **bisection** (or the **bisection method**).

Example 6.3.5 Applying the Bisection Method to Approximate a Solution of an Equation

In the introduction to this subsection, we studied the equation

$$f(x) = 3.2xe^{-x} + 6.5 - x = 0$$

and determined that it has a solution in the interval [5, 7]. We now apply the bisection method to obtain a more precise estimate for the solution; see Table 6.3.2.

Our analysis shows that the solution lies in the interval [6.5, 6.53125]. For the record, a more accurate approximation of the solution is 6.530475. ▲

Table 6.3.2

Midpoint	Value of f	Interval Containing the Solution
	$f(5) \approx 1.60781 > 0$	[5, 7]
	$f(7) \approx -0.47957 < 0$	
6	$f(6) \approx 0.54759 > 0$	[6, 7]
6.5	$f(6.5) \approx 0.03127 > 0$	[6.5, 7]
6.75	$f(6.75) \approx -0.22471 < 0$	[6.5, 6.75]
6.625	$f(6.625) \approx -0.09687 < 0$	[6.5, 6.625]
6.5625	$f(6.5625) \approx -0.03284 < 0$	[6.5, 6.5625]
6.53125	$f(6.53125) \approx -0.00079 < 0$	[6.5, 6.53125]

The bisection method is one of many numerical procedures developed to solve equations. In Section 6.6 we discuss a procedure based on derivatives that usually works better than bisection (in the sense that it creates approximations that approach the solution faster than those generated by the bisection method).

Final remark: In the bisection method, we keep calculating the value of the function at the midpoint of a shrinking sequence of intervals, keeping track of whether it's positive or negative. If the value of the function turns out to be zero, then we are done—we have found a solution!

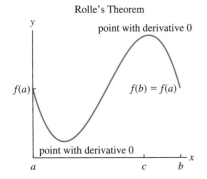

FIGURE 6.3.46
Rolle's theorem

Rolle's Theorem and the Mean Value Theorem

The Intermediate Value Theorem and the Extreme Value Theorem guarantee that a continuous function must take on particular values. **Rolle's Theorem** and the **Mean Value Theorem** guarantee that the *derivative* must take on particular values.

Rolle's theorem is closely related to the Extreme Value Theorem. It states that a differentiable function that takes on equal values at its endpoints must have a derivative equal to zero at some point in between (i.e., it must have a critical number).

Theorem 6.3.2 **Rolle's Theorem**

If $f(x)$ is continuous on the closed interval $[a, b]$ and differentiable on the open interval (a, b), and $f(a) = f(b)$, then there exists a number c such that $a < c < b$ and $f'(c) = 0$. ▲

As with the Intermediate Value Theorem, this theorem guarantees that there is at least one number at which the derivative is zero (see Figure 6.3.46). It does not tell us how many numbers possess that property, nor how to find them.

Example 6.3.6 Application of Rolle's Theorem

The function $f(x) = e^{0.2x^2} \sin 3x$ takes on the value zero for both $x = 0$ and $x = \pi$. As a product of two continuous functions, it is continuous for all real numbers, and in

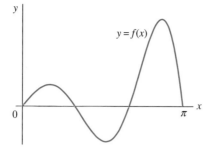

FIGURE 6.3.47
The graph of $f(x) = e^{0.2x^2} \sin 3x$

particular on the interval $[0, \pi]$. As well, both factors are differentiable, so $f(x)$ is differentiable on $(0, \pi)$. Rolle's Theorem guarantees that there must be at least one point between 0 and π where the derivative is zero.

Note that this result is by no means trivial. To find the critical points algebraically, we compute

$$f'(x) = 0.4xe^{0.2x^2} \sin 3x + 3e^{0.2x^2} \cos 3x = 0$$
$$e^{0.2x^2}(0.4x \sin 3x + 3 \cos 3x) = 0$$
$$0.4x \sin 3x + 3 \cos 3x = 0$$

(since $e^{0.2x^2} > 0$). This equation cannot be solved algebraically (we must use a numeric method, such as bisection). As well, it is not at all clear (without further analysis) that this equation does have a solution.

The graph of $f(x)$ in Figure 6.3.47 shows that there are three points in $(0, \pi)$ where $f'(x) = 0$.

Example 6.3.7 Using Rolle's Theorem

Consider the function

$$f(x) = \frac{x^3(3-x)^2}{\sqrt{x+4}+1}$$

Looking at its numerator, we see that $f(0) = f(3) = 0$. $f(x)$ is continuous on $[0, 3]$, as the quotient of two continuous functions (with nonzero denominator). It is differentiable on $(0, 3)$, since both numerator and denominator are differentiable. Thus, f satisfies all assumptions of Rolle's Theorem, and we conclude that there must be a number, c, in $(0, 3)$ where $f' = 0$.

By the Extreme Value Theorem, f must have absolute maximum and absolute minimum values in $[0, 3]$. Because f is positive on $(0, 3)$, it follows that its absolute maximum must be positive. Thus, one of the numbers guaranteed to exist by Rolle's Theorem is where f assumes its absolute maximum value.

The proof of Rolle's Theorem uses the Extreme Value Theorem to show that the function must have a minimum or a maximum in the interior of the interval (unless the function is constant) and the requirement that an interior minimum or maximum must occur at a critical point. If the function is differentiable, then the critical point must be the one where the derivative is equal to zero.

The Mean Value Theorem is a "tilted" version of Rolle's Theorem (see Exercises 17 and 18).

Theorem 6.3.3 Mean Value Theorem

If $f(x)$ is continuous on the closed interval $[a, b]$ and differentiable on the open interval (a, b), then there exists a number c such that $a < c < b$ and

$$f'(c) = \frac{f(b) - f(a)}{b - a}$$

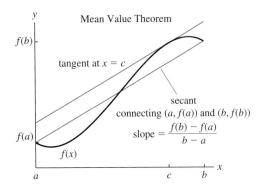

FIGURE 6.3.48

The Mean Value Theorem

Note that
$$\frac{f(b) - f(a)}{b - a}$$
is the slope of the secant line joining $(a, f(a))$ and $(b, f(b))$; see Figure 6.3.48. The Mean Value Theorem claims that there is a number c in (a, b) where the tangent line has the same slope as this secant line, i.e., where it is parallel to the secant line. Alternatively, if the average rate of change during the interval from a to b is A, then there is a point in the interval when the instantaneous rate of change is equal to A.

Example 6.3.8 Application of the Mean Value Theorem to Speed

The most popular application of this theorem involves speed. If a car travels 140 km in 2 hours, the average rate of change over this time is 70 km/h. A graph of position versus time must pass through the two points $(0, 0)$ and $(2, 140)$, and the line connecting these is a secant line of slope 70. The Mean Value Theorem guarantees that the instantaneous speed (on the speedometer) must have been exactly 70 km/h at some time c (Figure 6.3.49).

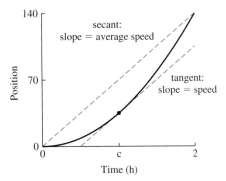

FIGURE 6.3.49

The Mean Value Theorem applied to velocity

Example 6.3.9 Finding the Point Guaranteed by the Mean Value Theorem

The Mean Value Theorem guarantees that for some x with $1 < x < 2$, the slope of the function $f(x) = \ln x$ must exactly match the slope of the secant line connecting $(1, f(1))$ and $(2, f(2))$. The slope of the secant is
$$\text{slope of secant} = \frac{f(2) - f(1)}{2 - 1} = \frac{\ln 2 - \ln 1}{1} = \ln 2$$
Where is $f'(x) = \ln 2$? We recall that
$$\frac{d}{dx}(\ln x) = \frac{1}{x}$$

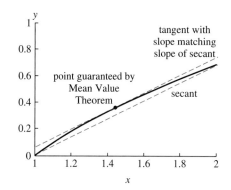

FIGURE 6.3.50

The Mean Value Theorem applied to ln x

so the value is the solution of

$$f'(x) = \frac{1}{x} = \ln 2$$

Therefore, $x = \frac{1}{\ln 2} \approx 1.443$ (Figure 6.3.50).

Example 6.3.10 Estimating Growth Using the Mean Value Theorem

Suppose that $f(x)$ is a differentiable function such that $f(3) = 4$ and $f'(x) \leq 2.5$ for all x. What is the largest possible value of $f(6)$?

We can rewrite the Mean Value Theorem,

$$\frac{f(b) - f(a)}{b - a} = f'(c)$$

as

$$f(b) - f(a) = f'(c)(b - a)$$

where $a < c < b$. Let $a = 3$ and $b = 6$ (so that the interval on which we apply the Mean Value Theorem is [3, 6]). Thus

$$f(6) - f(3) = f'(c)(6 - 3)$$
$$f(6) - 4 = 3f'(c)$$
$$f(6) = 3f'(c) + 4$$

where c is a number between 3 and 6. It is given that $f'(x) \leq 2.5$ for all x; thus, in particular, $f'(c) \leq 2.5$ and

$$f(6) \leq 3(2.5) + 4 = 11.5$$

Thus, the largest possible value for $f(6)$ is 11.5.

Continuity and Differentiability

We now state two important (and useful!) theorems that relate continuity and differentiability.

Discussing differentiability in Example 4.5.4 in Section 4.5, we discovered that a function that is not continuous at a number cannot be differentiable there either. Rewording our conclusion using positive statements (continuous, differentiable) rather than negative (not continuous, not differentiable) we obtain the following theorem.

Theorem 6.3.4 Differentiability Implies Continuity

Assume that a function $f(x)$ is differentiable at a. Then $f(x)$ is continuous at a.

Proof: We assume that

$$f'(a) = \lim_{h \to 0} \frac{f(a+h) - f(a)}{h}$$

is a real number, we need to prove that

$$\lim_{x \to a} f(x) = f(a)$$

The proof is technical, and not intuitive at all. First, we compute

$$\lim_{h \to 0} (f(a+h) - f(a))$$

Divide and multiply by h (note that $h \to 0$ and thus $h \neq 0$), then calculate the limit as the product of the limits:

$$\lim_{h \to 0} (f(a+h) - f(a)) = \lim_{h \to 0} \frac{f(a+h) - f(a)}{h} \cdot h$$

$$= \lim_{h \to 0} \frac{f(a+h) - f(a)}{h} \cdot \lim_{h \to 0} h$$

$$= f'(a) \cdot 0 = 0$$

Next, we compute $\lim_{h \to 0} f(a+h)$; start by adding and subtracting $f(a)$, then use the limit we just calculated:

$$\lim_{h \to 0} f(a+h) = \lim_{h \to 0} [f(a) + (f(a+h) - f(a))]$$

$$= \lim_{h \to 0} f(a) + \lim_{h \to 0} (f(a+h) - f(a))$$

$$= f(a) + 0 = f(a)$$

Almost done! We just need to relate h to x. Take $h = x - a$ (so that $x = a + h$). Then $h \to 0$ implies that $x - a \to 0$ and $x \to a$. Now we rewrite

$$\lim_{h \to 0} f(a+h) = f(a)$$

as

$$\lim_{x \to a} f(x) = f(a)$$

and the proof is complete.

You might ask yourself—why are the statements "if f is not continuous then f is not differentiable" (Example 4.5.4) and "if f is differentiable then f is continuous" (Theorem 6.3.4) the same? Because one is the contrapositive of the other (see the subsection Implications in Section 1.5).

By the way, note that the converse statement "if f is continuous then f is differentiable" is false: for instance, $f(x) = |x|$ is continuous at $x = 0$, but it is not differentiable there because it has a corner.

Our next theorem is a consequence of the Mean Value Theorem. It will help us answer the following question: If we know that $g'(x) = h'(x)$ for all x in an interval (a, b), what can we say about $g(x)$ and $h(x)$? Do they have to be equal, or not?

Part of the answer is easy: $g(x)$ and $h(x)$ do not have to be equal. For instance, $g(x) = x^2$ and $h(x) = x^2 + 3$ satisfy $g'(x) = h'(x)$ but $g(x) \neq h(x)$. So now the question is—how much different can $g(x)$ and $h(x)$ be?

Theorem 6.3.5 Assume that $f(x)$ is differentiable and $f'(x) = 0$ for all x in an interval (a, b). Then $f(x)$ is constant on (a, b).

With this theorem, we can answer our question. Assume that $g'(x) = h'(x)$ and form the new function

$$f(x) = g(x) - h(x)$$

Now
$$f'(x) = g'(x) - h'(x) = 0$$
for all x. The theorem tell us that $f(x)$ must be a constant function, i.e.,
$$f(x) = C$$
for some real number C. Thus
$$f(x) = g(x) - h(x) = C$$
and
$$g(x) = h(x) + C$$

In words, if two functions have equal derivatives, then they can differ (at most) by a constant (geometrically, they must be vertical shifts of each other).

To prove Theorem 6.3.5 we will show that $f(x_1) = f(x_2)$ for any x_1 and x_2 in (a, b). By assumption, f is differentiable on (a, b), and thus it is differentiable on a subset (x_1, x_2) of (a, b). By Theorem 6.3.4, f is also continuous on $[x_1, x_2]$. So, all assumptions of the Mean Value Theorem are satisfied. We conclude that there is a number c in (x_1, x_2) such that
$$\frac{f(x_2) - f(x_1)}{x_2 - x_1} = f'(c)$$
i.e.,
$$f(x_2) - f(x_1) = f'(c)(x_2 - x_1)$$

Because $f'(x) = 0$ for all x in (a, b), it follows that $f'(c) = 0$ and therefore
$$f(x_2) - f(x_1) = 0$$
and $f(x_2) = f(x_1)$.

Summary With a minimum of algebra, we can use theorems about continuous and differentiable functions to deduce mathematical conclusions. The **Intermediate Value Theorem** guarantees that a continuous function takes on all values between those at its endpoints. Using **Rolle's Theorem,** we argued that a differentiable function that takes on equal values at its endpoints must have a critical point in between where the derivative is equal to zero. The generalization of Rolle's Theorem, the **Mean Value Theorem,** states that the derivative of a differentiable function must at some point match the slope of the secant. Finally, we proved that a differentiable function must be continuous. As a consequence of the Mean Value Theorem, we argued that a function whose derivative is zero on some interval must be constant on that interval.

6.3 Exercises

Mathematical Techniques

1–6 ▪ Use the Intermediate Value Theorem to show that each equation has a solution for $0 \leq x \leq 1$.

1. $e^x + x^2 - 2 = 0$
2. $e^x - 3x^2 = 0$
3. $e^x + x^2 - 2 = x$
4. $e^x + x^2 - 2 = \cos(2\pi x) - 1$
5. $xe^{-3(x-1)} - 2 = 0$ (you will need to check an intermediate point)
6. $x^3 e^{-4(x-1)} - 1.1 = 0$ (you will need to check an intermediate point)

7–10 ▪ Show that each function has a positive maximum on the interval $0 \leq x \leq 1$.

7. $f(x) = x(x-1)e^x$

8. $f(x) = ex - xe^x$

9. $f(x) = 5x(1-x)(2-x) - 1$

10. $f(x) = 8x(1-x)^2 - 1$

11–14 ▪ Find the points guaranteed by the Mean Value Theorem and sketch the associated graph.

11. The graph of the function $f(x) = x^2$ must match the slope of the secant connecting $x = 0$ and $x = 1$.

12. The graph of the function $f(x) = x^2$ must match the slope of the secant connecting $x = 0$ and $x = 2$.

13. The graph of the function $g(x) = \sqrt{x}$ must match the slope of the secant connecting $x = 0$ and $x = 1$. (The Mean Value Theorem still applies even though $\sqrt{x}$ is not differentiable at $x = 0$.)

14. The graph of the function $g(x) = \sqrt{x}$ must match the slope of the secant connecting $x = 0$ and $x = 2$.

15–16 ▪ Check whether the conclusions of the Intermediate Value Theorem and the Mean Value Theorem fail in the following cases where the function is not continuous.

15. Consider the Heaviside function (Section 4.4, Exercises 44–45) defined by

$$H(x) = \begin{cases} 0 & \text{if } x < 0 \\ 1 & \text{if } x \geq 0 \end{cases}$$

Show that there is no solution to the equation $H(x) = 1/2$ and that there is no tangent that matches the slope of the secant connecting $(-1, H(-1))$ and $(1, H(1))$.

16. Consider the absolute value function $g(x) = |x|$. Does this satisfy the conditions for the Intermediate Value Theorem? Show that there is no tangent whose slope matches the slope of the secant connecting $(-1, g(-1))$ and $(2, g(2))$.

17–18 ▪ There is a clever proof of the Mean Value Theorem from Rolle's Theorem. The idea is to tilt the function f so that it takes on the same values at the endpoints a and b. In particular, we apply Rolle's Theorem to the function

$$g(x) = f(x) - (x-a)\frac{f(b) - f(a)}{b - a}$$

For the following functions, show that $g(a) = g(b)$, apply Rolle's Theorem to the function g, and find the derivative of f at a point where $g'(x) = 0$.

17. $f(x) = x^2$, $a = 1$ and $b = 2$.

18. In general, without assuming a particular form for $f(x)$ or values for a and b.

Applications

19–22 ▪ Try to apply the Intermediate Value Theorem to the following problems.

19. The price of gasoline rises from \$1.199 to \$1.279. Why is it not necessarily true that the price was exactly \$1.25 at some time?

20. A pot is dropped from the top of a 150-m building exactly 60 m above your office. Must it have fallen right past your office window?

21. A cell takes up $1.5 \cdot 10^{-9}$ mL of water in the course of an hour. Must the cell have taken up exactly $1 \cdot 10^{-9}$ mL at some time? Is it possible that the cell took up exactly $2 \cdot 10^{-9}$ mL at some time?

22. The population of bears in Yellowstone Park has increased from 100 to 1000. Must it have been exactly 314 at some time? What additional assumption would guarantee this?

23–25 ▪ Use the bisection method with the suggested interval and number of steps to find an approximate solution to each equation.

23. $\sin x = \frac{1}{3}x$, [2, 3], 4 steps (keep in mind that x is in radians).

24. $\ln x - 0.01x^2 = 0$, [0.8, 1.6], 3 steps.

25. $x^5 - 3x^4 - 5x^2 = -2$, [0, 1], 4 steps.

26. Find an approximation of the solution of the equation $f'(x) = 0$ from Example 6.1.12. using five steps of the bisection method starting with the interval [0, 1].

27. Consider the function $f(x) = \arcsin x + \arccos x$. Using the fact that $(\arccos x)' = -1/\sqrt{1-x^2}$, prove that $f'(x) = 0$ for all x in $(-1, 1)$. Use Theorem 6.3.5 to show that $\arcsin x + \arccos x = \pi/2$ for all x in $(-1, 1)$. By checking the formula directly, show that $\arcsin x + \arccos x = \pi/2$ when $x = 1$ and $x = -1$.

28–31 ▪ The Intermediate Value Theorem can often be used to prove that complicated discrete-time dynamical systems have equilibria.

28. Prove that the dynamical system $x_{t+1} = \cos x_t$ has an equilibrium between 0 and $\pi/2$.

29. A lung follows the discrete-time dynamical system $c_{t+1} = 0.25e^{-3c_t}c_t + 0.75\gamma$, where $\gamma = 5$. Show that there is an equilibrium between 0 and γ.

30. A lung follows the discrete-time dynamical system $c_{t+1} = 0.75\alpha(c_t)c_t + 0.25\gamma$, where $\gamma = 5$, the function $\alpha(c_t)$ is positive and decreasing, and $\alpha(0) = 1$. Show that there is an equilibrium between 0 and γ.

31. A lung follows the discrete-time dynamical system $c_{t+1} = f(c_t)$. We know only that neither c_{t+1} nor c_t can exceed 1 mol/L. Use the Intermediate Value Theorem to show that this discrete-time dynamical system must have an equilibrium.

32–33 ▪ Apply the Intermediate Value Theorem to the following situations.

32. A farmer sets off on Saturday morning at 6 A.M. to bring a crop to market, arriving in town at noon. On Sunday she sets off in the opposite direction at 6 A.M. and returns home along the same route, arriving once again at noon. Use the Intermediate Value Theorem to show that at some point along the path, her watch must have read exactly the same time on both of the two days.

33. Suppose instead that the farmer sets off one morning at 6 A.M. to bring a crop to market and arrives in town at noon. Having received a great price for her crop, she buys a new car and drives home the next day along the same route, leaving at 10 A.M. and arriving home at 11 A.M. Is it still true that at some point along the route her watch must read exactly the same time on both of the two days? If so, must that time occur between 10 and 11 A.M.?

34–37 ▪ An organism grows from 4 kg to 60 kg in 14 years. Suppose that mass is a differentiable function of time.

34. Why must the mass have been exactly 10 kg at some time?

35. Why must the rate of increase have been exactly 4 kg/year at some time?

36. Draw a graph of mass against time where the mass is increasing, is equal to 10 kg at 13 years, and has a growth rate of exactly 4 kg/year after 1 year.

37. Draw a graph of mass against time where the organism reaches 10 kg at 1 year and has a growth rate of exactly 4 kg/year at 13 years.

38–40 ▪ Draw the positions of cars from the following descriptions of one-hour trips. What speed must the car achieve according to the Mean Value Theorem? What speeds must the car achieve according to the Intermediate Value Theorem?

38. A car starts at 100 km/h and slows down to 0 km/h. The average speed is 40 km/h after one hour.

39. A car starts at 100 km/h, slows down to 40 km/h, and then speeds up to 80 km/h by the end of one hour. The average speed over the whole time is 60 km/h.

40. A car drives 100 km in one hour and never varies speed by more than 15 km/h.

6.4 Leading Behaviour and L'Hôpital's Rule

Finding limits at infinity gives general information about the behaviour of functions. In particular, we learned how to compare the ratios of different functions by describing which increased to infinity or decreased to zero faster. We now study a much larger class of functions, **sums** of functions and the ratios of sums. The technique, called the **method of leading behaviour,** consists of focusing on the largest piece of the function. By determining the leading behaviour of a function at both infinity and zero, we can deduce a great deal about the *shape* of the function by using the technique of **matched leading behaviours.** When the method of leading behaviour fails, **L'Hôpital's rule** provides an alternative way to compare the behaviour of functions.

Leading Behaviour of Functions at Infinity

Suppose we wish to describe how the sum of several functions, such as

$$f(x) = 5e^{2x} + 34e^x + 45x^5 + 56 \ln x + 10$$

behaves for large values of x. We might suspect that this function will be dominated by a single term that approaches infinity more quickly than any other term in the sense of the concepts that we developed in Section 4.3. The **method of leading behaviour** is based on this idea.

We study two cases: the leading behaviour of a function at infinity (i.e., as the argument approaches ∞) and the leading behaviour of a function at zero (i.e., as the argument approaches 0). As we will soon witness, not every function can be reduced to a single term that is representative of its behaviour. For the leading behaviour to make sense, we consider

(1) functions that contain at least one term that approaches ∞ or $-\infty$, and

(2) functions all of whose terms approach real numbers, but in this case, we further assume that at most one term approaches a nonzero number.

Definition 6.4.1 The **leading behaviour** of a function at infinity is the term that is largest in absolute value as the argument approaches infinity. We write f_∞ to represent the leading behaviour of the function f at infinity.

In applications, we often refer to the leading behaviour at infinity as long-term behaviour or asymptotic behaviour.

The largest term in the function $f(x)$ is $5e^{2x}$ because the exponential term with the largest parameter in the exponent grows to infinity most quickly. Therefore,

$$f_\infty(x) = 5e^{2x}$$

Likewise, if $g(x) = 3x^2 - x^3 + 2\sqrt{x} - 3.7$, then $g_\infty(x) = -x^3$; the leading behaviour of the function $h(x) = e^{x/10} + 150e^{-x} + x^7 - 14$ is $h_\infty(x) = e^{x/10}$; if $k(x) = 451 - 3\ln x$, then $k_\infty(x) = -3\ln x$.

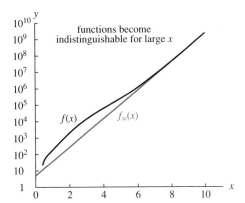

FIGURE 6.4.51

A comparison of a function and its leading behaviour in a semilog plot

In what sense does the leading behaviour describe a function? Consider again the function $f(x) = 5e^{2x} + 34e^x + 45x^5 + 56\ln x + 10$ from the start of this section. On a semilog plot, the graph of the leading behaviour looks indistinguishable from the graph of the original function when x is large, even though the two functions are quite different for small x (Figure 6.4.51). If we divide a function by its leading behaviour, the limit is 1 (i.e., f and f_∞ approach infinity at the same rate). For example,

$$\lim_{x \to \infty} \frac{f(x)}{f_\infty(x)} = \lim_{x \to \infty} \frac{5e^{2x} + 34e^x + 45x^5 + 56\ln x + 10}{5e^{2x}}$$
$$= \lim_{x \to \infty} \left[1 + \frac{34e^x}{5e^{2x}} + \frac{45x^5}{5e^{2x}} + \frac{56\ln x}{5e^{2x}} + \frac{10}{5e^{2x}} \right]$$
$$= 1 + 0 + 0 + 0 + 0 = 1$$

All terms after the first approach 0 because $5e^{2x}$ approaches infinity most quickly. Alternatively,

$$\frac{34e^x}{5e^{2x}} = \frac{34}{5}e^{-x}$$

approaches 0 as $x \to \infty$. Likewise,

$$\frac{10}{5e^{2x}} = 2e^{-2x} \to 0$$

as $x \to \infty$. Once we introduce L'Hôpital's rule, we will be able to verify algebraically that

$$\frac{x^5}{e^{2x}} \to 0 \quad \text{and} \quad \frac{\ln x}{e^{2x}} \to 0$$

as $x \to \infty$.

Although f and f_∞ become similar relatively (their ratio approaches 1), they do not become similar absolutely (their difference does not approach 0). See Table 6.4.1.

Example 6.4.1 Comparing Functions with the Method of Leading Behaviour

Complicated functions can be compared by comparing their leading behaviours. To find whether the function $f(x) = 5e^{2x} + 34e^x + 45x^5 + 56\ln x + 10$ from above increases to infinity more quickly than the function $g(x) = 23e^{2.5x} + 3e^{2x} + 2x^6$ we need only

Table 6.4.1

x	f(x)	$f_\infty(x)$	$\dfrac{f(x)}{f_\infty(x)}$	$f(x) - f_\infty(x)$
5	$2.559 \cdot 10^5$	$1.101 \cdot 10^5$	2.324	$1.458 \cdot 10^5$
10	$2.431 \cdot 10^9$	$2.426 \cdot 10^9$	1.002	$5.249 \cdot 10^6$
20	$1.177 \cdot 10^{18}$	$1.177 \cdot 10^{18}$	1.000	$1.7 \cdot 10^{10}$
30	$5.710 \cdot 10^{26}$	$5.710 \cdot 10^{26}$	1.000	$3.633 \cdot 10^{14}$

compare the leading behaviour of $f(x)$ with that of $g(x)$. The leading behaviour of the function $g(x)$ happens to be its first term, as the term with the largest parameter in the exponent, so

$$g_\infty(x) = 23e^{2.5x}$$

To see whether $f(x)$ approaches infinity more quickly than $g(x)$, we compute the limit of the ratio of the functions as x approaches infinity. Because each function is well represented by its leading behaviour, we have

$$\lim_{x\to\infty} \frac{f(x)}{g(x)} = \lim_{x\to\infty} \frac{f_\infty(x)}{g_\infty(x)}$$
$$= \lim_{x\to\infty} \frac{5e^{2x}}{23e^{2.5x}}$$
$$= \lim_{x\to\infty} \frac{5}{23} e^{-0.5x}$$
$$= 0$$

so the function $g(x)$ approaches infinity more quickly than $f(x)$.

The leading behaviour gives more information than the limit. For large x,

$$\frac{f(x)}{g(x)} \approx \frac{f_\infty(x)}{g_\infty(x)}$$
$$= \frac{5e^{2x}}{23e^{2.5x}}$$
$$= \frac{5}{23} e^{-0.5x}$$

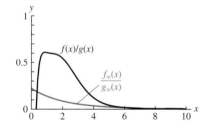

FIGURE 6.4.52
The ratio of two functions compared with the ratio of their leading behaviours

This computation not only tells us that $g(x)$ approaches infinity more quickly than $f(x)$ but also gives us an idea of how much more quickly. The ratio of the functions behaves much like the ratio of the leading behaviours for large x (Figure 6.4.52).

The same definition of leading behaviour works for sums of functions that approach 0. Remember that the largest term is the term approaching 0 the *most slowly* in the sense of the concepts that we defined in Section 4.3.

Example 6.4.2 Computing the Leading Behaviour at Infinity of a Function that Approaches Zero

The leading behaviour of the function

$$f(x) = 5e^{-2x} + 34e^{-x} + 45x^{-5}$$

is the power term $45x^{-5}$ because it approaches 0 the most slowly and is therefore the largest for large x, so

$$f_\infty(x) = 45x^{-5}$$

FIGURE 6.4.53

Power functions $f(x) = x^r$, with $r < 0$

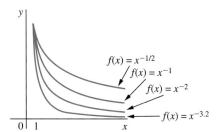

In Figure 6.4.53 we show graphs of several power functions $f(x) = x^r$ for negative values of r. We see that the leading behaviour at infinity of a collection of functions $f(x) = x^r$ (with $r < 0$) is the function with the smallest negative power.

For instance, the leading behaviour of

$$f(x) = 3x^{-2} + x^{-1} - 5.3x^{-4} + 7x^{-0.6}$$

is

$$f_\infty(x) = 7x^{-0.6}$$

Example 6.4.3 The Leading Behaviour at Infinity of a Sum That Includes a Term That Does Not Approach Zero

Consider the function

$$f(x) = 5e^{-2x} + 34e^{-x} + 45x^{-5} + 5$$

Note that $e^{-2x} \to 0$, $e^{-x} \to 0$, and $x^{-5} \to 0$ as $x \to \infty$. Thus, $f(x)$ has leading behaviour

$$f_\infty(x) = 5$$

because the constant 5 does not approach 0 and is therefore the largest (Figure 6.4.54).

All three terms in the function $g(x) = x^{-2} + 4 + \arctan x$ approach real numbers as $x \to \infty$. Since the last two terms approach nonzero numbers, the leading behaviour $g_\infty(x)$ is not defined.

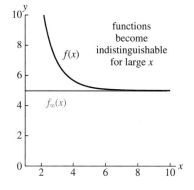

FIGURE 6.4.54

A function that approaches a finite value and its leading behaviour

Example 6.4.4 Application to the Absorption Function Shown in Figure 4.3.44b

We can apply the idea of leading behaviour to supplement our reasoning about the absorption functions in Figure 4.3.44b. The absorption function with saturation has formula

$$\alpha(c) = \frac{Ac}{k+c}$$

where c is the concentration and A and k are constant parameters. Both numerator and denominator increase to infinity. The leading behaviour of the denominator is c, the larger of the two terms. Replacing the denominator by the leading behaviour, we have

$$\alpha_\infty(c) = \frac{Ac}{c} = A$$

for large values of c (Figure 6.4.55). In the figure, both k and A have been set to 1. In the long term, absorption saturates at $A = 1$.

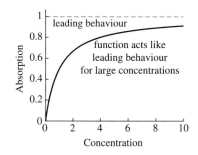

FIGURE 6.4.55

An absorption function with saturation: $\alpha(c) = \dfrac{c}{1+c}$

Note The method of leading behaviour tells us nothing new about the absorption function

$$\alpha(c) = Ace^{-\beta c}$$

We can rewrite this function as a ratio by placing the negative power in the denominator, giving $\alpha(c) = \dfrac{Ac}{e^{\beta c}}$. The limit as c approaches infinity is zero because an exponential function grows faster than a linear function. Because the numerator and denominator each have only a single term, we cannot simplify further.

Example 6.4.5 Application to the Absorption Function Shown in Figure 4.3.44e

The absorption function with overcompensation (Figure 4.3.44e),

$$\alpha(c) = \frac{Ac}{k + c^2}$$

can be simplified with the method of leading behaviour. In this case, the leading behaviour of the denominator is the quadratic term c^2, so

$$\alpha_\infty(c) = \frac{Ac}{c^2} = \frac{A}{c}$$

Absorption decreases to 0 and does so like the function $\dfrac{A}{c}$ (Figure 6.4.56).

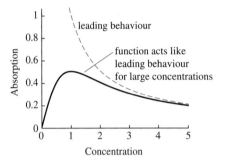

FIGURE 6.4.56

An absorption function with overcompensation

Leading Behaviour of Functions at Zero

In each comparison of an absorption function with its leading behaviour, we have done well for large values of the concentration and poorly for small values. To complete our analysis of the absorption functions, we need a better sense of what is happening near 0. Once again, the method of leading behaviour can be used to identify which small terms can be ignored.

First, we define the leading behaviour of a function at zero.

Definition 6.4.2 The **leading behaviour** of a function at zero is the term that is largest in absolute value as the argument approaches 0. We write f_0 to represent the leading behaviour of the function f at zero.

In applications, f_0 is sometimes referred to as the initial behaviour; as well, f is said to start like f_0.

In Figure 6.4.57, we show several power functions $f(x) = x^r$, for $r < 0$. Note that

$$\lim_{x \to 0^+} x^r = \infty$$

and that the largest function (i.e., the leading behaviour of a collection of functions $f(x) = x^r$ for $r < 0$) is the one with the most negative power (i.e., the one with the smallest r).

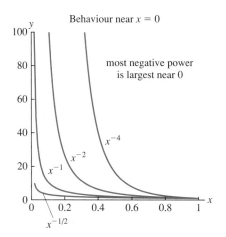

FIGURE 6.4.57

Power functions with negative powers

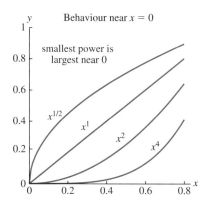

FIGURE 6.4.58

Power functions with positive powers

For example, the leading behaviour of $f(x) = x^{-1} + 3x^{-4}$ at zero is given by $f_0(x) = 3x^{-4}$. Likewise, if $g(x) = x^{-2} - 2x^{-3}$, then $g_0(x) = -2x^{-3}$, and if $h(x) = x^{-2.3} + 4x^{-2.4} + 5x^{-2.2} + 25$, then $h_0(x) = 4x^{-2.4}$.

All power functions of the form

$$f(x) = x^r$$

for positive values of r approach 0 as x approaches 0 (Figure 6.4.58). The power function with the smallest power is the leading behaviour of all $f(x) = x^r$ for $r > 0$.

It follows that the leading behaviour of $f(x) = x^2 + 4x - 12x^{2.5}$ at zero is given by $f_0(x) = 4x$. Likewise, if $g(x) = 3x^{0.6} - 2\sqrt{x} + x$, then $g_0(x) = -2\sqrt{x}$.

Note that the leading behaviour of $f(x) = 5x - x^2 + 4x^3 + 2$ at zero is $f_0(x) = 2$ (since all terms with x approach zero, the term 2 is the largest). As well, if $g(x) = 2x - x^4 + e^{-3x}$, then $g_0(x) = e^{-3x}$, since the first two terms approach 0 and $e^{-3x} \to 1$ as $x \to 0$.

The leading behaviour $f_0(x)$ of $f(x) = 3x^2 - x - 1 + 4e^x$ is not defined—the first two terms approach zero, but the last two terms approach nonzero real numbers (-1 and 4, respectively) as $x \to 0$. Likewise, neither term in $g(x) = 2 + \cos x$ can be isolated as the leading behaviour $g_0(x)$.

Example 6.4.6 The Leading Behaviour of a Polynomial at Zero

Consider the polynomial $f(x) = 4x - 2x^2 - x^3 + x^4$. The leading behaviour at zero is given by the term with the smallest power, i.e., $f_0(x) = 4x$ (Figure 6.4.59a). The leading behaviour for large x, on the other hand, is given by the last term, $f_\infty(x) = x^4$, because it has the largest power (Figure 6.4.59b).

The Method of Matched Leading Behaviours

The idea of studying a function for both large and small values of x can be formalized as the **method of matched leading behaviours.**

▶▶ **Algorithm 6.4.1** The Method of Matched Leading Behaviours

1. Find the leading behaviour at zero and at infinity.

2. Sketch the graph of each leading behaviour.

3. Sketch the graph that matches the leading behaviour at zero and approaches the leading behaviour at infinity.

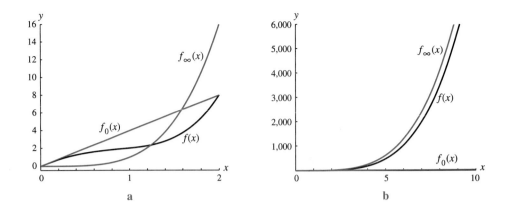

FIGURE 6.4.59

Polynomial $f(x)$ and its leading behaviours

How do we identify the leading behaviour of a fraction? It is not at all clear what "the term largest in absolute value" means when we consider, say, the function

$$f(x) = \frac{3x^2 + 2x}{x^2 + 4}$$

Without proof, we state the fact that the leading behaviour of a fraction is calculated by dividing the leading behaviour of the numerator by the leading behaviour of the denominator. That this makes sense is illustrated in the following example.

Example 6.4.7 Leading Behaviour of a Rational Function

Consider the function

$$f(x) = \frac{3x^2 + 2x}{x^2 + 4}$$

At infinity, its leading behaviour is

$$f_\infty(x) = \frac{3x^2}{x^2} = 3$$

and at zero, the leading behaviour is

$$f_0(x) = \frac{2x}{4} = \frac{1}{2}x$$

In Figure 6.4.60, we plot the function $f(x)$ and its leading behaviours, $f_\infty(x)$ and $f_0(x)$. Near zero, $f(x)$ looks like a line of slope $1/2$. As $x \to \infty$, the graph of $f(x)$ flattens out and looks like the horizontal line $y = 3$.

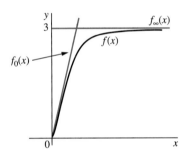

FIGURE 6.4.60

The function $f(x)$ and leading behaviours $f_\infty(x)$ and $f_0(x)$

Example 6.4.8 Applying Matched Leading Behaviours to Absorption Functions

Consider the absorption function with saturation

$$\alpha(c) = \frac{Ac}{k+c}$$

where $A, k > 0$. Since the term k is larger than c for small values of c, we conclude that

$$\alpha_0(c) = \frac{Ac}{k} = \frac{A}{k}c$$

For large values of c, the denominator becomes $(k+c)_\infty = c$ and thus

$$\alpha_\infty(c) = \frac{Ac}{c} = A$$

The graph of $\alpha(c)$ starts as the linear function $\alpha_0(c) = \frac{A}{k}c$ (of positive slope); in the long term, it flattens out and approaches $\alpha_\infty(c) = A$. Figure 6.4.61 shows the graph of $\alpha(c)$ with parameter values $A = 1$ and $k = 1$. This kind of absorption is called **saturated absorption.**

FIGURE 6.4.61

Saturated absorption

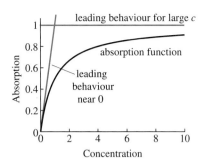

For the absorption function

$$\alpha(c) = \frac{Ac^2}{k+c^2}$$

using the method of matched leading behaviours we find

$$\alpha_0(c) = \frac{Ac^2}{k} = \frac{A}{k}c^2 \quad \text{and} \quad \alpha_\infty(c) = \frac{Ac^2}{c^2} = A$$

Approximating these portions of the graph with a smooth curve shows that the graph is concave up for small c and concave down for large c (sketched with parameter values $A = 1$ and $k = 1$ in Figure 6.4.62). This kind of absorption is called **absorption with saturation and a threshold.**

FIGURE 6.4.62

Saturated absorption with threshold

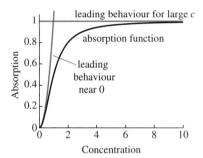

For the **saturated absorption with overcompensation**

$$\alpha(c) = \frac{Ac}{k+c^2}$$

the method of matched leading behaviours gives

$$\alpha_0(c) = \frac{Ac}{k} = \frac{A}{k}c \quad \text{and} \quad \alpha_\infty(c) = \frac{Ac}{c^2} = \frac{A}{c}$$

The curve begins by increasing like a line with slope A/k but eventually begins decreasing like the function A/c. As a consequence, this curve (being continuous) must have a maximum (Figure 6.4.63).

FIGURE 6.4.63

Saturated absorption with overcompensation

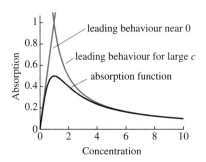

Chapter 6 Applications of Derivatives

L'Hôpital's Rule

Suppose we wish to find

$$\lim_{x \to 0} \frac{e^{2x}-1}{x} \qquad (6.4.1)$$

Both numerator and denominator approach 0. We cannot simplify the numerator with the method of leading behaviour because both terms approach a limit of 1 rather than 0 or ∞.

There is a general and powerful rule for dealing with cases like this. Instead of comparing the functions, we compare their derivatives. This is known as L'Hôpital's rule. First, we define the cases where this rule can be applied, called indeterminate forms.

Definition 6.4.3 The limit of the ratio

$$\lim_{x \to a} \frac{f(x)}{g(x)}$$

(where a could be ∞ or $-\infty$) is called an **indeterminate form of type 0/0** if

$$\lim_{x \to a} f(x) = 0 \text{ and } \lim_{x \to a} g(x) = 0$$

and an **indeterminate form of type ∞/∞** if

$$\lim_{x \to a} f(x) = \pm\infty \text{ and } \lim_{x \to a} g(x) = \pm\infty$$

Theorem 6.4.1 L'Hôpital's Rule

Suppose that f and g are differentiable functions such that

$$\lim_{x \to a} \frac{f(x)}{g(x)}$$

is an indeterminate form of type 0/0 or ∞/∞. If $g'(x) \neq 0$ near a (could be 0 at a) then

$$\lim_{x \to a} \frac{f(x)}{g(x)} = \lim_{x \to a} \frac{f'(x)}{g'(x)}$$

provided that the limit on the right side exists (i.e., is equal to a real number), or is equal to ∞ or $-\infty$.

In words, if the ratio of functions is an indeterminate form, then the limit of the ratio of functions is equal to the limit of the ratio of their derivatives.

L'Hôpital's rule applies to one-sided limits as well as to limits at infinity. Thus, $x \to a$ in the theorem can be replaced by any of $x \to a^+$, $x \to a^-$, $x \to \infty$, and $x \to -\infty$.

As with any other theorem, if we wish to use L'Hôpital's rule, we need to make sure that all assumptions are satisfied.

Example 6.4.9 Comparing Functions Using L'Hôpital's Rule

With L'Hôpital's rule we can prove that exponentials grow more quickly than power functions and that power functions grow more quickly than logarithms as x approaches infinity (Table 4.3.7).

Consider

$$\lim_{x \to \infty} \frac{e^x}{x}$$

As $x \to \infty$, both the numerator and the denominator approach infinity; thus, the above limit is an indeterminate form. We usually indicate it by writing

$$\lim_{x \to \infty} \frac{e^x}{x} = \frac{\infty}{\infty}$$

Using L'Hôpital's rule, we get

$$\lim_{x\to\infty} \frac{e^x}{x} = \frac{\infty}{\infty} = \lim_{x\to\infty} \frac{(e^x)'}{(x)'}$$
$$= \lim_{x\to\infty} \frac{e^x}{1}$$
$$= \infty$$

Example 6.4.10 Comparing the Logarithmic Function with a Linear Function

To compare the behaviour of x and $\ln x$ at infinity using L'Hôpital's rule, we compute

$$\lim_{x\to\infty} \frac{x}{\ln x} = \frac{\infty}{\infty} = \lim_{x\to\infty} \frac{(x)'}{(\ln x)'}$$
$$= \lim_{x\to\infty} \frac{1}{1/x}$$
$$= \lim_{x\to\infty} x$$
$$= \infty$$

If a single application of L'Hôpital's rule results in another indeterminate form, a second or third application might make it possible to evaluate the limit.

Example 6.4.11 Cases Requiring Multiple Applications of L'Hôpital's Rule

To show that e^x increases more quickly than x^3, we must take the derivative three times:

$$\lim_{x\to\infty} \frac{e^x}{x^3} = \lim_{x\to\infty} \frac{e^x}{3x^2} \quad \text{indeterminate form, take derivatives}$$
$$= \lim_{x\to\infty} \frac{e^x}{6x} \quad \text{still indeterminate, take derivatives}$$
$$= \lim_{x\to\infty} \frac{e^x}{6} = \infty \quad \text{limit of the exponential function is infinity}$$

Likewise,

$$\lim_{x\to\infty} \frac{e^{0.5x}}{x^{3/2}} = \frac{\infty}{\infty} = \lim_{x\to\infty} \frac{0.5 e^{0.5x}}{\frac{3}{2} x^{1/2}}$$
$$= \frac{1}{3} \lim_{x\to\infty} \frac{e^{0.5x}}{x^{1/2}}$$
$$= \frac{\infty}{\infty} = \frac{1}{3} \lim_{x\to\infty} \frac{0.5 e^{0.5x}}{\frac{1}{2} x^{-1/2}}$$
$$= \frac{1}{3} \lim_{x\to\infty} e^{0.5x} x^{1/2} = \infty$$

In the same way we can prove that

$$\lim_{x\to\infty} \frac{a e^{\beta x}}{c x^n} = \infty$$

for $a > 0$, $c > 0$, $\beta > 0$, and $n > 0$. This means that $a e^{\beta x}$ approaches infinity more quickly than $c x^n$, as claimed in Section 4.3 (see Table 4.3.7 and the text that follows).

Example 6.4.12 L'Hôpital's Rule Can Succeed When the Method of Leading Behaviours Fails

L'Hôpital's rule is indispensable when the method of leading behaviours does not apply. The expression

$$\lim_{x\to 0} \frac{e^{2x} - 1}{x}$$

is an indeterminate form of type 0/0. We apply L'Hôpital's rule, finding that

$$\lim_{x \to 0} \frac{e^{2x} - 1}{x} = \lim_{x \to 0} \frac{2e^{2x}}{1} = 2$$

Example 6.4.13 L'Hôpital's Rule Used Incorrectly

L'Hôpital's rule generally gives the wrong answer if applied to an expression that is not an indeterminate form. Here is an example. By direct substitution, we compute

$$\lim_{x \to 0} \frac{e^{2x} - 1}{x + 1} = \frac{1 - 1}{0 + 1} = 0$$

If we apply L'Hôpital's rule, we get

$$\lim_{x \to 0} \frac{e^{2x} - 1}{x + 1} = \lim_{x \to 0} \frac{2e^{2x}}{1} = 2$$

which is incorrect.

Note Why does L'Hôpital's rule work? To indicate reasons (this is not going to be a complete proof) we consider the case

$$\lim_{x \to a} \frac{f(x)}{g(x)} = \frac{0}{0}$$

i.e., $\lim_{x \to a} f(x) = 0$ and $\lim_{x \to a} g(x) = 0$.

Assume that $f(x)$ and $g(x)$ are differentiable. Near a, $f(x)$ can be approximated by the tangent

$$f(x) \approx f(a) + f'(a)(x - a)$$

Since $f(x)$ is differentiable, it is also continuous and thus

$$f(a) = \lim_{x \to a} f(x) = 0$$

So the tangent line approximation becomes

$$f(x) \approx f'(a)(x - a)$$

Likewise

$$g(x) \approx g'(a)(x - a)$$

and therefore

$$\frac{f(x)}{g(x)} \approx \frac{f'(a)(x - a)}{g'(a)(x - a)} = \frac{f'(a)}{g'(a)}$$

assuming that $x - a \neq 0$ and $g'(a) \neq 0$. Taking the limit, we obtain

$$\lim_{x \to a} \frac{f(x)}{g(x)} = \frac{f'(a)}{g'(a)}$$

If $f'(x)$ is continuous, then

$$\lim_{x \to a} f'(x) = f'(a)$$

As well, $\lim_{x \to a} g'(x) = g'(a)$, assuming that $g'(x)$ is continuous. So

$$\lim_{x \to a} \frac{f(x)}{g(x)} = \frac{f'(a)}{g'(a)} = \lim_{x \to a} \frac{f'(x)}{g'(x)}$$

Keep in mind that this is neither a rigid nor a complete proof (we can find one in more advanced calculus or analysis texts).

Example 6.4.14 Using a Linear Approximation to Calculate a Limit

Consider again the limit
$$\lim_{x \to 0} \frac{e^{2x}-1}{x} = \frac{0}{0}$$

Let $f(x) = e^{2x} - 1$. From $f(0) = 0$, $f'(x) = 2e^{2x}$, and $f'(0) = 2$, we construct the linear approximation
$$L(x) = f(0) + f'(0)(x-0) = 2x$$

Near 0,
$$f(x) = e^{2x} - 1 \approx L(x) = 2x$$

and therefore
$$\lim_{x \to 0} \frac{e^{2x}-1}{x} = \lim_{x \to 0} \frac{2x}{x} = 2$$

The limits leading to 0/0 or ∞/∞ are not the only indeterminate forms. In total, there are seven indeterminate forms:
$$\frac{0}{0}, \frac{\infty}{\infty}, 0 \cdot \infty, \infty - \infty, 0^0, \infty^0, \text{ and } 1^\infty$$

What is an indeterminate form? What makes the above seven limits indeterminate?

Indeterminate forms appear in the context of limits only, when the functions involved approach the expressions listed above, such as
$$\lim_{x \to 0} \frac{\sin x}{x} = \frac{0}{0}$$

What is 0/0? It turns out that, unlike 4/2 = 2 or 1/5 = 0.2, the fraction 0/0 *does not have a unique numerical value*—and it might not have one at all. No matter how we calculate 4/2, no matter what context (or application) it came from, the answer is 2. However, to say what 0/0 is, we need to know *exactly what limit it came from*.

Using L'Hôpital's rule, we compute several limits that involve 0/0:
$$\lim_{x \to 0} \frac{\sin x}{x} = \frac{0}{0} = \lim_{x \to 0} \frac{\cos x}{1} = 1$$

Can we say that the value of 0/0 is 1? No, because if we consider the limit
$$\lim_{x \to 1} \frac{x^2-1}{x-1} = \frac{0}{0} = \lim_{x \to 1} \frac{2x}{1} = 2$$

then we need to make 0/0 equal to 2.

As a matter of fact, we can make 0/0 equal to any real number we wish! For instance, to make it equal to 34.1, we could use
$$\lim_{x \to 0} \frac{34.1 \sin x}{x} = \frac{0}{0} = \lim_{x \to 0} \frac{34.1 \cos x}{1} = 34.1$$

As well, we can make 0/0 equal to ∞:
$$\lim_{x \to 0} \frac{x}{x^3} = \frac{0}{0} = \lim_{x \to 0} \frac{1}{x^2} = \infty$$

or make it not exist, as in
$$\lim_{x \to 0} \frac{x}{x^2} = \frac{0}{0} = \lim_{x \to 0} \frac{1}{x}$$

(since the left limit is $-\infty$ and the right limit is $+\infty$).

What we said here for 0/0 can be repeated for any of the remaining six indeterminate forms.

Chapter 6 Applications of Derivatives

Since L'Hôpital's rule (Theorem 6.4.1) involves fractions, all other types of indeterminate forms must be reduced to a fraction.

In a sequence of exercises, we will explore methods of calculating limits involving indeterminate forms.

Example 6.4.15 Indeterminate Product, $0 \cdot \infty$

Compute
$$\lim_{x \to 0^+} x^3 \ln x$$

As $x \to 0^+$, $x^3 \to 0$ and $\ln x \to -\infty$; thus, the given limit is an indeterminate form. To convert the product into a quotient, we use this simple algebraic trick:
$$A \cdot B = \frac{A}{1/B} = \frac{B}{1/A}$$

We need to decide which of the two factors to keep on top, and which one goes into the denominator (we try one way, and see how it goes; if we see that the calculations become more and more complicated, or seem to be going nowhere, we try the other way around).

Keeping $\ln x$ in the numerator, we compute
$$\lim_{x \to 0^+} x^3 \ln x = \lim_{x \to 0^+} \frac{\ln x}{1/x^3} = \frac{-\infty}{\infty}$$
$$= \lim_{x \to 0^+} \frac{(\ln x)'}{(x^{-3})'}$$
$$= \lim_{x \to 0^+} \frac{1/x}{-3x^{-4}}$$
$$= -\frac{1}{3} \lim_{x \to 0^+} x^3 = 0$$

Note that to simplify the fraction we wrote
$$\frac{1/x}{-3x^{-4}} = -\frac{1}{3}x^{-1} \cdot x^4 = -\frac{1}{3}x^3$$

If we try to start the limit as
$$\lim_{x \to 0^+} x^3 \ln x = \lim_{x \to 0^+} \frac{x^3}{1/\ln x}$$
we soon realize (after calculating the derivative in the denominator) that we have ended up with a limit that is more complicated than the one we started with. ▲

Example 6.4.16 Indeterminate Difference, $\infty - \infty$

Compute
$$\lim_{x \to 1^+} \left(\frac{1}{\ln x} - \frac{1}{x-1} \right)$$

As $x \to 1^+$, $\ln x \to 0$, and (since $\ln x > 0$ for $x > 1$) it follows that $\frac{1}{\ln x} \to +\infty$. Likewise, because x approaches 1 from the right, $x - 1 > 0$, and so $\frac{1}{x-1} \to +\infty$. So the given limit is an indeterminate form of type $\infty - \infty$.

We calculate the common denominator and then use L'Hôpital's rule:
$$\lim_{x \to 1^+} \left(\frac{1}{\ln x} - \frac{1}{x-1} \right) = \lim_{x \to 1^+} \frac{x - 1 - \ln x}{(x-1)\ln x} = \frac{0}{0}$$

$$= \lim_{x \to 1^+} \frac{1 - \frac{1}{x}}{\ln x + (x-1)\frac{1}{x}}$$

$$= \lim_{x \to 1^+} \frac{x - 1}{x \ln x + x - 1}$$

In the last step, we multiplied both the numerator and the denominator by x. Note that the limit is still an indeterminate form, so we apply L'Hôpital's rule once again:

$$\lim_{x \to 1^+} \frac{x - 1}{x \ln x + x - 1} = \frac{0}{0} = \lim_{x \to 1^+} \frac{1}{\ln x + x\frac{1}{x} + 1}$$

$$= \lim_{x \to 1^+} \frac{1}{\ln x + 2}$$

$$= \frac{1}{2}$$

Example 6.4.17 Indeterminate Difference, $\infty - \infty$

Find the value of

$$\lim_{x \to \pi/2} (\sec x - \tan x)$$

As $x \to (\pi/2)^+$, $\cos x \to 0$ and $\cos x < 0$; thus both

$$\sec x = \frac{1}{\cos x} \quad \text{and} \quad \tan x = \frac{\sin x}{\cos x}$$

approach $-\infty$. When $x \to (\pi/2)^-$, $\cos x \to 0$ and $\cos x > 0$; both $\sec x$ and $\tan x$ approach $+\infty$. Therefore, in either case, the limit is an indeterminate form.

We simplify first and then use L'Hôpital's rule:

$$\lim_{x \to \pi/2} (\sec x - \tan x) = \lim_{x \to \pi/2} \left(\frac{1}{\cos x} - \frac{\sin x}{\cos x} \right)$$

$$= \lim_{x \to \pi/2} \frac{1 - \sin x}{\cos x} = \frac{0}{0}$$

$$= \lim_{x \to \pi/2} \frac{-\cos x}{-\sin x}$$

$$= 0$$

by direct substitution.

Example 6.4.18 Indeterminate Power, 0^0

Calculate

$$\lim_{x \to 0^+} x^x$$

Using the fact that $A = e^{\ln A}$, we rewrite the limit as

$$\lim_{x \to 0^+} x^x = \lim_{x \to 0^+} e^{\ln x^x} = \lim_{x \to 0^+} e^{x \ln x}$$

Next, we need to interchange the exponential function and the limit. Recall that

$$\lim_{x \to a} f(g(x)) = f\left(\lim_{x \to a} g(x) \right)$$

as long as the function f is continuous.

To write the limit in the exponent of e would require a very small font size; as an alternative, we use exp notation for the exponential function, and write $\exp(A)$ when we mean e^A. Thus

$$\lim_{x \to 0^+} e^{x \ln x} = \lim_{x \to 0^+} \exp(x \ln x) = \exp\left(\lim_{x \to 0^+} x \ln x \right)$$

Now we calculate
$$\lim_{x\to 0^+} x\ln x$$
using L'Hôpital's rule:
$$\lim_{x\to 0^+} x\ln x = 0 \cdot (-\infty)$$
$$= \lim_{x\to 0^+} \frac{\ln x}{\frac{1}{x}}$$
$$= \lim_{x\to 0^+} \frac{\frac{1}{x}}{-\frac{1}{x^2}}$$
$$= \lim_{x\to 0^+} (-x)$$
$$= 0$$

Finally, gathering together important steps,
$$\lim_{x\to 0^+} x^x = \lim_{x\to 0^+} e^{x\ln x}$$
$$= \exp\left(\lim_{x\to 0^+} x\ln x\right)$$
$$= \exp(0) = e^0 = 1$$

Example 6.4.19 Indeterminate Power, 1^∞

All indeterminate limits that involve powers are done in the same way as in the previous example.

To calculate
$$\lim_{x\to\infty}\left(1+\frac{1}{x}\right)^x = 1^\infty$$
we rewrite the expression using the exponential function and replace e^A by $\exp(A)$:
$$\lim_{x\to\infty}\left(1+\frac{1}{x}\right)^x = \lim_{x\to\infty} e^{\ln(1+1/x)^x}$$
$$= \lim_{x\to\infty} e^{x\ln(1+1/x)}$$
$$= \lim_{x\to\infty} \exp\left[x\ln\left(1+\frac{1}{x}\right)\right]$$
$$= \exp\left\{\lim_{x\to\infty}\left[x\ln\left(1+\frac{1}{x}\right)\right]\right\}$$

In the last line, we interchanged the exponential function and the limit (which we are allowed to do since e^x is continuous). Using L'Hôpital's rule,
$$\lim_{x\to\infty}\left[x\ln\left(1+\frac{1}{x}\right)\right] = \lim_{x\to\infty}\frac{\ln\left(1+\frac{1}{x}\right)}{\frac{1}{x}} = \frac{0}{0}$$
$$= \lim_{x\to\infty}\frac{\frac{1}{1+1/x}\left(-\frac{1}{x^2}\right)}{-\frac{1}{x^2}}$$
$$= \lim_{x\to\infty}\frac{1}{1+\frac{1}{x}}$$
$$= 1$$

Therefore,
$$\lim_{x\to\infty}\left(1+\frac{1}{x}\right)^x = \exp\left\{\lim_{x\to\infty}\left[x\ln\left(1+\frac{1}{x}\right)\right]\right\} = \exp(1) = e$$

Summary We have learned two ways to compute the behaviour of complicated functions. The **method of leading behaviour** is a way to examine sums of functions and determine which piece increases to infinity *most quickly* or to zero *most slowly*. By graphing the leading behaviour of functions at both zero and infinity, we can use the **method of matched leading behaviours** to sketch an accurate graph. In other cases, we can evaluate **indeterminate forms** with **L'Hôpital's rule,** which says that a ratio of functions that both approach zero or both approach infinity has the same limit as the ratio of their derivatives.

6.4 Exercises

Mathematical Techniques

1–6 ■ Find the leading behaviour of each function at zero and infinity.

1. $f(x) = 1 + x$
2. $g(y) = y + y^3$
3. $h(z) = z + e^z$
4. $F(x) = 1 + 2x + 3e^x$
5. $m(a) = 100a + 30a^2 + \dfrac{1}{a}$
6. $G(c) = e^{-4c} + \dfrac{5}{c^2} + \dfrac{3}{c^5} + 10e^{-3c}$

7–16 ■ For each pair of functions, use the basic functions (when possible) to say which approaches its limit more quickly, and then check with L'Hôpital's rule.

7. x^2 and e^{2x} as $x \to \infty$.
8. x^2 and $1000x$ as $x \to \infty$.
9. $0.1x^{0.5}$ and $30 \ln x$ as $x \to \infty$.
10. x and $(\ln x)^2$ as $x \to \infty$.
11. e^{-2x} and x^{-2} as $x \to \infty$.
12. $1/\ln x$ and $30x^{-0.1}$ as $x \to 0$.
13. x^{-1} and $-\ln x$ as $x \to 0$. Use your result to figure out $\lim_{x \to 0}[x \ln x]$.
14. x^{-1} and $\dfrac{1}{e^x - 1}$ as $x \to 0$.
15. x^2 and x^3 as $x \to 0$.
16. x^2 and $e^x - x - 1$ as $x \to 0$.

17–39 ■ Find each limit using L'Hôpital's rule (if possible; check assumptions before you try to use it!) or otherwise.

17. $\lim_{x \to 2} \dfrac{x^3 - x^2 - x - 2}{x - 2}$
18. $\lim_{x \to 1} \dfrac{x^{10} - 1}{x - 1}$
19. $\lim_{x \to 2} \dfrac{x^{10} - 1}{x^3 - 1}$
20. $\lim_{x \to 0^+} \dfrac{1 - \sqrt{1-x}}{x}$
21. $\lim_{x \to 0} \dfrac{x \cos x}{\sin x}$
22. $\lim_{x \to \pi/2} \dfrac{x \cos x}{\sin x}$
23. $\lim_{x \to 0} \dfrac{\tan x}{x}$
24. $\lim_{x \to 0} \dfrac{\arctan x}{x}$
25. $\lim_{x \to 1} \dfrac{\sin x - 1}{x - 1}$
26. $\lim_{x \to 0^+} \dfrac{\ln x}{x^{0.68}}$
27. $\lim_{x \to \infty} \dfrac{\ln(\ln x)}{3x}$
28. $\lim_{x \to \infty} \dfrac{e^x - 2}{3 - 2e^x}$
29. $\lim_{x \to \infty} \dfrac{x^{100}}{e^{0.003x}}$
30. $\lim_{x \to 0} \dfrac{4^x - 1}{7^x - 1}$
31. $\lim_{x \to 0} \dfrac{\cos x - 1 + \frac{1}{2}x^2}{x^4}$
32. $\lim_{x \to \infty} (x^2 e^{-0.2x})$
33. $\lim_{x \to 0^+} (x^{0.5} \ln x)$
34. $\lim_{x \to \infty} [x \sin(1/x)]$
35. $\lim_{x \to 0^+} \left(\dfrac{1}{x} - \dfrac{1}{\sin x}\right)$
36. $\lim_{x \to 0^+} (\cot x - \csc x)$
37. $\lim_{x \to 0^+} x^{4x}$
38. $\lim_{x \to \infty} x^{1/x}$
39. $\lim_{x \to \infty} (1 + x)^{1/x}$

40–45 ■ For each function, find the leading behaviour of the numerator, the denominator, and the whole function at both zero and infinity. Find the limit of the function at zero and infinity (and check with L'Hôpital's rule when appropriate). Use the method of matched leading behaviours to sketch the graph of each function.

40. $\alpha(c) = \dfrac{2c^2}{1+c}$

41. $\alpha(c) = \dfrac{c^2}{1+2c}$

42. $\alpha(c) = \dfrac{1+c+c^2}{1+c}$

43. $\alpha(c) = \dfrac{1+c}{1+c+c^2}$

44. $\alpha(c) = \dfrac{3c}{1+\ln(1+c)}$

45. $\alpha(c) = \dfrac{e^c + 1}{e^{2c} + 1}$

46–49 ▪ Write the tangent line approximation for the numerator and denominator of each function, and show that the result of applying L'Hôpital's rule matches that of comparing the linear approximations.

46. $f(x) = \dfrac{2x + x^2}{3x + 2x^2}$ at $x = 0$.

47. $f(x) = \dfrac{\ln(1+x)}{e^{2x} - 1}$ at $x = 0$.

48. $f(x) = \dfrac{\ln x}{x^2 - 1}$ at $x = 1$.

49. $f(x) = \dfrac{\cos x + 1}{\sin x}$ at $x = \pi$.

Applications

50–55 ▪ Find the leading behaviours $\alpha_0(c)$ and $\alpha_\infty(c)$ of the given absorption functions. Sketch $\alpha_0(c)$ and $\alpha_\infty(c)$ and draw a possible graph of $\alpha(c)$.

50. $\alpha(c) = \dfrac{5c}{1+c}$

51. $\alpha(c) = \dfrac{c}{5+c}$

52. $\alpha(c) = \dfrac{5c^2}{1+c^2}$

53. $\alpha(c) = \dfrac{5c}{e^{2c}}$

54. $\alpha(c) = \dfrac{5c}{1+c^2}$

55. $\alpha(c) = 5c(1+c)$

56–59 ▪ Use the method of matched leading behaviours to graph each Hill function or variant.

56. $h_3(x) = \dfrac{x^3}{1+x^3}$

57. $g_3(x) = \dfrac{x^3}{10+x^3}$

58. $h_{10}(x) = \dfrac{x^{10}}{1+x^{10}}$

59. $g_{10}(x) = \dfrac{x^{10}}{0.1+x^{10}}$

60–63 ▪ The following discrete-time dynamical systems describe the populations of two competing strains of bacteria:

$$a_{t+1} = s a_t$$
$$b_{t+1} = r b_t$$

For the following values of the initial conditions a_0 and b_0, and the per capita production s and r:

a. Find the number of each type as a function of time.

b. Find the fraction of type a as a function of time.

c. Use leading behaviour or L'Hôpital's rule to find the limit of the fraction as $t \to \infty$.

d. Compute the fraction at $t = 0$, 10, 20, and 50, and compare with your limit.

60. $a_0 = 10^4$, $b_0 = 10^6$, $s = 2$, $r = 1.5$

61. $a_0 = 10^4$, $b_0 = 10^6$, $s = 1.5$, $r = 2$

62. $a_0 = 10^4$, $b_0 = 10^5$, $s = 0.8$, $r = 1.2$

63. $a_0 = 10^4$, $b_0 = 10^5$, $s = 0.5$, $r = 0.3$

64–65 ▪ Many absorption equations are of the form

$$\alpha(c) = A \dfrac{r(c)}{k + r(c)}$$

where A and k are positive parameters, and where $r(0) = 0$, $\lim_{c \to \infty} r(c) = \infty$, and $r'(c) > 0$.

64. In each case, identify $r(c)$ and show that $\alpha(c)$ is increasing. Use L'Hôpital's rule to find the limit as $c \to \infty$, and use the method of leading behaviour to describe absorption near $c = 0$ and $c = \infty$.

a. $\alpha(c) = \dfrac{Ac}{k+c}$ (from Table 4.3.10 in Chapter 4).

b. $\alpha(c) = \dfrac{Ac^2}{k+c^2}$ (from Table 4.3.10 in Chapter 4).

c. $\alpha(c) = \dfrac{Ac^n}{k+c^n}$, where n is a positive integer.

65. Answer the same questions as in Exercise 64, but without plugging in a particular form for $r(c)$.

Computer Exercises

66. Consider the functions

$$f(c) = \dfrac{c^2}{1+c^2}$$

$$g(c) = \dfrac{c}{1+c^2}$$

where $c \geq 0$. Find the leading behaviour of each at zero and infinity. Suppose we approximated each by a function defined in pieces:

$$\check{f}(c) = \begin{cases} f_0(c) & \text{if } f_0(c) \leq f_\infty(c) \\ f_\infty(c) & \text{if } f_0(c) > f_\infty(c) \end{cases}$$

and similarly for $\check{g}(c)$. Plot this approximation in each case. Find and plot the ratios

$$\frac{\check{f}(c)}{f(c)} \quad \text{and} \quad \frac{\check{g}(c)}{g(c)}$$

When is the approximation best? When is it worst?

67. Consider the function

$$f(x) = \frac{1 + x + x^2 + x^3 + x^4}{5 + 4x + 3x^2 + 2x^3 + x^4}$$

Find the leading behaviour for large x. Next find a function that keeps both the largest and the second-largest term from the numerator and denominator. How much better is this new approximation? How much improvement do you get by adding more and more terms?

6.5 Graphing Functions: A Summary

So far we have drawn the graphs of many functions, but in most cases it was to illustrate a particular feature—such as a horizontal asymptote, or some limit, or a relative extreme value. In this section we collect all information about a function that we can obtain using calculus to produce an accurate graph, in order to show its interesting and important features.

We could use a graphing device or software to obtain the graph of a function much more quickly than when we do it by hand. However, we need to be careful as we might miss important features of the graph. As well, no matter how much we zoom in, we can only get approximations for the values of the variables.

As an illustration, consider the graph of the polynomial

$$f(x) = (x - 1)^2(x - 2)^3 + 1 = x^5 - 8x^4 + 25x^3 - 38x^2 + 28x - 7$$

drawn for $0 \leq x \leq 3$ and $-2 \leq y \leq 4$ (Figure 6.5.64a). It looks like the graph is almost horizontal from $x = 1$ to a bit beyond $x = 2$. However, zooming in on the range ($0.5 \leq y \leq 1.5$ in Figure 6.5.64b and $0.9 \leq y \leq 1.1$ in Figure 6.5.64c) reveals a more complex situation. We see that $f(x)$ has a relative maximum—but is it at $x = 1$ or at some number near it? The maximum value is around 1, but what is its exact value? As well, $f(x)$ has a relative minimum around $x = 1.5$ (where exactly, and what is its exact value?). We can only guess what is going on near $x = 2$.

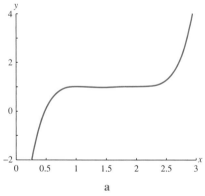
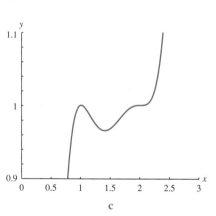

FIGURE 6.5.64
Computer-generated plots of $f(x)$

On the other hand, a short calculation,

$$f'(x) = 2(x - 1)(x - 2)^3 + (x - 1)^2 3(x - 2)^2$$
$$= (x - 1)(x - 2)^2 [2(x - 2) + 3(x - 1)]$$
$$= (x - 1)(x - 2)^2 (5x - 7)$$

proves (among other things) that $x = 7/5$ is a critical point; next, the first derivative test would confirm that it is a relative minimum; we calculate the minimum value to be

$$f(7/5) = \left(\frac{7}{5} - 1\right)^2 \left(\frac{7}{5} - 2\right)^3 + 1 = \frac{4}{25}\left(-\frac{27}{125}\right) + 1 = \frac{3017}{3125} = 0.96544$$

This is a result that we cannot obtain by manipulating the viewing window of a computer-generated graph.

And what is going on near $x=2$? A closer blowup (Figure 6.5.65) suggests that there might be an inflection point there (and again—if so, where exactly is it, and what is its y-coordinate? Check Exercise 20).

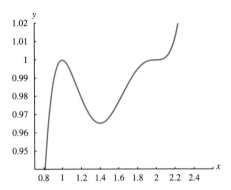

FIGURE 6.5.65

Zooming in on the graph of $f(x)$

In order to draw an accurate graph of a function, we collect information from three sources. From the formula for the function we identify its domain and range, x- and y-intercepts, and horizontal and vertical asymptotes. As well, we might test the function for symmetry, or recognize its special values, or check whether or not it is periodic.

From the first derivative we find intervals where the function is increasing and where it is decreasing, and identify relative extreme values.

Using the second derivative, we detect intervals where the function is concave up and where it is concave down, and find inflection points. As well, we might use the second derivative as an alternative to testing critical points with the first derivative test.

Example 6.5.1 Graph of a Rational Function

Sketch the graph of the function $f(x) = \dfrac{2x}{4+x^2}$.

The function $f(x)$ is defined for all x because the denominator is never equal to zero. Since $4 + x^2 > 0$ for all x, we conclude that $f(x) < 0$ for $x < 0$ and $f(x) > 0$ for $x > 0$. The y-intercept of $f(x)$ is $f(0) = 0/4 = 0$.

Using the method of matched leading behaviours, we find that $f_0(x) = 2x/4 = x/2$ and $f_\infty(x) = 2x/x^2 = 2/x$. Thus, near zero, $f(x)$ looks like a line of slope $1/2$, and in the long term (as x grows large) $f(x)$ decreases like $y = 2/x$.

From

$$\lim_{x \to \pm\infty} \frac{2x}{4+x^2} = \lim_{x \to \pm\infty} \frac{2x}{x^2} = \lim_{x \to \pm\infty} \frac{2}{x} = 0$$

we conclude that $y = 0$ is a horizontal asymptote of the graph of $f(x)$. Since it is defined for all x, $f(x)$ has no vertical asymptotes. Note that

$$f(-x) = \frac{2(-x)}{4+(-x)^2} = -\frac{2x}{4+x^2} = -f(x)$$

Thus, the graph of $f(x)$ is a mirror image of itself if we mirror it with respect to *both* the x-axis and the y-axis (functions with this property are called **odd functions**).

From

$$f'(x) = \frac{2(4+x^2) - 2x(2x)}{(4+x^2)^2} = \frac{8 - 2x^2}{(4+x^2)^2} = 2\frac{4-x^2}{(4+x^2)^2} = 0$$

it follows that $4 - x^2 = 0$, $x^2 = 4$, and so $x = \pm 2$ are the critical numbers. In the table below we analyze $f(x)$ based on its first derivative:

6.5 Graphing Functions: A Summary

x	$(-\infty, -2)$	-2	$(-2, 2)$	2	$(2, \infty)$
$f'(x)$	negative	0	positive	0	negative
$f(x)$	decreasing	relative minimum	increasing	relative maximum	decreasing

The second derivative of $f(x)$ is

$$f''(x) = 2\frac{(-2x)(4+x^2)^2 - (4-x^2)\,2(4+x^2)(2x)}{(4+x^2)^4}$$

$$= 2\frac{(-2x)(4+x^2) - 4x(4-x^2)}{(4+x^2)^3}$$

$$= 2\frac{(-2x)(4+x^2+8-2x^2)}{(4+x^2)^3}$$

$$= -4\frac{x(12-x^2)}{(4+x^2)^3}$$

From $f''(x) = 0$ we get $x = 0$ and $x = \pm\sqrt{12} = \pm\sqrt{4 \cdot 3} = \pm 2\sqrt{3}$. Using the table below, we investigate the concavity of the graph of $f(x)$:

x	$(-\infty, -2\sqrt{3})$	$-2\sqrt{3}$	$(-2\sqrt{3}, 0)$	0	$(0, 2\sqrt{3})$	$2\sqrt{3}$	$(2\sqrt{3}, \infty)$
$f''(x)$	negative	0	positive	0	negative	0	positive
$f(x)$	concave down	inflection	concave up	inflection	concave down	inflection	concave up

Collecting all information we have calculated, we sketch the graph of $f(x)$; see Figure 6.5.66.

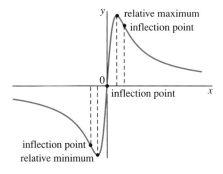

FIGURE 6.5.66

The graph of $f(x) = 2x/(4 + x^2)$

The function in this example appears in applications. For instance, $f(x)$ is an example of saturated absorption with overcompensation (see Section 4.3; in that case, x represents the concentration of a chemical, $x \geq 0$). The function $f(x)$ could represent the yield of some agricultural crop; in that case x (assuming that $x \geq 0$) is the level of nitrogen in the soil. ▲

Example 6.5.2   Bell-Shaped Curve

Sketch the graph of the function $f(x) = e^{-x^2}$.

The domain of $f(x)$ consists of all real numbers. Exponential functions are positive, so $f(x) > 0$ for all x; thus, $f(x)$ has no x-intercepts. The y-intercept is $f(0) = e^0 = 1$.

Since $f(x)$ is defined for all x, it has no vertical asymptotes. From

$$\lim_{x \to \infty} e^{-x^2} = e^{-\infty} = 0 \quad \text{and} \quad \lim_{x \to -\infty} e^{-x^2} = e^{-\infty} = 0$$

we conclude that $y = 0$ is a horizontal asymptote of the graph of $f(x)$.

Note that $f(-x) = e^{-(-x)^2} = e^{-x^2} = f(x)$; thus, $f(x)$ is its own reflection across the y-axis. In other words, the graph of $f(x)$ is symmetric with respect to the y-axis. (Functions with this property are called **even functions.**)

From

$$f'(x) = -2xe^{-x^2} = 0$$

we conclude that $x = 0$ is the only critical point of $f(x)$. When $x < 0$, $f'(x) > 0$ and $f(x)$ is increasing; when $x > 0$, $f'(x) < 0$ and $f(x)$ is decreasing. We conclude that $f(0) = 1$ is a relative maximum.

The second derivative is computed to be

$$f''(x) = -2e^{-x^2} - 2xe^{-x^2}(-2x) = -2e^{-x^2} + 4x^2 e^{-x^2} = 2e^{-x^2}(-1 + 2x^2)$$

Thus, $f''(x) = 0$ when $-1 + 2x^2 = 0$, i.e., when $x = \pm\sqrt{1/2} = \pm 1/\sqrt{2}$. The analysis of the behaviour of $f(x)$ based on $f''(x)$ is given in the table below:

x	$(-\infty, -1/\sqrt{2})$	$-1/\sqrt{2}$	$(-1/\sqrt{2}, 1/\sqrt{2})$	$1/\sqrt{2}$	$(1/\sqrt{2}, \infty)$
$f''(x)$	positive	0	negative	0	positive
$f(x)$	concave up	inflection point	concave down	inflection point	concave up

Collecting all information, we draw the graph (Figure 6.5.67). The relative maximum is at $(0, 1)$, and the inflection points are $(1/\sqrt{2}, e^{-1/2})$ and $(-1/\sqrt{2}, e^{-1/2})$.

A modification of $f(x) = e^{-x^2}$ given by

$$s(x) = \frac{1}{\sqrt{2\pi}} e^{-x^2/2}$$

determines the **standard normal distribution,** one of the most important functions in probability (see Exercises 21 and 22). Another modification appears in the model for diffusion (see Exercise 30).

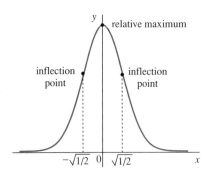

FIGURE 6.5.67

The graph of $f(x) = e^{-x^2}$

Example 6.5.3 Logistic Curve

Sketch the graph of the function $P(t) = \dfrac{10}{1 + 9e^{-2t}}$ for $t \geq 0$. This curve is an example of a **logistic curve,** which often appears in population modelling.

The function $P(t)$ is defined for all $t \geq 0$; it is positive because $e^{-2t} > 0$ (and thus has no t-intercepts). Actually, we can do better: since $1 + 9e^{-2t} > 1$, it follows that $P(t) < 10$ for all $t \geq 0$. As well, the initial population is $P(0) = 10/(1 + 9) = 1$.

6.5 Graphing Functions: A Summary

The calculation

$$\lim_{t \to \infty} P(t) = \lim_{t \to \infty} \frac{10}{1 + 9e^{-2t}} = \frac{10}{1 + 9(0)} = 10$$

proves that $P = 10$ is a horizontal asymptote of the graph of $P(t)$. When $P(t)$ models population dynamics, the value (sometimes referred to as the **plateau**) $P = 10$ is called the **carrying capacity.**

Rewriting the function as $P(t) = 10(1 + 9e^{-2t})^{-1}$ and using the chain rule, we obtain

$$P'(t) = 10(-1)(1 + 9e^{-2t})^{-2} 9e^{-2t}(-2) = 180e^{-2t}(1 + 9e^{-2t})^{-2}$$

Since $P'(t) > 0$, it follows that $P(t)$ is increasing for all $t > 0$. The second derivative is

$$P''(t) = 180 \left[e^{-2t}(-2)(1 + 9e^{-2t})^{-2} + e^{-2t}(-2)(1 + 9e^{-2t})^{-3} 9e^{-2t}(-2) \right]$$
$$= -360e^{-2t}(1 + 9e^{-2t})^{-3} \left[(1 + 9e^{-2t}) - 18e^{-2t} \right]$$
$$= -360e^{-2t}(1 + 9e^{-2t})^{-3}(1 - 9e^{-2t})$$

All terms except the last one are nonzero; thus, $P''(t) = 0$ implies $1 - 9e^{-2t} = 0$ and

$$e^{-2t} = 1/9$$
$$-2t = \ln(1/9) = \ln 1 - \ln 9 = -\ln 9$$
$$t = \frac{\ln 9}{2} \approx 1.0986$$

It follows that if $t < \ln 9/2$, then $P''(t) > 0$ and $P(t)$ is concave up; if $t > \ln 9/2$, then $P''(t) < 0$ and $P(t)$ is concave down. When $t = \ln 9/2$, then $e^{-2t} = 1/9$ and so

$$P(\ln 9/2) = \frac{10}{1 + 9(1/9)} = \frac{10}{2} = 5$$

We conclude that $(\ln 9/2, 5)$ is the point of inflection. (That the value of $P(t)$ at the inflection is one half of the carrying capacity is not a coincidence; see Exercise 26.) The graph of $P(t)$ is shown in Figure 6.5.68.

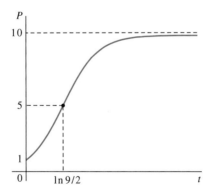

FIGURE 6.5.68
Logistic curve

Example 6.5.4 Graph of a Logarithmic Function

Sketch the graph of the function $f(x) = \dfrac{\ln x}{x^2}$.

The domain of $f(x)$ is $x > 0$. Setting $f(x) = 0$, we get $\ln x = 0$, so $x = 1$ is the (only) x-intercept. Using L'Hôpital's rule, we find

$$\lim_{x \to \infty} \frac{\ln x}{x^2} = \frac{\infty}{\infty} = \lim_{x \to \infty} \frac{1/x}{2x} = \lim_{x \to \infty} \frac{1}{2x^2} = 0$$

It follows that $y=0$ is a horizontal asymptote. Checking for vertical asymptotes, we compute

$$\lim_{x \to 0^+} \frac{\ln x}{x^2} = \lim_{x \to 0^+} \ln x \, \frac{1}{x^2} = (-\infty)(\infty) = -\infty$$

Thus, the graph of $f(x)$ approaches the negative y-axis as x approaches 0 from the right. Solving

$$f'(x) = \frac{\frac{1}{x}x^2 - \ln x(2x)}{x^4} = \frac{x - 2x \ln x}{x^4} = \frac{1 - 2\ln x}{x^3} = 0$$

we obtain $1 - 2\ln x = 0$, $\ln x = 1/2$, and so $x = e^{1/2}$ is the (only) critical point. If $x < e^{1/2}$, then $f'(x) > 0$ and $f(x)$ is increasing; when $x > e^{1/2}$, then $f'(x) < 0$ and $f(x)$ is decreasing. Consequently, $f(x)$ has a local maximum at $x = e^{1/2}$, and the maximum value is

$$f\left(e^{1/2}\right) = \frac{\ln\left(e^{1/2}\right)}{\left(e^{1/2}\right)^2} = \frac{1/2}{e} = \frac{1}{2e}$$

Next, we find the second derivative:

$$f''(x) = \frac{-2\frac{1}{x}x^3 - (1 - 2\ln x)(3x^2)}{x^6} = \frac{-2x^2 - 3(1 - 2\ln x)x^2}{x^6} = \frac{-5 + 6\ln x}{x^4}$$

Setting $f''(x) = 0$, we obtain $-5 + 6\ln x = 0$ and $x = e^{5/6}$. When $x < e^{5/6}$, then $f''(x) < 0$ and $f(x)$ is concave down; when $x > e^{5/6}$, then $f''(x) > 0$ and $f(x)$ is concave up. By computing

$$f\left(e^{5/6}\right) = \frac{\ln\left(e^{5/6}\right)}{\left(e^{5/6}\right)^2} = \frac{5/6}{e^{5/3}} = \frac{5}{6e^{5/3}}$$

we identify the point of inflection: $(e^{5/6}, 5/6e^{5/3})$. The graph of $f(x)$ is drawn in Figure 6.5.69.

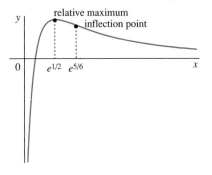

FIGURE 6.5.69

The graph of $f(x) = \ln x / x^2$

Example 6.5.5 Hill Functions

Functions of the form $f(x) = \dfrac{x^n}{1 + x^n}$, where n is a positive integer, are called **Hill functions.** They are used to describe all kinds of biological phenomena, including certain activities of enzymes and feeding patterns of animals. In the same coordinate system, sketch the graphs of the Hill function for $x \geq 0$ when $n = 1$, $n = 2$, and $n = 3$.

We keep in mind that $x \geq 0$. In that case, $x^n \geq 0$ and therefore the denominator $1 + x^n$ is never zero—so $f(x)$ is defined for all x in $[0, \infty)$. As well, $f(x) \geq 0$ and $f(0) = 0$. Since $x^n = 0$ only when $x = 0$, there are no other x-intercepts.

Note that $f(1) = 1/2$ no matter what n is. Since $f(x)$ is defined for all $x \geq 0$, it has no vertical asymptotes. From

$$\lim_{x \to \infty} \frac{x^n}{1+x^n} = \lim_{x \to \infty} \frac{x^n}{x^n} = 1$$

we conclude that $y = 1$ is a horizontal asymptote of the graph of $f(x)$ for all n.

By the method of matched leading behaviours, $f_\infty(x) = x^n/x^n = 1$ (confirming the calculation of the horizontal asymptote) and $f_0(x) = x^n/1 = x^n$. Thus, the graph of $f(x)$ starts as a power of x and flattens out for large x.

Differentiating $f(x)$ using the quotient rule, we obtain

$$f'(x) = \frac{nx^{n-1}(1+x^n) - x^n nx^{n-1}}{(1+x^n)^2} = \frac{nx^{n-1}}{(1+x^n)^2}$$

When $n = 1$, the numerator in $f'(x)$ is 1, and $f(x)$ has no critical points. When $n > 1$, then $f'(x) = 0$ implies $nx^{n-1} = 0$ and thus $x = 0$ is the only critical point. In any case, $f'(x) > 0$ for all $x > 0$, and we conclude that $f(x)$ is an increasing function on $(0, \infty)$ for all positive integers n.

When $n = 1$, $f'(x) = 1/(1+x)^2$ and therefore $f''(x) = -2/(1+x)^3$. We conclude that the Hill function $f(x) = x/(1+x)$ is concave down for all $x > 0$.

In general, getting the second derivative takes a bit more work

$$f''(x) = \frac{n(n-1)x^{n-2}(1+x^n)^2 - nx^{n-1}2(1+x^n)nx^{n-1}}{(1+x^n)^4}$$

$$= \frac{n(n-1)x^{n-2}(1+x^n) - 2n^2 x^{2n-2}}{(1+x^n)^3}$$

$$= \frac{nx^{n-2}[(n-1)(1+x^n) - 2nx^n]}{(1+x^n)^3}$$

$$= \frac{nx^{n-2}[(n-1) - x^n(1+n)]}{(1+x^n)^3}$$

Thus, $f''(x) = 0$ when $x = 0$ or when $(n-1) - x^n(1+n) = 0$, in which case

$$x = \sqrt[n]{\frac{n-1}{n+1}}$$

Next, we analyze $f''(x)$ for concavity:

x	$\left(0, \sqrt[n]{(n-1)/(n+1)}\right)$	$\sqrt[n]{(n-1)/(n+1)}$	$\left(\sqrt[n]{(n-1)/(n+1)}, \infty\right)$
$f''(x)$	positive	0	negative
$f(x)$	concave up	inflection point	concave down

When $n = 2$, the inflection point occurs at $x_2 = \sqrt{1/3} \approx 0.577$; for $n = 3$, the inflection point occurs at $x_3 = \sqrt[3]{1/2} \approx 0.794$. See Figure 6.5.70.

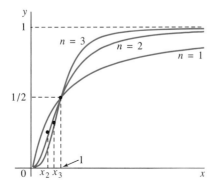

FIGURE 6.5.70

Hill functions

Example 6.5.6 Graph Involving a Trigonometric Function

Sketch the graph of the function $f(x) = x + 2\sin x$.

The domain of $f(x)$ consists of all real numbers (so $f(x)$ has no vertical asymptotes). Since $-1 \leq \sin x \leq 1$, it follows that $-2 \leq 2\sin x \leq 2$, and

$$x - 2 \leq x + 2\sin x \leq x + 2$$

Thus, the graph of $f(x)$ lies in the region between the parallel lines $y = x - 2$ and $y = x + 2$ (and thus has no horizontal asymptotes). The y-intercept of the graph of $f(x)$ is $f(0) = 0$.

Note that

$$f(-x) = -x + 2\sin(-x) = -x - 2\sin x = -f(x)$$

(recall that $\sin(-A) = -\sin A$ for any real number A). Thus, $f(x)$ is an odd function; i.e., the graph of $f(x)$ is a mirror image of itself when we mirror it with respect to both the x-axis and the y-axis.

From $f'(x) = 1 + 2\cos x = 0$ we get $\cos x = -1/2$, and the critical numbers are (see Example 2.3.9) $x = 2\pi/3 + 2\pi k$ and $x = 4\pi/3 + 2\pi k$, where k is an integer. Note that although $f(x)$ is not a periodic function, its derivative is periodic (with period 2π); thus, it suffices to analyze $f'(x)$ on the interval $[0, 2\pi]$:

x	$[0, 2\pi/3)$	$2\pi/3$	$(2\pi/3, 4\pi/3)$	$4\pi/3$	$(4\pi/3, 2\pi]$
$f'(x)$	positive	0	negative	0	positive
$f(x)$	increasing	relative maximum	decreasing	relative minimum	increasing

To obtain information for all x, we need to add $2\pi k$ to the values in the table: for instance, $f(x)$ has relative maxima at $2\pi/3 + 2\pi k$ (so they occur with period 2π), and the corresponding maximum value—which changes with k—is

$$f(2\pi/3 + 2\pi k) = \frac{2\pi}{3} + 2\pi k + 2\sin(2\pi/3 + 2\pi k) = \frac{2\pi}{3} + \sqrt{3} + 2\pi k$$

For instance, the first positive maximum ($k = 0$) occurs at $f(2\pi/3) = 2\pi/3 + \sqrt{3} \approx 3.826$; the next positive maximum ($k = 1$) occurs at $f(2\pi/3 + 2\pi) = f(8\pi/3) = 8\pi/3 + \sqrt{3} \approx 10.110$; and so on.

Next, $f''(x) = -2\sin x$ (so $f''(x)$ is periodic with period 2π), and $f''(x) = 0$ implies $x = \pi k$. If x is in $(-\pi, 0)$, $f''(x) > 0$ and $f(x)$ is concave up; if x is in $(0, \pi)$, $f''(x) < 0$ and $f(x)$ is concave down; for x in $(\pi, 2\pi)$, $f''(x) > 0$ and $f(x)$ is concave up. (Again, we add $2\pi k$ to obtain information for all x.) It follows that $x = \pi k$ is a point of inflection, and $f(\pi k) = \pi k + 2\sin(\pi k) = \pi k$.

The graph of $f(x)$ is shown in Figure 6.5.71. (The x-intercept $x = 2$ of the line $y = x - 2$ is very close to $x = 2\pi/3 \approx 2.09$.)

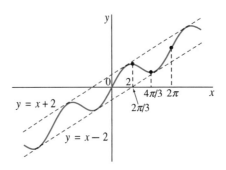

FIGURE 6.5.71

The graph of $f(x) = x + 2\sin x$

6.5 Exercises

Mathematical Techniques

1–19 ■ For each function find or discuss (as appropriate) the domain, range, asymptotes, intercepts, symmetry, periodicity, intervals of increase and intervals of decrease, relative extreme values, concavity, and inflection points. Based on the information you obtained, sketch the graph.

1. $f(x) = x^3 - 2x^2 - 4x + 4$
2. $f(x) = x^5 - 5x$
3. $f(x) = (x^2 - 1)(x+2)^2$
4. $f(x) = x(x-2)^3$
5. $f(x) = \dfrac{3x}{x+4}$ (this is an example of saturated absorption if $x \geq 0$)
6. $f(x) = \dfrac{x^2 - 1}{x^2 + 1}$
7. $f(x) = \dfrac{1}{x^2 - 16}$
8. $f(x) = \dfrac{x+5}{x^2}$
9. $\alpha(c) = \dfrac{c^2}{2 + c^2}$ (this is an example of saturated absorption with threshold if $c \geq 0$)
10. $\alpha(c) = \dfrac{c}{10 + c^2}$ (this is an example of saturated absorption with overcompensation if $c \geq 0$)
11. $f(x) = e^{-x} \sin x$, for x in $[0, 2\pi]$
12. $f(x) = xe^{-2x}$
13. $f(x) = x^2 e^x$
14. $g(x) = 2e^{-0.2x} - e^{-0.4x}$
15. $s(x) = e^{3x} - e^x$
16. $g(x) = x \ln x$
17. $f(x) = \dfrac{\ln x}{x}$
18. $f(x) = \dfrac{x}{2} + \cos x$
19. $f(x) = \sin x + \cos x$

20. Consider the function $f(x) = (x-1)^2 (x-2)^3 + 1$ from the introduction. Show that $f''(x) = 2(x-2)(10x^2 - 28x + 19)$. Find intervals where $f(x)$ is concave up and where it is concave down and all inflection points.

Applications

21. Explain how to obtain the graph of the function $s(x) = \dfrac{1}{\sqrt{2\pi}} e^{-x^2/2}$ from the graph of $f(x) = e^{-x^2}$ in Example 6.5.2. Sketch the graph of $s(x)$ and label its relative extreme values and inflection points.

22. The function $n(x) = \dfrac{1}{\sigma \sqrt{2\pi}} e^{-(x-\mu)^2/2\sigma^2}$ determines the general **normal distribution** in probability (σ and μ are positive constants). Starting from the graph of $f(x) = e^{-x^2}$, sketch the graph of $n(x)$ and label its relative extreme values and inflection points.

23. What happens to the inflection points of Hill functions $f(x) = x^n/(1 + x^n)$ as n keeps increasing? Can you prove it?

24. Ignore the requirement $t \geq 0$ used in Example 6.5.3 and sketch the graph of $P(t) = 10/(1 + 9e^{-2t})$ for all t.

25. Sketch the graph of the function $P(t) = 10/(1 - 0.5 e^{-2t})$ for $t \geq 0$.

26. In Example 5.6.11 we showed that the graph of the function $P(t) = L/(1 + Ce^{-kt})$ has an inflection point when $t_i = (\ln C)/k$. Show that $P(t_i) = L/2$; i.e., show that the value of the function at the inflection point is equal to one-half the carrying capacity.

27. Show that the graph of the function $P(t) = L/(1 + Ce^{-kt})$ has no inflection points if $C < 0$ (as usual, $L > 0$ and $k > 0$).

28. Sketch the graph of the function $L(t) = M(1 - e^{-kt})$ for $t \geq 0$, where M and k are positive constants. Now assume that $k = 1$. Find t so that $L(t)$ reaches 50% and 90% of its maximum value. This function is called the **learning curve**; explain what kind of learning pattern is implied by this model.

29. A commonly used model for reproduction is the **Ricker model**, $P(x) = rxe^{-x}$, where $r > 0$ (see Section 6.8). Sketch the graph of $P(x)$ for $r = 1$ and $x \geq 0$.

Computer Exercises

30. The concentration of a substance diffusing from a source (for instance, a pollutant in the air) at a location x units away from the source at time t changes according to the formula

$$c(x) = \dfrac{N}{\sqrt{4\pi kt}} e^{-x^2/4kt}$$

where the quantities N and k characterize the substance involved. The same model is used to describe the way heat flows from a source toward colder regions.

Take $N = 10$ and $k = 1$. Plot the graphs of $c(x)$ on $[0, 6]$ for $t = 1$, $t = 2$, and $t = 3$. Describe what they represent and how they compare to each other. Does it make sense?

31. Consider the logistic curve defined by $P(t) = L/(1 + Ce^{-2t})$. Recall (Example 5.6.11) that $C = (L - P_0)/P_0$, where P_0 is the initial population. Take $L = 10$.

   a. Sketch the graphs of $P(t)$ on $[0, 6]$ for $P_0 = 0.01$, $P_0 = 0.1$, and $P_0 = 1$. Compare the three graphs in terms of concavity. What seems to be the value of P at the inflection points?

   b. Sketch the graphs of $P(t)$ on $[0, 3]$ for $P_0 = 6$, $P_0 = 9$, $P_0 = 12$, and $P_0 = 20$. Explain why they make sense in the context of population modelling.

6.6 Newton's Method

With powerful calculators and computers, it might seem unnecessary to approximate a function with a tangent line. On a calculator, exponentiating is no harder than multiplying or adding. But although computing specific functional values is easy, *solving equations* for specific values can be difficult. We have seen how to use the Intermediate Value Theorem to show that an equation *has* a solution. As well, we found a way to obtain approximations of a solution to an equation using the **bisection method.** When it can be used, **Newton's method**—which we discuss in this section—is superior to bisection. Newton's method replaces the original equation with the tangent line approximation and derives a recursion that converges with remarkable speed to the solution.

Solving Equations

Finding inverse functions, computing critical numbers and points of inflection, and locating equilibrium points of a dynamical system are only a few instances of where we need to **solve equations.** Actually, no matter what branch of mathematics (not just applied mathematics!) we work in, we are very likely to encounter a problem that requires that we solve an equation or a system of equations.

Some equations, such as linear ($ax + b = 0$) or quadratic ($ax^2 + bx + c = 0$), are easy to solve because we have formulas for their solutions. There exist formulas for solutions of polynomial equations of degree three and four. And that's it! There are no formulas for the solutions of a general polynomial equation of degree five or higher. As well, most equations that involve transcendental functions (such as $\sin x = x$ and $x^2 = e^x - 2$) cannot be solved algebraically.

Solving equations is of great importance in math, so numerous methods and algorithms have been developed for the purpose. We have already studied one—the bisection method—in the context of the Intermediate Value Theorem.

The bisection method is an example of a **numerical algorithm.** In this section, we study another algorithm, called **Newton's method** or the **Newton–Raphson method.** Newton's method, its variations, and some other algorithms are built into every piece of software that enables calculators or computers to solve equations.

Besides algebraic techniques and numerical algorithms, we can use **graphing** to solve equations. To solve the equation $f(x) = 0$ for x means to find the x-intercepts of the graph of f. By zooming in on the graph, we can get better and better approximations for the location of the intercept(s).

All numeric and geometric approaches work on the same principle, called **iteration**. Iteration requires that we provide a starting point, such as a guess for what a solution to the equation we are trying to solve could be (recall that the Intermediate Value Theorem produces an interval where we look for solution(s)—that's a good start; we pick a number from that interval!). Then, a certain procedure (a sequence of steps, also called an algorithm) is applied to the starting value to produce a number that is—we hope—a better approximation of the solution than our initial guess. Next, the same procedure is applied to the number just obtained, giving an even better approximation. (In other words, the output of one iteration step becomes the input for the next step.)

In this way, we create a sequence of approximations, with each term in the sequence being closer to the actual solution than the previous term.

Newton's Method

Suppose that we want to solve the equation

$$f(x) = 0$$

If we have some idea that a solution is near the value x_0 (initial guess in Figure 6.6.72a), we can replace the original equation by the approximate equation

$$L(x) = 0$$

where $L(x)$ is the tangent line approximation at x_0 (Figure 6.6.72b). The equation for the tangent line $L(x)$ is

$$L(x) = f(x_0) + f'(x_0)(x - x_0)$$

so our approximate equation $L(x) = 0$ is

$$f(x_0) + f'(x_0)(x - x_0) = 0$$

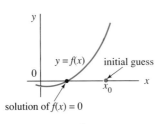
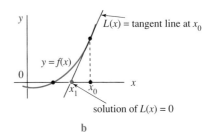

FIGURE 6.6.72

Newton's method: first step

a   b

As long as we can compute the derivative of $f(x)$, we can solve this equation for x. Graphically, the solution of the original equation is the point where the curve intersects the x-axis. The solution of the approximate equation is the point where the tangent line intersects the x-axis.

Example 6.6.1   Computing a Square Root with Newton's Method: First Step

Suppose we wish to solve the equation

$$f(x) = x^2 - 3 = 0$$

The solution of this equation for $x \geq 0$ is $\sqrt{3}$, a numerical value that is hard to compute by hand. The first step in Newton's method is to take a guess. We might begin with a rather poor guess of $x_0 = 3$. Next, we need to find the tangent line approximation by using the derivative

$$f'(x) = 2x$$

The tangent line approximation is

$$\begin{aligned} L(x) &= f(3) + f'(3)(x - 3) \\ &= 6 + 6(x - 3) \\ &= 6x - 12 \end{aligned}$$

We now replace the given equation

$$f(x) = x^2 - 3 = 0$$

with the approximate (linear!) equation

$$L(x) = 6x - 12 = 0$$

whose solution is $x = 2$.

For our initial guess, $x_0 = 3$, the value of the function is $f(3) = 6$ (keep in mind that we are trying to find an x that will make the function equal to zero). When $x = 2$, $f(2) = 1$. We see that the value of f at $x = 2$ is closer to zero than at $x_0 = 3$. So $x = 2$ is indeed closer to the solution of the given equation.

By using a guess, x_0, for the solution (Figure 6.6.73a), we replaced the difficult equation $f(x) = 0$ with the linear equation $L(x) = 0$. It is possible to solve this linear equation for x (as long as $f'(x_0) \neq 0$). We find

$$f(x) = 0 \quad \text{original equation}$$
$$f(x_0) + f'(x_0)(x - x_0) = 0 \quad \text{replace by tangent line approximation}$$
$$f'(x_0)(x - x_0) = -f(x_0)$$
$$x - x_0 = \frac{-f(x_0)}{f'(x_0)}$$
$$x = x_0 - \frac{f(x_0)}{f'(x_0)}$$

(Figure 6.6.73). The value x is the point where the tangent line intersects the horizontal axis. Our hope is that this point is closer to the unknown exact answer than the original guess was. If so, we can use x as a new guess, x_1, which we compute with the formula

$$x_1 = x_0 - \frac{f(x_0)}{f'(x_0)} \quad (6.6.1)$$

Starting from the point x_1, we can follow the same steps to find the tangent line and solve for the intersection with the horizontal axis (Figure 6.6.73). The new guess, x_2, will have the same formula but with x_1 substituted for x_0, or

$$x_2 = x_1 - \frac{f(x_1)}{f'(x_1)}$$

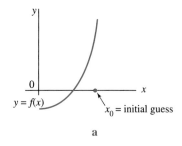
a
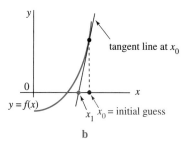
b
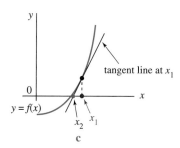
c

FIGURE 6.6.73

Newton's method: first two steps

Example 6.6.2 Computing a Square Root with Newton's Method: Second Step

In the example with $f(x) = x^2 - 3 = 0$, our first guess was $x_0 = 3$. By finding the tangent line and solving the equation, we found $x_1 = 2$. Alternatively, we could use Equation 6.6.1 with $f'(x) = 2x$, $x_0 = 3$, $f(x_0) = f(3) = 6$, and $f'(x_0) = f'(3) = 6$, to find

$$x_1 = 3 - \frac{f(3)}{f'(3)}$$
$$= 3 - \frac{6}{6} = 2$$

Starting from the new value, $x_1 = 2$, we get

$$x_2 = x_1 - \frac{f(x_1)}{f'(x_1)}$$
$$= 2 - \frac{f(2)}{f'(2)}$$
$$= 2 - \frac{2^2 - 3}{2 \cdot 2}$$

$$= 2 - \frac{1}{4} = 1.75$$

This is much closer to the exact answer because $f(1.75) = 1.75^2 - 3 = 0.0625$ (which is close to 0).

At each step, we applied the **Newton's method recursive formula** (or the **Newton's method discrete-time dynamical system**)

$$x_{t+1} = x_t - \frac{f(x_t)}{f'(x_t)}$$

This leads to the following algorithm.

▶▶ **Algorithm 6.6.1** Newton's Method for Solving a Nonlinear Equation

To solve the equation $f(x) = 0$:

1. Come up with an initial guess called x_0.

2. Use the Newton's method recursive formula

$$x_{t+1} = x_t - \frac{f(x_t)}{f'(x_t)}$$

to find x_1, x_2, and so forth until the answer converges.

By "until the answer converges" we mean to say that we stop the process when the differences between the successive terms in the sequence of approximations (also called iterations),

$$x_0, x_1, x_2, x_3, x_4, \ldots, x_t, x_{t+1}, \ldots$$

become very small (see Table 6.6.1 in Example 6.6.3 below). In other words, we stop when

$$x_{t+1} \approx x_t$$

i.e., when the terms get close enough to each other.

OK, so we stop when $x_{t+1} \approx x_t$. But why is x_t an approximate solution; that is, why is $f(x_t) \approx 0$? From the Newton's method formula

$$x_{t+1} = x_t - \frac{f(x_t)}{f'(x_t)} \approx x_t$$

(assuming that $f'(x_t) \neq 0$), we get

$$-\frac{f(x_t)}{f'(x_t)} \approx 0$$

and $f(x_t) \approx 0$.

Thus, if the sequence stabilizes at some number x_t, then x_t is an approximate solution of the given equation!

Note that

$$x_{t+1} = x_t - \frac{f(x_t)}{f'(x_t)} \approx x_t$$

states that x_t is (an approximate) equilibrium point of the Newton's method dynamical system.

Example 6.6.3 Computing a Square Root with Newton's Method

With the function $f(x) = x^2 - 3$, the Newton's method discrete-time dynamical system is

$$x_{t+1} = x_t - \frac{f(x_t)}{f'(x_t)}$$
$$= x_t - \frac{x_t^2 - 3}{2x_t}$$

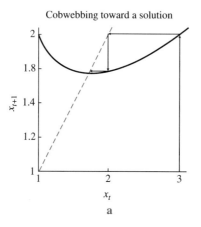
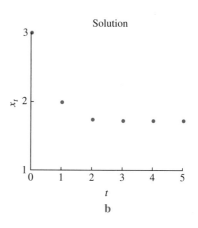

FIGURE 6.6.74

The Newton's method discrete-time dynamical system

The results of cobwebbing the equation $f(x) = x^2 - 3 = 0$ using the Newton's method dynamical system are shown in Figure 6.6.74. Note that the graph in Figure 6.6.74a represents the updating function of the system

$$x_{t+1} = x_t - \frac{x_t^2 - 3}{2x_t}$$

and *not* the function $f(x)$.

Table 6.6.1 shows the values of the first few iterations of Newton's method, as well as the absolute value of the difference between consecutive iterations.

Table 6.6.1

Iteration	Approximation	Difference
x_0	3	
x_1	2	1
x_2	1.75	0.25
x_3	1.732142857	0.017857143
x_4	1.732050810	0.000092047
x_5	1.732050808	0.000000002
x_6	1.732050808	0.000000000

The zeros in the last row mean that the difference between the fifth and the sixth iterations cannot be detected using nine decimal places (i.e., it is smaller than 10^{-9}). Requesting higher precision, we calculate

$$x_5 = 1.7320508075688772953$$
$$x_6 = 1.7320508075688772935$$

whose difference is

$$|x_6 - x_5| \approx 1.8 \cdot 10^{-18}$$

Newton's method converges to the answer very quickly and is correct to eight decimal places after only five steps. And, besides the derivative, the calculation involved nothing more complicated than multiplication, subtraction, and division. ▲

Example 6.6.4 Using Newton's Method

In Example 6.3.5 in Section 6.3, we used the bisection method to obtain an approximation of the solution of the equation

$$f(x) = 3.2xe^{-x} + 6.5 - x = 0$$

Now we apply Newton's method. We compute

$$f'(x) = 3.2e^{-x} - 3.2xe^{-x} - 1$$

and so the Newton's method recursive formula is

$$x_{t+1} = x_t - \frac{f(x_t)}{f'(x_t)}$$
$$= x_t - \frac{3.2x_t e^{-x_t} + 6.5 - x_t}{3.2e^{-x_t} - 3.2x_t e^{-x_t} - 1}$$

Take $x_0 = 4$ as our initial guess. The table below shows the values of the first few iterations. The "Difference" column keeps track of the absolute value of the difference between the present iteration and the one obtained in the previous step:

Iteration	Approximation	Difference
x_0	4	
x_1	6.325540144	2.325540144
x_2	6.529995822	0.204455677
x_3	6.530475119	0.000479298
x_4	6.530475122	0.000000002

In Example 6.3.5, after six steps, we concluded that the solution must be in the interval [6.5, 6.53125]. Note that after only four steps of Newton's method we reached the approximation 6.530475122, which is far better (for the record, a close approximation of the solution is 6.530475121663287). ▲

Example 6.6.5 Using Newton's Method to Find Equilibrium

Suppose we have to find an equilibrium point of the dynamical system

$$x_{t+1} = x_t(1 - x_t e^{-0.9x_t}) + 0.2$$

One way to do this is to start somewhere, say $x_0 = 1$, and keep calculating the values $x_1, x_2, x_3, \ldots$ in the hope that they will bring us closer and closer to the equilibrium. If we do so, we will obtain the values shown in Table 6.6.2.

Alternatively, to find equilibrium point(s), we need to solve—somehow—the equation

$$z(1 - ze^{-0.9z}) + 0.2 = z$$

(we have already used x, so we introduce the new symbol z for the variable and for the iterations below). Since we cannot solve the equation algebraically, we use Newton's

Table 6.6.2

Iteration	Approximation	Iteration	Approximation
x_0	1		
x_1	0.7934303403	x_{10}	0.5811411316
x_2	0.6851871929	x_{15}	0.5808020169
x_3	0.6317894405	x_{20}	0.5807923122
x_4	0.6057416195	x_{25}	0.5807920345
x_5	0.5930200101	x_{30}	0.5807920265
x_6	0.5867918774	x_{35}	0.5807920263

method. Define

$$f(z) = z(1 - ze^{-0.9z}) + 0.2 - z$$
$$= -z^2 e^{-0.9z} + 0.2$$

(so we need to solve $f(z) = 0$).

Then

$$f'(z) = -2ze^{-0.9z} - z^2 e^{-0.9z}(-0.9)$$
$$= -(2ze^{-0.9z} - 0.9z^2 e^{-0.9z})$$

and the Newton's method recursive formula is

$$z_{t+1} = z_t - \frac{f(z_t)}{f'(z_t)}$$
$$= z_t - \frac{-z_t^2 e^{-0.9z_t} + 0.2}{-(2z_t e^{-0.9z_t} - 0.9z_t^2 e^{-0.9z_t})}$$
$$= z_t - \frac{z_t^2 e^{-0.9z_t} - 0.2}{2z_t e^{-0.9z_t} - 0.9z_t^2 e^{-0.9z_t}}$$

(In the last step we multiplied both the numerator and the denominator by -1.) The first few iterations are recorded in the table below.

Iteration	Approximation
z_0	1
z_1	0.5381096566
z_2	0.5810776419
z_3	0.5807920349
z_4	0.5807920263
z_5	0.5807920263

As we can see, Newton's method gave us a better approximation a lot more quickly (to compare: a very good approximation is 0.58079202630476).

It is beyond the scope of this book to give a detailed analysis and criticism of Newton's method. (Does it always work? If not, what assumptions will make it work? How quickly do the iterations converge? Is the method capable of capturing all solutions of an equation, or only some?)

Here we study two examples that show certain features of Newton's method.

Example 6.6.6 A Case When Newton's Method Does Not Work

We now construct the iteration procedure to solve the equation
$$\sqrt[3]{x} = 0$$
using Newton's method (we know that $x = 0$ is the only solution).

Define
$$f(x) = \sqrt[3]{x} = x^{1/3}$$

Then
$$f'(x) = \frac{1}{3}x^{-2/3}$$

and the Newton's method recursive formula is
$$\begin{aligned} x_{t+1} &= x_t - \frac{f(x_t)}{f'(x_t)} \\ &= x_t - \frac{x_t^{1/3}}{\frac{1}{3}x_t^{-2/3}} \\ &= x_t - 3x_t \\ &= -2x_t \end{aligned}$$

Suppose we start at $x_0 = 1$. The sequence of iterations
$$\begin{aligned} x_1 &= -2(1) = -2 \\ x_2 &= -2(-2) = 4 \\ x_3 &= -2(4) = -8 \\ x_4 &= -2(-8) = 16 \end{aligned}$$
...

moves farther and farther away from the solution $x = 0$! As a matter of fact, the Newton's method system will push any nonzero initial guess farther and farther away from $x = 0$. Thus, in this case Newton's method fails to provide a solution.

Example 6.6.7 Solving Equations with Multiple Solutions

Consider the equation
$$f(x) = x^4 - 8x^3 + 21x^2 - 19x + 3$$

Note that f is continuous for all real numbers. Since
$$f(0) = 3 > 0$$
and
$$f(1) = -2 < 0$$
the Intermediate Value Theorem guarantees that there is a solution of the equation $f(x) = 0$ somewhere in the interval $[0, 1]$.

Let's try to find it using Newton's method. Computing
$$f'(x) = 4x^3 - 24x^2 + 42x - 19$$
we construct the Newton's method iteration formula
$$\begin{aligned} x_{t+1} &= x_t - \frac{f(x_t)}{f'(x_t)} \\ &= x_t - \frac{x_t^4 - 8x_t^3 + 21x_t^2 - 19x_t + 3}{4x_t^3 - 24x_t^2 + 42x_t - 19} \end{aligned}$$

Let's pick a starting value in the interval $[0, 1]$, say $x_0 = 0.8$. We compute the first few iterations

$x_1 = 2.699378882$

$x_2 = 2.957177257$

$x_3 = 2.996046754$

$x_4 = 2.999954676$

$x_5 = 2.999999994$

$\ldots$

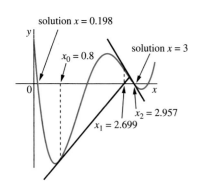

FIGURE 6.6.75
Newton's method misses the solution

The iterations seem to be converging toward 3. We can easily check that $x = 3$ is indeed a solution of the given equation—just not the one we were looking for!

Using an equation solver, we find out that the solutions of the given equation are $x \approx 0.198$ (the one we were looking for!), $x \approx 1.555$, $x = 3$ (the one we found instead), and $x \approx 3.250$.

So what happened? The graph of the function (Figure 6.6.75) helps us understand the situation. The tangent at $x_0 = 0.8$ crosses the x-axis at $x_1 \approx 2.699$, far from the interval [0, 1]. The tangent at x_1 then moves even farther away, toward the solution $x = 3$.

Summary   Solving equations is often impossible to do algebraically. Newton's method is a technique that can be implemented on the computer. By replacing the original equation with the tangent line approximation, we derived the **Newton's method recursive formula,** which usually converges very quickly to the solution.

6.6 Exercises

Mathematical Techniques

1–4 ▪ Try Newton's method graphically for two steps, starting from the given points on the figure.

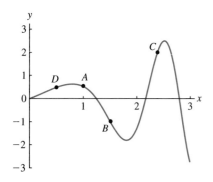

1. The point marked A.
2. The point marked B.
3. The point marked C.
4. The point marked D.

5–8 ▪ Use Newton's method for three steps to find the following. Find and sketch the tangent line for the first step, and then find the Newton's method recursive formula to check your answer for the first step and compute the next two values.

5. $x^2 - 5x + 1 = 0$ (this can be solved exactly with the quadratic formula).

6. $\sqrt[3]{20}$ (solve $f(x) = x^3 - 20 = 0$).

7. A positive solution of $e^{x/2} = x + 1$ (solve $h(x) = e^{x/2} - x - 1 = 0$).

8. The point where $\cos x = x$ (in radians, of course).

9–12 ▪ Many equations can also be solved by repeatedly applying a discrete-time dynamical system. Compare the following discrete-time dynamical systems with the Newton's method discrete-time dynamical system. Show that each has the same equilibrium, and see how close you get in three steps with each method.

9. Use the discrete-time dynamical system $x_{t+1} = e^{x_t} - 2$ to solve $e^x = x + 2$.

10. Use the discrete-time dynamical system $x_{t+1} = x_t^3 + x_t - 20$ (based on Exercise 6).

11. Use the discrete-time dynamical system $x_{t+1} = e^{x_t/2} - 1$ (based on Exercise 7).

12. Use the discrete-time dynamical system $x_{t+1} = \cos x_t$ (based on Exercise 8).

13–14 ▪ Find the value of the parameter r for which the given discrete-time dynamical system will converge most rapidly to its positive equilibrium. Follow the system for four steps, starting from the given initial condition.

13. The logistic dynamical system $x_{t+1} = rx_t(1-x_t)$. Start from $x_0 = 0.75$.

14. The Ricker dynamical system $x_{t+1} = rx_t e^{-x_t}$. Start from $x_0 = 0.75$.

15–18 ▪ As mentioned in the text, although Newton's method works incredibly well most of the time, it can fail or work less well in many circumstances. For each of the following exercises, graph the function to illustrate the failures.

15. Find all initial values from which Newton's method fails to solve $x(x-1)(x+1) = 0$. Which starting points converge to a negative solution?

16. Find two initial values from which Newton's method fails to solve $x^3 - 6x^2 + 9x - 1 = 0$. Graphically indicate a third such value.

17. Use Newton's method to solve $x^2 = 0$ (the solution is $x = 0$). Why does it approach the solution so slowly?

18. Use Newton's method to solve $\sqrt{|x|} = 0$ (the solution is $x = 0$). This is the square root of the absolute value of x. Why does the method fail?

19. Find an approximation of the solution of the equation $f'(x) = 0$ from Example 6.1.12 using Newton's method for three steps starting at $x_0 = 1$.

20. Find the critical point of the function $f(x) = x^2 - \sin x$ using Newton's method for three steps starting at $x_0 = 1$.

21–22 ▪ Suppose we wish to solve the equation $f(x) = 0$ but cannot compute the derivative $f'(x)$. (This kind of problem arises when the function $f(x)$ must be evaluated with a complicated computer program.) An alternative approach approximates the derivative $f'(x)$ with $f(x+1) - f(x)$ (the secant line). For each case, write an approximate Newton's method recursive formula, illustrate the procedure with a diagram, and try it for five steps to see how quickly it approaches the solution.

21. $f(x) = e^x - x - 2$

22. $f(x) = x^3 - 20$.

23–24 ▪ There is an alternative way to approximate the slope of the function at the value x_t:

$$f'(x_t) \approx \frac{f(x_t) - f(x_{t-1})}{x_t - x_{t-1}}$$

For each equation, use this estimate to write an approximate Newton's method recursive formula and illustrate the idea with a diagram. How is it different from an ordinary discrete-time dynamical system? Run it for a few steps, starting from x_0 and x_1 from Exercises 21–22. Does it converge faster than the earlier approximation? Why? How does it compare with Newton's method itself?

23. $f(x) = e^x - x - 2$ (from Exercise 21).

24. $f(x) = x^3 - 20$ (from Exercise 22).

Applications

25–26 ▪ Suppose the total amount of nectar that comes out of a flower after time t follows

$$F(t) = \frac{t^2}{1+t^2}$$

After noting how this function differs from the forms studied in Section 6.2, write the equation used to find the optimal time to remain, and then solve it with Newton's method (or algebraically) if the travel time τ takes on the following values. Draw the associated Marginal Value Theorem diagram.

25. $\tau = 0$

26. $\tau = 1$

27–28 ▪ Suppose a fish population follows the discrete-time dynamical system

$$N_{t+1} = rN_t e^{-N_t} - hN_t$$

For the following values of r, find the equilibrium N^* as a function of h; write the equation for the critical point of the payoff function, $P(h) = hN^*$; and use Newton's method to find the best h.

27. $r = 2.5$

28. $r = 1.5$

29–30 ▪ Consider a variant of the medication discrete-time dynamical system

$$M_t = p(M_t)M_t + 1$$

where the function $p(M_t)$ represents the fraction used (see Section 3.4, Exercise 39). Suppose that $p(M_t) = \alpha e^{-0.1M_t}$. For the following values of α, show that there is an equilibrium by using the Intermediate Value Theorem, follow the solution of the discrete-time dynamical system until it gets close to the equilibrium (about three decimal places), and find the equilibrium with Newton's method.

29. $\alpha = 0.5$

30. $\alpha = 0.9$

31–32 ▪ Thomas Malthus predicted doom for the human species when he argued that populations grow exponentially but their resources grow only linearly. Find the time when the population runs out of resources in each case.

31. The population grows according to $b(t) = 100e^{0.1t}$, and resources grow according to $R(t) = 400 + 100t$. The population starves when $b(t) = R(t)$.

32. The population grows according to $b(t) = 100e^{0.1t}$, and resources grow according to $R(t) = 4000 + 500t$. The population starves when $b(t) = R(t)$.

33–34 ▪ A lung follows the discrete-time dynamical system

$$c_{t+1} = 0.75\alpha(c_t)c_t + 0.25\gamma$$

where , the function $\alpha(c_t)$ is positive and decreasing, and $\alpha(0) = 1$. We used the Intermediate Value Theorem (Exercise 30 in

Section 6.3) to show that there is an equilibrium for any such function $\alpha(c)$. Use Newton's method to solve for the equilibrium for each form of $\alpha(c)$.

33. $\alpha(c) = e^{-c}$

34. $\alpha(c) = e^{-0.1c}$

Computer Exercises

35. Use bisection, which is based on the Intermediate Value Theorem, to solve $g(x) = e^x - x - 2 = 0$.

    a. We know there is a solution between $x = 0$ and $x = 2$. Show that there is a solution between $x = 1$ and $x = 2$.

    b. By computing $g(1.5)$, show that there is a solution between 1 and 1.5.

    c. Compute $g(1.25)$. There is a solution either between 1 and 1.25 or between 1.25 and 1.5. Which is it?

    d. Compute the value of g at the midpoint of the previous interval you determind in part (c), and find an interval half as big that contains a solution.

    e. Continue bisecting the interval until your answer is right to three decimal places.

    f. About how many more steps would it take to reach six decimal places?

36. Solve the equation $e^x - x - 2 = 0$ by using the quadratic approximation (Section 5.7) to the function. Compare how fast it converges with Newton's method. Which method do you think is better?

6.7 Stability of Discrete-Time Dynamical Systems

In our initial study of discrete-time dynamical systems, we observed that equilibria can be **stable** or **unstable**. By examining the graph and cobweb of a discrete-time dynamical system, we will develop a method for evaluating the stability of an equilibrium by computing the **derivative of the updating function**. This method explains why some solutions, such as that for a growing bacterial population, move away from equilibrium, whereas others, like that of the administration of pain medication, move toward their equilibrium.

Motivation

Figures 6.7.76–6.7.78 review the cobwebbing diagrams for five of the discrete-time dynamical systems we have studied, and the following list reviews the terminology of discrete-time dynamical systems:

- A **discrete-time dynamical system** is a rule (described by its **updating function**) that takes a measurement at one time step as an input and returns the measurement at the next time step as an output (see Section 3.1 for definitions and examples).

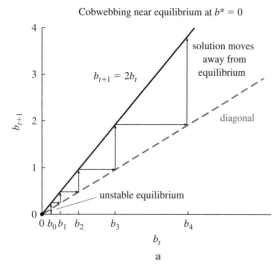
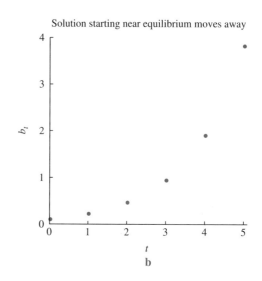

FIGURE 6.7.76

Bacterial population growth discrete-time dynamical system $b_{t+1} = 2b_t$

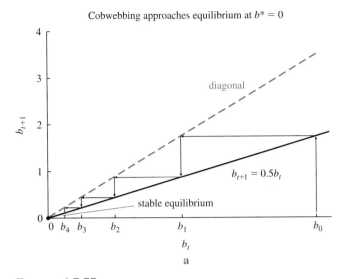

FIGURE 6.7.77

Bacterial population growth discrete-time dynamical system $b_{t+1} = 0.5b_t$

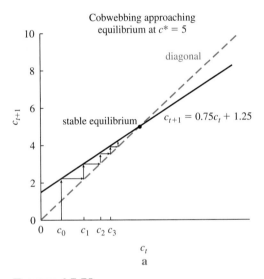
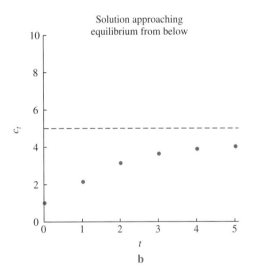

FIGURE 6.7.78

Absorption discrete-time dynamical system $c_{t+1} = 0.75c_t + 1.25$

- **Cobwebbing** is a graphical method for finding solutions of discrete-time dynamical systems (see Section 3.2 to recall how this is done).

- A **solution** gives the values of the measurement as a function of time.

- An **equilibrium** is a value that the discrete-time dynamical system does not change. Graphically, an equilibrium is a point where the graph of the updating function crosses the diagonal line.

- An equilibrium is **stable** if solutions that start near the equilibrium stay near, or move closer to the equilibrium. An equilibrium is **unstable** if solutions that start near the equilibrium move away from the equilibrium.

The figures show four stable equilibria: $b^* = 0$ in Figure 6.7.77, $c^* = 5$ in Figure 6.7.78, $p^* = 1$ in Figure 6.7.79, and $p^* = 0$ in Figure 6.7.80. There are three unstable equilibria: $b^* = 0$ in Figure 6.7.76, $p^* = 0$ in Figure 6.7.79, and $p^* = 1$ in Figure 6.7.80. What is the pattern? How can we recognize which equilibria are stable?

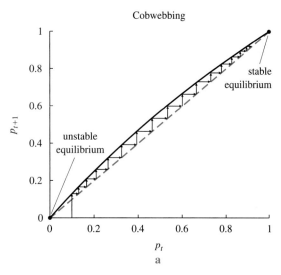
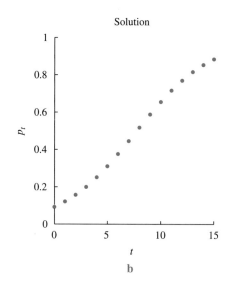

FIGURE 6.7.79
Bacterial selection discrete-time dynamical system $p_{t+1} = \frac{2p_t}{2p_t+1.5(1-p_t)}$

Note In this section, we study dynamical systems whose *updating functions are increasing* (i.e., have positive slope). In the next section, we will develop criteria for the stability of a general dynamical system.

Think about cobwebbing near a stable equilibrium, such as the one in Figure 6.7.78. If the initial condition is slightly less than the equilibrium, the solution *increases* because the graph of the updating function lies above the diagonal. Similarly, if we start slightly above the equilibrium, the solution *decreases* because the graph of the updating function lies below the diagonal. In other words, if the graph of the updating function crosses from above the diagonal to below the diagonal, the equilibrium is stable. Does this work for the other stable equilibria? It does if we imagine extending the graphs beyond the biologically meaningful realm. For example, in Figure 6.7.79, the updating function intersects the diagonal from above near $p^* = 1$. If we extend the curve beyond $p^* = 1$, it crosses from above to below (Figure 6.7.81).

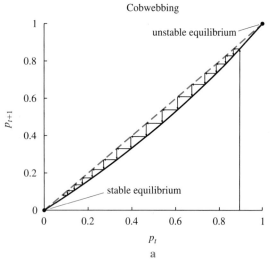
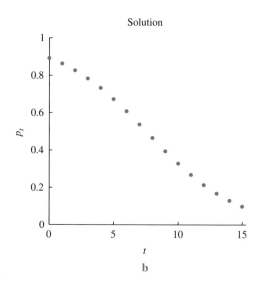

FIGURE 6.7.80
Bacterial selection discrete-time dynamical system $p_{t+1} = \frac{1.5p_t}{1.5p_t+2(1-p_t)}$

6.7 Stability of Discrete-Time Dynamical Systems   427

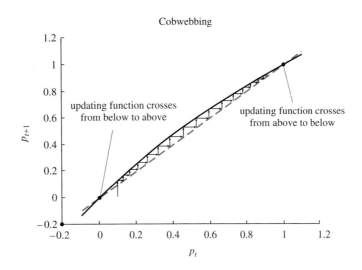

FIGURE 6.7.81
Bacterial selection discrete-time dynamical system with extended graph

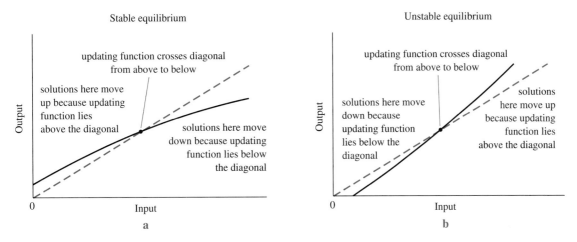

FIGURE 6.7.82
Graphical criterion for stability

Unstable equilibria, in contrast, are points where the updating function crosses the diagonal from below to above. In this case, a solution that starts below the equilibrium will decrease further, moving away from the equilibrium. A solution that starts above the equilibrium will increase further, again moving away from the equilibrium. At the unstable equilibrium at $p^* = 0$ in Figure 6.7.79, the updating function crosses from below to above if we extend it below $p^* = 0$ (Figure 6.7.81).

We can summarize our observations with the following condition (Figure 6.7.82):

Graphical Criterion for Stability of an Equilibrium for a Dynamical System whose Updating Function Is Increasing

- An equilibrium is stable if the graph of the updating function crosses the diagonal from above to below.

- An equilibrium is unstable if the graph of the updating function crosses the diagonal from below to above.

Stability and the Slope of the Updating Function

When does an increasing updating function cross the diagonal from below to above? The diagonal is the graph of $y = x$, a line with a slope equal to 1. The graph of the updating function will cross from below to above when its slope is *greater than 1*. Conversely, a curve will cross from above to below when its slope is *less than 1*.

We can therefore translate the graphical criterion for stability into a condition about the **derivative of the updating function,** because the derivative is equal to the slope (Figure 6.7.83).

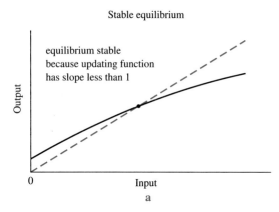
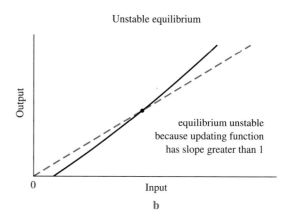

FIGURE 6.7.83
Slope criterion for stability

> **Slope Criterion for Stability of an Equilibrium for a Dynamical System Whose Updating Function Is Increasing**
>
> - An equilibrium is stable if the value of the derivative of the updating function at the equilibrium is less than 1.
> - An equilibrium is unstable if the value of the derivative of the updating function at the equilibrium is greater than 1.

A nonlinear discrete-time dynamical system can have many equilibria. With the aid of the graphical and slope criteria for stability, we can recognize stable and unstable equilibria by examining the slope of the updating function (Figure 6.7.84). We will see in the next section, however, that the dynamics can be much more complicated when the updating function has a negative slope.

The slope criterion for stability says nothing about what happens when the slope is exactly 1, i.e., when the diagonal is tangent to the updating function at the equilibrium point. Figure 6.7.85 shows an equilibrium where the updating function does not cross the diagonal but, rather, remains below the diagonal both to the left and to the right of the equilibrium. A solution that starts from any point below the equilibrium will decrease and move away from the equilibrium, as in the unstable case. A solution that starts from any point above the equilibrium will decrease and move toward the equilibrium, as in the stable case. Figure 6.7.86 shows other cases of tangency: the updating function has slope 1 at the equilibrium point. In Figure 6.7.86a, it crosses the diagonal from above to below, so the equilibrium is stable. The updating function

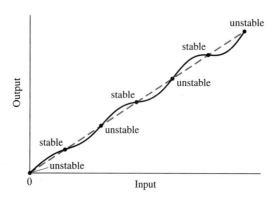

FIGURE 6.7.84
Recognizing the stability of a dynamical system with many equilibria

6.7 Stability of Discrete-Time Dynamical Systems

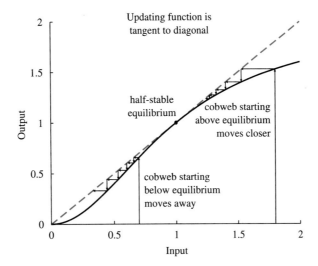

FIGURE 6.7.85
A half-stable equilibrium

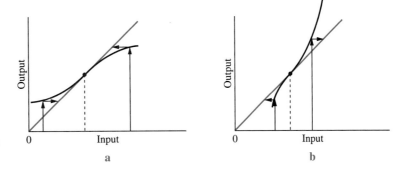

FIGURE 6.7.86
Cases of stable and unstable equilibria when the slope of the updating function is 1

in Figure 6.7.86b crosses the diagonal from below to above, resulting in an unstable equilibrium.

Evaluating Stability with the Derivative

We can test the slope criterion for stability by computing the derivatives of the discrete-time dynamical systems pictured in Figures 6.7.76–6.7.80.

Example 6.7.1 Using the Derivative to Check Stability: Bacterial Population Growth

In Figure 6.7.76, the discrete-time dynamical system is

$$b_{t+1} = 2b_t$$

with the updating function

$$f(b) = 2b$$

The only equilibrium is $b^* = 0$, corresponding to extinction. The updating function is a line with slope 2. More formally,

$$f'(b) = 2$$

Because this slope is greater than 1, the equilibrium is unstable. This makes biological sense because any nonzero population doubles each generation and grows away from a population size of zero.

Example 6.7.2 Using the Derivative to Check Stability: Bacterial Population Decline

In Figure 6.7.77, the discrete-time dynamical system is

$$b_{t+1} = 0.5b_t$$

with the updating function

$$f(b) = 0.5b$$

Again, the only equilibrium is $b^* = 0$. This updating function, however, has a slope of 0.5 because

$$f'(b) = 0.5$$

The slope is less than 1, and this equilibrium is stable. This makes biological sense because this population is being halved each generation and hence is decreasing toward extinction. ▲

Example 6.7.3 Using the Derivative to Check Stability: The Absorption Model

Consider the absorption discrete-time dynamical system

$$c_{t+1} = 0.75c_t + 1.25$$

(Figure 6.7.78). If we designate the updating function as $f(c) = 0.75c + 1.25$, we find

$$f'(c) = 0.75$$

because the updating function is linear with slope 0.75. This slope is less than 1, implying that the single equilibrium is stable. ▲

Example 6.7.4 Using the Derivative to Check Stability: The Selection Model

What happens when the discrete-time dynamical system is nonlinear? Figure 6.7.79 uses the bacterial selection discrete-time dynamical system

$$p_{t+1} = \frac{2p_t}{2p_t + 1.5(1 - p_t)}$$

where p_t represents the proportion of mutant bacteria at time t, 2 is the per capita production of mutant bacteria, and 1.5 is the per capita production of wild-type bacteria. There are two equilibria, at $p^* = 0$ and $p^* = 1$ (Section 3.4). The first corresponds to extinction of the mutant and the second to extinction of the wild type.

By writing the updating function as

$$f(p) = \frac{2p}{2p + 1.5(1 - p)} = \frac{2p}{0.5p + 1.5}$$

we compute

$$f'(p) = \frac{(0.5p + 1.5)(2) - 2p(0.5)}{(0.5p + 1.5)^2}$$

$$= \frac{3}{(0.5p + 1.5)^2}$$

At the equilibrium $p^* = 0$,

$$f'(0) = \frac{3}{(1.5)^2} \approx 1.333$$

which is greater than 1. This equilibrium is unstable. At the equilibrium $p^* = 1$,

$$f'(1) = \frac{3}{(2)^2} = 0.75$$

which is less than 1. This equilibrium is stable (Figure 6.7.87). Biologically, any population that begins with a positive fraction of mutants will be taken over by mutants. ▲

6.7 Stability of Discrete-Time Dynamical Systems 431

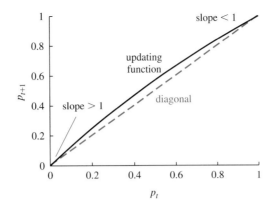

FIGURE 6.7.87
The discrete-time dynamical system for the fraction of mutants

Example 6.7.5 Stability of the Pain Medication Discrete-Time Dynamical System

Consider the discrete-time dynamical system for medication concentration in the bloodstream,

$$M_{t+1} = 0.5M_t + 1$$

describing a patient who is administered one unit of medication each day but also, each day, uses up half of the previous amount (Example 3.1.4). The equilibrium is calculated to be

$$M^* = 0.5M^* + 1$$

i.e., $0.5M^* = 1$ and $M^* = 2$. Because the updating function is linear with slope $0.5 < 1$, this equilibrium is stable (Figure 6.7.88), as shown by cobwebbing.

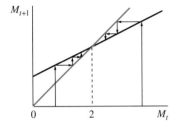

FIGURE 6.7.88
Stability of the pain medication discrete-time dynamical system

Example 6.7.6 Stability of the Pain Medication Discrete-Time Dynamical System if Run Backwards

What happens if we run this system backward in time? The backward discrete-time dynamical system can be found by solving for M_t as

$$M_t = 2(M_{t+1} - 1) = 2M_{t+1} - 2$$

The backward updating function is

$$f(M) = 2M - 2$$

It shares the equilibrium at 2 (Figure 6.7.89). However, because the graph of the backward updating function (the inverse function) is the mirror image across the diagonal, the slope at the equilibrium is the reciprocal of 0.5, i.e., is equal to 2 (alternatively, compute the derivative of $f(M) = 2M - 1$). An equilibrium that is stable when time runs forward is *unstable* when time runs backwards.

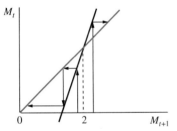

FIGURE 6.7.89
Stability of the pain medication discrete-time dynamical system if run backwards

Example 6.7.7 Stability of Limited Population

Consider a population where the per capita production is a decreasing function of the population size, x, according to

$$\text{per capita production} = \frac{2}{1 + 0.001x}$$

The per capita production is 2 when $x = 0$ and decreases as x becomes larger (Figure 6.7.90a). For example, when $x = 500$,

$$\text{per capita production} = \frac{2}{1 + 0.001 \cdot 500} \approx 1.333$$

Using the fact that the updated population is the per capita production (offspring per individual) times the old population (the number of individuals), the discrete-time dynamical system for this population is

$$x_{t+1} = \text{(per capita production)} \cdot x_t$$

$$x_{t+1} = \frac{2x_t}{1 + 0.001x_t}$$

(Figure 6.7.90b).

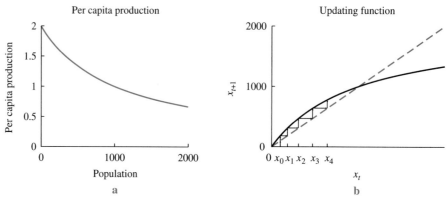

FIGURE 6.7.90

A discrete-time dynamical system describing competition

To find the equilibria, we solve the equation

$$x^* = \frac{2x^*}{1 + 0.001x^*}$$
$$x^*(1 + 0.001x^*) = 2x^*$$
$$x^*(1 + 0.001x^*) - 2x^* = 0$$
$$x^*(1 + 0.001x^* - 2) = 0$$
$$x^*(0.001x^* - 1) = 0$$

Thus, $x^* = 0$ or

$$0.001x^* - 1 = 0$$
$$0.001x^* = 1$$
$$x^* = \frac{1}{0.001} = 1000$$

The equilibrium $x^* = 0$ corresponds to extinction. The equilibrium $x^* = 1000$ is the point where

$$\text{per capita production} = \frac{2}{1 + 0.001(1000)} = 1$$

The per capita production of 1 means that a parent is replaced by one offspring, so the population breaks even (i.e., neither increases nor decreases).

The equilibrium at $x^* = 0$ appears to be unstable, and the equilibrium at $x^* = 1000$ appears to be stable (Figure 6.7.90b). In accordance with the graphical criterion for stability, the graph of the updating function crosses the diagonal from below to above at $x^* = 0$ and from above to below at $x^* = 1000$. We therefore expect that the derivative of the updating function is greater than 1 at $x^* = 0$ and less than 1 at $x^* = 1000$.

In functional notation, the updating function is

$$f(x) = \frac{2x}{1 + 0.001x}$$

The derivative of the updating function, found with the quotient rule, is

$$f'(x) = \frac{(1 + 0.001x)(2) - 2x(0.001)}{(1 + 0.001x)^2}$$

$$= \frac{2}{(1 + 0.001x)^2}$$

Evaluating at the equilibria yields

$$f'(0) = \frac{2}{(1+0.001 \cdot 0)^2} = 2$$

$$f'(1000) = \frac{2}{(1+0.001 \cdot 1000)^2} = 0.5$$

As indicated by the graph, the equilibrium at $x^* = 0$ is unstable and the equilibrium at $x^* = 1000$ is stable. If reduced to a low level, this population would increase back toward the stable equilibrium at $x^* = 1000$. If artificially increased to a high level (greater than 1000), the population would decrease toward the stable equilibrium at $x^* = 1000$.

Example 6.7.8 Stability of a Limited Population with Lower Production

Consider a variant of the model in Example 6.7.7 where the per capita production is a decreasing function of the population size, x, according to

$$\text{per capita production} = \frac{1}{1+0.001x}$$

The per capita production is 1 when $x = 0$ and decreases as x becomes larger (Figure 6.7.91a). The discrete-time dynamical system for this population is

$$x_{t+1} = (\text{per capita production}) \cdot x_t$$

$$x_{t+1} = \frac{x_t}{1+0.001x_t}$$

(Figure 6.7.91b).

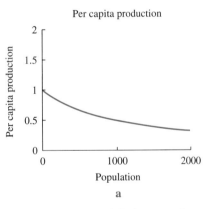
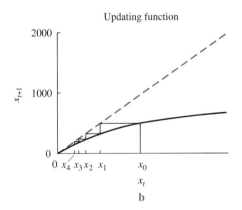

FIGURE 6.7.91
A discrete-time dynamical system with reduced per capita production

To find the equilibria, we solve

$$x^* = \frac{x^*}{1+0.001x^*}$$

$$(1+0.001x^*)x^* - x^* = 0$$

$$x^*(1+0.001x^* - 1) = 0$$

$$x^* \cdot 0.001x^* = 0$$

This system has only a single equilibrium at $x^* = 0$, corresponding to extinction.

The updating function is

$$f(x) = \frac{x}{1+0.001x}$$

and its derivative is

$$f'(x) = \frac{(1+0.001x) - x(0.001)}{(1+0.001x)^2}$$

$$= \frac{1}{(1 + 0.001x)^2}$$

Evaluating at the equilibrium yields

$$f'(0) = \frac{1}{(1 + 0.001 \cdot 0)^2} = 1$$

The updating function is tangent to the diagonal at the equilibrium $x^* = 0$, meaning that we cannot determine the stability from the slope criterion. By more carefully examining the function, we see that the graph of the updating function lies below the diagonal for $x > 0$. Algebraically,

$$f(x) = \frac{x}{1 + 0.001x} < x$$

because the denominator is greater than 1 for $x > 0$. Therefore, any positive population decreases, and the equilibrium $x^* = 0$ is stable.

Summary We used cobwebbing and logic to derive two criteria for the stability of equilibria for those dynamical systems whose updating function is increasing (i.e., has positive slope). When the updating function crosses the diagonal from below to above, initial conditions both to the left and to the right of the equilibrium are pushed away, making the equilibrium unstable (graphical criterion for stability). For the updating function to cross from below to above, the slope at the equilibrium must be greater than 1 (slope criterion for stability). The opposite occurs at stable equilibria, where the updating function crosses from above to below and has slope less than 1. In the special case that the slope is exactly equal to 1, the equilibrium might be neither stable nor unstable.

6.7 Exercises

Mathematical Techniques

1–4 ■ Find the equilibria of each discrete-time dynamical system from its graph and apply the graphical criterion for stability to find which are stable. Check by cobwebbing.

1.
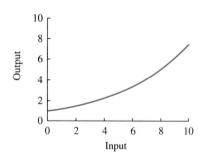

2.
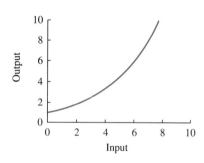

3.
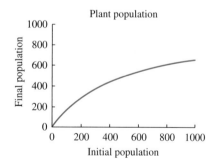

4.

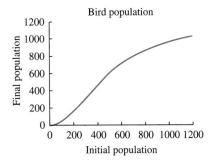

5–12 ■ Graph each discrete-time dynamical system, find the equilibria algebraically, and check whether the stability derived from the slope criterion for stability matches that found with cobwebbing.

5. $c_{t+1} = 0.5c_t + 8$ for $0 \leq c_t \leq 30$.

6. $b_{t+1} = 3b_t$ for $0 \leq b_t \leq 10$.

7. $b_{t+1} = 0.3b_t$ for $0 \leq b_t \leq 10$.

8. $b_{t+1} = 2b_t - 5$ for $0 \leq b_t \leq 10$.

9. The updating function is $f(x) = x^2$ for $0 \leq x \leq 2$.

10. The updating function is $g(y) = y^2 - 1$ for $0 \leq y \leq 2$.

11. $x_{t+1} = \dfrac{x_t}{1 + x_t}$

12. $x_{t+1} = -1 + 4x_t - 3x_t^2 + x_t^3$ (the only equilibrium is $x^* = 1$).

13–18 ■ The unusual equilibrium in the text has an updating function that lies below the diagonal both to the left and to the right of the equilibrium. There are several other ways in which an updating function can intersect the diagonal at an equilibrium. In each case, cobweb starting from points to the left and to the right of the equilibrium, and describe the stability.

13. Graph an updating function that lies above the diagonal both to the left and to the right of an equilibrium.

14. Graph an updating function that is tangent to the diagonal at an equilibrium but crosses from below to above. Show by cobwebbing that the equilibrium is unstable. What is the second derivative at the equilibrium?

15. Graph an updating function that is tangent to the diagonal at an equilibrium but crosses from above to below. Show by cobwebbing that the equilibrium is stable. What is the second derivative at the equilibrium?

16. Sketch the graph of an updating function that has a corner at an equilibrium and is stable.

17. Sketch the graph of an updating function that has a corner at an equilibrium and is unstable.

18. Sketch the graph of an updating function that has a corner at an equilibrium and is neither stable nor unstable.

19–20 ■ Another peculiarity of an updating function that is tangent to the diagonal at an equilibrium is that slight changes in the graph can produce big changes in the number of equilibria. The following exercises are based on Figure 6.7.85, repeated here.

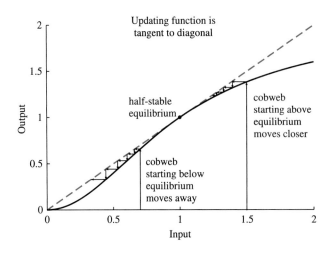

19. Move the curve slightly down (while keeping the diagonal in the same place). How many equilibria are there now? What happens when you cobweb starting from a point at the right-hand edge of the figure?

20. Move the curve slightly up (again keeping the diagonal in the same place). How many equilibria are there? Describe their stability.

21–22 ■ Find the inverse of each updating function, and compute the slope of both the original updating function and the derivative at the equilibrium.

21. The updating function $f(x) = \dfrac{x}{1+x}$ (compare with Exercise 11).

22. The updating function $f(x) = \dfrac{x}{x-1}$.

Applications

23–26 ■ Recall the updating function for the fraction, p, of mutant bacteria given by

$$f(p) = \frac{sp}{sp + r(1-p)}$$

where s is the per capita production of the mutant and r is the per capita production of the wild type. Find the derivative in each case, and evaluate at the equilibria, $p^* = 0$ and $p^* = 1$. Are the equilibria stable?

23. $s = 1.2$, $r = 2$

24. $s = 3$, $r = 1.2$

25. $s = 1.5$, $r = 1.5$

26. In general (without substituting numerical values for s and r). Can both equilibria be stable? What happens if $r = s$?

27–28 ■ Find the equilibrium population of bacteria with supplementation. Graph the updating function, and use the slope criterion for stability to check the stability.

27. A population of bacteria has per capita production $r = 0.6$, and 10^6 bacteria are added each generation (as in Section 3.5, Exercise 31).

28. A population of bacteria has per capita production $r = 0.2$, and $5 \cdot 10^6$ bacteria are added each generation (as in Section 3.5, Exercise 32).

29–30 ▪ A lab is growing and harvesting a culture of valuable bacteria described by the discrete-time dynamical system

$$b_{t+1} = rb_t - h$$

The bacteria have per capita production r, and h are harvested each generation (as in Section 3.5, Exercises 44 and 45). Graph the updating function for each, and use the slope criterion for stability to check the stability.

29. Suppose that $r = 1.5$ and $h = 10^6$.

30. Without setting r and h to particular values, find the equilibrium algebraically. When is the equilibrium stable?

31–32 ▪ The model describing the dynamics of the concentration of pain medication in the bloodstream,

$$M_{t+1} = 0.5M_t + 1$$

becomes nonlinear if the fraction of medication used is a function of the concentration. In each case, use the slope criterion for stability to check the stability of the equilibrium.

31. The nonlinear discrete-time dynamical system

$$M_{t+1} = M_t - \frac{0.5}{1 + 0.1M_t} M_t + 1$$

(studied in Section 3.4, Exercise 39).

32. The nonlinear discrete-time dynamical system

$$M_{t+1} = M_t - \frac{1}{1 + 0.1M_t} M_t + 1$$

How does this differ from the model in Exercise 31? Why is the equilibrium smaller?

33–36 ▪ An equilibrium that is stable when time goes forward should be unstable when time goes backwards. Find the inverse of the updating function associated with each discrete-time dynamical system, and find the derivative at the equilibria.

33. $c_{t+1} = 0.5c_t + 8$ (compare with Exercise 5)

34. $b_{t+1} = 3b_t$ (compare with Exercise 6)

35. $x_{t+1} = \dfrac{2x_t}{1 + 0.001x_t}$ (compare with Example 6.7.7)

36. $p_{t+1} = \dfrac{2p_t}{2p_t + (1 - p_t)}$ (the selection system with $s = 2$ and $r = 1$)

37–38 ▪ Consider a population, x_t, with per capita production $r \dfrac{x_t}{1 + x_t^2}$. After writing the discrete-time dynamical system, do the following for each value of the parameter r.

a. Find the equilibria.
b. Graph the updating function.
c. Indicate which equilibria are stable and which are unstable, and check via the slope criterion for stability.
d. Describe how the population would behave.

37. $r = 1$
38. $r = 2.5$

Computer Exercises

39. Consider the discrete-time dynamical system found in Section 3.4, Exercise 48. In that exercise, there were two cultures, 1 and 2. In culture 1, the mutant does better than the wild type, and in culture 2, the wild type does better than the mutant. In particular, suppose that $s = 2$ and $r = 0.3$ in culture 1, and that $s = 0.6$ and $r = 2$ in culture 2. Define updating functions f_1 and f_2 to describe the dynamics in the two cultures. The overall updating function after mixing equal amounts from the two is

$$f(p) = \frac{f_1(p) + f_2(p)}{2}$$

a. Write the updating function explicitly.
b. Find the equilibria.
c. Find the derivative of the updating function.
d. Evaluate the stability of the equilibrium.

40. Consider again each discrete-time dynamical system (Section 3.5, Exercise 46). Check the stability of the equilibria.

a. $x_{t+1} = \cos x_t$
b. $y_{t+1} = \sin y_t$
c. $z_{t+1} = \sin z_t + \cos z_t$

41. Consider the discrete-time dynamical system

$$z_{t+1} = a^{z_t}$$

(thanks to Larry Okun). We will study this for different values of a.

a. Follow the dynamics starting from initial condition $z_0 = 1$ for $a = 1, 1.1, 1.2, 1.3,$ and 1.4. Keep running the system until it seems to reach an equilibrium.

b. Do the same, but increase a slowly past a critical value of about 1.4446679. Solutions should creep up for a while and then increase very quickly. Graph the updating function for values above and below the critical value and try to explain what is going on.

c. At the critical value, the slope of the updating function is 1 at the equilibrium. Show that this occurs with $a = e^{1/e}$. What is the equilibrium?

42. Consider a population with

$$\text{per capita production} = r\frac{x_t}{1 + x_t^2}$$

a. Write the discrete-time dynamical system.
b. Set $r = 3$ and $x_0 = 2$. Compute the solution until it gets close to an equilibrium, and record this value.

c. Decrease r to 2.9. Use the equilibrium found with $r = 3$ as an initial condition and compute the solution until it approaches an equilibrium. Record this value.

d. Continue decreasing r down to $r = 1$, using the last equilibrium as the initial condition. What happens when r crosses 2?

e. If this were a real population and r were a measure of the quality of habitat, how would you interpret this behaviour?

f. Follow the same procedure, but start with $r = 1$ and increase r to 3.

g. Can you explain what is going on?

6.8 The Logistic Dynamical System and More Complex Dynamics

The derivative or slope of an updating function determines whether an equilibrium is stable or unstable. We have yet, however, to consider cases where the slope of the updating function is **negative** at an equilibrium. By studying several such cases, we will see that rather exotic behaviours are possible. Using the idea of **qualitative dynamical systems,** where we approximate a nonlinear discrete-time dynamical system with its **tangent line approximation** at an equilibrium, we will find a general condition that includes these new cases.

The Logistic Dynamical System

In the previous section, we looked at a model where large population size reduced per capita production according to

$$\text{per capita production} = \frac{2}{1 + 0.001x}$$

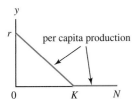

FIGURE 6.8.92

Per capita production for the logistic growth model

(Example 6.7.7 and Figure 6.7.90a). Another widely studied model is the **logistic dynamical system.** In this model, the per capita production of a population decreases linearly with population size according to

$$\text{per capita production} = r\left(1 - \frac{N}{K}\right)$$

where N represents population size. The parameter r is the greatest possible production, and K is the maximum possible population. For populations greater than K, the per capita production is assumed to be zero; see Figure 6.8.92. Using the fact that the new population is the per capita production times the old population, we find that the discrete-time dynamical system for the logistic dynamical system is

$$N_{t+1} = r\left(1 - \frac{N_t}{K}\right) N_t$$

To make calculations simpler, we define the new variable

$$x_t = \frac{N_t}{K}$$

to represent the fraction of the maximum possible population. If K is 1000, a population of $N_t = 500$ corresponds to the fraction $x_t = 0.5$. We can write the discrete-time dynamical system for x_t as

$$x_{t+1} = \frac{N_{t+1}}{K}$$

$$= \frac{r\left(1 - \frac{N_t}{K}\right) N_t}{K}$$

$$= r\left(1 - \frac{N_t}{K}\right)\frac{N_t}{K}$$

$$= r(1 - x_t)x_t$$

The factors on the right-hand side are usually written in a different order, giving the **logistic dynamical system**

$$x_{t+1} = rx_t(1 - x_t) \tag{6.8.1}$$

We begin by finding the equilibria:

$$x^* = rx^*(1 - x^*)$$

$$x^* - rx^*(1 - x^*) = 0$$

$$x^*[1 - r(1 - x^*)] = 0$$

$$x^* = 0 \quad \text{or} \quad 1 - r(1 - x^*) = 0$$

$$x^* = 0 \quad \text{or} \quad 1 - x^* = \frac{1}{r}$$

$$x^* = 0 \quad \text{or} \quad x^* = 1 - \frac{1}{r}$$

Do these solutions make sense? The first, $x^* = 0$, is the extinction equilibrium that shows up in all population models we have studied so far. (However, this is not true in general—a population model with immigration does not show the same dynamics. "Immigration" means an increase in population due to the introduction of new members from outside the population.) The second equilibrium depends on the maximum per capita production, r. If $r < 1$, this equilibrium is negative and biologically impossible. A population with a *maximum* per capita production less than 1 cannot replace itself and will go extinct. For larger values of r, the equilibrium becomes larger, a result consistent with the fact that a population with higher potential production grows to a larger value.

We will examine four cases: $r = 0.5$, $r = 1.5$, $r = 2.5$, and $r = 3.5$. The equilibria and their behaviour as deduced from cobwebbing (see Figure 6.8.93a–d) are summarized in the following table:

r	x^*	Stability
0.5	0	stable
1.5	0	unstable
1.5	$1 - \frac{1}{1.5} \approx 0.333$	stable
2.5	0	unstable
2.5	$1 - \frac{1}{2.5} = 0.6$	stable (oscillates)
3.5	0	unstable
3.5	$1 - \frac{1}{3.5} \approx 0.714$	unstable (oscillates)

The results match the graphical criterion for stability pretty well until $r = 3.5$ (Figure 6.8.93d). The updating function crosses the diagonal from above to below at the positive equilibrium, so we expect this equilibrium to be stable. However, a solution starting near the positive equilibrium $x^* \approx 0.714$ moves away, and it does so by jumping back and forth. Closer examination of the case with $r = 2.5$ (Figure 6.8.93c) reveals that solutions jump back and forth as they approach the equilibrium at $x^* = 0.6$. What is going on?

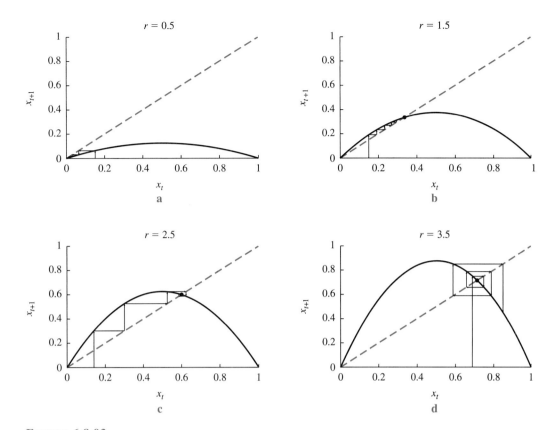

FIGURE 6.8.93

The behaviour of the logistic dynamical system

Qualitative Dynamical Systems

An equilibrium where the updating function has a positive slope is stable precisely when that slope is less than 1. These results match the behaviour of the bacterial growth model

$$b_{t+1} = rb_t$$

at the equilibrium $b^* = 0$. This linear system is stable precisely when the slope r is less than 1.

What happens if $r < 0$? Of course, this case does not make sense for populations, but it will provide insight into what happens near equilibria with negative slopes, as found in Figure 6.8.93c and d.

Consider a nonlinear discrete-time dynamical system, $x_{t+1} = f(x_t)$, and its tangent line, $L(x_t)$, at an equilibrium, x^* (Figure 6.8.94a and b). The two systems share the equilibrium at x^* and have the same slope. Furthermore, solutions remain very close to each other (Figure 6.8.94c). According to the slope criterion for stability, the equilibrium of L should be stable precisely when the equilibrium of f is stable. Starting from a point near x^*, the solutions are nearly identical because the tangent line lies close to the curve (Figure 6.8.95).

The original updating function and the tangent line are not exactly equal. The exact values in the solution will therefore also be different. When we ask about stability, however, we are concerned with general behaviour rather than exact values. Studying general aspects of dynamical systems without requiring exact measurements is the realm of **qualitative dynamical systems.** The approach is essential in biology, where measurements often include a great deal of noise. Nonetheless, we would like to use the linear model to make predictions. Some predictions are **qualitative,** verbal descriptions of behaviour such as "the concentration of pain medication will approach

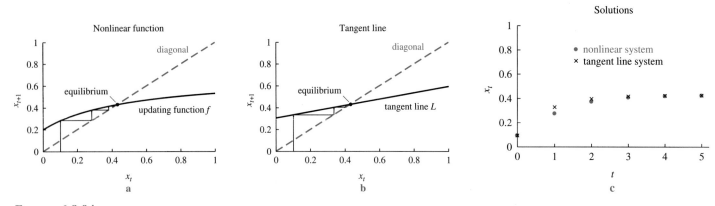

FIGURE 6.8.94

Comparing an updating function with its tangent line

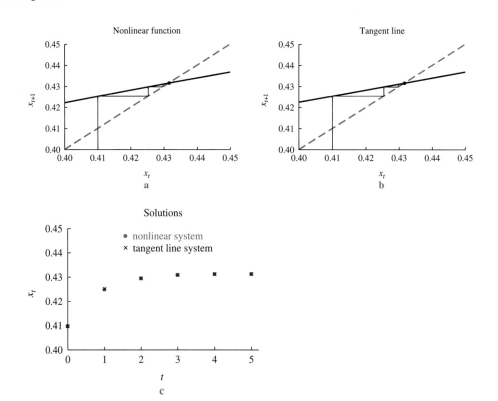

FIGURE 6.8.95

Zooming in on an updating function and its tangent line

an equilibrium." Others are **quantitative,** numerical descriptions of behaviour such as "the concentration of pain medication will reach exactly 25 units in 3 hours."

The qualitative theory of dynamical systems requires comparing the dynamics produced by similar discrete-time dynamical systems. If two nearly indistinguishable discrete-time dynamical systems produce qualitatively different dynamics, the underlying biological system might be highly sensitive to small changes in conditions. Such situations do occur. In this chapter, we restrict our attention to cases where similar discrete-time dynamical systems produce qualitatively similar dynamics.

Using this philosophy, we can try to figure out what happens at an equilibrium where the slope is negative by studying the linear bacterial growth model

$$b_{t+1} = rb_t$$

with unrealistic negative values for the per capita production r.

Example 6.8.1 The Consequences of "Negative" per Capita Production: $r = -0.5$

If $r = -0.5$, the discrete-time dynamical system is

$$b_{t+1} = -0.5b_t$$

Starting from an initial condition of $b_0 = 0.5$, we find that the solution is

$$b_1 = -0.25$$
$$b_2 = 0.125$$
$$b_3 = -0.0625$$
$$b_4 = 0.03125$$
$$b_5 = -0.015625$$
$$\ldots$$

The absolute value decreases while the sign switches back and forth (Figure 6.8.96).

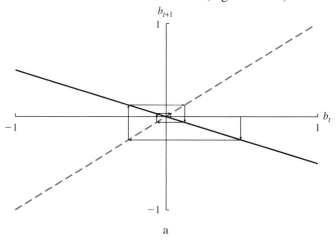
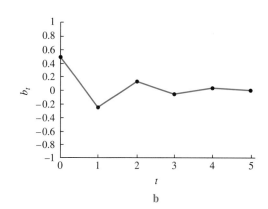

FIGURE 6.8.96
Cobwebbing and solution with $r = -0.5$

Example 6.8.2 The Consequences of "Negative" per Capita Production: $r = -0.9$

If $r = -0.9$, the discrete-time dynamical system is

$$b_{t+1} = -0.9 b_t$$

Starting from an initial condition of $b_0 = 0.5$, we find that the solution is

$$b_1 = -0.45$$
$$b_2 = 0.405$$
$$b_3 = -0.3645$$
$$b_4 = 0.32805$$
$$b_5 = -0.29524$$
$$\ldots$$

The absolute value decreases slowly while the sign switches back and forth (Figure 6.8.97).

Example 6.8.3 The Consequences of "Negative" per Capita Production: $r = -2$

Suppose that $r = -2$. Starting from an initial condition of $b_0 = 0.02$, we find that the solution is

$$b_1 = -0.04$$
$$b_2 = 0.08$$

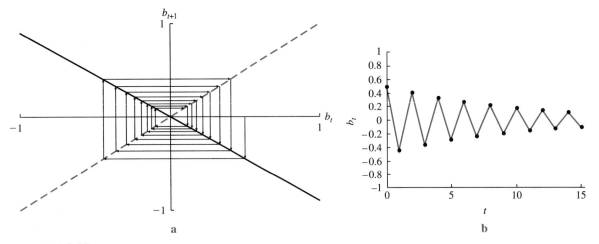

FIGURE 6.8.97
Cobwebbing and solution with $r = -0.9$

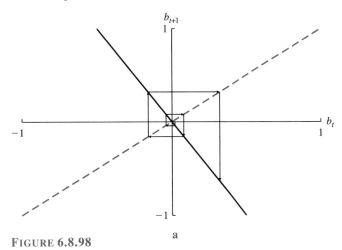

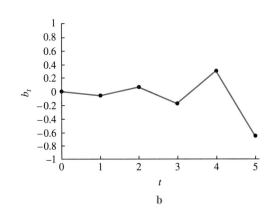

FIGURE 6.8.98
Cobwebbing and solution with $r = -2$

$$b_3 = -0.16$$
$$b_4 = 0.32$$
$$b_5 = -0.64$$
$$\ldots$$

Again the sign switches back and forth, but the absolute value now increases (Figure 6.8.98).

Example 6.8.4 The Consequences of "Negative" per Capita Production: $r = -1.1$

Finally, suppose that $r = -1.1$. Starting from an initial condition of $b_0 = 0.1$, we find that the solution is

$$b_1 = -0.11$$
$$b_2 = 0.121$$
$$b_3 = -0.1331$$
$$b_4 = 0.14641$$
$$b_5 = -0.16105$$
$$b_6 = 0.17716$$
$$\ldots$$

The sign switches back and forth, and the absolute value increases slowly (Figure 6.8.99).

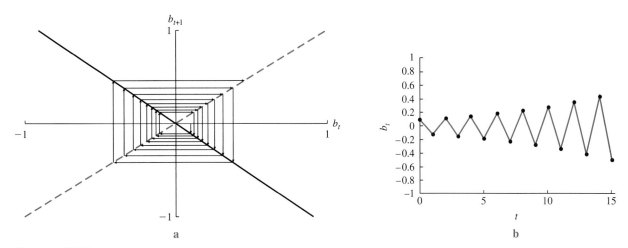

FIGURE 6.8.99

Cobwebbing and solution with $r = -1.1$

We see that, for negative r, the system is stable when $r > -1$ and unstable when $r < -1$. Putting this together with the results with positive r gives the following results:

Behaviour of Solutions of $b_{t+1} = rb_t$

r	Stability	Behaviour
$r > 1$	unstable	moves away from equilibrium
$0 < r < 1$	stable	moves toward equilibrium
$-1 < r < 0$	stable	oscillates toward equilibrium
$r < -1$	unstable	oscillates away from equilibrium

In other words, an equilibrium is stable if r is less than 1 in **absolute value,** or $|r| < 1$, and is unstable if r is greater than 1 in absolute value, or $|r| > 1$.

Because the behaviour of a nonlinear dynamical system resembles that of its tangent line, we can extend these results to the following theorem.

Theorem 6.8.1 **Stability Theorem for Discrete-Time Dynamical Systems**

Suppose that the discrete-time dynamical system

$$x_{t+1} = f(x_t)$$

has an equilibrium at x^*. The equilibrium at x^* is stable if

$$|f'(x^*)| < 1$$

and unstable if

$$|f'(x^*)| > 1$$

The proof depends on the Mean Value Theorem (Section 6.3) and will not be given here.

Using this method, we can read off the stability of equilibria of more complicated discrete-time dynamical systems (Figure 6.8.100). By including lines with slope -1 on our graph in addition to the diagonal, we can see that the equilibrium at 0.5 is stable because the slope is between 0 and -1, whereas the equilibrium at 1.5 is unstable because the slope is less than -1.

444 **Chapter 6** Applications of Derivatives

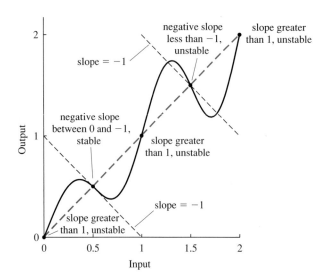

FIGURE 6.8.100

Applying the stability theorem for discrete-time dynamical systems

Example 6.8.5 Using the Stability Theorem: Alcohol Absorption Model

In Example 3.3.6 in Section 3.3 we studied the alcohol absorption model

$$a_{t+1} = a_t - \frac{10.1 a_t}{4.2 + a_t} + 7$$

where a_t is the amount of alcohol present at time t. In the same example we calculated the equilibrium $a^* \approx 9.5$; now we check its stability using Theorem 6.8.1.

The updating function is

$$f(a) = a - \frac{10.1 a}{4.2 + a} + 7$$

and so

$$f'(a) = 1 - \frac{(4.2 + a)(10.1) - 10.1 a}{(4.2 + a)^2} = 1 - \frac{42.42}{(4.2 + a)^2}$$

Since

$$|f'(a^*)| = \left| 1 - \frac{42.42}{(4.2 + 9.5)^2} \right| \approx 0.77$$

is smaller than 1, the equilibrium $a^* \approx 9.5$ is stable.

Analysis of the Logistic Dynamical System

We now investigate whether these results help make sense of the behaviour of the logistic dynamical system

$$x_{t+1} = r x_t (1 - x_t)$$

The derivative of the updating function is

$$\begin{aligned} f'(x) &= [rx(1-x)]' \\ &= r[1 - x + x(-1)] \\ &= r(1 - 2x) \end{aligned}$$

We evaluate the derivative at the two equilibria. At $x^* = 0$,

$$f'(0) = r(1 - 2 \cdot 0) = r$$

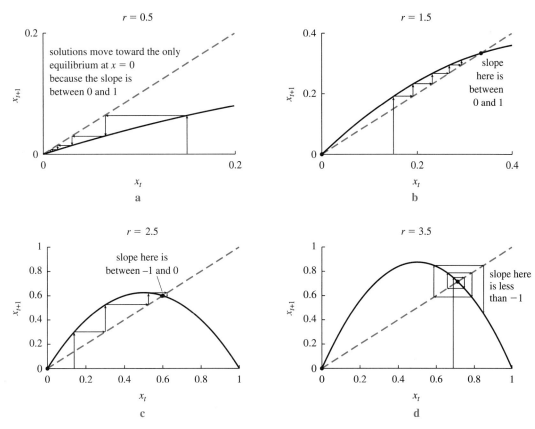

FIGURE 6.8.101
The behaviour of the logistic dynamical system explained

(Keep in mind that $r > 0$.) According to the stability theorem for discrete-time dynamical systems, the equilibrium at $x^* = 0$ is stable if $r < 1$ and unstable if $r > 1$. When $r > 1$, the derivative at the positive equilibrium $x^* = 1 - \frac{1}{r}$ is

$$f'\left(1 - \frac{1}{r}\right) = r\left[1 - 2\left(1 - \frac{1}{r}\right)\right]$$
$$= r\left(-1 + \frac{2}{r}\right)$$
$$= -r + 2$$

The results are summarized in the following table and in Figure 6.8.101. (Recall that for $x^* \neq 0$ the entries in the table have been calculated from $x^* = 1 - \frac{1}{r}$ and $f'(x^*) = -r + 2$. At the beginning of this subsection, we calculated that $f'(x^*) = r$ for $x^* = 0$.)

r	x^*	$f'(x^*)$	Stability
0.5	0	0.5	stable
1.5	0	1.5	unstable
1.5	0.333	0.5	stable
2.5	0	2.5	unstable
2.5	0.600	−0.5	stable (oscillates)
3.5	0	3.5	unstable
3.5	0.714	−1.5	unstable (oscillates)

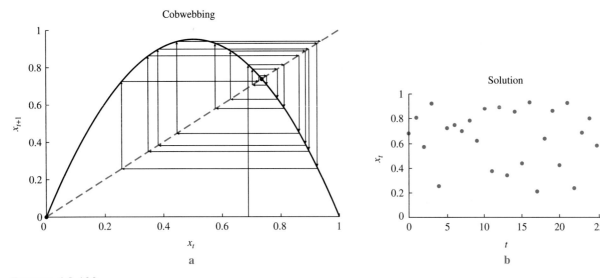

FIGURE 6.8.102

The long-term behaviour of the logistic dynamical system with $r = 3.8$

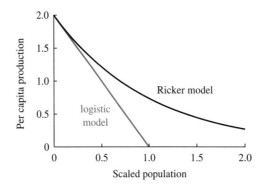

FIGURE 6.8.103

The per capita production in the Ricker and logistic models with $r = 2$

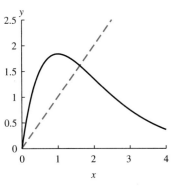

FIGURE 6.8.104

The Ricker discrete-time dynamical system with $r = 5$

When $r = 3.5$, neither equilibrium is stable. This population has nowhere to settle down and will continue jumping around indefinitely. In Figure 6.8.102 we show the behaviour of the system

$$x_{t+1} = rx_t(1 - x_t)$$

for $r = 3.8$. The solution (Figure 6.8.102b) no longer reveals a recognizable pattern (such as approaching a particular value or oscillation). This kind of behaviour is an example of **chaotic behaviour** or **chaos.**

Ricker Model

A commonly used model for reproduction in fisheries is the **Ricker model,** defined by

$$x_{t+1} = rx_t e^{-x_t}$$

The per capita production, re^{-x_t}, declines exponentially from a maximum of r as the population becomes larger. As in the logistic dynamical system, production decreases rapidly, but this model is a bit more realistic because per capita production never reaches zero (Figure 6.8.103). If we write the updating function (see Figure 6.8.104) as

$$f(x) = rxe^{-x}$$

the first derivative is

$$f'(x) = r[e^{-x} + xe^{-x}(-1)]$$
$$= r(1 - x)e^{-x}$$

Where are the equilibria of the Ricker model? We solve

$$x^* = rx^*e^{-x^*}$$
$$x^* - rx^*e^{-x^*} = 0$$
$$x^*\left(1 - re^{-x^*}\right) = 0$$

Thus, $x^* = 0$ or

$$1 - re^{-x^*} = 0$$
$$e^{-x^*} = \frac{1}{r}$$
$$-x^* = \ln\frac{1}{r} = \ln r^{-1} = -\ln r$$
$$x^* = \ln r$$

At the equilibrium $x^* = 0$, $f'(0) = r$. This equilibrium is stable when the maximum per capita production is $r < 1$ and unstable when $r > 1$. The equilibrium $x^* = \ln r$ is positive when $r > 1$. At this point, the derivative of the updating function is

$$f'(\ln r) = r(1 - \ln r)e^{-\ln r}$$
$$= r(1 - \ln r)\frac{1}{r}$$
$$= 1 - \ln r$$

Results for various values of r are summarized in the following table:

r	x^*	$f'(x^*)$	Stability
0.5	0	0.5	stable
2	0	2	unstable
2	$\ln 2 \approx 0.693$	$1 - \ln 2 \approx 0.307$	stable
4	0	4	unstable
4	$\ln 4 \approx 1.386$	$1 - \ln 4 \approx -0.386$	stable (oscillates)
8	0	8	unstable
8	$\ln 8 \approx 2.079$	$1 - \ln 8 \approx -1.079$	unstable (oscillates)
13	0	8	unstable
13	$\ln 13 \approx 2.565$	$1 - \ln 13 \approx -1.565$	unstable (oscillates)

As long as $\ln r < 2$, or $r < e^2 \approx 7.39$, the positive equilibrium is stable (Figure 6.8.105a and b). Like the logistic model, this discrete-time dynamical system produces unstable dynamics when the maximum possible per capita production is large (Figure 6.8.105c and d). In this case, fish can produce so many offspring that they destroy their resource base and induce a population crash in the following year.

Summary   The approach of **qualitative dynamical systems** lets us think of behaviour in general terms without depending on quantitative details about the discrete-time dynamical system. As an example, we compare the dynamics generated by a nonlinear discrete-time dynamical system with those generated by the tangent line at the equilibrium. Because stability of equilibria can be understood by studying linear discrete-time dynamical systems, we examined linear systems with negative slopes, finding that such systems oscillate. On the basis of these results, we stated the **stability theorem for discrete-time dynamical systems:** an equilibrium is stable if the *absolute value* of the slope is less than 1. We applied this condition to the **logistic dynamical system** and the **Ricker model** and found that there are parameter values with no stable equilibrium.

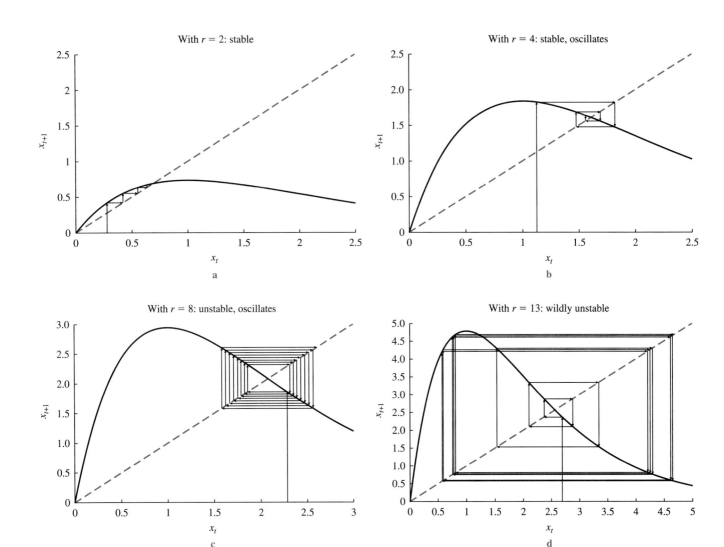

FIGURE 6.8.105
The dynamics of the Ricker model with various values of r

6.8 Exercises

Mathematical Techniques

1–4 ■ Draw the tangent line approximating the given system at the specified equilibrium, and compare the cobweb diagrams. Use the stability theorem to check whether the equilibrium is stable.

1. The bacterial selection equation $p_{t+1} = \dfrac{1.5 p_t}{1.5 p_t + 2(1 - p_t)}$ at the equilibrium $p^* = 0$.

2. As in Exercise 1, but at the equilibrium $p^* = 1$.

3. $x_{t+1} = 1.5 x_t (1 - x_t)$ at the equilibrium $x^* = 0$.

4. $x_{t+1} = 1.5 x_t (1 - x_t)$ at the equilibrium $x^* = 1/3$.

5–8 ■ Starting from the given initial condition, find the solution for five steps of each of the following.

5. $y_{t+1} = 1.2 y_t$ with $y_0 = 2$. When will the value exceed 100?

6. $y_{t+1} = -1.2 y_t$ with $y_0 = 2$. When will the value exceed 100?

7. $y_{t+1} = 0.8 y_t$ with $y_0 = 2$. When will the value be less than 0.2?

8. $y_{t+1} = -0.8 y_t$ with $y_0 = 2$. When will the value be between 0 and 0.2?

9–12 ■ Consider the linear discrete-time dynamical system $y_{t+1} = 1 + m(y_t - 1)$. For each of the following values of m:

a. Find the equilibrium.

b. Graph and cobweb.

c. Compare your results with the stability condition.

9. $m = 0.9$

10. $m = 1.5$

11. $m = -0.5$

12. $m = -1.5$

13–16 ▪ The following discrete-time dynamical systems have slope of exactly -1 at the equilibrium. Check this, and then iterate the function for a few steps starting from near the equilibrium to see whether it is stable, unstable, or neither.

13. $x_{t+1} = 4 - x_t$

14. $x_{t+1} = \dfrac{x_t}{x_t - 1}$ for $x_t > 1$.

15. $x_{t+1} = 3x_t(1 - x_t)$, the logistic system with $r = 3$.

16. $x_{t+1} = \dfrac{2}{1 + x_t^2}$ (the equilibrium is at $x^* = 1$)

17–18 ▪ Equilibria where the slope of the tangent line is exactly zero are also special. Show that the following systems satisfy this special relationship. What does this say about the stability of the equilibrium? (Think about a linear dynamical system with a slope of zero. How quickly do solutions approach the equilibrium?) Draw a cobweb diagram to illustrate these results.

17. The logistic dynamical system with $r = 2$.

18. The dynamical system $x_{t+1} = x_t e^{1-x_t}$ (equivalent to the Ricker model with $r = e$).

Applications

19–22 ▪ We have studied several systems where the fraction of medication absorbed depends on the concentration of medication in the bloodstream. These take the form

$$M_{t+1} = M_t - f(M_t)M_t + 1$$

where $f(M_t)$ is the fraction absorbed and 1 is the supplement. If the fraction absorbed increases, it seems possible that the equilibrium level will become unstable (high levels are rapidly reduced). For each of the following forms of $f(M_t)$, the equilibrium level is $M^* = 2$. Find the slope of the updating function at the equilibrium and check whether it is stable. In which cases does the solution oscillate?

19. $f(M_t) = \dfrac{M_t}{2 + M_t}$

20. $f(M_t) = \dfrac{M_t^2}{4 + M_t^2}$

21. $f(M_t) = \dfrac{M_t^4}{16 + M_t^4}$

22. $f(M_t) = \dfrac{M_t^8}{256 + M_t^8}$

23–26 ▪ Find values of r that satisfy the following conditions for the Ricker model. In each case, graph and cobweb.

23. The value of r where the positive equilibrium switches from having a positive to having a negative slope.

24. One value of r between 1 and the value found in Exercise 23.

25. One value of r between the value found in Exercise 23 and $r = e^2$.

26. One value of r greater than $r = e^2$.

27–28 ▪ The logistic model quantifies a competitive interaction, where per capita production is a decreasing function of population size. In some situations, per capita production is enhanced by population size. For each of the following such cases:

a. Write the updating function.

b. Find the equilibria and their stability.

c. Graph the updating function and cobweb. Which equilibrium is stable?

d. Explain what this population is doing.

27. Per capita production $= 0.5 + 0.5 b_t$

28. Per capita production $= 0.5 + 0.5 b_t^2$

29–30 ▪ Consider a modified version of the logistic dynamical system $x_{t+1} = rx_t(1 - x_t^n)$. For each value of n:

a. Sketch the updating function with $r = 2$.

b. Find the equilibria.

c. Find the derivative of the updating function at the equilibria.

d. For what values of r is the equilibrium at $x = 0$ stable? For what values of r is the positive equilibrium stable?

29. $n = 2$

30. $n = 3$

31–32 ▪ Expanding oscillations can result from improperly tuned feedback systems. Suppose that a thermostat is supposed to keep a room at 20°C. Take the following steps to figure out what is happening in each of the given cases.

a. Suppose the temperature produced is a linear function of the temperature on the thermometer. Find this function.

b. You continue to respond to a temperature of x degrees Celsius above 20°C by setting the thermostat to $(20 - x)$ degrees Celsius and to a temperature of x degrees Celsius below 20°C by setting the thermostat to $(20 + x)$ degrees Celsius. Find the temperature for the next few days.

c. Denote the temperature on day t by T_t. Find a formula for the thermostat setting, z_t, in response.

d. Use the answer to part (d) to find T_{t+1}. Write the updating function.

e. Use the stability condition to describe what will happen in this room.

31. You come in one morning and find that the temperature is 21°C. To correct this, you move the thermostat down by 1°C to 19°C. But the next day the temperature has dropped to 18°C.

32. You come in one morning and find that the temperature is 21°C. To correct this, you move the thermostat down by 1°C to 19°C. But the next day the temperature has dropped to 18.5°C.

33–34 ■ The model of bacterial selection includes no **frequency dependence,** meaning that the per capita production of the different types does not depend on the fraction of types in the population. Each of the following discrete-time dynamical systems for the number of mutants, a_t, and the number of wild type, b_t, depends on the fraction of mutants, p_t. For each, explain how each type is affected by the frequency of the mutants, find the discrete-time dynamical system for p_t, find the equilibria, evaluate their stability, and plot a cobweb diagram. Do any of them oscillate?

33. $a_{t+1} = 2(1 - p_t)a_t$ and $b_{t+1} = (1 + p_t)b_t$.

34. $a_{t+1} = 2(1 - p_t)^2 a_t$ and $b_{t+1} = (1 + p_t)b_t$.

35–38 ■ Crowded plants grow to smaller size. Smaller plants make fewer seeds. The following exercises describe the dynamics of a population described by n, the total number of seeds, and s, the size of the adult produced. Assume that adult plants die after producing seeds. In each case:

a. Start from $n = 20$ and find the total number of seeds for the next two years. (If the number of seeds per plant is a fraction, don't worry. Just think of it as an average.)

b. Write the discrete-time dynamical system for the number of seeds.

c. Find the equilibrium number of seeds.

d. Graph the updating function and cobweb.

e. How is this result related to the stability condition?

35. If there are n seeds, each sprouts and grows to a size of $s = \dfrac{100}{n}$. An adult of size s produces $s - 1$ seeds (because it must use one unit of energy to survive).

36. If there are n seeds, each sprouts and grows to a size of $s = \dfrac{100}{n}$. An adult of size s produces $s - 0.5$ seeds.

37. If there are n seeds, each sprouts and grows to a size of $s = \dfrac{100}{n}$. Suppose that an adult of size s produces $s - 2$ seeds.

38. If there are n seeds, each sprouts and grows to a size of $s = \dfrac{100}{n+5}$. An adult of size s produces $s - 1$ seeds.

Computer Exercises

39. Study the behaviour of the logistic dynamical system first for values of r near 3 and then for values between 3.5 and 4. Try the following with 5 values of r near 3 (such as 2.9, 2.99, 3, 3.01, and 3.1) and 10 values of r between 3.5 and 4.

    a. Use a computer to find solutions for 100 steps.

    b. Look at the last 50 or so points on the solution and try to describe what is going on.

    c. The case with $r = 4$ is rather famously chaotic. One of the properties of chaotic systems is sensitivity to initial conditions. Run the system for 100 steps from one initial condition, and then run it again from an initial condition that is very close. If you compare your two solutions, they should be similar for a while but should eventually become completely different. What if a real system had this property?

40. Consider a population following
$$x_{t+1} = rx_t^2 e^{-x_t}$$

    a. Graph the updating function for the following values of r: $r = 1$, $r = 2.6$, $r = e$, $r = 2.8$, $r = 3.6$, $r = \dfrac{e^2}{2}$, $r = 3.8$, $r = 6.6$, $r = \dfrac{e^3}{3}$, $r = 6.8$, and $r = 10$.

    b. Find the equilibria for these values of r (the equation cannot be solved in general, but a computer should have a routine for solving, or just guess). Make sure you find them all.

    c. Find the derivative of the updating function at each equilibrium.

    d. Find which of the equilibria are stable.

    e. When all of the positive equilibria are unstable, how might this model behave differently from the Ricker model with $r > e^2$? Can you explain why?

Chapter Summary: Key Terms and Concepts

Define or explain the meaning of each term and concept.
Extreme values: relative (local) minimum, relative (local) maximum, absolute (global) minimum, absolute (global) maximum; critical number (critical point); Fermat's Theorem, Extreme Value Theorem, first derivative test, second derivative test
Continuity and differentiability: Intermediate Value Theorem, Rolle's Theorem, Mean Value Theorem; differentiability implies continuity
Limits of functions: leading behaviour at infinity, leading behaviour at zero, method of matched leading behaviours; indeterminate form, L'Hôpital's rule
Solving equations: bisection method, Newton's method
Discrete-time dynamical systems: equilibrium, stable equilibrium, unstable equilibrium; graphical criterion for stability, slope criterion for stability (Stability Theorem); logistic dynamical system

Concept Check: True/False Quiz

Determine whether each statement is true or false. Give a reason for your choice.

1. If $f(x) = (x^2 - x - 2)(2x - 1)^{-1}$, then $f_\infty(x) = x/2$.
2. If $f'(a) = 0$, then $f(x)$ has a relative extreme value at a.
3. There is a continuous function $f(x)$ defined on the interval $(0, 5)$ that has an absolute minimum and an absolute maximum in $(0, 5)$.
4. Every continuous function defined on a closed interval $[a, b]$ has an absolute minimum and an absolute maximum in $[a, b]$.
5. If a continuous function $f(x)$ has a relative extreme value at a, then $f'(a) = 0$.
6. If a differentiable function $f(x)$ has a relative extreme value at a, then $f'(a) = 0$.
7. If $f(x)$ is continuous on $[-1, 1]$ and $f(-1) = f(1)$, then there is a point c in $(-1, 1)$ where $f'(c) = 0$.
8. By L'Hôpital's rule, $\lim\limits_{x \to 2} \dfrac{x^2 - 4}{x + 2} = \lim\limits_{x \to 2} \dfrac{2x}{1} = 2$.
9. The same value, m^*, is a stable equilibrium for both the dynamical system $m_{t+1} = f(m_t)$ and the backward dynamical system $m_t = f^{-1}(m_{t+1})$.
10. Both equilibria of the dynamical system $x_{t+1} = 2x_t(1 - x_t)$ are unstable.

Supplementary Problems

1–2 ▪ For the functions shown:
a. Sketch the derivative.
b. Label relative and absolute maxima.
c. Label relative and absolute minima.
d. Find subsets of the domain with positive second derivative.

1.
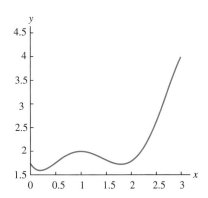

2.
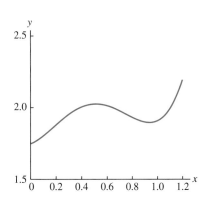

3–4 ▪ Write the tangent line approximation for each function, and estimate the requested values.

3. $f(x) = \dfrac{1+x}{1+e^{3x}}$. Estimate $f(-0.03)$.

4. $g(y) = (1 + 2y)^4 \ln y$. Estimate $g(1.02)$.

5–6 ▪ Sketch graphs of the following functions. Find all critical points, and state whether they are minima or maxima. Find the limit of the function as $x \to \infty$.

5. $(x^2 + 2x)e^{-x}$ for positive x.

6. $\ln x/(1 + x)$ for positive x. Do not solve for the maximum; just show that there must be one.

7–8 ▪ Find the Taylor polynomial of degree 2 approximating each absorption function.

7. $g(x) = \dfrac{1+x}{1+e^x}$ for x near 0.

8. $h(x) = \dfrac{x}{2 - e^x}$ for x near 0.

9–10 ▪ Combine the Taylor polynomials from the previous set of exercises with the leading behaviour of the functions for large x to sketch graphs

9. $g(x) = \dfrac{1+x}{1+e^x}$ (Exercise 7)

10. $h(x) = \dfrac{x}{2 - e^x}$ (Exercise 8)

11. Between days 0 and 150 (measured from November 1), the snow at a certain ski resort is given by

$$S(t) = -\dfrac{1}{4}t^4 + 60t^3 - 4000t^2 + 96{,}000t$$

where S is measured in microns (one micron, μm, is 10^{-4} cm).

a. A yeti tells you that this function has critical points at $t = 20$, $t = 40$, and $t = 120$. Confirm this assertion.

b. Find the absolute maximum and absolute minimum amounts of snow in centimetres.

c. Use the second derivative test to identify the critical points as relative maxima and minima.

d. Sketch the function.

12. Let $r(x)$ be the function giving the per capita production as a function of population size, x:

$$r(x) = \frac{4x}{1 + 3x^2}$$

a. Find the population size that produces the highest per capita production.

b. Find the highest per capita production.

c. Check with the second derivative test.

13. a. Using L'Hôpital's rule, calculate $\lim_{x \to 0} \frac{e^x - 1 - x}{x^2}$.

b. Using L'Hôpital's rule, calculate $\lim_{x \to 0} \frac{e^x - 1 - x - \frac{1}{2}x^2}{x^3}$.

c. Can you guess what $\lim_{x \to 0} \frac{e^x - 1 - x - \frac{1}{2}x^2 - \frac{1}{6}x^3}{x^4}$ is? Confirm your answer using L'Hôpital's rule.

d. Explain how the limits in parts (a) to (c) were constructed and where the answer comes from.

14. Show that $\lim_{x \to \infty} \left(1 + \frac{a}{x}\right)^{bx} = e^{ab}$ for real numbers a and b. Explain how this result illustrates the fact that 1^∞ is an indeterminate form.

15. Suppose the volume of a plant cell follows $V(t) = 1000(1 - e^{-t})$ cubic microns (μm^3) for t measured in days. Suppose the fraction of the cell in a vacuole (a water-filled portion of the cell) is $H(t) = e^t/(1 + e^t)$.

a. Sketch the graph of the total size of the cell as a function of time.

b. Find the volume of the cell outside the vacuole.

c. Find and interpret the derivative of this function. Don't forget the units.

d. Find when the volume of the cell outside the vacuole reaches a maximum.

16. Consider the function

$$F(t) = \frac{\ln(1 + t)}{t + t^2}$$

a. What is $\lim_{t \to 0} F(t)$?

b. What is $\lim_{t \to 0} F'(t)$?

c. What is $\lim_{t \to 0} F''(t)$?

d. Sketch the graph of this function.

17. During Thanksgiving dinner, the table is replenished with food every five minutes. Let F_t represent the fraction of the table laden with food.

$$F_{t+1} = F_t - \text{amount eaten} + \text{amount replenished}$$

Suppose that

$$\text{amount eaten} = \frac{bF_t}{1 + F_t}$$

$$\text{amount replenished} = a(1 - F_t)$$

and that $a = 1$ and $b = 1.5$.

a. Explain the terms that describe the amount eaten and the amount replenished.

b. If the table starts out empty, how much food is there after 5 minutes? How much is there after 10 minutes?

c. Use the quadratic formula to find the equilibria.

d. How much food will there be five minutes after the table is 60% full? Sketch the solution.

18. Let N_t represent the difference between the sodium concentration inside and outside a cell at some time. After one second, the value of N_{t+1} is given by

$$N_{t+1} = \begin{cases} 0.5N_t & \text{if } N_t \leq 2 \\ 4N_t - 7 & \text{if } 2 < N_t \leq 4 \\ -0.25N_t + 10 & \text{if } 4 < N_t \end{cases}$$

a. Graph the updating function and show that it is continuous.

b. Find the equilibria and their stability.

c. Find all initial conditions for which solutions approach $N = 0$.

19. Consider the problem of finding a positive solution of the equation

$$e^x = 2x + 1$$

a. Draw a graph and pick a reasonable starting value.

b. Write the Newton's method recursive formula for this equation.

c. Find your next guess.

d. Show explicitly that the slope of the updating function for this iteration is 0 at the solution.

20. Consider the problem of solving the equation

$$\ln x = \frac{x}{3}$$

a. Convince yourself that there is indeed a solution and find a reasonable guess.

b. Use Newton's method to update your guess twice.

c. What would be a bad choice of an initial guess?

21. Suppose a bee gains an amount of energy

$$F(t) = \frac{3t}{1 + t}$$

after it has been on a flower for time t but that it uses $2t$ energy units in that time (it has to struggle with the flower).

a. Find the net energy gain as a function of t.
b. Find when the net energy gain per flower is a maximum.
c. Suppose the travel time between flowers is $\tau = 1$. Find the time spent on the flower that maximizes the rate of energy gain.
d. Draw a diagram illustrating the results of parts (b) and (c). Why is the answer to part (c) smaller?

22. Consider the function for net energy gain from the previous exercise.

    a. Use the Extreme Value Theorem to show that there must be a maximum.
    b. Use the Intermediate Value Theorem to show that there must be a residence time t that maximizes the rate of energy gain.

23. A peculiar variety of bacterium enhances its own per capita production. In particular, the number of offspring per bacterium increases according to the function

    $$\text{per capita production} = r\left(1 + \frac{b_t}{K}\right)$$

    Suppose that $r = 0.5$ and $K = 10^6$.

    a. Graph per capita production as a function of population size.
    b. Find and graph the updating function for this population.
    c. Find the equilibria.
    d. Find their stability.

24. A type of butterfly has two morphs, a and b. Each type reproduces annually after predation. Twenty percent of type a are eaten, and 10% of type b are eaten. Each type doubles its population when it reproduces. However, the types do not breed true. Only 90% of the offspring of type a are of type a, the rest being of type b. Only 80% of the offspring of type b are of type b, the rest being of type a.

    a. Suppose there are 10,000 of each type before predation and reproduction. Find the number of each type after predation and reproduction.
    b. Find the updating function for types a and b.
    c. Find the updating function for the fraction, p, of type a.
    d. Find the equilibria.
    e. Find their stability.

25. A population of size x_t follows the rule

    $$\text{per capita production} = \frac{4x_t}{1 + 3x_t^2}$$

    a. Find the updating function for this population.
    b. Find the equilibria.
    c. What is the stability of each equilibrium?
    d. Find the equation of the tangent line at each equilibrium.
    e. What is the behaviour of the approximate dynamical system defined by the tangent line at the middle equilibrium?

26. Consider a population following the updating function

    $$N_{t+1} = \frac{rN_t}{1 + N_t^2}$$

    a. What is the per capita production?
    b. Find the equilibrium as a function of r.
    c. Find the stability of the equilibrium as a function of r.
    d. Does this positive equilibrium become unstable as r becomes large?

Projects

1. Consider the following alternative version of the Ricker model for a fishery:

   $$x_{t+1} = rx_t e^{-x_{t-1}}$$

   The idea is that the per capita production this year (year t) is a decreasing function of the population size last year (year $t - 1$).

   a. Discuss why this model might make more sense than the basic Ricker model.
   b. Find the equilibrium of this equation (set $x_{t-1} = x_t = x_{t+1} = x^*$).
   c. Use a computer to study the stability of the equilibrium. Do you have any idea why it might be different from that of the basic Ricker model?
   d. Write a model in which the per capita production depends on the population size two years ago. Find the equilibrium and use a computer to test stability.
   e. Subtract a harvest, h. Find the maximum sustained yield for different values of h. Is the equilibrium still stable?
   f. Models with an extra delay are supposed to be simplified versions of models with two variables. The per capita production is an increasing function of the amount of food available in that year, whereas the amount of food available is a decreasing function of the number of fish the previous year. Try to write a pair of equations describing this situation. Try to find equilibria, and discuss whether your results make sense.

2. Many bees collect both pollen and nectar. Pollen is used for protein, and nectar is used for energy. Suppose the amount of nectar harvested during t seconds on a flower is

   $$F(t) = \frac{t}{1+t}$$

and that the amount of pollen harvested during t seconds on a flower is

$$G(t) = \frac{t}{2+t}$$

The bee collects pollen and nectar simultaneously. Travel time between flowers is $\tau = 1$ second.

a. What is the optimal time for the bee to leave one flower for the next in order to collect nectar at the maximum rate?

b. What is the optimal time for the bee to leave one flower for the next in order to collect pollen at the maximum rate? Why are the two times different?

c. Suppose that the bee values pollen twice as much as nectar. Find a single function, $V(t)$, that gives the value of resources collected by time t. What is the optimal time for the bee to leave? (Solving the equation is easiest with Newton's method.)

d. Suppose that the bee values pollen k times as much as nectar. Compute the optimal time to leave, and graph it as a function of k. Do the values at $k=0$ and the limit as k approaches infinity make sense?

e. Suppose that the bee first collects nectar and then switches to pollen. Assume it spends 1 second collecting nectar. How long should it spend on pollen?

f. Suppose again that the bee values pollen k times as much as nectar. Experiment to try to find a solution that is best when nectar and pollen are harvested sequentially.

3. We know enough to study the solutions of the logistic dynamical system for values of r between 3 and 3.5. Define the updating function to be

$$f(x) = rx(1-x)$$

a. Find the two-step updating function $g = f \circ f$.

b. Graph the function g along with the diagonal for values of r ranging from 2.8 to 3.6.

c. Write the equation for the equilibria of g.

d. Two of the equilibria, $x^* = 0$ and $x^* = 1 - \frac{1}{r}$, match those for f. Why?

e. The terms x and $x - x^*$ factor out of the equilibrium equation $g(x) - x = 0$. Why?

f. Factor $g(x) - x$ and find the other two equilibria, x_1 and x_2. For what values of r do they make sense? What do they mean? (Think about $f(x_1)$ and $f(x_2)$.)

g. For what values of r are x_1 and x_2 stable?

h. Describe the dynamics of f and g in these cases.

i. What happens for values of r just above the point where x_1 and x_2 become unstable?

Chapter 7

Integrals and Applications

Biological systems are constantly changing. Describing this change and deducing its consequences constitute the dynamical approach to the understanding of life. In several chapters of this book, we studied problems that could be treated in discrete time. Given the state of a system (such as a population size) at one time, the discrete-time dynamical system gives the state of the system at a later time. Next, we developed the idea of the derivative, which describes both the *slope* of a graph and the *instantaneous rate of change* of a measurement.

With the derivative, we can study a different kind of dynamical system, the **differential equation**. The instantaneous rate of change of a measurement provides the information needed to deduce the future state of a system from its present state. Because the system can be measured at any time, these systems are a type of **continuous-time dynamical system**.

In order to solve differential equations, we need to know how to reverse differentiation. We study **antiderivatives** from the graphical, numeric, and algebraic viewpoints, developing a variety of techniques that allow us to effectively calculate them. Independently, we define **area** and the **definite integral** as limits of approximating sums (Riemann sums). The **Fundamental Theorem of Calculus** brings these concepts together and provides us with a deeper understanding of the processes of differentiation and integration. Near the end, we study several **applications** of integration, such as **volume, average value,** and **mass**.

7.1 Differential Equations

When we have a measurement, we can differentiate to find the rate of change (Figure 7.1.1). What if we know the rate of change and want to compute the measurement (Figure 7.1.2)? If we know, for example, that the velocity of an object is $v(t)$, a function of t, then the position $p(t)$ must satisfy the **differential equation**

$$\frac{dp(t)}{dt} = v(t)$$

For example,

$$\frac{dp(t)}{dt} = 4 - 2t$$

Sometimes we drop the independent variable from the notation for the function and write

$$\frac{dp}{dt} = 4 - 2t$$

Of course, we will also continue using the prime notation for derivatives:

$$p'(t) = 4 - 2t \quad \text{or} \quad p' = 4 - 2t$$

This equation relates two quantities, saying that one—the velocity, $v(t)$—is the derivative of the other—the position, $p(t)$. Until now, we have usually thought of ourselves as knowing the position and taking the derivative to find the velocity. With a differential equation, we know the velocity and wish to find the position.

Value (measured)	Differentiate	Rate of Change (computed)
position	→	speed
mass	→	rate of change of mass (growth rate)
amount of sodium	→	rate sodium enters a cell
population size	→	rate of change of population size

FIGURE 7.1.1
Finding the rate of change with the derivative

Rate of Change (measured)	Solve Differential Equation	Value (computed)
speed	→	position
rate of change of mass (growth rate)	→	mass
rate sodium enters a cell	→	amount of sodium
rate of change of population size	→	population size

FIGURE 7.1.2
Finding a measurement from the rate of change

Consider another example: according to Statistics Canada, the population of Canada in 2011 was 33,477 thousand (33.477 million). We cannot predict what the population will be in 2021 *unless we know how the population changes* (i.e., what the rate of change of the population is).

Denote by $P(t)$ the total population of Canada in thousands and assume that the *relative rate of change* is 0.0092. "Relative rate of change" means that every group of 1000 Canadians will (through births and immigration) increase to about 1009 Canadians in a year.

The derivative, $P'(t)$, is the rate of change. Dividing by $P(t)$, we obtain the relative rate of change

$$\frac{P'(t)}{P(t)} = 0.0092$$

and so

$$P'(t) = 0.0092 P(t)$$

We will learn how to solve this differential equation, i.e., how to obtain a formula for $P(t)$. That formula will allow us to calculate a prediction for the population of Canada in 2021.

There are situations where measuring the rate of change of a quantity is easier or more practical than measuring the quantity directly. For instance, it might be easier to measure how many ions enter and leave a cell each second by measuring changes in electrical charge than to track down and count every sodium ion floating around in a cell.

There are several types of differential equations. The simplest is of the form

$$y' = G(x)$$

where y is the unknown function, x is the independent variable, and $G(x)$ is a given function of x. For example,

$$y' = 3 - 2\ln x$$

(here $G(x) = 3 - 2\ln x$) and

$$\frac{dy}{dt} = \sin t - 2\cos t$$

in which case $G(t) = \sin t - 2\cos t$, are of this form.

We obtain this form of differential equation in applications when we measure the rate of change. We write

derivative of unknown quantity (function) = measured rate of change

By "measured" we mean to say that the rate of change is a function of the independent variable (in applications, it is usually the time, t). Such a differential equation is called a **pure-time differential equation.**

The general form of a pure-time differential equation is

$$\frac{df(t)}{dt} = G(t)$$

where $f(t)$ is the unknown function and $G(t)$ is the measured rate of change. Nothing but the independent variable appears on the right side. For example,

$$\frac{df(t)}{dt} = e^t$$

and

$$g'(t) = 3t + \sin t$$

and the equation from the start of the section,

$$\frac{dp(t)}{dt} = 4 - 2t$$

are pure-time differential equations. Even when we use another symbol for the independent variable, such as x in

$$\frac{df(x)}{dx} = G(x)$$

we still call it a pure-time differential equation.

There is another way to arrive at a differential equation. On the basis of biological principles, we might know a *rule* describing how a measurement changes, just as we did when deriving discrete-time dynamical systems. For example, when resources are not limiting, the rate of population growth is proportional to population size. In these cases, we can write down an **autonomous differential equation:**

derivative of unknown quantity = some function of unknown quantity

In symbols, an autonomous differential equation is of the form

$$\frac{df(t)}{dt} = \text{function of } f(t)$$

i.e., the independent variable does not appear explicitly on the right side. Examples of autonomous differential equations are

$$\frac{df(t)}{dt} = 3f(t) - 1.2$$

$$\frac{df}{dt} = \sin(f) + f^2$$

(using either f or $f(t)$ is common practice), and the population of Canada model,

$$P'(t) = 0.0092 P(t)$$

When we are not discussing applications, we usually use x and y to denote the independent variable and the function. The equations

$$y' = 3y^2 - 4$$

and

$$\frac{dy}{dx} = \sin y$$

are autonomous differential equations.

Some differential equations are neither pure-time nor autonomous, such as

$$\frac{df}{dt} = t^3 f - 4$$

and

$$f'(t) = t + f(t)$$

and

$$y' = 2xy + y^2$$

A differential equation that has a constant on the right side, such as

$$f'(t) = 2$$

can be interpreted as either pure-time or autonomous.

No matter what the form, a differential equation gives information about the rate of change of an unknown function. To solve a differential equation means to find that function. The methods that we use (in this chapter and the next) to find solution(s) depend on the form of the equation.

A function can be represented algebraically (formula), geometrically (graph), or numerically (table of values). We will learn how to obtain solutions of differential equations in all three representations.

Differential Equations: Examples and Terminology

Suppose that 1 μm^3 of water enters a cell each second. The volume, V, of the cell is thus increasing by 1 $\mu m^3/s$. Mathematically, the rate of change of V is 1 $\mu m^3/s$, or

$$\frac{dV}{dt} = 1 \tag{7.1.1}$$

The quantity being differentiated (V in this case) is called the **unknown function** or the **state variable.** Because we *measured* the rate of change, this is a **pure-time differential equation.**

Our goal is to find the volume, V, of the cell as a function of time. Such a function is called a **solution.** In this case, we try by *guessing*. What function has a constant rate of change equal to 1? Geometrically, what function has a constant slope of 1? The answer is a line with slope 1. One such line is

$$V(t) = t$$

(Figure 7.1.3). We can check whether this guess solves the differential equation by taking the derivative:

$$\frac{dV}{dt} = 1$$

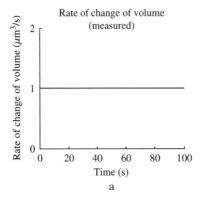
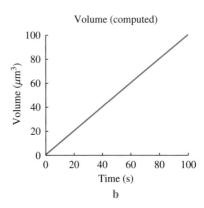

FIGURE 7.1.3

A function that solves the differential equation $\frac{dV}{dt} = 1$

Suppose we know that the volume of the cell at time $t = 0$ is 300 μm^3. Although the guess $V(t) = t$ is one solution of the differential equation, it does not match this **initial condition** since $V(0) = 0$. As with discrete-time dynamical systems, a solution also depends on where something starts. From our knowledge of lines, we realize that any function $V(t)$ of the form

$$V(t) = t + C$$

where C is a real number, has a constant slope of 1. We find the derivative to check this: $V'(t) = (t + C)' = 1$. The value of C can be chosen to match the initial condition. In this case, because

$$V(0) = 300 = 0 + C$$

we know that $C = 300$. The solution of the differential equation

$$\frac{dV}{dt} = 1$$

with the initial condition $V(0) = 300$ is

$$V(t) = t + 300$$

(Figure 7.1.4). From knowledge of the initial condition and a measurement of the rate of change, we have a formula giving the volume at any time. For instance, when $t = 50$, the volume is 350 μm^3.

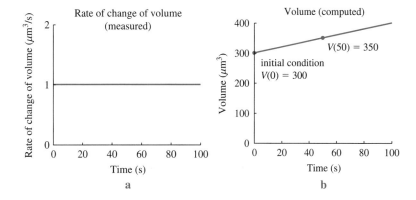

FIGURE 7.1.4
Volume as a function of time

A function f is a **solution** of a differential equation if the equation is true when f and f' are substituted into it. We illustrate this concept in our next example.

Example 7.1.1 Solution of a Differential Equation

Show that the function

$$y = \frac{\ln x + 6}{x}$$

is a solution of the differential equation

$$y' = \frac{1 - xy}{x^2}$$

with the initial condition $y(1) = 6$.

Using the quotient rule to compute the derivative of y, we get

$$y' = \frac{x\frac{1}{x} - (\ln x + 6)}{x^2} = \frac{1 - \ln x - 6}{x^2} = \frac{-\ln x - 5}{x^2}$$

The right side is equal to

$$\frac{1-xy}{x^2} = \frac{1-x\frac{\ln x+6}{x}}{x^2} = \frac{1-\ln x - 6}{x^2} = \frac{-\ln x - 5}{x^2}$$

Thus, the two sides of the given differential equation are identical. Furthermore,

$$y(1) = \frac{\ln 1 + 6}{1} = 6$$

and therefore $y = \frac{\ln x + 6}{x}$ is a solution of the given equation that satisfies the given initial condition $y(1) = 6$.

A differential equation, together with an initial condition, is called an **initial value problem** (often abbreviated as IVP). To solve an IVP means to find a function that satisfies the given differential equation *and* the initial condition.

Example 7.1.2 Solution of an Initial Value Problem

Prove that the function

$$P(t) = M - Me^{-kt}$$

is a solution of the initial value problem

$$P'(t) = k[M - P(t)], P(0) = 0$$

where k and M are constants.

We substitute $P(t) = M - Me^{-kt}$ into both sides of the equation to show that they are equal:

$$P'(t) = (M - Me^{-kt})' = 0 - Me^{-kt}(-k) = kMe^{-kt}$$

and

$$k[M - P(t)] = k[M - (M - Me^{-kt})] = kMe^{-kt}$$

To check that $P(t)$ satisfies the initial condition, we compute

$$P(0) = M - Me^{-k(0)} = M - M = 0$$

Example 7.1.3 A Pure-Time Differential Equation for Chemical Production

By $p(t)$ we denote the quantity of some chemical that is produced (say, in the human body). Assume that the *rate* of production of $p(t)$ is e^{-t}, where t is time in minutes. In other words, $p(t)$ satisfies the pure-time differential equation

$$p'(t) = e^{-t}$$

What can we say about $p(t)$? Since $p'(t) > 0$ for all t (recall that the exponential function is always positive), we conclude that $p(t)$ is an increasing function. Since e^{-t} is decreasing, it follows that the *rate at which $p(t)$ increases* is decreasing. Thus, $p(t)$ increases more and more slowly.

Alternatively, we compute the derivative of $p'(t)$ to get

$$p''(t) = -e^{-t}$$

Since $p''(t)$ is negative, we conclude that $p(t)$ is concave down for all t.

Let's try to find a formula for $p(t)$. Suppose that, initially, there was no chemical present, so that $p(0) = 0$. By guessing, we see that if

$$p(t) = -e^{-t} + C$$

(C is an arbitrary constant), then

$$p'(t) = -(-e^{-t}) + 0 = e^{-t}$$

exactly as needed. Substituting $t=0$ and $p(0)=0$ into $p(t)=-e^{-t}+C$, we find
$$0=-e^{-0}+C$$
and so $C=1$. Thus,
$$p(t)=-e^{-t}+1=1-e^{-t}$$
The rate, $p'(t)$, and the solution, $p(t)$, are shown in Figure 7.1.5. ▲

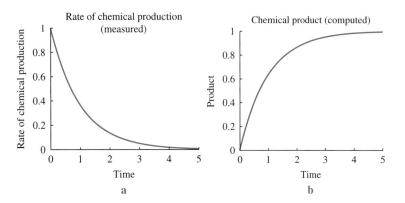

FIGURE 7.1.5

Product as a function of time

Example 7.1.4 Solving an Autonomous Differential Equation

Consider the differential equation from the population of Canada model mentioned earlier in this section:
$$P'(t)=0.0092P(t)$$

It is an autonomous differential equation, since the function $P(t)$ appears on the right side.

By t we denote the time in years, measured from 2011 (thus, $t=0$ represents the year 2011, $t=1$ is the year 2012, and so forth). The population of 33,477 thousand in 2011 is the initial condition:
$$P(0)=33{,}477$$

Before we find a formula for $P(t)$, let us think a bit about what $P(t)$ should look like. Because $P'(t)$ is positive, we see that $P(t)$ must be an increasing function. Moreover,
$$\begin{aligned}P''(t)&=0.0092P'(t)\\&=0.0092(0.0092P(t))\\&=0.0092^2P(t)>0\end{aligned}$$

so $P(t)$ must be concave up. (We could have reached the same conclusion in a slightly different way: we know that $P(t)$ is increasing, and therefore $0.0092P(t)$ is increasing as well; i.e., $P'(t)=0.0092P(t)$ is increasing. Thus, the rate at which $P(t)$ increases is getting larger and larger.)

Now we look for a formula for $P(t)$. The equation
$$P'(t)=0.0092P(t)$$

means that we are looking for a function whose derivative is 0.0092 times the function. We know that the exponential function $f(t)=Ce^t$ is equal to its derivative:
$$f'(t)=(Ce^t)'=Ce^t=f(t)$$

(C can be any real number). However, if $f(t)=Ce^{5t}$, then
$$f'(t)=(Ce^{5t})'=Ce^{5t}(5)=5Ce^{5t}=5f(t)$$

i.e., the derivative is five times the original function. Thus, if we adjust the exponent, we will get what we need—if

$$P(t) = Ce^{0.0092t}$$

then

$$P'(t) = C(0.0092)e^{0.0092t}$$
$$= 0.0092 \cdot Ce^{0.0092t}$$
$$= 0.0092 \cdot P(t)$$

so $P(t) = Ce^{0.0092t}$ satisfies the given equation—we have found the solution! To compute C, we use the initial condition:

$$P(t) = Ce^{0.0092t}$$
$$P(0) = Ce^{0.0092(0)}$$
$$33{,}477 = Ce^{0}$$
$$33{,}477 = C$$

The solution to the given differential equation is

$$P(t) = 33{,}477e^{0.0092t}$$

Using the solution, we can now calculate our prediction for the Canadian population in 2021:

$$P(\text{in year } 2021) = P(t = 10)$$
$$= 33{,}477e^{0.0092(10)}$$
$$= 33{,}477e^{0.092} \approx 36{,}703.01$$

i.e., about 36.7 million.

In the next section, we will start solving differential equations algebraically using **antiderivatives** (**indefinite integrals**). We first focus on differential equations of the form

$$f'(x) = \text{function of } x \text{ only}$$

(i.e., pure-time differential equations). In Chapter 8, we will learn how to solve other types of equations.

In the remaining part of this section, we will discuss solving differential equations graphically and numerically.

Graphical Solution of Pure-Time Differential Equations

By checking whether the graph of a function is increasing or decreasing, we can sketch the graph of the derivative. Solving a pure-time differential equation reverses this process. From the graph of the derivative, we wish to sketch the graph of the function itself.

The guideposts for this process are the reverse of those for reading and interpreting a graph, as summarized in the following table:

What the Derivative Does	What the Graph Does
derivative positive	graph increasing
derivative negative	graph decreasing
derivative zero	graph horizontal
derivative increasing	graph concave up
derivative decreasing	graph concave down

Example 7.1.5 Graphical Solution of a Pure-Time Differential Equation I

Suppose we wish to solve the pure-time differential equation

$$\frac{dp}{dt} = 4 - 2t$$

with initial condition $p(0) = 10$. The rate of change function, $F(t) = 4 - 2t$, is positive for $t < 2$, is zero at $t = 2$, and is negative for $t > 2$ (Figure 7.1.6a). The solution begins at $p(0) = 10$, increases until $t = 2$, has a derivative of zero at $t = 2$, and decreases thereafter (Figure 7.1.6b). Since p' is decreasing (and thus $p'' < 0$), the graph of $p(t)$ must be concave down. The graph of $p'(t)$ is a line, and therefore the graph of $p(t)$ is a parabola.

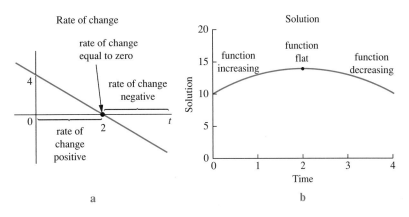

FIGURE 7.1.6

Graphical solution of a pure-time differential equation

Example 7.1.6 Graphical Solution of a Pure-Time Differential Equation II

Suppose we wish to find a measurement $M(t)$ with the rate of change given in Figure 7.1.7a, knowing that $M(5) = 10$.

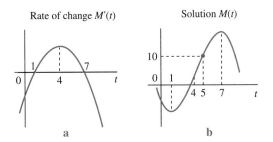

FIGURE 7.1.7

Finding a solution from the graph of the rate of change

Looking at the graph in Figure 7.1.7a, we see that the rate of change of $M(t)$ is positive when $1 < t < 7$ and negative when $t < 1$ and $t > 7$. Thus, $M(t)$ is decreasing on $(-\infty, 1)$, increasing on $(1, 7)$, and then decreasing again on $(7, \infty)$. We conclude that $M(1)$ is a relative minimum and $M(7)$ is a relative maximum.

To satisfy the initial condition, the graph of $M(t)$ must go through $(5, 10)$.

Furthermore, the derivative $M'(t)$ is increasing on $(-\infty, 4)$ and decreasing on $(4, \infty)$. Thus, the graph of $M(t)$ is concave up on $(-\infty, 4)$ and concave down on $(4, \infty)$. It has an inflection point at $t = 4$.

As well, $M'(t)$ is largest at $t = 4$, which means that $M(t)$ increases most rapidly around $t = 4$.

One possible graph of $M(t)$ is drawn in Figure 7.1.7b.

Euler's Method for Solving Differential Equations

How do we solve a differential equation? As we have just seen, we can sketch solutions of pure-time differential equations using our understanding of the derivative. In the next section, we will learn how to use **integration** to solve pure-time differential equations. Even that method, however, does not work for all equations. An alternative approach, called **Euler's method,** is a numerical approach that can be implemented on the computer. Like Newton's method (Section 6.6), Euler's method begins with the *tangent line approximation,* replacing a problem about *curves* with a problem about *lines.*

Example 7.1.7 Applying the Tangent Line Approximation to a Pure-Time Differential Equation

Consider the differential equation

$$f'(t) = 2t + 1$$

with initial condition $f(0) = 5$. Thinking about a numerical method, how can we use this information to find an approximate solution for $f(t)$?

Note that besides $f(0) = 5$ we also know (from the given equation) that $f'(0) = 1$. So we can compute the linear approximation (tangent) of $f(t)$ at $t = 0$:

$$\begin{aligned} L_1(t) &= f(0) + f'(0)(t - 0) \\ &= 5 + 1(t - 0) \\ &= 5 + t \end{aligned}$$

The function $L_1(t) = 5 + t$ is an approximation of the solution $f(t)$ we are looking for! Near $t = 0$, $L_1(t)$ approximates $f(t)$ well, but as we move away from $t = 0$, we can no longer be sure about it (actually it's very likely that $f(t)$ gets farther and farther away from $L_1(t)$); see Figure 7.1.8. So we cannot accept $L_1(t)$ as an approximation of $f(t)$ *for all t*—we must do better.

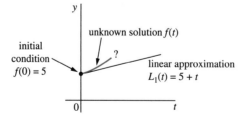

FIGURE 7.1.8

Linear approximation to the solution of the given differential equation

The idea (introduced by Swiss mathematician Leonhard Euler) is to start at the initial condition, $(0, 5)$, and walk along the tangent for some short time Δt (Δt is called the step size) and then stop, recalculate the tangent, and start walking in the direction of that new tangent for the same time, Δt. We then repeat this process for as many steps as needed.

Let's take the step size $\Delta t = 0.2$. So the t-coordinate of the point where we stop is 0.2. Since

$$L_1(0.2) = 5 + 0.2 = 5.2$$

the coordinates of the point at which we arrived are $(0.2, 5.2)$; see Figure 7.1.9a.

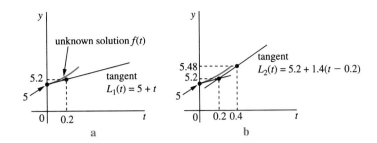

FIGURE 7.1.9

Euler's method, first two steps

Now we construct the linear approximation (tangent) at (0.2, 5.2). We know that this point is very likely *not on the graph of $f(t)$*, but somewhere near it (as shown in Figure 7.1.9a). Nevertheless, we use the given equation, $f'(t) = 2t + 1$, and take

$$f'(0.2) = 2(0.2) + 1 = 1.4$$

to be the slope of the tangent, and compute its equation

$$L_2(t) = f(0.2) + f'(0.2)(t - 0.2)$$
$$= 5.2 + 1.4(t - 0.2)$$

Now that we have corrected the direction, we walk along this new tangent for $\Delta t = 0.2$, arriving at the point whose coordinates are $(0.4, L_2(0.4))$. Since

$$L_2(0.4) = 5.2 + 1.4(0.4 - 0.2)$$
$$= 5.2 + 0.28 = 5.48$$

that point is (0.4, 5.48); Figure 7.1.9b.

Continuing this process, we arrive at the sequence of approximations shown in Table 7.1.1. This table is a table of values for our *approximation* of the solution, $f(t)$.

Table 7.1.1

t	Approximation of $f(t)$
0.2	5.2
0.4	5.48
0.6	5.84
0.8	6.28
1	6.80

In Figure 7.1.10a, we show the points that we calculated (so our approximation of the solution is not a curve, but a sequence of points). How close is our approximation to the actual solution? In Figure 7.1.10b, we compare our values with the graph of $f(t)$. By guessing, or by using the formulas introduced in the next section, we can show that $f(t) = t^2 + t + 5$, and so its actual value at $t = 1$ is $f(1) = 7$. Our approximation for $f(1)$ is 6.80.

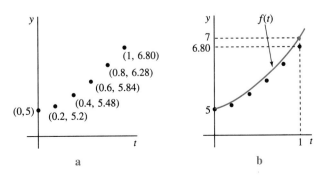

FIGURE 7.1.10

Euler's method compared with actual solution

Errors accumulate—the more steps we do, the less accurate our estimate will be. Running the above algorithm for 50 steps (so we move 50 times the step size of $\Delta t = 0.2$, i.e., 10 units away from 0), we obtain the approximation of 113 for the actual value $f(10) = 115$.

To improve our approximation of $f(1)$, we decrease the step size: let $\Delta t = 0.1$ (thus, we will compute the approximations for $f(t)$ at $t = 0.1, 0.2, 0.3, 0.4$, etc.).

Recall that the tangent at $(0, 5)$ was computed to be $L_1(t) = 5 + t$. Following it for $\Delta t = 0.1$, we arrive at the point

$$(0.1, L_1(0.1)) = (0.1, 5.1)$$

This is where we correct our direction. The tangent at $(0.1, 5.1)$ is

$$L_2(t) = f(0.1) + f'(0.1)(t - 0.1)$$
$$= 5.1 + 1.2(t - 0.1)$$

(since $f'(0.1) = 2(0.1) + 1 = 1.2$). Computing

$$L_2(0.2) = 5.1 + 1.2(0.2 - 0.1) = 5.22$$

we obtain the point $(0.2, 5.22)$, which is an approximation for the value of $f(t)$ at $t = 0.2$. The first 10 steps in Euler's method are shown in Table 7.1.2.

Table 7.1.2

t	Approximation of $f(t)$	t	Approximation of $f(t)$
0.1	5.1	0.6	5.90
0.2	5.22	0.7	6.12
0.3	5.36	0.8	6.36
0.4	5.52	0.9	6.62
0.5	5.70	1	6.90

Note that we managed to improve our estimate. This time, our approximation of 6.90 is closer to the true value $f(1) = 7$ than it was before.

Decreasing the step size makes our approximations more accurate. In Table 7.1.3, we show several approximations for the true value $f(1) = 7$ based on running Euler's method with different step sizes.

Table 7.1.3

Step size, Δt	Estimate for $f(1)$
0.5	6.50
0.25	6.75
0.2	6.80
0.1	6.90
0.05	6.95
0.01	6.99
0.001	6.999

Of course, the increase in accuracy comes with a price—the calculations become longer and longer. To calculate the estimates in Table 7.1.3, we needed 10 steps when $\Delta t = 0.1$, 100 steps when $\Delta t = 0.01$, and 1000 steps when $\Delta t = 0.001$.

If we have a few steps to do, we can do it by hand, using a calculator. However, for any calculation that requires a larger number of steps, we need a programmable calculator or math software (such as Maple).

By generalizing the process that we described in the example, we now derive a recursive formula (i.e., a discrete-time dynamical system) for Euler's method.

Assume that we wish to find a numeric approximation of the solution of the differential equation

$$f'(t) = G(t)$$

with initial condition $f(t_0) = y_0$. (In the example we just finished, $G(t) = 2t + 1$, $t_0 = 0$, and $y_0 = 5$.)

Choose a time step, Δt. Thus, we will approximate the values of $f(t)$ at

$$t_1 = t_0 + \Delta t$$
$$t_2 = t_1 + \Delta t = (t_0 + \Delta t) + \Delta t = t_0 + 2\Delta t$$
$$t_3 = t_2 + \Delta t = (t_0 + 2\Delta t) + \Delta t = t_0 + 3\Delta t$$

and so on. We label the corresponding approximations of $f(t)$ by y_1, y_2, y_3, and so forth (thus y_n will approximate $f(t_n)$).

In order to describe the dynamical system, we need to calculate one step. Suppose that we calculated the approximation y_n for t_n; i.e., we obtained the point (t_n, y_n) that approximates the value of the graph of $f(t)$ at $t = t_n$; see Figure 7.1.11. How do we obtain the next point, (t_{n+1}, y_{n+1})?

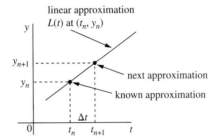

FIGURE 7.1.11
Euler's method, general step

Clearly, $t_{n+1} = t_n + \Delta t$. The equation of the tangent line at (t_n, y_n) is

$$L(t) = y_n + f'(t_n)(t - t_n)$$
$$= y_n + G(t_n)(t - t_n)$$

The value

$$L(t_{n+1}) = y_n + G(t_n)(t_{n+1} - t_n)$$
$$= y_n + G(t_n)\Delta t$$

is the desired approximation y_{n+1} of $f(t_{n+1})$. Thus,

$$y_{n+1} = y_n + G(t_n)\Delta t$$

▸▸ **Algorithm 7.1.1** Euler's Method for Solving a Pure-Time Differential Equation

Assume that we need to solve the differential equation

$$f'(t) = G(t)$$

with initial condition $f(t_0) = y_0$.

1. Choose a time step Δt (the length of time between approximations).

2. Use the formulas

$$t_{n+1} = t_n + \Delta t$$
$$y_{n+1} = y_n + G(t_n)\Delta t$$

to find approximations $y_1, y_2, y_3, \ldots$ of the solution $f(t)$ at $t_1, t_2, t_3, \ldots$.

Example 7.1.8 Using Euler's Method

Consider the differential equation

$$f'(t) = \frac{1}{3t-4}$$

with initial condition $f(2) = 1$. We use Euler's method with step size $\Delta t = 0.3$ and 10 steps to construct a table of values for an approximation of the solution, $f(t)$, of the given differential equation.

In our case, $t_0 = 2$, $y_0 = 1$, and $G(t) = \frac{1}{3t-4}$. Using the Euler's method formulas, we compute

$$t_1 = t_0 + \Delta t = 2 + 0.3 = 2.3$$
$$y_1 = y_0 + G(t_0)\Delta t$$
$$= 1 + \frac{1}{3(2)-4}(0.3)$$
$$= 1 + \frac{1}{2}(0.3) = 1.15$$

So we obtain the approximation $y_1 = 1.15$ for the value $f(t_1) = f(2.3)$. The next step is

$$t_2 = t_1 + \Delta t = 2.3 + 0.3 = 2.6$$
$$y_2 = y_1 + G(t_1)\Delta t$$
$$= 1.15 + \frac{1}{3(2.3)-4}0.3$$
$$= 1.15 + \frac{1}{2.9}0.3 \approx 1.25345$$

Proceeding in the same way, we calculate the values in Table 7.1.4.

Table 7.1.4

t	Approximation of $f(t)$	t	Approximation of $f(t)$
$t_0 = 2$	$y_0 = 1$		
$t_1 = 2.3$	$y_1 = 1.15$	$t_6 = 3.8$	$y_6 = 1.49595$
$t_2 = 2.6$	$y_2 = 1.25345$	$t_7 = 4.1$	$y_7 = 1.53649$
$t_3 = 2.9$	$y_3 = 1.33240$	$t_8 = 4.4$	$y_8 = 1.57264$
$t_4 = 3.2$	$y_4 = 1.39623$	$t_9 = 4.7$	$y_9 = 1.60524$
$t_5 = 3.5$	$y_5 = 1.44980$	$t_{10} = 5$	$y_{10} = 1.63495$

Euler's method gives the approximation $y_{10} = 1.63495$ for $f(5)$. For the record, a very close approximation is $f(5) \approx 1.56825$.

Example 7.1.9 Applying Euler's Method to a Differential Equation for Chemical Production

Consider the chemical production differential equation

$$p'(t) = e^{-t}$$

with initial condition $p(0) = 0$ from Example 7.1.3. Recall that by $G(t)$ we denote the right side of the equation, so $G(t) = e^{-t}$. By guessing, we obtained the exact solution

$$p(t) = 1 - e^{-t}$$

Using Euler's method, we now compute an approximation for $p(3)$ and compare it with the true value,

$$p(3) = 1 - e^{-3} \approx 0.95021$$

Take the step size to be $\Delta t = 0.25$. We are given $t_0 = 0$, $y_0 = 0$, and $G(t) = e^{-t}$, and so the formulas for Euler's method are

$$t_{n+1} = t_n + \Delta t = t_n + 0.25$$
$$\begin{aligned} y_{n+1} &= y_n + G(t_n)\Delta t \\ &= y_n + e^{-t_n}\Delta t \\ &= y_n + 0.25e^{-t_n} \end{aligned}$$

Starting with $t_0 = 0$ and $y_0 = 0$, we compute the first step:

$$t_1 = t_0 + 0.25 = 0.25$$
$$\begin{aligned} y_1 &= y_0 + 0.25e^{-t_0} \\ &= 0 + 0.25e^0 = 0.25 \end{aligned}$$

The next step is

$$t_2 = t_1 + 0.25 = 0.5$$
$$\begin{aligned} y_2 &= y_1 + 0.25e^{-t_1} \\ &= 0.25 + 0.25e^{-0.25} \approx 0.44470 \end{aligned}$$

To reach $t = 3$ using a step size of $\Delta t = 0.25$, we need 12 steps. The calculations are shown in Table 7.1.5.

Table 7.1.5

t	Approximation of $p(t)$	t	Approximation of $p(t)$
$t_0 = 0$	$y_0 = 1$		
$t_1 = 0.25$	$y_1 = 0.25$	$t_7 = 1.75$	$y_7 = 0.93380$
$t_2 = 0.5$	$y_2 = 0.44470$	$t_8 = 2$	$y_8 = 0.97725$
$t_3 = 0.75$	$y_3 = 0.59633$	$t_9 = 2.25$	$y_9 = 1.01108$
$t_4 = 1$	$y_4 = 0.71442$	$t_{10} = 2.5$	$y_{10} = 1.03743$
$t_5 = 1.25$	$y_5 = 0.80639$	$t_{11} = 2.75$	$y_{11} = 1.05795$
$t_6 = 1.5$	$y_6 = 0.87802$	$t_{12} = 3$	$y_{12} = 1.07393$

Euler's method gives $p(3) \approx y_{12} = 1.07393$, which is an overestimate of the true value of $p(3)$. Had we used a step size of $\Delta t = 0.1$, we would have obtained $p(3) \approx 0.99852$.

Euler's method applies not only to pure-time differential equations but also to a much broader class of differential equations. For instance, we can use it with autonomous differential equations, as our next example illustrates. We will explore this further in Chapter 8.

Example 7.1.10 Applying Euler's Method to the Population of Canada Model

To model the population of Canada, we used the differential equation $P'(t) = 0.0092P(t)$ with initial condition $P(0) = 33,477$ (Example 7.1.4). Recall that $P(t)$ is measured in thousands and the time is scaled so that $t = 0$ represents 2011. We now use Euler's method with $\Delta t = 1$ to find an approximation for $P(10)$, the population of Canada in 2021.

As usual, by p_0, p_1, p_2, etc., we denote the sequence of approximations of $P(t)$ at times t_0, t_1, t_2, etc. (because the step size is $\Delta t = 1$, each approximation advances one year into the future).

In writing the recursive formula for Euler's method, we keep in mind that this time the right side of the equation is $0.0092P(t)$ (so we replace $P(t)$ by p_n); thus,

$$t_{n+1} = t_n + \Delta t = t_n + 1$$
$$p_{n+1} = p_n + 0.0092 p_n \Delta t = 1.0092 p_n$$

where $t_0 = 0$ and $p_0 = P(0) = 33{,}477$.

As a matter of fact, there is no need to run the Euler's method algorithm for 10 steps. We got lucky—the iteration formula $p_{n+1} = 1.0092 p_n$ has solution $p_n = p_0 1.0092^t = 33{,}477 \cdot 1.0092^t$ (see Basic Exponential Discrete-Time Dynamical System in Section 3.1). Thus, the approximation for the population in 2021 is

$$P(10) \approx p_{10} = 33{,}477 \cdot 1.0092^{10} \approx 36{,}687.57$$

i.e., about 36.7 million.

Summary A **differential equation** expresses the rate of change of a quantity, the **state variable**, as a function of time or of the quantity itself. If the rate of change has been measured as a function of time, the equation is a **pure-time differential equation.** If the rate of change has been derived from a rule, the equation is an **autonomous** differential equation. A **solution** gives the value of the quantity as a function of the independent variable. The solution depends on the **initial condition. Euler's method,** which is based on the tangent line approximation, can be used to convert a differential equation into a discrete-time dynamical system.

7.1 Exercises

Mathematical Techniques

1–9 ■ Identify each of the following as a pure-time differential equation, an autonomous differential equation, or neither.

1. $\dfrac{dx}{dt} = t$
2. $\dfrac{dy}{dt} = 2y$
3. $\dfrac{dw}{dt} = \dfrac{2}{1+t}$
4. $y' = 2xe^x$
5. $y' = 2x + y$
6. $\dfrac{df}{dx} = \ln x + x - 1$
7. $\dfrac{df(t)}{dt} = 3f(t) - 4$
8. $\dfrac{df(t)}{dt} = 3t^3 f(t)$
9. $\dfrac{dz}{dt} = 2\sqrt{z}$

10–13 ■ Use the graph of the rate of change to sketch the graph of the function, starting from the given initial condition.

10. Start from $x(0) = 1$.

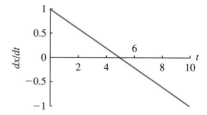

11. Start from $y(0) = -1$.

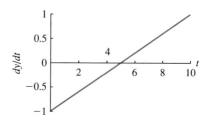

12. Start from $z(0) = 2$.

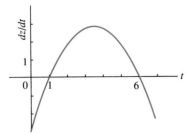

13. Start from $w(0) = 1$.

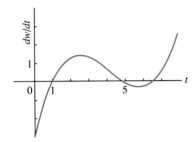

14. Consider the differential equation $y' = y^2$. Prove that the solution $y = y(x)$ is an increasing function if $y(x) \neq 0$. Discuss how the concavity of the graph of y depends on the values of y.

15. Consider the differential equation $y' = y^3 - y$.
    a. Find all constant solutions (i.e., solutions that are constant functions).
    b. Identify the values of y for which y is increasing.
    c. Identify the values of y for which y is decreasing.

16-19 ▪ Check that each of the following is a solution of the given differential equation. What was the initial condition (the value of the state variable at $t = 0$)?

16. $\dfrac{dx}{dt} = t$ has solution $x(t) = 1 + \dfrac{t^2}{2}$.

17. $\dfrac{dw}{dt} = \dfrac{2}{1+t}$ has solution $w(t) = 2\ln(1+t) + 3$.

18. $\dfrac{dy}{dt} = 2y$ has solution $y(t) = 4e^{2t}$.

19. $\dfrac{dz}{dt} = 2\sqrt{z}$ has solution $z(t) = (t+2)^2$.

20-23 ▪ Check that each of the following is a solution of the given differential equation.

20. $y' = \dfrac{1}{2}y^2 - \dfrac{1}{2}$ has solution $y = \dfrac{1+e^x}{1-e^x}$.

21. $y' = \dfrac{y+1}{x}$ has solution $y = x - 1$.

22. $y' = \dfrac{y}{ax}$ has solution $y = x^{1/a}$.

23. $y' = 2(y-3)$ has solution $y = 3 + e^{2x}$.

24. Show that $y = 2 + 3e^{-x^3}$ is a solution of the initial value problem $y' = 6x^2 - 3x^2 y$ with $y(0) = 5$.

25. Show that $f(t) = \sqrt{t^2 + 3}$ is a solution of the initial value problem $f'(t) = \dfrac{t}{f(t)}$ with $f(1) = 2$.

26. Prove that $y(x) = e^{1-\sqrt{1+x^2}} - 1$ is a solution of the IVP $y'(x) = -\dfrac{x(1+y)}{\sqrt{1+x^2}}$ with $y(0) = 0$.

27. Verify that $P(t) = \dfrac{7}{1 - 7\ln t}$ is a solution of the initial value problem $\dfrac{dP}{dt} = \dfrac{P^2}{t}$, $t > 0$, $P(1) = 7$.

28-29 ▪ Apply Euler's method to each differential equation to estimate the solution at $t = 1$ starting from the given initial condition. First use one step with $\Delta t = 1$, and then use two steps with $\Delta t = 0.5$. Compare with the exact result from the earlier problem.

28. $\dfrac{dx}{dt} = t$ with initial condition $x(0) = 1$ (as in Exercise 16).

29. $\dfrac{dw}{dt} = \dfrac{2}{1+t}$ with initial condition $w(0) = 3$ (as in Exercise 17).

30. Consider the differential equation $y' = 2(t - 3)$ with initial condition $y(0) = 3$. Use Euler's method with step size $\Delta t = 0.4$ to estimate the value, $y(2)$, of the solution at $t = 2$.

31. Consider the differential equation $y' = \dfrac{1}{2t}$ with initial condition $y(1) = 1$. Use Euler's method with step size $\Delta t = 0.25$ to estimate the value, $y(2)$, of the solution at $t = 2$.

32. Consider the initial value problem $c'(t) = e^{-0.8t^2}$ with $c(0) = 2.4$. Using Euler's method with step size $\Delta t = 0.25$, estimate the value, $c(1)$, of the solution at $t = 1$.

33. Consider the initial value problem $c'(t) = (0.2 + 0.1t^3)^{-1}$ with $c(3) = 1$. Using Euler's method with step size $\Delta t = 0.2$, estimate the value, $c(4)$, of the solution at $t = 4$.

Applications

34-37 ▪ For each description of the volume of a cell, write a differential equation, find and graph the solution, and indicate whether the solution makes sense for all time.

34. A cell starts at a volume of 600 μm^3 and loses volume at a rate of 2 $\mu m^3/s$.

35. A cell starts at a volume of 400 μm^3 and gains volume at a rate of 3 $\mu m^3/s$.

36. A cell starts at a volume of 900 μm^3 and loses volume at a rate of $2t$ micrometres per second.

37. A cell starts at a volume of 1000 μm^3 and loses volume at a rate of $3t^2$ micrometres per second.

38. A population, $P(t)$, of caribou is modelled by the autonomous differential equation $P'(t) = 2P(t)\left(1 - \dfrac{P(t)}{2500}\right)$, $P(t) > 0$.
    a. Find a constant solution (i.e., a solution that is a constant function).
    b. Show that the population is decreasing if $P(t)$ is larger than 2500.
    c. Show that the population is increasing if $P(t)$ is smaller than 2500.

d. Interpret your answers. What is the meaning of the number 2500 for the population of caribou?

39. A population, $P(t)$, of salmon in a river is modelled by the autonomous differential equation $P'(t) = 0.7P(t)\left(1 - \dfrac{1200}{P(t)}\right)$, $P(t) > 0$.

   a. Find a constant solution (i.e., a solution that is a constant function).
   b. Show that the population is decreasing if $P(t)$ is smaller than 1200.
   c. Show that the population is increasing if $P(t)$ is larger than 1200.
   d. Interpret your answers. What is the meaning of the number 1200 for the salmon population?

▼ 40–41 ■ The following describe the velocities of different animals. In each case:

a. Draw a graph of the velocity as a function of time.
b. Write a differential equation for the position.
c. Guess the solution of this equation.
d. Draw the graph of the position as a function of time.
e. Determine how long it will take the animal to reach its goal.

40. A snail starts crawling across a narrow path, trying to reach the other side, which is 50 cm away. The velocity of the snail t minutes after it starts is t centimetres per minute.

41. A cheetah is standing 1 m from the edge of the jungle. It starts sprinting across the savannah to attack a zebra that is 200 m from the edge of the jungle. After t seconds, the velocity of the cheetah is e^t metres per second.

▼ 42–47 ■ Apply Euler's method with the given value of Δt to the differential equation. Compare the approximate result with the exact result from the earlier problem.

42. The cell in Exercise 34. Use a step size of $\Delta t = 10$ to estimate the volume at $t = 40$.
43. The cell in Exercise 35. Use a step size of $\Delta t = 5$ to estimate the volume at $t = 30$.
44. The cell in Exercise 36. Use a step size of $\Delta t = 10$ to estimate the volume at $t = 30$.
45. The cell in Exercise 37. Use a step size of $\Delta t = 2$ to estimate the volume at $t = 10$.
46. The snail in Exercise 40. Use a step size of $\Delta t = 2$ to estimate the position at $t = 10$.

47. The cheetah in Exercise 41. Use a step size of $\Delta t = 1$ to estimate the position at $t = 5$.

▼ 48–49 ■ Use the hints to guess the solution of each differential equation describing the rate of production of some chemical. In each case, check your solution, and graph the rate of change and the solution.

48. $\dfrac{dP}{dt} = e^{-t+1}$ with initial condition $P(0) = 0$. Start by finding the derivative of e^{-t+1}, correct it by multiplying by some constant, and then add an appropriate value to match the initial condition.

49. $\dfrac{dP}{dt} = e^{-2t}$ with initial condition $P(0) = 0$. Start by finding the derivative of e^{-2t}, correct it by multiplying by some constant, and then add an appropriate value to match the initial condition.

▼ 50–53 ■ For each measurement, give circumstances under which you could measure the following:

a. The value but not the rate of change.
b. The rate of change but not the value.

50. Position (rate of change is velocity).
51. Mass (rate of change is growth rate).
52. Sodium concentration (rate of change is the rate at which sodium enters and leaves).
53. Total chemical (rate of change is the chemical production rate).

Computer Exercise

54. Apply Euler's method to solve the differential equation

$$\frac{db}{dt} = \frac{e^t}{t+1} + t$$

with the initial condition $b(0) = 1$. Compare with the sum of the results of

$$\frac{db_1}{dt} = \frac{e^t}{t+1}$$

and

$$\frac{db_2}{dt} = t$$

Do you think there is a sum rule for differential equations? If the computer has a method for solving differential equations, find the solution and compare it with your approximate solution from Euler's method.

7.2 Antiderivatives

In order to solve a pure-time differential equation, we have to recover the function from its rate of change (and initial condition, if given). In the previous section, we approached the problem from geometric and numeric viewpoints (Euler's method). Now we approach it algebraically by introducing the concept of the **antiderivative** or **indefinite integral.**

Antiderivatives

In order to solve the (pure-time) differential equation

$$\frac{dF}{dx} = 3x^2 - x - 1$$

we have to find all functions $F(x)$ whose derivative is equal to the given function $3x^2 - x - 1$.

In the language of applications, we are given the rate of change dF/dx of a function and are asked to find $F(x)$.

Definition 7.2.1 A function $F(x)$ defined on an interval (a, b) is called an **antiderivative** of a function $f(x)$ if $F'(x) = f(x)$ for all x in (a, b).

The interval (a, b) in the definition allows infinity; i.e., it could be $(-\infty, b)$, (a, ∞), or $(-\infty, \infty)$.

To say that $F(x)$ is an antiderivative of $f(x)$, we write

$$\int f(x)dx = F(x)$$

The integral sign $\int$ comes from a stretched uppercase letter S, used to denote the sum (we will soon learn about the connection between sums and integrals). The symbol dx is part of the notation. It is called the **differential,** and its meaning will be revealed as we continue developing the concept of the integral in this chapter.

Remember the following:

$F(x)$ Is an Antiderivative of $f(x)$
$\int f(x)dx = F(x) \quad \Leftrightarrow \quad F'(x) = f(x)$

So, it's like inverse functions:

If the derivative of A is B, then an antiderivative of B is A.

In other words, to prove that an antiderivative formula

$$\int f(x)dx = F(x)$$

is correct, we differentiate the function on the right side and show that it is equal to the function on the left side.

Example 7.2.1 Antiderivatives

a. Because $(\sin x)' = \cos x$, an antiderivative of $\cos x$ is $\sin x$; we write

$$\int \cos x \, dx = \sin x$$

b. From $(e^x)' = e^x$, we conclude that

$$\int e^x dx = e^x$$

c. Likewise,
$$\int \frac{1}{1+x^2} dx = \arctan x$$
is true because
$$(\arctan x)' = \frac{1}{1+x^2}$$

Since $(x^4)' = 4x^3$, it follows that an antiderivative of $4x^3$ is x^4:
$$\int 4x^3 dx = x^4$$
But $(x^4 + 13)' = 4x^3$, and so
$$\int 4x^3 dx = x^4 + 13$$
As well,
$$\int 4x^3 dx = x^4 - 3.7$$
since $(x^4 - 3.7)' = 4x^3$. So a given function has many antiderivatives. As a matter of fact, there are infinitely many of them: pick any real number C; from
$$(x^4 + C)' = 4x^3$$
we conclude that
$$\int 4x^3 dx = x^4 + C$$

Are there any more? Is there a function other than $F(x) = x^4 + C$ such that $F'(x) = 4x^3$?

The answer is no. As a consequence of Theorem 6.3.5 in Section 6.3 we proved that any two functions that have the same derivative differ *at most by a constant*.

Thus, if we find one antiderivative of a given function, we have found them all—since they differ from the one we found by a constant term only. Thus,
$$\int 4x^3 dx = x^4 + C$$
gives *all* antiderivatives of $4x^3$.

Theorem 7.2.1 If $F(x)$ is an antiderivative of $f(x)$, then the most general antiderivative (also called the **indefinite integral**) of $f(x)$ is $F(x) + C$; i.e.,
$$\int f(x) dx = F(x) + C$$
where C is a real number.

Geometrically, all antiderivatives $F(x) + C$ of a function $f(x)$ are vertical shifts of each other. Because $(F(x) + C)' = f(x)$, it follows that, at any number x (where they are defined), they all have equal slopes; see Figure 7.2.12.

In light of what we just said, we rewrite the antidifferentiation formulas of Example 7.2.1 as
$$\int \cos x \, dx = \sin x + C$$
$$\int e^x dx = e^x + C$$
$$\int \frac{1}{1+x^2} dx = \arctan x + C$$

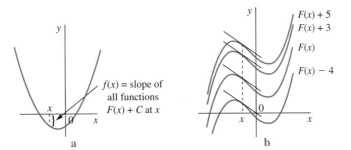

FIGURE 7.2.12
Antiderivatives of a function $f(x)$

(In all questions related to integration, C denotes an arbitrary constant.)
When we write

$$\int dx$$

we mean that the function we are antidifferentiating is the constant function $f(x) = 1$. Thus,

$$\int dx = \int 1\, dx = x + C$$

Example 7.2.2 Checking an Antiderivative by Differentiating

Show that

$$\int x \cos 4x\, dx = \frac{1}{16} \cos 4x + \frac{1}{4} x \sin 4x + C$$

Note that—using the definition of the antiderivative—all we need to do is to differentiate the function on the right side and check that we obtain the function on the left.
Using the product rule and the chain rule, we get

$$\left(\frac{1}{16} \cos 4x + \frac{1}{4} x \sin 4x + C \right)' = \frac{1}{16}(-\sin 4x)(4) + \frac{1}{4}[\sin 4x + x(\cos 4x)(4)]$$

$$= -\frac{1}{4} \sin 4x + \frac{1}{4} \sin 4x + \frac{4}{4} x \cos 4x$$

$$= x \cos 4x$$

Note Recall that by definition

$$\int f(x)\, dx = F(x) + C \quad \Leftrightarrow \quad (F(x) + C)' = f(x)$$

(the notation A ⇔ B says that the statements A and B are equivalent; i.e., A and B convey the same information, as in $y = \ln x \Leftrightarrow e^y = x$). Replacing $f(x)$ by $(F(x) + C)'$ in the indefinite integral, we obtain

$$\int (F(x) + C)'\, dx = F(x) + C$$

and, after differentiating the left side,

$$\int F'(x)\, dx = F(x) + C$$

So, if we start with $F(x)$, differentiate it, and then antidifferentiate the resulting function, we get $F(x)$ back.
Differentiating both sides of

$$\int f(x)\, dx = F(x) + C$$

and then using the fact that $F'(x) = f(x)$, we obtain

$$\left(\int f(x)dx\right)' = (F(x) + C)'$$

$$\left(\int f(x)dx\right)' = F'(x)$$

$$\left(\int f(x)dx\right)' = f(x)$$

The two formulas that we derived,

$$\int F'(x)dx = F(x) + C \quad \text{and} \quad \left(\int f(x)dx\right)' = f(x) \qquad (7.2.1)$$

are precise translations of the fact that differentiation and antidifferentiation are inverse operations.

Rules for Antiderivatives

To find the indefinite integrals (antiderivatives) of interesting functions, we begin by using three of the basic rules of differentiation: the power rule, the constant multiple rule, and the sum rule.

The Power Rule for Integrals The power rule for derivatives says that

$$(x^{n+1})' = (n+1)x^n$$

Dividing both sides by $n + 1$, we get

$$\frac{1}{n+1}(x^{n+1})' = x^n$$

$$\left(\frac{1}{n+1}x^{n+1}\right)' = x^n$$

We have found a function with derivative equal to x^n, which gives the **power rule** for indefinite integrals.

Theorem 7.2.2 **The Power Rule for Integrals**

Suppose $n \neq -1$. Then

$$\int x^n dx = \frac{x^{n+1}}{n+1} + C$$

Example 7.2.3 Applying the Power Rule for Integrals to a Quadratic

Applying the power rule for integrals with $n = 2$, we see that the indefinite integral of x^2 is

$$\int x^2 dx = \frac{x^{2+1}}{2+1} + C = \frac{x^3}{3} + C$$

Geometrically, we are given the graph of $f(x) = x^2$ (Figure 7.2.13a) and need to draw the graph of a function $F(x)$ whose slope at x is x^2. In other words, what is given is the derivative, $F'(x)$, and we need to draw $F(x)$ from it.

Since all slopes x^2 are positive (except at zero, where the slope is zero), we know that $F(x)$ must be an increasing function. As x moves away from the origin in either direction, the slopes increase. Thus, $F(x)$ increases more and more quickly as x becomes larger and larger positive or larger and larger negative; see Figure 7.2.13b.

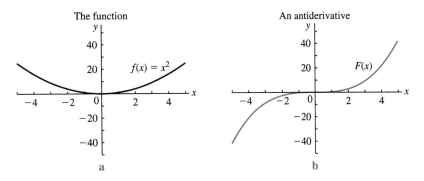

FIGURE 7.2.13
A quadratic function and its integral

Example 7.2.4 Applying the Power Rule for Integrals with a Negative Power

With $n = -2$, the indefinite integral of $g(x) = \dfrac{1}{x^2}$ is

$$g(x) = \int \frac{1}{x^2} dx = \int x^{-2} dx = \frac{x^{-2+1}}{-2+1} = -x^{-1} + C = -\frac{1}{x} + C$$

Example 7.2.5 Applying the Power Rule for Integrals with Fractional Powers

The power rule also works with fractional powers. When $n = 0.5$,

$$\int \sqrt{t}\, dt = \int t^{0.5} dt = \frac{t^{0.5+1}}{0.5+1} + C = \frac{t^{1.5}}{1.5} + C$$

and when $n = -0.5$,

$$\int \frac{1}{\sqrt{t}} dt = \int t^{-0.5} dt$$
$$= \frac{t^{-0.5+1}}{-0.5+1} + C$$
$$= \frac{t^{0.5}}{0.5} + C$$
$$= 2t^{0.5} + C = 2\sqrt{t} + C$$

(see Figure 7.2.14). The slopes of the graph in Figure 7.2.14b are shown in Figure 7.2.14a. As t approaches 0, the graph of $F(t)$ becomes steeper and steeper, and thus the slopes, $f(t)$, become larger and larger. As t approaches ∞, the graph of $F(t)$ becomes more and more flat, and the slopes get smaller and smaller, approaching the value 0.

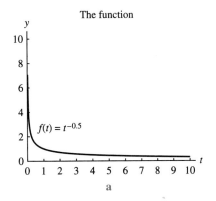
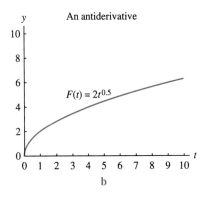

FIGURE 7.2.14
An indefinite integral of $f(t) = t^{-0.5}$

Example 7.2.6 Antiderivative of $f(x) = 1/x$

Because it would cause the appearance of zero in the denominator, we cannot use the formula

$$\int x^n dx = \frac{x^{n+1}}{n+1} + C$$

with $n = -1$, i.e., we cannot calculate $\int \frac{1}{x} dx$ from it. So how do we calculate $\int \frac{1}{x} dx$?

We know that

$$(\ln x)' = \frac{1}{x}$$

for $x > 0$ (we need this so that $\ln x$ is defined). So at least for $x > 0$, we get

$$\int \frac{1}{x} dx = \ln x + C$$

The function $f(x) = 1/x$ is defined for negative x, so what is its antiderivative? It cannot be $\ln x$, because it's not defined for $x < 0$. Let's try $\ln |x|$.

By the definition of absolute value, $|x| = -x$ if $x < 0$, and thus $\ln |x| = \ln(-x)$. We differentiate using the chain rule:

$$(\ln |x|)' = (\ln(-x))' = \frac{1}{-x}(-1) = \frac{1}{x}$$

This shows that for negative values of x,

$$\int \frac{1}{x} dx = \ln |x| + C$$

Merging the two cases ($x > 0$ and $x < 0$) together, we write

$$\int \frac{1}{x} dx = \ln |x| + C$$

for all $x \neq 0$. Equivalently,

$$(\ln |x|)' = \frac{1}{x}$$

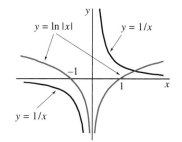

FIGURE 7.2.15

Comparing the graphs of $y = 1/x$ and $y = \ln |x|$

for all $x \neq 0$. Figure 7.2.15 shows the graphs of both $y = 1/x$ and $y = \ln |x|$. Note that $y = \ln |x|$ is increasing if $x > 0$ (and so its derivative, $y = 1/x$, is above the x-axis) and decreasing if $x < 0$ (its derivative is negative there).

The Constant Multiple Rule for Integrals The constant multiple rule for derivatives states that

$$(af(x))' = af'(x)$$

for any real number a. The derivative of a constant times a function is the constant times the derivative. Indefinite integrals work in the same way.

Theorem 7.2.3 The Constant Multiple Rule for Integrals

If a is a real number, then

$$\int af(x) dx = a \int f(x) dx$$

To prove this theorem, we differentiate the right side and show that it is equal to the function under the integral sign on the left side:

$$\left(a \int f(x)dx\right)' = a \left(\int f(x)dx\right)' = af(x)$$

(in the last step we used the inverse relationship between differentiation and antidifferentiation, stated in Equation 7.2.1). We're done!

Example 7.2.7 The Constant Multiple Rule for Integrals

a. The antiderivative of $5x^2$ is

$$\int 5x^2 dx = 5 \int x^2 dx = 5\frac{x^3}{3} + C$$

b. Similarly,

$$\int \frac{-3}{x^2} dx = -3 \int x^{-2} dx = -3\frac{x^{-1}}{-1} + C = \frac{3}{x} + C$$

c. As well,

$$\int 8.8\, t^{-0.12}\, dt = 8.8 \frac{t^{0.88}}{0.88} + C = 10\, t^{0.88} + C$$

The Sum Rule for Integrals The sum rule for derivatives says that the derivative of the sum is the sum of the derivatives, or

$$\frac{d(f+g)}{dx} = \frac{df}{dx} + \frac{dg}{dx}$$

Indefinite integrals work in the same way.

Theorem 7.2.4 The Sum Rule for Integrals

For two functions, $f(x)$ and $g(x)$,

$$\int (f(x) \pm g(x))dx = \int f(x)dx \pm \int g(x)dx$$

Again, to prove this statement, we differentiate the right side and use Equation 7.2.1:

$$\left(\int f(x)dx \pm \int g(x)dx\right)' = \left(\int f(x)dx\right)' \pm \left(\int g(x)dx\right)' = f(x) \pm g(x)$$

Example 7.2.8 The Sum Rule Applied to Power Functions

We have found that

$$\int 5x^2 dx = \frac{5x^3}{3} + C$$

$$\int (-3x^{-2})\, dx = \frac{3}{x} + C$$

Therefore,

$$\int (5x^2 - 3x^{-2})\, dx = \frac{5x^3}{3} + \frac{3}{x} + C$$

By reversing the differentiation formulas from Chapters 4 and 5 and adding the formulas we have derived in this section, we obtain the list of antiderivatives in Table 7.2.1.

Table 7.2.1

Function	Antiderivative	Function	Antiderivative
$f(x) = k$ (k is a constant)	$F(x) = kx + C$	$f(x) = \cos x$	$F(x) = \sin x + C$
$f(x) = x^n$ if $n \neq -1$	$F(x) = \dfrac{1}{n+1}x^{n+1} + C$	$f(x) = \sec^2 x$	$F(x) = \tan x + C$
$f(x) = x^{-1} = \dfrac{1}{x}$	$F(x) = \ln|x| + C$	$f(x) = \sec x \tan x$	$F(x) = \sec x + C$
$f(x) = e^x$	$F(x) = e^x + C$	$f(x) = \dfrac{1}{1+x^2}$	$F(x) = \arctan x + C$
$f(x) = \sin x$	$F(x) = -\cos x + C$	$f(x) = \dfrac{1}{\sqrt{1-x^2}}$	$F(x) = \arcsin x + C$

As we develop techniques of integration, we will be able to add more functions to this list.

Example 7.2.9 Antiderivative of $f(x) = a^x$

To find an antiderivative of a^x means to find a function whose derivative is a^x. From

$$(a^x)' = a^x \ln a \neq a^x$$

we see that a^x does not work. If we could get rid of $\ln a$, it would work. So we divide a^x by $\ln a$ and differentiate:

$$\left(\frac{1}{\ln a}a^x\right)' = \frac{1}{\ln a}(a^x)' = \frac{1}{\ln a}(a^x \ln a) = a^x$$

It works! Thus,

$$\int a^x dx = \frac{1}{\ln a}a^x + C$$

Solving Simple Differential Equations

Using antiderivatives, we can now solve some pure-time differential equations.

Example 7.2.10 Solving a Differential Equation

Find $f(x)$ if $f'(x) = 4\sec^2 x + e^x - 1$ and $f(0) = 7$.

First we find the antiderivative:

$$f(x) = \int (4\sec^2 x + e^x - 1)dx$$
$$= 4\tan x + e^x - x + C$$

To find the value of C, we use the initial condition $f(0) = 7$:

$$7 = 4\tan 0 + e^0 - 0 + C$$
$$7 = 1 + C$$

and so $C = 6$. We are done—the solution is

$$f(x) = 4\tan x + e^x - x + 6$$

Example 7.2.11 Solving a Differential Equation

Solve the differential equation

$$\frac{df}{dx} = x - (1 - x^2)^{-1/2}$$

with initial condition $f(1) = 2$.

We start by finding the general antiderivative:

$$f(x) = \int \left(x - (1-x^2)^{-1/2}\right) dx = \frac{1}{2}x^2 - \arcsin x + C$$

Now we substitute the initial condition to find C:

$$f(1) = \frac{1}{2}(1)^2 - \arcsin 1 + C = 2$$

$$\frac{1}{2} - \frac{\pi}{2} + C = 2$$

$$C = \frac{3}{2} + \frac{\pi}{2} = \frac{3+\pi}{2}$$

The solution is

$$f(x) = \frac{1}{2}x^2 - \arcsin x + \frac{3+\pi}{2}$$

Example 7.2.12 A Differential Equation for AIDS

During the early years of the AIDS epidemic, workers at the U.S. Centers for Disease Control (CDC) found that the number of new AIDS cases per year followed the formula

$$\text{rate at which new AIDS cases were reported} \approx 523.8t^2 \qquad (7.2.2)$$

where t is measured in years after the beginning of 1981. The CDC's formula for the number of AIDS cases, $A(t)$, can be written as a polynomial pure-time differential equation:

$$\frac{dA}{dt} = 523.8t^2$$

To solve this equation for the total number of AIDS cases, we require an initial condition. Surveys indicated that about 340 people had been infected at the beginning of 1981 ($t = 0$ in the model), so $A(0) = 340$. We can now solve for $A(t)$, finding

$$A(t) = \int 523.8t^2 \, dt = 523.8 \frac{t^3}{3} = 174.6t^3 + C$$

The constant C is the solution of the equation

$$A(0) = 174.6 \cdot 0^3 + C = 340$$

so $C = 340$. The solution is

$$A(t) = 174.6t^3 + 340$$

(Figure 7.2.16).

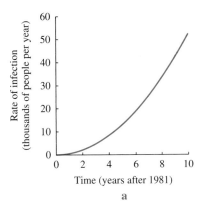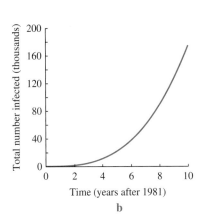

FIGURE 7.2.16
The early course of the AIDS epidemic in the United States

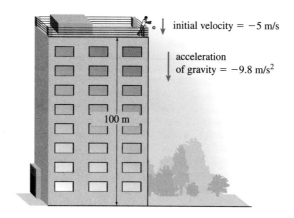

FIGURE 7.2.17
A falling rock

Problems that involve finding the indefinite integrals of polynomials arise when we study the motion of objects with **constant acceleration.** Suppose we are told that a rock is thrown downward from a building 100 m tall with initial downward velocity of 5 m/s. The acceleration of gravity is about −9.8 m/s² (Figure 7.2.17). How long will it take for the rock to hit the ground? How fast will it be going?

Let a represent the acceleration, v the velocity, and p the position of the rock. The basic differential equations from physics are

$$\frac{dv}{dt} = a \qquad \text{acceleration is the rate of change of velocity}$$

$$\frac{dp}{dt} = v \qquad \text{velocity is the rate of change of position}$$

We can use these equations to solve for the position of the rock.

Example 7.2.13 The Physics of Constant Acceleration: Finding the Velocity

To find the velocity, we know from physics that its rate of change, the acceleration, is $a = -9.8$. We are also given the initial condition $v(0) = -5$. Both values are negative because both point downward. We therefore have enough information to solve the differential equation

$$\frac{dv}{dt} = -9.8$$

with the initial condition $v(0) = -5$. Thus,

$$v(t) = \int (-9.8) dt = -9.8t + C$$

We find the constant C with the initial condition by solving

$$v(0) = -9.8 \cdot 0 + C = -5$$

so $C = -5$. Therefore, the equation for the velocity is

$$v(t) = -9.8t - 5$$

As time passes, the velocity becomes more and more negative as the rock falls more and more quickly (Figure 7.2.18a).

FIGURE 7.2.18
The velocity and position of a falling rock

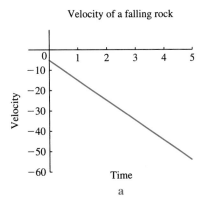

Example 7.2.14 The Physics of Constant Acceleration: Finding the Position

We now have enough information to solve for the position, p. We have the pure-time differential equation

$$\frac{dp}{dt} = -9.8t - 5$$

and the initial condition $p(0) = 100$. We first solve the differential equation with the indefinite integral, finding

$$p(t) = \int (-9.8t - 5)dt$$
$$= \int -9.8t\, dt - \int 5\, dt$$
$$= -9.8 \int t\, dt - 5 \int dt$$
$$= -9.8 \frac{t^2}{2} - 5t + C = -4.9t^2 - 5t + C$$

We use the initial condition to find the constant C by solving

$$p(0) = -4.9 \cdot 0^2 - 5 \cdot 0 + C = 100$$

to obtain $C = 100$. Therefore, the equation for the position is

$$p(t) = -4.9t^2 - 5t + 100$$

(Figure 7.2.18b).

The solution tells when the rock will hit the ground and how fast it will be going. It hits the ground at the time when $p(t) = 0$, which is the solution of

$$-4.9t^2 - 5t + 100 = 0$$

For convenience, we multiply by -1, so

$$4.9t^2 + 5t - 100 = 0$$

With the quadratic formula, we get

$$t = \frac{-5 \pm \sqrt{5^2 - 4 \cdot 4.9 \cdot (-100)}}{2 \cdot 4.9}$$
$$= \frac{-5 \pm \sqrt{1985}}{9.8}$$
$$\approx \frac{-5 \pm 44.55}{9.8}$$
$$\approx 4.036\, \text{s} \quad \text{or} \quad -5.056$$

The positive solution is the one that makes sense. How fast will the rock be going when it hits the ground? We use the formula for $v(t)$ to find

$$v(4.036) = -9.8 \cdot 4.036 - 5 \approx -44.55 \text{ m/s}$$

Summary Solving pure-time differential equations requires undoing the derivative. We defined the **antiderivative** of a function f as any function whose derivative is f. Unlike the derivative, the antiderivative includes an **arbitrary constant** to indicate that a whole family of functions have the same slope. From the rules for computing derivatives, we found the **power rule,** the **constant multiple rule,** and the **sum rule** for integrals. With these rules, we can find the indefinite integral of any polynomial. We applied this method to solve several pure-time differential equations.

7.2 Exercises

Mathematical Techniques

1–6 ■ From the graph, sketch the antiderivative of the function that passes through the given point.

1. The antiderivative that passes through the point (0, 500).

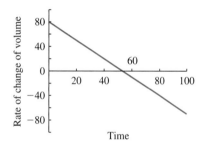

2. The antiderivative that passes through (50, 5000).

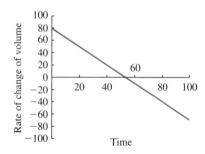

3. The antiderivative that passes through (100, 3000).

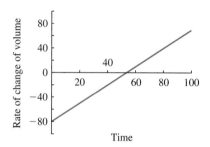

4. The antiderivative that passes through the point (0, 2500).

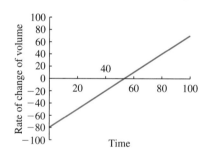

5. The antiderivative that passes through (100, 1000).

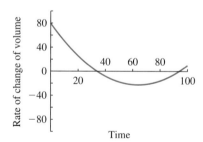

6. The antiderivative that passes through the point (0,1000).

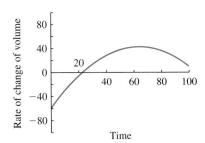

7–24 ▪ Find the general antiderivative of each function.

7. $f(x) = 4 - x - 5x^4$
8. $f(t) = 2$
9. $f(x) = x^{15} - 12x^{11} - 6$
10. $f(x) = x^{-2} - x^4$
11. $f(t) = 3t^{2.4} + e^t$
12. $f(x) = \sqrt{x} - \sqrt[4]{x^3}$
13. $g(x) = \dfrac{x - \sqrt{x} + 2}{x^2}$
14. $f(x) = \dfrac{11}{x} - \dfrac{12}{x^{1/3}} + \dfrac{13}{\sqrt{x}}$
15. $f(t) = \sin t - t^{3/4}$
16. $f(x) = 3\cos x - \sec^2 x$
17. $g(t) = 2 - \dfrac{3}{\sqrt{1 - t^2}}$
18. $f(x) = x + 3(1 + x^2)^{-1}$
19. $f(x) = 4\sec x \tan x - x^{-1}$
20. $f(t) = t - \sin t - \cos t$
21. $f(x) = \dfrac{1 + x^{1.5} - 3x^{4.3}}{x^{2.1}}$
22. $f(x) = 2^x + e^x$
23. $f(x) = 12^x + x^{12}$
24. $f(x) = \sqrt{\ln 2}$

25–30 ▪ Use antiderivatives to solve each differential equation. Sketch a graph of the rate of change and the solution on the given domain.

25. $\dfrac{dV}{dt} = 2t^2 + 5$ with $V(1) = 19$. Sketch the rate of change and solution for $0 \le t \le 5$.
26. $\dfrac{dV}{dt} = 2t^2 + 5$ with $V(0) = 19$. Sketch the rate of change and solution for $0 \le t \le 5$.
27. $\dfrac{df}{dt} = 5t^3 + 5t$ with $f(0) = -12$. Sketch the rate of change and solution for $0 \le t \le 2$.
28. $\dfrac{dg}{dt} = -3t + t^2$ with $g(0) = 10$. Sketch the rate of change and solution for $0 \le t \le 5$.
29. $\dfrac{dM}{dt} = t^2 + \dfrac{1}{t^2}$ with $M(3) = 10$. Sketch the rate of change and solution for $0 \le t \le 3$.
30. $\dfrac{dp}{dt} = 5t^3 + \dfrac{5}{t^2}$ with $p(1) = 12$. Sketch the rate of change and solution for $1 \le t \le 3$.

31–36 ▪ Find the function f.

31. $f'(x) = 4\sqrt{x} - \dfrac{2}{\sqrt{x}}, f(1) = 2$
32. $f'(t) = e^t - 2t + 1, f(0) = -4$
33. $f'(x) = 2(1 + x^2)^{-1}, f(1) = \pi$
34. $f'(x) = \dfrac{4}{x}, x < 0, f(-2) = 4$
35. $f'(\theta) = 4\sin\theta + \sec\theta\tan\theta, f(0) = -1$
36. $f'(x) = 6^x - 4, f(3) = -12$

37–38 ▪ There are no simple integral versions of product and quotient rules for derivatives. Use the given functions to show that the proposed rule does not work.

37. Use the functions $f(x) = x^2$ and $g(x) = x^3$ to show that the integral of a product is *not* equal to the product of the integrals.
38. Use the functions $f(x) = x^2$ and $g(x) = x^3$ to show that
$$\int f(x)g(x)dx \ne g(x)\int f(x)dx + f(x)\int g(x)dx$$

Applications

39–40 ▪ Suppose a cell is taking water into two vacuoles. Let V_1 denote the volume of the first vacuole and V_2 the volume of the second. In each of the following cases:

a. Solve the given differential equations for $V_1(t)$ and $V_2(t)$.
b. Write a differential equation for $V = V_1 + V_2$, including the initial condition.
c. Show that the solution of the differential equation for V is the sum of the solutions for V_1 and V_2.

39.
$$\dfrac{dV_1}{dt} = 2t + 5$$
$$\dfrac{dV_2}{dt} = 5t + 2$$
with initial conditions $V_1(0) = V_2(0) = 10$.

40.
$$\dfrac{dV_1}{dt} = 3.6t^2 + 5t$$
$$\dfrac{dV_2}{dt} = 5.2t^3 + 2$$
with initial conditions $V_1(0) = 5$ and $V_2(0) = 10$.

41–42 ▪ Suppose organisms grow in mass according to the differential equation

$$\frac{dM}{dt} = \alpha t^n$$

where M is measured in grams and t is measured in days. For each of the following values for n and α:

a. Find the units of α.

b. Suppose that $M(0) = 5$ g. Find the solution.

c. Sketch the graphs of the rate of change and the solution.

d. Describe your results in words.

41. $n = 1$, $\alpha = 2$

42. $n = -1/2$, $\alpha = 2$

43–46 ▪ In a new program, NASA sends astronaut-free expeditions to astronomical bodies. One of the more fascinating experiments involves having the robot release an object from a height $h = 100$ m with velocity $v = 5$ m/s (upward) to find its trajectory in the local gravitational field of strength a. For the following values of a:

a. Find the velocity and position of the object as functions of time.

b. How high will the object get?

c. How long will it take to pass the robot on the way down? How fast will it be moving?

d. How long will it take to hit the ground? How fast (in metres per second) will it be moving? How fast is this in kilometres per hour?

e. Graph the velocity and position as functions of time.

43. A practice test on Earth, where $a = -9.8$ m/s².

44. On the Moon, where $a = -1.62$ m/s².

45. On Jupiter, where $a = -22.88$ m/s².

46. On Mars's moon Deimos, where $a = 2.15 \cdot 10^{-3}$ m/s².

47–50 ▪ The velocities of four objects are measured at discrete times.

Time	Velocity of Object 1	Velocity of Object 2	Velocity of Object 3	Velocity of Object 4
0	1	9	25	1
1	3	7	16	3
2	5	5	9	6
3	7	3	4	10
4	9	1	1	15

Use Euler's method with $\Delta t = 1$ to estimate the position at $t = 4$ starting from the initial condition $p(0) = 10$. Next find a simple formula for the velocities, and use it to find the exact position at $t = 4$. Graph your results.

47. Object 1. To find the formula for the velocities, note that they increase linearly in time.

48. Object 2. To find the formula for the velocities, note that they decrease linearly in time.

49. Object 3. To find the formula for the velocities, compare them with the perfect square numbers.

50. Object 4. The velocities follow a quadratic equation of the form $v(t) = \frac{t^2}{2} + at + b$ for some values of a and b.

51–54 ▪ Consider again the velocities of the four objects used in the previous set of exercises. There is a more accurate variant of Euler's method that approximates the rate of change during a time interval as the average of the rate of change at the beginning of the interval and the rate of change at the end of the interval. For example, if $v(0) = 1$ and $v(1) = 3$, we approximate the rate of change for $0 \leq t \leq 1$ as 2. Use this variant of Euler's method with initial condition $p(0) = 10$ to estimate the position at $t = 4$ and compare your estimate with the exact position at $t = 4$. Graph your results.

51. Object 1.

52. Object 2.

53. Object 3.

54. Object 4.

7.3 Definite Integral and Area

In this section we discuss the definite integral, which is one of the fundamental concepts in mathematics. The construction involves approximating sums (**Riemann sums**) and limits. The definite integral is closely related to the area under the graph of a function: if the function is positive, the two are the same. Otherwise, the definite integral represents the **signed area,** i.e., the difference between the area of the region(s) above the x-axis and the area of the region(s) below the x-axis. We learn how to work with Riemann sums, **evaluate** some simple definite integrals, and interpret the definite integral as the **total change.** In the next section, we relate the antiderivative (indefinite integral) to the definite integral.

Area

In order to figure out the effects of pollutants in a lake, we need to know its surface area. Estimating the number of bacteria in a colony is based on knowing the area that the colony occupies in a Petri dish. By comparing the combined area of clouds with the area of clear skies in a photograph, we determine the cloud cover. The size of a two-dimensional image of a tumour (say, obtained by a mammogram) helps us estimate the size of the tumour. In numerous other situations, we need to know the area of some object.

How do we calculate area? The area of a rectangle is the product of the base length and the height. Because we know the area of a triangle (one half the base length times the height) we can calculate the area of any polygon (by dividing it up into triangles). We know the area of a circle (π times the square of the radius) and the area of an ellipse (π times the semi-major axis times the semi-minor axis); see Figure 7.3.19.

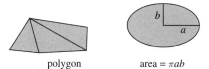

FIGURE 7.3.19

Polygon and ellipse

polygon   area = πab

And that's it! There are no formulas for areas of regions bounded by any other curve.

However—as we will soon see—there is a procedure we can use to approximate the area of a complicated region or, in some cases, to obtain its exact value.

We now state the fundamental area question: Consider a continuous function $f(x)$ defined on a closed interval $[a, b]$. Assume that $f(x) \geq 0$ for all x in $[a, b]$. The graph of $f(x)$, the vertical lines $x = a$ and $x = b$, and the x-axis define the region called the **region under f on $[a, b]$**; see Figure 7.3.20. Find its area.

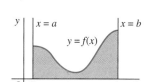

FIGURE 7.3.20

Region under f on $[a, b]$

In the process of answering this question, we will actually define what the area of a region is based on the fact that we know the area of a rectangle. The concept of area might feel intuitively clear—however, we need to have a precise mathematical definition so that we can work with it.

The idea of approximating area—to use simple objects to approximate a more complicated region—is definitely not new. Greek mathematician Archimedes used it to figure out areas of regions bounded by some special curves. The simple objects that we use are rectangles. We explain the idea in an example first.

Example 7.3.1   Estimating the Area Under a Parabola

We approximate the area, A, of the region under $f(x) = x^2$ on $[0, 1]$, shown in Figure 7.3.21a.

First, we divide the interval $[0, 1]$ into a certain number—say, five—of subintervals of equal length:

$$[0, 0.2], [0.2, 0.4], [0.4, 0.6], [0.6, 0.8], [0.8, 1]$$

The length of each subinterval is $\Delta x = 1/5 = 0.2$.

Next, we build rectangles whose bases are these subintervals. In order to do that, we need to know where to measure their heights—so we adopt the following rule: the height is the value of the function $f(x) = x^2$ measured at the right end of each subinterval. In this way, we obtain our first approximation of the region under $f(x) = x^2$ by rectangles; see Figure 7.3.21b. The area, A, under $f(x) = x^2$ will be approximated by the sum of the areas of the five rectangles.

FIGURE 7.3.21

First approximation of the area

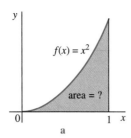
a

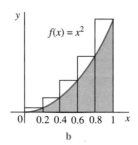
b

The height of the rectangle over the subinterval [0, 0.2] is the value of $f(x)=x^2$ at the right end, i.e., $f(0.2)=(0.2)^2=0.04$. So the area of the rectangle is (base length times height)

$$(0.2)(0.04) = 0.008$$

The height of the rectangle over the subinterval [0.2, 0.4] is $f(0.4)=(0.4)^2=0.16$. Its area is

$$(0.2)(0.16) = 0.032$$

The remaining three rectangles have heights equal to $f(0.6)=(0.6)^2$, $f(0.8)=(0.8)^2$, and $f(1)=(1)^2$. We now calculate the sum of the areas of the five rectangles:

$$R_5 = 0.2(0.2)^2 + 0.2(0.4)^2 + 0.2(0.6)^2 + 0.2(0.8)^2 + 0.2(1)^2$$
$$= 0.008 + 0.032 + 0.072 + 0.128 + 0.2$$
$$= 0.44$$

Thus, $R_5 = 0.44$ is an approximation for the area A. Since the rectangular regions (combined) are larger than the region whose area we want to find (see Figure 7.3.21b), R_5 is an overestimate; i.e., $A < 0.44$.

The symbol R_5 keeps track of what we have done: the subscript 5 stands for the number of rectangles that we used, and R says that we calculated their heights at the right endpoint of each subinterval. The number R_5 is called the **right sum with five rectangles.**

In order to improve our estimate, we do the obvious—we use more rectangles.

To calculate R_{10}, the right sum with 10 rectangles, we divide the given interval [0, 1] into 10 subintervals (Figure 7.3.22):

$$[0, 0.1], [0.1, 0.2], [0.2, 0.3], \ldots, [0.8, 0.9], [0.9, 1]$$

The length of each subinterval is $\Delta x = 1/10 = 0.1$. The sum, R_{10}, consists of 10 terms, each of the form

$$\text{(base length)(height)}$$
$$= (\Delta x)(\text{the value of } f(x)=x^2 \text{ at the right end of each subinterval})$$

Thus,

$$R_{10} = 0.1(0.1)^2 + 0.1(0.2)^2 + 0.1(0.3)^2 + 0.1(0.4)^2 + 0.1(0.5)^2$$
$$+ 0.1(0.6)^2 + 0.1(0.7)^2 + 0.1(0.8)^2 + 0.1(0.9)^2 + 0.1(1)^2$$
$$= 0.385$$

Although R_{10} is an overestimate of the area A (see Figure 7.3.22), it is closer to its exact value than R_5.

If instead of measuring at the right end, we measure the heights at the left end of each subinterval, we obtain the **left sum.**

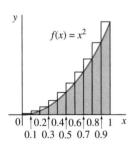

FIGURE 7.3.22

Using R_{10} to estimate the area

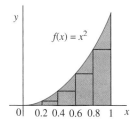

FIGURE 7.3.23

Using the left sum L_5 to estimate the area

Let's calculate L_5 (the left sum with five rectangles). We divide the interval $[0, 1]$ into five subintervals of length $\Delta x = 0.2$:

$$[0, 0.2], [0.2, 0.4], [0.4, 0.6], [0.6, 0.8], [0.8, 1]$$

(see Figure 7.3.23).

The rectangle over $[0, 0.2]$ has collapsed to a line segment and has area 0. The height of the rectangle over $[0.2, 0.4]$ is $f(0.2) = (0.2)^2$, so its area is

$$\Delta x f(0.2) = 0.2 f(0.2) = (0.2)(0.04) = 0.08$$

In the same way, we compute the remaining areas, thus obtaining

$$L_5 = 0.2(0)^2 + 0.2(0.2)^2 + 0.2(0.4)^2 + 0.2(0.6)^2 + 0.2(0.8)^2$$
$$= 0 + 0.08 + 0.032 + 0.072 + 0.128$$
$$= 0.24$$

From Figure 7.3.23, we see that $L_5 = 0.24$ is an underestimate for the area. Combining both R_5 and L_5, we obtain

$$0.24 < A < 0.44$$

Calculating

$$L_{10} = 0.1(0)^2 + 0.1(0.1)^2 + 0.1(0.2)^2 + 0.1(0.3)^2 + 0.1(0.4)^2$$
$$+ 0.1(0.5)^2 + 0.1(0.6)^2 + 0.1(0.7)^2 + 0.1(0.8)^2 + 0.1(0.9)^2$$
$$= 0.285$$

and recalling that we calculated $R_{10} = 0.385$, we now improve our estimate to

$$0.285 < A < 0.385$$

Using software that can calculate left and right sums, or writing our own computer code, we obtain the values in Table 7.3.1.

Table 7.3.1

n	L_n	R_n
10	0.285	0.385
100	0.32835	0.33835
1,000	0.33283	0.33383
5,000	0.33323	0.33343
10,000	0.33328	0.33338

Looking at the last row, we get a fairly close estimate:

$$0.33328 < A < 0.33338$$

The left sum and the right sum are also called **Riemann sums** (after German mathematician Bernhard Riemann). Another example of a Riemann sum is the **midpoint sum, M_n**.

Example 7.3.2 Calculating a Midpoint Sum

We approximate the area, A, of the region under $f(x) = x^2$ on $[0, 1]$ using the midpoint sum M_5. We use the same subintervals

$$[0, 0.2], [0.2, 0.4], [0.4, 0.6], [0.6, 0.8], [0.8, 1]$$

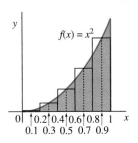

FIGURE 7.3.24

Using the midpoint sum M_5 to estimate the area

as in the calculation of R_5 and L_5, but this time we compute the height at the midpoint of each subinterval (i.e., at 0.1, 0.3, 0.5, 0.7, and 0.9); see Figure 7.3.24. We get

$$M_5 = 0.2(0.1)^2 + 0.2(0.3)^2 + 0.2(0.5)^2 + 0.2(0.7)^2 + 0.2(0.9)^2$$
$$= 0.002 + 0.018 + 0.05 + 0.098 + 0.162$$
$$= 0.33$$

For the record (using software),

$$M_{5000} = 0.3333333300$$
$$M_{10,000} = 0.3333333325$$

The three sums R_n, L_n, and M_n seem to be approaching the same number as n grows larger. That this must be so is intuitively clear: as n increases, the base length of each subinterval shrinks, and so the left end, the right end, and the midpoint get very close to each other. Because f is continuous, its values at these three numbers must be close to each other as well.

Let's try to figure out what number these sums will approach as $n \to \infty$.

Example 7.3.3 Calculating a Riemann Sum with n Subintervals

We compute the right sum R_n for the function $f(x) = x^2$ on the interval $[0, 1]$. First, we divide $[0, 1]$ into subintervals of length $\Delta x = 1/n$:

$$\left[0, \frac{1}{n}\right], \left[\frac{1}{n}, \frac{2}{n}\right], \left[\frac{2}{n}, \frac{3}{n}\right], \ldots, \left[\frac{n-2}{n}, \frac{n-1}{n}\right], \left[\frac{n-1}{n}, \frac{n}{n}\right]$$

Keep in mind that the area of each rectangle is calculated by multiplying the base length, $\Delta x = 1/n$, by the value of the function at the right end. Thus,

$$R_n = \Delta x f\left(\frac{1}{n}\right) + \Delta x f\left(\frac{2}{n}\right) + \Delta x f\left(\frac{3}{n}\right) + \cdots + \Delta x f\left(\frac{n}{n}\right)$$
$$= \frac{1}{n}\left(\frac{1}{n}\right)^2 + \frac{1}{n}\left(\frac{2}{n}\right)^2 + \frac{1}{n}\left(\frac{3}{n}\right)^2 + \cdots + \frac{1}{n}\left(\frac{n}{n}\right)^2$$
$$= \frac{1}{n}\frac{1^2}{n^2} + \frac{1}{n}\frac{2^2}{n^2} + \frac{1}{n}\frac{3^2}{n^2} + \cdots + \frac{1}{n}\frac{n^2}{n^2}$$
$$= \frac{1}{n}\frac{1}{n^2}(1^2 + 2^2 + 3^2 + \cdots + n^2)$$

We are interested in finding out what happens to R_n as $n \to \infty$. We don't know how, because of the sum involved (as n increases, so does the number of terms in the brackets). Luckily, there is a formula that helps us simplify the expression for R_n:

$$1^2 + 2^2 + 3^2 + \cdots + n^2 = \frac{n(n+1)(2n+1)}{6}$$

Thus,

$$R_n = \frac{1}{n^3}(1^2 + 2^2 + 3^2 + \cdots + n^2)$$
$$= \frac{1}{n^3}\frac{n(n+1)(2n+1)}{6}$$
$$= \frac{(n+1)(2n+1)}{6n^2}$$

Check: For $n = 5$, we get

$$R_5 = \frac{6 \cdot 11}{6 \cdot 25} = 0.44$$

and, if $n = 10$,

$$R_{10} = \frac{11 \cdot 21}{6 \cdot 100} = 0.385$$

which is what we obtained earlier.

Taking the limit of R_n as the number of subintervals, n, approaches infinity, we get

$$\lim_{n\to\infty} R_n = \lim_{n\to\infty} \frac{(n+1)(2n+1)}{6n^2}$$
$$= \lim_{n\to\infty} \frac{2n^2 + 3n + 1}{6n^2}$$
$$= \lim_{n\to\infty} \frac{2n^2}{6n^2}$$
$$= \frac{1}{3}$$

(Recall that the leading behaviour of a polynomial at infinity is its largest power term!)
Similarly, we could calculate the left sum, L_n, and show that

$$\lim_{n\to\infty} L_n = \frac{1}{3}$$

With a little more work, we could obtain

$$\lim_{n\to\infty} M_n = \frac{1}{3}$$

for the midpoint sum (for details, see Exercises 22 and 23; for extra practice in working with Riemann sums, see Exercises 24–27).

Thus, we can *define* the area of the region under $f(x) = x^2$ on [0, 1] to be the limit of *any* of the left, the right, or the midpoint sums as n approaches infinity. Next—for practice—we calculate a few more Riemann sums. After that, we generalize the calculations and come back to the issue of defining the area.

Example 7.3.4 Calculating Riemann Sums for $f(x) = \ln x$

Calculate the Riemann sums R_8, L_8, and M_8 for the function $f(x) = \ln x$ defined on the interval [1, 3].

We divide the given interval into eight subintervals, with the length of each equal to

$$\Delta x = \frac{3-1}{8} = \frac{1}{4} = 0.25$$

The subintervals are

$$[x_0, x_1], [x_1, x_2], [x_2, x_3], \ldots, [x_7, x_8]$$

where

$$x_0 = 1,\ x_1 = 1.25,\ x_2 = 1.5,\ x_3 = 1.75,\ x_4 = 2,$$
$$x_5 = 2.25,\ x_6 = 2.5,\ x_7 = 2.75,\ x_8 = 3$$

Let's compute R_8 first. Each term in R_8 is of the form

(base length)(height)
$= (\Delta x)$(the value of $f(x) = \ln x$ at the right end of the subinterval)

Thus,

$$R_8 = \Delta x f(x_1) + \Delta x f(x_2) + \Delta x f(x_3) + \Delta x f(x_4)$$
$$+ \Delta x f(x_5) + \Delta x f(x_6) + \Delta x f(x_7) + \Delta x f(x_8)$$
$$= 0.25 \ln 1.25 + 0.25 \ln 1.5 + 0.25 \ln 1.75 + 0.25 \ln 2$$
$$+ 0.25 \ln 2.25 + 0.25 \ln 2.5 + 0.25 \ln 2.75 + 0.25 \ln 3$$
$$\approx 1.42970$$

Likewise,

$$L_8 = \Delta x f(x_0) + \Delta x f(x_1) + \Delta x f(x_2) + \Delta x f(x_3)$$
$$+ \Delta x f(x_4) + \Delta x f(x_5) + \Delta x f(x_6) + \Delta x f(x_7)$$
$$= 0.25 \ln 1 + 0.25 \ln 1.25 + 0.25 \ln 1.5 + 0.25 \ln 1.75$$
$$+ 0.25 \ln 2 + 0.25 \ln 2.25 + 0.25 \ln 2.5 + 0.25 \ln 2.75$$
$$\approx 1.1505$$

The midpoints of the eight subintervals are

$$\frac{x_0 + x_1}{2} = 1.125, \quad \frac{x_1 + x_2}{2} = 1.375,$$
$$\frac{x_2 + x_3}{2} = 1.625, \; 1.875, \; 2.125, \; 2.375, \; 2.625, \; 2.875$$

and the midpoint sum is

$$M_8 = \Delta x f(1.125) + \Delta x f(1.375) + \Delta x f(1.625) + \Delta x f(1.875)$$
$$+ \Delta x f(2.125) + \Delta x f(2.375) + \Delta x f(2.625) + \Delta x f(2.875)$$
$$= 0.25 \ln 1.125 + 0.25 \ln 1.375 + 0.25 \ln 1.625 + 0.25 \ln 1.875$$
$$+ 0.25 \ln 2.125 + 0.25 \ln 2.375 + 0.25 \ln 2.625 + 0.25 \ln 2.875$$
$$\approx 1.2976$$

Using software or a programmable calculator, we obtain the values of the three sums for large values of n; see Table 7.3.2.

Table 7.3.2

n	L_n	R_n	M_n
100	1.28483	1.30680	1.29585
1,000	1.29474	1.29694	1.29584
5,000	1.29562	1.29606	1.29584
10,000	1.29573	1.29595	1.29584
100,000	1.29583	1.29585	1.29584

All three sums seem to be converging toward the same number.

Sigma Notation

The notation that we now introduce is a convenient way to write a sum. For $m \leq n$, we define

$$\sum_{i=m}^{n} a_i = a_m + a_{m+1} + a_{m+2} + \cdots + a_n$$

The uppercase Greek letter sigma (Σ) denotes the fact that we are adding terms of the form a_i. The numbers below and above Σ tell us where to start and end: the sum starts at the value $i = m$ and ends at $i = n$. In other words, the summation index i starts at m and advances by 1 until it reaches n. For example,

$$\sum_{i=0}^{3} a_i = a_0 + a_1 + a_2 + a_3$$

$$\sum_{j=5}^{7} f(x_j) = f(x_5) + f(x_6) + f(x_7)$$

$$\sum_{i=2}^{5} e^{-i} = e^{-2} + e^{-3} + e^{-4} + e^{-5} \approx 0.21018$$

$$\sum_{j=1}^{3} j^3 = 1^3 + 2^3 + 3^3 = 36$$

$$\sum_{i=5}^{11} \frac{1}{i} = \frac{1}{5} + \frac{1}{6} + \frac{1}{7} + \frac{1}{8} + \frac{1}{9} + \frac{1}{10} + \frac{1}{11} \approx 0.93654$$

The formula

$$1^2 + 2^2 + 3^2 + \cdots + n^2 = \frac{n(n+1)(2n+1)}{6}$$

that we used in Example 7.3.3 can be written as

$$\sum_{i=1}^{n} i^2 = \frac{n(n+1)(2n+1)}{6}$$

The expression

$$R_8 = \Delta x f(x_1) + \Delta x f(x_2) + \Delta x f(x_3) + \Delta x f(x_4) \\ + \Delta x f(x_5) + \Delta x f(x_6) + \Delta x f(x_7) + \Delta x f(x_8)$$

for the right sum from the previous example can be shortened using sigma notation to

$$R_8 = \sum_{i=1}^{8} \Delta x f(x_i) = \sum_{i=1}^{8} f(x_i) \Delta x$$

Likewise, we have

$$L_8 = \sum_{i=0}^{7} f(x_i) \Delta x$$

M_8 is bit trickier:

$$M_8 = \sum_{i=0}^{7} f\left(\frac{x_i + x_{i+1}}{2}\right) \Delta x$$

Area and Definite Integral

Assume that $y = f(x)$ is a continuous function defined on $[a, b]$ and is such that $f(x) \geq 0$ for all x in $[a, b]$. We now construct Riemann sums in general for an arbitrary number of subintervals.

The length of $[a, b]$ is $b - a$. We divide it into n subintervals of length

$$\Delta x = \frac{b-a}{n}$$

The subintervals are

$$[x_0, x_1], [x_1, x_2], [x_2, x_3], \ldots, [x_{n-2}, x_{n-1}], [x_{n-1}, x_n]$$

where

$$x_0 = a$$
$$x_1 = a + \Delta x$$
$$x_2 = a + 2\Delta x$$
$$\cdots$$
$$x_{n-1} = a + (n-1)\Delta x$$
$$x_n = a + n\Delta x = b$$

The right sum, R_n, with n rectangles is

$$R_n = f(x_1)\Delta x + f(x_2)\Delta x + f(x_3)\Delta x + \cdots + f(x_n)\Delta x$$
$$= \sum_{i=1}^{n} f(x_i)\Delta x$$

and the left sum is

$$L_n = f(x_0)\Delta x + f(x_1)\Delta x + f(x_2)\Delta x + \cdots + f(x_{n-1})\Delta x$$
$$= \sum_{i=0}^{n-1} f(x_i)\Delta x$$

Figure 7.3.25 shows rectangles belonging to R_n for $n = 10$, $n = 20$, and $n = 30$. In Figure 7.3.26 we show rectangles belonging to the left sum, L_n, for $n = 10$, $n = 20$, and $n = 30$.

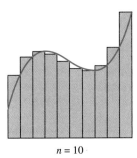
$n = 10$
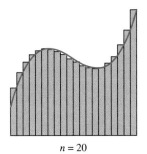
$n = 20$
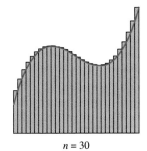
$n = 30$

FIGURE 7.3.25
Rectangles for the right Riemann sums, R_n

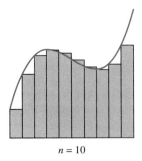
$n = 10$
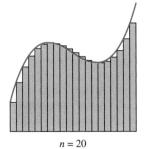
$n = 20$
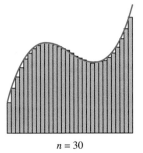
$n = 30$

FIGURE 7.3.26
Rectangles for the left Riemann sums, L_n

It can be proven that, for continuous functions, as n increases, both R_n and L_n converge—and *converge to the same number*. Moreover, we can form the sum

$$S_n = \sum_{i=1}^{n} f(x_i^*)\Delta x$$

where x_i^* is any point in a subinterval (to calculate R_n, we used x_i^* = right endpoint; for the left sum, L_n, we let x_i^* = left endpoint; and for the midpoint sum, M_n, we used x_i^* = midpoint; we form other Riemann sums by changing the rule for x_i^*; as well, we can choose x_i^* randomly within each subinterval if we wish to do so). S_n is the **general form of a Riemann sum.** The sum S_n, *no matter how the x_i^* are chosen,* will converge to the common limit of all Riemann sums. Thus, the following definition makes sense.

Definition 7.3.1 Let $f(x)$ be a continuous function defined on $[a, b]$ such that $f(x) \geq 0$ for all x in $[a, b]$. The **area, A, of the region under $f(x)$ on $[a, b]$** is

$$A = \lim_{n \to \infty} S_n = \lim_{n \to \infty} \sum_{i=1}^{n} f(x_i^*)\Delta x$$

where S_n is any of the left sum, the right sum, the midpoint sum, and any other Riemann sum, provided that the limit exists. The **definite integral of $f(x)$ on $[a, b]$**, denoted by

$$\int_a^b f(x)dx$$

is defined by

$$\int_a^b f(x)dx = \lim_{n \to \infty} S_n = \lim_{n \to \infty} \sum_{i=1}^{n} f(x_i^*)\Delta x$$

if the limit exists.

It can be proven that the limits in the definition exist (i.e., are real numbers) whenever $f(x)$ is continuous on $[a, b]$.

The concept of area in Definition 7.3.1 is an extension of our common geometric understanding of area (as in "the area of a triangle is one-half times the base length times the height," or "the area of a rectangle is the product of the length and the width").

The reason we use the integral sign $\int$ for both the antiderivative and the definite integral will become transparent in the next section, where we discover how they relate to each other.

The numbers a and b in the definite integral $\int_a^b f(x)dx$ are called the **limits of integration.**

In light of Definition 7.3.1, we write the result of Example 7.3.3 and the subsequent calculation as

$$\int_0^1 x^2 dx = \frac{1}{3}$$

Looking at Example 7.3.4, we conclude that (based on the left, right, and midpoint sums with 100,000 rectangles)

$$\int_1^3 \ln x\, dx \approx 1.29584$$

496 Chapter 7 Integrals and Applications

Let us repeat that when $f(x) \geq 0$ on $[a, b]$, then

$$\int_a^b f(x)dx$$

is equal to the area of the region below $f(x)$ on $[a, b]$ (and is therefore a positive number—or zero if $f(x) = 0$ for all x in $[a, b]$).

Usually, calculating the value of a definite integral using Riemann sums is messy, hard, and/or impossible to do. The good news is that—in some situations—we will be able to avoid it. In the next section, we will introduce an important algebraic way of calculating definite integrals.

However, that will not make the concept of Riemann sums obsolete—as a matter of fact, any computer software that calculates definite integrals uses Riemann sums, or a similar numeric approach.

In the next few examples, we explore the connection between definite integral and area. These are cases that involve geometrically simple regions that will allow us to calculate definite integrals very quickly.

Example 7.3.5 Calculating Definite Integrals

Compute

$$\int_{-2}^{3} 4\, dx$$

Let's interpret this integral geometrically. The function involved is the constant function $f(x) = 4$, and the interval is $[-2, 3]$. The region under $f(x)$ is a rectangle (see Figure 7.3.27a), and therefore

$$\int_{-2}^{3} 4\, dx = \text{area of the shaded rectangle} = 5 \cdot 4 = 20$$

To strengthen this message, we discuss another example. The region corresponding to the integral

$$\int_0^8 x\, dx$$

is the triangle bounded by the graph of $f(x) = x$ on the interval $[0, 8]$; see Figure 7.3.27b. We conclude that

$$\int_0^8 x\, dx = \text{area of the shaded triangle} = \frac{1}{2}(8 \cdot 8) = 32$$

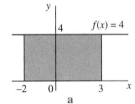
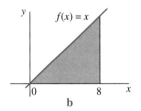

FIGURE 7.3.27

Interpreting the definite integral as an area

Example 7.3.6 Calculating Definite Integrals II

Compute

$$\int_{-1}^{4} |x|\, dx$$

Because $f(x) = |x| \geq 0$ for all x in $[-1, 4]$, we interpret the given integral as an area. The region in question is bounded from above by the graph of $f(x) = |x|$, from below by the x-axis, and from the sides by the vertical lines $x = -1$ and $x = 4$; see Figure 7.3.28. The region consists of two triangles, and

$$\int_{-1}^{4} |x|\, dx = \text{sum of the areas of the two triangles} = \frac{1}{2}(1 \cdot 1) + \frac{1}{2}(4 \cdot 4) = \frac{17}{2}$$

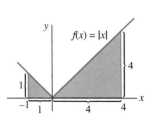

FIGURE 7.3.28

Definite integral as an area

Example 7.3.7 Calculating Definite Integrals III

Compute

$$\int_{0}^{2} \sqrt{4 - x^2}\, dx$$

In this case, the function is $f(x) = \sqrt{4 - x^2}$, defined on $[0, 2]$. Since $f(x) \geq 0$, we know that the graph of $f(x)$ lies above the x-axis, or touches it. Squaring $y = \sqrt{4 - x^2}$ and rearranging the terms, we get

$$y^2 = 4 - x^2$$
$$x^2 + y^2 = 4$$

which represents the circle centred at the origin of radius 2. The graph of $f(x)$ is the upper half of that circle; see Figure 7.3.29. It follows that

$$\int_{0}^{2} \sqrt{4 - x^2}\, dx = \frac{1}{4} \text{ of the area of the circle of radius } 2 = \frac{1}{4}\pi(2)^2 = \pi$$

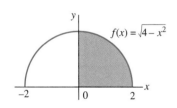

FIGURE 7.3.29

Interpreting a definite integral as an area

The Definite Integral

So far, we have learned what the definite integral

$$\int_{a}^{b} f(x)\, dx$$

is, and what it represents, for *positive continuous functions* defined on $[a, b]$. Since area cannot be negative, the definite integral satisfies the following:

If $f(x) \geq 0$ on $[a, b]$, then $\int_{a}^{b} f(x)\, dx \geq 0$.

How do we define the definite integral for general functions that are not necessarily positive? The answer is—in much the same way. However, we will have to be a bit more careful about how we interpret it.

Given any continuous function $f(x)$ on $[a, b]$, we form Riemann sums S_n (left sum, right sum, midpoint sum, and so forth) in exactly the same way as before. So,

$$S_n = \sum_{i=1}^{n} f(x_i^*) \Delta x$$

where the interval $[a, b]$ has been divided into n subintervals of length Δx, and x_i^* is the point (left endpoint, right endpoint, midpoint, etc.) at which f is evaluated.

It can be proven that if $f(x)$ is continuous, then all Riemann sums *converge to the same real number*.

Definition 7.3.2 Let $f(x)$ be a continuous function defined on an interval $[a, b]$. The **definite integral of $f(x)$ on $[a, b]$** is the real number

$$\int_a^b f(x)dx = \lim_{n \to \infty} S_n$$

There are some discontinuous functions for which the above limit exists as well (we will say a bit more about this later in this section).

In general, a function $f(x)$ defined on $[a, b]$ for which

$$\lim_{n \to \infty} S_n$$

exists (i.e., is a real number) is called **integrable** on $[a, b]$. (Thus, we see that all continuous functions are integrable.)

The symbol $\int$ in

$$\int_a^b f(x)dx$$

is called an **integral sign**. The numbers a and b are the **limits of integration** (a is the **upper limit** and b is the **lower limit**), and $f(x)$ is called the **integrand**. The symbol dx indicates that the variable involved is x. A good way of thinking about $f(x)dx$ is to think of a rectangle in a Riemann sum (the value of the function times the shrinking base length of the rectangle).

Keep in mind that the antiderivative

$$\int f(x)dx$$

is a function (actually a family of functions whose members differ by a constant), whereas the definite integral

$$\int_a^b f(x)dx$$

is a real number. So, in definite integrals, it does not matter what variable we use, i.e.,

$$\int_a^b f(x)dx = \int_a^b f(t)dt = \int_a^b f(u)du$$

7.3 Definite Integral and Area

Can we interpret the definite integral geometrically, as we did for positive functions?

Let us consider an example. We compute Riemann sums R_{10} and L_{10} for the function $f(x) = x^2 - 2x$ defined on $[-1, 4]$.

The length of $[-1, 4]$ is 5, and so $\Delta x = 5/10 = 0.5$. The subintervals of $[a, b] = [-1, 4]$ are

$$[x_0, x_1], [x_1, x_2], [x_2, x_3], \ldots, [x_8, x_9], [x_9, x_{10}]$$

The numbers x_i, together with the corresponding values $f(x_i)$, are given in the table below.

x_i	$f(x_i)$	x_i	$f(x_i)$
$x_0 = a = -1$	3		
$x_1 = -0.5$	1.25	$x_6 = 2$	0
$x_2 = 0$	0	$x_7 = 2.5$	1.25
$x_3 = 0.5$	-0.75	$x_8 = 3$	3
$x_4 = 1$	-1	$x_9 = 3.5$	5.25
$x_5 = 1.5$	-0.75	$x_{10} = b = 4$	8

Thus

$$R_{10} = \sum_{i=1}^{10} f(x_i) \Delta x$$
$$= f(x_1)\Delta x + f(x_2)\Delta x + f(x_3)\Delta x + \cdots + f(x_{10})\Delta x$$
$$= (1.25)(0.5) + (0)(0.5) + (-0.75)(0.5) + (-1)(0.5) + (-0.75)(0.5)$$
$$+ (0)(0.5) + (1.25)(0.5) + (3)(0.5) + (5.25)(0.5) + (8)(0.5)$$
$$= 8.125$$

and

$$L_{10} = \sum_{i=0}^{9} f(x_i) \Delta x$$
$$= f(x_0)\Delta x + f(x_1)\Delta x + f(x_2)\Delta x + \cdots + f(x_9)\Delta x$$
$$= (3)(0.5) + (1.25)(0.5) + (0)(0.5) + (-0.75)(0.5) + \cdots + (5.25)(0.5)$$
$$= 5.625$$

If we ignore the minus signs in R_{10}, we see that all terms in it represent the areas of the approximating rectangles drawn in Figure 7.3.30a. Where the terms differ is how

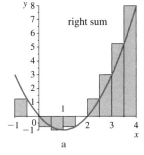
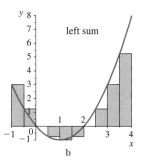

FIGURE 7.3.30
Interpreting Riemann sums as a signed area

they contribute to R_{10}: the areas of those rectangles that are above the x-axis are added to R_{10}, whereas the areas of those that are below the x-axis are subtracted from R_{10}. Thus,

$R_{10} =$ (sum of the areas of the rectangles above the x-axis)
$\quad -$ (sum of the areas of the rectangles below the x-axis)

The same can be said for the left sum L_{10} (Figure 7.3.30b), as well as for any sum with any number of rectangles. Since the definite integral is the limit of approximating (Riemann) sums, this means that

$$\int_a^b f(x)dx$$

is the *difference of areas*, i.e.,

$$\int_a^b f(x)dx = \text{(sum of the areas of the regions above the } x\text{-axis)}$$
$\quad -$ (sum of the areas of the regions below the x-axis)

or, in short,

$$\int_a^b f(x)dx = \text{``area above''} - \text{``area below''}$$

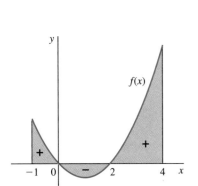

FIGURE 7.3.31

Interpreting the definite integral as a signed area

This difference in areas is sometimes called the **signed area.** For the function in Figure 7.3.31,

$$\int_{-1}^{4} f(x)dx = \text{(area from } -1 \text{ to } 0) - \text{(area from 0 to 2)} + \text{(area from 2 to 4)}$$

Example 7.3.8 Evaluating Integrals Using Signed Area I

Find the value of

$$\int_0^6 (2x - 4)dx$$

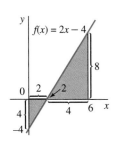

FIGURE 7.3.32

Calculating the definite integral using signed area

The graph of the function $f(x) = 2x - 4$ is the line of slope 2 that intersects the x-axis at $x = 2$ (Figure 7.3.32). It is both positive and negative on $[0, 6]$, and thus

$$\int_0^6 (2x - 4)dx = \text{(area of the triangle above the } x\text{-axis)}$$
$\quad -$ (area of the triangle below the x-axis)

The areas of the two triangles are computed to be 16 and 4, and so

$$\int_0^6 (2x - 4)dx = 16 - 4 = 12$$

Example 7.3.9 Evaluating Integrals Using Signed Area II

The region corresponding to the integral

$$\int_0^6 (-4)dx$$

is the rectangle over the interval [0, 6] of height 4 that lies entirely below the *x*-axis. The area of the rectangle is 24, and therefore

$$\int_0^6 (-4)dx = -24$$

Example 7.3.10 Evaluating Integrals Using Signed Area III

Find

$$\int_0^\pi \cos x\, dx$$

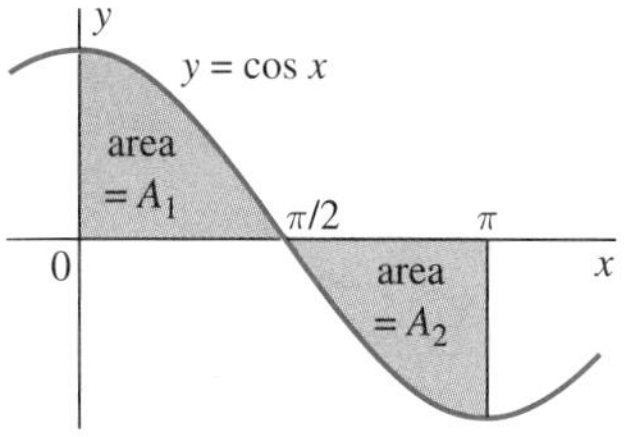

Looking at the graph in Figure 7.3.33, we see that

$$\int_0^\pi \cos x\, dx = \text{area above} - \text{area below} = A_1 - A_2$$

Due to symmetry, the two shaded regions have the same area. It follows that

$$\int_0^\pi \cos x\, dx = 0$$

FIGURE 7.3.33

The definite integral as a signed area

Example 7.3.11 Interpreting the Value of a Definite Integral

What can we say about a continuous function $f(x)$ if we know that

$$\int_a^b f(x)dx = -5$$

Because the integral is negative, we know that $f(x)$ has to be negative on at least part of the interval [*a*, *b*].

Think about the region between the graph of $f(x)$ and the *x*-axis bounded by $x = a$ and $x = b$. If the whole region is below the *x*-axis, then its area is 5. Otherwise, all we can say is that more of the region (5 area units) lies below the *x*-axis.

Recall that a function is called integrable if the limit of all of its Riemann sums is the same real number. We mentioned the fact—without providing a proof—that all continuous functions defined on an interval [*a*, *b*] are integrable. In our next example, we show that a discontinuous function can be integrable as well. We will not prove it using Riemann sums, but instead we will use the interpretation of a definite integral as a signed area.

Example 7.3.12 An Integrable Discontinuous Function

Consider the function $f(x)$ defined on [0, 5] whose graph is drawn in Figure 7.3.34. It has two points of discontinuity, at $x = 2$ and at $x = 4$. However, we can still use the interpretation

$$\int_0^5 f(x)dx = \text{area above} - \text{area below}$$

Thus,

$$\int_0^5 f(x)dx = (\text{area from 0 to 2}) - (\text{area from 2 to 4}) + (\text{area from 4 to 5})$$

$$= 2 \cdot 1 - 2 \cdot 3 + 1 \cdot 2 = -2 \qquad \blacktriangle$$

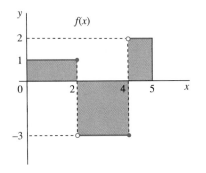

FIGURE 7.3.34

An integrable function that is not continuous

Properties of Integrals

We now list several properties of definite integrals that we will use. First of all, we recall

$$\int_a^b f(x)dx = \lim_{n \to \infty} (\text{any Riemann sum})$$

so, when appropriate, or needed, we think of using rectangles (Riemann sums) to get an approximate value for the integral.

We define

$$\int_a^a f(x)dx = 0$$

This makes sense—the region under $f(x)$ on $[a, a]$ is the vertical line segment from the x-axis to $f(a)$. Since a line segment is not a two-dimensional object, it has no area.

Changing the order of the upper and lower limits of integration results in

$$\int_b^a f(x)dx = -\int_a^b f(x)dx$$

The minus sign accounts for the fact that the orientation within the interval has been reversed from positive (direction of the x-axis) to negative.

If c is a constant, how does $\int_a^b cf(x)dx$ relate to $\int_a^b f(x)dx$? Assume that $c > 0$, and think of how the Riemann sums for the two integrals compare. An approximating rectangle belonging to a Riemann sum for $\int_a^b cf(x)dx$ has the same base length Δx as the corresponding rectangle in the Riemann sum for $\int_a^b f(x)dx$; see Figure 7.3.35. However, its height is scaled by the factor of c, and so its area is equal to c times the area of the approximating rectangle for $\int_a^b f(x)dx$. Taking the limit (and keeping in mind that we can factor a constant out of the limit), we get

$$\int_a^b cf(x)dx = c\int_a^b f(x)dx$$

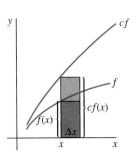

FIGURE 7.3.35

Comparing the Riemann sums of $f(x)$ and $cf(x)$

The formula is true for $c = 0$ (since both sides are zero). Arguing similarly as above, we could prove that it holds for negative values of c as well.

FIGURE 7.3.36

Properties of the definite integral

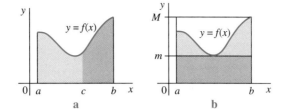

Using the Riemann sums argument again, we could prove that

$$\int_a^b (f(x) \pm g(x))dx = \int_a^b f(x)dx \pm \int_a^b g(x)dx$$

Generalizing our calculations in Examples 7.3.5 and 7.3.9, we write

$$\int_a^b c\,dx = c(b-a)$$

The **summation property** for definite integrals states that

$$\int_a^b f(x)dx = \int_a^c f(x)dx + \int_c^b f(x)dx$$

for any numbers a, b, and c. The general case (any arrangement of the three numbers, such as $b < c$ or $a > b$) is not easy to prove; however, if we think of a positive function $f(x)$, and if c is inside the interval $[a, b]$ (as in Figure 7.3.36a), then we can use the area argument:

$$\int_a^b f(x)dx = \text{area from } a \text{ to } b$$

$$= (\text{area from } a \text{ to } c) + (\text{area from } c \text{ to } b)$$

$$= \int_a^c f(x)dx + \int_c^b f(x)dx$$

Finally, assume that $m \leq f(x) \leq M$ for all x in $[a, b]$. Then

$$m(b-a) \leq \int_a^b f(x)dx \leq M(b-a)$$

Again, if $f(x)$ is positive on $[a, b]$, this is easy to see (Figure 7.3.36b): the region under $f(x)$ on $[a, b]$ contains the rectangle over $[a, b]$ whose height is m, and is contained in the rectangle over the same interval whose height is M. Thus, the area under $f(x)$ is between the areas of the two rectangles.

To summarize, we list the properties of the definite integral that we have discussed so far.

Assume that $f(x)$ and $g(x)$ are continuous functions and a, b, and c are real numbers such that $a < b$. Then we have the following:

(1) $\int_a^a f(x)dx = 0$

(2) $\int_b^a f(x)dx = -\int_a^b f(x)dx$

(3) $\int_a^b cf(x)dx = c\int_a^b f(x)dx$ for any real number c.

(4) $\int_a^b (f(x) \pm g(x))dx = \int_a^b f(x)dx \pm \int_a^b g(x)dx$

(5) $\int_a^b c\,dx = c(b-a)$

(6) $\int_a^b f(x)dx = \int_a^c f(x)dx + \int_c^b f(x)dx$

(7) If $m \leq f(x) \leq M$ for all x in $[a, b]$, then $m(b-a) \leq \int_a^b f(x)dx \leq M(b-a)$.

Example 7.3.13 Estimating the Definite Integral

Using minimum and maximum values of the function $f(x) = \dfrac{1}{1+x^3}$ on the interval $[2, 5]$, find an upper bound and a lower bound for the integral

$$\int_2^5 \frac{1}{1+x^3}\,dx$$

The function $f(x)$ is continuous on $[2, 5]$ and thus, by the Extreme Value Theorem, must have an absolute minimum and an absolute maximum—and these are the numbers m and M from the estimate

$$m(5-2) \leq \int_2^5 \frac{1}{1+x^3}\,dx \leq M(5-2)$$

that we will use. Since $f(x)$ is decreasing, we get

$$m = f(5) = \frac{1}{126} \quad \text{and} \quad M = f(2) = \frac{1}{9}$$

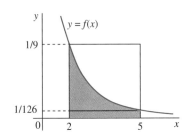

FIGURE 7.3.37

Estimating the definite integral

(see Figure 7.3.37) and therefore

$$\frac{1}{126}(5-2) \leq \int_2^5 \frac{1}{1+x^3}\,dx \leq \frac{1}{9}(5-2)$$

$$\frac{3}{126} \leq \int_2^5 \frac{1}{1+x^3}\,dx \leq \frac{3}{9}$$

$$0.02381 \leq \int_2^5 \frac{1}{1+x^3} dx \leq 0.33333$$

Note that using minimum and maximum values of $f(x)$ on the whole interval $[2, 5]$ gives a wide range for the value of the integral. In general, better techniques of estimation are needed. ◢

Example 7.3.14 Riemann Sums, Definite Integral, and Total Change

Suppose we wish to find the amount of chemical, p, produced after 2 seconds in a reaction that is modelled by

$$\frac{dp}{dt} = e^{-t}$$

We are in a situation where we know how the function changes (instantaneously) at any time t—that's what dp/dt represents—and are asked to calculate the total (net) change in p. In other words, we have to figure out the difference

$$p(2) - p(0)$$

that will tell us how much p changed from $t = 0$ to $t = 2$.

The idea is to look at the changes in p over small intervals. We divide the interval $[0, 2]$ into 10 subintervals, each of length

$$\Delta t = \frac{2}{10} = 0.2$$

Look at the first subinterval, $[0, 0.2]$: the rate of change of p, given by e^{-t}, is *not constant* on $[0, 0.2]$, but we *approximate it by a constant function*. We take the value of the approximation to be equal to the value of e^{-t} at the start of the interval, i.e., $e^{-0} = 1$. We do the same for all remaining subintervals—we approximate the function $f(t) = e^{-t}$ by its value at the start of each subinterval. In this way, we obtain a step function (Figure 7.3.38a).

FIGURE 7.3.38

Step function approximation for the rate of change

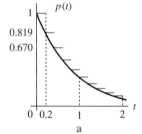
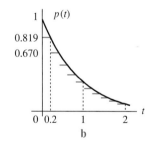

Over the interval $[0, 0.2]$ the rate of change of p is approximated by the constant function 1; thus, the total change in p on $[0, 0.2]$ is

$$(\text{rate of change})(\text{time}) = 1 \cdot 0.2 = 0.2$$

Over $[0.2, 0.4]$, the rate of change of p is again assumed constant, approximated by the value

$$e^{-0.2} \approx 0.819$$

Thus, on the interval [0.2, 0.4], p changes by

$$(\text{rate of change})(\text{time}) = 0.819 \cdot 0.2 = 0.164$$

Note that we started calculating the left sum of e^{-t} with 10 subintervals. This time, instead of interpreting each term of the form

$$(\text{value of the function})(\text{the length of the subinterval})$$

as

$$(\text{height})(\text{base length})$$

(i.e., area of a rectangle) we interpret it as

$$(\text{rate of change of the function})(\text{time interval during which the function changed at that rate})$$

(i.e., total change over the subinterval).

In this context, L_{10} is the sum of the approximate changes in p over all subintervals, which represents (an approximation for) the total change in p from $t=0$ to $t=2$. We compute

$$L_{10} = e^{-0} \cdot 0.2 + e^{-0.2} \cdot 0.2 + e^{-0.4} \cdot 0.2 + e^{-0.6} \cdot 0.2 + e^{-0.8} \cdot 0.2$$
$$+ e^{-1} \cdot 0.2 + e^{-1.2} \cdot 0.2 + e^{-1.4} \cdot 0.2 + e^{-1.6} \cdot 0.2 + e^{-1.8} \cdot 0.2$$
$$\approx 0.2(1 + 0.819 + 0.670 + 0.549 + 0.449$$
$$+ 0.368 + 0.301 + 0.247 + 0.202 + 0.165)$$
$$\approx 0.954$$

Similarly, we could approximate the rate of change of p on each subinterval by a constant function whose value is the value of e^{-t} at the end of the subinterval (Figure 7.3.38b). In this way, we would obtain the right sum approximation $R_{10} \approx 0.781$.

Using software that calculates Riemann sums, we obtain values of L_n and R_n for large n. For example, the sums

$$L_{1000} = 0.8655297$$
$$R_{1000} = 0.8638003$$
$$L_{5000} = 0.8648377$$
$$R_{5000} = 0.8644918$$

approximate the total change in p.

As $n \to \infty$, the Riemann sums approach the definite integral

$$\int_0^2 e^{-t} dt \approx 0.8646647$$

Thus, the total change in p from $t=0$ to $t=2$ is

$$p(2) - p(0) = \int_0^2 e^{-t} dt \approx 0.8646647$$

The conclusion of our previous example is worth remembering:

> **Assume that $f(x)$ represents the *rate of change* of a quantity $p(x)$ on $[a, b]$. Then we have the following:**
>
> A Riemann sum of $f(x)$ on $[a, b]$ approximates the *total change* $p(b) - p(a)$.
>
> The definite integral $\int_a^b f(x)\, dx$ is equal to the *total change* $p(b) - p(a)$.

Summary We defined the **definite integral** as a limit of approximating sums, called **Riemann sums**. The definite integral is related to the **area:** if a function is positive, then the definite integral is equal to the area below the function. In general, the definite integral represents the **signed area,** i.e., the difference between the area of the region(s) above the x-axis and the area of the region(s) below the x-axis. Near the end of the section, we listed the important properties of definite integrals.

7.3 Exercises

Mathematical Techniques

1–4 ▪ Estimate the area under the graph of $f(x)$ on the given interval using the specified Riemann sum. In each case, sketch the graph of the function and the approximating rectangles. State whether your approximation is an underestimate or an overestimate of the actual area.

1. $f(x) = \dfrac{1}{x}$, from $x = 1$ to $x = 4$, using R_6.
2. $f(x) = e^{-x}$, from $x = 0$ to $x = 1$, using R_5.
3. $f(x) = e^{-x}$, from $x = 0$ to $x = 1$, using L_5.
4. $f(x) = 16 - x^2$, from $x = 3$ to $x = 4$, using M_2.

5–12 ▪ Write each sum in expanded form. If possible, evaluate it.

5. $\sum_{i=2}^{5} i^2$
6. $\sum_{j=0}^{4} (1 + 2j)$
7. $\sum_{i=7}^{10} (2f(x_i) - 3)$
8. $\sum_{i=3}^{4} \dfrac{1}{i}$
9. $\sum_{j=0}^{4} 3^j$
10. $\sum_{i=1}^{4} (f(a_i) - f(a_{i-1}))$
11. $\sum_{i=0}^{4} f(x_j) \Delta x_j$
12. $\sum_{i=6}^{11} 5$

13–16 ▪ Write each sum using sigma notation (do not evaluate the sum).

13. $2 + 4 + 6 + \cdots + 2n$
14. $\dfrac{1}{2} + \dfrac{2}{3} + \dfrac{3}{4} + \cdots + \dfrac{9}{10}$
15. $\sin \dfrac{\pi}{2} + \sin \dfrac{\pi}{3} + \sin \dfrac{\pi}{4} + \cdots + \sin \dfrac{\pi}{12}$
16. $\sqrt{2} + \sqrt{4} + \sqrt{6} + \cdots + \sqrt{122}$

17–21 ▪ For the given integral, write the required approximating sum using sigma notation.

17. $\displaystyle\int_0^1 e^x\, dx$; find R_n.
18. $\displaystyle\int_0^{10} e^{-x}\, dx$; find L_n.
19. $\displaystyle\int_0^1 x^4\, dx$; find L_n.
20. $\displaystyle\int_1^5 \ln x\, dx$; find R_n.
21. $\displaystyle\int_1^2 \dfrac{1}{x}\, dx$; find L_n.

22–23 ▪ Consider the function $f(x) = x^2$ on the interval $[0, 1]$ from Example 7.3.3.

22. By imitating the calculation done there, show that the left sum is given by $L_n = \dfrac{1}{n^3}\left(0^2 + 1^2 + 2^2 + \cdots + (n-1)^2\right)$. Use the summation formula from the example to obtain $L_n = \dfrac{1}{6n^2}(n-1)(2n-1)$ and compute the limit of L_n as $n \to \infty$.

23. Find the midpoints of all intervals listed at the start of the example (recall that the midpoint of the interval $[a,b]$ is $(a+b)/2$). Then show that $M_n = \frac{1}{4n^3}(1^2 + 3^2 + 5^2 + \cdots + (2n-1)^2)$. Use the formula $1^2 + 3^2 + 5^2 + \cdots + (2n-1)^2 = \frac{1}{3}n(2n-1)(2n+1)$ to obtain a formula for M_n. Compute the limit of M_n as $n \to \infty$.

24–27 ▪ In each case:

a. Find the formula for the Riemann sum of the function $f(x)$ on the given interval.

b. Compute the limit of the Riemann sum in (a) as $n \to \infty$.

c. What definite integral does your answer in (b) represent?

d. Use a different argument to confirm that you answer in (b) is correct.

In your calculations, you will need the summation formula $1 + 2 + 3 + \cdots + m = \frac{1}{2}m(m+1)$ or $1 + 3 + 5 + \cdots + (2m-1) = m^2$.

24. $f(x) = x$, $[0, 1]$; R_n

25. $f(x) = x$, $[0, 2]$; M_n

26. $f(x) = 4x$, $[0, 1]$; L_n

27. $f(x) = x + 2$, $[0, 1]$; R_n

28–40 ▪ Evaluate each definite integral by interpreting it as an area or as a difference of areas (signed area).

28. $\int_{-10}^{20} 6\,dx$

29. $\int_{-6}^{4} x\,dx$

30. $\int_0^4 (1 - 2x)\,dx$

31. $\int_0^4 |x - 2|\,dx$

32. $\int_0^{10} 3|x|\,dx$

33. $\int_0^4 \sqrt{16 - x^2}\,dx$

34. $\int_{-2}^{2} \sqrt{4 - x^2}\,dx$

35. $\int_{-1}^{1} \tan x\,dx$

36. $\int_0^{2\pi} \cos x\,dx$

37. $\int_0^2 (3 + \sqrt{4 - x^2})\,dx$

38. $\int_{-1}^{4} (6 - |x|)\,dx$

39. $\int_{-2}^{2} \arctan x\,dx$

40. $\int_{-2}^{3} (1 - |x|)\,dx$

41–44 ▪ Find the definite integral of each discontinuous function.

41. Find $\int_0^5 f(x)\,dx$ for $f(x) = \begin{cases} -2 & \text{if } 0 \leq x \leq 2 \\ 4 & \text{if } 2 < x \leq 3 \\ -1 & \text{if } 3 < x \leq 5 \end{cases}$

42. Find $\int_{-2}^{1} f(x)\,dx$ for $f(x) = \begin{cases} x + 2 & \text{if } -2 \leq x \leq 0 \\ 1 & \text{if } 0 < x \leq 1 \end{cases}$

43. Find $\int_{-1}^{5} f(x)\,dx$ for $f(x) = \begin{cases} x - 1 & \text{if } -1 \leq x \leq 2 \\ x - 2 & \text{if } 2 < x \leq 5 \end{cases}$

44. Find $\int_{-1}^{2} f(x)\,dx$ for $f(x) = \begin{cases} \sqrt{1 - x^2} & \text{if } -1 \leq x \leq 0 \\ \sqrt{4 - x^2} & \text{if } 0 < x \leq 2 \end{cases}$

45–50 ▪ Determine whether each integral is positive, negative, or zero. Justify your answer.

45. $\int_0^3 \cos x\,dx$

46. $\int_0^4 \cos x\,dx$

47. $\int_0^{3\pi} \sin 2x\,dx$

48. $\int_0^{3\pi} \sin x\,dx$

49. $\int_{-\pi/4}^{0} \arctan x\,dx$

50. $\int_{-4}^{4} \arctan 2x\,dx$

51–54 ▪ Using Property (7) in the subsection Properties of Integrals, find an upper bound and a lower bound for each integral.

51. $\int_0^1 e^{-x}\,dx$

52. $\int_0^2 \frac{1}{1 + x^5}\,dx$

53. $\int_e^{10} \ln x\,dx$

54. $\int_0^1 \arctan x \, dx$

55–59 ■ Explain why each inequality is true. (Hint: Compare the region under the curve with a rectangle.)

55. $\int_0^1 e^{-x} dx < 1$

56. $\int_1^2 \frac{1}{x} dx > \frac{1}{2}$

57. $\int_0^\pi \sin x \, dx < \pi$

58. $\int_1^e \ln x \, dx < e - 1$

59. $\int_2^3 \frac{1}{1+x^2} dx < \frac{1}{5}$

Applications

60–63 ■ Use summation notation and find the total number of offspring for each of the following organisms.

60. The organism has two offspring in year 1, three offspring in year 2, five offspring in year 3, four offspring in year 4, and one offspring in year 5.

61. The organism has 0 offspring in year 1, 8 offspring in year 2, 15 offspring in year 3, 24 offspring in year 4, 31 offspring in year 5, 11 offspring in year 6, and 3 offspring in year 7.

62. The organism has $B_i = i(6 - i)$ offspring in years $i = 0$ through $i = 6$.

63. The organism has $B_i = \frac{i(i+1)}{2} + 4$ offspring in years $i = 0$ through $i = 7$.

64–67 ■ Find the left- and right-hand estimates for the definite integral of each function. Think of $f(t)$ as a rate of influx.

64. $f(t) = 2t$, limits of integration 0 to 1, $n = 5$.

65. $f(t) = 2t$, limits of integration 0 to 2, $n = 5$.

66. $f(t) = t^2$, limits of integration 0 to 2, $n = 5$.

67. $f(t) = 1 + t^3$, limits of integration 0 to 1, $n = 5$.

68–71 ■ Use Euler's method to estimate the solution of each differential equation with the given parameters. Suppose that $V(0) = 0$ in each case. Your answer should exactly match one of the estimates in Exercise 64–67. Can you explain why?

68. $\frac{dV}{dt} = 2t$; estimate $V(1)$ using $\Delta t = 0.2$ (as in Exercise 64).

69. $\frac{dV}{dt} = 2t$; estimate $V(2)$ using $\Delta t = 0.4$ (as in Exercise 65).

70. $\frac{dV}{dt} = t^2$; estimate $V(2)$ using $\Delta t = 0.4$ (as in Exercise 66).

71. $\frac{dV}{dt} = 1 + t^3$; estimate $V(1)$ using $\Delta t = 0.2$ (as in Exercise 67).

72–75 ■ Suppose the speed of a bee is given in the following table:

Time (s)	Speed (cm/s)
0	127
1	122
2	118
3	115
4	113
5	112
6	112
7	113
8	116
9	120
10	125

72. Using the measurements on even-numbered seconds, find the left-hand and right-hand estimates for the distance the bee moved during the experiment.

73. Using all of the measurements, find the left-hand and right-hand estimates for the distance the bee moved during the experiment.

74. Use a midpoint sum to estimate the distance moved (using all of the measurements).

75. Figure out a way to use the measurements on odd-numbered seconds to estimate the distance the bee moved during the experiment.

76–79 ■ Biologists measured the number of aspen that germinate in four sites over eight years but could measure only two sites per year. In the following table, NA indicates that no measurement was made in that year:

Year	Site 1	Site 2	Site 3	Site 4
2007	12	23	NA	NA
2008	NA	NA	34	10
2009	16	NA	NA	15
2010	17	21	NA	NA
2011	NA	NA	40	18
2012	NA	23	31	NA
2013	NA	27	NA	8
2014	13	NA	37	NA

The goal is to estimate the total number of aspen that germinated in each of the four sites during all eight years.

76. In site 1, estimate the total number of aspen using a modification of the left-hand estimate. How does this compare

with finding the total number that germinated in the four years with measurements and multiplying by 2? Why are they different?

77. In site 2, estimate the total number of aspen using a modification of the left-hand estimate. How does this compare with finding the total number that germinated in the four years with measurements and multiplying by 2?

78. In site 3, estimate the total number of aspen using a modification of the right-hand estimate. How does this compare with finding the total number that germinated in the four years studied and multiplying by 2?

79. In site 4, no measurement was made in the first and last years. Come up with some variation on the left-hand or right-hand estimate to estimate the total number of aspen. Compare the result with finding the total number that germinated in the four years studied and multiplying by 2.

Computer Exercises

80. We will compare various methods used to estimate the solution of
$$\frac{dV}{dt} = t^2$$
with $V(0) = 0$. We wish to find $V(2)$.

 a. Graph the rate of change as a function of t.
 b. Use the right-hand estimate with $\Delta t = 0.2, 0.1$, and 0.02.
 c. Use the left-hand estimate with $\Delta t = 0.2, 0.1$, and 0.02.
 d. Find the average of these two estimates for each Δt.

81. We will compare various methods used to estimate the solution of
$$\frac{dp}{dt} = \ln(1 + \sqrt{t} - t^3)$$
with $p(0) = 0$. We wish to find $p(1)$.

 a. Graph the rate of change as a function of t.
 b. Use the right-hand estimate with $\Delta t = 0.2, 0.1$, and 0.02.
 c. Use the left-hand estimate with $\Delta t = 0.2, 0.1$, and 0.02.
 d. Try to figure out from your graph why the two estimates are the same.

7.4 Definite and Indefinite Integrals

We now have two types of integrals. The **antiderivative** or the **indefinite integral**
$$\int f(x)\,dx$$
is any function $F(x) + C$ such that $F'(x) = f(x)$. If we are given an initial condition, we can calculate the value of the constant C.

The **definite integral**
$$\int_a^b f(x)\,dx$$
is a real number, defined as the limit of Riemann sums.

What is the relationship between these two mathematical approaches? In this section, we derive the connection, which is called the **Fundamental Theorem of Calculus.**

The Fundamental Theorem of Calculus: Computing Definite Integrals with Indefinite Integrals

First, we take another look at Riemann sums.

Example 7.4.1 Riemann Sums and the Total Change of a Function

A cell is modelled as a sphere of radius x (x is measured in micrometres (μm); 1 μm $= 10^{-6}$ m). In order to grow, the cell needs to absorb the material from its neighbourhood. It is known that the rate of absorption is proportional to the surface area of the cell, i.e.,
$$V'(x) = k \cdot 4\pi x^2$$

where $V(x)$ is the volume of the material absorbed and k is a constant. (Recall that the surface area of a sphere of radius x is $4\pi x^2$.) For simplicity, we assume that $4\pi k = 1$, so that
$$V'(x) = x^2$$
(We interpret $V'(x) = dV/dx$ as the rate of change in the volume of absorbed material with respect to the radius of the cell.)

We need to estimate the amount of material absorbed as the cell grows from $x = 0$ (just created) to $x = 1$.

As in Example 7.3.14 in the previous section, we divide the interval $[0, 1]$ into subintervals (say, five) and approximate the rate of change on each subinterval by a constant function; see Table 7.4.1 and Figure 7.4.39.

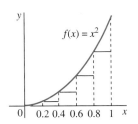

FIGURE 7.4.39

Estimating total change

Table 7.4.1

Subinterval (growth in radius)	Approximation of the Rate of Change $V'(x) = x^2$
[0, 0.2]	$0^2 = 0$
[0.2, 0.4]	$0.2^2 = 0.04$
[0.4, 0.6]	$0.4^2 = 0.16$
[0.6, 0.8]	$0.6^2 = 0.36$
[0.8, 1]	$0.8^2 = 0.64$

Keep in mind that the rate of change (i.e., volume per unit length) times the change in the radius of the cell gives the total change in the volume corresponding to the given growth in radius.

Thus, the change in the volume as the radius grows from 0 μm to 0.2 μm (i.e., over $[0, 0.2]$) is $0^2 \cdot 0.2 = 0$. Over the interval $[0.2, 0.4]$, the volume increases by $0.2^2 \cdot 0.2 = 0.04 \cdot 0.2 = 0.008$ units. The change in the volume as the cell grows in radius from 0.4 μm to 0.6 μm is $0.4^2 \cdot 0.2 = 0.16 \cdot 0.2 = 0.032$, and so forth. Adding up all the changes,
$$0^2 \cdot 0.2 + 0.2^2 \cdot 0.2 + 0.4^2 \cdot 0.2 + 0.6^2 \cdot 0.2 + 0.8^2 \cdot 0.2 = 0.24$$
we get an approximation for the total change in V as x changes from $x = 0$ to $x = 1$. At the same time, we have calculated the left Riemann sum with five rectangles, L_5, for the function $V(x) = x^2$ (as in Example 7.3.1 in the previous section).

We conclude that the Riemann sum L_5 is an approximation for the total change in V as x changes from $x = 0$ to $x = 1$. By computing $R_5 = 0.44$ (see Example 7.3.1), we obtain another estimate for the total change. In general,

Riemann sum $\approx$ total change in V from $x = 0$ to $x = 1$

and, by taking the limit as the number of subintervals goes to infinity,
$$\int_a^b x^2 dx = \text{total change in } V \text{ from } x = 0 \text{ to } x = 1$$

What we discovered in the previous example (also in Example 7.3.14 in the previous section) is important, so we repeat it: if $f(x)$ represents the *rate of change* of some quantity, then
$$\int_a^b f(x) dx = \textit{total change} \text{ in the quantity from } x = a \text{ to } x = b$$

512 **Chapter 7** Integrals and Applications

Let's go back to the equation $V'(x) = x^2$ from the previous example. Subtraction provides another way to express the total change, i.e.,

(total change in V from $x = 0$ to $x = 1$) $= V(1) - V(0)$

(Figure 7.4.40). Therefore, if V is a solution of the differential equation, then

$$\int_0^1 x^2 dx = V(1) - V(0)$$

We can thus compute the value of the definite integral without any Riemann sums if we can use the indefinite integral to compute the solution $V(x)$ and then the values $V(1)$ and $V(0)$. Now we show how this is done by solving for $V(x)$ based on different initial conditions (Examples 7.4.2–7.4.4).

Example 7.4.2 Finding the Total Change in Volume

Suppose that $V(0) = 0$. We solve the differential equation $V'(x) = x^2$ (i.e., we find an antiderivative of x^2) using the power rule

$$V(x) = \int x^2 dx = \frac{x^3}{3} + C$$

Substituting the initial condition $V(0) = 0$, we get

$$0 = \frac{(0)^3}{3} + C$$

i.e., $C = 0$; thus, the solution is

$$V(x) = \frac{x^3}{3}$$

With this formula, we can find the total change by subtracting:

(total change between $x = 0$ and $x = 1$) $= V(1) - V(0)$

$$= \frac{1^3}{3} - \frac{0^3}{3} = \frac{1}{3} \approx 0.3333$$

(see Figure 7.4.40a). This is the value we attempted to approximate in Example 7.4.1.

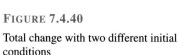

Figure 7.4.40

Total change with two different initial conditions

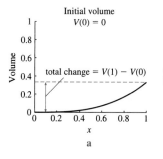

a

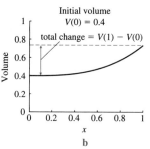

b

Example 7.4.3 Finding the Total Change: Different Initial Condition

Now we consider the same differential equation, $V'(x) = x^2$, but change the initial condition.

Suppose that $V(0) = 0.4$. This time,

$$V(x) = \frac{x^3}{3} + C$$

where C is determined from

$$0.4 = \frac{(0)^3}{3} + C$$

Thus, $C = 0.4$ and the solution is

$$V(x) = \frac{x^3}{3} + 0.4$$

We can again find the total change by subtracting:

(total change between $x = 0$ and $x = 1$) $= V(1) - V(0)$

$$= \left(\frac{1^3}{3} + 0.4\right) - \left(\frac{0^3}{3} + 0.4\right) = \frac{1}{3} \approx 0.3333$$

Although the two solutions to the differential equation are not equal, they are *parallel* (Figure 7.4.40). Both increase by exactly the same amount. △

Example 7.4.4   Finding the Total Change: Irrelevance of Initial Condition

In general, suppose that $V(0) = V_0$, where $V_0 \geq 0$. The solution of the differential equation is still

$$V(x) = \frac{x^3}{3} + C$$

but now the constant C must satisfy

$$V(0) = \frac{(0)^3}{3} + C = V_0$$

and so $C = V_0$. The solution is

$$V(x) = \frac{x^3}{3} + V_0$$

and the total change in volume is

$$V(1) - V(0) = \left(\frac{1^3}{3} + V_0\right) - \left(\frac{0^3}{3} + V_0\right) = \frac{1}{3} \approx 0.3333 \quad △$$

The total change does not depend on the initial condition. The initial condition is required to answer a question such as "Where are you after driving 5 km due north?" But the *change* in position is clear; you are 5 km north of where you started.

This idea is the essence of the **Fundamental Theorem of Calculus.** A definite integral can be evaluated in two ways:

1. by evaluating Riemann sums and taking the limit, and
2. by finding an antiderivative and subtracting.

When we can find an antiderivative, the second method is much simpler. However, when the indefinite integral is difficult or impossible to find algebraically, Riemann sums provide a guaranteed, if laborious, alternative that usually requires a calculator or a computer.

It is convenient to write the Fundamental Theorem of Calculus with one bit of new notation. To represent the change in the value of $F(x)$ between $x = a$ and $x = b$, we use the shorthand

$$F(x)\big|_a^b = F(b) - F(a)$$

Instead of reading this notation in a new way, we say "$F(b)$ minus $F(a)$."
We can now state the theorem.

Theorem 7.4.1 The Fundamental Theorem of Calculus (Part 1)

If $f(x)$ is continuous on $[a, b]$ and $F(x)$ is any antiderivative of $f(x)$, i.e., $F'(x) = f(x)$, then

$$\int_a^b f(x)dx = F(x)\big|_a^b = F(b) - F(a)$$

This theorem is called *fundamental* because it establishes a relationship between definite integrals and antiderivatives (indefinite integrals). Replacing $f(x)$ by $F'(x)$ in the statement of the theorem, we obtain

$$\int_a^b F'(x)dx = F(b) - F(a)$$

In words, the integral of the rate of change of a function is equal to the total change in the function.

In light of the theorem, we write

$$\int_0^1 x^2 dx = \frac{x^3}{3}\bigg|_0^1 = \frac{(1)^3}{3} - \frac{(0)^3}{3} = \frac{1}{3}$$

At the end of this section, we will sketch the proof of the theorem. First, we learn how to use it.

Example 7.4.5 Using the Fundamental Theorem of Calculus

To find $\int_1^2 \frac{1}{x} dx$, we find any indefinite integral of $\frac{1}{x}$, valid for $1 \leq x \leq 2$, such as

$$\int \frac{1}{x} dx = \ln x$$

Note that we did not write $\ln |x|$ but $\ln x$. The absolute value is not required since $1 \leq x \leq 2$, and so x is always positive. We compute

$$\int_1^2 \frac{1}{x} dx = \ln x \big|_1^2 = \ln 2 - \ln 1 \approx 0.693$$

Example 7.4.6 Using the Fundamental Theorem of Calculus to Evaluate a Definite Integral

Find

$$\int_2^4 (x^3 - x - 1)dx$$

We find an antiderivative and then evaluate:

$$\int_2^4 (x^3 + x - 1)dx = \left(\frac{x^4}{4} + \frac{x^2}{2} - x\right)\bigg|_2^4$$

$$= \left(\frac{(4)^4}{4} + \frac{(4)^2}{2} - 4\right) - \left(\frac{(2)^4}{4} + \frac{(2)^2}{2} - 2\right)$$

$$= 68 - 4 = 64$$

7.4 Definite and Indefinite Integrals

The discussion leading to the Fundamental Theorem of Calculus shows that we do not have to include the constant of integration when we evaluate definite integrals. If we do include it, we will see that—every time—it cancels out; for example,

$$\int_2^4 (x^3 + x - 1)dx = \left(\frac{x^4}{4} + \frac{x^2}{2} - x + C\right)\Big|_2^4$$

$$= \left(\frac{(4)^4}{4} + \frac{(4)^2}{2} - 4 + C\right) - \left(\frac{(2)^4}{4} + \frac{(2)^2}{2} - 2 + C\right)$$

$$= (68 + C) - (4 + C) = 64$$

Example 7.4.7 Using the Fundamental Theorem of Calculus

Find the area of the region under $f(x) = \frac{1}{x^2}$ and above the x-axis, from $x = 1$ to $x = 4$.

Since $f(x) = \frac{1}{x^2} \geq 0$ on $[1, 4]$ and f is continuous,

$$\text{area} = \int_1^4 \frac{1}{x^2}dx = \int_1^4 x^{-2}dx$$

$$= \frac{x^{-1}}{-1}\Big|_1^4 = -\frac{1}{x}\Big|_1^4 = \left(-\frac{1}{4}\right) - \left(-\frac{1}{1}\right) = \frac{3}{4}$$

Example 7.4.8 Calculating Definite Integrals

a. Find

$$\int_{-1}^1 (3e^x + 2x)dx$$

The function $f(x) = 3e^x + 2x$ is continuous on $[-1, 1]$; by Theorem 7.4.1,

$$\int_{-1}^1 (3e^x + 2x)dx = (3e^x + x^2)\Big|_{-1}^1$$

$$= [3e^1 + (1)^2] - [3e^{-1} + (-1)^2]$$

$$= 3e - 3e^{-1} \approx 7.05121$$

b. The calculation

$$\int_0^\pi \sin x\, dx = (-\cos x)\Big|_0^\pi$$

$$= (-\cos \pi) - (-\cos 0)$$

$$= -(-1) + 1 = 2$$

shows that the area under $y = \sin x$ from $x = 0$ to $x = \pi$ is equal to 2.

c. To find

$$\int_0^1 \frac{1}{1+x^2}dx$$

we recall that

$$(\arctan x)' = \frac{1}{1+x^2}$$

Therefore,

$$\int_0^1 \frac{1}{1+x^2} dx = \arctan x \Big|_0^1$$
$$= (\arctan 1) - (\arctan 0)$$
$$= \frac{\pi}{4} - 0 = \frac{\pi}{4}$$

Example 7.4.9 Calculating a Definite Integral I

Find

$$\int_1^2 \frac{x^2 + \sqrt{x} - 2}{x} dx$$

In order to find the antiderivative, we simplify the integrand first:

$$\int_1^2 \frac{x^2 + \sqrt{x} - 2}{x} dx = \int_1^2 \left(\frac{x^2}{x} + \frac{\sqrt{x}}{x} - \frac{2}{x} \right) dx$$
$$= \int_1^2 \left(x + x^{-1/2} - \frac{2}{x} \right) dx$$
$$= \left(\frac{x^2}{2} + 2x^{1/2} - 2\ln x \right) \Big|_1^2$$
$$= \left(\frac{(2)^2}{2} + 2(2)^{1/2} - 2\ln 2 \right) - \left(\frac{(1)^2}{2} + 2(1)^{1/2} - 2\ln 1 \right)$$
$$= (2 + 2\sqrt{2} - 2\ln 2) - \left(\frac{1}{2} + 2 - 0 \right)$$
$$= -\frac{1}{2} + 2\sqrt{2} - 2\ln 2 \approx 0.94213$$

Example 7.4.10 Calculating a Definite Integral II

Find

$$\int_0^2 e^{3x} dx$$

We need an antiderivative of e^{3x} first; i.e., we need to find a function $F(x)$ such that $F'(x) = e^{3x}$.

If $F(x) = e^{3x}$, then $F'(x) = 3e^{3x}$. Although $F'(x)$ is not equal to e^{3x}, we see that it differs from it by a constant factor only. So we try $F(x) = \frac{1}{3} e^{3x}$; this time

$$F'(x) = \frac{1}{3} e^{3x} \cdot 3 = e^{3x}$$

It works! Thus,

$$\int_0^2 e^{3x} dx = \left(\frac{1}{3} e^{3x} \right) \Big|_0^2$$
$$= \frac{1}{3} e^{3 \cdot 2} - \frac{1}{3} e^{3 \cdot 0}$$
$$= \frac{1}{3} e^6 - \frac{1}{3} \approx 134.14$$

The substitution method for integration that we discuss in the next section will free us from the need to guess an antiderivative. However, some simple integrands might need no more than an adjustment by a constant (such as the division by 3 that we had to do in the previous example), and guessing an antiderivative might be the most efficient way to solve an integral. For example, if a is a nonzero constant, then

$$\int e^{ax} dx = \frac{1}{a} e^{ax} + C$$

$$\int \cos(ax) dx = \frac{1}{a} \sin(ax) + C$$

$$\int \sin(ax) dx = -\frac{1}{a} \cos(ax) + C$$

We need to keep in mind that the Fundamental Theorem of Calculus applies to continuous functions. If we try to integrate a discontinuous function, things could go wrong, as our next example shows.

Example 7.4.11 Definite Integral of a Discontinuous Function

The function $f(x) = \frac{1}{x^2}$ is positive on $[-10, 2]$ and thus

$$\int_{-10}^{2} \frac{1}{x^2} dx$$

must be positive as well (the definite integral of a positive function is positive; see Section 7.3). However, if we try to evaluate the integral using the fundamental theorem, we get

$$\int_{-10}^{2} \frac{1}{x^2} dx = \left(-\frac{1}{x}\right)\Big|_{-10}^{2}$$

$$= -\frac{1}{2} + \frac{1}{10}$$

$$= -\frac{2}{5} < 0$$

The problem is that $f(x) = \frac{1}{x^2}$ is not continuous at $x = 0$, which is within the interval of integration $[-10, 2]$. Thus, the answer from the Fundamental Theorem of Calculus is meaningless.

As we will see in Section 7.7, the integral

$$\int_{-10}^{2} \frac{1}{x^2} dx$$

is not equal to a real number.

Example 7.4.12 Using the Fundamental Theorem of Calculus to Find How Far a Rock Falls

Suppose that the position of a rock, $p(t)$, follows the differential equation

$$\frac{dp}{dt} = v(t) = -9.8t - 5$$

where t is measured in seconds and p is measured in metres. How far does the rock fall between $t = 1$ and $t = 3$? The total change in position is given by the definite integral of the rate of change of position,

$$\int_1^3 v(t)dt$$

According to the Fundamental Theorem of Calculus, we can compute this value by finding *any* indefinite integral of $v(t)$ and subtracting the value at $t = 1$ from the value at $t = 3$. One indefinite integral is

$$\int v(t)dt = \int (-9.8t - 5)dt$$

$$= -9.8\frac{t^2}{2} - 5t$$

$$= -4.9t^2 - 5t$$

where we set the arbitrary constant to $C = 0$ for convenience. Then

$$(\text{total change in position between } t=1 \text{ and } t=3) = \left(-4.9t^2 - 5t\right)\Big|_1^3$$

$$= \left(-4.9 \cdot 3^2 - 5 \cdot 3\right)$$

$$- \left(-4.9 \cdot 1^2 - 5 \cdot 1\right)$$

$$= -49.2$$

The rock will have fallen 49.2 m during this time.

Instead of figuring out the indefinite integral and evaluating it in two separate calculations, we can write (as we have done before in similar cases)

$$\int_1^3 v(t)dt = \int_1^3 (-9.8t - 5)dt$$

$$= \left(-4.9t^2 - 5t\right)\Big|_1^3$$

$$= [-4.9(3)^2 - 5(3)] - [-4.9(1)^2 - 5(1)]$$

$$= -49.2$$

Example 7.4.13 Using the Fundamental Theorem of Calculus to Find How Much a Fish Grows

Suppose the change in length of a fish follows the equation

$$\frac{dL}{dt} = 6.48e^{-0.09t}$$

with t measured as age in years and L measured in centimetres. How much does the fish grow between ages 2 and 5? The total change is

$$\int_2^5 6.48e^{-0.09t}dt$$

By guessing, or by using the formula we mentioned after Example 7.4.10, we obtain

$$\int e^{-0.09t}dx = \frac{1}{-0.09}e^{-0.09t} + C$$

and

$$\int_2^5 6.48e^{-0.09t}\,dt = 6.48\left(\frac{1}{-0.09}e^{-0.09t}\right)\bigg|_2^5$$
$$= (-72e^{-0.09t})\big|_2^5$$
$$= (-72e^{-0.09\cdot 5}) - (-72e^{-0.09\cdot 2})$$
$$\approx (-45.9) - (-60.1) = 14.2$$

Note that while neither of the component terms -45.9 and -60.1 makes biological sense, their difference gives the correct answer (this is because the integration constant C does not influence the difference $F(b) - F(a)$ in the fundamental theorem, since it cancels out).

However, there is a way to do a biologically sound calculation. From

$$\frac{dL}{dt} = 6.48e^{-0.09t}$$

we get

$$L(t) = \int 6.48e^{-0.09t}\,dt$$
$$= 6.48 \cdot \frac{1}{-0.09}e^{-0.09t} + C$$
$$= -72e^{-0.09t} + C$$

Using a realistic initial condition, $L(0) = 0$, we get

$$0 = -72e^{-0.09\cdot 0} + C$$

and so $C = 72$. Thus

$$L(t) = -72e^{-0.09t} + 72 = 72(1 - e^{-0.09t})$$

The change is

$$L(t)\big|_2^5 = L(5) - L(2)$$
$$= 72(1 - e^{-0.09\cdot 5}) - 72(1 - e^{-0.09\cdot 2})$$
$$\approx 26.1 - 11.9 = 14.2$$

The difference is now found by subtracting the actual lengths at ages 5 and 2, which are both positive, as we would expect (Figure 7.4.41). ▲

The Integral Function and the Proof of the Fundamental Theorem of Calculus

For completeness of exposition, we present the proof of the Fundamental Theorem of Calculus. To do so, we need to introduce a new object—the integral function of a given function. Although the integral function will not appear again in the calculus that we discuss in this book, it is quite important. For instance, the cumulative density function (one of the key functions in probability) is an integral function.

Assume that f is a continuous function defined on an interval $[a, b]$. For x in $[a, b]$, we define the **integral function** of f by

$$g(x) = \int_a^x f(t)\,dt$$

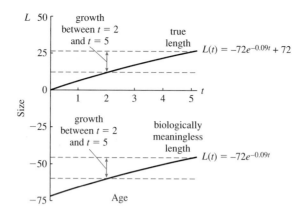

FIGURE 7.4.41
The growth of fish

Recall that the definite integral

$$\int_a^b f(t)\,dt$$

is a real number. What makes $g(x)$ a function is the x in the upper limit of the integral (and so g is a function of x and not of t). We interpret g as a signed area, i.e.,

$$g(x) = \text{"area above"} - \text{"area below" from } a \text{ to } x$$

Example 7.4.14 Simple Integral Function

Find the integral function $g(x)$ for the function $f(t) = 2$ defined on $[0, 10]$.
Because $f(t) \geq 0$ for all t in $[0, 10]$, it follows that

$$g(x) = \int_0^x 2\,dt$$

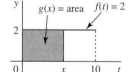

is the area under $f(t) = 2$ on $[0, x]$. The region in question is a rectangle (Figure 7.4.42) and so

$$g(x) = \int_0^x 2\,dt = 2x$$

FIGURE 7.4.42
The integral function of a constant function

for $0 \leq x \leq 10$.

Example 7.4.15 Understanding the Integral Function

Consider the function $f(t)$, defined on $[0, 4]$, whose graph is shown in Figure 7.4.43. Define the integral function

$$g(x) = \int_0^x f(t)\,dt$$

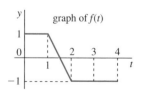

for x in $[0, 4]$. Compute $g(0)$, $g(1)$, $g(2)$, $g(3)$, and $g(4)$.
When $x = 0$, we get

$$g(0) = \int_0^0 f(t)\,dt = 0$$

FIGURE 7.4.43
Graph of $f(t)$

(recall that the definite integral with equal lower and upper limits is zero).
For $x = 1$, we have

$$g(1) = \int_0^1 f(t)dt = \text{area} = 1 \cdot 1 = 1$$

To calculate $g(2)$, we note that f is both positive and negative on $[0, 2]$:

$$g(2) = \int_0^2 f(t)dt = \text{area from 0 to 1.5} - \text{area from 1.5 to 2}$$
$$= \text{area from 0 to 1} = 1$$

since the contributions from the two triangular regions cancel each other.
Next,

$$g(3) = \int_0^3 f(t)dt = 0$$

because the regions over $[0, 1.5]$ and $[1.5, 3]$ are exactly the same (i.e., same shape), with one lying above the x-axis and the other below it. Finally, using the summation property of the definite integral,

$$g(4) = \int_0^4 f(t)dt = \int_0^3 f(t)dt + \int_3^4 f(t)dt = 0 + (-1) = -1$$

Of course, we could have used the "area above minus area below" argument to arrive at the same answer.

The graph of $g(x)$ is given in Figure 7.4.44. An alternative to plotting points is suggested in Exercise 65.

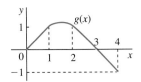

FIGURE 7.4.44
Graph of the integral function $g(x)$

Example 7.4.16 Integral Function

If $f(t) = t$ on $[0, 20]$, then

$$g(x) = \int_0^x t\,dt = \text{area of the triangle} = \frac{1}{2}x \cdot x = \frac{1}{2}x^2$$

for $0 \leq x \leq 20$. See Figure 7.4.45.

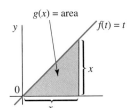 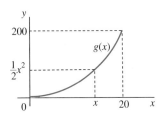

FIGURE 7.4.45
Integral function of $f(t) = t$

The most important property of the integral function is contained in the following theorem.

Theorem 7.4.2 **Fundamental Theorem of Calculus (Part 2)**

Assume that a function f is continuous on an interval $[a, b]$. The function

$$g(x) = \int_a^x f(t)\,dt$$

is differentiable on (a, b) and

$$g'(x) = f(x)$$

The message of the theorem is very important. In words, Theorem 7.4.2 claims that *every continuous function has an antiderivative.*

Substituting the formula for $g(x)$ into $g'(x) = f(x)$, we write

$$\left(\int_a^x f(t)\,dt \right)' = f(x)$$

Thus, if we take a function f, integrate (in this special way, from a to x), and then differentiate what we get, we get the function f back. For instance, differentiating

$$g(x) = \frac{1}{2}x^2$$

from Example 7.4.16, we get

$$g'(x) = \frac{1}{2}(2x) = x$$

which is equal to $f(x)$.

Example 7.4.17 Differentiating an Integral Function

a. Consider the function $f(t) = t^3 e^{-t^2}$ defined on the interval $[-2, 5]$. Note that $f(t)$ is continuous at all real numbers, and in particular on $[-2, 5]$. The integral function of f is given by

$$g(x) = \int_{-2}^x t^3 e^{-t^2}\,dt$$

where x is in $[-2, 5]$. Theorem 7.4.2 tells us that $g(x)$ is a differentiable function, and

$$g'(x) = \left(\int_{-2}^x t^3 e^{-t^2}\,dt \right)' = x^3 e^{-x^2}$$

b. Consider the function $f(t) = \sin(t^2 - 1) + t^3 - 2$ on $[0, 10]$. Since f is continuous, it must have an antiderivative. According to Theorem 7.4.2, if we define

$$g(x) = \int_0^x \left(\sin(t^2 - 1) + t^3 - 2 \right) dt$$

for all x in $[0, 10]$, then

$$g'(x) = \sin(x^2 - 1) + x^3 - 2 = f(x)$$

In words, we have just shown that g is an antiderivative of f.

c. Using the Fundamental Theorem, we find that

$$\left(\int_1^x (6-\cos 5t)^{2/3}\, dt\right)' = (6-\cos 5x)^{2/3}$$

d. We use the chain rule as usual. For instance, to find $g'(x)$ when

$$g(x) = \int_1^{x-2e^x} (6-\cos 5t)^{2/3}\, dt$$

we multiply the derivative of the integral function (calculated using Theorem 7.4.2) by the derivative of the function in the upper limit:

$$g'(x) = \left(\int_1^{x-2e^x} (6-\cos t)^{2/3}\, dt\right)'$$
$$= (6-\cos(x-2e^x))^{2/3}\, (x-2e^x)'$$
$$= (6-\cos(x-2e^x))^{2/3}\, (1-2e^x)$$

e. Likewise,

$$\left(\int_0^{x^3} e^{-2t}\, dt\right)' = e^{-2x^3}\, (x^3)' = 3x^2 e^{-2x^3}$$

Now we prove Theorem 7.4.2. By definition,

$$g'(x) = \lim_{h \to 0} \frac{g(x+h) - g(x)}{h}$$

First, we work on the difference quotient

$$\frac{1}{h}[g(x+h) - g(x)] = \frac{1}{h}\left(\int_a^{x+h} f(t)dt - \int_a^x f(t)dt\right)$$
$$= \frac{1}{h}\left(\int_a^{x+h} f(t)dt + \int_x^a f(t)dt\right)$$
$$= \frac{1}{h}\left(\int_x^a f(t)dt + \int_a^{x+h} f(t)dt\right)$$
$$= \frac{1}{h}\int_x^{x+h} f(t)dt \qquad (7.4.1)$$

by the summation property of the definite integral.

Since $f(t)$ is continuous on a closed interval $[x, x+h]$, the Extreme Value Theorem implies that there are numbers c_1 and c_2 in $[x, x+h]$ such that $f(c_1) = m$ and $f(c_2) = M$, where m is the absolute minimum and M is the absolute maximum of $f(t)$ on $[x, x+h]$. In other words,

$$m \leq f(t) \leq M$$

for all t in $[x, x+h]$. Now (see Properties of Integrals in the previous section, where we integrated a constant function)

$$\int_x^{x+h} m\,dt = m(x+h-x) = mh \quad \text{and} \quad \int_x^{x+h} M\,dt = M(x+h-x) = Mh$$

and, integrating the above inequality, we get

$$mh \leq \int_x^{x+h} f(t)\,dt \leq Mh$$

$$f(c_1)h \leq \int_x^{x+h} f(t)\,dt \leq f(c_2)h$$

Assuming that $h > 0$, we divide all sides by h:

$$f(c_1) \leq \frac{1}{h}\int_x^{x+h} f(t)\,dt \leq f(c_2)$$

Combining this inequality with Equation 7.4.1 gives

$$f(c_1) \leq \frac{1}{h}(g(x+h) - g(x)) \leq f(c_2) \tag{7.4.2}$$

If $h < 0$, then the inequality signs in Equation 7.4.2 are reversed. Nevertheless, the argument proceeds in exactly the same way.

As $h \to 0$, the numbers c_1 and c_2, which are in $[x, x+h]$, will approach x. Since $f(t)$ is continuous, $f(c_1) \to f(x)$ and $f(c_2) \to f(x)$. Thus, taking the limit of Equation 7.4.2 as $h \to 0$, we obtain

$$\lim_{h \to 0} f(c_1) \leq \lim_{h \to 0} \frac{1}{h}(g(x+h) - g(x)) \leq \lim_{h \to 0} f(c_2)$$
$$f(x) \leq g'(x) \leq f(x)$$

i.e.,

$$g'(x) = f(x)$$

and the proof is complete.

Now recall Part 1 of the Fundamental Theorem of Circulus, which we reproduce here using the same theorem number.

Theorem 7.4.1 **The Fundamental Theorem of Calculus (Part 1)**

If $f(x)$ is continuous on $[a, b]$ and $F'(x) = f(x)$, then

$$\int_a^b f(x)\,dx = F(x)\big|_a^b = F(b) - F(a)$$

To prove the theorem, we consider the integral function

$$g(x) = \int_a^x f(t)\,dt$$

We have just proven (Theorem 7.4.2) that

$$g'(x) = f(x)$$

i.e., $g(x)$ is an antiderivative of $f(x)$. Since $F(x)$ is an antiderivative of $f(x)$ as well, we conclude that

$$F(x) = g(x) + C$$

(recall that any two antiderivatives of the same function differ by a constant). Now

$$\begin{aligned} F(b) - F(a) &= (g(b) + C) - (g(a) + C) \\ &= g(b) - g(a) \\ &= \int_a^b f(t)dt - \int_a^a f(t)dt \\ &= \int_a^b f(t)dt \end{aligned}$$

since $\int_a^a f(t)dt = 0$. Reading what we proved from end to start, we get

$$\int_a^b f(t)dt = F(b) - F(a)$$

Thus (the variable name in the definite integral does not matter),

$$\int_a^b f(x)dx = F(b) - F(a)$$

and the proof is complete.

As we have seen, the Fundamental Theorem of Calculus has two parts, establishing the relationship between the definite integral (integral function) and the derivative. Keeping in mind the precise meaning given in the statements of the two parts of the theorem (and summarized in the table below), we can say that definite integration and differentiation are inverse operations.

Fundamental Theorem of Calculus and the Relationship between the Derivative and the Definite Integral

If f is continuous on $[a, b]$ and F is an antiderivative of f, i.e., $F' = f$, then

$$\int_a^b f(x)dx = F(b) - F(a) \qquad \int_a^b F'(x)dx = F(b) - F(a)$$

Assume that f is continuous on $[a, b]$. The integral function $g(x) = \int_a^x f(t)dt$ is differentiable on (a, b) and

$$g'(x) = f(x) \qquad \left(\int_a^x f(t)dt\right)' = f(x)$$

Summary The **Fundamental Theorem of Calculus** describes the connection between definite and indefinite integrals; the definite integral is equal to the difference between the values of the indefinite integral at the limits of integration. The Fundamental Theorem simplifies calculations of total change for pure-time differential equations by eliminating the need to solve for the arbitrary constant.

7.4 Exercises

Mathematical Techniques

1. Explain how the definite integral connects the concepts of rate of change and total change of a function.
2. What is the difference between the definite integral and the integral function?
3. What is the integral function of a constant function $f(t) = 1$ defined on $[0, 1]$?
4. Identify the error in the following calculation:

$$\int_0^\pi \sec^2 x \, dx = \tan x \Big|_0^\pi = \tan \pi - \tan 0 = 0$$

5–37 ■ Compute each definite integral.

5. $\int_0^1 7x^2 \, dx$
6. $\int_0^1 (10t^9 + 6t^5) \, dt$
7. $\int_{-1}^2 (72t + 5) \, dt$
8. $\int_{-2}^2 (y^4 + 5y^3) \, dy$
9. $\int_1^2 \frac{5}{x^3} \, dx$
10. $\int_1^4 3z^{\frac{3}{7}} \, dz$
11. $\int_1^8 \left(\frac{2}{\sqrt[3]{t}} + 3\right) dt$
12. $\int_{1.2}^{2.4} (5z^{-1.2} - 1.2) \, dz$
13. $\int_2^3 \left(\frac{3}{z^2} + \frac{z^2}{3}\right) dz$
14. $\int_0^1 (3e^x + 2x^3) \, dx$
15. $\int_1^4 \left(e^x + \frac{1}{x}\right) dx$
16. $\int_{-3}^{-1} \left(\frac{2}{t} + \frac{t}{2}\right) dt$
17. $\int_0^\pi (2\sin x + 3\cos x) \, dx$
18. $\int_{-\pi/2}^{\pi/2} [x^2 - 20 \sin x] \, dx$
19. $\int_0^5 3e^{\frac{x}{5}} \, dx$
20. $\int_1^2 \frac{3x - 4}{x^2} \, dx$
21. $\int_0^1 \frac{6}{1 + x^2} \, dx$
22. $\int_0^1 \frac{x^2}{1 + x^2} \, dx$ (Hint: Start by using long division.)
23. $\int_0^4 \frac{x - 6}{x + 1} \, dx$ (Hint: Start by using long division.)
24. $\int_{-2}^0 (x - 2)(x + 4) \, dx$
25. $\int_0^1 \sqrt[3]{x}(x^2 - 1) \, dx$
26. $\int_1^2 \frac{3 - 5x}{\sqrt{x}} \, dx$
27. $\int_1^2 \frac{x^2 + x + 1}{4x} \, dx$
28. $\int_{-1}^1 \frac{e^x - e^{-x}}{2} \, dx$
29. $\int_0^2 \frac{2e^x - e^{2x}}{e^{3x}} \, dx$
30. $\int_{-1/2}^1 \frac{1}{\sqrt{1 - x^2}} \, dx$
31. $\int_e^{e^3} \frac{1}{x} \, dx$
32. $\int_0^4 3^x \, dx$

33. $\int_{1/4}^{4} \frac{1}{\sqrt{x}} dx$

34. $\int_{0}^{\pi/4} \sec x \tan x \, dx$

35. $\int_{0}^{\pi/4} \sec^2 x \, dx$

36. $\int_{-3}^{0} \frac{1}{4+t} dt$

37. $\int_{0}^{2} \frac{1}{1+4t} dt$

38–42 ▪ Evaluate each integral, sketch the graph of the function involved, and interpret as an area or a difference of areas.

38. $\int_{0}^{\pi} \sin x \, dx$

39. $\int_{-2}^{1} x^3 \, dx$

40. $\int_{-1}^{1} \frac{1}{1+x^2} dx$

41. $\int_{\pi/2}^{\pi} \cos x \, dx$

42. $\int_{-1}^{1} (e^x - 1) dx$

43–48 ▪ Compute the definite integrals of the following functions from $t = 1$ to $t = 2$, from $t = 2$ to $t = 3$, and finally from $t = 1$ to $t = 3$ to check the summation property of definite integrals.

43. $g(t) = t^2$

44. $h(t) = 1 + t^3$

45. $L(t) = \frac{5}{t^3}$

46. $B(t) = 3t^{\frac{3}{7}}$

47. $F(t) = e^t + \frac{1}{t}$

48. $G(t) = \frac{2}{t} + \frac{t}{2}$

49–52 ▪ For the function $f(t)$, consider the integral function $g(x) = \int_{a}^{x} f(t) dt$.

a. Sketch the graph of $f(t)$.

b. Calculate the values $g(0)$, $g(1)$, $g(3)$, $g(4)$, and $g(6)$.

c. Find the intervals on which $g(x)$ is increasing and the intervals on which it is decreasing.

d. Identify numbers where $g(x)$ has a maximum or a minimum value.

e. Sketch the graph of $g(x)$.

49. $f(t) = 5$ for t in $[0, 6]$.

50. $f(t) = |t - 3| - 2$ for t in $[0, 6]$.

51. $f(t) = t - 4$ for t in $[0, 6]$.

52. $f(t) = \begin{cases} t - 1 & \text{if } 0 \leq t \leq 2 \\ 1 & \text{if } 2 < t \leq 4 \\ 5 - t & \text{if } 4 < t \leq 6 \end{cases}$

53–60 ▪ Find the derivative of each integral function.

53. $g(x) = \int_{0}^{x} (te^t - t + 4) dt$

54. $g(x) = \int_{3}^{x} \left(\ln t - \frac{1}{t} \right) dt$

55. $g(x) = \int_{0}^{x} \arctan t^4 \, dt$

56. $g(x) = \int_{x}^{3} \sin t \cos t \, dt$ (Hint: Reverse the limits of integration first.)

57. $g(x) = \int_{2}^{3x-5} e^t \, dt$ (Hint: Use the chain rule.)

58. $g(x) = \int_{2}^{\sin x} e^t \, dt$ (Hint: Use the chain rule.)

59. $g(x) = \int_{2}^{x^3} \arctan t^2 \, dt$ (Hint: Use the chain rule.)

60. $g(x) = \int_{6}^{e^x} \frac{\sin t}{\sqrt{t-4}} dt$ (Hint: Use the chain rule.)

61–64 ▪ The Fundamental Theorem of Calculus (Part 2, (Theorem 7.4.2)) states that

$$\frac{d}{dx} \left(\int_{a}^{x} f(t) \, dt \right) = f(x)$$

for any value of a. Check this in each case by computing the definite integral and then taking its derivative.

61. $f(x) = x^2$ with $a = 0$.

62. $f(x) = 1 + x^3$ with $a = 1$.

63. $f(x) = \left(1 + \frac{x}{2}\right)^4$ with $a = -1$.

64. $f(x) = (1 + 2x)^{-4}$ with $a = 0$.

65. We can find the formula for the integral function $g(x)$ from Example 7.4.15 using the Fundamental Theorem of Calculus.

a. Represent $f(t)$ algebraically as a function defined piecewise.

b. Show that $g(x) = x$ if $0 \leq x \leq 1$.

c. Show that $g(x) = -x^2 + 3x - 1$ if $1 \leq x \leq 2$.

d. Show that $g(x) = 3 - x$ if $2 \leq x \leq 4$.

Applications

66–71 ■ Find the change in the state variable between the given times first by solving the differential equation with the given initial conditions and then by using the definite integral.

66. The change of position of a rock between times $t = 1$ and $t = 5$ with position following the differential equation $\frac{dp}{dt} = -9.8t - 5$ and initial condition $p(0) = 200$. Here t is measured in seconds and p in metres.

67. The amount a fish grows between ages $t = 1$ and $t = 5$ if it follows the differential equation $\frac{dL}{dt} = 6.48e^{-0.09t}$ with initial condition $L(0) = 5$. Here t is measured in years and L in centimetres.

68. The amount a fish grows between ages $t = 0.5$ and $t = 1.5$ if it follows the differential equation $\frac{dL}{dt} = 64.3e^{-1.19t}$ with initial condition $L(0) = 5$. Here t is measured in years and L in centimetres.

69. The number of new AIDS cases between 1985 and 1987 if the number of AIDS cases follows $\frac{dA}{dt} = 523.8(t - 1981)^2$ with initial condition $A(1981) = 340$. Here t is measured in years.

70. The amount of chemical produced between times $t = 5$ and $t = 10$ if the amount P follows $\frac{dP}{dt} = \frac{5}{1 + 2t}$ with initial condition $P(0) = 2$. Here t is measured in minutes and P in moles.

71. The amount of chemical produced between times $t = 5$ and $t = 10$ if the amount P follows $\frac{dP}{dt} = 5e^{-2t}$ with initial condition $P(0) = 2$. Here t is measured in minutes and P in moles.

72–77 ■ Check the summation property for the solution of each differential equation by showing that the change in value of the whole interval is equal to the change during the first half of the interval plus the change during the second half of the interval.

72. The position of a rock obeys the differential equation $\frac{dp}{dt} = -9.8t - 5$ with initial condition $p(0) = 200$. Show that the distance moved between times $t = 1$ and $t = 5$ is equal to the sum of the distance moved between $t = 1$ and $t = 3$ and the distance moved between $t = 3$ and $t = 5$.

73. The growth of a fish obeys the differential equation $\frac{dL}{dt} = 6.48e^{-0.09t}$ with initial condition $L(0) = 5$ (as in Exercise 67). Show that the growth between times $t = 1$ and $t = 5$ is equal to the sum of the growth between $t = 1$ and $t = 3$ and the growth between $t = 3$ and $t = 5$.

74. The growth of a fish obeys the differential equation $\frac{dL}{dt} = 64.3e^{-1.19t}$ with initial condition $L(0) = 5$ (as in Exercise 68). Show that the growth between times $t = 0.5$ and $t = 1.5$ is equal to the sum of the growth between $t = 0.5$ and $t = 1$ and the growth between $t = 1$ and $t = 1.5$.

75. The number of AIDS cases obeys $\frac{dA}{dt} = 523.8(t - 1981)^2$ with initial condition $A(1981) = 340$ (as in Exercise 69). Show that the number of new cases between times $t = 1985$ and $t = 1987$ is equal to the sum of the number of new cases between $t = 1985$ and $t = 1986$ and the number of new cases between $t = 1986$ and $t = 1987$.

76. The amount of chemical obeys $\frac{dP}{dt} = \frac{5}{1 + 2t}$ with initial condition $P(0) = 2$ (as in Exercise 70). Show that the amount produced between times $t = 5$ and $t = 10$ is equal to the sum of the amount produced between $t = 5$ and $t = 7.5$ and the amount produced between $t = 7.5$ and $t = 10$.

77. The amount of chemical obeys $\frac{dP}{dt} = 5e^{-2t}$ with initial condition $P(0) = 2$ (as in Exercise 71). Show that the amount produced between times $t = 5$ and $t = 10$ is equal to the sum of the amount produced between $t = 5$ and $t = 7.5$ and the amount produced between $t = 7.5$ and $t = 10$.

78–79 ■ Two rockets are shot from the ground. Each has a different upward acceleration and a different amount of fuel. After the fuel runs out, each rocket falls with an acceleration of -9.8 m/s^2. For each rocket:

a. Write and solve differential equations describing the velocity and position of the rocket while it still has fuel.

b. Find the velocity and height of the rocket when it runs out of fuel.

c. Write and solve differential equations describing the velocity and position of the rocket after it has run out of fuel. What is the initial condition for each?

d. Find the maximum height reached by the rocket. Does it rise more while it has fuel or after the fuel has run out? Why?

e. Find the velocity when it hits the ground.

78. The upward acceleration is 12 m/s^2 and it has 10 s worth of fuel.

79. The upward acceleration is 2 m/s^2 and it has 60 s worth of fuel.

Computer Exercises

80. Imagine that toward the end of the universe, acceleration due to gravity begins to break down. Suppose that

$$a = -9.8\left(\frac{1}{1+t}\right)$$

where time is measured in seconds after the beginning of the end. An object begins falling from 10 m above the ground.

a. Find the velocity at time t.

b. Find the position at time t.

c. Graph acceleration, velocity, and position on the same graph. Which of these measurements are integrals of each other?

d. When will this object hit the ground?

81. Suppose the volume of water in a vessel obeys the differential equation

$$\frac{dV}{dt} = f(t) = 1 + 3t + 3t^2$$

with $V(0) = 0$.

a. Graph the functions f and V for $t = 0$ to $t = 10$ and label the curves.

b. Find the volume at time $t = 10$. What is the definite integral that has the same value?

c. Define a function $A(T)$ that gives the *average* rate of change of volume as a function of time (the total volume added between times $t = 0$ and $t = T$ divided by the elapsed time). Graph this on the same graph as $f(t)$. Label the curves (and write the formula for $A(t)$ as a definite integral).

d. Graph $f(t)$ between $t = 0$ and $t = 10$ and the constant function with rate equal to the average at time $T = 10$. What is the area under the line? Does it match what you found in part (b)? Why should it?

7.5 Techniques of Integration: Substitution and Integration by Parts

By the Fundamental Theorem of Calculus, computing the definite integral

$$\int_a^b f(x)dx$$

can be broken down into two steps: finding an antiderivative $F(x)$ of $f(x)$ and evaluating the difference $F(b) - F(a)$. The latter step is easy, but the former could be quite difficult, or impossible, to do algebraically.

In this section, we investigate methods that will help us find the antiderivatives of some functions. Based on the chain rule, we derive the **substitution** (also known as the change of variables) method. Using the product rule for derivatives, we develop the **integration by parts** formula. Near the end, we show how to use **Taylor polynomials** to obtain approximations for definite integrals that we cannot calculate using algebraic means.

A Useful Shortcut

Recall that, given a continuous function $f(x)$,

$$\int f(x)dx = F(x) + C \text{ is equivalent to } F'(x) = f(x)$$

The function $F(x)$ is called an antiderivative of $f(x)$.

Example 7.5.1 Finding Antiderivatives by Guess-and-Check

Find the indefinite integral

$$\int \frac{1}{2x+4}dx$$

Recalling that

$$\int \frac{1}{x}dx = \ln|x| + C$$

we try to guess the answer:

$$F(x) = \ln|2x+4|$$

Let's check: the derivative

$$F'(x) = \frac{1}{2x+4} \cdot 2 \neq \frac{1}{2x+4}$$

is not equal to the integrand; however, it is not far from it—the difference is in the factor of 2. If we can remove it, we'll be done. So we try

$$F(x) = \frac{1}{2} \ln |2x+4|$$

This time

$$F'(x) = \frac{1}{2}\left(\frac{1}{2x+4}\right)(2) = \frac{1}{2x+4}$$

It works! Therefore,

$$\int \frac{1}{2x+4} dx = \frac{1}{2} \ln |2x+4| + C$$

For obvious reasons, the method we used in the previous example is sometimes called guess-and-check. Recall that in the previous section we used the same approach to show that

$$\int e^{ax} dx = \frac{1}{a} e^{ax} + C$$

Here is one more example.

Example 7.5.2 Finding Antiderivatives by Guess-and-Check

Find

$$\int \frac{1}{\sqrt{3x-4}} dx$$

We know that

$$\int \frac{1}{\sqrt{x}} dx = \int x^{-1/2} dx = \frac{x^{1/2}}{1/2} = 2\sqrt{x}$$

Based on this, our guess for the antiderivative of $f(x) = \frac{1}{\sqrt{3x-4}}$ is

$$F(x) = 2\sqrt{3x-4}$$

Differentiate $F(x)$:

$$F'(x) = 2\frac{1}{2}(3x-4)^{-1/2}\, 3 = \frac{3}{\sqrt{3x-4}} \neq \frac{1}{\sqrt{3x-4}}$$

It's not exactly $f(x)$, but close—we need to get rid of the factor of 3. So we try

$$F(x) = \frac{1}{3} \cdot 2\sqrt{3x-4} = \frac{2}{3}\sqrt{3x-4}$$

This time,

$$F'(x) = \frac{2}{3} \cdot \frac{1}{2}(3x-4)^{-1/2} \cdot 3 = (3x-4)^{-1/2} = \frac{1}{\sqrt{3x-4}}$$

and we are done. The answer is

$$\int \frac{1}{\sqrt{3x-4}} dx = \frac{2}{3}\sqrt{3x-4} + C$$

The Chain Rule and Integration by Substitution

Why is integration (in general) difficult? Here is one reason.
We know that

$$\int x^3 dx = \frac{1}{4}x^4$$

(for simplicity, we take the integration constant C to be zero). The calculation

$$\left(\int x\,dx\right)\left(\int x^2\,dx\right) = \frac{1}{2}x^2 \cdot \frac{1}{3}x^3 = \frac{1}{6}x^5 \neq \frac{1}{4}x^4$$

shows that we *cannot* integrate the product of two functions by multiplying the integral of the first function by the integral of the second function. That is,

$$\int x^3\,dx \neq \left(\int x\,dx\right)\left(\int x^2\,dx\right)$$

In general, there is no formula that we can use to integrate the product (or the quotient) of two functions—this is why various **methods of integration** have been developed.

In this section, we explore the two most commonly used methods, **substitution** and **integration by parts,** which will allow us to integrate *some* products and quotients. First, we discuss the integration by substitution method.

In order to calculate

$$\int 2xe^{x^2}\,dx$$

we need to find a function whose derivative is equal to the product $2x \cdot e^{x^2}$. How do we obtain a product when we differentiate a function? One way is to use the chain rule ("the derivative of the outer function *times* the derivative of the inner function"). So, thinking about the chain rule, we recognize $2x \cdot e^{x^2}$ as the derivative of e^{x^2}:

$$(e^{x^2})' = e^{x^2} \cdot 2x$$

Thus,

$$\int 2xe^{x^2}\,dx = e^{x^2} + C$$

Let's try another example. To find

$$\int 3x^2 \cos x^3\,dx$$

we note that the integrand contains x^3 as well as its derivative, $3x^2$ (so x^3 must be the inner function in the composition). So we try $\sin x^3$:

$$(\sin x^3)' = \cos x^3 \cdot 3x^2$$

It works—thus,

$$\int 3x^2 \cos x^3\,dx = \sin x^3 + C$$

It seems that if an *integrand comes from the chain rule, then we can calculate the integral*. That this is really so is confirmed by the following theorem.

Theorem 7.5.1 Substitution Rule for Indefinite Integrals

Assume that $f(x)$ and $g(x)$ are such that the composition $f(g(x))$ is defined, and that $f(x)$ and $g'(x)$ are continuous. If $u = g(x)$, then

$$\int f(g(x))g'(x)\,dx = \int f(u)\,du$$

Note that the assumptions guarantee that the integrand on the left side is a continuous function. At the end of this subsection, we will prove the theorem.

532   **Chapter 7**   Integrals and Applications

The substitution rule tells us that if the integrand came from the chain rule, then we can find its antiderivative. As a new variable u we select the "inner" function in the composition.

If $u = g(x)$, then

$$\frac{du}{dx} = g'(x)$$

If we view du and dx as ordinary variables, inside the integral, we can write

$$du = g'(x)dx$$

(That we are allowed to do so follows from the statement of Theorem 7.5.1.)

Example 7.5.3   Using the Substitution Rule I

Find

$$\int x^2 \sin(1 + x^3)dx$$

Let $u = 1 + x^3$. (Why? We will explain after the calculation.) Then

$$du = (1 + x^3)'dx$$
$$du = 3x^2 dx$$
$$\frac{1}{3}du = x^2 dx$$

Using the substitution rule, we write

$$\int x^2 \sin(1 + x^3)dx = \int \sin(1 + x^3) \cdot x^2 dx$$
$$= \int \sin u \cdot \frac{1}{3}du$$
$$= \frac{1}{3}\int \sin u \, du$$
$$= -\frac{1}{3}\cos u + C$$
$$= -\frac{1}{3}\cos(1 + x^3) + C$$

Note that in the last step of the example we go back to the original variable, x. We check that our answer is correct by differentiating (using the chain rule):

$$\left(-\frac{1}{3}\cos(1 + x^3) + C\right)' = -\frac{1}{3}[-\sin(1 + x^3)] \cdot 3x^2 = x^2 \sin(1 + x^3)$$

Why did we take $u = 1 + x^3$? Look at the substitution rule formula: the "inner" function $u = g(x)$ appears in the integral twice: as itself, in one factor, and as the derivative of itself in the other factor. Hence a very helpful hint: if we can recognize a *function and its derivative* in the integrand (ignoring possible constants), then u is made equal to that function.

Example 7.5.4   Using the Substitution Rule II

Find

$$\int \frac{(\ln x)^2}{x}dx$$

We see that both $\ln x$ and its derivative $\frac{1}{x}$ are present as factors in the integrand—so we take $u = \ln x$. Then

$$du = \frac{1}{x}dx$$

and

$$\int \frac{(\ln x)^2}{x} dx = \int (\ln x)^2 \cdot \frac{1}{x} dx$$

$$= \int u^2 du$$

$$= \frac{1}{3}u^3 + C$$

$$= \frac{1}{3}(\ln x)^3 + C$$

Example 7.5.5 Using the Substitution Rule III

Find the indefinite integral

$$\int \frac{e^{2x}}{(1+e^{2x})^2} dx$$

The derivative of $1 + e^{2x}$ is $2e^{2x}$. Ignoring the constant factor of 2, we see that both the function $1 + e^{2x}$ and its derivative appear as factors in the integrand.

Let $u = 1 + e^{2x}$. Then

$$du = 2e^{2x} dx$$

$$\frac{1}{2} du = e^{2x} dx$$

and, by the substitution rule,

$$\int \frac{e^{2x}}{(1+e^{2x})^2} dx = \int \frac{1}{(1+e^{2x})^2} \cdot e^{2x} dx$$

$$= \int \frac{1}{u^2} \cdot \frac{1}{2} du$$

$$= \frac{1}{2} \int u^{-2} du$$

$$= \frac{1}{2} \frac{u^{-1}}{-1} + C$$

$$= -\frac{1}{2u} + C$$

$$= -\frac{1}{2(1+e^{2x})} + C$$

We now show how to evaluate definite integrals using the substitution rule. The two approaches that we discuss differ in how they treat the limits of integration.

Example 7.5.6 Evaluating a Definite Integral Using the Substitution Rule

Find

$$\int_0^1 x^3 (x^4 + 2)^4 dx$$

We evaluate this integral in two steps: first, we ignore the limits of integration and find an antiderivative of the given function. Then, using the Fundamental Theorem of Calculus, we evaluate the given definite integral.

Let $u = x^4 + 2$. Then

$$du = 4x^3 dx$$
$$\frac{1}{4}du = x^3 dx$$

and

$$\int x^3(x^4+2)^4 dx = \int (x^4+2)^4 \cdot x^3 dx$$
$$= \int u^4 \cdot \frac{1}{4} du$$
$$= \frac{1}{4}\frac{u^5}{5} + C$$
$$= \frac{1}{20}(x^4+2)^5 + C$$

By the Fundamental Theorem of Calculus,

$$\int_0^1 x^3(x^4+2)^4 dx = \frac{1}{20}(x^4+2)^5 \Big|_0^1$$
$$= \frac{1}{20}(3)^5 - \frac{1}{20}(2)^5 = 10.55$$

The alternative approach to evaluating

$$\int_0^1 x^3(x^4+2)^4 dx$$

consists of changing the limits of integration as the new variable u is introduced. We start as before: take $u = x^4 + 2$ and compute

$$du = 4x^3 dx$$
$$\frac{1}{4}du = x^3 dx$$

Next, we calculate the limits of integration for u: when $x = 0$, $u = (0)^4 + 2 = 2$. When $x = 1$, $u = (1)^4 + 2 = 3$. Thus

$$\int_0^1 x^3(x^4+2)^4 dx = \int_0^1 (x^4+2)^4 \cdot x^3 dx$$
$$= \int_2^3 u^4 \cdot \frac{1}{4} du$$
$$= \frac{1}{4}\frac{u^5}{5}\Big|_2^3$$
$$= \frac{1}{20}(3)^5 - \frac{1}{20}(2)^5 = 10.55$$

The second method that we used in Example 7.5.6 is correct by the following theorem.

Theorem 7.5.2 **Substitution Rule for Definite Integrals**

Assume that $f(x)$ and $g(x)$ are such that the composition $f(g(x))$ is defined and that $f(x)$

and $g'(x)$ are continuous. If $u = g(x)$, then

$$\int_a^b f(g(x))g'(x)dx = \int_{g(a)}^{g(b)} f(u)du$$

Proof: Assume that F is an antiderivative of f; i.e., $F' = f$. Compute the left side: by the chain rule,

$$f(g(x))g'(x) = F'(g(x))g'(x) = (F(g(x)))'$$

and, using the Fundamental Theorem of Calculus,

$$\int_a^b f(g(x))g'(x)dx = \int_a^b (F(g(x)))'dx = F(g(x))|_a^b = F(g(b)) - F(g(a))$$

For the right side we use the Fundamental Theorem again:

$$\int_{g(a)}^{g(b)} f(u)du = F(u)|_{g(a)}^{g(b)} = F(g(b)) - F(g(a))$$

Thus, both sides are equal.

Example 7.5.7   Evaluating a Definite Integral Using the Substitution Rule

Find the numerical value of the integral

$$\int_0^{\sqrt{\pi}} x \cos x^2 \, dx$$

Let $u = x^2$; then

$$du = 2x\,dx$$
$$\frac{1}{2}du = x\,dx$$

Now transform the limits of integration: when $x = 0$, then $u = (0)^2 = 0$. When $x = \sqrt{\pi}$, then $u = (\sqrt{\pi})^2 = \pi$. Thus

$$\int_0^{\sqrt{\pi}} x \cos x^2 \, dx = \int_0^{\sqrt{\pi}} \cos x^2 \cdot x\,dx$$
$$= \int_0^{\pi} \cos u \cdot \frac{1}{2} du$$
$$= \frac{1}{2} \sin u \Big|_0^{\pi}$$
$$= \frac{1}{2} \sin \pi - \frac{1}{2} \sin 0 = 0$$

Example 7.5.8   Applying Substitution to Fish Growth

Suppose fish length begins at $L(0) = 0$ (measured from fertilization) and follows the differential equation

$$\frac{dL}{dt} = 6.48 e^{-0.09t}$$

FIGURE 7.5.46

Growth of the fish

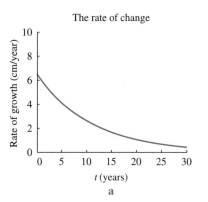
a

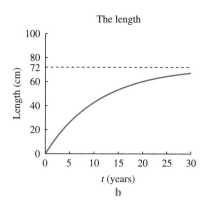
b

This equation expresses the fact that fish grow more and more slowly, but never stop growing, throughout their lives (Figure 7.5.46a). Growth data for walleye in North Caribou Lake, Ontario, are consistent with this equation. To solve it, we need to find the indefinite integral of the rate of change, $6.48e^{-0.09t}$, and then compute the constant. Thus,

$$L(t) = \int 6.48 e^{-0.09t} dt = 6.48 \int e^{-0.09t} dt$$

To find the integral—as extra practice—we use the substitution (alternatively, we could guess-and-check). Let $u = -0.09t$. Then

$$du = -0.09 dt$$

$$\frac{1}{-0.09} du = dt$$

and by the substitution rule

$$\int e^{-0.09t} dt = \int e^u \frac{1}{-0.09} du$$

$$= \frac{1}{-0.09} e^u$$

$$= \frac{1}{-0.09} e^{-0.09t}$$

Therefore,

$$L(t) = 6.48 \, \frac{1}{-0.09} e^{-0.09t} + C = -72 e^{-0.09t} + C \qquad (7.5.1)$$

To finish solving the differential equation, we must find the integration constant. The equation is

$$L(0) = 0 = -72 e^{-0.09 \cdot 0} + C = -72 + C$$

so $C = 72$. The solution is

$$L(t) = 72 - 72 e^{-0.09t} = 72 \left(1 - e^{-0.09t}\right) \qquad (7.5.2)$$

This is called the **von Bertalanffy growth equation** (Figure 7.5.46). The size approaches a limit of 72 cm, although the fish never stops growing.

Example 7.5.9 Finding the Age When a Fish Matures

We can use the solution to the previous example to find the age at which fish mature. Walleye begin to reproduce when they reach about 45 cm in length (but continue to grow after that). How long will it take this fish to mature?

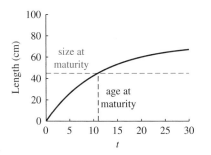

FIGURE 7.5.47
Finding the age of maturity of a walleye

We must solve

$$L(t) = 45$$

or

$$72\left(1 - e^{-0.09t}\right) = 45$$
$$1 - e^{-0.09t} = \frac{45}{72} = 0.625$$
$$e^{-0.09t} = 0.375$$
$$-0.09t = \ln 0.375$$
$$t = \frac{\ln 0.375}{-0.09} \approx 10.9$$

So, the fish will take about 11 years to mature (Figure 7.5.47). ▲

Proof of Theorem 7.5.1 We now articulate why Theorem 7.5.1 is true. The argument is technical, and not at all intuitive; to make it more transparent, we illustrate the steps in the proof with an example.

Suppose we wish to find

$$\int 2xe^{x^2}\,dx$$

If we let $f(u) = e^u$ and $g(x) = x^2$ (so that $g'(x) = 2x$), we can rewrite this integral as

$$\int 2xe^{x^2}\,dx = \int g'(x)e^{g(x)}\,dx$$
$$= \int g'(x)f(g(x))\,dx$$
$$= \int f(g(x))g'(x)\,dx$$

If f has an antiderivative, F, then $F' = f$ and the integrand is equal to

$$f(g(x))g'(x) = F'(g(x))g'(x)$$

By the chain rule,

$$(F(g(x)))' = F'(g(x))g'(x)$$

and thus, putting it all together,

$$\int f(g(x))g'(x)\,dx = \int F'(g(x))g'(x)\,dx$$
$$= \int (F(g(x)))'\,dx$$
$$= F(g(x)) + C \qquad (7.5.3)$$

Going back to our example, the antiderivative of $f(u) = e^u$ is $F(u) = e^u$ and, therefore, using Equation 7.5.3, we have

$$\int 2xe^{x^2}\,dx = F(g(x)) + C = e^{g(x)} + C = e^{x^2} + C$$

By differentiating

$$(e^{x^2} + C)' = e^{x^2} \cdot 2x = 2xe^{x^2}$$

we verify that our answer is correct.

If we introduce the new variable $u = g(x)$, then we write Equation 7.5.3 as

$$\int F'(g(x))g'(x)\,dx = F(g(x)) + C = F(u) + C = \int F'(u)\,du$$

i.e. (since $f = F'$),

$$\int f(g(x))g'(x)dx = \int f(u)du$$

This is the statement of Theorem 7.5.1.

Integration by Parts

The sum rule and constant multiple rule for derivatives translate directly into similar rules for integrals. The chain rule for derivatives translates into integration by substitution. The product rule for derivatives also leads to a corresponding method for integrals, called **integration by parts.** Like substitution, it provides a method that can sometimes transform unfamiliar integrals into familiar forms but is not guaranteed to work.

Recall the product rule for derivatives, written for the functions $u(x)$ and $v(x)$,

$$[u(x)v(x)]' = u'(x)v(x) + u(x)v'(x)$$

Solving for $u(x)v'(x)$,

$$u(x)v'(x) = [u(x)v(x)]' - u'(x)v(x)$$

and integrating both sides gives

$$\int u(x)v'(x)dx = \int [u(x)v(x)]'dx - \int u'(x)v(x)dx$$

$$\int u(x)v'(x)dx = u(x)v(x) - \int u'(x)v(x)dx$$

We have thus obtained the **integration by parts formula**

$$\int u(x)v'(x)dx = u(x)v(x) - \int v(x)u'(x)dx \qquad (7.5.4)$$

This formula expresses the integral on the left side as the difference of the function $u(x)v(x)$ (which needs no integration!) and another integral. How can we use it?

If we can recognize the given integrand as the product

$$u(x)v'(x)$$

and are able to integrate $v'(x)$ to get $v(x)$, and the remaining integral

$$\int v(x)\,u'(x)\,dx$$

is easier than the one we started with, then the method will work.

Example 7.5.10 An Example of Integration by Parts

Suppose we wish to find the indefinite integral

$$\int xe^x dx$$

Note that this function does not have a simple antiderivative. We cannot use the substitution method because xe^x cannot be obtained by differentiating a single function using the chain rule. The integrand is a product, so we choose $u(x) = x$ and $v'(x) = e^x$. Then

$$v(x) = \int e^x dx = e^x$$

(where we chose $C = 0$ for the arbitrary constant) and $u'(x) = 1$. Substituting into the formula for integration by parts gives

$$\int xe^x dx = u(x)v(x) - \int v(x)u'(x)\,dx$$
$$= xe^x - \int e^x dx$$
$$= xe^x - e^x + C$$

Checking, we use the product rule to get

$$\frac{d}{dx}(xe^x - e^x + C) = e^x + xe^x - e^x = xe^x$$

Let's see what happens if we switch the choices for $u(x)$ and $v'(x)$ in the previous example. Suppose we take $u(x) = e^x$ and $v'(x) = x$.
Then $u'(x) = e^x$ and $v(x) = \frac{1}{2}x^2$ and thus

$$\int xe^x dx = u(x)v(x) - \int v(x)u'(x)\,dx$$
$$= \frac{1}{2}x^2 e^x - \frac{1}{2}\int x^2 e^x dx$$

We don't care how complicated the first term is, since we no longer need to integrate it. However, the remaining integral is more complicated than the original integral, as x gained an extra power: instead of xe^x in the original integrand we now have $x^2 e^x$. This indicates that our first choice for $u(x)$ and $v'(x)$ was correct in this case.

Looking back at our calculation in Example 7.5.10, we see that after the integration by parts formula was applied, the integrand xe^x was replaced by e^x. What made it simpler, i.e., what made the x disappear?

It is because we took $u(x) = x$. Note that in the integral that is left after the integration by parts formula is used,

$$\int v(x)u'(x)\,dx$$

$u(x)$ appears as its derivative and, in our case, $u'(x) = 1$.
Let us try this again.

Example 7.5.11 Using the Integration by Parts Formula

Find

$$\int x^2 \cos x\,dx$$

We take $u(x) = x^2$ and $v'(x) = \cos x$. Then $u'(x) = 2x$ and $v(x) = \sin x$ and

$$\int x^2 \cos x\,dx = u(x)v(x) - \int v(x)u'(x)\,dx$$
$$= x^2 \sin x - \int \sin x \cdot 2x\,dx$$
$$= x^2 \sin x - 2\int x \sin x\,dx$$

The integrand $x \sin x$ is simpler than the original integrand $x^2 \cos x$, since the power of x has decreased by one. To complete the calculation, we integrate

$$\int x \sin x\,dx$$

by parts again. Let $u(x) = x$ and $v'(x) = \sin x$. Then $u'(x) = 1$ and $v(x) = -\cos x$ and

$$\int x \sin x \, dx = u(x)v(x) - \int v(x) u'(x) \, dx$$

$$= -x \cos x - \int (-\cos x) \, dx$$

$$= -x \cos x + \int \cos x \, dx$$

$$= -x \cos x + \sin x$$

Putting the pieces together, we get

$$\int x^2 \cos x \, dx = x^2 \sin x - 2 \int x \sin x \, dx$$

$$= x^2 \sin x - 2(-x \cos x + \sin x) + C$$

$$= x^2 \sin x + 2x \cos x - 2 \sin x + C$$

Example 7.5.12 Integration by Parts Applied to $\ln x$

Find the indefinite integral

$$\int \ln x \, dx$$

The trick is to think of the integrand $\ln x$ as the product $\ln x \cdot 1$. While we do not know the integral of $\ln x$, we *do know its derivative*, which gives us a hint that we can try integration by parts. So we take $u(x) = \ln x$ and $v'(x) = 1$. To use the integration by parts formula, we compute $u'(x) = \frac{1}{x}$ and $v(x) = x$, obtaining

$$\int \ln x \, dx = u(x)v(x) - \int v(x) u'(x) \, dx$$

$$= x \ln x - \int x \frac{1}{x} \, dx$$

$$= x \ln x - \int 1 \, dx$$

$$= x \ln x - x + C$$

Checking yields

$$\frac{d}{dx}(x \ln x - x + C) = \ln x + 1 - 1 = \ln x$$

Integration by parts can fail, as the following example shows.

Example 7.5.13 Failure of Integration by Parts

Suppose we wish to find the indefinite integral

$$\int \frac{e^x}{x} \, dx$$

This can be thought of as the product of e^x and $\frac{1}{x}$. We could set $u(x) = \frac{1}{x}$ and $v'(x) = e^x$. Then $v(x) = \int e^x \, dx = e^x$ and $u'(x) = -\frac{1}{x^2}$. Substituting, we get

$$\int \frac{e^x}{x} \, dx = \frac{e^x}{x} + \int \frac{e^x}{x^2} \, dx$$

The new integral is worse than the one we started with.

Alternatively, we could set $u(x) = e^x$ and $v'(x) = \frac{1}{x}$. Then

$$v(x) = \int \frac{1}{x}\, dx = \ln|x|$$

and $u'(x) = e^x$. It follows that

$$\int \frac{e^x}{x}\, dx = e^x \ln|x| - \int e^x \ln|x|\, dx$$

Unfortunately, this new integral is no easier than the original, so integration by parts has not helped us.

Example 7.5.14 Evaluating a Definite Integral

Using integration by parts, evaluate

$$\int_0^1 xe^{-0.2x}\, dx$$

The technique for evaluating definite integrals by parts is the same as for indefinite integrals, with the only difference lying in the fact that we need to evaluate after integration.

Choose $u(x) = x$ and $v'(x) = e^{-0.2x}$. Then $u'(x) = 1$ and

$$v(x) = \int e^{-0.2x}\, dx = \frac{1}{-0.2} e^{-0.2x} = -5e^{-0.2x}$$

Thus,

$$\int_0^1 xe^{-0.2x}\, dx = u(x)v(x)\Big|_0^1 - \int_0^1 v(x)u'(x)\, dx$$

$$= -5xe^{-0.2x}\Big|_0^1 - \int_0^1 (-5e^{-0.2x})\, dx$$

$$= (-5(1)e^{-0.2 \cdot 1}) - (-5(0)e^{-0.2 \cdot 0}) + 5 \int_0^1 e^{-0.2x}\, dx$$

$$= -5e^{-0.2} + 5\left(\frac{1}{-0.2}\right) e^{-0.2x}\Big|_0^1$$

$$= -5e^{-0.2} - 25e^{-0.2x}\Big|_0^1$$

$$= -5e^{-0.2} - 25(e^{-0.2} - e^0)$$

$$= -30e^{-0.2} + 25$$

Sometimes, a bit more work is needed to calculate an integral.

Example 7.5.15 Calculating a Definite Integral

Evaluate

$$\int_0^1 \arctan x\, dx$$

As in Example 7.5.12, where we calculated the integral of $\ln x$ (which we did not know) by using its *derivative* (which we did know), we recognize that we know the derivative of arctan x, but not its integral. So we use integration by parts with $u(x) = \arctan x$ and $v'(x) = 1$. Then

$$u'(x) = \frac{1}{1+x^2}$$

and $v(x) = x$, so

$$\int_0^1 \arctan dx = u(x)v(x)\Big|_0^1 - \int_0^1 v(x)u'(x)dx$$

$$= x \arctan x\Big|_0^1 - \int_0^1 \frac{x}{1+x^2}dx$$

We evaluate the first term:

$$x \arctan x\Big|_0^1 = 1 \arctan 1 - 0 \arctan 0 = \arctan 1 = \frac{\pi}{4}$$

and note that the second term is still an integral:

$$\int_0^1 \frac{x}{1+x^2}dx$$

With the combination of $1+x^2$ and its derivative x (actually it is $2x$, but we can easily handle the constant), the integrand is perfect for evaluation by substitution. Let $w = 1 + x^2$ (having already used u in this example, we use w instead). Then

$$dw = 2xdx$$
$$\frac{1}{2}dw = xdx$$

When $x = 0$, $w = 1 + (0)^2 = 1$; when $x = 1$, $w = 1 + (1)^2 = 2$; thus, the limits of integration for w are $w = 1$ and $w = 2$. We compute

$$\int_0^1 \frac{x}{1+x^2}dx = \int_0^1 \frac{1}{1+x^2} \cdot xdx$$

$$= \int_1^2 \frac{1}{w} \cdot \frac{1}{2}dw$$

$$= \frac{1}{2} \ln|w|\Big|_1^2$$

$$= \frac{1}{2}\ln|2| - \frac{1}{2}\ln|1| = \frac{1}{2}\ln 2$$

We conclude that

$$\int_0^1 \arctan dx = x \arctan x\Big|_0^1 - \int_0^1 \frac{x}{1+x^2}dx = \frac{\pi}{4} - \frac{1}{2}\ln 2 \approx 0.439$$

There are no theorems that tell us which method to use to find an antiderivative of a given function. But there is a strategy—shown in Table 7.5.1—that will assist us in picking the appropriate method. The table lists all integration techniques we have discussed, from the simplest to the most advanced (i.e., requiring the most work). Given a function, we start at the top of the table. If the method suggested does not apply, we move to the line below.

Table 7.5.1

Method	Applies When ...
basic antiderivative	... the integrand is recognized as the reversal of a differentiation formula, such as $\int x^n dx = \frac{1}{n+1}x^{n+1}$, $\int \frac{1}{1+x^2}dx = \arctan x$, and $\int \cos x\, dx = \sin x$.
guess-and-check	... the integrand differs from a basic antiderivative in that x is replaced by $ax + b$, such as $\int \cos 6x\, dx$, $\int \frac{1}{x-7}dx$, and $\int e^{4x+5}dx$.
substitution	... both a function and its derivative (up to a constant) appear in the integrand, i.e., the integrand has been obtained from the chain rule, such as $\int 4x^3 \cos(x^4 + 1)dx$, $\int \frac{\ln x}{x}dx$, and $\int xe^{4x^2}dx$.
integration by parts	... the integrand is the product of a power of x and one of $\sin x$, $\cos x$, and e^x, such as $\int x \cos 3x\, dx$, $\int x^3 e^x dx$, and $\int x^2 \sin 4x\, dx$.
	... the integrand contains a single function whose derivative we know, such as $\int \ln x\, dx$ or $\int \arctan x\, dx$.

Of course, it could happen that no method mentioned in Table 7.5.1 applies. This just means that we have to use other integration techniques and/or resources. Substitution and integration by parts are not the only algebraic methods; there are many more—some can be found in various calculus textbooks. Then, there are tables of integrals (both in written form and on the Internet) as well as mathematics software (Maple, WolframAlpha, Mathematica, etc.) capable of calculating antiderivatives symbolically.

Integration Using Taylor Polynomials

What happens if we cannot integrate a function exactly, i.e., when substitution, integration by parts, or any other method we try fails to produce an answer? One thing we can try is to approximate a function by its Taylor polynomial and then integrate.

Example 7.5.16 Integration Using Taylor Polynomials

To find the indefinite integral

$$\int \frac{e^x}{x}dx$$

we recall that we can approximate $f(x) = e^x$ by its Taylor polynomial (say, of degree five)

$$T_5(x) = 1 + x + \frac{1}{2}x^2 + \frac{1}{6}x^3 + \frac{1}{24}x^4 + \frac{1}{120}x^5$$

Thus,

$$\int \frac{e^x}{x}dx \approx \int \frac{T_5(x)}{x}dx$$

$$= \int \frac{1 + x + \frac{1}{2}x^2 + \frac{1}{6}x^3 + \frac{1}{24}x^4 + \frac{1}{120}x^5}{x}dx$$

$$= \int \left(\frac{1}{x} + 1 + \frac{1}{2}x + \frac{1}{6}x^2 + \frac{1}{24}x^3 + \frac{1}{120}x^4\right)dx$$

$$= \ln|x| + x + \frac{1}{2 \cdot 2}x^2 + \frac{1}{6 \cdot 3}x^3 + \frac{1}{24 \cdot 4}x^4 + \frac{1}{120 \cdot 5}x^5 + C$$

$$= \ln|x| + x + \frac{1}{4}x^2 + \frac{1}{18}x^3 + \frac{1}{96}x^4 + \frac{1}{600}x^5 + C$$

544 Chapter 7 Integrals and Applications

This looks messy, but at least we've got something! Keep in mind, though, that this approximation works *only near 0*, because this is where $T_5(x)$ approximates e^x well. Away from 0, it does not make any sense (look at the next example). To improve the approximation (and keep it accurate for a larger interval around 0) we need to take a higher-degree Taylor polynomial. (It can be proven that if we take infinitely many terms in the Taylor polynomial—thus obtaining a Taylor series—our calculation will give the exact answer!) ▲

Example 7.5.17   Evaluating a Definite Integral Using Taylor Polynomials

Let's compute

$$\int_{0.1}^{1} \frac{e^x}{x} dx$$

Now that we have an (approximate) antiderivative, all we need to do is to evaluate

$$\int_{0.1}^{1} \frac{e^x}{x} dx \approx \left(\ln|x| + x + \frac{1}{4}x^2 + \frac{1}{18}x^3 + \frac{1}{96}x^4 + \frac{1}{600}x^5 \right) \Big|_{0.1}^{1}$$

$$= \left(\ln|1| + 1 + \frac{1}{4}(1)^2 + \frac{1}{18}(1)^3 + \frac{1}{96}(1)^4 + \frac{1}{600}(1)^5 \right)$$

$$- \left(\ln|0.1| + 0.1 + \frac{1}{4}(0.1)^2 + \frac{1}{18}(0.1)^3 + \frac{1}{96}(0.1)^4 + \frac{1}{600}(0.1)^5 \right)$$

$$\approx 3.51767$$

This is quite close to the true value

$$\int_{0.1}^{1} \frac{e^x}{x} dx \approx 3.51793$$

If we move farther from 0, we cannot use $T_5(x)$. For instance

$$\int_{0.1}^{10} \frac{e^x}{x} dx \approx 2493.85179$$

whereas the Taylor approximation gives 365.89150. ▲

Example 7.5.18   Evaluating a Definite Integral Using Taylor Polynomials

Let us apply the same idea to obtain an approximation for

$$\int_{0}^{1} e^{-x^2} dx$$

Instead of calculating the Taylor polynomial of e^{-x^2}, we use what we already have: the Taylor polynomial $T_3(x)$ of e^x based at $x=0$ gives an approximation

$$e^x \approx T_3(x) = 1 + x + \frac{1}{2}x^2 + \frac{1}{6}x^3$$

(for simplicity, we take T_3 instead of some higher-order polynomial). Replacing x by $-x^2$, we obtain

$$e^{-x^2} \approx 1 + (-x^2) + \frac{1}{2}(-x^2)^2 + \frac{1}{6}(-x^2)^3$$

$$= 1 - x^2 + \frac{1}{2}x^4 - \frac{1}{6}x^6$$

Now we approximate the definite integral

$$\int_0^1 e^{-x^2} dx \approx \int_0^1 \left(1 - x^2 + \frac{1}{2}x^4 - \frac{1}{6}x^6\right) dx$$

$$= \left(x - \frac{1}{3}x^3 + \frac{1}{10}x^5 - \frac{1}{42}x^7\right)\Big|_0^1$$

$$= 1 - \frac{1}{3} + \frac{1}{10} - \frac{1}{42} = 0.74286$$

This is a fairly good approximation of the value

$$\int_0^1 e^{-x^2} dx \approx 0.74682$$

Summary To evaluate integrals, we developed **integration by substitution,** the integral version of the chain rule of differentiation. This technique can be used to integrate a few complicated combinations of functions. The method of **integration by parts,** the integral version of the product rule, can be used to compute a few additional integrals. In some cases, we use **Taylor polynomials** to obtain an approximation for an integral.

7.5 Exercises

Mathematical Techniques

1–14 ■ Attempt to find each integral using guess-and-check. Alternatively, use substitution.

1. $\int (x-3)^{3.4} dx$

2. $\int (2x+4)^4 dx$

3. $\int \frac{1}{3x-11} dx$

4. $\int \left(1 + \frac{x}{4}\right)^{1/3} dx$

5. $\int e^{-6.7x} dx$

6. $\int 3^{x-5} dx$

7. $\int_0^1 (1 - 0.5x)^4 dx$

8. $\int_3^5 \sqrt{2x+3} \, dx$

9. $\int_{-2}^{1/3} e^{1-3x} dx$

10. $\int_0^{\pi/8} \cos 4x \, dx$

11. $\int \sin(x/3) dx$

12. $\int \sin(2\pi(x-1)) dx$

13. $\int \frac{1}{1+4x^2} dx$

14. $\int \frac{1}{\sqrt{1-9x^2}} dx$

15–22 ■ Use substitution to find the indefinite integral of each function.

15. $f(x) = \dfrac{e^x}{1+e^x}$

16. $f(t) = e^t(1+e^t)^4$

17. $f(y) = 2y\sqrt{1+y^2}$

18. $f(x) = \cos x \, e^{\sin x}$

19. $f(\theta) = \tan \theta$ (Write it as $\dfrac{\sin \theta}{\cos \theta}$ and use substitution for the denominator.)

20. $f(t) = \dfrac{t}{1+t}$

21. $f(x) = \dfrac{1}{x \ln x}$

22. $f(x) = \dfrac{\ln x}{x}$

23–35 ▪ Find each integral.

23. $\displaystyle\int_{-1}^{1} (1 - x^4) x^3 \, dx$

24. $\displaystyle\int_{0}^{\sqrt{\pi}} x \sin x^2 \, dx$

25. $\displaystyle\int_{1}^{4} \dfrac{1}{x^2} \sqrt{1 - \dfrac{1}{x}} \, dx$

26. $\displaystyle\int_{0}^{1/2} e^{2x} \sqrt{1 + e^{2x}} \, dx$

27. $\displaystyle\int_{0}^{2\pi} \dfrac{\sin x}{3 + \cos x} \, dx$

28. $\displaystyle\int \dfrac{e^{\sqrt{x}}}{\sqrt{x}} \, dx$

29. $\displaystyle\int x e^{-x^2} \, dx$

30. $\displaystyle\int x^{-2} \cos(1/x) \, dx$

31. $\displaystyle\int_{0}^{\pi/2} \sin x \cos^2 x \, dx$

32. $\displaystyle\int_{2}^{3} \dfrac{1}{x \ln^2 x} \, dx$

33. $\displaystyle\int_{0}^{\pi/4} \dfrac{\sin x}{\cos^3 x} \, dx$

34. $\displaystyle\int_{-1}^{0} e^{4x} \sin e^{4x} \, dx$

35. $\displaystyle\int_{0}^{\sqrt{\pi/2}} x \sec x^2 \tan x^2 \, dx$

36–47 ▪ Use integration by parts to evaluate each integral. Check your answer by taking the derivative.

36. $\displaystyle\int x e^{2x} \, dx$

37. $\displaystyle\int x \cos 3x \, dx$

38. $\displaystyle\int x^2 e^x \, dx$ (You will need to integrate by parts twice.)

39. $\displaystyle\int_{0}^{\pi/4} x \cos 2x \, dx$

40. $\displaystyle\int_{1}^{9} \sqrt{x} \ln x \, dx$

41. $\displaystyle\int \ln^2 x \, dx$

42. $\displaystyle\int \arcsin x \, dx$

43. $\displaystyle\int x \arctan x \, dx$

44. $\displaystyle\int_{0}^{1} x 3^x \, dx$

45. $\displaystyle\int x^2 e^{-x} \, dx$ (You will need to integrate by parts twice.)

46. $\displaystyle\int \sin \sqrt{x} \, dx$ (Hint: Start with substitution, and then use integration by parts.)

47. $\displaystyle\int x^5 e^{x^3} \, dx$ (Hint: Start by substituting $u = x^3$.)

48–69 ▪ Using an appropriate technique (consult Table 7.5.1 if you need it), find each integral.

48. $\displaystyle\int \dfrac{x + 1}{x^2 + 1} \, dx$

49. $\displaystyle\int_{0}^{3} \dfrac{1}{9 + x^2} \, dx$

50. $\displaystyle\int_{0}^{3} \dfrac{3x}{9 + x^2} \, dx$

51. $\displaystyle\int_{1}^{2} \dfrac{x - 2}{2x + 3} \, dx$

52. $\displaystyle\int \dfrac{x - 1}{\sqrt{1 - x^2}} \, dx$

53. $\displaystyle\int \dfrac{1}{\sqrt{x}(\sqrt{x} + 4)} \, dx$

54. $\displaystyle\int_{-2}^{0} \sqrt{3 - 4x} \, dx$

55. $\displaystyle\int_{0}^{\pi/4} \tan x \, dx$

56. $\displaystyle\int x^{-2} \sin(1/x) \, dx$

57. $\displaystyle\int \cot 4x \, dx$

58. $\displaystyle\int \cos 4x \, dx$

59. $\int \sin^4 x \cos x \, dx$

60. $\int_0^\pi x \sin(x/3) \, dx$

61. $\int_0^1 x^2 \, 5^x \, dx$

62. $\int_0^1 x \, 5^{x^2} \, dx$

63. $\int_0^1 x^\pi \, dx$

64. $\int \left(\ln x^3\right)^2 dx$

65. $\int \dfrac{2}{x \ln x^3} \, dx$

66. $\int_e^{e^2} \dfrac{\ln x}{10x} \, dx$

67. $\int x \log_{10} x \, dx$

68. $\int_0^1 \dfrac{e^x - e^{-x}}{e^x + e^{-x}} \, dx$

69. $\int x\left(e^x - e^{-x}\right) dx$

70. Using the Taylor polynomial T_5 for $f(x) = \sin x$ based at $x = 0$, find an approximation for $\int_{0.1}^1 \dfrac{\sin x}{x} \, dx$.

71. Using the Taylor polynomial T_6 for $f(x) = \cos x$ based at $x = 0$, find an approximation for $\int \dfrac{\cos x}{x^2} \, dx$.

72. Using the Taylor polynomial T_4 for $f(x) = \cos x$ based at $x = 0$, find an approximation for $\int \cos x^3 \, dx$.

73. Using the Taylor polynomial T_3 for $f(x) = \sin x$ based at $x = 0$, find an approximation for $\int_0^1 \sin x^2 \, dx$.

74. Find $\int_{0.9}^{1.1} \ln x \, dx$ in two ways, as suggested.
    a. Use integration by parts.
    b. Use the Taylor polynomial T_2 for $f(x) = \ln x$ based at $x = 1$ to find an approximation of the given integral. Compare your answers to (a) and (b).

75. Find $\int_0^{0.2} \arctan x \, dx$ in two ways, as suggested.
    a. Use integration by parts.
    b. Use the Taylor polynomial T_3 for $f(x) = \arctan x$ based at $x = 0$ to find an approximation of the given integral. Compare your answers to (a) and (b).

76–77 ▪ In Examples 7.5.10 and 7.5.12, we chose the constant $C = 0$ when finding $v(x)$. Follow the steps for integration by parts, but leave C as an arbitrary constant. Do you get the same answer?

76. Find the indefinite integral $\int xe^x dx$ as in Example 7.5.10.

77. Find the indefinite integral $\int \ln x \, dx$ as in Example 7.5.12.

78–79 ▪ Sometimes integrating by parts seems to lead in a circle, but the answer can still be found. Try the following.

78. Find the indefinite integral $e^x \sin x$ using integration by parts twice.

79. Find the indefinite integral $\dfrac{\ln x}{x}$ using integration by parts.

Applications

80–83 ▪ The following differential equations for the production of a chemical P share the properties that the rate of change at $t = 0$ is 5 and the limit as $t \to \infty$ is 0. For each, find the solution starting from the initial condition $P(0) = 0$, sketch the solution, and indicate what happens to $P(t)$ as $t \to \infty$. Compute $P(10)$ and $P(100)$. Why do some increase to infinity whereas others do not?

80. $\dfrac{dP}{dt} = \dfrac{5}{1 + 2t}$

81. $\dfrac{dP}{dt} = 5e^{-0.2t}$

82. $\dfrac{dP}{dt} = 2.5\left(\dfrac{1}{1+t} + e^{-t}\right)$

83. $\dfrac{dP}{dt} = \dfrac{5}{(1+t)^2}$

84–85 ▪ Use integration by parts to find the solution of each differential equation.

84. Suppose the mass, M, of a toad grows according to the differential equation $\dfrac{dM}{dt} = (t + t^2)e^{-2t}$ with $M(0) = 0$. When does this toad grow fastest? Find $M(1)$. How much larger would the toad be at time $t = 1$ if it always grew at the maximum rate?

85. Suppose the mass, W, of a worm grows according to the differential equation $\dfrac{dW}{dt} = (4t - t^2)e^{-3t}$ with $W(0) = 0$. When does this worm grow fastest? Find $W(2)$. How much larger would the worm be at $t = 2$ if it always grew at the maximum rate?

86–87 ▪ The following problems give the parameters for walleye in a variety of locations. For each location, the differential equation has the form $\dfrac{dL}{dt} = \alpha e^{-\beta t}$.

a. Find the solution of the differential equation if $L(0) = 0$.
b. Find the limit of size as t approaches infinity.
c. Assume that all walleye mature at 45 cm in length. How old are these walleye when they mature?
d. Graph the size and compare with Ontario walleye (Figure 7.5.47).

86. In Texas, where $\alpha = 64.3$ and $\beta = 1.19$.

87. In Saskatchewan, where $\alpha = 6.48$ and $\beta = 0.06$.

88–91 ■ A population of leopards, $L(t)$, is increasing according to the formula given in each of the following problems. However, poaching is decreasing the population at a rate equal to $0.1L$.

a. Write a pure-time differential equation for the number of leopards, $B(t)$, being poached.

b. Solve using the initial condition $B(0) = 0$.

c. Graph the number of leopards left and the number poached.

d. Find the limit of the ratio $\dfrac{B(t)}{L(t)}$ as t approaches infinity.

88. $L(t) = 1000e^{0.2t}$

89. $L(t) = 1000e^{0.05t}$

90. $L(t) = 100t$

91. $L(t) = 100 + 100e^{0.5t}$

92–95 ■ Growth rates of insects depend on the temperature, T. Suppose that the length of an insect, L, follows the differential equation

$$\frac{dL}{dt} = 0.001 T(t)$$

with t measured in days starting from January 1 ($t = 0$) and temperature measured in degree Celsius. Insects hatch with an initial size of 0.1 cm. For each of the following equations for $T(t)$:

a. Sketch the graph of the temperature over the course of a year.

b. Suppose an insect starts growing on January 1. How big will it be after 30 days?

c. Suppose an insect starts growing on June 1 (day 151). How big will it be after 30 days?

92. $T(t) = 0.001t(365 - t)$

93. $T(t) = 40 - 5 \times 10^{-6} t^2 (365 - t)$

94. $T(t) = 20 + 10 \cos\left[\dfrac{2\pi(t - 190)}{365}\right]$

95. $T(t) = 20 + 10 \cos\left[\dfrac{2\pi(t - 90)}{182.5}\right]$

Computer Exercises

96. Consider again the fish in Caribou Lake, growing according to

$$\frac{dL}{dt} = \alpha e^{-\beta t}$$

where $\alpha = 6.48$ and $\beta = 0.09$. However, suppose there is some variability among fish in the values of these two parameters.

a. Find the growth trajectories of five fish with values of α evenly spread from 10% below to 10% above 6.48.

b. Find the growth trajectories of five fish with values of β evenly spread from 10% below to 10% above 0.09.

c. Which parameter has a greater effect on the size of the fish?

97. Clever genetic engineers design a set of fish that grow according to the following equations:

$$\frac{dL_1}{dt} = 1$$

$$\frac{dL_2}{dt} = \frac{1}{1 + \sqrt{t}}$$

$$\frac{dL_3}{dt} = \frac{1}{1 + t}$$

$$\frac{dL_4}{dt} = \frac{1}{1 + t^2}$$

$$\frac{dL_5}{dt} = \frac{1}{1 + t^3}$$

$$\frac{dL_6}{dt} = e^{-t}$$

$$\frac{dL_7}{dt} = e^{-t^2}$$

Suppose they all start at size 0. Use a computer to sketch the growth of these fish for $0 \leq t \leq 5$. Then zoom in near $t = 0$. Could you tell which fish was which?

7.6 Applications

In this section, we introduce several remarkable applications of the definite integral. First, we use the fact that the graph describing the Riemann sum can be interpreted geometrically as a way to approximate the **area under a curve.** In fact, geometric problems of this sort provided the motivation for Archimedes' near discovery of the integral about two thousand years ago, long before Newton introduced the study of differential equations. The idea of chopping quantities into small bits, adding them up with Riemann sums, and computing exact answers with the definite integral has many other applications, including finding the **volume** of a solid, the **length** of a curve, the **average value** of a function, and the total mass from a **density.**

Integrals and Areas

Recall that if $f(x) \geq 0$ on $[a, b]$, then

$$\int_a^b f(x)\,dx = \text{area of the region between the } x\text{-axis and}$$
$$\text{the graph of } f(x) \text{ over the interval } [a, b]$$

(Figure 7.6.48a). In words, for positive functions, the definite integral and the area are equal.

Suppose that $f(x) \leq 0$ for all x in $[a, b]$. Then

$$\int_a^b f(x)\,dx = \text{area above the } x\text{-axis} - \text{area below the } x\text{-axis}$$
$$= -\text{area below the } x\text{-axis}$$

since no part of the region is above the x-axis (Figure 7.6.48b). In this case, the integral

$$\int_a^b f(x)\,dx$$

is negative, and

$$-\int_a^b f(x)\,dx = \text{area of the region between the } x\text{-axis and } f(x) \text{ on } [a, b]$$

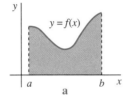
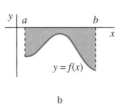

FIGURE 7.6.48
Definite integral and area

Example 7.6.1 Calculating Area

Find the area, A, of the region between the graph of the function $f(x) = x^2 - 9$ and the x-axis, bounded by the vertical lines $x = 0$ and $x = 4$.

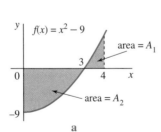
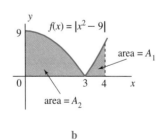

FIGURE 7.6.49
The region between $f(x) = x^2 - 9$ and the x-axis on $[0, 4]$

Looking at Figure 7.6.49a, we see that one part of the graph of $f(x) = x^2 - 9$ lies below the x-axis and the other lies above the x-axis. Since $f(x) \geq 0$ on $[3, 4]$, the area under $f(x)$ on $[3, 4]$ is

$$A_1 = \int_3^4 (x^2 - 9)\,dx$$

$$= \left(\frac{x^3}{3} - 9x\right)\bigg|_3^4$$

$$= \left(\frac{(4)^3}{3} - 9(4)\right) - \left(\frac{(3)^3}{3} - 9(3)\right) = \frac{10}{3}$$

Next, we evaluate

$$\int_0^3 (x^2 - 9) dx = \left(\frac{x^3}{3} - 9x\right)\Big|_0^3$$

$$= \left(\frac{(3)^3}{3} - 9(3)\right) - \left(\frac{(0)^3}{3} - 9(0)\right) = -18$$

Thus, the area of the region between the x-axis and $f(x)$ on $[0, 3]$ is $A_2 = 18$. The area of the whole region is

$$A = A_1 + A_2 = 18 + \frac{10}{3} = \frac{64}{3}$$

Note that

$$\int_0^4 (x^2 - 9)\, dx = \left(\frac{x^3}{3} - 9x\right)\Big|_0^4 = \left(\frac{64}{3} - 36\right) - (0) = -\frac{44}{3}$$

does not give the sum of the areas of the two shaded regions in Figure 7.6.49a. Recalling the definition, we know that this integral actually *subtracts* the area of the region below the x-axis from the area of the region above the x-axis; i.e.,

$$\int_0^4 (x^2 - 9)\, dx = A_1 - A_2 = \frac{10}{3} - 18 = -\frac{44}{3}$$

Thus, to find the *total area* of the shaded regions, we need to integrate the *absolute value* of the function:

$$\int_0^4 |x^2 - 9|\, dx$$

(Figure 7.6.49b). Of course, in order to integrate absolute value, we need to find where the integrand is positive and where it is negative (so, one way or another we end up with a calculation as in Example 7.6.1). Here is another example.

Example 7.6.2 Calculating Area

Find the area, A, of the region bounded by the graph of the function $f(x) = \ln x$, the x-axis, and the vertical lines $x = 1/2$ and $x = 2$.

The area is given by

$$A = \int_{1/2}^2 |\ln x|\, dx$$

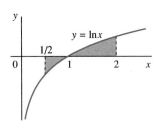

Figure 7.6.50

Region between $y = \ln x$ and the x-axis on $[1/2, 2]$

(Figure 7.6.50). The function $\ln x$ is negative for $x < 1$, and thus $|\ln x| = -\ln x$ when $x < 1$. If $x \geq 1$, then $\ln x \geq 0$, and $|\ln x| = \ln x$. We conclude that

$$A = \int_{1/2}^2 |\ln x|\, dx = -\int_{1/2}^1 \ln x\, dx + \int_1^2 \ln x\, dx$$

In Example 7.5.12 we used integration by parts to show that $\int \ln x\, dx = x \ln x - x + C$. Thus

$$A = -(x \ln x - x)\Big|_{1/2}^1 + (x \ln x - x)\Big|_1^2$$

$$= -(\ln 1 - 1) + \left(\frac{1}{2}\ln(1/2) - \frac{1}{2}\right) + (2\ln 2 - 2) - (\ln 1 - 1)$$

$$= 1 - \frac{1}{2}\ln 2 - \frac{1}{2} + 2\ln 2 - 2 + 1 = \frac{3}{2}\ln 2 - \frac{1}{2} \approx 0.5397$$

(recall that $\ln(1/2) = \ln 1 - \ln 2 = -\ln 2$). So the area of the shaded region in Figure 7.6.50 is approximately 0.5397. ▲

Next, we learn how to calculate areas of certain more complicated regions.

In particular, we consider regions in the *xy*-plane that are **bounded** and lie between the graphs of continuous functions. A region is called **bounded** if it is contained within some disk; in other words, the coordinates of points in a bounded region cannot grow arbitrarily large. (We will meet unbounded regions in Section 7.7.)

Four regions (A, B, C, and D) lie between the graphs $y = f(x)$ and $y = g(x)$ in Figure 7.6.51a. However, C and D are not bounded, and hence the region between $f(x)$ and $g(x)$ consists of the two shaded regions A and B. The *x*-coordinates of the points in A and B are determined by the points of intersection of $f(x)$ and $g(x)$. For instance, for all points (x, y) in A, $x_1 \le x \le x_2$. Sometimes we impose a condition on the range of values for *x*. For instance, the region between $f(x)$ and $g(x)$ and between $x = a$ and $x = b$ in Figure 7.6.51b consists of the three shaded regions labelled A, B, and C.

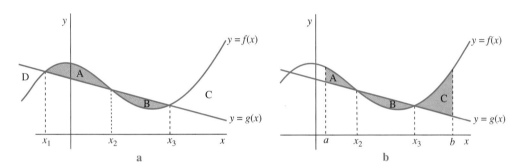

FIGURE 7.6.51

Region between $f(x)$ and $g(x)$

How do we calculate the area in this case?

Consider the region that lies between the graphs of two continuous functions, $y = f(x)$ and $y = g(x)$, and between the vertical lines $x = a$ and $x = b$. Assume that $f(x) \ge g(x)$ for all x in $[a, b]$; see Figure 7.6.52. We would like to compute the area, A, of the region.

We consider several cases, depending on the location of the two functions. First, assume that both f and g lie above the *x*-axis, as shown in Figure 7.6.52a. Then

$$A = \text{area under } f \text{ on } [a, b] - \text{area under } g \text{ on } [a, b]$$

$$= \int_a^b f(x)dx - \int_a^b g(x)dx$$

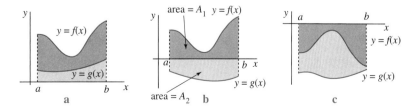

FIGURE 7.6.52

The region between two curves

$$= \int_a^b [f(x) - g(x)]dx$$

Next, assume that $f(x) \geq 0$ and $g(x) \leq 0$ on the whole interval $[a, b]$, as in Figure 7.6.52b. In that case, since $f(x) \geq 0$,

$$A_1 = \text{area under } f \text{ on } [a, b] = \int_a^b f(x)dx$$

Because $g(x) \leq 0$, we get

$$A_2 = \text{area between } x\text{-axis and } g \text{ on } [a, b] = -\int_a^b g(x)dx$$

Therefore,

$$A = A_1 + A_2$$
$$= \int_a^b f(x)dx + \left(-\int_a^b g(x)dx\right)$$
$$= \int_a^b [f(x) - g(x)]dx$$

the same as above. Finally, assume that both f and g lie below the x-axis, as in Figure 7.6.52c. Then

$$A_1 = \text{area between the } x\text{-axis and } f \text{ on } [a, b] = -\int_a^b f(x)dx$$

and

$$A_2 = \text{area between the } x\text{-axis and } g \text{ on } [a, b] = -\int_a^b g(x)dx$$

and therefore

$$A = A_2 - A_1$$
$$= -\int_a^b g(x)dx - \left(-\int_a^b f(x)dx\right)$$
$$= \int_a^b [f(x) - g(x)]dx$$

Thus, the area of the region between the graphs of continuous functions $f(x)$ and $g(x)$, bounded by the vertical lines $x = a$ and $x = b$, where $f(x) \geq g(x)$ for all x in $[a, b]$, is given by

$$A = \int_a^b [f(x) - g(x)]dx$$

regardless of the signs of $f(x)$ and $g(x)$.

Example 7.6.3 The Area of a Region between Two Curves

The region between the curves $y = \dfrac{1}{x}$ and $y = x$ on $[1, 2]$ is shown in Figure 7.6.53. Since

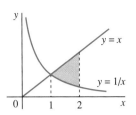

FIGURE 7.6.53
The region between $y = \dfrac{1}{x}$ and $y = x$

$x \geq \dfrac{1}{x}$ over the entire interval $[1, 2]$, the area of the region is

$$\int_1^2 \left(x - \dfrac{1}{x}\right) dx = \left(\dfrac{x^2}{2} - \ln x\right)\bigg|_1^2$$

$$= \left(\dfrac{(2)^2}{2} - \ln 2\right) - \left(\dfrac{(1)^2}{2} - \ln 1\right)$$

$$= \dfrac{3}{2} - \ln 2 \approx 0.807$$

Example 7.6.4 The Area of a Region between Two Curves That Intersect I

Find the area, A, of the region between $y = x$ and $y = \sqrt{x}$ and between $x = 0$ and $x = 1.5$.

The two curves intersect at $x = 0$ and $x = 1$ (see Figure 7.6.54). Note that $\sqrt{x} \geq x$ for x in $[0, 1]$, and $x \geq \sqrt{x}$ if x is in $[1, 1.5]$. Therefore,

$$A = \int_0^1 (\sqrt{x} - x)dx + \int_1^{1.5} (x - \sqrt{x})dx$$

$$= \left(\dfrac{2}{3}x^{3/2} - \dfrac{1}{2}x^2\right)\bigg|_0^1 + \left(\dfrac{1}{2}x^2 - \dfrac{2}{3}x^{3/2}\right)\bigg|_1^{1.5}$$

$$= \left(\dfrac{2}{3} - \dfrac{1}{2}\right) + \left(\dfrac{1}{2}(1.5)^2 - \dfrac{2}{3}(1.5)^{3/2}\right) - \left(\dfrac{1}{2} - \dfrac{2}{3}\right)$$

$$\approx 0.234$$

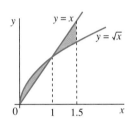

FIGURE 7.6.54
The region between $y = x$ and $y = \sqrt{x}$

Example 7.6.5 The Area of a Region between Two Curves That Intersect II

Find the area of the region between $y = \dfrac{2}{1 + x^2}$ and $y = 1$ and between $x = 0$ and $x = 3$.

Solving

$$\dfrac{2}{1 + x^2} = 1$$

for x we get $1 + x^2 = 2$, and thus $x^2 = 1$ and $x = \pm 1$. Within the interval $[0, 3]$, the two curves intersect at $x = 1$; see Figure 7.6.55. On the interval $[0, 1]$,

$$\dfrac{2}{1 + x^2} \geq 1$$

whereas on the interval $[1, 3]$,

$$1 \geq \dfrac{2}{1 + x^2}$$

The area is

$$A = \int_0^1 \left(\dfrac{2}{1 + x^2} - 1\right) dx + \int_1^3 \left(1 - \dfrac{2}{1 + x^2}\right) dx$$

$$= (2 \arctan x - x)\big|_0^1 + (x - 2 \arctan x)\big|_1^3$$

$$= (2 \arctan 1 - 1) - (2 \arctan 0 - 0) + (3 - 2 \arctan 3) - (1 - 2 \arctan 1)$$

$$= 2\left(\dfrac{\pi}{4}\right) - 1 + 3 - 2 \arctan 3 - 1 + 2\left(\dfrac{\pi}{4}\right)$$

$$= \pi + 1 - 2 \arctan 3 \approx 1.644$$

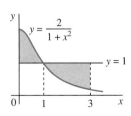

FIGURE 7.6.55
The region between $y = \dfrac{2}{1 + x^2}$ and $y = 1$

Example 7.6.6 Estimating the Surface Area of Lake Ontario

We use the idea of approximating rectangles to estimate the surface area of Lake Ontario. Using a satellite image of the lake (Figure 7.6.56a), we drew an outline and measured the distances across the lake at equally spaced locations (Figure 7.6.56b; note that the satellite image has a bar in the lower left corner indicating distance units; we measured the distances in the image and then converted to the actual distance). Expecting a reasonable degree of precision, we took 16 rectangles.

Figure 7.6.56a

Satellite image of Lake Ontario
Courtesy NOAA

In the language of Riemann sums, we now calculate the midpoint sum with 16 rectangles (Figure 7.6.56c). The base length of each rectangle is 18.76 km, and the heights are shown in Figures 7.6.56b and c. Thus,

$$M_{16} = 18.76 \cdot 18.76 + 18.76 \cdot 38.86 + 18.76 \cdot 52.26 + \cdots + 18.76 \cdot 53.60$$
$$= 19{,}054.91$$

Our estimate of about 19,055 km^2 is within the estimates provided by various sources (most sources—Government of Canada, Environmental Protection Agency, Encyclopaedia Britannica, National Oceanographic and Atmospheric Administration—place their estimates around 19,000 km^2).

Is this how it is really done? Yes—in the sense of the idea of using simple geometric objects to approximate the shape of the lake. A commonly used method (because it can be easily implemented on a computer) replaces rectangles with squares.

A square grid is placed over an image of the lake. The squares whose centres fall within the lake or sit on its boundary are counted in. In practice, the centres of these squares are marked by a dot (we marked a few in Figure 7.6.57), and then all we need to do is to count the number of dots and multiply by the area of one square.

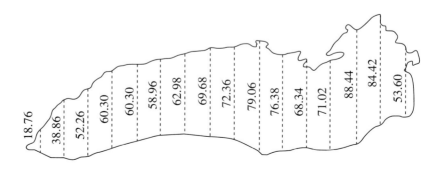

Figure 7.6.56b

Distances across Lake Ontario

FIGURE 7.6.56c

Approximating rectangles

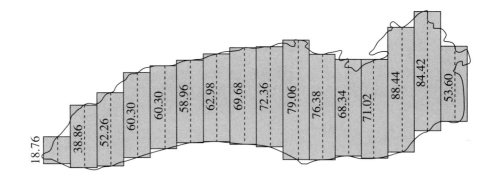

FIGURE 7.6.57

Counting dots to estimate the area of Lake Ontario

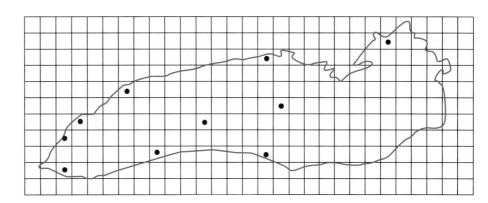

In our case, we counted 137 squares. The size of the grid (in this case) is 12 km, so the estimate for the area is

$$137 \cdot 12^2 = 137 \cdot 144 = 19{,}728 \text{ km}^2$$

Note that every square that is counted contributes 144 km² to the total area. So our decision on which squares to count (or not) around the edges could significantly affect the estimate. ▲

Integrals and Volumes

We *approximated* the area of a (two-dimensional) region between the graph of a function and the x-axis by using simple geometric objects (rectangles). Then, by computing the limit of approximations by rectangles we arrived at the *exact* value for the area (Section 7.3).

We compute the volume of a three-dimensional solid in space in essentially the same way. This time, the simple three-dimensional objects we use are cylinders. A **right cylinder** is a solid bounded by a two-dimensional region (called the **base**) and an identical region (called the **top**) in a plane parallel to the base. The cylinder contains all points on the line segments perpendicular to the base (hence "right") that connect the base and the top. The distance between the base and the top is the **height** of the cylinder (labelled h in Figure 7.6.58). In Figure 7.6.58a we show a right circular cylinder (since its base is a disk). Two more cylinders are drawn in Figures 7.6.58b and c. The volume of a cylinder is computed by multiplying the base area by the height. Thus, the volume of the right circular cylinder of radius r and height h is $\pi r^2 h$.

Let's see how cylinders are used to approximate three-dimensional regions, thus producing an approximation for their volume.

Example 7.6.7 Volume of a Heart Chamber

In Figure 7.6.59a we see a sonogram (an ultrasound image) of a human heart, showing its structures—in particular, the four chambers. The information about the volume of

FIGURE 7.6.58
Cylinders

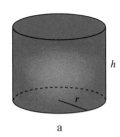
a

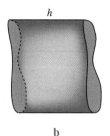
b

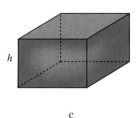
c

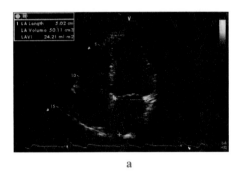
a

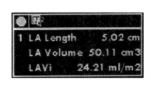
b

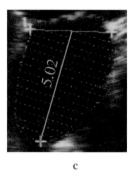
c

FIGURE 7.6.59
Ultrasound image of a human heart
Miroslav Lovric

the chambers is essential in determining certain features of the blood flow in and out of the heart. The zoom-in of the upper left corner of the sonogram reads "LA Volume 50.11 cm^3" (Figure 7.6.59b; LA stands for left atrium chamber, shown on the bottom right of the sonogram image). How was the volume calculated?

With a computer mouse, an ultrasound technician outlines the shape of the chamber. (Note that it is not possible to identify where exactly the boundary of the chamber is located.) When the mouse button is released, the computer draws the grid shown in Figures 7.6.59a and c. The measurement "LA Length 5.02 cm" (Figure 7.6.59b) is the distance across the chamber shown in Figure 7.6.59c.

The grid consists of 15 strips of thickness $5.02/15 \approx 0.335$ cm. By using the lengths of the parallel dashed lines as diameters, we draw right circular cylinders (Figures 7.6.60a, b), obtaining an approximation of the chamber by circular cylinders (Figure 7.6.60c). The height of each cylinder is 0.335 cm.

The diameter of the cylinder at the bottom of the chamber (Figure 7.6.60a) is measured to be 2 cm; thus, its volume is $\pi(1)^2(0.335) \approx 1.052$ cm^3. The diameter of the cylinder above it is 2.5 cm, and its volume is $\pi(1.25)^2(0.335) \approx 1.644$ cm^3. The remaining diameters (from bottom to top) are 3.0, 3.21, 3.43, 3.57, 3.93, 4.07, 4.29, 4.29, 4.29, 4.14, 4.07, 3.43, and 2.14 cm, respectively. Computing the corresponding volumes and then adding up the volumes of all 15 cylinders, we obtain 50.342 cm^3 as an approximation of the volume of the heart chamber.

Our estimate is very close to the approximation of 50.11 cm^3 produced by the computer and shown in the sonogram. The discrepancy is due to the errors we made in measuring the diameters.

We use the same idea to find the volume of any three-dimensional solid S (Figure 7.6.61). First we divide S into n pieces and approximate each piece with a cylinder. By adding up the volumes of all cylinders (thus forming a Riemann sum!), we obtain an approximation for the volume of S. Increasing the number of cylinders (analogous to the case of the area in Section 7.3) generates better and better approximations. Finally, in the limit as $n \to \infty$, we obtain the exact value of the volume.

7.6 Applications 557

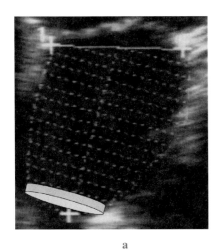
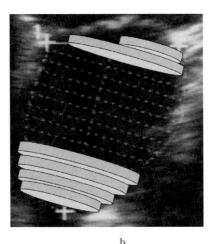
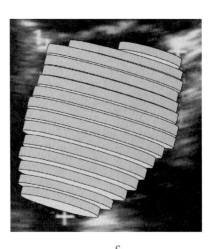

a b c

FIGURE 7.6.60
Approximating cylinders
Miroslav Lovric

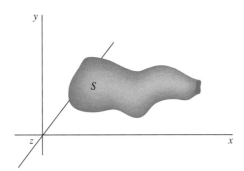

FIGURE 7.6.61
Three-dimensional solid S

Now the details. Assume that S lies between two parallel planes that are perpendicular to the x-axis and intersect it at $x = a$ and $x = b$ (Figure 7.6.62a). Pick any location x between a and b, and cut S with a plane perpendicular to the x-axis passing through x. This way, we obtain a two-dimensional region R_x called the *cross-section* of the solid S (Figure 7.6.62a). Denote by $A(x)$ the area of R_x (we learned how to calculate areas earlier in this chapter). As cross-sections vary with x, so does their area—thus A is indeed a function of x; it is defined on $[a, b]$.

Divide the interval $[a, b]$ into n subintervals (n is a positive integer)

$$[a = x_1, x_2], [x_2, x_3], \ldots, [x_i, x_{i+1}], \ldots, [x_n, x_{n+1} = b]$$

of length $\Delta x = (b - a)/n$ and form the cross-sections of S at points $x_1, x_2, \ldots, x_n$. "Thicken" each cross-section to make a cylinder whose base is the cross-section and whose height is Δx; see Figure 7.6.62b, where we thickened the cross-section at x_i. The volume of the cylinder thus obtained is the base area, $A(x_i)$, multiplied by the height, Δx. By adding up the volumes of all thickened cylinders,

$$V_n = A(x_1)\Delta x + A(x_2)\Delta x + A(x_3)\Delta x + \cdots + A(x_n)\Delta x = \sum_{i=1}^{n} A(x_i)\Delta x$$

we obtain an approximation of the volume, V, of the solid S. (Note that V_n is the left Riemann sum for the function $A(x)$ on the given interval.)

It can be proven that, as n increases (i.e., as the approximating cylinders become thinner and thinner), the approximations V_n become better and better. We now define the volume as the limit of V_n. Keep in mind that, at the same time, we are computing the limit of Riemann sums—thus obtaining a definite integral.

558 Chapter 7 Integrals and Applications

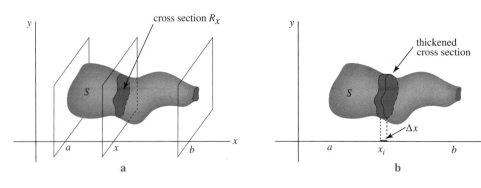

FIGURE 7.6.62

Approximating the solid S with cylinders

Definition 7.6.1 Assume that S is a solid three-dimensional region that lies between $x=a$ and $x=b$. Denote by $A(x)$ the area of the cross-section of S by the plane perpendicular to the x-axis that passes through x. Assume that $A(x)$ is continuous on $[a, b]$. Then the **volume** V of S is given by

$$V = \lim_{n \to \infty} V_n = \lim_{n \to \infty} \sum_{i=1}^{n} A(x_i) \Delta x = \int_a^b A(x) dx$$

provided that the limit exists (i.e., is equal to a real number). ▲

Note that Definition 7.6.1 is essentially—with necessary technical adjustments—the definition of the area (Definition 7.3.1) moved one dimension higher.

In conclusion, to compute the volume of a solid S that lies between $x=a$ and $x=b$ we integrate the areas of its cross-sections from a to b. We now illustrate this idea in several examples.

Example 7.6.8 Volume of a Cylinder

Assume that the cross-sectional area of the cylinder in Figure 7.6.58b is A. In Figure 7.6.63a we redrew the cylinder, placing it so that we can calculate its volume by integration. Since all cross-sections are identical, we find

$$V = \int_0^h A(x) dx = \int_0^h A dx = Ax \Big|_0^h = Ah$$

as claimed at the start of this section. ▲

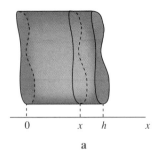
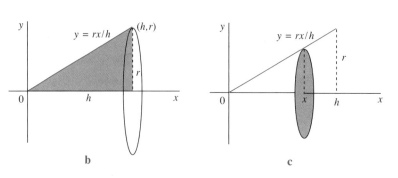

FIGURE 7.6.63

A cylinder and a cone

Example 7.6.9 Volume of a Cone

Find the volume of a cone of radius r and height h.

Look at Figure 7.6.63b. The line joining the point (h, r) to the origin has slope r/h, and so its equation is $y = rx/h$. The cone is now obtained by rotating the shaded triangular region about the x-axis. At a location x, the cross section is a disk of radius rx/h (Figure 7.6.63c). By Definition 7.6.1, the volume of the cone is

$$V = \int_0^h A(x)dx = \int_0^h \pi \left(\frac{rx}{h}\right)^2 dx = \pi \frac{r^2}{h^2} \int_0^h x^2 dx = \pi \frac{r^2}{h^2}\left(\frac{x^3}{3}\bigg|_0^h\right) = \pi \frac{r^2}{h^2} h^3 = \pi r^2 h \quad \blacktriangle$$

The cone is an example of a **solid of revolution.** In general, we obtain a solid of revolution by rotating a two-dimensional region about an axis in space.

As a special case, the region R between the graph of a continuous function $f(x)$ and the x-axis, and between the vertical lines $x = a$ and $x = b$, is rotated about the x-axis (Figure 7.6.64a). The resulting solid of revolution, S, is shown in Figure 7.6.64b.

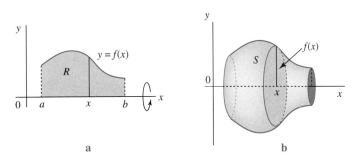

FIGURE 7.6.64

Solid of revolution

At a location x, the cross-section of S is a disk of radius $|f(x)|$ (in the case drawn in Figure 7.6.64a, $f(x) \geq 0$; in general, $f(x)$ could be negative). The area of the cross-section is $A(x) = \pi |f(x)|^2 = \pi [f(x)]^2$, and the volume of S is

$$V = \int_a^b A(x)dx = \pi \int_a^b [f(x)]^2 \, dx$$

Example 7.6.10 Volume of a Sphere

Find the volume of a sphere of radius r.

From $x^2 + y^2 = r^2$ we obtain the equation of the semicircle $y = \sqrt{r^2 - x^2}$ in the upper half-plane. Denote by R the region enclosed by this semicircle and the x-axis. The rotation of R about the x-axis generates the sphere (Figure 7.6.65; so a sphere can be viewed as a solid of revolution). Its volume is

$$V = \pi \int_{-r}^r [f(x)]^2 \, dx = \pi \int_{-r}^r (r^2 - x^2) \, dx$$

$$= \pi \left(r^2 x - \frac{x^3}{3}\right)\bigg|_{-r}^r$$

$$= \pi \left(r^3 - \frac{r^3}{3}\right) - \pi \left(-r^3 + \frac{r^3}{3}\right) = \frac{4}{3}\pi r^3 \quad \blacktriangle$$

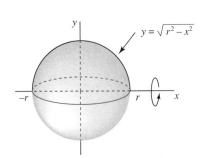

FIGURE 7.6.65

Sphere as a solid of revolution

Example 7.6.11 Volume of a Solid of Revolution Obtained by Rotation about the y-axis

Denote by R the bounded region in the first quadrant between the curves $y = e^x$ and $y = e$. Find the volume of the solid, S, obtained by rotating R about the y-axis.

We sketch the region R in Figure 7.6.66a. Because R is rotated about the y-axis, we consider the cross-sections perpendicular to the y-axis—and therefore to obtain the volume we need to integrate along the y-axis. For that reason, we view the curve $y = e^x$ as $x = \ln y$.

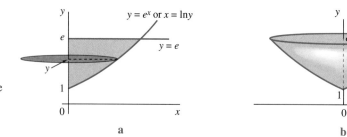

FIGURE 7.6.66

Integrating along the y-axis to calculate the volume

The cross-section at a location y is a disk of radius $\ln y$. Its area is $A(y) = \pi (\ln y)^2$, and the volume of S is

$$V = \pi \int_1^e A(y)\, dy = \pi \int_1^e (\ln y)^2\, dy$$

Using integration by parts (see Exercise 41 in Section 7.5), we obtain

$$\int (\ln y)^2\, dy = y(\ln y)^2 - 2y \ln y + 2y$$

The volume of S (shown in Figure 7.6.66b) is computed to be

$$V = \pi \left[y(\ln y)^2 - 2y \ln y + 2y \right]\Big|_1^e = \pi \left[(e - 2e + 2e) - 2 \right] = \pi(e - 2) \quad \blacktriangle$$

Example 7.6.12 Volume of a Solid of Revolution

Find the volume of a solid S generated by rotating the region R from Example 7.6.11 about the x-axis.

This time, the cross-section at a location x is not a disk, but a washer (a disk from which a smaller concentric disk has been removed); see Figure 7.6.67a and b. The radius of the larger (outer) disk is constant and equal to e, and the radius of the smaller (inner) disk is e^x. The area of the washer is obtained by subtracting the area of the inner disk from the area of the outer disk:

$$A(x) = \pi(e)^2 - \pi(e^x)^2 = \pi \left(e^2 - e^{2x} \right)$$

The volume of S is

$$V = \pi \int_0^1 A(x)\, dx = \pi \int_0^1 \left(e^2 - e^{2x} \right) dx = \pi \left[e^2 x - \frac{1}{2} e^{2x} \right]\Big|_0^1 = \pi \left[\frac{1}{2} e^2 + \frac{1}{2} \right] = \frac{\pi}{2} \left(e^2 + 1 \right)$$

The solid S is drawn in Figure 7.6.67c. $\quad \blacktriangle$

Integrals and Lengths

Area is a measurement of the size of two-dimensional regions, whereas volume determines the size of three-dimensional solids. **Length** measures the size of one-dimensional objects—line segments and curves.

The length of a straight line segment joining two points (x_1, y_1) and (x_2, y_2) in a plane is given by the **distance formula**

$$d = \sqrt{(x_2 - x_1)^2 + (y_2 - y_1)^2}$$

7.6 Applications   561

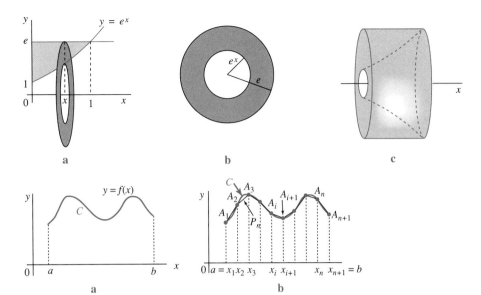

FIGURE 7.6.67
Finding the volume

FIGURE 7.6.68
Approximating a curve by a polygon

By adding up the lengths of all line segments (sides) that form a polygon, we find its length.

How do we find the length of a curve, such as the one in Figure 7.6.68a? More than two thousand years ago, mathematicians calculated (actually estimated) the perimeter of a circle by inscribing polygons with an increasing number of sides (for instance, Archimedes studied polygons with 12, 24, 48, and 96 sides to obtain a good estimate for π). In general, to find the length of any curve, the same idea will work—we construct approximations of the curve using polygons (this will feel like a déjà vu for a reason).

Consider a curve C defined as the graph of a continuous function $f(x)$ on an interval $[a, b]$, shown in Figure 7.6.68a. Divide $[a, b]$ into n subintervals (n is a positive integer) of length $\Delta x = (b - a)/n$ and label their endpoints as $a = x_1, x_2, \ldots, x_i, x_{i+1}, \ldots, x_n, x_{n+1} = b$. The corresponding points on the graph of $f(x)$,

$$A_1 = (x_1, f(x_1)), A_2 = (x_2, f(x_2)), \ldots, A_i = (x_i, f(x_i)), \ldots, A_{n+1} = (x_{n+1}, f(x_{n+1}))$$

are taken as the vertices of a polygon (call it P_n). This polygon has n sides and is an approximation of the curve C (Figure 7.6.68b).

We will define the length of C in essentially the same way as we defined the area and the volume: it will be the limit of the lengths of approximating polygons.

First we find the length of the polygon P_n. The length of the side from $A_i = (x_i, f(x_i))$ to $A_{i+1} = (x_{i+1}, f(x_{i+1}))$ is obtained from the distance formula:

$$d_i = \sqrt{(x_{i+1} - x_i)^2 + (f(x_{i+1}) - f(x_i))^2}$$

We simplify the expression under the square root. The Mean Value Theorem (Theorem 6.3.3), applied to $f(x)$ on the interval $[x_i, x_{i+1}]$, implies that there is a number x_i^* in (x_i, x_{i+1}) such that

$$f(x_{i+1}) - f(x_i) = f'(x_i^*)(x_{i+1} - x_i)$$

For this to work (recall the statement of the Mean Value Theorem), we must assume that $f(x)$ is differentiable on (a, b). Thus,

$$d_i = \sqrt{(x_{i+1} - x_i)^2 + (f'(x_i^*)(x_{i+1} - x_i))^2}$$
$$= \sqrt{(1 + [f'(x_i^*)]^2)(x_{i+1} - x_i)^2}$$
$$= \sqrt{(1 + [f'(x_i^*)]^2)} \, |x_{i+1} - x_i|$$
$$= \sqrt{(1 + [f'(x_i^*)]^2)} \, (x_{i+1} - x_i) = \sqrt{(1 + [f'(x_i^*)]^2)} \Delta x$$

(in this calculation we applied the formula $\sqrt{A^2} = |A| = A$ to $A = x_{i+1} - x_i > 0$). As usual, $\Delta x = x_{i+1} - x_i$. The length of the polygon P_n is the sum of the lengths of all its sides:

$$\text{length}(P_n) = \sum_{i=1}^{n} \sqrt{(1 + [f'(x_i^*)]^2)} \Delta x$$

The length of C is the limit of lengths of polygons (constructed in the same way for every $n \geq 1$):

$$\text{length}(C) = \lim_{n \to \infty} \text{length}(P_n) = \lim_{n \to \infty} \sum_{i=1}^{n} \sqrt{(1 + [f'(x_i^*)]^2)} \Delta x$$

The expression on the right side is, by definition, equal to the definite integral

$$\int_a^b \sqrt{1 + [f'(x)]^2} \, dx$$

as long as the integrand is continuous. So we need to assume that the derivative $f'(x)$ is continuous.

Definition 7.6.2 Assume that $f'(x)$ is a continuous function defined on an interval $[a, b]$. The **length** of the curve (i.e., of the graph of the function) $y = f(x)$ from $x = a$ to $x = b$ is given by

$$L = \int_a^b \sqrt{1 + [f'(x)]^2} \, dx$$

Note that the integrand $\sqrt{1 + [f'(x)]^2}$ could easily get messy (and it does). In some cases we can simplify and integrate using the formulas and the methods we discussed earlier in this chapter. In general, however, we need to use alternative techniques (such as numeric integration or software).

To illustrate how Definition 7.6.2 works, we check the formula for the circumference of a circle.

Example 7.6.13 Circumference of a Circle

Find the length of the graph of the function $f(x) = \sqrt{r^2 - x^2}$ for $-r \leq x \leq r$.

The graph of $f(x)$ is the semicircle of radius r in the upper half-plane. We start by simplifying the integrand. From

$$f'(x) = \frac{1}{2}\left(r^2 - x^2\right)^{-1/2}(-2x) = -\frac{x}{\sqrt{r^2 - x^2}}$$

we find

$$1 + [f'(x)]^2 = 1 + \frac{x^2}{r^2 - x^2} = \frac{r^2 - x^2 + x^2}{r^2 - x^2} = \frac{r^2}{r^2 - x^2}$$

$$\sqrt{1 + [f'(x)]^2} = \sqrt{\frac{r^2}{r^2 - x^2}} = \frac{r}{\sqrt{r^2 - x^2}}$$

The length of the semicircle is

$$L = \int_{-r}^{r} \sqrt{1 + [f'(x)]^2} \, dx = \int_{-r}^{r} \frac{r}{\sqrt{r^2 - x^2}} \, dx = r \int_{-r}^{r} \frac{1}{\sqrt{r^2 - x^2}} \, dx$$

To compute the antiderivative we write

$$\int \frac{1}{\sqrt{r^2-x^2}}\,dx = \int \frac{1}{\sqrt{r^2(1-x^2/r^2)}}\,dx = \frac{1}{r}\int \frac{1}{\sqrt{1-(x/r)^2}}\,dx$$

$$= \frac{1}{r}\Big(r\arcsin(x/r)\Big)$$

$$= \arcsin(x/r)$$

(see Exercise 75 for details). As expected

$$L = r\left(\arcsin(x/r)\Big|_{-r}^{r}\right) = r[\arcsin 1 - \arcsin(-1)] = r\pi$$

(recall that $\arcsin 1 = \pi/2$ and $\arcsin(-1) = -\pi/2$).

Example 7.6.14 Length of a Curve

Find the length of the curve $f(x) = \dfrac{1}{2a}\left(e^{ax}+e^{-ax}\right)$ from $x=-1$ to $x=1$. Assume that $a > 0$.

As in the previous example, we simplify the integrand first. Since

$$f'(x) = \frac{1}{2a}\left(ae^{ax}-ae^{-ax}\right) = \frac{1}{2}\left(e^{ax}-e^{-ax}\right)$$

it follows that

$$1 + [f'(x)]^2 = 1 + \frac{1}{4}\left(e^{ax}-e^{-ax}\right)^2 = 1 + \frac{1}{4}\left(e^{2ax}-2+e^{-2ax}\right)$$

$$= \frac{1}{4}\left(e^{2ax}+2+e^{-2ax}\right)$$

$$= \frac{1}{4}\left(e^{ax}+e^{-ax}\right)^2$$

$$\sqrt{1+[f'(x)]^2} = \frac{1}{2}\left(e^{ax}+e^{-ax}\right)$$

The length of the curve is

$$L = \int_{-1}^{1} \frac{1}{2}\left(e^{ax}+e^{-ax}\right)\,dx = \frac{1}{2}\left(\frac{1}{a}e^{ax}-\frac{1}{a}e^{-ax}\right)\Big|_{-1}^{1}$$

$$= \frac{1}{2}\left(\frac{1}{a}e^{a}-\frac{1}{a}e^{-a}\right) - \frac{1}{2}\left(\frac{1}{a}e^{-a}-\frac{1}{a}e^{a}\right)$$

$$= \frac{1}{a}e^{a}-\frac{1}{a}e^{-a} = \frac{1}{a}\left(e^{a}-e^{-a}\right)$$

This curve is called **catenary;** see Figure 7.6.69. Catenaries are used to model the shape of cables or flexible chains hanging between two poles of equal height and acted upon by gravity only.

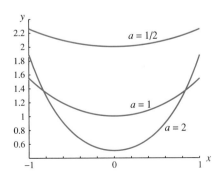

FIGURE 7.6.69

Three catenary curves

FIGURE 7.6.70

Analyzing the average thickness of ice

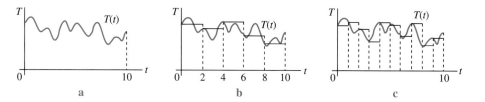

Integrals and Averages

The **average value**, $\bar{m}$, of a finite number of measurements $m_1, m_2, m_3, \ldots, m_n$ is calculated using the formula

$$\bar{m} = \frac{m_1 + m_2 + m_3 + \cdots + m_n}{n}$$

Often, when we need to represent a set of data by a single number, we use the average value (also called the **mean value**). For instance, we talk about average daily temperature, average blood pressure within a certain population (say, people over 55), average number of offspring per bacterium, and so forth.

Now suppose that we have measured the thickness, $T(t)$, of ice at some location in the Arctic over a period of 10 months using an instrument that returns a plot rather than a set of discrete measurements (Figure 7.6.70a). How do we figure out the average thickness? We do not have a finite set of data, but rather an infinite set represented by the points on the graph.

We use the usual trick: divide the interval [0, 10] into a number of subintervals (for example, five), and assume that, on each subinterval, the thickness has a constant value equal to the value of $T(t)$ at the start (left end) of each subinterval; see Figure 7.6.70b. In this way, we obtain five (finitely many) numbers, $T(0), T(2), T(4), T(6)$, and $T(8)$. Their average is

$$\frac{T(0) + T(2) + T(4) + T(6) + T(8)}{5}$$

To improve accuracy, we repeat this calculation with a larger number of subintervals. So, take n subintervals (in which case each will have length $\Delta t = 10/n$); see Figure 7.6.70c. We now have n values of the thickness:

$$T(t_1), T(t_2), T(t_3), \ldots, T(t_n)$$

and their average is

$$\frac{T(t_1) + T(t_2) + T(t_3) + \cdots + T(t_n)}{n}$$

Since $\Delta t = \frac{10}{n}$, it follows that $n = \frac{10}{\Delta t}$, and thus

$$\frac{T(t_1) + T(t_2) + \cdots + T(t_n)}{n} = \frac{T(t_1) + T(t_2) + \cdots + T(t_n)}{10/\Delta t}$$

$$= \frac{1}{10}(T(t_1) + T(t_2) + \cdots + T(t_n))\Delta t$$

$$= \frac{1}{10}(T(t_1)\Delta t + T(t_2)\Delta t + \cdots + T(t_n)\Delta t)$$

Note that the term in brackets is a Riemann sum corresponding to the integral of the function $T(t)$ on [0, 10]. Therefore, we can obtain the *exact* value of the average thickness using the limit as $n \to \infty$ (we use $\bar{T}$ to denote the average value of $T(t)$):

$$\bar{T} = \frac{1}{10}\int_0^{10} T(t)\,dt$$

7.6 Applications

In general, for a continuous function f on $[a, b]$, we define the **average value $\bar{f}$ of f on $[a, b]$** by

$$\bar{f} = \frac{1}{b-a} \int_a^b f(x) dx \qquad (7.6.1)$$

Example 7.6.15 Finding an Average Rate

Suppose that water is flowing into a vessel at a rate of $(1 - e^{-t})$ cubic centimetres per second for the 2 s between $t = 0$ and $t = 2$. What is the *average* rate at which water enters during this time? The average is the total amount of water that enters divided by the time, or

$$\text{average rate} = \frac{\text{total water entering}}{\text{total time}}$$

The total amount of water that enters is

$$\text{total water entering} = \int_0^2 (1 - e^{-t}) dt$$

$$= (t + e^{-t})\big|_0^2$$

$$= 2 + e^{-2} - (0 + e^{-0}) \approx 1.135$$

The average rate is

$$\text{average rate} = \frac{\text{total water entering}}{\text{total time}} \approx \frac{1.135}{2} \approx 0.568$$

(Figure 7.6.71). If water enters at a constant rate of 0.568 cm³/s for 2 s, then 1.135 cm³ would enter, equal to the amount of water that entered at the variable rate $1 - e^{-t}$ during those same 2 s. Geometrically, the area under the horizontal line at 0.568 between $t = 0$ and $t = 2$ is equal to the area under the curve $1 - e^{-t}$.

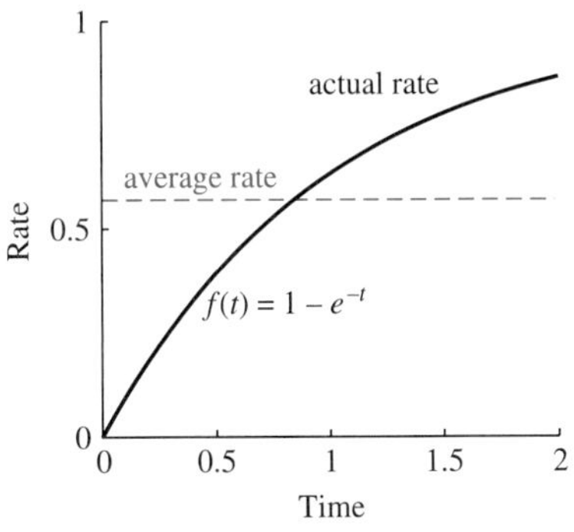

FIGURE 7.6.71
The average value of a rate

From the definition in Equation 7.6.1 of average value,

$$\bar{f} = \frac{1}{b-a} \int_a^b f(x) dx$$

we obtain

$$\bar{f} \cdot (b - a) = \int_a^b f(x) dx$$

The left side of this equation is the area of the rectangle on $[a, b]$ of height $\bar{f}$. So the area under the horizontal line representing the average is equal to the area under the curve (Figure 7.6.72). In other words, if we could cut up the region under f on $[a, b]$ into pieces and reassemble them in the shape of a rectangle on $[a, b]$ (in such a way that there are no gaps and no overlaps), then the height of that rectangle would be $\bar{f}$.

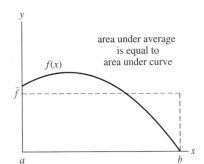

FIGURE 7.6.72
The average value in general

Integrals and Mass

Integration can be used to find the mass of an object with known **density**. Here we consider only the one-dimensional case, such as a thin rod. Suppose the density of the bar, measured in grams per centimetre, is $\rho(x)$ at x (Figure 7.6.73). To estimate

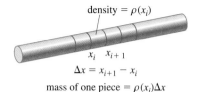

Figure 7.6.73
The mass of a bar

the mass of the bar, we break it into n small pieces of length Δx. The mass of the piece between x_i and $x_i + \Delta x$ is approximately $\rho(x_i)\Delta x$, the density at the left end of the piece times the length. Adding up all the pieces, we get

$$\text{mass of the bar} \approx \sum_{i=0}^{n-1} \rho(x_i) \Delta x$$

This has the exact form of a Riemann sum. The limit as Δx approaches zero and n approaches infinity is equal to the definite integral, so

$$\text{mass of the bar} = \int_a^b \rho(x) dx \qquad (7.6.2)$$

Example 7.6.16 Finding the Mass of a Bar

Consider a 100-cm vertical bar composed of a substance that has settled and become denser near the ground. Let z denote the height above ground, and suppose that the density is given by

$$\rho(z) = e^{-0.01z}$$

in grams per centimetre. The mass is

$$\int_0^{100} e^{-0.01z} dz = \frac{1}{-0.01} \left(e^{-0.01z}\right)\Big|_0^{100}$$
$$= -100 \left(e^{-0.01z}\right)\Big|_0^{100}$$
$$= -100 \left(e^{-0.01 \cdot 100}\right) - \left[-100 \left(e^{-0.01 \cdot 0}\right)\right]$$
$$= 100 \left(1 - e^{-1}\right) \approx 63.21$$

Does this result make sense? The density at the bottom of the bar is $\rho(0) = 1$ g/cm. If the entire bar had this maximum density, the mass would be 100 g. The density at the top of the bar is $\rho(100) = e^{-1} \approx 0.368$ g/cm, so the mass of the bar would be 36.8 g if the entire bar had this minimum density. Our result lies between these extremes. Furthermore,

$$\text{average density} = \frac{\text{total mass}}{\text{total length}} = \frac{\int_0^{100} \rho(x) dx}{100} \approx \frac{63.2}{100} = 0.632$$

The result lies between the minimum density of 0.368 and the maximum of 1 (Figure 7.6.74).

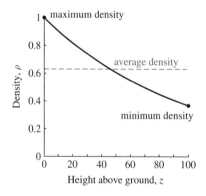

Figure 7.6.74
The average density of a bar

Example 7.6.17 Counting Otters with Integration

The same technique can be used to find totals when density is measured in other units. Suppose the density of otters along the coast of British Columbia is

$$f(x) = 0.0003x(1000 - x) = 3 \cdot 10^{-4} x(1000 - x)$$

in otters per kilometre, where x is measured in kilometres from the U.S. border and can take on values between 0 and 1000. The population density takes on a maximum value of 75 otters/km halfway up the coast at $x = 500$ and a minimum value of 0 at $x = 0$ and $x = 1000$ (Figure 7.6.75).

The total number T is the definite integral of the density, or

FIGURE 7.6.75
The average density of otters

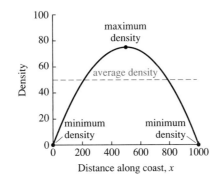

$$T = \int_0^{1000} 0.0003x(1000 - x)\,dx$$

$$= \int_0^{1000} (0.3x - 0.0003x^2)\,dx$$

$$= (0.15x^2 - 0.0001x^3)\Big|_0^{1000}$$

$$= 1.5 \cdot 10^5 - 10^5$$

$$= 0.5 \cdot 10^5 = 50{,}000 \text{ otters}$$

The average density is

$$\text{average density} = \frac{\text{total number}}{\text{total distance}}$$

$$= \frac{50{,}000 \text{ otters}}{1000 \text{ km}} = 50 \frac{\text{otters}}{\text{km}}$$

This value lies between the maximum density of 75 otters/km and the minimum density of 0 otters/km. The average value is *not* the average of these minimum and maximum values.

Summary The definite integral can be used to find the **area between two curves**, the **volume of a solid** (in some cases), and the **length of a curve.** We used the definite integral to calculate **average** values of functions by dividing the integral (the total amount) by the length of the interval. Similarly, we computed masses or total numbers from **densities.** In each case, the underlying idea is that of the Riemann sum: chopping things up into small pieces and adding the results.

7.6 Exercises

Integrals and Areas

1–6 ▪ Find the area under each curve.

1. Area under $f(x) = 3x^3$ from $x = 0$ to $x = 3$.
2. Area under $g(x) = e^x$ from $x = 0$ to $x = \ln 2$.
3. Area under $h(x) = e^{x/2}$ from $x = 0$ to $x = \ln 2$.
4. Area under $f(t) = (1 + 3t)^3$ from $t = 0$ to $t = 2$.
5. Area under $G(y) = (3 + 4y)^{-2}$ from $y = 0$ to $y = 2$.
6. Area under $s(z) = \sin(z + \pi)$ from $z = 0$ to $z = \pi$.

7–12 ▪ The definite integral can be used to find the area between two curves. In each case:

a. Sketch the graphs of the two functions over the given range, and shade the region between the curves.

b. Sketch the graph of the difference between the two curves. The area *under* this curve matches the area *between* the original curves.

c. Find the area of the region between the two curves.

7. Find the area between $f(x) = 2x$ and $g(x) = x^2$ for $0 \le x \le 2$.
8. Find the area between $f(x) = e^x$ and $g(x) = x + 1$ for $-1 \le x \le 1$.
9. Find the area between $f(x) = 2x$ and $g(x) = x^2$ for $0 \le x \le 3$.
10. Find the area between $f(x) = x^2$ and $g(x) = x^3$ for $0 \le x \le 2$.
11. Find the area between $f(x) = e^x$ and $g(x) = \dfrac{e^{2x}}{2}$ for $0 \le x \le 1$.
12. Find the area between $f(x) = \sin 2x$ and $g(x) = \cos 2x$ for $0 \le x \le \pi$.
13. Find the area between $y = \sin x$ and $y = \cos x$ from $x = 0$ to $x = \pi$.
14. Find the area between $y = \sin 2x$ and $y = \cos x$ from $x = \pi/2$ to $x = \pi$.
15. Find the area between $y = x^{-1}$ and $y = x^{-2}$ from $x = 1$ to $x = 2$.
16. Find the area between the functions $y = |x|$ and $y = 2 - x^2$.
17. Find the area between the functions $y = \sqrt{x}$ and $y = 4x$ from $x = 0$ to $x = 1$.

18–21 ▪ Estimate the area of each region as suggested. In Exercises 18 and 21 your answer will depend on which sum (left, right, midpoint) you decide to use. Since the left-sum approximation misses the leftmost end of the region, use an appropriate rectangle to include that part. Likewise, add a rectangle to the right-sum approximation to account for the rightmost end of the region.

18. Use approximating rectangles. The widths (in metres) were measured at 3-m intervals.

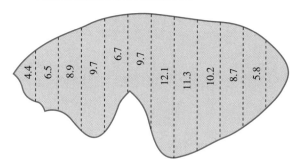

19. Count the squares. The side of the square forming the grid measures 2 km. (Your answer will depend on how you treat the squares that contain the boundary.)

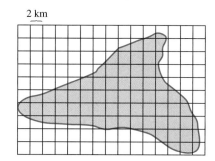

20. Count the squares. The side of the square forming the grid measures 5 m. (Your answer will depend on how you treat the squares that contain the boundary.)

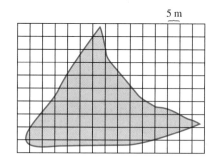

21. Use approximating rectangles. The widths (in kilometres) from left to right are 6.6, 8.2, 8.2, 7.1, 4.5, 2.6, 2.5, 4.0, 6.1, 8.5, 10.1, 11.5, and 8.6. They were measured at 3-km intervals.

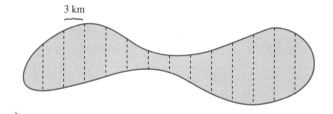

22–23 ■ Use integration by parts to evaluate each of the following as a definite integral.

22. Find the area under the curve $g(x) = x \ln x$ for $1 \leq x \leq 2$. Sketch a graph to see whether your answer makes sense.

23. Find the integral of $g(x) = x \sin(2\pi x)$ for $0 \leq x \leq 2$. Sketch a graph to see whether your answer makes sense.

24–25 ■ We have used small vertically oriented rectangles to compute areas. There is no reason why small horizontal rectangles cannot be used. Here are the steps to find the area under the curve $y = f(x)$ from $x = 0$ to $x = 1$ by using horizontal rectangles.

a. Draw a picture with five horizontal rectangles, each of height 0.2, approximately filling the region to the right of the curve and to the left of the line $x = 1$.

b. Calculate an upper and a lower estimate of the length of each rectangle.

c. Use the areas of these rectangles to find upper and lower estimates of the area.

d. Think now of a very thin rectangle of height y. How long is the rectangle?

e. Write a definite integral expression for the area.

f. Evaluate the integral and check that the answer is correct.

24. With $f(x) = x^2$.

25. With $f(x) = \sqrt{x}$.

26–27 ■ Archimedes developed the basic idea of integration to find the areas of geometric figures. Often, this involves building regions out of small pieces with shapes more complicated than rectangles.

26. Use the fact that the perimeter of a circle of radius r is $2\pi r$ to find the area of a circle with radius 1. Think of the circular region as being built out of little rings with some small width Δr.

27. Use the fact that the area of a circle of radius r is πr^2 to find the volume of a cone of height one that has radius r at a height r. Think of the cone as being built of a stack of little circular disks with some small thickness Δr.

28–31 ■ Some books define the natural log function with a definite integral:

$$l(a) = \int_1^a \frac{1}{x} dx$$

Using this definition, we can prove the laws of logarithms.

28. Show that $l(6) - l(3) = l(2)$. (Use the summation property of the definite integral to write the difference as an integral, and then use the substitution $y = \frac{x}{3}$.)

29. Find the integral from a to $2a$ by following the same steps as in Exercise 28 (make the substitution $y = \frac{x}{a}$).

30. Show that $l(10^2) = 2 \cdot l(10)$. (Try the substitution $y = \sqrt{x}$ in $\int_1^{10^2} \frac{1}{x} dx$.)

31. Show that $l(a^b) = b \cdot l(a)$. (Try the substitution $y = \sqrt[b]{x}$ in $\int_1^{a^b} \frac{1}{x} dx$.)

Integrals and Volumes

32–51 ■ Find the volume of the solid obtained by rotating the region R enclosed (bounded) by the given curves about the given axis.

32. $y = 4 - 2x$, $y = 0$, and $x = 0$; about the x-axis.

33. $y = 4 - x^2$ and $y = 0$; about the x-axis.

34. $y = 4 - 2x$, $y = 0$, and $x = 0$; about the y-axis.

35. $x = 1 - y^2$ and $x = 0$; about the y-axis.

36. $y = \sqrt{\sin x}$, $y = 0$, $x = 0$, and $x = \pi/2$; about the x-axis.

37. $y = \sec x$, $y = 0$, $x = 0$, and $x = \pi/4$; about the x-axis.

38. $y = e^{-x}$, $y = 0$, $x = -2$, and $x = 4$; about the x-axis.

39. $y = \cos x$, $y = 0$, $x = 0$, and $x = \pi/2$; about the x-axis. (Use $\cos^2 x = (1 + \cos 2x)/2$ to simplify before integration.)

40. $y = x$ and $y = x^3$ in the first quadrant; about the x-axis.

41. $y = x^{-1}$, $y = \sqrt{x}$, and $x = 4$; about the x-axis.

42. $y = x$ and $y = x^3$ in the first quadrant; about the y-axis.

43. $xy = 1$, $y = x^2$, and $y = 2$; about the y-axis.

44. $y = 8 - x$, $y = 1$, $x = 2$, and $x = 5$; about the x-axis.

45. $y = x^2$ and $y = 18 - x^2$; about the x-axis.

46. $y = 8 - x$, $y = 1$, $x = 2$, and $x = 5$; about the y-axis.

47. $x = y^2 + 2$ and $x = 6$; about the y-axis.

48. $x + y = 1$, $x = 0$, and $y = 0$; about the line $x = 4$.

49. $y = x^3$, $x = 1$, and $y = 0$; about the line $x = -2$.

50. $x + y = 1$, $x = 0$, and $y = 0$; about the line $y = -2$.

51. $y = x^3$, $x = 1$, and $y = 0$; about the line $y = 4$.

52–55 ■ Describe the solid of revolution whose volume is given by each integral.

52. $\pi \int_0^4 x^2 \, dx$

53. $\int_0^4 x^2 \, dx$

54. $\pi \int_{1/2}^1 \left(x^{-1} - x^3 \right) dx$

55. $\pi \int_{-2}^2 \left(\left(3 + \sqrt{4 - x^2}\right)^2 - 9 \right) dx$

56. An ellipsoid of revolution is obtained by rotating about the x-axis the part of the ellipse $x^2/a^2 + y^2/b^2 = 1$ that lies above the x-axis. Find its volume.

57. Let S be the solid obtained by rotating about the y-axis the part of the ellipse $x^2/a^2 + y^2/b^2 = 1$ that lies to the right of the y-axis. Find the volume of S.

58. Find the volume of a pyramid of height h whose base is a square with side a.

59. The upper (pointed) part of a cone of radius r and height h is cut off parallel to the base of the cone. If the height of the remaining part (called the frustum of a cone) is $h/2$, find its volume.

Integrals and Lengths

60. Find the length of a straight line segment joining the points $(-2, 1)$ and $(2, 9)$ in two ways:
    a. using the distance formula.
    b. using Definition 7.6.2 applied to $f(x) = 2x + 5$, $-2 \leq x \leq 2$.

61. Find the length of a straight line segment joining the origin and the point (a, b), $a \neq 0$, in two ways:
    a. using the distance formula.
    b. applying Definition 7.6.2 to a suitably chosen function.

62–69 ▪ Find the length of each curve.

62. $y = 2x^{3/2}$, from $x = 1$ to $x = 2$.

63. $y = \dfrac{x^3}{6} + \dfrac{1}{2x}$, from $x = 1$ to $x = 2$.

64. $y = \dfrac{1}{3}(x^2 + 2)^{3/2}$, from $x = 0$ to $x = 1$.

65. $y = \dfrac{2}{3}(x - 1)^{3/2}$, from $x = 3$ to $x = 4$.

66. $y = \ln(\sec x)$, from $x = 0$ to $x = \pi/4$.

67. $y = \ln(\cos x)$, from $x = -\pi/4$ to $x = \pi/4$.

68. $y = \dfrac{1}{4}(e^{2x} + e^{-2x})$, from $x = 0$ to $x = 1$.

69. $y = \dfrac{x^2}{8} - \ln x$, from $x = 1$ to $x = 4$.

70–73 ▪ Set up the integral for the length of each curve. Do not evaluate it.

70. $y = x^2$, from $x = -3$ to $x = 3$.

71. $y = x^{-1}$, from $x = 1/2$ to $x = 2$.

72. $y = e^{-x}$, from $x = -1$ to $x = 1$.

73. $y = 2 \ln x$, from $x = 1$ to $x = e$.

74. Inscribe a regular polygon P_n with n sides into a circle of radius r. Show that the length of a side of P is $2r \sin(\pi/n)$ and thus the length of P is $2rn \sin(\pi/n)$. Using L'Hôpital's rule, show that $\lim\limits_{n \to \infty} 2rn \sin(\pi/n) = 2\pi r$.

75. Using the substitution $u = x/r$, show that $\int \dfrac{1}{\sqrt{1-(x/r)^2}} dx = r \arcsin(x/r)$. To practise the technique used in Example 7.6.13, find $\int \dfrac{1}{a^2+x^2} dx$ where $a > 0$.

Integrals and Averages

76–79 ▪ Find the average value of each function over the given range. Sketch the graph of the function along with a horizontal line at the average to make sure that your answer makes sense.

76. $f(x) = x^2$ for $0 \leq x \leq 3$.

77. $f(x) = \dfrac{1}{x}$ for $0.5 \leq x \leq 2$.

78. $f(x) = x - x^3$ for $-1 \leq x \leq 1$.

79. $f(x) = \sin 2x$ for $0 \leq x \leq \pi/2$.

80–81 ▪ The average of a step function computed with the definite integral matches the average computed in the usual way. Test this in the following situations by finding the average of the values, first directly and then as the integral of a step function.

80. Suppose a math class has four equally weighted tests. A student gets 60 on the first test, 70 on the second, 80 on the third, and 90 on the last.

81. A math class has 20 students. In a quiz worth 10 points, four students get 6 points, seven students get 7, five students get 8, three students get 9, and one student gets 10.

82–85 ▪ Suppose water is entering a tank at a rate of $g(t) = 360t - 39t^2 + t^3$, where g is measured in litres per hour and t is measured in hours. The rate is 0 at times 0, 15, and 24.

82. Find the total amount of water entering during the first 15 h, from $t = 0$ to $t = 15$. Find the average rate at which water entered during this time.

83. Find the total amount and average rate from $t = 15$ to $t = 24$.

84. Find the total amount and average rate from $t = 0$ to $t = 24$.

85. Suppose that energy is produced at a rate of

$$E(t) = |g(t)|$$

in joules per hour. Find the total energy generated from $t = 0$ to $t = 24$. Find the average rate of energy production.

Integrals and Mass, Various Applications

86–87 ▪ Several very thin 2-m-long snakes are collected in the Amazon. Each has density of $\rho(x)$ given by the following formulas, where ρ is measured in grams per centimetre and x is measured in centimetres from the tip of the tail.

a. Find the minimum and maximum density of each snake. Where does the maximum occur?

b. Find the total mass of each snake.

c. Find the average density of each snake. How does this compare with the minimum and maximum?

d. Graph the density and average for each snake.

86. $\rho(x) = 1 + 2 \cdot 10^{-8} x^2 (300 - x)$

87. $\rho(x) = 1 + 2 \cdot 10^{-8} x^2 (240 - x)$

88–91 ■ Suppose water is entering a series of vessels at the given rate. In each case, find the total amount of water entering during the first second and the average rate during that time. Compare the average rate with the rate at the "average time," at $t = 0.5$ halfway through the time period from 0 to 1. In which case is the average rate greater than the rate at the average time? Graph the flow rate function, and mark the flow rate at the average time. Can you guess what it is about the shape of the graph that determines how the average rate compares with the rate at the average time?

88. Water is entering at a rate of t^3 cubic centimetres per second.

89. Water is entering at a rate of $\sqrt{t}$ cubic centimetres per second.

90. Water is entering at a rate of t cubic centimetres per second.

91. Water is entering at a rate of $4t(1 - t)$ cubic centimetres per second.

Computer Exercise

92. Find the area between the two curves $f(x) = \cos x$ and $g(x) = 0.1x$ for $0 \leq x \leq 10$ (see Exercises 7–17).

    a. Graph the two functions. There should be four separate regions between them.

    b. Have a computer find where each region begins and ends.

    c. Integrate to find the area of each region.

    d. Add the areas.

7.7 Improper Integrals

So far, we have considered definite integrals of functions that do not approach infinity between finite limits of integration. Integrals of this sort are called **proper integrals**. **Improper integrals** are of two types: integrals with infinite limits of integration and integrals of functions that approach infinity somewhere within the limits of integration. We now learn how to compute and apply improper integrals.

Infinite Limits of Integration

"Infinite" measurements cannot crop up in biological experiments. Nonetheless, infinity is a useful mathematical abstraction of "very long" or "long term" or "very far." Consider again the equation for chemical production with exponentially declining rate,

$$\frac{dP}{dt} = e^{-t}$$

in moles per second. The amount of product produced between $t = 0$ and $t = T$ is given by the definite integral

$$\text{production between 0 and } T = \int_0^T e^{-t} dt$$

The longer we wait, the more product has been produced. Let P_∞ denote the amount that would be produced if the experiment ran forever. We would like to write

$$P_\infty = \int_0^\infty e^{-t} dt$$

but we need to define what it means to integrate a function over the infinite interval from zero to infinity.

We *defined* the definite integral to be the limit of Riemann sums. Computing Riemann sums requires that we divide the interval of integration into n subintervals of equal and *finite* size (the subintervals are not allowed to be of infinite length). However, the infinite interval $(0, \infty)$ cannot be divided up that way and so the Riemann sums approach does not work.

Instead, we can think of this **improper integral** as a *limit*, as follows:

$$\int_0^\infty e^{-t} dt = \lim_{T \to \infty} \int_0^T e^{-t} dt$$

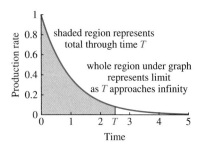

FIGURE 7.7.76
Computing an integral with infinite limits of integration

This formula captures the spirit of what we want. The proper integral gives the amount of production up to time T, and the limit enables us to make T "very large" (Figure 7.7.76). In this case,

$$\int_0^T e^{-t}\,dt = -e^{-t}\Big|_0^T = -e^{-T} + 1$$

Therefore,

$$\int_0^\infty e^{-t}\,dt = \lim_{T \to \infty}\left(1 - e^{-T}\right)$$
$$= \lim_{T \to \infty} 1 - \lim_{T \to \infty} e^{-T}$$
$$= 1 - 0 = 1$$

Exactly 1 mol of the product would be created after an infinite amount of time. If we wait for a "long time," such as 10 s, the exact amount is

$$1 - e^{-10} \approx 0.99995$$

which is quite close to the limit.

Formally, we have the following definition.

Definition 7.7.1 Assume that the definite integral $\int_a^T f(x)\,dx$ exists (i.e., is equal to a real number) for every $T \geq a$. Then we define the **improper integral of $f(x)$ on (a, ∞)** by

$$\int_a^\infty f(x)\,dx = \lim_{T \to \infty} \int_a^T f(x)\,dx$$

provided that the limit on the right side exists.

Recall that the definite integral

$$\int_a^T f(x)\,dx$$

is a limit (of Riemann sums of $f(x)$ constructed over the interval $[a, T]$)—this explains the opening line in the definition. With this in mind, we see that

$$\lim_{T \to \infty} \int_a^T f(x)\,dx$$

is a limit of limits!

If the definite integral $\int_T^a f(x)\,dx$ exists for every $T \leq a$, we define the **improper integral of $f(x)$ on $(-\infty, a)$** by

$$\int_{-\infty}^a f(x)\,dx = \lim_{T \to -\infty} \int_T^a f(x)\,dx$$

assuming that the limit on the right side exists.

We say that an improper integral is **convergent** if its corresponding limit exists (i.e., is equal to a real number). Otherwise, the integral is called **divergent.**

Consider the improper integral

$$\int_{-\infty}^\infty f(x)\,dx$$

If both improper integrals, $\int_a^\infty f(x)\, dx$, and $\int_{-\infty}^a f(x)\, dx$, converge for some real number a, then we say that the improper integral $\int_{-\infty}^\infty f(x)\, dx$ converges, and

$$\int_{-\infty}^\infty f(x)\, dx = \int_{-\infty}^a f(x)\, dx + \int_a^\infty f(x)\, dx$$

Otherwise, we say that $\int_{-\infty}^\infty f(x)\, dx$ diverges.

For positive functions, a *convergent* improper integral can still be interpreted as an area; thus, the calculation before the definition shows that the area under $f(t) = e^{-t}$ on $[0, \infty)$ is equal to 1. If we accept this, then we have to accept the fact that an unbounded region (such as the one in Figure 7.7.76) might have *finite or infinite area*.

Improper Integrals: Examples

As implied by the definition, the limit used to calculate an improper integral need not be finite. Nothing prevents the definite integral from 0 to T from getting larger and larger as T approaches infinity. Consider a different chemical produced at the diminishing rate

$$\frac{dQ}{dt} = \frac{1}{1+t}$$

again in moles per second. This rate decreases to zero more slowly than the exponential function (Section 6.4). How much product would this reaction produce after a long time?

The total product produced, Q_∞, is computed with the improper integral

$$Q_\infty = \int_0^\infty \frac{1}{1+t}\, dt$$

$$= \lim_{T \to \infty} \int_0^T \frac{1}{1+t}\, dt$$

Recall that

$$\int \frac{1}{1+t}\, dt = \ln|1+t| + C$$

(we guessed the answer; alternatively, we could have used the substitution $u = 1 + t$).

Thus (because $1 + t \geq 0$ we drop the absolute value sign),

$$Q_\infty = \lim_{T \to \infty} \left[\ln(1+t) \Big|_0^T \right]$$

$$= \lim_{T \to \infty} [\ln(1+T) - \ln 1]$$

$$= \lim_{T \to \infty} [\ln(1+T)] = \infty$$

If this rule were followed forever, the amount of product would be infinite (Figure 7.7.77). In this case, the integral diverges. Such a result might seem absurd and irrelevant. But this very absurdity provides a valuable negative result: no real system could follow this law indefinitely (i.e., for all time). Even though the rate gets smaller and smaller, the total production increases without bound.

What laws can be maintained indefinitely without producing an infinite amount of product? Any decreasing positive function that does not decrease to zero has an infinite improper integral (Figure 7.7.78).

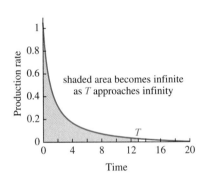

FIGURE 7.7.77

A divergent integral where the integrand decreases to zero slowly

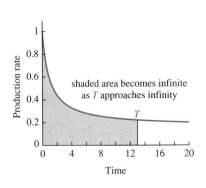

FIGURE 7.7.78

A divergent integral where the integrand does not decrease to zero

The region under the curve can be thought of as including a rectangle with positive height and an infinite length.

Example 7.7.1   Calculating Improper Integrals

Compare the convergence of the improper integrals $\int_1^\infty \frac{1}{x}\,dx$ and $\int_1^\infty \frac{1}{x^2}\,dx$.
By definition,

$$\int_1^\infty \frac{1}{x}\,dx = \lim_{T\to\infty} \int_1^T \frac{1}{x}\,dx$$

$$= \lim_{T\to\infty}\left[\ln|x|\Big|_1^T\right]$$

$$= \lim_{T\to\infty}(\ln T - \ln 1)$$

$$= \lim_{T\to\infty}(\ln T) = \infty$$

Thus, the integral $\int_1^\infty \frac{1}{x}\,dx$ is divergent. Using the same definition,

$$\int_1^\infty \frac{1}{x^2}\,dx = \lim_{T\to\infty}\int_1^T \frac{1}{x^2}\,dx$$

$$= \lim_{T\to\infty}\left[\left(-\frac{1}{x}\right)\Big|_1^T\right]$$

$$= \lim_{T\to\infty}\left(-\frac{1}{T}+\frac{1}{1}\right) = 1$$

Since the limit is a real number, we conclude that $\int_1^\infty \frac{1}{x^2}\,dx$ converges, and its value is equal to 1.   ▲

Having examined two special cases in the previous example, we now calculate the improper integral

$$\int_1^\infty \frac{1}{x^p}\,dx$$

for an arbitrary power $p > 1$:

$$\int_1^\infty \frac{1}{x^p}\,dx = \lim_{T\to\infty}\int_1^T x^{-p}\,dx$$

$$= \lim_{T\to\infty}\left(\frac{x^{-p+1}}{-p+1}\Big|_1^T\right)$$

$$= \lim_{T\to\infty}\left(\frac{T^{-p+1}}{-p+1} - \frac{1^{-p+1}}{-p+1}\right)$$

$$= \lim_{T\to\infty}\left(\frac{T^{1-p}}{-p+1} - \frac{1}{-p+1}\right)$$

When $p > 1$, the power $1 - p$ is negative. Therefore, T^{1-p} approaches zero as T approaches infinity. The improper integral is

$$\int_1^\infty \frac{1}{x^p}\,dt = \lim_{T\to\infty}\left(\frac{T^{1-p}}{1-p} - \frac{1}{1-p}\right)$$

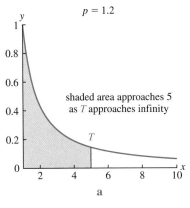
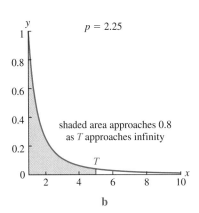

FIGURE 7.7.79

Reciprocal power functions with $p > 1$

$$= 0 - \frac{1}{1-p}$$
$$= \frac{1}{p-1}$$

This integral converges. We examine the case $p < 1$ after the following example.

Example 7.7.2 Improper Integrals of Reciprocal Power Functions with $p > 1$

The improper integral of the function $\frac{1}{x^p}$ for $p = 1.2$ is

$$\int_1^\infty \frac{1}{x^{1.2}} dx = \frac{1}{1.2 - 1} = 5$$

(Figure 7.7.79a). With the larger value $p = 2.25$,

$$\int_1^\infty \frac{1}{x^{2.25}} dx = \frac{1}{2.25 - 1} = 0.8$$

(Figure 7.7.79b). The value of the improper integral (the total area under the curve, or the limiting amount of product produced) becomes smaller as the value of p becomes larger. ▲

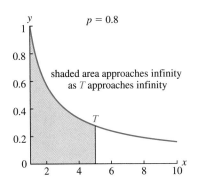

FIGURE 7.7.80

Reciprocal power function with $p < 1$

With $p < 1$, the integration is the same. But when we take the limit, we find

$$\int_1^\infty \frac{1}{x^p} dx = \lim_{T \to \infty} \left(\frac{T^{1-p}}{1-p} - \frac{1}{1-p} \right)$$
$$= \infty$$

because T is taken to the *positive* power $1 - p$ when $p < 1$. This integral diverges. Recall that $y = 1/x^p$ decreases to zero faster for larger values of p. If the value of p is sufficiently large (greater than 1), the area under the curve is finite. When p is small (less than or equal to 1), the function decreases to zero more slowly and the area is infinite (Figure 7.7.80).

To summarize,

$$\int_1^\infty \frac{1}{x^p} dx \text{ is convergent when } p > 1 \text{ and divergent when } p \leq 1. \tag{7.7.1}$$

When the rate decreases exponentially according to $e^{-\alpha x} (\alpha > 0)$, the improper integral is

$$\int_0^\infty e^{-\alpha x} dx = \lim_{T \to \infty} \int_0^T e^{-\alpha x} dx$$

$$= \lim_{T \to \infty} \left[\left(-\frac{1}{\alpha} e^{-\alpha x} \right) \Big|_0^T \right]$$

$$= \lim_{T \to \infty} \left(-\frac{e^{-\alpha T}}{\alpha} + \frac{1}{\alpha} \right)$$

$$= \frac{1}{\alpha} \qquad (7.7.2)$$

This exponential integral converges for every positive value of α, but the integral takes on a larger value for smaller α, as is consistent with the fact that $e^{-\alpha x}$ decreases to zero more slowly when α is small (Example 7.7.3 and Figure 7.7.81).

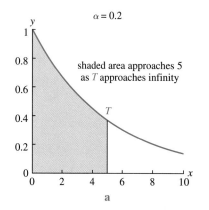
a

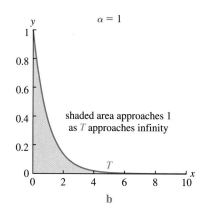
b

FIGURE 7.7.81
Improper integrals with different parameters in the exponent

Example 7.7.3 Improper Integrals of Exponential Functions

The integral of the function $y = e^{-0.2x}$ from zero to infinity is

$$\int_0^\infty e^{-0.2x}\, dx = \frac{1}{0.2} = 5$$

With the larger value $\alpha = 1$,

$$\int_0^\infty e^{-x}\, dx = \frac{1}{1} = 1$$

The value of the improper integral (the total area under the curve, or the limiting amount of product produced) becomes smaller as the value of α becomes larger.

Example 7.7.4 Calculating an Improper Integral from $-\infty$ to ∞

Determine whether the improper integral $\int_{-\infty}^{\infty} xe^{-x^2}\, dx$ converges or not.

According to the text following Definition 7.7.1, we need to investigate the improper integrals

$$\int_{-\infty}^{a} xe^{-x^2}\, dx \quad \text{and} \quad \int_{a}^{\infty} xe^{-x^2}\, dx$$

where a is a real number.

Note that, using substitution with $u = -x^2$, $du = -2x\, dx$, and $x\, dx = -du/2$, we get

$$\int xe^{-x^2}\, dx = -\frac{1}{2} \int e^u\, du = -\frac{1}{2} e^u = -\frac{1}{2} e^{-x^2} + C$$

By definition

$$\int_a^\infty xe^{-x^2}dx = \lim_{T\to\infty}\int_a^T xe^{-x^2}dx$$

$$= \lim_{T\to\infty}\left[\left(-\frac{1}{2}e^{-x^2}\right)\Big|_a^T\right]$$

$$= \lim_{T\to\infty}\left(-\frac{1}{2}e^{-T^2}+\frac{1}{2}e^{-a^2}\right) = \frac{1}{2}e^{-a^2}$$

Likewise,

$$\int_{-\infty}^a xe^{-x^2}dx = \lim_{T\to-\infty}\int_T^a xe^{-x^2}dx$$

$$= \lim_{T\to-\infty}\left[\left(-\frac{1}{2}e^{-x^2}\right)\Big|_T^a\right]$$

$$= \lim_{T\to-\infty}\left(-\frac{1}{2}e^{-a^2}+\frac{1}{2}e^{-T^2}\right) = -\frac{1}{2}e^{-a^2}$$

Since both integrals are convergent, we conclude that $\int_{-\infty}^\infty xe^{-x^2}dx$ is convergent and

$$\int_{-\infty}^\infty xe^{-x^2}dx = \int_{-\infty}^a xe^{-x^2}dx + \int_a^\infty xe^{-x^2}dx = \frac{1}{2}e^{-a^2} - \frac{1}{2}e^{-a^2} = 0$$

Applying the Method of Leading Behaviour to Improper Integrals The **method of leading behaviour** (Section 6.4) can be used to analyze more complicated functions. The idea is that the behaviour of the integral is determined by the leading behaviour of the integrand, the piece that decreases to zero most slowly. An alternative approach, called the **comparison test,** compares a given integrand with a known integrand directly.

Example 7.7.5   Using Leading Behaviour to Establish That an Integral Diverges

Suppose we wish to know whether the integral

$$\int_1^\infty \left(\frac{1}{x}+\frac{1}{e^x}\right)dx$$

converges. The leading behaviour of the integrand is the largest part, namely, the part that decreases most slowly. Because power functions always grow more slowly than exponential functions, the leading behaviour of $f(x) = \frac{1}{x} + \frac{1}{e^x}$ is

$$f_\infty(x) = \frac{1}{x}$$

Now we use the fact that the original integral converges (diverges) if the integral of the leading behaviour converges (diverges). (We do not prove this fact here.) In this case,

$$\int_1^\infty f_\infty(x)dx = \int_1^\infty \frac{1}{x}dx = \infty$$

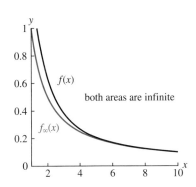

FIGURE 7.7.82
A divergent improper integral checked with the method of leading behaviour

by Equation 7.7.1; see Figure 7.7.82. Therefore, the original integral also diverges.

Example 7.7.6 Using Leading Behaviour to Establish That an Integral Converges

Now let's consider the integral

$$\int_1^\infty \frac{1}{x+e^x}dx$$

The leading behaviour of the integrand can be found by computing the leading behaviour of the denominator, $x+e^x$. Because the exponential is larger, we replace $x+e^x$ with e^x. The integral of the leading behaviour is

$$\int_1^\infty \frac{1}{e^x}dx = \int_1^\infty e^{-x}dx$$

which converges by Equation 7.7.2; see Figure 7.7.83. Thus, the given integral converges.

It is, however, impossible to compute the value

$$\int_1^\infty \frac{1}{x+e^x}dx$$

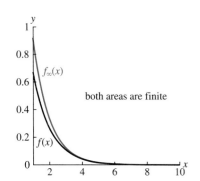

Figure 7.7.83
A convergent improper integral checked with the method of leading behaviour

exactly without using numerical methods.

The **comparison test** provides an alternative method. If we can compare our function with a simple function, we can often establish whether the integral converges or diverges.

The Comparison Test Assume that $f(x)$ and $g(x)$ are continuous functions such that $f(x) \geq g(x) \geq 0$ for all $x \geq a$. Then we have the following:

a. If $\int_a^\infty f(x)\,dx$ is convergent, then $\int_a^\infty g(x)\,dx$ is convergent.

b. If $\int_a^\infty g(x)\,dx$ is divergent, then $\int_a^\infty f(x)\,dx$ is divergent.

The statements remain true if $\int_a^\infty$ is replaced by $\int_{-\infty}^a$ or by $\int_{-\infty}^\infty$.

In words, if a positive function $f(x)$ lies above the positive function $g(x)$ and the area under $f(x)$ is finite, then the area under $g(x)$ must be finite, i.e., $\int_a^\infty g(x)\,dx$ is convergent (Figure 7.7.84a shows the case $a=0$). If $\int_a^\infty g(x)\,dx$ is divergent, then the area under $f(x)$ cannot be finite and so $\int_a^\infty f(x)\,dx$ is divergent as well (Figure 7.7.84b shows the case $a=0$).

Example 7.7.7 Using the Comparison Test to Establish That an Integral Converges

Consider again the integral

$$\int_1^\infty \frac{1}{x+e^x}dx$$

from Example 7.7.6. Because $x > 0$, we get $x+e^x > e^x$ and thus

$$0 < \frac{1}{x+e^x} < \frac{1}{e^x} = e^{-x}$$

Since $\int_1^\infty e^{-x}dx$ converges, we conclude that the integral $\int_1^\infty \frac{1}{x+e^x}dx$ converges.

7.7 Improper Integrals

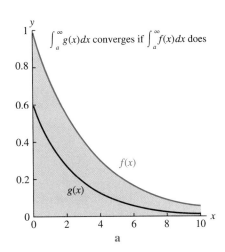
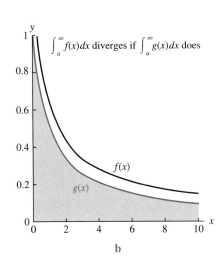

FIGURE 7.7.84

The comparison test with infinite limits of integration

Example 7.7.8 Using the Comparison Test to Establish That an Integral Diverges

Consider evaluating

$$\int_1^\infty \frac{1}{x+\sqrt{x}}\,dx$$

For $x > 1$, $\sqrt{x} < x$, so

$$\frac{1}{x+\sqrt{x}} > \frac{1}{x+x} = \frac{1}{2x}$$

Since $\int_1^\infty \frac{1}{2x}\,dx$ diverges, the original integral diverges as well.

Infinite Integrands

Although functions with infinite integrands crop up less frequently in biological problems, it is useful to be familiar with their behaviour. Consider trying to find the area under the curve $f(x) = \frac{1}{\sqrt{x}}$ between $x = 0$ and $x = 1$ (Figure 7.7.85). This function approaches infinity as $x \to 0^+$.

Although one can, with some care, define this integral as a limit of right-hand Riemann sums, we instead use the limit to define and compute this second type of improper integral.

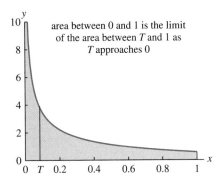

FIGURE 7.7.85

Computing an integral with an infinite integrand

Definition 7.7.2 Assume that $f(x)$ is continuous on $[a, b)$ but not continuous at $x = b$. We define

$$\int_a^b f(x)dx = \lim_{T \to b^-} \int_a^T f(x)dx$$

if the limit exists. If $f(x)$ is continuous on $(a, b]$, but not continuous at $x = a$, then

$$\int_a^b f(x)\,dx = \lim_{T \to a^+} \int_T^b f(x)\,dx$$

if the limit exists.

If the corresponding limit exists, then we say that the **improper integral** $\int_a^b f(x)\,dx$ is **convergent**. Otherwise, it is called **divergent**.

If $f(x)$ is continuous on $[a, b]$ except at $x = c$, $a < c < b$, and if the improper integrals $\int_a^c f(x)\,dx$ and $\int_c^b f(x)\,dx$ converge, then the improper integral $\int_a^b f(x)\,dx$ converges and

$$\int_a^b f(x)\,dx = \int_a^c f(x)\,dx + \int_c^b f(x)\,dx$$

Example 7.7.9 A Finite Area Even When a Function Approaches Infinity

With this definition, we have

$$\int_0^1 \frac{1}{\sqrt{x}}dx = \lim_{T \to 0^+} \int_T^1 \frac{1}{\sqrt{x}}dx$$

$$= \lim_{T \to 0^+} \left(2\sqrt{x}\Big|_T^1\right)$$

$$= \lim_{T \to 0^+} (2 - 2\sqrt{T}) = 2$$

Although the function is not continuous at 0, the area is perfectly well defined.

Example 7.7.10 An Infinite Area under a Function That Approaches Infinity

If $g(x) = \frac{1}{x^2}$, we have

$$\int_0^1 \frac{1}{x^2}dx = \lim_{T \to 0^+} \int_T^1 \frac{1}{x^2}dx$$

$$= \lim_{T \to 0^+} \left[\left(-\frac{1}{x}\right)\Big|_T^1\right]$$

$$= \lim_{T \to 0^+} \left(-1 + \frac{1}{T}\right) = \infty$$

For power functions of the form $1/x^p$, where $p > 0$ and $p \neq 1$, we can derive a general result for convergence over the interval from 0 to a:

$$\int_0^a \frac{1}{x^p}dx = \lim_{T \to 0^+} \int_T^a x^{-p}\,dx$$

$$= \lim_{T \to 0^+} \left[\left(\frac{x^{1-p}}{1-p}\right)\Big|_T^a\right]$$

$$= \lim_{T \to 0^+} \left(\frac{a^{1-p}}{1-p} - \frac{T^{1-p}}{1-p}\right)$$

If $0 < p < 1$, then
$$\lim_{T \to 0^+} T^{1-p} = 0$$
and the integral converges and is equal to $\frac{a^{1-p}}{1-p}$. If $p > 1$, then
$$\lim_{T \to 0^+} T^{1-p} = \infty$$
and the integral diverges.

If $p = 1$, then
$$\int_0^1 \frac{1}{x} dx = \lim_{T \to 0^+} \int_T^1 \frac{1}{x} dx$$
$$= \lim_{T \to 0^+} \left[\ln |x| \Big|_T^1 \right]$$
$$= \lim_{T \to 0^+} (\ln 1 - \ln T)$$
$$= \lim_{T \to 0^+} (-\ln T) = \infty$$

To summarize:
$$\int_0^1 \frac{1}{x^p} dx \text{ is convergent when } 0 < p < 1 \text{ and divergent when } p \geq 1 \quad (7.7.3)$$

Example 7.7.11 Evaluating an Improper Integral

Determine whether the improper integral $\int_0^5 \frac{1}{x-2} dx$ is convergent or not.

Note that the function $f(x) = \frac{1}{x-2}$ has a vertical asymptote at $x = 2$; i.e., it is not continuous there. Definition 7.7.2 tells us that we need to examine the improper integrals
$$\int_0^2 \frac{1}{x-2} dx \quad \text{and} \quad \int_2^5 \frac{1}{x-2} dx$$

The calculation
$$\int_0^2 \frac{1}{x-2} dx = \lim_{T \to 2^-} \int_0^T \frac{1}{x-2} dx$$
$$= \lim_{T \to 2^-} \left[\ln |x-2| \Big|_0^T \right]$$
$$= \lim_{T \to 2^-} \left[\ln |T-2| - \ln 2 \right] = -\infty$$

shows that the integral $\int_0^2 \frac{1}{x-2} dx$ is divergent. We conclude that the improper integral $\int_0^5 \frac{1}{x-2} dx$ is divergent. (Note that there is no need to examine the integral $\int_2^5 \frac{1}{x-2} dx$ for convergence, as it will not change our answer.) ▲

Applying the Method of Leading Behaviour to Improper Integrals As before, these rules can be extended with the method of leading behaviour. The portion that counts is the part that increases most quickly.

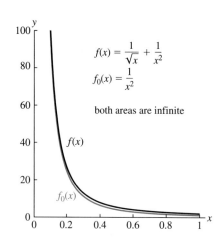

FIGURE 7.7.86
A divergent improper integral checked with the method of leading behaviour

Example 7.7.12 Using the Method of Leading Behaviour When the Integrand Approaches Infinity I

The integrand of the integral

$$\int_0^1 \left(\frac{1}{\sqrt{x}} + \frac{1}{x^2}\right) dx$$

has two components, the larger of which is $\frac{1}{x^2}$. Therefore, the leading behaviour is

$$\int_0^1 \frac{1}{x^2} dx = \infty$$

because the power is greater than 1, meaning that the original integral diverges by Equation 7.7.3; see Figure 7.7.86.

Example 7.7.13 Using the Method of Leading Behaviour When the Integrand Approaches Infinity II

The leading behaviour at zero of the denominator of the integrand of

$$\int_0^1 \frac{1}{\sqrt{x} + x^2} dx$$

is $\sqrt{x}$ (recall that $\sqrt{x} \geq x^2$ if $0 \leq x \leq 1$). The integral of the leading behaviour is

$$\int_0^1 \frac{1}{\sqrt{x}} dx$$

which converges by Equation 7.7.3 because the power is less than 1 (Figure 7.7.87).

Summary We have introduced **improper integrals** of two types: integrals with infinite limits of integration and integrals with infinite integrands. The former are defined as the limit when one limit of integration increases to infinity and the second as the limit when one limit of integration approaches a point where the function itself approaches infinity. If the limit exists, we say the integral **converges;** if not, we say the integral **diverges.** We found conditions for convergence or divergence of integrals of power and exponential functions and then extended these results with the method of leading behaviour and the **comparison test.**

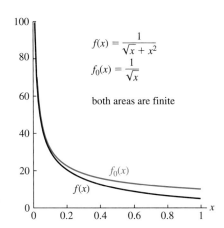

FIGURE 7.7.87

A convergent improper integral checked with the method of leading behaviour

7.7 Exercises

Mathematical Techniques

1–4 ■ Many improper integrals can be evaluated by comparing functions with the method of leading behaviour. State which of the given pair of functions approaches its limit more quickly, and demonstrate the result with L'Hôpital's rule when needed.

1. Which function approaches zero faster as x approaches infinity: $f(x) = e^{-x}$ or $f(x) = \frac{1}{x}$?

2. Which function approaches zero faster as x approaches infinity: $f(x) = \frac{1}{1+x^2}$ or $f(x) = \frac{1}{1+x}$?

3. Which function approaches infinity faster as x approaches zero: $f(x) = \frac{1}{x^2}$ or $f(x) = \frac{1}{x}$?

4. Which function approaches infinity faster as x approaches zero: $f(x) = \frac{1}{x}$ or $f(x) = \frac{1}{\sqrt{x}}$?

5–10 ■ Evaluate each improper integral or explain why it doesn't converge.

5. $\int_0^\infty e^{-3x}\,dx$

6. $\int_0^\infty e^x\,dx$

7. $\int_1^\infty \frac{1}{\sqrt{x}}\,dx$

8. $\int_5^\infty \frac{1}{x^2}\,dx$

9. $\int_0^\infty \frac{1}{(1+3x)^{3/2}}\,dx$

10. $\int_0^\infty \frac{1}{(2+5x)^4}\,dx$

11–14 ■ Evaluate each improper integral or explain why it doesn't converge.

11. $\int_0^1 \frac{1}{x}\,dx$

12. $\int_0^{0.001} \frac{1}{x^2}\,dx$

13. $\int_0^{0.001} \frac{1}{\sqrt[3]{x}}\,dx$

14. $\int_0^\infty \frac{1}{\sqrt[3]{x}}\,dx$

15–25 ■ Determine whether each integral is convergent or not (calculations of integrals, in some cases, require substitution or integration by parts).

15. $\int_2^\infty e^{-x}\,dx$

16. $\int_0^\infty xe^{-x}\,dx$

17. $\int_1^\infty \frac{\ln x}{x}\,dx$

18. $\int_0^\infty \frac{1}{1+x^2}\,dx$

19. $\int_{-\infty}^\infty \frac{1}{1+x^2}\,dx$

20. $\int_{-\infty}^\infty \frac{x}{1+x^2}\,dx$

21. $\int_{-\infty}^2 \frac{3x}{x^2+4}\,dx$

22. $\int_1^\infty \frac{\ln x}{x^4}\,dx$

23. $\int_0^1 \frac{\ln x}{\sqrt{x}}\,dx$

24. $\int_0^{\pi/2} \sec^2 x \, dx$

25. $\int_{\pi/2}^{\pi} \tan x \, dx$

26–29 ▪ Use the method of leading behaviour to deduce whether each integral converges.

26. $\int_0^1 \frac{1}{\sqrt[3]{x} + x^3} dx$

27. $\int_0^1 \frac{1}{x^2 + e^x} dx$

28. $\int_1^{\infty} \frac{1}{\sqrt[3]{x} + x^3} dx$

29. $\int_1^{\infty} \frac{1}{\sqrt{x} + e^x} dx$

30–33 ▪ Use the comparison test to deduce whether each integral converges. If it does, find an upper bound on the value.

30. $\int_0^1 \frac{1}{\sqrt[3]{x} + x^3} dx$

31. $\int_0^1 \frac{1}{x^2 + e^x} dx$

32. $\int_1^{\infty} \frac{1}{\sqrt[3]{x} + x^3} dx$

33. $\int_1^{\infty} \frac{1}{\sqrt{x} + e^x} dx$

Applications

34–37 ▪ Write a pure-time differential equation to describe each situation and determine the long-term behaviour of its solution (i.e., calculate the corresponding improper integral). Given this long-term behaviour, could this process continue indefinitely in the real world?

34. The volume of a cell is increasing at a rate of $\frac{100}{(1+t)^2}$ cubic micrometres per second, starting from a size of 500 μm^3.

35. The concentration of a toxin in a cell is increasing at a rate of $50e^{-2t}$ micromoles per litre per second, starting from a concentration of 10 μmol/L. If the cell is poisoned when the concentration exceeds 30 μmol/L, could this cell survive?

36. A population of bacteria is increasing at a rate of $\frac{1000}{(2+3t)^{0.75}}$ bacteria per hour, starting from a population of 10^6. Could this sort of growth be maintained indefinitely? When would the population reach $2 \cdot 10^6$? Would you say that this population is growing quickly?

37. A population of bacteria is increasing at a rate of $\frac{1000}{(2+3t)^{1.5}}$ bacteria per hour, starting from a population of 1000. Could this sort of growth be maintained indefinitely? Would the population reach 2000?

Computer Exercise

38. Use a computer to find the values of the integrals in Exercises 26 and 27 to eight decimal places and compare with the integrals of the leading behaviour.

Chapter Summary: Key Terms and Concepts

Define or explain the meaning of each term and concept.
Differential equations: pure-time differential equation, autonomous differential equation; solution, initial condition, initial value problem; Euler's method for solving differential equations
Antiderivatives and techniques of integration: power rule, constant product rule, sum rule; substitution, integration by parts, integration using Taylor polynomials
Definite integrals: region under the graph of a function, left sum, right sum, midpoint sum, Riemann sum; definite integral, limits of integration, integrand, integrable function; area, signed area; integral of rate of change, total change; Fundamental Theorem of Calculus, integral function, evaluation of definite integrals
Applications: area between curves, cross-section, volume of a solid, solid of revolution, length of a curve, average value, density, mass
Improper integrals: infinite limits of integration, integral of a discontinuous function; convergent integral, divergent integral; comparison test

Concept Check: True/False Quiz

Determine whether each statement is true or false. Give a reason for your choice.

1. $P'(t) = 2P(t) - P(t)^2 + t$ is an autonomous differential equation.

2. $y = -2e^{3t}$ is the solution of the initial value problem $y' = 3y$, $y(0) = -2$.

3. The left sum L_{10} is an underestimate of the area of the region under the curve $y = x^3$ on $[0, 4]$.

4. The function $g(x) = 1/x$ is an antiderivative of $h(x) = \ln x$.

5. The formula $\int [af(x) + g(x)] dx = a \int f(x) dx + \int g(x) dx$ holds for all real numbers a.

6. The area under the graph of $y = e^{-x}$ and above the x-axis and between $x = 0$ and $x = 1$ is greater than 1.

7. By the Fundamental Theorem of Calculus,

$$\int_{-1}^{3} \frac{1}{x} dx = \ln |x| \Big|_{-1}^{3} = \ln |3| - \ln |-1| = \ln 3$$

8. The average value of $f(x) = x$ on $[0, 10]$ is 5.

9. The improper integral $\int_0^1 x^{-0.99} dx$ is convergent.

10. The improper integral $\int_1^{\infty} x^{-0.99} dx$ is convergent.

Supplementary Problems

1. The voltage, v, of a neuron follows the differential equation

$$\frac{dv}{dt} = 1 + \frac{1}{1 + 0.02t} - e^{0.01t}$$

over the course of 100 ms, where t is measured in milliseconds and v in millivolts. We start at $v(0) = -70$ mV.

   a. Sketch the graph of the rate of change. Indicate on your graph the times when the voltage reaches minima and maxima (you don't need to solve for the numerical values).

   b. Sketch the graph of the voltage as a function of time.

   c. What is the voltage after 100 ms?

2. Consider again the differential equation in the previous problem,

$$\frac{dv}{dt} = 1 + \frac{1}{1 + 0.02t} - e^{0.01t}$$

with $v(0) = -70$.

   a. Use Euler's method to estimate the voltage after 1 ms and again 1 ms after that.

   b. Estimate the voltage after 2 ms using left-hand and right-hand Riemann sums.

   c. Which of your estimates matches Euler's method and why?

3. A neuron in your brain sends a charge down an 80-cm axon (a long skinny thing) toward your hand at a speed of 10 m/s. At the time when the charge reaches your elbow, the voltage in the axon is -70 mV, except on the 6-cm piece between 47 cm and 53 cm from your brain. On this piece, the voltage is

$$v(x) = -70 + 10 \left[9 - (x - 50)^2 \right] \text{ millivolts}$$

where $v(x)$ is the voltage at a distance of x centimetres from the brain.

   a. How long will it take the information to get to your hand? How long did it take the information to reach your elbow?

   b. Sketch the graph of the voltage along the whole axon.

   c. Find the average voltage of the 6-cm piece.

   d. Find the average voltage of the whole axon.

4. The charge in a dead neuron decays according to

$$\frac{dv}{dt} = \frac{1}{\sqrt{1 + 4t}} - \frac{2}{(1 + 4t)^{\frac{3}{2}}}$$

starting from $v(0) = -70$ mV at $t = 0$.

   a. Is the voltage approaching zero as $t \to \infty$? How do you know that it will eventually reach zero?

   b. Write an equation (but don't solve it) for the time when the voltage reaches zero.

   c. What is wrong with this model?

5. Consider the differential equations

$$\frac{db}{dt} = 2b$$

and

$$\frac{dB}{dt} = 1 + 2t$$

   a. Which of these is a pure-time differential equation? Describe circumstances in which you might find each of these equations.

   b. Suppose $b(0) = B(0) = 1$. Use Euler's method with $\Delta t = 0.1$ to find estimates for $b(0.1)$ and $B(0.1)$.

   c. Use Euler's method with $\Delta t = 0.1$ to find estimates for $b(0.2)$ and $B(0.2)$.

6. Consider the differential equation

$$\frac{dp}{dt} = e^{-4t}$$

where $p(t)$ is product in moles at time t, and t is measured in seconds.

   a. Explain what is going on.

   b. Suppose $p(0) = 1$. Find $p(1)$.

7. Consider the differential equation

$$\frac{dV}{dt} = 4 - t^2$$

where $V(t)$ is volume in litres at time t, and t is measured in minutes.

   a. Explain what is going on.
   b. At what time is the volume a maximum?
   c. Break the interval from $t=0$ to $t=3$ into three parts, and find the left-hand and right-hand estimates of the volume at $t=3$ (assume $V(0)=0$).
   d. Write the definite integral expressing volume at $t=3$ and evaluate.

8. Find the area under the curve $f(x) = 3 + \left(1 + \frac{x}{3}\right)^2$ between $x=0$ and $x=3$.

9. The population density of trout in a stream is

$$\rho(x) = |-x^2 + 5x + 50|$$

where ρ is measured in trout per kilometre and x is measured in kilometres. x runs from 0 to 20.

   a. Graph $\rho(x)$ and find the minimum and maximum.
   b. Find the total number of trout in the stream.
   c. Find the average density of trout in the stream.
   d. Indicate on your graph how you would find where the actual density is equal to the average density.

10. The amount of product is described by the differential equation

$$\frac{dp}{dt} = \frac{1}{\sqrt{1+3t}}$$

starting at time $t=0$. Suppose p is measured in moles, t is measured in hours, and $p(0) = 0$.

   a. Find the limiting amount of product.
   b. Find the average rate at which product is produced on the interval from 0 to t, and compute the limit as $t \to \infty$.
   c. Find the limit as $t \to 0$ of the average rate at which product is produced.

11. A student is hooked up to an EEG during a test, and her α brain wave power follows

$$A(t) = \frac{50}{2 + 0.3t} + 10e^{0.0125t}$$

where t runs from 0 to 120 min.

   a. Convince yourself that brain wave power has a minimum value sometime during the test. Sketch the graph of the function. Find the maximum of $A(t)$ on $[0, 120]$.
   b. Find the total brain wave energy (the integral of brain wave power) during the test.
   c. Find the average brain wave power during the test. Sketch the corresponding line on your graph.
   d. Draw a graph showing how you would estimate the total brain wave energy using the right-hand approximation with $n = 6$. Write the associated sum. Do you think your estimate is high or low?

12. Consider the function $G(h)$ giving the density of nutrients in a plant stem as a function of the height, h,

$$G(h) = 5 + 3e^{-2h}$$

where G is measured in moles per metre and h is measured in metres.

   a. Find the total amount of nutrient if the stem is 2 m tall.
   b. Find the average density in the stem.
   c. Find the exact and approximate amount between 1 and 1.01 m. To find the approximate amount, use the left sum with one rectangle.

Projects

1. As noted in the text, Euler's method is not a very good way to solve differential equations numerically. In this project, you will compare Euler's method with the **midpoint** (or **second-order Runge-Kutta**) **method** and the **implicit Euler method.**

   Suppose we know (or estimate) that the solution of the general differential equation

$$\frac{dy}{dt} = f(t, y)$$

takes on some value $y = y_0$ at time $t = t_0$. Our goal is to estimate the value of $y_1 = y(t_0 + h)$ with *step size h*.

   We have seen that Euler's method uses the tangent line to estimate

$$y_1 = y(t_0) + hf(t_0, y_0)$$

This method has the drawback that it uses the derivative only at the beginning of the interval between times t_0 and $t_0 + h$. If the derivative changes significantly during this interval, the method can be very inaccurate.

   An alternative called the midpoint method uses an estimate of the derivative at time $t_0 + \frac{h}{2}$ instead. The next step is

$$k = hf(t_0, y_0)$$

$$y_1 = y(t_0) + hf\left(t_0 + \frac{h}{2}, y_0 + \frac{k}{2}\right)$$

where k is the estimated change computed by Euler's method, and we use it to guess the value of y halfway through the interval. This method is more accurate than Euler's method.

Both Euler's method and the midpoint method are **unstable,** meaning that the approximate solutions they produce can fail to approach a stable equilibrium if the step size, h, is too large. The **implicit Euler scheme** is stable for linear equations. The idea is to use the derivative at the end of the interval instead of the beginning. That is,

$$y_1 = y(t_0) + hf(t_0 + h, y_1)$$

Because we do not know the value of y_1, we have to solve for it.

This project asks you to compare these three methods on four differential equations:

$$\frac{db}{dt} = b$$
$$\frac{dx}{dt} = -x$$
$$\frac{dV}{dt} = -e^{-t}$$
$$\frac{dy}{dt} = -e^{-t} + e^{-y}$$

Suppose the state variable in each equation has the initial value of 1 at $t = 0$.

a. First, find the solution of each at $t = 1$. Try each method (if you can figure out how to get the implicit method to work on the last equation) with values of h ranging from 1 down to 0.001.

b. The equations for x and V should both approach zero exponentially as t becomes large. Use large values of h (such as 10 or 100) in each of the three methods. How well do they do for large t?

c. The midpoint method is related to approximating the solution with a quadratic Taylor polynomial (Section 5.7). Develop an extension of Euler's method based on the quadratic approximation and compare with the midpoint method.

Press, W. H., Teukolsky, S. A., Vettering, W. T., and Flannery, B.P. *Numerical Recipes in FORTRAN: The Art of Scientific Computing,* 2nd Edition, Cambridge University Press, Cambridge, U.K., 1992.

2. In ancient times, the Greeks were fascinated by geometric problems such as finding the area of a circle. Archimedes, perhaps the greatest mathematician in the ancient world, came up with an idea closely related to the Riemann sum to solve this problem. Using some modern tools, we can apply his ideas to compute the value of π.

   a. Break a circle of radius 1 into n wedges, each with angle $\theta = 2\pi/n$. Show that the area of a right triangle inside each wedge is $\sin\theta/2$. Be sure to draw a picture.

   b. Approximate the area of a circle using $n = 8$. Look up (or remember) the half-angle formula giving $\sin(\theta/2)$ in terms of $\cos\theta$ to approximate the area with $n = 16$.

   c. Continue in this way to approximate the area with $n = 32$, and so forth.

   d. Use the same approach, but approximate each wedge with a right triangle that lies outside the circle. Show that the area of the triangle is $\tan\theta/2$.

   e. Approximate the area of a circle using $n = 8$ (what happens when $n = 4$?). Look up (or remember) the half-angle formula giving $\tan(\theta/2)$ in terms of $\tan\theta$ to approximate the area with $n = 16$.

   f. Continue in this way to approximate the area with $n = 32$, and so forth.

   g. Can you think of a better method that Archimedes could have used to compute the value of π?

For more information, consult Beckmann, P. *A History of π,* 2nd Edition, Golem Press, Boulder, CO, 1971.

Chapter 8

Differential Equations

We have developed the indefinite integral and definite integral as tools to solve **pure-time differential equations** and have linked them with the Fundamental Theorem of Calculus. We will now use integrals and several graphical tools to study **autonomous differential equations,** where the rate of change is a **rule** relating change to the current state. In particular, we will use the **phase-line diagram** to find **equilibria** and their **stability,** much as we used cobwebbing to find the equilibria and stability of discrete-time dynamical systems.

Because most biological systems involve several interacting measurements, we will extend our methods to address **systems of autonomous differential equations,** where the rate of change of each measurement depends on the value of the others. Generalization of the phase line to the **phase plane** makes possible the study of two-dimensional systems.

8.1 Basic Models with Differential Equations

In this section, we examine models involving **autonomous differential equations,** including several models of **population growth,** a model for **temperature change** and **diffusion** across a membrane, and a **model of selection.** We discuss the equations from various viewpoints, thus building our understanding and gaining important insights. In the next two sections, we further our **qualitative study,** and in Section 8.4 we learn how to **solve** (some) differential equations **algebraically.**

Autonomous Differential Equations

We have developed techniques to solve pure-time differential equations with the general form

$$\frac{df}{dt} = F(t) \tag{8.1.1}$$

In these equations, the rate of change is a function of the independent variable (in this case, t) and not of the unknown function, $f(t)$. (The function $f(t)$ is also called the state variable.) Change is imposed from outside the system; consider the control of a lake level by the weather or by an engineer. In most biological systems, however, the rate of change depends on the current state of the system as well as on external factors. A lake that experienced increased evaporation at high lake levels would change as a function of the level itself, not just as a function of the weather or the decisions of engineers.

The class of differential equations where the rate of change depends only on the state of the system and not on external circumstances is particularly important. These equations are called **autonomous differential equations.** We have already studied one autonomous differential equation, which describes the growth of a bacterial population, $b(t)$, as follows:

$$\frac{db}{dt} = 2b$$

The rate of change of the population size b depends only on the population size, b, and not on the time, t.

The general autonomous differential equation for a measurement f is of the form

$$\frac{df}{dt} = \text{function of } f \qquad (8.1.2)$$

When the rate of change depends on both the state variable and the time, the equation is called a **nonautonomous differential equation.** The general form is

$$\frac{df}{dt} = \text{function of both } f \text{ and } t \qquad (8.1.3)$$

These equations are generally more difficult to solve than either pure-time or autonomous differential equations.

Keep in mind that although the *rate of change* in an autonomous differential equation does not depend explicitly on time, the *solution* does.

Example 8.1.1   The Basic Exponential Model Revisited

Consider a population, $P(t)$, that changes according to

$$P'(t) = kP(t)$$

where k is a constant, t is time, and $P(0) = P_0$ is the initial population. We have already met this model in numerous situations, such as bacterial growth, the growth of the Canadian population, radioactive decay, and the metabolic breakdown (dissolution) of drugs (caffeine).

Dividing by $P(t)$, we get

$$\frac{P'(t)}{P(t)} = k$$

The expression on the left side (the ratio of the rate of change to the total population) is called the **relative rate of change,** the **per capita production (reproduction) rate,** or the **per capita growth rate.**

Thus, the population is characterized by the fact that *the relative rate of change is constant* (equal to k).

If $k > 0$, then

$$P'(t) = kP(t) > 0$$

and the population increases. If $k < 0$, then

$$P'(t) = kP(t) < 0$$

and the population decreases exponentially.

We can check (we will prove this in Section 8.4) that $P(t)$ is given by

$$P(t) = P(0)e^{kt}$$

The graph of $P(t)$ for several values of k is shown in Figure 8.1.1.

When $k > 0$, the model predicts unlimited growth, which is certainly not realistic. Nevertheless, for a short time it can predict the behaviour of many systems quite accurately (for instance, the initial growth of some bacterial or animal populations, or the initial spread of an infection). A reminder:

Basic Exponential Model

If $P'(t) = kP(t)$, then $P(t) = P(0)e^{kt}$, where $P(0)$ is the initial population.

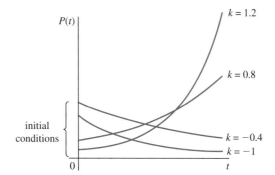

FIGURE 8.1.1

Exponential growth and decay model $P'(t) = kP(t)$

Example 8.1.2 Logistic Differential Equation

The exponential model cannot accurately predict the behaviour of a population for a long time. Limitations on resources (food, space) will force the population to level off, i.e., reach a size where it can no longer grow.

Consider the **logistic differential equation**

$$\frac{dP}{dt} = kP\left(1 - \frac{P}{L}\right) \tag{8.1.4}$$

where k and L are positive constants and $P = P(t)$ is the population at time t. The initial condition is $P(0) = P_0$.

Let's try to understand what this equation says.

(1) If $P < L$, then $P/L < 1$ and

$$\frac{dP}{dt} = kP\left(1 - \frac{P}{L}\right) > 0$$

and the population increases. However, if $P > L$, then $P/L > 1$ and

$$\frac{dP}{dt} = kP\left(1 - \frac{P}{L}\right) < 0$$

and the population decreases. The constant L is called the **carrying capacity** of the population (within an ecosystem). If the population is below L, it will increase. If the population manages to exceed L (for instance, as a result of immigration), then it will decrease.

(2) If $P = L$, then $dP/dt = 0$ and the population does not change. This is an example of an **equilibrium** (also called **ecological balance**). When a population begins at an equilibrium, or reaches it, it stays there (at least in theory). The rate at which members die is equal to the rate at which they are replaced.

(3) If the population, P, is small compared to the carrying capacity, L, then P/L is small (close to zero) and

$$1 - \frac{P}{L} \approx 1$$

Hence an important conclusion: for small populations, the logistic model

$$\frac{dP}{dt} = kP\left(1 - \frac{P}{L}\right)$$

can be approximated by the exponential model

$$\frac{dP}{dt} = kP$$

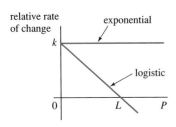

FIGURE 8.1.2

Comparison of relative rates of change

We could have arrived at the same conclusion by writing

$$kP\left(1 - \frac{P}{L}\right) = kP - \frac{k}{L}P^2$$

and using the fact that its leading behaviour at zero (i.e., at $P = 0$) is kP.

(4) Dividing

$$P' = kP\left(1 - \frac{P}{L}\right)$$

by P, we get

$$\frac{P'}{P} = k\left(1 - \frac{P}{L}\right) = k - \frac{k}{L}P$$

Thus, the per capita production rate (or the relative rate of change) of P is *not* constant (it is linear, with negative slope equal to $-k/L$). So it decreases as P increases, which makes sense: the larger the population, the smaller the number of new individuals per individual that will be produced. The relative rates of change for exponential and logistic models are compared in Figure 8.1.2.

(5) What does the graph of the solution of the logistic equation (see Equation 8.1.4) look like? In Section 8.4, we will solve the equation algebraically and obtain a formula. In Figure 8.1.3, we summarize what we have learned so far about it.

(6) To gain a bit more insight, let's consider the equation

$$\frac{dP}{dt} = 0.09P\left(1 - \frac{P}{2000}\right)$$

with $P(0) = 350$ and apply Euler's method to obtain a numerical approximation of its solution.

Recall that Euler's method is given by

$$t_{n+1} = t_n + \Delta t$$
$$y_{n+1} = y_n + G(t_n)\Delta t$$

where $P(t_0) = y_0$ is the initial condition, Δt is the time step, $G(t_n)$ is the slope at t_n, and y_n is the approximation for the solution, $P(t_n)$. With time step $\Delta t = 0.5$, the system is

$$t_{n+1} = t_n + 0.5$$
$$y_{n+1} = y_n + 0.09y_n\left(1 - \frac{y_n}{2000}\right)0.5$$
$$= y_n + 0.045y_n\left(1 - \frac{y_n}{2000}\right)$$

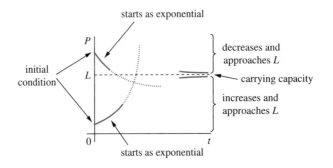

FIGURE 8.1.3

Sketching the solution to the logistic equation

Table 8.1.1

t	Approximation of $P(t)$	t	Approximation of $P(t)$
$t_0 = 0$	$y_0 = 350$		
$t_1 = 0.5$	$y_1 = 363$	$t_{50} = 25$	$y_{50} = 1332$
$t_2 = 1$	$y_2 = 376$	$t_{100} = 50$	$y_{100} = 1903$
$t_{10} = 5$	$y_{10} = 497$	$t_{150} = 75$	$y_{150} = 1990$
$t_{20} = 10$	$y_{20} = 681$	$t_{200} = 100$	$y_{200} = 1999$
$t_{30} = 15$	$y_{30} = 894$	$t_{250} = 125$	$y_{250} = 1999.9 \approx 2000$

Starting with $t_0 = 0$ and $y_0 = 350$, we compute

$$t_1 = t_0 + \Delta t = 0 + 0.5 = 0.5$$
$$y_1 = y_0 + 0.045 y_0 \left(1 - \frac{y_0}{2000}\right)$$
$$= 350 + 0.045 \cdot 350 \cdot \left(1 - \frac{350}{2000}\right) \approx 363$$

The next step is

$$t_2 = t_1 + \Delta t = 1$$
$$y_2 = y_1 + 0.0045 y_1 \left(1 - \frac{y_1}{2000}\right)$$
$$\approx 363 + 0.0045 \cdot 363 \cdot \left(1 - \frac{363}{2000}\right) \approx 376$$

Proceeding in the same way, we obtain the values in Table 8.1.1.

We see how the values approach the carrying capacity of 2000; see also Figure 8.1.4a, where we show the approximations. In Figure 8.1.4b, we show the approximation of the solution using Euler's method with initial condition $P(0) = 2500$ (and the same step size).

The first few calculations in Euler's method can be done by hand (using a calculator). However, the approximations in Table 8.1.1, as well as the graphs in Figure 8.1.4, required a computer (we used Maple). ▲

Example 8.1.3 Modified Logistic Differential Equation (the Allee Effect)

As good as the logistic differential equation model is, it has a flaw: it predicts that any population, *no matter how small,* will grow and sooner or later reach the carrying

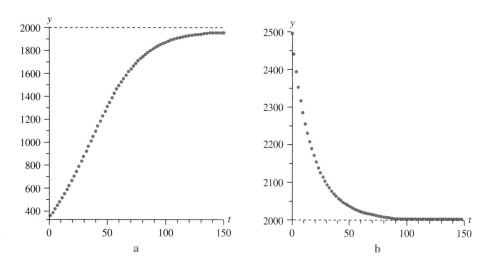

FIGURE 8.1.4
Solutions of the logistic equation using Euler's method

capacity. As is well known, that is not true: if a population (such as a sexually reproducing species) falls to a certain minimum, it will keep decreasing (due to the negative effects of inbreeding and/or the inability to adjust to climate and other changes) and eventually become extinct. This phenomenon is known as the **Allee effect** (after W. C. Allee, the American zoologist and ecologist who first identified and studied it).

To build a model that accounts for this fact, we adjust Equation 8.1.4; consider

$$P' = kP\left(1 - \frac{P}{L}\right)\left(1 - \frac{m}{P}\right)$$

where $P(0) = P_0$ is the initial condition; k, m, and L are positive constants; and $m < L$.

How does the extra factor

$$1 - \frac{m}{P}$$

affect the solution, P?

Assume that P is such that $m < P < L$. Then

$$1 - \frac{P}{L} > 0 \quad \text{and} \quad 1 - \frac{m}{P} > 0$$

and

$$P' = kP\left(1 - \frac{P}{L}\right)\left(1 - \frac{m}{P}\right) > 0$$

i.e., if P is above m (and below L), then it increases over time. However, if $P < m < L$, then

$$1 - \frac{P}{L} > 0 \quad \text{and} \quad 1 - \frac{m}{P} < 0$$

and

$$P' = kP\left(1 - \frac{P}{L}\right)\left(1 - \frac{m}{P}\right) < 0$$

Thus, if P is below m, then P decreases over time.

Thus, m represents that existential threshold: if P falls below m, the population can no longer recover and will die out. ▲

Note Certain small populations (such as a tumour) will grow if they start from a single cell. Thus, both logistic and modified logistic (Allee) models are relevant for studying population growth.

Both the concentration of a substance in a fluid and the temperature change continuously in time, making them appropriate quantities to model with continuous functions. The rules describing the movement of fluids and heat can be described compactly and intuitively with autonomous differential equations. We explore these two phenomena in the following examples.

Example 8.1.4 Newton's Law of Cooling

The basic idea of the movement (transfer) of heat is expressed by **Newton's law of cooling,** which states that

> the rate at which an object loses or gains heat is proportional to the difference between the temperature of the object and the ambient temperature

Denote the temperature of an object by $T(t)$ and the fixed ambient temperature by A. Newton's law of cooling can be written as the differential equation

$$\frac{dT}{dt} = \alpha(A - T) \tag{8.1.5}$$

where α is a positive constant with dimensions of 1/time. This is an **autonomous differential equation** because it describes a *rule* and because the rate of change is a function of the state variable, T, and not of the time, t. Remember that T is a function of time but that α and A are constants.

If the temperature of the object is higher than the ambient temperature ($T > A$), the rate of change of temperature is negative and the object cools. If the temperature of the object is lower than the ambient temperature ($T < A$), the rate of change of temperature is positive and the object warms up (Figure 8.1.5). If the temperature of the object is equal to the ambient temperature, the rate of change of temperature is zero, and the temperature of the object does not change.

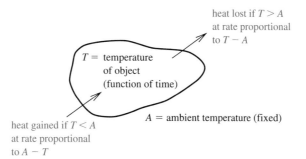

FIGURE 8.1.5
Newton's law of cooling

The parameter α, with dimensions of 1/time, depends on the **specific heat** and shape of the material. Materials with high specific heat, such as water, retain heat for a long time, have low values of α, and experience small rates of heat change for a given temperature difference. Materials with low specific heat, such as metals, lose heat rapidly, have large values of α, and experience high rates of heat change for a given temperature difference. The parameter α also depends on the object's ratio of surface area to volume. An object with a large exposed surface relative to its volume will heat or cool rapidly, as a shallow puddle freezes quickly on a cold evening. An object with a relatively small surface area heats or cools more slowly.

As with all equations we discuss in this section, we will solve Newton's law of cooling explicitly in Section 8.4. ▲

Example 8.1.5 Diffusion across a Membrane

The processes of heat exchange and chemical exchange have many parallels. The underlying mechanism is that of **diffusion.** Substances leave the cell at a rate proportional to their concentration inside and enter the cell at a rate proportional to their concentration outside. The constants of proportionality depend on the properties of the substance and on the properties of the membrane separating the two regions (Figure 8.1.6).

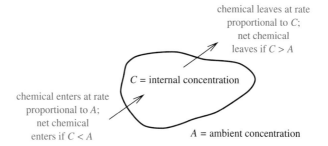

FIGURE 8.1.6
Diffusion between two regions

Denote the concentration inside the cell by $C = C(t)$, the concentration outside by A, and the constant of proportionality by β. The rate at which the chemical leaves the cell is βC, and the rate at which it enters is βA. It follows that

$$\frac{dC}{dt} = \text{rate at which chemical enters} - \text{rate at which chemical leaves}$$
$$= \beta A - \beta C = \beta(A - C) \tag{8.1.6}$$

This is exactly the same as Newton's law of cooling! The rate of change of concentration is proportional to the difference in concentration inside and outside the cell. ▲

Because this model is identical to Newton's law of cooling, their solutions must have the same form (and predict the same behaviour). The concentration of chemical will get closer and closer to the ambient concentration, just as it did with the discrete-time dynamical system (see Exercises 41–44 in Section 8.4).

So Newton's law of cooling and diffusion are represented by the same equation. Furthermore, that equation appears in other domains, with the same mathematics but a different interpretation. One of the simplest models used to describe fish growth claims that the length, $L = L(t)$, of a fish at age t changes according to

$$\frac{dL}{dt} = k(A - L)$$

where A is the maximum length the fish could reach. (This model is called the **von Bertalanffy limited growth model**.) See Examples 7.5.8 and 7.5.9, Exercise 29 in this section, and Exercises 35–36 in Section 8.4.

Example 8.1.6   A Continuous-Time Model for Selection

Suppose we have two strains of bacteria, with population sizes $a = a(t)$ and $b = b(t)$, with rate of change proportional to population size. If the per capita production rate of strain a is μ and that of b is λ, they follow the equations

$$\frac{da}{dt} = \mu a$$

$$\frac{db}{dt} = \lambda b$$

If $\mu > \lambda$, type a has a higher per capita production rate than b, and we expect it to take over the population. However, we can often measure only the *fraction* of bacteria of type a and not the total number. We would like to write a differential equation for the fraction, and we can do so by following steps much like those used to find the discrete-time dynamical system for competing bacteria (Section 3.4).

The fraction, p, of type a is

$$p = \frac{a}{a+b}$$

and the fraction of type b is

$$\frac{b}{a+b} = 1 - p$$

We compute the derivative of p:

$$\frac{dp}{dt} = \frac{d}{dt}\left(\frac{a}{a+b}\right)$$

$$= \frac{(a+b)\frac{da}{dt} - a\frac{d(a+b)}{dt}}{(a+b)^2}$$

$$= \frac{a\frac{da}{dt} + b\frac{da}{dt} - a\frac{da}{dt} - a\frac{db}{dt}}{(a+b)^2}$$

$$= \frac{b\frac{da}{dt} - a\frac{db}{dt}}{(a+b)^2}$$

$$= \frac{\mu ab - \lambda ab}{(a+b)^2}$$

$$= \frac{(\mu - \lambda)ab}{(a+b)^2}$$

We are not yet finished, however. To write this as an autonomous differential equation, we must express the derivative as a function of p, not of the unmeasurable total populations a and b. We write

$$\frac{dp}{dt} = \frac{(\mu - \lambda)ab}{(a+b)^2}$$

$$\frac{dp}{dt} = (\mu - \lambda)\left(\frac{a}{a+b}\right)\left(\frac{b}{a+b}\right)$$

$$= (\mu - \lambda)p(1 - p)$$

Note that this is a logistic differential equation with $k = \mu - \lambda$ and $L = 1$ (this time, however, k can be negative).

Although our derivation was similar, this differential equation looks nothing like the discrete-time dynamical system for this system,

$$p_{t+1} = \frac{sp_t}{sp_t + r(1 - p_t)}$$

(Equation 3.4.7). For instance, the parameters in the two equations have different meanings. The parameters s and r in the discrete-time dynamical system represent the *per capita production*, the number of bacteria produced per bacterium in one generation, whereas the parameters μ and λ in the differential equation represent the *per capita production rates*, the rates at which one bacterium produces new bacteria.

The differential equation

$$\frac{dp}{dt} = (\mu - \lambda)p(1 - p)$$

is **nonlinear** because the state variable p appears (on the right side) inside a nonlinear function, in this case a quadratic.

We begin by solving the differential equations for a and b. Suppose the initial number of type a bacteria is a_0 and the initial number of type b is b_0 (even though these values cannot be measured). The solutions are

$$a(t) = a_0 e^{\mu t}$$
$$b(t) = b_0 e^{\lambda t}$$

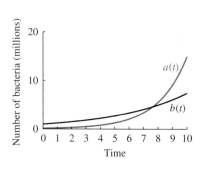

FIGURE 8.1.7

Exponential growth of two bacterial populations: $a_0 = 0.1$, $b_0 = 1$, $\mu = 0.5$, $\lambda = 0.2$

(Figure 8.1.7).

We can now find the equation for the fraction p as a function of t as

$$p(t) = \frac{a(t)}{a(t) + b(t)}$$

$$= \frac{a_0 e^{\mu t}}{a_0 e^{\mu t} + b_0 e^{\lambda t}}$$

The initial conditions, a_0 and b_0, cannot be measured; we can measure only the initial fraction, $p(0) = p_0$. However, we do know that

$$p_0 = \frac{a_0}{a_0 + b_0}$$

Dividing the numerator and denominator of our expression for $p(t)$ by $a_0 + b_0$ yields

$$p(t) = \frac{\frac{a_0}{a_0 + b_0} e^{\mu t}}{\frac{a_0}{a_0 + b_0} e^{\mu t} + \frac{b_0}{a_0 + b_0} e^{\lambda t}}$$

$$= \frac{p_0 e^{\mu t}}{p_0 e^{\mu t} + (1 - p_0) e^{\lambda t}} \quad (8.1.7)$$

Example 8.1.7 Behaviour of the Solution of the Selection Model

Suppose we use the parameter values $\mu = 0.5$ and $\lambda = 0.2$ and the initial condition

$$p_0 = \frac{a_0}{a_0 + b_0} = \frac{0.1}{0.1 + 1} \approx 0.091$$

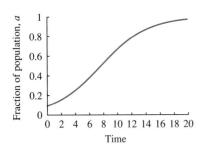

Figure 8.1.8
Solution for the fraction of type a bacteria

(as in Figure 8.1.7). The solution is

$$p(t) = \frac{0.091 e^{0.5t}}{0.091 e^{0.5t} + (1 - 0.091) e^{0.2t}}$$

$$= \frac{0.091 e^{0.5t}}{0.091 e^{0.5t} + 0.909 e^{0.2t}}$$

Because $0.5 > 0.2$, the leading behaviour of the denominator for large values of t is the term $0.091 e^{0.5t}$ (Section 6.4). Therefore, for large t

$$p(t) \approx \frac{0.091 e^{0.5t}}{0.091 e^{0.5t}} = 1$$

The fraction of type a approaches 1 as t approaches infinity (Figure 8.1.8).

Summary The rate of change of an **autonomous differential equation** depends on the state variable and not on the time. The rate of change of a **nonautonomous differential equation** depends on both the state variable and the time. We have introduced autonomous differential equations describing population dynamics, cooling, diffusion, and selection, finding that some processes are governed by the same equation. The equation for selection, derived by returning to the underlying model, is a **nonlinear** equation, but it can be solved by combining the solutions for each separate population.

8.1 Exercises

Mathematical Techniques

1–4 ■ Identify each of the following as a pure-time, autonomous, or nonautonomous differential equation. In each case, identify the state variable.

1. $\dfrac{dF}{dt} = F^2 + kt$

2. $\dfrac{dx}{dt} = \dfrac{x^2}{x - \lambda}$

3. $\dfrac{dy}{dt} = \mu e^{-t} - 1$

4. $\dfrac{dm}{dt} = \dfrac{e^{\alpha m} m^2}{mt - \lambda}$

5–8 ■ For the given time, value of the state variable, and values of the parameters, indicate whether the state variable is increasing, decreasing, or remaining unchanged.

5. $t = 0$, $F = 1$, and $k = 1$ in the differential equation in Exercise 1.

6. $t = 0$, $x = 1$, and $\lambda = 2$ in the differential equation in Exercise 2.

7. $t = 1$, $y = 1$, and $\mu = 2$ in the differential equation in Exercise 3.

8. $t = 2$, $m = 0$, $\alpha = 2$, and $\lambda = 1$ in the differential equation in Exercise 4.

9–16 ■ Describe each situation using a differential equation. In each case:

a. Identify the dependent variable (unknown function) and the independent variable. Pick the symbols for both variables, and identify their units (if possible).

b. Write the differential equation and the initial condition (i.e., describe the given event as an initial value problem).

c. Identify the differential equation as pure-time, autonomous, or nonautonomous.

Note: Some differential equations will involve a constant of proportionality whose value cannot be determined from the given information.

9. A population of amoebas starts with 120 amoebas and grows with a constant per capita reproduction rate of 1.3 per member per day.

10. The relative rate of change of the population of wild foxes in an ecosystem is 0.75 baby foxes per fox per month. Initially, the population is 74,000.

11. Experimental evidence suggests that—assuming constant temperature—the rate of change of the atmospheric pressure with respect to the altitude (height above sea level) is proportional to the pressure. The pressure at sea level is p_0 atmospheres (atm).

12. The rate of change of the thickness of the ice on a lake is inversely proportional to the square root of the thickness. Initially, the ice is 3 mm thick.

13. The population of an isolated island in the Pacific Ocean is 7500. Initially, 13 people are infected with a flu virus. The rate of change of the number of infected people is proportional to the product of the number of people who are infected and the number of people who are not yet infected.

14. Volcanic mud, initially at 125°C, cools in contact with the surrounding air, whose temperature is 10°C. The rate of change of the temperature of the mud is proportional to the difference between the temperature of the mud and the temperature of the surrounding air.

15. The rate of change of the biomass of plankton is proportional to the fourth root of the biomass. (Biomass is the total mass of living organisms in a given area at a given time.) The initial biomass of plankton is estimated to be 3 million tonnes.

16. The volume of the brain of an adult monkey is V cubic centimetres, and the volume of the brain of a newborn monkey is V_0 cubic centimetres. The rate of change in the volume of a growing monkey's brain is proportional to the difference between the volume of the brain of an adult monkey and the volume of the growing monkey's brain.

17–22 ■ The following exercises compare the behaviour of two similar-looking differential equations: the pure-time differential equation $\frac{dp}{dt} = t$ and the autonomous differential equation $\frac{db}{dt} = b$.

17. Use integration to solve the pure-time differential equation starting from the initial condition $p(0) = 1$, find $p(1)$, and sketch the solution.

18. Solve the pure-time differential equation starting from the initial condition $p(1) = 1$, find $p(2)$, and add the curve to your graph from Exercise 17.

19. Check that the solution of the autonomous differential equation starting from the initial condition $b(0) = 1$ is $b(t) = e^t$. Find $b(1)$ and sketch the solution.

20. Check that the solution of the autonomous differential equation starting from the initial condition $b(1) = 1$ is $b(t) = e^{t-1}$. Find $b(2)$ and add it to the sketch of the solution.

21. Exercises 17 and 18 give the value of p one time unit after it took on the value 1. Why don't the two answers match? (This behaviour is typical of pure-time differential equations.)

22. Exercises 19 and 20 give the value of b one time unit after it took on the value 1. Why do the two answers match? (This behaviour is typical of autonomous differential equations.)

23–26 ■ Check the solution of each autonomous differential equation, making sure that it also matches the initial condition.

23. Check that $x(t) = -\frac{1}{2} + \frac{3}{2}e^{2t}$ is the solution of the differential equation $\frac{dx}{dt} = 1 + 2x$ with initial condition $x(0) = 1$.

24. Check that $b(t) = 10e^{3t}$ is the solution of the differential equation $\frac{db}{dt} = 3b$ with initial condition $b(0) = 10$.

25. Check that $G(t) = 1 + e^t$ is the solution of the differential equation $\frac{dG}{dt} = G - 1$ with initial condition $G(0) = 2$.

26. Check that $z(t) = 1 + \sqrt{1 + 2t}$ is the solution of the differential equation $\frac{dz}{dt} = \frac{1}{z - 1}$ with initial condition $z(0) = 2$.

27. Consider the logistic equation

$$P' = 0.13P\left(1 - \frac{P}{1200}\right)$$

a. For which values of P is the population increasing?

b. For which values of P is the population decreasing?

c. Check that the constant function $P(t) = 1200$ is a solution of the equation. What is special about it?

d. Make a rough sketch of the solutions with initial populations $P(0) = 300$ and $P(0) = 1800$.

28. Consider the modified logistic equation

$$P' = 0.09P\left(1 - \frac{P}{2000}\right)\left(1 - \frac{120}{P}\right)$$

In the same coordinate system, sketch what the solutions with initial conditions $P(0) = 80$, $P(0) = 600$, and $P(0) = 2350$ might look like. To your graph, add exact solutions with initial conditions $P(0) = 120$ and $P(0) = 2000$. What do the latter two solutions look like? Why?

29. Consider von Bertalanffy's limited growth model

$$\frac{dL}{dt} = k(A - L)$$

introduced after Example 8.1.5. Explain why the name of the model includes the words *limited* and *growth*. Show that the solution, L, must be concave down if $L < A$.

30–33 ▪ Use Euler's method to estimate the solution of the differential equation at the given time, and compare with the value given by the exact solution. Sketch the graph of the solution along with the lines predicted by Euler's method.

30. Estimate $x(2)$ if x obeys the differential equation $\frac{dx}{dt} = 1 + 2x$ with initial condition $x(0) = 1$. Use Euler's method with $\Delta t = 1$ for two steps. Compare with the exact answer in Exercise 23.

31. Estimate $b(1)$ if b obeys the differential equation $\frac{db}{dt} = 3b$ with initial condition $b(0) = 10$. Use Euler's method with $\Delta t = 0.5$ for two steps. Compare with the exact answer in Exercise 24.

32. Estimate $G(1)$ if G obeys the differential equation $\frac{dG}{dt} = G - 1$ with initial condition $G(0) = 2$. Use Euler's method with $\Delta t = 0.2$ for five steps. Compare with the exact answer in Exercise 25.

33. Estimate $z(4)$ if z obeys the differential equation $\frac{dz}{dt} = \frac{1}{z-1}$ with initial condition $z(0) = 2$. Use Euler's method with $\Delta t = 1$ for four steps. Compare with the exact answer in Exercise 26.

34–37 ▪ The derivation of the differential equation for p in the text requires combining two differential equations for a and b. Often, one can find a differential equation for a new variable derived from a single equation. In the following cases, use the chain rule to derive a new differential equation.

34. Suppose $\frac{dx}{dt} = 2x - 1$. Set $y = 2x - 1$ and find a differential equation for y. The end result should be simpler than the original equation.

35. Suppose $\frac{db}{dt} = 4b + 2$. Set $z = 4b + 2$ and find a differential equation for z. The end result should be simpler than the original equation.

36. Suppose $\frac{dx}{dt} = x + x^2$. Set $y = \frac{1}{x}$ and find a differential equation for y. This transformation changes a nonlinear differential equation for x into a linear differential equation for y (this is called a Bernoulli differential equation).

37. Suppose $\frac{dx}{dt} = 2x + \frac{1}{x}$. Set $y = x^2$ and find a differential equation for y. This is another example of a Bernoulli differential equation.

Applications

38–41 ▪ The simple model of bacterial growth assumes that the per capita production rate does not depend on population size. The following problems help you derive models of the form

$$\frac{db}{dt} = \lambda(b)b$$

where the per capita production rate, λ, is a function of the population size, b.

38. One widely used nonlinear model of competition is the logistic model, where the per capita production rate is a linearly decreasing function of population size. Suppose that the per capita production rate has a maximum of $\lambda(0) = 1$ and that it decreases with a slope of -0.002. Find $\lambda(b)$ and the differential equation for b. Is $b(t)$ increasing when $b = 10$? Is $b(t)$ increasing when $b = 1000$?

39. Suppose that the per capita production rate decreases linearly from a maximum of $\lambda(0) = 4$ with slope -0.001. Find $\lambda(b)$ and the differential equation for b. Is $b(t)$ increasing when $b = 1000$? Is $b(t)$ increasing when $b = 5000$?

40. In some circumstances, individuals reproduce better when the population size is large and fail to reproduce when the population size is small (the Allee effect). Suppose that the per capita production rate is an increasing linear function with $\lambda(0) = -2$ and a slope of 0.01. Find $\lambda(b)$ and the differential equation for b. Is $b(t)$ increasing when $b = 100$? Is $b(t)$ increasing when $b = 300$?

41. Suppose that the per capita production rate increases linearly with $\lambda(0) = -5$ and a slope of 0.001. Find $\lambda(b)$ and the differential equation for b. Is $b(t)$ increasing when $b = 1000$? Is $b(t)$ increasing when $b = 3000$?

42–47 ▪ The derivation of the movement of chemical assumed that chemical moved as easily into the cell as out of it. If the membrane can act as a filter, the rates at which chemical enters and leaves might differ or might depend on the concentration itself. In each of the following cases, draw a diagram illustrating the situation, and write the associated differential equation. Let C be the concentration inside the cell, A the concentration outside, and β the constant of proportionality relating the concentration and the rate.

42. Suppose that no chemical re-enters the cell. This should look like the differential equation for a population. What would happen to the population?

43. Suppose that no chemical leaves the cell. What would happen to the concentration?

44. Suppose that the constant of proportionality governing the rate at which chemical enters the cell is three times as large as the constant governing the rate at which it leaves. Would the concentration inside the cell be increasing or decreasing if $C = A$? What would this mean for the cell?

45. Suppose that the constant of proportionality governing the rate at which chemical enters the cell is half as large as the constant governing the rate at which it leaves. Would the concentration inside the cell be increasing or decreasing if $C = A$? What would this mean for the cell?

46. Suppose that the constant of proportionality governing the rate at which chemical enters the cell is proportional to $1 + C$ (because the chemical helps to open special channels). Would the concentration inside the cell be increasing or decreasing if $C = A$? What would this mean for the cell?

47. Suppose that the constant of proportionality governing the rate at which chemical enters the cell is proportional to $1 - C$ (because the chemical helps to close special channels). Would the concentration inside the cell be increasing or decreasing if $C = A$? What would this mean for the cell?

48–49 ■ The model of selection includes no interaction between bacterial types a and b (the per capita production rate of each type is a constant). Write a pair of differential equations for a and b with the following forms for the per capita production rate, and derive an equation for the fraction, p, of type a. Assume that the basic per capita production rate for type a is $\mu = 2$, and that for type b is $\lambda = 1.5$.

48. The per capita production rate of each type is reduced by a factor of $1 - p$ (so that the per capita production rate of type a is $2(1 - p)$). This is a case where a large proportion of type a reduces the production of both types. Will type a take over?

49. The per capita production rate of type a is reduced by a factor of $1 - p$, and the per capita production rate of type b is reduced by a factor of p. This is a case where a large proportion of type a reduces the production rate of type a, and a large proportion of type b reduces the production of type b. Do you think that type a will still take over?

50–51 ■ We will find later (with separation of variables) that the solution for Newton's law of cooling with initial condition $T(0)$ is

$$T(t) = A + (T(0) - A)e^{-\alpha t}$$

For each set of given parameter values:

a. Write and check the solution.
b. Find the temperature at $t = 1$ and $t = 2$.
c. Sketch the graph of your solution. What happens as t approaches infinity?

50. Set $\alpha = 0.2$/min, $A = 10°C$, and $T(0) = 40°C$.

51. Set $\alpha = 0.02$/min, $A = 30°C$, and $T(0) = 40°C$.

52–53 ■ Use Euler's method to estimate the temperature for the following cases of Newton's law of cooling. Compare with the exact answer.

52. $\alpha = 0.2$/min and $A = 10°C$, and $T(0) = 40°C$. Estimate $T(1)$ and $T(2)$ using $\Delta t = 1$. Compare with Exercise 50.

53. $\alpha = 0.02$/min and $A = 30°C$, and $T(0) = 40°C$. Estimate $T(1)$ and $T(2)$ using $\Delta t = 1$. Compare with Exercise 51a. Why is the result so close?

54–55 ■ Use the solution for Newton's law of cooling (Exercise 50 and 51) to find the solution expressing the concentration of chemical inside a cell as a function of time in the following examples. Find the concentration after 10, 20, and 60 s. Sketch your solutions for the first minute. (Recall that β has the same meaning as α in the Newton's law of cooling model.)

54. $\beta = 0.01$/s, $C(0) = 5$ mmol/cm^3, and $A = 2$ mmol/cm^3.

55. $\beta = 0.1$/s, $C(0) = 5$ mmol/cm^3, and $A = 2$ mmol/cm^3.

56–59 ■ Recall that the solution of the discrete-time dynamical system $b_{t+1} = rb_t$ is $b_t = r^t b_0$. This is closely related to the differential equation $\dfrac{db}{dt} = \lambda b$.

56. For what values of b_0 and r does this solution match $b(t) = 10^6 e^{2t}$ (the solution of the differential equation with $\lambda = 2$ and $b(0) = 10^6$) for all values of t?

57. For what values of b_0 and r does this solution match $b(t) = 100e^{-3t}$ (the solution of the differential equation with $\lambda = -3$ and $b(0) = 100$) for all values of t?

58. For what values of λ do solutions of the differential equation grow? For what values of r do solutions of the discrete-time dynamical system grow?

59. What is the relation between r and λ? That is, what value of r produces the same growth as a given value of λ?

60–61 ■ Use Euler's method to estimate the value of $p(t)$ from the selection differential equation for the given parameter values. Compare with the exact answer using the equation for the solution. Graph the solution, including the estimates from Euler's method.

60. Suppose $\mu = 2$, $\lambda = 1$, and $p(0) = 0.1$. Estimate the proportion after 2 min using a time step of $\Delta t = 0.5$.

61. Suppose $\mu = 2.5$, $\lambda = 3$, and $p(0) = 0.6$. Estimate the proportion after 1 min using a time step of $\Delta t = 0.25$.

62–63 ■ The rate of change in the differential equation

$$\frac{dp}{dt} = (\mu - \lambda)p(1 - p)$$

is a measure of the strength of selection. For each value of λ, graph the rate of change as a function of p with $\mu = 2$. For what value of p is the rate of change greatest?

62. $\lambda = 1$

63. $\lambda = 0.1$

Computer Exercises

64. Suppose that the ambient temperature oscillates with period a according to

$$A(t) = 20 + \cos\left(\frac{2\pi t}{a}\right)$$

The differential equation is

$$\frac{dT}{dt} = \alpha(A(t) - T)$$

which has a solution with $A(0) = 20$ of

$$T(t) = 20$$
$$+ \frac{\alpha^2 a^2 \cos\left(\frac{2\pi t}{a}\right) + 2\alpha\pi a \sin\left(\frac{2\pi t}{a}\right) - \alpha^2 a^2 e^{-\alpha t}}{\alpha^2 a^2 + 4\pi^2}$$

a. Use a computer algebra system to check that this answer works.

b. Plot $T(t)$ for five periods using values of a ranging from 0.1 to 10 when $\alpha = 1$. Compare $T(t)$ with $A(t)$. When does the temperature of the object most closely track the ambient temperature?

65. Apply Euler's method to solve the differential equation

$$\frac{db}{dt} = b + t$$

with the initial condition $b(0) = 1$. Compare with the sum of the results of

$$\frac{db_1}{dt} = b_1$$

and

$$\frac{db_2}{dt} = t$$

Do you think there is a sum rule for differential equations? Compare with Section 7.1, Exercise 54. If the computer has a method for solving differential equations, find the solution and compare with your approximate solution from Euler's method.

8.2 Equilibria and Display of Autonomous Differential Equations

Euler's method provides an algorithm for computing approximate solutions of autonomous differential equations. As with discrete-time dynamical systems, much can be deduced about the dynamics by algebraically solving for **equilibria** and plotting them on a graphical summary of the dynamics, called a **phase-line diagram.**

Equilibria

One of the most important tools for analyzing discrete-time dynamical systems is the equilibrium, a state of the system that remains unchanged by the dynamical system. Solutions starting from an equilibrium do not change over time. For example, solutions of the bacterial selection model (Equation 3.4.5) approach an equilibrium at $p = 1$ when the mutants are superior to the wild type whether the model is a discrete-time dynamical system (Figure 8.2.9a) or an autonomous differential equation (Figure 8.2.9b; see also Figure 8.1.8). In either case, if we started with all mutant bacteria, the population would never change.

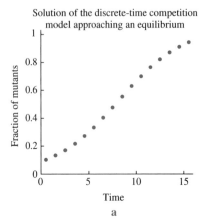
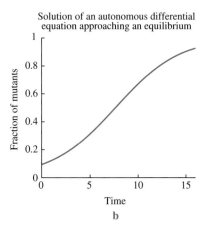

FIGURE 8.2.9
Solutions and equilibria: discrete-time dynamical systems and differential equations

8.2 Equilibria and Display of Autonomous Differential Equations

Equilibria are as useful in analyzing autonomous differential equations as they are in analyzing discrete-time dynamical systems. As before, an equilibrium is a value of the state variable that does not change.

Definition 8.2.1 A value, m^*, of the state variable is called an **equilibrium** of the autonomous differential equation

$$\frac{dm}{dt} = f(m)$$

if

$$f(m^*) = 0$$

At the value where $f(m^*) = 0$, the rate of change is zero. A rate of change of zero indicates precisely that the value of the measurement remains the same.

The steps for finding equilibria closely resemble those for finding equilibria of a discrete-time dynamical system.

▶▶ Algorithm 8.2.1 Finding Equilibria of an Autonomous Differential Equation

1. Make sure that the differential equation is autonomous.
2. Write the equation for the equilibria and solve it.
3. Think about the results. Do they make sense in the context of the given biological (or other) realization?

Example 8.2.1 The Equilibria for the Bacterial Population Model

The simplest autonomous differential equation we have considered is the equation

$$\frac{db}{dt} = \lambda b$$

where $b = b(t)$ is the number of bacteria and λ is a constant. What are the equilibria of this differential equation? The rate of change depends only on b and not on the time, t, so this equation is autonomous. The algebraic steps are

$$\lambda b^* = 0$$
$$\lambda = 0 \quad \text{or} \quad b^* = 0$$

If $\lambda = 0$, every value of b is an equilibrium because the population does not change. This equilibrium corresponds to the case $r = 1$ of the related discrete-time dynamical system, $b_{t+1} = rb_t$. The other equilibrium at $b^* = 0$ indicates extinction, again matching the results for the discrete-time dynamical system. A solution starting from $b = 0$ remains there forever (Figure 8.2.10).

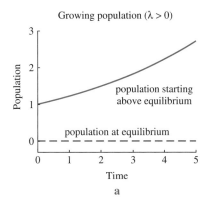
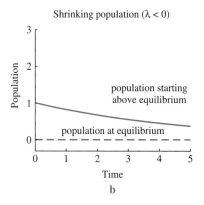

FIGURE 8.2.10
Solutions of the bacterial growth model

Example 8.2.2 The Equilibria for Newton's Law of Cooling

For Newton's law of cooling, the differential equation is

$$\frac{dT}{dt} = \alpha(A - T)$$

where $T = T(t)$ is the temperature of an object, A is the ambient temperature (assumed constant), and $\alpha \geq 0$. The rate of change depends only on T and not on the time, t, so this equation is autonomous. What are the equilibria? Let's think about it—how can it happen that an object placed in an environment does not experience any change in temperature? One possibility is obvious—if the temperature of the object is exactly the same as the ambient temperature, the object will not become warmer or colder. Are there any other possibilities?

To find the equilibria, T^*, algebraically, we write

$$\alpha(A - T^*) = 0$$
$$\alpha = 0 \quad \text{or} \quad A - T^* = 0$$
$$\alpha = 0 \quad \text{or} \quad T^* = A$$

The solutions are $\alpha = 0$ and $T^* = A$. If $\alpha = 0$, no heat is exchanged with the outside world. Such an object is always at equilibrium. The other equilibrium is $T^* = A$. The object is in equilibrium when its temperature is equal to the ambient temperature. A solution with initial condition $T(0) = A$ remains there forever.

Example 8.2.3 Equilibria of the Logistic and Modified Logistic Differential Equations

To find the equilibria of

$$P'(t) = kP(t)\left(1 - \frac{P(t)}{L}\right)$$

we solve

$$kP^*\left(1 - \frac{P^*}{L}\right) = 0$$
$$k = 0 \quad \text{or} \quad P^* = 0 \quad \text{or} \quad 1 - \frac{P^*}{L} = 0$$
$$k = 0 \quad \text{or} \quad P^* = 0 \quad \text{or} \quad P^* = L$$

If $k = 0$, then any constant solution is an equilibrium (not an interesting case). The case $P^* = 0$ states that a population of zero will never change (on its own). The equilibrium $P^* = L$ represents the carrying capacity (so a population starting at L will remain at L).

The modified logistic equation

$$P'(t) = kP(t)\left(1 - \frac{P(t)}{L}\right)\left(1 - \frac{m}{P(t)}\right)$$

that we discussed in Example 8.1.3 has an extra equilibrium, $P^* = m$, that corresponds to the minimum number of individuals needed for the population to survive. A population starting at $P(0) = m$ will remain there for all time.

Example 8.2.4 The Equilibria for the Bacterial Selection Model

We follow the same algorithm to find the equilibria of the selection model

$$\frac{dp}{dt} = (\mu - \lambda)p(1 - p)$$

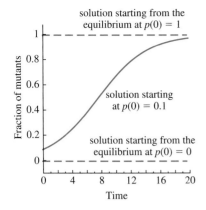

FIGURE 8.2.11

Three solutions of the bacterial selection equation

The equation is autonomous because the rate of change is not a function of t. The equation for the equilibria is

$$(\mu - \lambda)p^*(1 - p^*) = 0$$
$$\mu - \lambda = 0 \quad \text{or} \quad p^* = 0 \quad \text{or} \quad 1 - p^* = 0$$
$$\mu = \lambda \quad \text{or} \quad p^* = 0 \quad \text{or} \quad p^* = 1$$

If $\mu = \lambda$, the per capita production rates of the two types match and the fraction of mutants does not change. The other two equilibria correspond to extinction. If $p^* = 0$, there are no mutants, and if $p^* = 1$, there are no wild type. Solutions starting from either of these points remain there (Figure 8.2.11). Now assume that the initial condition satisfies $0 < p(0) < 1$. If $\mu > \lambda$, then the solution starting at $p(0)$ approaches the equilibrium $p^* = 1$ (see Example 8.1.7, where we took $p(0) = 0.1$, and also Figure 8.2.11). If $\mu < \lambda$, any solution starting at $p(0)$ decreases toward the equilibrium $p^* = 0$. Although the differential equation describing these dynamics looks very different from the discrete-time dynamical system model describing the same process (Equation 3.4.5), the equilibria match.

Graphical Display of Autonomous Differential Equations

The graphical technique of cobwebbing helped us sketch solutions of discrete-time dynamical systems. A similarly useful graphical way to display autonomous differential equations is called the **phase-line diagram.** The diagram summarizes where the state variable is increasing, where it is decreasing, and where it is unchanged.

According to the interpretation of the derivative, the state variable is increasing when the rate of change is positive, decreasing when the rate of change is negative, and unchanging when the rate of change is zero. By figuring out where the rate of change is positive, where it is negative, and where it is zero, we can get a good idea of how solutions behave; recall that we have already used this strategy to study some differential equations in Section 8.1.

The rate of change for Newton's law of cooling model

$$T' = \alpha(A - T)$$

(where $\alpha > 0$) is graphed as a function of the state variable T. Its graph is a line of slope $-\alpha$ (top diagram in Figure 8.2.12). The rate of change is positive when $T < A$, zero at $T = A$, and negative when $T > A$. The temperature is therefore increasing when $T < A$, constant at the equilibrium $T = A$, and decreasing when $T > A$.

This description can be translated into a one-dimensional drawing of the dynamics, the phase-line diagram, as shown at the bottom of Figure 8.2.12. The line represents the state variable, in this case the temperature, sometimes referred to as the **phase** of the system. To construct the diagram, draw rightward-pointing arrows at points where the temperature is increasing, leftward-pointing arrows at points where the temperature is decreasing, and big dots where the temperature is fixed. The directions of the arrows come from the values of the rate of change. When the rate of change is positive, the temperature is increasing and the arrow points to the right.

Solutions follow the arrows. Starting below the equilibrium, the arrows push the temperature up toward the equilibrium. Starting above the equilibrium, the arrows push the temperature down toward the equilibrium.

More information can be encoded on the diagram by making the size of the arrows correspond to the magnitude of the rate of change. In Figure 8.2.12, the arrows are larger when T is farther from the equilibrium where the absolute value of the rate of change of temperature is larger. The solutions (Figure 8.2.13 shows the case $A = 20$) change rapidly far from the equilibrium, A, and more slowly near A.

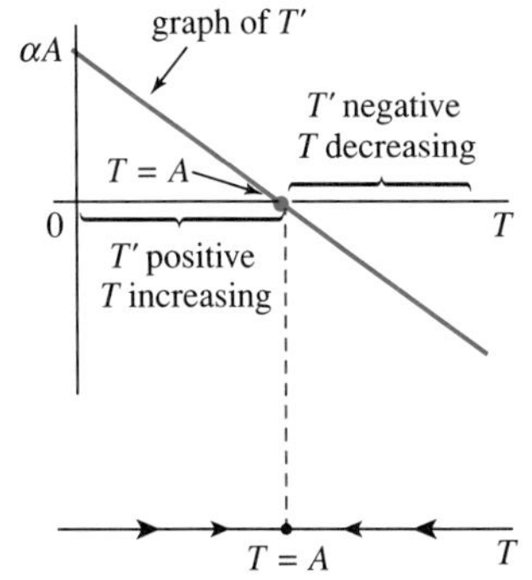

FIGURE 8.2.12

The rate of change and phase-line diagram for Newton's law of cooling

FIGURE 8.2.13

Two solutions of Newton's law of cooling

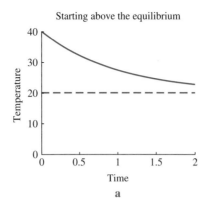
Starting above the equilibrium

a

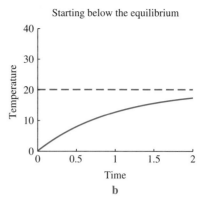
Starting below the equilibrium

b

Example 8.2.5 Phase-Line Diagram

Draw the phase-line diagram for the autonomous differential equation

$$y' = (y-2)^2 - 1$$

where y is a function of t.

The equilibria of the differential equation are the solutions of the equation

$$(y^* - 2)^2 - 1 = 0$$

i.e., $y^* = 1$ and $y^* = 3$. The graph of the rate of change, $y' = (y-2)^2 - 1$, is the parabola shown in Figure 8.2.14.

When $y < 1$, the graph of y' is positive, and thus y is increasing. This fact is indicated using rightward-pointing arrows in the phase-line diagram. The rate of change when $y = a$ is larger than when $y = b$, which means that the function y increases faster when it reaches the value a than when it reaches the value b. Thus, the arrow at a is larger than the arrow at b.

When the state variable (or the phase) y is between 1 and 3, the rates of change y' are negative. Thus, y decreases and the arrows in the phase-line diagram are drawn pointing to the left. The magnitude of the rate of change is larger when $y = c$ than when $y = d$ (the function y decreases more rapidly when $y = c$ than when $y = d$), which is why the arrow at c is larger than the arrow at d.

When $y > 3$, y' is positive and increasing; thus, y increases more and more rapidly (y is concave up!). This fact is indicated in the phase-line diagram by drawing rightward-pointing arrows whose size increases as y increases.

FIGURE 8.2.14

Phase-line diagram

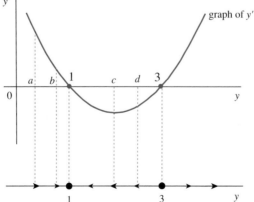

Example 8.2.6 Phase-Line Diagram for the Selection Model: $\mu > \lambda$

The rate of change of the selection model,

$$\frac{dp}{dt} = (\mu - \lambda)p(1-p)$$

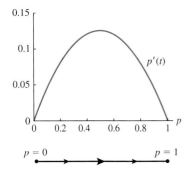

FIGURE 8.2.15
The rate of change and phase-line diagram for the selection model

as a function of the state variable p in the case $\mu > \lambda$ is positive except at the equilibria (Figure 8.2.15). All the arrows on the phase-line diagram point to the right, pushing solutions toward the equilibrium at $p = 1$. The rate of change takes on its maximum value at $p = 0.5$, which means that the largest arrow is at $p = 0.5$ (i.e, at the vertex of the parabola). A solution starting near $p = 0$ begins by increasing slowly, increases faster as it passes $p = 0.5$, and then slows down as it approaches $p = 1$ (Figure 8.2.11).

The phase-line diagram for the selection model (Figure 8.2.15) has two different kinds of equilibria. The arrows push the solution away from the equilibrium at $p = 0$ and toward the equilibrium at $p = 1$. We propose an informal definition of stable and unstable equilibria similar to that for discrete-time dynamical systems.

Definition 8.2.2 An equilibrium of an autonomous differential equation is **stable** if solutions that begin near the equilibrium stay near or approach the equilibrium. An equilibrium of an autonomous differential equation is **unstable** if solutions that begin near the equilibrium move away from the equilibrium.

In the theory of autonomous differential equations, the stability as we just defined it is called **asymptotic stability.** The adjective *asymptotic* refers to the long-term behaviour of the solution (i.e., as the time increases toward infinity). We study stability in the next section.

Summary **Equilibria** of an autonomous differential equation are found by setting the rate of change equal to zero and solving for the state variable. By graphing the rate of change as a function of the state variable, we can identify values where the state variable is increasing or decreasing. This information can be translated into a **phase-line diagram.**

8.2 Exercises

Mathematical Techniques

▼1–4 ▪ Find the equilibria of each autonomous differential equation.

1. $\frac{dx}{dt} = 1 - x^2$

2. $\frac{dx}{dt} = 1 - e^x$

3. $\frac{dy}{dt} = y \cos y$

4. $\frac{dz}{dt} = \frac{1}{z} - 3$

▼5–8 ▪ Find the equilibria of each autonomous differential equation with parameters.

5. $\frac{dx}{dt} = 1 - ax$

6. $\frac{dx}{dt} = cx + x^2$

7. $\frac{dW}{dt} = \alpha e^{\beta W} - 1$

8. $\frac{dy}{dt} = ye^{-\beta y} - ay$

▼9–10 ▪ From each graph of the rate of change as a function of the state variable, draw the phase-line diagram.

9.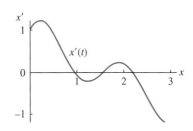

10.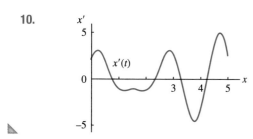

11–12 ▪ From each phase-line diagram, sketch the solution starting from the specified initial condition (at $x = 5$).

13–14 ▪ From the given phase-line diagram, sketch a possible graph of the rate of change of x as a function of x.

13. The phase line in Exercise 11.

14. The phase line in Exercise 12.

15–18 ▪ Graph the rate of change as a function of the state variable, and draw the phase-line diagram for each differential equation.

15. $\frac{dx}{dt} = 1 - x^2$. Graph for $-2 \leq x \leq 2$.

16. $\frac{dx}{dt} = 1 - e^x$. Graph for $-2 \leq x \leq 2$.

17. $\frac{dy}{dt} = y \cos y$. Graph for $-2 \leq y \leq 2$.

18. $\frac{dz}{dt} = \frac{1}{z} - 3$. Graph for $0 < z \leq 1$.

Applications

19–20 ▪ Suppose that a population is growing at constant per capita production rate, λ, but that individuals are harvested at a rate of h. The differential equation describing such a population is

$$\frac{db}{dt} = \lambda b - h$$

For each of the following values of λ and h, find the equilibrium, draw the phase-line diagram, and sketch one solution with initial condition below the equilibrium and another with initial condition above the equilibrium. Explain your result.

19. $\lambda = 2$, $h = 1000$

20. $\lambda = 0.5$, $h = 1000$

21–24 ▪ Find the equilibria, graph the rate of change, $\frac{db}{dt}$, as a function of b, and draw a phase-line diagram for the following models describing bacterial population growth.

21. $\frac{db}{dt} = (1 - 0.002b)b$. Check that your arrows are consistent with the behaviour of $b(t)$ at $b = 10$ and $b = 1000$.

22. $\frac{db}{dt} = (4 - 0.001b)b$. Check that your arrows are consistent with the behaviour of $b(t)$ at $b = 1000$ and $b = 5000$.

23. $\frac{db}{dt} = (-2 + 0.01b)b$. Check that your arrows are consistent with the behaviour of $b(t)$ at $b = 100$ and $b = 300$.

24. $\frac{db}{dt} = (-5 + 0.001b)b$. Check that your arrows are consistent with the behaviour of $b(t)$ at $b = 1000$ and $b = 3000$.

25–26 ▪ Find the equilibria; graph the rate of change $\frac{dC}{dt}$ as a function of C; and draw a phase-line diagram for the following models describing chemical diffusion.

25. $\frac{dC}{dt} = -\beta C + 3\beta A$. Check that the direction arrow is consistent with the behaviour of $C(t)$ at $C = A$.

26. $\frac{dC}{dt} = -\beta C + 0.5\beta A$. Check that the direction arrow is consistent with the behaviour of $C(t)$ at $C = A$.

27–28 ▪ Find the equilibria; graph the rate of change $\frac{dp}{dt}$ as a function of p; and draw a phase-line diagram for the following models describing selection.

27. $\frac{dp}{dt} = 0.5p(1 - p)^2$. What happens to a solution starting from a small, but positive, value of p?

28. $\frac{dp}{dt} = (2(1 - p) - 1.5p)p(1 - p)$. What happens to a solution starting from a small, but positive, value of p?

29–36 ▪ Find the equilibria and draw the phase-line diagram for the following differential equations, in addition to answering the questions.

29. Suppose the population size of some species of organism follows the model

$$\frac{dN}{dt} = \frac{3N^2}{2 + N^2} - N$$

where N is measured in hundreds. Why might this population behave as it does at small values? This is another example of the Allee effect discussed in Section 8.1.

30. Suppose the population size of some species of organism follows the model

$$\frac{dN}{dt} = \frac{5N^2}{1 + N^2} - 2N$$

where N is measured in hundreds. What is the critical value below which this population is doomed to extinction (as in Exercise 29)?

31. The drag on a falling object is proportional to the square of its speed. The differential equation is

$$\frac{dv}{dt} = a - Dv^2$$

where v is speed, a is acceleration, and D is drag. Suppose that $a = 9.8$ m/s^2 and that $D = 0.0032$ per metre (values for a falling skydiver). Check that the units in the differential equation are consistent. What does the equilibrium speed mean?

32. Consider the same situation as in Exercise 31 but for a skydiver diving head down with her arms against her sides and her toes pointed, thus minimizing drag. The drag is reduced to $D = 0.00048/m$. Find the equilibrium speed. How does it compare to the ordinary skydiver?

33. According to Torricelli's law of draining, the rate at which a fluid flows out of a cylinder through a hole at the bottom is proportional to the square root of the depth of the water. Let y represent the depth of the water in centimetres. The differential equation is

$$\frac{dy}{dt} = -c\sqrt{y}$$

where $c = 2\sqrt{cm}$/s. Show that the units are consistent. Use a phase-line diagram to sketch solutions starting from $y = 10$ and $y = 1$.

34. Write a differential equation describing the depth of water in a cylinder where water enters at a rate of 4 cm/s but drains out as in Exercise 33. Use a phase-line diagram to sketch solutions starting from $y = 10$ and $y = 1$.

35. One of the most important differential equations in chemistry uses the **Michaelis-Menten** or **Monod** equation. Suppose S is the concentration of a substrate that is being converted into a product. Then

$$\frac{dS}{dt} = -k_1 \frac{S}{k_2 + S}$$

describes how the substrate is used. Set $k_1 = k_2 = 1$. How does this equation differ from Torricelli's law of draining (Exercise 33)?

36. Write a differential equation describing the amount of substrate if substrate is added at rate R but is converted into product as in Exercise 35. Find the equilibrium. Draw the phase-line diagram and a representative solution with $R = 0.5$ and $R = 1.5$. Can you explain your results?

37–38 ▪ Small organisms such as bacteria take in food at rates proportional to their surface area but use energy at higher rates.

37. Suppose that energy is used at a rate proportional to the mass. In this case,

$$\frac{dV}{dt} = a_1 V^{2/3} - a_2 V$$

where V represents the volume in cubic centimetres and t is time measured in days. The first term says that surface area is proportional to volume to the 2/3 power. The constant a_1 gives the rate at which energy is taken in and has units of centimetres per day. a_2 is the rate at which energy is used and has units of 1/day. Check the units. Find the equilibrium. What happens to the equilibrium as a_1 becomes smaller? Does this make sense? What happens to the equilibrium as a_2 becomes smaller? Does this make sense?

38. Suppose that energy is used at a rate proportional to the mass to the 3/4 power (closer to what is observed). In this case,

$$\frac{dV}{dt} = a_1 V^{2/3} - a_2 V^{3/4}$$

Find the units of a_2 if V is measured in cubic centimetres and t is measured in days. (They should look rather strange.) Find the equilibrium. What happens to the equilibrium as a_1 becomes smaller? Does this make sense? What happens to the equilibrium as a_2 becomes smaller? Does this make sense?

Computer Exercises

39. In Section 8.1, Exercise 64, we considered the equation

$$\frac{dT}{dt} = \alpha(A(t) - T)$$

where

$$A(t) = 20 + \cos\left(\frac{2\pi t}{a}\right)$$

Using either the solution given in that problem or a computer system that can solve the equation, show that the solution always approaches the "equilibrium" at $T = A(t)$. Use the parameter value $\alpha = 1$ and values of a ranging from 0.1 to 10. For which value of a does the solution get closest to $A(t)$? Can you explain why?

8.3 Stability of Equilibria

Phase-line diagrams for autonomous differential equations have two types of equilibria: stable and unstable. Solutions starting near a stable equilibrium move closer to the equilibrium. The carrying capacity at $P = L$ in a logistic differential equation and the equilibrium at $p = 1$ of the selection equation with superior mutants (Figure 8.2.15) are stable in this sense. Solutions starting near an unstable equilibrium move farther from the equilibrium. The equilibria at $p = 0$ of the selection equation (Figure 8.2.15) and $P = m$ for the modified logistic differential equation are unstable. By reasoning about the graph of the rate of change, we can find an algebraic way to recognize stable and unstable equilibria.

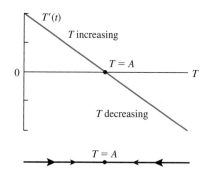

FIGURE 8.3.16
The rate of change and phase-line diagram for Newton's law of cooling revisited

Recognizing Stable and Unstable Equilibria

Consider again Newton's law of cooling

$$\frac{dT}{dt} = \alpha(A - T)$$

(Figure 8.3.16). Our phase-line diagram and our intuition suggest that the equilibrium $T = A$ is stable. Arrows to the left of the equilibrium point right, pushing solutions up toward the equilibrium (a cool object warms up). Arrows to the right of the equilibrium point left, pushing solutions down toward the equilibrium (a hot object cools off). The equilibrium must be stable.

The rate of change is positive to the left and negative to the right of a stable equilibrium. Therefore, the rate-of-change function must be **decreasing** at the equilibrium, implying that the **derivative** of the rate-of-change function is negative. An equilibrium is stable if the derivative of the rate of change *with respect to the state variable* is negative at the equilibrium. The right-hand equilibrium in Figure 8.3.17 satisfies this criterion. As well, the equilibrium $y^* = 1$ in Example 8.2.5 is stable.

Example 8.3.1 The Derivative of the Rate-of-Change Function for Newton's Law of Cooling

The function $\alpha(A - T)$ gives the rate of change of T in Newton's law of cooling. If T is slightly larger than the equilibrium value A, the rate of change is negative. If T is slightly smaller than A, the rate of change is positive. To check stability with the derivative criterion, we compute the derivative of the rate of change with respect to the state variable, T,

$$\frac{d}{dT}[\alpha(A - T)] = -\alpha < 0$$

The rate of change is a decreasing function at the equilibrium, and the equilibrium is therefore stable (Figure 8.3.16).

At an unstable equilibrium, the arrows push solutions away. Solutions starting above the equilibrium increase, and those starting below decrease. In terms of the rate of change, an equilibrium is unstable if the rate of change is negative below the equilibrium and positive above it. This implies that the rate-of-change function is increasing and that the derivative of the rate-of-change function is positive. An equilibrium is unstable if the derivative of the rate of change with respect to the state variable is positive at the equilibrium. The left-hand equilibrium in Figure 8.3.17 satisfies this criterion. As well, the equilibrium $y^* = 3$ in Example 8.2.5 is unstable.

Example 8.3.2 The Rate of Change for a Reverse Version of Newton's Law of Cooling

Imagine a reverse version of Newton's law of cooling given by

$$\frac{dT}{dt} = \alpha(T - A)$$

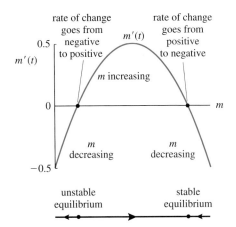

FIGURE 8.3.17

Behaviour at equilibria with positive and negative derivatives

with $\alpha > 0$. Objects cooler than A decrease in temperature, and objects warmer than A increase in temperature. The rate of change and the phase-line diagram are shown in Figure 8.3.18. The derivative of the rate of change with respect to T at $T = A$ is

$$\frac{d}{dT}[\alpha(T - A)] = \alpha > 0$$

The rate of change is an increasing function, and the equilibrium is unstable.

We summarize these results in the following powerful theorem.

Theorem 8.3.1 Stability Theorem for Autonomous Differential Equations

Suppose

$$\frac{dm}{dt} = f(m)$$

is an autonomous differential equation with an equilibrium at m^*. The equilibrium at m^* is stable if

$$f'(m^*) < 0$$

and unstable if

$$f'(m^*) > 0$$

Whenever $f'(m^*) = 0$ at an equilibrium m^*, the stability theorem does not apply, and we have to use other means; see Example 8.3.7.

An outline of the proof of Theorem 8.3.1 is given at the end of this section.

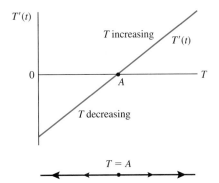

FIGURE 8.3.18

A reverse version of Newton's law of cooling

Applications of the Stability Theorem

We now apply the Stability Theorem for Autonomous Differential Equations to our models.

Example 8.3.3   Applying the Stability Theorem to Bacterial Population Dynamics

The equation for population growth is

$$\frac{db}{dt} = \lambda b$$

If $\lambda \neq 0$, the only equilibrium is $b^* = 0$. The derivative of the rate of change, $f(b) = \lambda b$, is

$$f'(b) = \lambda$$

According to the Stability Theorem for Autonomous Differential Equations, the equilibrium is unstable if the derivative is positive and stable if the derivative is negative. If $\lambda > 0$, the equilibrium is unstable (exponential growth; see Figures 8.3.19a and 8.2.10a), and if $\lambda < 0$ the equilibrium is stable (exponential decay; see Figures 8.3.19b and 8.2.10b).

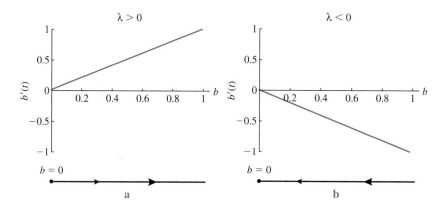

FIGURE 8.3.19

Phase-line diagrams for the bacterial growth model

There are two important differences between using the Stability Theorem for Autonomous Differential Equations and solving the equation to figure out whether the equilibrium is stable. The stability theorem helps us recognize stable and unstable equilibria with a minimum of work; we solve for the equilibria and compute the derivative of the rate of change. Finding what happens near an equilibrium by computing the solution requires more work: finding the solution and then evaluating its limit as t approaches infinity. The other difference between the two methods is that the simpler calculation based on the stability theorem gives less information about the solutions, telling us only what happens to solutions that start *near* the equilibrium.

Example 8.3.4   Applying the Stability Theorem to the Logistic Equation

In Example 8.2.3 in the previous section, we showed that the equation

$$P' = f(P) = kP\left(1 - \frac{P}{L}\right)$$

has equilibria at $P^* = 0$ and $P^* = L$ (recall that k is a positive constant). From

$$f(P) = kP - \frac{k}{L}P^2$$

we compute

$$f'(P) = k - \frac{2k}{L}P$$

Since $f'(0) = k > 0$, we conclude that $P^* = 0$ is unstable. From

$$f'(L) = k - \frac{2k}{L}L = -k < 0$$

we conclude that $P^* = L$ is stable.

Example 8.3.5 Applying the Stability Theorem to the Modified Logistic Equation

For the modified logistic equation

$$P' = f(P) = kP\left(1 - \frac{P}{L}\right)\left(1 - \frac{m}{P}\right)$$

we simplify:

$$f(P) = \left(kP - \frac{k}{L}P^2\right)\left(1 - \frac{m}{P}\right)$$

$$= kP - \frac{k}{L}P^2 - mk + \frac{mk}{L}P$$

$$= -\frac{k}{L}P^2 + \frac{mk}{L}P + kP - mk$$

and then differentiate:

$$f'(P) = -\frac{2k}{L}P + \frac{mk}{L} + k$$

$$= k\left(-\frac{2}{L}P + \frac{m}{L} + 1\right)$$

In Example 8.2.3, we found three equilibria: $P^* = 0$, $P^* = L$, and $P^* = m$. The equilibrium $P^* = 0$ is unstable because

$$f'(0) = k\left(\frac{m}{L} + 1\right) > 0$$

When $P^* = L$,

$$f'(L) = k\left(-2 + \frac{m}{L} + 1\right) = k\left(\frac{m}{L} - 1\right) < 0$$

(because $m < L$), and so the equilibrium is stable. Finally,

$$f'(m) = k\left(-\frac{2m}{L} + \frac{m}{L} + 1\right) = k\left(1 - \frac{m}{L}\right) > 0$$

tells us that $P^* = m$ is unstable.

Example 8.3.6 Applying the Stability Theorem to the Selection Equation

The selection equation,

$$\frac{dp}{dt} = f(p) = (\mu - \lambda)p(1 - p)$$

has equilibria at $p = 0$ and $p = 1$ (Section 8.2, Example 8.2.4). The derivative of the rate of change $f(p)$ with respect to p is

$$f'(p) = (\mu - \lambda)(1 - 2p)$$

Then

$$f'(0) = \mu - \lambda$$
$$f'(1) = \lambda - \mu$$

If $\mu > \lambda$, the equilibrium $p = 0$ has a positive derivative and is unstable, and the equilibrium $p = 1$ has a negative derivative and is stable, matching Figure 8.3.20. These results make biological sense; the mutant is superior in this case.

If $\mu < \lambda$, the mutant is at a disadvantage and the $p = 0$ equilibrium is stable and the $p = 1$ equilibrium is unstable.

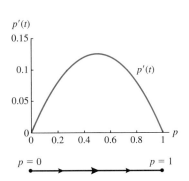

FIGURE 8.3.20

The rate of change and phase-line diagram for the selection model with $\mu - \lambda = 0.5$

Example 8.3.7 A Case Where the Stability Theorem Does Not Apply

Consider the differential equation

$$\frac{db}{dt} = f(b) = b^2$$

that we will revisit in Example 8.4.9 in Section 8.4. From $f(b) = b^2 = 0$, it follows that $b^* = 0$ is the only equilibrium point. Because $f'(b) = 2b$ gives $f'(b^*) = 0$, we cannot use Theorem 8.3.1 to figure out the stability of $b^* = 0$.

From $db/dt = b^2 > 0$ if $b \neq 0$, we conclude that any solution that starts near 0 will move away from it, since it must be increasing (in Example 8.4.9 we will show that it will actually approach infinity in finite time). Thus, $b^* = 0$ is an unstable equilibrium.

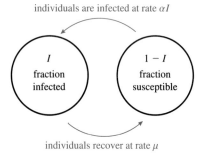

Figure 8.3.21

Factors involved in disease dynamics

A Model for a Disease

Suppose a disease is circulating in a population. Individuals recover from this disease unharmed but are susceptible to reinfection. How many people will be sick at any given time? Is there any way that such a disease will die out? The factors affecting the dynamics are sketched in Figure 8.3.21.

Let I denote the fraction of infected individuals in the population. Each uninfected, or susceptible, individual has a chance of getting infected when she encounters an infectious individual (depending, perhaps, on whether she gets sneezed on). It seems plausible that a susceptible individual will run into infectious individuals at a rate proportional to the number of infectious individuals. Then

$$\text{per capita rate at which a susceptible individual is infected} = \alpha I$$

The positive parameter α combines the rate at which people are encountered and the probability that an encounter produces an infection.

What fraction of individuals are susceptible? Every individual is either infected or not, so a fraction $1 - I$ are susceptible. The overall rate at which new individuals are infected is the per capita rate times the fraction, or

$$\text{rate at which susceptible individuals are infected}$$
$$= \text{per capita rate} \cdot \text{fraction of individuals}$$
$$= \alpha I(1 - I)$$

Individuals recover from the disease at a rate proportional to the fraction of infected individuals, or

$$\text{rate at which infected individuals recover} = \mu I$$

(where $\mu > 0$). Putting the two processes together by adding the rates, we have

$$\frac{dI}{dt} = \alpha I(1 - I) - \mu I \qquad (8.3.1)$$

Suppose we start with a few infected individuals, a small value of I. Will the disease persist? The equilibria and their stability can provide the answer. To find the equilibria, we solve

$$\alpha I^*(1 - I^*) - \mu I^* = 0$$
$$I^* [\alpha(1 - I^*) - \mu] = 0$$
$$I^* = 0 \quad \text{or} \quad \alpha(1 - I^*) - \mu = 0$$
$$I^* = 0 \quad \text{or} \quad I^* = 1 - \frac{\mu}{\alpha}$$

Do these make sense? There are no sick people if $I^* = 0$. In the absence of some process bringing in infected people (such as migration), a population without illness will remain so. The other equilibrium depends on the parameters α and μ. There are two cases, depending upon which parameter is larger:

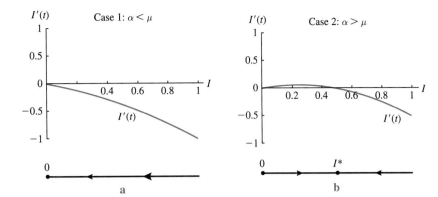

FIGURE 8.3.22
Phase-line diagrams for the disease model

Case 1: If $\alpha < \mu$, $1 - \frac{\mu}{\alpha} < 0$, which is nonsense. When the recovery rate is large, the only biologically plausible equilibrium is $I^* = 0$.

Case 2: If $\alpha > \mu$, $0 < 1 - \frac{\mu}{\alpha} < 1$, which is biologically possible. There are two equilibria if the infection rate is larger than the recovery rate.

To draw the phase-line diagrams for the two cases, we must compute the stability of the equilibria. The derivative of the rate of change with respect to the state variable I is

$$\frac{d}{dI}[\alpha I(1-I) - \mu I] = \alpha - 2\alpha I - \mu \qquad (8.3.2)$$

Case 1: If $\alpha < \mu$, the derivative at $I^* = 0$ is

$$\alpha - 2\alpha \cdot 0 - \mu = \alpha - \mu < 0$$

The single equilibrium is stable.

Case 2: If $\alpha > \mu$, the derivative at $I^* = 0$ is

$$\alpha - 2\alpha \cdot 0 - \mu = \alpha - \mu > 0$$

and this equilibrium is unstable. At $I^* = 1 - \frac{\mu}{\alpha}$, the derivative of the rate of change is

$$\alpha - 2\alpha \cdot \left(1 - \frac{\mu}{\alpha}\right) - \mu = \mu - \alpha < 0$$

The positive equilibrium is stable. The phase-line diagrams for the two cases are drawn in Figure 8.3.22.

From our phase-line diagram, we can deduce the behaviour of solutions. In case 1, with $\alpha < \mu$, all solutions converge to the equilibrium at $I^* = 0$ (Figure 8.3.23a). If control measures could be implemented to increase the recovery rate, μ, or reduce the transmission, α, the disease could be completely eliminated from a population. The transmission need not be reduced to zero to eliminate the disease. In case 2, with $\mu < \alpha$, the equilibrium, I^*, is positive and stable. Solutions starting near 0 increase to I^* and the disease remains present in the population (Figure 8.3.23b). Such a disease is called **endemic.**

Outline of the Proof of Theorem 8.3.1

To start the proof, we need a fact about continuous functions that we have not mentioned before: if a continuous function $f(x)$ is positive (negative) at some value x, then it is positive (negative) on some open interval containing x. (We present this without

FIGURE 8.3.23

Solutions of the disease equation in two cases

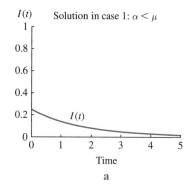
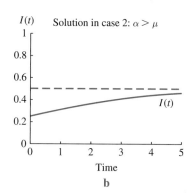

proof here; although the statement might seem plausible (convince yourself by drawing a graph), it still needs to be proven.)

Consider the case $f'(m^*) < 0$. We need to show that m^* is stable (the case $f'(m^*) > 0$ is argued analogously).

Assuming that f' is continuous, we can find an interval, (m_1, m_2), containing m^* such that $f'(m) < 0$ for all m in (m_1, m_2). Thus, f is decreasing on (m_1, m_2), and since $f(m^*) = 0$, we conclude that $f(m) > 0$ on (m_1, m^*) and $f(m) < 0$ on (m^*, m_2); see Figure 8.3.24a.

FIGURE 8.3.24

Analysis of stability

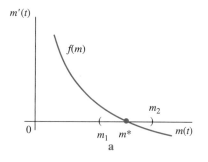
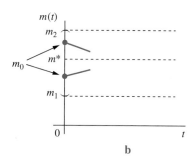

To prove that m^* is a stable equilibrium, we need to show that a solution that starts close to m^* will get closer to it (i.e., will approach m^*).

Consider the solution $m(t)$ such that $m(0) = m_0$ is in (m_1, m^*); see Figure 8.3.24b. Its tangent line approximation is

$$m(t) \approx m(0) + m'(0)t = m_0 + f(m_0)t$$

Since m_0 is in (m_1, m^*), $f(m_0) > 0$, i.e., the tangent line has positive slope. Thus, starting below m^*, $m(t)$ will increase.

If $m(t)$ starts above m^* (i.e., $m(0) = m_0$ is in (m^*, m_2)), then $f(m_0) < 0$. The slope of the tangent is negative and $m(t)$ has to decrease.

Summary

By examining the phase-line diagram near an equilibrium, we found the stability theorem for autonomous differential equations. An equilibrium of an autonomous differential equation is stable if the derivative of the rate of change with respect to the state variable is negative and unstable if this derivative is positive. Calculating stability in this way is easier than solving the equation but does not provide exact information about the behaviour of solutions far from the equilibrium. We applied this method to several familiar models and to a model of a disease, showing that a disease could be eradicated without entirely stopping transmission.

8.3 Exercises

Mathematical Techniques

1–2 ■ From the following graphs of the rate of change as a function of the state variable, identify stable and unstable equilibria by checking whether the rate of change is an increasing or a decreasing function of the state variable.

1.
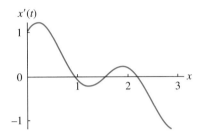

2.
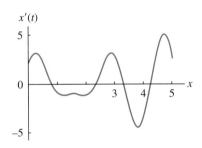

3–6 ■ Use the stability theorem to evaluate the stability of the equilibria of each autonomous differential equation.

3. $\frac{dx}{dt} = 1 - x^2$. Compare your results with the phase-line in Section 8.2, Exercise 15.

4. $\frac{dx}{dt} = 1 - e^x$. Compare your results with the phase-line in Section 8.2, Exercise 16.

5. $\frac{dy}{dt} = y \cos y$. Compare your results with the phase-line in Section 8.2, Exercise 17.

6. $\frac{dz}{dt} = \frac{1}{z} - 3$. Compare your results with the phase-line in Section 8.2, Exercise 18.

7–10 ■ Find the stability of the equilibria of each autonomous differential equation with parameters.

7. $\frac{dx}{dt} = 1 - ax$. Suppose that $a > 0$.

8. $\frac{dx}{dt} = cx + x^2$. Suppose that $c > 0$.

9. $\frac{dW}{dt} = \alpha e^{\beta W} - 1$. Suppose that $\alpha > 0$ and $\beta < 0$.

10. $\frac{dy}{dt} = ye^{-\beta y} - ay$. Suppose that $\beta < 0$ and $a > 1$.

11–14 ■ As with discrete-time dynamical systems, equilibria can act strangely when the slope of the rate of change function is exactly equal to the critical value of 0.

11. Consider the differential equation $\frac{dx}{dt} = x^2$. Find the equilibrium, graph the rate of change as a function of x, and draw the phase-line diagram. Would you consider the equilibrium to be stable or unstable?

12. Consider the differential equation $\frac{dx}{dt} = -(1-x)^4$. Find the equilibrium, graph the rate of change as a function of x, and draw the phase-line diagram. Would you consider the equilibrium to be stable or unstable?

13. Graph a rate-of-change function that has a slope of 0 at a stable equilibrium. What is the sign of the second derivative at the equilibrium? What is the sign of the third derivative at the equilibrium?

14. Graph a rate-of-change function that has a slope of 0 at an unstable equilibrium. What is the sign of the second derivative at the equilibrium? What is the sign of the third derivative at the equilibrium?

15–16 ■ The fact that the rate-of-change function is continuous means that many behaviours are impossible for an autonomous differential equation.

15. Try to draw a phase-line diagram with two stable equilibria in a row. Use the Intermediate Value Theorem to sketch a proof of why this is impossible.

16. Why is it impossible for a solution of an autonomous differential equation to oscillate?

17–20 ■ When parameter values change, the number and stability of equilibria sometimes change. Such changes are called bifurcations, and they play a central role in the study of differential equations. The following exercises illustrate several of the more important bifurcations. Graph the equilibria as functions of the parameter value, using a solid line when an equilibrium is stable and a dashed line when an equilibrium is unstable. This picture is called a **bifurcation diagram.**

17. Consider the equation
$$\frac{dx}{dt} = ax - x^2$$
for both positive and negative values of x. Find the equilibria as functions of a for values of a between -1 and 1. Draw a bifurcation diagram and describe what happens at $a = 0$. The change that occurs at $a = 0$ is called a **transcritical bifurcation.**

18. Consider the equation
$$\frac{dx}{dt} = a - x^2$$
for both positive and negative values of x. Find the equilibria as functions of a for values of a between -1 and 1. Draw a bifurcation diagram and describe what happens at $a = 0$. The change that occurs at $a = 0$ is called a **saddle-node bifurcation.**

19. Consider the equation

$$\frac{dx}{dt} = ax - x^3$$

for both positive and negative values of x. Find the equilibria as functions of a for values of a between -1 and 1. Draw a bifurcation diagram and describe what happens at $a = 0$. The change that occurs at $a = 0$ is called a **pitchfork bifurcation.**

20. Consider the equation

$$\frac{dx}{dt} = ax + x^3$$

for both positive and negative values of x. Find the equilibria as functions of a for values of a between -1 and 1. Draw a bifurcation diagram and describe what happens at $a = 0$. The change that occurs at $a = 0$ is a slightly different type of **pitchfork bifurcation** (Exercise 19) called a **subcritical bifurcation** (Exercise 19 is **supercritical**). How does your picture differ from a simple mirror image of that in Exercise 19?

Applications

21–24 ▪ Use Theorem 8.3.1 to determine the stability of equilibria for each bacterial population model.

21. $\dfrac{db}{dt} = (1 - 0.002b)b$

22. $\dfrac{db}{dt} = (4 - 0.001b)b$

23. $\dfrac{db}{dt} = (-2 + 0.01b)b$

24. $\dfrac{db}{dt} = (-5 + 0.001b)b$

25–26 ▪ Use Theorem 8.3.1 to determine the stability of equilibria for each selection model.

25. $\dfrac{dp}{dt} = 0.5p(1 - p)^2$

26. $\dfrac{dp}{dt} = (2(1 - p) - 1.5p)p(1 - p)$

27–30 ▪ A **reaction-diffusion equation** describes how chemical concentration changes in response to two factors simultaneously, reaction and movement. A simple model has the form

$$\frac{dC}{dt} = \beta(A - C) + R(C)$$

The first term describes diffusion, and the second term, $R(C)$, is the reaction, which could have a positive or a negative sign (depending on whether chemical is being created or destroyed). Suppose that $\beta = 1/\text{min}$ and $A = 5$ mol/L. For each form of $R(C)$:

a. Describe how the reaction rate depends on the concentration.
b. Find the equilibria and their stability.
c. Describe how absorption changes the results.

27. $R(C) = -C$

28. $R(C) = 0.5C$

29. $R(C) = \dfrac{C}{2 + C}$

30. $R(C) = -\dfrac{C}{2 + C}$

31–38 ▪ Apply Theorem 8.3.1 to determine the stability of equilibria for each model. As well, draw the phase-line diagram. Check that your algebraic and geometric answers match. Give a biological interpretation of the stability of each equilibrium.

31. The population size of some species of organism follows the model $\dfrac{dN}{dt} = \dfrac{3N^2}{2 + N^2} - N$, where N is measured in hundreds.

32. The population size of some species of organism follows the model $\dfrac{dN}{dt} = \dfrac{5N^2}{1 + N^2} - 2N$, where N is measured in hundreds.

33. The drag of a falling object is proportional to the square of its speed. The differential equation is $v'(t) = a - Dv^2$, where v is the speed, $a = 9.8$ m/s^2, and $D = 0.0032$ per metre.

34. The drag of a falling object is proportional to the square of its speed. The differential equation is $v'(t) = a - Dv^2$, where v is the speed, $a = 9.8$ m/s^2, and $D = 0.00048$ per metre.

35. Assume that S is the concentration of the substrate that is being converted into a product. The **Michaelis–Menten** or **Monod** equation states that $\dfrac{dS}{dt} = -k_1 \dfrac{S}{k_2 + S}$. Assume that $k_1 = k_2 = 1$.

36. Assume that S is the concentration of the substrate that is being converted into a product. To account for the addition of the substrate at the rate R, the Michaelis–Menten equation is adjusted in the following way: $\dfrac{dS}{dt} = R - k_1 \dfrac{S}{k_2 + S}$. Assume that $k_1 = k_2 = 1$ and $R = 0.5$.

37. The volume of a small growing organism changes according to $\dfrac{dV}{dt} = a_1 V^{2/3} - a_2 V$, where $a_1, a_2 > 0$. The constant a_1 gives the rate at which energy is taken in, and a_2 gives the rate at which energy is used. (For details about the context of this exercise, see Exercise 37 in Section 8.2.)

38. The volume of a small growing organism changes according to $\dfrac{dV}{dt} = a_1 V^{2/3} - a_2 V^{3/4}$, where $a_1, a_2 > 0$. The constant a_1 gives the rate at which energy is taken in, and a_2 gives the rate at which energy is used. (For details about the context of this exercise, see Exercises 37 and 38 in Section 8.2.)

39–42 ▪ Exercises 17–20 show how the number and stability of equilibria can change when a parameter changes. Often, bifurcations have important biological applications, and bifurcation diagrams help explain how the dynamics of a system can suddenly change when a parameter changes only slightly. In each case, graph the equilibria against the parameter value, using a solid line when an equilibrium is stable and a dashed line when an equilibrium is unstable to draw the bifurcation diagram.

39. Consider the logistic differential equation (Section 8.1, Exercise 38) with harvesting proportional to population size, or $\frac{db}{dt} = b(1 - b - h)$, where h represents the fraction harvested. Graph the equilibria as functions of h for values of h between 0 and 2, using a solid line when an equilibrium is stable and a dashed line when an equilibrium is unstable. Even though they do not make biological sense, include negative values of the equilibria on your graph. You should find a *transcritical bifurcation* (Exercise 17) at $h = 1$.

40. Suppose $\mu = 1$ in the basic disease model $\frac{dI}{dt} = \alpha I(1 - I) - \mu I$. Graph the two equilibria as functions of α for values of α between 0 and 2, using a solid line when an equilibrium is stable and a dashed line when an equilibrium is unstable. Even though they do not make biological sense, include negative values of the equilibria on your graph. You should find a *transcritical bifurcation* (Exercise 17) at $\alpha = 1$.

41. Consider a version of the equation in Section 8.2, Exercise 29, that includes the parameter r:

$$\frac{dN}{dt} = \frac{rN^2}{1 + N^2} - N$$

Graph the equilibria as functions of r for values of r between 0 and 3, using a solid line when an equilibrium is stable and a dashed line when an equilibrium is unstable. The algebra for checking stability is messy, so check stability only at $r = 3$. You should find a *saddle-node bifurcation* (Exercise 18) at $r = 2$.

42. Consider a variant of the basic disease model given by

$$\frac{dI}{dt} = \alpha I^2(1 - I) - I$$

Graph the equilibria as functions of α for values of α between 0 and 5, using a solid line when an equilibrium is stable and a dashed line when an equilibrium is unstable. The algebra for checking stability is messy, so check stability only at $\alpha = 5$. You should find a *saddle-node bifurcation* (Exercise 18) at $\alpha = 4$.

43–44 ■ Right at a bifurcation point, the stability theorem fails because the slope of the rate of change function at the equilibrium is exactly 0. In each of the following cases, check that the stability theorem fails, and then draw a phase-line diagram to find the stability of the equilibrium.

43. Analyze the stability of the positive equilibrium in the model from Exercise 42 when $\alpha = 4$, the point where the bifurcation occurs.

44. Analyze the stability of the disease model when $\alpha = \mu = 1$, the point where the bifurcation occurs in 40.

8.4 Separable Differential Equations

In this section, we develop the **separation of variables** technique, which will help us solve autonomous differential equations such as the ones that appear in the models we studied in previous sections. As it involves integration, separation of variables can be hard to do or, worse yet, can fail to produce a solution in the form of an explicit formula.

Separation of Variables

A **separable differential equation** is a differential equation in which we can separate the unknown function (also called the state variable) from the independent variable. The general form of a separable equation is

$$f(y)y' = g(x) \quad \text{or} \quad f(y)\frac{dy}{dx} = g(x)$$

where $f(y)$ is a function of y only, and $g(x)$ is a function of x only. For instance,

$$\frac{dy}{dx} = 3xe^x(y^2 + 1)$$

is separable because we can write

$$\frac{1}{y^2 + 1}\frac{dy}{dx} = 3xe^x$$

(so $f(y) = \frac{1}{y^2 + 1}$ and $g(x) = 3xe^x$). Likewise,

$$y' = \frac{yt^3}{4 - y}$$

is separable:
$$\frac{4-y}{y}y' = t^3$$

However, it is impossible to separate x from y in the equation
$$\frac{dy}{dx} = xy + 1$$

What is important for us is that *all autonomous differential equations are separable.* For instance:

$$\text{Basic exponential model for } P(t): \frac{1}{P}P' = k$$

$$\text{Logistic equation for } P(t): \frac{1}{P(1-P/L)}P' = k$$

$$\text{Diffusion equation for } C(t): \frac{1}{A-C}\frac{dC}{dt} = \alpha$$

Once we learn how to solve separable equations, we will be able to obtain algebraic solutions for the equations that we studied in previous sections.

To solve a separable differential equation,
$$f(y)\frac{dy}{dx} = g(x)$$

we integrate both sides with respect to the independent variable
$$\int f(y)\frac{dy}{dx}\,dx = \int g(x)\,dx$$

By the substitution rule for indefinite integrals (see Theorem 7.5.1 in Section 7.5), the left side transforms to
$$\int f(y)\,dy = \int g(x)\,dx$$

So, if we can integrate the functions on both sides, we will be able to solve the equation.

Example 8.4.1 Solving a Differential Equation by Separation of Variables

Solve the differential equation
$$\frac{dy}{dx} = e^y x^2$$

with $y(0) = 4$.

We divide the given equation by e^y:
$$e^{-y}\frac{dy}{dx} = x^2$$

and realize that it is separable. Integrate
$$\int e^{-y}\frac{dy}{dx}\,dx = \int x^2\,dx$$
$$\int e^{-y}\,dy = \int x^2\,dx$$
$$-e^{-y} + C_1 = \frac{1}{3}x^3 + C_2$$

where C_1 and C_2 are constants. We move the constants to the same side

$$-e^{-y} = \frac{1}{3}x^3 + C_2 - C_1$$

and write

$$-e^{-y} = \frac{1}{3}x^3 + C$$

where $C = C_2 - C_1$. (Thus, there is no need to add a constant to both sides when we integrate a separable equation.)

We need to do two more things: solve for C (since we have an initial condition) and try to get an explicit formula for the solution (which will work in this case but is not always possible).

Substituting $y(0) = 4$ into the solution, we get

$$-e^{-4} = \frac{1}{3}(0)^3 + C$$

and so $C = -e^{-4}$. The solution is

$$-e^{-y} = \frac{1}{3}x^3 - e^{-4}$$

$$e^{-y} = -\frac{1}{3}x^3 + e^{-4}$$

$$-y = \ln\left(-\frac{1}{3}x^3 + e^{-4}\right)$$

$$y = -\ln\left(-\frac{1}{3}x^3 + e^{-4}\right)$$

Example 8.4.2 Separation of Variables

Solve the differential equation

$$y' = \frac{\cos t - 2t}{\ln y}$$

with initial condition $y(0) = 1$.

The equation is separable:

$$\ln y \, y' = \cos t - 2t$$

and we solve it by integration:

$$\int \ln y \, y' \, dt = \int (\cos t - 2t) \, dt$$

$$\int \ln y \, dy = \int (\cos t - 2t) \, dt$$

To integrate $\ln y$, we use integration by parts (see Example 7.5.12 in Section 7.5) to get

$$y \ln y - y = \sin t - t^2 + C$$

Substituting $t = 0$ and $y = 1$, we get

$$\ln 1 - 1 = \sin 0 - (0)^2 + C$$

and $C = -1$. The solution

$$y \ln y - y = \sin t - t^2 - 1$$

cannot be solved explicitly for y as a function of t.

Now we turn to the models from previous sections.

Chapter 8 Differential Equations

Example 8.4.3 Solving the Basic Exponential Model Equation

After using it for a long time, we can finally compute systematically the solution of the differential equation

$$P'(t) = kP(t)$$

where $P(0) = P_0$.
  Separate the variables:

$$\frac{1}{P(t)} P'(t) = k$$

and integrate

$$\int \frac{1}{P} P' dt = \int k dt$$

$$\int \frac{1}{P} dP = \int k dt$$

$$\ln |P| = kt + C$$

Since $P > 0$, we do not need the absolute value. When $t = 0$, $P = P_0$ and

$$\ln P_0 = k(0) + C$$

i.e., $C = \ln P_0$. Therefore, the solution is

$$\ln P = kt + \ln P_0$$
$$e^{\ln P} = e^{kt + \ln P_0}$$
$$e^{\ln P} = e^{kt} e^{\ln P_0}$$
$$P = P(t) = P_0 e^{kt}$$

To simplify notation, we abbreviated $P(t)$ to P while integrating and simplifying. ▲

Example 8.4.4 Solving the Newton's Law of Cooling Equation

Consider the law of cooling differential equation model (see Example 8.1.4)

$$\frac{dT(t)}{dt} = \alpha(A - T(t))$$

where $T(t)$ is the temperature at time t and $T(0) = T_0$.
  To solve the equation, we separate variables:

$$\frac{1}{A - T} \frac{dT}{dt} = \alpha$$

integrate and simplify:

$$\int \frac{1}{A - T} \frac{dT}{dt} dt = \int \alpha dt$$
$$\int \frac{1}{A - T} dT = \int \alpha dt$$
$$-\ln |A - T| = \alpha t + C$$
$$\ln |A - T| = -\alpha t - C$$
$$|A - T| = e^{-\alpha t} e^{-C} \qquad (8.4.1)$$

If $T < A$ (i.e., the object is cooler than the ambient temperature), then $|A - T| = A - T$ and

$$A - T = e^{-\alpha t} e^{-C}$$
$$T = A - Ke^{-\alpha t}$$

where $K = e^{-C}$ is a constant whose value we now determine. From $T(0) = T_0$, we get

$$T_0 = A - Ke^{-\alpha \cdot 0}$$

and $K = A - T_0$. The solution is

$$T(t) = A - (A - T_0)e^{-\alpha t} \tag{8.4.2}$$

If the object is warmer than the ambient temperature A, then $|A - T| = -(A - T) = T - A$ and Equation 8.4.1 yields

$$T - A = e^{-\alpha t}e^{-C}$$
$$T = A + Ke^{-\alpha t}$$

Using $T(0) = T_0$, we compute $K = T_0 - A$. In this case, the solution is

$$T(t) = A + (T_0 - A)e^{-\alpha t} \tag{8.4.3}$$

Note that, in either case,

$$\lim_{t \to \infty} T(t) = A$$

For example, if $\alpha = 0.1, A = 40°C$, and $T(0) = T_0 = 10°C$, then Equation 8.4.2 gives

$$T(t) = 40 - 30e^{-0.1t}$$

If $\alpha = 0.1, A = 40°C$, and $T(0) = T_0 = 60°C$, then from Equation 8.4.3

$$T(t) = 40 + 20e^{-0.1t}$$

(see Figure 8.4.25).

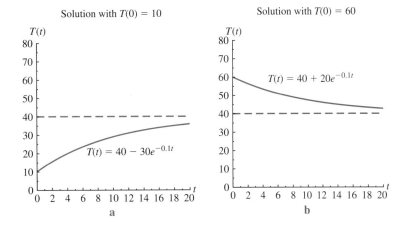

FIGURE 8.4.25
Solutions of Newton's law of cooling

We solve the diffusion equation (Example 8.1.5) and the von Bertalanffy limited growth model (mentioned after Example 8.1.5) in exactly the same way. Of course, the interpretations differ, depending on the nature of the application. ▲

Example 8.4.5 Logistic Differential Equation

Now we solve the logistic differential equation (from Example 8.1.2)

$$P'(t) = kP(t)\left(1 - \frac{P(t)}{L}\right)$$

where k and L are positive constants and $P(t)$ is the population at time t. The initial condition is $P(0) = P_0$.

Multiply by L (to simplify fractions) and separate variables:

$$LP' = kP(L - P)$$

$$\frac{1}{P(L-P)}P' = \frac{k}{L}$$

and integrate:

$$\int \frac{1}{P(L-P)}P'\,dt = \int \frac{k}{L}\,dt$$

$$\int \frac{1}{P(L-P)}\,dP = \int \frac{k}{L}\,dt$$

To compute the integral on the left side, we use the following formula

$$\int \frac{1}{P(L-P)}\,dP = \frac{1}{L}\ln\left|\frac{P}{L-P}\right| + C_1$$

(In Exercise 52, we suggest how to calculate it, using the method of partial fractions; alternatively, we could use software that does symbolic integration.)

Thus,

$$\frac{1}{L}\ln\left|\frac{P}{L-P}\right| = \frac{k}{L}t + C_1$$

$$\ln\left|\frac{P}{L-P}\right| = kt + LC_1$$

$$\left|\frac{P}{L-P}\right| = e^{kt+LC_1} = e^{LC_1}e^{kt}$$

$$\frac{P}{L-P} = \pm e^{LC_1}e^{kt}$$

$$\frac{P}{L-P} = Ce^{kt} \qquad (8.4.4)$$

where $C = \pm e^{LC_1}$. Next, we solve for an explicit formula for P:

$$\frac{P}{L-P} = Ce^{kt}$$

$$LCe^{kt} - PCe^{kt} = P$$

$$P(1 + Ce^{kt}) = LCe^{kt}$$

$$P = \frac{LCe^{kt}}{1 + Ce^{kt}}$$

Finally, dividing by Ce^{kt}, we get

$$P = \frac{L}{1 + \frac{1}{C}e^{-kt}}$$

To find C, we substitute $P(0) = P_0$ into Equation 8.4.4 and get

$$C = \frac{P_0}{L - P_0}$$

We are done. The solution of the logistic equation with the given initial condition is

$$P(t) = \frac{L}{1 + \frac{L - P_0}{P_0}e^{-kt}} \qquad (8.4.5)$$

Note that, as $t \to \infty$, $e^{-t} \to 0$, and

$$\lim_{t \to \infty} P(t) = L$$

Thus, we have confirmed that, in the long run, the population will approach the carrying capacity. In Figure 8.4.26, we show several logistic curves corresponding to the same values of k and L but with different initial conditions.

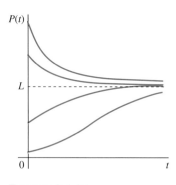

FIGURE 8.4.26

Logistic curves corresponding to different initial conditions

Example 8.4.6 Calculations with the Logistic Differential Equation

The solution of the equation

$$\frac{dP}{dt} = 0.09P\left(1 - \frac{P}{2000}\right)$$

with initial condition $P(0) = 350$ that we studied in Example 8.1.2 is given by

$$P(t) = \frac{2000}{1 + \frac{2000 - 350}{350} e^{-0.09t}}$$

$$\approx \frac{2000}{1 + 4.71 e^{-0.09t}}$$

(see Figure 8.4.27).

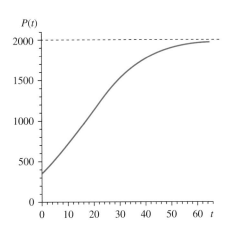

FIGURE 8.4.27
Logistic curve

Example 8.4.7 Solution of the Bacteria Selection Model

In Example 8.1.6 we studied the differential equation

$$p'(t) = (\mu - \lambda) p(t)[1 - p(t)]$$

for the fraction of type a bacteria. This is a logistic differential equation with $L = 1$, $k = \mu - \lambda$, and $p(0) = p_0$. Applying Equation 8.4.5, we write the solution

$$p(t) = \frac{1}{1 + \frac{1 - p_0}{p_0} e^{-(\mu - \lambda)t}}$$

In particular, when $\mu = 0.5$, $\lambda = 0.2$, and $p_0 = 0.1/1.1$ (as in Example 8.1.7), we obtain

$$p(t) = \frac{1}{1 + 10 e^{-0.3t}}$$

Dividing the formula we obtained in Example 8.1.7 by $0.091 e^{0.5t}$, we obtain the same answer.

Example 8.4.8 Gompertz Growth Model

Initially used to study human mortality, the **Gompertz differential equation**

$$P'(t) = kP(t)[\ln L - \ln P(t)]$$

has been successfully used in various growth models. The function $P(t)$ represents the population (or length, or weight, or other variables used in describing growth) at time

t; L is the theoretically largest possible size (depends on the context); and $k > 0$ is the growth rate. Another way to describe L is to say that, when $P(t)$ reaches L, the growth is zero. Assume that the initial population is $P(0) = P_0$.

Note that the per capita growth rate

$$\frac{P'(t)}{P(t)} = k\left[\ln L - \ln P(t)\right]$$

decreases as $P(t)$ increases—as is often the case in nature.

We now solve this autonomous differential equation. Separate the variables and integrate (write P instead of $P(t)$ to simplify notation):

$$\frac{P'}{P(\ln L - \ln P)} = k$$

$$\int \frac{P'}{P(\ln L - \ln P)} dt = \int k\, dt$$

$$\int \frac{1}{P(\ln L - \ln P)} dP = \int k\, dt$$

Using substitution with $u = \ln L - \ln P$, $du/dP = -1/P$, and $dP/P = -du$, we write the integral on the left side as

$$\int \frac{1}{P(\ln L - \ln P)} dP = -\int \frac{1}{u} du = -\ln|u| = -\ln|\ln L - \ln P|$$

Since $L > P$, we drop the absolute value sign. Going back to solving the equation, we write

$$-\ln(\ln L - \ln P) = kt + C$$
$$\ln(\ln L - \ln P) = -kt - C$$
$$\ln L - \ln P = e^{-kt-C} = e^{-kt}e^{-C} = e^{-kt}K$$

where K is a constant. Using the initial condition $P(0) = P_0$, we find $\ln L - \ln P_0 = K$, i.e., $K = \ln(L/P_0)$. It follows that

$$\ln L - \ln P = e^{-kt} \ln(L/P_0)$$
$$\ln L - \ln P = \ln(L/P_0)^{e^{-kt}}$$
$$\ln P = \ln L - \ln(L/P_0)^{e^{-kt}}$$
$$\ln P = \ln \frac{L}{(L/P_0)^{e^{-kt}}}$$
$$P(t) = \frac{L}{(L/P_0)^{e^{-kt}}} = L\left(\frac{P_0}{L}\right)^{e^{-kt}} \tag{8.4.6}$$

This solution can also be written as

$$P(t) = L\left(e^{\ln(P_0/L)}\right)^{e^{-kt}} = L e^{\ln(P_0/L)e^{-kt}}$$

The **Gompertz curve** (Figure 8.4.28a) looks similar to the logistic curve. In Figure 8.4.28b we graphed both the logistic curve (Equation 8.4.5) and the Gompertz curve (Equation 8.4.6) with $k = 0.09$, $L = 2000$, and $P_0 = 350$. We see that the Gompertz curve grows faster and reaches the plateau value of 2000 faster than the logistic curve (see Exercises 37 and 38).

Example 8.4.9   A Population Explosion

As a more unusual example, suppose that the per capita production rate (also known as the per capita reproduction rate) of some population with size $b = b(t)$ is

$$\text{per capita production rate} = b$$

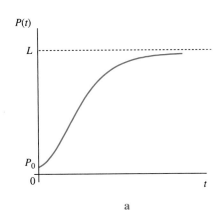
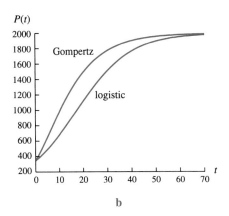

FIGURE 8.4.28
Gompertz and logistic growth curves

This means that individuals reproduce more and more rapidly the more of them there are. We expect that this population will grow very quickly, and we can use separation of variables to compute how quickly. The differential equation for growth is

$$\frac{db}{dt} = \text{per capita production rate} \cdot b$$
$$= b \cdot b = b^2$$

Separation of variables proceeds as follows:

$$\frac{1}{b^2}\frac{db}{dt} = 1$$

$$\int \frac{1}{b^2}\frac{db}{dt}dt = \int dt$$

$$\int \frac{1}{b^2}db = \int dt$$

$$-\frac{1}{b} = t + C$$

$$b = -\frac{1}{t+C}$$

Suppose that $b(0) = 10$. We solve for C using the initial condition, finding

$$10 = -\frac{1}{C}$$

and so $C = -0.1$. Therefore,

$$b = b(t) = -\frac{1}{t - 0.1}$$

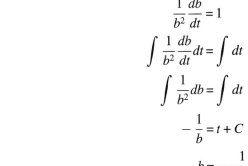

FIGURE 8.4.29
Semilog plot of a population with self-enhancing growth

(Figure 8.4.29). The population blasts off to infinity at time $t = 0.1$. What happens at $t = 0.11$? The solution does not exist. As far as this differential equation is concerned, the world comes to an end at $t = 0.1$. Models that include self-enhancing growth tend to have this sort of "chain reaction" or "blow-up" property.

Summary

We developed the method of **separation of variables** to solve autonomous differential equations. In this method, we isolate the state variable on the left-hand side and time on the right-hand side and then integrate to find a solution. Autonomous differential equations describing population growth and Newton's law of cooling can be solved with this technique.

8.4 Exercises

Mathematical Techniques

1–5 ■ Use separation of variables to solve each autonomous differential equation. Check your answers by differentiating.

1. $\dfrac{db}{dt} = 0.01b$, $b(0) = 1000$

2. $\dfrac{db}{dt} = -3b$, $b(0) = 10^6$

3. $\dfrac{dN}{dt} = 1 + N$, $N(0) = 1$

4. $\dfrac{dN}{dt} = 4 + 2N$, $N(0) = 1$

5. $\dfrac{db}{dt} = 1000 - b$, $b(0) = 500$

6–9 ■ Solve each autonomous differential equation.

6. $\dfrac{dP}{dt} = 1 + P^2$, $P(0) = 1$

7. $P' = \dfrac{1}{2}P^3$, $P(1) = 4$

8. $\dfrac{dy}{dt} = 0.4e^{-y}$, $y(0) = 1$

9. $\dfrac{dy}{dt} = 2\sec y$, $y(0) = 0$

10–13 ■ Use separation of variables to solve each pure-time differential equation. Check your answers by differentiating.

10. $\dfrac{dP}{dt} = \dfrac{5}{1+2t}$ with $P(0) = 0$.

11. $\dfrac{dP}{dt} = 5e^{-2t}$ with $P(0) = 0$.

12. $\dfrac{dL}{dt} = 1000e^{0.2t}$ with $L(0) = 1000$.

13. $\dfrac{dL}{dt} = 64.3e^{-1.19t}$ with $L(0) = 0$.

14–17 ■ Use separation of variables to solve each differential equation.

14. $\dfrac{dy}{dt} = -8ty^2$, $y(0) = 3$

15. $y' - (x-3)(1+y^2) = 0$, $y(0) = -1$

16. $P'(t) - 2P(t)\sqrt{t} = 0$, $P(0) = 250$

17. $\dfrac{dP}{dt} = 2t^3 e^P$, $P(0) = 1$

18–23 ■ Write the solution of the logistic differential equation

$$\dfrac{dP}{dt} = kP\left(1 - \dfrac{P}{L}\right)$$

in each case. Make a rough sketch of the solution. Find the time needed for the population to reach 95% of its carrying capacity.

18. $k = 0.8$, $L = 4000$, and $P(0) = 500$.

19. $k = 0.08$, $L = 4000$, and $P(0) = 500$.

20. $k = 0.02$, $L = 4000$, and $P(0) = 20$.

21. $k = 0.02$, $L = 4000$, and $P(0) = 1000$.

22. $k = 0.06$, $L = 4000$, and $P(0) = 1000$.

23. $k = 0.12$, $L = 4000$, and $P(0) = 1000$.

24. The rate of spread of a virus is proportional to the product of the fraction, $I(t)$, of the population of people who are infected and the fraction of those who are not yet infected, where t is time in days.

   a. Write a differential equation for $I(t)$, assuming that initially 15 people of the total population of 500 were infected.

   b. Solve the equation (look at Example 8.4.5). It is known that 3 days later, 125 people were infected.

   c. How long will it take before 50% of the whole population is infected?

25–28 ■ Separation of variables can help to solve some nonautonomous differential equations. For example, suppose that the per capita production rate is $\lambda(t)$, a function of time, so

$$\dfrac{db}{dt} = \lambda(t)b$$

This equation can be separated into parts depending only on b and on t by dividing both sides of the equation by b and multiplying by dt. For the following functions $\lambda(t)$, interpret and solve the equation with the initial condition $b(0) = 10^6$. Sketch each solution. Check your answer by substituting into the differential equation.

25. $\lambda(t) = t$

26. $\lambda(t) = \dfrac{1}{1+t}$

27. $\lambda(t) = \cos t$

28. $\lambda(t) = e^{-t}$

29–32 ■ Find the solution of the differential equation $\dfrac{db}{dt} = b^p$ in each case. At what time does it approach infinity? Sketch the graph.

29. $p = 2$ (as in Example 8.4.9) and $b(0) = 100$.

30. $p = 2$ and $b(0) = 0.1$.

31. $p = 1.1$ and $b(0) = 100$.

32. $p = 1.1$ and $b(0) = 0.1$.

33–34 ■ Each autonomous differential equation can be solved except for the step of writing x explicitly as a function of t. Use the steps to figure out how the solution behaves.

a. Solve the equation using separation of variables.

b. It is impossible to solve algebraically for x in terms of t. However, you can still find the arbitrary constant. Find it.

c. Although you cannot find x as a function of t, you can find t as a function of x. Graph this function for $1 \leq x \leq 10$.

d. Sketch the solution for x as a function of t.

33. The autonomous differential equation $\frac{dx}{dt} = \frac{x}{1+x}$ with $x(0) = 1$. This describes a population with a per capita production rate that decreases like $\frac{1}{1+x}$.

34. The autonomous differential equation $\frac{dx}{dt} = \frac{x}{1+x^2}$ with $x(0) = 1$. This describes a population with a per capita production rate that decreases like $\frac{1}{1+x^2}$. Describe how the solution differs from that in Exercise 33.

Applications

35. Solve the differential equation $L' = k(A - L)$, where $L = L(t)$ represents the length of a fish at time t, A is the maximum length, and $k > 0$ is the growth rate. Assume that $L(0) = 0$.

36. Solve the differential equation $L' = 0.09(72 - L)$, where $L = L(t)$ represents the length of a fish at time t. Assume that $L(0) = 0$. Compare your answer with Example 7.5.8.

37. Write down the solutions of the logistic and the Gompertz differential equations in the case $k = 0.09$, $L = 2000$, and $P_0 = 350$. In each case, find the time needed for the solution to reach 90% of the carrying capacity.

38. Write down the solutions of the logistic and the Gompertz differential equations in the case $k = 0.15$, $L = 800$, and $P_0 = 20$. In each case, find the time needed for the solution to reach 95% of the carrying capacity.

39. By differentiating the Gompertz differential equation, find the point of inflection of the Gompertz curve. Compare with Exercise 40.

40. By differentiating Equation 8.4.6 twice, find the point of inflection of the Gompertz curve. Compare with Exercise 39.

41–44 ▪ Using the separation of variables method, find the solution of the chemical diffusion equation $\frac{dC}{dt} = \beta(A - C)$ with the following parameter values and initial conditions. Find the concentration after 10 s. How long would it take for the concentration to get halfway to the equilibrium value?

41. $\beta = 0.01/s$, $C(0) = 5$ mmol/cm^3, and $A = 2$ mmol/cm^3.

42. $\beta = 0.01/s$, $C(0) = 1$ mmol/cm^3, and $A = 2$ mmol/cm^3. Why do you think the time matches that in Exercise 41?

43. $\beta = 0.1/s$, $C(0) = 5$ mmol/cm^3, and $A = 2$ mmol/cm^3.

44. $\beta = 0.1/s$, $C(0) = 1$ mmol/cm^3, and $A = 2$ mmol/cm^3.

45–46 ▪ Consider Torricelli's law of draining, $\frac{dy}{dt} = -2\sqrt{y}$ (for the context, see Section 8.2, Exercise 33) with the constant set to 2.

45. Suppose the initial condition is $y(0) = 4$. Find the solution with separation of variables, and graph the result. What really happens at time $t = 2$? And what happens after this time? How does this differ from the solution of the equation $\frac{dy}{dt} = -2y$?

46. Suppose the initial condition is $y(0) = 16$. Find the solution with separation of variables, and graph the result. When does the solution reach 0? What would the depth be at this time if draining followed the equation $\frac{dy}{dt} = -2y$?

47–48 ▪ Suppose that a population is growing at constant rate, λ, but that individuals are harvested at a rate of h, following the differential equation $\frac{db}{dt} = \lambda b - h$. For each of the following values of λ and h, use separation of variables to find the solution, and compare graphs of the solution with those found earlier from a phase-line diagram.

47. $\lambda = 2$, $h = 1000$ (as in Section 8.2, Exercise 19).

48. $\lambda = 0.5$, $h = 1000$ (as in Section 8.2, Exercise 20).

49–50 ▪ Use separation of variables to solve for C in the following models describing chemical diffusion, and find the solution starting from the initial condition $C = A$.

49. $dC/dt = -\beta C + 3\beta A$

50. $dC/dt = -\beta C + 0.5\beta A$

51. Separation of variables and a trick known as **integration by partial fractions** can be used to solve the selection equation. For simplicity, we will consider the case $\frac{dp}{dt} = p(1-p)$, where $\mu - \lambda = 1$.

    a. Separate variables.

    b. Show that
    $$\frac{1}{p(1-p)} = \frac{1}{p} + \frac{1}{1-p}$$
    and rewrite the left-hand side of the equation you obtained in (a).

    c. Integrate the rewritten left-hand side.

    d. Combine the two natural logarithm terms into one using a law of logarithms.

    e. Write the equation for the solution with a single constant, C.

    f. Exponentiate both sides and solve for p.

    g. Using the initial condition $p(0) = 0.01$, find the value of the constant. Evaluate the limit of the solution as t approaches infinity.

    h. Using the initial condition $p(0) = 0.5$, find the value of the constant. Evaluate the limit of the solution as t approaches infinity.

52. Consider calculating the indefinite integral $\int \frac{1}{x(a+bx)} dx$.

    a. We refine the idea presented in Exercise 51. We break down the integrand into simpler fractions,
    $$\frac{1}{x(a+bx)} = \frac{A}{x} + \frac{B}{a+bx}$$
    where A and B are constants whose value we need to find. Compute the common denominator on the right side, and then compare with the original fraction to conclude that $Aa = 1$ and $Ab + B = 0$.

b. Solve for A and B to show that

$$\frac{1}{x(a+bx)} = \frac{1/a}{x} - \frac{b/a}{a+bx}$$

This expression is called a partial fraction decomposition.

c. Integrate the given function and simplify to show that

$$\int \frac{1}{x(a+bx)} dx = \frac{1}{a} \ln\left|\frac{x}{a+bx}\right| + C$$

53. There are many important differential equations for which separation of variables fails that can be solved with other techniques. An important category involves Newton's law of cooling when the ambient temperature is changing. Consider, in particular, the case where $A(t) = e^{\beta t}$. Assume that the constant α is 1, so the differential equation is $\frac{dT}{dt} = -T + A(t)$. It is impossible to separate variables in this equation.

   a. Create the new variable $y = e^t T$ and find a differential equation for y.

   b. Identify the type of differential equation, and solve it with the initial condition $T(0) = 0$.

   c. Graph your solution and the ambient temperature when β is small, say $\beta = 0.1$. Describe the result.

   d. Graph your solution and the ambient temperature when β is large, say $\beta = 1$. Why are the two curves so much farther apart?

Computer Exercises

54. We have seen that solutions of the differential equation

$$\frac{db}{dt} = b^2$$

approach infinity in a finite amount of time. We will try to stop $b(t)$ from approaching infinity by multiplying the rate of change by a decreasing function of t in the nonautonomous equation

$$\frac{db}{dt} = g(t)b^2$$

Try the following three functions for $g(t)$:

$$g_1(t) = e^{-t}$$
$$g_2(t) = \frac{1}{1+t^2}$$
$$g_3(t) = \frac{1}{1+t}$$

   a. Which of these functions decreases most quickly and should best be able to stop $b(t)$ from approaching infinity?

   b. Have a computer solve the equation in each case with initial conditions ranging from $b(0) = 0.1$ to $b(0) = 5$. Which solutions approach infinity?

   c. Use separation of variables to try to find the solution in each of these cases (have the computer help with the integral of $g_2(t)$). Can you figure out when the solutions approach infinity? How well does this match your results from part (b)?

8.5 Systems of Differential Equations; Predator–Prey Model

We now begin the study of problems in two dimensions. The discrete-time dynamical systems and differential equations we have considered so far have described the dynamics of a single state variable: a population, a concentration, a fraction, and so forth. Most biological systems cannot be fully described without multiple measurements and multiple state variables. The tools for studying these systems, differential equations and discrete-time dynamical systems, remain the same, as does the goal of figuring out what will happen. The methods for getting from the problem to the answer, however, are more complicated. In this section, we introduce two important equations for population dynamics and an extension of Newton's law of cooling that takes into account the changing temperature of the room. These are called **systems of autonomous differential equations** or **coupled autonomous differential equations**. Euler's method can be applied to find approximate solutions of these systems of equations.

Predator–Prey Dynamics

Our basic model of population growth followed a single species that existed undisturbed in isolation. Life is rarely so peaceful. Imagine a bacterial population disrupted by the arrival of a predator, perhaps some sort of amoeba. We expect the bacteria to do worse when more predatory amoebas are around. Conversely, we expect the amoebas to do better when more of their bacterial prey are around. We now translate this intuition into differential equations.

8.5 Systems of Differential Equations; Predator–Prey Model

Denote the population of bacteria at time t by $b(t)$ and the population of amoebas by $p(t)$ to represent predation. We can build the equations by considering the factors affecting the population of each type. Suppose that the bacteria would grow exponentially in the absence of predators according to

$$\frac{db}{dt} = \lambda b$$

How might predation affect the bacterial population? One ecologically naive but mathematically convenient approach is to assume that the organisms obey the principle of mass action.

The Principle of Mass Action: Individual bacteria encounter amoebas at a rate proportional to the number of amoebas.

Doubling the number of predators therefore doubles the rate at which bacteria run into predators. If running into an amoeba spells doom, this doubles the rate at which each bacterium risks being eaten.

In equation form,

$$\text{rate at which an individual bacterium is eaten} = \epsilon p$$

where ϵ is the constant of proportionality. Therefore,

$$\text{per capita growth rate of bacteria} = \lambda - \epsilon p$$

This separates the growth rate into the usual per capita production rate, λ, and the predation rate, ϵp. This equation is illustrated with $\lambda = 1$ and $\epsilon = 0.001$ in Figure 8.5.30a. The growth rate of a population is the product of the per capita growth rate and the population size,

$$\frac{db}{dt} = \text{per capita growth rate} \cdot \text{population size}$$
$$= (\lambda - \epsilon p)b$$

Suppose that the predators have a negative per capita rate of $-\delta$ in the absence of their prey, so

$$\frac{dp}{dt} = -\delta p$$

Solutions of this equation converge to zero. Predators must eat to live. How might predation affect the amoeba population? As above, we assume mass action: doubling the number of bacteria doubles the rate at which predators run into bacteria and thus doubles the rate at which they eat:

$$\text{rate at which an amoeba eats bacteria} = \eta b$$

(η is the Greek letter eta). So the per capita production rate is increased by the eating rate:

$$\text{per capita growth rate of predators} = -\delta + \eta b$$

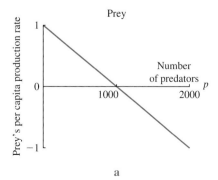
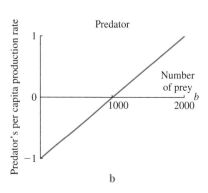

FIGURE 8.5.30
Per capita growth rates of prey and predator species

These per capita growth rates are illustrated with $\delta = 1$ and $\eta = 0.001$ in Figure 8.5.30b. Multiplying the per capita growth rate of the amoebas by their population size gives

$$\frac{dp}{dt} = \text{per capita growth rate} \cdot \text{population size}$$
$$= (-\delta + \eta b)p$$

We combine the differential equations into the **system of autonomous differential equations**

$$\frac{db}{dt} = (\lambda - \epsilon p)b$$
$$\frac{dp}{dt} = (-\delta + \eta b)p$$
(8.5.1)

(where $\lambda, \varepsilon, \delta, \eta > 0$). These are also called **coupled autonomous differential equations** because the rate of change of the bacterial population depends on both their own population size and that of the amoebas, and the rate of change of the amoebas depends on both their own population and that of the bacteria. Two separate measurements are required to find the rate of change of either population. These equations are **autonomous** because neither rate of change depends explicitly on time.

Dynamics of Competition

We can use similar reasoning to describe the competitive interaction between two populations. Our basic model of selection is based on the *uncoupled* pair of differential equations

$$\frac{da}{dt} = \mu a$$
$$\frac{db}{dt} = \lambda b$$

Although there are two measurements, the rate of change of each type does not depend on the population size of the other. This lack of interaction is an idealization. Two populations in a single vessel will probably interact. Suppose that the per capita growth rate of each type declines as a linear function of the total number, $a + b$, according to

$$\text{per capita growth rate of type } a = \mu\left(1 - \frac{a+b}{K_a}\right)$$
$$\text{per capita growth rate of type } b = \lambda\left(1 - \frac{a+b}{K_b}\right)$$

We have written the growth rates in this form to separate out the maximum per capita production μ of type a and λ of type b. Growth of type a becomes negative when the total population exceeds K_a, and growth of type b becomes negative when the total population exceeds K_b. The values K_a and K_b are sometimes called the **carrying capacities** of types a and b, respectively.

Example 8.5.1 Per Capita Growth Rates with Particular Parameter Values

With $\mu = 2$, $\lambda = 2$, $K_a = 1000$, and $K_b = 500$,

$$\text{per capita growth rate of type } a = 2\left(1 - \frac{a+b}{1000}\right)$$

per capita growth rate of type $b = 2\left(1 - \dfrac{a+b}{500}\right)$

See Figure 8.5.31; the horizontal axis represents the total population, $a + b$.

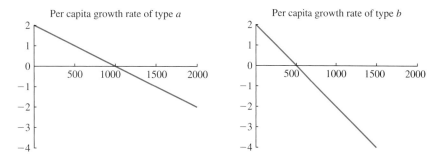

FIGURE 8.5.31

Per capita growth rates of competing species

We find the rates of change of the populations by multiplying the per capita growth rates by the population sizes, finding the coupled system of differential equations

$$\begin{aligned}\dfrac{da}{dt} &= \mu\left(1 - \dfrac{a+b}{K_a}\right)a \\ \dfrac{db}{dt} &= \lambda\left(1 - \dfrac{a+b}{K_b}\right)b\end{aligned} \quad (8.5.2)$$

The behaviour of each population depends both on its own size and on that of the other species. Neither depends explicitly on time, so this system of equations is autonomous.

Example 8.5.2   Competition Model with Parameter Values

With $\mu = 2$, $\lambda = 2$, $K_a = 1000$, and $K_b = 500$ (as in Example 8.5.1), the model becomes

$$\dfrac{da}{dt} = 2a\left(1 - \dfrac{a+b}{1000}\right)$$

$$\dfrac{db}{dt} = 2b\left(1 - \dfrac{a+b}{500}\right)$$

Newton's Law of Cooling

Both Newton's law of cooling and the model of chemical diffusion across a membrane ignored the fact that the ambient temperature or concentration might also change. A small hot object placed in a large room will have little effect on the room temperature, but a large hot object not only will cool off itself but also will heat up the room. We require a system of autonomous differential equations to describe both temperatures simultaneously (Figure 8.5.32). We will now derive the coupled differential equations that describe this situation.

Newton's law of cooling expresses the rate of change of the temperature, T, of an object as a function of the ambient temperature, A, by the equation

$$\dfrac{dT}{dt} = \alpha(A - T)$$

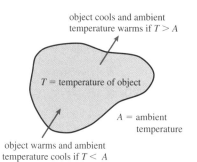

FIGURE 8.5.32

Newton's law of cooling revisited

This equation is valid even if the ambient temperature, A, is itself changing. If $A < T$, heat is leaving the object and warming the room. The room follows the same law as the object itself, but we expect that the factor α, which depends on the size, shape, and material of the object, will be different from that of the object, or that

$$\frac{dA}{dt} = \alpha_2(T - A)$$

The rate of change of temperature of each object depends on the temperature of the other. Together, these equations give the following system of coupled autonomous differential equations:

$$\frac{dT}{dt} = \alpha(A - T)$$

$$\frac{dA}{dt} = \alpha_2(T - A)$$

What is the relation between α and α_2? In general, α_2 will be smaller as the room becomes larger. If the "object" is made of the same material as the "room," the ratio of α to α_2 is equal to the ratio of their sizes.

Example 8.5.3 Newton's Law of Cooling When the Object Is Smaller Than the Room

If the room is three times as big as the object but is made of the same substance (such as a balloon of air), then

$$\alpha_2 = \frac{\alpha}{3}$$

(Figure 8.5.33).

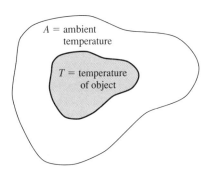

if the ambient space is three times as large as the object, the rate of change of the ambient temperature is one third that of the object

FIGURE 8.5.33
Newton's law of cooling applied to objects of different sizes

Example 8.5.4 Newton's Law of Cooling When the Object Cools More Rapidly Than the Room

Suppose that the specific heat of the object in Example 8.5.3 is 10 times that of the room (meaning that a small amount of heat warms the room a great deal). Then α_2 is smaller by a factor of 3 because the room is larger, but larger by a factor of 10 because the specific heat is smaller. Thus,

$$\alpha_2 = 10\frac{\alpha}{3} \approx 3.33\alpha$$

Applying Euler's Method to Systems of Autonomous Differential Equations

How do we find solutions of systems of autonomous differential equations? In general, this is a very difficult problem. In the next section, we will extend the method of equilibria and phase-line diagrams to sketch solutions. Surprisingly, perhaps, Euler's method for finding approximate solutions with the tangent line approximation works in exactly the same way for systems as for single equations. We will apply the method to figure out how solutions behave for systems describing predator–prey dynamics and competition.

Consider the predator–prey equations in Equation 8.5.1 with the parameter values $\lambda = 1$, $\delta = 1$, $\epsilon = 0.001$, and $\eta = 0.001$:

$$\frac{db}{dt} = (1 - 0.001p)b$$

$$\frac{dp}{dt} = (-1 + 0.001b)p$$

Suppose the initial condition is $b(0) = 800$ and $p(0) = 200$, so, initially, there are 800 prey and 200 predators. What will happen to these populations after 2 time units?

Let's set up Euler's method: as usual, Δt is the time step. We will calculate approximations to the solutions $b = b(t)$ and $p = p(t)$ at times

$$t_0, t_1 = t_0 + \Delta t, t_2 = t_0 + 2\Delta t, t_3 = t_0 + 3\Delta t, \ldots$$

By b_n we denote the approximations for $b(t_n)$, and by p_n the approximations for $p(t_n)$. The initial conditions state that when $t_0 = 0$, then $b_0 = 800$ and $p_0 = 200$.

The Euler's method recursive formula is given by

$$t_{n+1} = t_n + \Delta t$$
$$b_{n+1} = b_n + (1 - 0.001p_n)b_n \Delta t \qquad (8.5.3)$$
$$p_{n+1} = p_n + (-1 + 0.001b_n)p_n \Delta t$$

In the first step, using a time step of $\Delta t = 0.2$, we compute

$$t_1 = 0 + 0.2 = 0.2$$
$$b_1 = b_0 + \Delta t(1 - 0.001p_0)b_0$$
$$= 800 + 0.2(1 - 0.001 \cdot 200)800 = 928$$
$$p_1 = p_0 + \Delta t(-1 + 0.001b_0)p_0$$
$$= 200 + 0.2(-1 + 0.001 \cdot 800)200 = 192$$

In the next step,

$$t_2 = 0.2 + 0.2 = 0.4$$
$$b_2 = b_1 + \Delta t(1 - 0.001p_1)b_1$$
$$= 928 + 0.2(1 - 0.001 \cdot 192)928 = 1078$$
$$p_2 = p_1 + \Delta t(-1 + 0.001b_1)p_1$$
$$= 192 + 0.2(-1 + 0.001 \cdot 928)192 = 189$$

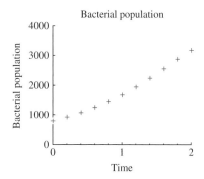
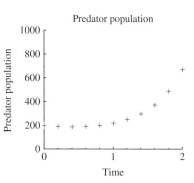

FIGURE 8.5.34

Euler's method applied to the predator–prey equations

Proceeding in the same way (using a programmable calculator or software), we obtain the values recorded in Table 8.5.1; see also Figure 8.5.34.

As we can see, the bacterial population increases. After a small initial dip, the amoeba population recovers and starts increasing.

This is only a small glimpse into the complex dynamics of the two populations. As we will see in the remaining sections of this chapter, the two populations change periodically (oscillate). That is logical: as the bacterial population increases, there is more food for the amoebas, so the amoebas will increase in number. As the amoeba population grows, more and more bacteria are consumed, and the bacterial population will start decreasing. As that happens, there will be less food for the amoeba population, which will force it to decrease. As the amoeba population decreases, there are fewer predators, so the bacterial population will increase, and the cycle is complete.

Table 8.5.1

t	Approximation for $b(t)$	Approximation for $p(t)$
$t_0 = 0$	$b_0 = 800$	$p_0 = 200$
$t_1 = 0.2$	$b_1 = 928$	$p_1 = 192$
$t_2 = 0.4$	$b_2 = 1078$	$p_2 = 189.2$
$t_3 = 0.6$	$b_3 = 1252.8$	$p_3 = 192.2$
$t_4 = 0.8$	$b_4 = 1455.2$	$p_4 = 201.9$
$t_5 = 1$	$b_5 = 1687.4$	$p_5 = 220.3$
$t_6 = 1.2$	$b_6 = 1950.6$	$p_6 = 250.6$
$t_7 = 1.4$	$b_7 = 2242.9$	$p_7 = 298.2$
$t_8 = 1.6$	$b_8 = 2557.8$	$p_8 = 372.3$
$t_9 = 1.8$	$b_9 = 2878.8$	$p_9 = 488.3$
$t_{10} = 2$	$b_{10} = 3173.4$	$p_{10} = 671.8$

Further calculations using Euler's method reveal some of these dynamics (Table 8.5.2). Keep in mind that we are using tangent line approximations, so the farther we go, the more errors accumulate. Beyond a certain point, Euler's method does not give meaningful results.

Table 8.5.2

t	Approximation for $b(t)$	Approximation for $p(t)$
$t_{12} = 2.4$	$b_{12} = 3406.2$ maximum	$p_{12} = 1423$
$t_{13} = 2.6$	$b_{13} = 3118$	$p_{13} = 2107.8$
$t_{16} = 3.2$	$b_{16} = 624$	$p_{16} = 4208.9$ maximum
$t_{17} = 4.4$	$b_{17} = 223.5$	$p_{17} = 3892.4$
$t_{24} = 4.8$	$b_{24} = 19.5$ minimum	$p_{24} = 916.4$
$t_{25} = 5$	$b_{25} = 19.8$	$p_{25} = 736.7$
$t_{51} = 10.2$	$b_{51} = 1173.5$	$p_{51} = 9$ minimum
$t_{52} = 10.4$	$b_{52} = 1406$	$p_{52} = 9.3$

There is no difference between using Euler's method to solve one equation, or two, or more. Nevertheless, we review it here—but instead of using formulas, we do it in words.

▶▶ Algorithm 8.5.1 Euler's Method for Solving Coupled Autonomous Differential Equations

1. Choose a time step; set up the formula

    new time step = previous time step + step size

2. Build the discrete-time dynamical system for approximations of each unknown function (measurement) in the following way:

    next approximation = previous approximation
    + rate of change of the function times the step size

All expressions on the right side refer to the previous situation (past).

Revisit the previous example to see that this is how the formulas were set up. As well, look at the following example.

Example 8.5.5 Applying Euler's Method to the Differential Equations Describing Competition

Suppose we wish to apply Euler's method to the coupled differential equations for competition using the parameter values in Example 8.5.2,

$$\frac{da}{dt} = 2a\left(1 - \frac{a+b}{1000}\right)$$

$$\frac{db}{dt} = 2b\left(1 - \frac{a+b}{500}\right)$$

with initial condition $a(0) = 750$ and $b(0) = 750$ to estimate $a(0.3)$ and $b(0.3)$.

Pick the time step $\Delta t = 0.1$. By a_n we denote the approximations for $a(t)$, and by b_n the approximations for $b(t)$. We need to go for three steps to obtain a_3 (to approximate $a(0.3)$) and b_3 (approximation for $b(0.3)$). The initial conditions state that when $t_0 = 0$, then $a_0 = 750$ and $b_0 = 750$.

The Euler's method recursive formula is given by

$$t_{n+1} = t_n + \Delta t$$

$$a_{n+1} = a_n + 2a_n\left(1 - \frac{a_n + b_n}{1000}\right)\Delta t$$

$$b_{n+1} = b_n + 2b_n\left(1 - \frac{a_n + b_n}{500}\right)\Delta t$$

In the first step, we compute

$$t_1 = 0 + 0.1 = 0.1$$

$$a_1 = a_0 + 2a_0\left(1 - \frac{a_0 + b_0}{1000}\right)\Delta t$$

$$= 750 + 2 \cdot 750\left(1 - \frac{750 + 750}{1000}\right)(0.1) = 675$$

$$b_1 = b_0 + 2b_0\left(1 - \frac{a_0 + b_0}{500}\right)\Delta t$$

$$= 750 + 2 \cdot 750\left(1 - \frac{750 + 750}{500}\right)(0.1) = 450$$

In the next step,

$$t_2 = 0.1 + 0.1 = 0.2$$

$$a_2 = a_1 + 2a_1\left(1 - \frac{a_1 + b_1}{1000}\right)\Delta t$$

$$= 675 + 2 \cdot 675\left(1 - \frac{675 + 450}{1000}\right)(0.1) = 658.1$$

$$b_2 = b_1 + 2b_1\left(1 - \frac{a_1 + b_1}{500}\right)\Delta t$$

$$= 450 + 2 \cdot 450\left(1 - \frac{675 + 450}{500}\right)(0.1) = 337.5$$

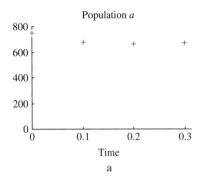
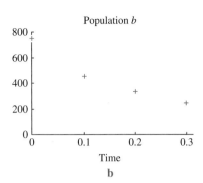

FIGURE 8.5.35
Euler's method applied to the competition equations

As a result of the next step, we would obtain the desired approximations

$$a(0.3) \approx a_3 \approx 658.7$$
$$b(0.3) \approx b_3 \approx 270.6$$

These values are plotted in Figure 8.5.35.

Euler's method is an effective way to compute solutions but requires a great deal of numerical calculation. In the next section, we will begin to develop methods to predict how solutions will behave using graphical techniques.

Summary We have introduced three **coupled autonomous differential equations,** pairs of differential equations in which the rate of change of each state variable depends on its own value and on the value of the other state variable. We derived models of a predator and its prey, a competitive analogue of the system studied to describe selection, and a version of Newton's law of cooling that keeps track of the change in room temperature. Each system is **autonomous** because the rates of change depend only on the state variables and not on time. **Euler's method** can be used to compute approximate solutions of these equations using the tangent line approximation.

8.5 Exercises

Mathematical Techniques

1–4 ▪ Consider the following special cases of the predator–prey equations:

$$\frac{db}{dt} = (\lambda - \epsilon p)b \quad \text{and} \quad \frac{dp}{dt} = (-\delta + \eta b)p$$

Write the differential equations and tell what they mean.

1. $\epsilon = \eta = 0$
2. $\delta = 0$
3. $\eta = 0$
4. $\epsilon = 0$

5–6 ▪ If each state variable in a system of autonomous differential equations does not respond to changes in the value of the other, but depends only on a constant value, the two equations can be considered separately. For example, in the competition equation

$$\frac{da}{dt} = \mu\left(1 - \frac{a+b}{K_a}\right)a$$

we could treat b as a constant value. In each case, find the equilibrium and draw a phase-line diagram for a. Set the parameters to $\mu = 2$ and $K_a = 1000$.

5. Suppose that types a and b do not interact (equivalent to setting $b = 0$ in the differential equation for a).

6. Suppose that types a and b interact with a fixed population of 500 of the other (set $b = 500$ in the differential equation for a).

7–8 ▪ Apply Euler's method to the competition equations

$$\frac{da}{dt} = \mu\left(1 - \frac{a+b}{K_a}\right)a$$

$$\frac{db}{dt} = \lambda\left(1 - \frac{a+b}{K_b}\right)b$$

starting from the given initial conditions. Assume that $\mu = 2$, $\lambda = 2$, $K_a = 1000$, and $K_b = 500$.

7. Use $a(0) = 750$ and $b(0) = 500$. Take two steps, with a step length of $\Delta t = 0.1$.

8. Use $a(0) = 750$ and $b(0) = 500$. Take four steps, with a step length of $\Delta t = 0.05$. How do your results compare with those in Exercise 7?

9–10 ▪ Apply Euler's method to Newton's law of cooling,

$$\frac{dT}{dt} = \alpha(A - T)$$

$$\frac{dA}{dt} = \alpha_2(T - A)$$

with the given parameter values and starting from the given initial conditions.

9. Suppose $\alpha = 0.3$ and $\alpha_2 = 0.1$. Use $T(0) = 60$ and $A(0) = 20$. Take two steps, with a step length of $\Delta t = 0.1$.

10. Suppose $\alpha = 3$ and $\alpha_2 = 1$. Use $T(0) = 60$ and $A(0) = 20$. Take two steps, with a step length of $\Delta t = 0.25$. Do the results look reasonable?

Applications

11–14 ▪ Consider the following types of predator–prey interactions. Graph the per capita rates of change and write the associated system of autonomous differential equations.

11. per capita growth of prey $= 1 - 0.05p$
     per capita growth of predators $= -1 + 0.02b$

12. per capita growth of prey $= 2 - 0.01p$
     per capita growth of predators $= 1 + 0.01b$

    How does this differ from the basic predator–prey system (Equation 8.5.1)?

13. per capita growth of prey $= 2 - 0.0001p^2$
     per capita growth of predators $= -1 + 0.01b$

14. per capita growth of prey $= 2 - 0.01p$
     per capita growth of predators $= -1 + 0.0001b^2$

15–18 ▪ Write systems of differential equations describing the following situations. Feel free to make up parameter values as needed.

15. Two predators that must eat each other to survive.

16. Two predators that must eat each other to survive, but with the per capita growth rate of each reduced by competition with its own species.

17. Two competitors where the per capita growth rate of a is decreased by the total population, and the per capita growth rate of b is decreased by the population of b.

18. Two competitors where the per capita growth rate of each type is affected only by the population size of the other type.

19–20 ▪ Follow these steps to derive the equations for chemical exchange between two adjacent cells of different size. Suppose the concentration in the first cell is designated by the variable C_1 and the concentration in the second cell is designated by the variable C_2. In each case:

a. Write an expression for the total amount of chemical A_1 in the first cell and of chemical A_2 in the second cell.

b. Suppose that the amount of chemical moving from the first cell to the second cell is β times C_1 and that the amount of chemical moving from the second cell to the first cell is β times C_2. Write equations for the rates of change of A_1 and A_2.

c. Divide by the volumes to find differential equations for C_1 and C_2.

d. In which cell is the concentration changing more rapidly?

19. Suppose that the size of the first cell is 2 μL and the size of the second cell is 5 μL.

20. Suppose that the size of the first cell is 5 μL and the size of the second cell is 12 μL.

21–22 ▪ Write systems of autonomous differential equations describing the temperature of an object and the temperature of the room in each case.

21. The size of the room is 10 times that of the object, but the specific heat of the room is 0.2 times that of the object (meaning that a small amount of heat produces a large change in the temperature of the room).

22. The size of the room is five times that of the object, but the specific heat of the room is twice that of the object (meaning that a large amount of heat produces only a small change in the temperature of the room).

23–28 ▪ There are many important extensions of the basic disease model (Equation 8.3.1) that include more categories of people and model processes of birth, death, and immunity. The simplest two-dimensional model involves the same assumptions as the basic model but explicitly tracks the number of susceptible individuals (S) and the number of infected individuals (I) rather than the fraction. Individuals become infected (move from the S class into the I class) at a rate proportional to the product of the number of infected individuals and the number of susceptible people. Individuals recover (move from the I class into the S class) at a rate proportional to the number of infected individuals. Equations modelling this situation are

$$\frac{dI}{dt} = \alpha IS - \mu I$$

$$\frac{dS}{dt} = -\alpha IS + \mu I$$

23. Suppose that individuals who leave the infected class through recovery become permanently immune rather than becoming susceptible again. Write the differential equations, and compare your model with the predator-prey model (Equation 8.5.1).

24. Suppose that one-third of the individuals who leave the infected class through recovery become permanently immune and that the other two-thirds become susceptible again.

25. Suppose that all individuals become susceptible upon recovery (as in the basic model) but that there is a source of mortality, so both infected and susceptible individuals die at per capita rate k.

26. Suppose that all individuals become susceptible upon recovery (as in the basic model) but that there is a source

of mortality, whereby susceptible individuals die at per capita rate k but infected individuals die at a per capita rate that is twice as large.

27. Suppose that all individuals become susceptible upon recovery (as in the basic model) but that all individuals give birth at rate b. The offspring of susceptible individuals are susceptible, and the offspring of infected individuals are infected.

28. Suppose that all individuals become susceptible upon recovery (as in the basic model) but that all individuals give birth at rate b and that all offspring are susceptible.

29–30 ▪ Write the following models as systems of autonomous differential equations for a and b.

29. The situation in Section 8.1, Exercise 48.

30. The situation in Section 8.1, Exercise 49.

Computer Exercises

31. Consider Euler's method for the predator–prey interaction given by Equation 8.5.3. Using a programmable calculator or appropriate software, investigate the system as suggested.

    Starting from the initial condition $(b(0), p(0)) = (800, 200)$, follow this system until it loops around near its initial condition. Use the following values of Δt.

    a. $\Delta t = 1$

    b. $\Delta t = 0.2$

    c. $\Delta t = 0.1$

    d. $\Delta t = 0.01$

32. Follow the same steps as in the previous problem for the spring equations

$$\frac{dx}{dt} = v$$

$$\frac{dv}{dt} = -x$$

How close is your estimated solution to the exact solution? What happens if you keep running for many cycles?

8.6 The Phase Plane

Systems of autonomous differential equations are generally impossible to solve exactly. We have seen how to use Euler's method to find approximate solutions. As with autonomous differential equations and discrete-time dynamical systems, we can deduce a great deal about the behaviour of solutions from an appropriate graphical display. For systems of autonomous differential equations, the tool is the **phase-plane** diagram, an extension of the phase-line diagram. Our goal is again to find **equilibria,** points where each of the state variables remains unchanged. Finding these points on the phase plane requires a new tool, the **nullcline,** a graph of the set of points where each state variable separately remains unchanged.

Equilibria and Nullclines: Predator–Prey Equations

A single autonomous differential equation has an equilibrium where the rate of change of the state variable is zero. An **equilibrium** of a two-dimensional system of autonomous differential equations is a point where the rate of change of all state variables is zero.

Consider again a predator and its prey described by the system of autonomous differential equations

$$\frac{db}{dt} = (1 - 0.001p)b$$

$$\frac{dp}{dt} = (-1 + 0.001b)p$$

(Equation 8.5.1 with $\lambda = 1$, $\delta = 1$, $\epsilon = 0.001$, and $\eta = 0.001$). The rates of change of both b and p are equal to zero when the following equations are satisfied:

$$\frac{db}{dt} = (1 - 0.001p)b = 0$$

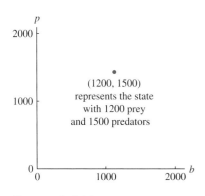

FIGURE 8.6.36

The phase plane with p on the vertical axis and b on the horizontal axis

$$\frac{dp}{dt} = (-1 + 0.001b)p = 0$$

To graph the solutions of

$$\frac{db}{dt} = (1 - 0.001p)b = 0$$

we must pick one of the variables b and p to place on the vertical axis. We can pick either one, and here we choose p. Our next step is to plot an empty graph with the vertical axis labelled p and the horizontal axis labelled b (Figure 8.6.36). This is a graph of the **phase plane.** Like the phase-line diagram, a phase-plane diagram is a picture of all possible values the state variables can take. For example, the point (1200, 1500) represents the system with 1200 prey and 1500 predators.

Now we can solve the equation

$$\frac{db}{dt} = (1 - 0.001p)b = 0$$

to find all values of b and p where the state variable b does not change. We obtain

$$1 - 0.001p = 0 \quad \text{and} \quad b = 0$$
$$p = \frac{1}{0.001} = 1000 \quad \text{and} \quad b = 0$$

In the phase plane, the graph of $p = 1000$ is a horizontal line crossing the p-axis at 1000 (Figure 8.6.37a). The equation $b = 0$ represents a vertical line coinciding with the p-axis (Figure 8.6.37b).

FIGURE 8.6.37

The two components of the b-nullcline

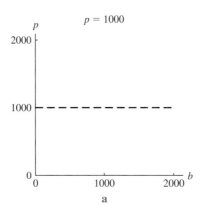
a

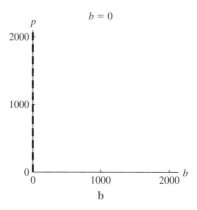
b

These solutions represent the set of values where the rate of change of b is zero. This entire set is called the b-**nullcline** and is graphed in the phase plane in Figure 8.6.38. This might look like a strange graph, consisting as it does of two distinct pieces. Although no function could have a graph like this, it is typical of the behaviour of nullclines.

We use the same method to find the p-nullcline, the set of points where the rate of change of p is zero. The equation

$$\frac{dp}{dt} = (-1 + 0.001b)p = 0$$

has a solution where either of the factors is equal to zero. Thus,

$$-1 + 0.001b = 0 \quad \text{or} \quad p = 0$$

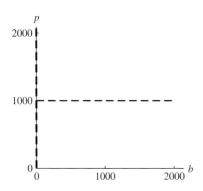

FIGURE 8.6.38

The b-nullcline: points in the phase plane where b does not change

Thus, $b = 1000$ or $p = 0$. The equation $b = 1000$ represents the vertical line crossing the b-axis at 1000 (Figure 8.6.39a). The equation $p = 0$ is the b-axis (Figure 8.6.39b). The entire p-nullcline is the combination of these two pieces (Figure 8.6.40).

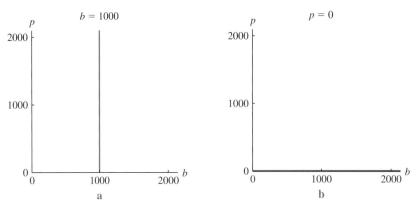

FIGURE 8.6.39

The two components of the p-nullcline

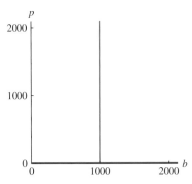

FIGURE 8.6.40

The p-nullcline: points in the phase plane where p does not change

The state variable b does not change on the b-nullcline. The state variable p does not change on the p-nullcline. Therefore, neither b nor p changes at any point where the two nullclines intersect. These intersections, therefore, are the equilibria. To find equilibria, plot both nullclines in the same phase plane, being careful to distinguish which piece belongs to which nullcline (Figure 8.6.41). There are two intersections of the nullclines, at $(0, 0)$ and $(1000, 1000)$. At the first equilibrium, both populations are extinct. At the second, both populations are positive and the system is balanced.

The points $(0, 1000)$ and $(1000, 0)$ are not equilibria because they do not lie at the intersection of the two nullclines. The point $(0, 1000)$ lies on both pieces of the b-nullcline but not on the p-nullcline. We can check whether a point is an equilibrium by substituting into the system of differential equations. At $b = 0$ and $p = 1000$,

$$\frac{db}{dt} = (1 - 0.001p)b = (1 - 0.001 \cdot 1000) \cdot 0 = 0$$

$$\frac{dp}{dt} = (-1 + 0.001b)p = (-1 + 0.001 \cdot 0) \cdot 1000 = -1000$$

Because the rate of change of the predator population p is not equal to zero, this is not an equilibrium. At this point, there are no prey and the predator population declines.

Similarly, the point $(1000, 0)$ lies on both pieces of the p-nullcline but not on the b-nullcline. At $b = 1000$ and $p = 0$,

$$\frac{db}{dt} = (1 - 0.001p)b = (1 - 0.001 \cdot 0) \cdot 1000 = 1000$$

$$\frac{dp}{dt} = (-1 + 0.001b)p = (-1 + 0.001 \cdot 1000) \cdot 0 = 0$$

Again, the rate of change of one state variable is not zero, and this point is not an equilibrium.

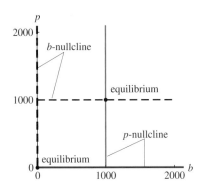

FIGURE 8.6.41

The nullclines of the predator–prey system

The following algorithm gives the steps to find the nullclines and equilibria of a system of autonomous differential equations.

▶▶ **Algorithm 8.6.1** Finding the Nullclines and Equilibria of Coupled Autonomous Differential Equations

1. Decide which variable is represented by the horizontal axis and which one by the vertical axis in the phase plane.

2. Write the equations for the nullclines, and solve them.

3. Graph each solution in the phase plane.

4. Identify the intersections of nullclines belonging to different variables, as these are the equilibria of the system.

Solutions that begin at an equilibrium remain there. To figure out what other solutions do, we would need to assess the stability of the equilibria. The techniques to do this, extensions of the methods used in one dimension, involve *linear algebra*, a subject that is beyond the scope of this book. In the next section, however, we will learn to analyze at least some situations by drawing direction arrows on a phase-plane diagram.

Equilibria and Nullclines: Competition Equations

We can use Algorithm 8.6.1 to find the equilibria of competing bacterial types that follow the equations

$$\frac{da}{dt} = 2\left(1 - \frac{a+b}{1000}\right)a$$

$$\frac{db}{dt} = 2\left(1 - \frac{a+b}{500}\right)b$$

(Equation 8.5.2 with $\mu = 2$, $\lambda = 2$, $K_a = 1000$, and $K_b = 500$).

1. We decide to place a on the horizontal axis and b on the vertical axis.

2. To find the a-nullcline, we solve

$$\frac{da}{dt} = 2\left(1 - \frac{a+b}{1000}\right)a = 0$$

$$2\left(1 - \frac{a+b}{1000}\right) = 0 \quad \text{or} \quad a = 0$$

We solve the first equation for b, finding

$$2\left(1 - \frac{a+b}{1000}\right) = 0$$

$$1 - \frac{a+b}{1000} = 0$$

$$\frac{a+b}{1000} = 1$$

$$a + b = 1000$$

$$b = 1000 - a$$

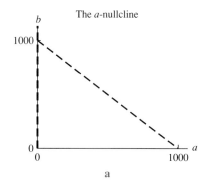

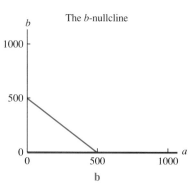

FIGURE 8.6.42
The nullclines of the competition system

This is a line with slope -1 and b-intercept 1000. The second factor represents the vertical line at $a = 0$; see Figure 8.6.42a.

To find the b-nullcline, we solve

$$\frac{db}{dt} = 2\left(1 - \frac{a+b}{500}\right)b = 0$$

$$2\left(1 - \frac{a+b}{500}\right) = 0 \quad \text{or} \quad b = 0$$

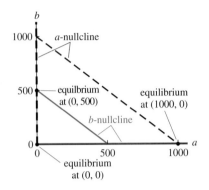

FIGURE 8.6.43

The nullclines and equilibria of the competition system

We solve the first equation:

$$1 - \frac{a+b}{500} = 0$$
$$\frac{a+b}{500} = 1$$
$$a + b = 500$$
$$b = 500 - a$$

This is a line with slope -1 and b-intercept 500. The second factor is a horizontal line at $b = 0$; see Figure 8.6.42b.

3. All nullclines are shown in Figure 8.6.43.

4. To find the equilibria, we plot both nullclines on the same graph of the phase plane (Figure 8.6.43). There are three intersections and thus three equilibria: at $(0, 0)$, $(1000, 0)$, and $(0, 500)$. Both types are extinct at $(0, 0)$, type a dominates the population at $(1000, 0)$, and type b dominates the population at $(0, 500)$.

The points $(0, 1000)$ and $(500, 0)$ are not equilibria. At $(0, 1000)$,

$$\frac{db}{dt} = 2\left(1 - \frac{1000}{500}\right)(1000) = -2000 < 0$$

The point lies on both pieces of the a-nullcline but not on the b-nullcline. Similarly, at $(500, 0)$,

$$\frac{da}{dt} = 2\left(1 - \frac{500}{1000}\right)(500) = 500 > 0$$

This point lies on both branches of the b-nullcline but not on the a-nullcline.

Equilibria and Nullclines: Newton's Law of Cooling

We use the same steps to find the equilibria for Newton's law of cooling:

$$\frac{dT}{dt} = \alpha(A - T)$$
$$\frac{dA}{dt} = \alpha_2(T - A)$$

(assume that $\alpha > 0$ and $\alpha_2 > 0$).

1. We choose the temperature, T, to be on the vertical axis.

2. The A-nullcline is the set of points where

$$\frac{dA}{dt} = \alpha_2(T - A) = 0$$

Thus, $T - A = 0$ and $T = A$. The graph is the diagonal in the first quadrant; see Figure 8.6.44a.

The T-nullcline is the set of points where

$$\frac{dT}{dt} = \alpha(A - T) = 0$$

Thus, $A - T = 0$ and $T = A$, the same as before; see Figure 8.6.44b.

3. Both nullclines are shown in Figure 8.6.45.

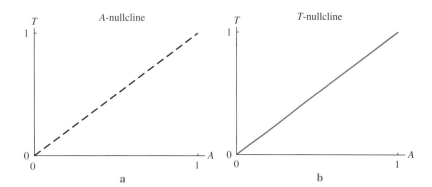

FIGURE 8.6.44
The nullclines of Newton's law of cooling

4. The two nullclines exactly overlap. Because all points where the two nullclines intersect are equilibria, every point on the line $T = A$ is an equilibrium (Figure 8.6.45).

This result makes sense. When the two temperatures are the same, there is no further change in temperature by either the object or the room.

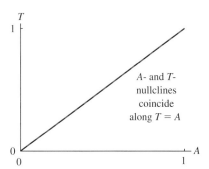

FIGURE 8.6.45
The nullclines and equilibria of Newton's law of cooling

Summary   We have introduced the **phase plane, nullclines,** and **equilibria** as tools to study autonomous systems of differential equations. Equilibria occur where the rate of change of all state variables is zero. They can be found in the phase plane (the Cartesian plane with axes labelled by the state variables) by graphing the two nullclines. The nullcline associated with a state variable is the set of values where that state variable remains unchanged.

8.6 Exercises

Mathematical Techniques

▼1–4 ■ Finding equilibria of coupled differential equations requires solving simultaneous equations. The following are linear equations, where the only possibilities are no solutions, one solution, or a whole line of solutions. For each pair, solve each equation for y in terms of x, set the two equations for y equal, and solve for x. Check that both equations give the same value for y. Sketch the graph with y on the vertical axis.

1.
$-3 - y + 3x = 0$
$-2 + 2y - 4x = 0$

2.
$-6 + 3y + 3x = 0$
$-2 - 2y - 6x = 0$

3.
$3 - 3y + 3x = 0$
$-2 + 2y - 2x = 0$

What goes wrong? Use your graph with y on the vertical axis to explain the problem.

4.
$8 - 2y + 4x = 0$
$-2 + y - 2x = 0$

What goes wrong? Use your graph with y on the vertical axis to explain the problem.

5–10 ■ Finding equilibria of nonlinear coupled differential equations requires solving nonlinear simultaneous equations, which can have any number of solutions. For each pair, solve each equation for y in terms of x, set the two equations for y equal, and solve for x. Check that both equations give the same value for y. Sketch the graph with y on the vertical axis.

5. $$-5 - y + 3x^2 + 2x = 0$$
   $$-2 + 2y - 4x = 0$$

6. $$-3 - y + 3x^2 + 2x = 0$$
   $$-2 + 2y - 4x + 2x^2 = 0$$

7. $$-y^2 + x^2 = 0$$
   $$-2 + 2y - 4x = 0$$
   Solving the first equation for y in terms of x does not give a function. Graph the relation and find the solutions.

8. $$y(y - x^2) = 0$$
   $$-6 + 2y - 4x = 0$$
   Solving the first equation for y in terms of x does not give a function. Graph the relation and find the solutions.

9. $$x(y - x) = 0$$
   $$-6 + 2y - 4x = 0$$
   Solving the first equation for y in terms of x includes a vertical section. Graph the relation and find the solutions.

10. $$(x - 1)(y^2 - x^2) = 0$$
    $$-2 + 2y - 6x = 0$$
    Solving the first equation for y in terms of x does not give a function and includes a vertical section. Graph the relation and find the solutions.

11–16 ■ Graph the nullclines in the phase plane and find the equilibria of each of the following.

11. Predator–prey model
    $$\frac{db}{dt} = (\lambda - \epsilon p)b$$
    $$\frac{dp}{dt} = (-\delta + \eta b)p$$
    (Equation 8.5.1) with $\lambda = 1$, $\delta = 3$, $\epsilon = 0.002$, and $\eta = 0.005$.

12. Predator–prey model with $\lambda = 1$, $\delta = 3$, $\epsilon = 0.005$, and $\eta = 0.002$.

13. Newton's law of cooling
    $$\frac{dT}{dt} = \alpha(A - T)$$
    $$\frac{dA}{dt} = \alpha_2(T - A)$$
    with $\alpha = 0.01$ and $\alpha_2 = 0.1$.

14. Newton's law of cooling with $\alpha = 0.1$ and $\alpha_2 = 0.5$.

15. Competition model
    $$\frac{da}{dt} = \mu\left(1 - \frac{a+b}{K_a}\right)a$$
    $$\frac{db}{dt} = \lambda\left(1 - \frac{a+b}{K_b}\right)b$$
    (Equation 8.5.2) with $\lambda = 2$, $\mu = 1$, $K_a = 10^6$, and $K_b = 10^7$.

16. Competition model with $\lambda = 1$, $\mu = 2$, $K_a = 10^6$, and $K_b = 10^7$. How do the results compare with those in Exercise 15? Why?

17–20 ■ Redraw the phase planes for the following problems, but make the other choice for the vertical variable. Check that you get the same equilibrium.

17. The equations in Exercise 11.
18. The equations in Exercise 12.
19. The equations in Exercise 15.
20. The equations in Exercise 16.

21–22 ■ If each state variable in a system of autonomous differential equations does not respond to changes in the value of the other, but depends only on a constant value, the two equations can be considered separately. In this case, the phase plane is particularly simple. Find the nullclines and equilibria in the following cases.

21. System 8.5.2: in the equation for da/dt set $b = 0$; in the equation for db/dt set $a = 0$. Use the values $\mu = 2$, $\lambda = 3$, $K_a = 1000$, and $K_b = 300$.

22. System 8.5.2: in the equation for da/dt set $b = 500$; in the equation for db/dt set $a = 500$. Use the values $\mu = 2$, $\lambda = 3$, $K_a = 1000$, and $K_b = 300$.

Applications

23–26 ■ Find the nullclines and equilibria for the following predator–prey models.

23. $\frac{db}{dt} = (1 - 0.05p)b$, $\frac{dp}{dt} = (-1 + 0.02b)p$

24. $\frac{db}{dt} = (2 - 0.01p)b$, $\frac{dp}{dt} = (1 + 0.01b)p$

25. $\frac{db}{dt} = (2 - 0.0001p^2)b$, $\frac{dp}{dt} = (-1 + 0.01b)p$

26. $\frac{db}{dt} = (2 - 0.01p)b$, $\frac{dp}{dt} = (-1 + 0.0001b^2)p$

27–30 ■ Find and graph the nullclines, and find the equilibria for the following models.

27. $\frac{dp_1}{dt} = (-1 + 0.001p_2)p_1$, $\frac{dp_2}{dt} = (-1 + 0.001p_1)p_2$

28. $\frac{dp_1}{dt} = (-1 + 0.001p_2 - 0.001p_1)p_1$
    $$\frac{dp_2}{dt} = (-1 + 0.001p_1 - 0.001p_2)p_2$$

29. $\frac{da}{dt} = \mu\left(1 - \frac{a+b}{K_a}\right)a$, $\frac{db}{dt} = \lambda\left(1 - \frac{b}{K_b}\right)b$

30. $\frac{da}{dt} = \mu\left(1 - \frac{b}{K_a}\right)a$, $\frac{db}{dt} = \lambda\left(1 - \frac{a}{K_b}\right)b$

31–32 ■ The models of diffusion derived in Section 8.5, Exercises 19 and 20, assume that the membrane between the vessels is equally permeable in both directions. Suppose instead that the constant of proportionality governing the rate at which chemical moves differs in the two directions. In each of the following cases:

a. Find the rate at which chemical moves from the smaller to the larger vessel.

b. Find the rate at which chemical moves from the larger to the smaller vessel.

c. Find the rate of change of the amount of chemical in each vessel.

d. Divide by the volumes, V_1 and V_2, to find the rate of change of concentration.

e. Find and graph the nullclines.

f. What are the equilibria? Do they make sense?

31. The constant of proportionality governing the rate at which chemical enters the cell is three times as large as the constant governing the rate at which it leaves (as in Section 8.1, Exercise 44).

32. The constant of proportionality governing the rate at which chemical enters the cell is half the constant governing the rate at which it leaves (as in Section 8.1, Exercise 45).

33–34 ■ In our model of competition, the per capita growth rates of types a and b are functions only of the total population size. This means that reproduction is reduced just as much by an individual of type a as by an individual of type b. In many systems, each type interferes differently with type a than with type b. Check that the given set of equations matches the assumptions in each case, and find and graph the equilibria and nullclines.

33. Suppose that individuals of type b reduce the per capita growth rate of type a by half as much as individuals of type a, and that individuals of type a reduce the per capita growth rate of type b by twice as much as individuals of type b. The equations are

$$\frac{da}{dt} = \left(1 - \frac{a+b/2}{1000}\right)a$$

$$\frac{db}{dt} = \left(1 - \frac{2a+b}{1000}\right)b$$

34. Suppose that individuals of type b reduce the per capita growth rate of type a by half as much as individuals of type a, and that individuals of type a reduce the per capita growth rate of type b by half as much as individuals of type b. The equations are

$$\frac{da}{dt} = \left(1 - \frac{a+b/2}{1000}\right)a$$

$$\frac{db}{dt} = \left(1 - \frac{a/2+b}{1000}\right)b$$

(There should be four equilibria).

35–40 ■ Draw the nullclines and find equilibria of the following extensions of the basic disease model (for the context, see Section 8.5, Exercises 23–28).

35. The model $I' = \alpha IS - \mu I$, $S' = -\alpha IS$. Find the nullclines and equilibria of this model when $\alpha = 2$ and $\mu = 1$.

36. The model $I' = \alpha IS - \mu I$, $S' = -\alpha IS + 2\mu I/3$. Find the nullclines and equilibria of this model when $\alpha = 2$ and $\mu = 1$.

37. The model $I' = \alpha IS - \mu I - kI$, $S' = -\alpha IS + \mu I - kS$. Find the nullclines and equilibria of this model when $\alpha = 2$, $\mu = 1$, and $k = 0.5$.

38. The model $I' = \alpha IS - \mu I - 2kI$, $S' = -\alpha IS + \mu I - kS$. Find the nullclines and equilibria of this model when $\alpha = 2$, $\mu = 1$, and $k = 4$.

39. The model $I' = bI + \alpha IS - \mu I$, $S' = bS - \alpha IS + \mu I$. Find the nullclines and equilibria of this model when $\alpha = 2$, $\mu = 1$, and $b = 2$.

40. The model $I' = \alpha IS - \mu I$, $S' = bI + bS - \alpha IS + \mu I$. Find the nullclines and equilibria of this model when $\alpha = 2$, $\mu = 1$, and $b = 1$.

Computer Exercises

41. One complicated equation for chemical kinetics is the Schnakenberg reaction. Let A and B denote the concentrations of two chemicals, A and B. A is added at constant rate k_1, B is added at constant rate k_4, A breaks down at rate k_2, and B is converted into A with an **autocatalytic reaction** (a reaction where the product increases the reaction rate). The equations are

$$\frac{dA}{dt} = k_1 - k_2 A + k_3 A^2 B$$

$$\frac{dB}{dt} = k_4 - k_3 A^2 B$$

The final term is somewhat like the term αIS in epidemic equations, but it differs in that the rate of the reaction becomes faster the larger the concentration of A. Suppose that $k_2 = k_3 = 1$.

a. Have a computer draw the nullclines and find the equilibria in the case $k_1 = 0.2$ and $k_4 = 2$.

b. Do the same with $k_1 = -0.2$ and $k_4 = 2$.

c. Try to explain your results.

42. A variant of the Schnakenberg reaction has the chemical A inhibiting its own production. In particular, assume that we replace $k_3 A^2 B$ with

$$k_3 \frac{AB}{1+A}$$

Suppose that $k_2 = k_3 = 1$.

a. Write the equations describing this system.

b. Have a computer draw the nullclines and find the equilibria in the case $k_1 = 0.2$ and $k_4 = 2$.

c. Do the same with $k_1 = -0.2$ and $k_4 = 2$.

d. Try to explain your results.

8.7 Solutions in the Phase Plane

We have seen how to find nullclines and equilibria in the phase plane. Our real goal is to find *solutions,* descriptions of how the state variables change over time. We begin by graphing the results from Euler's method in the phase plane. To deduce the behaviour of solutions without all the calculations necessary for Euler's method, we will add **direction arrows** to the phase-plane diagram. Like the arrows that appear in phase-line diagrams, they indicate where solutions are increasing or decreasing and can be used to sketch **phase-plane trajectories,** or solutions in the phase plane.

Euler's Method in the Phase Plane

We applied Euler's method to the predator–prey equations

$$\frac{db}{dt} = (1 - 0.001p)b$$

$$\frac{dp}{dt} = (-1 + 0.001b)p$$

in Section 8.5 and computed the values for 10 steps of length $\Delta t = 0.2$ starting from the initial condition $b(0) = 800$ and $p(0) = 200$:

t	Approximation for b	Approximation for p
0	800	200
0.2	928	192
0.4	1078	189.2
0.6	1252.8	192.2
0.8	1455.2	201.9
1	1687.4	220.3
1.2	1950.6	250.6
1.4	2242.9	298.2
1.6	2557.8	372.3
1.8	2878.8	488.3
2	3173.4	671.8

We can plot these values in the phase plane (Figure 8.7.46). The value at $t = 0$ is plotted as the point (800, 200) in the phase plane, the value at $t = 0.2$ is plotted as (928, 192), and so forth. Following the points through time, we see that the prey population increases steadily and that the predator population begins by decreasing and then increases.

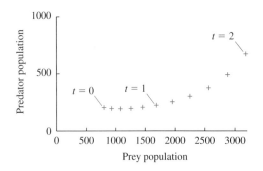

FIGURE 8.7.46

Results from Euler's method plotted in the phase plane

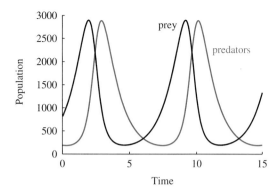

FIGURE 8.7.47

Solutions of the predator–prey system

Euler's method does not provide an exact solution, however, and takes a lot of calculation. More precise techniques correct the errors produced by using the tangent line approximation and can be programmed on a computer to generate solutions accurate to any desired level. A solution generated with one such method (called the Runge–Kutta method; we do not give details here) is plotted in Figure 8.7.47. The initial condition in this graph is $b(0) = 800$ and $p(0) = 200$. The solution consists of two curves, one for each of the state variables. At any time t, the values of b and p can be read from the graphs of b and p.

The populations oscillate, with the peak predator population occurring after the peak prey population.

As with the results from Euler's method, we can graph these solutions in the phase plane as a **phase-plane trajectory** (Figure 8.7.48). The initial condition, $(800, 200)$, is plotted at the point $(800, 200)$ (labelled $t = 0$). At $t = 2$, the population of prey is 2894 and the population of predators is 1051, so the point $(2894, 1051)$ is plotted in the phase plane (the values 2894 and 1051 were obtained using the aforementioned Runge–Kutta method). Because time does not appear explicitly as part of a phase-plane trajectory, we have labelled several points with the time. The initial condition, (b_0, p_0), is labelled with $t = 0$. We have graphed three things on one graph: the time, the prey population, and the predator population.

How can we translate back and forth between the solutions plotted as functions of time and the phase-plane trajectory? Starting from the solution, we can plot the number of predators against the number of prey at several times (such as $t = 0$, $t = 1$, etc.) and connect the dots. Starting from the phase-plane trajectory, we can sketch the solutions by tracing along the graph at a constant speed. The horizontal location of our pencil gives the prey population, or the height of the graph of $b(t)$. The vertical location of our pencil gives the predator population, or the height of the graph of $p(t)$. In our example, we see that the prey population begins by increasing, reaches a maximum value of

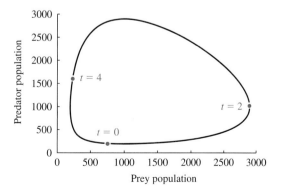

FIGURE 8.7.48

Solution of the predator–prey equations in the phase plane

FIGURE 8.7.49

Graphing a solution and a phase-plane trajectory

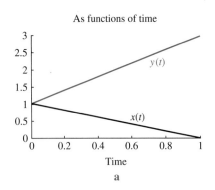
a

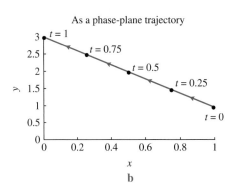
b

nearly 3000, and then decreases to a value of about 200 before beginning to increase again. The predator population decreases to slightly below 200 before beginning an increase to nearly 3000. The predators reach their maximum after the prey and then decrease again.

Example 8.7.1 Graphing a Solution and a Phase-Plane Trajectory

Suppose that $x(t) = 1 - t$ and $y(t) = 1 + 2t$ are the solutions of two coupled differential equations for $0 \leq t \leq 1$. Both $x(t)$ and $y(t)$ are linear functions of time. We can plot $x(t)$ by computing $x(0) = 1$ and $x(1) = 0$ and connecting the points $(0, 1)$ and $(1, 0)$ with a line. Similarly, we can plot $y(t)$ by computing $y(0) = 1$ and $y(1) = 3$ and connecting the points $(0, 1)$ and $(1, 3)$ with a line (Figure 8.7.49a). To plot the phase-plane trajectory, we can make a chart of values:

t	x(t)	y(t)
0	1	1
0.25	0.75	1.5
0.5	0.5	2
0.75	0.25	2.5
1	0	3

The values $x(t)$ and $y(t)$ are plotted as the points $(x(t), y(t))$ in Figure 8.7.49b, labelled by the corresponding values of t.

Example 8.7.2 Translating a Graph of a Solution into a Phase-Plane Trajectory

Suppose that we are given the graph of a solution in Figure 8.7.50a. To plot the phase-plane trajectory, we identify the values of t for which the state variables $A(t)$ and $B(t)$ are increasing and decreasing.

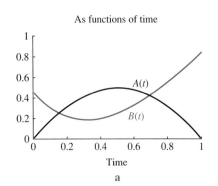
a

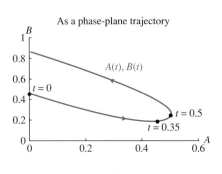
b

FIGURE 8.7.50

Translating the graph of a solution into a phase-plane trajectory

From $t=0$ until about $t=0.35$, $A(t)$ is increasing and $B(t)$ is decreasing. From $t=0.35$ until $t=0.5$, $A(t)$ is increasing and $B(t)$ is increasing. From $t=0.5$ on, $A(t)$ is decreasing and $B(t)$ is increasing (see Figure 8.7.50b):

t	Behaviour of $A(t)$	Behaviour of $B(t)$
0–0.35	increasing	decreasing
0.35–0.5	increasing	increasing
0.5–1	decreasing	increasing

Example 8.7.3 Translating a Phase-Plane Trajectory into the Graph of a Solution

Suppose instead that we are given the graph of the phase-plane trajectory in Figure 8.7.51a. To plot the solution, we use the phase-plane trajectory to identify

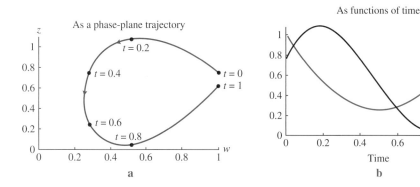

FIGURE 8.7.51

Translating a phase-plane trajectory into the graph of a solution

values of t for which the state variables w and z are increasing and decreasing. From $t=0$ until about $t=0.2$, $w(t)$ is decreasing and $z(t)$ is increasing. From $t=0.2$ until $t=0.5$, $w(t)$ is decreasing and $z(t)$ is decreasing. From $t=0.5$ until $t=0.75$, $w(t)$ is increasing and $z(t)$ is decreasing. From $t=0.75$ on, both $w(t)$ and $z(t)$ are increasing; see Figure 8.7.51b:

t	Behaviour of $w(t)$	Behaviour of $z(t)$
0–0.2	decreasing	increasing
0.2–0.5	decreasing	decreasing
0.5–0.75	increasing	decreasing
0.75–1	increasing	increasing

Direction Arrows: Predator–Prey Equations

The numerical solutions (Figures 8.7.47 and 8.7.48) give an accurate description of the dynamics. Our goal, however, is to understand the behaviour of the populations without solving the equations. For one-dimensional systems, we sketched solutions from the direction arrows and equilibria on a phase-line diagram. We know how to draw nullclines and find equilibria on a phase-plane diagram. The next step is to figure out how to draw the direction arrows.

We have redrawn the nullclines and equilibria in the phase plane for the predator–prey system (Figure 8.7.52). The nullclines break the phase plane into four regions, labelled I, II, III, and IV. In each, we wish to determine whether the populations of predators and prey are increasing or decreasing.

FIGURE 8.7.52

Regions of the phase plane for the predator–prey equations

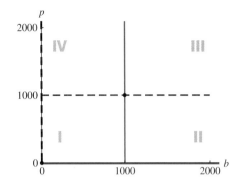

There are three different approaches to finding this out.

- Method 1: Pick a pair of values, (b, p), in the region and substitute into the differential equation. Check whether $\frac{db}{dt}$ and $\frac{dp}{dt}$ are positive or negative.
- Method 2: Manipulate the inequalities algebraically.
- Method 3: Reason about the equations.

Which method is best depends on the equations. We will apply all three to the predator–prey phase plane.

In region I, the predator population is below the b-nullcline $p = 1000$, and the prey population is below the p-nullcline $b = 1000$. Method I requires picking a pair of values in this region. One point in this region is $(500, 500)$. We can substitute this into the differential equations to check whether the populations are increasing or decreasing. With these values,

$$\frac{db}{dt} = (1 - 0.001 \cdot 500) \cdot 500 = 250 > 0$$

$$\frac{dp}{dt} = (-1 + 0.001 \cdot 500) \cdot 500 = -250 < 0$$

The prey population is increasing because $\frac{db}{dt} > 0$, and the predator population is decreasing because $\frac{dp}{dt} < 0$. We indicate this by an arrow pointing toward larger values of the prey (to the right) and toward smaller values of the predator (down) (Figure 8.7.53). This is a **direction arrow.**

Method 2 is algebraic. If we are in region I, then

$$0 < b < 1000$$
$$0 < p < 1000$$

FIGURE 8.7.53

Direction arrow in region I of the predator–prey phase plane

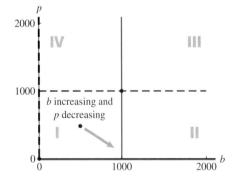

FIGURE 8.7.54

Direction arrows for the predator–prey equations

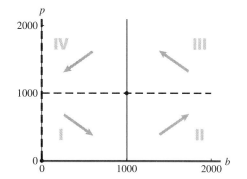

Then

$$\frac{db}{dt} = (1 - 0.001p)b > 0$$

because both $1 - 0.001p > 0$ and $b > 0$. The prey population increases in this region. Similarly,

$$\frac{dp}{dt} = (-1 + 0.001b)p < 0$$

because $-1 + 0.001b < 0$ and $p > 0$. The predator population decreases. Again, the direction arrow points to the right and down.

Method 3 uses reasoning about the equations. In region I, the prey and predator populations are both low. This means that the prey are happy (few predators to eat them) and the predators are sad (too few prey to eat). The prey population will increase and the predator population will decrease.

We can use any of the three methods to find the direction arrows in the three remaining regions (Figure 8.7.54). Using the reasoning method, in region II, we find that the prey population is large and the predator population is small. Both species should be happy and should increase, generating a direction arrow that points up and to the right. In region III, both populations are large. This is bad for the prey and good for the predators. The direction arrow therefore points toward lower values of prey (to the left) and larger values of predators (up). Finally, in region IV, the prey population is small and the predator population is large. Neither species does well under these circumstances, and the direction arrow points left and down.

As on a phase-line diagram, solutions follow the arrows (Figure 8.7.55). Starting in region I, where predator and prey populations are small, the arrow points down and to the right, pushing the solution into region II. Both populations then increase, following the arrows up and to the right into region III. The predators then increase while the prey decrease, moving the population into region IV, from which both populations decrease into region I. We cannot tell from this description whether the phase-plane trajectory circles around, spirals toward, or spirals away from the equilibrium.

FIGURE 8.7.55

Direction arrows and solution for the predator–prey equations

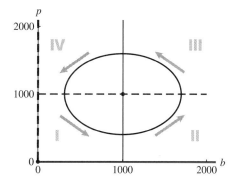

FIGURE 8.7.56

Direction arrows on nullclines

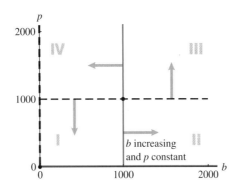

What happens to direction arrows right on the nullclines? For example, what happens to a population with $b = 1000$ and $p < 1000$? This point lies on the p-nullcline, meaning that the population of predators remains unchanged. If an increasing predator population is associated with an upward-pointing arrow and a decreasing predator population is associated with a downward-pointing arrow, an unchanging predator population must be associated with an arrow that points neither up nor down. Such an arrow is horizontal. The population of prey, however, is changing. Because the number of predators is small, the prey population will increase and the arrow will point to the right (Figure 8.7.56). Alternatively, we use the equations: when $b = 1000$ and $p < 1000$, then

$$\frac{db}{dt} = (1 - 0.001p)b > 0$$

and so the prey population increases. Because we are on the p-nullcline, $dp/dt = 0$ and therefore the arrow must be horizontal. The other four direction arrows on nullclines can be found in a similar way.

Finding the direction arrows on the nullclines is useful for two reasons. First, it provides a useful check on the rest of the direction arrows. The directions can only change one at a time; when we move from region I to region II, the arrow switches from pointing down and to the right to pointing up and to the right. The only change was the vertical direction, and this change happens right at the nullcline where the arrow is horizontal. Second, these direction arrows can help in sketching more accurate phase-plane trajectories. Because solutions must follow the arrows, solutions must be horizontal when they cross the p-nullcline and vertical when they cross the b-nullcline. The solution sketched in Figure 8.7.55 satisfies these criteria.

Direction Arrows: Competition Equations

For the competition model, the nullclines break the phase plane into three regions (Figure 8.7.57). We have again used the parameter values $K_a = 1000$ and $K_b = 500$, and we assume that μ and λ are positive. The a-nullcline lies above the b-nullcline. We can determine the direction arrows by using each of the three methods: substituting values,

FIGURE 8.7.57

Direction arrows for the competition system

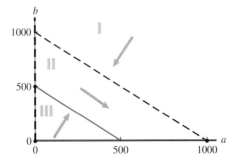

using algebra, and reasoning. With the first method, a point in region I is (1000, 1000). Then

$$\frac{da}{dt} = \mu \left(1 - \frac{1000 + 1000}{1000}\right) \cdot 1000 = -1000\mu < 0$$

$$\frac{db}{dt} = \lambda \left(1 - \frac{1000 + 1000}{500}\right) \cdot 1000 = -3000\lambda < 0$$

Because both derivatives are negative, both a and b are decreasing, meaning that the direction arrow points down and to the left. Region II is defined by

$$500 < a + b < 1000$$

so one point in the region is (400, 400). Then

$$\frac{da}{dt} = \mu \left(1 - \frac{400 + 400}{1000}\right) \cdot 400 = 80\mu > 0$$

$$\frac{db}{dt} = \lambda \left(1 - \frac{400 + 400}{500}\right) \cdot 400 = -240\lambda < 0$$

The direction arrow points down and to the right. The point (100, 100) lies in region III, where

$$\frac{da}{dt} = \mu \left(1 - \frac{100 + 100}{1000}\right) \cdot 100 = 80\mu > 0$$

$$\frac{db}{dt} = \lambda \left(1 - \frac{100 + 100}{500}\right) \cdot 100 = 60\lambda > 0$$

The arrow points up and to the right, because the rate of change of each state variable is positive.

Algebraically, in region I, $a + b > 1000$, so

$$1 - \frac{a+b}{1000} < 0$$

$$1 - \frac{a+b}{500} < 0$$

Therefore,

$$\frac{da}{dt} = \mu \left(1 - \frac{a+b}{1000}\right) a < 0$$

$$\frac{db}{dt} = \lambda \left(1 - \frac{a+b}{500}\right) b < 0$$

Both populations decrease, and the direction arrow points down and to the left. In region II, $500 < a + b < 1000$ and

$$\frac{da}{dt} = \mu \left(1 - \frac{a+b}{1000}\right) a > 0$$

$$\frac{db}{dt} = \lambda \left(1 - \frac{a+b}{500}\right) b < 0$$

The population of a increases, the population of b decreases, and the direction arrow points down and to the right. In region III, $a + b < 500$, so

$$\frac{da}{dt} = \mu \left(1 - \frac{a+b}{1000}\right) a > 0$$

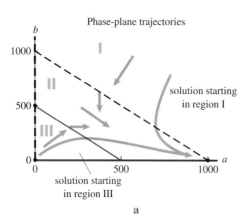
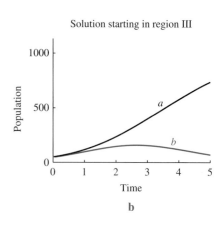

Figure 8.7.58
Solutions of the competition system

$$\frac{db}{dt} = \lambda \left(1 - \frac{a+b}{500}\right) b > 0$$

Both populations increase, and the direction arrow points up and to the right.

With the reasoning method, think of $K_a = 1000$ and $K_b = 500$ as the largest total populations that types a and b can tolerate. In region I, the total population exceeds both K_a and K_b. Both types suffer from overpopulation and have shrinking populations, generating a direction arrow that points down and to the left. In region II, the total population lies between 500 and 1000. Type b cannot withstand competition that type a can tolerate, producing a direction arrow that points down and to the right. In region III, the total population is less than both K_a and K_b. Both types can grow, generating a direction arrow pointing up and to the right.

The two phase-plane trajectories plotted in Figure 8.7.58a follow the arrows and are forced toward the equilibrium at $(1000, 0)$, where type b has been driven extinct. This figure includes the direction arrows on the nullclines. On the a-nullcline, the population of b is decreasing because the total population is 1000, exceeding the tolerance of type b. The direction arrow therefore points straight down. On the b-nullcline, the population of a is increasing because the total population is 500, less than K_a for type a. The direction arrow points straight to the right. The solution starting in region III is plotted as a function of time in Figure 8.7.58b. Even though type b is doomed to extinction, it has an initial period of growth. The solution for b reaches a maximum when the trajectory crosses the b-nullcline.

This model differs in several ways from the differential equations

$$\frac{da}{dt} = \mu a$$

$$\frac{db}{dt} = \lambda b$$

(which we reduced to a single equation for the fraction p of type a in Section 8.1). When $\lambda < \mu$, type b grows more slowly and goes extinct. In the competition model, the type better able to withstand competition (the one with the larger value of K) eventually wins out, even if its growth rate is lower.

Direction Arrows: Newton's Law of Cooling

The direction arrows are simpler for Newton's law of cooling. The nullclines coincide and break the plane into only two regions (Figure 8.7.59). In region I, $T > A$. Therefore,

$$\frac{dT}{dt} = \alpha(A - T) < 0$$

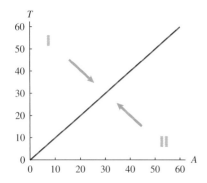

Figure 8.7.59
Direction arrows for Newton's law of cooling

$$\frac{dA}{dt} = \alpha_2(T - A) > 0$$

meaning that T decreases and A increases. The direction arrow points down and to the right. In region II, $T < A$. Therefore,

$$\frac{dT}{dt} = \alpha(A - T) > 0$$

$$\frac{dA}{dt} = \alpha_2(T - A) < 0$$

meaning that T increases and A decreases. The direction arrow points up and to the left. Solutions are pushed toward the line of equilibria. Physically, this means that the object and the room will tend to approach the same temperature.

Summary We have seen how to plot solutions of two-dimensional differential equations as functions of time and as **phase-plane trajectories.** The nullclines break the phase plane into regions, in each of which we can find a **direction arrow** indicating whether the state variables are increasing or decreasing. For additional information on the behaviour of the phase-plane trajectories, we can draw vertical or horizontal direction arrows on the nullclines. Phase-plane trajectories follow the direction arrows.

8.7 Exercises

Mathematical Techniques

1–4 ■ Suppose that each function is a solution of some differential equation. Graph these as functions of time and as phase-plane trajectories for $0 \leq t \leq 2$. Mark the position at $t = 0$, $t = 1$, and $t = 2$.

1. $x(t) = t$, $y(t) = 3t$
2. $a(t) = 2e^{-t}$, $b(t) = e^{-2t}$
3. $f(t) = 1 + t$, $g(t) = e^{-t}$
4. $x(t) = 1 + t(t - 2)$, $y(t) = t(3 - t)$

5–6 ■ From each graph of solutions of differential equations as functions of time, graph the matching phase-plane trajectory.

5.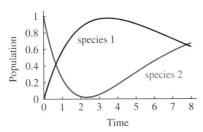

6.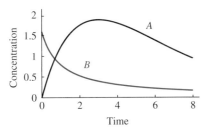

7–8 ■ From each graph of a phase-plane trajectory, graph the matching solutions of differential equations as functions of time.

7.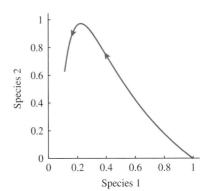

8.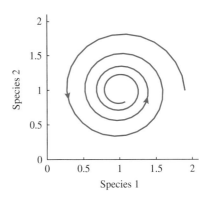

9–10 ■ On each phase-plane diagram, use the direction arrows to sketch phase-plane trajectories starting from two different initial conditions.

9.
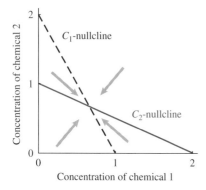

10.
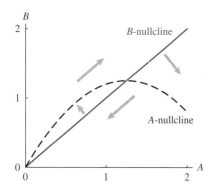

11–12 ■ Use the information in the phase-plane diagram to draw direction arrows on the nullclines.

11. The diagram in Exercise 9.

12. The diagram in Exercise 10.

13. Compare solutions estimated with Euler's method with the phase-plane diagram and direction arrows found in the text for the competition equations ((Figure 8.7.58)

$$\frac{da}{dt} = \mu \left(1 - \frac{a+b}{K_a}\right) a$$

$$\frac{db}{dt} = \lambda \left(1 - \frac{a+b}{K_b}\right) b$$

starting from the given initial conditions. Assume that $\mu = 2$, $\lambda = 2$, $K_a = 1000$, and $K_b = 500$. Use $a(0) = 750$ and $b(0) = 500$. Take two steps, with a step length of $\Delta t = 0.1$, as in Section 8.5, Exercise 7.

14–15 ■ Compare solutions estimated with Euler's method with the phase-plane diagram and direction arrows found in the text for Newton's law of cooling (Figure 8.7.59),

$$\frac{dT}{dt} = \alpha(A - T)$$

$$\frac{dA}{dt} = \alpha_2(T - A)$$

with the given parameter values and starting from the given initial conditions.

14. Suppose $\alpha = 0.3$ and $\alpha_2 = 0.1$. Use $T(0) = 60$ and $A(0) = 20$. Take two steps, with a step length of $\Delta t = 0.1$, as in Section 8.5, Exercise 9.

15. Suppose $\alpha = 3$ and $\alpha_2 = 1$. Use $T(0) = 60$ and $A(0) = 20$. Take two steps, with a step length of $\Delta t = 0.25$, as in Section 8.5, Exercise 10. Does this diagram help explain what went wrong?

Applications

16–29 ■ For each system, add direction arrows to the phase plane.

16. $\frac{db}{dt} = (1 - 0.05p)b$, $\frac{dp}{dt} = (-1 + 0.02b)p$

17. $\frac{db}{dt} = (2 - 0.01p)b$, $\frac{dp}{dt} = (1 + 0.01b)p$

18. $\frac{dp_1}{dt} = (-1 + 0.001p_2)p_1$, $\frac{dp_2}{dt} = (-1 + 0.001p_1)p_2$

19. $\frac{dp_1}{dt} = (-1 + 0.001p_2 - 0.001p_1)p_1$

    $\frac{dp_2}{dt} = (-1 + 0.001p_1 - 0.001p_2)p_2$

20. $\frac{da}{dt} = \mu\left(1 - \frac{a+b}{K_a}\right)a$, $\frac{db}{dt} = \lambda\left(1 - \frac{b}{K_b}\right)b$

21. $\frac{da}{dt} = \mu\left(1 - \frac{b}{K_a}\right)a$, $\frac{db}{dt} = \lambda\left(1 - \frac{a}{K_b}\right)b$

22. $\frac{da}{dt} = \left(1 - \frac{a+b/2}{1000}\right)a$, $\frac{db}{dt} = \left(1 - \frac{2a+b}{1000}\right)b$

23. $\frac{da}{dt} = \left(1 - \frac{a+b/2}{1000}\right)a$, $\frac{db}{dt} = \left(1 - \frac{a/2+b}{1000}\right)b$

24. $I' = 2IS - I$, $S' = -2IS$

25. $I' = 2IS - I$, $S' = -2IS + 2I/3$

26. $I' = 2IS - I - 0.5I$, $S' = -2IS + I - 0.5S$

27. $I' = 2IS - I - 8I$, $S' = -2IS + I - 4S$

28. $I' = 2I + 2IS - I$, $S' = 2S - 2IS + I$

29. $I' = 2IS - I$, $S' = I + S - 2IS + I$

30–37 ■ For each problem, use the direction arrows on your phase plane to sketch a solution starting from the given initial condition.

30. The model $\frac{dp_1}{dt} = (-1 + 0.001p_2)p_1$, $\frac{dp_2}{dt} = (-1 + 0.001p_1)p_2$ starting from $(1500, 200)$. Is there another path for the solution that is consistent with the direction arrows?

31. The model $\frac{dp_1}{dt} = (-1 + 0.001p_2 - 0.001p_1)p_1$, $\frac{dp_2}{dt} = (-1 + 0.001p_1 - 0.001p_2)p_2$ starting from $(1500, 200)$.

32. The model $\frac{da}{dt} = \mu\left(1 - \frac{a+b}{K_a}\right)a$, $\frac{db}{dt} = \lambda\left(1 - \frac{b}{K_b}\right)b$ starting from $(200, 300)$.

33. The model $\dfrac{da}{dt} = \mu\left(1 - \dfrac{b}{K_a}\right)a$, $\dfrac{db}{dt} = \lambda\left(1 - \dfrac{a}{K_b}\right)b$ starting from (200, 300).

34. The model $I' = 2IS - I$, $S' = -2IS$ starting from (0.5, 1).

35. The model $I' = 2IS - I$, $S' = -2IS + 2I/3$ starting from (0.5, 1).

36. The model $I' = 2I + 2IS - I$, $S' = 2S - 2IS + I$ starting from (0.5, 1).

37. The model $I' = 2IS - I$, $S' = I + S - 2IS + I$ starting from (0.5, 0.5). Can you be sure that the solution behaves exactly like your picture?

Computer Exercises

38. Consider the following differential equations describing the diffusion and utilization of a chemical:

$$\dfrac{dC}{dt} = \alpha(A - C) - \dfrac{\delta C}{1 + C}$$

$$\dfrac{dA}{dt} = \dfrac{\alpha}{K}(C - A) + S$$

The parameters have the following meanings:

Name	Meaning	Values to Use
α	diffusion rate	1
δ	use efficiency	4 and 1
K	ratio of volumes	2
S	supplementation rate	1

a. Set $\delta = 4$ and the rest of the parameters to their designated values. Plot the nullclines and find the equilibrium.

b. Follow the same steps with $\delta = 1$. Is there an equilibrium? Can you say why not? (No math jargon allowed.) Sketch C and A as functions of time.

c. Try to figure out the critical value of δ where the behaviour changes.

39. Many biological systems need to be able to respond to changes in the level of some signal (such as a hormone) without responding to the actual level. For example, a cell might have no response to a low level of hormone. If the hormone level rapidly increases, the cell responds. But if the hormone level then remains constant at the higher level, the cell again stops responding. This process is sometimes called *adaptation*.

One mechanism for this process is summarized in the following model. Internal response is a function of the fraction, p, of cell surface receptors that are bound by the hormone. This fraction increases when the hormone level, H, is high. However, hormone also dissociates from bound receptors. Assume this happens at a rate A but that this rate is controlled by the cell. One possible set of equations is

$$\dfrac{dp}{dt} = k_1 H(1 - p) - Ap$$

$$\dfrac{dA}{dt} = \epsilon(H - A)$$

Suppose that $k_1 = 0.5$ and that ϵ is a small value (such as 0.1 or 0.01). The value of H is determined by conditions external to the cell and does not have its own differential equation.

a. Find the nullclines and equilibria of this model, assuming that H is a constant. Does H appear in your final results? Explain why the cell should respond in the same way to any constant level of H.

b. Use a computer to simulate the response when the level of H jumps quickly from $H = 1$ to $H = 10$. One way to do this is to solve the equations with $H = 10$, using as initial conditions the equilibrium values of p and A when $H = 1$. Draw graphs of p and A in the phase plane and as functions of time. Explain what is happening.

c. Do part (b) assuming that the level of H drops rapidly from $H = 10$ to $H = 1$.

Chapter Summary: Key Terms and Concepts

Define or explain the meaning of each term and concept.
Differential equations: autonomous differential equation; relative rate of change, per capita production (growth) rate; logistic equation, Allee effect, carrying capacity, Newton's law of cooling, diffusion equation, continuous selection model; separable differential equation, Euler's method
Analysis of differential equations: equilibrium solution, stable equilibrium, unstable equilibrium, Stability Theorem for Autonomous Differential Equations; phase-line diagram
Systems of differential equations: predator–prey model, competition model, Newton's law of cooling; Euler's method for systems of differential equations; equilibria, phase plane, nullclines, solution (phase-plane trajectory), direction arrow

Concept Check: True/False Quiz

Determine whether each statement is true or false. Give a reason for your choice.

1. $P'(t) = 2P(t)[1 - 10/P(t)]$ is an example of a logistic equation.
2. The solution of the differential equation $P'(t) = P(t)[1 - P(t)]$ starting at $P(0) = 0.7$ is increasing.
3. In the Newton's law of cooling equation $T'(t) = 2[45 - T(t)]$, the ambient temperature is $90°$.
4. $x = -c$ is an equilibrium of the autonomous differential equation $x'(t) = cx + x^2$.
5. The differential equation $p' = p^3 - 2p + 1$ has four equilibria.
6. The differential equation $y' = e^{x-y}$ is separable.
7. All solutions of the logistic differential equation are increasing.
8. The system $m' = m - 0.5mn$, $n' = -n - 0.1mn$ could describe a predator–prey interaction.
9. The system $m' = m - 0.5mn$, $n' = -n + 0.1mn$ has an equilibrium for which $m \neq 0$ and $n \neq 0$.
10. If a solution in a phase plane is a circle, then the variables involved change periodically.

Supplementary Problems

1. Consider the differential equation
$$\frac{dC}{dt} = 3(A - C) + 1$$
where C is the concentration of some chemical in a cell, measured in moles per litre, and A is a constant.
   a. What kind of differential equation is this? Explain the terms in the equation.
   b. Draw the phase-line diagram.
   c. Verify the stability of the equilibrium by using the derivative.
   d. Sketch solutions as functions of time starting from two initial conditions: $C(0) = 0$ and $C(0) = A + 1$.

2. Consider the differential equation
$$\frac{dC}{dt} = 3(A - C) + 1$$
where A is a constant.
   a. Solve the equation when $C(0) = 0$.
   b. Check your answer.
   c. Find $C(0.4)$.
   d. After what time will the solution be within 5% of its limit?

3. Consider the system of equations
$$\frac{dA}{dt} = (C - A) - \frac{A^2}{3}$$
$$\frac{dC}{dt} = 3(A - C) + 1$$
where C is the internal concentration of a chemical and A is the external concentration.

   a. Describe the two processes affecting concentration in the external environment. How big is the external environment relative to the cell?
   b. Draw a phase plane, including nullclines, equilibria, and direction arrows. (Hint: It is easier to put C on the vertical axis.)

4. Give conditions wherein you might observe population growth described by the following:
   a. One-dimensional autonomous differential equation.
   b. One-dimensional nonautonomous differential equation.
   c. One-dimensional pure-time differential equation.
   d. Two-dimensional autonomous differential equation.

5. Consider the differential equation
$$\frac{dV}{dt} = 12 - t^2$$
where $V(t)$ is volume in litres at time t in seconds.
   a. What kind of differential equation is this?
   b. Graph the rate of change and use it to sketch the graph of the solution.
   c. At what time does V take on its maximum?
   d. Suppose $V(0) = 0$. Use Euler's method to estimate $V(0.1)$.
   e. Suppose $V(0) = 0$. Find the time, T, when $V(t)$ is zero again.
   f. What is the average volume between zero and T?

6. Consider the differential equation
$$\frac{dx}{dt} = 3x(x - 1)^2$$

a. Draw the phase-line diagram of this equation.

b. Find the stability of the equilibria using the derivative.

c. Sketch trajectories of x as a function of time for initial conditions $x(0) = -0.5$, $x(0) = 0.5$, and $x(0) = 1.5$.

7. Suppose the per capita production rate of a bacterial population is given by

$$\text{per capita production rate} = \frac{1}{\sqrt{b}}$$

where $b(t)$ is the population size at time t in hours.

a. Find the differential equation describing this population.

b. Solve the equation and check your answer.

c. What is the population after 2 h if the population starts at $b(0) = 10{,}000$?

d. Does this population grow more quickly or more slowly than one growing exponentially? Why?

8. Differential equations to describe an epidemic are sometimes given as

$$\frac{dS}{dt} = \beta(S+I) - cSI$$

$$\frac{dI}{dt} = cSI - \delta I$$

where S measures the number of susceptible people and I the number of infected people.

a. Compare these equations with the predator–prey equations. What is different in these equations? What biological process does each term on the right-hand side describe?

b. Sketch the nullclines and find the equilibria if $\beta = 1$ and $c = \delta = 2$. (Draw only the parts where S and I are positive.)

c. Sketch direction arrows on your phase-plane diagram.

9. Consider the following differential equation, which describes the concentration of sodium ions in a cell following consumption of a bag of potato chips at time $t = 0$ s:

$$\frac{dN}{dt} = 2 - 10t$$

Suppose $N(0) = 50$ mmol/cm^3.

a. What kind of differential equation is this?

b. Sketch the graph of the rate of change as a function of time.

c. Sketch the graph of the concentration as a function of time.

d. Use Euler's method to estimate $N(0.1)$.

e. Find $N(0.1)$ exactly.

f. At what times is $N(t) = 50$?

10. Consider the following differential equation, which describes the concentration of sodium ions in a cell:

$$\frac{dN}{dt} = 2(N - 50) - (N - 50)^2$$

Assume this equation only works for $45 \leq N \leq 55$.

a. What kind of differential equation is this? What does each term mean?

b. Sketch the graph of the rate of change as a function of concentration.

c. Draw the phase-line diagram.

d. Sketch solutions starting from $N(0) = 48$, $N(0) = 51$, and $N(0) = 54$.

e. Suppose $N(0) = 51$. Estimate $N(0.1)$.

f. What method would you use to solve this equation?

11. Two types of bacteria with populations a and b are living in a culture. Suppose

$$\frac{da}{dt} = 2a\left(1 - \frac{a}{500} - \frac{b}{200}\right)$$

$$\frac{db}{dt} = 3b\left(1 + \frac{a}{1000} - \frac{b}{100}\right)$$

a. What kind of differential equations are these? Explain the terms.

b. Draw the phase plane, including nullclines, equilibria, and direction arrows.

12. The density of sugar in a hummingbird's 20-mm tongue is

$$s(x) = \frac{1.2}{1 + 0.2x}$$

where x is measured in millimetres from the end of the tongue and s is measured in moles per metre.

a. Find the total amount of sugar in the hummingbird's tongue.

b. Find the average density of sugar in the tongue.

c. Compare the average with the minimum and maximum densities. Does your answer make sense?

13. The length, L, of a microtubule is found to follow the differential equation

$$\frac{dL}{dt} = -L(2-L)(1-L)$$

where L is measured in micrometres (μm) and t is measured in seconds.

a. Draw the phase-line diagram.

b. Check the stability of the equilibria using the derivative.

c. Sketch trajectories starting from $L(0) = 0.5$, $L(0) = 1.5$, and $L(0) = 2.5$.

14. A lab finds that

$$\frac{dL}{dt} = -2L + 5.4LE$$

$$\frac{dE}{dt} = 3.5 - LE$$

where E is the level of some component of a certain microtubule and L is the length of the microtubule.

a. Explain the terms in these equations.

b. Draw the phase-plane diagram, including nullclines, equilibria, and direction arrows.

15. Consider the following differential equation describing a population of mathematically sophisticated bacteria:

$$\frac{db}{dt} = b \ln\left(\frac{2b+1}{2+b}\right)$$

a. What kind of differential equation is this?

b. Draw the derivative as a function of the state variable, and draw the phase-line diagram.

c. Sketch trajectories starting from $b(0) = 0.8$ and $b(0) = 1.2$.

d. Check the stability of the equilibrium at $b = 0$ by taking the derivative of the rate of change.

e. Use Euler's method to estimate $b(0.01)$ if $b(0) = 0.5$.

Projects

1. Computer Exercise 39 in Section 8.7 presents one possible model of adaptation. This project studies a simpler, alternative model (proposed by H. G. Othmer). Because this model is so simple, we can compare the results of phase-plane analysis with actual solutions of the equations.
   Consider the equations

   $$\frac{dp}{dt} = k(H - A - p)$$

   $$\frac{dA}{dt} = \epsilon(H - A)$$

   where p represents the response of the cell and H the external condition driving the response. The variable A describes some internal state of the cell. First, suppose that H is constant.

   a. Find the nullclines and equilibria and draw the phase plane, including direction arrows.

   b. The equation for A is the same as for Newton's law of cooling. Write down the solution for A with an arbitrary initial condition.

   c. Substitute this solution into the equation for p. There is a clever way to solve the resulting nonautonomous equation. Make up a new variable, $q(t) = e^{kt}p(t)$. With a bit of manipulation, you can write a pure-time differential equation for q. Solve this equation and find the solution for $p(t)$.

   d. Plot this solution as a phase-plane trajectory for several different initial conditions. Are the results consistent with the phase plane?

   Using these results, we can study what happens when H changes. Suppose first that the cell is at equilibrium with $H = 1$ and then that H jumps up to 10. Set $k = 1$ and $\epsilon = 0.1$.

   e. How long will it take before A has roughly reached its new equilibrium value?

   f. What is the value of p at this time? How long will it take p to nearly reach its equilibrium value again?

   g. Suppose now that $H = 1$ for a time T, jumps to $H = 10$ for the same time T, jumps back to 1, and so forth. Experiment with different values of T. How large must T be before the cell produces a healthy response to each change? What happens when T is much smaller than this value? What might the cell do to be able to respond more quickly?

2. Consider an interaction between two mutually inhibiting proteins with concentrations x and y, given by the differential equations

   $$\frac{dx}{dt} = f(y) - x$$

   $$\frac{dy}{dt} = g(x) - y$$

   Both $f(y)$ and $g(x)$ are decreasing functions.

   a. Explain each term in these equations.

   b. Try to imagine a biological situation they might describe.

   c. Sketch the nullclines (remember that both f and g are decreasing).

   d. Show that equilibria occur where $f(g(x)) = x$. What discrete-time dynamical system shares the equilibria of the system of differential equations?

   Next try the following steps for functions of three different forms:

   Case 1: $f(y) = \dfrac{1}{1+\alpha y}$, $g(x) = \dfrac{1}{1+\alpha x}$

   Case 2: $f(y) = e^{-\alpha y}$, $g(x) = e^{-\alpha x}$

   Case 3: $f(y) = \dfrac{1}{1+\alpha y^2}$, $g(x) = \dfrac{1}{1+\alpha x^2}$

   Experiment with different values of α.

   e. Find the function $f(g(x))$ and graph it on a cobwebbing diagram. Does the equilibrium look stable?

f. Draw the nullclines and direction arrows for the differential equation. Is the equilibrium stable?

g. Why does the stability you found in part (e) match that in part (f)?

h. Try to figure out whether it is possible to have three equilibria (it is possible in cases 2 and 3).

Why might it be important for a biological system to have three equilibria? How could it operate as a switch?

Cherry, J. L., and Adler, F. R. How to make a biological switch. *Journal of Theoretical Biology* 203: 117–133, 2000.

Chapter 1

Section 1.3

1. The variables are the altitude and the wombat density, which we can call a and w, respectively. The parameter is the rainfall, which we can call R.

3. $f(x)$ is neither increasing nor decreasing.

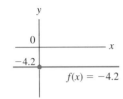

5. $f(x)$ is decreasing on $x < 0$ and on $x > 0$.

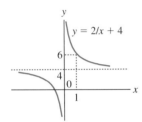

7. $f(x)$ is increasing on $x < 0$ and decreasing on $x > 0$.

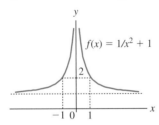

9. $(2, 2)$

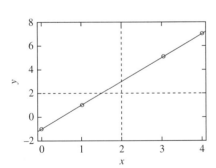

11. $(4, 12)$

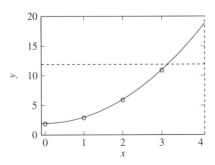

13. $f(a) = a + 5, f(a + 1) = a + 6, f(4a) = 4a + 5$

15. $h\left(\dfrac{c}{5}\right) = \dfrac{1}{c}, h\left(\dfrac{5}{c}\right) = \dfrac{c}{25}, h(c + 1) = \dfrac{1}{5c + 5}$

17. $\mathbb{R} = (-\infty, \infty)$

19. $\mathbb{R} = (-\infty, \infty)$

21. $x \neq 0$

23. $(2, \infty)$

25. $x \neq 0$

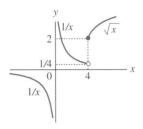

27. All real numbers

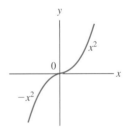

29. $\mathbb{R}$

31. The range is $y \geq 3$. See the graph below.

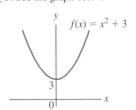

33. $[0, \infty)$

35. $[0, \infty)$

37.

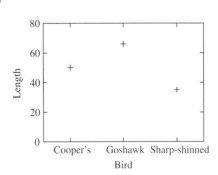

39. The fish population is steadily declining between 1950 and 1990.

41. The stock increases, crashes, increases, crashes, increases, crashes, and levels out.

43.

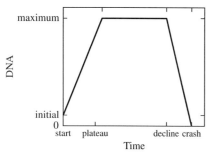

45.

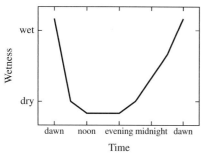

47. If $r=0, c=0$; if $r=4, c=0$; if $r=6, c=2$; if $r=10, c=6$. It looks like these cells can tolerate up to 5 rad before they begin to become cancerous.

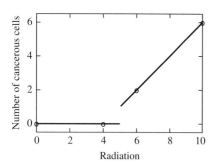

49. If $a=0, h=0$; if $a=100, h=50$; if $a=500, h=83.3$; if $a=1000$, $h=90.9$; if $a=10,000, h=99.0$. It looks like it would reach 100 m.

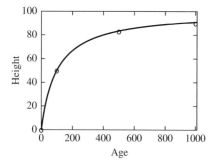

Section 1.4

1.

x	f(x)	g(x)	(f + g)(x)
−2	−1	−11	−12
−1	1	−8	−7
0	3	−5	−2
1	5	−2	3
2	7	1	8

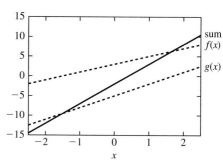

3.

x	F(x)	G(x)	(F+G)(x)
−2	5	−1	4
−1	2	0	2
0	1	1	2
1	2	2	4
2	5	3	8

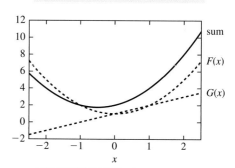

5.

x	f(x)	g(x)	(f · g)(x)
−2	−1	−11	11
−1	1	−8	−8
0	3	−5	−15
1	5	−2	−10
2	7	1	7

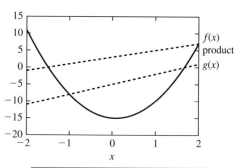

7.

x	F(x)	G(x)	(F · G)(x)
−2	5	−1	−5
−1	2	0	0
0	1	1	1
1	2	2	4
2	5	3	15

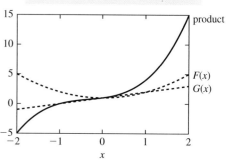

9. $f \cdot g = x^3 - x^4$, domain is $\mathbb{R} = (-\infty, \infty)$; $\frac{f}{g} = \frac{x^3}{1-x}$, domain is $x \neq 1$

11. $f \cdot g = \sqrt{x - x^2}$, domain is $x \geq 1$; $\frac{f}{g} = \sqrt{\frac{x}{x-1}}$, domain is $x > 1$

13. $(f \circ g)(x) = 64 - 96x + 48x^2 - 8x^3$, $(g \circ f)(x) = 4 - 2x^3$

15. $(f \circ g)(x) = \frac{1}{x-3}$, $(g \circ f)(x) = \frac{1}{x} - 3$

17. $(f \circ g)(x) = \frac{1-x}{1+x}$, $(g \circ f)(x) = \frac{x+1}{x-1}$

19. $(f \circ g)(x) = \frac{\sqrt{x}}{1+x}$, $(g \circ f)(x) = \sqrt{\frac{x}{1+x^2}}$

21. $(f \circ g)(x) = f(3x - 5) = 2 \cdot (3x - 5) + 3 = 6x - 7$ and
    $(g \circ f)(x) = g(2x + 3) = 3 \cdot (2x + 3) - 5 = 6x + 4$.

    These do not match, so the functions do not commute.

23. $(F \circ G)(x) = F(x + 1) = (x + 1)^2 + 1 = x^2 + 2x + 2$ and
    $(G \circ F)(x) = G(x^2 + 1) = x^2 + 2$.

    These do not match, so the functions do not commute.

25. $f^{-1}(x) = \frac{x-3}{2}$. Also, $f(1) = 5$ and $f^{-1}(5) = 1$.

27. The function $F(x)$ fails the horizontal line test because, for example, $F(-1) = F(1) = 2$. Therefore, it has no inverse.

29.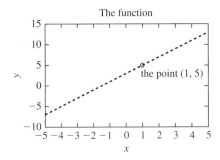
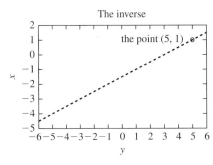

31. This function does not have an inverse because it fails the horizontal line test. From the graph, we could not tell whether $F^{-1}(2)$ is 1 or -1.

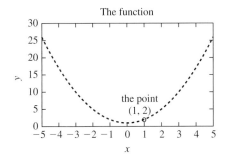

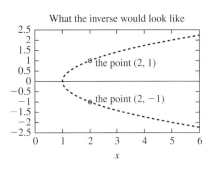

33. Domain is $\mathbb{R}$; range is $\mathbb{R}$; f has inverse; $f^{-1}(x) = (x - 4)^3$

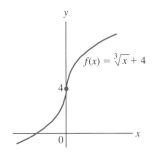

35. Domain consists of two sets, $[-1, 0)$ and $(0, 1]$; range is $[1, \infty)$; no inverse

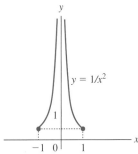

37. Domain is $x \neq 3$; range is $y \neq 0$; g has inverse; $y^{-1}(x) = \frac{13}{x} + 3$

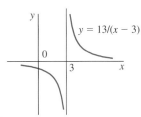

39. $f^{-1}(x) = \frac{(x+1)^3 + 11}{2}$

41. $g^{-1}(x) = \frac{1}{2x+1}$

43. $m(h) = \frac{h^3}{a^3}$

45. $m(p) = \frac{3p}{2-p}$

47. The vertical axis is shifted by a value less than 0, moving it down.

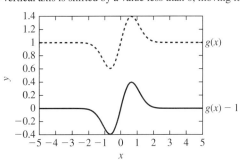

49. The horizontal axis is shifted by a value greater than 0, moving it to the left.

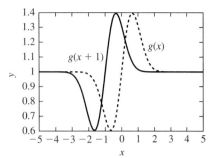

51. Shift the graph of $y = \frac{1}{x^2}$ to the right 3 units.

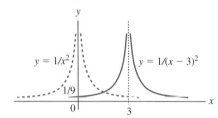

53. Shift the graph of $y = \sqrt{x}$ to the right 3 units.

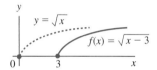

55. Start with the graph of $y = \sqrt{x}$, shift it right 2 units, then reflect that graph across the x-axis, and finally move the resulting graph 3 units up.

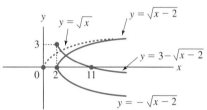

57. Shift the graph of $y = x^2$ 1/2 units to the left and then move the resulting graph 23/4 units up.

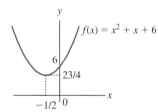

59. Move the graph of $y = |x|$ 2 units to the right and then 3 units down.

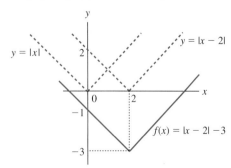

61.

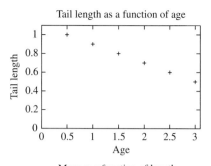

63.

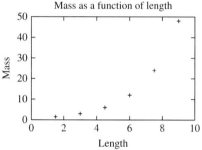

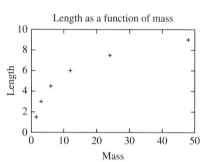

The graph of length as a function of mass looks like the graph of mass as a function of length turned on its side.

65. $S(T(W)) = S(40 - 0.2W) = 6 + 0.04W$. Plugging in $W = 10$ gives 6.4 hours of sleep.

67. $M(V(L(T))) = M(V(10 + T/10)) = M(2(10 + T/10)^3) = 2.6(10 + T/10)^3$. At $T = 25$, $M = 5.08 \times 10^3$ mg $= 5.08$ g.

69. The formula is $V(t) = \left(3t + 20 + \frac{t^2}{2}\right) + (t + 40) = 2t + 60 + \frac{t^2}{2}$.

t	$V_a(t)$	$V_b(t)$	$V(t)$
0	20	40	60
0.5	21.625	39.5	61.125
1	23.5	39	62.5
1.5	25.625	38.5	64.125
2	28	38	66
2.5	30.625	37.5	68.125
3	33.5	37	70.5

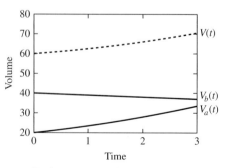

V_a increases, V_b decreases, and the total, V, increases.

71. Volume never has the same value twice and therefore V has an inverse that contains sufficient information to find the age.

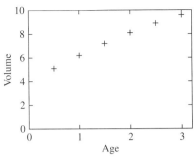

73. There are two different values of G (9.4 and 8.2) when M is 5.6. This is not a function because it fails the vertical line test. Furthermore, we could not tell if M was 5.1 or 5.6 when $G = 8.2$.

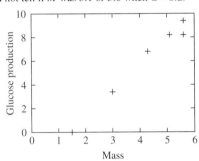

75. We denote the total mass by $T(t)$. Then $T(t) = P(t)W(T) = (2 \times 10^6 - 2 \times 10^4 t)(80 + 0.5t)$.

t	$P(t)$	$W(t)$	$T(t)$
0	2	80	160
20	1.6	90	144
40	1.2	100	120
60	0.8	110	88
80	0.4	120	48
100	0	130	0

The population decreases, the mass per individual increases, and the total mass decreases.

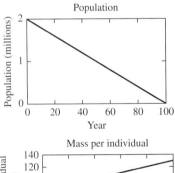

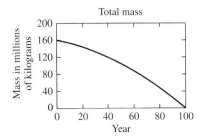

77. We denote the total mass by $T(t)$. Then $T(t) = P(t)W(T) = (2 \times 10^6 + 2 \times 10^4 t)(80 - 0.005t^2)$.

t	$P(t)$	$W(t)$	$T(t)$
0	2	80	160
20	2.4	78	187.2
40	2.8	72	201.6
60	3.2	62	198.4
80	3.6	48	172.8
100	4	30	120

The population decreases, the mass per individual increases, and the total mass increases and then decreases.

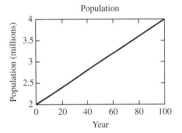

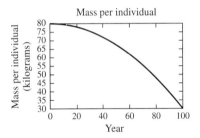

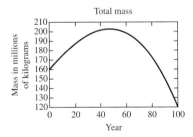

79. Move $y = 1/c$ left by k units (thus getting $y = 1/(k+c)$), then scale vertically by a factor of Ak (to obtain $y = Ak/(k+c)$), then reflect across the x-axis (to obtain $y = -Ak/(k+c)$), and finally move up A units.

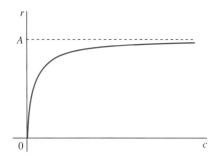

True/False Quiz

1. FALSE. $(g \circ f)(x) = g(f(x)) = g(x^2) = 7$.
3. TRUE. Reflecting $f(x)$ with respect to the x-axis, we obtain $-f(x)$. Horizontal compression by a factor of 3 corresponds to replacing x by $x/3$, so the resulting function is $-f(x/3)$.
5. FALSE. $f(g(x)) = f(2x - 1) = 2(2x - 1) + 1 = 4x - 1$ is not equal to x (so f and g do not "cancel each other").
7. TRUE. Pick $c \geq 0$. From $\sqrt{5x - 2} = c$ we get $5x - 2 = c^2$ and $x = (c^2 + 2)/5$. For that value of x, $f(x) = c$.
9. TRUE, by the definition of the inverse function.

Supplementary Problems

1. Divide both the numerator and the denominator by c^2 to get $A(c) = \frac{n/c}{k/c^2+1}$. As c grows larger and larger, the values of $A(c)$ approach $0/(0 + 1) = 0$.
3.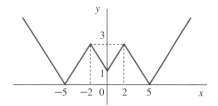
5. $(-\infty, 3)$ and $(3, 7/2]$
7. Approximately 1880 deer
9. $4x$
11. Approximately 466.7 watts

Chapter 2

Section 2.1

1. The points are (1, 5) and (3, 9). The change in input is 2, the change in output is 4, and the slope is 2. This is not a proportional relation because the ratio of output to input changes from 5 at the first point to 3 at the second point. This relation is increasing because larger values of x lead to larger values of y (and the slope is positive).

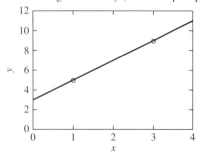

3. The points are (1, 3) and (3, 13). The change in input is 2, the change in output is 10, and the slope is 5. This is a not a proportional relation because the ratio of output to input changes from 3 at the first point to 4.333 at the second point. This relation is increasing because larger values of w lead to larger values of z (and the slope is positive).

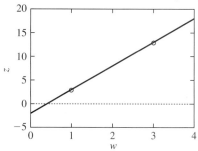

5. The point lies on the line because $f(2) = 2 \cdot 2 + 3 = 7$. The point-slope form is $f(x) = 2(x - 2) + 7$. Multiplying out gives $f(x) = 2x - 4 + 7 = 2x + 3$ as it should.

7. Multiplying out, we find that $f(x) = 2x + 1$. The slope is 2 and the y-intercept is 1. The original point is (1, 3).

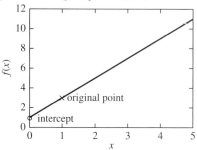

9. In point-slope form, this line has equation $f(x) = -2(x - 1) + 6$. Multiplying out, we find that $f(x) = -2x + 8$. The slope is -2 and the y-intercept is 8.

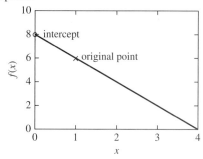

11. The slope between the two points is

$$\text{slope} = \frac{\text{change in output}}{\text{change in input}} = \frac{3 - 6}{4 - 1} = -1.$$

In point-slope form, the line has equation $f(x) = -1 \cdot (x - 1) + 6$. In slope-intercept form, it is $f(x) = -x + 7$. This line has slope -1 and y-intercept 7.

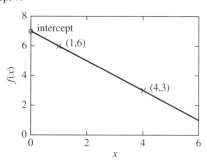

13. This is not linear because the input z appears in the denominator.

15. This is linear because the input q is only multiplied by constants and has constants added to it.

17. $h(1) = \frac{1}{5}$, $h(2) = \frac{1}{10}$, $h(4) = \frac{1}{20}$. Slope between $z = 1$ and $z = 2$ is $-1/10$, and between $z = 2$ and $z = 4$ it is $-1/40$.

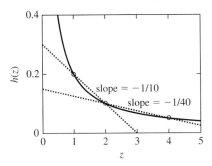

19. $2x = 7 - 3 = 4$, so $x = 4/2 = 2$. Plugging in, $2 \cdot 2 + 3 = 7$.

21. $2x - 3x = 7 - 3 = 4$, so $-x = 4$ or $x = -4$. Plugging in, $2 \cdot (-4) + 3 = -5 = 3 \cdot (-4) + 7$.

23. Multiplying out, we get $10x - 4 = 10x + 5$. This has no solution.

25. $2x = 7 - b$, so $x = \frac{7-b}{2}$.

27. $(2 - m)x = 7 - b$, so $x = \frac{7-b}{2-m}$. There is no solution if $m = 2$. However, if $m = 2$ and $b = 7$, both sides are identical and any value of x works.

29. 1 in. = 2.54 cm. The slope is 2.54 $\frac{\text{cm}}{\text{in.}}$.

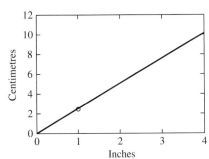

31. 1 g $\approx \frac{1}{453.6}$ lb ≈ 0.0022 lb. The slope is 0.0022 $\frac{\text{lb}}{\text{g}}$.

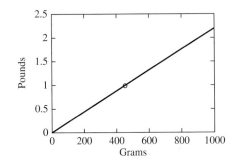

33. $(f \circ g)(x) = (mx + b) + 1$ and $(g \circ f)(x) = m(x + 1) + b = mx + m + b$. These match only if the intercepts are equal, or $b + 1 = m + b$. This is true for any b as long as $m = 1$. In this case, both f and g have slope 1, meaning that each just adds a constant to its input. The order cannot matter because addition is commutative.

35. $y = \frac{7}{6}x$

37. $y = 4\sqrt{x}$

39. $a = \frac{3}{b^3}$

41. $y = \frac{98}{x^2}$

43. $M = aB^{1.03}$, where $a = \frac{40.5}{1000^{1.03}} \approx 0.0329$.

45. $f(x) = x$ increases faster.

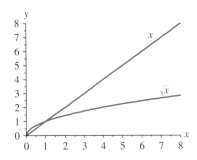

47. $f(x) = x^{1.1}$ increases faster.

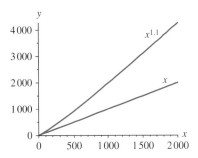

49. $f(x) = \frac{1}{x}$ decreases faster.

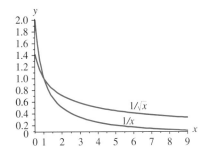

51. $f(x) = x^{-1.1}$ decreases faster.

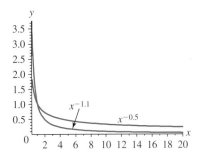

53. The slope is 7 cm^2, and the equation is $V = 7T$.

55. The slope is 10^6, and the equation is $M = 10^6 m$.

57. The slope has no units, so it will not change. The altitude at the top is 3000 m = $3000/0.3048 = 9842.52$ ft, so the equation of the line is $a = -0.2d + 9842.52$.

59. Solve $2600 = -0.2d + 3000$ for d to find $d = 2000$ m.

61. The surface elevation changed linearly from 1980 to 1990 (the slope is 0.8 m per 5 years) and during 2000 to 2010 (the slope is −0.8 m per 5 years).

63. Between 2000 and 2010 the slope was $(1750 - 1751.6)/10 = -0.16$ (i.e., the elevation was decreasing at a rate of 0.16 m per year). Based on the fact that the elevation in 2010 was 1750, the formula for the elevation is (point-slope form)

$$E(y) = 1750 - 0.16(y - 2010)$$

where y represents the year. If it had changed at the same rate, the elevation in 1980 would have been

$$E(1980) = 1750 - 0.16(1980 - 2010) = 1750 + 4.8 = 1754.8$$

metres.

65. Using the first two rows, we find a slope of

$$\text{slope} = \frac{\text{change in volume}}{\text{change in age}} = \frac{6.2 - 5.1}{1 - 0.5} = 2.2$$

Using $(1, 6.2)$ as the base point, we find

$$V = 2.2(a - 1) + 6.2 = 2.2a + 4$$

Therefore, $V(2.75) = 10.05$. The y-intercept gives the initial volume of 4.

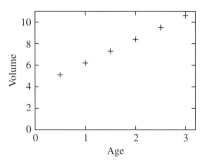

67. The volume changes by 0.9 when the mass changes by 1.5, giving a slope of 0.6. The point-slope form of the line, using the last data point, is $V = 0.6(m - 10) + 10.6$, which simplifies to $V = 0.6m + 4.6$ in slope-intercept form. When $m = 30$, $V = 22.6$. The density is then $\rho = 30/22.6 \approx 1.33$. The density at $a = 0.5$ is $\rho = 2.5/5.1 \approx 0.49$.

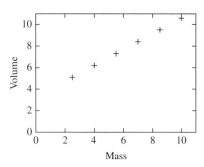

69. a. Let $t = 0$ represent the year 2001. Then the data points are $(0, 135, 294)$ and $(5, 135, 851)$. The slope is $m = 111.4$. The expression $p(t) = 135,294 + 111.4t$ represents the population of Prince Edward Island, $p(t)$, as a function of time, t, in years, measured from 2001.

b. The population is increasing at a rate of about 111 people per year.

c. The projected population for 2011 is

$$p(t) = 135,294 + 111.4(10) = 136,408$$

The model underestimates the population in 2011. Some time after 2006 the population started increasing at a rate larger than 111.4 people/year.

71. a. Let $t = 0$ represent the year 2001. Then the data points are $(0, 729, 498)$ and $(5, 729, 997)$. The slope is $m = 99.8$. The expression $p(t) = 729,498 + 99.8t$ represents the population of New Brunswick, $p(t)$, as a function of time, t, in years, measured from 2001.

b. The population is increasing at a rate of about 99 people per year.

c. The projected population for 2011 is

$$p(t) = 729,498 + 99.8(10) = 730,496$$

The model underestimates the population in 2011 by a large number. Some time after 2006 the population started increasing at a rate much larger than 99 people/year.

73. The line has equation $N = 70(W - 0.5) + 80$.

75. Predict $N = 70(0 - 0.5) + 80 = 45$. This seems rather unlikely, because these wasps had nothing to eat. I would not expect any to develop.

Section 2.2

1. Law 6: $43.2^0 = 1$

3. Law 3: $43.2^{-1} = 1/43.2 \approx 0.023$

5. Law 4: $43.2^{7.2}/43.2^{6.2} = 43.2^{7.2-6.2} = 43.2^1 = 43.2$

7. Law 2: $(3^4)^{0.5} = 3^{4 \cdot 0.5} = 3^2 = 9$

9. To use law 1, we need to first multiply out the exponents, finding $2^{2^3} \cdot 2^{2^2} = 2^8 \cdot 2^4 = 2^{12} = 4096$.

11. Law 6: $\ln 1 = 0$

13. Law 5: $\log_{43.2} 43.2 = 1$

15. Law 1: $\log_{10} 5 + \log_{10} 20 = \log_{10}(5 \cdot 20) = \log_{10} 100 = \log_{10}(10^2) = 2$

17. Law 4: $\log_{10} 500 - \log_{10} 50 = \log_{10}(500/50) = \log_{10} 10 = 1$

19. Law 2: $\log_{43.2}(43.2^7) = 7$

21. Law 3: $\log_7\left(\frac{1}{43.2}\right) = -\log_7 43.2 \approx -1.935$

23. $e^{3x} = \frac{21}{7} = 3$. Taking logs of both sides, $3x = \ln 3 \approx 1.099$, and $x \approx 0.366$. Checking, $7e^{3 \cdot 0.366} \approx 21.0$.

25. Taking logs of both sides gives $\ln 4 - 2x + 1 = \ln 7 + 3x$. Moving the x's to one side gives $5x = \ln 4 + 1 - \ln 7 = 1 + \ln\left(\frac{4}{7}\right) \approx 0.440$, and $x \approx 0.088$. Checking, $4e^{-2 \cdot 0.088 + 1} = 7e^{3 \cdot 0.088} \approx 9.11$.

27. $e^{2x} = 7$ gives $2x = \ln 7 \approx 1.946$ or $x \approx 0.973$. This function is increasing, and doubles after a "time" of $x = \ln 2/2 \approx 0.346$.

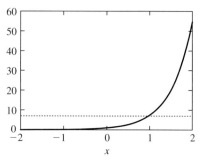

Therefore, $e^{2x} = 14$ when $x \approx 0.973 + 0.346 = 1.319$, and $e^{2x} = 3.5$ when $x \approx 0.973 - 0.346 = 0.627$.

29. $5e^{0.2x} = 7$ when $e^{0.2x} = 1.4$, or $0.2x = \ln 1.4 \approx 0.336$ or $x \approx 1.68$. This function is increasing, and doubles after a "time" of $x = \frac{\ln 2}{0.2} \approx 3.46$.

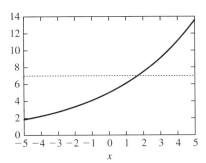

Therefore, $5e^{0.2x} = 14$ when $x \approx 1.68 + 3.46 = 5.14$, and $5e^{0.2x} = 3.5$ when $x \approx 1.68 - 3.46 = -1.78$.

31. Write $y = 1.4^x = e^{\ln 1.4^x} = e^{x \ln 1.4} \approx e^{0.336x}$. So, $a = \ln 1.4 \approx 0.336$. Thus, to obtain the graph of $y = 1.4^x$ we expand the graph of $y = e^x$ by a factor of $1/\ln 1.4 \approx 2.972$.

33. $f(g(x)) = 4x^2$

35. $f(g(x)) = \frac{e^{6x}}{16}$

37. $f(g(x)) = \log_{10} 4 - 2 \log_{10} x - x$

39. $x = e^{1/4} + 1 \approx 2.28403$

41. $x = 1099$

43. $x = 10^{10}$

45. $B = \frac{T^4}{17.73^4}$. If we know the blood circulation time, T, of a mammal, this formula tells us how to figure out its body mass.

47. $D = \frac{-\alpha + \sqrt{\alpha^2 - 4\beta \ln S}}{2\beta}$. If we know that S percent of cancer cells survived, we can compute the dose, D, that was applied in the treatment.

49. The graph of $y = \ln x^2$ for $x > 0$ is obtained by vertically expanding the graph of $y = \ln x$ by a factor of 2. The graph of $y = \ln x^2$ for $x < 0$ is the mirror image of the graph of $y = \ln x^2$, $x > 0$, across the y-axis.

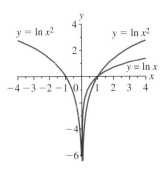

51. Shift the graph of $y = \log_{10} x$ to the right 2 units and then reflect across the x-axis.

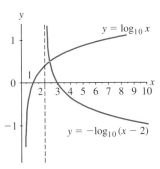

53. $t_d \approx 0.6931/1.0 = 0.6931$ d. It will take twice as long as this to quadruple, or 1.386 d.

55. $t_d \approx 0.6931/0.1 = 6.931$ h. It will quadruple in 13.86 h.

57. Converting to base e, we have that $S(t) = 2.34e^{\ln(10) \cdot 0.5t} \approx 2.34e^{1.151t}$. The doubling time is $t_d \approx \frac{0.6931}{1.151} = 0.602$ d. It will have increased by a factor of 10 when $\alpha t = 1$, which occurs when $t = 2$.

59. $Q(50000) = 6 \cdot 10^{10} e^{-0.000122 \cdot 50000} \approx 1.34 \cdot 10^8$. This is equal to $\frac{1.34 \cdot 10^8}{6 \cdot 10^{10}} \approx 0.00223$ of the original amount.

61. $t_h \approx \frac{0.693}{0.000122} \approx 5680$ years

63. It will have doubled twice and will be 2000.

65. If the doubling time is 24 years, the parameter α is $\frac{0.693}{24} \approx 0.0289$. Therefore, $P(t) \approx 500 e^{0.0289t}$.

67. It will have halved twice, to become 400.

69. About 65 million years

71. We plot the line $\ln(Q(t)) = \ln\left(6 \cdot 10^{10} e^{-0.000122t}\right) \approx 24.82 - 0.000122t$.

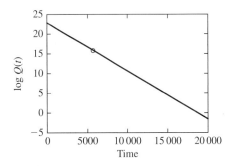

73. $\ln(P(t)) = \ln\left(1600 e^{-0.0161t}\right) \approx 7.38 - 0.0161t$

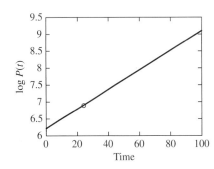

75. We compute $\ln f(t) = \ln 11 - \frac{1}{3} \ln t$. The graph of $f(t)$ in the log–log plot is a line of slope $-1/3$ with a vertical intercept of $\ln 11 \approx 2.398$.

77. We find $\log_{10} S = \log_{10} 4.84 + \frac{2}{3} \log_{10} V$. The graph of S in the log–log plot is a line of slope $2/3$ with a vertical intercept of $\log_{10} 4.84 \approx 0.685$.

Section 2.3

1. $300°$

3. $5940°$

5. $112.5°$

7. $\frac{3\pi}{2}$ rad

9. $\frac{\pi}{5}$ rad

11. 0.855 rad

13. $\sin(5\pi/4) = \cos(5\pi/4) = -1/\sqrt{2}$; $\tan(5\pi/4) = 1$; $\sec(5\pi/4) = -\sqrt{2}$

15. $\sin(5\pi/6) = 1/2$, $\cos(5\pi/6) = -\sqrt{3}/2$, $\tan(5\pi/6) = -1/\sqrt{3}$, $\sec(5\pi/6) = -2/\sqrt{3}$

17. $\sin(-3\pi/4) = \cos(-3\pi/4) = -1/\sqrt{2}$, $\tan(-3\pi/4) = 1$, $\sec(-3\pi/4) = -\sqrt{2}$

19. $\sin\theta = 3/\sqrt{10}$, $\cos\theta = 1/\sqrt{10}$, $\sec\theta = \sqrt{10}$, $\csc\theta = \sqrt{10}/3$, $\cot\theta = 1/3$

21. $\sin\theta = -3/5$, $\tan\theta = -3/4$, $\sec\theta = 5/4$, $\csc\theta = -5/3$, $\cot\theta = -4/3$

23. Expand the graph of $y = \sin x$ horizontally by a factor of 3.

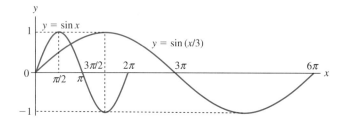

25. Shift the graph of $y = \cos x$ to the right π units and then shift it vertically 1 unit.

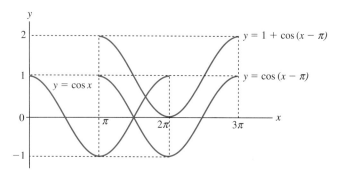

27. $\cos(0) = \sqrt{1}$, $\cos\left(\dfrac{\pi}{4}\right) = \dfrac{\sqrt{2}}{2} = \sqrt{\dfrac{1}{2}}$, $\cos\left(\dfrac{\pi}{2}\right) = 0 = \sqrt{\dfrac{0}{2}}$

29. $\cos(0 - \pi) = \cos(-\pi) = -1 = -\cos(0)$,
$\cos(\pi/4 - \pi) = \cos(-3\pi/4) = -\sqrt{2}/2 = -\cos(\pi/4)$,
$\cos(\pi/2 - \pi) = \cos(-\pi/2) = 0 = -\cos(\pi/2)$,
$\cos(\pi - \pi) = \cos(0) = 1 = -\cos(\pi)$

31. $\cos(2 \cdot 0) = \cos^2(0) - \sin^2(0) = 1 - 0 = 1 = \cos(0)$,
$\cos(2 \cdot \pi/4) = \cos^2(\pi/4) - \sin^2(\pi/4) = 1/2 - 1/2 = 0 = \cos(\pi/2)$,
$\cos(2 \cdot \pi/2) = \cos^2(\pi/2) - \sin^2(\pi/2) = 0 - 1 = -1 = \cos(\pi)$,
$\cos(2 \cdot \pi) = \cos^2(\pi) - \sin^2(\pi) = 1 - 0 = 1 = \cos(2\pi)$

33. Replacing y with $-y$ in the angle addition formula for the sine function, we obtain

$$\sin(x - y) = \sin(x + (-y))$$
$$= \sin x \cos(-y) + \cos x \sin(-y)$$
$$= \sin x \cos y - \cos x \sin y$$

since $\sin(-y) = -\sin y$ and $\cos(-y) = \cos y$.

35. Use the definition $\tan x = \sin x / \cos x$, and compute the common denominator.

37. Use the definition of the tangent and then the double-angle formulas.

$$\tan 2x = \frac{\sin 2x}{\cos 2x} = \frac{2 \sin x \cos x}{\cos^2 x - \sin^2 x}$$

Then divide both the numerator and the denominator by $\cos^2 x$.

39. Use the double-angle formulas and then factor out $\cos x$.

$\sin x \sin 2x + \cos x \cos 2x$
$= \sin x (2 \sin x \cos x) + \cos x \left(\cos^2 x - \sin^2 x\right)$
$= \cos x \left(2 \sin^2 x + \cos^2 x - \sin^2 x\right)$
$= \cos x \left(\sin^2 x + \cos^2 x\right) = \cos x$

41. Multiplying the factor of 5 through gives $r(t) = 10 + 5\cos(2\pi t)$ with average 10, amplitude 5, period 1, and phase 0.

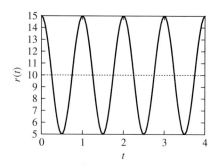

43. Use the fact that $\cos(t - \pi) = -\cos(t)$. Then $f(t) = 2 + 1\cos(t - \pi)$ with average 2, amplitude 1, period 2π, and phase π.

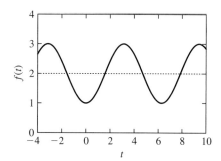

45. $\pi/6$ and $5\pi/6$
47. $\pi/3$ and $5\pi/3$
49. $\pi/4$ and $5\pi/4$
51. $x = \pi k$, $x = \pi/3 + 2\pi k$, $x = 5\pi/3 + 2\pi k$, where k is an integer.
53. $x = 2\pi/3 + 2\pi k$, $x = 4\pi/3 + 2\pi k$, where k is an integer.
55. $-\pi/6$
57. $\pi/3$
59. $-\pi/4$
61. Reflect the graph of $y = \arcsin x$ with respect to the x-axis and then move up 4 units. The domain of $y = -\arcsin x + 4$ is $[-1, 1]$.

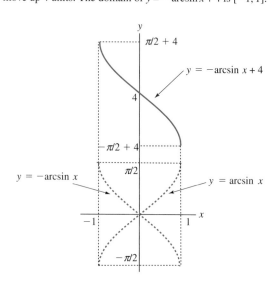

63. Stretch the graph of $y = \arctan x$ vertically by a factor of 3 and then reflect with respect to the x-axis. The domain of both functions is $\mathbb{R}$.

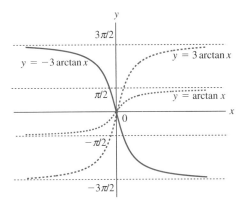

65. Average is 2, minimum is -4, maximum is 8, amplitude is 6, period is 7, and phase is 4.

67. Average is 2, minimum is 1.8, maximum is 2.2, amplitude is 0.2, period is 0.2, and phase is 0.1.

69. Average is 4, amplitude is 3, maximum is 7, minimum is 1, period is 1, and phase is 5 (which is the same as 0).

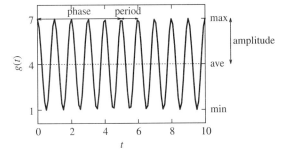

71. Average is -2, amplitude is 3, maximum is 1, minimum is -5, period is 0.2, and phase is -0.1 (which is the same as 0.1).

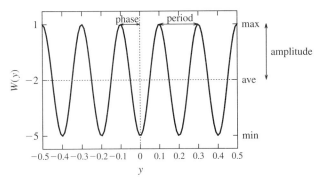

73. $p(t) = 2.8 + 1 \cdot \cos\left(\frac{\pi}{3}(t-1)\right)$ (in millions), where t is time in months

75. $p(t) = 100 + 20\cos(144\pi t)$, where t is time in minutes

77. The population $p(t)$ exhibits periodic behaviour, with period $2\pi/2 = \pi$. It oscillates between a low of $p(0)e^{-1.4} \approx 0.247 p(0)$ and a high of $p(0)e^{1.4} \approx 4.055 p(0)$. This population undergoes high fluctuations, which range from about 25% of the initial population count to about 4 times the initial population count.

79. This function increases overall, but wiggles around a bit. It might describe the size of an organism that grows on average, but grows quickly during the day and shrinks down a tiny bit at night.

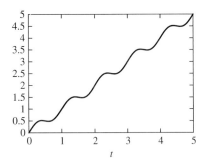

81. This function wiggles up and down with decreasing amplitude. This graph could represent a sound wave produced by a dolphin; the oscillations dampen over time.

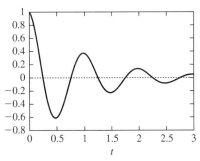

83. This function simply increases and does not oscillate at all. It could describe just about anything.

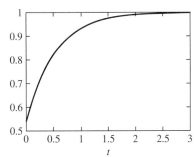

85. Let $S_u(t)$ represent the ultradian rhythm. Then $S_u(t) = 0.4\cos\left(\frac{2\pi t}{4}\right)$.

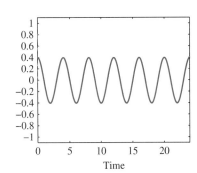

87. Most sleepy right at midnight, when both cycles peak together. Least sleepy at 10:00 A.M. and 2:00 P.M. There is not a minimum at noon because the ultradian rhythm and the circadian rhythm are opposed.

89. Oscillations with amplitudes that increase and decrease periodically. Might sound like Japanese opera.

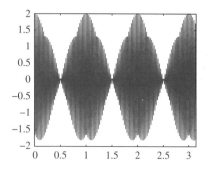

True/False Quiz

1. TRUE. From $A = c\frac{1}{B}$, it follows that $B = c\frac{1}{A}$.

3. FALSE. After three months, the quantity doubles, and then doubles again (after another three months). So, the quantity quadruples in six months.

5. FALSE. This fact is true only for lines through the origin.

7. FALSE. A 75% loss occurs over the time of two half-lives (i.e., 100% to 50% decrease and then further decrease to 25%). So the half-life is 1.5 hours.

9. TRUE. See formula 2.3.1.

Supplementary Problems

1. a. $3 \cdot 10^{10}$ bacteria
   b. $2 \cdot 10^{10} \leq$ number $\leq 5 \cdot 10^{10}$

3. a. $f^{-1}(y) = -\ln y/2$, $g^{-1}(y) = \sqrt[3]{y-1}$. $f(x) = 2$ when $x = f^{-1}(2) = -\frac{\ln(2)}{2} \approx -0.345$. Similarly, we find $g(1) = 2$.
   b. $(f \circ g)(x) = e^{-2(x^3+1)}$, $(g \circ f)(x) = e^{-6x} + 1$, $(f \circ g)(2) = e^{-18} = 1.5 \times 10^{-8}$, $(g \circ f)(2) = 1 + e^{-12} = 1.00006$
   c. $(g \circ f)^{-1}(y) = -\ln(y-1)/6$, domain is $y > 1$.

5. a. $\pi/3$ radians. $\sin\theta = \sqrt{3}/2$, $\cos\theta = 1/2$.

   b. $-\pi/3$ radians. $\sin\theta = -\sqrt{3}/2$, $\cos\theta = 1/2$.

   c. 1.919 radians. $\sin\theta \approx 0.9397$, $\cos\theta = -0.3420$.

   d. -3.316 radians. $\sin\theta \approx 0.1736$, $\cos\theta = -0.9848$.

   e. 20.25 radians. $\sin\theta \approx 0.9848$, $\cos\theta = 0.1736$.

7. The domain of all three functions is the set of real numbers. The graphs of $\sin x$, $\sin 2x$, and $\sin 4x$ oscillate about the x-axis: $\sin x$ with a period of 2π, $\sin 2x$ twice as fast as $\sin x$ (the period of $\sin 2x$ is π), and $\sin 4x$ four times as fast as $\sin x$ (the period of $\sin 4x$ is $\pi/2$). Adding x means that the graphs of $x + \sin x$, $x + \sin 2x$, and $x + 4\sin x$ oscillate about the line $y = x$. As well,

$$x - 1 \leq x + \sin x, \; x + \sin 2x, \; x + \sin 4x \leq x + 1$$

i.e., the graphs of all three functions are contained within the slanted strip bounded by the lines $y = x - 1$ and $y = x + 1$. From $f(x) = x + \sin x = x$ we conclude that $\sin x = 0$. Thus, it intersects the line $y = x$ at the points where the x-coordinates are the same as those of $\sin x = 0$; i.e., $x = 0, \pi, 2\pi, 3\pi$, etc. The analogous statements hold for $f(x) = x + \sin 2x$ and $f(x) = x + \sin 4x$.

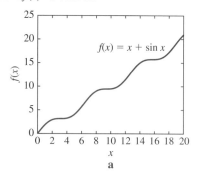

a

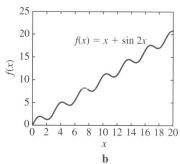

b

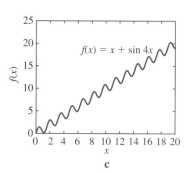

c

9. a. To obtain the graph of $|f(x)|$, we keep the parts of the graph of $f(x)$ that lie above the x-axis (or touch the x-axis). The parts that lie below the x-axis are reflected with respect to the x-axis.
To draw the graph of $f(|x|)$, we keep the part of the graph of $f(x)$ that corresponds to $x \geq 0$. For $x < 0$, we reflect the graph of $f(x)$ for $x > 0$ across the y-axis.

b.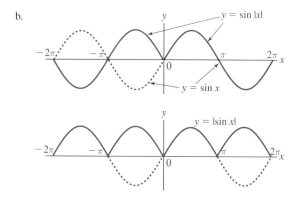

11. a. The size at $t = 30$ is $S(30) = 1e^{0.1 \cdot 30} \approx 20.08$. The size after treatment, which we can denote as $S_T(t)$, is the line through the point $(30, 20.08)$ with slope -0.4, so

$$S_T(t) = -0.4(t - 30) + 20.08$$

b.

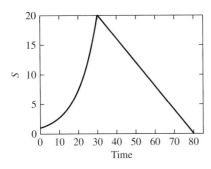

c. We solve $S_T(t) = 0$, and obtain

$$t = 30 + \frac{20.08}{0.4} \approx 80.2$$

13. a.

t	P(t)
0	20
1	32.22025
2	41.56266
3	46.52545
4	48.66306
5	49.49971
8	49.97485
10	49.99659

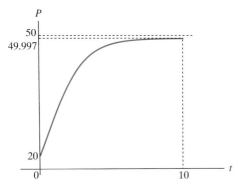

b. The initial population is $P(0) = 20$. After the initial fast rate of increase, the population slows down, increasing at a slower and slower rate. As t approaches 10, the population seems to be approaching 50.

15. a. We need to solve for the time when $b(t) = 3b(0)$. The tripling time t_t is

$$t_t = \frac{\ln 3}{0.333} \approx 3.30 \text{ hours}$$

b. The value of α will be

$$\alpha = \frac{\ln 3}{t_t} \approx 0.033$$

The equation is $b(t) = 3 \cdot 10^3 e^{0.033t}$

Chapter 3

Section 3.1

1. The updating function is $f(p_t) = p_t - 2$, and $f(5) = 3$, $f(10) = 8$, $f(15) = 13$. This is a linear function.

3. The updating function is $f(x_t) = x_t^2 + 2$, and $f(0) = 2$, $f(2) = 6$, $f(4) = 18$. This is not a linear function because the input x_t is squared.

5. Denote the updating function by $f(v) = 1.5v$. Then $(f \circ f)(v) = f(1.5v) = 1.5(1.5v) = 2.25v$, so $v_{t+2} = 2.25v_t$. Applying f to the initial condition twice gives $f(1220) = 1830$ and $f(1830) = 2745$, which is equal to $2.25 \cdot 1220$.

7. Denote the updating function by $h(n) = 0.5n$, then $(h \circ h)(n) = h(0.5n) = 0.5(0.5n) = 0.25n$ and $n_{t+2} = 0.25n_t$. Applying the updating function to the initial condition twice gives $h(1200) = 600$ and $h(600) = 300$, matching $(h \circ h)(1200) = 0.25 \cdot 1200 = 300$.

9. Solving for v_t gives $v_t = \frac{v_{t+1}}{1.5}$. Then $v_0 = \frac{1220}{1.5} = 813.3$.

11. Solving for n_t gives $n_t = 2n_{t+1}$. Then $n_0 = 2 \cdot 1200 = 2400$.

13. From $p_{t+1} = \frac{4}{p_t^3}$ we get $p_t^3 = \frac{4}{p_{t+1}}$ and $P_t = \sqrt[3]{\frac{4}{p_{t+1}}}$. Thus, $p_0 = \sqrt[3]{\frac{4}{p_1}} = \sqrt[3]{\frac{4}{32}} = \frac{1}{2}$.

15. $(f \circ f)(x) = f(f(x)) = f\left(\frac{x}{1+x}\right) = \frac{\frac{x}{1+x}}{1 + \frac{x}{1+x}} = \frac{x}{1+2x}$

To find the inverse, set $y = f(x)$ and solve

$$y = \frac{x}{1+x}$$
$$(1+x)y = x$$
$$y + xy = x$$
$$y = x - xy$$
$$\frac{y}{1-y} = x$$

Therefore, $f^{-1}(y) = \frac{y}{1-y}$.

17. $v_t = 1.5^t \cdot 1220 \, \mu\text{m}^3$

$$v_1 = 1.5 \cdot 1220 = 1830$$
$$v_2 = 1.5 \cdot 1830 = 2745$$
$$v_3 = 1.5 \cdot 2745 = 4117.5$$
$$v_4 = 1.5 \cdot 4117.5 = 6176.25$$
$$v_5 = 1.5 \cdot 6176.25 = 9264.375$$

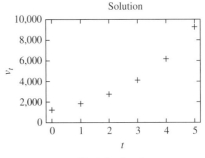

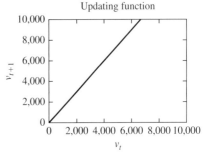

19. $n_t = 0.5^t \cdot 1200$

$$n_1 = 0.5 \cdot 1200 = 600$$
$$n_2 = 0.5 \cdot 600 = 300$$
$$n_3 = 0.5 \cdot 300 = 150$$
$$n_4 = 0.5 \cdot 150 = 75$$
$$n_5 = 0.5 \cdot 75 = 37.5$$

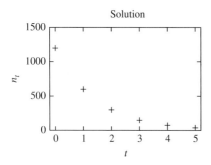

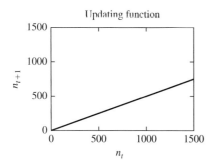

21. Plugging $t = 20$ into the solution $v_t = 1.5^t \cdot 1220 \ \mu m^3$, we get $v_{20} = 4.05 \cdot 10^6 \ \mu m^3$. This might be reasonable.

23. Plugging $t = 20$ into the solution $n_t = 0.5^t \cdot 1200$, we get $n_{20} = 0.0011$. This is an unreasonably small population.

25. $b_t = 20 \cdot 0.8^t - 10$

27. If $b \neq 1$, then $a_t = \frac{b+c-1}{b-1} b^t - \frac{c}{b-1}$; if $b = 1$, then $a_t = 1 + tc$.

29. $x_1 = 2/3, x_2 = 2/5, x_3 = 2/7, x_4 = 2/9$. It looks like $x_t = \frac{2}{1+2t}$.

31. $x_1 = 3/2, x_2 = 3, x_3 = 3/2, x_4 = 3$. It seems to be jumping back and forth between 3 and 3/2. If I start at $x_0 = 5$, the solution jumps between 5 and 5/4. But if I start at $x_0 = 2$, the solution just sits there.

33. If you started with 100, doubled (giving 200), and then divided by 4, you would end up with 50. If you first divided by 4 (giving 25) and then doubled, you would again end up with 50. This works in general, because the final population is $P_{t+1} = \frac{P_t}{2}$. These do commute.

35. Starting from $100, half gives $50, add $10 to get $60. Starting from $100, add $10 to get $110, lose half to get $55. These do not commute.

37. 80 m, which is a bit tall, even for a redwood.

39. $1.1 \cdot 10^{12}$ million bacteria. This population would weigh over one gram, which is a lot.

41. The solution goes $x_0 = 10, x_1 = 40, x_2 = 100, x_3 = 220$. Adding 20 gives $x_t = 30 \cdot 2^t - 20$.

43.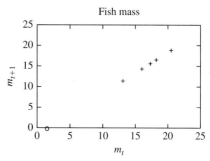

The discrete-time dynamical system is $m_{t+1} = l_t - 1.7$ and the missing value is -0.2. This does not make much sense as a mass, so the discrete-time dynamical system only makes sense when $m_t > 1.7$.

45.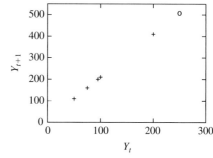

The discrete-time dynamical system is $Y_{t+1} = 2Y_t + 10$ and the missing value is 510.

47. The tail length decreases by 0.1 cm each half day, so $T_{t+1} = T_t - 0.1$.

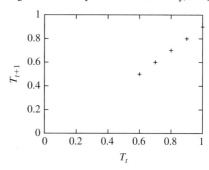

49. Age, not surprisingly, increases by one half day each half day, so $a_{t+1} = a_t + 0.5$ days.

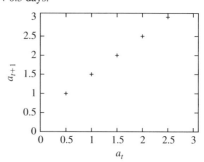

51. Let a_t and a_{t+1} be the total area before and after the experiment. Then

$$a_t = \frac{v_t}{20} \text{ and } a_{t+1} = \frac{v_{t+1}}{20}$$

The original discrete-time dynamical system is

$$v_{t+1} = 2v_t$$

Therefore,
$$a_{t+1} = \frac{v_{t+1}}{20} = \frac{2v_t}{20} = 2\frac{v_t}{20} = 2a_t$$

53. Let V_t and V_{t+1} be the initial and final volumes, respectively. We know that
$$V_t = \frac{4\pi}{3}r_t^3$$
$$V_{t+1} = \frac{4\pi}{3}r_{t+1}^3$$

Therefore,
$$V_{t+1} = \frac{4\pi}{3}(r_t + 0.8)^3$$

We would like to express the right-hand side in terms of V_t. We can solve for r_t in terms of V_t as
$$r_t = \sqrt[3]{\frac{3V_t}{4\pi}}$$

Substituting, we get
$$V_{t+1} = \frac{4\pi}{3}\left(\sqrt[3]{\frac{3V_t}{4\pi}} + 0.8\right)^3$$

55. The points for the second patient are $(0, 2)$, $(2, 3.2)$, and $(3.2, 3.92)$.

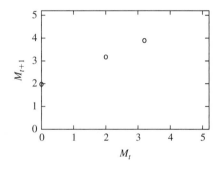

Let the level be M_t at the beginning of the day and M_{t+1} at the end of the day. Using the first two lines in the table, we find the two points to be $(0, 2)$ and $(2, 3.2)$. The slope is
$$\text{slope} = \frac{\text{change in output}}{\text{change in input}} = \frac{3.2 - 2}{2 - 0} = 0.6$$

In point-slope form, the discrete-time dynamical system is
$$M_{t+1} = 0.6(M_t - 0) + 2 = 0.6M_t + 2$$

57. The first has height $h_t = 10 + 0.8t$, and the second has height $h_t = 2 + 0.8t$. The difference is always 8 metres. The ratio is $(10 + 0.8t)/(2 + 0.8t)$, which is 5 at $t = 0$, 2.33 at $t = 5$, 1.44 at $t = 20$, and 1.09 at $t = 100$. The ratio gets closer to 1 as the trees get larger because the 8-metre height difference becomes more and more insignificant.

59. a. $b_1 = 4 \cdot 10^6$, $b_2 = 6 \cdot 10^6$, $b_3 = 10 \cdot 10^6$
    b. $b_{t+1} = 2(b_t - 10^6) = 2b_t - 2 \cdot 10^6$
    c. The population is smaller because we are harvesting before reproduction.

61. a. The ratio is 1.01.
    b. 4.0804
    c. $T_{t+1} = 1.01T_t$. The tones heard are $T_0 = 400$, $T_1 = 404$, $T_2 = 408.04$, $T_3 = 412.12$, $T_4 = 416.24$, and $T_5 = 420.40$.

d. The ratio is 1.00125. The tones are $T_0 = 400$, $T_1 = 400.5$, $T_2 = 401$, $T_3 = 401.5$, $T_4 = 402$, and $T_5 = 402.51$.

63. We have that $b_{t+1} = 2b_t$ and $m_{t+1} = m_t + 1$ (measured in nanograms). Then the total mass $M_{t+1} = m_{t+1}b_{t+1} = (m_t + 1) \cdot 2b_t = 2M_t + 2b_t$. This is not a proper discrete-time dynamical system because M_{t+1} depends on b_t in addition to M_t. Knowing the total mass before is not enough information to find the total mass after, because a colony with a few large bacteria will grow differently from one with many small bacteria.

Section 3.2

1.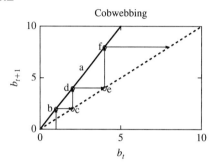

3. The solution is $v_t = 1.5^t \cdot 1220 \; \mu\text{m}^3$, consistent with a cobweb diagram that predicts a solution that increases faster and faster.

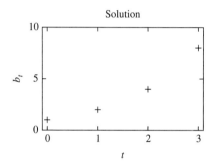

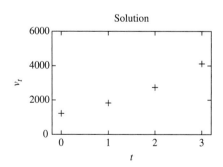

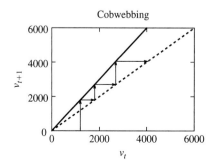

5. The solution is $n_t = 0.5^t \cdot 1200$, consistent with a cobweb diagram that decays toward 0.

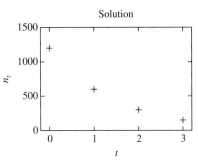

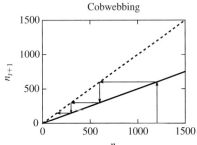

7.

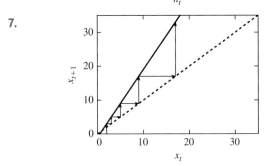

9.

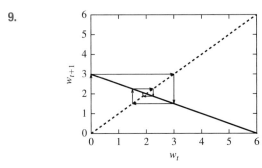

11.

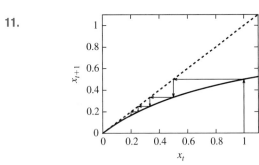

13. The equilibrium seems to be at about 1.3. It is stable.

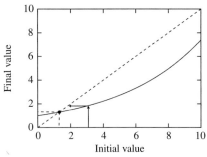

15. The equilibria seem to be at about 0 and about 7.5. It is not clear what the cobwebbing yields for either equilibrium.

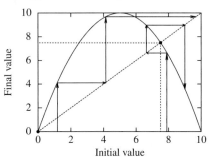

17. $f(x) = x$ when $x^2 = x$, or $x^2 - x = 0$, or $x(x-1) = 0$, which has solutions at $x = 0$ and $x = 1$.

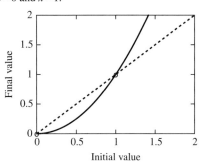

19. The equilibrium is where $c^* = 0.5c^* + 8$ or $c^* = 16$.

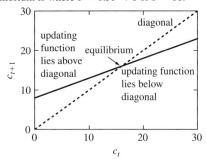

21. The equilibrium is $b^* = 0$.

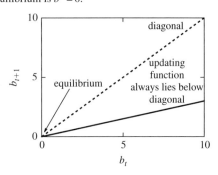

23. $v^* = 1.5v^*$ implies that $v^* = 0$. The equilibrium is unstable.

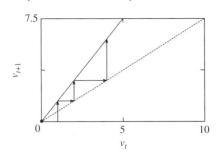

25. $x^* = 2x^* - 1$ has solution $x^* = 1$. The equilibrium is unstable.

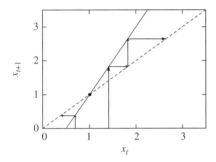

27. $w^* = -0.5w^* + 3$ has solution $w^* = 2$. The equilibrium is stable.

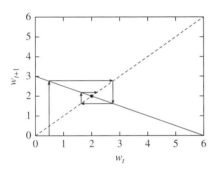

29. $x^* = \frac{x^*}{1+x^*}$ has solution $x^* = 0$. The equilibrium is stable from the left side and unstable from the right side.

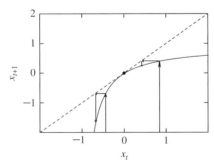

31.
$$w^* = aw^* + 3$$
$$w^* = \frac{3}{1-a}$$

This solution does not exist if $a = 1$ and is negative if $a > 1$.

33.
$$x^* = \frac{ax^*}{1+x^*}$$
$$x^*(1+x^*) = ax^*$$
$$x^*(1+x^*) - ax^* = 0$$
$$x^*(1+x^* - a) = 0$$

There are two solutions, $x^* = 0$ and $x^* = a - 1$. The second is negative if $a < 1$. The two solutions are equal (leaving only one) when $a = 1$.

35.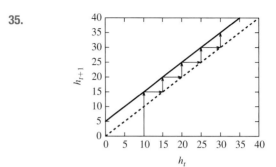

37.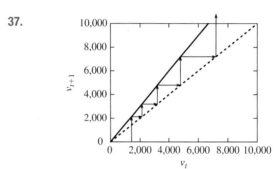

39.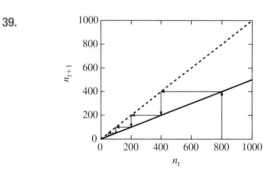

41.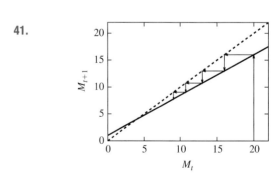

The equilibrium is
$$M^* = 0.75M^* + 1$$
$$0.25M^* = 1$$
$$M^* = 4$$

43.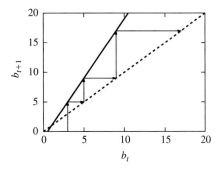

The equilibrium is
$$b^* = 2b^* - 10^6$$
$$b^* = 10^6$$

The population grows and seems to be moving away from the equilibrium.

45. The equilibrium is
$$b^* = 2b^* - h$$
$$b^* = h$$

It is strange that the equilibrium gets larger as the harvest gets larger. However, the cobwebbing diagram indicates that only populations above the equilibrium will grow, and those below it will shrink. The equilibrium in this case is the minimum population required for the population to survive.

Section 3.3

1. $b_8 = 12{,}000 \cdot 1.3^8 \approx 97887.7$

3. $b_{500} = 100 \cdot 0.99^{500} \approx 0.66$

5. $r = 3/2$ and $b_0 = 8{,}000/9 \approx 888.9$

7. $r = \left(\frac{1000}{3}\right)^{1/25} \approx 1.26$

9. The per capita production $r(1 - p_t) = -rp_t + r$ is the line of slope $-r$ whose y-intercept is equal to r; the equilibria of the given system are $p^* = 0$ and $p^* = 1 - 1/r$.

11. $b_1 = 6{,}000$, $b_2 = 10{,}285.71$, $b_3 = 10{,}936.71$, $b_4 = 10{,}994.70$, $b_5 = 10{,}999.56$; the numbers seem to be approaching $11{,}000$.

13. There are two equilibrium points, $b^* = 0$ and $b^* = 11{,}000$; the population that starts at either $b^* = 0$ or $b^* = 11{,}000$ will remain at its starting value for all time.

15. $a_5 = 31.58$; if you start with three rapid drinks (i.e., three times 14 g of alcohol) and then consume half a drink every hour, the amount of alcohol in your body will decrease and reach 31.58 g after 5 hours.

17. $a_5 = 52.66$; a person that starts with two rapid drinks and then consumes one drink per hour will see an increase in the level of alcohol in his or her body, which will reach 52.66 g in 5 hours.

19. There are three equilibrium points, $p^* = 0$ and $p^* = \frac{r \pm \sqrt{r^2 - 4}}{2}$.

21. $b_t = 10^6 \cdot 1.2^t$; the population will reach 10^7 between $t = 12$ and $t = 13$.

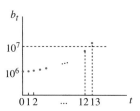

23. $b_t = 10{,}000 \cdot 0.9^t$; the population will reach $1{,}000$ between $t = 21$ and $t = 22$.

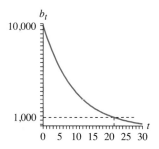

25. a. From $\frac{b_0}{b_1} = \frac{1}{r}$ we get $\frac{b_0}{b_1} = \alpha b_0 + \beta = \frac{1}{r}$; from $\frac{b_N}{b_{N+1}} = 1$ we get $\frac{b_N}{b_{N+1}} = \alpha b_N + \beta = 1$. Solving this system gives $\alpha = \frac{r-1}{r(b_N - b_0)}$ and $\beta = \frac{b_N - rb_0}{r(b_N - b_0)}$.

b. From $\frac{b_t}{b_{t+1}} = \alpha b_t + \beta$ we compute

$$b_{t+1} = \frac{b_t}{\alpha b_t + \beta} = \frac{b_t}{\frac{r-1}{r(b_N - b_0)} b_t + \frac{b_N - rb_0}{r(b_N - b_0)}}$$
$$= \frac{b_t}{\frac{(r-1)b_t + b_N - rb_0}{r(b_N - b_0)}}$$
$$= \frac{b_t r(b_N - b_0)}{(r-1)b_t + b_N - rb_0}$$
$$= \frac{r(K - b_0)b_t}{(r-1)b_t + K - rb_0}$$

27. $p_t = 2500 \cdot 0.99667^t$; the population will reach $m = 500$ after about 483 years.

29. $p_t = 25{,}000 \cdot 0.99187^t$; the population will reach $m = 200$ after about 592 years.

31. $c_t = 600 \cdot 0.87^t$; about 5 hours (half-life!)

33. It will take between 15 and 16 hours for the level of caffeine to reach 90% of the equilibrium (the equilibrium value is 923.08 mg).

Section 3.4

1. There are now 400 red birds and 800 blue birds, or $1/3$ red birds and $2/3$ blue birds. These fractions add to 1.

3. There are now $200r$ red birds and 800 blue birds. The fraction of red birds is $200r$ out of $200r + 800$ or a fraction $\frac{r}{r+4}$. There are 800 blue birds out of $200r + 800$ or a fraction $\frac{4}{r+4}$. These fractions add to 1 no matter what r is.

5.

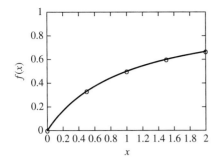

7.
$$p_t = \frac{m_t}{m_t + b_t} = \frac{1.2 \cdot 10^5}{1.2 \cdot 10^5 + 3.5 \cdot 10^6} \approx 0.033$$
$$m_{t+1} = 1.2 m_t = 1.44 \cdot 10^5$$
$$b_{t+1} = 2 b_t = 7 \cdot 10^6$$
$$p_{t+1} = \frac{m_{t+1}}{m_{t+1} + b_{t+1}} = \frac{1.44 \cdot 10^5}{1.44 \cdot 10^5 + 7 \cdot 10^6} \approx 0.020$$

9. $p_t \approx 0.033$, $m_{t+1} = 0.36 \cdot 10^5$, $b_{t+1} = 1.75 \cdot 10^6$, $p_{t+1} \approx 0.020$

11. The equation for the equilibria is $p^* = \dfrac{p^*}{p^* + 2(1 - p^*)}$. Then

$$p^*(p^* + 2(1 - p^*)) = p^*$$
$$p^*(p^* + 2(1 - p^*)) - p^* = 0$$
$$p^*(p^* + 2(1 - p^*) - 1) = 0$$
$$p^*(1 - p^*) = 0$$
$$p^* = 0 \text{ or } p^* = 1$$

13. The equation for the equilibria is $x^* = \dfrac{x^*}{1 + ax^*}$. Solving,

$$x^*(1 + ax^*) = x^*$$
$$x^*(1 + ax^*) - x^* = 0$$
$$x^*(1 + ax^* - 1) = 0$$
$$x^*(ax^*) = 0$$

The only equilibrium is at $x^* = 0$, as long as $a \neq 0$. If $a = 0$, the system is $x_{t+1} = x_t$, which has every value of x^* as an equilibrium.

15. The equilibrium seems to be at about 1.3.

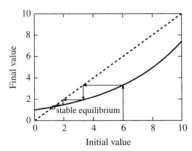

17. The equilibria seem to be at about 0 and 7.5.

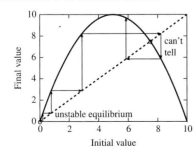

19. The discrete-time dynamical system is
$$p_{t+1} = \frac{1.2 p_t}{1.2 p_t + 2(1 - p_t)}$$

The equilibrium at $p^* = 0$ is stable and the one at $p^* = 1$ is unstable. This makes sense because the wild type are reproducing faster than the mutants ($r > s$) and should dominate the population.

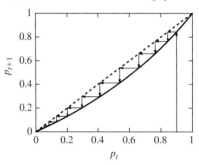

21. The discrete-time dynamical system is
$$p_{t+1} = \frac{0.3 p_t}{0.3 p_t + 0.5(1 - p_t)}$$

The equilibrium at $p^* = 0$ is stable and the one at $p^* = 1$ is unstable. The picture looks the same as in Exercise 19 because the ratio of r to s is the same. But in this case, both populations are decreasing.

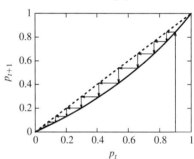

23.

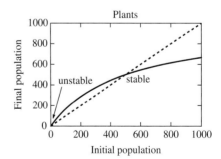

25.
a. $2 \cdot 10^5$ mutate and 10^4 revert.
b. There are $10^6 - 2 \cdot 10^5 + 10^4 = 8.1 \cdot 10^5$ wild type, and $10^5 - 10^4 + 2 \cdot 10^5 = 2.9 \cdot 10^5$ mutants.
c. The total number before and after is $1.1 \cdot 10^6$. It does not change because the bacteria are not reproducing or dying, just changing their type.
d. The fraction before is $10^5 / 1.1 \cdot 10^6 = 0.091$. The fraction after is $2.9 \cdot 10^5 / 1.1 \cdot 10^6 \approx 0.264$.

27.
a. $0.2 b_t$ mutate and $0.1 m_t$ revert.
b. $b_{t+1} = b_t - 0.2 b_t + 0.1 m_t = 0.8 b_t + 0.1 m_t$; $m_{t+1} = m_t - 0.1 m_t + 0.2 b_t = 0.9 m_t + 0.2 b_t$

c. The total number before is $b_t + m_t$. The total number after is

$$b_{t+1} + m_{t+1} = (0.8b_t + 0.1m_t) + (0.9m_t + 0.2b_t)$$
$$= 0.8b_t + 0.2b_t + 0.1m_t + 0.9m_t$$
$$= b_t + m_t$$

d. Divide the discrete-time dynamical system for m_{t+1} by $b_{t+1} + m_{t+1}$,

$$p_{t+1} = \frac{m_{t+1}}{b_{t+1} + m_{t+1}} = \frac{0.9m_t + 0.2b_t}{b_{t+1} + m_{t+1}} = \frac{0.9m_t + 0.2b_t}{b_t + m_t}$$
$$= \frac{0.9m_t}{b_t + m_t} + \frac{0.2b_t}{b_t + m_t} = 0.9p_t + 0.2(1 - p_t) = 0.2 + 0.7p_t$$

e. The equilibrium solves $p^* = 0.2 + 0.7p^*$, or $p^* \approx 0.667$.

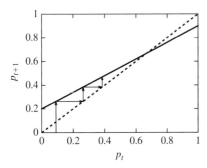

f. The equilibrium seems to be stable. The fraction of mutants will increase until it reaches 66.7%.

29. a. 10^5 mutate.
    b. There are $10^6 - 10^5 = 9 \cdot 10^5$ wild type, and $10^5 + 10^5 = 2 \cdot 10^5$ mutants.
    c. There are $1.8 \cdot 10^6$ wild type and $3 \cdot 10^5$ mutants.
    d. The total number after is $2.1 \cdot 10^6$.
    e. The fraction before is $10^5/1.1 \cdot 10^6 \approx 0.091$. The fraction after is $3 \cdot 10^5/2.1 \cdot 10^6 \approx 0.143$.

31. a. $0.1b_t$ mutate.
    b. There are $0.9b_t$ wild type and $m_t + 0.1b_t$ mutants after mutation.
    c. There are $b_{t+1} = 1.8b_t$ wild type and $m_{t+1} = 0.5m_t + 0.15b_t$ mutants after reproduction.
    d. The total number after is $1.95b_t + 1.5m_t$.
    e. Divide the discrete-time dynamical system for m_{t+1} by $b_{t+1} + m_{t+1}$,

$$p_{t+1} = \frac{m_{t+1}}{b_{t+1} + m_{t+1}} = \frac{1.5m_t + 0.15b_t}{1.95b_t + 1.5m_t} = \frac{\frac{1.5m_t}{m_t + b_t} + \frac{0.15b_t}{m_t + b_t}}{\frac{1.95b_t}{m_t + b_t} + \frac{1.5m_t}{m_t + b_t}}$$
$$= \frac{1.5p_t + 0.15(1 - p_t)}{1.5p_t + 1.95(1 - p_t)}$$

f. The equilibrium solves

$$p^* = \frac{1.5p^* + 0.15(1 - p^*)}{1.5p^* + 1.95(1 - p^*)}$$

Following the algebra gives

$$p^* (1.5p^* + 1.95(1 - p^*)) = 1.5p^* + 0.15(1 - p^*)$$
$$p^* (1.5p^* + 1.95(1 - p^*)) - 1.5p^* + 0.15(1 - p^*) = 0$$
$$0.15 - 0.6p^* + 0.45(p^*)^2 = 0$$
$$0.45(p^* - 1/3)(p^* - 1) = 0$$

Therefore, $p^* = 1/3$ or $p^* = 1$.

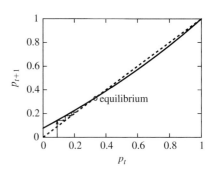

g. The equilibrium at 1/3 seems to be stable. The fraction of mutants will end up at about 33.3%.

33. $x_1 = 100 - 20 + 30 = 110$; $y_1 = 100 - 30 + 20 = 90$; $x_2 = 115$ and $y_2 = 85$

35. $x_{t+1} = x_t - 0.2x_t + 0.3y_t = 0.8x_t + 0.3y_t$;
    $y_{t+1} = y_t - 0.3y_t + 0.2x_t = 0.7y_t + 0.2x_t$

37. a. Consider the first island. After migration, there are 80 butterflies that started on the first island and 30 that started on the second. The 80 reproduce, making a total of 190. On the second island, there are 20 from the first and 70 from the second after migration. The 20 reproduce, making a total of 110. Following the same reasoning, $x_2 = 337$ and $y_2 = 153$.
    b. $x_{t+1} = 2(x_t - 0.2x_t) + 0.3y_t = 1.6x_t + 0.3y_t$;
       $y_{t+1} = y_t - 0.3y_t + 2 \cdot 0.2x_t = 0.7y_t + 0.4x_t$
    c. $x_{t+1} + y_{t+1} = 2x_t + y_t$. Then

$$p_{t+1} = \frac{x_{t+1}}{x_{t+1} + y_{t+1}} = \frac{1.6x_t + 0.3y_t}{x_{t+1} + y_{t+1}}$$
$$= \frac{1.6x_t + 0.3y_t}{2x_t + y_t}$$
$$= \frac{\frac{1.6x_t}{x_t + y_t} + \frac{0.3y_t}{x_t + y_t}}{\frac{2x_t}{x_t + y_t} + \frac{y_t}{x_t + y_t}}$$
$$= \frac{1.6p_t + 0.3(1 - p_t)}{2p_t + 1 - p_t}$$

d. We must solve the equation

$$p^* = \frac{1.6p^* + 0.3(1 - p^*)}{2p^* + 1 - p^*} = \frac{0.3 + 1.3p^*}{1 + p^*}$$

Multiplying out and using the quadratic formula, we find $p^* \approx 0.718$. This is larger than the result in Exercise 3.4.36 because the butterflies from the first island reproduce.

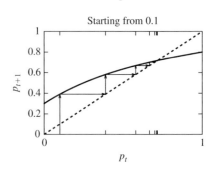

Starting from 0.1

39. The discrete-time dynamical system is

$$M_{t+1} = M_t - \frac{0.5}{1 + 0.1M_t} M_t + 1$$

To find the equilibrium,

$$M^* = M^* - \frac{0.5}{1 + 0.1M^*} M^* + 1$$

$$\frac{0.5}{1 + 0.1M^*} M^* = 1$$

$$0.5 M^* = 1 + 0.1 M^*$$

$$0.4 M^* = 1$$

The equilibrium is therefore 2.5. It is larger because the fraction absorbed is always less than 0.5.

41. The discrete-time dynamical system is

$$M_{t+1} = M_t - \frac{\beta}{1 + 0.1M_t} M_t + 1$$

To find the equilibrium,

$$M^* = M^* - \frac{\beta}{1 + 0.1M^*} M^* + 1$$

$$\frac{\beta}{1 + 0.1M^*} M^* = 1$$

$$\beta M^* = 1 + 0.1 M^*$$

$$(\beta - 0.1) M^* = 1$$

The equilibrium is therefore

$$M^* = \frac{1}{\beta - 0.1}$$

This becomes smaller as β becomes larger. Large values of β indicate that the fraction absorbed is large, which makes sense. The fact that the equilibrium is negative when $\beta \leq 0.1$ indicates that there is no equilibrium. The body cannot use up the dose of 1 each day, and the level just keeps building up.

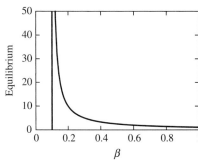

The second diagram looks a lot like the tree growth model. Once the concentration becomes large, it simply increases by 1 per day.

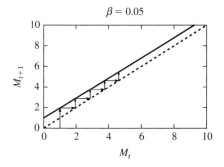

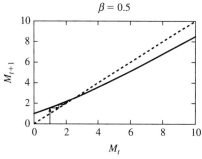

43. a.

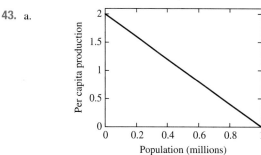

b. $b_{t+1} = 2b_t \left(1 - \frac{b_t}{10^6}\right)$
c. The equilibria are $b^* = 0$ and $b^* = 5 \cdot 10^5$.
d. $b^* = 0$ is unstable; $b^* = 5 \cdot 10^5 = 0.5 \cdot 10^6$ is stable.

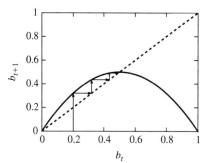

45. a.

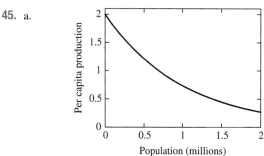

b. $b_{t+1} = 2b_t e^{-b_t/10^6}$
c. The equilibria are $b^* = 0$ and $b^* = \ln(2) \cdot 1 \cdot 10^6 \approx 6.93 \cdot 10^5$.
d. $b^* = 0$ is unstable; $b^* = 10^6 \ln 2 \approx 0.693 \cdot 10^6$ is stable.

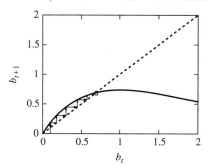

Section 3.5

1. $2/3$ of the water is at $100°C$, and $1/3$ is at $30°C$. The final temperature is then the weighted average, $T = \frac{1}{3} \cdot 30°C + \frac{2}{3} \cdot 100°C \approx 76.7°C$.

3. $1/3$ of the water is at T_1, and $2/3$ is at T_2. The final temperature is then the weighted average, $T = \frac{1}{3} \cdot T_1 + \frac{2}{3} \cdot T_2$. If $T_1 = 30$ and $T_2 = 100$, we get $T = \frac{1}{3} \cdot 30 + \frac{2}{3} \cdot 100 = 76.7$, as before.

5. A fraction $V_1/(V_1 + V_2)$ is at T_1, and a fraction $V_2/(V_1 + V_2)$ is at T_2. The final temperature is the weighted average

$$T = T_1 \frac{V_1}{V_1 + V_2} + T_2 \frac{V_2}{V_1 + V_2}$$

7. The $100°C$ water cools to $50°C$, so $2/3$ of the water is at $50°C$, and $1/3$ is at $15°C$. The final temperature is then the weighted average, $T = \frac{1}{3} \cdot 15°C + \frac{2}{3} \cdot 50°C \approx 38.3°C$. This is indeed half the value in Exercise 3.5.1.

9. a. amount = volume times concentration, or $Vc_0 = 2 \cdot 1 = 2$ mmol.
   b. 0.5 L at 1 mmol/L = 0.5 mmol
   c. 1.5 L at 1 mmol/L = 1.5 mmol
   d. 0.5 L at 5 mmol/L = 2.5 mmol
   e. $1.5 + 2.5 = 4$ mmol
   f. 4 mmol/2 L = 2 mmol/L
   g. $q = 0.5/2 = 0.25$. Then $c_{t+1} = (1-q)c_t + q\gamma = 0.75 \cdot c_t + 0.25 \cdot 5$. When $c_0 = 1$, $c_1 = 0.75 \cdot 1 + 0.25 \cdot 5 = 2$ mmol/L.

11. Start with 9 mmol, breathe out 8.1 mmol, leaving 0.9 mmol, breathe in 4.5 mmol, ending with 5.4 mmol, and a concentration of 5.4 mmol/L. This checks with the discrete-time dynamical system. In this case, $q = 0.9$ and $\gamma = 5$, so $c_{t+1} = 0.1c_t + 0.9 \cdot 5$. Substituting $c_0 = 9$, we find $c_1 = 5.4$.

13. The discrete-time dynamical system has $q = 0.5/2 = 0.25$ and $\gamma = 5$ and thus has formula $c_{t+1} = (1-0.25)c_t + 0.25 \cdot 5 = 0.75c_t + 1.25$. We want to start from 1 mmol/L.

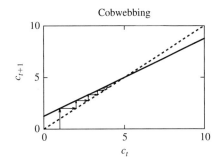

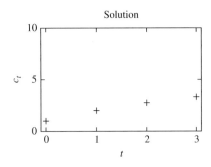

15. The discrete-time dynamical system is $c_{t+1} = 0.1c_t + 4.5$, starting from $c_0 = 9$.

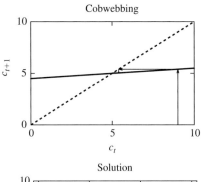

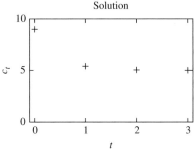

17. The discrete-time dynamical system is $c_{t+1} = 0.75 \cdot c_t + 1.25$. Solving for the equilibrium, we find $c^* = 0.75 \cdot c^* + 1.25$, or $0.25c^* = 1.25$ or $c^* = 5$. This matches the value of γ.

19. The discrete-time dynamical system is $c_{t+1} = 0.1 \cdot c_t + 4.5$. Solving for the equilibrium, we find $c^* = 0.1 \cdot c^* + 4.5$, or $0.9c^* = 4.5$ or $c^* = 5$. This matches the value of γ.

21. Using the equation $c^* = \frac{q\gamma}{1 - (1-q)(1-\alpha)}$ for the equilibrium, we find that $c^* = \frac{0.4 \cdot 0.21}{1 - 0.6 \cdot 0.9} \approx 0.183$. The concentration is higher because more of the air in the lung at any one time comes from outside.

23. The concentration after absorption is $c_t - 0.02$. Using the weighted average idea,

$$c_{t+1} = (1-q)(c_t - 0.02) + q\gamma = 0.8(c_t - 0.02) + 0.042$$
$$= 0.8c_t + 0.026$$

The equilibrium solves

$$c^* = 0.8c^* + 0.026$$
$$c^* = 0.13$$

This function does not really make sense if $c_t < 0.02$ because there would not be enough there to absorb.

25. The concentration after absorption is $c_t - 0.2(c_t - 0.05) = 0.8c_t + 0.01$. Then

$$c_{t+1} = (1-q)(0.8c_t + 0.01) + q\gamma = 0.8(0.8c_t + 0.01) + 0.042$$
$$= 0.64c_t + 0.05.$$

The equilibrium is then

$$c^* = 0.64c^* + 0.05$$
$$c^* \approx 0.0139$$

27. The concentration after absorption is $c_t - A$. Then

$$c_{t+1} = (1-q)(c_t - A) + q\gamma = 0.8(c_t - A) + 0.042$$

The equilibrium solves

$$c^* = 0.8c^* + 0.042 - 0.8A$$

$$0.2c^* = 0.042 - 0.8A$$
$$c^* = 0.21 - 4A$$

Then $c^* = 0.15$ if $0.21 - 4A = 0.15$ or $A = 0.015$. The lung must reduce the oxygen concentration by 1.5%. In Example 3.5.9, we found that the lung absorbs 10% of the equilibrium concentration of 0.15, which is equivalent to $A = 0.15$.

29. The concentration before exchanging air is $c_t + 0.001$, so the discrete-time dynamical system is the weighted average

$$c_{t+1} = (1-q)(c_t + 0.001) + q\gamma$$
$$c_{t+1} = 0.8(c_t + 0.001) + 0.2 \cdot 0.0004 = 0.8c_t + 0.00088$$

The equilibrium is

$$c^* = 0.8c^* + 0.00088$$
$$c^* = 0.0044$$

This is about 11 times as high as the external concentration.

31. a. Population after reproduction is $0.6 \cdot 3 \cdot 10^6 = 1.8 \cdot 10^6$.
b. Population after supplementation is $1.8 \cdot 10^6 + 10^6 = 2.8 \cdot 10^6$.
c. $b_{t+1} = 0.6b_t + 10^6$

33. The discrete-time dynamical system is $b_{t+1} = 0.6b_t + 10^6$. The equilibrium satisfies $b^* = 0.6b^* + 10^6$, or $0.4b^* = 10^6$ or $b^* = 2.5 \cdot 10^6$.

35. The discrete-time dynamical system is $b_{t+1} = rb_t + 10^6$. The equilibrium satisfies $b^* = rb^* + 10^6$, or $(1-r)b^* = 10^6$ or $b^* = \frac{10^6}{1-r}$. If $r = 0$, $b^* = 10^6$. This makes sense because all the old bacteria die so the population will reach an equilibrium equal to the number supplemented. If r is close to 1, b^* becomes very large. This makes sense because the old bacteria could almost maintain their population, so deaths will not balance out supplementation until there are a lot of bacteria.

37. There are $3.3 \cdot 10^7 s_t + 3 \cdot 10^3$ cubic metres of salt in $3.6 \cdot 10^7$ cubic metres of water. After evaporation, there are $3.45 \cdot 10^7$ cubic metres of water, in which the concentration is

$$s_{t+1} = \frac{3.3 \cdot 10^7 s_t + 3 \cdot 10^3}{3.45 \cdot 10^7} \approx 0.957 s_t + 8.69 \cdot 10^{-5}$$

Because water that flows out is well mixed, this gives the discrete-time dynamical system.

39. There are $3.3 \cdot 10^7 s_t + 3 \cdot 10^3$ cubic metres of salt in $3.6 \cdot 10^7$ cubic metres of water. After evaporation, there are $3.4 \cdot 10^7$ cubic metres of water, in which the concentration is

$$s_{t+1} = \frac{3.3 \cdot 10^7 s_t + 3 \cdot 10^3}{3.4 \cdot 10^7} \approx 0.971 s_t + 8.82 \cdot 10^{-5}$$

41. The discrete-time dynamical system is $s_{t+1} = 0.957 s_t + 8.69 \cdot 10^{-5}$. The equilibrium solves

$$s^* = 0.957 s^* + 8.69 \cdot 10^{-5}$$
$$0.043 s^* = 8.69 \cdot 10^{-5}$$
$$s^* \approx 0.002$$

The water ends up twice as salty as the water that flows in, perhaps because only half the water flows out.

43. The discrete-time dynamical system is $s_{t+1} = 0.971 s_t + 8.82 \cdot 10^{-5}$. The equilibrium solves

$$s^* = 0.971 s^* + 8.82 \cdot 10^{-5}$$
$$0.029 s^* = 8.82 \cdot 10^{-5}$$
$$s^* \approx 0.003$$

The water ends up three times as salty as the water that flows in, perhaps because 2/3 of the water that enters evaporates.

45. $b^* = \frac{h}{r-1}$. Oddly, the equilibrium is larger when the harvest h is larger, and smaller when the per capita production r is larger. This occurs because the equilibrium is unstable and acts as a threshold between populations that go extinct and populations that grow. A larger equilibrium means that the population is more susceptible to extinction.

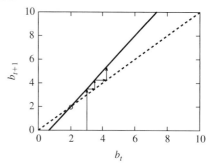

True/False Quiz

1. FALSE. From $m^* = f(m^*)$ it follows that $f^{-1}(m^*) = f^{-1}(f(m^*)) = m^*$. Thus, both systems have the same equilibrium points.

3. FALSE. $h_{t+1} = 1000 H_{t+1} = 1000(2H_t + 3) = 2(1000H_t) + 3000 = 2h_t + 3000$

5. TRUE. When $m_t = 2$, then $m_{t+1} = e^{2-2} + 1 = 1 + 1 = 2$.

7. FALSE. Since the slope of the updating function is larger than 1, any starting value m_0 will be increased by cobwebbing.

9. FALSE. The equation for the equilibrium points of this system is a degree 3 polynomial equation—thus, it has at most three solutions.

Supplementary Problems

1. a. $f(b) = 3b$
b. $1.62 \cdot 10^9$
c. $B(t) = 2 \cdot 10^7 \cdot 3^{t/2}$ with t measured in hours
d. $t = 2\ln\left(\frac{10^9}{2 \cdot 10^7}\right) / \ln 3 = 2\ln 50 / \ln 3$

3. $p^* = 0$ and $p^* = 1$

5. $M^* = S/\alpha$; to increase M^* (equilibrium for the medication), we can either increase S (daily dosage) or decrease α (fraction of the drug absorbed by the body during a day).

7. 1.486 moles/L; it is exactly twice the concentration found earlier.

9. If $r = 0$ or $r = 1$, the system has one equilibrium, $b^* = 0$. If $r \neq 0$ and $r \neq 1$, the system has two equilibria, $b^* = 0$ and $b^*_* = (r-1)/3$.

11. a. Let B_t be the number of butterflies in the late summer. There are then $1.2B_t$ eggs, leading to $0.6B_t$ new butterflies from reproduction plus 1000 from immigration. The discrete-time dynamical system is $B_{t+1} = 0.6B_t + 1000$.

b.

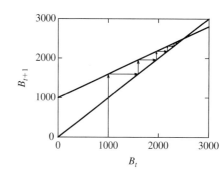

c. Equilibrium has 2500 butterflies.

13.
a. 2.66 cm²
b. The discrete-time dynamical system is $A_{t+1} = 1.1 A_t$.
c. 1.82 cm²
d. 0.55
e. When $2 \cdot 1.1^t = 10$ or in 16.9 hours

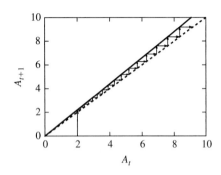

15.
a. The higher the concentration of chemical, the deeper it breathes.
b. $c_{t+1} = (1-q)c_t + q\gamma = \left(1 - \dfrac{c_t}{c_t + \gamma}\right)c_t + \dfrac{c_t}{c_t + \gamma}\gamma = \dfrac{2\gamma c_t}{c_t + \gamma}$
c. $c_t = 0$ or $c_t = \gamma$

17.
a. First line: 10 people served, 20 join, 10 switch out, 10 switch in, so there are $100 - 10 + 20 - 10 + 10 = 110$ at the end of the minute. Second line: 30 people served, 20 join, 10 switch out, 10 switch in, so there are $100 - 30 + 20 - 10 + 10 = 90$ at the end of the minute.
b. Let a be the number in the first line and b the number in the second line. Then
$$a_{t+1} = 0.9a_t + 0.1(a_t + b_t) - 0.1a_t + 0.1b_t = 0.9a_t + 0.2b_t$$
Similarly,
$$b_{t+1} = 0.2a_t + 0.7b_t$$
c. $p_{t+1} = \dfrac{a_{t+1}}{a_{t+1} + b_{t+1}} = \dfrac{0.9a_t + 0.2b_t}{1.1a_t + 0.9b_t}$
$= \dfrac{0.9p_t + 0.2(1 - p_t)}{1.1p_t + 0.9(1 - p_t)} = \dfrac{0.7p_t + 0.2}{0.2p_t + 0.9}$

19.
a. $S = \dfrac{V}{4}$
b. Both R and α have units of moles per square centimetre.
c. $R = 1.2$ moles/cm², so the total absorbed is $1.2 \cdot 10^2$ moles
d. $T = 0.3V$

21.
a. Spend $60 million on operations, leaving $280 million. Get $28 million in interest and $50 million in donations for a total of $358 million.
b. $M_{t+1} = 0.825 M_t + 77.5$
c.

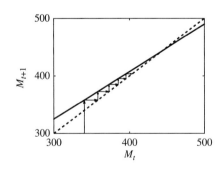

23.
a.

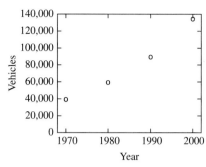

b. It is increasing by 50% per decade, so
$$T_{t+1} = 1.5 T_t$$
where T is traffic and t is time in decades.
c. We divide by 1.5 to find 26,667.
d. After 5 more decades, we would find $135{,}000 \cdot 1.5^5$ cars.
e. To find the doubling time, we write the solution in exponential form at $e^{\ln(1.5)t}$. The doubling time is $\ln(2)/\ln(1.5) \approx 1.71$ decades, or 17.1 years.

Chapter 4

Section 4.1

1. With $\Delta t = 1$, $\Delta f = f(2) - f(1) = 3$, so $\dfrac{\Delta f}{\Delta t} = 3$.
With $\Delta t = 0.5$, $\Delta f = f(1.5) - f(1) = 1.5$, so $\dfrac{\Delta f}{\Delta t} = 3$.
With $\Delta t = 0.1$, $\Delta f = f(1.1) - f(1) = 0.3$, so $\dfrac{\Delta f}{\Delta t} = 3$.
With $\Delta t = 0.01$, $\Delta f = f(1.01) - f(1) = 0.03$, so $\dfrac{\Delta f}{\Delta t} = 3$.

3. With $\Delta t = 1$, $\Delta G = G(1) - G(0) \approx 6.389$, so $\dfrac{\Delta G}{\Delta t} \approx 6.389$.
With $\Delta t = 0.5$, $\Delta G = G(0.5) - G(0) \approx 1.718$, so $\dfrac{\Delta G}{\Delta t} \approx 3.436$.
With $\Delta t = 0.1$, $\Delta G = G(0.1) - G(0) \approx 0.221$, so $\dfrac{\Delta G}{\Delta t} \approx 2.21$.
With $\Delta t = 0.01$, $\Delta G = G(0.01) - G(0) \approx 0.0202$, so $\dfrac{\Delta G}{\Delta t} \approx 2.02$.

5. Each secant line has equation $f_s(t) = 2 + 3t$.

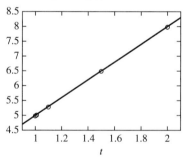

7. The coordinates of the base point are $(0, 1)$, so the secant lines are as follows: with $\Delta t = 1$, $G_s(t) = 1 + 6.389t$, with $\Delta t = 0.5$, $G_s(t) = 1 + 3.436t$, with $\Delta t = 0.1$, $G_s(t) = 1 + 2.21t$, with $\Delta t = 0.01$, $G_s(t) = 1 + 2.02t$.

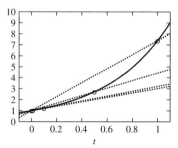

Answers to Odd-Numbered Questions

9. The slope is 3, so the tangent line is $y = L(t) = 2 + 3t$.

11. It looks like the slopes are getting close to 2, so the tangent line is $y = L(t) = 1 + 2t$.

13. The slopes are given in the table below.

Δx	Slope of Secant Line $\arctan \Delta x / \Delta x$	Δx	Slope of Secant Line $\arctan \Delta x / \Delta x$
−0.5	0.92730	0.5	0.92730
−0.1	0.99669	0.1	0.99669
−0.005	0.99999	0.01	0.99997
		0.005	0.99999

Our guess for the slope of the tangent is 1; the equation of the tangent is $y = x$.

15. The quantity $m(6) - m(2)$ represents the change in the number of monkeys between February and June. The quantity $\frac{m(6)-m(2)}{6-2} = \frac{m(6)-m(2)}{4}$ represents the average rate of change in the number of monkeys between February and June. The instantaneous rate of change when $t = 6$ represents the rate of change in the number of monkeys in June. The units of the instantaneous rate of change are number of monkeys per month.

17.

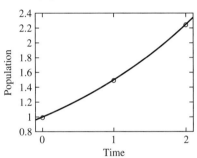

  a. $b(0) = 1$, $b(1) = 1.5$, $b(2) = 2.25$.
  b. $\Delta b = 1.5 - 1 = 0.5$, so $\Delta b / \Delta t = 0.5$.
  c. $\Delta b = 2.25 - 1.5 = 0.75$, so $\Delta b / \Delta t = 0.75$.

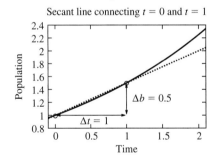

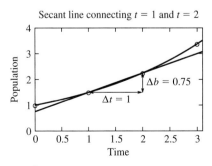

19. a. i. $\Delta b = 1.5^1 - 1 = 0.5$, and $\Delta b / \Delta t = 0.5$.

  ii. $\Delta b = 1.5^{0.1} - 1 \approx 0.0413$, and $\Delta b / \Delta t \approx 0.414$.

  iii. $\Delta b = 1.5^{0.01} - 1 \approx 0.00406$, and $\Delta b / \Delta t \approx 0.406$.

  iv. $\Delta b = 1.5^{0.001} - 1 \approx 0.000405$, and $\Delta b / \Delta t \approx 0.405$.

  b. The limit looks like 0.405.
  c.

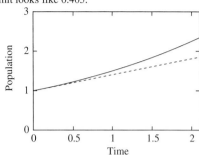

21. a. i. The slope is $(5 \cdot 1^2 - 0)/1 = 5$.

  ii. The slope is $(5 \cdot 0.1^2 - 0)/0.1 = 0.5$.

  iii. The slope is $(5 \cdot 0.01^2 - 0)/0.01 = 0.05$.

  iv. The slope is $(5 \cdot 0.001^2 - 0)/0.001 = 0.005$.

  b. The slope gets close to 0.
  c.

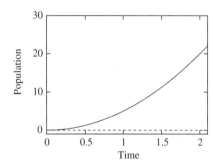

23. During the first hour, 3 bacteria/h; during the first half hour, 2.485 bacteria/h; during the second half hour, 3.515 bacteria/h. The population changes faster during the second half hour.

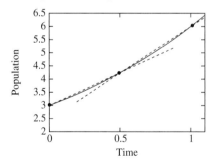

25. During the first hour, −0.79 bacteria/h; during the first half hour, −0.88 bacteria/h; during the second half hour, −0.69 bacteria/h. The population changes faster during the first half hour.

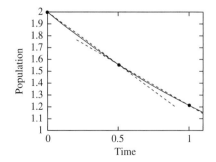

27. a. Using the previous measurement as the point,

Age	Rate of Change (metres per year)	Rate Divided by Height
1	1.07	0.0957
2	1.22	0.0984
3	1.34	0.0975
4	1.27	0.0846
5	1.60	0.0963
6	1.66	0.0909
7	1.90	0.0942
8	1.84	0.0836
9	2.44	0.0998
10	2.40	0.0894

b.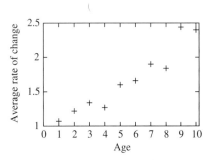

c.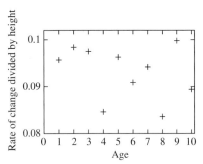

d. Both seem to jump around a bit, but the tree increases its height by about 9% per year. The approximate differential equation is

$$\frac{dh}{dt} = 0.09h$$

Section 4.2

1. $\lim_{x \to 0^-} f(x) = -2$ means that we can make the values $f(x)$ to be as close to -2 as desired by taking the numbers x close enough to 0, and negative.
$\lim_{x \to 0^+} f(x) = 4$ means that we can make the values $f(x)$ as close to 4 as desired by taking positive values of x that are close enough to 0.

3.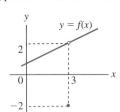

5. a. $\lim_{x \to 1^-} f(x) = 2$ b. $\lim_{x \to 1^+} f(x) = 2$ c. $\lim_{x \to 1} f(x) = 2$
d. $\lim_{x \to 4^-} f(x) = 6$ e. $\lim_{x \to 4^+} f(x) = 6$ f. $\lim_{x \to 4} f(x) = 6$ g. $f(1) = 2$
h. $f(4) = 5$

7. a. $\lim_{x \to 3^-} f(x) = 3$ b. $\lim_{x \to 3^+} f(x) = 1$ c. $\lim_{x \to 3} f(x)$ does not exist
d. $\lim_{x \to -1.3^-} f(x) = 2$ e. $\lim_{x \to -1.3^+} f(x) = 2$ f. $\lim_{x \to -1.3} f(x) = 2$
g. $f(3) = 1$ h. $f(-1.3)$ is not defined

9.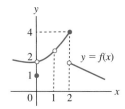

11. $\lim_{x \to \pi^-} f(x) = \lim_{x \to \pi^+} f(x) = 0$; $\lim_{x \to \pi} f(x) = 0$

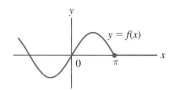

13. $\lim_{x \to 1^-} f(x) = \lim_{x \to 1^+} f(x) = 1$; $\lim_{x \to 1} f(x) = 1$

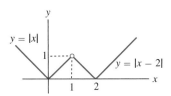

15. At $x = 0.1$, value is 0.998; at $x = 0.01$ it is 0.999. The limit seems to be 1.

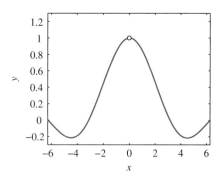

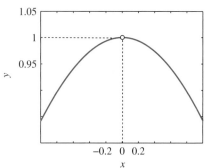

17. $0.1^{0.1} \approx 0.794, 0.01^{0.01} \approx 0.955, 0.001^{0.001} \approx 0.993$. The limit seems to be 1.

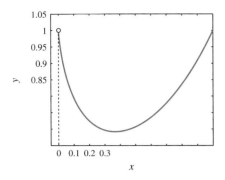

19. Denoting the function by $g(x)$, we find that $g(0.1) \approx 2.214$, $g(0.01) \approx 2.0201$, $g(0.001) \approx 2.002$. This seems to approach 2.

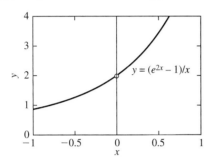

21. $\sqrt{\ln(1.1)} \approx 0.309$, $\sqrt{\ln(1.01)} \approx 0.099$, $\sqrt{\ln(1.001)} \approx 0.032$. It seems to be approaching 0, but rather slowly.

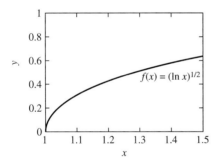

23. This will be 3 times the previous limit plus 4, or $3 \cdot 1 + 4 = 7$ using Theorem 2.2c and 2.2a.

25. This will be the ratio of two earlier limits, or $\frac{e}{1} = e$, using Theorem 2.2d.

27. $f_2(x) < 0.1$ if $x < 0.1$; $f_2(x) < 0.01$ if $x < 0.01$. This approaches 0 at a medium rate. Small values of the input produce small values of the output.

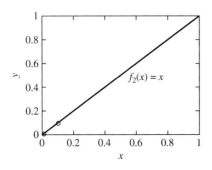

29. $f_4(x) < 0.1$ if $x < 0.562$; $f_4(x) < 0.01$ if $x < 0.316$. This approaches 0 very quickly because small values of the input produce tiny values of the output. Graphically, the value of the function becomes tiny even when x is not that small.

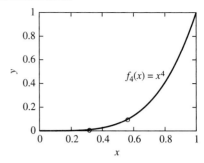

31. -3

33. $1/2$

35. -3

37. $27/2$

39. Does not exist

41. Does not exist

43. $-1/2$

45. Does not exist

47. 1

49. $-1/4$

51. $-\pi/2$

53. Does not exist

55. 0

57. We find

$$\lim_{x \to 0} \frac{1 - \cos x}{x} = \lim_{x \to 0} \frac{1 - \cos x}{x} \cdot \frac{1 + \cos x}{1 + \cos x}$$
$$= \lim_{x \to 0} \frac{1 - \cos^2 x}{x(1 + \cos x)}$$
$$= \lim_{x \to 0} \frac{-\sin^2 x}{x(1 + \cos x)}$$
$$= -\lim_{x \to 0} \frac{\sin x}{x} \cdot \frac{\sin x}{1 + \cos x} = -1 \cdot \frac{0}{2} = 0$$

59. $\lim_{x \to 0} \frac{\sin x^2}{2x} = \frac{1}{2} \lim_{x \to 0} \frac{\sin x^2}{x^2} \cdot x = \frac{1}{2} \lim_{x^2 \to 0} \frac{\sin x^2}{x^2} \cdot x = \frac{1}{2} \cdot 1 \cdot 0 = 0$

61. $-\infty$

63. The average rate of change is 5 (unless $\Delta x = 0$). The limit is 5.

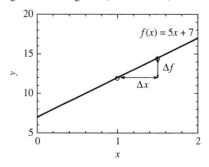

65. $\Delta f = 5(1 + \Delta x)^2 - 5$, so the average rate of change is $10 + 5\Delta x$ (unless $\Delta x = 0$). The value gets closer and closer to 10 for small values of Δx, so the limit must be 10.

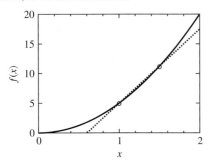

67. $\Delta f = 5(2 + \Delta x)^2 + 7(2 + \Delta x) - 37$, so the average rate of change is $27 + 5\Delta x$ (unless $\Delta x = 0$). The value gets closer and closer to 27 for small values of Δx, so the limit must be 27.

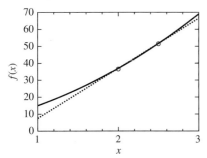

69. From $\cos(\pi/x) = 0$ we conclude that $\frac{\pi}{x} = \frac{\pi}{2} + \pi k$ and so $x = \frac{2}{1+2k}$. This calculation proves that the function $f(x) = \cos(\pi/x)$ takes on the value 0 at infinitely many values of x that approach zero. Similarly, from $\cos(\pi/x) = 1$ we conclude that $\frac{\pi}{x} = 2\pi k$, $\frac{1}{x} = 2k$, and so $x = \frac{1}{2\pi k}$. This calculation implies that $f(x)$ takes on the value 1 at infinitely many values of x that (as in the previous case) approach zero. Thus, the limit of $f(x) = \cos(\pi/x)$ as $x \to 0$ does not exist (as the function keeps oscillating between different values).

71. Since the limits of the two ends of the given inequality do not approach the same number, the Squeeze Theorem does not apply. In particular,

$$\lim_{x \to \pi/4} \sin x = \sin(\pi/4) = \frac{\sqrt{2}}{2} \text{ and } \lim_{x \to \pi/4} \tan x = \tan(\pi/4) = 1$$

The only thing we can say is that if the limit exists, then $\sqrt{2}/2 \leq \lim_{x \to \pi/4} f(x) \leq 1$.

73. Start with $-1 \leq \sin(1/x) \leq 1$. Then,

$$-3x^2 \leq 3x^2 \sin(1/x) \leq 3x^2$$
$$-\lim_{x \to 0} 3x^2 \leq \lim_{x \to 0} 3x^2 \sin(1/x) \leq \lim_{x \to 0} 3x^2$$
$$0 \leq \lim_{x \to 0} 3x^2 \sin(1/x) \leq 0$$

Therefore, $\lim_{x \to 0} 3x^2 \sin(1/x) = 0$.

75. a. $\lim_{T \to 0} H(T) = 10$
b. $H(2) = 3.33$
c. $H(1) = 5$
d. We need to solve $H(T) = 9.9$, finding $T = 0.01$ K.

77. The average rate of change between $t = 0$ and $t = \Delta t$ is

$$\frac{b(\Delta t) - b(0)}{\Delta t} = \frac{\Delta t + 0.1 \Delta t^2 - 0}{\Delta t} = 1 + 0.1 \Delta t$$

So Δt would have to be less than 0.1 for the average rate of change to be within 1% of the instantaneous rate of change.

79. The average rate of change between $t = 0$ and $t = \Delta t$ is

$$\frac{b(\Delta t) - b(0)}{\Delta t} = \frac{\sin(\Delta t) - 0}{\Delta t}$$

Plugging in small values of Δt, the average rate of change is 0.99 when $\Delta t = 0.25$.

81. a. This would cost \$10,000.
b. This would cost \$1 million.
c. This would cost \$10 billion.

83. a. The radioactivity would be 5 rad and would cost \$1000 to detect.
b. The radioactivity would be 2.5 rad and would cost \$2000 to detect.
c. The radioactivity would be reduced by a factor of $2^{3.3} \approx 9.85$, so the radioactivity would be 1.01 rad and would cost \$4924 to detect.

Section 4.3

1. $\lim_{x \to 2^-} f(x) = -\infty$ means that we can make the values of $f(x)$ as small as desired (i.e., as large negative as desired) by taking x close enough to 2, and at the same time smaller than 2; $\lim_{x \to 0^+} f(x) = \infty$ means that we can make the values of $f(x)$ as large as desired by taking positive numbers x to be close enough to 0.

3. $f(x) = -\frac{1}{x-1}$

5. $f(x) = -\frac{1}{x^2} + \frac{1}{(x-1)^2}$

7. $\lim_{x \to -\pi^-} \cot x = -\infty$, $\lim_{x \to -\pi^+} \cot x = \infty$, $\lim_{x \to \pi^-} \cot x = -\infty$, $\lim_{x \to \pi^+} \cot x = \infty$, $\lim_{x \to 2\pi^-} \cot x = -\infty$, $\lim_{x \to 2\pi^+} \cot x = \infty$

9. $-\infty$

11. $-\infty$

13. $+\infty$

15. Does not exist

17. $+\infty$

19. $-\infty$

21. $x = 0$ is a vertical asymptote. $x = 1$ is not a vertical asymptote. $x = -1$ is not a vertical asymptote.

23. $x = 2$ is not a vertical asymptote. $x = -2$ is a vertical asymptote.

25. No vertical asymptotes

27. $g_1(x) > 10$ if $x < 0.01$. $g_1(x) > 100$ if $x < 0.0001$. This approaches infinity slowly because tiny values of the input are required to produce large values of the output. Graphically, the value of the function only becomes large when x is very small.

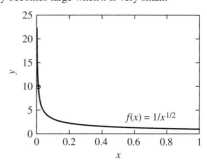

29. $g_3(x) > 10$ if $x < 0.316$. $g_3(x) > 100$ if $x < 0.1$. This approaches infinity quickly because small values of the input produce huge values of the output. Graphically, the value of the function becomes large even when x is not too small.

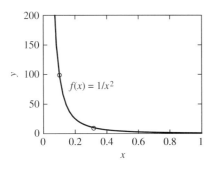

31. 0
33. 0
35. 0
37. ∞
39. ∞
41. 0
43. -2
45. $1/2$
47. ∞
49. ∞
51. 0
53. $-\pi/2$
55. 0

57. Dividing the numerator and the denominator by e^{-x} (equivalently, multiplying by e^x), we obtain

$$\lim_{x \to -\infty} \frac{e^x - 2e^{-x}}{3e^x + e^{-x}} = \lim_{x \to -\infty} \frac{e^x \cdot e^x - 2e^{-x} \cdot e^x}{3e^x \cdot e^x + e^{-x} \cdot e^x} = \lim_{x \to -\infty} \frac{e^{2x} - 2}{3e^{2x} + 1}$$
$$= \frac{0 - 2}{0 + 1} = -2$$

since $e^{2x} \to 0$ as $x \to -\infty$.

59. $y = 1$
61. $y = \pi/2$ is a horizontal asymptote at $+\infty$.
63. $y = -1$ and $y = 0$
65. $y = 0$
67. Start with $-1 \leq \cos x \leq 1$. Then

$$0 \leq \cos^2 x \leq 1$$
$$0 \leq e^{-7x} \cos^2 x \leq e^{-7x}$$
$$0 \leq \lim_{x \to \infty} e^{-7x} \cos^2 x \leq \lim_{x \to \infty} e^{-7x}$$
$$0 \leq \lim_{x \to \infty} e^{-7x} \cos^2 x \leq 0$$

Therefore, $\lim_{x \to \infty} e^{-7x} \cos^2 x = 0$.

69. x^3 approaches infinity faster because the power is larger. For x^3, the values are 1, 1000, and 1,000,000. For $1000x$, the values are 1000, 10,000, and 100,000, which are always larger. x^3 passes $1000x$ between $x = 10$ and $x = 100$.

71. e^{5x} approaches infinity faster because the coefficient in the exponent is larger. For $5e^x$, the values are 13.6, $1.1 \cdot 10^5$, and $1.3 \cdot 10^{44}$. For e^{5x}, the values are 148.4, $5.2 \cdot 10^{21}$, and $1.4 \cdot 10^{217}$. These are much larger.

73. $x^{0.5}$ approaches infinity faster because the power is larger. For $10x^{0.1}$, the values are 10, 12.59, and 15.85. For $x^{0.5}$, the values are 1, 3.16, and 10.0. The two functions do not get into the right order until $x > 316.2$.

75. The log–log graphs of the four functions are given by

$$\log_{10} y = \log_{10}(100x^5) = \log_{10} 100 + \log_{10} x^5 = 2 + 5 \log_{10} x$$
$$\log_{10} y = \log_{10}(x^6) = 6 \log_{10} x$$
$$\log_{10} y = \log_{10}(0.1x^7) = \log_{10} 0.1 + \log_{10} x^7 = -1 + 7 \log_{10} x$$
$$\log_{10} y = \log_{10}(0.01x^8) = \log_{10} 0.01 + \log_{10} x^8 = -2 + 8 \log_{10} x$$

Since all four graphs are lines, ordering in terms of growth means comparing their slopes. From the one approaching ∞ most slowly to the one approaching ∞ most quickly, the order is as follows: $2 + 5\log_{10} x$, $6 \log_{10} x$, $-1 + 7 \log_{10} x$, $-2 + 8 \log_{10} x$, i.e., $100x^5$, x^6, $0.1x^7$, $0.01x^8$.

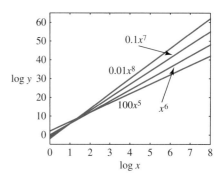

77. $10e^{-x}$ approaches zero faster because the parameter in the exponent is larger. For $10e^{-x}$, the values are 3.68, $4.5 \cdot 10^{-4}$, and $3.72 \cdot 10^{-41}$. For $0.1e^{-0.2x}$, the values are 0.82, 0.0135, and $2.06 \cdot 10^{-10}$. The two functions do not get into the right order until $x = 100$.

79. $25x^{-0.2}$ approaches zero faster because the power is more negative. For $x^{-0.1}$, the values are 1, 0.79, and 0.63. For $25x^{-0.2}$, the values are 25, 15.77, and 9.95. The two functions do not get into the right order until x is about 10^{14}.

81. $30x^{-0.1}$ approaches zero faster because power functions are faster than natural logs. For $30x^{-0.1}$, the values are 30, 23.83, and 18.93. For $1/\ln x$, the values are undefined (divide by 0), 0.434, and 0.217. The two functions do not get into the right order until x is about 10^{34}.

83. The log–log graphs of the four functions are given by

$$\log_{10} y = \log_{10}(x^{-1}) = -\log_{10} x$$
$$\log_{10} y = \log_{10}(10x^{-2}) = \log_{10} 10 + \log_{10} x^{-2} = 1 - 2 \log_{10} x$$
$$\log_{10} y = \log_{10}(100x^{-3}) = \log_{10} 100 + \log_{10} x^{-3} = 2 - 3 \log_{10} x$$
$$\log_{10} y = \log_{10}(1000x^{-4}) = \log_{10} 1000 + \log_{10} x^{-4} = 3 - 4 \log_{10} x$$

We need to arrange the functions by how quickly they approach $-\infty$ as $x \to \infty$. Since all four graphs are lines, ordering in terms of growth means comparing their slopes. The four functions, ordered from the one approaching $-\infty$ most slowly to the one approaching $-\infty$ most quickly, are as follows: $-\log_{10} x$, $1 - 2 \log_{10} x$, $2 - 3 \log_{10} x$, $3 - 4 \log_{10} x$, i.e., x^{-1} (approaches 0 most slowly), $10x^{-2}$, $100x^{-3}$, $1000x^{-4}$ (approaches 0 most quickly).

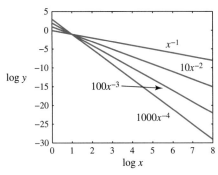

85. Approaches ∞ because the denominator is a natural log function, which grows more slowly than the linear function in the numerator.

87. Approaches ∞ because the denominator is a linear function, which grows more slowly than the quadratic function in the numerator.

89. (a) The limit is 0. (b) The function has larger and larger oscillations as $x \to \infty$.

91. From
$$x^{-2} < 10^{-7}$$
$$x^2 > 10^7$$
$$x > \sqrt{10^7} = 10^3\sqrt{10} \approx 3162.28$$

we see that we can take $N > 10^3\sqrt{10}$.

93. From
$$x^{-1/2} < 0.01$$
$$x^{1/2} > 100$$
$$x > 100^2 = 10,000$$

we see that we can take $N > 10,000$.

95. This population increases to infinity because $r > 1$. $b_t = 10^8(1.5)^t > 10^{10}$ if

$$1.5^t > 10^2$$
$$e^{\ln(1.5)t} > 10^2$$
$$\ln(1.5)t > \ln(10^2)$$
$$t > \ln(10^2)/\ln(1.5) \approx 11.35$$

It takes 11.35 generations to reach 10^{10}.

97. This population decreases to zero because $r < 1$. Solving $b_t = 10^3$, gives that it takes 109.3 generations to reach 10^3.

99. The ratio is $\frac{2t}{2^{t+1}+2t}$, which approaches a limit of zero because the denominator increases exponentially and the numerator increases only linearly. Plugging in numbers, it would take 25 generations for less than one in a million to be too long.

101. The 0.5^t term approaches 0, so the limit is 2.

103. We could show that the slope of the updating function at the equilibrium is less than 1 (which it is), or find the solution and prove it approaches the equilibrium as a limit (which it does).

105. The lung has a limited capacity to take up chemical.

107. When chemical concentration becomes high, new absorption pathways are turned on.

109. a. The temperature is 1 million degrees Celsius, and would cost $1 million.
   b. The temperature is 10 million degrees Celsius, and would cost $1000 million or $1 billion.
   c. The temperature is 20 million degrees Celsius, and would cost $8000 million or $8 billion.

Section 4.4

1. $\lim_{x \to -2} f(x) = f(-2)$

3.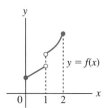

5. There are many functions with this property, such as $f(x) = \frac{1}{x+2} + \frac{1}{(x+4)^2}$.

7. The only numbers x where $1/f(x)$ is not continuous are those for which $f(x) = 0$.

9. The product of

$$f(x) = \begin{cases} 1 & \text{if } x \leq 0 \\ 0 & \text{if } x > 0 \end{cases} \quad \text{and} \quad g(x) = \begin{cases} 0 & \text{if } x \leq 0 \\ 1 & \text{if } x > 0 \end{cases}$$

is the constant function $f(x) \cdot g(x) = 0$, which is continuous everywhere.

11. This is a polynomial and is continuous everywhere.

13. This is the product of a continuous polynomial (y^2) and a natural logarithm. The natural logarithm is the composition of a linear function ($y - 1$) with the natural log. The only potential trouble point is where $y - 1 = 0$, or at $y = 1$. The function is not defined for $y \leq 1$.

15. This is a triple composition, of cosine with the exponential with a polynomial. Because each piece is continuous, the combination is also.

17. The two pieces of the function are a polynomial (t^2) and a linear function (0), both of which are continuous. The only trouble spot is where the pieces of the definition must match up. In this case, however, both pieces are equal to 0, so the function is continuous.

19. This is the quotient of the constant 1 and the function $(1 + z^2)^2$, which is a polynomial. This is guaranteed to be continuous, except where the denominator is 0. But the denominator is always positive, so this function must be continuous everywhere.

21. $p(2) = 87, p(2.1) \approx 103.4, p(2.01) \approx 88.53, p(1.9) \approx 72.9, p(1.99) \approx 85.49$

23. $h(1)$ cannot be computed. $h(1.1) \approx -2.786, h(1.01) \approx -4.698$. Numbers less than 1 are not in the domain of the function. It looks like $\lim_{y \to 1+} = -\infty$.

25. $h(2) = 0$, $h(2.1) = 0.42$, $h(2.01) = 0.04$, $h(1.9) = -0.38$, $h(1.99) = -0.040$

27. $F(2) = -0.37$, $F(2.1) = 0.832$, $F(2.01) = 0.96$, $F(1.9) = 0.743$, $F(1.99) = -0.586$. These do not look very close. But $F(2.00001) = -0.369$. We must be very close to 2.

29. $a(0) = 0, a(0.1) = 0.01, a(0.01) = 0.0001, a(-0.1) = 0, a(-0.01) = 0$

31. $f(x) = 3.1$ if $x = 1.05$ and $f(x) = 2.9$ if $x = 0.95$, so $0.95 \leq x \leq 1.05$.

33. $f(x) = 20.1$ if $x = 2.005$ and $f(x) = 19.9$ if $x = 1.995$, so $1.995 \leq x \leq 2.005$.

35. $f(x)$ is continuous for all real numbers x.

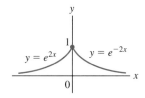

37. $f(x)$ is not continuous at $x = 4$ only.

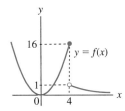

39. $g(x)$ is not continuous when $x < 0$.

41. $f(x)$ is not continuous at $x = 2$.

43. $f(x)$ is not continuous at $x = 4$.

45. Any value of $x < 0$ gives a value within the tolerance, but no positive value, however small, produces an output within the given tolerance.

47. a.

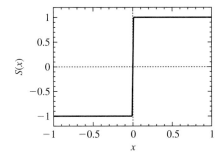

b. The slope on the central part is 100.

$$S(x) = \begin{cases} -1 & \text{if } x \leq -0.01 \\ -1 + 100(x + 0.01) & \text{if } 0.01 < x < 0.01 \\ 1 & \text{if } x \geq 0.01 \end{cases}$$

c. The input would have to be between -0.001 and 0.001.

49.
$$4\pi - 0.5 < A < 4\pi + 0.5$$
$$12.07 < \pi r^2 < 13.07$$
$$3.84 < r^2 < 4.16$$
$$1.96 < r < 2.04$$

The radius must be within 0.04 of 2.0 to get within 0.5 of the desired area.

51.
$$2.618 < S < 2.818$$
$$2.618 < e^{0.001t} < 2.818$$
$$0.963 < 0.001t < 1.036$$
$$963 < t < 1036$$

The time must be within about 37 seconds of 1000.

53. The tolerance is $3.12 \cdot 10^6$. The tolerance becomes smaller because a small difference in initial conditions is amplified as the population moves away from its unstable equilibrium.

55. Between $3.52 \cdot 10^5$ and $2.88 \cdot 10^5$, there is a tolerance of $3.2 \cdot 10^4$.

57. $T_{10} = 0.52$ requires $T_5 = \frac{0.52}{0.5^5} \approx 16.64$, and $T_{10} = 0.48$ requires $T_5 = \frac{0.48}{0.5^5} \approx 15.36$. The tolerance is 0.64 g/L. The tolerances become larger because a difference in initial conditions is decreased as the concentration moves toward its stable equilibrium.

59. T_9 would have to be between 4.8 and 5.2, a tolerance of 0.2 g/l. T_5 would have to be between $4.8 \cdot 10^4$ and $5.2 \cdot 10^4$, a tolerance of $2 \cdot 10^3$. T_0 would have to be between $4.8 \cdot 10^9$ and $5.2 \cdot 10^9$, a tolerance of $2 \cdot 10^8$. That's a huge tolerance. The tolerances are larger because the system zooms toward the stable equilibrium more quickly in this case.

61. a.

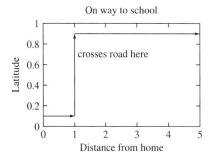

b.

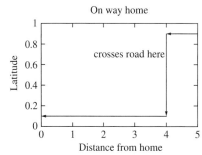

c. Yes. Knowing his distance from home is not enough to guess which side of the road he is on.

Section 4.5

1. $f'(-2) = \lim_{h \to 0} \frac{f(-2+h) - f(-2)}{h}$

3. The derivative $f'(x) = 2ax + b$ is a real number for all x.

5.

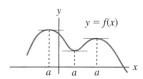

7. The derivative is $f'(x) = 2ax + b > 0$ when $x < -b/2a$ and $f'(x) = 2ax + b < 0$ when $x > -b/2a$.

9. See the figure below for labels. The function is not continuous at b; it is not differentiable at b (discontinuity), at d (vertical tangent), and at a (corner); the derivative is zero at c.

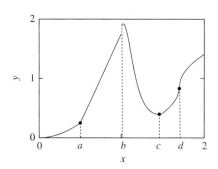

11. There are many correct answers.

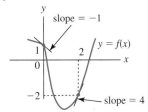

13.

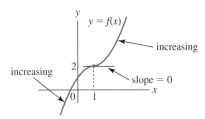

15.

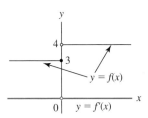

17.

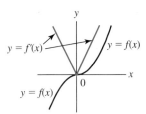

19.

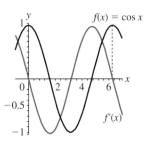

21.

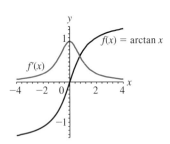

23. $f(x) = 3x^4$ and $a = 1$

25. $f(x) = x^2$ and $a = 10$

27. $f(x) = e^x$ and $a = 0$

29. $M'(x) = 0.5$, $\dfrac{dM}{dx} = 0.5$. This function is increasing.

31. $g'(y) = -3$, $\dfrac{dg}{dy} = -3$. This function is decreasing.

33. $f(1) = 3$, $f(1 + \Delta x) = 3 - 2\Delta x - \Delta x^2$, $\Delta f = -2\Delta x - \Delta x^2$, so the slope of the secant is $\dfrac{\Delta f}{\Delta x} = -2 - \Delta x$, if $\Delta x \neq 0$. Taking the limit of this constant function by plugging in $\Delta x = 0$ we find that the slope of the tangent is $f'(1) = -2$.

35. $f(x + \Delta x) = 4 - x^2 - 2x\Delta x - \Delta x^2$, $\Delta f = -2x\Delta x - \Delta x^2$, so the slope of the secant is $\dfrac{\Delta f}{\Delta x} = -2x - \Delta x$, if $\Delta x \neq 0$. Taking the limit, the derivative is $f'(x) = -2x$, or $\dfrac{df}{dx} = -2x$.

37. The derivative is $f'(x) = -2x$, so the critical point occurs at $x = 0$. For $x < 0$, the derivative is positive, while for $x > 0$ it is negative. The function thus switches from increasing to decreasing at $x = 0$.

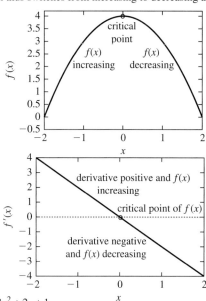

39. $f'(x) = 3x^2 + 2x + 1$

41. $f'(x) = -\dfrac{2}{(x-1)^2}$

43. $f'(x) = -\dfrac{8}{x^3}$

45. $y = -\dfrac{1}{16}x + \dfrac{1}{2}$

47. $y = -1$

49.

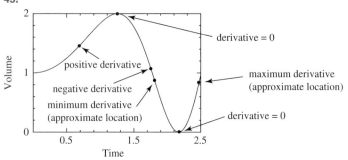

51.

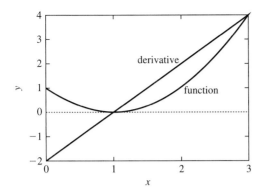

53.

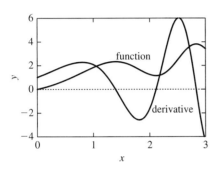

55. With $\Delta x = 1.0$, the slope of the secant is 1. With $\Delta x = 0.1$, the slope of the secant is 3.16. With $\Delta x = 0.01$, the slope of the secant is 10. With $\Delta x = 0.0001$, the slope of the secant is 100. The limit seems to be infinity. The tangent line is a vertical line with equation $x = 0$.

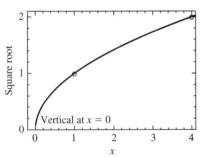

57. The slope of any secant with $\Delta x > 0$ is $1/\Delta x$, which has a limit of infinity. The slope of any secant with $\Delta x < 0$ is $-1/\Delta x$, which has a limit of negative infinity. There is nothing even resembling a tangent line for this function.

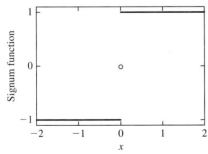

59. Forward: $P'(2001) \approx \frac{P(2006)-P(2001)}{2006-2001} = \frac{913{,}462 - 908{,}007}{5} = 1091$

Backward: $P'(2001) \approx \frac{P(1996)-P(2001)}{1996-2001} = \frac{909{,}282-908{,}007}{-5} = -255$

Central: $P'(2001) \approx \frac{1091 - 255}{2} = 418$

61. Forward: $f'(0) \approx \frac{f(1)-f(0)}{1-0} = \frac{6.4-8.8}{1} = -2.4$

Backward: $f'(0) \approx \frac{f(-1)-f(0)}{-1-0} = \frac{8.2-8.8}{-1} = 0.6$

Central: $f'(0) \approx \frac{-2.4+0.6}{2} = -0.9$

63. Forward: $f'(3) \approx \frac{f(4)-f(3)}{4-3} = \frac{6.7-4.2}{1} = 2.5$

Backward: $f'(3) \approx \frac{f(2)-f(3)}{2-3} = \frac{3.1-4.2}{-1} = 1.1$

Central: $f'(2) \approx \frac{2.5+1.1}{2} = 1.8$

65.

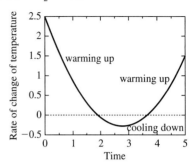

67.

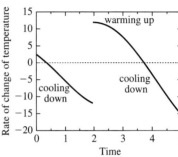

69.

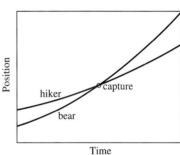

71.

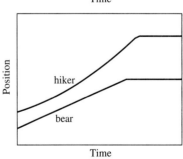

73. The time solves $M(t) = 0$, or $100 = \frac{1.62}{2}t^2$, or $t^2 \approx 123.4$ or $t \approx 11.1$ seconds. The speed is the derivative of the position, or $M'(t) = at = 1.62t$. The speed when it hits the ground is 17.98 m/s.

75. When $M(t) = 0$, then $100 = \frac{2.15 \cdot 10^{-3}}{2}t^2$, or $t \approx 305$ seconds. The speed is the derivative of the position, or $M'(t) = at = 2.15 \cdot 10^{-3} t$. The speed when it hits the ground is 0.65 m/s.

True/False Quiz

1. FALSE. The limit of $f(x)$ could be equal to 4 even if $f(x)$ is not defined at $x = 5$.

3. TRUE. $g(x)$ is a composition of continuous functions.
5. FALSE. Take $f(x) = 5/x$ and $g(x) = 1/x$. Then $f(x) > g(x)$ for all $x > 0$; however, both functions approach 0 as $x \to \infty$.
7. TRUE. This is the definition of continuity.
9. FALSE. Consider the function defined piecewise by $f(x) = 1/x$ if $x \neq 0$ and $f(0) = 1$. The domain of $f(x)$ consists of all real numbers, but $x = 0$ is its vertical asymptote.

Supplementary Problems

1. We can plug in because the function is continuous. The limit is 1.
3. The right limit is infinity; the left limit is negative infinity. The limit does not exist.
5. This increases to infinity.
7. 1
9. 0
11. -2
13. $1/15$
15. Limit is 0; near $x = 0$, the trigonometric function $f(x) - \sin x$ is approximately equal to the polynomial $g(x) = x - \frac{1}{6}x^3$.
17. $g(x)$ is continuous at all real numbers x such that $x \neq 1$.
19. $g(x)$ is continuous at all numbers $x \neq 0$.
21. $g(x)$ is continuous on $(-\infty, -\sqrt{3})$ and $(\sqrt{3}, \infty)$.
23. y is continuous when $x \neq 1$.
25. $g(x)$ is continuous on $[0 + \pi k, \pi/2 + \pi k)$, where k is an integer.
27. $\ln f(x) + \sqrt{f(x)}$ is continuous at all x for which $f(x) > 0$.
29. $f'(x) = -\dfrac{25}{(x-6)^2}$
31. $f'(x) = \dfrac{1}{2\sqrt{x-1}} + \dfrac{1}{2\sqrt{x+1}}$
33. $f'(x) = \pi$
35. $y = \dfrac{2}{\sqrt{3}}x + \sqrt{3}$
37.

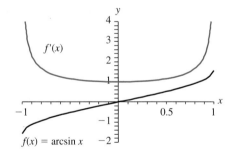

39. a.

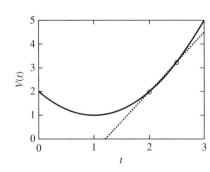

b. The secant line has slope $(V(2.5) - V(2))/0.5 = 2.5$. Because $V(2) = 2$, the equation is $V_s(t) = 2.5(t-2) + 2$.
c. $V'(t) = 2t - 2$, so $V'(2) = 2$ in units of thousands of cubic microns per minute.
d. At $t = 2$, cell volume is increasing by 2 thousand cubic microns per minute.
e.

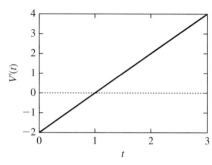

41. a.

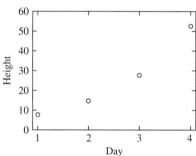

b. I picked days 3 and 4 because they are closest to day 5. The secant line has equation $\hat{h}(t) = 28.0 + 25.0(t - 3)$, and $\hat{h}(5) = 78.0$.
c. On day 4, the tangent line is $\hat{h}(t) = 53.0 + 34.6(t - 4)$, and $\hat{h}(5) = 77.6$.
d. Probably the tangent line because it uses the latest information.
e. Maybe they went out and measured the change in plant height over one hour.

43. a.

With increasing temperature

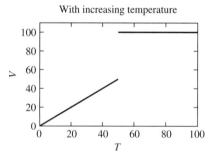

With decreasing temperature

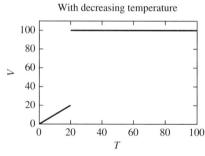

b. $\lim\limits_{T \to 50^+} V_i(T) = 100$; $\lim\limits_{T \to 50^-} V_i(T) = 50$;
$\lim\limits_{T \to 50^+} V_d(T) = \lim\limits_{T \to 50^-} V_d(T) = 100$

c. $\lim_{T \to 20^+} V_i(T) = \lim_{T \to 20^-} V_i(T) = 20$ $\lim_{T \to 20^+} V_d(T) = 100$
   $\lim_{T \to 20^-} V_d(T) = 20$

Chapter 5

Section 5.1

1. $5x^4$

3. $0.2x^{-0.8}$

5. $p = e$ gives ex^{e-1}.

7. $p = 1/e$ gives $\frac{1}{e}x^{\frac{1}{e}-1}$.

9. $f'(x) = \frac{d(3x^2)}{dx} + \frac{d(3x)}{dx} + \frac{d1}{dx}$
   $= 3\frac{d(x^2)}{dx} + 3\frac{dx}{dx}$
   $= 3 \cdot 2x + 3 \cdot 1 = 6x + 3$

11. $g'(z) = \frac{d(3z^3)}{dz} + \frac{d(2z^2)}{dz}$
    $= 3\frac{d(z^3)}{dz} + 2\frac{dz^2}{dz}$
    $= 3 \cdot 3z^2 + 2 \cdot 2z = 9z^2 + 4z$

13. $x^3 + 3x^2 + 3x + 1$

15. $8x^3 + 12x^2 + 6x + 1$

17. $f'(x) = 3x^2 - 12$

19. $f'(x) = 4\pi x^3$

21. $h'(x) = 1/4 + 3e^x$

23. $f'(x) = 4.5x^{0.5} + 3.2x^{-4.2}$

25. $m'(t) = -\frac{3}{4}t^{-7/4} - 1.08t^{-1/2}$

27. $A'(r) = 2\pi r$

29. $f'(x) = 4x + 1$

31. $g'(t) = -(1/2)t^{-3/2} - 5 - t^{-2}$

33. Using the binomial formula, $f(x) = 81x^4 - 216x^3 + 216x^2 - 96x + 16$. Thus, $f'(x) = 324x^3 - 648x^2 + 432x - 96$.

35. Using the binomial formula, $f(x) = m^3x^3 + 3m^2nx^4 + 3mn^2x^5 + n^3x^6$. Thus, $f'(x) = 3m^3x^2 + 12m^2nx^3 + 15mn^2x^4 + 6n^3x^5$.

37. $u'(x) = -\frac{a}{x^2} - \frac{2b}{x^3} - \frac{c}{2\sqrt{x^3}}$

39. Write $f(z) = 3e^ze^{12} - 1 = 3e^{12}e^z - 1$. Then $f'(z) = 3e^{12}e^z - 0 = 3e^{z+12}$.

41. Write $f(x) = \frac{9}{10}e^x - \frac{1}{5}x^{1/3} + \frac{2}{5}$. Then $f'(x) = \frac{9}{10}e^x - \frac{1}{15}x^{-2/3}$.

43. $-x + 12y - 16 = 0$

45. $y = -2x - 2$

47. We are looking for all x such that $y' = 3e^x - 1 = 0$. Solving, we get $x = -\ln 3$.

49. Using the definition,

$$f'(x) = \lim_{h \to 0} \frac{f(x+h) - f(x)}{h} = \lim_{h \to 0} \frac{5^{x+h} - 5^x}{h}$$
$$= \lim_{h \to 0} \frac{5^x 5^h - 5^x}{h} = 5^x \lim_{h \to 0} \frac{5^h - 1}{h}$$

To find an approximation for the limit, we use a table of values (values depend on the precision of the calculator/software used):

h	$\frac{5^h - 1}{h}$	h	$\frac{5^h - 1}{h}$
0.1	1.74619	−0.1	1.48660
0.01	1.62246	−0.01	1.59656
0.001	1.61073	−0.001	1.60814
0.0001	1.60957	−0.0001	1.60930
0.00001	1.60950	−0.00001	1.60942

Thus, $\lim_{h \to 0} \frac{5^h - 1}{h} \approx 1.609$, and we conclude that $(5^x)' \approx 1.609 \cdot 5^x$.

51. $f'(x) = -2 + 2x$, which is negative for $x < 1$, zero at $x = 1$, and positive for $x > 1$.

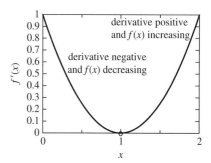

53. $h'(x) = 3x^2 - 3$, which is negative for $x < 1$, zero at $x = 1$, and positive for $x > 1$.

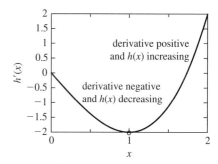

55. The constant 2 is the derivative of a linear function with slope 2. One such function is $2x$.

57. This looks just like the derivative of x^{15}.

59. x^{-1}

61. $A'(t) = 525t^2$, with units of people per year. New cases went up quickly.

63. $A'(r) = 2\pi r$. The units are centimetres. The derivative is equal to the perimeter of the circle, corresponding to the area of the little ring around the circle.

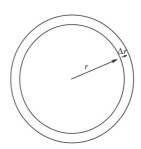

65.

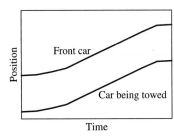

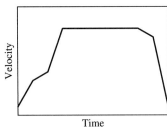

67. Subtract the 15 km/h for the passenger from the 120 km/h for the train to get 105 km/h (west).

69. Relative to passenger: $40 - 10 = 30$ km/h, relative to train: $30 - 15 = 15$ km/h, relative to ground: $120 + 15 = 135$ km/h.

71. a. $P(t) = 2 - 2t + 2t^2$
b. $a'(t) = 2t$ and $b'(t) = -2 + 2t$. The units are millions of bacteria per hour.
c. $P'(t) = -2 + 4t$, which is equal to $a'(t) + b'(t) = 2t - 2 + 2t$.
d. The population of a is always growing, the population of b shrinks before it grows, and the whole population also shrinks before it grows.
e.

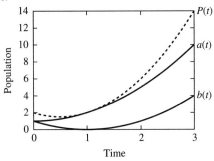

73. The speed is $M'(t) = 10 - at = 10 - 9.78t$. The critical point is then when $t \approx 1.02$ s. Plugging in, $M(1.02) \approx 105.11$. It hits the ground when $M(t) = 0$, or at $t \approx 5.66$. The speed is $M'(5.66) \approx 45.35$ m/s.

75. The speed is $M'(t) = 10 - at = 10 - 22.88t$. The critical point is then when $t \approx 0.44$ s. Plugging in, $M(0.44) \approx 102.18$. It hits the ground when $M(t) = 0$, or at $t \approx 3.43$. The speed is $M'(3.43) \approx 68.48$ m/s.

77. $S(t) = 3t^2$ works. $S(1) = 3$ and $S(2) = 12$. That seems pretty fast.

79. $S(t) = t^6$ because $\frac{dS}{dt} = 6t^5 = 6\frac{S(t)}{t}$. $S(1) = 1$ and $S(2) = 64$. That is quite fast, because the derivative gets bigger the larger the organism is.

Section 5.2

1. $f'(x) = -12x - 5$

3. $r'(y) = 15y^2 - 6y - 5$

5. $f'(x) = (x^3 + 3x^2 - x - 1)e^x$

7. $h'(x) = -e^{-x}(2x^{-3} + x^{-2})$

9. $g'(t) = \frac{14}{3}t^{2/3} - \frac{2}{3} + 7t^{2/3} - \frac{1}{3} = \frac{35}{3}t^{2/3} - 1$

11. $u'(z) = -4ABz^{-5} + 5B^2z^{-6} + ABz^{-2}$

13. Write $f(x) = \frac{3}{5}x^{1/2}(x^2 - 2x + 1)$ and use the product rule:

$$f'(x) = \frac{3}{5}\left(\frac{5}{2}x^{3/2} - 3x^{1/2} + \frac{1}{2}x^{-1/2}\right)$$

15. $h'(x) = (2x + 3)(-3x + 2) - (12x + 5)(x + 2)$

17. $f'(x) = 0.5e^x(7x^6 - 3x^2 + x^7 - x^3)$

19. $f'(x) = \frac{-x}{(2+x)^2}$

21. $g'(z) = \frac{(1 + 2z^3)2z - (1 + z^2)6z^2}{(1 + 2z^3)^2} = \frac{-2z(-1 + 3z + z^3)}{(1 + 2z^3)^2}$

23. $F'(x) = \frac{(1)(2 + x)(3 + x) - (1 + x)(5 + 2x)}{(2 + x)^2(3 + x)^2} = \frac{-x^2 - 2x + 1}{(2 + x)^2(3 + x)^2}$

25. $f'(x) = \frac{e^x(x-1)}{7x^2}$

27. $g'(t) = \frac{3e^t(-1+t)}{(e^t + t)^2}$

29. $g'(x) = \frac{(bc - ad)e^x}{(c + de^x)^2}$

31. Write $h(x) = \frac{1}{2}(e^x e^5 + e^{-x}e^{-5})$; then $h'(x) = \frac{1}{2}(e^{x+5} - e^{-x-5})$.

33. $f'(1) = 1$

35. $f'(0) = 0$

37. $f'(1) = 9$

39. $f'(1) = -4$

41. $f'(1) = 0$

43. $g'(1) = 4e$

45. $\Delta f = 0.21$, $\Delta g = 0.63$, and $\Delta(fg) = 2.4423$. This is equal to $g(x_0)\Delta f + f(x_0)\Delta g + \Delta f \Delta g = 0.42 + 1.89 + 0.1323$. With $\Delta x = 0.01$, $\Delta f = 0.0201$, $\Delta g = 0.0603$, and $\Delta(fg) = 0.22231203$. This is equal to $g(x_0)\Delta f + f(x_0)\Delta g + \Delta f \Delta g = 0.0402 + 0.1809 + 0.00121203$. The last term is now much smaller than the rest.

47. With this incorrect rule, we would get that

$$\frac{d1 \cdot x^3}{dx} = \frac{d1}{dx}\frac{dx^3}{dx} = 0 \cdot 3x^2 = 0$$

However, the product $f(x)g(x) = 1 \cdot x^3 = x^3$ really has derivative $3x^2$ by the power rule.

49. The derivative $g'(x) = (1 - x)e^x$ is positive whenever $x < 1$. Thus, the function is increasing on the given interval. To obtain the graph, plot several points:

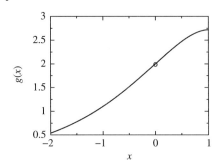

51. The derivative

$$F'(z) = \frac{3z^2 e^z - z^3 e^z}{(e^z)^2} = \frac{3z^2 - z^3}{e^z} = \frac{z^2}{e^z}(3 - z)$$

is positive when $z < 3$ (so that's where the function is increasing) and negative when $z > 3$ (function is decreasing). Plot points to obtain an accurate graph:

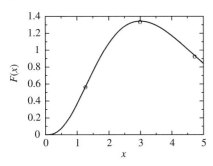

53. $\dfrac{d}{dx}\dfrac{1}{f(x)} = \dfrac{f(x) \cdot 0 - 1 \cdot f'(x)}{f(x)^2} = \dfrac{-f'(x)}{f(x)^2} < 0$

because $f'(x) > 0$ and $f(x) > 0$. Therefore, $\dfrac{1}{f(x)}$ is decreasing.

55. $\dfrac{dx^2}{dx} = x \cdot \dfrac{dx}{dx} + \dfrac{dx}{dx} \cdot x = 2x$. It worked.

57. Suppose the power rule works for a particular value of n, so that

$$\dfrac{d(x^n)}{dx} = nx^{n-1}$$

Then, writing $x^{n+1} = x \cdot x^n$, we have

$$\begin{aligned}\dfrac{dx^{n+1}}{dx} &= \dfrac{dx}{dx}x^n + \dfrac{d(x^n)}{dx}x \\ &= 1 \cdot x^n + nx^{n-1} \cdot x \\ &= 1 \cdot x^n + nx^n = (n+1)x^n\end{aligned}$$

which matches the rule for $n + 1$. By the principle of mathematical induction, the power rule works for all positive integer values of n.

59. We denote the total mass by $T(t)$.
   a. $T(t) = P(t)W(T) = (2 \cdot 10^6 - 2 \cdot 10^4 t)(80 + 0.5t)$
   b. $T'(t) = 6 \cdot 10^5 - 2 \cdot 10^4 t$
   c. $T'(t) = 0$ when $-6 \cdot 10^5 - 2 \cdot 10^4 t = 0$ or when $t = -30$ yr. At time -30, the population is $2.6 \cdot 10^6$, the weight per person is 65 kg, and the total weight is $1.69 \cdot 10^8$ kg.
   d.

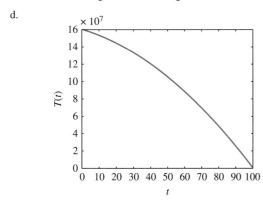

61. We denote the total mass by $T(t)$.
   a. $T(t) = P(t)W(T) = (2 \cdot 10^6 + 2 \cdot 10^4 t)(80 - 0.005 t^2)$
   b. $T'(t) = 1.6 \cdot 10^6 - 20{,}000t - 300t^2$
   c. $T'(t) = 0$ at $t \approx 46.94$, by solving the quadratic with the quadratic formula. At this time, the population is $2.93 \cdot 10^6$, the weight per person is 68.98 kg, and the total weight is $2.03 \cdot 10^8$ kg.
   d.

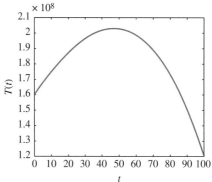

63. Total mass below ground is $(-1t + 40)(1.8 + 0.02t)$. The derivative is $-0.04t - 1$, which is always negative.

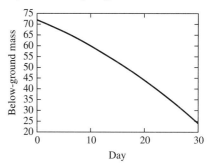

65. $S(1) = 0.84$, $S(5) = 1$, $S(10) = -6.0$, $S'(N) = 1 - 0.32N$; the bird does best by laying about 3 eggs.

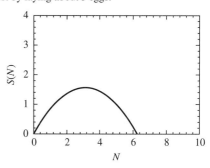

67. $S(1) = 0.91$, $S(5) = 1.43$, $S(10) = 0.91$

$$S'(N) = \dfrac{1 - 0.1N^2}{(1 + 0.1N^2)^2}$$

This bird does best by laying about 3 eggs.

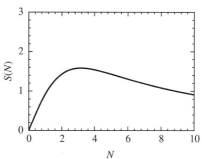

69. The derivative simplifies to $\dfrac{1.44}{(1.8p + 0.8(1-p))^2}$.

71. a. $\rho(t) = \dfrac{1 + t^2}{1 + 2t}$
   b. $\rho'(t) = \dfrac{2t^2 + 2t - 2}{(1 + 2t)^2}$

c. This is positive when $2t^2 + 2t - 2 > 0$. This occurs for t larger than the solution of $t^2 + t - 1 = 0$, which can be found with the quadratic formula to be 0.618. After that, the density is increasing.

d.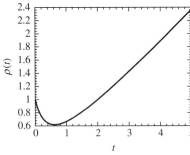

73. The final population is the product of the per capita production and the initial population, b_t, or $b_{t+1} = \dfrac{2b_t}{1+\frac{b_t}{1000}}$. The derivative of $f(b) = \dfrac{2b}{1+\frac{b}{1000}}$ is $f'(b) = \dfrac{200{,}000}{(1000+b)^2}$, which is always positive.

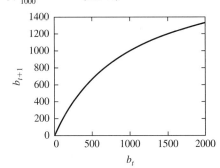

75. a. $h_{10}(0) = 0$, $h_{10}(1) = 0.5$, $h_{10}(2) \approx 0.999$
b. Using the quotient rule with $u(x) = x^{10}$ and $v(x) = 1 + x^{10}$ gives $h'_{10}(x) = \dfrac{10x^9}{(1+x^{10})^2}$. Evaluating, we find $h'_{10}(0) = 0$, $h'_{10}(1) = 2.5$, $h'_{10}(2) \approx 0.0048$.

c.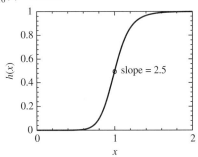

d. This gives a nice sharp response. Inputs much less than 1 produce a very small output and inputs much greater than 1 produce a large output.

77. The total mass is $M(t) = 10(1-t)e^t$. Then $M'(t) = -10te^t$, which is negative for $t > 0$. The total mass decreases from the start, and is never greater than the initial value of 10. It reaches zero when

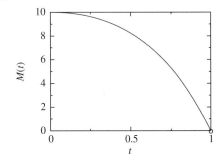

79. The total mass is $M(t) = 10\left(1 - \frac{t^2}{4}\right)e^t$. Then $M'(t) = 10(1 - t/2 - t^2/4)e^t$, which is negative for $t > \sqrt{5} - 1$ and positive for $t < \sqrt{5} - 1$. The total mass is greater than the initial value until $t = \sqrt{5} - 1$. It reaches zero when $t = 2$.

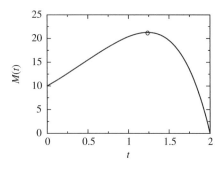

81. This equation says that the rate of change of the population is equal to the population size. A solution is a function that is equal to its own derivative, such as $b(t) = e^t$. This solution is increasing.

Section 5.3

1. $f(x) = \sqrt{x}$ and $g(x) = e^{x+1}$; $F'(x) = \dfrac{\sqrt{e^{x+1}}}{2}$

3. $f(x) = x^4$ and $g(x) = x - 4 - \dfrac{2}{x}$; $F'(x) = 4\left(x - 4 - \dfrac{2}{x}\right)^3\left(1 + \dfrac{2}{x^2}\right)$

5. $f(x) = 1/x$ and $g(x) = 4e^x + 3$; $F'(x) = -\dfrac{4e^x}{(4e^x + 3)^2}$

7. $f'(x) = 1 + \dfrac{4}{x}$

9. $g'(z) = \ln z + 1 + \dfrac{4}{z}$

11. $F'(w) = e^w \ln w + \dfrac{e^w}{w}$

13. $\ln x^2 = 2\ln x$, so $s'(x) = \dfrac{2}{x}$

15. $M'(x) = \ln(x) + 1 + \dfrac{2}{x}$ which is always positive. This function is increasing for $x > 2$. For an accurate graph, plot points.

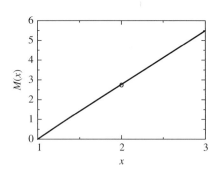

17. $g'(x) = 6(1 + 3x)$

19. $f'_1(t) = 90(1+3t)^{29}$

21. $r'(x) = -\dfrac{6x(1+3x)}{(1+2x)^4}$

23. Write $f(x) = (1 - x - 2x^3)^{1/2}$. Then $f'(x) = \dfrac{-1 - 6x^2}{2(1-x-2x^3)^{1/2}}$

25. $f'(x) = \frac{1}{(x-2)^2} \cdot \left(2 - \frac{x}{x-2}\right)^{-1/2}$

27. $F'(z) = -6\left(1 + \frac{2}{1+z}\right)^2 (1+z)^{-2}$

29. $f'(x) = -3e^{-3x}$

31. $f'(y) = \frac{1}{1+y}$

33. $f'(x) = \frac{3}{x \ln 10}$

35. $f'(x) = \frac{2(2 + \log_{10} x)}{x \ln 10}$

37. $f'(x) = \frac{1}{2}(\ln x + x - 4)^{-1/2}\left(\frac{1}{x} + 1\right)$

39. $G'(x) = 16x e^{x^2}$

41. Simplify first: $L(x) = \ln\sqrt{\ln x} = \ln(\ln x)^{1/2} = \frac{1}{2}\ln(\ln x)$. Then, $L'(x) = \frac{1}{2x \ln x}$.

43. $y = 0$

45. $F'(6) = -6$

47. Write $f(x) = (\ln x)^x = e^{\ln((\ln x)^x)} = e^{x \ln(\ln x)}$. Then $f'(x) = (\ln x)^x \left(\ln(\ln x) + \frac{1}{\ln x}\right)$.

49. Write $f(x) = (1+x)^{2+x} = e^{\ln((1+x)^{2+x})} = e^{(2+x)\ln(1+x)}$. Then $f'(x) = (1+x)^{2+x}\left(\ln(1+x) + \frac{2+x}{1+x}\right)$.

51. With the quotient rule, set $u(y) = 1$ and $v(y) = 1 + y^3$. Then

$$H'(y) = \frac{v(y)u'(y) - u(y)v'(y)}{v^2(y)} = \frac{(1+y^3) \cdot 0 - 1 \cdot 3y^2}{(1+y^3)^2} = -\frac{3y^2}{(1+y^3)^2}$$

With the chain rule, set $H(y) = r(p(y))$, where $p(y) = 1 + y^3$ and $r(p) = 1/p$. Then $p'(y) = 3y^2$ and $r'(p) = \frac{-1}{p^2}$. Thus,

$$H'(y) = r'(p)p'(y) = \frac{-1}{p^2} \cdot 3y^2 = \frac{-3y^2}{(1+y^3)^2}$$

53. Simplify: $\ln x^3 = 3 \ln x$. Then $\frac{d \ln(x^3)}{dx} = 3\frac{d \ln x}{dx} = \frac{3}{x}$.
With the chain rule, set $f(x) = x^3$ and $l(f) = \ln(f)$. Then

$$(l \circ f)'(x) = l'(f(x))f'(x) = \frac{1}{f(x)}3x^{3-1} = \frac{1}{x^3}3x^2 = \frac{3}{x}$$

55. $F(x) = 1 + 6x + 12x^2 + 8x^3$, and $F'(x) = 6 + 24x + 24x^2$. With the chain rule, we think of F as the composition $F(x) = g(h(x))$, where $h(x) = 1 + 2x$ and $g(h) = h^3$. Then $h'(x) = 2$ and $g'(h) = 3h^2$, so $F'(x) = g'(h)h'(x) = 3h^2 \cdot 2 = 6(1+2x)^2$. If multiplied out, this matches the result found directly.

57. $F'(x) = -5x^{-6}$. We can write $F(x) = x^{-5} = e^{\ln x^{-5}} = e^{-5 \ln x}$, a composition $f(g(x))$ where $f(g) = e^g$ and $g(x) = -5 \ln x$. Then $f'(g) = e^g$ and $g'(x) = -5/x$, so

$$F'(x) = f'(g)g'(x) = e^g \cdot \frac{-5}{x} = e^{-5 \ln x}\frac{-5}{x} = x^{-5}\frac{-5}{x} = -5x^{-6}$$

59. Using the chain rule,

$$(x^n)' = \left(e^{n \ln x}\right)' = e^{n \ln x}(n \ln x)' = e^{n \ln x}\left(n\frac{1}{x}\right) = x^n n\frac{1}{x} = nx^{n-1}$$

61. The radius connects the points $(0, 0)$ and $\left(x, \sqrt{1-x^2}\right)$ and therefore its slope is $\frac{\sqrt{1-x^2}-0}{x-0} = \frac{\sqrt{1-x^2}}{x}$. Because a line perpendicular to a line with slope m has slope $-1/m$, the tangent has slope $-\frac{x}{\sqrt{1-x^2}}$. This matches the result using the chain rule.

63. Write $c(x) = \frac{N}{\sqrt{4\pi kt}}e^{-(1/4kt)x^2}$. Then,

$$c'(x) = \frac{N}{\sqrt{4\pi kt}}e^{-(1/4kt)x^2}\left(-\frac{1}{4kt}\right)2x = -\frac{N}{\sqrt{4\pi kt}} \cdot \frac{1}{2kt}e^{-(1/4kt)x^2}x$$

The derivative $c'(x)$ represents the rate of change of the concentration of the pollutant with respect to the distance from the source at a fixed time t; thus, its units are ppm/distance (i.e., parts per million per distance unit). All factors except for the minus sign are positive, and therefore $c'(x) < 0$. So, at a fixed time t, the concentration is a decreasing function of the distance from the source; i.e., it decreases as we move away from the source.

65. $S'(T) = -1/5$ and $T'(W) = -0.2$, so the derivative of the composition is the product 0.04.

67. $M'(V) = 1.3$, $V'(L) = 6L^2$, $L'(T) = 0.1$. The derivative of the composition is the product of the derivatives or $0.78L^2$. In terms of the temperature, this is $0.78(10 + T/10)^2$.

69. $Q'(0) = -7.32 \cdot 10^6$. It would take $3.0 \cdot 10^{10}/7.32 \cdot 10^6 \approx 4098.3$ years. The half-life is 5681 years, which is greater because the decay gets slower as time passes.

71. The derivative $\alpha'(c) = \frac{5}{(1+c)^2}$ so $\alpha'(0) = 5$ and the limit as c approaches infinity is 0. This is consistent with a graph that starts with a positive slope and eventually flattens out.

73. The derivative $\alpha'(c) = \frac{5-10c}{e^{2c}}$ so $\alpha'(0) = 5$ and the limit as c approaches infinity is 0, but the derivative is negative. This is consistent with a graph that starts out increasing but then decreases to 0.

75. The derivative is $\alpha'(c) = 5 + 10c$ so $\alpha'(0) = 5$ and the limit as c approaches infinity is ∞ consistent with a graph that starts out increasing and then increases at an increasing rate.

77. The derivative is $\frac{db}{dt} = -20e^{-2t} = -2 \cdot 10e^{-2t} = -2b(t)$. This solution is decreasing.

79. The derivative is $\frac{db}{dt} = -40e^{-2t} = 10 - 2b(t)$. This solution is decreasing.

Section 5.4

1. $f'(x) = \cos x + 3 \sec x \tan x - \sec^2 x$

3. $g'(x) = 3 \sec 3x \tan 3x$

5. $y' = -a\sin(ax + b)$

7. $g'(x) = 2x \cos x - x^2 \sin x$

9. Using the quotient rule with $u(\theta) = \sin\theta$ and $v(\theta) = 1 + \cos\theta$ gives

$$q'(\theta) = \frac{(1 + \cos\theta)\cos\theta + \sin\theta \sin\theta}{(1 + \cos(\theta))^2}$$

$$= \frac{\cos\theta + (\cos\theta)^2 + (\sin\theta)^2}{(1 + \cos(\theta))^2}$$

$$= \frac{\cos\theta + 1}{(1 + \cos\theta)^2}$$

$$= \frac{1}{1 + \cos\theta}$$

11. $h'(x) = 4x^3 \sin(x^3) + 3x^6 \cos(x^3)$

13. $y' = -\dfrac{(x-\pi)\csc^2 x + \cot x}{(x-\pi)^2}$

15. $h'(x) = -4x^3 \sin x^4 - 4\sin x \cos^3 x$

17. $f'(x) = -\dfrac{\sin(\ln x)}{x} - \tan x$

19. $g'(x) = 2x \cdot \sin\dfrac{1}{x} - \cos\dfrac{1}{x}$

21. $f'(x) = \dfrac{1}{\sec x + \tan x} \cdot \sec x(\tan x + \sec x) = \sec x$

23. Write $y = \sec(t+1)^{1/2}$; $y' = \dfrac{1}{2\sqrt{t+1}} \sec\sqrt{t+1}\tan\sqrt{t+1}$.

25. $G'(t) = -\dfrac{4\pi}{5} \sin\left[\dfrac{2\pi}{5}(t-3)\right]$

27. $f'(x) = -e^x \sin(e^x)$

29. $f'(x) = 3\ln 2 \cdot 2^{3\tan x} \sec^2 x$

31. $f'(x) = -(1 + 10^x \ln 10)\csc^2(x + 10^x)$

33. $(\cot x)' = \left(\dfrac{\cos x}{\sin x}\right)' = \dfrac{(-\sin x)\sin x - \cos x \cos x}{\sin^2 x} = \dfrac{-1}{\sin^2 x} = -\csc^2 x$

35. $(\csc x)' = \left(\dfrac{1}{\sin x}\right)' = -\dfrac{\cos x}{\sin^2 x} = -\dfrac{\cos x}{\sin x}\dfrac{1}{\sin x} = -\cot x \csc x$

37. The double-angle formula is $\sin(2x) = 2\sin x \cos x$, with derivative
$$\dfrac{d}{dx} 2\sin x \cos x = 2\left((\cos x)^2 - (\sin x)^2\right) = 2\cos(2x)$$
This matches the result with the chain rule.

39. $y = -x + \dfrac{\pi}{4} - \dfrac{1}{2}\ln 2$

41. The angle addition formula says that $\cos(x+y) = \cos x \cos y - \sin x \sin y$. Taking the derivative,
$$\dfrac{d}{dx}\cos(x+y) = -\sin x \cos y - \cos x \sin y = -\sin(x+y)$$
This matches the result with the chain rule.

43. $a'(x) = 3 - \sin x$. $a(0) = 1$, $a'(0) = 3$, $a(\pi/2) = 3\pi/2$, $a'(\pi/2) = 2$, $a(\pi) = 3\pi - 1$, $a'(\pi) = 3$. The function is increasing, but with a slight slowing at around $x = \pi$.

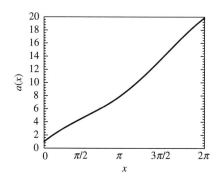

45. $c'(z) = (\cos z - \sin z)/e^z$; $c(0) = 0$, $c'(0) = 1$, $c(\pi/2) = e^{-\pi/2}$, $c'(\pi/2) = -e^{-\pi/2}$, $c(\pi) = 0$, $c'(\pi) = -e^{-\pi}$. The function zips up to a maximum at around $x = \pi/2$, dips down to zero at $x = \pi$, becomes negative, and then returns to zero at $x = 2\pi$.

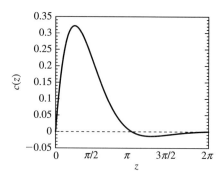

47. $s'(t) = 0.2e^{0.2t}\cos t - e^{0.2t}\sin t$; $s(0) = 1$, $s'(0) = 0.2$, $s(\pi/2) = 0$, $s'(\pi/2) = -1.369$, $s(\pi) = -1.874$, $s'(\pi) = -0.375$. This oscillation expands exponentially.

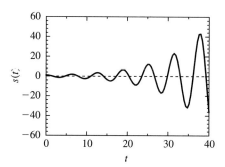

49. $f'(x) = \dfrac{2}{\sqrt{1-4x^2}}$

51. $g'(z) = (\arctan z)^{-1} - \dfrac{z}{1+z^2}(\arctan z)^{-2}$

53. $f'(x) = \dfrac{6}{\sqrt{1-9x^2}}(1 + \arctan 3x)$

55. $F'(t) = \dfrac{\frac{t}{1+t^2} - \arctan t}{t^2}$

57. Start by differentiating $\cos(\arccos x) = x$ to get $(\arccos x)' = -\dfrac{1}{\sin(\arccos x)}$; show that $\sin(\arccos x) = \sqrt{1-x^2}$.

59. Use the formula from Exercise 57 to differentiate $\arccos x$. Then use the fact that $f(x)$ is a constant function.

61. $g'(t) = -6\pi \sin(2\pi(t-5))$. $g'(0) = 0$, consistent with the fact that this oscillation starts at a peak.

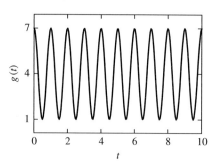

63. $W'(y) = -30\pi \sin\left(2\pi\dfrac{y+0.1}{0.2}\right)$. $W'(0) = -30\pi \sin\left(2\pi\dfrac{0.1}{0.2}\right) = 0$, consistent with the fact that this oscillation starts at a minimum.

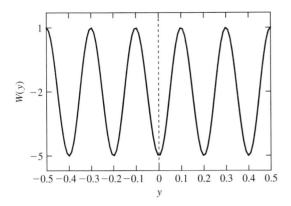

65. The monthly oscillation is almost unnoticeable.

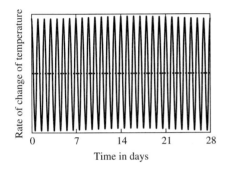

67. We can find the derivative of $b(t)$ with the chain rule by noting that $b(t) = f(g(t))$, where $f(g) = e^g$ and $g(t) = 0.1t + 0.125\sin(0.8t)$. Then $f'(g) = e^g$ and $g'(t) = 0.1 + 0.1\cos(0.8t)$, so

$$b'(t) = (0.1 + 0.1\cos(0.8t))e^{0.1t+0.125\sin(0.8t)}$$
$$= 0.1(1 + \cos(0.8t))b(t)$$

69. The oscillation in the per capita production is larger for larger A. When A is larger than 1, there are times when the population decreases. See Exercise 71.

71.

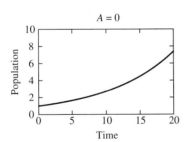

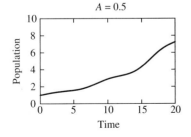

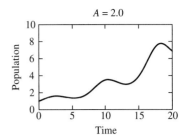

73. a. The average is 0 and the amplitude is 212.0 L/s.
b. The period is 10 seconds. The number of beats per minute is 6.
c. The flow is zero at $t = 2.5$, $t = 7.5$, and so forth. At these times, there is no flow, meaning that the blood has stopped and is turning around.
d. The rate of change is $F'(t) = -212\frac{2\pi}{10}\sin\left(\frac{2\pi t}{10}\right)$. This is zero at $t = 0$, $t = 5$, and so forth. At $t = 0$, the flow is 212 L/s, the maximum in one direction. At $t = 5$, the flow is -212 L/s, the maximum in the other direction.

Section 5.5

1. $y' = -\frac{3x^2}{20y^9 - 4}$

3. $y' = \frac{e^y - ye^x}{e^x - xe^y}$

5. $y' = -\frac{y}{x - xy - 1}$

7. $y' = \frac{3 - \sec x \tan x \tan y}{\sec x \sec^2 y}$

9. $y'(1) = 13/51$

11. $y = \frac{2}{9}x + 1$

13. $y' = 6x^3 5^{x^2}\left(\frac{3}{x} + 2x\ln 5\right)$

15. $y' = \sin x^{\tan x}\left(\sec^2 x \ln(\sin x) + 1\right)$

17. $f'(x) = \left(1 + \frac{1}{x}\right)^x\left[\ln\left(1 + \frac{1}{x}\right) - \frac{1}{x+1}\right]$

19. $y' = \sqrt{\frac{x^4 3^x}{(x^2+5)^3}}\left[\frac{4}{x} + \ln 3 - \frac{6x}{x^2+5}\right]$

21. a. Write $\ln y = \ln\left(f(x)^{g(x)}\right) = g(x)\ln f(x)$. Then

$$\frac{1}{y}y' = g'(x)\ln f(x) + g(x)\frac{1}{f(x)}f'(x)$$

$$y' = f(x)^{g(x)}\left(g'(x)\ln f(x) + g(x)\frac{f'(x)}{f(x)}\right)$$

$$y' = f(x)^{g(x)}g'(x)\ln f(x) + f(x)^{g(x)}g(x)\frac{f'(x)}{f(x)}$$

$$y' = f(x)^{g(x)}g'(x)\ln f(x) + g(x)f(x)^{g(x)-1}f'(x)$$

b. The derivative, assuming that the base $f(x)$ is constant:

$$\left(f(x)^{g(x)}\right)' = f(x)^{g(x)}\ln f(x) \cdot g'(x)$$

The derivative, assuming that the exponent $g(x)$ is constant:

$$\left(f(x)^{g(x)}\right)' = g(x)f(x)^{g(x)-1}f'(x)$$

Adding up the two results, we obtain

$$f(x)^{g(x)}\ln f(x) \cdot g'(x) + g(x)f(x)^{g(x)-1}f'(x)$$

which is the correct formula for the derivative, as shown in (a).

23. $\frac{1}{P(t)}P'(t) = 0.7\frac{1}{Q(t)}Q'(t)$

25. $\frac{1}{y(t)}y'(t) = 4\frac{1}{p(t)}p'(t)$

27. Differentiating $E = 120.7M^{0.74}$, we obtain

$$\frac{dE}{dt} = 120.7(0.74)M^{-0.26}\frac{dM}{dt} = 89.318M^{-0.26}\frac{dM}{dt}$$

It is given that $M = 45$ kg and $dM/dt = 0.2$ kg/day. With these values,

$$\frac{dE}{dt} = 89.318(45)^{-0.26}(0.2) \approx 6.639$$

Thus, the metabolic rate of a 45-kg dog, at the moment when it gains mass at a rate of 0.2 kg/day, increases at a rate of approximately 6.639 kcal/day per day.

29. From $P(t) = aQ(t)^b$ we obtain $\ln P(t) = \ln(aQ(t)^b) = \ln a + b\ln Q(t)$ and therefore (by implicit differentiation) $\frac{1}{P(t)}P'(t) = b\frac{1}{Q(t)}Q'(t)$.

31. $\frac{dP}{dt} = -\frac{3}{8}$

33. $\frac{dS}{dt} \approx -0.0102$ m²/month

35. It decreases at a rate of approximately 5.0396 units per hour.

37. The area is increasing at a rate of approximately 3.39m²/h.

39. The ladder is travelling toward the wall at a rate of approximately 0.85 m/s.

41. The volume is decreasing at a rate of approximately 311.83 cm³/h.

Section 5.6

1.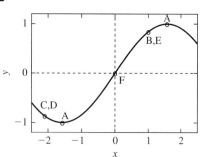

3.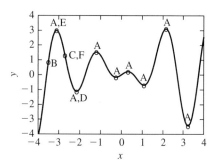

5. Positive increasing derivative
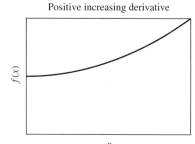

7. Negative increasing derivative
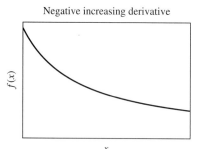

9. The point of inflection at $x = 0$ has a negative third derivative because the second derivative changes from positive to negative values and is therefore decreasing.

11. $s'(x) = -1 + 2x - 3x^2 + 4x^3$; $s''(x) = \frac{ds'(x)}{dx} = 2 - 6x + 12x^2$

13. $h'(y) = 10y^9 - 9y^8$; $h''(y) = 90y^8 - 72y^7$

15. $F'(z) = (1+z)(2+z) + z(2+z) + z(1+z)$;
$F''(z) = (1+z) + (2+z) + z + (2+z) + z + (1+z) = 6z + 6$

17. $f'(x) = \frac{2x - 2(3+x)}{4x^2} = \frac{-3}{2x^2}$; $f''(x) = \frac{3}{x^3}$

19. $f'(x) = 3x^2 \ln x + x^2$, $f''(x) = 6x\ln x + 5x$

21. $f'(x) = e^{-x^2}(-2x) = -2xe^{-x^2}$, $f''(x) = -2e^{-x^2} + 4x^2e^{-x^2}$

23. $f'(x) = 2\cos 2x + 2\sin x \cos x = 2\cos 2x + \sin 2x$,
$f''(x) = -4\sin 2x + 2\cos 2x$

25. $f^{(4)}(x) = 24$

27. $f^{(3)}(x) = 64e^{4x}$

29. $f^{(14)}(x) = -\cos x$

31. $f^{(5)}(x) = 24x^{-5}$

33. If $n \le 5$, then $f^{(n)}(0) = n!$, and if $n > 5$, then $f^{(n)}(x) = 0$.

35. $f^{(n)}(x) = (-1)^n e^{-x}(x - n)$

37. $g'(z) = 1 - z^{-2}$, so there is a critical point at $z = 1$. $g''(z) = 2z^{-3} > 0$, so the function is always concave up.

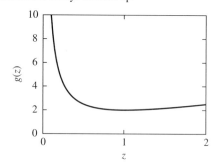

39. We can use the quotient rule with $u(t) = t$ and $v(t) = 1 + t$ to find

$$M'(t) = \frac{(1+t) \cdot 1 - t \cdot 1}{(1+t)^2} = \frac{1}{(1+t)^2}$$

This is positive, meaning that the function is increasing. By using the quotient rule again, we compute $M''(t) = -\frac{2}{(1+t)^3} < 0$, so the graph is concave down.

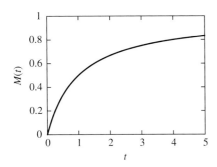

41. Applying the power rule with $p = -2$ gives $f'(x) = -2x^{-3}$. The first derivative is negative for $x > 0$. Again applying the power rule, this time with $p = -3$, along with the constant product rule gives $f''(x) = 6x^{-4}$. The second derivative is always positive.

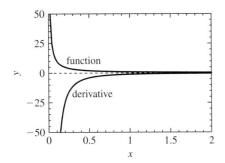

43. The derivative of this quadratic function is $f'(x) = 1 - 2x$. The first derivative is positive when $x < 0.5$ and negative when $x > 0.5$. The second derivative is $f''(x) = -2$, which is always negative.

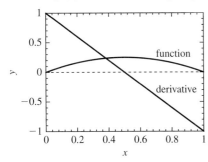

45. Look for those x-values where one graph has the largest or the smallest values (i.e., where it has a horizontal tangent) and some other graph crosses the x-axis. For example, the graphs at x_1 show that $b' = a$; the graphs at x_2 show that $c' = b$. Thus, $c = f$, $b = c' = f'$, and $a = b' = c'' = f''$.

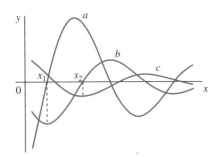

47. $G'(x) = -\frac{1+2x}{2x^{3/2}}e^{-x}$, which is negative. It also says that the curve is infinitely steep at $x = 0$. $G''(x) = \frac{3+4x+4x^2}{4x^{5/2}}e^{-x}$, which is positive. This function is decreasing and concave up.

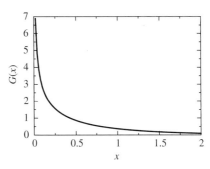

49. Constant function $y = 120$

51. 120

53. The derivative is $g'(x) = 1 - x^{-2}$ and the second derivative is $g''(x) = 2x^{-3}$.

a. Because $g(1) = 2$ and $g'(1) = 0$, the tangent line is $L(x) = 2$.
b. Because $g''(1) = 2$, the quadratic $T_2(x) = 2 + (x-1)^2$ matches both the first and second derivatives at $x = 1$.
c.

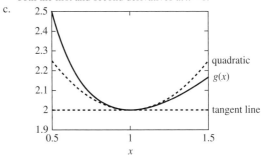

55. $(x + 11)e^x$

57. $h(x) = (ax + b)e^x = 0$ at $x^* = \frac{-b}{a}$. The critical point solves $h'(x) = (ax + a + b)e^x = 0$, or $x = -1 + \frac{-b}{a} = x^* - 1$. The point of inflection solves $h''(x) = (ax + 2a + b)e^x = 0$, or $x = -2 + \frac{-b}{a} = x^* - 2$.

59. From

$$f'(0) \approx \frac{f(-1) - f(0)}{-1 - 0} = \frac{8.2 - 8.8}{-1} = 0.6$$

$$f'(-1) \approx \frac{f(-2) - f(-1)}{-2 - (-1)} = \frac{12.8 - 8.2}{-1} = -4.6$$

we find

$$f''(0) \approx \frac{f'(-1) - f'(0)}{-1 - 0} \approx \frac{-4.6 - 0.6}{-1} = 5.2$$

61. From

$$f'(0) \approx \frac{f(-1) - f(0)}{-1 - 0} = \frac{8.2 - 8.8}{-1} = 0.6$$

$$f'(1) \approx \frac{f(0) - f(1)}{0 - 1} = \frac{8.8 - 6.4}{-1} = -2.4$$

we find

$$f''(1) \approx \frac{f'(0) - f'(1)}{0 - 1} \approx \frac{0.6 - (-2.4)}{-1} = -3$$

63. a. Velocity is $p'(t) = -274t + 20$. Acceleration is $p''(t) = -274$.
b. The position at $t = 3$ is

$$p(3) = -137 \cdot 9 + 20 \cdot 9 + 500 = -553$$

far less than 0, so the object has already hit the ground.

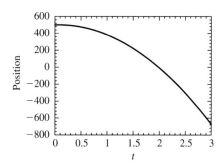

c. The tower was 500 metres high, and the object was thrown upward at 20 m/s. The acceleration due to gravity on the Sun is huge.

65. a. Velocity is $p'(t) = -3.7t + 20$. Acceleration is $p''(t) = -3.7$.

b.
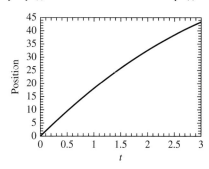

c. The object was thrown upward from ground level at 20 m/s. The acceleration of gravity on Mercury is smaller than on Earth, so the object has not yet begun to fall after three seconds.

67. The second derivative is −20,000, matching the graph, which is always concave down.

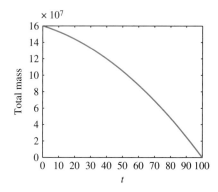

69. The second derivative is $-600t - 20{,}000$, which is always negative. This matches graph, which is always concave down.

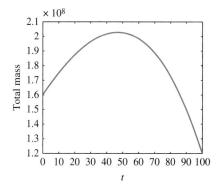

71.
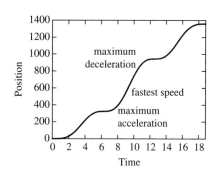

73. The derivative of $f(b) = 2b^2\left(1 - \dfrac{b}{1000}\right)$ is $f'(b) = 4b\left(1 - \dfrac{b}{1000}\right) - \dfrac{b^2}{500}$, and the second derivative is $4 - \dfrac{3b}{250}$, which is positive for $b \le 333.33$ and negative otherwise.

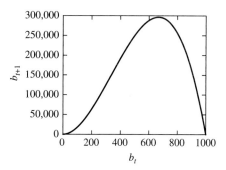

75. The first derivative can be found with the quotient rule to be

$$h'(x) = \dfrac{2x}{(1+x^2)^2}$$

The second derivative can also be found with the quotient rule as

$$h''(x) = \dfrac{2(1-3x^2)}{(1+x^2)^3}$$

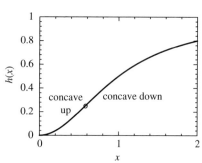

This is positive when $x < \sqrt{1/3}$ and negative when $x > \sqrt{1/3}$. The function is concave up for $x < \sqrt{1/3}$ and concave down when $x > \sqrt{1/3}$.

77. We want to maximize $F(p) - cp$. The derivative of this function is $F'(p) - c$ and is equal to 0 when $F'(p) = c$.

79.

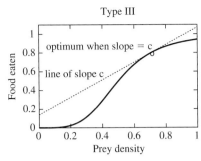

Type III

81. The function is $F(p) = \dfrac{p}{1+p} - cp$ with derivative

$$F'(p) = \dfrac{1}{(1+p)^2} - c$$

Because the first term is always less than 1, this derivative is negative if $c > 1$. The function is thus decreasing, and the optimal prey density is 0. If $c < 1$, we can solve $F'(p) = 0$ for $p = \dfrac{1}{\sqrt{c}} - 1$. This is the maximum because the function switches from increasing to decreasing at this point.

Section 5.7

1. $T_1(x) = 1$, $T_2(x) = 1 - \dfrac{1}{2}\left(x - \dfrac{\pi}{2}\right)^2$

3. $T_1(x) = x$, $T_2(x) = x$

5. $T_3(x) = 1 + x + x^2 + x^3$, $T_4(x) = 1 + x + x^2 + x^3 + x^4$

7. $T_1(x) = 1$, $T_2(x) = 1 + \dfrac{1}{2}x^2$

9. Let $f(x) = x^2$ with base point $a = 3$. Then $f'(x) = 2x$ and $f(3) = 9$ and $f'(3) = 6$. $L(x) = 9 + 6(x - 3)$ and $L(3.03) = 9 + 6(3.03 - 3) = 9.18$. To find the secant line, we evaluate $f(4) = 16$, so the secant line has slope 7. Therefore, $f_s(t) = 9 + 7(x - 3)$ and $f_s(3.03) = 9 + 7 \cdot 0.03 = 9.21$. The exact answer is $T_2(3.03) = 9.1809$, pretty close to the tangent line approximation.

11. Let $f(x) = \sqrt{x}$ with base point 4. $f(4) = 2$, $f'(x) = \dfrac{1}{2}x^{-1/2}$, and $f'(4) = 0.25$. So $L(x) = 2 + 0.25(x - 4)$ and $L(6) = 2.5$. To find the secant line, we evaluate $f(9) = 3$, so the secant line has slope 0.2. Therefore, $f_s(t) = 2 + 0.2(x - 4)$ and $f_s(6) = 2 + 0.2 \cdot 2 = 2.4$. The exact answer is 2.44949 to five decimal places.

13. Let $f(x) = \cos x$. $f(0) = \cos 0 = 1$. $f'(x) = -\sin x$ and $f'(0) = 0$. So $L(x) = 1 + 0(x - 0)$ and $L(-0.02) = 1$. The secant line does not help because there is no easy value of x to evaluate this function. The exact answer is 0.99980 to five decimal places.

15. Let $f(x) = x^2$ with base point $a = 3$. Because $f'(x) = 2x$, $f''(x) = 2$. Therefore, $f''(3) = 2$. $T_2(x) = 9 + 6(x - 3) + (x - 3)^2$ and $T_2(3.03) = 9.1809$. This is exactly right because the original function is quadratic already.

17. Let $f(x) = \sqrt{x}$ with base point 4. We have that $f'(x) = \dfrac{1}{2}x^{-1/2}$ and $f''(x) = \dfrac{-1}{4}x^{-3/2}$. Then $f''(4) = -1/32$, so $T_2(x) = 2 + 0.25(x - 4) - (x - 4)^2/64$ and $T_2(6) = 2.4375$. The exact answer is 2.44949 to five decimal places.

19. Let $f(x) = \cos x$ with base point 0. Then $f'(x) = -\sin x$ and $f''(x) = -\cos x$, so $f''(0) = -1$. Then $T_2(x) = 1 + 0(x - 0) - (x - 0)^2/2$ and $T_2(-0.02) = 0.9998$. The exact answer is 0.9998000067 to 10 decimal places. This is really close.

21. This is a composition $f(g(x))$, where $g(x) = \sqrt{x}$ and $f(g) = \ln g$. These have derivatives $g'(x) = \dfrac{1}{2\sqrt{x}}$ and $f'(g) = \dfrac{1}{g}$. Near the base point $x = 1$,

we have that $f(g(1)) = 0$ and $(f \circ g)'(x) = \dfrac{1}{2x}$, so $(f \circ g)'(1) = 1/2$. The tangent line approximation is $0 + 1/2(x - 1)$, which has value 0.01 at $x = 0.98$. [To distinguish linear approximations, instead of using L we place the hat above the symbol for the function.] In steps, the tangent line to g at $x = 1$ is $\hat{g}(x) = 1 + 1/2(x - 1)$, so $\hat{g}(0.98) = 0.99$. We then evaluate f near 1, finding $f(1) = 0$ and $f'(1) = 1$. Then $\hat{f}(x) = 0 + 1(x - 1)$ and $\hat{f}(0.99) = -0.01$, exactly as before.

23. This is a composition of $f(g(h(x)))$ where $h(x) = (1 + x)^3$, $g(h) = \ln h$ and $f(g) = \sin g$. These have derivatives $h'(x) = 3(1 + x)^2$, $g'(h) = 1/h$, and $f'(g) = \cos g$. Near the base point $x = 0$, we have that $h(0) = 1$ and $g(1) = 0$ and $f(0) = 0$, so $f(g(h(0))) = 0$. By the chain rule, $(f \circ g \circ h)'(0) = 3$. The tangent line approximation is $0 + 3x$ which has value 0.3 at $x = 0.1$. [To distinguish linear approximations, instead of using L we place the hat above the symbol for the function.] In steps, the tangent line to h at $x = 0$ is $\hat{h}(x) = 1 + 3x$, so $\hat{h}(0.1) = 1.3$. The tangent line to g near $h = 1$ is $\hat{g}(h) = 0 + (h - 1)$ so $\hat{g}(1.3) = 0.3$. We then evaluate f near 0, finding $f(0) = 0$ and $f'(0) = 1$. Then $\hat{f}(g) = g$ and $\hat{f}(0.3) = 0.3$, exactly as before.

25. $\ln 1.1 = 0.095 < 0.1$. $\ln 0.9 = -0.105 < -0.1$. The estimates are high because the graph of $\ln x$ is concave down and lies below the tangent.

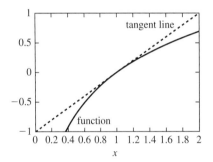

27. $\sqrt{1.1} \approx 1.049 < 1.05$. $\sqrt{0.9} \approx 0.949 < 0.95$. The estimates are high because the graph of $\sqrt{x}$ is concave down and lies below the tangent.

29. $T_3(x) = 1 + 3x + 4x^2 + x^3$

31. $T_3(x) = 13 + 30(x - 1) + 31(x - 1)^2 + 17(x - 1)^3$

33. $T_3(x) = x - x^3/6$

35. 6.0691

37. The Taylor polynomials are

$$T_1(x) = x$$
$$T_2(x) = x - \dfrac{1}{2}x^2$$
$$T_3(x) = x - \dfrac{1}{2}x^2 + \dfrac{1}{3}x^3$$
$$T_4(x) = x - \dfrac{1}{2}x^2 + \dfrac{1}{3}x^3 - \dfrac{1}{4}x^4$$

and so forth. Because the Taylor polynomials approximate the function f, the sum of the series is $f(1) = \ln 2$. I added up the first 10 terms (up to $-\dfrac{1}{10}$) and got 0.645. This is not that close to $\ln 2 = 0.693$, but the values are jumping around it and apparently getting closer.

39. $\alpha(0) = 0$ and $\alpha'(c) = \dfrac{5}{(5 + c)^2}$, so $\alpha'(0) = 1/5$. Therefore, the tangent line is $L(c) = c/5$, matching the result found with leading behaviour.

41. $\alpha(0) = 0$ and $\alpha'(c) = \dfrac{5 - 10c}{e^{2c}}$, so $\alpha'(0) = 5$. Therefore, the tangent line is $L(c) = 5c$, matching the result found with leading behaviour.

43. $\alpha(0) = 0$ and $\alpha'(c) = 5 + 10c$, so $\alpha'(0) = 5$. Therefore, the tangent line is $L(c) = 5c$, matching the result found with leading behaviour.

45. The tangent at $t = 0$ gives 0.5, the tangent at $t = 1$ gives 0.625, and the secant gives 0.75. The exact answer is $\frac{2}{3} \approx 0.667$, closest to the tangent at $t = 1$. The secant is in the right ball-park, and the tangent at $t = 0$ is farther off.

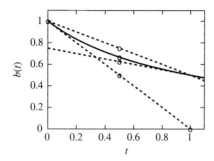

47. The tangent at $t = 0$ gives -0.1, the tangent at $t = 1$ gives 0.475, and the secant gives 0.45. To three decimal places, the exact answer is 0.476, closest to the tangent at $t = 1$. The secant is fairly close and the tangent at $t = 0$ is negative, an impossible result.

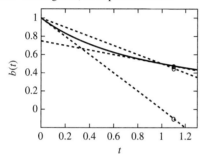

49. Using the base point $a = 1.5$, $M(1.5) = 3.375$. $M'(a) = 3a^2$, so $M'(1.5) = 6.75$. Then the tangent line is $T_1(a) = 3.375 + 6.75(a - 15)$, and $T_1(1.45) = 3.0375$. The secant line connecting $a = 1$ and $a = 15$ is $M_s(a) = 3.375 + 4.75(a - 1.5)$, and $M_s(1.45) = 3.1375$. The exact value is 3.0486 to five decimal places. The tangent line is quite close, but only the secant line would be possible if we did not know the formula.

51. First, we can use the equation $T(t) = t + t^2$ to find the tangent line at $t = 2$ as $T_1(t) = 6 + 5(t - 2)$, and estimate $T(3) = 11$. We can find the tangent line at $t = 4$ as $T_1(t) = 20 + 9(t - 4)$, and again estimate $T(3) = 11$. The secant line between the actual data points has slope $\frac{20.24 - 6.492}{2} = 6.874$, giving the line $T_s(t) = 6.492 + 6.874(t - 2)$ with the value $T_s(3) = 13.366$. None of these is very close to the correct answer.

True/False Quiz

1. TRUE. $\left(2\sqrt{g(x)}\right)' = 2\frac{1}{2}(g(x))^{-1/2}g'(x) = \frac{g'(x)}{(g(x))^{1/2}} = \frac{g'(x)}{\sqrt{g(x)}}$.

3. FALSE. From $(fg)' = f'g + fg'$ using the product rule again, we obtain

$$(fg)'' = f''g + f'g' + f'g' + fg'' = f''g + 2f'g' + fg''$$

5. TRUE. The derivatives repeat in a cycle of 4. Thus, $f^{(4)}(x) = f^{(8)}(x) = \cos x$ and therefore $f^{(9)}(x) = -\sin x$ and $f^{(10)}(x) = -\cos x$.

7. TRUE. From $A = cB$ for some constant c we obtain $\ln A = \ln c + \ln B$ and therefore, by the chain rule, $A'/A = 0 + B'/B = B'/B$.

9. TRUE. The tangent line approximation of $f(x) = \ln x$ at $x = 1$ is

$$L(x) = f(1) + f'(1)(x - 1) = 0 + 1(x - 1) = x - 1$$

Therefore, $\ln 1.2 = f(1.2) \approx L(1.2) = 1.2 - 1 = 0.2$.

Supplementary Problems

1. $F'(y) = 4y^3 + 10y$

3. $H'(c) = \frac{2c(1+c)}{(1+2c)^2}$ for $c \neq -1/2$.

5. $b'(y) = -0.75y^{-1.75}$ for $y > 0$.

7. $g'(x) = 60x(4 + 5x^2)^5$

9. $s'(t) = 3/t$

11. $f'(x) = \cos(\sin(\sin x)) \cdot \cos(\sin x) \cdot \cos x$

13. If $\sin x > 0$, then $h'(x) = -1$. If $\sin x < 0$, then $h'(x) = 1$.

15. $f'(x) = \frac{1}{3}\left(2 + \frac{x}{e^{3x}}\right)^{-2/3} \frac{1 - 3x}{e^{3x}}$

17. $g'(y) = (9y^2 + 4y + 1)e^{3y^3 + 2y^2 + y}$

19. $g'(x) = \frac{2x}{1+x^2}$; $g(x)$ is decreasing when $x < 0$.

21. $c'(z) = \frac{e^{2z} - 1 - 2ze^{2z}}{z^2}$; the limit is 2.

23. $g'(x) = -4x^3 e^{-x^4}$, which is 0 at $x = 0$. $g''(x) = (16x^6 - 12x^2)e^{-x^4}$, which is 0 at $x = 0$ and $x = \pm 0.931$.

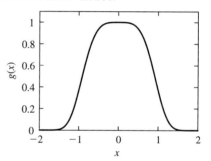

25. $F'(c) = -2e^{-2c} + e^c$, which is 0 where $c = \ln 2/3$. $F''(c) = 4e^{-2c} + e^c$, which is always positive. There are no points of inflection.

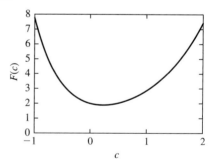

27. $f'(x) = \frac{4}{x \ln 2} + \frac{\cot x}{\ln 2} - \log_2 e$

29. $y = -2x + 3$

31. a. The average rate of change is

$$\frac{\Delta P}{\Delta t} = \frac{P(2) - P(1)}{2 - 1} = \frac{2/5 - 1/3}{1} = 1/15$$

b. The secant line passes through the point $(1, 1/3)$ with slope $1/15$ and has equation

$$P_s(t) = \frac{1}{3} + \frac{1}{15}(t - 1)$$

c.

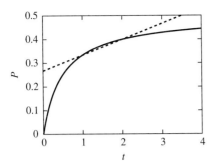

d. I'd say it is less steep because the curve gets less and less steep for larger values of t.

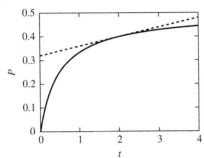

e. $\lim_{\Delta t \to 0} \dfrac{P(2 + \Delta t) - P(2)}{\Delta t}$

33.

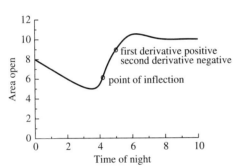

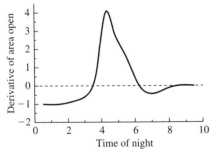

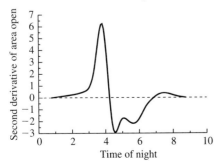

Chapter 6

Section 6.1

1. For example: $f(x) = (x-1)^2$ or $f(x) = \sqrt[3]{x-1}$

3.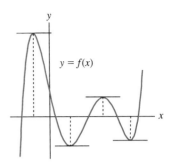

5. For example: $f(x) = (x-3)^3$

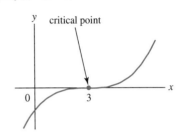

7. $f'(x) = 2 - 4x$. The only point where $f'(x) = 0$ occurs where $2 - 4x = 0$ or $x = 1/2$.

9. $g'(y) = \dfrac{1 + y^2 - 2y^2}{(1+y^2)^2} = \dfrac{1 - y^2}{(1+y^2)^2}$. The equation $g'(y) = 0$ has solutions where $1 - y^2 = 0$, or at $y = -1$ and $y = 1$.

11. $c'(\theta) = 2\pi \sin(2\pi\theta)$. This is 0 when θ is an integral multiple of $1/2$, such as $\theta = 0, \theta = 1/2, \theta = 1, \theta = 3/2$, etc.

13. $x = \sqrt[4]{3}$ and $x = -\sqrt[4]{3}$.

15. $x = \dfrac{\pi}{2} + k\pi$, where k is an integer.

17. $x = -1/4$

19. $x = -\dfrac{\pi}{2} + 2k\pi$ where k is an integer.

21. $x = 5/3$

23. $f(x)$ is increasing on $(-1, 0), (1, \infty)$ and decreasing on $(-\infty, -1), (0, 1)$; relative maximum: $f(0) = 1$; relative minimum: $f(1) = f(-1) = 0$.

25. $f(x)$ is increasing on $(-\sqrt{3} - 1, 0)$ and $(0, \sqrt{3} - 1)$ and decreasing on three intervals: $(-\infty, -\sqrt{3} - 1), (\sqrt{3} - 1, 2)$, and $(2, \infty)$; relative maximum: $f(\sqrt{3} - 1) \approx -1.866$; relative minimum: $f(-\sqrt{3} - 1) \approx -0.134$.

27. $f(x)$ is increasing on $(0, 1)$ and decreasing on $(-\infty, 0)$ and $(1, \infty)$; relative maximum: $f(1) = e^{-2} \approx 0.135$; relative minimum: $f(0) = 0$.

29. $f(x)$ is increasing on $(2\pi k, \pi/4 + 2\pi k)$ and $(5\pi/4 + 2\pi k, 2\pi + 2\pi k)$ and decreasing on $(\pi/4 + 2\pi k, 5\pi/4 + 2\pi k)$; relative maximum: $f(\pi/4 + 2\pi k) = \sqrt{2}$; relative minimum: $f(5\pi/4 + 2\pi k) = -\sqrt{2}$ (k is an integer).

31. The maximum is at $x = 0.5$, and the minimum is at $x = 2$.

33. The maximum is at $y = 1$, and the minimum at $y = 0$.

35. The maximum is at $x = 3$ and the minimum is at $x = 1$.

37. Absolute maximum value: $f(0) = 4$; absolute minimum value: $f(4) = 0$

39. Absolute maximum value: $f(\pi/2) = 1$; absolute minimum value: $f(\pi) = 0$

41. Absolute maximum value: $f(1) = e^{-1}$; absolute minimum value: $f(0) = 0$

43. $f(-\pi) = -\pi$ is the absolute minimum and $f(6\pi) = 6\pi$ is the absolute maximum.

45.

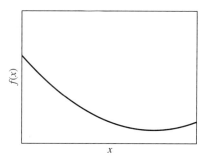

47.

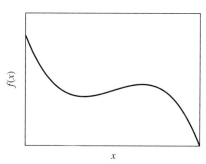

49. The function $f(x) = \frac{1}{x^2}$ defined for $-1 \le x \le 1$, but with $f(0) = 0$

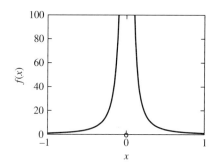

51. $f''(x) = -4 < 0$. The graph has a maximum at $x = 0.5$, consistent with the fact that it is concave down.

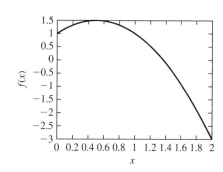

53. $g''(y) = \frac{2y(y^2 - 3)}{(1 + y^2)^3}$. Therefore, $g''(1) = -1/2$ and $g''(-1) = 1/2$, consistent with the local maximum at $y = 1$ and the local minimum at $y = -1$.

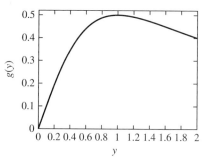

55. $c''(\theta) = -(2\pi)^2 \cos(2\pi\theta)$. At the critical point $\theta = -1$, the second derivative is negative, so it is a maximum. At the critical point $\theta = -1/2$, the second derivative is positive, so it is a minimum. At the critical point $\theta = 0$, the second derivative is negative, so it is a maximum. At the critical point $\theta = 1/2$, the second derivative is positive, so it is a minimum. At the critical point $\theta = 1$, the second derivative is negative, so it is a maximum.

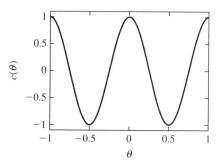

57. a. $h(x) = 1 - f(x)$, so $h'(x) = -f'(x)$ by the constant multiple rule. If $f'(x) = 0$ then $h'(x) = 0$.

b. $h''(x) = -f''(x)$

c. If $f(x)$ has a local maximum, $h(x)$ has a local minimum. This makes sense; when $f(x)$ is large, $h(x)$ is small and vice versa.

d. The function is $h(x) = 1 - xe^{-x}$. The derivative is $h'(x) = (x - 1)e^{-x}$, which has a critical point at $x = 1$. The second derivative is $h''(x) = (2 - x)e^{-x}$, which is positive at $x = 1$. This is indeed a minimum.

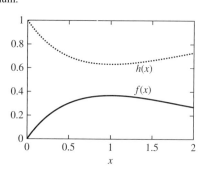

59. a. $h'(x) = f'(x) - 2f'(x)f(x)$ by the chain rule. If $f'(x) = 0$, then $h'(x) = 0$.

b. $h''(x) = f''(x) - 2f''(x)f(x) - 2f'(x)^2$. If $f'(x) = 0$, then $h''(x) = f''(x)(1 - 2f(x))$. This has the same sign as $f''(x)$ if $f(x) < 1/2$, and the opposite sign if $f(x) > 1/2$.

c. If $f(x)$ has a local maximum that is less than $1/2$, $h(x)$ does also, but if $f(x)$ has a local maximum that is greater than $1/2$, $h(x)$ has a local minimum. This is strange.

d. The function is $h(x) = xe^{-x} - x^2 e^{-2x}$. The derivative is $h'(x) = (1-x)e^{-x} + 2x(x-1)e^{-2x}$, which has a critical point at $x = 1$. The second derivative is $h''(x) = (x-2)e^{-x} + (8x - 2 - 4x^2)e^{-2x}$, which is negative at $x = 1$. This is indeed a maximum. This is consistent with our calculation because $f(1) = e^{-1} = 0.37 < 0.5$.

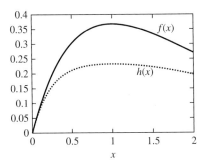

61. By about 416%

63. We find that $C'(t) = 3t^2 - 60t$, with critical points at $t = 0$ and $t = 20$. Evaluating at the critical points and endpoints, we find $C(0) = 6000$, $C(20) = 2000$, and $C(25) = 2875$, giving a maximum at $t = 0$ and a minimum at $t = 20$.

65. a. $H(a) = -a \ln a - (1-a) \ln(1-a)$

b. The critical point $a = 1/2$ yields the maximum value for H. In that case, $b = 1/2$. The maximum occurs when the two species are represented in equal proportion.

67. $(2, 2)$ and $(-2, -2)$

69. $(1/2, 0)$

71. The rectangle with maximum area is the square of side $R\sqrt{2}$.

73. $\alpha = \pi/2$.

75. The side of the rectangle is the interval $[0, 1]$ on the x-axis.

77. The base is the square of side $\sqrt{S}/2$; the height is $\sqrt{S}/4$.

79. Side of the square: $\frac{6}{\pi+4}$; radius of the disk: $\frac{3}{\pi+4}$.

81. Consider the following diagram:

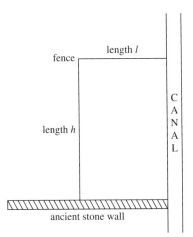

The total length of the fence is $l + h = 1000$, and the area enclosed is lh. We can solve for h as $h = 1000 - l$, so the area is $A(l) = l(1000 - l)$. Then $A'(l) = 1000 - 2l$, which has a critical point at $l = 500$ (meaning that $h = 500$). This is a global maximum because the area at the endpoints $l = 0$ and $l = 500$ is 0. The maximum area is then $500 \cdot 500 = 25,000$ m^2.

Section 6.2

1. We find
$$r'(m) = \left(\frac{M}{K}\right)^{1/3} \left(-\frac{1}{3}\right) \left(1 - m^4\right)^{-4/3} \left(-4m^3\right)$$
$$= \frac{4}{3} \left(\frac{M}{K}\right)^{1/3} \frac{m^3}{\left(1 - m^4\right)^{4/3}}$$

and thus

$$r'(0.1) = \frac{4}{3}\left(\frac{M}{K}\right)^{1/3} \frac{(0.1)^3}{\left(1 - (0.1)^4\right)^{4/3}} \approx 0.00133 \left(\frac{M}{K}\right)^{1/3}$$

$$r'(0.5) = \frac{4}{3}\left(\frac{M}{K}\right)^{1/3} \frac{(0.5)^3}{\left(1 - (0.5)^4\right)^{4/3}} \approx 0.18164 \left(\frac{M}{K}\right)^{1/3}$$

$$r'(0.9) = \frac{4}{3}\left(\frac{M}{K}\right)^{1/3} \frac{(0.9)^3}{\left(1 - (0.9)^4\right)^{4/3}} \approx 4.03419 \left(\frac{M}{K}\right)^{1/3}$$

When a bone is thin ($m = 0.9$), making it thinner (by increasing m a bit) increases the radius at a fast rate. However, when a bone is thick ($m = 0.1$), making it a bit thinner does not cause a large change in its radius.

3. The mass of the bone is (as in the text) $\rho\pi \left(\frac{M}{K}\right)^{2/3} \left(1 - m^2\right) \left(1 - m^4\right)^{-2/3}$. The mass of the marrow is (replace $1/2$ by 0.55) $0.55\rho\pi \left(\frac{M}{K}\right)^{2/3} m^2 \left(1 - m^4\right)^{-2/3}$. The total mass is

$$T(m) = \rho\pi \left(\frac{M}{K}\right)^{2/3} \left(1 - m^4\right)^{-2/3} \left(1 - 0.45m^2\right)$$

Thus, we need to minimize the function
$f(m) = \left(1 - 0.45m^2\right) \left(1 - m^4\right)^{-2/3}$. We compute

$$f'(m) = -m \left(1 - m^4\right)^{-5/3} \frac{1}{30} \left[9m^4 - 80m^2 + 27\right]$$

Solving $9m^4 - 80m^2 + 27 = 0$ for m, we get $m = \sqrt{\frac{80 - \sqrt{5428}}{18}} \approx 0.59278$. Thus, of all bones that are capable of withstanding the same moment, the one whose marrow cavity radius is approximately 59.2 percent of the total radius is the lightest. This value is close to some values in the table (humerus for a fox or femur for a hare).

5. a. Following the steps in the examples in the text gives an optimum of $t = 1.0$. At this point, the derivative is

$$F'(1.0) = \frac{0.5}{(1.0 + 0.5)^2} \approx 0.222$$

The tangent line is $\hat{F}(t) = F(1) + F'(1)(t - 1) = 0.667 + 0.222(t - 1)$. We can check directly that $\hat{F}(-2.0) = 0$.

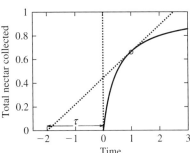

b. $t = 0.5$. The tangent line is $\hat{F}(t) = \frac{1}{2} + \frac{1}{2}(t - 0.5)$. It is true that $\hat{F}(-0.5) = 0$.

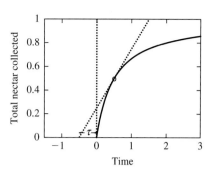

c. $t = \sqrt{0.05} \approx 0.223$. The tangent line is $\hat{F}(t) = 0.309 + 0.955(t - 0.223)$. It is true that $\hat{F}(-0.1) = 0$.

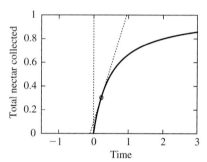

7. a. $t = \sqrt{2} \approx 1.414$

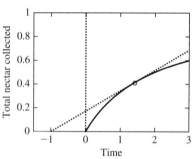

b. $t = 1.0$

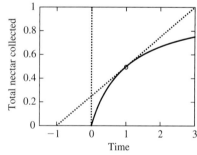

c. $t = \sqrt{0.1} \approx 0.316$

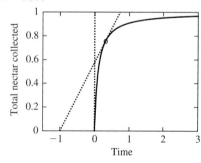

9. a. The travel time must be 2.0 minutes.
   b. $\tau = 0.02$
   c. $\tau = 32.0$
11. $R(t) = \dfrac{t}{(1+t)^2}$, so
$$R''(t) = \dfrac{2(t-2)}{(t+1)^4}$$
which is negative at $t = 1$, the optimal solution. Therefore, this is at least a local maximum.
13. The bees are trying to maximize $R(n) = \dfrac{n}{P(n)} = \dfrac{n}{1+n^2}$. Taking the derivative, this has a maximum at $n = 1$.
15. $R(n) = \dfrac{n}{P(n)}$, so $R'(n) = \dfrac{P(n) - nP'(n)}{P(n)^2}$. The critical point is where $P(n) = nP'(n)$ or $P'(n) = \dfrac{P(n)}{n}$. If $P(n) = 1 + n^2$, then $P'(n) = 2n$, so the condition is $2n = \dfrac{1+n^2}{n}$ or $2n^2 = 1 + n^2$ or $n = 1$. If $P(n) = 1 + n$, then $P'(n) = 1$, so the condition is $1 = \dfrac{1+n}{n}$, which has no solution.
17. a. $N^* = 0$ or $N^* = 1 - \dfrac{1+h}{2.0}$. The largest possible h is 1.
    b. $P(h) = h\left(1 - \dfrac{1+h}{2.0}\right)$
    c. $P'(h) = 0$ at $h = 0.5$. This is a maximum because $P''(h) = -1$.
    d. $P(0.5) = 0.125$
19. The derivative of the updating function is $2.5(1 - 2N_t) - h$. The equilibrium is $N^* = 1 - \dfrac{1+h}{2.5}$. Plugging into the derivative gives $h - 0.5$ (after some algebra). The equilibrium is stable as long as $h < 1.5$. At $h = 0.75$, the slope is 0.25, indicating stability.

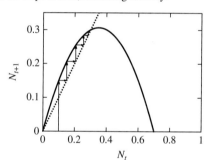

21. a. $N^* = \dfrac{2.5}{1+h} - 1$
    b. $h = 1.5$
    c. $P(h) = hN^*$. The maximum is at $h = \sqrt{2.5} - 1 \approx 0.58$. This takes on the value of approximately 0.58 for $r = 2.5$.
    d. With $r = 2.5$, $P(0.58) \approx 0.338$.

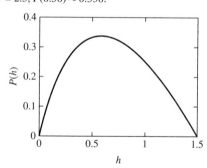

23. a. Since $f'(x) = 1 - \cos x > 0$ (except at $x = 2k\pi$, where it is zero) we conclude that $f(x)$ is increasing. Therefore, whenever $x > 0$, then

$$f(x) > f(0)$$
$$x - \sin x > 0$$
$$x > \sin x$$

b. Use the definition of cosecant and cotangent and then multiply by the positive quantity $\sin^2(\pi/x)$:

$$\pi \csc^2(\pi/x) - x\cot(\pi/x) > 0$$
$$\pi \frac{1}{\sin^2(\pi/x)} - x\frac{\cos(\pi/x)}{\sin(\pi/x)} > 0$$
$$\pi - x\cos(\pi/x)\sin(\pi/x) > 0$$

c. We compute

$$\pi - x\cos(\pi/x)\sin(\pi/x) > 0$$
$$\pi - x\frac{1}{2}\sin(2\pi/x) > 0$$
$$x\frac{1}{2}\sin(2\pi/x) < \pi$$
$$\sin(2\pi/x) < \frac{2\pi}{x}$$

d. In (a) we showed that $\sin A < A$ for $A > 0$. Apply this to $A = 2\pi/x$ to prove that (c) is true. Thus, (b) is true and we are done.

Section 6.3

1. Let $f(x) = e^x + x^2 - 2$. Then $f(0) = -1 < 0$ and $f(1) = e - 1 > 0$. By the Intermediate Value Theorem, there must be a solution in between.

3. To get this into the right form, subtract x from both sides to give the equation $e^x + x^2 - 2 - x = 0$. Let $f(x) = e^x + x^2 - 2 - x$. Then $f(0) = -1 < 0$ and $f(1) = e - 2 > 0$. By the Intermediate Value Theorem, there must be a solution where $f(x) = 0$, or where $e^x + x^2 - 2 - x = 0$. This point must also solve the original equation.

5. Let $f(x) = xe^{-3(x-1)} - 2$. Then $f(0) = -2 < 0$ and $f(1) = -1 < 0$. The Intermediate Value Theorem tells us nothing. However, $f(1/2) = 0.24 > 0$. The Intermediate Value Theorem guarantees (because f is continuous) solutions between 0 and 1/2 and also between 1/2 and 1.

7. $f(0) = f(1) = 0$, and $f(x) > 0$ for $0 \leq x \leq 1$. As well, $f(x)$ is continuous on $[0, 1]$ and differentiable on $(0, 1)$. Therefore, there must be a positive maximum in that range.

9. $f(0) = f(1) = -1$. As well, $f(x)$ is continuous on $[0, 1]$ and differentiable on $(0, 1)$. There must be a maximum in between. Also, $f(0.5) = 0.875$, which is positive; therefore, there must be a positive maximum.

11. $f'(x) = 2x$. The slope of the secant is $\frac{f(1) - f(0)}{1 - 0} = 1$. $f'(x) = 1$ at $x = 0.5$.

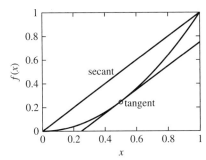

13. $g'(x) = \frac{1}{2\sqrt{x}}$. The slope of the secant is 1, and $g'(x) = 1$ at $x = 1/4$.

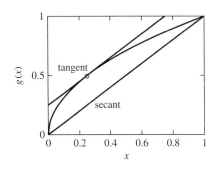

15. The function never takes on any values other than 0 and 1, so can never equal 1/2. Also, the slope of the secant connecting $x = -1$ and $x = 1$ is 1/2, but the tangent at every point (except at the point of discontinuity $x = 0$) has slope 0.

17. With these values, $f(b) = f(2) = 4$ and $f(a) = f(1) = 1$, so

$$g(x) = x^2 - 3(x - 1)$$

Then $g(1) = g(2) = 1$. Therefore, there must be some value c between $x = 1$ and $x = 2$ where $g'(c) = 0$. But $g'(x) = 2x - 3 = f'(x) - 3$. The point where $g'(x) = 0$ is then a point where $f'(x) = 3$. This is the point guaranteed by the Mean Value Theorem, because the slope of the secant connecting $x = 1$ and $x = 2$ is 3.

19. The price of gasoline does not change continuously and therefore need not take on all intermediate values.

21. The Intermediate Value Theorem guarantees this crossing. It is possible that it crosses the larger value, but it need not.

23. A solution is within $[2.25, 2.3125]$.

25. A solution is within $[0.5625, 0.625]$.

27. From the fact that

$$f'(x) = (\arcsin x)' + (\arccos x)' = \frac{1}{\sqrt{1-x^2}} - \frac{1}{\sqrt{1-x^2}} = 0$$

we conclude that $f(x) = C$ for some real number C. Since

$$f(0) = \arcsin 0 + \arccos 0 = 0 + \frac{\pi}{2} = \frac{\pi}{2}$$

it follows that $C = \pi/2$. Therefore,

$$f(x) = \arcsin x + \arccos x = \frac{\pi}{2}$$

for all x in $(-1, 1)$.

29. $c_{t+1} > c_t$ when $c_t = 0$, and $c_{t+1} < c_t$ when $c_t = \gamma$. Because this discrete-time dynamical system is continuous, there must be a crossing in between.

31. Any point where $f(c) = c$ is an equilibrium, where $g(c) = f(c) - c = 0$. Because $f(0) \geq 0$, $g(0) \geq 0$. If $f(0) = 0$, this is our equilibrium, so we can assume $f(0) > 0$ and $g(0) > 0$. Because $f(1) \leq 1$, $g(1) \leq 0$. If $f(1) = 1$, this is our equilibrium, so we can assume $f(1) < 1$ and $g(1) < 0$. By the Intermediate Value Theorem, $g(c)$ must cross 0, and there must be an equilibrium.

33. Again, we plot the times on her watch during the two trips and observe that the difference in times must switch from positive at $t = 10$ to negative at $t = 11$.

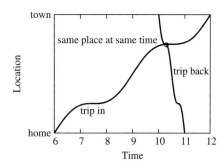

35. The average rate of increase is the slope of the secant connecting age 0 and age 14, which has a slope of 4.0 kg/year. Therefore, the Mean Value Theorem guarantees such a point.

37.

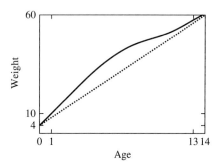

39. The Mean Value Theorem guarantees that the speed at some instant is equal to the average speed of 60 km/h. The Intermediate Value Theorem guarantees that the speed must hit *every* value between 40 and 100.

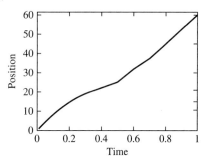

Section 6.4

1. $f_0(x) = 1$; $f_\infty(x) = x$

3. $h_0(z) = e^z$; $h_\infty(z) = e^z$

5. $m_0(a) = \dfrac{1}{a}$; $m_\infty(a) = 30a^2$

7. The exponential function e^{2x} approaches infinity faster. By L'Hôpital's rule,
$$\lim_{x \to \infty} \frac{e^{2x}}{x^2} = \lim_{x \to \infty} \frac{2e^{2x}}{2x} = \lim_{x \to \infty} \frac{4e^{2x}}{2} = \infty$$

9. The power function $0.1x^{0.5}$ approaches infinity faster. By L'Hôpital's rule,
$$\lim_{x \to \infty} \frac{0.1x^{0.5}}{30 \ln x} = \lim_{x \to \infty} \frac{0.05x^{-0.5}}{30x^{-1}} = \lim_{x \to \infty} \frac{1}{600} x^{0.5} = \infty$$

11. The exponential function e^{-2x} approaches zero faster. L'Hôpital's rule does not really make things simpler directly, but
$$\lim_{x \to \infty} \frac{e^{-2x}}{x^{-2}} = \lim_{x \to \infty} \frac{x^2}{e^{2x}} = \lim_{x \to \infty} \frac{2x}{2e^{2x}} = \lim_{x \to \infty} \frac{2}{4e^{2x}} = 0.$$

13. I would guess that the power function approaches infinity faster
$$\lim_{x \to 0} \frac{x^{-1}}{-\ln x} = \lim_{x \to 0} \frac{-x^{-2}}{-x^{-1}} = \lim_{x \to 0} x^{-1} = \infty$$
Therefore, $\lim_{x \to 0} \dfrac{1}{x \ln x} = \infty$, and $\lim_{x \to 0} x \ln x = 0$.

15. The power function with the larger power, x^3, approaches 0 more quickly:
$$\lim_{x \to 0} \frac{x^3}{x^2} = \lim_{x \to 0} \frac{3x^2}{2x} = \lim_{x \to 0} \frac{6x}{2} = 0$$

17. 7

19. $\dfrac{1023}{7}$

21. 1

23. 1

25. Does not exist

27. 0

29. 0

31. $\dfrac{1}{24}$

33. 0

35. 0

37. 1

39. 1

41. The numerator has only one term, so the leading behaviour is c^2 at both 0 and ∞. The denominator has leading behaviour 1 for c near 0 and $2c$ for c large. Therefore,
$$\alpha_0(c) = \frac{c^2}{1} = c^2$$
$$\alpha_\infty(c) = \frac{c^2}{2c} = \frac{c}{2}$$
$$\lim_{c \to 0} \alpha(c) = 0$$
$$\lim_{c \to \infty} \alpha(c) = \infty$$

L'Hôpital's rule is not appropriate at $c = 0$, because the denominator approaches 1. This limit can be found by plugging in. As $c \to \infty$, both the numerator and the denominator approach infinity, so we can use L'Hôpital's rule to check:
$$\lim_{c \to \infty} \frac{c^2}{1 + 2c} = \lim_{c \to \infty} \frac{2c}{2} = \infty$$

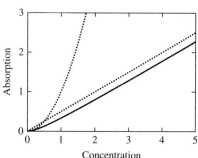

43. The numerator has leading behaviour 1 near 0 and c for c large. The denominator has leading behaviour 1 for c near 0 and c^2 for c large. Therefore,

$$\alpha_0(c) = \frac{1}{1} = 1$$
$$\alpha_\infty(c) = \frac{c}{c^2} = \frac{1}{c}$$
$$\lim_{c \to 0} \alpha(c) = 1$$
$$\lim_{c \to \infty} \alpha(c) = 0$$

L'Hôpital's rule is not appropriate at $c = 0$, because both the numerator and denominator approach 1. This limit can be found by plugging in. As $c \to \infty$, both the numerator and denominator approach infinity, so we can use L'Hôpital's rule to check:

$$\lim_{c \to \infty} \frac{1+c}{1+c+c^2} = \lim_{c \to \infty} \frac{1}{1+2c} = 0$$

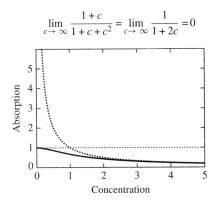

45. Neither the numerator nor the denominator can be simplified for c near 0. The leading behaviour of the numerator is e^c and the leading behaviour of the denominator is e^{2c} for c large. Therefore,

$$\alpha_0(c) = \alpha(c)$$
$$\alpha_\infty(c) = \frac{e^c}{e^{2c}} = e^{-c}$$
$$\lim_{c \to 0} \alpha(c) = 1$$
$$\lim_{c \to \infty} \alpha(c) = 0$$

L'Hôpital's rule is not appropriate at $c = 0$, because both the numerator and denominator approach 2. This limit can be found by plugging in. As $c \to \infty$, both the numerator and denominator approach infinity, so we can use L'Hôpital's rule to check:

$$\lim_{c \to \infty} \frac{e^c + 1}{e^{2c} + 1} = \lim_{c \to \infty} \frac{e^c}{e^{2c}} \lim_{c \to \infty} e^{-c} = 0$$

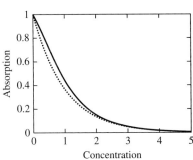

47. The tangent line to $\ln(1+x)$ near $x=0$ is x. The tangent line to e^{2x} near $x = 0$ is $1 + 2x$. Therefore, $e^{2x} - 1 \approx 2x$. For small x, $f(x) \approx \frac{x}{2x} =$ 1/2. With L'Hôpital's rule, we find

$$\lim_{x \to 0} \frac{\ln(1+x)}{(e^{2x}-1)} = \lim_{x \to 0} \frac{1/(1+x)}{2e^{2x}} = \frac{1}{2}$$

49. The tangent line to $\cos x + 1$ near $x = \pi$ is 0. The tangent line to $\sin x$ near $x = \pi$ is $-(x - \pi)$. For small x, $f(x) \approx \frac{0}{-(x-\pi)} = 0$. With L'Hôpital's rule,

$$\lim_{x \to \pi} \frac{\cos x + 1}{\sin x} = \lim_{x \to \pi} \frac{-\sin x}{\cos x} = 0$$

51. This function acts like $c/5$ for small c and like 1 for large c.

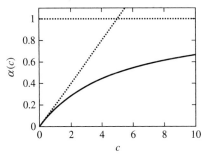

53. This function acts like $5c$ for small c and like 0 for large c.

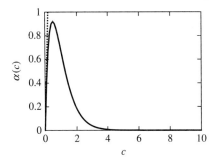

55. The first term decreases to zero more slowly for c near zero, to $\alpha_0(c) = 5c$. The second term grows more quickly for large c, so $\alpha_\infty(c) = 5c^2$.

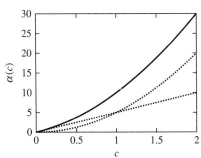

57. $g_3(x)_0 = \dfrac{x^3}{10}$; $g_3(x)_\infty = 1$

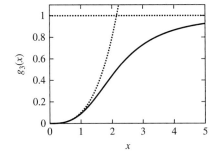

59. $g_{10}(x)_0 = 10x^{10}$; $g_{10}(x)_\infty = 1$

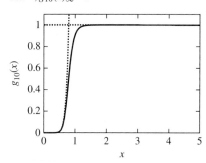

61. a. Using the general solution of the bacterial discrete-time dynamical system we find $a_t = 10^4 \cdot 1.5^t$, $b_t = 10^6 \cdot 2.0^t$.

b. $p_t = \dfrac{10^4 \cdot 1.5^t}{10^4 \cdot 1.5^t + 10^6 \cdot 2.0^t}$.

c. In exponential notation, we can rewrite the denominator as $10^4 e^{\ln(1.5)t} + 10^6 e^{\ln(2.0)t}$, with leading behaviour $10^6 e^{\ln(2.0)t}$ because the parameter in the exponent is larger. Therefore, $\lim_{t \to 0} p_t = 0$.

d. $p_0 = 0.01$, $p_{10} \approx 0.0005$, $p_{20} \approx 0.00003$, $p_{50} \approx 5.6 \cdot 10^{-9}$. This is very close to the limit.

63. a. Using the general solution of the bacterial discrete-time dynamical system, we find $a_t = 10^4 \cdot 0.5^t$, and $b_t = 10^5 \cdot 0.3^t$.

b. $p_t = \dfrac{10^4 \cdot 0.5^t}{10^4 \cdot 0.5^t + 10^5 \cdot 0.3^t}$.

c. In exponential notation, the denominator is $10^4 e^{\ln(0.5)t} + 10^5 e^{\ln(0.3)t} \approx 10^4 e^{-0.693t} + 10^5 e^{-1.204t}$. The leading behaviour is $10^4 e^{-0.693t}$ because the parameter in the exponent is less negative. Therefore, $\lim_{t \to 0} p_t = 1$.

d. $p_0 = 0.09$, $p_{10} \approx 0.94$, $p_{20} \approx 0.9996$, $p_{50} \approx 0.999999999$. This is incredibly close to the limit.

65.
$$\alpha'(c) = A \dfrac{(k + r(c))r'(c) - r(c)r'(c)}{(k + r(c))^2}$$
$$= A \dfrac{kr'(c)}{(k + r(c))^2} > 0$$

If $r(c) \to \infty$, the limit is indeterminate. The derivatives of the numerator and denominator of $\alpha(c)$ are both equal to $r'(c)$, so the limit of $\alpha(c)$ is 1. Because $r(0) = 0$, the leading behaviour of the denominator is k near 0 and the function acts like $r(c)/k$ for c near 0. It acts like α for large values of c.

Section 6.5

1. $f(x)$ is a polynomial, hence continuous for all real numbers; it has no asymptotes; y-intercept $f(0) = 4$; $f'(x) = 3x^2 - 4x + 4$; increasing on $(-\infty, -2/3)$ and $(2, \infty)$; decreasing on $(-2/3, 2)$; relative maximum $f(-2/3) = 148/27 \approx 5.48$; relative minimum $f(2) = -4$; $f''(x) = 6x - 4$; concave up on $(2/3, \infty)$; concave down on $(-\infty, 2/3)$.

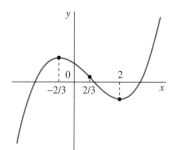

3. $f(x)$ is a polynomial, hence continuous for all real numbers; it has no asymptotes; y-intercept $f(0) = -4$; x-intercepts $x = -2$ and $x = \pm 1$; $f'(x) = 2(x + 2)(2x^2 + 2x - 1)$; increasing on $\left(-2, \left(-\sqrt{3} - 1\right)/2\right)$ and $\left(\left(\sqrt{3} - 1\right)/2, \infty\right)$; decreasing on $(-\infty, -2)$ and $\left(\left(-\sqrt{3} - 1\right)/2, \left(\sqrt{3} - 1\right)/2\right)$; relative maximum $f\left(\left(-\sqrt{3} - 1\right)/2\right) \approx 0.348$; relative minima $f(-2) = 0$ and $f\left(\left(\sqrt{3} - 1\right)/2\right) \approx -4.848$; $f''(x) = 12x^2 + 24x + 6 = 0$ for $x_1 = -1 - \sqrt{2}/2$ and $x_2 = -1 + \sqrt{2}/2$; concave up on $(-\infty, x_1)$ and (x_2, ∞); concave down on (x_1, x_2).

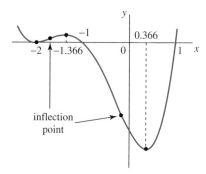

5. Domain: $x \neq -4$; horizontal asymptote $y = 3$; vertical asymptote $x = -4$; y-intercept $f(0) = 0$; x-intercept $x = 0$; $f'(x) = \dfrac{12}{(x+4)^2}$; no critical points; $f(x)$ is increasing on $(-\infty, -4)$ and $(-4, \infty)$; no relative extreme values; $f''(x) = -\dfrac{24}{(x+4)^3}$; concave up on $(-\infty, -4)$; concave down on $(-4, \infty)$; no inflection points.

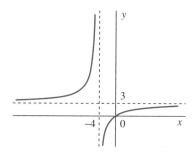

7. Domain: $x \neq \pm 4$; $y = 0$ is a horizontal asymptote; vertical asymptotes are $x = \pm 4$; y-intercept $f(0) = -1/16$; no x-intercepts; from $f(-x) = \dfrac{1}{(-x)^2 - 16} = \dfrac{1}{x^2 - 16} = f(x)$, we conclude that the graph of $f(x)$ is symmetric with respect to y-axis; $f'(x) = -2x/(x^2 - 16)^2$; increasing on $(-\infty, -4)$ and $(-4, 0)$; decreasing on $(0, 4)$ and $(4, \infty)$; relative maximum: $f(0) = -1/16$; relative minimum: none; $f''(x) = \dfrac{6x^2 + 32}{(x^2 - 16)^3}$; concave up on $(-\infty, -4)$ and $(4, \infty)$; concave down on $(-4, 4)$; no inflection points.

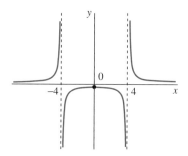

9. Domain: all real numbers; note that $\alpha(c) \geq 0$ for all c; $\alpha = 1$ is a horizontal asymptote; no vertical asymptotes; intercept $\alpha(0) = 0$; because $\alpha(-c) = \alpha(c)$, the graph is symmetric with respect to the vertical axis; $\alpha'(c) = \dfrac{4c}{(c^2 + 2)^2}$; increasing on $(0, \infty)$ and decreasing on $(-\infty, 0)$; relative maximum: none; relative minimum: $\alpha(0) = 0$; $\alpha''(c) = -\dfrac{12c^2 - 8}{(c^2 + 2)^3}$; concave up on $\left(-\sqrt{6}/3, \sqrt{6}/3\right)$; concave down on $\left(-\infty, -\sqrt{6}/3\right)$ and $\left(\sqrt{6}/3, \infty\right)$; inflection points at $c = \pm\sqrt{6}/3$.

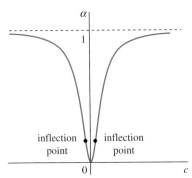

11. Domain $[0, 2\pi]$; no asymptotes; y-intercept $f(0) = 0$; x-intercepts $x = 0, \pi, 2\pi$; no asymptotes; $f'(x) = e^{-x}(\cos x - \sin x)$; critical points $x = \pi/4$ and $x = 5\pi/4$; increasing on $(0, \pi/4)$ and $(5\pi/4, 2\pi)$; decreasing on $(\pi/4, 5\pi/4)$; relative maximum $f(\pi/4) = e^{-\pi/4}/\sqrt{2} \approx 0.3223$; relative minimum $f(5\pi/4) = -e^{-5\pi/4}/\sqrt{2} \approx -0.0139$; $f''(x) = -2e^{-x}\cos x$; zeros $x = \pi/2$ and $x = 3\pi/2$; concave up on $(\pi/2, 3\pi/2)$; concave down on $(0, \pi/2)$ and $(3\pi/2, 2\pi)$.

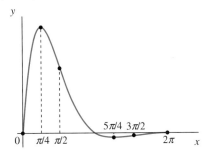

13. Domain: all real numbers; range: $f(x) \geq 0$; $y = 0$ is a horizontal asymptote at the left end; no vertical asymptotes; intercepts: $f(0) = 0$; $f'(x) = xe^x(2 + x)$; critical points: $x = 0$ and $x = -2$; increasing on $(-\infty, -2)$ and $(0, \infty)$; decreasing on $(-2, 0)$; relative maximum: $f(-2) = 4e^{-2} \approx 0.5413$; relative minimum: $f(0) = 0$; $f''(x) = e^x(x^2 + 4x + 2) = 0$ when $x = -2 \pm \sqrt{2}$; concave up on $\left(-\infty, -2 - \sqrt{2}\right)$ and $\left(-2 + \sqrt{2}, \infty\right)$; concave down on $\left(-2 - \sqrt{2}, -2 + \sqrt{2}\right)$.

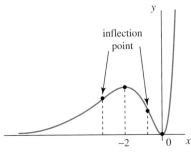

15. Domain: all real numbers; $y = 0$ is a horizontal asymptote at the left end; no vertical asymptotes; intercepts: $s(0) = 0$; $s'(x) = e^x(3e^{2x} - 1)$; critical point: $x = -\frac{1}{2}\ln 3 \approx -0.5493$; $s(x)$ is increasing on $(-\ln 3/2, \infty)$ and decreasing on $(-\infty, -\ln 3/2)$; relative minimum: $f(-\ln 3/2) \approx -0.3849$; relative maximum: none; $f''(x) = e^x(9e^{2x} - 1)$; the second derivative is zero when $x = -\ln 3$; $s(x)$ is concave up on $(-\ln 3, \infty)$ and concave down on $(-\infty, -\ln 3)$; inflection point at $x = -\ln 3 \approx 1.0986$.

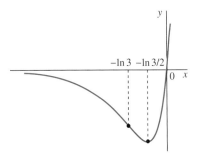

17. Domain: $x > 0$; $y = 0$ is a horizontal asymptote; $x = 0$ is a vertical asymptote; no y-intercepts; x-intercept occurs when $\ln x = 0$, i.e., when $x = 1$; $f'(x) = \frac{1 - \ln x}{x^2}$; critical point: $x = e$; $f(x)$ is increasing on $(0, e)$ and decreasing on (e, ∞); relative minimum: none; relative maximum: $f(e) = \ln e/e = 1/e$; $f''(x) = \frac{-3 + 2\ln x}{x^3}$; $f''(x) = 0$ implies that $x = e^{3/2}$; $f(x)$ is concave up on $(e^{3/2}, \infty)$ and concave down on $(0, e^{3/2})$; inflection point at $x = e^{3/2} \approx 4.4817$.

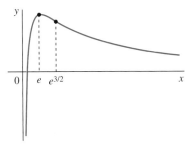

19. Domain: all real numbers; because $-1 \leq \sin x, \cos x \leq 1$ it follows that $-2 \leq f(x) \leq 2$; no asymptotes; y-intercept: $f(0) = 1$; x-intercepts: $x = 3\pi/4 + \pi k$; the function is periodic with period 2π; $f'(x) = \cos x - \sin x$; Critical points: $x = \pi/4 + 2\pi k$ and $x = 5\pi/4 + 2\pi k$; increasing on $(0 + 2\pi k, \pi/4 + 2\pi k)$ and $(5\pi/4 + 2\pi k, 2\pi + 2\pi k)$; decreasing on $(\pi/4 + 2\pi k, 5\pi/4 + 2\pi k)$; relative minimum: $f(5\pi/4 + 2\pi k) = -\sqrt{2}$; relative maximum: $f(\pi/4 + 2\pi k) = \sqrt{2}$; $f''(x) = -\sin x - \cos x$; the zeros of the second derivative are the same as the x-intercepts: $x = 3\pi/4 + 2\pi k$ and $x = 7\pi/4 + 2\pi k$; concave up on $(3\pi/4 + 2\pi k, 7\pi/4 + 2\pi k)$; concave down on $(0 + 2\pi k, 3\pi/4 + 2\pi k)$ and $(7\pi/4 + 2\pi k, 2\pi + 2\pi k)$.

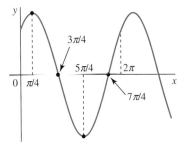

21. Given $f(x) = e^{-x^2}$, to obtain $s(x)$ we replace x by $\left(1/\sqrt{2}\right)x \approx 0.707x$ and then multiply the function by $1/\sqrt{2\pi} \approx 0.399$. Thus, we expand the graph of $f(x)$ horizontally by a factor of $\sqrt{2}$ and then scale it vertically by a factor of $1/\sqrt{2\pi}$. The maximum of $s(x)$ is the scaled maximum of $f(x)$:

$$s(0) = \frac{1}{\sqrt{2\pi}}f(0) = \frac{1}{\sqrt{2\pi}} \cdot 1 = \frac{1}{\sqrt{2\pi}}$$

The inflection points move farther away from the y-axis: from $x = -1/\sqrt{2}$ for $f(x)$ to $x = -1$ for $s(x)$ and from $x = 1/\sqrt{2}$ for $f(x)$ to $x = 1$ for $s(x)$.

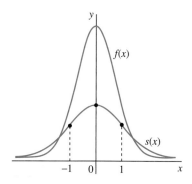

23. The inflection points approach the point $(1, 1/2)$. The inflection point of the nth Hill function is located at $x = \sqrt[n]{\frac{n-1}{n+1}} = \left(\frac{n-1}{n+1}\right)^{1/n}$. To compute the limit as $n \to \infty$ use L'Hôpital's rule.

25. Domain (given): $t \geq 0$; $P = 10$ is a horizontal asymptote; there are no vertical asymptotes; intercept: $f(0) = 20$; $P'(t) = -\frac{10e^{-2t}}{(1-0.5e^{-2t})^2}$; $P(t)$ has no critical points; because $P'(t) < 0$ for all t, $P(t)$ is a decreasing function for all t, so it has no relative extreme values; $P''(t) = \frac{20e^{-2t}(1+0.5e^{-2t})}{(1-0.5e^{-2t})^3}$ is positive for all t; $P(t)$ is concave up for all t, and there are no points of inflection.

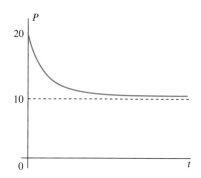

27. We find $P'(t) = kCL\frac{e^{-kt}}{(1+Ce^{-kt})^2}$ and $P''(t) = -k^2 CL\frac{e^{-kt}(1-Ce^{-kt})}{(1+Ce^{-kt})^3}$. $P''(t) = 0$ implies that $e^{-kt} = \frac{1}{C}$, which has no solutions when $C < 0$.

29. Domain (given): $x \geq 0$; range: $P(x) \geq 0$; $P = 0$ is a horizontal asymptote; no vertical asymptotes; intercepts: $f(0) = 0$; $P'(x) = e^{-x} + xe^{-x}(-1) = e^{-x}(1-x)$; There is one critical point, $x = 1$; $P(x)$ is decreasing on $(1, \infty)$ and increasing on $(0, 1)$; relative maximum: $P(1) = e^{-1} \approx 0.367$; no other extreme values; $P''(x) = e^{-x}(-2+x) = 0$ when $x = 2$; $P(x)$ is concave up on $(2, \infty)$ and concave down on $(0, 2)$; the only inflection point is $x = 2$.

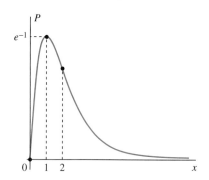

Section 6.6

1.

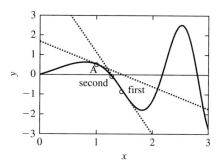

3.

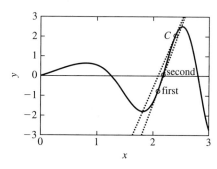

5. Set $f(x) = x^2 - 5x + 1$. Then $f(0) = 1$ and $f(1) = -3$, so there must be a solution in between by the Intermediate Value Theorem. If we start with a guess of $x = 0$, we find $f'(0) = -5$, so the tangent line is $L(x) = 1 - 5x$, which intersects the horizontal axis at $x = 0.2$. The Newton's method discrete-time dynamical system is

$$x_{t+1} = x_t - \frac{f(x_t)}{f'(x_t)} = x_t - \frac{x_t^2 - 5x_t + 1}{2x_t - 5}$$

If $x_0 = 0$, then $x_1 = 0.2$, $x_2 = 0.2086956522$, and $x_3 = 0.2087121525$. The exact answer to 10 decimal places is 0.2087121525.

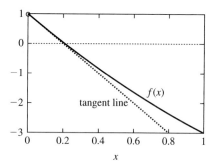

7. Suppose the initial guess is $x_0 = 1$. Because $h'(x) = 0.5e^{x/2} - 1$, $h(1) = -0351$, and $h'(3) = -0.1756$, the tangent line is $L(x) = -0.351 - 0.1756(x - 1)$, which intersects the horizontal axis at $x = -1$. The Newton's method discrete-time dynamical system is

$$x_{t+1} = x_t - \frac{h(x_t)}{h'(x_t)} = x_t - \frac{e^{x_t/2} - x_t - 1}{0.5e^{x_t/2} - 1}$$

If $x_0 = 1$, then $x_1 = -100$, $x_2 = -0.1294668027$, $x_3 = -0.0037771286$. This seems to be approaching 0. If we start from $x_0 = 2$, we get $x_1 = 2.7844$, $x_2 = 2.5479$, and $x_3 = 2.51355$. After many steps, the answer converges to 2.512862414.

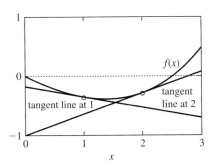

9. It is true that $x = e^x - 2$ if $e^x = x + 2$. But starting from $x_0 = 1$, solutions are $x_1 = 0.718$, $x_2 = 0.051$, and $x_3 = -0.947$. After a while, it seems to converge to another equilibrium at -1.84.

11. It is true that $e^{x/2} = x + 1$, if $h(x) = 0$. Starting from a guess $x_0 = 2$, we find $x_1 = 1.718$, $x_2 = 1.361$, and $x_3 = 0.975$. This seems to be going very slowly to $x = 0$.

13. It should converge most rapidly when the equilibrium is superstable, or the slope is 0. The slope at the equilibrium is $2 - r$, so the most rapid convergence should be when $r = 2$. Starting from $x_0 = 0.75$, we get $x_1 = 0.375$, $x_2 = 0.46875$, $x_3 = 0.4980468750$, and $x_4 = 0.4999923706$.

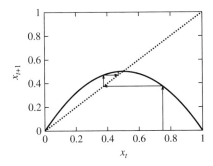

15. The method fails if we start at one of the critical points of the function, which occur where $x = \pm 0.577$. All values below the lower critical point converge to the negative solution.

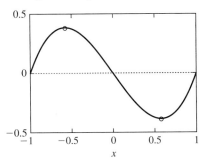

17. The Newton's method discrete-time dynamical system is

$$x_{t+1} = x_t - \frac{x_t^2}{2x_t} = \frac{x_t}{2}$$

This converges to 0 rather slowly, because the derivative of the function x^2 is 0 at the solution.

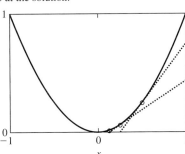

19. We are asked to approximate the solution of $f'(x) = -\frac{x}{1+x} + \ln\frac{3}{1+x} = 0$ in the interval $[0, 1]$. The Newton's method iteration formula reads

$$x_{n+1} = x_n - \frac{f'(x_n)}{f''(x_n)} = x_n + \frac{-\frac{x_n}{1+x_n} + \ln\frac{3}{1+x_n}}{\frac{2+x_n}{(1+x_n)^2}}$$

where $x_0 = 1$. The first three iterations are shown in the table below.

Iteration	Approximation
x_0	1
x_1	0.873953
x_2	0.879077
x_3	0.879087

21. $f'(x) \approx e^{x+1} - (x+1) - 2 - (e^x - x - 2) = e^{x+1} - 1 - e^x$. The approximate Newton's method discrete-time dynamical system is

$$x_{t+1} = x_t - \frac{e^{x_t} - x_t - 2}{e^{x_t+1} - 1 - e^{x_t}}$$

Starting from an initial guess of $x_0 = 10$, I got $x_1 \approx 1.0767$, $x_2 \approx 1.1118$, $x_3 \approx 1.1288$, $x_4 \approx 1.1373$, $x_5 \approx 1.1419$. After 5 steps, it matches the first couple of decimal places in the exact answer, to 10 decimal places, of 1.146193221.

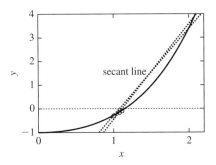

23. The discrete-time dynamical system is

$$x_{t+1} = x_t - \frac{f(x_t)(x_t - x_{t-1})}{f(x_t) - f(x_{t-1})}$$

which depends on both x_t and x_{t-1}, unlike an ordinary discrete-time dynamical system. Starting from $x_0 = 10$ and

$$x_1 = x_0 - \frac{e^{x_0} - x_0 - 2}{e^{x_0+1} - 1 - e^{x_0}}$$

from Exercise 21, we find $x_1 = 1.076746253$, $x_2 = 1.154339800$, $x_3 = 1.145768210$, $x_4 = 1.146190691$, and $x_5 = 1.146193221$, which is correct to 9 decimal places. This method is better because it uses a more accurate version of the secant line approximation.

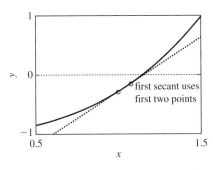

25. This function differs in that it is concave up for small t. The nectar comes out slowly when the bee first arrives. The derivative is

$F'(t) = \dfrac{2t}{(1+t^2)^2}$, so the Marginal Value Theorem equation is

$$F'(t) = \dfrac{F(t)}{t}$$
$$\dfrac{2t}{(1+t^2)^2} = \dfrac{t}{1+t^2}$$
$$\dfrac{2}{1+t^2} = 1$$
$$1+t^2 = 2$$
$$t^2 = 1$$
$$t = 1$$

It turned out this could be solved algebraically.

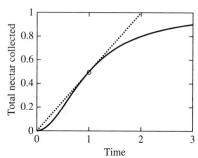

27. $N^* = \ln\left(\dfrac{2.5}{1+h}\right)$. Therefore, $P(h) = hN^* = h\ln\left(\dfrac{2.5}{1+h}\right)$. Then

$$P'(h) = \ln\left(\dfrac{2.5}{1+h}\right) - \dfrac{h}{1+h} = 0$$

We can use $P(h)$ and $P'(h)$ in Newton's method, finding the discrete-time dynamical system

$$h_{t+1} = h_t - \dfrac{P(h_t)}{P'(h_t)}$$

we used an initial guess of $h_0 = 0.75$ because this is the solution we found when the fish followed the logistic model. After five steps, it converges to 1.5.

29. Let G be the updating function. Then $G(0) = 1$ and $G(5) = 2.516$. There must be an equilibrium in between. Starting from $M_0 = 2$, get to 1.726 after about 6 steps. Solving the equation $0.5e^{-.01x}x + 10 - x = 0$ with Newton's method, it reaches 1.726 after 2 steps.

31. We need to solve the equation $100e^{01t} = 400 + 100t$ or $100e^{01t} - 400 - 100t = 0$. The Newton's method discrete-time dynamical system is

$$x_{t+1} = x_t - \dfrac{100e^{0.1x_t} - 400 - 100x_t}{10e^{0.1x_t} - 100}$$

If food resources were constant at the initial value of 400, the population would run out of food when $b(t) = 400$, or at time $10\ln 4 \approx 138$. Using an initial guess of $x_0 = 138$, the solution shoots off to a negative value. Using an initial guess of $x_0 = 30$ instead we find the solution as $t = 37.18$ after three steps.

33. We will start with the guess $c_0 = 50$, because the equilibrium is between 0 and γ. We need to solve the equation $c^* = 0.75e^{-c^*}c^* + 1.25$ or $f(c) = 0$, where $f(c) = c - 0.75e^{-c^*}c^* - 1.25$. Then $f'(c) = 1 + 0.75e^{-c}c - 0.75e^{-c}$, and the Newton's method discrete-time dynamical system is

$$c_{t+1} = c_t - \dfrac{c_t - 0.75e^{-c_t}c_t - 1.25}{1 + 0.75e^{-c_t}c_t - 0.75e^{-c_t}}$$

With $c_0 = 5.0$, $c_1 = 1.349066687$, $c_2 = 1.502145217$, and $c_3 = 1.500942584$.

Section 6.7

1. The updating function crosses from above to below and is stable.

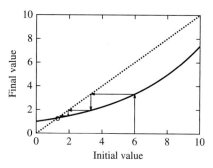

3. At the lower equilibrium the updating function crosses the diagonal from below to above and is therefore unstable. At the upper equilibrium, the updating function crosses from above to below and is stable.

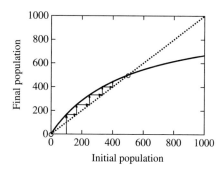

5. The equilibrium satisfies $c^* = 0.5c^* + 8.0$ or $c^* = 16.0$. The slope of the updating function $f(c) = 0.5c + 8.0$ is $f'(c) = 0.5 < 1$. The equilibrium is stable.

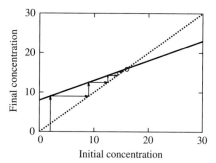

7. The equilibrium is $b^* = 0$. The slope of the updating function $f(b) = 0.3b$ is $f'(b) = 0.3 < 1$. The equilibrium is stable.

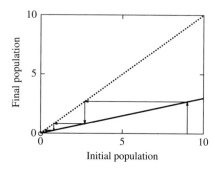

9. $f(x) = x$ when $x^2 = x$, or $x^2 - x = 0$, or $x(x-1) = 0$, which has solutions at $x = 0$ and $x = 1$. The derivative is $f'(x) = 2x$, so $f'(0) = 0 < 1$ (stable) and $f'(1) = 2 > 1$ (unstable).

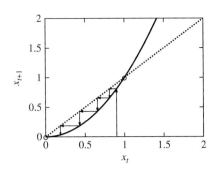

11. $x^* = \dfrac{x^*}{1+x^*}$ has solution $x^* = 0$. Also, if $f(x) = \dfrac{x}{1+x}$, then $f'(x) = \dfrac{1}{(1+x)^2}$, so $f'(0) = 1$. We can not tell if this is stable or not using the Slope Criterion for stability. However, the graph of the updating function lies below the diagonal for all $x > 0$, meaning that the equilibrium is stable.

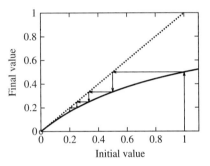

13.

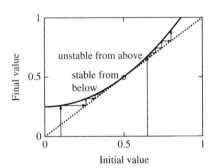

15. The second derivative is 0 at the equilibrium.

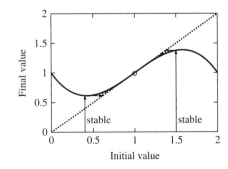

17.

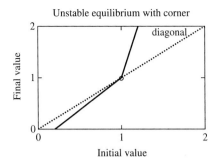

19. There is no equilibrium, and the cobwebbing creeps slowly past the point where the equilibrium used to be.

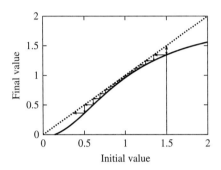

21. The inverse is $f^{-1}(x) = \dfrac{x}{1-x}$. The only equilibrium is $x = 0$. The slopes of both the original updating function and the inverse are 1 at this point.

23. $f'(p) = \dfrac{2.4}{(1.2p + 2.0(1-p))^2}$; $f'(0) = 0.6$; $f'(1) \approx 1.667$. The equilibrium at $p = 0$ is stable; the equilibrium at $p = 1$ is unstable.

25. $f'(p) = \dfrac{2.25}{(1.5p + 1.5(1-p))^2} = 1$; $f'(0) = f'(1) = 1$. Both derivatives are exactly 1, so we cannot tell. This updating function exactly matches the diagonal, meaning that solutions move neither toward nor away from equilibria.

27. The discrete-time dynamical system is $b_{t+1} = 0.6b_t + 10^6$. The equilibrium satisfies $b^* = 0.6b^* + 10^6$ or $0.4b^* = 10^6$ or $b^* = 2.5 \cdot 10^6$. The derivative of the updating function $f(b) = 0.6b + 10^6$ is $f'(b) = 0.6$, which is always less than 1. The equilibrium is stable.

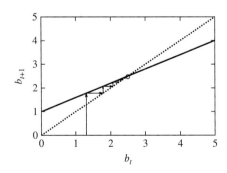

29. The updating function is $b_{t+1} = 1.5b_t - 10^6$, and the equilibrium is $b^* = 2 \cdot 10^6$. The derivative of the updating function is $f'(b) = 1.5 > 1$, so the equilibrium is unstable.

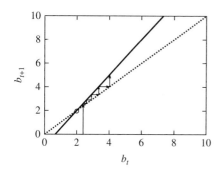

31. The equilibrium concentration of medication is 2.5. The derivative of the updating function is

$$f'(M) = 1 - \frac{0.5}{(1+0.1M)^2}$$

so $f'(2.5) \approx 0.68$. The equilibrium is stable.

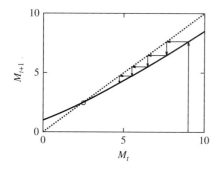

33. The inverse is $c_t = 2(c_{t+1} - 8) = f^{-1}(c_{t+1})$. The equilibrium is $c^* = 16$. The derivative of the backward updating function is $(f^{-1})'(c) = 2$, so the equilibrium is unstable. The same equilibrium was stable in the forward direction.

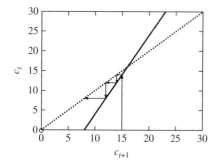

35. The inverse is

$$x_t = \frac{x_{t+1}}{2 - 0.001 x_{t+1}} = f^{-1}(x_{t+1})$$

The equilibria are $x^* = 0$ and $x^* = 1000$. The derivative of the backward updating function is $(f^{-1})'(x) = \frac{2}{(2-0.001x_{t+1})^2}$. Therefore, $(f^{-1})'(0) = 0.5$ and $(f^{-1})'(1000) = 2$. The equilibrium at $x = 0$ is stable and the equilibrium at $x = 1000$ is unstable.

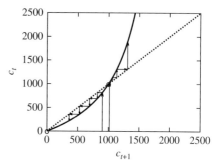

37. $x_{t+1} = \dfrac{x_t^2}{1.0 + x_t^2}$. The equilibria satisfy

$$x = \frac{x^2}{1+x^2}$$
$$x(1+x^2) = x^2$$
$$x(1+x^2 - x) = 0$$
$$x = 0 \text{ or } (1+x^2 - x) = 0$$

Because $1 + x^2 - x = 0$ has no real solution (the quadratic formula gives the square root of a negative number), the only equilibrium is $x = 0$.

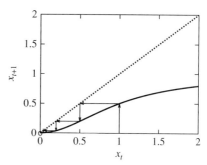

The derivative of the updating function is

$$f'(x) = \frac{2x}{(1+x^2)^2}$$

so $f'(0) = 0$, and the equilibrium is stable. This population is going extinct.

Section 6.8

1. The derivative of the updating function $f(p) = \dfrac{1.5p}{1.5p + 2(1-p)}$ is $f'(p) = \dfrac{3}{(1.5p + 2(1-p))^2}$, so $f'(0) = 0.75$. Because $f(0) = 0$, the tangent line is $L(p) = f(0) + f'(0)p = 0.75p$ and the equilibrium is stable.

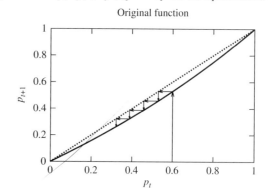

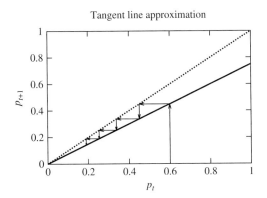

3. The derivative of the updating function $f(x) = 1.5x(1 - x)$ is $f'(x) = 1.5(1 - 2x)$, so $f'(0) = 1.5$. The tangent line is $L(x) = f(0) + f'(0)x = 1.5x$ and the equilibrium is unstable.

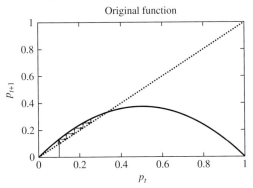

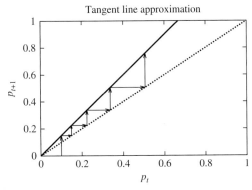

5. The solution is $y_t = 2 \cdot (1.2)^t$ with values 2, 2.4, 2.88, 3.456, 4.147, and 4.977. Solving $y_t = 100$ gives $1.2^t = 50$ or $t = \ln 50 / \ln 1.2 = 21.45$. It would cross 100 at time step 22.

7. The solution is $y_t = 2 \cdot (0.8)^t$ with values 2, 1.6, 1.28, 1.024, 0.819, and 0.655. Solving $y_t = 0.2$ gives $0.8^t = 0.1$ or $t = \ln 0.1 / \ln 0.8 = 10.31$. It would cross 0.2 at time step 11.

9. The equilibrium is $y^* = 1$.

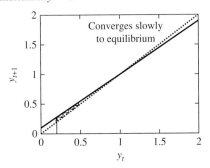

It is stable because the slope m is less than 1.

11. The equilibrium is $y^* = 1$.

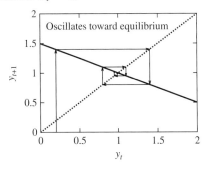

It is stable but oscillatory because $-1 < m < 0$.

13. The slope of the updating function is exactly -1 everywhere. Solutions jump back and forth and are neither stable nor unstable.

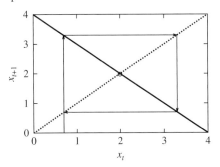

15. $x^* = 3x^*(1 - x^*)$ has solutions $x^* = 0$ and $x^* = 2/3$. The derivative of the updating function $f(x) = 3x(1 - x)$ is $f'(x) = 3(1 - 2x)$, so $f'(0) = 3$ and $f'(2/3) = -1$. The zero equilibrium is unstable, but a solution starting at $x_0 = 0.6$ gets slowly closer to x^*, with $x_2 \approx 0.6048$ and $x_4 \approx 0.6087$. The positive equilibrium is stable.

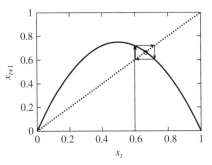

17. $x^* = 0.5$, $f'(x) = 2 - 4x$, and $f'(0.5) = 0$. The equilibrium is highly stable. The solutions in this case move toward the equilibrium very quickly.

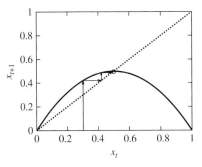

19. The derivative of the updating function $g(M) = M - \dfrac{M^2}{2+M} + 1$ is $g'(M) = \dfrac{4}{(2+M)^2}$, so $g'(2) = 1/4$. The equilibrium is stable and does not oscillate.

21. The derivative of the updating function $g(M) = M - \dfrac{M^5}{16+M^4} + 1$ is $g'(M) = \dfrac{16(3M^4 - 16)}{(16+M^4)^2}$, and $g'(2) = -1/2$. The equilibrium is stable, but solutions oscillate.

23. The slope at the equilibrium is $1 - \ln r$, so it switches sign when $\ln r = 1$ or $r = e$.

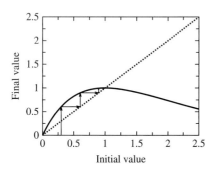

25. We chose the value $r = e^{1.5} \approx 4.482$.

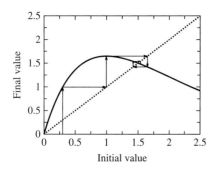

27. a. The discrete-time dynamical system is $b_{t+1} = b_t(0.5 + 0.5b_t)$.

b. The equilibria are $b^* = 0$ and $b^* = 1$. The derivative of the updating function $f(b) = (0.5 + 0.5b)b$ is $f'(b) = 0.5 + b$, so $f'(0) = 0.5$ and $f'(1) = 1.5$. The equilibrium $b = 0$ is stable, and $b = 1$ is unstable.

c.

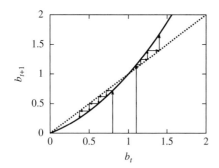

d. Populations starting below 1 die out, while those starting above 1 blast off to infinity. Members of this species do well with a little help from their friends.

29. a.

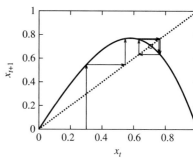

b. The equilibria are $x = 0$ and $x^* = \sqrt{1 - \dfrac{1}{r}}$ (when $r \geq 1$).

c. $f'(x) = r - 3rx^2 \cdot f'(0) = r$ and $f'(x^*) = 3 - 2r$.

d. The first equilibrium is stable when $r < 1$ as usual, and the positive equilibrium is stable when $r < 2$.

31. a. This line passes through the points (20, 21) and (19, 18). Let z be the setting on the thermostat and T be the temperature. This line has slope
$$m = \dfrac{21 - 18}{20 - 19} = 3$$
Because it passes through the point (20, 21), we can write the equation in point-slope form as $T = 3(z - 20) + 21$.

b. When it is 18°, you set the thermometer to 22°. This results in a temperature of 27°. Setting the thermometer to 13° then results in a temperature of 0°. Things are getting pretty chilly.

c. $z_t = 40 - T_t$

d. $T_{t+1} = 3(z_t - 20) + 21 = 3(40 - T_t - 20) + 21 = 81 - 3T_t$

e. The slope of -3 means that the temperatures will oscillate more and more widely. The system needs a better correction mechanism. I would have it estimate T as a function of z and correct on that basis.

33. The mutants do worse when mutants are common, and the wild type do better. The discrete-time dynamical system is
$$p_{t+1} = \dfrac{a_{t+1}}{a_{t+1} + b_{t+1}} = \dfrac{2(1 - p_t)a_t}{2(1 - p_t)a_t + (1 + p_t)b_t}$$
$$= \dfrac{2(1 - p_t)p_t}{2(1 - p_t)p_t + (1 + p_t)(1 - p_t)} = \dfrac{2p_t}{2p_t + (1 + p_t)}$$

The equilibria are at $p = 0$ and $p = 1/3$. The derivative of the updating function is $f'(p) = \dfrac{2}{(1 + 3p)^2}$, so $f'(0) = 2 > 1$ and $f'(1/3) = 1/2 < 1$.

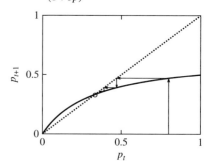

35. a. There are 20 plants; each grows to size $100/20 = 5$ and makes 4 seeds. This gives a total of 80. These 80 plants grow to size $\dfrac{100}{80} = 1.25$, so each makes 0.25 seeds (or one in four plants makes a seed). The total number of seeds is then 20 in the next generation. The values just keep jumping back and forth.

b. There are n_t plants of size $\dfrac{100}{n_t}$. Each of the n_t plants makes $\dfrac{100}{n_t} - 1$ seeds, for a total of $n_{t+1} = 100 - n_t$.

c. $n^* = 50$
d.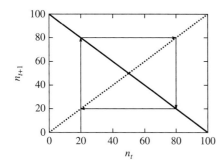

e. The slope is -1, and the equilibrium is neither stable nor unstable.

37. a. There are 20 plants; each makes 3 seeds, or a total of 60. These 60 plants grow to size 1.667, each making a negative number of seeds. This population just went extinct.
b. The discrete-time dynamical system is $n_{t+1} = 100 - 2.0n_t$.
c. There is an equilibrium of $n^* \approx 333$.

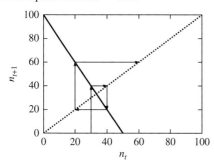

d. The slope is -2, and the equilibrium is unstable and oscillatory.

True/False Quiz

1. TRUE. The leading behaviour of a polynomial at infinity is determined by its highest term.

3. TRUE. For instance, consider $f(x) = \sin x$ on the interval $(0, 5)$: it has an absolute maximum of 1 at $x = \pi/2$ and an absolute minimum of -1 at $x = 3\pi/2$.

5. FALSE. It could happen that $f(x)$ has a corner.

7. FALSE. $f(x)$ needs to be differentiable for this statement to hold. Counterexample: $f(x) = |x|$ is continuous on $[-1, 1]$ and $f(-1) = f(1) = 1$, but there is no point c in $(-1, 1)$ where $f'(c) = 0$.

9. FALSE. The value $m^* = 0$ is an equilibrium of both $m_{t+1} = 0.5m_t$ and its corresponding backward system $m_t = 2m_{t+1}$. However, $m^* = 0$ is stable for $m_{t+1} = 0.5m_t$ and unstable for $m_t = 2m_{t+1}$.

Supplementary Problems

1.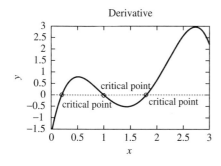

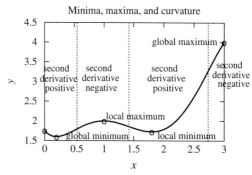

3. $L(x) = 0.5 - 0.25x$; $L(-0.03) = 0.5075$. In this case, $f(-0.03) \approx 0.5068$.

5. $f'(x) = (2 - x^2)e^{-x}$, which is 0 at $x = \sqrt{2}$. Because $f(0) = 0$ and $\lim_{x \to \infty} f(x) = 0$ (the exponential function declines faster than the quadratic function increases), this must be a maximum.

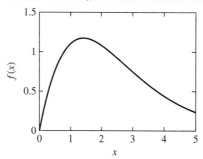

7. $T_2 = \dfrac{1}{2} + \dfrac{1}{4}x - \dfrac{1}{4}x^2$.

9.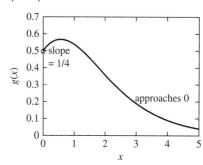

11. a. $S'(t) = -t^3 + 180t^2 - 8000t + 96{,}000$, which is 0 at $t = 20$, $t = 40$, and $t = 120$.
b. Substituting the endpoints ($t = 0$ and $t = 150$) and the critical points into the function $S(t)$, we find a maximum of 576,000 at $t = 120$.
c. $S''(t) = -3t^3 - 360t - 8000$. Then $S''(20) = -2000$, $S''(40) = 1600$, and $S''(120) = -8000$. The first and last are maxima and the middle one is a minimum.
d.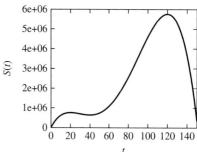

13. a. $1/2$
b. $1/6$
c. $1/24$

d. The limits are of the form $\lim_{x \to 0} \dfrac{e^x - T_n(x)}{x^{n+1}}$, where $T_n(x)$ is the Taylor polynomial of $f(x) = e^x$ near $x = 0$; the answer is the coefficient of x^{n+1} in the Taylor polynomial of $f(x) = e^x$ near $x = 0$.

15. a.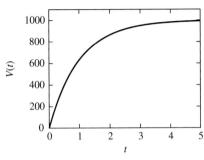

   b. The fraction outside is $1 - H(t) = 1/(1 + e^t)$, so the total volume outside is
   $$\dfrac{1000(1 - e^{-t})}{1 + e^t}$$

   c. Call this function $V_o(t)$:
   $$V'_o(t) = \dfrac{1000(2 + e^{-t} - e^t)}{(1 + e^t)^2}$$

   d. This is a bit tricky. The maximum is the critical point where $V'_o(t) = 0$, or where $2 + e^{-t} - e^t = 0$. Letting $x = e^t$, this is $2 + 1/x - x = 0$. Multiplying both sides by x, we get the quadratic $2x + 1 - x^2 = 0$, which can be solved with the quadratic formula to give $x = 1 + \sqrt{2}$, so that $t = \ln(1 + \sqrt{2}) \approx 0.88$ days.

17. a. The more food there is, the more is eaten, up to a limit of b. The replenishment is enough to refill the table.
   b. Full after 5 min; 25% full after 10 min.
   c. The equilibrium is at $F = 0.5$.
   d. $G(06) = 0.4375$.

   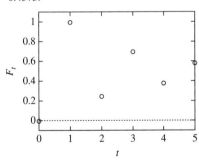

19. a. It looks like $x_0 = 1$ is a good guess.

   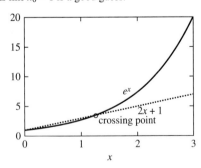

   b. We need to solve $g(x) = e^x - 2x - 1 = 0$, so
   $$x_{t+1} = x_t - \dfrac{g(x_t)}{g'(x_t)} = x_t - \dfrac{e^x - 2}{e^x - 2x - 1}$$

   c. $x_1 = 1.3922$.

d. The derivative of the updating function is
   $$\dfrac{e^x(e^x - 2x - 1)}{(e^x - 2)^2}$$
   which is 0 when $e^x - 2x - 1 = 0$.

21. a. Let $N(t)$ be the net energy gain. Then $N(t) = F(t) - 2t$.
   b. At $t = \sqrt{3/2} - 1 \approx 0.225$.
   c. We need to maximize $\dfrac{N(t)}{1 + t}$, which occurs when $t = 0.2$.
   d. The answer to part (c) is smaller because the bee has other options besides sucking as much nectar as possible out of the flower.

   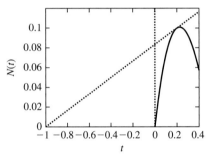

23. a.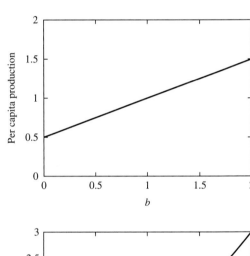

   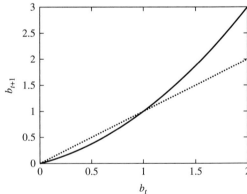

   b. The discrete-time dynamical system is $b_{t+1} = rb_t\left(1 + \dfrac{b_t}{K}\right)$.
   c. Equilibria at $b^* = 0$ and $b^* = 10^6$.
   d. Equilibrium at $b^* = 0$ is stable and equilibrium at $b^* = 10^6$ is unstable.

25. a. The updating function f is $f(x) = 4x^2/(1 + 3x^2)$, multiplying the per capita production by x, the number of individuals.
   b. First, find that 0 is a solution. The rest is a quadratic, which has solutions at $x = 1/3$ and $x = 1$.
   c. We find that $f'(x) = 8x/(1 + 3x^2)^2$. Then $f'(0) = 0$, $f'(1/3) = 3/2$, and $f'(1) = 1/2$. Therefore, the equilibria at 0 and 1 are stable and the one at 1/3 is unstable.

d. The tangent line at 0 is $L(x) = 0$. The tangent line at $x = 1/3$ is $L(x) = 1/3 + 3/2(x - 1/3)$. The tangent line at $x = 1$ is $L(x) = 1 + 1/2(x - 1)$.

e. This dynamical system shoots off to infinity for starting points greater than 1/3, and off to negative infinity for starting points less than 1/3. This is different from the behaviour of the original system, in which such solutions approach $x = 1$ and $x = 0$, respectively.

Chapter 7

Section 7.1

1. Pure-time
3. Pure-time
5. Neither pure-time nor autonomous
7. Autonomous
9. Autonomous
11.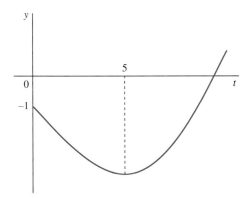
13.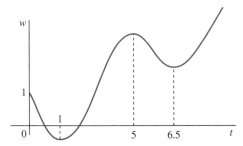
15. a. $y(x) = 0$, $y(x) = 1$, and $y(x) = -1$
    b. $(-1, 0)$ and $(1, \infty)$
    c. $(-\infty, -1)$ and $(0, 1)$
17. Taking the derivative, we find that $\dfrac{dw}{dt} = \dfrac{2}{1+t}$, as required by the differential equation. The initial condition must have been $w(0) = 3$.
19. Taking the derivative, we find $\dfrac{dz}{dt} = 2(t + 2)$. Because $z = (t + 2)^2$, we can rewrite this as $\dfrac{dz}{dt} = 2\sqrt{z}$, as required by the differential equation. The initial condition must have been $z(0) = (0 + 2)^2 = 4$.
21. Both sides are equal to 1.
23. Both sides are equal to $2e^{2x}$.
25. Both sides are equal to $t(t^2 + 3)^{-1/2}$; the initial condition is satisfied.
27. Both sides are equal to $\dfrac{49}{t(1 - 7\ln t)^2}$; the initial condition is satisfied.
29. With $\Delta t = 1$, $w(1) \approx w_1 = 3 + w'(0) \cdot 1 = 3 + 2 \cdot 1 = 5$. With $\Delta t = 0.5$, the first step is $w_1 = 3 + w'(0) \cdot 0.5 = 3 + 2 \cdot 0.5 = 4$. The second step is $w(1) \approx w_2 = 4 + 1.33 \cdot 0.5 = 4.67$. The exact answer, using the solution $w(t) = 2\ln(1 + t) + 3$, is $w(1) = 4.386$.
31. $y(2) \approx y_4 = 1.37976$
33. $c(4) \approx 1.251183$
35. The differential equation is $\dfrac{dV}{dt} = 3$. The function $V(t) = 400 + 3t$ has derivative 3 and satisfies the initial condition $V(0) = 400$. Mathematically, the solution makes sense for all time, but eventually the cell becomes unrealistically large.

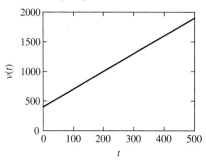

37. The differential equation is $\dfrac{dV}{dt} = -3t^2$. The function $V(t) = 1000 - t^3$ has derivative $-3t^2$ and satisfies the initial condition $V(0) = 1000$. The solution stops making sense after $t = 10$ when the volume becomes negative.

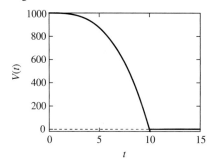

39. a. $P(t) = 1200$
    b. Show that $P'(t) < 0$.
    c. Show that $P'(t) > 0$.
    d. A population that starts with 1200 salmon remains there for all time. If there are more than 1200 salmon, the population increases. If there are less than 1200 salmon, the population decreases. The number 1200 represents the minimum number needed for survival.

41. a.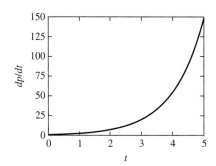

    b. $\dfrac{dp}{dt} = e^t$
    c. The function $p(t) = e^t$ has derivative e^t. Also, $p(0) = 1$, which matches the initial condition.

d.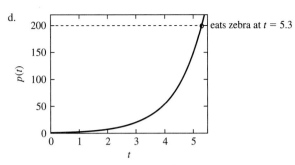

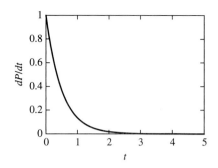

e. We have to find when $p(t) = 200$. $e^t = 200$ when $t = \ln 200 \approx 5.30$.

43. This cell follows the differential equation $\dfrac{dV}{dt} = 3$ with initial condition $V(0) = 400$. The step size is $\Delta t = 5$, so we need to find V_6. We compute

$$\begin{aligned} V_1 &= V(0) + V'(0)\Delta t = 400 + 3 \cdot 5 = 415 \\ V_2 &= V_1 + V'(5)\Delta t = 415 + 3 \cdot 5 = 430 \\ V_3 &= V_2 + V'(10)\Delta t = 430 + 3 \cdot 5 = 445 \\ V_4 &= V_3 + V'(15)\Delta t = 445 + 3 \cdot 5 = 460 \\ V_5 &= V_4 + V'(20)\Delta t = 460 + 3 \cdot 5 = 475 \\ V(30) &\approx V_6 = V_5 + V'(25)\Delta t = 475 + 3 \cdot 5 = 490 \end{aligned}$$

The solution we guessed was $V(t) = 400 + 3t$, and $V(30) = 490$. In this case, Euler's method gives exactly the right answer.

45. This cell follows the differential equation $\dfrac{dV}{dt} = -3t^2$ with initial condition $V(0) = 1000$. The step size is $\Delta t = 2$, so we need to find V_5. We compute

$$\begin{aligned} V_1 &= V(0) + V'(0)\Delta t = 1000 - 3 \cdot 0^2 \cdot 2 = 1000 \\ V_2 &= V_1 + V'(2)\Delta t = 1000 - 3 \cdot 2^2 \cdot 2 = 976 \\ V_3 &= V_2 + V'(4)\Delta t = 976 - 3 \cdot 4^2 \cdot 2 = 880 \\ V_4 &= V_3 + V'(6)\Delta t = 880 - 3 \cdot 6^2 \cdot 2 = 664 \\ V(10) &\approx V_5 = V_4 + V'(8)\Delta t = 664 - 3 \cdot 8^2 \cdot 2 = 280 \end{aligned}$$

The solution we guessed was $V(t) = 1000 - t^3$ and $V(10) = 0$. Euler's method is pretty far off.

47. This cheetah follows the differential equation $\dfrac{dp}{dt} = e^t$ with initial condition $p(0) = 1$. The step size is $\Delta t = 1$, so we need to find p_5. We compute

$$\begin{aligned} p_1 &= p(0) + p'(0)\Delta t = 1 + e^0 \cdot 1 = 2 \\ p_2 &= p_1 + p'(1)\Delta t = 2 + e^1 \cdot 1 \approx 4.72 \\ p_3 &= p_2 + p'(2)\Delta t = 4.72 + e^2 \cdot 1 \approx 12.11 \\ p_4 &= p_3 + p'(3)\Delta t = 12.11 + e^3 \cdot 1 \approx 32.19 \\ p(5) &\approx p_5 = p_4 + p'(4)\Delta t = 32.19 + e^4 \cdot 1 \approx 86.79 \end{aligned}$$

The solution we guessed was $p(t) = e^t$, and $p(5) \approx 148.4$. Euler's method is pretty far off.

49. The derivative of e^{-2t} is $-2e^{-2t}$. We need to multiply by $-1/2$, so the derivative of $-\dfrac{e^{-2t}}{2}$ is e^{-2t}. If $P(t) = -\dfrac{e^{-2t}}{2}$, then $P(0) = -1/2$, which is $1/2$ too low. We need to add $1/2$, getting $P(t) = 1/2 - \dfrac{e^{-2t}}{2}$. The derivative is $\dfrac{dP}{dt} = e^{-2t}$ and $P(0) = 1/2 - 1/2 = 0$. This checks.

51. a. You could measure mass with a balance at any time.
    b. You could measure growth rate but not total mass if you checked how much carbon a growing plant absorbed from the atmosphere.

53. a. You could measure total chemical with a destructive device that separated out the chemical.
    b. You could measure the rate of chemical production but not the total if the reaction produced an easily measured by-product such as heat or carbon dioxide.

Section 7.2

1.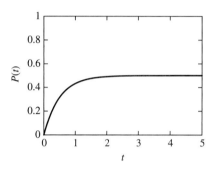

3.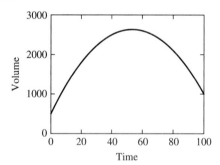

5.

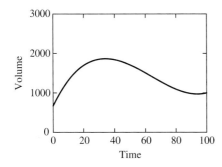

7. $4x - \frac{1}{2}x^2 - x^5 + C$

9. $\frac{1}{16}x^{16} - x^{12} - 6x + C$

11. $\frac{3}{3.4}t^{3.4} + e^t + C$

13. $\ln|x| + 2x^{-1/2} - 2x^{-1} + C$

15. $-\cos t - \frac{4}{7}t^{7/4} + C$

17. $2t - 3\arcsin t + C$

19. $4\sec x - \ln|x| + C$

21. $-\frac{1}{1.1}x^{-1.1} + 2.5x^{0.4} - \frac{3}{3.2}x^{3.2} + C$

23. $\frac{12^x}{\ln 12} + \frac{1}{13}x^{13} + C$

25. Integrating, we find that $V(t) = 2t^3/3 + 5t + c$. Substituting the initial condition, $V(1) = 2/3 + 5 + c = 19$, so $c = 40/3$. The solution is $V(t) = 2t^3/3 + 5t + 40/3$.

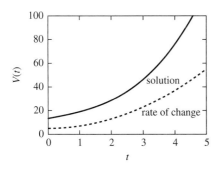

27. Integrating, we find that $f(t) = 1.25t^4 + 2.5t^2 + c$. Substituting the initial condition, $f(0) = c = -12$, so $c = -12$. The solution is $f(t) = 1.25t^4 + 2.5t^2 - 12$.

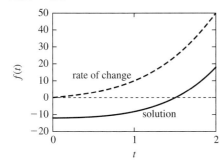

29. $M(t) = \frac{t^3}{3} - \frac{1}{t} + c$. Substituting the initial condition, we get $M(3) = \frac{27}{3} - \frac{1}{3} + c = 10$, so $c = \frac{4}{3}$.

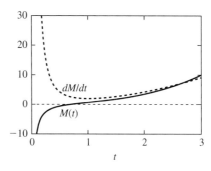

31. $f(x) = \frac{8}{3}x^{3/2} - 4x^{1/2} + \frac{10}{3}$

33. $f(x) = 2\arctan x + \frac{\pi}{2}$

35. $f(\theta) = -4\cos\theta + \sec\theta + 2$

37. Let $F(x) = \int f(x)dx = \frac{x^3}{3} + c$, and $G(x) = \int g(x)dx = \frac{x^4}{4} + c$. The product of the functions is $f(x)g(x) = x^5$, with integral $\int x^5 dx = \frac{x^6}{6} + c \ne F(x)G(x)$.

39. a. $V_1(t) = t^2 + 5.0t + 10$, $V_2(t) = 2.5t^2 + 2t + 10$.
b. $\frac{dV}{dt} = 7t + 7$ with initial condition $V(0) = V_1(0) + V_2(0) = 20$.
c. $V(t) = 3.5t^2 + 7t + 20$. This is indeed the sum of $V_1(t)$ and $V_2(t)$.

41. a. The units of α must be $\frac{\text{grams}}{\text{day}^2}$.
b. $M(t) = t^2 + c$. Substituting the initial condition, we have $M(0) = c = 5$, so $M(t) = t^2 + 5$.
c.

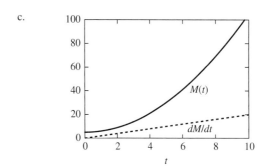

d. The mass increases more and more quickly.

43. a. $v(t) = -9.8t + 5$; $p(t) = -4.9t^2 + 5t + 100$.
b. The maximum height is when $v(t) = 0$, or at $t = \frac{5}{9.8} \approx 0.51$. At this time,

$$p(0.51) = -4.9(0.51)^2 + 5 \cdot 0.51 + 100 \approx 101.27.$$

c. It passes the robot when $p(t) = 100$, or at $t = 0$ and $t \approx 1.02$. The velocity -5 m/s.
d. It will take 5.06 s to hit the ground, and will be moving at -44.55 m/s.

e.

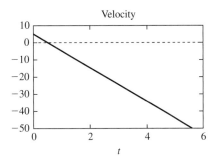

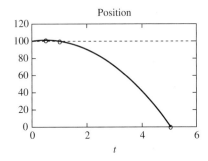

45.
a. $v(t) = -22.88t + 5$; $p(t) = -11.44t^2 + 5.0t + 100$.

b. The maximum height is when $v(t) = 0$, or at $t \approx 0.22$. The position is 100.55 m.

c. It passes the robot when $p(t) = 100$, or at $t = 0$ and $t \approx 0.44$. The velocity is -5 m/s.

d. It take 3.18 s to hit the ground, and will be moving at -67.83 m/s.

e.

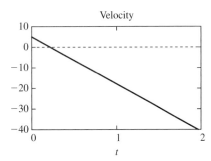

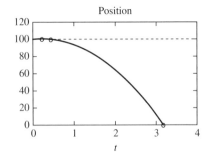

47. With Euler's method, $p_1 = 11$, $p_2 = 14$, $p_3 = 19$, and $p(4) \approx p_4 = 26$. The velocities fall on the line $v(t) = 1 + 2t$. Therefore, position satisfies the differential equation $\frac{dp}{dt} = 1 + 2t$. This has solution $p(t) = 1t + t^2 + 10$ when $p(0) = 10$. At $t = 4$, the exact solution is $p(t) = 30$.

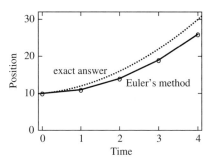

49. With Euler's method, $p_1 = 35$, $p_2 = 51$, $p_3 = 60$, and $p(4) \approx p_4 = 64$. The velocities fall according to the quadratic $v(t) = (t-5)^2 = t^2 - 10t + 25$. The position satisfies the differential equation $\frac{dp}{dt} = t^2 - 10t + 25$. This has solution $p(t) = \frac{t^3}{3} - 5t^2 + 25t + 10$ when $p(0) = 10$. At $t = 4$, the exact solution is $p(t) \approx 51.333$.

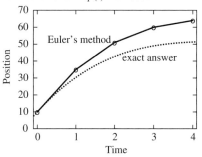

51. We estimate the velocity to be 2 during the first minute, 4 during the second, 6 during the third, and 8 during the fourth. Then $p_1 = 12$, $p_2 = 16$, $p_3 = 22$, and $p(4) \approx p_4 = 30$. At $t = 4$, this matches the exact solution of $p(4) = 30$.

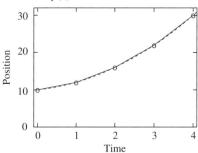

53. We estimate the velocity to be 20.5 during the first minute, 12.5 during the second, 6.5 during the third, and 2.5 during the fourth. Then $p_1 = 30.5, p_2 = 43, p_3 = 49.5$, and $p_4 \approx p(4) = 52$. At $t = 4$, the exact solution is $p(4) \approx 51.333$, so this is much closer.

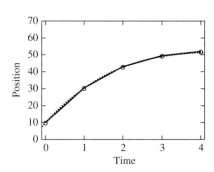

Section 7.3

1. $R_6 \approx 1.21786$; underestimate

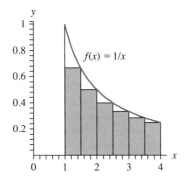

3. $L_5 \approx 0.69744$; overestimate

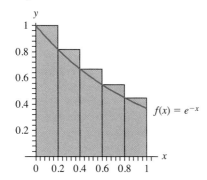

5. $2^2 + 3^2 + 4^2 + 5^2 = 54$

7. $(2f(x_7) - 3) + (2f(x_8) - 3) + (2f(x_9) - 3) + (2f(x_{10}) - 3)$

9. $3^0 + 3^1 + 3^2 + 3^3 + 3^4 = 121$

11. $f(x_0)\Delta x_0 + f(x_1)\Delta x_1 + f(x_2)\Delta x_2 + f(x_3)\Delta x_3 + f(x_4)\Delta x_4$

13. $\sum_{i=1}^{n} 2i$

15. $\sum_{j=2}^{12} \sin\frac{\pi}{j}$

17. $\sum_{i=1}^{n} \frac{1}{n} e^{i/n}$

19. $\sum_{i=0}^{n-1} \frac{i^4}{n^5}$

21. $\sum_{i=0}^{n-1} \frac{1}{n+i}$

23. Divide $[0, 1]$ into n subintervals of length $\Delta x = 1/n$: $[0, 1/n]$, $[1/n, 2/n], [2/n, 3/n], \ldots, [(n-1)/n, n/n = 1]$. The midpoints are $\frac{1}{2n}, \frac{3}{2n}, \frac{5}{2n}, \ldots, \frac{2n-1}{2n}$. The midpoint sum is

$$M_n = \Delta x f(1/2n) + \Delta x f(3/2n) + \cdots + \Delta x f((2n-1)/2n)$$
$$= \frac{1}{12n^2}(2n-1)(2n+1)$$

The limit is $\lim_{n \to \infty} M_n = \frac{1}{3}$.

25. a. $M_n = 2$
    b. $\lim_{n \to \infty} M_n = 2$
    c. $\int_0^2 x\,dx$

d. The region involved (under the graph of the line $y = x$ on the interval $[0, 2]$) is a triangle with base 2 and height 2; its area is 2.

27. a. $R_n = \frac{n+1}{2n} + 2$
    b. $\lim_{n \to \infty} R_n = \frac{5}{2}$
    c. $\int_0^1 (x+2)\,dx$
    d. The region involved (under the graph of the line $y = x + 2$ on the interval $[0, 1]$) consists of a rectangle of base 1 and height 2 and a triangle of base 1 and height 1 on top; the area of the region is $1 \cdot 2 + 1 \cdot 1/2 = 5/2$.

29. Difference of areas of two triangles; -10

31. Sum of areas of two triangles; 4

33. Quarter-circle of radius 4; 4π

35. Difference of areas of regions of the same shape; 0

37. The region is a quarter circle placed on top of a rectangle; $6 + \pi$.

39. The integral is zero; it is the difference of the areas of two regions of the same shape.

41. Think of the integral as a signed area: $\int_0^5 f(x)\,dx = -4 + 4 - 2 = -2$.

43. 3

45. Positive

47. Zero

49. Negative

51. $e^{-1} \leq \int_0^1 e^{-x}\,dx \leq 1$

53. $10 - e \leq \int_e^{10} \ln x\,dx \leq (10 - e)\ln 10$

55. The region is contained in the rectangle over $[0, 1]$ whose height is 1 (and whose area is 1).

57. The region is contained in the rectangle over $[0, \pi]$ whose height is 1 (and whose area is π).

59. The region is contained in the rectangle over $[2, 3]$ whose height is $1/5$ (and whose area is $1/5$).

61. Let B_i represent the number of offspring in year i. Then the total number of offspring, T, is $T = \sum_{i=1}^{7} B_i = 0 + 8 + 15 + 24 + 31 + 11 + 3 = 92$.

63. The total number of offspring T is $T = \sum_{i=0}^{7} B_i = \sum_{i=0}^{7} \frac{i(i+1)}{2} + 4 = 4 + 5 + 7 + 10 + 14 + 19 + 25 + 32 = 116$.

65. The left-hand estimate is 3.20; the right-hand estimate is 4.80.

67. The left-hand estimate is 1.16; the right-hand estimate is 1.36.

69.
$$V_1 = V(0) + 2 \cdot 0 \cdot 0.4 = 0$$
$$V_2 = V_1 + 2 \cdot 0.4 \cdot 0.4 = 0.32$$
$$V_3 = V_2 + 2 \cdot 0.8 \cdot 0.4 = 0.96$$
$$V_4 = V_3 + 2 \cdot 1.2 \cdot 0.4 = 1.92$$
$$V(2) \approx V_5 = V_4 + 2 \cdot 1.6 \cdot 0.4 = 3.2$$

Euler's method gives the same answer as the left-hand estimate.

71.
$$V_1 = V(0) + \left(1 + 0^3\right) \cdot 0.2 = 0.2$$
$$V_2 = V_1 + \left(1 + 0.2^3\right) \cdot 0.2 = 0.4016$$
$$V_3 = V_2 + \left(1 + 0.4^3\right) \cdot 0.2 = 0.6144$$

$$V_4 = V_3 + \left(1 + 0.6^3\right) \cdot 0.2 = 0.8576$$
$$V(1) \approx V_5 = V_4 + \left(1 + 0.8^3\right) \cdot 0.2 = 1.16$$

Euler's method gives the same answer as the left-hand estimate.

73.
$$I_l = (127 + 122 + 118 + 115 + 113 \\ + 112 + 112 + 113 + 116 + 120) \cdot 1 = 1168$$
$$I_r = (122 + 118 + 115 + 113 + 112 \\ + 112 + 113 + 116 + 120 + 125) \cdot 1 = 1166.$$

75. We could use just these measurements and treat the time intervals between them as being 2.0 seconds, giving $I_m = (122 + 115 + 112 + 113 + 120) \cdot 2 = 1164$.

77. $23 \cdot 3 + 21 \cdot 2 + 23 + 27 \cdot 2 = 188$. The total number counted in the four years is 94, so we also estimate 188 with this method.

79. We could fill in the first NA using the next measurement, and then use the left-hand estimate idea thereafter. We find $10 \cdot 2 + 15 \cdot 2 + 18 \cdot 2 + 8 \cdot 2 = 102$. The total number counted in the four years is 51, so we also estimate 102 with this method.

Section 7.4

1. The definite integral of the rate of change of a function $f(x)$ over $[a,b]$ is equal to the total change $f(b) - f(a)$ of $f(x)$ from $x = a$ to $x = b$.

3. The integral function $g(x) = x$ measures the area of the rectangle over $[0, x]$ of height 1.

5. $\frac{7}{3}$

7. 123

9. $\frac{15}{8}$

11. 30

13. $\left(\frac{-3}{z} + \frac{z^3}{9}\right)\Big|_2^3 \approx 2.61$

15. $(e^x + \ln |x|)|_1^4 \approx 53.27$

17. 4

19. $\int_0^5 3e^{x/5} dx = 15e^{x/5}\Big|_0^5 = 15e - 15 \approx 25.77$

21. $3\pi/2$

23. $4 - 7\ln 5$

25. $-9/20$

27. $\frac{5}{8} + \frac{1}{4}\ln 2$

29. $-e^{-4} + e^{-2}$

31. 2

33. 3

35. 1

37. $\int_0^2 \frac{1}{1+4t} dt = \frac{\ln(1+4t)}{4}\Big|_0^2 = \frac{\ln 9}{4} \approx 0.549$

39. Area above minus area is $-15/4$.

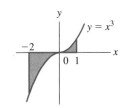

41. The integral is equal to -1; the area of the shaded region is 1.

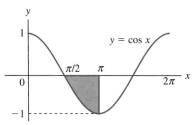

43.
$$\int_1^2 t^2 dt = \frac{t^3}{3}\Big|_1^2 = \frac{7}{3}$$
$$\int_2^3 t^2 dt = \frac{t^3}{3}\Big|_2^3 = \frac{19}{3}$$
$$\int_1^3 t^2 dt = \frac{t^3}{3}\Big|_1^3 = \frac{26}{3}$$

which is equal to $\frac{7}{3} + \frac{19}{3}$.

45. $\int_1^2 L(t)dt = 1.875$; $\int_2^3 L(t)dt \approx 0.347$; $\int_1^3 L(t)dt \approx 2.222 = 1.875 + 0.347$.

47. $\int_1^2 F(t)dt \approx 5.36$; $\int_2^3 F(t)dt \approx 13.10$; $\int_1^3 F(t)dt \approx 18.46 = 5.36 + 13.10$.

49. a. See figure below.
b. $g(0) = 0$, $g(1) = 5$, $g(3) = 15$, $g(4) = 20$, and $g(6) = 30$.
c. $g(x)$ is increasing on $[0, 6]$.
d. $g(x)$ has the (absolute) minimum value at $x = 0$ and the (absolute) maximum value at $x = 6$.
e. See figure below.

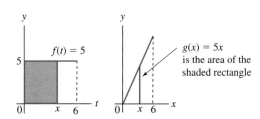

51. a. See the figure below.
b. $g(0) = 0$, $g(1) = -7/2$, $g(3) = -15/2$, $g(4) = -8$, and $g(6) = -6$.
c. and d.

Value(s) of x	(0, 4)	4	(4, 6)
g'	negative	zero	positive
g	decreasing	minimum	increasing

e. See the figure below

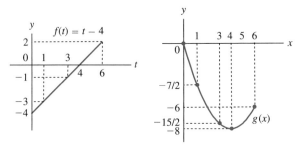

53. $g'(x) = xe^x - x + 4$

55. $g'(x) = \arctan x^4$.

57. $g'(x) = 3e^{3x-5}$

59. $g'(x) = 3x^2 \arctan x^6$

61. $\int_0^x s^2 ds = \frac{s^3}{3}|_0^x = \frac{x^3}{3}$. But $\frac{d}{dx}\left(\frac{x^3}{3}\right) = x^2 = f(x)$. It worked.

63. We find
$$\int_{-1}^x \left(1+\frac{t}{2}\right)^4 dt = 2\frac{(1+\frac{t}{2})^5}{5}|_{-1}^x = \frac{2}{5}\left(1+\frac{x}{2}\right)^5 - \frac{1}{80}$$
But $\frac{d}{dx}\left(\frac{2}{5}\left(1+\frac{x}{2}\right)^5 - \frac{1}{80}\right) = \left(1+\frac{x}{2}\right)^4$, if we use the chain rule.

65. a.
$$f(t) = \begin{cases} 1 & \text{if } 0 \le t \le 1 \\ 3-2t & \text{if } 1 < t \le 2 \\ -1 & \text{if } 2 < t \le 4 \end{cases}$$
b., c., and d. Use the Fundamental Theorem of Calculus to compute the integrals.

67. Using the indefinite integral and the substitution $u = -0.09t$, we find that
$$L(t) = \int 6.48e^{-0.09t} dt = -72.0e^{-0.09t} + c$$
The initial condition $L(0) = 5 = -72e^0 + c$ implies that $c = 77$, so the solution is $L(t) = 77 - 72e^{-0.09t}$. The change of size is $L(5) - L(1) \approx 31.09 - 11.20 = 19.89$. The definite integral is
$$\int_1^5 6.48e^{-0.09t} dt = -72e^{-0.09t}|_1^5 = -72e^{-0.09 \cdot 5} - 72e^{-0.09 \cdot 1}$$
$$\approx -45.91 - 65.8 = 19.89$$

69. The solution is $A(t) = 174.6(t-1981)^3 + 340$. The number of new cases is $A(1987) - A(1985) = 26,539$, which matches $\int_{1985}^{1987} 523.8(t-1981)^2 dt$.

71. The solution is $P(t) = 4.5 - 2.5e^{-2t}$. The amount produced is $P(10) - P(5) \approx 0.0001$, which matches $\int_5^{10} 5e^{-2.0t} dt$.

73. We found that $L(t) = 77 - 72e^{-0.09t}$. The change of size between times 1 and 3 is $L(3) - L(1) \approx 10.84$, while that between times 3 and 5 is $L(5) - L(3) \approx 9.05$. We found that $L(5) - L(1) \approx 19.89$, which is equal to $10.84 + 9.05$.

75. $A(1986) - A(1985) \approx 10,651$, $A(1987) - A(1986) \approx 15,889$. These add to approximately 26,539.

77. $P(7.5) - P(5) \approx 0.0001$, $P(10) - P(7.5) \approx 7.5 \cdot 10^{-7}$. These add to 0.0001; nothing much happens during the second half of the interval.

79. a. The velocity follows $\frac{dv}{dt} = 2$, which has solution $v(t) = 2t$ if $v(0) = 0$. Then $\frac{dp}{dt} = 2t$ is a differential equation for position. This has solution $p(t) = t^2$ using the initial condition $p(0) = 0$.
b. $v(60) = 120$ and $p(60) = 3600$.
c. The velocity follows $\frac{dv}{dt} = -9.8$, which has solution $v(t) = -9.8t + 708$, if $v(60) = 120$. Then $\frac{dp}{dt} = -9.8t + 708$, is a differential equation for position. This has solution $p(t) = -4.9t^2 + 708t + 21,240$ using the initial condition $p(60) = 3600$.
d. The maximum height is reached when $v = 0$ or when $-9.8t + 708 = 0$. This happens at $t \approx 72.24$. The height is then 4334. It rises less after running out of fuel, because the acceleration of gravity is stronger than the acceleration of the engine.

e. It hits the ground when $p(t) = 0$, or at $t \approx 101.99$. The velocity is -291.50 m/s.

Section 7.5

1. $\frac{1}{4.4}(x-3)^{4.4} + C$

3. $\frac{1}{3} \ln|3x-11| + C$

5. $-\frac{1}{6.7}e^{-6.7x} + C$

7. $\frac{2}{5}(1 - 0.5^5) = 0.3875$

9. $\frac{1}{3}(e^7 - 1)$

11. $-3\cos(x/3) + C$

13. $\frac{1}{2}\arctan(2x) + C$

15. Substitute $u = 1 + e^x$; $\ln(1+e^x) + C$.

17. Substitute $u = 1 + y^2$; $\frac{2}{3}(1+y^2)^{3/2} + C$.

19. Substitute $y = \cos\theta$; $-\ln|\cos\theta| + C$.

21. Substitute $y = \ln x$; $\ln|\ln x| + C$.

23. 0

25. $\frac{2}{3}\left(\frac{3}{4}\right)^{3/2} \approx 0.433$

27. 0

29. $-\frac{1}{2}e^{-x^2} + C$

31. $1/3$

33. $1/2$

35. $\frac{1}{2}(\sqrt{2} - 1)$

37. Let $u(x) = x$ and $v'(x) = \cos 3x$; $\frac{x \sin 3x}{3} + \frac{\cos 3x}{9} + C$.

39. $\frac{\pi}{8} - \frac{1}{4}$

41. $x \ln^2 x - 2(x \ln x - x) + C$

43. $\frac{1}{2}x^2 \arctan x - \frac{1}{2}(x - \arctan x) + C$

45. Integrate by parts twice, start with $u(x) = x^2$ and $v'(x) = e^{-x}$; $-x^2 e^{-x} - 2xe^{-x} - 2e^{-x} + C$.

47. $\frac{1}{3}e^{x^3}(x^3 - 1) + C$

49. $\pi/12$

51. Start by using long division; $\int_1^2 \frac{x-2}{2x+3} dx = \int_1^2 \frac{1}{2} + \frac{-7/2}{2x+3} dx = \frac{1}{2} + \frac{7}{4} \ln 5 - \frac{7}{4} \ln 7$.

53. Use the substitution $u = \sqrt{x} + 4$; $2 \ln(\sqrt{x} + 4) + C$.

55. Write $\tan x = \sin x/\cos x$ and use the substitution $u = \cos x$; $\int \tan x\, dx = -\ln|\cos x| + C$ and $\int_0^{\pi/4} \tan x\, dx = \frac{1}{2} \ln 2$.

57. Write $\cot 4x = \cos 4x/\sin 4x$ and use the substitution $u = \sin 4x$; $\frac{1}{4} \ln|\sin 4x| + C$.

59. Use the substitution $u = \sin x$; $\frac{1}{5} \sin^5 x + C$.

61. Integrate by parts twice, starting with $u = x^2, v' = 5^x$; $\frac{5}{\ln 5} - \frac{10}{(\ln 5)^2} + \frac{8}{(\ln 5)^3}$.

63. $\frac{1}{\pi+1}$

65. $\frac{2}{3} \ln |\ln x| + C$

67. Integrate by parts using $u = \log_{10} x, v' = x$; $\frac{x^2}{2} \log_{10} x - \frac{x^2}{4 \ln 10} + C$.

69. $x(e^x + e^{-x}) - (e^x - e^{-x}) + C$

71.
$$\int \frac{\cos x}{x^2} dx \approx \int \frac{T_6(x)}{x^2} dx = \int \frac{1 - \frac{1}{2}x^2 + \frac{1}{24}x^4 - \frac{1}{720}x^6}{x^2} dx$$
$$= -\frac{1}{x} - \frac{1}{2}x + \frac{1}{24} \cdot \frac{x^3}{3} - \frac{1}{720} \cdot \frac{x^5}{5} + C$$

73.
$$\int_0^1 \sin(x^2) dx \approx \int_0^1 \left(x^2 - \frac{1}{6}x^6\right) dx$$
$$= \frac{1}{3} - \frac{1}{42} \approx 0.309524$$

75. a. Let $u = \arctan x$, $v' = 1$; $\int \arctan x \, dx = x \arctan x - \frac{1}{2} \ln(1 + x^2) + C$, and $\int_0^{0.2} \arctan x \, dx \approx 0.019869$.

b.
$$\int_0^{0.2} \arctan x \, dx \approx \int_0^{0.2} T_3(x) dx = \int_0^{0.2} \left(x - \frac{1}{3}x^3\right) dx$$
$$\approx 0.019867$$

77. Set $u(x) = \ln x$ and $\frac{dv}{dx} = 1$. Then $v(x) = \int 1 \, dx = x + c$ and $\frac{du}{dx} = \frac{1}{x}$, giving
$$\int \ln x \, dx = \ln x(x + c) - \int \frac{x + c}{x} dx$$
$$= \ln x(x + c) - x - c \ln x + c_1 = x \ln x - x + c_1$$

The arbitrary constant c cancelled, leaving the original answer.

79. Set $u(x) = \ln x$ and $\frac{dv}{dx} = \frac{1}{x}$. Then $v(x) = \ln x$ and $\frac{du}{dx} = \frac{1}{x}$, giving
$$\int \frac{\ln x}{x} dx = (\ln x)^2 - \int \frac{\ln x}{x} dx$$

We can solve for the original integral, giving
$$2 \int \frac{\ln x}{x} dx = (\ln x)^2$$

so
$$\int \frac{\ln x}{x} dx = \frac{(\ln x)^2}{2} + c$$

81. Integrating, $P(t) = -25e^{-0.2t} + c$. Substituting the initial condition, we find $c = 25$. The limit is 25 as t approaches infinity. In fact, $P(10) \approx 21.62$ and $P(100) \approx 25$. This does not increase to infinity because the rate of chemical production decreases to 0 quickly (exponentially).

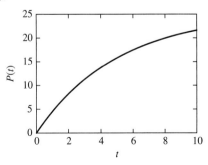

83. Integrating, $P(t) = \frac{-5}{1+t} + c$. Substituting the initial condition, we find $c = 5$. The limit is 5 as t approaches infinity. In fact, $P(10) \approx 4.54$ and $P(100) \approx 4.95$. The rate of chemical production decreases to zero moderately quickly.

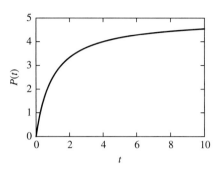

85. Let $g(t) = (4t - t^2)e^{-3t}$. To find the time when the worm grows fastest, we compute
$$g'(t) = (4 - 14t + 3t^2)e^{-3t}$$

Solving $g'(t) = 0$ requires solving the quadratic $4 - 14t + 3t^2 = 0$ because e^{-3t} is never zero. The solution is
$$t = \frac{14 \pm \sqrt{14^2 - 4 \cdot 3 \cdot 4}}{6}$$

The only value with $t < 2$ is $t \approx 0.306$. This must be a maximum because $g(0) = g(2) = 0$. The maximum growth rate is $g(0.306) \approx 0.451$. Using integration by parts, we find that
$$\int_0^2 (4t - t^2)e^{-3t} = \frac{-10 - 30t + 9t^2}{27} e^{-3t} \Big|_0^2 \approx 0.367$$

If the worm grew at its maximum rate, the size at $t = 2$ would be $2 \cdot 0.451 = 0.902$. $W(2)$ is only about 40% as large.

87. a. $L(t) = 108.0(1 - e^{-0.06t})$
b. The limit is 108.0 cm.
c. These walleye reach maturity when $L(t) = 45$. To solve,
$$108.0(1 - e^{-0.06t}) = 45.0$$
$$e^{-0.06t} = 1 - \frac{45.0}{108.0}$$
$$-0.06t = \ln\left(1 - \frac{45.0}{108.0}\right)$$
$$t = \frac{\ln(1 - \frac{45.0}{108.0})}{-0.06} \approx 8.983$$

d. These walleye reach maturity a bit earlier than the Ontario Walleye and grow quite a bit larger.

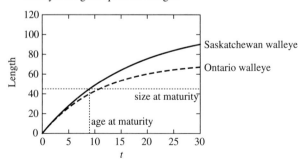

89. a. $\dfrac{dB}{dt} = 100e^{0.05t}$

b. Integrating, $B(t) = 2000e^{0.05t} + c$. The initial condition implies that $c = -2000$, so $B(t) = 2000e^{0.05t} - 2000$.

c.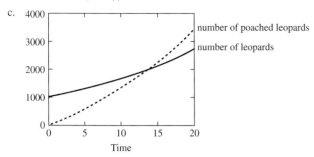

d. The limit of the ratio is 2.0.

91. a. $\dfrac{dB}{dt} = 10 + 10e^{0.5t}$

b. Integrating, $B(t) = 10t + 20e^{0.5t} + c$. The initial condition implies that $c = -20$, so $B(t) = 10t + 20e^{0.5t} - 20$.

c.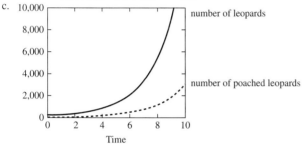

d. The limit of the ratio is 0.2.

93. a.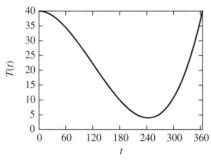

b. The solution is
$$L(t) = \int 0.04 - 5 \cdot 10^{-9}t^2(365 - t)\,dt$$
$$= \int 0.04 - 1.825 \cdot 10^{-6}t^2 - 5.0 \cdot 10^{-9}t^3\,dt$$
$$= 0.04t - 6.08 \cdot 10^{-7}t^3 - 1.25 \cdot 10^{-9}t^4 + c$$

Using the initial condition $T(0) = 0.1$, we find $c = 0.1$. The size at $t = 30$ is 1.28 cm.

c. Using the initial condition $T(150) = 0.1$, we find $c \approx -4.4797$. The size at $t = 180$ is 0.4847 cm. This bug is smaller because this place is cooler in the middle of the year.

95. a.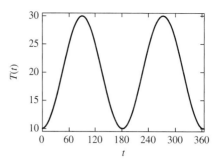

b. The solution is
$$L(t) = \int 0.02 + 0.01 \cos\left(\dfrac{2\pi(t - 190)}{182.5}\right) dt$$
$$= 0.02t + 0.01 \sin\left(\dfrac{2\pi(t - 190)}{182.5}\right) \dfrac{182.5}{2\pi} + c$$

Using the initial condition $T(0) = 0.1$, we find $c \approx 0.1742$. At $t = 30$, $L(30.0) \approx 0.9773$.

c. Using the initial condition $T(150) = 0.1$, we find $c \approx -2.615$. After 30 days, at $t = 180$, $L(180.0) \approx 0.887$. This bug is almost the same size because this climate has two warm spells.

Section 7.6

1. $\int_0^3 3x^3\,dx = 60.75$

3. $\int_0^{\ln 2} e^{x/2}\,dx = 2(\sqrt{2} - 1)$

5. $\int_0^2 (3 + 4y)^{-2}\,dy \approx 0.061$

7. a.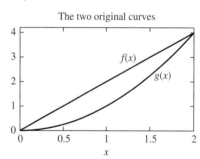

b.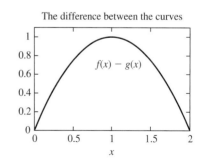

c. $2x - x^2 \geq 0$ for all $0 \leq x \leq 2$. Therefore,
$$\int_0^2 (2x - x^2)\,dx = \left(x^2 - \dfrac{x^3}{3}\right)\bigg|_0^2 = \dfrac{4}{3}$$

9. a.

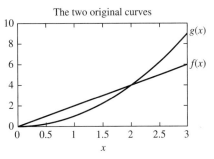

b.

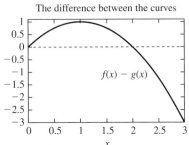

c. $2x - x^2 > 0$ for $0 \leq x < 2$, but $2x - x^2 < 0$ for $2 < x \leq 4$. Therefore, the area is $\int_0^2 (2x - x^2)dx + \int_2^3 (x^2 - 2x)dx = \frac{8}{3}$.

11. a.

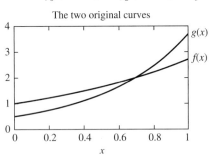

b.

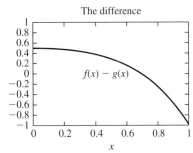

c. $e^x - \frac{e^{2x}}{2} > 0$ if $x < \ln 2$. Therefore, the area is $\int_0^{\ln 2} \left(e^x - \frac{e^{2x}}{2}\right) dx + \int_{\ln 2}^1 \left(\frac{e^{2x}}{2} - e^x\right)dx \approx 0.379$.

13. $2\sqrt{2}$

15. $\ln 2 - \frac{1}{2}$

17. 257/192

19. 65 squares, estimate: 260 km²; 67 squares, estimate: 268 km²

21. Right sum: 291.3 km²; left sum: 285.3 km²

23. Use $u(x) = x$ and $v'(x) = \sin(2\pi x)$ to get $\int_0^2 x\sin(2\pi x)dx = \frac{-1}{\pi} \approx -0.318$. The integral is negative, but this is consistent with the fact that the function gives expanding oscillations, with each negative region including more area than the preceding positive region.

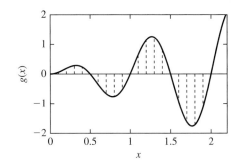

25. a.

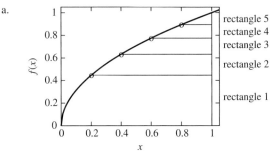

b. Rectangle 1: lower size estimate is $0.8 \cdot f(0.2) = 0.358$; upper size estimate is $1 \cdot f(0.2) = 0.447$. Rectangle 2: lower size estimate is $0.6 \cdot (f(0.4) - f(0.2)) = 0.111$; upper size estimate is $0.8 \cdot (f(0.4) - f(0.2)) = 0.148$. Rectangle 3: lower size estimate is $0.4 \cdot (f(0.6) - f(0.4)) = 0.0568$; upper size estimate is $0.6 \cdot (f(0.6) - f(0.4)) = 0.0853$. Rectangle 4: lower size estimate is $0.2 \cdot (f(0.8) - f(0.6)) = 0.024$, upper size estimate is $0.4 \cdot (f(0.8) - f(0.6)) = 0.048$. Rectangle 5: lower size estimate is 0.0; upper size estimate is $0.2 \cdot (f(1) - f(0.8)) = 0.021$.

c. Lower estimate is 0.550; upper estimate is 0.7493.

d. A rectangle at height y goes from the point where $\sqrt{x} = y$, or $x = y^2$, to $x = 1$. Its length is $1 - y^2$.

e. Area $= \int_0^1 (1 - y^2)\, dy$

f. $\int_0^1 (1 - y^2)\, dy = \left(y - \frac{y^3}{3}\right)\Big|_0^1 = \frac{2}{3}$. It checks.

27. A little disk at height r will have volume approximately equal to $\pi r^2 \Delta r$. If we pick n disks with thickness Δr (so that $n\Delta r = 1$), the total volume will be approximated by the Riemann sum $\sum_{i=0}^n \pi r^2 \Delta r$. In the limit, this is the definite integral $\int_0^1 \pi r^2 dr = \pi \frac{r^3}{3}\Big|_0^1 = \frac{\pi}{3}$.

29. Set $y = x/a$, so that $dx = a\, dy$. The integrand becomes $\frac{dy}{y}$ and the limits of integration go from 1 to 2. The area from a to $2a$ is equal to the area from 1 to 2, and thus is equal to $l(2)$.

31. We have that $\int_1^{a^b} \frac{1}{x} dx = l(a^b)$. Substituting $y = \sqrt[b]{x}$, we find that

$$\frac{dy}{dx} = \frac{1}{b}x^{\frac{1}{b}-1} = \frac{1}{b}\frac{y}{x}.$$

Then the integrand becomes

$$\frac{1}{x}dx = \frac{b}{y}dy$$

and the limits of integration go from 1 to a. So

$$\int_1^{a^b} \frac{1}{x} dx = \int_1^a \frac{b}{y} dy = bl(a).$$

This matches the law of logs.

33. $\frac{512\pi}{15}$

35. $\frac{16\pi}{15}$

37. The volume is $V = \pi \int_0^{\pi/4} \sec^2 x\, dx = \pi$.

39. $V = \pi \int_0^{\pi/2} (\cos x)^2 \, dx = \frac{\pi}{2} \int_0^{\pi/2} (1 + \cos 2x)\, dx = \frac{\pi^2}{4}$

41. $V = \pi \int_1^4 \left[(\sqrt{x})^2 - (1/x)^2 \right] dx = \frac{27\pi}{4}$

43. $V = \pi \int_1^2 \left[(\sqrt{y})^2 - (1/y)^2 \right] dy = \pi$

45. $V = \pi \int_{-3}^{3} \left[(18 - x^2)^2 - (x^2)^2 \right] dx = 1296\pi$

47. $V = \pi \int_{-2}^{2} \left[(6)^2 - (y^2 + 2)^2 \right] dy = \frac{1408}{15}\pi$

49. $V = \pi \int_0^1 \left[(3)^2 - (2 + y^{1/3})^2 \right] dy = \frac{7\pi}{5}$

51. $V = \pi \int_0^1 \left[(4)^2 - (4 - x^3)^2 \right] dx = \frac{13\pi}{7}$

53. The solid is obtained by rotating the triangular region between $y = 0$ and $y = x/\sqrt{\pi}$ and between $x = 0$ and $x = 4$ about the x-axis. It is a cone of radius $4/\sqrt{\pi}$ and height 4.

55. The solid is obtained by rotating the region under the upper semicircle of radius 2 shifted 3 units up about the x-axis.

57. $V = \pi \int_{-b}^{b} a^2 \left(1 - \frac{y^2}{b^2} \right) dy = \frac{4}{3} \pi a^2 b$

59. $V = \pi \int_{h/2}^{h} \left(\frac{rx}{h} \right)^2 dx = \frac{7}{24} \pi r^2 h$

61. a. $d = \sqrt{(a-0)^2 + (b-0)^2} = \sqrt{a^2 + b^2}$
    b. The line has slope b/a, and its equation is $f(x) = bx/a$. We find $f'(x) = b/a$ and $1 + [f'(x)]^2 = 1 + b^2/a^2$. When $x = 0$, then $y = f(x) = 0$, and when $x = a$, then $y = b$. Thus, $L = \int_0^a \sqrt{1 + b^2/a^2}\, dx = \sqrt{a^2 + b^2}$.

63. $L = \int_1^2 \left(\frac{x^2}{2} + \frac{1}{2x^2} \right) dx = \frac{17}{12}$

65. $L = \int_3^4 \sqrt{x}\, dx \approx 1.869$

67. We compute
$$L = \int_{-\pi/4}^{\pi/4} \sec x\, dx = (\ln |\sec x + \tan x|)\Big|_{-\pi/4}^{\pi/4}$$
$$= \ln \left(\sqrt{2} + 1 \right) - \ln \left(\sqrt{2} - 1 \right) \approx 1.763$$

Note: To compute the integral of $\sec x$, multiply and divide by $\sec x + \tan x$;
$$\int \sec x\, dx = \int \frac{\sec x (\sec x + \tan x)}{\sec x + \tan x}\, dx = \int \frac{\sec^2 x + \sec x \tan x}{\sec x + \tan x}\, dx$$

Use the substitution $u = \sec x + \tan x$; then $du = (\sec x \tan x + \sec^2 x)\, dx$, and
$$\int \sec x\, dx = \int \frac{1}{u}\, du = \ln |u| = \ln |\sec x + \tan x| + C$$

69. $L = \int_1^4 \left(\frac{x}{4} + \frac{1}{x} \right) dx = \frac{15}{8} + \ln 4$

71. $L = \int_{1/2}^{2} \sqrt{1 + x^{-4}}\, dx$

73. $L = \int_1^e \sqrt{1 + 4x^{-2}}\, dx$

75. From $u = x/r$ we get $du = dx/r$ and $dx = r\, du$. Thus,
$$\int \frac{1}{\sqrt{1 - (x/r)^2}}\, dx = \int \frac{1}{\sqrt{1 - u^2}}\, r\, du = r \arcsin u + C$$
$$= r \arcsin (x/r) + C$$

Similarly,
$$\int \frac{1}{a^2 + x^2}\, dx = \int \frac{1}{a^2 (1 + (x/a)^2)}\, dx = \frac{1}{a^2} \int \frac{1}{1 + (x/a)^2}\, dx$$

Use the substitution $u = x/a$; then $du = dx/a$ and $dx = a\, du$, and
$$\int \frac{1}{a^2 + x^2}\, dx = \frac{1}{a^2} \int \frac{1}{1 + u^2}\, a\, du = \frac{1}{a} \arctan u + C$$
$$= \frac{1}{a} \arctan (x/a) + C$$

77. $\int_{0.5}^{2.0} \frac{1}{x}\, dx = \ln x \big|_{0.5}^{2.0} = 2 \ln 2$. The average is therefore $\frac{2 \ln 2}{1.5} \approx 0.924$.

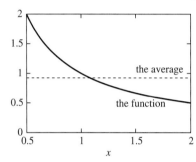

79. $\int_0^{\pi/2} \sin(2x)\, dx = -\frac{1}{2} \cos(2x) \big|_0^{\pi/2} = 1$. The average is therefore the total (1) divided by the width of the interval $\left(\frac{\pi}{2} \right)$ or $\frac{2}{\pi} \approx 0.637$.

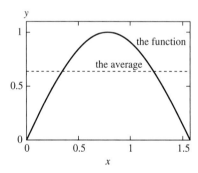

81. The function has regions of width 4 and value 6, width 7 and value 7, width 5 and value 8, width 3 and value 9, and width 1 and value 10.

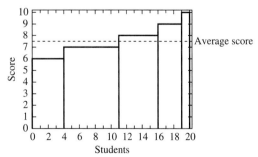

The average is $(4 \cdot 6 + 7 \cdot 7 + 5 \cdot 8 + 3 \cdot 9 + 1 \cdot 10)/20 = 7.5$. The function is

$$f(x) = \begin{cases} 6 & \text{for } 0 \leq x < 4 \\ 7 & \text{for } 4 \leq x < 11 \\ 8 & \text{for } 11 \leq x < 16 \\ 9 & \text{for } 16 \leq x < 19 \\ 10 & \text{for } 19 \leq x < 20 \end{cases}$$

$$\text{Total score} = \int_0^{20} f(x)dx = 150$$

$$\text{Average score} = \frac{\text{Total score}}{\text{width of interval}} = \frac{150}{20} = 7.5$$

83. $\int_{15}^{24}(360t - 39t^2 + t^3)\,dt = -2369.25$.

Over these 9 h, the average rate is -263.25 L/h. Water is leaving.

85. We must find where $g(t)$ is positive and negative. Solving the equation $g(t)=0$ gives $g(t)=t(360-39t+t^2)=t(t-15)(t-24)$, which is positive for $t<15$ and negative for $15<t<24$. Then $|g(t)|=g(t)$ for $t<15$ and $|g(t)|=-g(t)$ for $15<t<24$. Therefore

$$\int_0^{24}|g(t)|dt = \int_0^{15}g(t)dt - \int_{15}^{24}g(t)dt$$
$$= 9281.25 - (-2369.25) = 11{,}650.5$$

The average rate of energy production is the total energy divided by 24 hours or $\frac{11{,}650.5}{24} = 485.4$ J/h.

87. a. The critical points are $x=0$ and $x=160$. $x=0$ is the minimum (where $\rho(0)=1.0$) and $x=160$ is the maximum (where $\rho(160)\approx 1.0491$).
    b. $\int_0^{200}\rho(x)dx \approx 204.8$ g.
    c. The average is 1.024 g/cm, which lies between the minimum and the maximum.
    d.

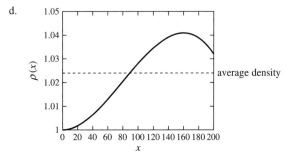

89. $\int_0^1 \sqrt{t}\,dt \approx 0.667$ cm³. The average rate is 0.667 cm³/s. The rate at time 0.5 is $\sqrt{0.5}\approx 0.707$, greater than the average rate during the first second. This seems to be because the function is concave down (has negative second derivative).

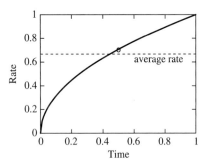

91. $\int_0^1 4t(1-t)dt \approx 0.667$ cm³. The average rate is 0.667 cm³/s. The rate at $t=0.5$ is 1.0 (the maximum). This is greater than the average rate, because the function is concave down.

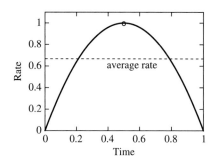

Section 7.7

1. We know that e^{-x} is faster because exponential functions are faster than power functions. With L'Hôpital's rule,

$$\frac{e^{-x}}{\frac{1}{x}} = \frac{x}{e^x}$$

and

$$\lim_{x\to\infty}\frac{x}{e^x} = \lim_{x\to\infty}\frac{1}{e^x} = 0$$

e^{-x} does indeed approach 0 faster than $\frac{1}{x}$ as x approaches infinity.

3. The power in the denominator of $\frac{1}{x^2}$ is larger, so it approaches infinity more quickly. Algebraically,

$$\frac{\frac{1}{x^2}}{\frac{1}{x}} = \frac{x}{x^2} = \frac{1}{x}$$

which approaches infinity. Therefore, $\frac{1}{x^2}$ does approach infinity more quickly than $\frac{1}{x}$ as x approaches 0.

5. $\int_0^{\infty} e^{-3t}dt = 1/3$

7. Does not converge because the integrand approaches zero too slowly.

9. Using the substitution $u=1+3x$, we find that this integral is $\frac{2}{3}$.

11. Diverges because the power is not less than 1.

13. $\int_0^{0.001}\frac{1}{\sqrt[3]{x}}dx = 0.015$

15. Convergent; e^{-2}

17. Divergent

19. Convergent; π

21. Divergent

23. Convergent; -4

25. Divergent

27. The leading behaviour of the denominator is e^x (because it does not even approach 0). The leading behaviour of the integrand is e^{-x}, which does not approach infinity anywhere. The integral converges.

29. The leading behaviour of the denominator is e^x. The leading behaviour of the integrand is e^{-x}, so the integral converges.

31. For $x < 1$, $x^2 + e^x > e^x$, so

$$\int_0^1 \frac{1}{x^2 + e^x}\,dx < \int_0^1 \frac{1}{e^x}\,dx = e^{-x}\Big|_0^1 = 1 - e^{-1} \approx 0.632$$

The value is less than 0.632.

33. For $x > 1$, $e^x > \sqrt{x}$, so

$$\int_1^\infty \frac{1}{\sqrt{x} + e^x}\,dx < \int_1^\infty \frac{1}{e^x}\,dx = -e^{-x}\Big|_1^\infty = 1/e.$$

The value is less than 1.

35. The differential equation is

$$\frac{dP}{dt} = 50e^{-2t}$$

The solution is

$$P(t) = 10 + \int_0^t 50e^{-2s}\,ds = 35 - 25e^{-2t}$$

In principle, this rule could be followed indefinitely, but the limit of the toxin is 35, greater than the tolerance of the cell.

37. The differential equation is

$$\frac{db}{dt} = \frac{1000}{(2 + 3t)^{1.5}}$$

The solution is

$$b(t) = 1000 + \int_0^t \frac{1000}{(2 + 3s)^{1.5}}\,ds$$

$$\approx 1000 - 667(2 + 3s)^{-0.5}\Big|_0^t$$

$$= 1000 - 667(2 + 3t)^{-0.5} + 667(2^{-0.5})$$

$$\approx 1471 - 667(2 + 3t)^{-0.5}$$

This rule could be followed indefinitely, and the population would approach a limiting size of 1471. It would never double.

True/False Quiz

1. FALSE. An autonomous function does not allow for the explicit appearance of the independent variable (without t it would have been autonomous).

3. TRUE. The function $y = x^3$ is increasing, and so all rectangles belonging to L_{10} are contained in the region under $y = x^3$.

5. TRUE. This formula is a combination of the sum and the constant multiple rules.

7. FALSE. The calculation makes no sense since the function $f(x) = 1/x$ is not continuous at $x = 0$. In other words, the Fundamental Theorem of calculus does not apply.

9. TRUE. Apply the general formula with $p = 0.99 < 1$.

Supplementary Problems

1. a.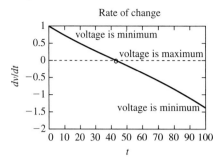

b.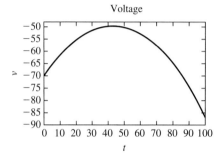

c. Solving with the indefinite integral, we get

$$v(t) = t + 50\ln(1 + 0.02t) - 100e^{0.01t} + c.$$

Substituting the initial condition, we have that $c = 30$. Therefore, $v(100) = 100 + 50\ln(3.0) - 100e^1 + 30 \approx -86.9$.

3. a. At a speed of $10\frac{\text{m}}{\text{s}} = 1000\frac{\text{cm}}{\text{s}}$, it takes $t = 80$ cm/(1000 cm/sec) $= 0.08$ sec to reach your hand. Similarly, your elbow seems to be 50 cm from the brain, so it takes the signal 0.05 s to get there.

b.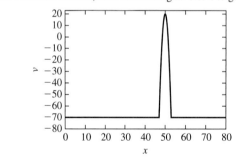

c. Average $= \frac{1}{6}\int_{47}^{53}\left[-70 + 10(9 - (x - 50)^2)\right]dx = -10$

d. The total voltage along the 6-cm piece is -60. Along the rest (74 cm) the total is $74 \cdot (-70) = -5180$. The total along the whole thing is $-5180 - 60 = -5240$, so the average is $-5240/80 = -65.6$.

5. a. The second is a pure-time differential equation. The first might describe a population of bacteria that are autonomously reproducing, and the second might describe a population being supplemented from outside at an ever-increasing rate.

b. $b(0.1) \approx b_1 = b(0) + b'(0) \cdot 0.1 = 1 + 2 \cdot 0.1 = 1.2$
Similarly,

$$B(0.1) \approx B_1 = B(0) + B'(0) \cdot 0.1 = 1 + 2 \cdot 0.1 = 1.2$$

c. $b(0.2) \approx b_2 = b_1 + 2b_1 \cdot 0.1 = 1.2 + 2 \cdot 1.2 \cdot 0.1 = 1.44$
Similarly,

$$B(0.2) \approx B_2 = B_1 + B'(0.1) \cdot 0.1 = 1.2 + 1.2 \cdot 0.1 = 1.32$$

7.
  a. The volume is increasing until time $t = 2$, and decreases thereafter.
  b. The volume is a maximum when the rate of change is 0, or at $t = 2$.
  c. LHE $= V(0) \cdot 1 + V(1) \cdot 1 + V(2) \cdot 1 = 7$ and RHE $= V(1) \cdot 1 + V(2) \cdot 1 + V(3) \cdot 1 = -2$.
  d. $V(3) = \int_0^3 (4 - t^2) dt = 4t - \frac{t^3}{3}\big|_0^3 = 3$

9.
  a. The function hits 0 when $x = 10$. The derivative is 0 at $x = 2.5$, where the density is 56.25. At the endpoints, we have densities of 50 (at $x = 0$) and 250 (at $x = 20$). The maximum is thus at $x = 20$, with the minimum at $x = 10$.

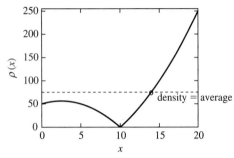

  b. Taking into account the absolute values, the total number is
  $$\text{total} = \int_0^{20} |-x^2 + 5x + 50| dx$$
  $$= \int_0^{10} \left(-x^2 + 5x + 50\right) dx + \int_{10}^{20} \left(x^2 - 5x - 50\right) dx = 1500$$

  c. The average density is 75.0.
  d. Shown in figure for a.

11.
  a. $A'(t) = -15/(2 + 0.3t)^2 + 0.125e^{0.125t}$, so $A'(0) = -3.625$. Also, $A(0) = 35.0$ and $A(120) \approx 46.13$. The graph thus begins decreasing and then increases. The maximum is at $t = 120$.

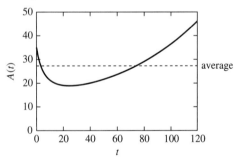

  b. $\int_0^{120} A(t) dt = \frac{50}{0.3} \ln(2 + 0.3t) + 800 e^{0.0125t} \big|_0^{120} \approx 3276$
  c. The average is $3276/120 \approx 27.3$.
  d. The minimum value must be less than 27.3.
  e. RHE $= \sum_{i=1}^{6} A(20i) 20$. I bet the estimate is high.

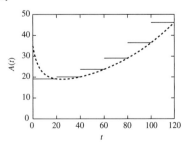

Chapter 8

Section 8.1

1. This differential equation is nonautonomous because the state variable F and the time t both appear on the right-hand side. The parameter k does not affect the kind of differential equation.

3. This is a pure-time differential equation because the only variable on the right-hand side is the time t (along with the parameter μ).

5. Substituting these values into the right-hand side gives $\frac{dF}{dt} = 1^2 + 1 \cdot 0 = 1 > 0$. Therefore, F is increasing.

7. Substituting these values into the right-hand side gives $\frac{dy}{dt} = 2e^{-1} - 1 = -0.26 < 0$. Therefore, y is decreasing.

9.
  a. Dependent variable: number of amoebas $A(t)$; independent variable: time t in days
  b. $A'(t) = 1.3 A(t), A(0) = 120$
  c. Autonomous

11.
  a. Dependent variable: atmospheric pressure $p(h)$ in atmospheres; independent variable: altitude h (measured in some units of length)
  b. $p'(h) = k \cdot p(h), p(0) = p_0$, and k is a constant.
  c. Autonomous

13.
  a. Dependent variable: number of infected people $I(t)$; independent variable: time t.
  b. $I'(t) = k \cdot I(t) \cdot (7{,}500 - I(t)), I(0) = 13$, and k is a constant.
  c. Autonomous

15.
  a. Dependent variable: biomass of plankton $B(t)$ in units of million tonnes; independent variable: time t.
  b. $B'(t) = k \cdot \sqrt[4]{B(t)}, B(0) = 3$, where k is a constant.
  c. Autonomous

17. $p(t) = \frac{t^2}{2} + 1$. Then $p(1) = 1.5$.

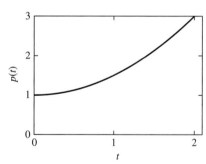

19. The derivative of e^t is e^t, and $e^0 = 1$, so this is the solution that matches the initial conditions. Also, $b(1) = e = 2.718$.

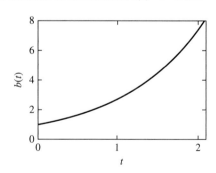

21. They do not match because the rate of change gets larger the longer you wait.

23. First, $x(0) = -\frac{1}{2} + \frac{3}{2}e^0 = 1$, so the initial condition matches. Next,
$$\frac{dx}{dt} = 3e^{2t} = 2\left(-\frac{1}{2} + \frac{3}{2}e^{2t}\right) + 1 = 1 + 2x$$
This checks.

25. First, $G(0) = 1 + e^0 = 2$, so the initial condition matches. Next, $\frac{dG}{dt} = e^t = (1 + e^t) - 1 = G - 1$. This checks.

27.
a. $P < 1200$
b. $P > 1200$
c. The population that starts at 1200 members will remain there for all time.
d. See figure below.

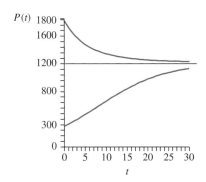

29. If $L < A$, then L is increasing ("growth"); if $L > A$, then L decreases ("limited"). Differentiating $\frac{dL}{dt} = k(A - L)$, we get
$$\frac{d^2L}{dt^2} = -k\frac{dL}{dt} = -k(k(A - L)) = -k^2(A - L)$$
So if $L < A$, the second derivative of L is negative; i.e., L is concave down.

31. $b_1 = b(0) + b'(0) \cdot 0.5 = 10 + 30 \cdot 0.5 = 25$. When $b = 25$, then $b' = 3 \cdot 25 = 75$, and $b(1) \approx b_2 = b_1 + b'(25) \cdot 0.5 = 25 + 75 \cdot 0.5 = 62.5$. The exact answer is $10e^{3.0} = 200.85$. Euler's method is way off.

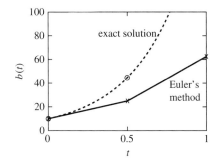

33.
$z_1 = z(0) + z'(0) \cdot 1 = 2 + 1 \cdot 1 = 3$
$z_2 = z_1 + z'(1) \cdot 1 = 3 + 0.5 \cdot 1 = 3.5$
$z_3 = z_2 + z'(2) \cdot 1 = 3.5 + 0.4 \cdot 1 = 3.9$
$z(4) \approx z_4 = z_3 + z'(3) \cdot 1 = 3.9 + 0.344 \cdot 1 = 4.244$.

The exact answer is $1 + \sqrt{1 + 2 \cdot 4} = 4$. Euler's method is pretty close.

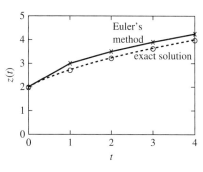

35. The derivative of z can be found with the chain rule, by thinking of $z(t)$ as the composition $F \circ b(t)$ where $F(b) = 4b + 2$. Then
$$\frac{dz}{dt} = \frac{dF}{db}\frac{db}{dt} = 4\frac{db}{dt} = 4(4b + 2)$$
We need to rewrite the right-hand side entirely in terms of z, and get
$$\frac{dz}{dt} = 4z$$
This is simpler because the +2 term has disappeared.

37. The derivative of y can be found with the chain rule, by thinking of $y(t)$ as the composition $F \circ x(t)$ where $F(x) = x^2$. Then
$$\frac{dy}{dt} = \frac{dF}{dx}\frac{dx}{dt} = 2x\frac{dx}{dt} = 2x\left(2x + \frac{1}{x}\right) = 4x^2 + 2$$
We need to rewrite the right-hand side entirely in terms of y, and get
$$\frac{dy}{dt} = 4y + 2$$
This differential equation is linear and the transformation succeeded in simplifying the equation.

39. A line with intercept 4 and slope -0.001 has equation $\lambda(b) = 4 - 0.001b$, so that the differential equation is $\frac{db}{dt} = (4 - 0.001b)b$. When $b = 1000$, $\frac{db}{dt} = 3000 > 0$, so this population would increase. When $b = 5000$, $\frac{db}{dt} = -5000 < 0$, so this population would decrease.

41. A line with intercept -5 and slope 0.001 has equation $\lambda(b) = -5 + 0.001b$, so that the differential equation is $\frac{db}{dt} = (-5 + 0.001b)b$. When $b = 1000$, $\frac{db}{dt} = -4000 < 0$, so this population would decrease. When $b = 3000$, $\frac{db}{dt} = -6000 < 0$, so this population would also decrease.

43.

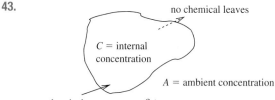

The equation is $\frac{dC}{dt} = \beta A$. The chemical concentration increases linearly without bound.

45.

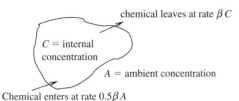

The equation is $\frac{dC}{dt} = -\beta C + 0.5\beta A$. If the concentrations were equal, the derivative would be negative, meaning that the internal concentration would decrease and become lower than the external concentration.

47.

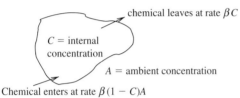

The equation is $\frac{dC}{dt} = -\beta C + \beta(1-C)A$. If the concentrations were equal, the derivative would be negative, meaning that the internal concentration would decrease and become smaller than the external concentration.

49. The differential equations are

$$\frac{da}{dt} = 2(1-p)a$$
$$\frac{db}{dt} = 1.5pb$$

Then

$$\frac{dp}{dt} = \frac{b\frac{da}{dt} - a\frac{db}{dt}}{(a+b)^2}$$
$$= \frac{2(1-p)(ab) - 1.5p(ab)}{(a+b)^2}$$
$$= (2(1-p) - 1.5p)p(1-p)$$

The rate of change is positive for small values of p, but negative for large values of p. Type a will thus decrease when it becomes common and not be able to take over.

51. a. $T(t) = 30 + 10e^{-0.02t}$. $T(0) = 40$, matching the initial condition:

$$\frac{dT}{dt} = -0.2e^{-0.02t}$$

This is supposed to match $\alpha(A - T(t))$, which is

$$\alpha(A - T(t)) = 0.02(30 - (30 + 10e^{-0.02t}))$$
$$= -0.2e^{-0.02t}$$

It checks.

b. $T(1) \approx 39.8$ and $T(2) \approx 39.6$.

c. The term $10e^{-0.02t}$ approaches 0, so $\lim_{t \to \infty} T(t) = 30$, the ambient temperature.

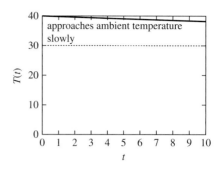

53. We find $T'(0) = 0.02(30 - 40) = -0.2$. Then $T_1 = T(0) + T'(0) \cdot 1 = 40 - 0.2 \cdot 1 = 39.8$. Next, $T'(1) = 0.02(30 - 39.8) = -0.196$, and so

$$T(2) \approx T_2 = T(1) + T'(1) \cdot 1 = 39.8 - 0.196 \cdot 1 = 39.604°C$$

This is right on. Euler's method is very accurate because the temperature is changing almost linearly.

55.
$$C(t) = A + e^{-\beta t}(C(0) - A) = 2.0 + e^{-0.1t}(5-2)$$
$$= 2 + 3e^{-0.1t}$$
$$C(10) = 2 + 3e^{-1.0} \approx 3.104$$
$$C(20) = 2 + 3e^{-2.0} \approx 2.406$$
$$C(60) = 2 + 3e^{-6.0} \approx 2.007$$

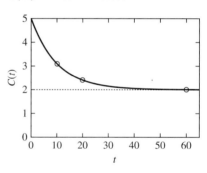

57. When $b_0 = 100$ and $\ln r = -3$ or $r \approx 0.05$

59. The growth matches when $\ln r = \lambda$ or $r = e^\lambda$.

61. We need to take four steps. Recall that $p' = (\mu - \lambda)p(1-p) = -0.5p(1-p)$:

$$p_1 = p(0) + p'(0) \cdot 0.25 = 0.6 - 0.5 \cdot 0.6 \cdot 0.4 \cdot 0.25 = 0.57$$
$$p_2 = p_1 + p'(0.5) \cdot 0.25 = 0.57 - 0.5 \cdot 0.57 \cdot 0.43 \cdot 0.25 \approx 0.539$$
$$p_3 = p_2 + p'(1) \cdot 0.25 = 0.539 - 0.5 \cdot 0.539 \cdot 0.461 \cdot 0.25 \approx 0.508$$
$$p(1) \approx p_4 = p_3 + p'(1.5) \cdot 0.25 = 0.508 - 0.5 \cdot 0.508 \cdot 0.492 \cdot 0.25 \approx 0.477$$

The exact solution is

$$p(t) = \frac{0.6e^{2.5t}}{0.6e^{2.5t} + 0.4e^{3.0t}}$$

so that $p(1) = 0.476$. Euler's method is extremely close.

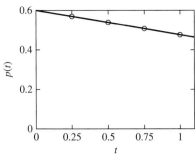

63.

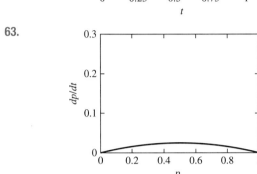

The rate of change is greatest when $p = 0.5$.

Section 8.2

1. This equation is autonomous because the only variable on the right-hand side is the state variable x. To find equilibria, we solve $1 - x^2 = (1-x)(1+x) = 0$; thus, $x = 1$ or $x = -1$.

3. This equation is autonomous because the only variable on the right-hand side is the state variable y. Equilibria: $y \cos y = 0$, i.e., $y = 0$ or $y = \frac{\pi}{2} + \pi n$ for any integer value of n.

5. This equation is autonomous. Equilibrium: $1 - ax = 0$, i.e., $x = \frac{1}{a}$.

7. This equation is autonomous. To find the equilibrium, we solve $\alpha e^{\beta W} - 1 = 0$: from $\alpha e^{\beta W} = 1$ we get $e^{\beta W} = \frac{1}{\alpha}$, thus $\beta W = \ln(1/\alpha) = -\ln \alpha$ and $W = -\frac{\ln \alpha}{\beta}$.

9.

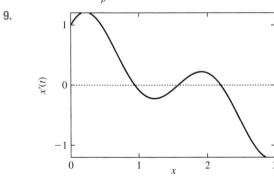

11.

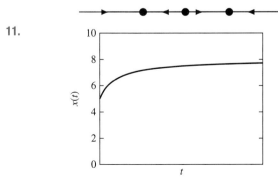

13.

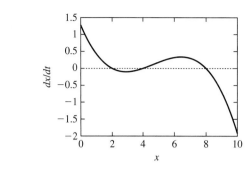

15.

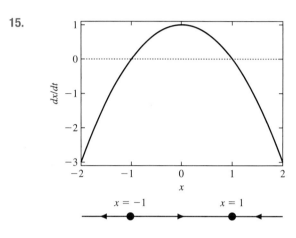

17.

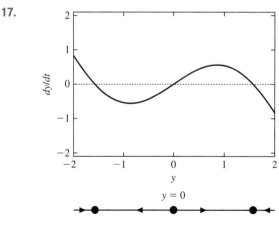

19. The equilibrium is $b^* = 500$.

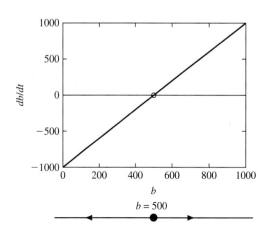

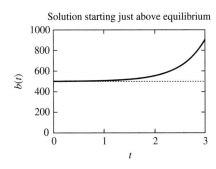

Solution starting just above equilibrium

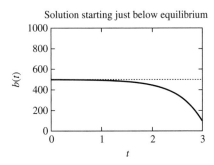

Solution starting just below equilibrium

This population can outgrow the harvest if it starts at a large enough value. If it starts too small, the harvest will drive it to extinction.

21. We found that the population obeys the autonomous differential equation $\frac{db}{dt} = (1 - 0.002b)\,b$. This is in factored form and has equilibria at $b = 500$ and at $b = 0$.

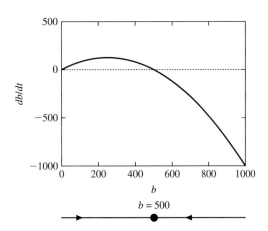

$b = 500$

The arrow points up at $b = 10$, consistent with an increasing population, and down at $b = 1000$, consistent with a decreasing population.

23. We found that the population obeys the autonomous differential equation $\frac{db}{dt} = (-2 + 0.01b)\,b$. This is in factored form and has equilibria at $b = 200$ and at $b = 0$.

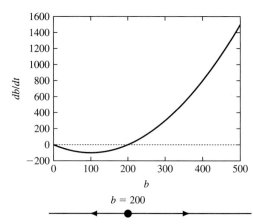

$b = 200$

The arrow points down at $b = 100$, consistent with a decreasing population, and up at $b = 300$, consistent with an increasing population.

25. We found that the population obeys the autonomous differential equation $\frac{dC}{dt} = -\beta C + 3\beta A$. As long as $\beta \neq 0$, this has equilibria at $C = 3A$.

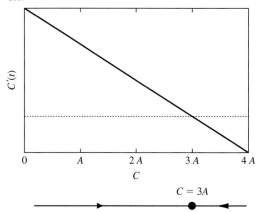

$C = 3A$

The arrow points up at $C = A$, consistent with an increasing concentration.

27. We found that the population obeys the autonomous differential equation $\frac{dp}{dt} = 0.5p\,(1 - p)^2$. This has equilibria at $p = 0$ and $p = 1$.

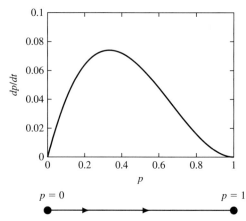

$p = 0$ $\qquad\qquad\qquad\qquad p = 1$

All the arrows point up, except at the equilibria, so the solution moves up to $p = 1$, meaning that a takes over.

29. The equilibria are $N=0$ and the solution of $\frac{3N}{2+N^2} - 1 = 0$, which occurs where $N^2 - 3N + 2 = 0$. This factors to have solutions at $N = 1$ and $N = 2$. To see whether N is increasing or decreasing between the equilibria, we need to check whether $\frac{dN}{dt}$ is positive or negative. We find that $f(1/2) = -1/6 < 0$, $f(3/2) = 3/34 > 0$, and $f(3) = -6/11 < 0$. Therefore, the graph of the rate of change and the phase-line diagram must be the following:

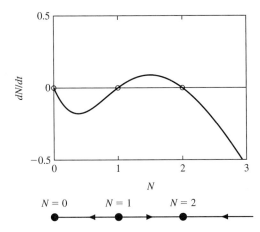

This population dies out if it drops below $N = 1$. Perhaps they cannot find mates when the population gets below one hundred.

31. Everything has units of metres per second squared. The equilibrium is the solution of $9.8 - 0.0032v^2 = 0$, or $v^* = \sqrt{9.8/0.0032} \approx 55.3$ m/s. This is the terminal velocity of a skydiver in free fall.

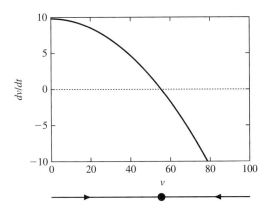

33. Both sides have units of centimeters per second. This checks. The equilibrium is $y^* = 0$, meaning that all water has drained out of the cylinder.

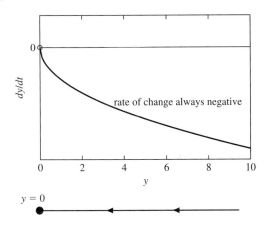

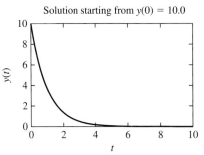

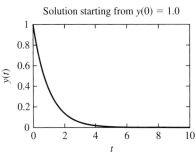

35. The equilibrium is at $S^* = 0$. Eventually, all substrate will be used. In both cases the rate is always negative. However, the graph of the rate for Torricelli's law of draining is much steeper near a value of 0 for the state variable.

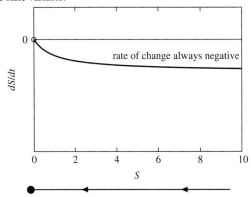

37. Everything has units of cubic centimeters per day. The equilibrium is

$$V^* = \left(\frac{a_1}{a_2}\right)^3$$

The equilibrium gets smaller for smaller values of a_1 because this animal is less effective at collecting food. The equilibrium gets larger when a_2 becomes smaller because this animal is more efficient at using energy.

Section 8.3

1.

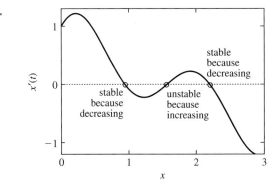

3. We found equilibria at $x=1$ and $x=-1$. The derivative of the rate of change is $-2x$, which is negative at $x=1$ and positive at $x=-1$. Therefore, $x=1$ is stable, consistent with the inward-pointing arrows on the phase-line diagram, and $x=-1$ is unstable, consistent with the outward-pointing arrows on the phase-line diagram.

5. We found equilibria at $y=0$ and $y=\frac{\pi}{2}+\pi n$. The derivative of the rate of change is $\cos y - y \sin y$. This is positive at $y=0$, negative at $y=\pi/2$, and negative at $y=-\pi/2$. Therefore, $y=0$ is unstable (outward-pointing arrows), $y=\pi/2$ is stable (inward-pointing arrows), and $y=-\pi/2$ is stable (inward-pointing arrows).

7. The derivative of the rate of change is $-a$, which is negative for all values of x when $a>0$. Therefore, the equilibrium must be stable.

9. The derivative of the rate of change is $\alpha\beta e^{\beta W}$, which is negative for all values of W when $\alpha>0$ and $\beta<0$. Therefore, the equilibrium must be stable.

11. The equilibrium occurs where $x^2 = 0$, or at $x=0$. The derivative of the rate of change function is $2x$, which is equal to 0 at the equilibrium. The Stability Theorem does not apply.

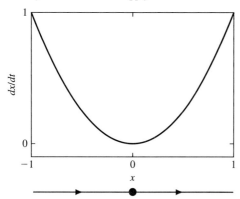

The rate of change is positive except at the equilibrium, so x is increasing except at the equilibrium. This equilibrium is half stable: stable from the left, and unstable to the right.

13. The second derivative is 0 at the equilibrium. The third derivative must be negative because the function switches from being concave up for values of x less than the equilibrium to concave down for values of x greater than the equilibrium.

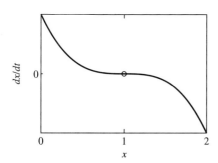

15. Suppose there are two stable equilibria in a row. Then the rate of change must cross from negative to positive at each. In particular, just above the lower one, the value of the rate of change is positive. Just below the upper one, the value of the rate of change is negative. By the Intermediate Value Theorem, there must be another crossing in between.

17. The equilibria are $x=0$ and $x=a$. The derivative of the rate of change function $f(x)=ax-x^2$ is $f'(x)=a-2x$. Then $f'(0)=a$, so the equilibrium at $x=0$ is stable if $a<0$ and unstable if $a>0$. $f'(a)=-a$, so the equilibrium at $x=a$ is stable if $a>0$ and unstable if $a<0$. At $a=0$ there is an exchange of stability when the two equilibria cross.

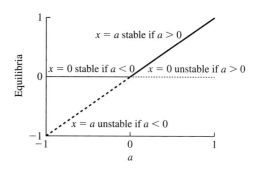

19. The equilibria are $x=0, x=\pm\sqrt{a}$. The last two equilibria do not exist if $a<0$. The derivative of the rate of change function $f(x) = ax - x^3$ is $f'(x) = a - 3x^2$. $f'(0) = a$, so the equilibrium at $x=0$ is stable if $a<0$ and unstable if $a>0$. $f'(\sqrt{a}) = -2a$, so the equilibrium at $x=\sqrt{a}$ is stable when $a>0$. $f'(-\sqrt{a}) = -2a$, so the equilibrium at $x=-\sqrt{a}$ is also stable when $a>0$.

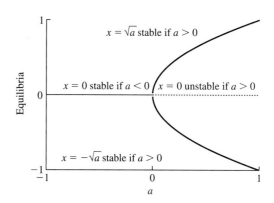

21. This population obeys the autonomous differential equation $\frac{db}{dt} = (1-0.002b)b$ and has equilibria at $b=500$ and at $b=0$. The derivative of the rate of change function $f(b)=(1-0.002b)b$ is $f'(b) = 1-0.004b$. Then $f'(0)=1>0$, so $b=0$ is unstable, and $f'(500) = -1 < 0$, so $b=500$ is stable.

23. This population obeys the differential equation $\frac{db}{dt} = (-2+0.01b)b$ and has equilibria at $b=200$ and at $b=0$. The derivative of the rate of change function $f(b)=(-2+0.01b)b$ is $f'(b)=-2+0.02b$. Then $f'(0)=-2<0$, so $b=0$ is stable, and $f'(200)=2>0$, so $b=200$ is unstable.

25. This population obeys the autonomous differential equation $\frac{dp}{dt} = 0.5p(1-p)^2$ and has equilibria at $p=0$ and at $p=1$. The derivative of the rate of change function $f(p)=0.5p(1-p)^2$ is $f'(p)= 0.5(1-p)^2 - p(1-p)$. Then $f'(0)=0.5>0$, so $p=0$ is unstable, and $f'(1)=0$, so we can not tell. However, the graph does indicate that this equilibrium is stable.

27. a. This is a case where chemical is used up at a rate proportional to its concentration.

   b. Solving $f(C) = (5-C) - C = 0$ gives $C=2.5$. The derivative of the rate of change function is $f'(C) = -2$, so the equilibrium is stable.

   c. Without the reaction, the equilibrium is $C=A=5$. Absorption decreases the equilibrium amount.

29. a. This is a case where chemical is created at a rate that gets larger as the concentration gets larger, but reaches a maximum of 0.5.
    b. Solving $f(C) = (5 - C) + \dfrac{C}{2+C} = 0$ gives $C = 2 \pm \sqrt{14}$. Only one of these values is positive, at $C = 2 + \sqrt{14} = 5.74$. The derivative of the rate of change function is
    $$f'(C) = -1 + \dfrac{1}{2+C} - \dfrac{C}{(2+C)^2}$$
    and $f'(5.74) = -0.97$, so the equilibrium is stable.
    c. Without the reaction, the equilibrium is $C = A = 5$. Chemical creation increases the equilibrium amount, but only slightly.

31. The derivative of the rate of change is
    $$\dfrac{d}{dN}\left(\dfrac{3N^2}{2+N^2} - N\right) = \dfrac{12N}{(2+N^2)^2} - 1$$
    At $N = 0$, this is -1, so the equilibrium at $N = 0$ is stable. At $N = 1$, this is $1/3$, so the equilibrium at $N = 1$ is unstable. At $N = 2$, this is $-1/3$, so the equilibrium at $N = 2$ is stable. This population of 1 acts as a threshold.

33. The derivative of the rate of change is
    $$\dfrac{d}{dv}(9.8 - 0.0032v^2) = -0.0064v$$
    which is negative for any positive speed, including the equilibrium at $v = 55.3$. The equilibrium is stable, consistent with the inward pointing arrows on the phase-line diagram. The falling object will approach its terminal velocity.

35. The derivative of the rate of change is
    $$\dfrac{d}{dS}\left(-\dfrac{S}{1+S}\right) = -\dfrac{1}{(1+S)^2}$$
    which is always negative. Any equilibrium must be stable. In particular, the equilibrium at $S = 0$ is stable, consistent with using up a substance.

37. The derivative of the rate of change is
    $$\dfrac{d}{dV}(a_1 V^{2/3} - a_2 V) = \dfrac{2}{3}a_1 V^{-1/3} - a_2$$
    At the equilibrium $V = \left(\dfrac{a_2}{a_1}\right)^3$, the value is
    $$\dfrac{2}{3}a_1 V^{-1/3} - a_2 = -\dfrac{1}{3}a_2 < 0$$
    The equilibrium is stable, consistent with the inward-pointing arrows on the phase-line diagram. This growing organism will reach a final size when it becomes too big to grow any more. The diagram below is drawn for $a_1 = a_2 = 1$, in which case $V^* = 1$.

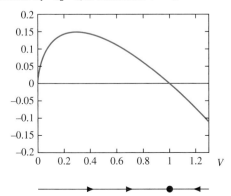

39. The equilibria are $b = 0$ and $b = 1 - h$. The second equilibrium is positive only if $h < 1$. The rate of change function $f(b) = b(1 - b - h)$ has derivative $f'(b) = 1 - h - 2b$. Then $f'(0) = 1 - h$, so $b = 0$ is stable if $h > 1$ and unstable if $h < 1$. Also $f'(1 - h) = h - 1$, so $b = 1 - h$ is stable if $h < 1$ and unstable if $h > 1$ (where it is negative and makes no biological sense).

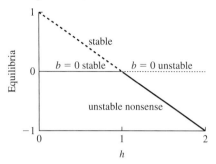

41. The equilibria are $N = 0, N = \dfrac{r + \sqrt{r^2 - 4}}{2}$, and $N = \dfrac{r - \sqrt{r^2 - 4}}{2}$. The last two only exist when $r \geq 2$. The rate of change function $f(N) = \dfrac{rN^2}{1+N^2} - N$ has derivative
    $$f'(N) = \dfrac{2rN}{1+N^2} - \dfrac{2rN^3}{(1+N^2)^2} - 1$$
    Then $f'(0) = -1$, so $I = 0$ is always stable. At $r = 3$, the equilibria are 2.618 and 0.382. Substituting in gives $f'(2.618) = -0.745$ and $f'(0.382) = 0.745$. The larger equilibrium is stable and the smaller one is unstable.

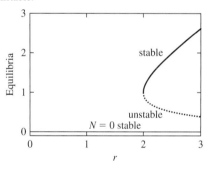

43. With $\alpha = 4$, the differential equation is $\dfrac{dI}{dt} = 4I^2(1 - I) - I$. The equilibria are $I = 0$ and $I = 1/2$.

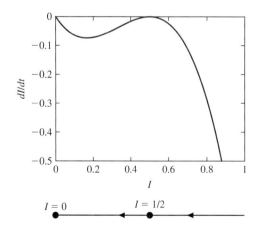

The derivative of the rate of change is $8I(1 - I) - 4I^2 - 1$, which is negative at $I = 0$ and equal to 0 at $I = 1/2$. From our phase-line diagram, however, we see that the rate of change is always negative, implying that the equilibrium at $I = 1/2$ is half-stable, stable from the right and unstable from the left.

Section 8.4

1. $\frac{db}{b} = 0.01 dt$, so $\ln b = 0.01t + c$ and $b(t) = Ke^{0.01t}$ with $K = 1000$.
Checking,
$$\frac{db}{dt} = \frac{d}{dt}(1000e^{0.01t}) = 10e^{0.01t} = 0.01 \cdot 1000e^{0.01t} = 0.01 b(t)$$

3. $\frac{dN}{1+N} = dt$, so $\ln(1+N) = t + c$, $1 + N(t) = Ke^t$, and $N(t) = Ke^t - 1$.
Substituting $t = 0$, we find $K = 2$, so $N(t) = 2e^t - 1$. Checking,
$$\frac{dN}{dt} = \frac{d}{dt}(2e^t - 1) = 2e^t = 1 + (2e^t - 1) = 1 + N(t)$$

5. $\frac{db}{1000 - b} = dt$, so $-\ln(1000 - b) = t + c$, $1000 - b(t) = Ke^{-t}$, and
$b(t) = 1000 - Ke^{-t}$. Substituting $t = 0$, we find $K = 500$, so $b(t) = 1000 - 500e^{-t}$. Checking,
$$\frac{db}{dt} = \frac{d}{dt}(1000 - 500e^{-t}) = 500e^{-t} = 1000 - (1000 - 500e^{-t})$$
$$= 1000 - b(t)$$

7. $P = \dfrac{4}{\sqrt{-16t + 17}}$

9. $y = \arcsin(2t)$

11. $dP = 5e^{-2t} dt$. Integrating with the substitution $s = -2t$, we find $P(t) = -2.5^{-2t} + c$. The constant is $c = 2.5$, giving a solution $P(t) = -2.5e^{-2t} + 2.5$. Checking,
$$\frac{dP}{dt} = \frac{d}{dt}(-2.5e^{-2t} + 2.5) = 5.0e^{-2t}$$

13. $dL = 64.3e^{-1.19t} dt$. Integrating with the substitution $s = -1.19t$, we find $L(t) = -54e^{-1.19t} + c$. The constant is $c = 54$, giving a solution $L(t) = -54e^{-1.19t} + 54$. Checking,
$$\frac{dL}{dt} = \frac{d}{dt}(-54e^{-1.19t} + 54) = 64.3e^{-1.19t}$$

15. $y = \tan\left(\dfrac{1}{2}x^2 - 3x - \dfrac{\pi}{4}\right)$

17. $P = -\ln\left(-\dfrac{1}{2}t^4 + e^{-1}\right)$

19. The solution is $P(t) = \dfrac{4000}{1 + 7e^{-0.08t}}$; from $P(t) = \dfrac{4000}{1 + 7e^{-0.08t}} = 0.95 \cdot 4000$ we find $t = -\dfrac{1}{0.08} \ln\left(\dfrac{1}{7} \cdot \dfrac{0.05}{0.95}\right) \approx 61.129$.

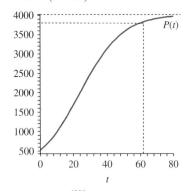

21. The solution is $P(t) = \dfrac{4000}{1 + 3e^{-0.02t}}$; from $P(t) = 0.95 \cdot 4000$ we find $t = -\dfrac{1}{0.02} \ln\left(\dfrac{1}{3} \cdot \dfrac{0.05}{0.95}\right) \approx 202.153$.

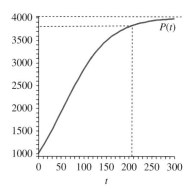

23. The solution is $P(t) = \dfrac{4000}{1 + 3e^{-0.12t}}$; from $P(t) = 0.95 \cdot 4000$ we find $t = -\dfrac{1}{0.12} \ln\left(\dfrac{1}{3} \cdot \dfrac{0.05}{0.95}\right) \approx 33.692$.

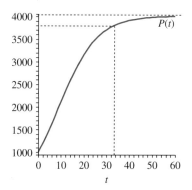

25. The per capita production is an increasing function of time, meaning that things are getting better and better for this population. $\dfrac{db}{b} = t\, dt$, so $\ln b = t^2/2 + c$ and $b = Ke^{t^2/2}$ with $K = 10^6$. This population grows faster than exponentially. The solution is $b(t) = 10^6 e^{t^2/2}$. To check, we must use the chain rule to find the derivative. If we write $b(t) = f(g(t))$, where $g(t) = t^2/2$ and $f(g) = 10^6 e^g$, then $g'(t) = t$ and $f'(g) = 10^6 e^g$, so $b'(t) = f'(g)g'(t) = 10^6 t e^g = 10^6 t e^{t^2/2} = t b(t)$.

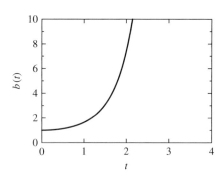

27. The per capita production is an oscillating function of time. $\dfrac{db}{b} = \cos t\, dt$, so $\ln b = \sin t + c$ and $b = Ke^{\sin t}$ with $K = 10^6$. This population oscillates around its initial value. The solution is $b(t) = 10^6 e^{\sin t}$. To check, we must use the chain rule to find the derivative. If we write $b(t) = f(g(t))$, where $g(t) = \sin t$ and $f(g) = 10^6 e^g$, then $g'(t) = \cos t$ and $f'(g) = 10^6 e^g$, so
$$b'(t) = f'(g)g'(t) = 10^6 \cos t\, e^g = 10^6 \cos t\, e^{\sin t} = \cos t\, b(t)$$

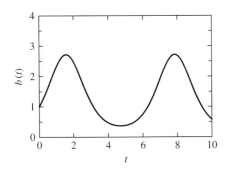

29. $b(t) = \dfrac{1}{0.01 - t}$. It blows up at time $t = 0.01$.

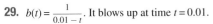

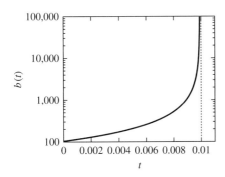

31. $b(t) = \left(\dfrac{10}{6.31 - t}\right)^{10}$. It blows up at time $t \approx 6.31$.

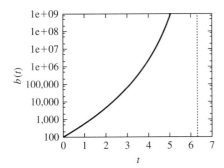

33. a.
$$\dfrac{1+x}{x} dx = dt$$
$$\ln x + x + c_1 = t + c_2$$
$$\ln x + x = t + c$$

b. Substitute $t = 0$ and $x = 1$, and find $c = 1$.

c. $t = \ln x + x - 1$

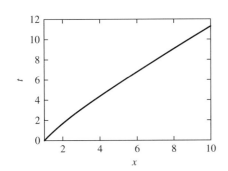

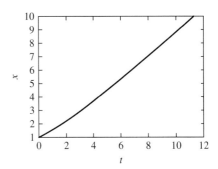

35. $L(t) = A - Ae^{-kt}$

37. Solution of the logistic equation: $P_L(t) = \dfrac{2000}{1 + \frac{33}{7} e^{-0.09t}}$
$P_L(t)$ will reach 90% of the carrying capacity when $t_L = \dfrac{1}{-0.09} \ln(7/297) \approx 41.64$.
Solution of the Gompertz equation: $P_G(t) = 2000 e^{\ln(7/40) \cdot e^{-0.09t}}$.
$P_G(t)$ will reach 90% of the carrying capacity when $t_G = -\dfrac{1}{0.09} \ln\left(\dfrac{\ln 0.9}{\ln(7/40)}\right) \approx 31.18$.

39. We find $P'' = k^2 P(\ln L - \ln P)(\ln L - \ln P - 1)$. Setting $P'' = 0$, we obtain $P = Le^{-1}$. By substituting this value into the solution of the Gompertz equation $P(t) = L e^{\ln(P_0/L) \cdot e^{-kt}}$ we can find the vlaue of t at the point of inflection: $t = \dfrac{1}{k} \ln(\ln(L/P_0))$.

41.
$$C(t) = A + e^{-\beta t}(C(0) - A)$$
$$= 2 + e^{-0.01t}(5 - 2)$$
$$= 2 + 3e^{-0.01t}$$
$$C(10) = 2 + 3e^{-0.1}$$
$$\approx 4.715$$

The equilibrium is 2, and it is halfway to the equilibrium when $C(t) = 3.5 \cdot 10^{-5}$. Solving for t,
$$3.5 = 2 + 3e^{-0.01t}$$
$$1.5 = 3e^{-0.01t}$$
$$0.5 = e^{-0.01t}$$
$$\ln 0.5 = -0.01t$$
$$-100 \cdot \ln 0.5 = t \approx 69.3$$

43. $C(t) = 2 + 3e^{-0.1t}$, $C(10) = 3.104$. The time to reach 3.5, halfway to the equilibrium, is 6.93 s.

45.
$$\dfrac{dy}{2\sqrt{y}} = -dt$$
$$\sqrt{y} + c_1 = -t + c_2$$
$$\sqrt{y} = -t + c$$
$$y = (-t + c)^2$$

The constant c solves $y(0) = 4 = (-0 + c)^2$, or $c = 2$.

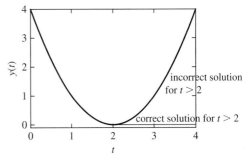

The cylinder is empty at $t = 2$, so the solution must stay at zero. The solution in this case actually reaches 0 at $t = 2$, unlike the exponentially decaying solution of the bacterial death equation, which only approaches 0 as a limit.

47.
$$\frac{db}{2b - 1000} = dt$$
$$\frac{1}{2}\ln(2b - 1000) + c_1 = t + c_2$$
$$\ln(2b - 1000) = 2t + c$$
$$2b - 1000 = Ke^{2t}$$
$$b = Ke^{2t} + 500$$

If the solution starts just above the equilibrium at 500, say with $b(0) = 501$, we find $K = 1$, and a solution $b(t) = 500 + e^{2t}$. This grows exponentially away from the equilibrium. If the solution starts just below the equilibrium at 500, say with $b(0) = 499$, we find $K = -1$, and a solution $b(t) = 500 - e^{2t}$. This declines exponentially away from the equilibrium.

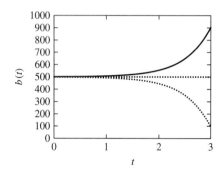

49. We found the differential equation $\frac{dC}{dt} = -\beta C + 3\beta A$. Separating variables,
$$\frac{dC}{C - 3A} = -\beta dt$$
$$\ln(C - 3A) + c_1 = -\beta t + c_2$$
$$C - 3A = Ke^{-\beta t}$$
$$C = Ke^{-\beta t} + 3A$$

With the initial condition $C(0) = A$, the constant K is $-2A$. The solution increases from this point up to the equilibrium at $C = 3A$.

51. a. $\frac{1}{p(1-p)} dp = dt$

b. Compute the common denominator to find $\frac{1}{p} + \frac{1}{1-p} = \frac{1-p}{p(1-p)} + \frac{p}{p(1-p)} = \frac{1}{p(1-p)}$.

c. $\int (\frac{1}{p} + \frac{1}{1-p}) dp = \ln(p) - \ln(1-p) + c_1$

d. The subtraction outside the logs turns into division inside, giving $\ln\left(\frac{p}{1-p}\right) + c_1$.

e. $\ln\left(\frac{p}{1-p}\right) = t + c$

f. Exponentiating both sides gives $\frac{p}{1-p} = e^{t+c} = Ke^t$. Solving, we get $p = \frac{Ke^t}{1+Ke^t}$.

g. $0.01 = \frac{K}{1+K}$, so $K = 0.0101$. The limit as t approaches infinity is $p = 1$.

h. $0.5 = \frac{K}{1+K}$, so $K = 1$. The limit as t approaches infinity is still $p = 1$.

53. a. We find
$$\frac{dy}{dt} = e^t T + e^t \frac{dT}{dt}$$
$$= e^t T + e^t(-T + A(t))$$
$$= e^t A(t) = e^{(1+\beta)t}$$

b. This is a pure-time differential equation, which can be solved by integration. We find that
$$y(t) = \frac{e^{(1+\beta)t}}{1+\beta} + c$$

Because $H(t) = e^{-t} y(t)$, then
$$H(t) = \frac{e^{\beta t}}{1+\beta} + e^{-t} c$$

To match the initial condition requires that $c = \frac{1}{1+\beta}$.

c. Using $\beta = 0.1$ gives the following graph.

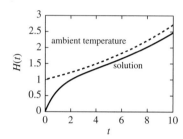

The solution does a pretty good job of tracking the ambient temperature.

d. Using $\beta = 1$ gives the following graph.

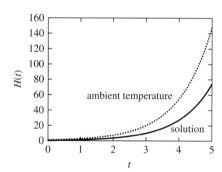

The solution lags behind because the temperature is changing too quickly for the object to keep up.

Section 8.5

1. The equations are
$$\frac{db}{dt} = \lambda b$$
$$\frac{dp}{dt} = -\delta p.$$

These two species ignore each other. The prey will grow exponentially and the predators will die out.

3. The equations are
$$\frac{db}{dt} = (\lambda + \varepsilon p)b$$
$$\frac{dp}{dt} = -\delta p$$

The predators will die out, even though they are eating some of the prey. Apparently, these prey are not at all nutritious.

5. The equation for a is

$$\frac{da}{dt} = 2\left(1 - \frac{a}{1000}\right)a.$$

with equilibria at $a = 0$ and $a = 1000$. The equilibrium at 0 is unstable and the equilibrium at 1000 is stable.

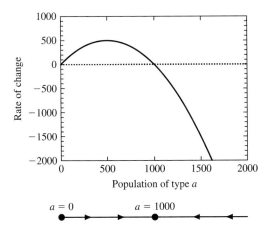

7.

t	$a'(t)$	$b'(t)$	Approximation for a	Approximation for b
0.0	−375	−1500	750	500
0.1	−89.06	−787.5	$750 − 375 \cdot 0.1 = 712.5$	$500 − 1500 \cdot 0.1 = 350$
0.2	35.4	−515.2	$712.5 − 89.06 \cdot 0.1 = 703.6$	$350 − 787.5 \cdot 0.1 = 271.3$

9.

t	$T'(t)$	$A'(t)$	Approximation for T	Approximation for A
0.0	−12	4	60	20
0.1	−11.5	3.84	$60 − 12 \cdot 0.1 = 58.8$	$20 + 4 \cdot 0.1 = 20.4$
0.2	−11.1	3.69	$58.8 − 11.5 \cdot 0.1 = 57.6$	$20.4 + 3.84 \cdot 0.1 = 20.8$

11.
$$\frac{db}{dt} = (1 - 0.05p)b$$
$$\frac{dp}{dt} = (-1 + 0.02b)p$$

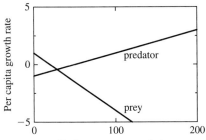

13.
$$\frac{db}{dt} = (2 - 0.0001p^2)b$$
$$\frac{dp}{dt} = (-1 + 0.01b)p$$

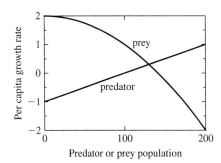

15. Call the predators p_1 and p_2. Then

$$\frac{dp_1}{dt} = (-1 + 0.001p_2)p_1$$
$$\frac{dp_2}{dt} = (-1 + 0.001p_1)p_2$$

17.
$$\frac{da}{dt} = \mu\left(1 - \frac{a+b}{K_a}\right)a$$
$$\frac{db}{dt} = \lambda\left(1 - \frac{b}{K_b}\right)b$$

19. a. $A_1 = C_1 V_1 = 2C_1$ and $A_2 = C_2 V_2 = 5C_2$.

b.
$$\frac{dA_1}{dt} = \beta(C_2 - C_1)$$
$$\frac{dA_2}{dt} = \beta(C_1 - C_2)$$

c. Then
$$\frac{dC_1}{dt} = \frac{\beta}{V_1}(C_2 - C_1) = \frac{\beta}{2.0}(C_2 - C_1)$$
$$\frac{dC_2}{dt} = \frac{\beta}{V_2}(C_1 - C_2) = \frac{\beta}{5.0}(C_1 - C_2)$$

d. The concentration changes faster in the smaller cell.

21. The general form is
$$\frac{dT}{dt} = \alpha(A - T)$$
$$\frac{dA}{dt} = \alpha_2(T - A)$$

α_2 will be 1/10 the size due to the size of the room, but 5 times as large due to the specific heat. Therefore, we expect that

$$\frac{dT}{dt} = \alpha(A - T)$$
$$\frac{dA}{dt} = \frac{\alpha}{2}(T - A)$$

23. The equations are
$$\frac{dI}{dt} = \alpha IS - \mu I$$
$$\frac{dS}{dt} = -\alpha IS$$

The diseased I individuals follow an equation that looks exactly like the predator equation (with μ instead of δ and α instead of η). The S equation is a little like the prey equation, but lacks any reproduction (the λ term), and the coefficient in front of the IS term is exactly equal to that in the I equation.

25. The equations are

$$\frac{dI}{dt} = \alpha IS - \mu I - kI$$

$$\frac{dS}{dt} = -\alpha IS + \mu I - kS$$

27. The equations are

$$\frac{dI}{dt} = bI + \alpha IS - \mu I$$

$$\frac{dS}{dt} = bS - \alpha IS + \mu I$$

29. In terms of p, we found differential equations

$$\frac{da}{dt} = 2(1-p)a$$

$$\frac{db}{dt} = 1.5(1-p)b$$

Substituting in $1 - p = \frac{b}{a+b}$ gives the equations

$$\frac{da}{dt} = 2\frac{b}{a+b}a$$

$$\frac{db}{dt} = 1.5\frac{b}{a+b}b$$

Section 8.6

1. Solving each equation for y gives

$$y = 3x - 3$$

$$y = 2x + 1$$

Setting the equations equal, we find $3x - 3 = 2x + 1$, so $x = 4$. Plugging in gives $y = 9$ in each equation.

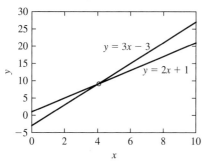

3. Solving each equation for y gives

$$y = x + 1$$

$$y = x + 1$$

Setting the equations equal, we find $x + 1 = x + 1$. Any value of x is a solution. This happens because the two lines are identical, and any value where $y = x + 1$ is a solution.

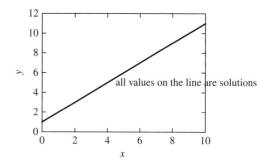

5. Solving each equation for y gives

$$y = 3x^2 + 2x - 5$$

$$y = 2x + 1$$

Combining the equations, we find $3x^2 + 2x - 5 = 2x + 1$, so $3x^2 = 6$, and $x = \pm\sqrt{2}$. With $x = \sqrt{2}$, the first equation gives $y = 1 + 2\sqrt{2}$, as does the second. With $x = -\sqrt{2}$, the first equation gives $y = 1 - 2\sqrt{2}$, as does the second.

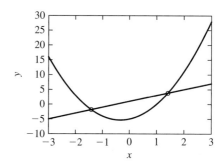

7. Solving the first equation for y gives $y = x$ and $y = -x$. The second gives $y = 2x + 1$. This gives two pairs of equations:

$$y = x$$
$$y = 2x + 1$$

and

$$y = -x$$
$$y = 2x + 1$$

Setting the first pair equal, we find $x = 2x + 1$, so $x = -1$ and $y = -1$. Setting the second pair equal, we find $-x = 2x + 1$, so $x = -1/3$ and $y = 1/3$.

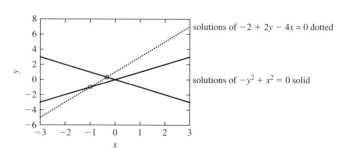

9. Solving the first equation for y gives $y = x$ and $x = 0$. The second gives $y = 2x + 3$. This gives two pairs of equations:

$$y = x$$
$$y = 2x + 3$$

and

$$x = 0$$
$$y = 2x + 3$$

Setting the first pair equal, we find $x = 2x + 3$, so $x = -3$ and $y = -3$. We can substitute $x = 0$ into the second equation of the second pair to find $y = 3$.

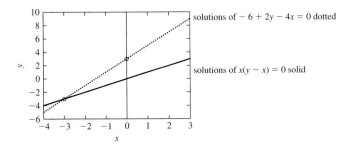

11. The p-nullcline consists of the two pieces $p = 0$ and $b = \delta/\eta = 600$. The b-nullcline consists of the two pieces $b = 0$ and $p = \lambda/\varepsilon = 500$. The equilibria are $(0, 0)$ and $(600, 500)$.

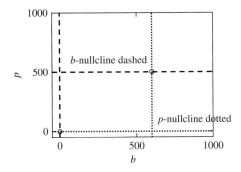

13. Both nullclines are the line $T = A$, which consists entirely of equilibria.

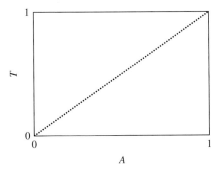

15. Place b on the vertical axis. The a-nullcline is the two pieces $a = 0$ and $b = 10^6 - a$. The b-nullcline is the two pieces $b = 0$ and $b = 10^7 - a$. The equilibria are $(0, 0)$, $(10^6, 0)$, and $(0, 10^7)$.

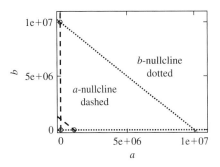

17. Place b on the vertical axis. The p-nullcline consists of the two pieces $p = 0$ and $b = 600$, and the b-nullcline consists of the two pieces $b = 0$ and $p = 500$ as before. The equilibria are $(0, 0)$ and $(500, 600)$.

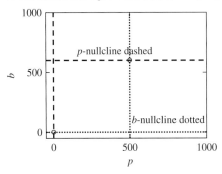

19. Place a on the vertical axis. The a-nullcline is the two pieces $a = 0$ and $a = 10^6 - b$. The b-nullcline is the two pieces $b = 0$ and $a = 10^7 - b$. The equilibria are $(0, 0)$, $(0, 10^6)$, and $(10^7, 0)$.

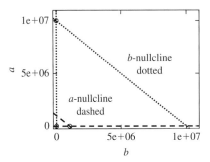

21. The equations are

$$\frac{da}{dt} = 2\left(1 - \frac{a}{1000}\right)a$$
$$\frac{db}{dt} = 3\left(1 - \frac{b}{300}\right)b$$

The a-nullcline is $a = 0$ and $a = 1000$. The b-nullcline is $b = 0$ and $b = 300$. The equilibria are $(0, 0)$ and $(1000, 300)$.

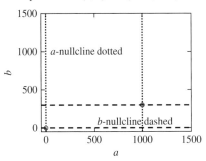

23. With p on the vertical axis, the b-nullcline is $b=0$ and $p=20$. The p-nullcline is $p=0$ or $p=50$. The only equilibria are $(0,0)$ and $(50, 20)$.

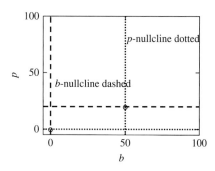

25. With p on the vertical axis, the b-nullcline is $b=0$ and $p=\sqrt{20{,}000} \approx 141$. The p-nullcline is $p=0$ and $b=100$. The equilibria are $(0,0)$ and $(100, 141)$.

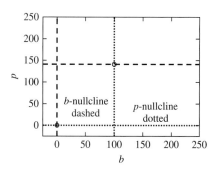

27. We found equations

$$\frac{dp_1}{dt} = (-1.0 + 0.001p_2)p_1$$

$$\frac{dp_2}{dt} = (-1.0 + 0.001p_1)p_2$$

Putting p_2 on the vertical axis, the p_1-nullcline is $p_1=0$ and $p_2=1000$. The p_2-nullcline is $p_2=0$ and $p_1=1000$. The equilibria are $(0,0)$ and $(1000, 1000)$.

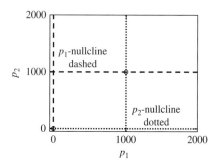

29. We found equations

$$\frac{da}{dt} = \mu\left(1 - \frac{a+b}{K_a}\right)a$$

$$\frac{db}{dt} = \lambda\left(1 - \frac{b}{K_b}\right)b$$

Putting b on the vertical axis, the a-nullcline is $a=0$ and $b = K_a - a$. The b-nullcline is $b=0$ and $a = K_b$. The equilibria are $(0,0)$, $(K_a, 0)$, and $(K_b, K_a - K_b)$ (if $K_a > K_b$).

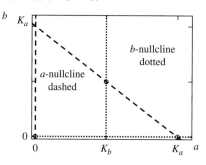

31. a. Let C_1 be the concentration in the first vessel. Then chemical moves from the first to the second at rate βC_1.

b. Let C_2 be the concentration in the second vessel. Then chemical moves from the second to the first at rate $\frac{\beta}{3}C_2$.

c. Let A_1 and A_2 be the total amounts. Then

$$\frac{dA_1}{dt} = \beta\left(\frac{C_2}{3} - C_1\right)$$

$$\frac{dA_2}{dt} = \beta\left(C_1 - \frac{C_2}{3}\right)$$

d.
$$\frac{dC_1}{dt} = \frac{\beta}{V_1}\left(\frac{C_2}{3} - C_1\right)$$

$$\frac{dC_2}{dt} = \frac{\beta}{V_2}\left(C_1 - \frac{C_2}{3}\right)$$

e. Place C_2 on the vertical axis. $C_2 = 3C_1$ is the nullcline for both C_1 and C_2.

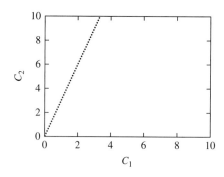

f. All points along the line $C_2 = 3C_1$ are equilibria. The equilibrium concentration in the second vessel is three times that in the first.

33. The equations look right because the per capita growth of a decreases half as quickly as a function of b as it does as a function of a. Therefore, individuals of type b decrease reproduction of individuals of type a only half as much. Similarly, the per capita growth of b decreases twice as quickly as a function of a as it does as a function of b. The a-nullcline is $a=0$ and $b=2(1000-a)$. The b-nullcline is $b=0$ and $b=1000-2a$.

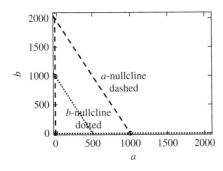

The two nullclines are parallel and do not intersect except at the boundaries.

35. The equations are

$$\frac{dI}{dt} = \alpha IS - \mu I = 2IS - I$$
$$\frac{dS}{dt} = -\alpha IS = -2IS$$

The I-nullcline is the two pieces $I=0$ and $S=1/2$. The S-nullcline is the two pieces $I=0$ and $S=0$. There are equilibria wherever $I=0$.

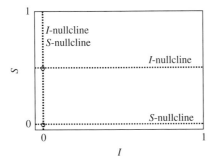

37. The equations are

$$\frac{dI}{dt} = 2IS - 1.5I$$
$$\frac{dS}{dt} = -2IS + I - 0.5S$$

The I-nullcline is the two pieces $I=0$ and $S=3/4$. The S-nullcline is $S = \frac{I}{2I+0.5}$. There is only one equilibrium, at $(0, 0)$, because $\frac{3}{4} = \frac{I}{2I+0.5}$ has no intersection.

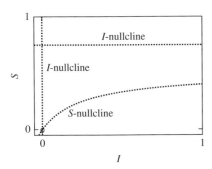

39. The equations are

$$\frac{dI}{dt} = I + 2IS$$
$$\frac{dS}{dt} = 2S - 2IS + I$$

The I-nullcline has only one reasonable piece at $I=0$ (along with $S=-1/2$). The S-nullcline is $S = \frac{I}{2(I-1)}$ (which is only defined for $I=0$ and $I>1$). There is only one equilibrium at $(0,0)$.

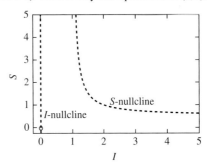

Section 8.7

1.

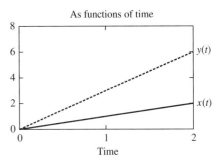

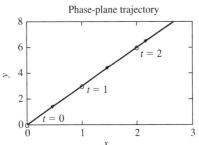

3.

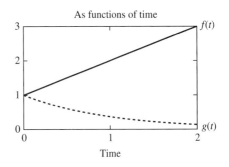

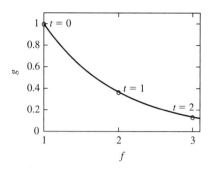

5.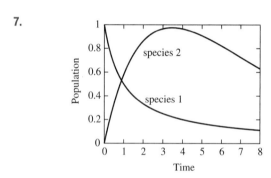

7.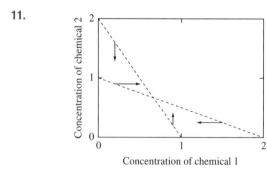

9.

11.

13. We use the hat symbol to denote the approximate values of the variables obtained using Euler's method. We found that $(\hat{a}(0.1), \hat{b}(0.1)) =$ (712.5, 350), with both values having decreased. This is consistent with the fact that our initial conditions lie in Region I (because 750 + 500 > 1000). Similarly, $(\hat{a}(0.2), \hat{b}(0.2)) = (703.6, 271.3)$, with both values having decreased. This is consistent with the fact that our initial conditions lie in Region I (because 712.5 + 350 > 1000). However, the solution has now moved into region II and a will begin to increase.

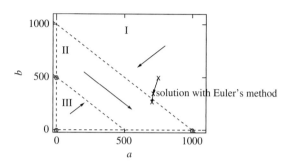

15. We use the hat symbol to denote the approximate values of the variables obtained using Euler's method. We found that $(\hat{T}(0.1), \hat{A}(0.1)) =$ (30, 30), with T having decreased and A increased. This is consistent with the fact that our initial conditions lie in Region II (because 60 > 20). However, $(\hat{T}(0.2), \hat{A}(0.2)) = (30, 30)$ because our first step landed right on the equilibrium. This overly long jump is a consequence of taking too big a step. A real solution could not cross the line of equilibria because it would get stuck.

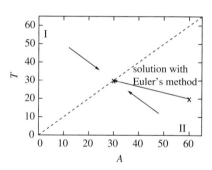

17.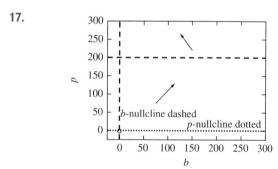

19.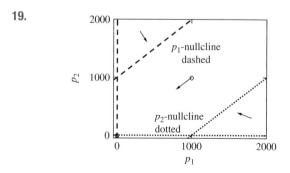

Answers to Odd-Numbered Questions   759

21.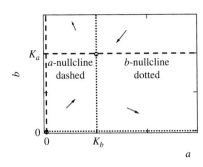

23.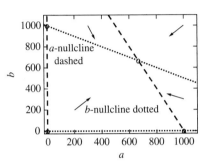

25.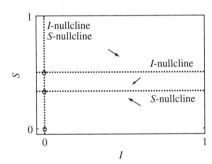

27.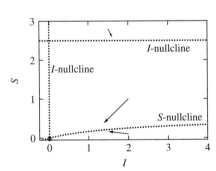

29.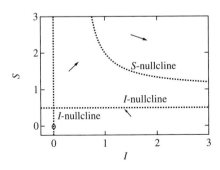

31.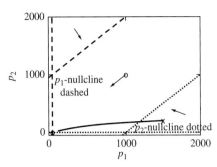

33.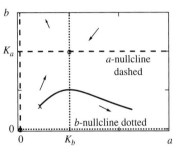

35.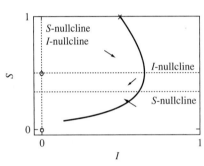

37.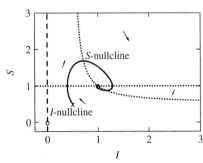

It is also possible that the solution winds outward away from the equilibrium.

True/False Quiz

1. FALSE. $P(t)$ should not be in the denominator in the last factor.

3. FALSE. The ambient temperature is $45°$.

5. FALSE. Setting $p' = 0$, we obtain a degree 3 polynomial equation—which has at most three real solutions.

7. FALSE. The solutions that start above the carrying capacity are decreasing. Or, the equilibrium solutions are constant functions (and not increasing).

9. TRUE. Solving the system $m' = 0$ and $n' = 0$, we obtain $m = n = 0$ and $m = 10, n = 2$.

Supplementary Problems

1. a. If A is constant, this is an autonomous differential equation. The first term represents diffusion of chemical, and the second term describes a constant increase from an outside source.

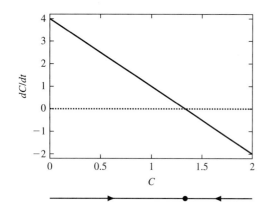

b.

c. The equilibrium is the solution of $3(A - C) + 1 = 0$, or $c = A + 1/3$. If $f(C) = 3(A - C) + 1$, then $f'(C) = -3$, which is clearly negative. Therefore, the equilibrium is stable.

d. With $A = 1$,

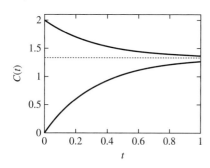

3. a. The first terms look like diffusion. The factor of 3 means that C changes more quickly than A, so it must have a smaller volume. The $A^2/3$ term means that A is being depleted the larger its concentration. The $+1$ is an outside supplement to C.

b.

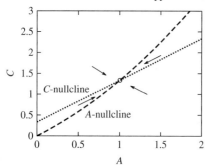

5. a. This is a pure-time differential equation because the rate of change depends only on the time.

b.

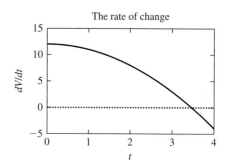

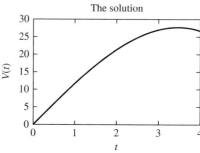

c. V takes on a maximum when $dV/dt = 0$, or at $t = \sqrt{12}$.
d. $V(0.1) \approx V_1 = V(0) + V'(0) \cdot (0.1) = 0 + 12 \cdot 0.1 = 1.2$.
e. Integrating, we find $V(t) = 12t - t^3/3 + c$, and the constant c must be 0. Solving $V(t) = 0$ gives $t = 6$.
f. The average volume is

$$\frac{1}{6} \int_0^6 \left(12t - \frac{t^3}{3}\right) dt = \frac{1}{6}\left(6t^2 - \frac{t^4}{12}\bigg|_0^6\right) = 18$$

7. a. $\dfrac{db}{dt} = \dfrac{1}{\sqrt{b}} \cdot b = \sqrt{b}$

b. Using separation of variables, we get

$$\frac{db}{\sqrt{b}} = dt$$
$$2\sqrt{b} = t + c$$
$$b = \left(\frac{t+c}{2}\right)^2$$

Checking, $b'(t) = (t+c)/2$, which is indeed $\sqrt{b}$.

c. If $b(0) = 10{,}000$, $(c/2)^2 = 10{,}000$, so $c = 200$. Substituting $t = 2$, we find $b = 101^2 = 10{,}201$.

d. A quadratic function increases more slowly than an exponential function, so this population grows more slowly. This occurs because per capita reproduction decreases with larger populations.

9. a. This is a pure-time differential equation because the rate of change depends only on time, not on N.

b.

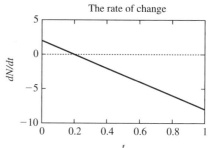

c.

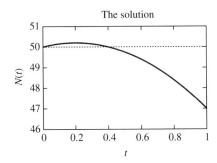

d. $N(0.1) \approx N_1 = N(0) + N'(0) \cdot 0.1 = 50 + 2 \cdot 0.1 = 50.2$.

e. Integrating, we get $N(t) = 2t - 5t^2 + c$. The constant is 50 from the initial condition. Substituting, we find that $N(1) = 47$.

f. $N(t) = 50$ at $t = 0$ and at $t = 0.4$.

11. a. This is a system of coupled autonomous differential equations. This looks like a system in which type a has a lower per capita production, and the per capita production of type a is reduced by its own presence and the presence of the other, and rather more by b, which probably eats a. The per capita production of type b is increased by type a and decreased by itself. Perhaps b's eat a's but fight with each other.

b. The a-nullcline is $a = 0$ and $b = 200 - 0.4a$. The b-nullcline is $b = 0$ and $b = 100 + 0.1a$. The equilibria are (0, 0), (0, 100), (500, 0), and (200, 120).

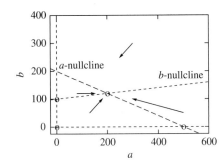

13. a. The equilibria are $L = 0$, $L = 1$, and $L = 2$. At $L = -0.5$, the derivative is positive, at $L = 0.5$, the derivative is negative, at $L = 1.5$ it is positive, and at $L = 2.5$ it is again negative.

$$L = 0 \qquad L = 1 \qquad L = 2$$

b. Taking the derivative of the rate of change with respect to L,

$$\frac{d}{dt}(-L(2-L)(1-L)) = -(2-L)(1-L) + L(1-L) + L(2-L)$$

At $L = 0$ this is -2, at $L = 1$ it is 1, and at $L = 2$ it is -2. Therefore, the equilibrium at $L = 1$ is unstable and the other two are stable.

c.

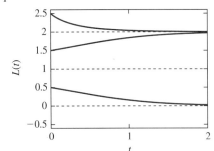

15. a. This is an autonomous differential equation.

b.

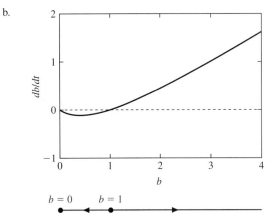

c.

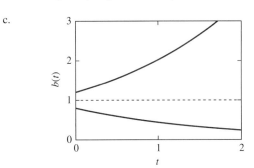

d. The derivative is $-\ln 2 < 0$, and the equilibrium is stable.

e. The rate of change if $b = 0.5$ is -0.111. Therefore,

$$\hat{b}(0.01) = 0.5 - 0.111 \cdot 0.01 \approx 0.49889$$

Index

Absolute (global) extreme values, **345–346**, 355–357
   finding, 357
Absolute maximum, **345**–346
Absolute minimum, **345**–346
Absolute value function, **20**
   continuity, 234–235
   graph, 20
   not differentiable, 245–246
Absorption
   of alcohol, 147–151
   of caffeine, 141–142
   lung dynamics with, 170–172, 222–223
   of pain medication, 115, 138
Absorption function(s), 222–224
   enhanced, 223
   leading behaviour, 394–395
   linear, 223
   with overcompensation, 223, 392
   saturated, 223, 394
   saturated with overcompensation, 223, 395
   saturated with threshold, 223, 395
Acceleration, 325–326
   constant, 482–483
   units, 10
Addition of functions, 25
AIDS
   spread of, 481, 528
Alcohol use
   dynamics, 147–151
   one drink, 147
Algebraic function(s), 37
   continuity, 235
   limit, **198**
Allee, W. C., 594
Allee effect, 163, **593–594**
Allometric relations (formulas), 2, **67**
Allometry, 2
Amplitude of an oscillation, 98–99
Angle addition formulas for sine and cosine, **97**, 297
Antiderivative, 462, **473**
   checking by differentiating, 475
   constant multiple rule, 478
   power rule, 476
   quadratic, 333–335
   sum rule, 479
   table of, 480
Approximating derivatives, 252–253, 325
Approximation of a function
   linear (linearization), 330
   secant line, 331–332
   tangent line, 330
   Taylor polynomial, 338–340
Archimedes, 487, 548, 561, 569, 587
Arcsin (arcsine function, inverse sine), 102–103
Arctan (inverse tangent), 103–104
Area, 455, **487**
   and definite integral, 493, **495**
   difference of, 500
   fixed while optimizing boundary, 360–361
   integral and, 549
   optimizing shape and, 372–375

Area (*Cont.*)
   region between two curves, 551–552
   region under a function on an interval, **495**, 548
   signed, 500
   surface, of Lake Ontario, 554–555
   under a curve, 495, 548
   under a parabola, 487–489
Area of convergence (blood splatters), 105
Asymptotes, **208**, **212**
Asymptotic behaviour, 389
Asymptotic stability, **607**. *See also* Stability
Autocatalytic reaction, **647**
Autonomous differential equation(s), **457**, **589**
   basic exponential model, 590–591
   continuous-time model of selection, 596–598
   coupled, 632
   diffusion, 595–596
   equilibrium (equilibria), 602–**603**, **607**
   Euler's method, 634–636
   graphical display, **605–606**
   logistic differential equation, 591–593
   modified logistic differential equation, 593–594
   Newton's laws of cooling, **594–595**, 596
   phase-line diagram, **605**
   population of Canada, 456–458, 461–462
   separation of variables, 619–620
   stability (asymptotic), **607**, 610, 611
   stability theorem, applications of, 611, 612–614
   stable equilibrium, 610
   systems of, **630**
   unstable equilibrium, 610
Average, **107**
   of an oscillation, 98–99
   rate of change of a function, 180
   value of a function, **564–565**
   weighted, **168–169**, 170
Average rate of change, 179–181
   of a function on an interval, **180**
   slope of a secant line, 181
Average speed (velocity), 184, 243
Average value of a function, **564–565**
Axes (coordinate), **11**

Backward difference quotient, **252**
Bacteria (bacterial population dynamics)
   discrete-time dynamical system, 114, 142–144
   estimation of a number, 9
   limited population, 145–147
   mass, 125
   model of selection, 153, 596
Base point, 58, **181**
Basic exponential model. *See also* Exponential decay; Exponential growth
   continuous, 78, 590, 622
   discrete, 117
Basic trigonometric identity, 93
Bees
   optimizing honeycomb shape, 372–375
   optimizing rate of food intake, 368–372
Bell-shaped curve, 407–408

Bifurcation, 617–619
   pitchfork, 618
   saddle-node, 617
   subcritical, 618
   supercritical, 618
   transcritical, 617
Bifurcation diagram, 617
Binomial theorem, 262
Bisection method, **381**, 414
Blood circulation time, 65–66
Blood flow, 307, 316, 361–362
Blood stains, analyzing, 105–106
Body mass index (BMI), 64–65
Bones
   growth, 68–69
   optimizing strength of, 365–368
Bounded region, **551**
Breathing, 164–171

Caffeine, 141–142
Cancellation formulas (inverse functions), 32, 76
Cancer
   linear-quadratic survival model, 87–88
   mammography, 84–85
   model of treatment, 232
Carbon-14 (radiocarbon) dating, 81
Carrying capacity, 3, 26, 85, 214, 409, 591–593, 632
Cartesian coordinate system, **11**
Catenary, **563**
Cell
   size in relation to shape, 360–361
   survival fraction, 286
Central difference quotient, **252**
Chain rule, **283–284**, 285
   implicit differentiation, 308
   integration by substitution, 530–531
   proof of, 292–293
Change, exponential, **75**
Change in input, **58**
Change in output, **58**
Change of variables (substitution) method, 529
Chaos (chaotic behaviour), **446**
Chemical diffusion. *See* Diffusion
Chemotherapy, 232
Circadian rhythm, 109
Circle, 19, 308
   area, 272, 360, 487
   circumference (perimeter), 374
   encloses largest area, 373
   semicircle, 187, 295, 308
   unit (definition of trig functions), 94
Closed interval, 9
Cobwebbing, **130–131**
   equilibria, 132, 135
Comparing functions at infinity, 219
Comparison test for improper integrals, **578**
Competition model (between populations), 632
   direction arrows, 654–655
   equilibrium, 643–644
   Euler's method, 637–638
   nullclines, 643–644

I-1

Composition (of functions), 27
    continuity, 234
    derivative. *See* Chain rule
    updating functions, 122
Concave down, 318
Concave up, 318
Constant acceleration
    velocity and position, 482–484
Constant function, 19
    antiderivative, 480
    continuity, 234
    derivative, 261
    limit, 195
Constant multiple (product) rule
    for derivatives, **266**, **275**
    for integrals, **478**
Constant of proportionality, **56**
Consumption of drugs. *See* Drug consumption
Continuity
    of composition of functions, 234
    and differentiability, 246–247, 384–385
    of an inverse function, 234–235
Continuous function(s), **230**
    at an endpoint, 234
    at a number, 230
    basic, 234
    composition of, 234, 236
    discontinuous, 230, 236
    Extreme Value Theorem, **356**
    from the left, 234
    from the right, 234
    interchanging limit and, 235
    Intermediate Value Theorem, **378**
    on an interval, 233
    list of, 235
Continuous-time dynamical system, **7–8**, 455
Continuous-time selection model, 596–598
Convergent improper integral, 572–573
Corner (in the graph of a function), **246**, 247, 248
    derivative at, 245
Cosine (function), 92, 95
    angle addition formula, 97
    antiderivative, 480
    derivative, 298
    double-angle formula, 97
    graph, 95
    is an even function, 95
    is periodic, 95
    main period, 95
    oscillations, 98. *See also* Oscillation(s)
Coupled (autonomous) differential equations, **632**
Critical number (point), 247–248, 251, **347**
Cube root function, **20**
Cubic function, **19**
Curvature of the graph of a function, **31**
Curve. *See also* Graph(s)
    bell-shaped, 407–408
    learning, 413
    length of, 562
    logistic, 214, 408–409
Cusp, **246**, 247, 248
Cylinder
    approximating (in finding volume), 556–557,
    right, 555
    use in finding volume,
    volume of, 558

Daily (monthly) temperature cycles, 101–102
Decreasing function, **22**, 60, 248
Definite integral, **495**, **498**
    area, **493**, **495**, 549
    average value of function, **564**
    difference of areas (signed area), 500
    Fundamental Theorem of Calculus, **514**, **522**, **524–525**
    left sum, 488
    length, 560
    mass, 565
    midpoint sum, 489
    properties of, 502–504
    Riemann sum, 487–492, 495, 498
    right sum, 488
    signed area, 500
    substitution rule, **531**–535
    summation property, 503
    total change, 505–507, 510–512
    volume, 555
Degree (angle measure), 92
Delta (Δ) notation, 56
Density, 9, 10
    bone (marrow), 366
    mass density, 57, 565
    mass density, units, 10
    of otter population, 566–567
    prey density, 329
Dependent variable, 11
Derivative, **244**
    approximating, **252–253**, 325
    Chain rule, **284–285**
    composition, 284–285
    constant function, 261
    constant product (multiple) rule, 266
    at a corner, 245–246
    cosine function, 296–298
    difference quotient, 244, 252
    difference rule, **265**
    at discontinuity, 246
    exponential function, 268–271
    first, 317
    general exponential function, 287
    graph, 247–251
    higher order, 319–320
    implicit, 308–309
    inverse function, of, 310
    inverse trigonometric function, 303–304
    linear function, 248–249
    logarithmic, 310–311
    oscillation(s), 299–300
    polynomial, 265–266
    power function, 262–263, 264
    power rule, 262–263, 264
    product rule, 273–275
    quotient rule, 277
    related rates, 312–314
    second, 316–317
    sine function, 296–298
    stability, evaluating with, 429–434, 443
    sum rule, 265
    trigonometric function(s), 296–301
    vertical tangent, 247
Derivative test for extreme values
    first, 349
    second, 352
Difference quotient, 244, 252

Difference rule for derivatives, 265
Differentiability, 245, 247
    and continuity, 246–247, 384–385
Differentiable (function)
    at a number, **245**
    on an open interval, **245**
Differential equation(s), 5–6, **8**, **252**, **270**, **455**. *See also* Autonomous differential equation(s)
    AIDS, for, 481
    autonomous, **457**, **489**
    Euler's method, 464–467
    initial condition(s), **459**
    initial value problem, 460
    logistic, 591–593
    modified logistic (Allee), 593–594
    nonlinear, 597
    pure-time, **457**, **458**
    separation of variables, 619
    solution, **458**, **459**
    state variable, **458**
Differential notation, 252, 263, 275
Diffusion, 85, 408, 413, 620
    across a membrane, 595
    chemical, 608
    equation, 314
    pollutant, 354–355
    reaction-diffusion equation, 618
Dimension
    discrete-time dynamical systems, 125
    dinosaurs, 1–2, 82
    skull length and spine length, 313
Direct substitution rule for limits, 197, 198
Direction arrows, **648**
    competition equations, 654–656
    Newton's law of cooling, 656–657
    predator–prey equations, 651–654
Discontinuity and derivative, 246–247, 347
Discontinuous function(s), 230
    cancer treatment model, 232
    in definite integrals, 517
    in improper integrals, 579–580
    integrable, 502
Discrete logistic model, 437–438
Discrete-time dynamical systems, 7, **113–114**
    additive (basic additive), 119
    backwards, 123
    bacterial population growth, 114, 116–118, 135–136
    Beverton-Holt, 152
    cobwebbing, 130
    consumption of drugs (caffeine and alcohol), 141
    dimensions, 125–126
    dynamical rule, **113**
    equilibrium (equilibria), **132–135**
    exponential (basic exponential), 117
    initial condition, **116**
    inverse (backward-time), 123
    limited population, 145–147
    logistic, 437–438
    lung, 164–172
    medication, 115–116
    mites 115, 122
    Newton's method, **417**
    nonlinear, 153
    oscillating, 123

Discrete-time dynamical systems (*Cont.*)
  pain medication, 115, 119–121, 131–132, 135, 138
  population, 142–147
  selection model, 153–159
  solution, **116**, 131
  tree growth, 115, 118–119
  units, 125–126
  updating function, **114**. *See also* Updating function
Discrete variable, **18**
Disease
  AIDS, 481
  endemic, **615**
  model for, 614–616
  spread of infectious disease, 322–323
Distance between two points, **560**
Divergent improper integral, 572
Domain of a function, **14**
  given, **14**
  natural, **14**
Double-angle formula for sine and cosine, 97
Double log (log-log) graph, 88–89
Doubling time, **78**
Drag on a falling object, 608
Drug consumption
  alcohol, 147–151
  caffeine, 141–142
  pain medication, 115, 119, 131, 135–138
DuBois and DuBois formula, **315**
Dynamical rule, **113**
Dynamical system. *See also* Discrete-time dynamical systems
  continuous-time, 8
  deterministic, 7
  discrete-time, 7
  probabilistic, 7

e (the number), 72, 269–270
Ecological balance, **591**
Elephants
  blood circulation time, 65–66
  heartbeat rate, 66–67
  relative surface area, 68
Endemic disease, 615
Equilibrium (equilibria) of an autonomous differential equation, 591, **603**, **640**
  Stability Theorem, 611
  of a system of autonomous equations, 640, 642–643
Equilibrium (equilibria) of a discrete-time dynamical system, 121, **132–133**, 135, 425
  algebraic approach, 135
  cobwebbing, 130–131, 132
  evaluating stability using derivative, 429–433
  graphical criterion for stability, 427
  half stable, 429
  slope criterion for stability, 428
  stable, 134, 225, 425
  Stability Theorem, 443
  unstable, 134, 425
Euler, Leonhard, 464
Euler's method, 464–466, **467**
  implicit, 586
  in the phase plane, 648–649
  system of autonomous differential equations, **633**, **636**

Even function, 95, 408
Exponential change, **75**
Exponential decay, **78**
Exponential function, **72**, **74**, 268–270
  with base a, 72
  with base e, 74
  cancellation law (with logarithms), 76
  continuity, 234–235
  derivative, 268–269, 287
  domain, 72–73
  graph, 72
  inverse, 75
  laws of exponents, 73, 74
  limits, 198
  models, 78–85, 590–591
  range, 7
Exponential growth, **78**
  population of Canada, 83–84
Exponentially changing quantity, 76
Exponential measurements, 290, 590–591
Exponential model, 75, 590–591
Extinction, 51–52, 152, 594
  extinction equilibrium, 146, 159, 430, 432
Extreme values of a function, 345
  on a closed interval, 349, **356**
  first derivative test, 349
  second derivative test, 353
Extreme Value Theorem, **356**

Factorial(s), 320, **338**
Fermat's Theorem, 348
First derivative, **317**
First derivative test, 349
Fish
  Beluga whales, 152
  growth model, 85, 214, 518
  population model, 137, 446–447
Flow of a liquid
  blood flow, 307, 316, 361–362
  resistance to, 361
  through a pipe, 242, 272
Food intake optimization, 368–372
Forward difference quotient, **252**
Fourier series, **110**
Function(s), 13
  absolute minimum and maximum, 345
  absolute value, 20
  absorption, 223
  addition, 25
  algebraic, 37
  algebraic operations, 25
  approaches infinity (at the same rate/more quickly/more slowly), 219–220
  approaches zero (at the same rate/more quickly/more slowly), 221
  average rate of change, **180**
  change in, 180
  commuting with respect to composition, 28
  composition of, **27**
  concave down, **318**
  concave up, **318**
  constant, 19
  continuous, **230**. *See also* Continuous function(s)
  critical point (number) of, **247–248**, 347
  cube root, 20
  cubic, 19

Function(s) (*Cont.*)
  decreasing, **22**, 60, 248
  defined piecewise, 20–21
  dependent variable, 14
  derivative, **244**
  difference of, 25
  differentiable at a number, **245**
  differentiable on an open interval, **245**
  discontinuous, **230**
  domain, 14
  domain, given, 14
  domain, natural, 14
  exponential, 72–75
  extreme values, 345–346
  graph, **16**
  Hill, 279
  horizontal line test, 32
  increasing, **22**, 60, 248
  independent variable, 14
  integrable, 498
  integral, 519
  inverse, 29–31, **32**
  inverse not possible to determine, 35
  limit(s), **188**, 193–194
  linear, 19, 55, **58–63**
  logarithmic, 75–78
  monod growth, 43
  multiplication, 25
  one-to-one, 32
  polynomial, 37
  power, 63
  product of, 25
  quadratic, 19
  quotient, 25
  range, **14**
  rational, 37
  reflecting, 42–43
  relation, 13
  relative minimum and maximum, 346
  scaling, 40–42
  shifting, 37–39
  sum of, 25
  updating, 114. *See also* Updating function
  vertical line test, **18**
Functional response, 223, 329
Fundamental Theorem of Calculus, **514**, **522**, **524–525**
  integral function, **519**–521
  proof, 523–525

Galois, Evariste, 35
Gamma distribution, 323–324
Generalized polynomials, **328**
General lung discrete-time dynamical system, 168
Geometric approach to investigating equilibrium, 132
Geometric interpretation of the derivative, 247–251
Given domain, **14**
Global (absolute) maximum, **345**
Global (absolute) minimum, **345**
Gompertz differential equation, **625–626**
Graph(s)
  of absolute value function, 20
  bell-shaped curve, 407–408

Graph(s) (*Cont.*)
　drawing, 11
　definition of, **16**
　double log, 88–89
　of exponential functions, 72, 74, 78
　of a function, 16
　of Hill functions, 410–411
　hyperbola, 19
　of a linear function, 58
　of logarithmic functions, 76
　logistic curve, **408**–409
　log-log, 88–89
　ordered pair, 16
　oscillation(s), 98–101
　parabola, 19
　plotting points, 16–17
　of power functions, 63
　of a proportional relation, 55–56
　reflecting, 42–43
　scaling, 40–42
　semilog, 86–87
　shifting, 37–39
　square root function, 20
　transformation of, 37–43
　trigonometric functions, 95–96
　from a verbal description, 18
Growth rate. *See* Per capita growth rate

Half life, 79–81
Half-saturation constant, **43**
Heartbeat frequency, 66–67
Heart, optimization of, 361–363
Hill function(s), **279**, 410–411
Hodgkin, Sir Alan, 7
Horizontal asymptote, **212**
Horizontal line test, 32
Huxley, Sir Andrew, 7
Hydrocephalus, 312
Hyperbola, 19
Hysteresis, **240**

Identity function, **195**
　continuity of, 234
Implicit differentiation, **308–310**
　comparison test, **578**
　convergent, 572–573
　divergent, 572–573
　improper integrals, **571**
　infinite integrand, 579–580
　infinite limits of integration, 571–572
　leading behaviour, 577, 581–582
Increasing function, **22**, 60, 248
Indefinite integral, **474**. *See also* Antiderivative; Integral
　link with definite integral, 510
Independent variable, 14
Indeterminate form, **218**, **396**
　limit(s), 396–402
　L'Hôpital's rule, **396**
Index (in summation notation), 493
Infectious disease. *See* Disease
Infinite integrand, 579–580
Infinite limits
　of integration, 571–572
　limit laws for, 218
Infinite series, **341**

Infinity, **206**
　comparing functions at, 219–220
　function approaching, at infinity, 219–220
　infinite limit(s) at, 215
　limit(s) at, 209–210
Inflection point, **318–319**, 322
Initial (population), **12**
Initial behaviour. *See* Leading behaviour
Initial condition(s)
　for a differential equation, **459**, 468
　for a discrete-time dynamical system, **116**
　for Euler's method, 467
　irrelevance of, in integration, 513
　sensitive dependence on, 163
Initial value. *See also* Initial condition(s)
　in exponential model, 78
　problem, 460
　problem, solution, 460
Initial value problem, 460
　solution of, 460
Inner function (in composition), 27, **283**
Input and output precision (tolerance), 238–240
Instantaneous rate of change, 182–185, **244**
　instantaneous speed, 184, 244
　tangent line, 185
Integrable function, **498**, 501
Integral. *See also* Antiderivative; Definite integral; Integration
　approximation. *See* Riemann sum
　connection between definite and indefinite integrals, 514
　constant multiple rule for, 478–479, 502
　definite, **493, 495**
　improper, 571
　indefinite, **474**
　properties of, 502–504
　Riemann sum, 496. *See also* Riemann sum
　summation property for, **503**
　sum rule, 479
Integral function, 519
Integral sign, **498**
Integrand, **498**
　infinite, 579–580
Integration
　constant multiple rule, 478–479, 502
　by guess-and-check, 530
　infinite limits of, 571–573
　methods, **529**
　by partial fractions, **629**
　by parts, **538–542**
　power rule, 476
　sum rule, 479
　using substitution, **530**–538
　using Taylor polynomials, 543–545
Intermediate Value Theorem, **378**
　bisection, used in, 380–381
Interpolation, **63**, 332
Inverse function, 29–31, **32**
　cancellation formulas, 32
　continuity of, 234–235
　derivative, 238, 310
　domain of, 32
　of the natural exponential function, 75
　range of, 32

Inverse function (*Cont.*)
　of trigonometric functions, 102
　updating function, 123
Inversely proportional relation (quantities), 63
Inverse trigonometric functions, **102–105**
　continuity of, 235
　derivative of, 296, 303–304

Lake Ontario, surface area, 554–555
Laws of exponents, 73
Laws of logarithms, 76
Leading behaviour, 220, **388, 392**
　applied to absorption function(s), 394–395
　of functions at infinity, **388–391**
　of functions at zero, **392–394**
　in improper integrals, 577, 581–582
　left-hand limit, **194**
　matched, 224, **393**–395
Left sum, **488**, 494
Length of a curve, **560**, **562**
L'Hôpital's rule, **396**
　comparing functions using, 396–397
　incorrect use of, 398
Limit(s), 183, **188**
　calculation of, 192
　comparing functions at infinity, 219
　of a constant function, 195
　in definition of continuity, 230
　in definition of derivative, 244
　direct substitution rule, 197, 198
　does not exist, **194, 215–216**
　geometric approach, 193
　of identity function, 195
　indeterminate form, **218**, **396**
　infinite, 205, 206
　infinite, at infinity, 215
　at infinity, 209–210
　L'Hôpital's rule, **396**
　left-hand, **194**
　long-term behaviour, 205
　numeric approach, 187, 192
　one-sided, 194
　polynomial, 196, 217
　properties of, 196
　right-hand, **193–194**
　scientific notion, 191
　of sequences, 224
　Squeeze (Sandwich) Theorem, 200
Limited population, 144–147
　cobwebbing, 147
　equilibria, 146
　stability, 431
Limit laws, **195**–196
　for infinite limits, 218
Limits of integration, **495, 498**
　infinite, 571–572
Limits of sequences and equilibria, 225
Line(s) **56, 58**. *See also* Linear function(s)
　graphing, 56, 58, 60
　point-slope form, **19**, **59**, 330
　slope, 56
　slope-intercept form, **59**
Linear approximation (linearization), **330**
Linear function(s), **19**, **55**, **58–63**
Linear interpolation, **332**
Linear model of population of Canada, 61–63
Linearization (linear approximation), **330**

Linear-quadratic survival model, 87–88
Loading dose (medication), 121
Local (relative) maximum, **346**
Local (relative) minimum, **346**
Logarithm(s), 75
  with base 10, 77
  converting between bases, 77
  laws of, 76
  natural, 75
Logarithmic differentiation, 310–311
Logarithmic function(s), **75**. *See also* Logarithm(s)
  cancellation laws, 76
  continuity, 234, 235, 236
  derivatives, 288–290
  domain, 75
  graph, 76
  range, 75
Logistic curve, **214**, **408**–409
Logistic differential equation, **591**–593
  equilibria, 604
  graph, 593
  modified, 593–594
  solution, 623–624
  stability, 612–613
Logistic dynamical system, **437**–439
  analysis of, 444–446
  per capita production for, 437
Logistic (growth) function, **85**, **322**–323
Logistic (population) model, 214
Log-log graph, 88–89
Long-term behaviour of a function, **205**
Lower limit of integration, **498**
Lung discrete-time dynamical system, 164
  absorption, with, 170–172, 222–223
  equilibria, 170
  general formulation, 167–170

Maintenance dose (medication), 121
Main period of a trigonometric function, **95**
Malaria, model, 6–7
Malthus, Thomas, 423
Mammography, 84–85, 343
Marginal Value Theorem (Property), **369**–370
Mass, 10, 274, 367
  of a bacterial population, 125–126
  of a bar, 566
  density, units, 10
  integrals and, 565–566
  proportional relation with volume, 57
  units, 10
Mass action, principle of, 631
Mathematical induction, 281
Mathematical model, 1, 3, 4–7. *See also* Model
Maximization. *See also* Optimization
  Extreme Value Theorem, 356
  Marginal Value Theorem, 369–370
Maximum
  absolute (global), 345
  of a definite integral, 503–504
  local (relative), **346**
  of an oscillation, 99
Mean value, 563. *See also* Average
Mean Value Theorem, **382**–383
  Rolle's Theorem, **381**–382

Medication. *See also* Pain medication
  absorption of, 115
  optimizing side effects, 353–354
Metabolic rate, 314
Method
  Euler's, 464–466, **467**
  of leading behaviour. *See* Leading behaviour
  of matched leading behaviours. *See* Leading behaviour
  Newton's (Newton–Raphson's), **414**–415
  Runge-Kutta, 586, 649
Michaelis constant, **43**
Michaelis-Menten model, **43**
Michaelis-Menten reaction kinetics, 223
Midpoint sum, 489
Milky Way Galaxy, 76
Minimum
  absolute (global), 345
  of a definite integral, 503–504
  local (relative), **346**
  of an oscillation, 99
Model, **1**, 3, 4–7
  absorption (substance) and replacement, 141
  Allee effect, 163, 593–594
  basic exponential, 78, 590
  Beverton-Holt, 152
  cancer, 232
  competition, 632
  disease, 614–615
  exponential, 78, 590
  gamma distribution, 323–324
  Gompertz growth, **625**–626
  limited population, 144–147
  linear-quadratic survival, 87–88
  logistic, 85, 214, 322–323, 591–593
  lung, 164–167
  malaria, 6–7
  Michaelis-Menten, 43
  modified logistic (Allee), 593–594, 604, 613
  monod growth, 43
  neurons, 6–7
  population of Canada. *See* Population of Canada
  population explosion, 627–628
  predator-prey, 630. *See also* Predator–prey model
  Ricker, 85, 413, 446–468
  selection (continuous), 596–598
  selection (discrete-time), 153–159
  spread of a pollutant, 85, 91, 295, 314, 354
  von Bertalanffy, **536**, 596
  world population, 106, 305
Model of selection. *See* Selection
Modified logistic differential equation, 593–594
  equilibria, 604
  stability, 613
Monkeys, 291–292
Monod equation, 224
Monod growth function, **43**
Monod reaction kinetics, 223
Monthly (daily) temperature cycles, 101–102
Mutant bacteria, 153
Mutation, 161–162, 177–178

Natural domain, **14**
Natural exponential function, **74**, **268**–270

Natural exponential function (*Cont.*)
  antiderivative, 480
  derivative, 269
  graph, 74
  inverse function, 75
Natural log (logarithmic) function, **75**, 76
  antiderivative, 540
  defined with a definite integral, 569
  derivative, 288
  graph, 76
Near a point, **188**
Negative angle, 92
Negative infinity, **206**
Neuron, 6–7, 240
  hysteresis, 240
Neuronal branching, 363
Newton-Raphson's method, 414. *See also* Newton's method
Newton's law of cooling, **594**
  autonomous differential equation, **594**–595, 596
  direction arrows, 656–657
  equilibria, 604, 644–645
  nullclines, 644–645
  phase-line diagram, 610
  reverse version, 610–611
  solution, 604–605, 622–623
  stability, 610
  system of autonomous differential equations, 633–634
Newton's method, **414**–415, 416, 417
  discrete-time dynamical system, **417**
  does not apply, 421
  equilibrium, 417
  multiple solutions, 421–422
  recursive formula, **417**
Newton's (second) law, **326**
Nonautonomous differential equation, **590**
Nonlinear
  differential equation, 597
  discrete-time dynamical system, 153, 428, 439, 443
Nonlinear equation
  Newton's Method, 417
Nonlinear function, recognizing, 59
Nullclines, **640**–641
  competition model, 643–644
  finding equilibrium, 642–643
  Newton's law of cooling, 644–645
  predator–prey, 640–642
Numerical algorithm, **414**

Odd function, **95**, 96
One-sided limit, **194**
One-to-one, 32
Open interval, 9
  function differentiable on an, **245**
Optical density, 28–29
Optimization. *See also* Maximum; Minimum
  bone strength, 365–368
  boundary of a rectangle, 359–361
  cell shape, 360–361
  food intake, 368–372, 453–454
  honeycomb shape, 372–375
  side effects of a medication, 353–354
  work of the heart, 361–363
Order of magnitude, 8

Ordered pair, 11, 13, 16
Ordering functions at infinity
    approaching infinity, 219–220
    approaching zero, 221–222
Origin (of a coordinate system), 11
Oscillating dynamical system, 123
Oscillation(s), **98**
    amplitude, 98
    average, 98
    daily and monthly temperature, 101–102
    derivative of, 299–300
    maximum, 98–99
    minimum, 98–99
    period, 98
    phase, 98
    scaling, 99–101
    shifting, 98–101
Outer function (in a composition), 27, 283
Overcompensation (absorption), 224

Pain medication, 115, 138
    absorption of, 115
    administration, 121
    discrete-time dynamical system, 115, 131–132
    equilibria, 135, 138
    loading dose, 121
    maintenance dose, 121
    solution, 119–121, 225
    stability of the model, 431
Parabola, **19**
    area under, 487–489
    graph, 19, 248–249
Parameter, **12**
Partial fractions, 629
Per capita growth rate, **43**, **590**
Per capita (re)production rate, 43, 143, **590**
Periodic function (sine and cosine), 95
Periodic hematopoiesis, 259
Period of oscillation, 98–99
Phase
    of an autonomous differential equation, **605**
    of an oscillation, 98–99
Phase-line diagram, **605**–606
Phase plane, **640**
    Euler's method, **648**
Phase-plane trajectory, **649**
    graphing, 649–651
Piecewise defined function, **20**
Plateau, 47, **214**, 409
Point of inflection, **318**–319, 322
Point-slope form of line, 19, 59, 330
Poiseuille's law, **361**
Pollutant, spread of, 85, 91, 295, 314, 354
Polymerase chain reaction (PCR), 229
Polynomial, 37
    approximating functions with, **330**
    continuity, 234, 235
    derivative, 265
    direct substitution rule, 197–198
    generalized, **328**
    limit(s), 196, 217
    quadratic approximation, 333
    Taylor polynomial, 338
Population, 3–6. *See also* Model
    Beverton-Holt, 152
    codfish in Eastern Canada, 137

Population (*Cont.*)
    competition, 632
    discrete-time dynamical system, 142–174
    explosion model, 626–627
    extinction, 51, 52, 152, 594
    final, 12, 114
    growth rate, 43
    initial, 12, 114
    limited, 144–147
    logistic, 214
    per capita growth rate, 43, 590
    per capita production rate 43, 143, 590
    predator–prey model, 630–632
    rate of change, 253, 291
    relative (specific) growth rate, 43
    relative rate of change, 291–292, 456, 592
    self-enhancing, 627
    world population model, 106, 305
Population of Canada
    autonomous differential equation, 456–458, 461–462
    census data, 29
    exponential growth model, 83–84
    Euler's method applied, 469–470
    linear model, 61–63
Positive angle, 92
Positive infinity, **206**
Power function(s), **63**
    antiderivative of, 476–478
    behaviour near infinity, 219–220
    behaviour near zero, 222
    derivative of, 261–264
    improper integrals, 574–575, 580–581
    power rule
        for derivatives, 261–264
        for integrals, 476–478
Precision (tolerance), 188
Predator–prey model, 648–650
    autonomous differential equations, 630–632
    direction arrows, 651–654
    Euler's method applied, 635–636, 648–649
    graph, 648–649
    nullclines and equilibria, 640–643
Prescribed precision (tolerance), **188**
Prime notation, 255, 263, 275, 280
    in the chain rule, 293
Principle of mass action, 631
Product rule for derivatives, 273–275
Proper integral(s), **571**
Proportionality constant, 56
Proportional relation, 55, 63
    between mass and volume, 57
Pure-time differential equation, **457**, 458
    Euler's method, 464–467
    graphical solution, 462–464
    solving using antiderivatives, **473**
Pythagorean Theorem, 94

Quadratic approximation, 333–335
    Taylor polynomials, 338
Quadratic function, **19**
    approximations using, 333
    graph of, 19
Qualitative dynamical systems, **439**–443
Qualitative reasoning about graphs, **18**
Quotient rule for derivatives, **277**

Radian (angle measure), 92
Radiocarbon dating, 81
    limits of, 82
Radiometric dating, 83
Range of a function, **14**
    exponential function(s), 72–73, 74
    inverse function, 32
    logarithmic function(s), 75
Rate of change, **291**
    approximating, of a population, 253
    average, 180
    instantaneous, 182–183
    relative, 291–292, 590
Rational function(s), **37**
    continuity, 234–235
    derivative, 277
    graphing, 406–407
    leading behaviour, 394
    limit(s), 196–197, 213, 217
Reaction-diffusion equation, **618**
Recursion (recursive relation), **114**, 116
Recursively defined sequence, **116**
Reflection(s) of a graph, 42–43
Region under a function and over an interval, **487**
    bounded, 551
Related rates, **312**–**314**
Relation, **13**
Relative (local) extreme value, 346
Relative (local) maximum and minimum, **346**
Relative rate of change, 291–292, 456
Relative (specific) growth rate, **43**
Relative surface area, 68
Ricker model, 85, 413, 446–468
Riemann, Bernhard, 489
Riemann sum, 487–492, 495, 498
    area under curve, 487–489
    definite integral, 495–496, 498
    general form of, 495, 498
    left sum, 488, 494
    midpoint sum, 489
    right sum, 488, 494
    signed area, 499
    total change, 505–507, 510–512
Right cylinder, 555
Right-hand limit, **194**–**195**
Right sum, 488, 494
Rolle's Theorem, **381**–**382**
Ross, Sir Ronald, 6
Runge-Kutta method, 586, 649

Saturation, **43**
    absorption with, 391, 394
    absorption with, and threshold, 395
Scaling graph(s), 39–42
Schnakenberg reaction, 647
Secant line, **181**, 331
    approximation, **331**–**332**
    average rate of change, **180**–181
    base point, 181
    difference quotient, 244
    instantaneous rate of change, 182–184
    slope, 181–182
Second derivative, **317**–**318**
    acceleration, 325–326
    approximating, 325
    concavity, 318, 322

Second derivative test, **352**
Second-order Runge-Kutta method, **586**
Selection (continuous model), 596–598
    equilibria, 604–605
    phase-line diagram, 607
    solution, 598, 625
    stability, 613
Selection (discrete-time dynamical system model), 153–159
    cobwebbing, 158
    equilibria, 156–159
    stability, 426–427, 430–431
Semilog graph, 86–88
Sensitive dependence on initial conditions, **163**
Sequence (of numbers), 116
    limit, 224–225, 226
    recursive (defined recursively), **116**
Shannon (Shannon-Wiener) index, **365**
Shifting graph(s) 37–39
Side effects, 50, 353–354
Sigma (Σ) notation, 492–493
Signed area, **500**
    and integral, 500
    Riemann sums interpreted as, 499
Signed vertical distance, **16**
Signum function, **233**
Simpson's diversity index, **365**
Simulation, 2
Sine function, 92, 95
    angle addition formula, 97
    antiderivative of, 480
    derivative of, 298
    double-angle formula, 97
    graph, 97
    inverse, 102–105
    main period, 95
    is an odd function, 95
    oscillations, 98. *See also* Oscillation(s)
    is periodic, 95
Sinusoidal function. *See* Oscillation(s)
Sinusoidal oscillation. *See* Oscillation(s)
Slope, 56
    as a proportionality constant, 58
    between two points, 59
    criterion for stability of an equilibrium, **428**
    delta (Δ) notation, 56
    infinite, 247
    of a line, 19, **56**, 59
    negative, 60, 248, 250
    positive, 60, 248, 250
    of a secant line, 181–182, 183
    of a tangent line, 183, 244
    zero, 60, 248, 250
Slope-intercept form of line, **59**
Solid of revolution, **559**
    volume, 560
Solution(s),
    of a differential equation, 458, **459**
    of a discrete-time dynamical system, 116, 131
    graphical, of a pure-time differential equation, 462
    graphical, of a system of autonomous differential equations, 640
    in a phase plane, 648–651
Specific growth rate. *See* Relative growth rate

Specific heat, **595**
Speed, 184, 244
    average, 184
    in Intermediate Value Theorem 378
    in Mean Value Theorem, 383
    units, 10
Square root function, **20**
    calculation using Newton's method, 415–417
    graph, **20**
Square wave, 110
Squeeze (Sandwich) Theorem, 200–**201**
Stability
    of an autonomous differential equation, 609–611
    of a discrete-time dynamical system, 134–136, **424**
    evaluating with the derivative, 429–434
Stability Theorem for autonomous differential equations, **611**
    applications, 612–615
    proof, outline, 615–616
Stability Theorem for discrete-time dynamical systems, **443**
    graphical criterion, 427
    slope criterion, 428
Stable equilibrium
    of an autonomous differential equation, 607
    of a discrete-time dynamical system, 134, 425
    and limits of sequences, 225
Standard normal distribution, **408**
State variable, **458**
Step function approximation, 505
Substitution method
    for definite integrals, 534–535
    for indefinite integrals, 531–534
    for integration, **530**
Summation (Σ) notation, 492–493
Summation property of indefinite integral, **503**, 504
Sum rule
    for antiderivatives, 479
    for definite integrals, 503–504
    for derivatives, 265
    for limits, 196
Surface area,
    of Lake Ontario, 554, 555
    relation with volume, and, 67–68, 266
    relative surface area, 266
System of autonomous differential equations, **630**
    direction arrows, 651–657
    equilibria, 640–645
    Euler's method, 634–638, 648–651
    nullclines, 640–645
    phase plane, **640**

Tangent function, 92, 96
    derivative of, 300–301
    graph of, 96
    inverse of, 103–104
    relation with secant function, 96
Tangent line, **183–184**, 244
    horizontal (critical point), 248
    slope, 183, 244
    vertical, 247

Tangent line approximation, **330**
    in Euler's method, 464, 634
    in Newton's method, 414–416
Taylor polynomials, **338**
    degree *n*, **338**
    infinite series, **341**
    integration using, 543–545
    logarithmic function, 338–339
Temperature
    converting between Celsius and Fahrenheit, 30–31, 60–61
    daily and monthly cycles, 101–102
Threshold
    absorption with saturation and, **395**
    behaviour, 240
Tolerance (precision), **188**
Torricelli's law of draining, 609
Total change, 505–507, 510–512
    definite integral as, 507, 511
Trajectory
    of motion, 231
    in a phase-plane, **648**–651
Transcendental functions, 37
Translation, of dimensions in a discrete-time dynamical system, 125
Tree growth discrete-time dynamical system, 115, 118–119, 125, 136
Triceratops, 1–2
Trigonometric equations, 97–98
Trigonometric function(s), **91**. *See also* Cosine function; Sine function; Tangent function
    continuity, 234, 235
    derivative, 296–301, 303–305
    graph, 95, 96, 301
    inverse, 102–105
    limit(s), 214, 215
Trigonometric identities, 93, 94, 95, 96, 97
Trigonometric ratios, 92–94

Ultradian rhythm, 109
Units, 10
    converting, 10
    in discrete-time dynamical systems, 125
Unstable equilibrium
    of an autonomous differential equation, 607
    of a discrete-time dynamical system, 134, 425
Updating function, **114, 424**
    backward, 123–124
    composition, 122
    converting units, 125
    equilibrium (equilibria), 133
    inverse, 123–124
    in Newton's method, 418
    repeated action of, 116
    slope, 428, 443
    stability criterion, 427, 428, 443
    units (dimensions), 125
Upper limit of integration, **498**

Variable, 11
    dependent, 11
    independent, 11
Velocity. *See* Speed
Vertical asymptote, **208**
Vertical line test, **18**
Vertical tangent, 247

Volume
    computing using integration, 555, 557–**558**
    of a cone, 559
    of a cylinder, 558
    of a heart chamber, 555–557
    proportional relation with mass, 57
    of a solid of revolution, 559–560, 561

Volume (*Cont.*)
    of a sphere, 559
    relation with surface area, 67–68, 266
Von Bertalanffy limited growth model, 536, 596

Walleye, 214, 536–537
Water pressure, 284
Weber-Fechner law, 129

Weighted average, 168–169, 170
    multicomponents, 170
Wild type bacteria, 153
World population model, 106–107, 305

Zero (number)
    continuity, division by, 236–237
    function approaching, at infinity, 221–222

Trigonometric functions of special angles: see table on page 94

Identities

$$\tan x = \frac{\sin x}{\cos x} \qquad \cot x = \frac{\cos x}{\sin x}$$

$$\sin^2 x + \cos^2 x = 1 \qquad 1 + \tan^2 x = \sec^2 x$$

$$\sin(-x) = -\sin x \qquad \cos(-x) = \cos x$$

$$\sin(x - \pi/2) = -\cos x \qquad \sin(x + \pi/2) = \cos x$$

$$\cos(x - \pi/2) = \sin x \qquad \cos(x + \pi/2) = -\sin x$$

$$\sin 2x = 2\sin x \cos x$$

$$\cos 2x = 2\cos^2 x - 1 = 1 - 2\sin^2 x \qquad \cos^2 x - \sin^2 x = \cos 2x$$

Inverse Trigonometric Functions

$\sin A = B$ and A in $[-\pi/2, \pi/2]$ is equivalent to $A = \arcsin B$

$\tan A = B$ and A in $(-\pi/2, \pi/2)$ is equivalent to $A = \arctan B$

$\sin(\arcsin x) = x$ if x is in $[-1, 1]$

$\arcsin(\sin x) = x$ if x is in $[-\pi/2, \pi/2]$

$\tan(\arctan x) = x$ for any real number x

$\arctan(\tan x) = x$ if x is in $(-\pi/2, \pi/2)$

EXPONENTIAL AND LOGARITHM FUNCTIONS

$\ln A = B$ is equivalent to $A = e^B$

$\log_a A = B$ is equivalent to $A = a^B$

$\ln x = \log_e x$

$\ln 1 = 0 \qquad \log_a 1 = 0$

$\ln e = 1 \qquad \log_a a = 1$

$\ln e^x = x \qquad \log_a a^x = x$

$e^{\ln x} = x \qquad a^{\log_a x} = x$

$\log_{10} 1 = 0 \qquad \log_{10} 10 = 1$

$\log_{10} 10^n = n$

$$\log_a x = \frac{\ln x}{\ln a} \qquad \ln x = \frac{\log_a x}{\log_a e}$$

Laws of Logarithms

$\log_a(x \pm y)$ does not simplify/expand

$\log_a(xy) = \log_a x + \log_a y$

$\log_a(x^r) = r \log_a x$

$\log_a \left(\dfrac{x}{y}\right) = \log_a x - \log_a y$

DIFFERENTIATION FORMULAS

Sum and difference rules: $(f(x) \pm g(x))' = f'(x) \pm g'(x)$

Constant multiple rule: $(cf(x))' = cf'(x)$

Product rule: $(f(x)g(x))' = f'(x)g(x) + f(x)g'(x)$

Quotient rule: $\left(\dfrac{f(x)}{g(x)}\right)' = \dfrac{f'(x)g(x) - f(x)g'(x)}{[g(x)]^2}$

Chain rule: $[f(g(x))]' = f'(g(x))g'(x)$

$(c)' = 0$ $\qquad (x^n)' = nx^{n-1}$

$(e^x)' = e^x$ $\qquad (a^x)' = a^x \ln a$

$(\ln x)' = \dfrac{1}{x}$ $\qquad (\log_a x)' = \dfrac{1}{x \ln a}$

$(\sin x)' = \cos x$ $\qquad (\csc x)' = -\csc x \cot x$

$(\cos x)' = -\sin x$ $\qquad (\sec x)' = \sec x \tan x$

$(\tan x)' = \sec^2 x$ $\qquad (\cot x)' = -\csc^2 x$

$(\arcsin x)' = \dfrac{1}{\sqrt{1 - x^2}}$ $\qquad (\arctan x)' = \dfrac{1}{1 + x^2}$

Approximate solutions of equations

An approximation of a solution $f(x) = 0$ is given by $x_{t+1} = x_t - \dfrac{f(x_t)}{f'(x_t)}$ where x_0 is an initial guess (Newton's method).

Approximations of functions

Linear approximation: $L(x) = f(a) + f'(a)(x - a)$

Quadratic approximation: $T_2(x) = f(a) + f'(a)(x - a) + \dfrac{f''(a)}{2}(x - a)^2$

Taylor polynomial: $T_n(x) = f(a) + f'(a)(x - a) + \dfrac{f''(a)}{2!}(x - a)^2 + \cdots + \dfrac{f^{(n)}(a)}{n!}(x - a)^n$

BASIC INTEGRATION FORMULAS

Antiderivatives

$\displaystyle\int (f(x) \pm g(x))\, dx = \int f(x)\, dx \pm \int g(x)\, dx$

$\displaystyle\int cf(x)\, dx = c \int f(x)\, dx$

Integration by parts: $\displaystyle\int u(x)v'(x)\, dx = u(x)v(x) - \int v(x)u'(x)\, dx$

$\displaystyle\int 0\, dx = C$ $\qquad \displaystyle\int dx = \int 1\, dx = x + C$

$\displaystyle\int x^n\, dx = \dfrac{1}{n+1} x^{n+1} + C$ if $n \neq -1$ $\qquad \displaystyle\int x^{-1}\, dx = \int \dfrac{1}{x}\, dx = \ln |x| + C$

$\displaystyle\int e^x\, dx = e^x + C$ $\qquad \displaystyle\int a^x\, dx = \dfrac{a^x}{\ln a} + C$

$\displaystyle\int \sin x\, dx = -\cos x + C$ $\qquad \displaystyle\int \cos x\, dx = \sin x + C$